# The Adams Spectral Sequence for Topological Modular Forms

Mathematical
Surveys
and
Monographs

Volume 253

# The Adams Spectral Sequence for Topological Modular Forms

Robert R. Bruner
John Rognes

AMERICAN MATHEMATICAL SOCIETY
Providence, Rhode Island

2020 *Mathematics Subject Classification.* Primary 55N34, 55P43, 55Q45, 55T15; Secondary 18G40, 55N35, 55P42, 55Q51, 55T05.

For additional information and updates on this book, visit
**www.ams.org/bookpages/surv-253**

**Library of Congress Cataloging-in-Publication Data**

Names: Bruner, R. R. (Robert Ray), 1950– author. | Rognes, John, author.
Title: The Adams spectral sequence for topological modular forms / Robert R. Bruner, John Rognes.
Description: Providence, Rhode Island : American Mathematical Society, 2021. | Series: Mathematical surveys and monographs, 0076-5376 ; Volume 253 | Includes bibliographical references and index.
Identifiers: LCCN 2021004452 | ISBN 9781470456740 (hardcover) | ISBN 9781470465636 (ebook)
Subjects: LCSH: Algebra, Homological. | Homology theory. | Adams spectral sequences. | AMS: Category theory; homological algebra – Spectral sequences, hypercohomology. | Algebraic topology – Homology and cohomology theories – Elliptic cohomology. | Algebraic topology – Homology and cohomology theories – Other homology theories in algebraic topology. | Algebraic topology – Homotopy theory – Stable homotopy theory, spectra. | Algebraic topology – Homotopy theory – Spectra with additional structure ($E_\infty$, $A_\infty$, ring spectra, etc.). | Algebraic topology – Homotopy groups – Stable homotopy of spheres. | Algebraic topology – Homotopy groups – $v_n$-periodicity. | Algebraic topology – Spectral sequences in algebraic topology – General. | Algebraic topology – Spectral sequences in algebraic topology – Adams spectral sequences.
Classification: LCC QA169 .B765 2021 | DDC 514/.23–dc23
LC record available at https://lccn.loc.gov/2021004452





∞ The paper used in this book is acid-free and falls within the guidelines established to ensure permanence and durability.
Visit the AMS home page at `https://www.ams.org/`

10 9 8 7 6 5 4 3 2 1 26 25 24 23 22 21

# Contents

# List of Figures

# List of Tables

# Preface

Inspired by earlier work exhibiting $v_1$-periodicity in the topological cyclic homology of the integers [**30**], [**31**], [**148**], [**149**], and subsequent work exhibiting $v_2$-periodicity in the topological cyclic homology of the connective complex $K$-theory ring spectrum and its Adams summand [**19**], [**18**], the authors started an investigation into the topological Hochschild homology and topological cyclic homology of the topological modular forms ring spectrum, aiming to study the $v_3$-action on $F_*TC(tmf)$ for suitable finite type 3 spectra $F$. In particular, at the prime $p = 2$ we can take $F$ to be the homotopy cofiber of a map $v_2^{32}\colon \Sigma^{192}M(1,4) \to M(1,4)$ as in [**26**], and then $F \wedge tmf \simeq tmf/(2, B, M)$ for certain Bott and Mahowald elements $B \in \pi_8(tmf)$ and $M \in \pi_{192}(tmf)$.

The Adams spectral sequence, in conjunction with the computer software package `ext` described in [**41**], provides a flexible and powerful tool for making calculations with *tmf*, $THH(tmf)$ and approximations to $TC(tmf)$. The additive structure of the Adams spectral sequence for *tmf*, and parts of its multiplicative structure, have been known to Mahowald and some other experts for many years [**76**], [**54**, Ch. 13], but for our project we expect to need full information about the multiplicative structure. Since we believe that this detailed information will be of use and interest also to other researchers in algebraic topology, we have composed the following account of the Adams spectral sequence for *tmf*, and related spectra such as $tmf/(2, B, M)$, aiming to give complete information and proofs of results that have otherwise mostly been available as folklore.

The first author would like to thank the Isaac Newton Institute for Mathematical Sciences for support and hospitality during the program *Homotopy Harnessing Higher Structures* during which some of the work in this book was done. This program was supported by EPSRC grant number EP/R014604/1. The first author also received funding from the Simons Foundation, project number 245786, the Research Council of Norway (RCN), project number 239015, and the Pure Mathematics in Norway project of the Trond Mohn Foundation (TMF).

# Introduction

In this book we study the graded ring $\pi_*(tmf)$ of homotopy groups of the connective ring spectrum of topological modular forms, by means of the classical Adams spectral sequence. We obtain precise information about the additive and multiplicative structure of this graded ring, in all degrees. As an application we calculate the full additive and multiplicative structure of $\pi_*(S)$, the stable homotopy groups of spheres, in degrees $* \le 44$.

In this introduction, we first review the context of topological modular forms and the Adams spectral sequence, and then turn to a discussion of the $E_2$-term, differential pattern and extension questions leading to $\pi_*(tmf)$ as a graded ring. Finally we outline our results about duality, the Adams spectral sequence for the sphere spectrum, and the case of odd primes.

## 0.1. Topological modular forms

The ring spectrum *tmf* is a connective form of a periodic ring spectrum $TMF$, first constructed as an $A_\infty$ ring spectrum (= $S$-algebra) by Mike Hopkins and Haynes Miller [**74**, §9], [**77**], [**146**], and then as an $E_\infty$ ring spectrum (= commutative $S$-algebra) by Paul Goerss and Hopkins [**65**], [**62**], [**54**, Ch. 12]. A different, but equivalent, construction was later developed by Jacob Lurie [**96**], [**97**], [**98**]. An elliptic cohomology theory is a Landweber exact cohomology theory associated to the formal group of an elliptic curve, and $TMF$ is in a sense the initial such theory, being defined as the global sections (or homotopy limit) of a sheaf of $E_\infty$ ring spectra over the moduli stack $\mathcal{M}_{ell}$ of elliptic curves. The sheaf extends over the Deligne–Mumford compactification $\overline{\mathcal{M}}_{ell}$ of this stack, allowing generalized elliptic curves with nodal singularities, and the global sections of the extended sheaf defines an intermediate $E_\infty$ ring spectrum *Tmf*, whose connective cover is *tmf*:

$$tmf = \tau_{\ge 0} Tmf \longrightarrow Tmf \longrightarrow TMF = tmf[1/\Delta]\,.$$

In particular, the topological modular forms spectrum *tmf* is itself an $E_\infty$ ring spectrum.

The natural transformation from the homotopy groups of a homotopy limit to the limit of the homotopy groups defines a ring homomorphism

$$e' \colon \pi_*(TMF) \longrightarrow MF_{*/2} = \mathbb{Z}[c_4, c_6, \Delta^{\pm 1}]/(c_4^3 - c_6^2 = 1728\Delta)$$

from the homotopy groups of $TMF$ to the graded ring of integral modular functions. Here $c_4$ and $c_6$ are multiples of the classical Eisenstein series, and $\Delta$ is the discriminant. More precisely, $e'$ is the edge homomorphism in a descent spectral sequence

$$H^s(\mathcal{M}_{ell}; \omega^{\otimes k}) \Longrightarrow \pi_{2k-s}(TMF)\,,$$

called the elliptic spectral sequence. The $E_2$-term, differential structure and additive extensions in this spectral sequence were determined by Hopkins and Mark Mahowald around 1994, see [**74**, §9], [**103**, §4] and [**76**]. It follows that both the kernel and the cokernel of the edge homomorphism are torsion groups annihilated by 24. In particular, $e'$ induces an isomorphism after inverting the primes 2 and 3. Localized at $p = 2$ or $p = 3$, however, $\pi_*(TMF)$ contains a rich pattern of torsion groups, which detects a large part of the known 2- and 3-power torsion in $\pi_*(S)$. Since $\Delta^8$ and $\Delta^3$ are infinite cycles in the 2- and 3-localized descent spectral sequences, respectively, there are invertible homotopy classes $M \in \pi_{192}(TMF)_{(2)}$ and $H \in \pi_{72}(TMF)_{(3)}$ that are detected by these powers of $\Delta$. Hence $\pi_*(TMF)$ repeats 192-periodically at $p = 2$ and 72-periodically at $p = 3$. Hopkins and Mahowald [**76**, §11] used this to exhibit many $v_2$-periodic families of elements in $\pi_*(S)$.

The sphere spectrum $S$ is connective, so the unit map $S \to \mathit{Tmf}$ factors through the connective cover $\mathit{tmf} \to \mathit{Tmf}$, and the edge homomorphism $e'$ restricts to a homomorphism

$$e\colon \pi_*(\mathit{tmf}) \longrightarrow \mathit{mf}_{*/2} = \mathbb{Z}[c_4, c_6, \Delta]/(c_4^3 - c_6^2 = 1728\Delta)$$

to the ring of integral modular forms, in which $\Delta$ is not inverted. In this framework, the calculation of a spectral sequence converging to $\pi_*(\mathit{tmf})$ was documented by Tilman Bauer [**23**], including the identification of the $E_2$-term as the cohomology of a Weierstrass curve Hopf algebroid $(A, \Gamma)$, the differential pattern, and the additive extensions. In particular, each homotopy group $\pi_n(\mathit{tmf})$ is finitely generated, so $\mathit{tmf}$ has finite type. Bauer also determined part of the multiplicative structure of $\pi_*(\mathit{tmf})$, including the products with the Hopf invariant one classes $\eta \in \pi_1(S) \cong \pi_1(\mathit{tmf})$ and $\nu \in \pi_3(S) \cong \pi_3(\mathit{tmf})$. It turns out that $\pi_7(\mathit{tmf}) = 0$, so the Hopf invariant one class $\sigma \in \pi_7(S)$ acts trivially on $\pi_*(\mathit{tmf})$. Inverting a power of $\Delta$ one recovers the elliptic spectral sequence studied by Hopkins and Mahowald, so Bauer's paper also serves to document the (unpublished) details of their calculation. Thereafter, most of the remaining multiplicative structure of $\pi_*(\mathit{tmf})$ was determined by Bauer and André Henriques, and concisely recorded by Henriques in [**54**, Ch. 13].

There is also a descent spectral sequence

$$H^s(\overline{\mathcal{M}}_{ell}; \omega^{\otimes k}) \Longrightarrow \pi_{2k-s}(\mathit{Tmf})$$

associated to the extended sheaf of $E_\infty$ ring spectra over $\overline{\mathcal{M}}_{ell}$, which is intermediate between Bauer's spectral sequence and the Hopkins–Mahowald elliptic spectral sequence. Its $E_2$-term, differential structure, additive extensions and most of the multiplicative structure were determined by Johan Konter [**89**], building on the work of Bauer. In particular, the computations of Konter prove the "Gap Theorem" that $\pi_n(\mathit{Tmf}) = 0$ for $-21 < n < 0$.

A major goal of the present work is to determine, with full proofs, the precise graded ring structure of $\pi_*(\mathit{tmf})$, together with substantial information about the ring homomorphisms $\iota\colon \pi_*(S) \to \pi_*(\mathit{tmf})$ and $e\colon \pi_*(\mathit{tmf}) \to \mathit{mf}_{*/2}$. After implicit completion at the prime 2, the additive structure of $\pi_*(\mathit{tmf})$ is given in Theorem 9.27 and Table 9.4, while the product structure is summarized in Theorem 9.54 and Tables 9.8 and 9.9. We pay particular attention to the coefficients of products landing in groups of order greater than 2; see Proposition 9.35 and Figure 9.5, which also specify the one bit of multiplicative information that we have left unresolved, regarding the sign $s \in \{\pm 1\}$ of a product $\nu_4 \cdot \nu_6$ in $\pi_{246}(\mathit{tmf})$.

In Corollary 9.55 we confirm and generalize an observation due to Mahowald, asserting that $\epsilon \in \pi_8(S) \to \pi_8(tmf)$ and certain related classes $\epsilon_k \in \pi_{8+24k}(tmf)$ have the same action on the $B$-power torsion in $\pi_*(tmf)$ as the Bott class $B$ and its relatives $B_k$, respectively. We determine the $tmf$-Hurewicz image of $\pi_n(S)$ in $\pi_n(tmf)$ for $n \leq 101$ in Proposition 11.83. The edge homomorphism to $mf_{*/2}$ is described in Proposition 9.19. As a consequence of these precise calculations, we show in Theorem 9.53 that, when viewed as a ring homomorphism to its image, the 2-completed edge homomorphism is split surjective in the sense that it admits a section $\mathrm{im}(e) \to \pi_*(tmf)$ that is also a ring homomorphism. Finally, in Remark 9.58 we give a detailed comparison of our results with those collected by Henriques, pointing out a short list of discrepancies.

At the prime 3, the corresponding results are given in Figure 13.2, Theorem 13.19, Table 13.2 and Proposition 13.29. There is one unresolved coefficient $t \in \{0, 1, 2\}$ in a product $B_2 \cdot B_2$ in $\pi_{112}(tmf)$, which, if nonzero, obstructs the existence of a ring homomorphism section to the 3-completed edge homomorphism $e\colon \pi_*(tmf) \to \mathrm{im}(e)$. At primes $p \geq 5$ the edge homomorphism is an isomorphism, so the coefficient $t$ is the only obstruction to the existence of an integrally defined section $\mathrm{im}(e) \to \pi_*(tmf)$ that respects the ring structures.

## 0.2. (Co-)homology and complex bordism of *tmf*

Let $n$ be a natural number. After inverting $n$, the moduli stack of elliptic curves admits an étale cover $\mathcal{M}(n) \to \mathcal{M}_{ell}$ classifying elliptic curves with level $n$ structure, and there is a corresponding étale extension $TMF[1/n] \to TMF(n)$ of $E_\infty$ ring spectra. Mike Hill and Tyler Lawson [**70**] extended the Goerss–Hopkins–Miller sheaf of $E_\infty$ ring spectra to a compactification $\overline{\mathcal{M}}(n)$ of $\mathcal{M}(n)$, with a log-étale map to $\overline{\mathcal{M}}_{ell}$, thereby obtaining extensions $Tmf[1/n] \to Tmf(n)$ of $E_\infty$ ring spectra. In particular, for $n = 1$ their construction provides one way of extending the Goerss–Hopkins–Miller sheaf from $\mathcal{M}_{ell}$ to $\overline{\mathcal{M}}_{ell}$. There are also $E_\infty$ ring spectra $Tmf_0(n)$ and $Tmf_1(n)$ corresponding to $\Gamma_0(n)$ and $\Gamma_1(n)$ level structures, respectively. Connective covers of these variants have proved useful in determining the mod $p$ cohomology and homology of $tmf$, as well as its complex bordism.

First, let $p = 2$ and let $A$ denote the mod 2 Steenrod algebra. It is generated by the Steenrod squaring operations $Sq^i$ for $i \geq 1$, subject to the Adem relations [**13**] [**160**, §I.1]. It is a cocommutative Hopf algebra over $\mathbb{F}_2$, and the structure of the dual Hopf algebra

$$A_* = \mathbb{F}_2[\xi_i \mid i \geq 1]$$

was determined by Milnor [**127**]. The coproduct is given by

$$\psi(\xi_k) = \sum_{i+j=k} \xi_i^{2^j} \otimes \xi_j$$

with $\xi_0 = 1$. The mod 2 cohomology $H^*(X) = H^*(X; \mathbb{F}_2)$ of any spectrum is naturally an $A$-module, and the mod 2 homology $H_*(X) = H_*(X; \mathbb{F}_2)$ is naturally an $A_*$-comodule. Let

$$A(1) = \langle Sq^1, Sq^2 \rangle$$
$$A(2) = \langle Sq^1, Sq^2, Sq^4 \rangle$$
$$E(2) = \langle Q_0, Q_1, Q_2 \rangle$$

be the subalgebras of $A$ generated by the listed elements, where $Q_0 = Sq^1$, $Q_1 = [Sq^2, Q_0]$ and $Q_2 = [Sq^4, Q_1]$. These are finite-dimensional of ranks 8, 64 and 8, respectively, and $E(2)$ is the exterior algebra on the three given generators. The $A(2)$-module $A(2)//E(2) = A(2) \otimes_{E(2)} \mathbb{F}_2$ is a "double" of $A(1)$, with $Sq^{2i}$ acting in $A(2)//E(2)$ as $Sq^i$ acts in $A(1)$, and can be realized as the cohomology of a 2-local 8-cell 12-dimensional CW spectrum $\Phi = \Phi A(1)$. (A more common notation for the double of $A(1)$ is $DA(1)$, but we prefer to reserve $DX$ to denote the Spanier–Whitehead dual $F(X, S)$ of a spectrum $X$.)

Akhil Mathew [**114**, Thm. 1.2] showed that $\mathit{Tmf} \wedge \Phi$ is 2-locally equivalent to the spectrum $\mathit{Tmf}_1(3)$ of topological modular forms for elliptic curves with $\Gamma_1(3)$ level structure, whose connective cover $\mathit{tmf}_1(3)$ is equivalent to a (generalized) truncated Brown–Peterson spectrum $BP\langle 2\rangle$ with cohomology $H^*(BP\langle 2\rangle) \cong A//E(2) = A \otimes_{E(2)} \mathbb{F}_2$. It follows from the Gap Theorem that

$$\mathit{tmf} \wedge \Phi \simeq_{(2)} \mathit{tmf}_1(3)\,,$$

and this in turn implies [**114**, Thm. 1.1] that

$$H^*(\mathit{tmf}) \cong A//A(2) = A \otimes_{A(2)} \mathbb{F}_2\,.$$

This will be a key input to our Adams spectral sequence computations. The surjection $A = H^*(H) \to H^*(\mathit{tmf})$ is induced by a unique $E_\infty$ ring spectrum map $\mathit{tmf} \to H = H\mathbb{F}_2$ to the mod 2 Eilenberg–Mac Lane spectrum, which also induces an injective algebra homomorphism $H_*(\mathit{tmf}) \to H_*(H) = A_*$, with image

$$H_*(\mathit{tmf}) \cong \mathbb{F}_2[\bar\xi_1^8, \bar\xi_2^4, \bar\xi_3^2, \bar\xi_i \mid i \ge 4] = A_* \,\square_{A(2)_*}\, \mathbb{F}_2\,.$$

Here $\bar\xi_i = \chi(\xi_i)$ denotes the Hopf algebra conjugate of the Milnor generator $\xi_i$, and $\square$ denotes the cotensor product.

Next, let $p = 3$ and let $A$ denote the mod 3 Steenrod algebra. It is generated by the Bockstein operation $\beta$ and the Steenrod power operations $P^i$ for $i \ge 1$, again subject to Adem relations [**160**, §VI.1]. The dual Hopf algebra is

$$A_* = \mathbb{F}_3[\xi_i \mid i \ge 1] \otimes E(\tau_i \mid i \ge 0)$$

with coproduct

$$\psi(\xi_k) = \sum_{i+j=k} \xi_i^{3^j} \otimes \xi_j$$

$$\psi(\tau_k) = \tau_k \otimes 1 + \sum_{i+j=k} \xi_i^{3^j} \otimes \tau_j\,,$$

where $\xi_0 = 1$. Let $P(0) = \langle P^1\rangle$ and $A(1) = \langle \beta, P^1\rangle$ be the subalgebras of $A$ generated by the listed elements. Here $P(0)$ is realized as the mod 3 cohomology of the 3-local 3-cell 8-dimensional CW spectrum $\Psi = S \cup_\nu e^4 \cup_\nu e^8$, and Mathew [**114**, Thm. 4.15] showed that $\mathit{Tmf} \wedge \Psi$ is 3-locally equivalent to $\mathit{Tmf}_0(2) = \mathit{Tmf}_1(2)$, whose connective cover $\mathit{tmf}_0(2)$ is equivalent to $BP\langle 2\rangle \vee \Sigma^8 BP\langle 2\rangle$. This leads to a calculation of the $A$-module coalgebra $H^*(\mathit{tmf})$ and the $A_*$-comodule algebra $H_*(\mathit{tmf})$. However, in this case it turns out to be more convenient to analyze $\pi_*(\mathit{tmf})$ using a variant of the Adams spectral sequence due to Andy Baker and Andrey Lazarev [**20**], namely one which is constructed entirely within the category of $\mathit{tmf}$-modules. The $E_2$-term of this $\mathit{tmf}$-module Adams spectral sequence is given by Ext over the $\mathit{tmf}$-module Steenrod algebra $A_{\mathit{tmf}} = H^*_{\mathit{tmf}}(H) = \pi_{-*}F_{\mathit{tmf}}(H, H)$, where $H = H\mathbb{F}_3$, rather than over the ordinary Steenrod algebra. Using the equivalence

$$tmf \wedge \Psi \simeq_{(3)} tmf_0(2)\,,$$

Hill and Henriques [**68**] show that $A_{tmf}$ is a quadratic extension of $A(1)$, dual to

$$A_*^{tmf} = H_*^{tmf}(H) = \pi_*(H \wedge_{tmf} H) \cong \mathbb{F}_3[\xi_1]/(\xi_1^3) \otimes E(\tau_0, \tau_1, \theta_2)\,,$$

where $|\theta_2| = 9$. We review this calculation in Chapter 13, see Theorem 13.6, and add the observation that this is a square–zero extension.

Mathew [**114**, §5] went on to determine the complex bordism $MU_*(tmf)$ as an $MU_*(MU)$-comodule, and to show that the $E_2$-term

$$\mathrm{Ext}^{s,t}_{MU_*(MU)}(MU_*, MU_*(tmf)) \Longrightarrow \pi_{t-s}(tmf)$$

of the Adams–Novikov spectral sequence for $tmf$ is isomorphic to the cohomology of the Weierstrass curve Hopf algebroid studied by Bauer. Hence the spectral sequence of [**23**] is in hindsight identical to this Adams–Novikov spectral sequence.

## 0.3. The Adams $E_2$-term for $S$

Let $p$ be any prime, and let $X/p^n = X \wedge Cp^n$ where $Cp^n = S \cup_{p^n} e^1$. We say that a spectrum $X$ has finite type mod $p$ if $\pi_*(X/p)$ is finite in each degree. For bounded below spectra $X$ this is equivalent to asking that $H_*(X) = H_*(X; \mathbb{F}_p)$ is finite in each degree, which in turn is equivalent to the condition that $H^*(X) = H^*(X; \mathbb{F}_p)$ is finite in each degree. If $X$ is bounded below and of finite type mod $p$, then the mod $p$ Adams spectral sequence for $X$ has $E_2$-term

$$E_2^{s,t}(X) = \mathrm{Ext}^{s,t}_A(H^*(X), \mathbb{F}_p)$$

and converges strongly to the homotopy groups

$$E_2^{s,t}(X) \Longrightarrow_s \pi_{t-s}(X_p^\wedge)$$

of the $p$-completion $X_p^\wedge = \mathrm{holim}_n X/p^n$ of $X$, cf. [**2**, Thm. 2.1]. (This reference assumes that $X$ is of finite type, not just mod $p$, but one can prove the same conclusion with the weaker hypotheses stated.) If $\pi_*(X)$ is finitely generated in each degree, then we say that $X$ is of finite type, and there are isomorphisms

$$\pi_*(X) \otimes \mathbb{Z}_p \cong \pi_*(X)_p^\wedge \cong \pi_*(X_p^\wedge)\,.$$

The same conclusion holds if $X$ is $p$-local and $\pi_*(X)$ is finitely generated over $\mathbb{Z}_{(p)}$ in each degree. The Adams $E_2$-term can also be expressed in terms of comodule Ext as

$$E_2^{s,t}(X) = \mathrm{Ext}^{s,t}_{A_*}(\mathbb{F}_p, H_*(X))\,.$$

If $X$ is a ring spectrum (up to homotopy), then $H_*(X)$ is an $A_*$-comodule algebra, $E_2(X) = \mathrm{Ext}_{A_*}(\mathbb{F}_p, H_*(X))$ is a bigraded $\mathbb{F}_p$-algebra, $\pi_*(X)$ is a graded ring, and the Adams spectral sequence for $X$ is an algebra spectral sequence. If $X$ is homotopy commutative, then $H_*(X)$, $E_2(X)$ and $\pi_*(X)$ are graded commutative. If $X$ is an $E_\infty$ ring spectrum, or more generally an $H_\infty$ ring spectrum [**45**, §I.3], then there are algebraic Steenrod operations acting on $E_2(X)$ and power operations acting on $\pi_*(X)$, and their compatibility forces certain relations to hold between the differentials in the Adams spectral sequence and the algebraic Steenrod operations [**45**, Ch. VI]. We shall make extensive use of these relations in this work, since they suffice to determine many of the more subtle Adams differentials.

The foremost example among the spectra relevant to stable homotopy theory is the sphere spectrum $S$, with $H^*(S) = \mathbb{F}_p$ and Adams spectral sequence

$$E_2^{s,t}(S) = \mathrm{Ext}_A^{s,t}(\mathbb{F}_p, \mathbb{F}_p) \Longrightarrow_s \pi_{t-s}(S)_p^\wedge .$$

The homotopy groups $\pi_*(S)$ are known as the stable homotopy groups of spheres, or as the "stable stems". The sphere spectrum is the initial commutative $S$-algebra, or $E_\infty$ ring spectrum, hence is also an $H_\infty$ ring spectrum. The bigraded cohomology algebra $E_2(S) = \mathrm{Ext}_A(\mathbb{F}_p, \mathbb{F}_p)$ of $A$ is only partially understood, and no viable explicit statement about its full structure is known to the authors, conjectural or not.

However, some features are understood. Let us concentrate on the case $p = 2$. In the $(t-s, s)$-plane, the Adams $E_2$-term has an $h_0$-tower along the vertical axis, and is otherwise concentrated within a triangular region with $s \geq 0$ and $t - s \geq 2s - 3$. A bird's-eye view for $t - s \leq 200$ is given in Figure 0.5. A machine computation for $t \leq 200$ was made using the first author's program package `ext`, which is available online and described in [**41**]. In this range of degrees we can also calculate the algebra structure on $\mathrm{Ext}_A(\mathbb{F}_2, \mathbb{F}_2)$, with the product given by Yoneda composition. The gray region with $t \geq 201$ does not indicate trivial groups, but rather the current limit of our detailed calculations. By Theorem 4.9, the Adams periodicity operator $\pi_5 \colon E_2^{s,t}(S) \to E_2^{s+32,t+96}(S)$ maps known calculations isomorphically onto the lighter gray region, while (by our approach) further machine computations would be needed to identify the groups in the darker gray region. More legible charts are shown in Figures 1.1 to 1.8. The algebra generators in topological degrees $t - s \leq 48$ are listed in Table 1.1 and labeled in Figures 1.9 and 1.10. An even larger chart, showing the region $t - s \leq 210$, can be found on the web page of Christian Nassau [**136**].

## 0.4. The Adams differentials for $S$

Let us review some of the results on the differentials and extensions in the mod 2 Adams spectral sequence for the sphere spectrum.

Starting from the horizontal $(t-s)$-axis and moving up, the first groups in the $E_2$-term are $E_2^{0,*}(S) = \mathbb{F}_2\{1\}$ and

$$E_2^{1,*}(S) = \mathbb{F}_2\{h_i \mid i \geq 0\} ,$$

with $h_i$ in topological degree $t-s = 2^i - 1$ corresponding to the primitive element $\xi_1^{2^i}$ in $A_*$, dual to the indecomposable class $Sq^{2^i}$ in $A$. These are tied together by the algebraic Steenrod operations: $Sq^0(h_i) = h_{i+1}$ for each $i \geq 0$. The classes $h_0$, $h_1$, $h_2$ and $h_3$ survive to $E_\infty(S)$ and detect the Hopf invariant one homotopy classes $2 \in \pi_0(S)$, $\eta \in \pi_1(S)$, $\nu \in \pi_3(S)$ and $\sigma \in \pi_7(S)$, respectively. Frank Adams [**3**] proved that the remaining classes $h_i$ do not detect homotopy classes: there are nonzero differentials $d_2(h_i) = h_0 h_{i-1}^2$ for all $i \geq 4$. As a consequence, the only spheres that are $H$-spaces are the unit spheres $S^0$, $S^1$, $S^3$ and $S^7$ in the classical real division algebras $\mathbb{R}$, $\mathbb{C}$, $\mathbb{H}$ and $\mathbb{O}$.

The Adams 2-line

$$E_2^{2,*}(S) = \mathbb{F}_2\{h_i h_j \mid i \leq j \neq i+1\}$$

is multiplicatively generated by the $h_i$, subject only to the relations $h_i h_{i+1} = 0$. Mahowald [**101**] showed that the classes $h_1 h_j$ survive to $E_\infty(S)$ and detect homotopy classes denoted $\eta_j \in \pi_{2^j}(S)$ (for $j \ge 3$). Mahowald and Martin Tangora [**107**, Thm. 8.1.1] proved that the classes $h_j^2$ for $j \le 4$ survive to $E_\infty(S)$. The corresponding result for $j = 5$ was obtained by Michael Barratt, John Jones and Mahowald [**21**, Thm. 2.1]. It then follows from the work of William Browder [**35**, Thm. 7.1] that these classes detect Kervaire invariant one homotopy classes $\theta_j \in \pi_{2^{j+1}-2}(S)$. More recently, Hill, Hopkins and Douglas Ravenel [**69**, Thm. 1.1] showed that none of the classes $h_j^2$ for $j \ge 7$ survive to detect homotopy classes. As a consequence, every closed framed $n$-manifold is framed cobordant to a homotopy sphere, unless $n = 2^{j+1} - 2$ with $j \le 6$. The case $j = 6$ remains open: it is not known whether there exists a class $\theta_6 \in \pi_{126}(S)$ detected by $h_6^2$. The only other products $h_i h_j$ that survive to $E_\infty(S)$ are $h_0h_2$, $h_0h_3$ and $h_2h_4$ detecting $2\nu$, $2\sigma$ and $\nu^*$, respectively, cf. the references to [**144**, Thm. 3.4.3].

John Wang [**176**, Thm. 2.11] showed that the Adams 3-line is spanned by classes $c_i$ in topological degree $t - s = 2^i \cdot 11 - 3$ for $i \ge 0$, together with the products $h_ih_jh_k$ for $i \le j \le k$. The latter are subject only to the relations $h_ih_{i+1} = 0$, $h_ih_{i+2}^2 = 0$ and $h_i^2h_{i+2} = h_{i+1}^3$ found by Adams [**3**, Thm. 2.5.1]. The indecomposable classes $c_i$ are connected by algebraic Steenrod operations: $Sq^0(c_i) = c_{i+1}$ for each $i \ge 0$. The classes $c_0$ and $c_1$ survive to $E_\infty(S)$ and detect homotopy classes denoted $\epsilon \in \pi_8(S)$ and $\bar\sigma \in \pi_{19}(S)$, respectively, whereas the remaining classes $c_i$ support differentials $d_2(c_i) = h_0f_{i-1}$ for $i \ge 2$, see [**45**, Prop. VI.1.16(i)], and Wen-Hsiung Lin [**93**, Thm. 1.4] proved that $h_0f_{i-1} \ne 0$. (These classes are unrelated to the modular forms $c_4$ and $c_6$.)

The paper [**93**] also describes the Adams 4-line $E_2^{4,*}(S)$, and the decomposable classes in $E_2^{5,*}(S)$. There are seven families of indecomposable classes on the 4-line, obtained by applying $Sq^0$ repeatedly to $d_0$, $e_0$, $f_0$, $g = g_1$, $p$, $D_3$ and $p'$ in topological degrees $t - s = 14$, 17, 18, 20, 33, 61 and 69, respectively. In particular, $d_0$, $g$, $p$ and $p'$ detect classes $\kappa \in \pi_{14}(S)$, $\bar\kappa \in \pi_{20}(S)$, $\nu\theta_4 \in \pi_{33}(S)$ and $\sigma\theta_5 \in \pi_{69}(S)$. The latter two claims are due to Barratt, Mahowald and Tangora [**22**, Prop. 3.3.7] and Daniel Isaksen, Guozhen Wang and Zhouli Xu [**83**, Table 21], respectively. On the other hand, $d_2(e_0) = h_1^2d_0$ and $d_2(f_0) = h_0^2e_0$, and Wang and Xu [**174**] recently showed that $d_3(D_3) = B_3$ is nonzero.

Starting instead from the vertical $s$-axis and moving to the right, the differential structure in the Adams spectral sequence was determined for $t-s \le 28$ by Richard Maunder [**116**] and Peter May [**117**], building on earlier calculations of unstable homotopy groups of spheres by Hirosi Toda [**171**] and Mamoru Mimura [**130**]. The stable calculations were extended to the range $t - s \le 45$ by Mahowald and Tangora [**107**]. In particular they used Mimura's result that $\epsilon\kappa \ne 0$ [**129**, Thm. B] to correct a mistake in the group structure of $\pi_{23}(S)$ related to a cluster of hidden 2-, $\eta$- and $\nu$-extensions landing in degrees 23, 22 and 23, respectively. Later papers by Barratt, Mahowald and Tangora [**22**] and the first author [**40**] corrected two other mistakes in the new range, finding nonzero differentials $d_3(h_2h_5) = h_0p$ and $d_3(e_1) = h_1t$, respectively. Thereafter, Barratt, Jones and Mahowald [**21**] gave complete information on the Adams spectral sequence differentials for $t - s \le 48$. (However, the argument given for $d_6(B_2) = 0$ appears to depend in a circular manner on a hidden $\eta$-extension in degree 47 found by Tangora [**166**, p. 582], as the latter reference applies Michael Moss' convergence theorem [**132**, Thm. 1.2] in a case that presumes the vanishing of $d_6(B_2)$. See Remark 11.60.)

Turning to higher degrees, Stanley Kochman [**87**] made a computer-assisted calculation of an Atiyah–Hirzebruch spectral sequence to calculate $\pi_*(S)$ for $* \le 64$. These results were transcribed as differentials in the Adams spectral sequence by Kochman and Mahowald [**88**], leading to several corrections in the range $54 \le * \le 64$. More recently, Isaksen [**82**] used a comparison of the classical mod 2 Adams spectral sequence with its motivic analogue, formed in Voevodsky's stable homotopy category of motives over $\mathrm{Spec}(\mathbb{C})$, and discovered a missing differential $d_3(Q_2) = gt$ affecting $\pi_{56}(S)$ and $\pi_{57}(S)$. With the aid of the motivic Adams spectral sequence, Isaksen obtained complete calculations in degrees $* \le 59$, with one differential ($d_2(D_1) = h_0^2 h_3 g_2$) being obtained jointly with Xu [**84**], and one additive extension being obtained by Wang and Xu [**175**]. Thereafter, Wang and Xu [**174**] calculated $\pi_{60}(S) \cong \mathbb{Z}/4\{\bar\kappa^3\}$ and $\pi_{61}(S) = 0$. As a consequence, the only odd-dimensional spheres with a unique smooth structure are now known to be $S^1$, $S^3$, $S^5$ and $S^{61}$.

In current work, Isaksen, Wang and Xu [**83**] combine comparisons of classical Adams and Adams–Novikov spectral sequences with motivic Adams and Adams–Novikov spectral sequences to obtain a nearly complete account of the first 90 stable stems. A key new input is the identification by Bogdan Gheorghe, Wang and Xu [**61**, Thm. 1.14] of the motivic Adams spectral sequence for the motivic spectrum $C\tau$ with the machine computable algebraic Novikov spectral sequence for the sphere spectrum.

To determine the full differential structure in the mod 2 Adams spectral sequence for *tmf* we shall use only a small part of the known information about the spectral sequence for $S$, all within the Toda–Mimura range. More precisely, we will use the fact that there is a hidden $\eta$-extension from $h_0^3 h_4$ detecting $\rho \in \pi_{15}(S)$ to $Pc_0$ detecting $\eta\rho$, and that there is a hidden $\eta$-extension from $h_1 g$ detecting $\eta\bar\kappa \in \pi_{21}(S)$ to $Pd_0$ detecting $\eta^2\bar\kappa$. The first of these is an easy consequence of the proven Adams conjecture, whereas the second is more subtle, and coincides with the mistake that was corrected in [**107**, Thm. 2.1.1]. We provide stable, i.e., spectrum level, proofs of these results in Chapter 11, benefiting from our easy access using `ext` to the action of $\mathrm{Ext}_A(\mathbb{F}_2, \mathbb{F}_2)$ on $\mathrm{Ext}_A(H^*(X), \mathbb{F}_2)$ for several small CW spectra $X$, in a moderate range of degrees. We shall also use the fact that $\eta^2\kappa = 0$ in $\pi_{16}(S)$. Once we have determined the differential structure on the spectral sequence for *tmf*, it becomes significantly easier to determine many of the remaining differentials in the mod 2 Adams spectral sequence for $S$. We take the opportunity to work some of this out in Chapter 11, obtaining the full differential structure of the latter spectral sequence in degrees $t - s \le 48$, and the full additive and multiplicative structure of its abutment $\pi_*(S)$ in degrees $* \le 44$. See Figures 11.10 to 11.14 and Remark 0.1.

## 0.5. The Adams $E_2$-term for *tmf*

The central object of study in this book is the classical mod 2 Adams spectral sequence for the $E_\infty$ ring spectrum *tmf* of topological modular forms. In Part I of our work, consisting of Chapters 1 to 4, we study the $E_2$-term of this Adams spectral sequence. We also determine the $E_2$-terms of the Adams spectral sequences for the *tmf*-modules $tmf/2 = tmf \wedge C2$, $tmf/\eta = tmf \wedge C\eta$ and $tmf/\nu = tmf \wedge C\nu$.

Our starting point will be that *tmf* is a connective $E_\infty$ ring spectrum of finite type with mod 2 cohomology $H^*(tmf) = A/\!/A(2)$. Its mod 2 Adams spectral sequence

$$E_2^{s,t}(tmf) = \mathrm{Ext}_A^{s,t}(H^*(tmf), \mathbb{F}_2) \Longrightarrow \pi_{t-s}(tmf)_2^\wedge$$

is an algebra spectral sequence, converging strongly to the graded commutative ring $\pi_*(tmf)_2^\wedge \cong \pi_*(tmf) \otimes \mathbb{Z}_2$. Since $A$ contains $A(2)$ as a sub Hopf algebra, $A$ is free over $A(2)$, so there is a change-of-algebra isomorphism

$$E_2(tmf) = \mathrm{Ext}_A(A//A(2), \mathbb{F}_2) \cong \mathrm{Ext}_{A(2)}(\mathbb{F}_2, \mathbb{F}_2)$$

taking the (graded) commutative algebra structure induced from the homotopy commutative ring structure on *tmf* to the (graded) commutative algebra structure on $\mathrm{Ext}_{A(2)}(\mathbb{F}_2, \mathbb{F}_2)$ induced from the cocommutative coproduct on $A(2)$, which in turn agrees with the product defined by Yoneda composition [**179**, Prop. 5.8].

The cohomology algebra $\mathrm{Ext}_{A(2)}(\mathbb{F}_2, \mathbb{F}_2)$ was obtained by May (unpublished), and by Nobuo Shimada and Akira Iwai [**155**, §8] using an injective resolution constructed as a twisted tensor product. They conclude, in the notation of [**54**, Ch. 13], that $\mathrm{Ext}_{A(2)}(\mathbb{F}_2, \mathbb{F}_2)$ is generated as an algebra by 13 indecomposable classes

| $x$ | $h_0$ | $h_1$ | $h_2$ | $c_0$ | $w_1$ | $\alpha$ | $d_0$ | $\beta$ | $e_0$ | $g$ | $\gamma$ | $\delta$ | $w_2$ |
|---|---|---|---|---|---|---|---|---|---|---|---|---|---|
| $t-s$ | 0 | 1 | 3 | 8 | 8 | 12 | 14 | 15 | 17 | 20 | 25 | 32 | 48 |
| $s$ | 1 | 1 | 1 | 3 | 4 | 3 | 4 | 3 | 4 | 4 | 5 | 7 | 8 |
| $d_2(x)$ | 0 | 0 | 0 | 0 | 0 | $h_2w_1$ | 0 | $h_0d_0$ | 0 | 0 | 0 | 0 | $\alpha\beta g$ |

subject to 54 relations

$$h_0h_1 = 0\,,\ h_0^2h_2 = h_1^3\,,\ h_1h_2 = 0\,,\ h_0h_2^2 = 0\,,\ \dots$$
$$\dots\,,\ \gamma^2 = h_1^2w_2 + \beta^2 g\,,\ \delta g = 0\,,\ \gamma\delta = h_1c_0w_2\,,\ \delta^2 = 0$$

that induce all other relations in $\mathrm{Ext}_{A(2)}(\mathbb{F}_2, \mathbb{F}_2)$. Furthermore, they note that this algebra is free as a module over the subalgebra $\mathbb{F}_2[w_1, w_2]$. A large-scale image of $E_2^{s,t}(tmf) = \mathrm{Ext}_{A(2)}^{s,t}(\mathbb{F}_2, \mathbb{F}_2)$ for $t - s \le 200$ is shown in Figure 0.6, repeated at a smaller scale in Figures 1.11 to 1.18. The 13 algebra generators are labeled in Figures 1.19 and 1.20. Note that the decomposable class $\alpha g$ lies in the same bidegree as $\delta$. On many occasions it will be convenient to work with the sum of these two classes, which we denote

$$\delta' = \delta + \alpha g\,.$$

The charts were obtained using `ext` to construct a minimal free resolution of $\mathbb{F}_2$ as an $A(2)$-module, in the finite range shown. See Table 3.3 for a dictionary relating the notational schemes used by [**155**], [**54**] and `ext` to identify the 13 algebra generators, and see Table 3.4 for the full list of 54 generating relations. It is straightforward for `ext` to verify that these relations hold in $\mathrm{Ext}_{A(2)}(\mathbb{F}_2, \mathbb{F}_2)$. The list of 54 relations is a minimal generating set for the ideal $I$ of relations satisfied by the 13 algebra generators, but it may be difficult to use this list to identify when two polynomial expressions are equal in $\mathrm{Ext}_{A(2)}(\mathbb{F}_2, \mathbb{F}_2)$. We therefore order the generators as follows

$$h_0 > h_1 > h_2 > c_0 > \alpha > \beta > d_0 > e_0 > \gamma > \delta > g > w_1 > w_2$$

and give a reduced Gröbner basis

$$h_0h_1\ ,\ h_1^3+h_0^2h_2\ ,\ h_0^3h_2\ ,\ h_1h_2\ ,\ \dots$$
$$\dots\ ,\ \alpha^3e_0+\gamma g w_1\ ,\ d_0e_0\gamma+\alpha^3 g\ ,\ \gamma\delta+h_1c_0w_2\ ,\ \delta^2$$

for the ideal $I$ in Table 3.5. Using this 77-term Gröbner basis there is a straightforward algorithm for bringing any polynomial in $P=\mathbb{F}_2[h_0,h_1,h_2,\dots,g,w_1,w_2]$ to an irreducible normal form, so that two polynomials have the same image in $P/I$ if and only if they have the same normal form. In the proof of Theorem 5.15 we give some worked examples of this reduction process.

Due to the scarcity of detail in the published references, we provide an independent proof of the theorem of Shimada–Iwai that the homomorphism $\phi\colon P/I\to \mathrm{Ext}_{A(2)}(\mathbb{F}_2,\mathbb{F}_2)$, sending the 13 algebra generators to the given Ext-classes, is an isomorphism. We do this by means of a spectral sequence due to Donald Davis and Mahowald [**52**], which is designed to calculate $\mathrm{Ext}_{A(n)}(M,\mathbb{F}_2)$ in terms of $\mathrm{Ext}_{A(n-1)}(N_\sigma\otimes M,\mathbb{F}_2)$, where the $N_\sigma$ for $\sigma\ge 0$ are a specific sequence of $A(n)$-modules. Davis and Mahowald applied this spectral sequence for $n=2$ to additively calculate $\mathrm{Ext}_{A(2)}(M,\mathbb{F}_2)$ for a number of $A(2)$-modules $M$. In Chapter 2 we rework their construction in comodule algebraic terms, so as to clarify the multiplicative aspects of their spectral sequence. We then apply this in Chapter 3 to calculate the Davis–Mahowald $E_\infty$-term, which is the associated graded of an exhaustive filtration of $\mathrm{Ext}_{A(2)}(\mathbb{F}_2,\mathbb{F}_2)$. By comparing this with the normal form generators of $P/I$, and a counting argument, we can conclude in Theorem 3.46 that $\phi$ is indeed an isomorphism. Along the way we verify, in Proposition 3.42, that the algebra given by the Shimada–Iwai presentation is free as a module over $\mathbb{F}_2[w_1,w_2]$. More precisely, we obtain a presentation for $\mathrm{Ext}_{A(2)}(\mathbb{F}_2,\mathbb{F}_2)$ as a direct sum of cyclic $R_0$-modules, where we use the notation $R_0=\mathbb{F}_2[g,w_1,w_2]$. While $R_0$ is not quite a system of parameters for $\mathrm{Ext}_{A(2)}(\mathbb{F}_2,\mathbb{F}_2)$, it serves a similar purpose, cf. Remark 3.44. The subalgebras $R_1$ and $R_2$ defined in Section 0.6 are then similarly relevant to the later stages of the Adams spectral sequence for *tmf*, as explained in that section.

Since $A$ and $A(2)$ are cocommutative Hopf algebras, there are compatible Steenrod operations

$$Sq^i\colon \mathrm{Ext}_A^{s,t}(\mathbb{F}_2,\mathbb{F}_2)\longrightarrow \mathrm{Ext}_A^{s+i,2t}(\mathbb{F}_2,\mathbb{F}_2)$$
$$Sq^i\colon \mathrm{Ext}_{A(2)}^{s,t}(\mathbb{F}_2,\mathbb{F}_2)\longrightarrow \mathrm{Ext}_{A(2)}^{s+i,2t}(\mathbb{F}_2,\mathbb{F}_2)$$

acting on their cohomology algebras. It is well known that $Sq^0(h_i)=h_{i+1}$ and $Sq^1(h_i)=h_i^2$, for all $i\ge 0$. We calculate all of these operations for $A(2)$ in Theorem 1.20, using explicit chain homotopies to handle the cases $Sq^1(c_0)$ and $Sq^2(c_0)$. Many of the operations for $A$ are calculated in [**133**], [**122**, §6] and [**45**, §VI.1], and we review and extend these results in Section 11.2.

In Chapter 4 we also determine the Adams $E_2$-terms for the *tmf*-module spectra $tmf/2$, $tmf/\eta$ and $tmf/\nu$, as modules over $E_2(tmf)$. For $i=2^j\in\{1,2,4\}$ we let $M_i=\mathbb{F}_2\{1,Sq^i\}$ denote a minimal $A(2)$-module with nontrivial action by $Sq^i$ on a class in degree 0. With this notation, the Adams spectral sequences for these *tmf*-modules take the forms

$$E_2^{s,t}(tmf/2)=\mathrm{Ext}_{A(2)}^{s,t}(M_1,\mathbb{F}_2)\Longrightarrow \pi_{t-s}(tmf/2)$$
$$E_2^{s,t}(tmf/\eta)=\mathrm{Ext}_{A(2)}^{s,t}(M_2,\mathbb{F}_2)\Longrightarrow \pi_{t-s}(tmf/\eta)^\wedge_2$$

$$E_2^{s,t}(tmf/\nu) = \mathrm{Ext}_{A(2)}^{s,t}(M_4, \mathbb{F}_2) \Longrightarrow \pi_{t-s}(tmf/\nu)_2^\wedge .$$

In each case the short exact sequence of $A(2)$-modules

$$0 \to \Sigma^i \mathbb{F}_2 \longrightarrow M_i \longrightarrow \mathbb{F}_2 \to 0$$

induces a long exact sequence

$$\dots \xrightarrow{h_j} \mathrm{Ext}_{A(2)}(\mathbb{F}_2, \mathbb{F}_2) \longrightarrow \mathrm{Ext}_{A(2)}(M_i, \mathbb{F}_2) \longrightarrow \mathrm{Ext}_{A(2)}(\Sigma^i \mathbb{F}_2, \mathbb{F}_2) \xrightarrow{h_j} \dots ,$$

and we use our $R_0$-module presentation of $\mathrm{Ext}_{A(2)}(\mathbb{F}_2, \mathbb{F}_2)$ to obtain $R_0$-module presentations of $\mathrm{Ext}_{A(2)}(M_i, \mathbb{F}_2)$ for $i \in \{1, 2, 4\}$ in Propositions 4.2, 4.11 and 4.15. In each case we calculate the kernel and cokernel of multiplication by $h_j$, and then identify the resulting extension of $R_0$-modules. The latter is determined by calculations in a finite range of degrees, which we perform using `ext`. For $i = 2^j \in \{1, 2, 4\}$ we write $\widetilde{x}$, $\widehat{x}$ and $\overline{x}$, respectively, for chosen lifts in $\mathrm{Ext}_{A(2)}(M_i, \mathbb{F}_2)$ of classes $x \in \ker(h_j) \subset \mathrm{Ext}_{A(2)}(\Sigma^i \mathbb{F}_2, \mathbb{F}_2)$. When multiple choices are possible, we specify our lifts in terms of the cocycles chosen by `ext`, as in Tables 4.2, 4.5 and 4.7. In particular, we find explicit generators for $E_2(tmf/2)$, $E_2(tmf/\eta)$ and $E_2(tmf/\nu)$ as modules over $E_2(tmf)$ in Corollaries 4.3, 4.13 and 4.16. Large-scale charts of these $E_2$-terms are shown in Figures 0.11, 0.14 and 0.17.

As an application of our calculation of $\mathrm{Ext}_{A(2)}(M_1, \mathbb{F}_2)$, we give a proof in Section 4.2 of May's improved version of the Adams periodicity theorem from [**7**]. Adams' original proof established periodicity above a line of slope 1/3 in the $(t-s, s)$-plane, while May's strengthened result gives periodicity above a line of slope 1/5.

## 0.6. The Adams differentials for *tmf*

In Part II of this book, consisting of Chapters 5 to 8 and Appendices A to D, we study the $d_r$-differentials for $r \geq 2$ in the mod 2 Adams spectral sequence for *tmf*, and for the closely related spectra $tmf/2$, $tmf/\eta$ and $tmf/\nu$.

The ring structure on *tmf* and its actions on $tmf/2$, $tmf/\eta$ and $tmf/\nu$ induce algebra and module structures in the respective Adams spectral sequences, leading to the Leibniz rule

$$d_r(xy) = d_r(x)y + xd_r(y)$$

in all cases. (There is no sign since we are working at $p = 2$.) Supplementing the usual multiplicative structure, our principal tool is the formula

$$d_*(Sq^i(x)) = Sq^{i+r-1}(d_r(x)) \dot{+} \begin{cases} 0 & \text{if } v > s-i+1, \\ \bar{a}\, x\, d_r(x) & \text{if } v = s-i+1, \\ \bar{a}\, Sq^{i+v}(x) & \text{if } v \leq \min\{s-i, 10\} \end{cases} \tag{0.1}$$

in the Adams spectral sequence for an $H_\infty$ ring spectrum $Y$, such as *tmf* or $S$. This result is due to Jukka Mäkinen [**109**] in the case $Y = S$, and to the first author [**45**, Thm. VI.1.1 and VI.1.2] for general $H_\infty$ ring spectra. Here $x \in E_2^{s,t}(Y) = \mathrm{Ext}_A^{s,t}(H^*(Y), \mathbb{F}_2)$ is an element that survives to the $E_r$-term, for some $r \geq 2$. Writing the 2-adic valuation of $t - i + 1$ as $4q + r$, with $0 \leq r \leq 3$, the "vector field number" is $v = 8q + 2^r$. If $v = 1$ then $\bar{a} = h_0$, while if $v \geq 2$ then $\bar{a} \in E_\infty(S)$ detects a generator of the image of the $J$-homomorphism in $\pi_{v-1}(S)$. The two summands in (0.1) are the leading contributions to an Adams differential on $Sq^i(x)$, and the symbol $\dot{+}$ indicates that if the terms have different Adams filtration, then only the term in lower Adams filtration appears in the differential. See Section 5.2 and Theorem 5.6 for further explanations in the context of *tmf*, and Section 11.1 and Theorem 11.22 for a full discussion in the context of $H_\infty$ ring spectra.

To determine the Adams $d_2$-differential and $E_3$-term for *tmf*, it suffices to determine $d_2(x)$ for each of the 13 algebra generators $x = h_0, h_1, h_2, \dots, g, w_1, w_2$ of $E_2(tmf)$. Due to the multiplicative structure, $d_2(x) = 0$ except for $x \in \{\alpha, \beta, w_2\}$. An application of equation (0.1) shows that $d_*(Sq^1(c_0)) = h_0 Sq^2(c_0)$, which evaluates to $d_2(h_2\beta) = h_0^2 e_0$. This readily implies that $d_2(\alpha) = h_2 w_1$ and $d_2(\beta) = h_0 d_0$.

It remains to determine $d_2(w_2)$. For this we make use of naturality with respect to the unit map $\iota\colon S \to tmf$, and a small piece of the known structure of $\pi_*(S)$, as discussed in Section 0.4. In Theorem 5.10 we use the hidden $\eta$-extension on $\rho \in \pi_{15}(S)$ and the fact that $\eta^2\kappa = 0$ to deduce that $d_3(e_0) = c_0 w_1$. Furthermore, in Theorem 5.12 we use the hidden $\eta$-extension on $\eta\bar\kappa \in \pi_{21}(S)$ to deduce that $d_4(e_0 g) = g w_1^2$. The multiplicative structure, including the relation $\gamma^2 = \beta^2 g + h_1^2 w_2$, then implies that $d_4(h_1^2 w_2) = \alpha^2 e_0 w_1$ is nonzero, which in turn implies that $d_3(h_1 w_2) = g^2 w_1$ and $d_2(w_2) = \alpha\beta g$ are nonzero. See Proposition 5.14 and Figures 1.19, 1.20 and 1.13.

We recall our presentation for $E_2(tmf)$ as a direct sum of cyclic $R_0$-modules, where $R_0 = \mathbb{F}_2[g, w_1, w_2]$, in Table 5.1. Since $g$, $w_1$ and $w_2^2$ are $d_2$-cycles, the $d_2$-differential is $R_1$-linear, where we let $R_1 = \mathbb{F}_2[g, w_1, w_2^2]$. Using the Leibniz rule, we can calculate $d_2$ on each $R_1$-module generator. It is then an algebraic exercise to calculate $E_3(tmf) = H(E_2(tmf), d_2)$ as an $R_1$-module, and we carry this out in Appendix A.1. The result is presented as a direct sum of mostly cyclic $R_1$-modules in Table 5.2, with the non-cyclic summands being made explicit in Table 5.3.

Next, we show that $E_3(tmf)$ is generated as an algebra by the 24 classes below.

| $x$ | $h_0$ | $h_1$ | $h_2$ | $c_0$ | $w_1$ | $h_0^3\alpha$ | $d_0$ | $e_0$ |
|---|---|---|---|---|---|---|---|---|
| $t-s$ | 0 | 1 | 3 | 8 | 8 | 12 | 14 | 17 |
| $s$ | 1 | 1 | 1 | 3 | 4 | 6 | 4 | 4 |
| $d_3(x)$ | 0 | 0 | 0 | 0 | 0 | 0 | 0 | $c_0 w_1$ |

| $x$ | $g$ | $\alpha^2$ | $\gamma$ | $\alpha\beta$ | $\beta^2$ | $\delta$ | $\alpha g$ | $h_0\alpha^3$ |
|---|---|---|---|---|---|---|---|---|
| $t-s$ | 20 | 24 | 25 | 27 | 30 | 32 | 32 | 36 |
| $s$ | 4 | 6 | 5 | 6 | 6 | 7 | 7 | 10 |
| $d_3(x)$ | 0 | $h_1 d_0 w_1$ | 0 | 0 | $h_1 g w_1$ | 0 | 0 | 0 |

| $x$ | $h_0 w_2$ | $h_1 w_2$ | $h_2 w_2$ | $c_0 w_2$ | $h_0^3 \alpha w_2$ | $\delta w_2$ | $h_0 \alpha^3 w_2$ | $w_2^2$ |
|---|---|---|---|---|---|---|---|---|
| $t-s$ | 48 | 49 | 51 | 56 | 60 | 80 | 84 | 96 |
| $s$ | 9 | 9 | 9 | 11 | 14 | 15 | 18 | 16 |
| $d_3(x)$ | 0 | $g^2 w_1$ | 0 | 0 | 0 | 0 | 0 | $\beta g^4$ |

Equation (0.1) shows that $d_3(\alpha^2) = h_0 \alpha d_2(\alpha) = h_1 d_0 w_1$, $d_3(\beta^2) = Sq^4(d_2(\beta)) = h_1 g w_1$ and $d_3(w_2^2) = Sq^9(d_2(w_2)) + h_0 w_2 d_2(w_2) = \beta g^4$. When combined with our earlier results and the multiplicative structure, this determines $d_3$ on the remaining algebra generators, see Theorem 5.18. Since $g$, $w_1$ and $w_2^4$ are $d_3$-cycles, the $d_3$-differential is $R_2$-linear, where $R_2 = \mathbb{F}_2[g, w_1, w_2^4]$. We calculate $d_3$ on each $R_2$-module generator using the Leibniz rule, and calculate $E_4(tmf) = H(E_3(tmf), d_3)$ in Appendix A.2. The resulting direct sum of mostly cyclic $R_2$-modules is presented in Tables 5.5 and 5.6.

Continuing, we check that $E_4(tmf)$ is generated as an algebra by the 52 classes below.

| $x$ | $h_0$ | $h_1$ | $h_2$ | $c_0$ | $w_1$ | $h_0^3\alpha$ | $d_0$ | $g$ | $h_0\alpha^2$ | $\gamma$ | $\alpha\beta$ | $d_0e_0$ | $\delta$ | $\alpha g$ |
|---|---|---|---|---|---|---|---|---|---|---|---|---|---|---|
| $t-s$ | 0 | 1 | 3 | 8 | 8 | 12 | 14 | 20 | 24 | 25 | 27 | 31 | 32 | 32 |
| $s$ | 1 | 1 | 1 | 3 | 4 | 6 | 4 | 4 | 7 | 5 | 6 | 8 | 7 | 7 |
| $d_4(x)$ | 0 | 0 | 0 | 0 | 0 | 0 | 0 | 0 | 0 | 0 | 0 | $d_0w_1^2$ | 0 | 0 |

| $x$ | $h_0\alpha^3$ | $e_0g$ | $\alpha^2g$ | $h_0w_2$ | $\alpha e_0g$ | $h_1^2w_2$ | $h_2w_2$ | $\beta g^2$ | $c_0w_2$ |
|---|---|---|---|---|---|---|---|---|---|
| $t-s$ | 36 | 37 | 44 | 48 | 49 | 50 | 51 | 55 | 56 |
| $s$ | 10 | 8 | 10 | 9 | 11 | 10 | 9 | 11 | 11 |
| $d_4(x)$ | 0 | $gw_1^2$ | $\alpha\beta w_1^2$ | $d_0\gamma w_1$ | $\delta' w_1^2$ | $\alpha^2e_0w_1$ | 0 | $\alpha d_0gw_1$ | 0 |

| $x$ | $h_0^3\alpha w_2$ | $h_0\alpha^2w_2$ | $\delta w_2$ | $h_0\alpha^3w_2$ | $h_0w_2^2$ | $h_1w_2^2$ | $h_2w_2^2$ | $c_0w_2^2$ | $w_1w_2^2$ |
|---|---|---|---|---|---|---|---|---|---|
| $t-s$ | 60 | 72 | 80 | 84 | 96 | 97 | 99 | 104 | 104 |
| $s$ | 14 | 15 | 15 | 18 | 17 | 17 | 17 | 19 | 20 |
| $d_4(x)$ | 0 | 0 | 0 | 0 | 0 | 0 | 0 | 0 | 0 |

| $x$ | $h_0^3\alpha w_2^2$ | $d_0w_2^2$ | $h_0\alpha^2w_2^2$ | $\alpha\beta w_2^2$ | $d_0e_0w_2^2$ | $\delta w_2^2$ | $\alpha gw_2^2$ | $h_0\alpha^3w_2^2$ |
|---|---|---|---|---|---|---|---|---|
| $t-s$ | 108 | 110 | 120 | 123 | 127 | 128 | 128 | 132 |
| $s$ | 22 | 20 | 23 | 22 | 24 | 23 | 23 | 26 |
| $d_4(x)$ | 0 | 0 | 0 | 0 | $d_0w_1^2w_2^2$ | 0 | 0 | 0 |

| $x$ | $e_0gw_2^2$ | $\alpha^2gw_2^2$ | $h_0w_2^3$ | $\alpha e_0gw_2^2$ | $h_1^2w_2^3$ | $h_2w_2^3$ | $c_0w_2^3$ |
|---|---|---|---|---|---|---|---|
| $t-s$ | 133 | 140 | 144 | 145 | 146 | 147 | 152 |
| $s$ | 24 | 26 | 25 | 27 | 26 | 25 | 27 |
| $d_4(x)$ | $gw_1^2w_2^2$ | $\alpha\beta w_1^2w_2^2$ | $d_0\gamma w_1w_2^2$ | $\delta' w_1^2w_2^2$ | $\alpha^2e_0w_1w_2^2$ | 0 | 0 |

| $x$ | $h_0^3\alpha w_2^3$ | $h_0\alpha^2w_2^3$ | $\delta w_2^3$ | $h_0\alpha^3w_2^3$ | $w_2^4$ |
|---|---|---|---|---|---|
| $t-s$ | 156 | 168 | 176 | 180 | 192 |
| $s$ | 30 | 31 | 31 | 34 | 32 |
| $d_4(x)$ | 0 | 0 | 0 | 0 | 0 |

Our earlier results and the multiplicative structure determine $d_4$ on all of these algebra generators, see Theorem 5.23. Since $g$, $w_1$ and $w_2^4$ are $d_4$-cycles, the $d_4$-differential is $R_2$-linear. We calculate $d_4$ on each $R_2$-module generator using the Leibniz rule, and then pass to homology to obtain $E_5(tmf) = H(E_4(tmf), d_4)$ in Appendix A.3. The resulting direct sum of mostly cyclic $R_2$-modules is presented in Tables 5.8 and 5.9.

Finally, we verify that $E_5(tmf)$ is generated as an algebra by the following 43 classes.

| $x$ | $h_0$ | $h_1$ | $h_2$ | $c_0$ | $w_1$ | $h_0^3\alpha$ | $d_0$ | $g$ | $h_0\alpha^2$ | $\gamma$ | $\alpha\beta$ | $\delta$ | $\delta'$ | $h_0\alpha^3$ |
|---|---|---|---|---|---|---|---|---|---|---|---|---|---|---|
| $t-s$ | 0 | 1 | 3 | 8 | 8 | 12 | 14 | 20 | 24 | 25 | 27 | 32 | 32 | 36 |
| $s$ | 1 | 1 | 1 | 3 | 4 | 6 | 4 | 4 | 7 | 5 | 6 | 7 | 7 | 10 |

| $x$ | $h_0^2w_2$ | $h_2w_2$ | $c_0w_2$ | $\alpha^3g + h_0w_1w_2$ | $h_0^3\alpha w_2$ | $h_0\alpha^2w_2$ | $\delta w_2$ | $h_0\alpha^3w_2$ |
|---|---|---|---|---|---|---|---|---|
| $t-s$ | 48 | 51 | 56 | 56 | 60 | 72 | 80 | 84 |
| $s$ | 10 | 9 | 11 | 13 | 14 | 15 | 15 | 18 |

| $x$ | $h_0w_2^2$ | $h_1w_2^2$ | $h_2w_2^2$ | $c_0w_2^2$ | $w_1w_2^2$ | $h_0^3\alpha w_2^2$ | $d_0w_2^2$ | $h_0\alpha^2w_2^2$ | $\alpha\beta w_2^2$ |
|---|---|---|---|---|---|---|---|---|---|
| $t-s$ | 96 | 97 | 99 | 104 | 104 | 108 | 110 | 120 | 123 |
| $s$ | 17 | 17 | 17 | 19 | 20 | 22 | 20 | 23 | 22 |

| $x$ | $\delta w_2^2$ | $\delta' w_2^2$ | $h_0\alpha^3w_2^2$ | $h_0^2w_2^3$ | $h_2w_2^3$ | $c_0w_2^3$ | $\alpha^3gw_2^2 + h_0w_1w_2^3$ | $h_0^3\alpha w_2^3$ |
|---|---|---|---|---|---|---|---|---|
| $t-s$ | 128 | 128 | 132 | 144 | 147 | 152 | 152 | 156 |
| $s$ | 23 | 23 | 26 | 26 | 25 | 27 | 29 | 30 |

| $x$ | $h_0\alpha^2w_2^3$ | $\delta w_2^3$ | $h_0\alpha^3w_2^3$ | $w_2^4$ |
|---|---|---|---|---|
| $t-s$ | 168 | 176 | 180 | 192 |
| $s$ | 31 | 31 | 34 | 32 |

We show in Theorem 5.27 that there is no room for any further differentials, so that $E_5(tmf) = E_\infty(tmf)$.

Tables 5.8 and 5.9 therefore also express $E_\infty(tmf)$ as a direct sum of $R_2$-modules. In particular, $E_\infty(tmf)$ is free as an $\mathbb{F}_2[w_2^4]$-module, but it has both $w_1$-periodic elements and $w_1$-power torsion elements. The latter are generated by classes in degrees $3 \le t-s \le 164$, repeating 192-periodically. The $E_\infty$-term is shown for $0 \le t-s \le 200$ in Figure 0.7, with the $w_1$-power torsion classes marked in red. The more interesting part is shown for $0 \le t-s \le 96$ and $96 \le t-s \le 192$ in Figure 0.8. More legible charts are provided in Figures 5.1 to 5.8.

Mahowald first calculated this Adams spectral sequence, as outlined in his paper [**76**, §9] with Hopkins, before the spectrum *tmf* with cohomology $A//A(2)$ was known to exist. Already in the 1998 version of that preprint, the authors wrote that this was a "calculation which has been known for about twenty years." Our computation confirms their outline, including the hidden 2- and $\eta$-extensions, except that the third differential in their Proposition 9.10 should be $d_3(v_1wg_{35,7}) = v_1^4g_{33,8}$, and in the chart of their Theorem 9.11 the classes in degrees $t-s = 70$ and 90 should be in Adams filtrations 14 and 18, respectively.

In Chapter 6 and Appendix B we determine the $d_2$-, $d_3$- and $d_4$-differentials in the Adams spectral sequence for *tmf*/2, as a module spectral sequence over the Adams spectral sequence for *tmf*, and calculate the resulting $E_3$-, $E_4$- and $E_5$-terms. All of these differentials follow algebraically from the known differentials for *tmf* and the module structure. The spectral sequence for *tmf*/2 collapses at the $E_5$-term, as we show in Theorem 6.13. The resulting $E_\infty$-term is presented as an $R_2$-module in Tables 6.10 and 6.11. It is free as an $\mathbb{F}_2[w_2^4]$-module, and is shown at large scale in Figures 0.12 and 0.13, and more legibly in Figures 6.1 to 6.8. See also Remark 0.1.

In Chapter 7 and Appendix C we determine the $d_2$- and $d_3$-differentials in the Adams spectral sequence for $tmf/\eta$, as a module spectral sequence over the Adams spectral sequence for *tmf*, and calculate the resulting $E_3$- and $E_4$-terms. The module structure determines almost all of the differentials, but for one exceptional differential (namely $d_3(h_2^2\widehat{\beta}) = i(d_0 w_1)$) we rely on the hidden $\eta$-extension from $h_1 g$ to $Pd_0$ in the Adams spectral sequence for $S$. Perhaps surprisingly, the Adams spectral sequence for $tmf/\eta$ collapses already at the $E_4$-term, as we show in Theorem 7.6. The resulting $E_\infty$-term is presented as an $R_2$-module in Tables 7.5 and 7.6. It is free as an $\mathbb{F}_2[w_2^4]$-module, and is shown at large scale in Figures 0.15 and 0.16, and more legibly in Figures 7.1 to 7.8.

In Chapter 8 and Appendix D we determine the $d_2$-, $d_3$- and $d_4$-differentials in the Adams spectral sequence for $tmf/\nu$, as a module spectral sequence over the Adams spectral sequence for *tmf*, and calculate the resulting $E_3$-, $E_4$- and $E_5$-terms. The module structure determines almost all of the differentials, except that for one differential (namely $d_2(\overline{\beta^2}) = i(h_1\delta)$) we rely on an ad hoc argument using the external pairing $tmf/\nu \wedge tmf/\nu \to tmf \wedge C\nu \wedge C\nu$. There are no further differentials, as we show in Theorem 8.12. The resulting $E_\infty$-term is presented as a direct sum of cyclic $R_2$-modules in Table 8.9. It is free as an $\mathbb{F}_2[w_2^4]$-module, and is shown at large scale in Figures 0.18 and 0.19, and more legibly in Figures 8.1 to 8.8.

The results on $E_\infty(tmf/2)$ and $E_\infty(tmf/\nu)$ give us a sufficiently good handle on $\pi_*(tmf/2)$ and $\pi_*(tmf/\nu)$ to determine the hidden 2- and $\nu$-extensions in $\pi_*(tmf)$. It turns out that all hidden $\eta$-extensions follow from these, mainly due to the relation $\eta^3 = 4\nu$, so the calculation of $E_\infty(tmf/\eta)$ is not strictly needed for our analysis of the ring structure on $\pi_*(tmf)$. We do, however, include this case for completeness, as a consistency check, and for future applications to other *tmf*-module spectra.

## 0.7. The graded homotopy ring of *tmf*

Part III commences with Chapter 9, which is the core of this book. Its aim is to calculate the graded homotopy ring $\pi_*(tmf)$, implicitly completed at $p = 2$.

Using the long exact sequence

$$\cdots \longrightarrow \pi_n(tmf) \xrightarrow{2} \pi_n(tmf) \xrightarrow{i} \pi_n(tmf/2) \xrightarrow{j} \pi_{n-1}(tmf) \longrightarrow \cdots$$

and our calculation of $E_\infty(tmf/2)$ we determine the hidden 2-extensions in the Adams spectral sequence for *tmf* in Theorem 9.8. This already determines the group structure of $\pi_n(tmf)$ in each degree $n$. In particular, there are hidden 2-extensions to $\alpha^3 g + h_0 w_1 w_2$ and to $\alpha^3 g w_2^2 + h_0 w_1 w_2^3$. It follows that $\pi_*(tmf)$ is generated as a graded ring by 40 homotopy classes, which are detected by the 40 out of 43 algebra generators of $E_\infty(tmf)$ that remain when $h_0$ and the two classes just mentioned are omitted.

For example, there are five classes $\eta$, $\nu$, $\epsilon$, $\kappa$ and $\bar{\kappa}$ in $\pi_*(tmf)$ that are detected by $h_1$, $h_2$, $c_0$, $d_0$ and $g$ in $E_\infty(tmf)$, respectively, and which are the images of the classes in $\pi_*(S)$ with the same names [**171**], [**130**]. (We note that this prescription only determines $\bar{\kappa}$ up to an odd multiple.) There are also two classes $B$ and $C$ in $\pi_*(tmf)$ that are detected by $w_1$ and $h_0^3\alpha$ in $E_\infty(tmf)$, respectively. We show in

Proposition 9.19 that they can be assumed to have images $c_4$ and $2c_6$, respectively, under the edge homomorphism $e\colon \pi_*(tmf) \to mf_{*/2}$, and that these conditions together uniquely determine these two classes. We refer to $B \in \pi_8(tmf)$ as the "Bott element", in part because $B$ and $C$ map to generators of $\pi_8(ko)$ and $\pi_{12}(ko)$ under the $E_\infty$ ring map $q_0\colon tmf \to ko$ constructed by Lawson and Niko Naumann [**91**, Thm. 1.2].

Together with the ring unit $D = 1$, these seven classes generate the remaining ring generators for $\pi_*(tmf)$ by "formal multiplication by powers of the discriminant $\Delta$, up to scalar multiples." This formal relationship can be expressed in terms of Massey products in $E_2(tmf) = \mathrm{Ext}_{A(2)}(\mathbb{F}_2, \mathbb{F}_2)$, as we discuss in Subsection 9.1.1, or in terms of modular form images. For $x \in \{\eta, \nu, \epsilon, \kappa, \bar{\kappa}, B, C, D\}$ and some or all $0 \le k \le 7$ we write $x_k$ for the $k$-th member of the family of ring generators for $\pi_*(tmf)$ that are related to $x = x_0$ through formal multiplication by powers of $\Delta$, up to scalars. More precisely, we have the following 40 ring generators.

| $x_k$ | $\eta$ | $\eta_1$ | $\eta_4$ |
|---|---|---|---|
| $n$ | 1 | 25 | 97 |
| $E_\infty(tmf)$ | $h_1$ | $\gamma$ | $h_1 w_2^2$ |

| $x_k$ | $\nu$ | $\nu_1$ | $\nu_2$ | $\nu_4$ | $\nu_5$ | $\nu_6$ |
|---|---|---|---|---|---|---|
| $n$ | 3 | 27 | 51 | 99 | 123 | 147 |
| $E_\infty(tmf)$ | $h_2$ | $\alpha\beta$ | $h_2 w_2$ | $h_2 w_2^2$ | $\alpha\beta w_2^2$ | $h_2 w_2^3$ |

| $x_k$ | $\epsilon$ | $\epsilon_1$ | $\epsilon_4$ | $\epsilon_5$ |
|---|---|---|---|---|
| $n$ | 8 | 32 | 104 | 128 |
| $E_\infty(tmf)$ | $c_0$ | $\delta'$ | $c_0 w_2^2$ | $\delta' w_2^2$ |

| $x_k$ | $\kappa$ | $\kappa_4$ |
|---|---|---|
| $n$ | 14 | 110 |
| $E_\infty(tmf)$ | $d_0$ | $d_0 w_2^2$ |

| $x_k$ | $\bar{\kappa}$ |
|---|---|
| $n$ | 20 |
| $E_\infty(tmf)$ | $g$ |

| $x_k$ | $B$ | $B_1$ | $B_2$ | $B_3$ | $B_4$ | $B_5$ | $B_6$ | $B_7$ |
|---|---|---|---|---|---|---|---|---|
| $n$ | 8 | 32 | 56 | 80 | 104 | 128 | 152 | 176 |
| $E_\infty(tmf)$ | $w_1$ | $\alpha g$ | $c_0 w_2$ | $\delta w_2$ | $w_1 w_2^2$ | $\alpha g w_2^2$ | $c_0 w_2^3$ | $\delta w_2^3$ |
| $mf_{*/2}$ | $c_4$ | $c_4\Delta$ | $c_4\Delta^2$ | $c_4\Delta^3$ | $c_4\Delta^4$ | $c_4\Delta^5$ | $c_4\Delta^6$ | $c_4\Delta^7$ |

| $x_k$ | $C$ | $C_1$ | $C_2$ | $C_3$ | $C_4$ | $C_5$ | $C_6$ | $C_7$ |
|---|---|---|---|---|---|---|---|---|
| $n$ | 12 | 36 | 60 | 84 | 108 | 132 | 156 | 180 |
| $E_\infty(tmf)$ | $h_0^3\alpha$ | $h_0\alpha^3$ | $h_0^3\alpha w_2$ | $h_0\alpha^3 w_2$ | $h_0^3\alpha w_2^2$ | $h_0\alpha^3 w_2^2$ | $h_0^3\alpha w_2^3$ | $h_0\alpha^3 w_2^3$ |
| $mf_{*/2}$ | $2c_6$ | $2c_6\Delta$ | $2c_6\Delta^2$ | $2c_6\Delta^3$ | $2c_6\Delta^4$ | $2c_6\Delta^5$ | $2c_6\Delta^6$ | $2c_6\Delta^7$ |

| $x_k$ | $D_1$ | $D_2$ | $D_3$ | $D_4$ | $D_5$ | $D_6$ | $D_7$ | $M$ |
|---|---|---|---|---|---|---|---|---|
| $n$ | 24 | 48 | 72 | 96 | 120 | 144 | 168 | 192 |
| $E_\infty(tmf)$ | $h_0\alpha^2$ | $h_0^2 w_2$ | $h_0\alpha^2 w_2$ | $h_0 w_2^2$ | $h_0\alpha^2 w_2^2$ | $h_0^2 w_2^3$ | $h_0\alpha^2 w_2^3$ | $w_2^4$ |
| $mf_{*/2}$ | $8\Delta$ | $4\Delta^2$ | $8\Delta^3$ | $2\Delta^4$ | $8\Delta^5$ | $4\Delta^6$ | $8\Delta^7$ | $\Delta^8$ |

See also Figure 9.1. More concisely, the $D$-family is characterized by $e(D_k) = d_k\Delta^k$, where the scalars $d_k$ are introduced in Definition 9.18. We call the final generator $M = D_8 \in \pi_{192}(tmf)$ the "Mahowald element". We show in Proposition 9.19 that we can choose the ring generators $B_k$, $C_k$, $D_k$ and $M$ to have the modular form images listed above, still subject to the constraint that they are detected by the given classes in $E_\infty(tmf)$. In the case of the $C$-family, our proof relies on the fact that the image of the edge homomorphism $e\colon \pi_n(tmf) \to mf_{n/2}$ is divisible by 2 for $n = 12 + 24k$, cf. [**75**, Prop. 4.6] and [**23**, §8]. The specified images in $E_\infty(tmf)$ and $mf_{*/2}$ suffice to determine most of the ring generators, but some ambiguity remains, especially in the $\nu$-family, which we discuss and almost completely eliminate in Definition 9.22. See Remark 9.24.

Multiplication by $M$ induces multiplication by $w_2^4$ in $E_\infty(tmf)$, hence acts freely on $\pi_*(tmf)$. Letting $N_* \subset \pi_*(tmf)$ denote the $\mathbb{Z}[B]$-submodule generated by the classes in degrees $0 \le * < 192$, we obtain a $\mathbb{Z}[B,M]$-module isomorphism $N_* \otimes \mathbb{Z}[M] \cong \pi_*(tmf)$. We summarize the $\mathbb{Z}[B,M]$-module structure on $\pi_*(tmf)$ in Theorem 9.27, by way of the $\mathbb{Z}[B]$-module structure on $N_*$. The $B$-power torsion submodule $\Gamma_B N_*$ is finite and concentrated in degrees $3 \le * \le 164$, see Table 9.4. There is a split extension of $\mathbb{Z}[B]$-modules

$$0 \to \Gamma_B N_* \longrightarrow N_* \longrightarrow N_*/\Gamma_B N_* \to 0\,.$$

The $B$-torsion free quotient $N_*/\Gamma_B N_*$ is a direct sum of eight *ko*-covers $ko[k]$, for $0 \le k \le 7$. Here $ko[k]$ is a $\mathbb{Z}[B]$-submodule of $\pi_*(ko)$ that starts in degree $24k$ and contains all classes in degrees $* > 4 + 24k$, see Theorem 9.26.

Turning to the multiplicative structure of $\pi_*(tmf)$, we show in Proposition 9.10 that there are no hidden $B$- or $M$-multiplications in $\pi_*(tmf)$, so that all of the $w_1$-power torsion in $E_\infty(tmf)$ is realized as $B$-power torsion, of the same exponent, in $\pi_*(tmf)$. It follows that the 2- and $B$-power torsion ideals in $\pi_*(tmf)$ are

$$\Gamma_2\pi_*(tmf) = (\eta_k, \nu_k, \epsilon_k, \kappa_k, \bar\kappa)$$
$$\Gamma_B\pi_*(tmf) = (\nu_k, \epsilon_k, \kappa_k, \bar\kappa)\,,$$

where in the latter case the $\nu$-family must be interpreted to include an "honorary" member $\nu_3 = \eta_1^3$.

We use the long exact sequence

$$\dots \longrightarrow \pi_{n-3}(tmf) \xrightarrow{\nu} \pi_n(tmf) \xrightarrow{i} \pi_n(tmf/\nu) \xrightarrow{j} \pi_{n-4}(tmf) \longrightarrow \dots$$

and our calculation of $E_\infty(tmf/\nu)$ to determine the hidden $\nu$-extensions in the Adams spectral sequence for *tmf* in Theorem 9.14, and from this we deduce the hidden $\eta$-extensions in Theorem 9.16. We then establish an interesting multiplicative relation in $\pi_{105}(tmf)$, namely

$$\nu^2\nu_4 = \eta\epsilon_4 + \eta_1\bar\kappa^4.$$

This exhibits a hidden $\nu$-extension from the $E_\infty$-class detecting $\nu\nu_4$ to the $E_\infty$-class detecting $\eta\epsilon_4$. However, this is not the whole relation in homotopy: there is also the higher filtration term $\eta_1\bar{\kappa}^4$. A hidden extension is simply the lowest filtration part of a nonzero product that is zero at $E_\infty$. Having determined the hidden 2-, $\eta$- and $\nu$-extensions, it is natural to consider $\pi_*(tmf)$ as a $T$-module, where

$$T = \mathbb{Z}[\eta, \nu, B, M]/(2\eta, \eta^3 + 4\nu, \eta\nu, 2\nu^2, \nu B, \nu^4)$$

is the (implicitly 2-completed) subring of $\pi_*(tmf)$ generated by $\eta$, $\nu$, $B$ and $M$. We produce a list of 58 $T$-module generators for $\pi_*(tmf)$ in Table 9.5. The structure of $\pi_*(tmf)$ as a graded abelian group, with all 2-, $\eta$-, $\nu$-, $B$- and $M$-multiplications, is shown at various scales in Figures 0.9, 0.10 and 9.6 through 9.13.

In Section 9.5 we undertake to compute the remaining products in $\pi_*(tmf)$. It suffices to calculate all products $xy$, where $x$ is one of 57 $T$-module generators of $\pi_*(tmf)$ (other than $x = 1$) and $y$ is one of 36 ring generators of $\pi_*(tmf)$ (other than $y \in \{\eta, \nu, B, M\}$). We achieve this for the 2-power torsion generators $y$ in Theorem 9.47, up to some signs $s$ and $s_i$ for $i \in \{0, 2, 4, 6\}$. The method of proof is principally to reason with the Adams filtration of $\pi_*(tmf)$, combined with previously established hidden extensions, and supplemented by the edge homomorphism $e$ to modular forms. At $p = 2$ the Adams filtration gives quite different information from that provided by the Adams–Novikov or descent filtration, and this makes the calculation possible. For example, the 2-torsion free classes lie in relatively high Adams filtration, but have Adams–Novikov filtration zero. We also perform this calculation for the 2-torsion free generators $y$, in Theorem 9.48. Again, the principal method is the use of the multiplicative Adams filtration, combined with previously established hidden extensions.

Our choices of ring generators $B_k$ for $\pi_*(tmf)$ were partially dictated by the need to reason, as outlined above, by means of the Adams filtration. For example, this is why we chose $B \in \pi_8(tmf)$ to be the class detected by $w_1$ in Adams filtration 4, rather than its sum $B + \epsilon$ in Adams filtration 3, even if both classes map to the usual Bott element in $\pi_8(ko)$. However, this has the effect that some of the multiplicative relations that hold in $mf_{*/2}$, such as $c_4\Delta^2 \cdot c_4\Delta^3 = c_4 \cdot c_4\Delta^5$, will only hold up to 2-torsion correction terms for our chosen lifts to $\pi_*(tmf)$. For instance, $B_2 \cdot B_3 = B \cdot B_5 + \eta\eta_1\kappa_4$ with $\eta\eta_1\kappa_4$ in Adams filtration 27.

Somewhat miraculously, it is possible to modify our choices of ring generators for $\pi_*(tmf)$ to eliminate these correction terms. The change amounts to replacing the $B$-family with a $\widetilde{B}$-family, as specified in Definition 9.50. (This decoration is unrelated to our notation $\tilde{x}$ for classes in $E_2(tmf/2)$.) The modular form images do not change, but the detecting classes in $E_\infty(tmf)$ are affected, so that $\widetilde{B}_k$ has Adams filtration $3 + 4k$ for all $0 \le k \le 7$. In particular, $\widetilde{B} = B + \epsilon$. A class is $\widetilde{B}$-power torsion if and only if it is $B$-power torsion.

| $x_k$ | $\widetilde{B}$ | $\widetilde{B}_1$ | $\widetilde{B}_2$ | $\widetilde{B}_3$ | $\widetilde{B}_4$ | $\widetilde{B}_5$ | $\widetilde{B}_6$ | $\widetilde{B}_7$ |
|---|---|---|---|---|---|---|---|---|
| $n$ | 8 | 32 | 56 | 80 | 104 | 128 | 152 | 176 |
| $E_\infty(tmf)$ | $c_0$ | $\delta$ | $c_0w_2$ | $\delta w_2$ | $c_0w_2^2$ | $\delta w_2^2$ | $c_0w_2^3$ | $\delta w_2^3$ |
| $mf_{*/2}$ | $c_4$ | $c_4\Delta$ | $c_4\Delta^2$ | $c_4\Delta^3$ | $c_4\Delta^4$ | $c_4\Delta^5$ | $c_4\Delta^6$ | $c_4\Delta^7$ |

By Theorem 9.53 the relations among the 2-torsion free generators $\widetilde{B}_k$, $C_k$ and $D_k$ in $\pi_*(tmf)$ are the same as among their images in $\mathrm{im}(e) \subset mf_{*/2}$. Hence the surjective ring homomorphism $\pi_*(tmf) \longrightarrow \mathrm{im}(e)$ mapping $x$ to $e(x)$ admits a multiplicative (and additive) section $\sigma\colon \operatorname{im}(e) \to \pi_*(tmf)$, given by the following table.

| $x$ | $c_4\Delta^k$ | $2c_6\Delta^k$ | $d_k\Delta^k$ | $\Delta^8$ |
|---|---|---|---|---|
| $\sigma(x)$ | $\widetilde{B}_k$ | $C_k$ | $D_k$ | $M$ |

It is not clear whether the existence of such a ring homomorphism $\sigma$ is part of the previous literature on the subject, in part due to the tendency to use ambiguous integral modular form notation, such as $c_4$, for topological modular forms, such as $B$ and $\widetilde{B}$.

With these changes, and our final normalization of the classes $\nu_k$, the products with 2-power torsion elements in $\pi_*(tmf)$ are somewhat more regular. Our conclusion is given in Theorem 9.54 and Tables 9.8 and 9.9. These give the products $xy$ for $x$ a $T$-module generator of $\pi_*(tmf)$ other than $x = 1$, replacing each $B_k$ with $\widetilde{B}_k$, and $y$ a 2-power torsion ring generator of $\pi_*(tmf)$, other than $y \in \{\eta, \nu, B, M\}$. Furthermore, the rows for $x \in \{\widetilde{B}_k, C_k, D_{2j+1}\}$ are omitted, because all products in these rows are zero, with the exception of

$$\eta_i \widetilde{B}_j = \eta \widetilde{B}_{i+j} \,.$$

Here $\widetilde{B}_{k+8}$ is interpreted as $\widetilde{B}_k M$, for $k \geq 0$. The remaining multiplication tables then "only" have 38 rows and 14 columns.

One bit of ambiguity remains: Having chosen $\nu_1$, $\nu_2$ and $\nu_4$, with $\nu D_4 = 2\nu_4$, there are unique choices for $\nu_5$ and $\nu_6$ satisfying $\nu_1\nu_5 = 2\nu\nu_6$ and $\nu_2\nu_4 = 3\nu\nu_6$. We then have $\nu_4\nu_6 = s\nu\nu_2 M$ in $\pi_{246}(tmf) \cong \mathbb{Z}/4$ for some sign $s \in \{\pm 1\}$. We have not determined this sign $s$, which is independent of the choice of $\nu_1$, $\nu_2$ and $\nu_4$. If $s = 1$, then the relation $\nu_i\nu_j = (i+1)\nu\nu_{i+j}$ holds for all $i$ and $j$.

Mahowald noted (cf. [**76**, Prop. 8.7]) that multiplication by $\epsilon$ agrees with multiplication by $B$ on the $B$-power torsion in $\pi_*(tmf)$. This is equivalent to the assertion that $\widetilde{B} \cdot y = 0$ for all $y \in \Gamma_B \pi_*(tmf) = (\nu_k, \epsilon_k, \kappa_k, \bar{\kappa})$. We prove in Corollary 9.55 that $\widetilde{B}_k \cdot y = 0$ for all $0 \leq k \leq 7$ and $y \in \Gamma_B \pi_*(tmf)$, thus generalizing Mahowald's assertion. (The honorary case $\widetilde{B}_k \cdot \nu_3 = 0$ is not made explicit in our tables, but $\eta_1^3 \widetilde{B}_k = \eta^3 \widetilde{B}_{k+3} = 4\nu \widetilde{B}_{k+3} = 0$.)

## 0.8. Duality

Working for a moment over $\mathbb{Z}[1/6]$, with 2 and 3 inverted, the compactified moduli stack $\overline{\mathcal{M}}_{ell}$ is equivalent to the weighted projective stack associated to the graded ring $\mathbb{Z}[1/6][c_4, c_6]$, cf. [**62**, §4.6]. Its cohomology satisfies Serre duality with respect to the dualizing sheaf $\Omega \cong \omega^{\otimes -10}$, corresponding to $1/c_4c_6$, meaning that there is a perfect pairing of finitely generated free $\mathbb{Z}[1/6]$-modules

$$H^s(\overline{\mathcal{M}}_{ell}; \omega^{\otimes k}) \otimes H^{1-s}(\overline{\mathcal{M}}_{ell}; \operatorname{Hom}(\omega^{\otimes k}, \Omega)) \longrightarrow H^1(\overline{\mathcal{M}}_{ell}; \Omega) \cong \mathbb{Z}[1/6] \,.$$

Hence the descent spectral sequence has $E_2$-term concentrated in the rows $s = 0$ and $s = 1$, with

$$E_2^{0,2k} = H^0(\overline{\mathcal{M}}_{ell}; \omega^{\otimes k})$$

linearly dual to

$$E_2^{1,-20-2k} = H^1(\overline{\mathcal{M}}_{ell}; \operatorname{Hom}(\omega^{\otimes k}, \Omega)) \,.$$

This implies that $\pi_n(Tmf)[1/6]$ is linearly dual to $\pi_{-21-n}(Tmf)[1/6]$, and can be refined to the spectrum level statement that $Tmf[1/6]$ is Anderson self-dual in the sense that

$$\Sigma^{21}Tmf[1/6] \simeq I_{\mathbb{Z}}(Tmf)[1/6]\,.$$

Here $I_{\mathbb{Z}}(X)$ denotes the Anderson dual of $X$, see Section 10.4. Vesna Stojanoska extended this result to the primes 3 and 2, by first establishing Anderson self-duality for covers $Tmf(2)$ of $Tmf[1/2]$ and $Tmf(3)$ of $Tmf[1/3]$, and then applying descent for the natural actions by the groups $GL_2(\mathbb{Z}/2)$ and $GL_2(\mathbb{Z}/3)$ of order 6 and 48, respectively. The argument for $p = 3$ appeared in [**161**], while part of the argument for $p = 2$ appeared in [**162**].

The computation of the sheaf cohomology of $\overline{\mathcal{M}}_{ell}$ can also be interpreted as saying that there is a homotopy fiber sequence

$$tmf \longrightarrow Tmf \longrightarrow \Sigma^{-1}tmf/(B^\infty, M^\infty)\,,$$

where $tmf/(B^\infty, M^\infty)$ is the iterated homotopy cofiber in the square

$$\begin{array}{ccc} tmf & \longrightarrow & tmf[1/B] \\ \downarrow & & \downarrow \\ tmf[1/M] & \longrightarrow & tmf[1/B, 1/M]\,. \end{array}$$

Formulated in terms of connective covers, Anderson duality implies an equivalence of $tmf$-modules

$$\Sigma^{20}tmf \simeq I_{\mathbb{Z}}(tmf/(B^\infty, M^\infty))\,.$$

In Chapter 10 we turn the argument around, and first establish the above equivalence after completion at 2, and then use an argument of John Greenlees and Stojanoska [**67**] to glue $tmf$ and its Anderson dual together to obtain Anderson self-duality for $Tmf$.

We obtain the equivalence above by descent along the map $\iota'\colon tmf \to tmf_1(3) \simeq BP\langle 2\rangle$, corresponding to a separable (in the sense of [**150**, §9.1]) extension $TMF \to TMF_1(3)$ of degree 8 inside the $GL_2(\mathbb{Z}/3)$-Galois extension $TMF \to TMF(3)$. The calculation of $\pi_*(tmf)$ as a $\mathbb{Z}[B, M]$-module from Theorem 9.27 shows that there is a top class $C_7/BM$ in $\pi_{-20}(tmf/(B^\infty, M^\infty))$, corresponding to a bottom class in $\pi_{20}$ of the Anderson dual. We can represent the latter homotopy class by a $tmf$-module map

$$a\colon \Sigma^{20}tmf \longrightarrow I_{\mathbb{Z}}(tmf/(B^\infty, M^\infty))\,.$$

We show in Theorem 10.6 that $a$ is a 2-adic equivalence. For the proof we use the finite CW spectrum $\Phi = \Phi A(1)$ from Lemma 1.42, which we may take to be Spanier–Whitehead self-dual, with mod 2 cohomology realizing the double $A(2)/\!/E(2)$ of $A(1)$. For any such spectrum $\Phi$ there is an equivalence $tmf \wedge \Phi \simeq BP\langle 2\rangle$. Smashing $a$ with the equivalence $\Sigma^{-12}\Phi \simeq D\Phi = F(\Phi, S)$ we obtain a map

$$\Sigma^8 BP\langle 2\rangle \longrightarrow I_{\mathbb{Z}}(BP\langle 2\rangle/(v_1^\infty, v_2^\infty))\,,$$

which can be verified to be an equivalence by an inspection of homotopy groups. This completes the proof of Anderson duality for $Tmf$ at $p = 2$. To be precise, we formulate our Theorem 10.6 in terms of the perhaps more familiar Brown–Comenetz duality functor $X \mapsto I(X)$, saying that there is a 2-adic equivalence of $tmf$-modules

$$\Sigma^{20}tmf \simeq I(tmf/(2^\infty, B^\infty, M^\infty))\,,$$

where $tmf/(2^\infty, B^\infty, M^\infty)$ is defined to be the iterated homotopy cofiber of a cubical diagram, similar to the square in the definition of $tmf/(B^\infty, M^\infty)$. However,

$I(X/2^\infty) \simeq I_{\mathbb{Z}}(X)$ after 2-adic completion for any spectrum $X$, by Lemma 10.10, so the two formulations are equivalent.

Recall that the $B$-power torsion in $\pi_*(tmf)$ repeats $M$-periodically, and is generated by finitely many classes of finite additive order in the range $0 \le * < 192$. The $B$-power torsion in $\pi_n(tmf)$ for $0 \le n < 192$ usually contributes classes of finite additive order in $\pi_{n-191}(tmf/(B^\infty, M^\infty))$, and appears in Pontryagin dual form as $B$-power torsion in $\pi_{190-n}(I_{\mathbb{Z}}(tmf/(B^\infty, M^\infty))) \cong \pi_{170-n}(tmf)$. Hence Anderson self-duality for *tmf* is visible as a Pontryagin self-duality in most of the $B$-power torsion $\Gamma_B N_* \subset \Gamma_B \pi_*(tmf)$, with the finite group in degree $n$ being Pontryagin dual to the finite group in degree $170 - n$.

However, there is one systematic family of exceptions. The $B$-power torsion classes $\langle \nu_k \rangle$ in degree $n = 3 + 24k$, for $0 \le k \le 6$, occur in $\pi_{n-191}(tmf/(B^\infty, M^\infty))$ as the quotients of torsion-free extensions by $\mathbb{Z}\{C_k/BM\}$. Hence these classes contribute under Anderson duality to the $B$-periodic, 2-torsion free part of

$$\pi_{191-n}(I_{\mathbb{Z}}(tmf/(B^\infty, M^\infty))) \cong \pi_{171-n}(tmf),$$

and are not visible in $\pi_{170-n}(tmf)$.

Conversely, the $B$-torsion free part of $N_*$ is a direct sum of $\mathbb{Z}[B]$-modules $ko[k]$, with bottom class $D_k$ in degree $n = 24k$. For $1 \le k \le 7$ the relation $B \cdot D_k = d_k B_k$ in $ko[k]$ implies that $\pi_{n-192}(tmf/(B^\infty, M^\infty))$ contains a finite group $\langle B_k/BM \rangle \cong \mathbb{Z}/d_k$. Its Pontryagin dual appears as $B$-power torsion in $\pi_{191-n}(I_{\mathbb{Z}}(tmf/(B^\infty, M^\infty))) \cong \pi_{171-n}(tmf)$, namely as the summand $\langle \nu_{7-k} \rangle$.

Thus, the duality of *tmf* is most visibly reflected in the part $\Theta\pi_*(tmf)$ of $\Gamma_B \pi_*(tmf)$ consisting of the $B$-power torsion classes that are not in degrees $* \equiv 3 \mod 24$. Here $\Theta\pi_*(tmf) = \Theta N_* \otimes \mathbb{Z}[M]$, with $\Theta N_*$ concentrated in degrees $* \not\equiv 3 \mod 24$, and there is a perfect pairing

$$(-,-)\colon \Theta N_{170-n} \otimes \Theta N_n \longrightarrow \mathbb{Q}/\mathbb{Z}$$

for all $n$. The remaining $B$-power torsion is not Pontryagin self-dual, but interacts as explained above with the non-free part of the $\mathbb{Z}[B, M]$-torsion free quotient of $\pi_*(tmf)$. This is illustrated in Figures 10.1 and 10.2, where $\Theta_n N_*$ and $\Theta_{170-n} N_*$ are shown above and below the "fold line", respectively. The exceptional classes $\nu_k$ in degrees $* \equiv 3 \mod 24$ appear just outside of the mirror symmetric parts of these pictures.

We introduce the notations $M/x^\infty$ and $\Gamma_x M$ in Section 10.2, and review the Brown–Comenetz duality functor $I$ and prove the 2-complete duality theorem for *tmf* in Section 10.3. We review Anderson duality and convert our duality theorem into a 2-complete self-duality theorem for *Tmf* in Section 10.4. In Section 10.5 we convert the spectrum level duality theorem for *tmf* into several algebraic duality statements, summarized in Theorem 10.26. In particular, we verify the claims made above about additive extensions in $\pi_*(tmf/(B^\infty, M^\infty))$, which lead to a less ad hoc Definition 10.18 of the $\pi_*(tmf)$-module $\Theta\pi_*(tmf) \cong \Theta N_* \otimes \mathbb{Z}[M]$, where $\Theta N_*$ is Pontryagin self-dual. Finally that self-duality is spelled out in Theorem 10.29 and Table 10.1. As a rule of thumb, cf. Remark 10.30, classes $x$ and $y$ of order 2 in $\Theta N_*$ are dual when they formally multiply to $(\eta\nu\epsilon\kappa)_6$.

## 0.9. The sphere spectrum

In Chapter 11, we discuss the mod 2 Adams spectral sequence for the sphere spectrum. Using the $H_\infty$ ring structure on $S$, and the proven Adams conjecture, we readily determine the full pattern of differentials originating in degrees $t - s \leq 29$. See Theorems 11.52, 11.54, 11.56 and 11.59. By means of a comparison of Adams spectral sequences along the maps

$$S \xrightarrow{i} C\eta \xrightarrow{1\wedge i} C\eta \wedge C\nu \,,$$

we show in Theorem 11.71 that there is a hidden $\eta$-extension from $h_1 g$ detecting $\eta\bar{\kappa}$ to $Pd_0$ detecting $\eta^2\bar{\kappa}$. This is equivalent to Mimura's result [**129**, Thm. B] that $\epsilon\kappa \neq 0$ in $\pi_{22}(S)$, but our proof is entirely stable. These methods quickly give the graded ring structure of $\pi_*(S)$ for $* \leq 28$, see Theorem 11.61.

We need some of this information about $\pi_*(S)$, in the smaller range $* \leq 22$, when we determine the Adams differentials for *tmf* in Chapter 5. Furthermore, once we have fully determined the differential structure for *tmf*, we can easily use naturality along

$$\iota\colon S \longrightarrow tmf$$

to determine the remaining Adams differentials for $S$ originating in the larger range $t-s \leq 48$. In particular, this leads to simple proofs of the differentials $d_2(v) = h_0 z$, $d_3(r) = h_1 d_0^2$, $d_3(d_0 e_0) = h_0^5 r$, $d_3(z) = 0$, $d_3(Ph_5 c_0) = 0$, $d_3(h_0^5 Q) = h_0 P^4 d_0$, $d_4(d_0 e_0 + h_0^7 h_5) = P^2 d_0$, $d_4(Pd_0 e_0) = P^3 d_0$, $d_4(P^2 d_0 e_0) = P^4 d_0$, $d_4(e_0 g) = d_0 P d_0$, $d_4(h_0 h_2 h_5) = 0$, $d_4(Ph_2 h_5) = 0$, $d_4(N) = 0$ and $d_5(f_1) = 0$. We therefore also document this extended calculation. This leads to a complete description of the $E_\infty$-term for $t - s \leq 48$, and of the graded ring $\pi_*(S)$ for $* \leq 44$, in the theorems referred to above. We choose to stop at this point because our methods do not seem to simplify the determination of the group structure of $\pi_{45}(S)$, which is due to Tangora [**166**, p. 583].

We also study the Adams spectral sequence for the homotopy cofiber *tmf*/*S* of $\iota$, using the long exact sequence of $E_2$-terms

$$\dots \longrightarrow \mathrm{Ext}_A(\mathbb{F}_2, \mathbb{F}_2) \xrightarrow{\iota} \mathrm{Ext}_{A(2)}(\mathbb{F}_2, \mathbb{F}_2) \longrightarrow \mathrm{Ext}_A(H^*(tmf/S), \mathbb{F}_2) \longrightarrow \dots$$

to obtain information about the *tmf*-Hurewicz homomorphism $\iota\colon \pi_*(S) \to \pi_*(tmf)$. In particular, we show $\iota(\{q\}) = \epsilon_1$ in degree 32 and $\iota(\eta\{u\}) = B\epsilon_1$ in degree 40, both of which involve a shift in Adams filtration. Finally, we show in Theorem 11.89 that for $* \leq 101$ (and for $* = 125$) the *tmf*-Hurewicz image of $\pi_*(S)$ in $\pi_*(tmf)$ equals the sum of the well-known *ko*-Hurewicz image in $\pi_*(ko)$, the group $\pi_3(S) \cong \pi_3(tmf)$, and the self-dual part $\Theta\pi_*(tmf)$ of the $B$-power torsion in $\pi_*(tmf)$. According to Mark Behrens, Mahowald and J.D. Quigley [**27**] this remains true in all degrees.

## 0.10. Finite coefficients

Having determined the differential structure in the Adams spectral sequences for $tmf/2 = tmf \wedge C2$, $tmf/\eta = tmf \wedge C\eta$ and $tmf/\nu = tmf \wedge C\nu$, it is relatively easy to determine the graded abelian group structure of $\pi_*(tmf/2)$, $\pi_*(tmf/\eta)$ and $\pi_*(tmf/\nu)$, together with the action of $\eta$, $\nu$, $B$ and $M$ on these $\pi_*(tmf)$-modules. With a few exceptions, we accomplish this in Chapter 12. At this point it is also relatively easy to calculate $\pi_*(tmf/B)$ and $\pi_*(tmf/(B, M))$, the latter being Anderson self-dual, and $\pi_*(tmf/(2, B))$ and $\pi_*(tmf/(2, B, M))$, where the latter is Brown–Comenetz self-dual. Here $tmf/(2, B) \simeq tmf \wedge M(1, 4)$ and $tmf/(2, B, M) \simeq tmf \wedge M(1, 4, 32)$, where $M(1, 4)$ and $M(1, 4, 32)$ are generalized Moore spectra of types 2 and 3, respectively. For these calculations it is convenient to use two modifications of the classical Adams spectral sequence, which we review in Section 12.6. The first, which we call the delayed sequence, also plays a role in the analysis in Chapter 11 of Steenrod operations in the Adams spectral sequence for an $H_\infty$ ring spectrum. The second, which we call the hastened sequence, was used by Behrens, Hill, Hopkins and Mahowald [**26**] in their construction of the self-map $v_2^{32}\colon \Sigma^{192}M(1,4) \to M(1,4)$ needed to construct $M(1,4,32)$.

## 0.11. Odd primes

We conclude Part III of this book with Chapter 13 on the case of odd primes, which essentially amounts to the case $p = 3$.

Following their construction of the Lubin–Tate spectrum $E_n$ as an $A_\infty$ ring spectrum, with an action by the extended Morava stabilizer group $\mathbb{G}_n$, Hopkins and Miller (ca. 1990) first calculated the homotopy fixed point spectral sequence for $EO_{p-1} = E_{p-1}^{hF}$, where $F$ is a maximal finite subgroup of $\mathbb{G}_{p-1}$. For $p = 3$ there is an equivalence $EO_2 \simeq L_{K(2)}TMF$, so the Hopkins–Miller calculation also amounts to the determination of the descent spectral sequence (and the Adams–Novikov spectral sequence) for $TMF$ at $p = 3$. These calculations were reviewed by Goerss, Hans–Werner Henn, Mahowald and Charles Rezk in [**64**, §3], and by Lee Nave in [**137**, §2.2].

We instead calculate the mod 3 Adams spectral sequence for *tmf*, formed in the category of *tmf*-modules, following Baker–Lazarev [**20**] and Hill [**68**]. We use the Davis–Mahowald spectral sequence from Chapter 2 to give a direct calculation of the Adams $E_2$-term, and use the $H_\infty$ ring structure to directly obtain the Adams differentials. We use the equivalence $tmf \wedge \Psi \simeq tmf_0(2)$ to determine the hidden $\nu$-extensions. Thereafter we determine the product structure on $\pi_*(tmf)$, establish the Brown–Comenetz and Anderson duality theorems, and discuss the *tmf*-Hurewicz image. The introduction to Chapter 13 gives a more detailed overview.

## 0.12. Adams charts

To round out this introduction we display the $(E_2, d_2)$-, $(E_3, d_3)$-, $(E_4, d_4)$- and $E_\infty$-terms of the mod 2 Adams spectral sequence for *tmf* in the range $t - s \le 48$, as well as some bird's-eye view charts of the spectral sequences for *tmf*, *tmf*/2, $tmf/\eta$ and $tmf/\nu$, giving $E_2$- and $E_\infty$-terms for $t - s \le 200$, and $E_\infty$-terms for $0 \le t - s \le 96$ and $96 \le t - s \le 192$. We also show the $E_2$-term for $S$ in the range $t \le 200$, as calculated by `ext`.

REMARK 0.1. We follow the standard convention of drawing Adams charts with the topological degree $t - s$ as the horizontal coordinate and the filtration degree $s$ as the vertical coordinate. The dots give a vector space basis (usually over $\mathbb{F}_2$) for the $E_r$-term shown. Solid lines increasing $(t - s, s)$-bidegrees by $(0, 1)$ and $(1, 1)$ indicate nonzero $h_0$- and $h_1$-multiplications, respectively, while dashed lines increasing bidegrees by $(3, 1)$ indicate nonzero $h_2$-multiplications. Nonzero $d_r$-differentials are shown as arrows of bidegree $(-1, r)$. We usually draw classes that support or are hit by differentials as open (white) circles, while classes that remain to the $E_{r+1}$-term are shown as filled (black) circles.

At the $E_\infty$-term for *tmf*-modules we usually show the $w_1$-power torsion classes in red, while the $w_1$-periodic classes are black. In general, we indicate hidden 2- and $\eta$-extensions by red dashed lines increasing $t - s$ by 0 and 1, respectively, while hidden $\nu$-extensions are shown by red dotted lines increasing $t - s$ by 3.

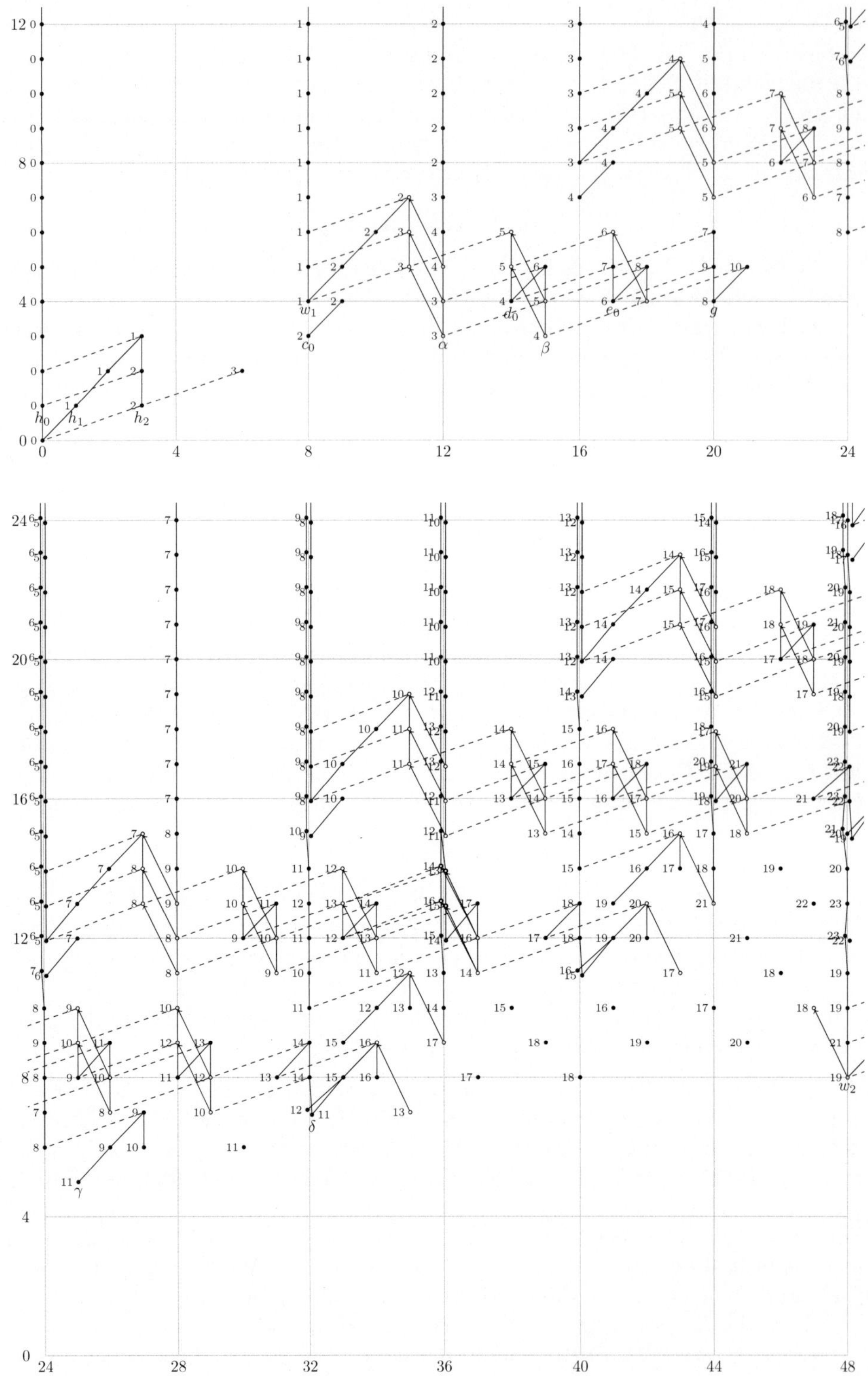

FIGURE 0.1. $(E_2(tmf), d_2)$ for $t - s \leq 48$

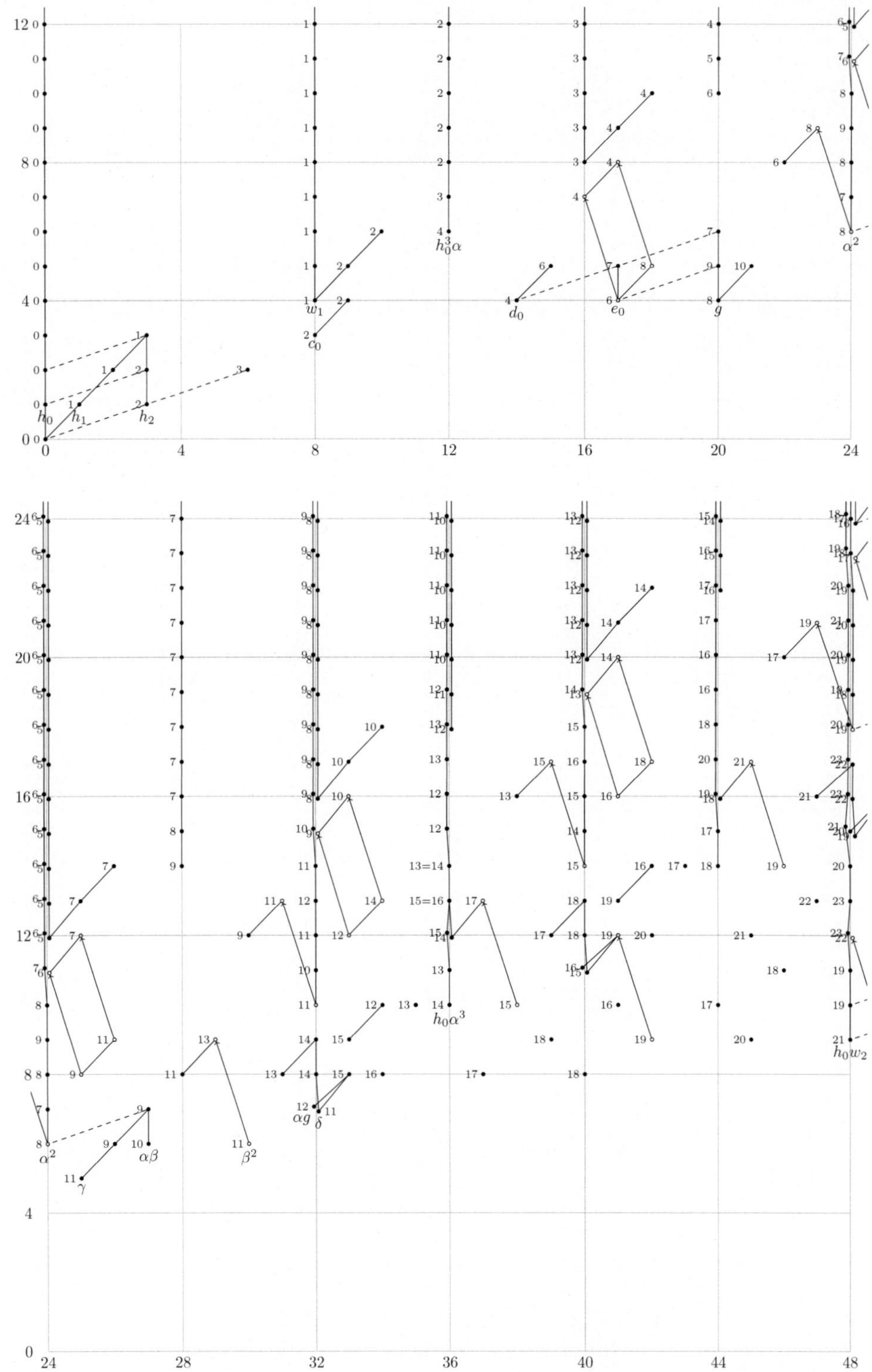

FIGURE 0.2. $(E_3(tmf), d_3)$ for $t - s \leq 48$

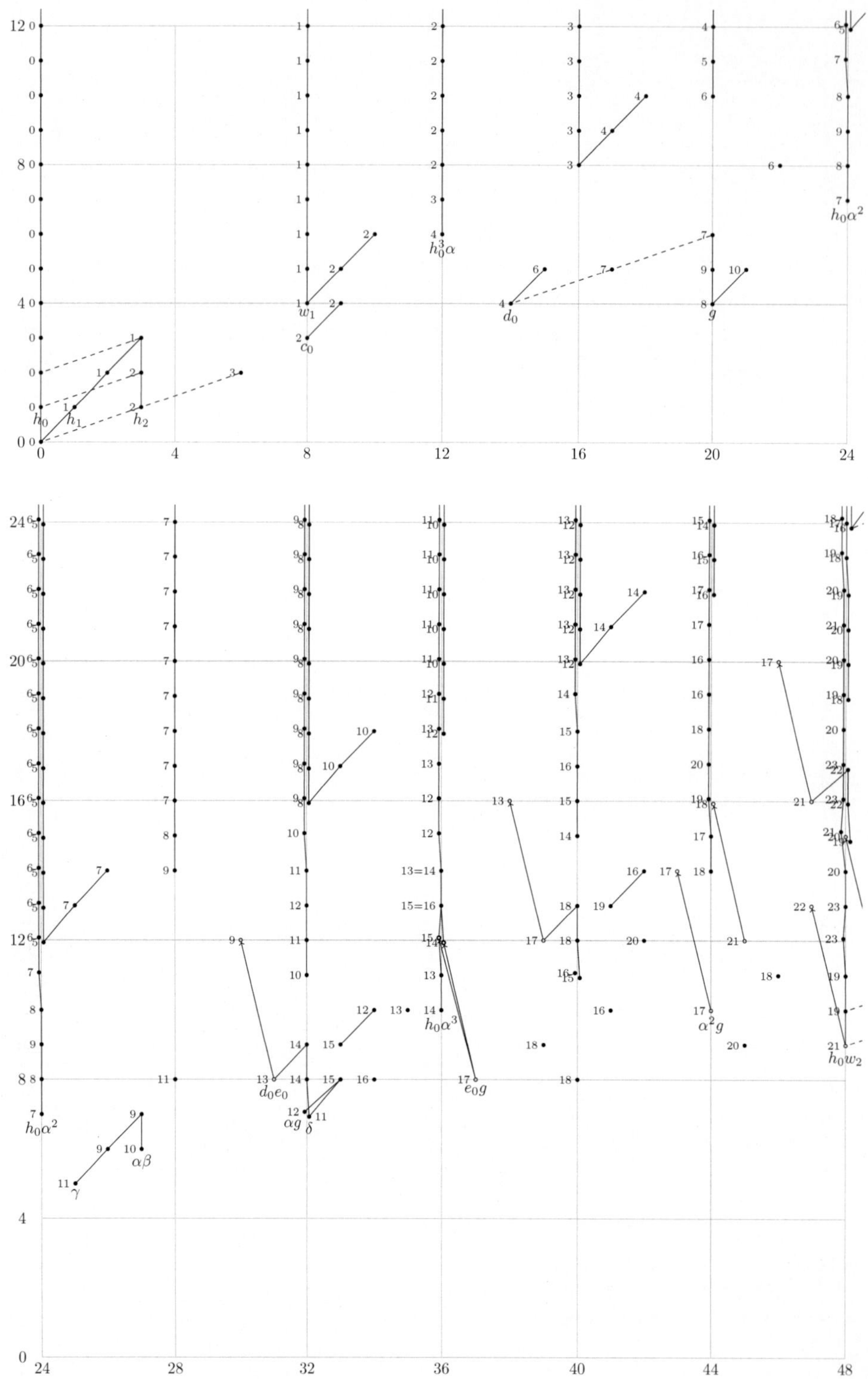

FIGURE 0.3. $(E_4(tmf), d_4)$ for $t - s \leq 48$

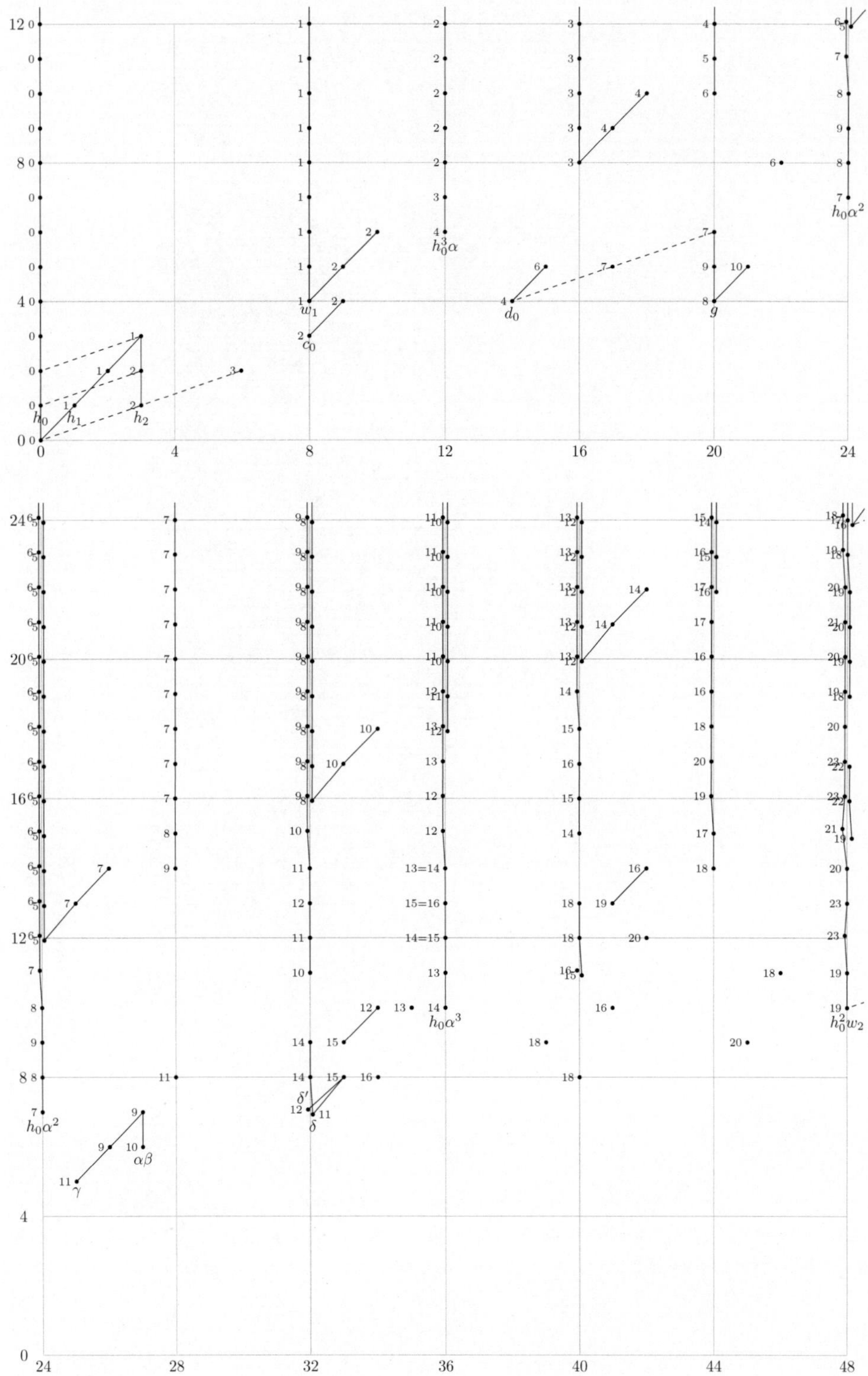


FIGURE 0.4. $E_5(tmf) = E_\infty(tmf)$ for $t - s \leq 48$

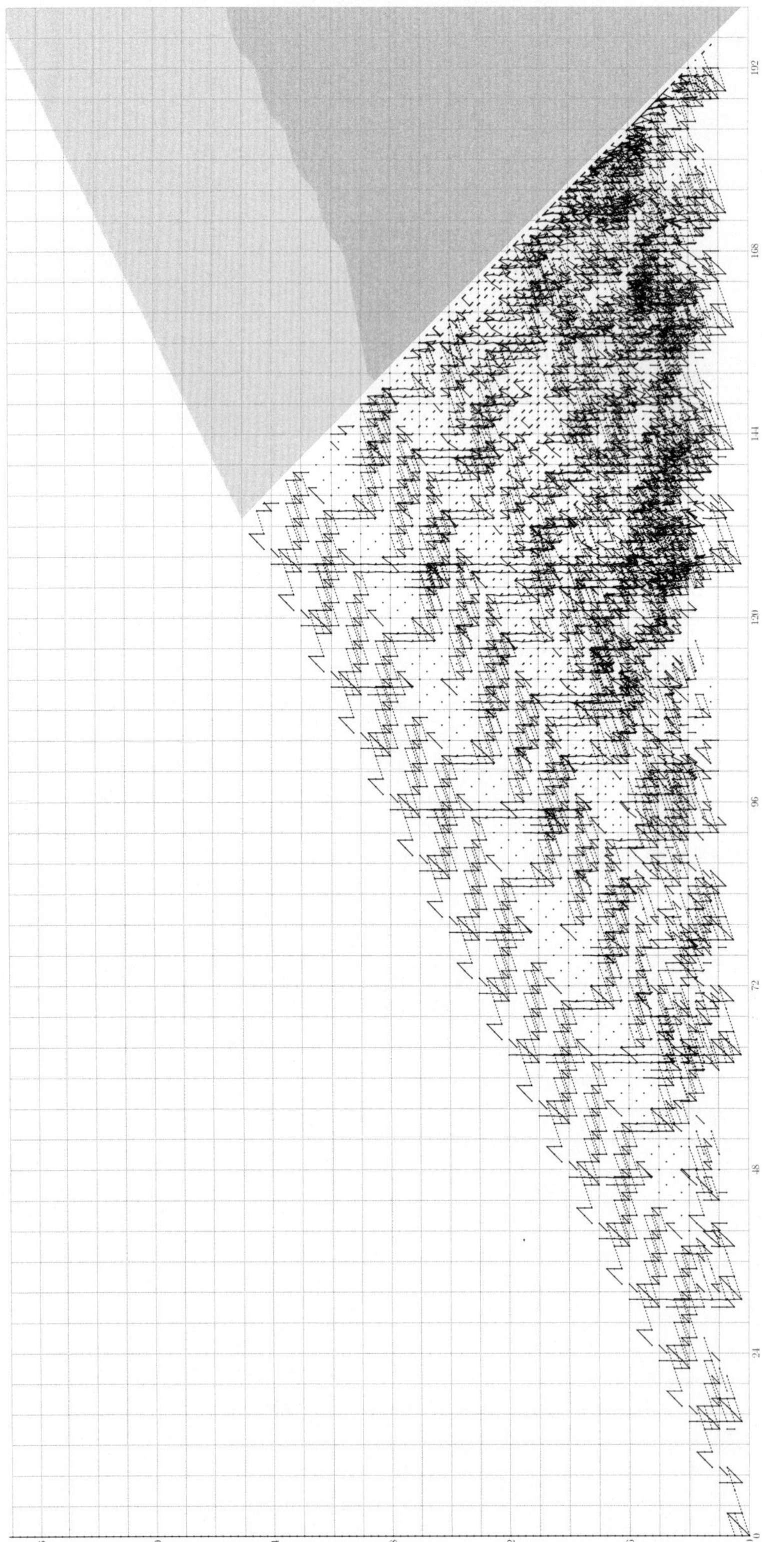


FIGURE 0.5. $E_2^{s,t}(S) = \mathrm{Ext}_A^{s,t}(\mathbb{F}_2, \mathbb{F}_2)$ for $t \leq 200$, cf. Section 0.3

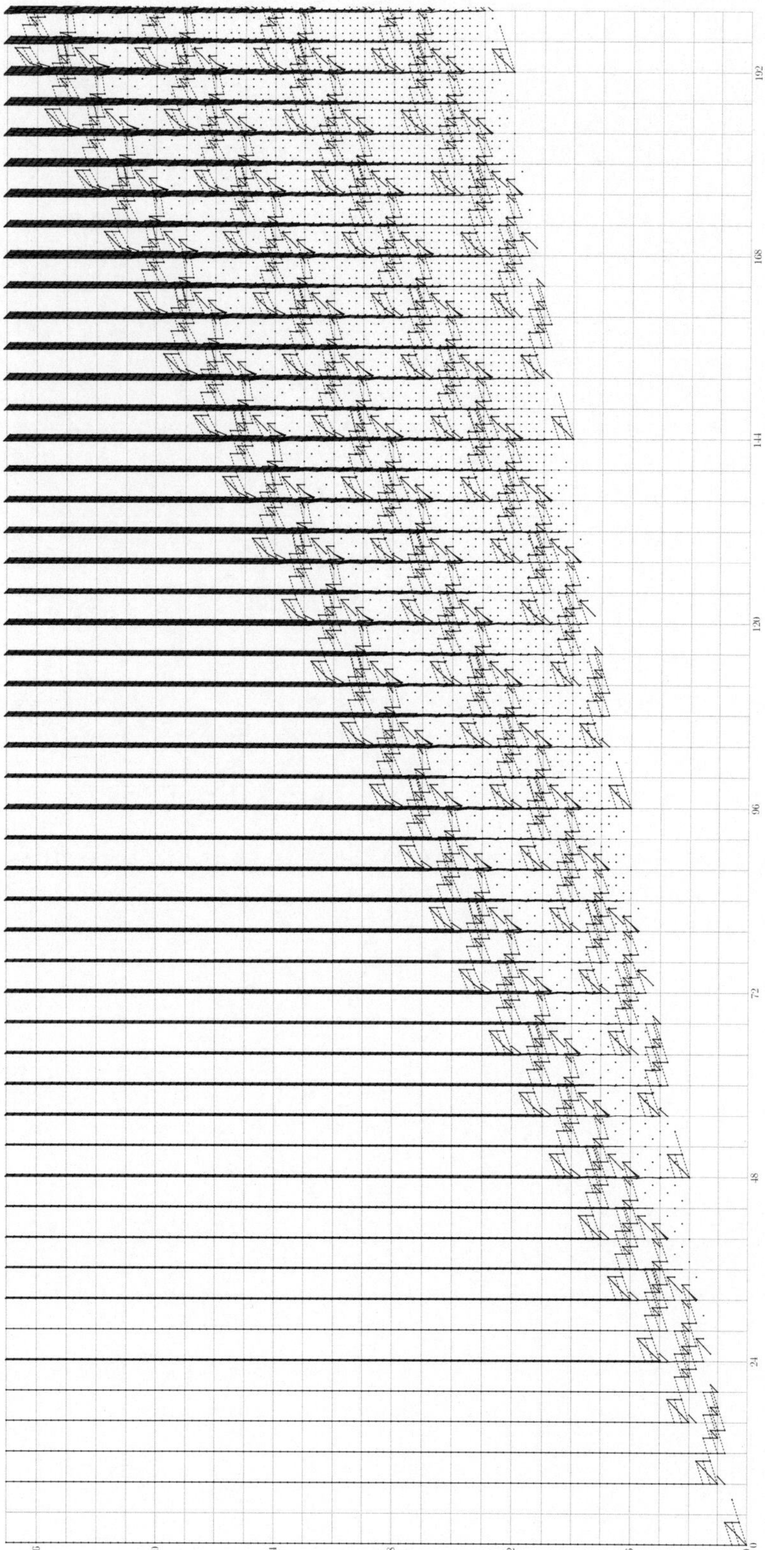


FIGURE 0.6. $E_2^{s,t}(tmf) = \mathrm{Ext}_{A(2)}^{s,t}(\mathbb{F}_2, \mathbb{F}_2)$ for $t - s \leq 200$

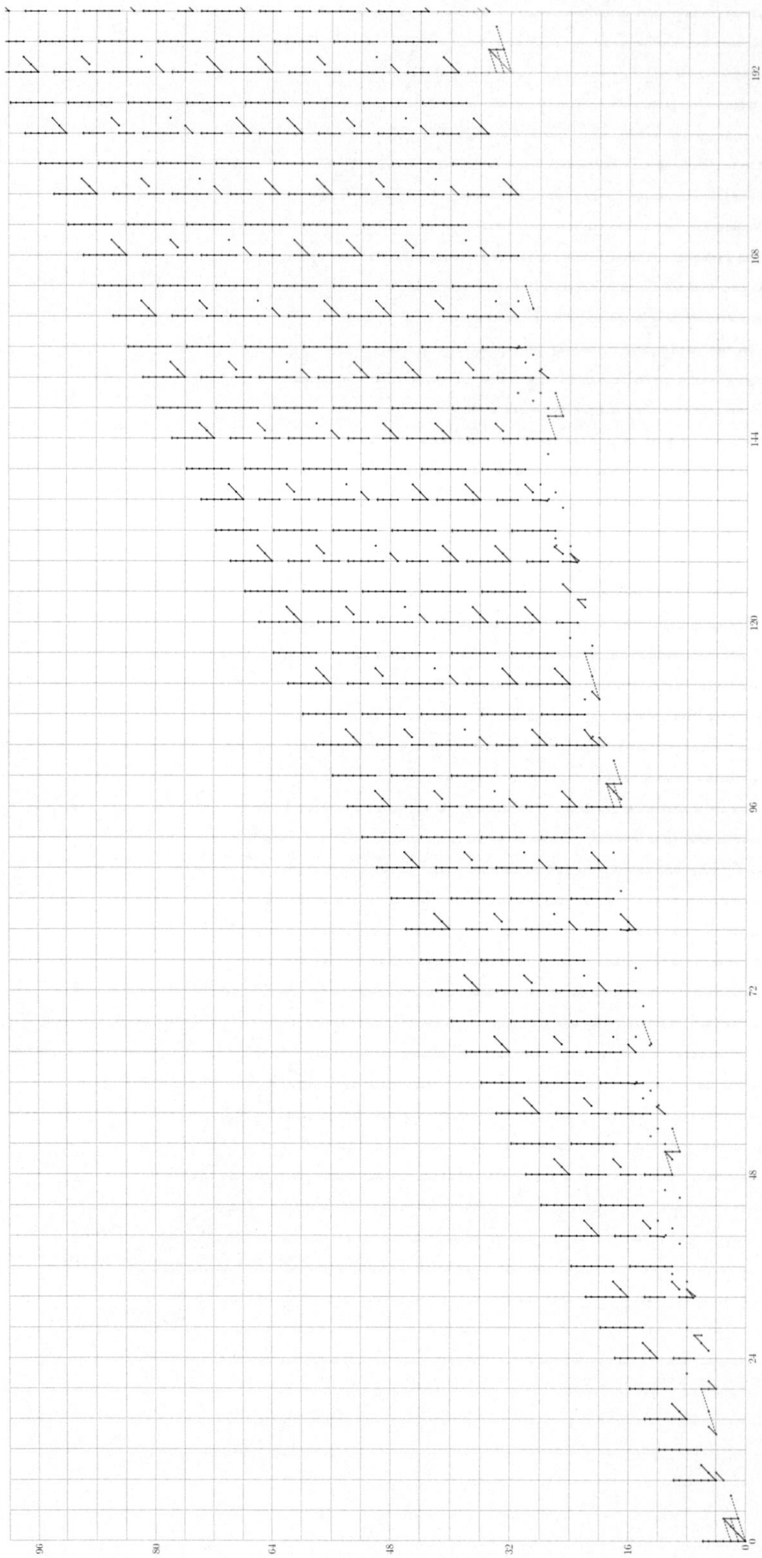


FIGURE 0.7. $E_\infty^{s,t}(tmf) \Longrightarrow \pi_{t-s}(tmf)$ for $0 \le t - s \le 200$

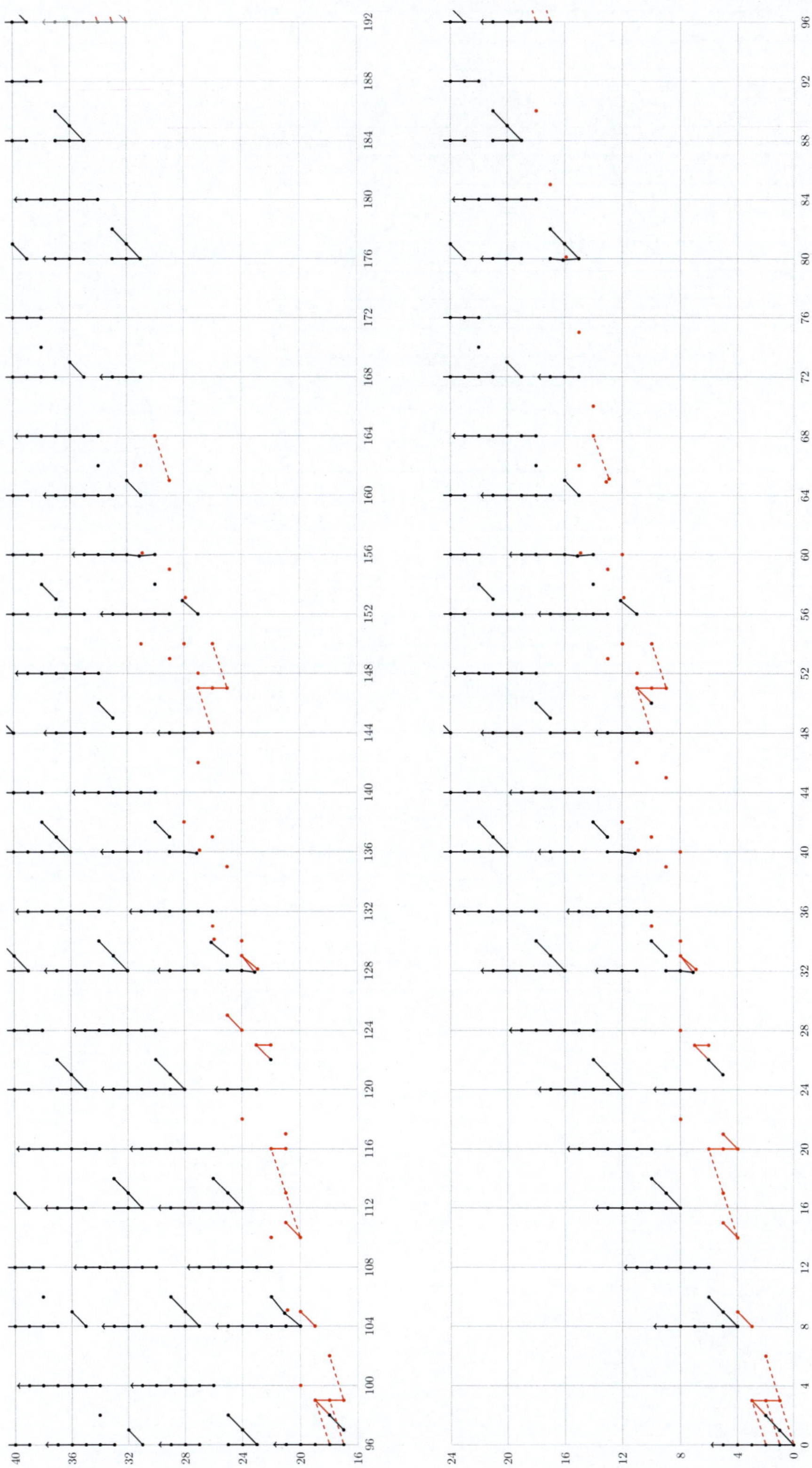

FIGURE 0.8. $E_\infty^{s,t}(tmf)$ for $0 \le t - s \le 96$ and $96 \le t - s \le 192$

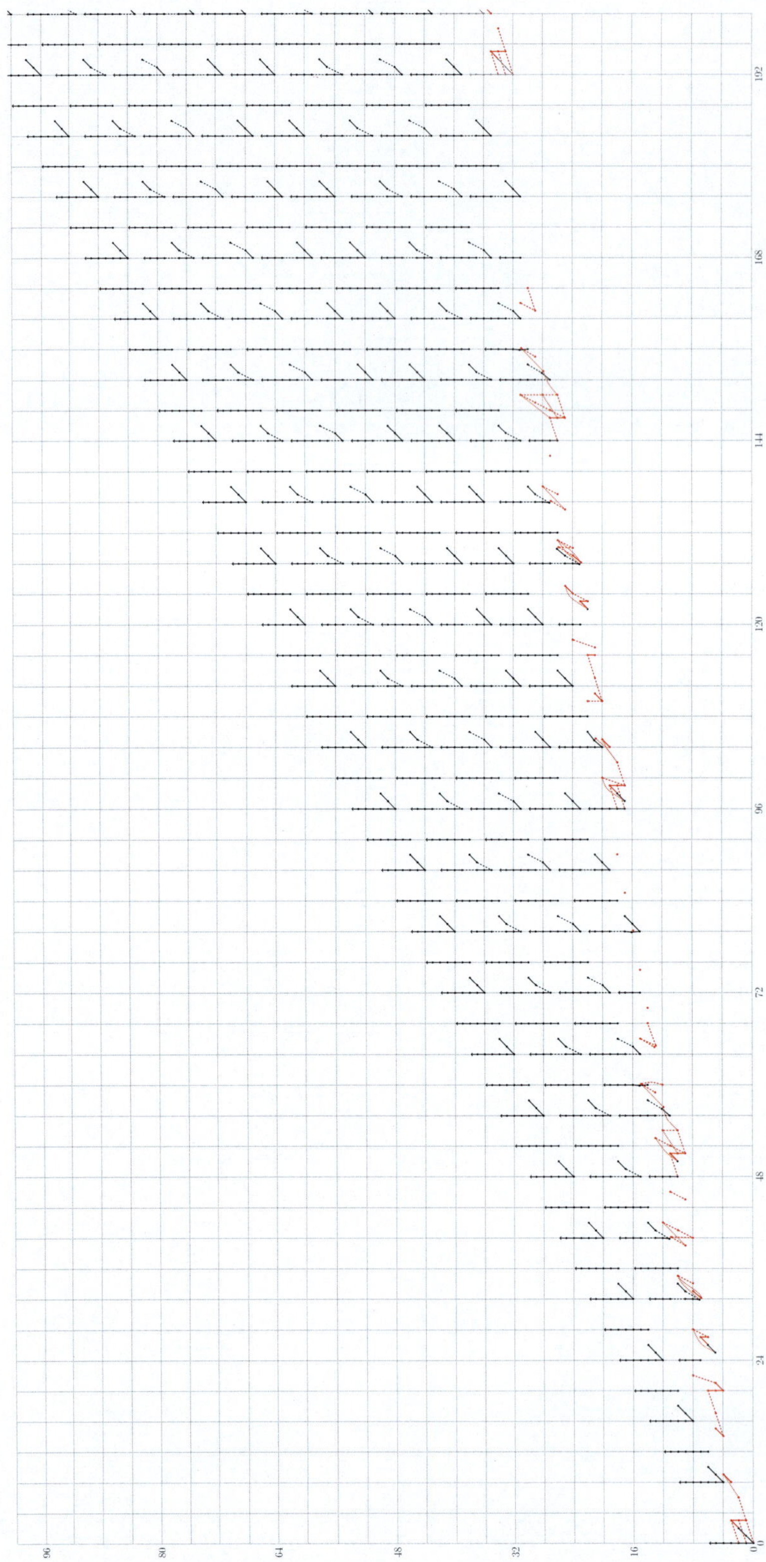


FIGURE 0.9. $\pi_n(tmf)$ for $0 \le n \le 200$

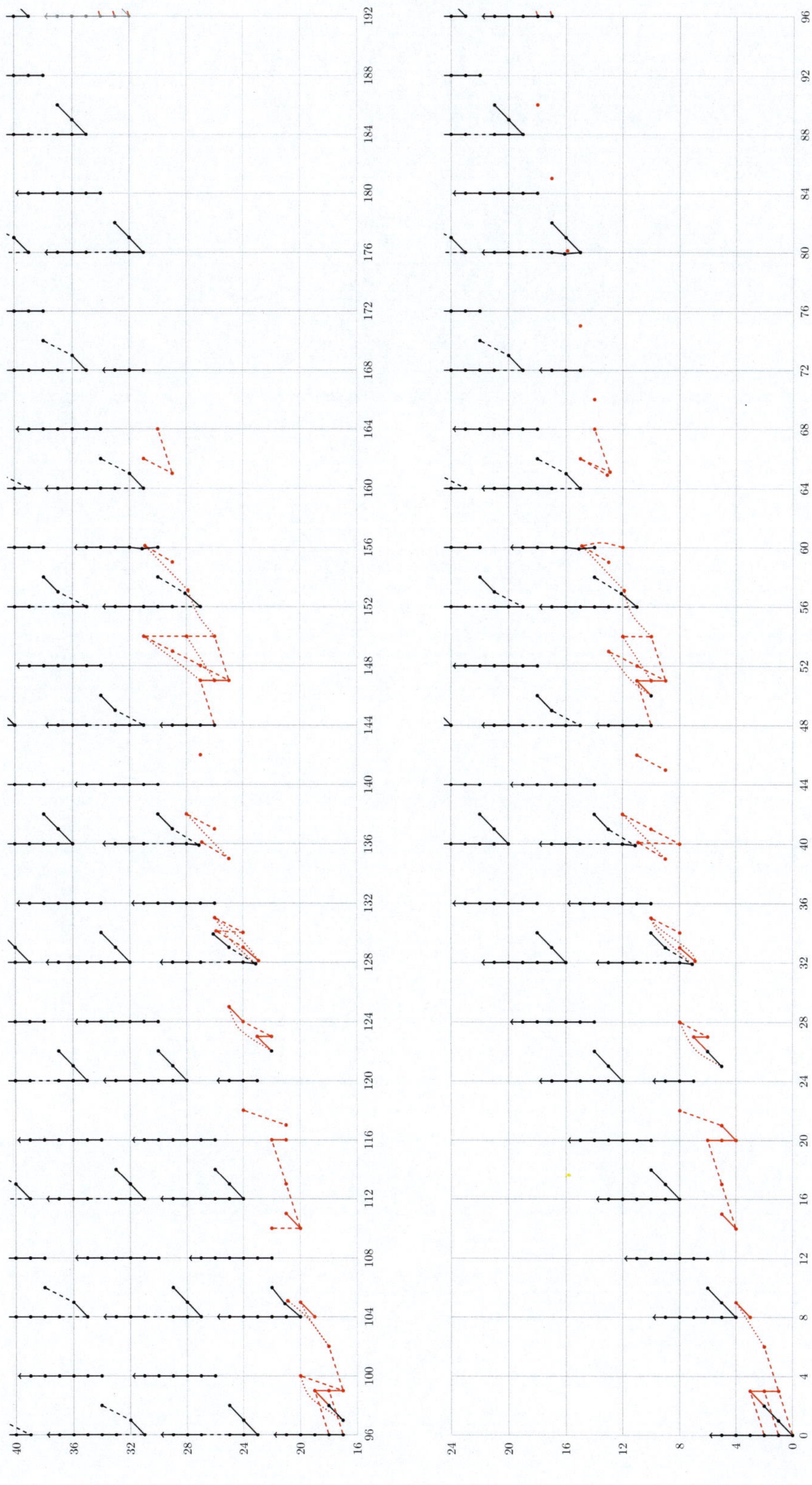

FIGURE 0.10. $\pi_n(tmf)$ for $0 \leq n \leq 96$ and $96 \leq n \leq 192$

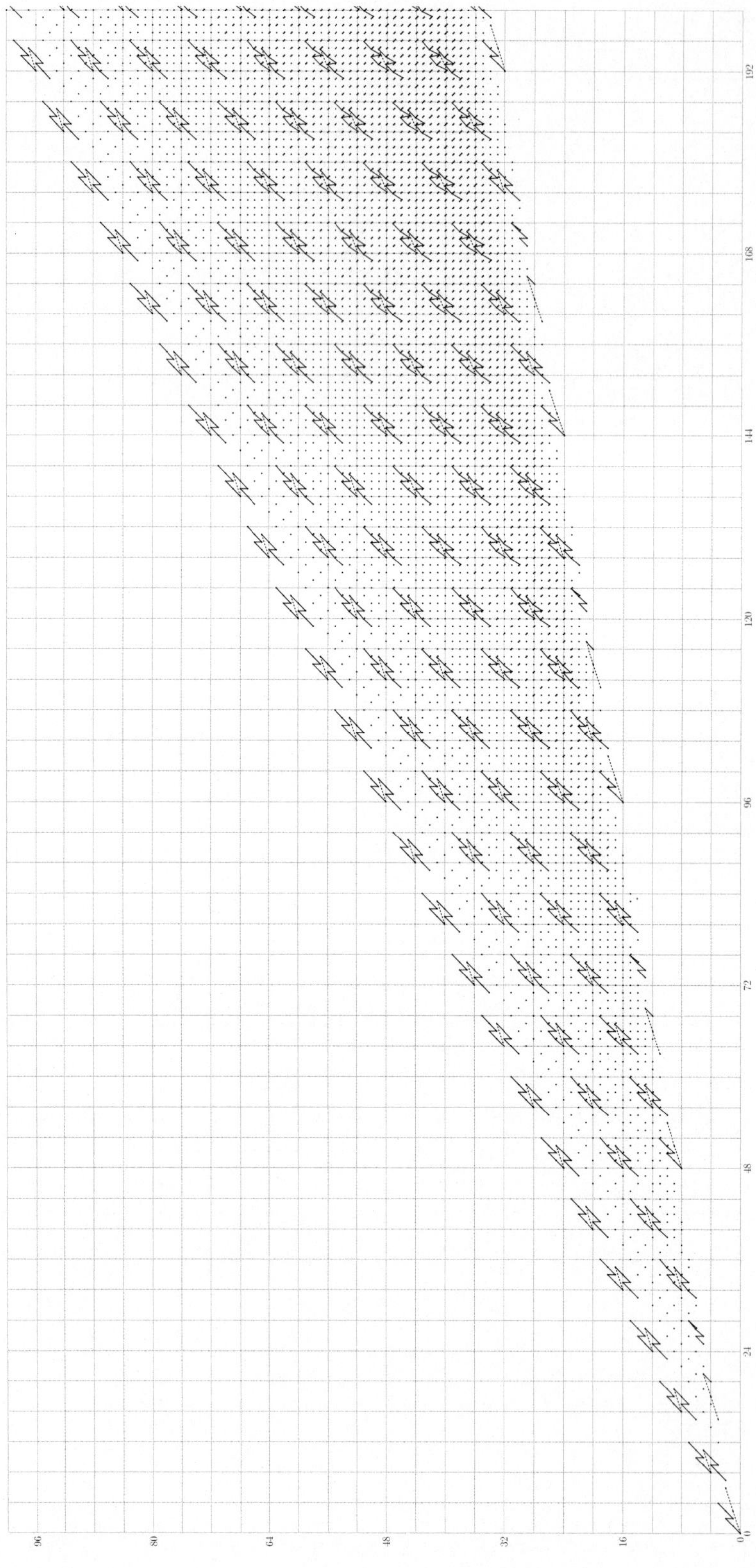


FIGURE 0.11. $E_2^{s,t}(tmf/2) = \mathrm{Ext}^{s,t}_{A(2)}(M_1, \mathbb{F}_2)$ for $t - s \leq 200$

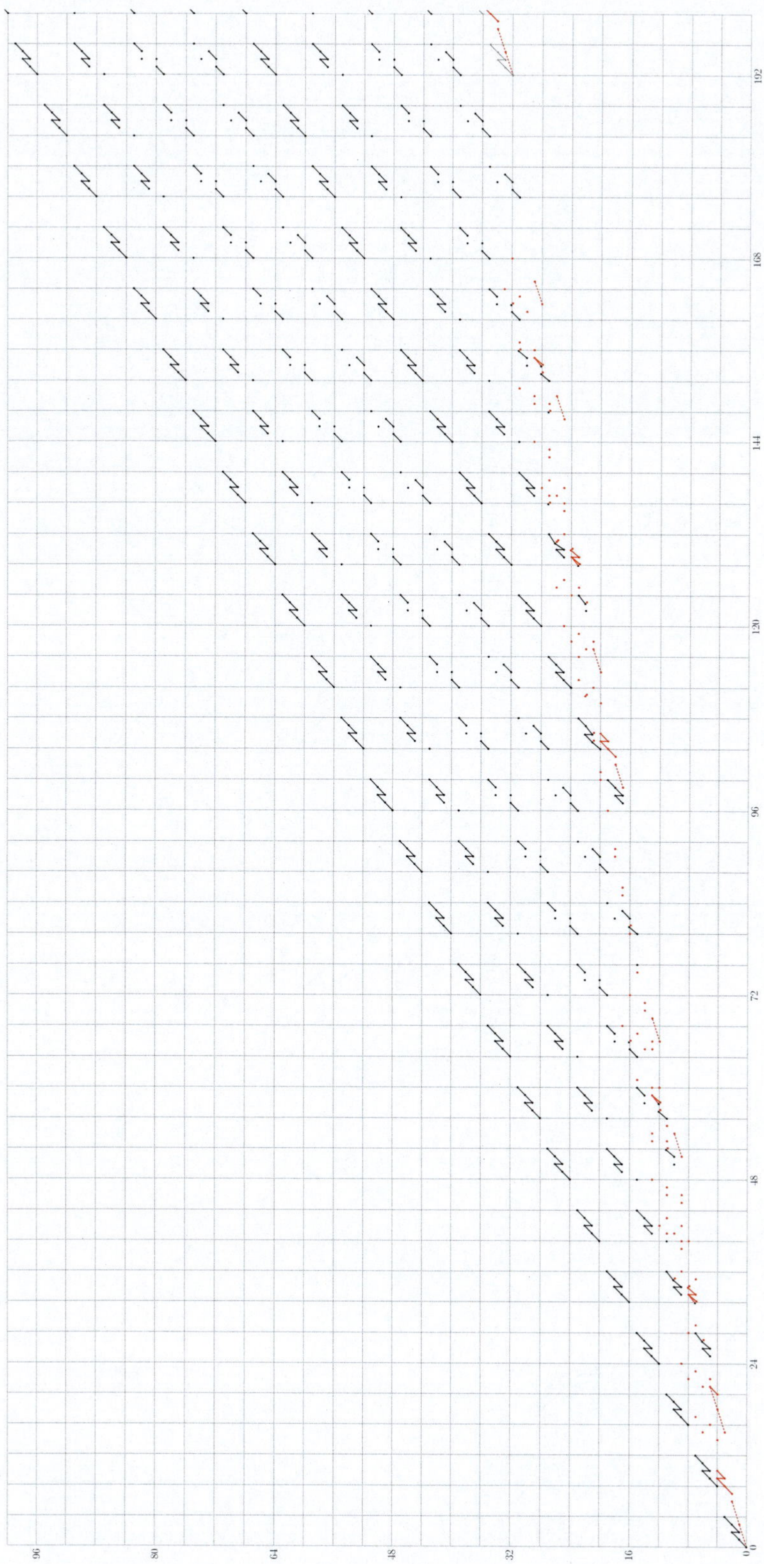


FIGURE 0.12. $E_\infty^{s,t}(tmf/2) \Longrightarrow \pi_{t-s}(tmf/2)$ for $0 \leq t-s \leq 200$

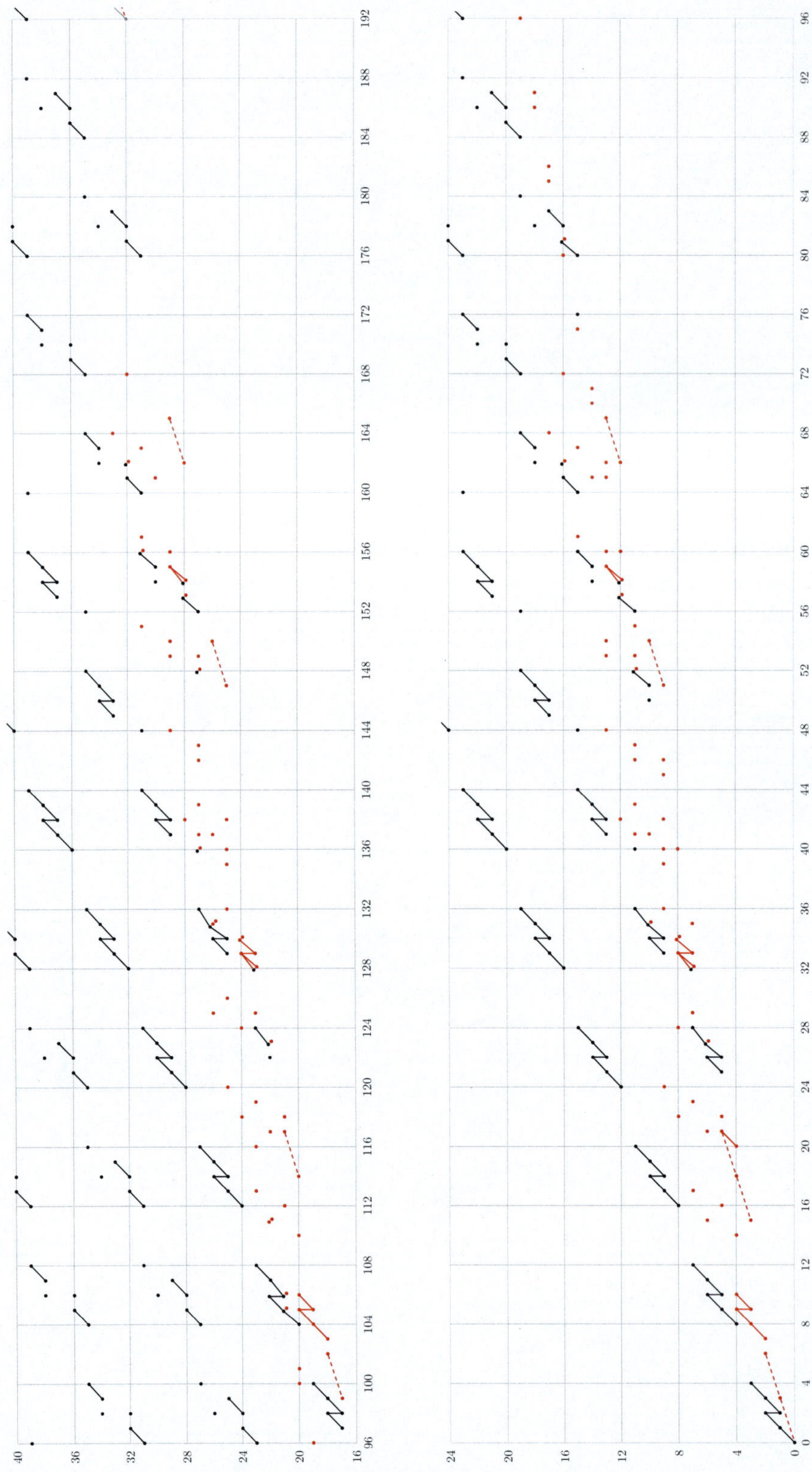


FIGURE 0.13. $E_\infty^{s,t}(tmf/2)$ for $0 \le t-s \le 96$ and $96 \le t-s \le 192$

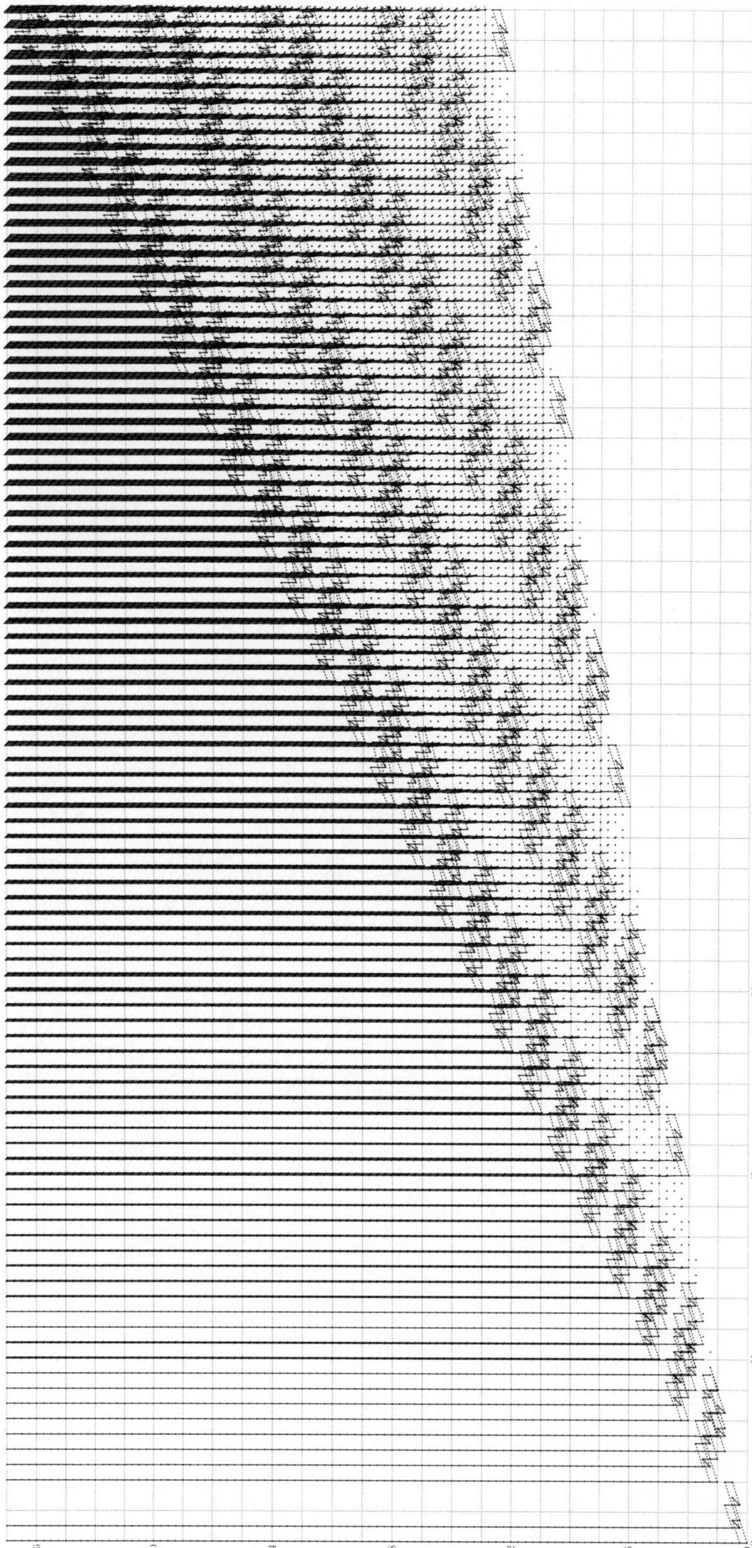


FIGURE 0.14. $E_2^{s,t}(tmf/\eta) = \mathrm{Ext}^{s,t}_{A(2)}(M_2, \mathbb{F}_2)$ for $t - s \leq 200$

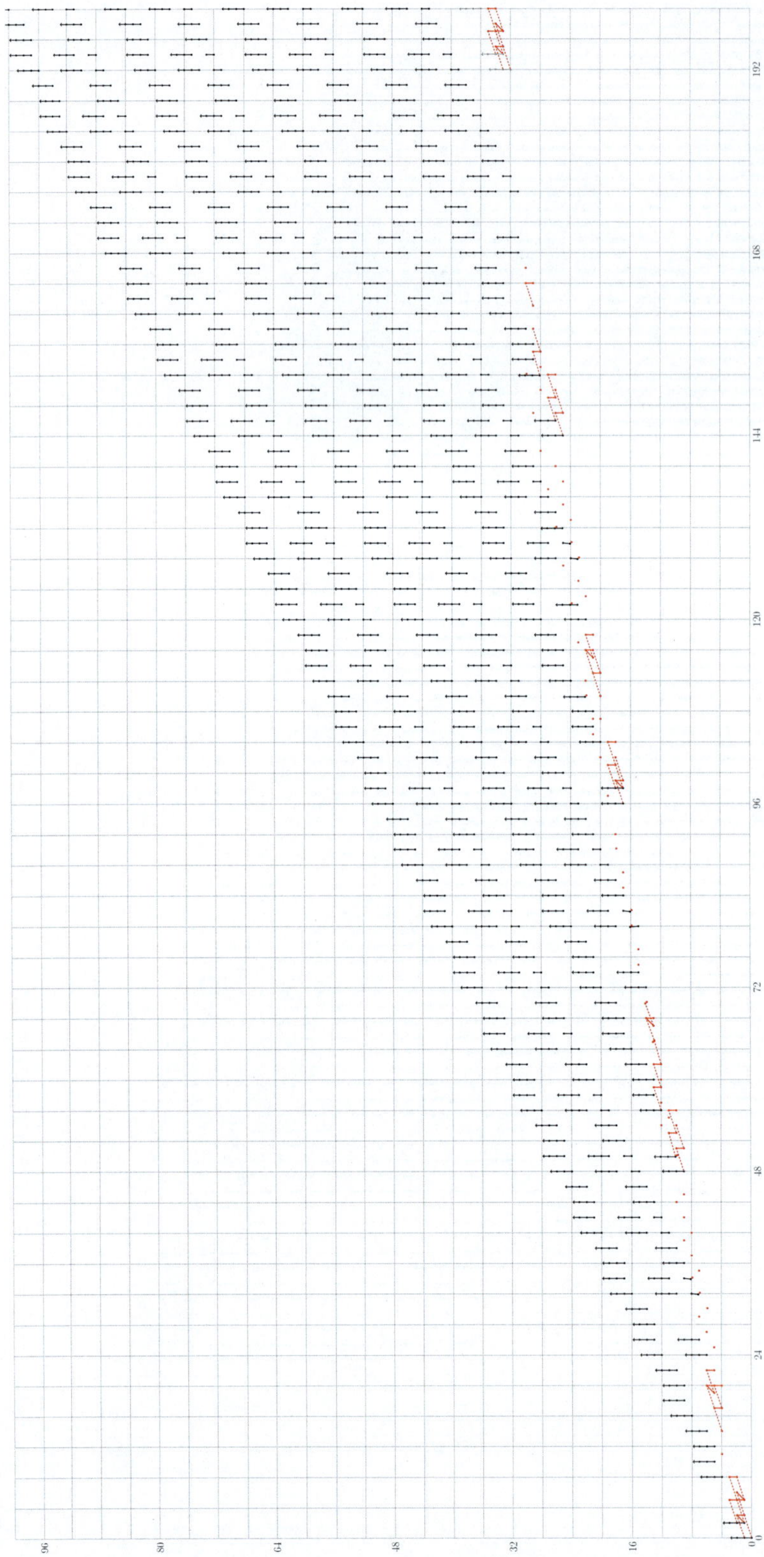


FIGURE 0.15. $E_\infty^{s,t}(tmf/\eta) \Longrightarrow \pi_{t-s}(tmf/\eta)$ for $0 \le t - s \le 200$

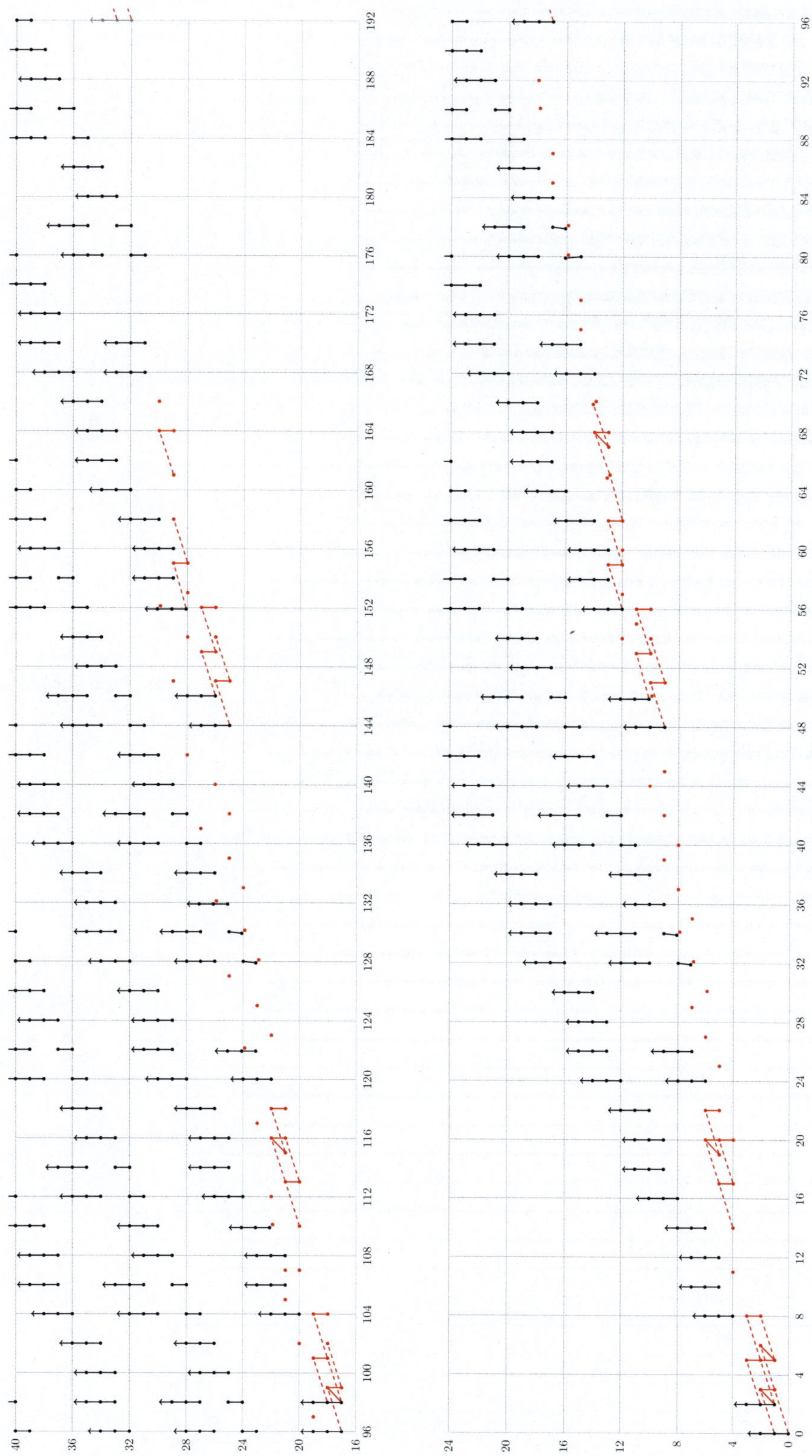

FIGURE 0.16. $E_\infty^{s,t}(tmf/\eta)$ for $0 \leq t-s \leq 96$ and $96 \leq t-s \leq 192$

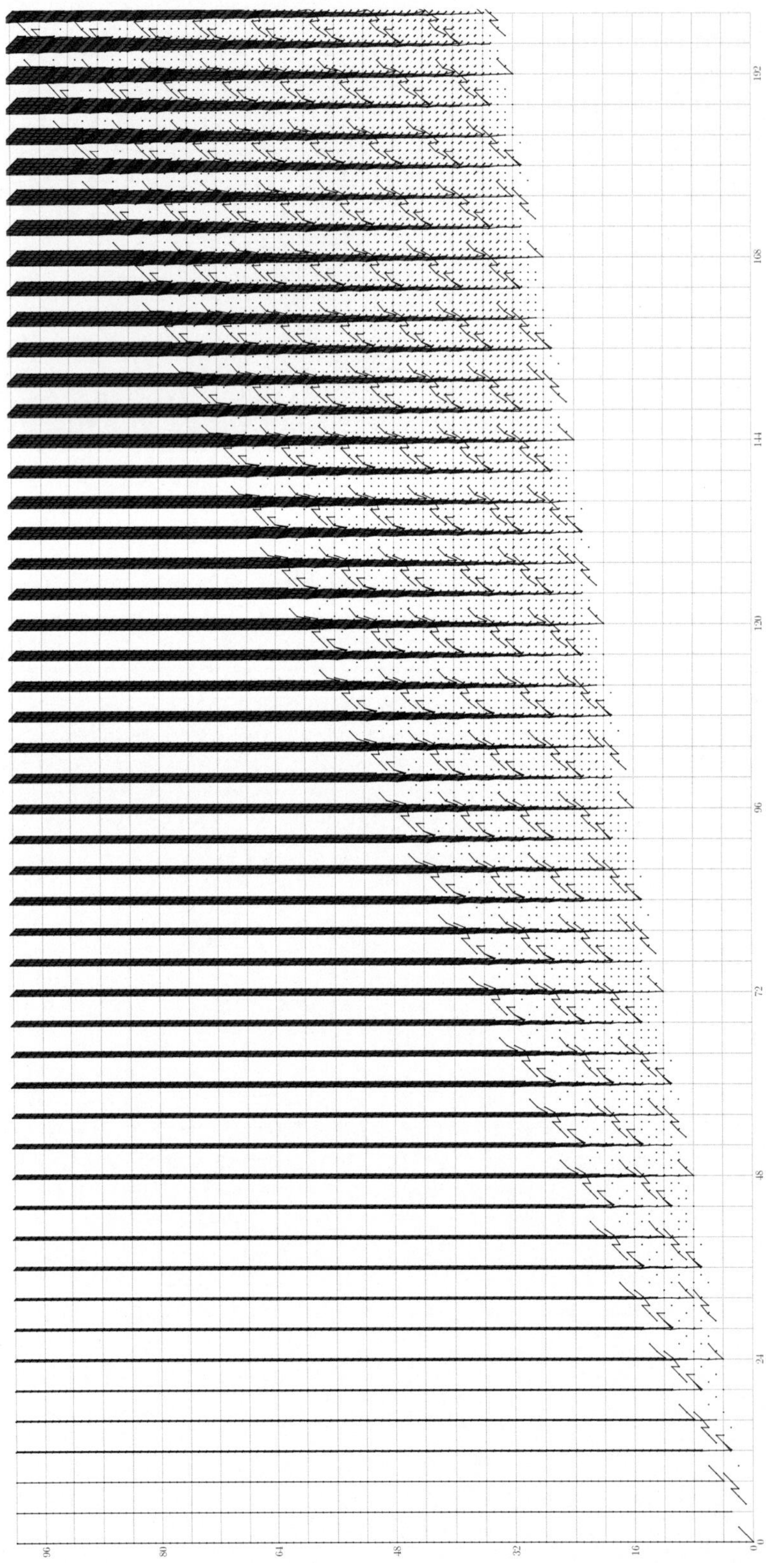


FIGURE 0.17. $E_2^{s,t}(tmf/\nu) = \mathrm{Ext}_{A(2)}^{s,t}(M_4, \mathbb{F}_2)$ for $t - s \leq 200$

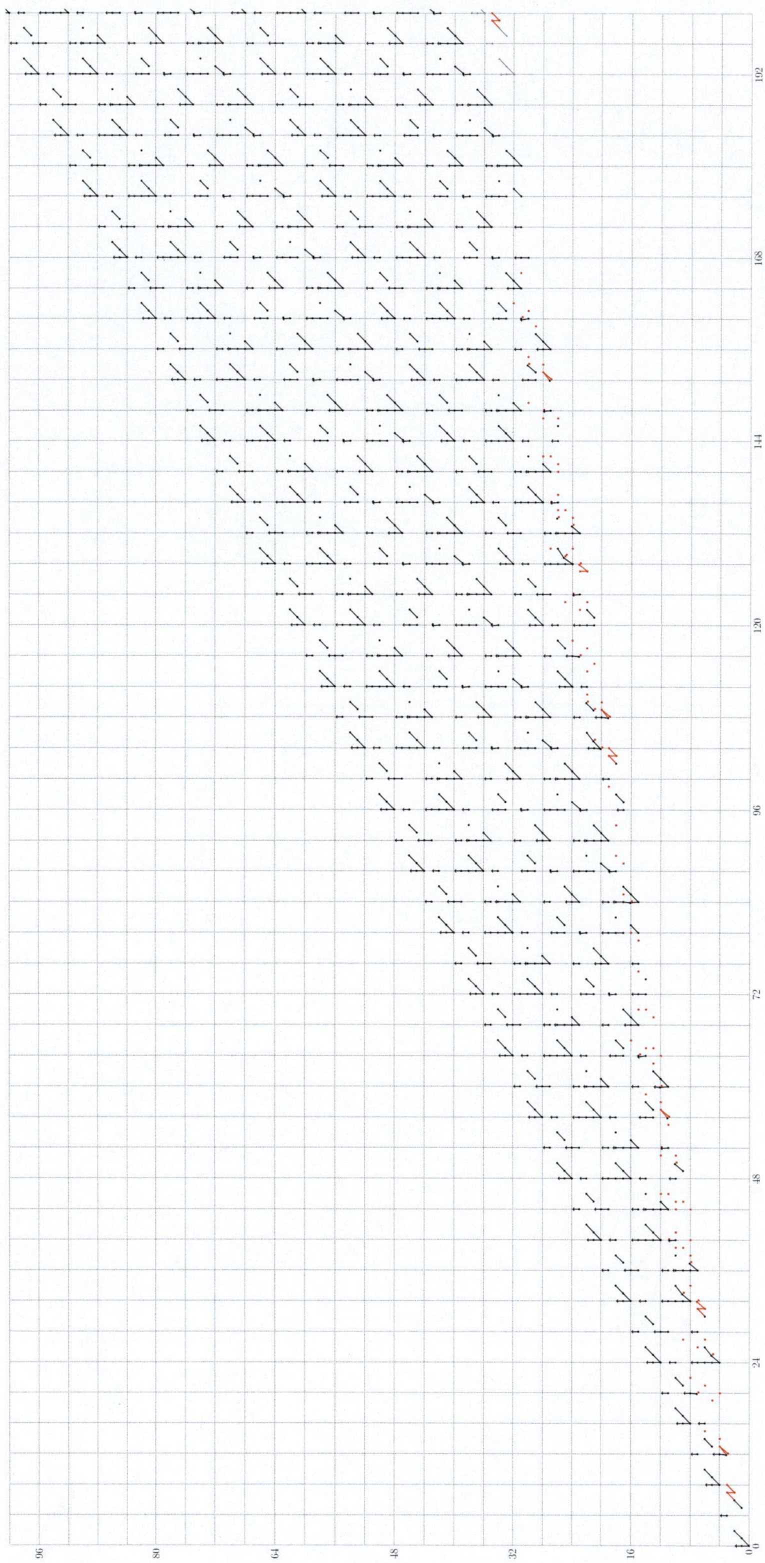


FIGURE 0.18. $E_\infty^{s,t}(tmf/\nu) \Longrightarrow \pi_{t-s}(tmf/\nu)$ for $0 \leq t-s \leq 200$

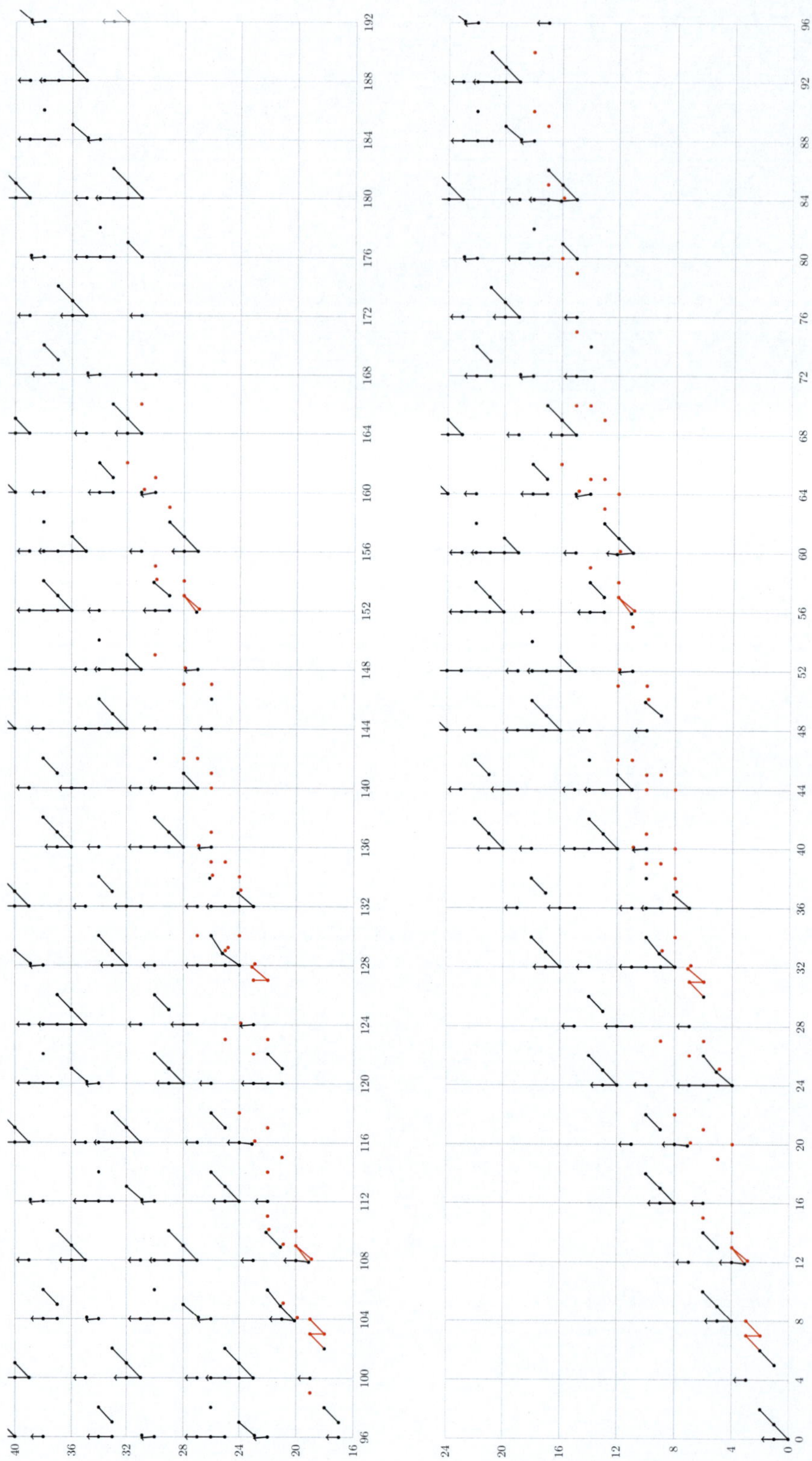


FIGURE 0.19. $E_\infty^{s,t}(tmf/\nu)$ for $0 \le t-s \le 96$ and $96 \le t-s \le 192$

# Part 1

# The Adams $E_2$-term

CHAPTER 1

# Minimal resolutions

The first author's computer program `ext` can calculate minimal resolutions and lift chain maps for finite modules, and for finitely presented modules, over the mod 2 Steenrod algebra $A$ and its subalgebra $A(2)$, in finite ranges of degrees.

## 1.1. The Adams $E_2$-term for $S$

The classical mod 2 Adams spectral sequence for the sphere spectrum $S$ is a strongly convergent algebra spectral sequence

$$E_2^{s,t}(S) = \mathrm{Ext}_A^{s,t}(\mathbb{F}_2, \mathbb{F}_2) \Longrightarrow_s \pi_{t-s}(S)_2^\wedge \,,$$

with $E_2$-term given by Ext over the Steenrod algebra $A$, and abutment the 2-completed homotopy groups of spheres.

The $A$-module component of the program `ext` will calculate a minimal resolution

$$\dots \xrightarrow{\partial} C_2 \xrightarrow{\partial} C_1 \xrightarrow{\partial} C_0 \xrightarrow{\epsilon} \mathbb{F}_2 \to 0$$

of $\mathbb{F}_2$ by free $A$-modules $C_s$, in a finite range of filtration degrees $s \ge 0$ and internal degrees $t \ge 0$. As part of the calculation it will choose a basis $\{s_g^*\}_g$ for each $A$-module $C_s$, indexed by non-negative integers $g \ge 0$, in a well-defined deterministic order of non-decreasing internal degrees. By minimality the coboundaries in the induced cocomplex

$$\dots \xleftarrow{\delta} \mathrm{Hom}_A(C_2, \mathbb{F}_2) \xleftarrow{\delta} \mathrm{Hom}_A(C_1, \mathbb{F}_2) \xleftarrow{\delta} \mathrm{Hom}_A(C_0, \mathbb{F}_2) \leftarrow 0$$

are zero, so $\mathrm{Ext}_A^s(\mathbb{F}_2, \mathbb{F}_2) = \mathrm{Hom}_A(C_s, \mathbb{F}_2)$.

DEFINITION 1.1. For $s, g \ge 0$, let $s_g \in \mathrm{Ext}_A^s(\mathbb{F}_2, \mathbb{F}_2) = \mathrm{Hom}_A(C_s, \mathbb{F}_2)$ be the cocycle that is dual to the $g$'th generator $s_g^*$ of $C_s$, i.e., the homomorphism that takes the value 1 on $s_g^*$ and maps the other basis elements to 0. The internal degree $t$ of $s_g$ is equal to the internal degree of the generator $s_g^*$.

The result of such a calculation for $s \le 100$ and $t \le 200$ is shown in Figures 1.1 to 1.8. The charts use Adams indexing, with the topological degree $t - s$ on the horizontal axis and the filtration degree $s$ on the vertical axis. The dot with label $g$ in bidegree $(t-s, s)$ corresponds to the generator $s_g \in \mathrm{Ext}_A^{s,t}(\mathbb{F}_2, \mathbb{F}_2)$, and these give a basis for $\mathrm{Ext}_A(\mathbb{F}_2, \mathbb{F}_2)$ as a bigraded $\mathbb{F}_2$-vector space, in this range of bidegrees.

A small part of the minimal resolution $(C_*, \partial)$, with $0 \le s \le 6$ and $0 \le t \le 22$, is shown in Table 1.2. Here we use the Milnor basis for $A$, with $Sq^{(i_1,\dots,i_r)}$ dual to $\xi_1^{i_1} \cdots \xi_r^{i_r}$ in the monomial basis for the dual Steenrod algebra, cf. Section 3.1.

EXAMPLE 1.2. The class $0_0 = 1$ in $\mathrm{Ext}_A^{0,0}(\mathbb{F}_2, \mathbb{F}_2)$ is the algebra unit. For each $i \ge 0$ the class $1_i = h_i$ in $\mathrm{Ext}_A^{1,2^i}(\mathbb{F}_2, \mathbb{F}_2)$ is dual to the algebra indecomposable $Sq^{2^i}$ in $A$. For each $s \ge 0$ the class $s_0 = h_0^s$ in $\mathrm{Ext}_A^{s,s}(\mathbb{F}_2, \mathbb{F}_2)$ detects $2^s \in \pi_0(S)_2^\wedge = \mathbb{Z}_2$.

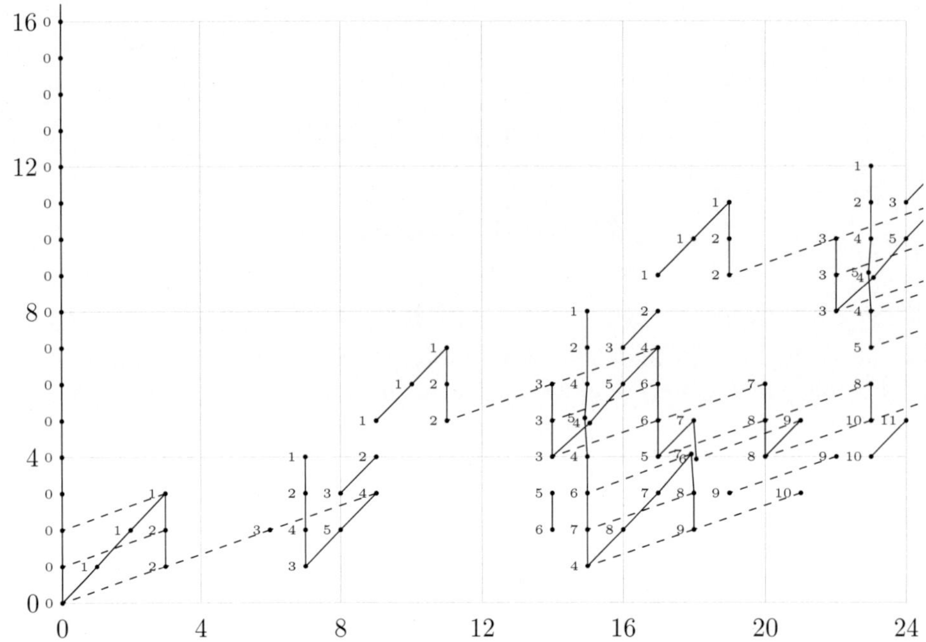

FIGURE 1.1. $\mathrm{Ext}_A^{s,t}(\mathbb{F}_2, \mathbb{F}_2)$ for $0 \le t - s \le 24$

The next algebra indecomposable in $\mathrm{Ext}_A(\mathbb{F}_2, \mathbb{F}_2)$ is $3_3 = c_0 \in \mathrm{Ext}_A^{3,11}(\mathbb{F}_2, \mathbb{F}_2)$, in Adams bidegree $(t - s, s) = (8, 3)$.

REMARK 1.3. To make these calculations, install `ext`, go to the directory `A`, and let `S.def` be a text file containing the numbers `1 0`. This defines the $A$-module with a single $\mathbb{F}_2$-generator in internal degree 0, necessarily with trivial action by each $Sq^i$ for $i \ge 1$. Use `newmodule S S.def` to create the module subdirectory `S`. Go to this subdirectory, and run `dims 0 75 &` (taking a couple of minutes) to calculate the minimal resolution for $0 \le s \le 40$ and $0 \le t \le 75$. The upper bound for $s$ is specified in the text file `MAXFILT`. A much higher upper bound for $t$ will take significantly longer to compute. When `dims` is finished, use `report` to extract the files `Shape`, `himults` and `lines` from the calculation. Thereafter use

```
chart 0 16 0 24 Shape himults Ext-A-0-24.tex Ext-A-F2
pdflatex Ext-A-0-24.tex
```

to obtain an Adams chart such as the one in Figure 1.1. Similarly, use

```
chart 0 24 24 48 Shape himults Ext-A-24-48.tex Ext-A-F2
pdflatex Ext-A-24-48.tex
```

to obtain an Adams chart such as the one in Figure 1.2.

At this stage, each file `Diff.s` for $0 \le s \le 40$ contains a description in internal degrees $0 \le t \le 75$ of the boundary homomorphism $\partial \colon C_s \to C_{s-1}$. More precisely, it contains a list of the internal degrees of the free $A$-module generators $s_0^*, s_1^*, \ldots$ of $C_s$, together with expressions for the boundaries $\partial(s_g^*)$ in $C_{s-1}$, as linear combinations of the corresponding free $A$-module generators of $C_{s-1}$. The coefficients of these linear combinations lie in $A$, and are encoded in an efficient machine readable

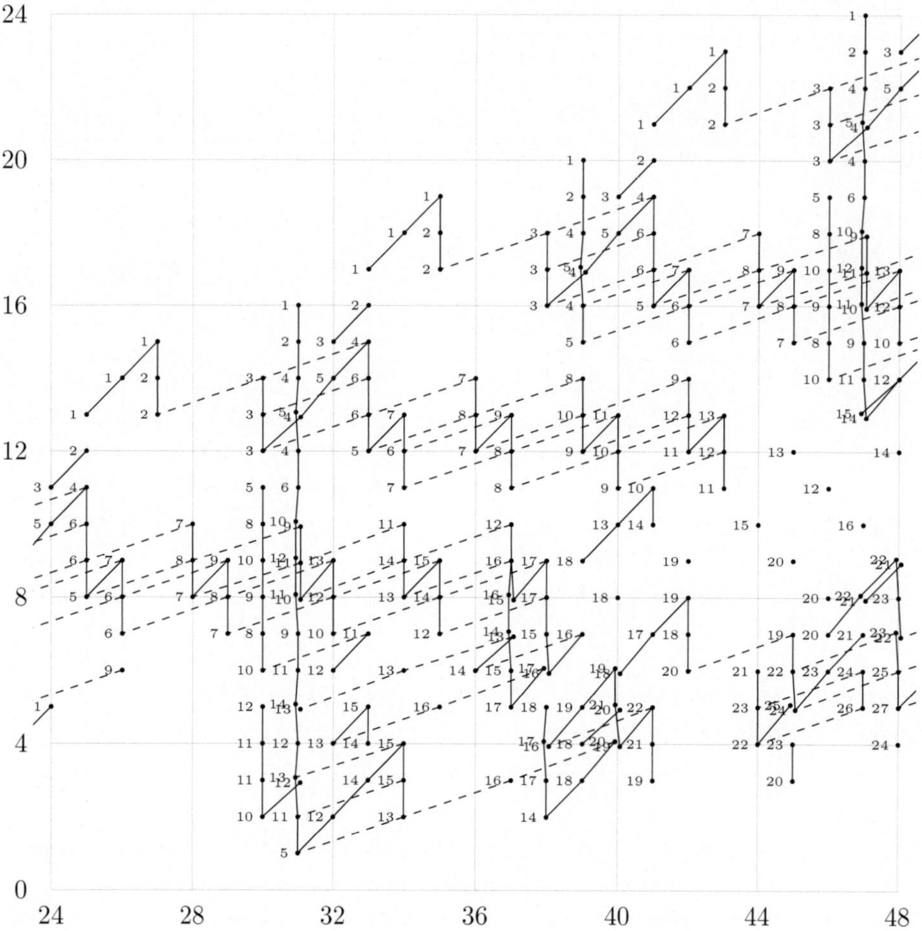


FIGURE 1.2. $\mathrm{Ext}_A^{s,t}(\mathbb{F}_2, \mathbb{F}_2)$ for $24 \leq t-s \leq 48$

format in the files `Diff.s`. They can, however, be converted to a humanly readable format using commands of the following form.

```
convert Diff.s hDiff.s 2 1 1 i
```

In the resulting file `hDiff.s` the coefficients in $A$ are expressed in terms of the Milnor basis. This can be done for all filtration degrees at once by running `seeres`, which creates the file `resolution`, giving humanly readable formulas for all of the boundary operators $\partial\colon C_s \to C_{s-1}$. The information in Table 1.2 was calculated in this way.

Yoneda composition of $s'$- and $s''$-fold $A$-module extensions defines a pairing

$$\mathrm{Ext}_A^{s',t'}(\mathbb{F}_2, \mathbb{F}_2) \otimes \mathrm{Ext}_A^{s'',t''}(\mathbb{F}_2, \mathbb{F}_2) \longrightarrow \mathrm{Ext}_A^{s'+s'',t'+t''}(\mathbb{F}_2, \mathbb{F}_2)$$

taking $x \otimes y$ to $xy$. For varying $s'$, $s''$, $t'$ and $t''$ these make $\mathrm{Ext}_A(\mathbb{F}_2, \mathbb{F}_2)$ a bigraded commutative algebra over $\mathbb{F}_2$. The Hopf algebra structure on $A$ also leads to a tensor product of $A$-modules and an induced pairing of Ext-groups, which we have already noted coincides with the Yoneda pairing.

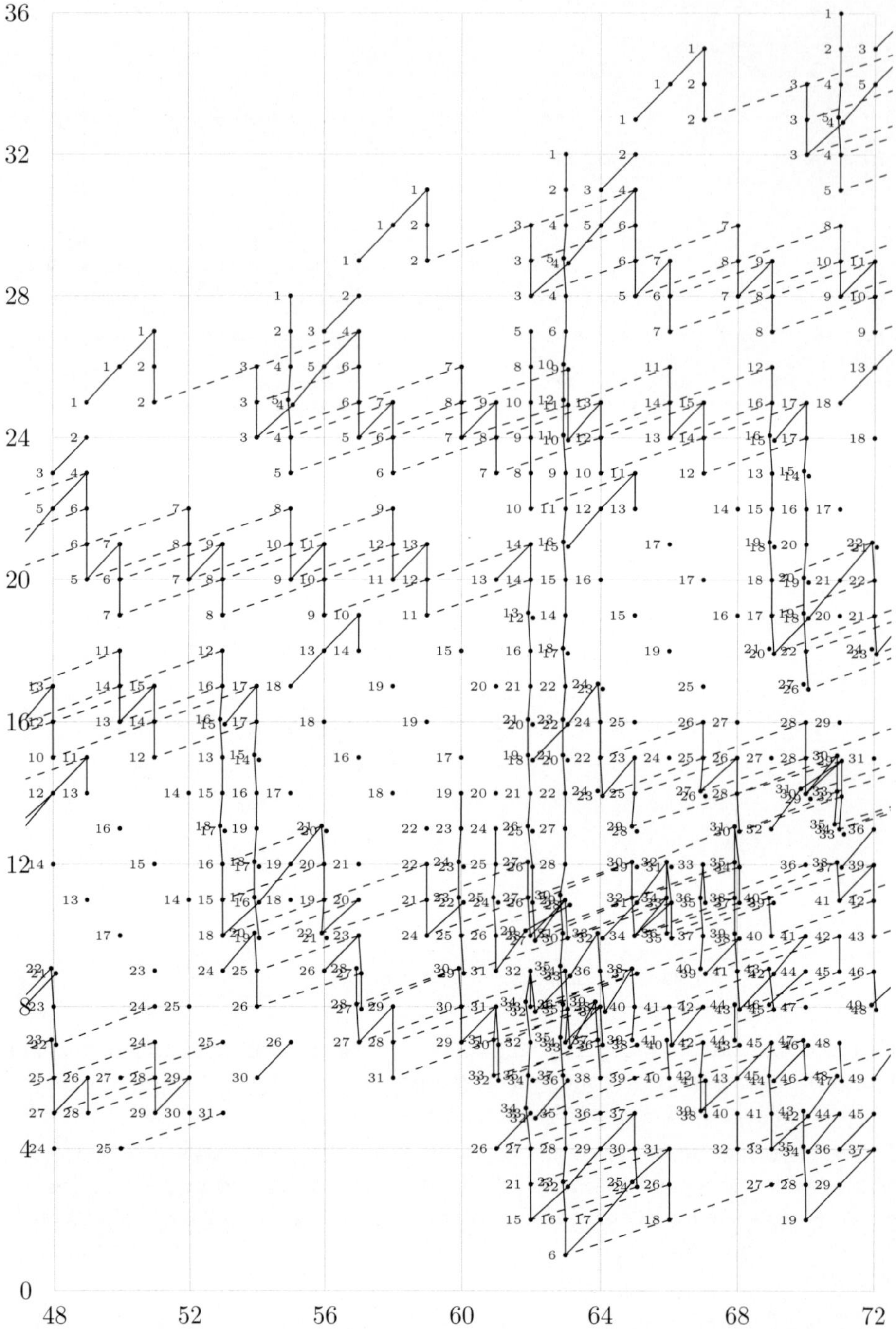

FIGURE 1.3. $\mathrm{Ext}_A^{s,t}(\mathbb{F}_2, \mathbb{F}_2)$ for $48 \le t - s \le 72$

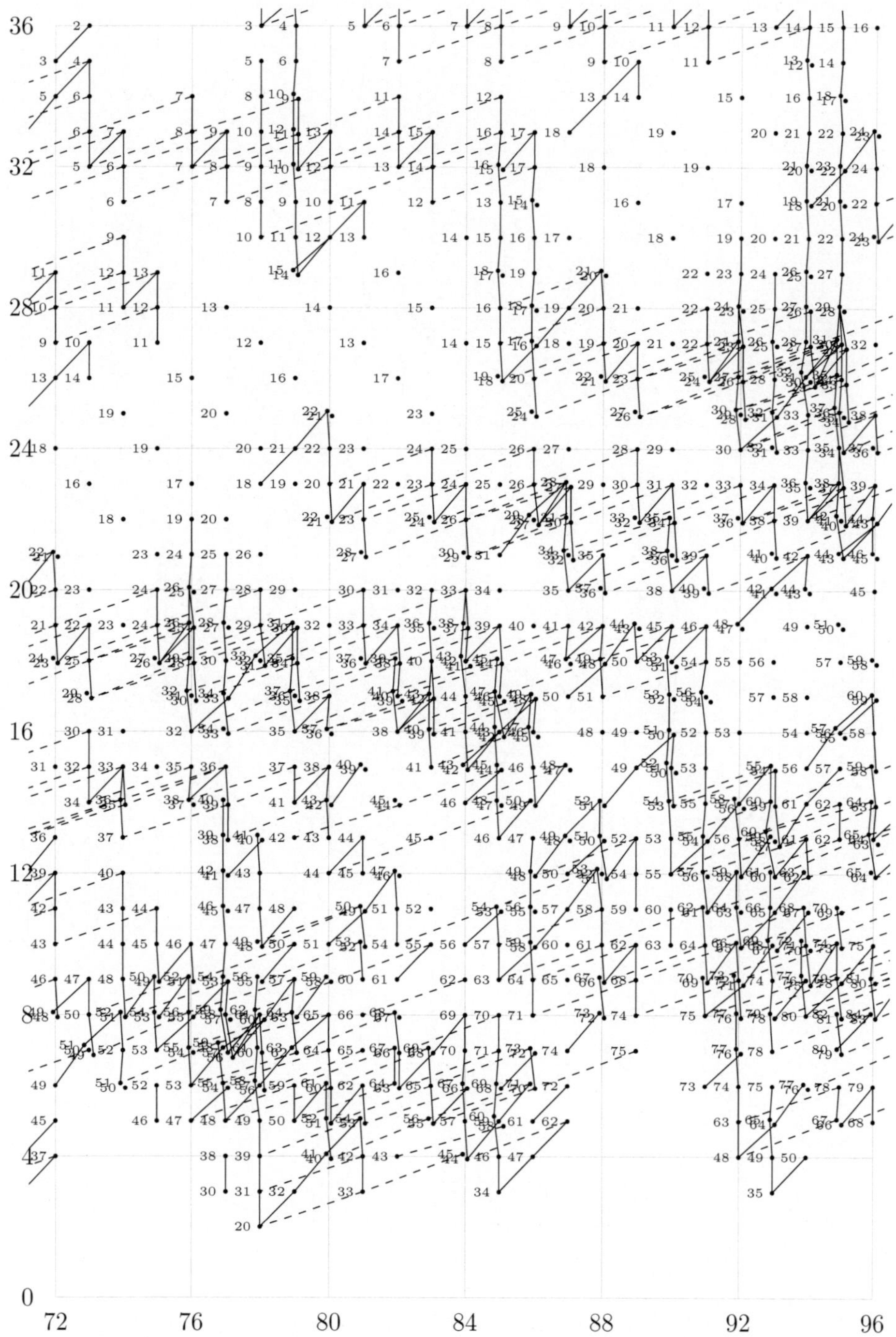

FIGURE 1.4. $\mathrm{Ext}_A^{s,t}(\mathbb{F}_2, \mathbb{F}_2)$ for $72 \le t - s \le 96$

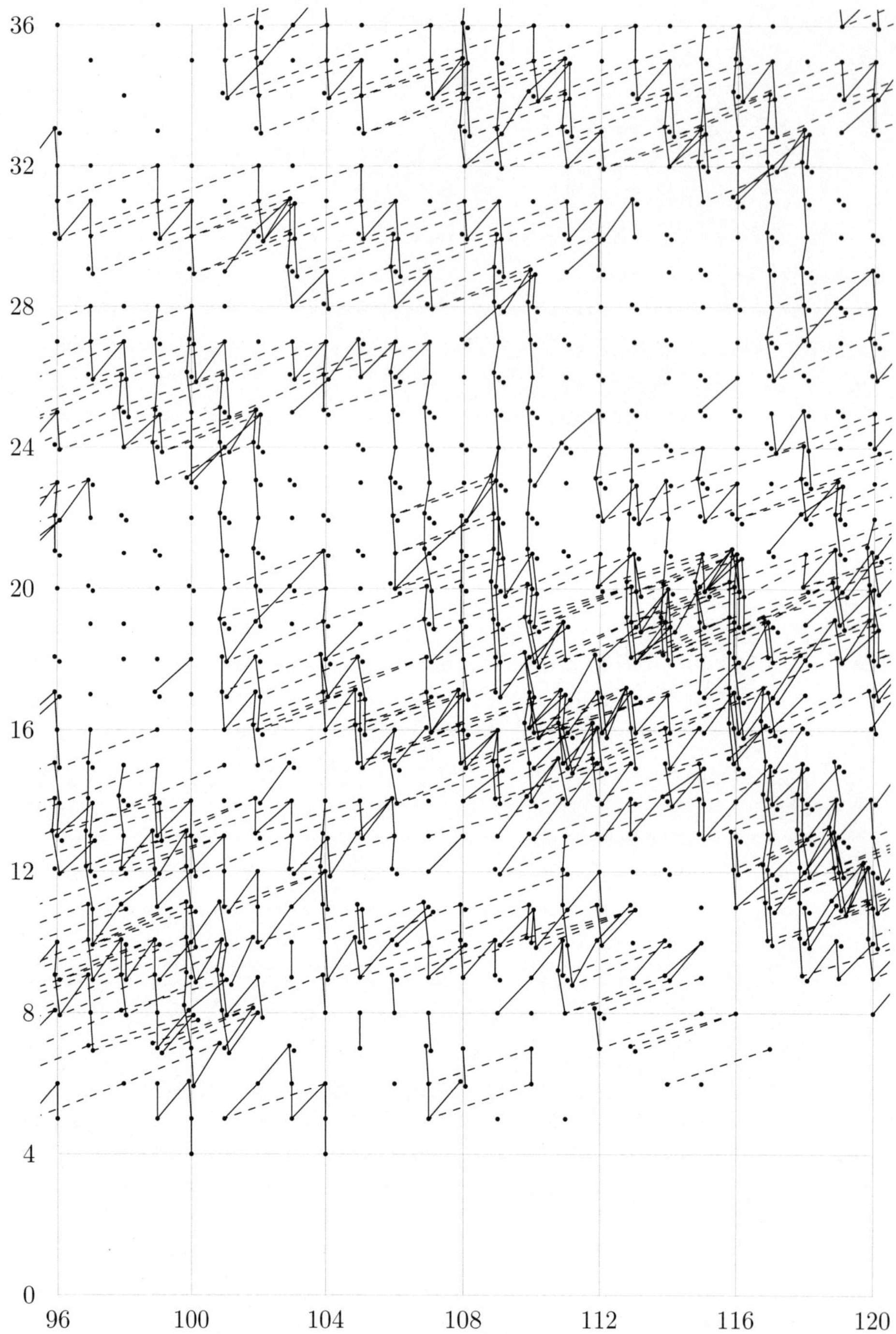

FIGURE 1.5. $\mathrm{Ext}_A^{s,t}(\mathbb{F}_2, \mathbb{F}_2)$ for $96 \leq t - s \leq 120$

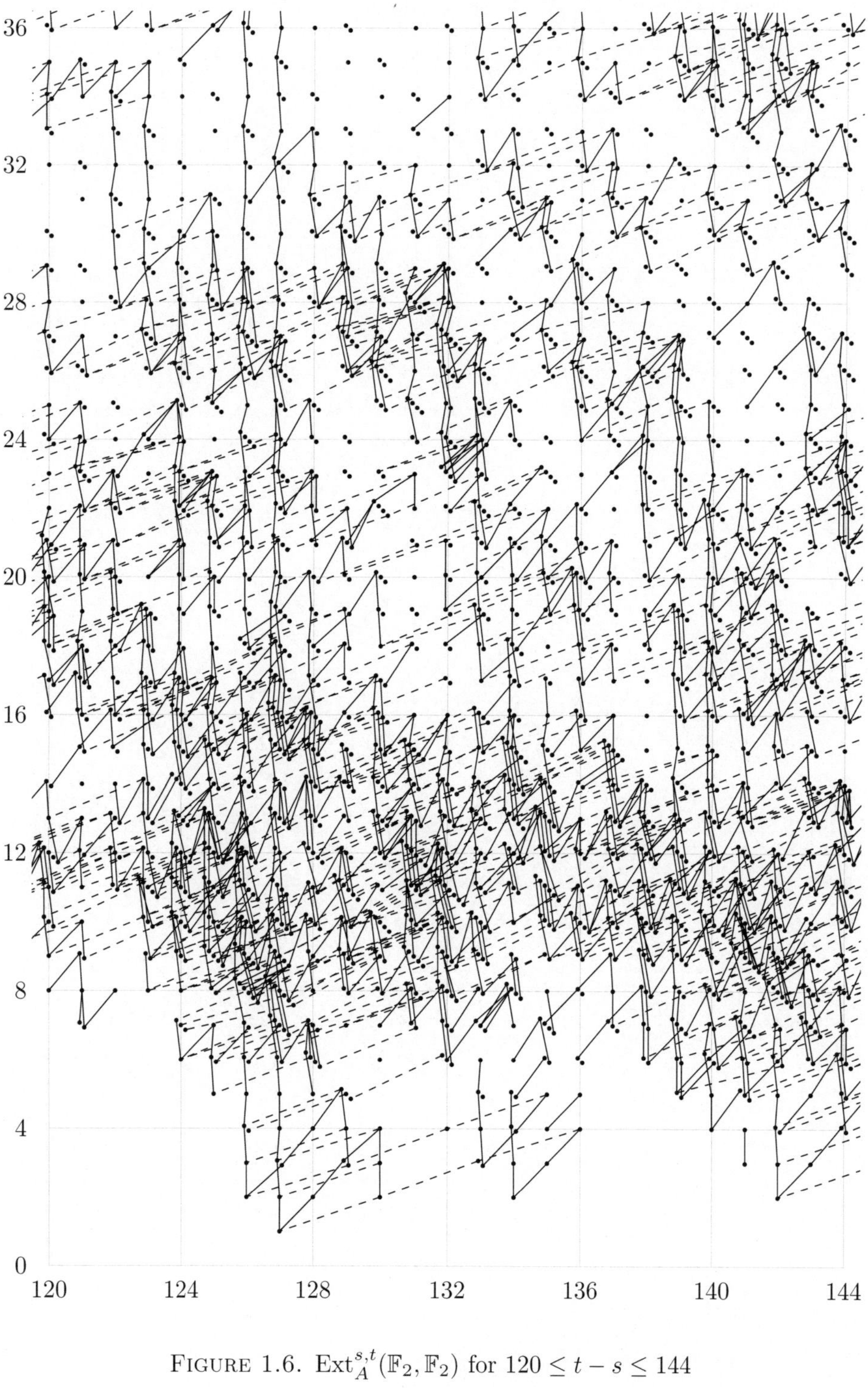


FIGURE 1.6. $\mathrm{Ext}_A^{s,t}(\mathbb{F}_2, \mathbb{F}_2)$ for $120 \le t - s \le 144$

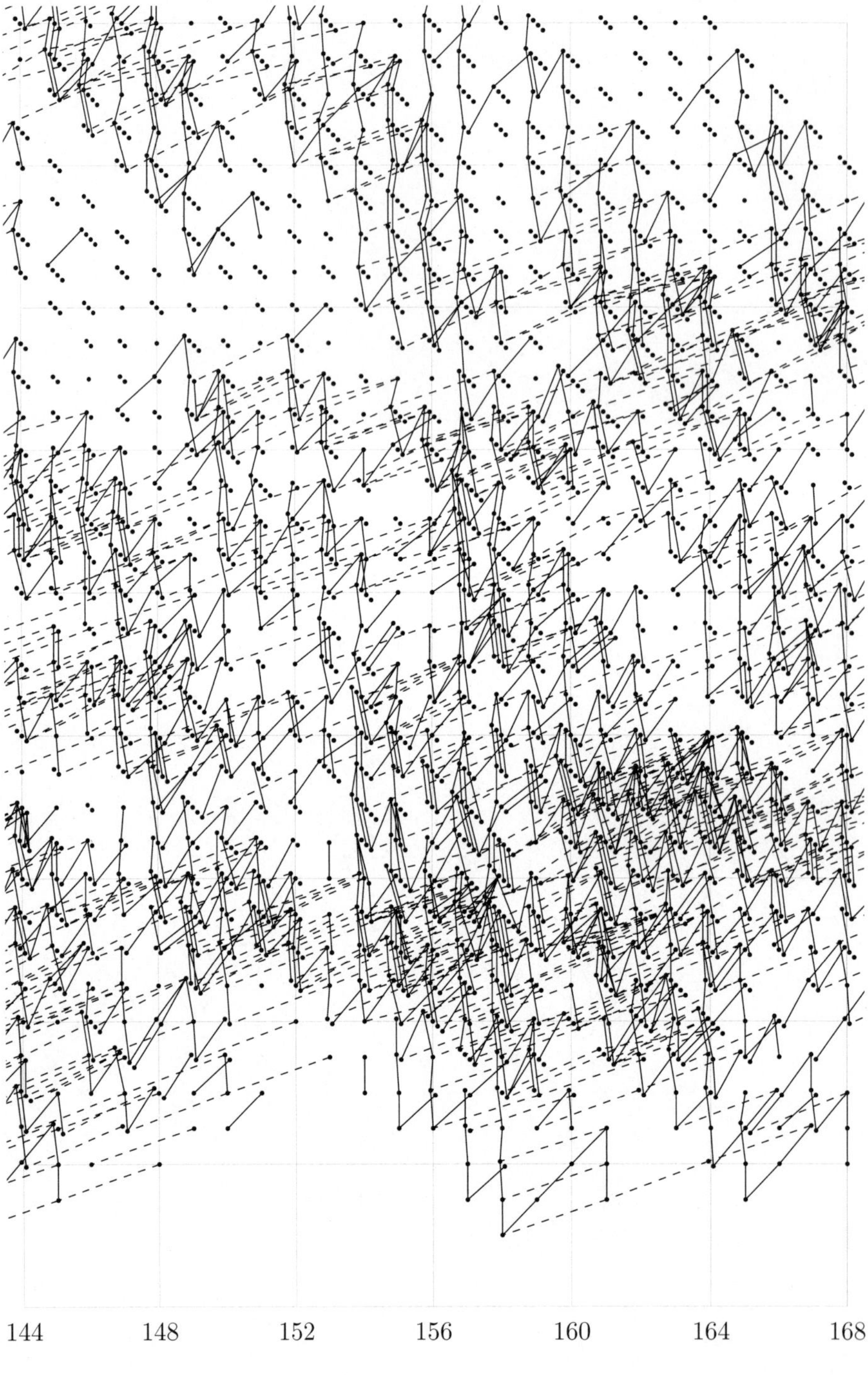


FIGURE 1.7. $\mathrm{Ext}_A^{s,t}(\mathbb{F}_2,\mathbb{F}_2)$ for $144 \le t-s \le 168$, $t \le 200$

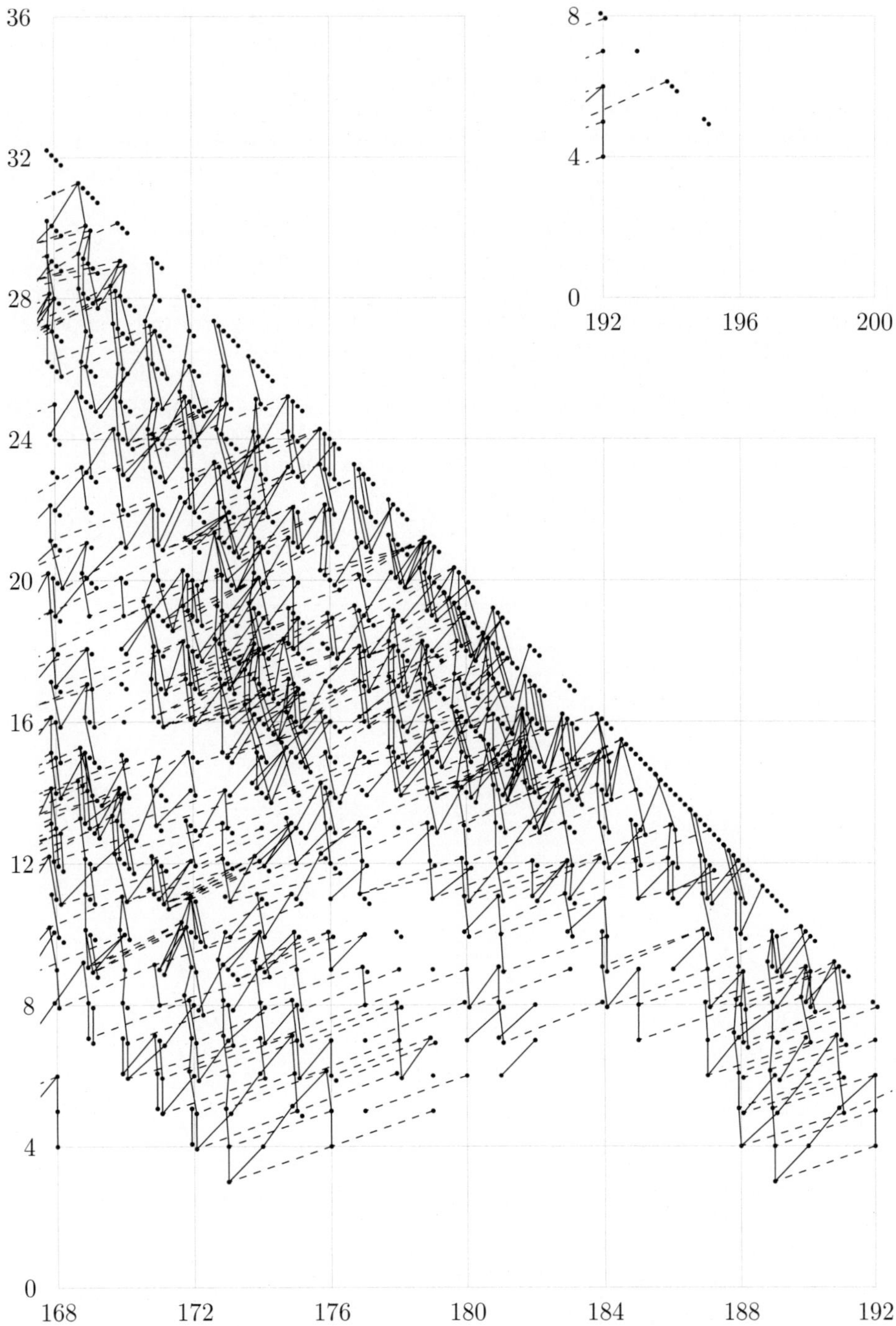

FIGURE 1.8. $\mathrm{Ext}_A^{s,t}(\mathbb{F}_2, \mathbb{F}_2)$ for $168 \le t - s \le 200$, $t \le 200$

The program `ext` can extract the Yoneda product $h_i y$ from the structure of the minimal resolution, for each $i \geq 0$ and for any cocycle $y = s_g$. In Figures 1.1 to 1.8 the nonzero multiplications by $h_0$ are shown as solid vertical lines from $y$ to $h_0 y$, the nonzero multiplications by $h_1$ are shown as solid lines of slope 1 from $y$ to $h_1 y$, and the nonzero multiplications by $h_2$ are shown as dashed lines of slope 1/3 from $y$ to $h_2 y$. We omit to show the $h_i$-multiplications for $i \geq 3$, as they would make the charts too crowded to be legible.

More generally, `ext` can calculate the Yoneda product $xy$ of two cocycles $x = s'_{g'} \colon C_{s'} \to \mathbb{F}_2$ and $y = s''_{g''} \colon C_{s''} \to \mathbb{F}_2$ by lifting $y$ to a chain map $\tilde{y} \colon C_{*+s''} \to C_*$, and then expressing the composite $x \circ \tilde{y} \colon C_{s'+s''} \to \mathbb{F}_2$ as a linear combination of cocycles $s_g$, with $s = s' + s''$. It is thereby possible to determine the indecomposable quotient of $\mathrm{Ext}_A(\mathbb{F}_2, \mathbb{F}_2)$, within the machine calculated range.

PROPOSITION 1.4. *In topological degrees $t - s \leq 48$, a basis for the algebra indecomposables in $\mathrm{Ext}_A(\mathbb{F}_2, \mathbb{F}_2)$ is given by the classes listed in Table 1.1. The same classes are labeled and emphasized in Figures 1.9 and 1.10.*

Table 1.1: Algebra indecomposables in $\mathrm{Ext}_A(\mathbb{F}_2, \mathbb{F}_2)$ for $t - s \leq 48$

| $t-s$ | $s$ | $g$ | $x$ | dec. | $\iota(x)$ | $d_2(x)$ |
|---|---|---|---|---|---|---|
| 0 | 1 | 0 | $h_0$ | | $h_0$ | 0 |
| 1 | 1 | 1 | $h_1$ | | $h_1$ | 0 |
| 3 | 1 | 2 | $h_2$ | | $h_2$ | 0 |
| 7 | 1 | 3 | $h_3$ | | 0 | 0 |
| 8 | 3 | 3 | $c_0$ | | $c_0$ | 0 |
| 9 | 5 | 1 | $Ph_1$ | | $h_1 w_1$ | 0 |
| 11 | 5 | 2 | $Ph_2$ | | $h_2 w_1$ | 0 |
| 14 | 4 | 3 | $d_0$ | | $d_0$ | 0 |
| 15 | 1 | 4 | $h_4$ | | 0 | $h_0 h_3^2$ |
| 16 | 7 | 3 | $Pc_0$ | | $c_0 w_1$ | 0 |
| 17 | 4 | 5 | $e_0$ | | $e_0$ | $h_1^2 d_0$ |
| 17 | 9 | 1 | $P^2 h_1$ | | $h_1 w_1^2$ | 0 |
| 18 | 4 | 6 | $f_0$ | $h_1^3 h_4$ | $h_2 \beta$ | $h_0^2 e_0$ |
| 19 | 3 | 9 | $c_1$ | | 0 | 0 |
| 19 | 9 | 2 | $P^2 h_2$ | | $h_2 w_1^2$ | 0 |
| 20 | 4 | 8 | $g$ | | $g$ | 0 |
| 22 | 8 | 3 | $Pd_0$ | | $d_0 w_1$ | 0 |
| 23 | 7 | 5 | $i$ | | $\beta w_1$ | $h_0 Pd_0$ |
| 24 | 11 | 3 | $P^2 c_0$ | | $c_0 w_1^2$ | 0 |
| 25 | 8 | 5 | $Pe_0$ | | $e_0 w_1$ | $h_1^2 Pd_0$ |

Table 1.1: Algebra indecomposables in $\mathrm{Ext}_A(\mathbb{F}_2, \mathbb{F}_2)$ (cont.)

| $t-s$ | $s$ | $g$ | $x$ | dec. | $\iota(x)$ | $d_2(x)$ |
|---|---|---|---|---|---|---|
| 25 | 13 | 1 | $P^3h_1$ | | $h_1w_1^3$ | 0 |
| 26 | 7 | 6 | $j$ | | $\alpha d_0$ | $h_0Pe_0$ |
| 27 | 13 | 2 | $P^3h_2$ | | $h_2w_1^3$ | 0 |
| 29 | 7 | 7 | $k$ | | $\alpha e_0$ | $h_0d_0^2$ |
| 30 | 6 | 10 | $r$ | | $\beta^2$ | 0 |
| 30 | 12 | 3 | $P^2d_0$ | | $d_0w_1^2$ | 0 |
| 31 | 1 | 5 | $h_5$ | | 0 | $h_0h_4^2$ |
| 31 | 5 | 13 | $n$ | $h_0^4h_5$ | 0 | 0 |
| 32 | 4 | 13 | $d_1$ | | 0 | 0 |
| 32 | 6 | 12 | $q$ | | 0 | 0 |
| 32 | 7 | 10 | $\ell$ | | $\alpha g$ | $h_0d_0e_0$ |
| 32 | 15 | 3 | $P^3c_0$ | | $c_0w_1^3$ | 0 |
| 33 | 4 | 14 | $p$ | | 0 | 0 |
| 33 | 12 | 5 | $P^2e_0$ | | $e_0w_1^2$ | $h_1^2P^2d_0$ |
| 33 | 17 | 1 | $P^4h_1$ | | $h_1w_1^4$ | 0 |
| 34 | 11 | 7 | $Pj$ | | $\alpha d_0w_1$ | $h_0P^2e_0$ |
| 35 | 7 | 12 | $m$ | | $\beta g$ | $h_0d_0g$ |
| 35 | 17 | 2 | $P^4h_2$ | | $h_2w_1^4$ | 0 |
| 36 | 6 | 14 | $t$ | | 0 | 0 |
| 37 | 5 | 17 | $x$ | | 0 | 0 |
| 38 | 4 | 16 | $e_1$ | $h_0^2h_3h_5$ | 0 | 0 |
| 38 | 6 | 16 | $y$ | $h_1x$ | 0 | $h_0^3x$ |
| 38 | 16 | 3 | $P^3d_0$ | | $d_0w_1^3$ | 0 |
| 39 | 9 | 18 | $u$ | | $d_0\gamma$ | 0 |
| 39 | 15 | 5 | $P^2i$ | | $\beta w_1^3$ | $h_0P^3d_0$ |
| 40 | 4 | 19 | $f_1$ | $h_1^2h_3h_5$ | 0 | 0 |
| 40 | 19 | 3 | $P^4c_0$ | | $c_0w_1^4$ | 0 |
| 41 | 3 | 19 | $c_2$ | | 0 | $h_0f_1$ |
| 41 | 10 | 14 | $z$ | | $\alpha^2e_0$ | 0 |
| 41 | 16 | 5 | $P^3e_0$ | | $e_0w_1^3$ | $h_1^2P^3d_0$ |
| 41 | 21 | 1 | $P^5h_1$ | | $h_1w_1^5$ | 0 |
| 42 | 9 | 19 | $v$ | | $e_0\gamma$ | $h_0z$ |

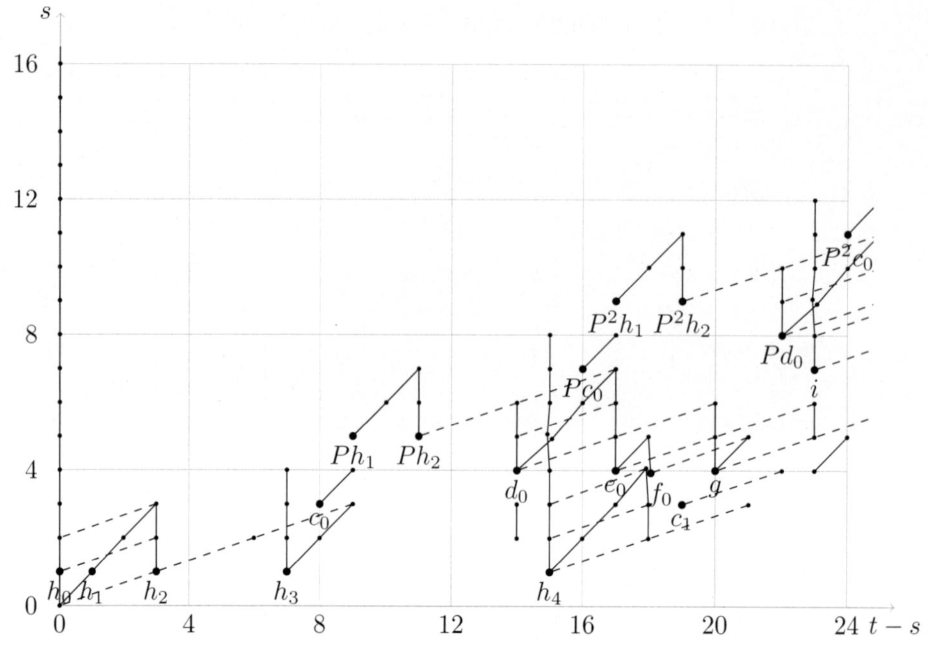


FIGURE 1.9. Indecomposables in $\mathrm{Ext}_A^{s,t}(\mathbb{F}_2, \mathbb{F}_2)$ for $0 \le t - s \le 24$

Table 1.1: Algebra indecomposables in $\mathrm{Ext}_A(\mathbb{F}_2, \mathbb{F}_2)$ (cont.)

| $t-s$ | $s$ | $g$ | $x$ | dec. | $\iota(x)$ | $d_2(x)$ |
|---|---|---|---|---|---|---|
| 42 | 15 | 6 | $P^2 j$ | | $\alpha d_0 w_1^2$ | $h_0 P^3 e_0$ |
| 43 | 21 | 2 | $P^5 h_2$ | | $h_2 w_1^5$ | 0 |
| 44 | 4 | 22 | $g_2$ | | 0 | 0 |
| 45 | 9 | 20 | $w$ | | $\gamma g$ | 0 |
| 46 | 7 | 20 | $B_1$ | | 0 | 0 |
| 46 | 8 | 20 | $N$ | | 0 | 0 |
| 46 | 20 | 3 | $P^4 d_0$ | | $d_0 w_1^4$ | 0 |
| 47 | 13 | 14 | $Q$ | | 0 | $h_0 i^2$ |
| 47 | 13 | 15 | $Pu$ | | $d_0 \gamma w_1$ | 0 |
| 48 | 7 | 22(?) | $B_2$ | $h_0^2 h_5 e_0$ | 0 | 0 |
| 48 | 23 | 3 | $P^5 c_0$ | | $c_0 w_1^5$ | 0 |

REMARK 1.5. In Table 1.1, the $(t-s)$- and $s$-columns give the Adams bigrading $(t-s, s)$ of the class $x$, while the $s$- and $g$-columns specify the cocycles $s_g$ corresponding to $x$ in the representation of $\mathrm{Ext}_A(\mathbb{F}_2, \mathbb{F}_2)$ given by the minimal resolution chosen by `ext`. In later tables the class $x$ will sometimes correspond to a

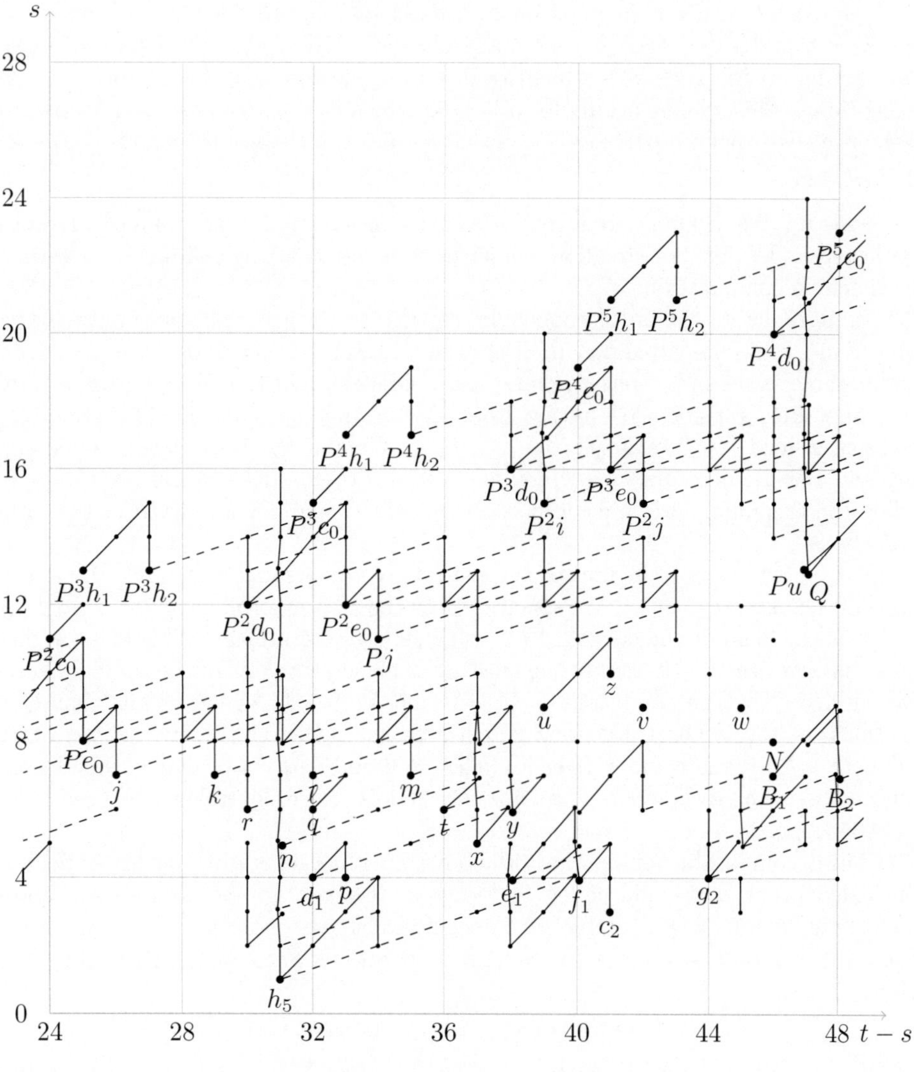


FIGURE 1.10. Indecomposables in $\mathrm{Ext}_A^{s,t}(\mathbb{F}_2, \mathbb{F}_2)$ for $24 \le t - s \le 48$

sum $s_{g_1} + s_{g_2}$ of cocycles, in which case we will write "$g_1 + g_2$" in the $g$-column. The dec.-column lists any decomposable class in the same bidegree. The rows are lexicographically ordered by topological degree $t - s$, by filtration $s$, and by the generator index $g$. The $\iota(x)$-column gives the restriction to $\mathrm{Ext}_{A(2)}(\mathbb{F}_2, \mathbb{F}_2)$ as will be explained in Lemma 1.15. The $d_2(x)$-column gives the Adams $d_2$-differentials on these algebra generators, which will be established in Theorem 11.52.

The names in the $x$-column are those inherited from the May spectral sequence calculations of May and Tangora [**165**, App. 1]. In particular, the Adams periodicity operator $P$ is given by the Massey product $Px \in \langle h_3, h_0^4, x\rangle$ when $h_0^4 x = 0$, and $P^2$ is given by $P^2x \in \langle h_4, h_0^8, x\rangle$ when $h_0^8 x = 0$. The program `ext` can evaluate Massey products of the form $\langle h_i, x, y\rangle$, once the cocycle $y$ has been lifted.

We will adopt the indexing scheme from [**165**] and [**45**, Def. VI.1.8], where we set $a_0 = a$ and $a_{i+1} = Sq^0(a_i)$ for many classes $a$. Here $Sq^0$ is a Steenrod operation acting on $\mathrm{Ext}_A(\mathbb{F}_2, \mathbb{F}_2)$, which we discuss in Section 1.3 and Chapter 11. An exception to this scheme occurs for $a = g$, in which case $g_0$ refers to a May spectral sequence class that supports a differential, so that the classes in $\mathrm{Ext}_A(\mathbb{F}_2, \mathbb{F}_2)$ start with $g = g_1$.

PROOF. We will make no formal use of this proposition, other than to introduce notation, and will therefore allow ourselves to assert that the claim can be verified by machine computation.

In more detail, the indecomposable quotient of $\mathrm{Ext}_A(\mathbb{F}_2, \mathbb{F}_2)$ can be calculated in a finite range as explained in Remarks 1.3 and 1.6. In most cases an indecomposable is the only nonzero class in its bidegree, and this lets us recognize its corresponding `ext`-cocycle directly from the minimal resolution. The remaining eight cases for $t - s \le 48$ are $f_0$, $n$, $e_1$, $y$, $f_1$, $Q$, $Pu$ and $B_2$. Six of these are defined modulo a single decomposable class, as indicated by the dec.-column in Table 1.1, while the remaining two indecomposable classes, $Q$ and $Pu$, are both in the same bidegree.

We specify $n$ to be the nonzero class in its bidegree satisfying $h_0 n = 0$, i.e., the class of the `ext`-cocycle $5_{13}$. We specify $Pu$ by the Massey product $Pu = \langle h_3, h_0^4, u\rangle$, which is the class of the cocycle $13_{15}$, with zero indeterminacy. This class is then also characterized by the conditions $h_0 Pu = 0$ and $h_1 Pu \neq 0$. In the same bidegree we specify $Q$ by the conditions $h_0 Q \neq 0$ and $h_1 Q \neq 0$, which means that $Q$ is the class of $13_{14}$. The third nonzero class in that bidegree is sometimes denoted $Q' = Q + Pu$. It is characterized by $h_0 Q' \neq 0$ and $h_1 Q' = 0$, and is the class of $13_{14} + 13_{15}$. These choices of classes $n$, $Pu$ and $Q$ are compatible with those made in [**165**, App. 1].

The decomposable ambiguity in the remaining five generators has little effect on our calculations, and could be left unspecified. However, for definiteness we choose to use the results of [**46**] to pin down specific `ext`-representatives for all but one of these indecomposables, using the Steenrod operations $Sq^i$ acting on $\mathrm{Ext}_A(\mathbb{F}_2, \mathbb{F}_2)$. Hence, we set

$$f_0 = Sq^1(c_0) \qquad \text{and} \qquad y = Sq^2(f_0)$$

as in [**45**, §VI.1], together with $e_1 = Sq^0(e_0)$ and $f_1 = Sq^0(f_0)$. Using direct cochain calculations, similar to the ones in the proof of Proposition 1.21, [**46**] show that with these choices $f_0$ is represented by the cocycle $4_6$, $e_1$ is represented by the cocycle $4_{16}$, $y$ is represented by the cocycle $6_{16}$, and $f_1$ is represented by the cocycle $4_{19}$.

The final indecomposable in this range of degrees is $B_2$ in bidegree $(t - s, s) = (48, 7)$. With our methods we can only specify it modulo the decomposable $h_0^2 h_5 e_0 = 7_{23}$, i.e., as $7_{22}$ or $7_{22} + 7_{23}$. This is equivalent to setting $B_2 = \langle h_2, h_0^3, g_2\rangle$, since this Massey product contains $7_{22}$ and has indeterminacy generated by $7_{23}$. For simplicity we will set $B_2$ to be the class of $7_{22}$, and indicate this uncertainty with a question-mark in the chart for $E_2(S)$. Note, however, that the indeterminacy in $B_2$ disappears at the $E_3$-term, due to an Adams differential $d_2(h_5 f_0) = h_0^2 h_5 e_0$, and therefore has no visible consequence after this point. □

REMARK 1.6. To make these calculations with `ext`, go to the directory `A` and use `cocycle S 1 0`, `cocycle S 1 1`, ..., `cocycle S 1 6` in turn. These create cocycle subdirectories 1_0, 1_1, ..., 1_6 in `A/S` and add their names to the list in `A/S/maps` of cocycles $y = s''_{g''}\colon C_{s''} \to \mathbb{F}_2$ that need to be lifted to chain maps $\tilde{y}\colon C_{*+s''} \to C_*$. Change directory to `A/S` and run `dolifts 0 40 maps` to calculate these lifts. Use `collect maps all` to extract the file `all`, which contains a row

```
s g (s' g' F2) s''_g''
```

for each summand $s_g$ in the product of $s'_{g'}$ and $s''_{g''}$. For example, the lines

```
3 1 (2 0 F2) 1_2
3 1 (2 1 F2) 1_1
3 1 (2 2 F2) 1_0
```

exhibit $3_1$ as $2_0 \cdot 1_2 = h_0^2 \cdot h_2$, as $2_1 \cdot 1_1 = h_1^2 \cdot h_1$, and as $2_2 \cdot 1_0 = h_0 h_2 \cdot h_0$. Each cocycle in filtration 2 is then seen to be decomposable, but in filtration 3 the cocycles $3_3$, $3_9$ and $3_{19}$ are seen to be indecomposable. Return to `A` and use `cocycle S 3 3`, `cocycle S 3 9` and `cocycle S 3 19` to create these cocycles, go to `A/S` and run `dolifts 0 40 maps` to lift them, and repeat.

REMARK 1.7. The classes $h_i$ and $P^i h_1$ are indecomposable for all $i \geq 0$, so the algebra $\mathrm{Ext}_A(\mathbb{F}_2, \mathbb{F}_2)$ is not finitely generated. This is in contrast to $\mathrm{Ext}_{A(2)}(\mathbb{F}_2, \mathbb{F}_2)$, which is finitely generated as an algebra.

Table 1.2: Minimal free $A$-module resolution $(C_*, \partial)$ of $\mathbb{F}_2$ with $C_s = A\{s_0^*, s_1^*, \dots\}$, for $s \le 6$ and $t \le 22$

| $t-s$ | $s$ | $x$ | $\partial(x)$ |
|---|---|---|---|
| 0 | 0 | $0_0^*$ | $1$ |
| 0 | 1 | $1_0^*$ | $Sq^1(0_0^*)$ |
| 1 | 1 | $1_1^*$ | $Sq^2(0_0^*)$ |
| 3 | 1 | $1_2^*$ | $Sq^4(0_0^*)$ |
| 7 | 1 | $1_3^*$ | $Sq^8(0_0^*)$ |
| 15 | 1 | $1_4^*$ | $Sq^{16}(0_0^*)$ |
| 0 | 2 | $2_0^*$ | $Sq^1(1_0^*)$ |
| 2 | 2 | $2_1^*$ | $Sq^3(1_0^*) + Sq^2(1_1^*)$ |
| 3 | 2 | $2_2^*$ | $Sq^4(1_0^*) + Sq^{(0,1)}(1_1^*) + Sq^1(1_2^*)$ |
| 6 | 2 | $2_3^*$ | $Sq^7(1_0^*) + Sq^6(1_1^*) + Sq^4(1_2^*)$ |
| 7 | 2 | $2_4^*$ | $(Sq^8 + Sq^{(2,2)})(1_0^*) + (Sq^7 + Sq^{(4,1)} + Sq^{(0,0,1)})(1_1^*) + Sq^1(1_3^*)$ |
| 8 | 2 | $2_5^*$ | $(Sq^9 + Sq^{(3,2)})(1_0^*) + (Sq^8 + Sq^{(5,1)})(1_1^*) + Sq^{(0,2)}(1_2^*) + Sq^2(1_3^*)$ |
| 14 | 2 | $2_6^*$ | $Sq^{15}(1_0^*) + Sq^{14}(1_1^*) + Sq^{12}(1_2^*) + Sq^8(1_3^*)$ |
| 15 | 2 | $2_7^*$ | $(Sq^{16} + Sq^{(10,2)} + Sq^{(7,3)} + Sq^{(4,4)} + Sq^{(2,0,2)})(1_0^*) + (Sq^{(12,1)} + Sq^{(3,4)} + Sq^{(0,5)} + Sq^{(8,0,1)} + Sq^{(0,0,0,1)})(1_1^*)$ |
| | | | $+ Sq^{13}(1_2^*) + Sq^1(1_4^*)$ |
| 16 | 2 | $2_8^*$ | $(Sq^{17} + Sq^{(11,2)} + Sq^{(3,0,2)})(1_0^*) + (Sq^{16} + Sq^{(4,4)} + Sq^{(1,5)})(1_1^*) + (Sq^{14} + Sq^{(8,2)} + Sq^{(0,0,2)})(1_2^*) + Sq^2(1_4^*)$ |
| 18 | 2 | $2_9^*$ | $(Sq^{19} + Sq^{(10,3)} + Sq^{(7,4)})(1_0^*) + (Sq^{18} + Sq^{(15,1)} + Sq^{(6,4)})(1_1^*) + (Sq^{16} + Sq^{(10,2)})(1_2^*) + Sq^{(0,4)}(1_3^*) + Sq^4(1_4^*)$ |

Table 1.2: Minimal free $A$-module resolution $(C_*, \partial)$ of $\mathbb{F}_2$, with $C_s = A\{s_0^*, s_1^*, \dots\}$, for $s \le 6$ and $t \le 22$ (cont.)

| $t-s$ | $s$ | $x$ | $\partial(x)$ |
|---|---|---|---|
| 0 | 3 | $3_0^*$ | $Sq^1(2_0^*)$ |
| 3 | 3 | $3_1^*$ | $Sq^4(2_0^*) + Sq^2(2_1^*) + Sq^1(2_2^*)$ |
| 7 | 3 | $3_2^*$ | $(Sq^8 + Sq^{(2,2)})(2_0^*) + (Sq^6 + Sq^{(0,2)})(2_1^*) + Sq^1(2_4^*)$ |
| 8 | 3 | $3_3^*$ | $(Sq^9 + Sq^{(3,2)})(2_0^*) + Sq^{(0,0,1)}(2_1^*) + Sq^6(2_2^*) + (Sq^3 + Sq^{(0,1)})(2_3^*)$ |
| 9 | 3 | $3_4^*$ | $Sq^{10}(2_0^*) + (Sq^8 + Sq^{(1,0,1)})(2_1^*) + Sq^4(2_3^*) + Sq^3(2_4^*) + Sq^2(2_5^*)$ |
| 14 | 3 | $3_5^*$ | $(Sq^{(9,2)} + Sq^{(6,3)})(2_0^*) + Sq^{(7,2)}(2_1^*) + Sq^{(0,4)}(2_2^*) + (Sq^9 + Sq^{(0,3)})(2_3^*) + (Sq^8 + Sq^{(2,2)})(2_4^*)$ |
| | | | $+ (Sq^7 + Sq^{(4,1)} + Sq^{(0,0,1)})(2_5^*) + Sq^1(2_6^*)$ |
| 15 | 3 | $3_6^*$ | $(Sq^{16} + Sq^{(10,2)} + Sq^{(7,3)} + Sq^{(4,4)} + Sq^{(1,5)} + Sq^{(2,0,2)})(2_0^*)$ |
| | | | $+ (Sq^{14} + Sq^{(11,1)} + Sq^{(8,2)} + Sq^{(2,4)} + Sq^{(7,0,1)} + Sq^{(4,1,1)} + Sq^{(0,0,2)})(2_1^*) + Sq^1(2_7^*)$ |
| 17 | 3 | $3_7^*$ | $(Sq^{18} + Sq^{(3,5)})(2_0^*) + (Sq^{16} + Sq^{(4,4)} + Sq^{(6,1,1)})(2_1^*) + (Sq^{15} + Sq^{(3,4)})(2_2^*)$ |
| | | | $+ (Sq^{12} + Sq^{(9,1)} + Sq^{(3,3)} + Sq^{(0,4)})(2_3^*) + Sq^3(2_7^*) + Sq^2(2_8^*)$ |
| 18 | 3 | $3_8^*$ | $(Sq^{19} + Sq^{(13,2)} + Sq^{(4,5)} + Sq^{(2,1,2)})(2_0^*) + (Sq^{17} + Sq^{(11,2)} + Sq^{(5,4)} + Sq^{(2,5)} + Sq^{(10,0,1)} + Sq^{(7,1,1)})(2_1^*)$ |
| | | | $+ (Sq^{16} + Sq^{(4,4)})(2_2^*) + (Sq^{13} + Sq^{(10,1)})(2_3^*) + (Sq^{12} + Sq^{(6,2)} + Sq^{(3,3)} + Sq^{(0,4)})(2_4^*)$ |
| | | | $+ (Sq^{(8,1)} + Sq^{(4,0,1)})(2_5^*) + Sq^5(2_6^*) + Sq^4(2_7^*) + Sq^{(0,1)}(2_8^*) + Sq^1(2_9^*)$ |
| 19 | 3 | $3_9^*$ | $(Sq^{(14,2)} + Sq^{(11,3)} + Sq^{(6,0,2)})(2_0^*) + (Sq^{(15,1)} + Sq^{(11,0,1)} + Sq^{(0,1,0,1)})(2_1^*)$ |
| | | | $+ (Sq^{(11,2)} + Sq^{(8,3)} + Sq^{(2,5)} + Sq^{(3,0,2)})(2_2^*) + (Sq^{14} + Sq^{(2,4)} + Sq^{(0,0,2)})(2_3^*) + (Sq^{13} + Sq^{(7,2)})(2_4^*)$ |
| | | | $+ Sq^{12}(2_5^*) + (Sq^6 + Sq^{(0,2)})(2_6^*)$ |

Table 1.2: Minimal free $A$-module resolution $(C_*, \partial)$ of $\mathbb{F}_2$, with $C_s = A\{s_0^*, s_1^*, \dots\}$, for $s \le 6$ and $t \le 22$ (cont.)

| $t-s$ | $s$ | $x$ | $\partial(x)$ |
|---|---|---|---|
| 0 | 4 | $4_0^*$ | $Sq^1(3_0^*)$ |
| 7 | 4 | $4_1^*$ | $Sq^8(3_0^*) + (Sq^5 + Sq^{(2,1)})(3_1^*) + Sq^1(3_2^*)$ |
| 9 | 4 | $4_2^*$ | $(Sq^{10} + Sq^{(4,2)})(3_0^*) + (Sq^7 + Sq^{(1,2)} + Sq^{(0,0,1)})(3_1^*) + Sq^2(3_3^*)$ |
| 14 | 4 | $4_3^*$ | $(Sq^{15} + Sq^{(9,2)} + Sq^{(6,3)} + Sq^{(0,5)})(3_0^*) + (Sq^{(9,1)} + Sq^{(6,2)})(3_1^*) + (Sq^7 + Sq^{(4,1)} + Sq^{(0,0,1)})(3_3^*) + Sq^{(3,1)}(3_4^*)$ |
| 15 | 4 | $4_4^*$ | $(Sq^{16} + Sq^{(7,3)} + Sq^{(4,4)} + Sq^{(1,5)})(3_0^*) + (Sq^{13} + Sq^{(10,1)} + Sq^{(7,2)} + Sq^{(4,3)} + Sq^{(1,4)} + Sq^{(6,0,1)} + Sq^{(0,2,1)})(3_1^*)$ $+ Sq^1(3_6^*)$ |
| 17 | 4 | $4_5^*$ | $(Sq^{(9,3)} + Sq^{(4,0,2)})(3_0^*) + (Sq^{(12,1)} + Sq^{(1,0,2)} + Sq^{(0,0,0,1)})(3_1^*) + Sq^{11}(3_2^*) + (Sq^{10} + Sq^{(7,1)})(3_3^*)$ $+ (Sq^{(6,1)} + Sq^{(3,2)} + Sq^{(0,3)} + Sq^{(2,0,1)})(3_4^*) + Sq^{(1,1)}(3_5^*)$ |
| 18 | 4 | $4_6^*$ | $(Sq^{(10,3)} + Sq^{(4,5)} + Sq^{(1,6)} + Sq^{(5,0,2)})(3_0^*) + (Sq^{(7,3)} + Sq^{(4,4)} + Sq^{(9,0,1)} + Sq^{(6,1,1)} + Sq^{(3,2,1)} + Sq^{(1,0,0,1)})(3_1^*)$ $+ Sq^{12}(3_2^*) + (Sq^{11} + Sq^{(4,0,1)})(3_3^*) + (Sq^{(7,1)} + Sq^{(0,1,1)})(3_4^*) + (Sq^5 + Sq^{(2,1)})(3_5^*)$ |
| 18 | 4 | $4_7^*$ | $(Sq^{19} + Sq^{(13,2)} + Sq^{(10,3)} + Sq^{(7,4)} + Sq^{(5,0,2)} + Sq^{(2,1,2)})(3_0^*) + (Sq^{16} + Sq^{(4,4)} + Sq^{(9,0,1)} + Sq^{(1,0,0,1)})(3_1^*)$ $+ (Sq^{12} + Sq^{(6,2)} + Sq^{(3,3)} + Sq^{(0,4)})(3_2^*) + Sq^{(4,0,1)}(3_3^*) + (Sq^{10} + Sq^{(4,2)} + Sq^{(1,3)} + Sq^{(0,1,1)})(3_4^*) + Sq^4(3_6^*)$ $+ Sq^2(3_7^*) + Sq^1(3_8^*)$ |

Table 1.2: Minimal free $A$-module resolution $(C_*, \partial)$ of $\mathbb{F}_2$, with $C_s = A\{s_0^*, s_1^*, \dots\}$, for $s \leq 6$ and $t \leq 22$ (cont.)

| $t-s$ | $s$ | $x$ | $\partial(x)$ |
|---|---|---|---|
| 0 | 5 | $5_0^*$ | $Sq^1(4_0^*)$ |
| 9 | 5 | $5_1^*$ | $Sq^{10}(4_0^*) + (Sq^3 + Sq^{(0,1)})(4_1^*)$ |
| 11 | 5 | $5_2^*$ | $Sq^{12}(4_0^*) + (Sq^5 + Sq^{(2,1)})(4_1^*) + Sq^3(4_2^*)$ |
| 14 | 5 | $5_3^*$ | $(Sq^{(9,2)} + Sq^{(6,3)} + Sq^{(3,4)})(4_0^*) + Sq^{(5,1)}(4_1^*) + (Sq^6 + Sq^{(0,2)})(4_2^*) + Sq^1(4_3^*)$ |
| 15 | 5 | $5_4^*$ | $(Sq^{(10,2)} + Sq^{(4,4)})(4_0^*) + (Sq^9 + Sq^{(6,1)})(4_1^*) + Sq^{(0,0,1)}(4_2^*) + Sq^2(4_3^*)$ |
| 15 | 5 | $5_5^*$ | $Sq^{16}(4_0^*) + (Sq^9 + Sq^{(6,1)})(4_1^*) + Sq^1(4_4^*)$ |
| 17 | 5 | $5_6^*$ | $(Sq^{(9,3)} + Sq^{(6,4)} + Sq^{(3,5)} + Sq^{(0,6)})(4_0^*) + (Sq^9 + Sq^{(6,1)})(4_2^*) + Sq^4(4_3^*) + Sq^1(4_5^*)$ |
| 0 | 6 | $6_0^*$ | $Sq^1(5_0^*)$ |
| 10 | 6 | $6_1^*$ | $Sq^{11}(5_0^*) + Sq^2(5_1^*)$ |
| 11 | 6 | $6_2^*$ | $Sq^{12}(5_0^*) + Sq^{(0,1)}(5_1^*) + Sq^1(5_2^*)$ |
| 14 | 6 | $6_3^*$ | $Sq^{15}(5_0^*) + Sq^6(5_1^*) + Sq^4(5_2^*) + Sq^1(5_3^*)$ |
| 15 | 6 | $6_4^*$ | $(Sq^{16} + Sq^{(10,2)})(5_0^*) + (Sq^7 + Sq^{(4,1)} + Sq^{(0,0,1)})(5_1^*) + Sq^1(5_5^*)$ |
| 16 | 6 | $6_5^*$ | $Sq^{(11,2)}(5_0^*) + (Sq^8 + Sq^{(5,1)})(5_1^*) + Sq^{(0,2)}(5_2^*) + Sq^3(5_3^*) + Sq^2(5_4^*)$ |

## 1.2. The Adams $E_2$-term for *tmf*

The topological modular forms spectrum *tmf* is an $E_\infty$ ring spectrum with mod 2 cohomology $H^*(tmf) = A/\!/A(2) = A \otimes_{A(2)} \mathbb{F}_2$, where $A(2) = \langle Sq^1, Sq^2, Sq^4 \rangle$ is the finite subalgebra of $A$ generated by the $Sq^{2^i}$ for $i \le 2$. The classical mod 2 Adams spectral sequence for *tmf* is an algebra spectral sequence

$$E_2^{s,t}(tmf) = \mathrm{Ext}_A^{s,t}(H^*(tmf), \mathbb{F}_2) \Longrightarrow_s \pi_{t-s}(tmf)_2^\wedge .$$

It is strongly convergent to the graded homotopy ring $\pi_*(tmf)_2^\wedge \cong \pi_*(tmf) \otimes \mathbb{Z}_2$, because *tmf* is connective and of finite type. The $E_\infty$ ring structure on *tmf* makes $H^*(tmf)$ a cocommutative $A$-module coalgebra, which in turn induces the bigraded commutative algebra structure on $\mathrm{Ext}_A(H^*(tmf), \mathbb{F}_2)$. It is this algebra structure on the $E_2$-term that carries over to the subsequent $E_r$-terms and makes $E_r(tmf)$ an algebra spectral sequence.

The $A$-module coalgebra structure on $H^*(tmf) = A \otimes_{A(2)} \mathbb{F}_2$ is induced from the evident $A(2)$-module coalgebra structure on $\mathbb{F}_2$. The change-of-algebras isomorphism

$$E_2(tmf) = \mathrm{Ext}_A(A \otimes_{A(2)} \mathbb{F}_2, \mathbb{F}_2) \cong \mathrm{Ext}_{A(2)}(\mathbb{F}_2, \mathbb{F}_2) ,$$

cf. Lemma 2.1, takes the algebra structure on the left hand side to the pairing on the right hand side that is induced by the tensor product of $A(2)$-modules. This is, in turn, equal to the Yoneda product in Ext over $A(2)$.

The $A(2)$-module component of the program `ext` will calculate a minimal free $A(2)$-module resolution

$$\dots \xrightarrow{\partial} C_2 \xrightarrow{\partial} C_1 \xrightarrow{\partial} C_0 \xrightarrow{\epsilon} \mathbb{F}_2 \to 0$$

of $\mathbb{F}_2$, in a finite range of filtration degrees $s \ge 0$ and internal degrees $t \ge 0$. As part of the calculation it will choose a basis $\{s_g^*\}_g$ indexed by non-negative integers $g \ge 0$ for each $A(2)$-module $C_s$. By minimality, $\mathrm{Ext}^s_{A(2)}(\mathbb{F}_2, \mathbb{F}_2) = \mathrm{Hom}_{A(2)}(C_s, \mathbb{F}_2)$.

Definition 1.8. For $s, g \ge 0$ let $s_g \in \mathrm{Ext}^{s,t}_{A(2)}(\mathbb{F}_2, \mathbb{F}_2) = \mathrm{Hom}^t_{A(2)}(C_s, \mathbb{F}_2)$ be the cocycle that is dual to the $g$'th generator $s_g^*$ of $C_s$. Here $t$ is the internal degree of that generator.

Adams-indexed charts of $\mathrm{Ext}^{s,t}_{A(2)}(\mathbb{F}_2, \mathbb{F}_2)$ for $0 \le t - s \le 192$ are shown in Figures 1.11 to 1.18. The dot with label $g$ in bidegree $(t-s, s)$ corresponds to the cocycle $s_g \in \mathrm{Ext}^{s,t}_{A(2)}(\mathbb{F}_2, \mathbb{F}_2)$. A small part of the minimal resolution $(C_*, \partial)$, with $0 \le s \le 6$ and $0 \le t \le 22$, is shown in Table 1.4.

Remark 1.9. To make these calculations, go to the directory `A2`, and let `tmf.def` be a text file containing the numbers `1 0`. This defines the $A(2)$-module with a single $\mathbb{F}_2$-generator in internal degree 0, necessarily with trivial action by each $Sq^i$. Use `newmodule tmf tmf.def` to create the module subdirectory `tmf`. Go to this subdirectory, and run `dims 0 240 &` (taking a couple of minutes) to calculate the minimal resolution for $0 \le s \le 40$ and $0 \le t \le 240$. When `dims` is finished, use `report` to extract data from the calculation. Thereafter use

```
chart 0 16 0 24 Shape himults Ext-A2-0-24.tex Ext-A2-F2
pdflatex Ext-A2-0-24.tex
```

to obtain the Adams chart in Figure 1.11. Then use

```
chart 4 20 24 48 Shape himults Ext-A2-24-48.tex Ext-A2-F2
pdflatex Ext-A2-24-48.tex
```

to obtain the Adams chart in Figure 1.12. Running the command `seeres` creates the file `resolution`, giving humanly readable formulas for the boundary operators $\partial\colon C_s \to C_{s-1}$, as shown in Table 1.4.

EXAMPLE 1.10. The class $0_0 = 1$ in $\mathrm{Ext}_{A(2)}^{0,0}(\mathbb{F}_2, \mathbb{F}_2)$ is the algebra unit. For $0 \le i \le 2$ the class $1_i = h_i$ in $\mathrm{Ext}_{A(2)}^{1,2^i}(\mathbb{F}_2, \mathbb{F}_2)$ is dual to $Sq^{2^i}$ in $A(2)$. For each $s \ge 0$ the class $s_0 = h_0^s$ in $\mathrm{Ext}_{A(2)}^{s,s}(\mathbb{F}_2, \mathbb{F}_2)$ detects $2^s \in \pi_0(tmf)_2^\wedge = \mathbb{Z}_2$. The next algebra indecomposable in $\mathrm{Ext}_{A(2)}(\mathbb{F}_2, \mathbb{F}_2)$ is $3_2 = c_0 \in \mathrm{Ext}_{A(2)}^{3,11}(\mathbb{F}_2, \mathbb{F}_2)$, in Adams bidegree $(t-s, s) = (8, 3)$.

REMARK 1.11. We use the same notation $s_g$ for `ext`-calculated cocycles in $\mathrm{Ext}_A^s(\mathbb{F}_2, \mathbb{F}_2)$ and in $\mathrm{Ext}_{A(2)}^s(\mathbb{F}_2, \mathbb{F}_2)$, so it must be understood from the context whether we are working over $A$ or over $A(2)$. The unit map $\iota\colon S \to tmf$ induces a morphism of Adams spectral sequences that is given at the $E_2$-term by the restriction homomorphism

$$\iota\colon \mathrm{Ext}_A(\mathbb{F}_2, \mathbb{F}_2) \longrightarrow \mathrm{Ext}_A(A/\!/A(2), \mathbb{F}_2) \cong \mathrm{Ext}_{A(2)}(\mathbb{F}_2, \mathbb{F}_2)$$

associated to the inclusion $A(2) \subset A$. This homomorphism takes $c_0 = 3_3$ in $\mathrm{Ext}_A^{3,11}(\mathbb{F}_2, \mathbb{F}_2)$ to $c_0 = 3_2$ in $\mathrm{Ext}_{A(2)}^{3,11}(\mathbb{F}_2, \mathbb{F}_2)$. The homomorphism $\iota$ preserves the filtration degree $s$, but does typically not preserve the generator index $g$.

PROPOSITION 1.12. *In topological degrees $t - s \le 200$, the algebra indecomposables in $\mathrm{Ext}_{A(2)}(\mathbb{F}_2, \mathbb{F}_2)$ are the classes listed in Table 1.3. The same classes are labeled and emphasized in Figures 1.19 and 1.20.*

Table 1.3: Algebra indecomposables in $\mathrm{Ext}_{A(2)}(\mathbb{F}_2, \mathbb{F}_2)$ (for $t-s \le 200$)

| $t-s$ | $s$ | $g$ | $x$ | dec. | $\iota'(x)$ |
|---|---|---|---|---|---|
| 0 | 1 | 0 | $h_0$ | | $v_0$ |
| 1 | 1 | 1 | $h_1$ | | 0 |
| 3 | 1 | 2 | $h_2$ | | 0 |
| 8 | 3 | 2 | $c_0$ | | 0 |
| 8 | 4 | 1 | $w_1$ | | $v_1^4$ |
| 12 | 3 | 3 | $\alpha$ | | $v_0v_2^2$ |
| 14 | 4 | 4 | $d_0$ | | 0 |
| 15 | 3 | 4 | $\beta$ | | 0 |
| 17 | 4 | 6 | $e_0$ | | 0 |
| 20 | 4 | 8 | $g$ | | 0 |
| 25 | 5 | 11 | $\gamma$ | | 0 |
| 32 | 7 | 11 | $\delta$ | $\alpha g$ | 0 |
| 48 | 8 | 19 | $w_2$ | | $v_2^8$ |

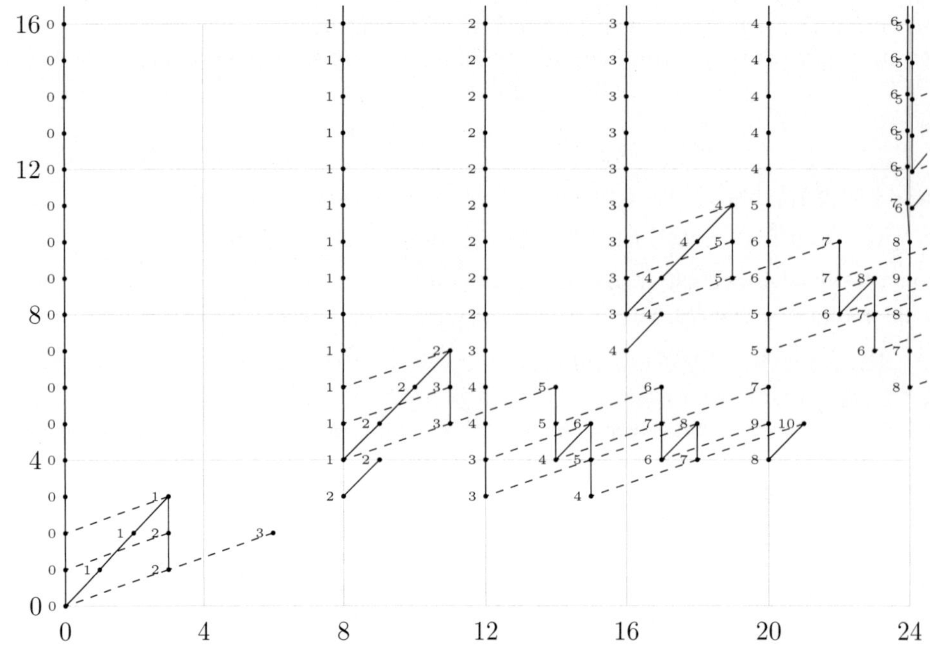

FIGURE 1.11. $\mathrm{Ext}^{s,t}_{A(2)}(\mathbb{F}_2, \mathbb{F}_2)$ for $0 \le t - s \le 24$

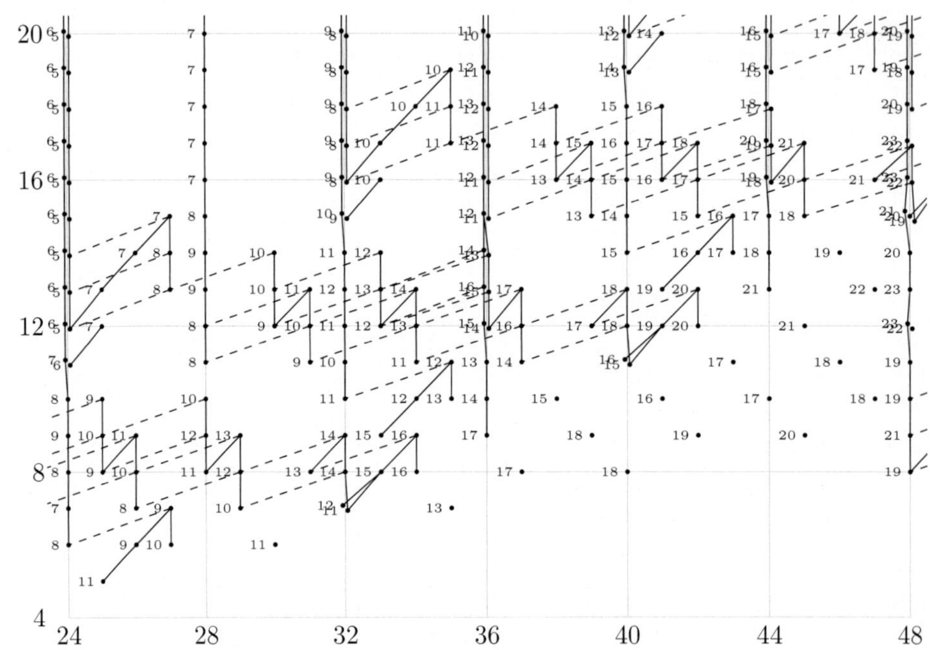

FIGURE 1.12. $\mathrm{Ext}^{s,t}_{A(2)}(\mathbb{F}_2, \mathbb{F}_2)$ for $24 \le t - s \le 48$

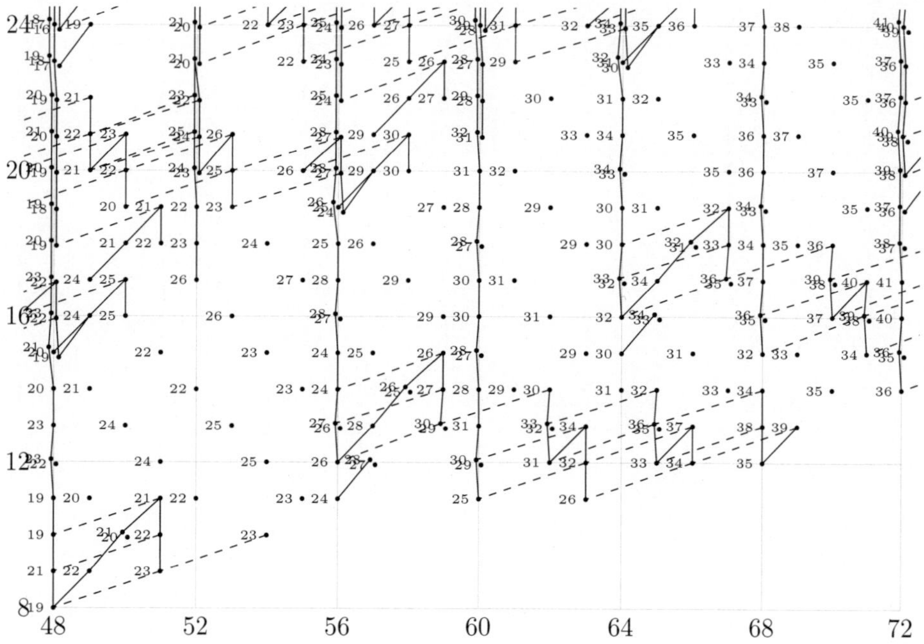

FIGURE 1.13. $\mathrm{Ext}_{A(2)}^{s,t}(\mathbb{F}_2, \mathbb{F}_2)$ for $48 \le t - s \le 72$

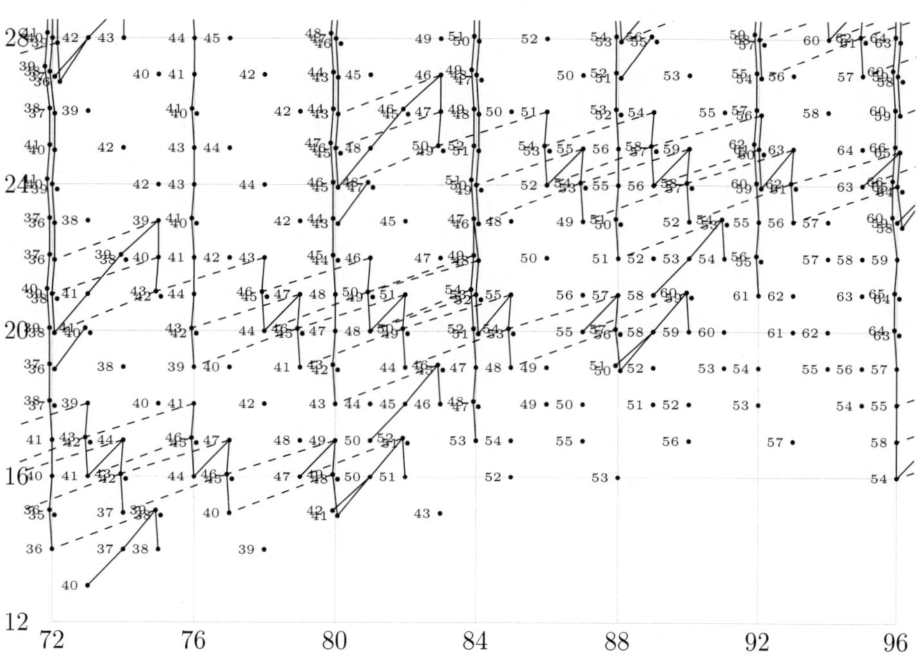

FIGURE 1.14. $\mathrm{Ext}_{A(2)}^{s,t}(\mathbb{F}_2, \mathbb{F}_2)$ for $72 \le t - s \le 96$

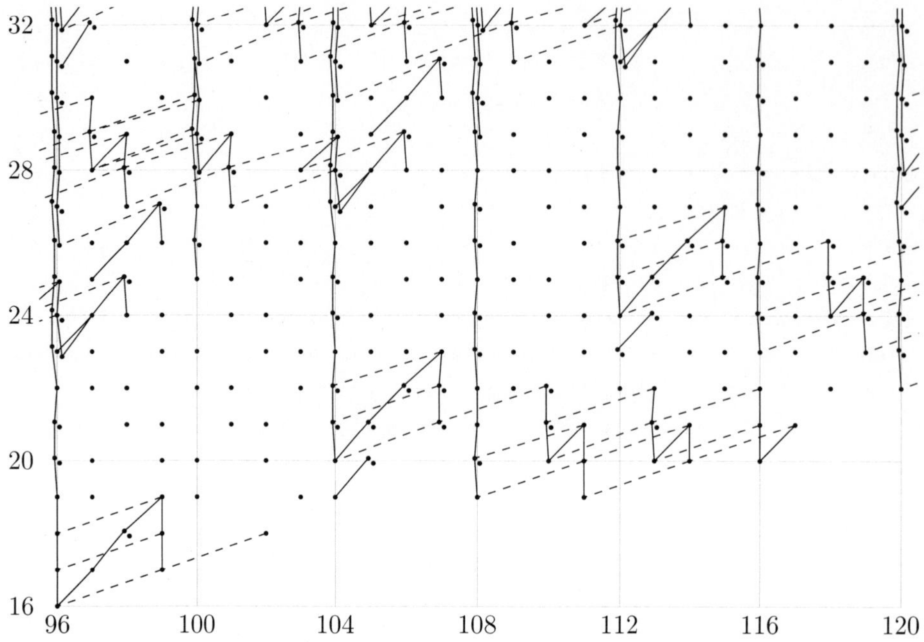

FIGURE 1.15. $\mathrm{Ext}_{A(2)}^{s,t}(\mathbb{F}_2, \mathbb{F}_2)$ for $96 \leq t - s \leq 120$

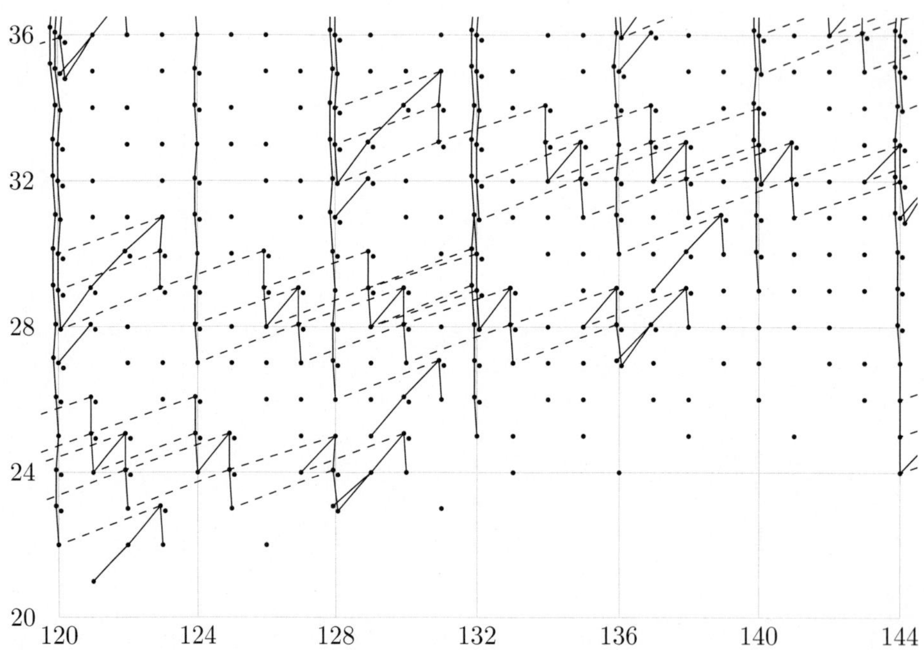

FIGURE 1.16. $\mathrm{Ext}_{A(2)}^{s,t}(\mathbb{F}_2, \mathbb{F}_2)$ for $120 \leq t - s \leq 144$

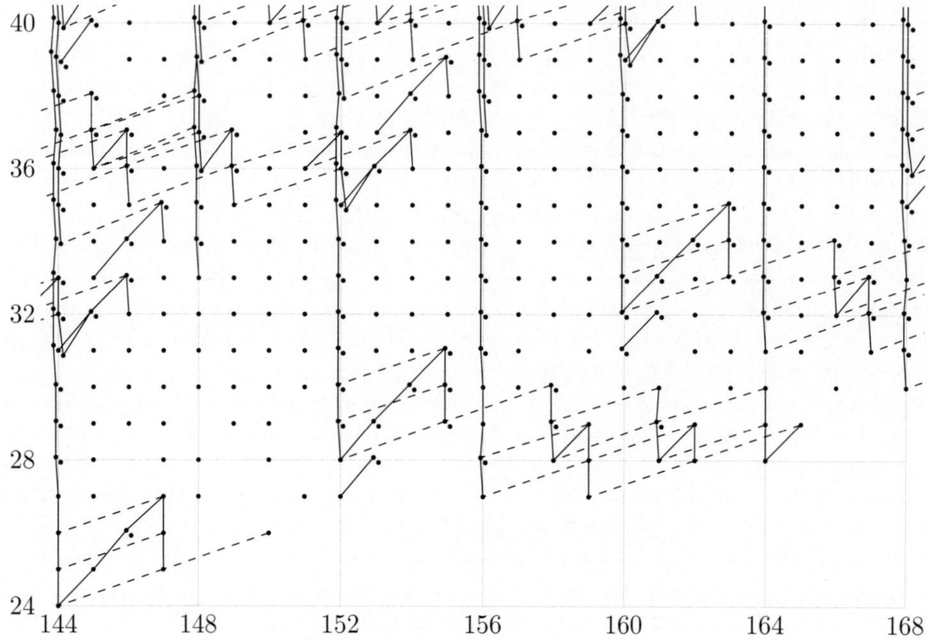

FIGURE 1.17. $\mathrm{Ext}_{A(2)}^{s,t}(\mathbb{F}_2, \mathbb{F}_2)$ for $144 \le t - s \le 168$

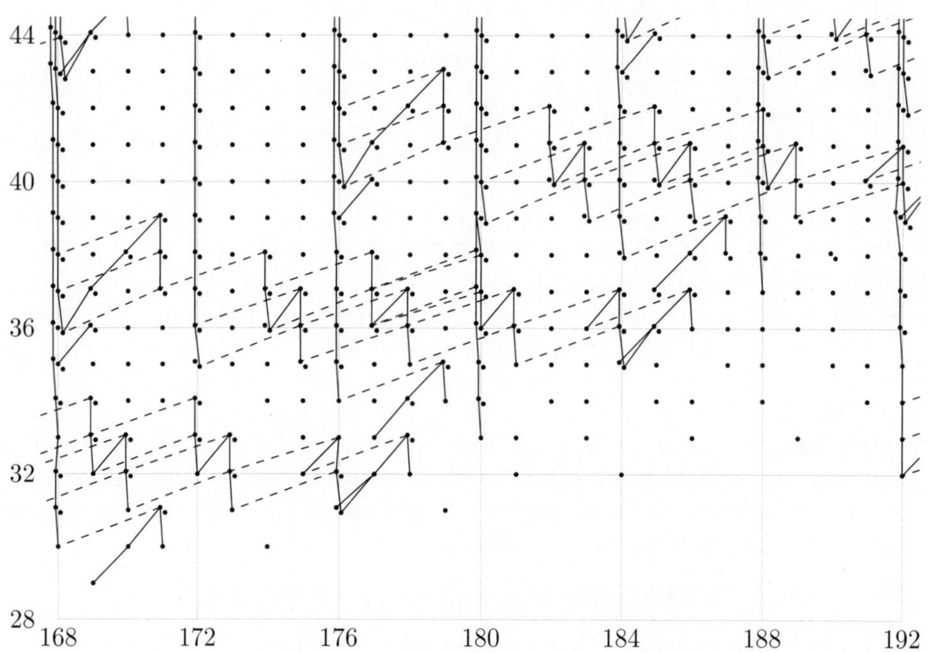

FIGURE 1.18. $\mathrm{Ext}_{A(2)}^{s,t}(\mathbb{F}_2, \mathbb{F}_2)$ for $168 \le t - s \le 192$

SKETCH PROOF. We will make no formal use of this proposition, other than to introduce notation, and will therefore allow ourselves to assert that the claim can be verified by machine computation. We will see later, in Theorem 3.46, that these classes generate all of $\mathrm{Ext}_{A(2)}(\mathbb{F}_2, \mathbb{F}_2)$ as an $\mathbb{F}_2$-algebra.

In more detail, the indecomposable quotient can be calculated in a finite range as explained in Remarks 1.9 and 1.13. Our notation for the 13 algebra generators $h_0, h_1, \dots, \delta, w_2$ follows Henriques [**54**, Ch. 13]. Each but one of the indecomposables is characterized by being the only nonzero class in its bidegree. The exceptional case is $\delta$ in bidegree $(t-s, s) = (32, 7)$, which also contains the decomposable class $\alpha g$. The third nonzero class in this bidegree, which we denote by $\delta' = \delta + \alpha g$, is thus also indecomposable. The class $\delta$ is characterized by the conditions $h_0\delta \neq 0$ and $h_1\delta \neq 0$, while $\alpha g$ satisfies $h_0\alpha g = h_0\delta$ and $h_1\alpha g = 0$, and $\delta'$ satisfies $h_0\delta' = 0$ and $h_1\delta' = h_1\delta$. As can be seen from Figure 1.12, this means that $\delta = 7_{11}$, $\alpha g = 7_{11} + 7_{12}$ and $\delta' = 7_{12}$ in the basis chosen by `ext`. □

REMARK 1.13. To make these calculations with `ext`, go to the directory `A2` and use `cocycle tmf 1 0`, `cocycle tmf 1 1` and `cocycle tmf 1 2` to create cocycle subdirectories `1_0`, `1_1` and `1_2` in `A2/tmf` and add their names to the list in `A2/tmf/maps` of cocycles that need to be lifted to chain maps. Go to `A2/tmf` and run `dolifts 0 40 maps` to calculate these lifts. Continue as in Remark 1.6, and repeat.

Whenever it is defined, the Adams periodicity operator $P$ in $\mathrm{Ext}_A(\mathbb{F}_2, \mathbb{F}_2)$ corresponds under $\iota$ to multiplication by $w_1$ in $\mathrm{Ext}_{A(2)}(\mathbb{F}_2, \mathbb{F}_2)$.

PROPOSITION 1.14 (Adams). *For $x \in \mathrm{Ext}_A(\mathbb{F}_2, \mathbb{F}_2)$ with $h_0^4 x = 0$,*

$$\iota(Px) = w_1 \cdot \iota(x)\,.$$

*More generally, for $i \geq 0$ and $h_0^{4\cdot 2^i} x = 0$,*

$$\iota(P^{2^i} x) = w_1^{2^i} \cdot \iota(x)\,.$$

PROOF. The first claim is a special case of [**7**, Lem. 4.5]. Using the description of $\mathrm{Ext}_A(\mathbb{F}_2, \mathbb{F}_2)$ as the cohomology of the cobar complex $(C^*_{A_*}(\mathbb{F}_2, \mathbb{F}_2), \delta)$, see Section 2.3, the classes $h_3$ and $h_0^4$ are represented by the cocycles $\xi = [\xi_1^8]$ and $\eta = [\xi_1|\xi_1|\xi_1|\xi_1]$, respectively. Let $\zeta$ be a cobar cocycle representing $x$ in bidegree $(t-s, s)$. Since $h_3 h_0^4 = 0$ and $h_0^4 x = 0$ we can write $\xi\eta = \delta(a)$ and $\eta\zeta = \delta(b)$, for cochains $a$ and $b$ in bidegrees $(8, 4)$ and $(t-s+1, s+3)$, respectively. By definition, $Px = \langle h_3, h_0^4, x\rangle$ is the class of the cocycle $a\zeta + \xi b$ in bidegree $(t-s+8, s+4)$, with indeterminacy $h_3 \mathrm{Ext}_A^{s+3,t+4}(\mathbb{F}_2, \mathbb{F}_2)$. (The group $\mathrm{Ext}_A^{4,12}(\mathbb{F}_2, \mathbb{F}_2)$ is trivial.) The restriction homomorphism $\iota$ is induced by the projection $A_* \to A(2)_*$, sending $\xi_1^8$, $\xi$ and $h_3$ to 0. Hence $a$ is sent to a cocycle, and Adams [**7**, Lem. 4.3] checks that this cocycle represents the nonzero class $w_1 \in \mathrm{Ext}_{A(2)}^{4,12}(\mathbb{F}_2, \mathbb{F}_2)$. Thus

$$\iota(Px) = \iota([a\zeta + \xi b]) = \iota([a])\iota([\zeta]) + 0 = w_1\iota(x)\,.$$

The cases $i \geq 1$ are similar, using [**7**, Lem. 4.4]. □

LEMMA 1.15. *In topological degrees $t - s \leq 48$ the values $\iota(x)$ of the restriction homomorphism $\iota\colon \mathrm{Ext}_A(\mathbb{F}_2, \mathbb{F}_2) \to \mathrm{Ext}_{A(2)}(\mathbb{F}_2, \mathbb{F}_2)$ on the algebra generators $x$ are as given in Table* 1.1.

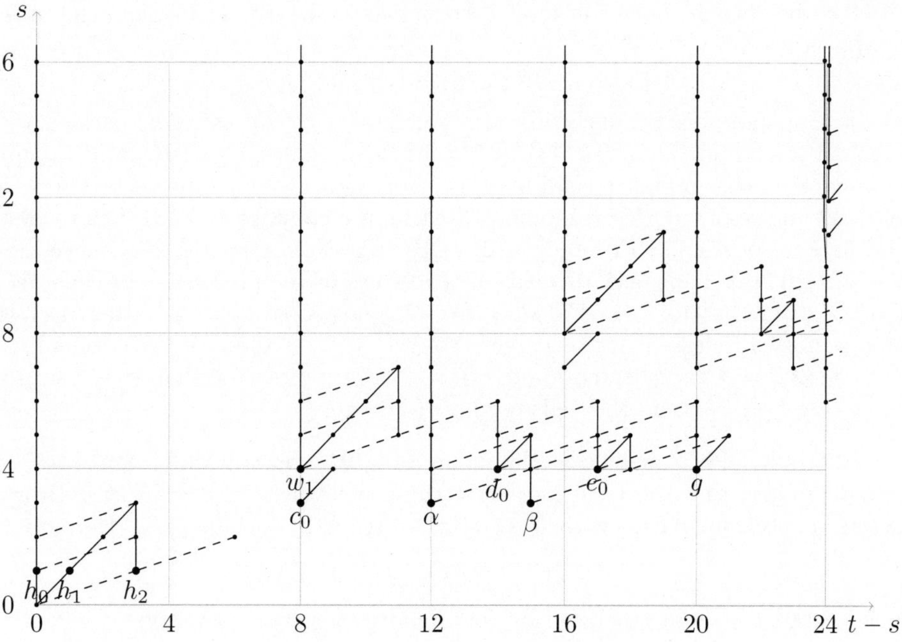


FIGURE 1.19. Indecomposables in $\mathrm{Ext}^{s,t}_{A(2)}(\mathbb{F}_2, \mathbb{F}_2)$ for $0 \le t - s \le 24$

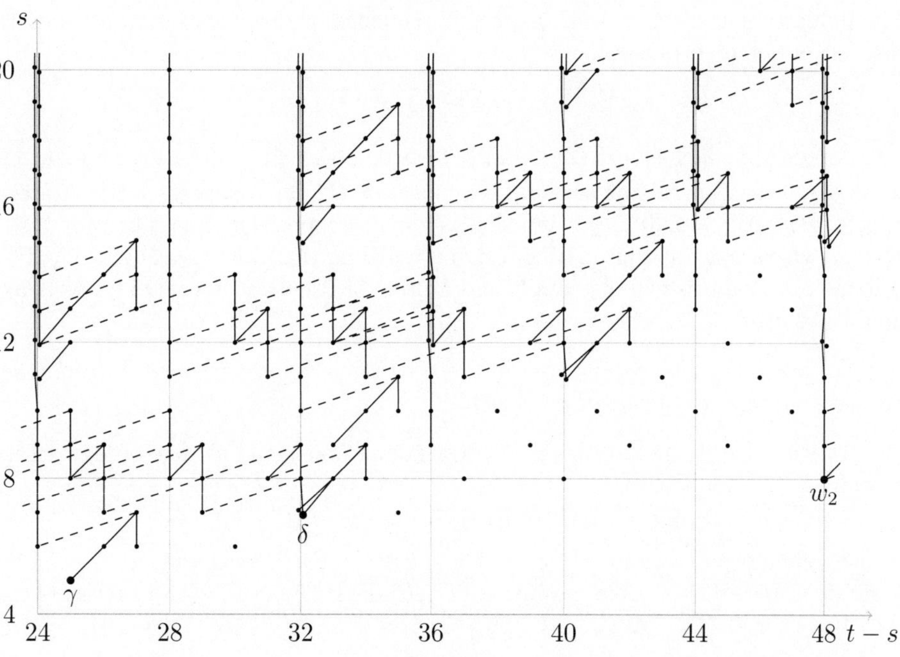


FIGURE 1.20. Indecomposables in $\mathrm{Ext}^{s,t}_{A(2)}(\mathbb{F}_2, \mathbb{F}_2)$ for $24 \le t - s \le 48$

PROOF. The homomorphism $\iota$ corresponds under the change-of-algebra isomorphism

$$\mathrm{Ext}_A(A//A(2), \mathbb{F}_2) \cong \mathrm{Ext}_{A(2)}(\mathbb{F}_2, \mathbb{F}_2)$$

to the homomorphism induced by $\epsilon\colon A//A(2) \to \mathbb{F}_2$. We use `ext` to calculate a minimal resolution $(D_*, \partial)$ of $A//A(2)$ by free $A$-modules, either by inducing up a minimal free $A(2)$-module resolution of $\mathbb{F}_2$, or by creating a module definition file for $A//A(2)$ and resolving this $A$-module. Next we use `cocycle` and `dolifts` to create and lift the cocycle $0_0\colon D_0 \to A//A(2) \to \mathbb{F}_2$ to a chain map $D_* \to C_*$ covering $\epsilon$. We use `collect maps all` to read off the values of the products $\iota(x) = x \cdot 0_0$ in $\mathrm{Ext}_A(A//A(2), \mathbb{F}_2)$ for $x$ in $\mathrm{Ext}_A(\mathbb{F}_2, \mathbb{F}_2)$. In most cases, $\iota(x)$ is either 0 or the unique nonzero class in its bidegree. For $t - s \le 48$ the only exceptional case is that of $x = \ell$ in bidegree $(t-s, s) = (32, 7)$, whose nonzero image satisfies $h_1\iota(\ell) = \iota(h_1\ell) = 0$, and this tells us that $\iota(\ell) = \alpha g$. □

REMARK 1.16. Lawson and Naumann [**90**], [**91**] constructed a map $\iota'\colon tmf \to tmf_1(3)$ of $E_\infty$ ring spectra, where $tmf_1(3)$ is equivalent to a truncated Brown–Peterson spectrum $BP\langle 2\rangle$ with $H^*(BP\langle 2\rangle) = A//E(2) = A \otimes_{E(2)} \mathbb{F}_2$. Here

$$E(2) = E(Q_0, Q_1, Q_2) \subset A(2)$$

is the exterior algebra generated by the Milnor (coalgebra) primitives

$$\begin{aligned} Q_0 &= Sq^1 \\ Q_1 &= [Sq^2, Q_0] = Sq^3 + Sq^2Sq^1 \\ Q_2 &= [Sq^4, Q_1] = Sq^7 + Sq^6Sq^1 + Sq^5Sq^2 + Sq^4Sq^2Sq^1 \,. \end{aligned}$$

The induced morphism of Adams spectral sequences is given at the $E_2$-term by the restriction homomorphism

$$\iota'\colon \mathrm{Ext}_{A(2)}(\mathbb{F}_2, \mathbb{F}_2) \longrightarrow \mathrm{Ext}_{E(2)}(\mathbb{F}_2, \mathbb{F}_2) = \mathbb{F}_2[v_0, v_1, v_2]$$

associated to the inclusion $E(2) \subset A(2)$. Here $v_i \in \mathrm{Ext}^{1,2^{i+1}-1}_{E(2)}(\mathbb{F}_2, \mathbb{F}_2)$ is dual to $Q_i$ for $0 \le i \le 2$. The Adams spectral sequence for $BP\langle 2\rangle$ collapses at the $E_2$-term, and $\pi_*(BP\langle 2\rangle)^\wedge_2 \cong \mathbb{Z}_2[v_1, v_2]$. We shall show in Proposition 1.44 that $tmf \wedge \Phi \simeq BP\langle 2\rangle$, where $\Phi$ is any finite CW spectrum realizing $A(2)//E(2) = A(2) \otimes_{E(2)} \mathbb{F}_2$ in cohomology, and this will play a role in our proof of Brown–Comenetz and Anderson duality for $tmf$, see Theorem 10.6.

LEMMA 1.17. *The values $\iota'(x)$ of the restriction homomorphism $\iota'$ on the algebra generators $x$ are as given in Table* 1.3.

PROOF. The homomorphism $\iota'$ corresponds under the change-of-algebra isomorphism

$$\mathrm{Ext}_{A(2)}(A(2)//E(2), \mathbb{F}_2) \cong \mathrm{Ext}_{E(2)}(\mathbb{F}_2, \mathbb{F}_2)$$

to the homomorphism induced by $\epsilon'\colon A(2)//E(2) \to \mathbb{F}_2$. We can use `ext` to calculate a minimal free $A(2)$-module resolution $(D'_*, \partial)$ of $A(2)//E(2)$, with generators dual to a basis for $\mathrm{Ext}^{s,t}_{A(2)}(A(2)//E(2), \mathbb{F}_2)$, for $s \le 8$ and $t \le 56$. Lifting the cocycle $0_0\colon D'_0 \to A(2)//E(2) \to \mathbb{F}_2$ gives a chain map $D'_* \to C_*$ covering $\epsilon'$, dual to the restriction homomorphism $\iota'$. From this we can read off that $\iota'(x)$ is nonzero for $x \in \{h_0, w_1, \alpha, w_2\}$ and zero for $x \in \{h_1, h_2, c_0, d_0, \beta, e_0, g, \gamma, \delta\}$. This determines $\iota'(x)$ in all but one case, that of $x = w_1$, for which $\iota'(w_1) \in \mathbb{F}_2\{v_0^2 v_1 v_2, v_1^4\}$.

$$\gamma_{0,0,0} \longmapsto 0_0^*$$

$$\gamma_{1,0,0} \longmapsto 1_0^*$$
$$\gamma_{0,1,0} \longmapsto Sq^2(1_0^*) + Sq^1(1_1^*)$$
$$\gamma_{0,0,1} \longmapsto Sq^4Sq^2(1_0^*) + Sq^4Sq^1(1_1^*) + Sq^{(0,1)}(1_2^*)$$

$$\gamma_{2,0,0} \longmapsto 2_0^*$$
$$\gamma_{1,1,0} \longmapsto Sq^2(2_0^*)$$
$$\gamma_{1,0,1} \longmapsto Sq^{(0,2)}(2_0^*) + Sq^3(2_2^*)$$
$$\gamma_{0,2,0} \longmapsto Sq^2(2_1^*)$$
$$\gamma_{0,1,1} \longmapsto (Sq^6 + Sq^{(0,2)})(2_1^*)$$

$$\gamma_{2,1,0} \longmapsto Sq^2(3_0^*)$$
$$\gamma_{2,0,1} \longmapsto Sq^{(0,2)}(3_0^*)$$
$$\gamma_{1,1,1} \longmapsto Sq^{(2,2)}(3_0^*) + Sq^5(3_1^*)$$
$$\gamma_{0,3,0} \longmapsto Sq^6(3_0^*) + Sq^2Sq^1(3_1^*)$$

$$\gamma_{2,1,1} \longmapsto Sq^{(2,2)}4_0^*$$
$$\gamma_{0,4,0} \longmapsto 4_1^*$$

FIGURE 1.21. Part of a chain map $E_* \to C_*$, showing that $\iota'(w_1) = v_1^4$

To settle that one case, we use the minimal free $E(2)$-module resolution $(E_*, \partial)$ of $\mathbb{F}_2$, with $E_s = E(2)\{\gamma_{i,j,k} \mid i+j+k = s\}$ and

$$\partial(\gamma_{i,j,k}) = Q_0\gamma_{i-1,j,k} + Q_1\gamma_{i,j-1,k} + Q_2\gamma_{i,j,k-1} .$$

Here $\gamma_{i,j,k}$ is dual to $v_0^i v_1^j v_2^k$, and is zero if $i < 0$, $j < 0$ or $k < 0$. Recall the minimal $A(2)$-free resolution $(C_*, \partial)$ of $\mathbb{F}_2$, given in Table 1.4 in the range $0 \le s \le 6$ and $0 \le t \le 22$. An $E(2)$-linear chain map $E_* \to C_*$ covering $\mathbb{F}_2$ is shown in Figure 1.21, on the subcomplex generated by the $\gamma_{i,j,k}$ with $(i,j,k) \le (2,1,1)$ or $(i,j,k) \le (0,4,0)$. In particular, the cocycle $w_1 = 4_1$ dual to $4_1^*$ restricts to the dual of $\gamma_{0,4,0}$, i.e., to $v_1^4$, with no contribution from $v_0^2 v_1 v_2$. □

REMARK 1.18. We shall see in Theorem 3.46 that there are no further algebra indecomposables in $\mathrm{Ext}_{A(2)}(\mathbb{F}_2, \mathbb{F}_2)$. The previous two lemmas show that the seven indecomposables $h_0$, $h_1$, $h_2$, $c_0$, $d_0$, $e_0$ and $g$ are the images of classes with the same names in $\mathrm{Ext}_A(\mathbb{F}_2, \mathbb{F}_2)$, and that the two indecomposables $w_1$ and $w_2$ map to powers of the classes $v_1$ and $v_2$ in $\mathrm{Ext}_{E(2)}(\mathbb{F}_2, \mathbb{F}_2)$. The Greek letters $\alpha$, $\beta$, $\gamma$ and $\delta$ are then used to denote the four remaining algebra generators.

Table 1.4: Minimal free $A(2)$-module resolution $(C_*, \partial)$ of $\mathbb{F}_2$ with $C_s = A(2)\{s_0^*, s_1^*, \dots\}$, for $s \le 6$ and $t \le 22$

| $t-s$ | $s$ | $x$ | $\partial(x)$ |
|---|---|---|---|
| 0 | 0 | $0_0^*$ | 1 |
| 0 | 1 | $1_0^*$ | $Sq^1(0_0^*)$ |
| 1 | 1 | $1_1^*$ | $Sq^2(0_0^*)$ |
| 3 | 1 | $1_2^*$ | $Sq^4(0_0^*)$ |
| 0 | 2 | $2_0^*$ | $Sq^1(1_0^*)$ |
| 2 | 2 | $2_1^*$ | $Sq^3(1_0^*) + Sq^2(1_1^*)$ |
| 3 | 2 | $2_2^*$ | $Sq^4(1_0^*) + Sq^{(0,1)}(1_1^*) + Sq^1(1_2^*)$ |
| 6 | 2 | $2_3^*$ | $Sq^7(1_0^*) + Sq^6(1_1^*) + Sq^4(1_2^*)$ |
| 0 | 3 | $3_0^*$ | $Sq^1(2_0^*)$ |
| 3 | 3 | $3_1^*$ | $Sq^4(2_0^*) + Sq^2(2_1^*) + Sq^1(2_2^*)$ |
| 8 | 3 | $3_2^*$ | $(Sq^{(6,1)} + Sq^{(3,2)})(2_0^*) + Sq^{(0,0,1)}(2_1^*) + Sq^6(2_2^*) + (Sq^3 + Sq^{(0,1)})(2_3^*)$ |
| 12 | 3 | $3_3^*$ | $(Sq^{(7,2)} + Sq^{(0,2,1)})(2_0^*) + (Sq^{(5,2)} + Sq^{(2,3)})(2_1^*) + Sq^7(2_3^*)$ |
| 15 | 3 | $3_4^*$ | $(Sq^{(7,3)} + Sq^{(3,2,1)})(2_0^*) + (Sq^{(5,3)} + Sq^{(4,1,1)})(2_1^*) + (Sq^{(7,2)} + Sq^{(0,2,1)})(2_2^*) + (Sq^{(4,2)} + Sq^{(1,3)})(2_3^*)$ |
| 0 | 4 | $4_0^*$ | $Sq^1(3_0^*)$ |
| 8 | 4 | $4_1^*$ | $Sq^{(6,1)}(3_0^*) + Sq^{(3,1)}(3_1^*)$ |
| 9 | 4 | $4_2^*$ | $Sq^{(4,2)}(3_0^*) + (Sq^7 + Sq^{(1,2)} + Sq^{(0,0,1)})(3_1^*) + Sq^2(3_2^*)$ |
| 12 | 4 | $4_3^*$ | $Sq^{(0,2,1)}(3_0^*) + Sq^{(7,1)}(3_1^*) + Sq^1(3_3^*)$ |

Table 1.4: Minimal free $A(2)$-module resolution $(C_*, \partial)$ of $\mathbb{F}_2$ with $C_s = A(2)\{s_0^*, s_1^*, \dots\}$, for $s \le 6$ and $t \le 22$ (cont.)

| $t-s$ | $s$ | $x$ | $\partial(x)$ |
|---|---|---|---|
| 14 | 4 | $4_4^*$ | $(Sq^{(6,3)} + Sq^{(2,2,1)})(3_0^*) + Sq^{(6,2)}(3_1^*) + (Sq^7 + Sq^{(4,1)} + Sq^{(0,0,1)})(3_2^*) + Sq^3(3_3^*)$ |
| 15 | 4 | $4_5^*$ | $Sq^{(7,3)}(3_0^*) + Sq^{(0,2,1)}(3_1^*) + Sq^4(3_3^*) + Sq^1(3_4^*)$ |
| 17 | 4 | $4_6^*$ | $Sq^{(5,2,1)}(3_0^*) + Sq^{(6,3)}(3_1^*) + (Sq^{(7,1)} + Sq^{(4,2)})(3_2^*) + (Sq^6 + Sq^{(0,2)})(3_3^*) + Sq^3(3_4^*)$ |
| 18 | 4 | $4_7^*$ | $Sq^{(6,2,1)}(3_0^*) + Sq^{(6,1,1)}(3_1^*) + (Sq^{(5,2)} + Sq^{(4,0,1)})(3_2^*) + Sq^4(3_4^*)$ |
| 0 | 5 | $5_0^*$ | $Sq^1(4_0^*)$ |
| 8 | 5 | $5_1^*$ | $Sq^{(6,1)}(4_0^*) + Sq^1(4_1^*)$ |
| 9 | 5 | $5_2^*$ | $Sq^2(4_1^*)$ |
| 11 | 5 | $5_3^*$ | $Sq^{(6,2)}(4_0^*) + Sq^4(4_1^*) + Sq^3(4_2^*)$ |
| 12 | 5 | $5_4^*$ | $Sq^{(0,2,1)}(4_0^*) + Sq^1(4_3^*)$ |
| 14 | 5 | $5_5^*$ | $Sq^{(6,3)}(4_0^*) + Sq^7(4_1^*) + (Sq^6 + Sq^{(0,2)})(4_2^*) + Sq^1(4_4^*)$ |
| 15 | 5 | $5_6^*$ | $Sq^{(3,2,1)}(4_0^*) + Sq^{(0,0,1)}(4_2^*) + Sq^4(4_3^*) + Sq^2(4_4^*) + Sq^1(4_5^*)$ |
| 17 | 5 | $5_7^*$ | $Sq^{(5,2,1)}(4_0^*) + Sq^{(6,1)}(4_2^*) + Sq^4(4_4^*) + Sq^1(4_6^*)$ |
| 0 | 6 | $6_0^*$ | $Sq^1(5_0^*)$ |
| 8 | 6 | $6_1^*$ | $Sq^{(6,1)}(5_0^*) + Sq^1(5_1^*)$ |
| 10 | 6 | $6_2^*$ | $Sq^3(5_1^*) + Sq^2(5_2^*)$ |
| 11 | 6 | $6_3^*$ | $Sq^{(6,2)}(5_0^*) + Sq^4(5_1^*) + Sq^{(0,1)}(5_2^*) + Sq^1(5_3^*)$ |
| 12 | 6 | $6_4^*$ | $Sq^{(0,2,1)}(5_0^*) + Sq^1(5_4^*)$ |
| 14 | 6 | $6_5^*$ | $Sq^{(6,3)}(5_0^*) + Sq^7(5_1^*) + Sq^6(5_2^*) + Sq^4(5_3^*) + Sq^1(5_5^*)$ |

## 1.3. Steenrod operations in $E_2(tmf)$

There are Steenrod operations

$$Sq^i \colon \operatorname{Ext}_\Gamma^{s,t}(\mathbb{F}_2, \mathbb{F}_2) \longrightarrow \operatorname{Ext}_\Gamma^{s+i,2t}(\mathbb{F}_2, \mathbb{F}_2)$$

acting on Ext over any cocommutative Hopf algebra $\Gamma$ defined over $\mathbb{F}_2$ (and similarly at odd primes), see [**95**, Ch. 2] and [**118**, §11]. Let $W_*$ be the standard free $\mathbb{F}_2[\Sigma_2]$-resolution of $\mathbb{F}_2$, with $W_i$ generated by $e_i$ for each $i \geq 0$, and let $C_*$ be a free $\Gamma$-module resolution of $\mathbb{F}_2$. There is a unique homotopy class of $\Sigma_2$-equivariant maps of $\Gamma$-module complexes

$$\Delta \colon W_* \otimes C_* \longrightarrow C_* \otimes C_*$$

covering $\mathbb{F}_2$, where $\Sigma_2$ acts freely on $W_*$ on the left hand side and by the symmetry isomorphism on the right hand side, while $\Gamma$ acts freely on $C_*$ on the left hand side and by the diagonal action on the right hand side. For each cocycle $x \colon C_s \to \Sigma^t \mathbb{F}_2$ the formula

$$a \longmapsto \langle x \otimes x, \Delta(e_i \otimes a) \rangle ,$$

where $\langle -, - \rangle$ denotes the (Kronecker) evaluation pairing, defines a cocycle $C_{2s-i} \to \Sigma^{2t}\mathbb{F}_2$. By construction, its cohomology class is $Sq^{s-i}(x) \in \operatorname{Ext}_\Gamma^{2s-i,2t}(\mathbb{F}_2, \mathbb{F}_2)$. These operations satisfy $Sq^s(x) = x^2$, and $Sq^i(x) = 0$ if $i < 0$ or $i > s$. Furthermore, the Cartan formula

$$Sq^k(xy) = \sum_{i+j=k} Sq^i(x) Sq^j(y) \tag{1.1}$$

and the Adem relations

$$Sq^a Sq^b = \sum_i \binom{b-i-1}{a-2i} Sq^{a+b-i} Sq^i \tag{1.2}$$

hold, where $a < 2b$. In particular, $Sq^0 Sq^i = Sq^i Sq^0$ for each $i \geq 0$.

DEFINITION 1.19. For $x \in \operatorname{Ext}_\Gamma^{s,t}(\mathbb{F}_2, \mathbb{F}_2)$ we let

$$Sq^*(x) = (x^2 = Sq^s(x), Sq^{s-1}(x), \dots, Sq^1(x), Sq^0(x))$$

be the total squaring operation on $x$.

When $\Gamma = A(2)$ we can completely determine the Steenrod operations in Ext. In contrast to the case $\Gamma = A$, there are only a few sequences $(h_0, h_1, h_2)$ and $(w_1, g)$ of generators connected by the $Sq^0$-operations, and $w_2 \neq Sq^0(w_1) = g$ deviates from the indexing scheme mentioned in Remark 1.5.

THEOREM 1.20. *The Steenrod operations in* $\operatorname{Ext}_{A(2)}(\mathbb{F}_2, \mathbb{F}_2)$ *are given by*

$$\begin{aligned}
Sq^*(h_0) &= (h_0^2, h_1) \\
Sq^*(h_1) &= (h_1^2, h_2) \\
Sq^*(h_2) &= (h_2^2, 0) \\
Sq^*(c_0) &= (0, h_0 e_0, h_2 \beta, 0) \\
Sq^*(\alpha) &= (\alpha^2, \gamma, 0, 0) \\
Sq^*(\beta) &= (\beta^2, 0, 0, 0) \\
Sq^*(d_0) &= (g w_1, 0, \beta^2, 0, 0) \\
Sq^*(e_0) &= (d_0 g, \beta g, 0, 0, 0)
\end{aligned}$$

$$
\begin{aligned}
Sq^*(\gamma) &= (\beta^2 g + h_1^2 w_2, h_2 w_2, 0,0,0,0) \\
Sq^*(\delta) &= (0, h_0 e_0 w_2, h_2\beta w_2, 0,0,0,0,0) \\
Sq^*(g) &= (g^2, 0,0,0,0) \\
Sq^*(w_1) &= (w_1^2, 0,0,0,g) \\
Sq^*(w_2) &= (w_2^2, 0,0,0,0,0,0,0,0)\,.
\end{aligned}
$$

PROOF. The products $Sq^s(x) = x^2$ are calculated with ext, cf. Table 3.5. The operations landing in trivial groups are obviously zero. It is well-known that $Sq^0(h_i) = h_{i+1}$ for each $i \geq 0$, see e.g. [**3**, p. 36] or [**118**, Def. 11.9]. This can also be verified directly for $i \in \{0,1\}$ by the method we use in Proposition 1.21 to calculate $Sq^*(c_0)$. The remaining operations are

$$
\begin{aligned}
Sq^2(\alpha) &\in \mathbb{F}_2\{\gamma\} \\
Sq^2(d_0) &\in \mathbb{F}_2\{\beta^2\} \\
Sq^3(d_0) &\in \mathbb{F}_2\{\alpha e_0\} \\
Sq^3(e_0) &\in \mathbb{F}_2\{\beta g\} \\
Sq^4(\gamma) &\in \mathbb{F}_2\{h_2 w_2\} \\
Sq^5(\delta) &\in \mathbb{F}_2\{h_2\beta w_2\} \\
Sq^6(\delta) &\in \mathbb{F}_2\{\gamma g^2, h_0 e_0 w_2\} \\
Sq^0(w_1) &\in \mathbb{F}_2\{g\}\,.
\end{aligned}
$$

As we now show, each of these can be determined by the Cartan formula and multiplicative relations that hold in $\mathrm{Ext}_{A(2)}(\mathbb{F}_2, \mathbb{F}_2)$, cf. Tables 3.4 and 3.5:

Applying $Sq^3$ to $h_1\alpha = 0$ gives $h_1^2 Sq^2(\alpha) = h_2\alpha^2 = h_1^2\gamma \neq 0$, so $Sq^2(\alpha) = \gamma$.

Applying $Sq^3$ to $h_1 d_0 = h_0 h_2 \alpha$ gives $h_1^2 Sq^2(d_0) = h_2 Sq^3(d_0)$. Here $h_1^2\beta^2 = 0$ and $h_2\alpha e_0 = h_0\alpha g \neq 0$, so $Sq^3(d_0) = 0$.

Applying $Sq^5$ to $h_0\gamma = 0$ gives $h_0^2 Sq^4(\gamma) = h_1(\beta^2 g + h_1^2 w_2) = h_0^2 h_2 w_2 \neq 0$, so $Sq^4(\gamma) = h_2 w_2$.

Applying $Sq^6$ to the relation $\beta d_0 = \alpha e_0$ gives $\alpha^2 Sq^3(e_0) = \gamma \cdot d_0 g = d_0\gamma g \neq 0$, so $Sq^3(e_0) = \beta g$.

Applying $Sq^5$ to the same relation gives $\beta^2 Sq^2(d_0) = \gamma \cdot \beta g = g^3 \neq 0$, so $Sq^2(d_0) = \beta^2$.

Applying $Sq^4$ to $d_0^2 = g w_1$ gives $g^2 Sq^0(w_1) = \beta^4 = g^3 \neq 0$, so $Sq^0(w_1) = g$.

Applying $Sq^5$ to the relation $c_0\gamma = h_1\delta$ gives $h_2 Sq^5(\delta) = h_2\beta \cdot h_2 w_2 = h_1 g w_2 \neq 0$, so $Sq^5(\delta) = h_2\beta w_2$.

Applying $Sq^6$ to the same relation gives $h_2 Sq^6(\delta) = h_0 e_0 \cdot h_2 w_2 = h_0^2 g w_2 \neq 0$, using $h_2\beta \cdot (\beta^2 g + h_1^2 w_2) = 0$ and $h_1^2 \cdot h_2\beta w_2 = 0$. Hence $Sq^6(\delta) \neq 0$.

Applying $Sq^8$ to $\alpha\delta = 0$ gives $\gamma Sq^6(\delta) = 0$, using $\alpha^2 \cdot h_2\beta w_2 = 0$. Here $\gamma \cdot \gamma g^2 = \beta^2 g^3 \neq 0$ and $\gamma \cdot h_0 e_0 w_2 = 0$, so $Sq^6(\delta) = h_0 e_0 w_2$. □

To calculate $Sq^i(c_0)$ we use a method suggested by Christian Nassau [**135**]. It goes back to Steenrod's second definition [**159**] of the squaring operations in terms of $\cup_i$-pairings giving chain homotopies between $\cup_{i-1}$ and $\cup_{i-1}\tau$, where $\tau$ denotes the symmetry isomorphism.

Consider a cocycle $x\colon C_s \to \Sigma^t \mathbb{F}_2$ that factors as $x = yf$, with $K_* \to \mathbb{F}_2$ a quasi-isomorphism, $f\colon C_* \to K_*$ a chain map over $\mathbb{F}_2$ and $y\colon K_s \to \Sigma^t \mathbb{F}_2$ a cocycle.

Here $C_* \to \mathbb{F}_2$ is the free resolution considered above, while $K_*$ will typically not consist of free modules. To evaluate $x \otimes x$ on $\Delta(e_i \otimes a)$ for $a \in C_{2s-i}$ we can instead evaluate $y \otimes y$ on $(f \otimes f)\Delta(e_i \otimes a)$:

$$\begin{array}{ccccc} W_* \otimes C_* & \xrightarrow{\Delta} & C_* \otimes C_* & & \\ & \searrow^{D} & \big\downarrow{\scriptstyle f\otimes f} & \searrow^{x\otimes x} & \\ & & K_* \otimes K_* & \xrightarrow[y\otimes y]{} & \Sigma^{2t}\mathbb{F}_2 . \end{array}$$

Any choice of $\Sigma_2$-equivariant chain map $D\colon W_* \otimes C_* \to K_* \otimes K_*$ covering $\mathbb{F}_2$ will make the left hand triangle commute up to chain homotopy, since $\Sigma_2$ acts freely on $W_*$ and $K_* \otimes K_* \to \mathbb{F}_2$ is a quasi-isomorphism.

Let $D_i\colon C_{*-i} \to K_* \otimes K_*$ be given by $D_i(a) = D(e_i \otimes a)$, for each $i \geq 0$. Then $D_0\colon C_* \to K_* \otimes K_*$ is a chain map over $\mathbb{F}_2$, and $(y \otimes y)D_0\colon C_{2s} \to \Sigma^{2t}\mathbb{F}_2$ represents $x^2 = Sq^s(x)$. Next, $D_1\colon C_{*-1} \to K_* \otimes K_*$ is a chain homotopy from $D_0$ to $\tau D_0$, in the sense that $\partial D_1 + D_1\partial = D_0 + \tau D_0$, and $(y \otimes y)D_1\colon C_{2s-1} \to \Sigma^{2t}\mathbb{F}_2$ represents $\cup_1(x) = Sq^{s-1}(x)$. Continuing, $D_2\colon C_{*-2} \to K_* \otimes K_*$ is a chain homotopy from $D_1$ to $\tau D_1$, in the sense that $\partial D_2 + D_2\partial = D_1 + \tau D_1$, and $(y \otimes y)D_2\colon C_{2s-2} \to \Sigma^{2t}\mathbb{F}_2$ represents $\cup_2(x) = Sq^{s-2}(x)$. In general, $(y \otimes y)D_i$ gives $Sq^{s-i}(x)$ for all $i \geq 0$.

Conversely, a diagonal approximation $D_0$ and a sequence of chain homotopies $D_i$ from $D_{i-1}$ to $\tau D_{i-1}$, for each $i \geq 1$, correspond precisely to a $\Sigma_2$-equivariant chain map $D$ as above. This process gives all the squaring operations on any cocycle $x$ for which we can write down a corresponding $s$-fold extension $K_*$. The computational efficacy of this process depends upon the size of the complex $K_* \otimes K_*$.

PROPOSITION 1.21. *$Sq^1(c_0) = h_2\beta$ and $Sq^2(c_0) = h_0e_0$ in* $\mathrm{Ext}_{A(2)}(\mathbb{F}_2, \mathbb{F}_2)$.

PROOF. The class $c_0 \in \mathrm{Ext}^{3,11}_{A(2)}(\mathbb{F}_2, \mathbb{F}_2)$ is represented by the 3-fold exact complex of $A(2)$-modules

$$0 \to K_3 \xrightarrow{\partial} K_2 \xrightarrow{\partial} K_1 \xrightarrow{\partial} K_0 \xrightarrow{\epsilon} \mathbb{F}_2 \to 0 \tag{1.3}$$

given in Figure 1.22, where we identify $K_3$ with $\Sigma^{11}\mathbb{F}_2$ by a cocycle $y\colon K_3 \to \Sigma^{11}\mathbb{F}_2$. Each $K_n$ is a cyclic $A(2)$-module generated by $k_n$, and $\epsilon(k_0) = 1$, $\partial(k_1) = Sq^1(k_0)$, $\partial(k_2) = Sq^4(k_1)$, $\partial(k_3) = Sq^6(k_2)$ and $y(k_3) = 1$.

Recall the minimal $A(2)$-free resolution $(C_*, \partial)$ of $\mathbb{F}_2$, given in Table 1.4 in the range $0 \leq s \leq 6$ and $0 \leq t \leq 22$. A chain map $f\colon C_* \to K_*$ covering $\mathbb{F}_2$ is given by

$$\begin{aligned} 0_0^* &\longmapsto k_0 \\ 1_0^* &\longmapsto k_1 \\ 2_2^* &\longmapsto k_2 \\ 3_2^* &\longmapsto k_3 , \end{aligned}$$

sending the remaining generators $s_g^*$ to zero. The composite $x = yf\colon C_3 \to \Sigma^{11}\mathbb{F}_2$ is then dual to $3_2^*$, which shows that (1.3) represents $3_2 = c_0$.

A chain map $D_0\colon C_* \to K_* \otimes K_*$ covering $\mathbb{F}_2$ is given as in Figure 1.23, sending the remaining $s_g^*$ to zero. In particular, $(y \otimes y)D_0\colon C_6 \to \mathbb{F}_2$ is zero, confirming that $Sq^3(c_0) = c_0^2 = 0$.

A chain homotopy $D_1\colon C_{*-1} \to K_* \otimes K_*$ from $D_0$ to $\tau D_0$ is given by

$$1_1^* \longmapsto k_1 \otimes k_1$$

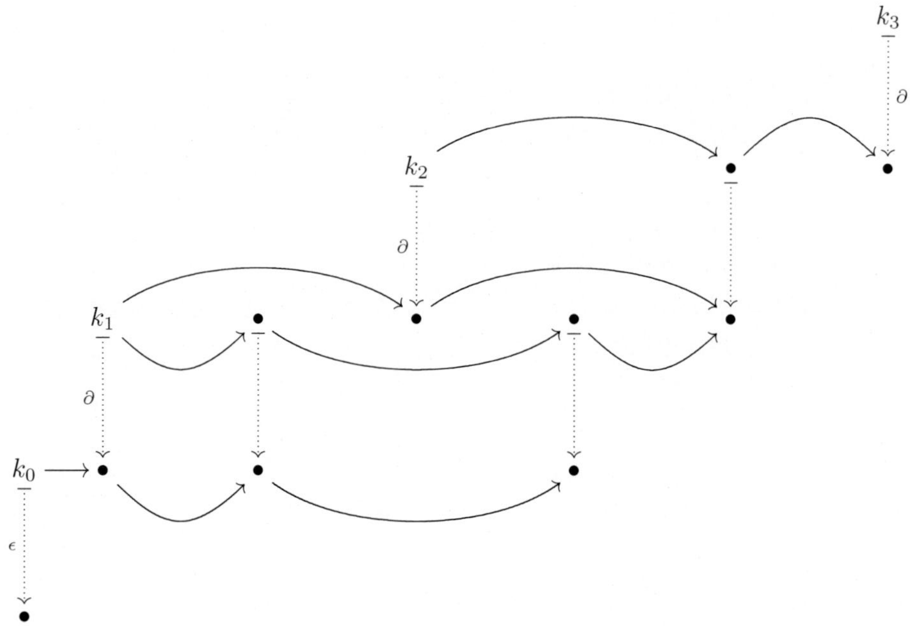


FIGURE 1.22. A 3-fold extension $K_*$ representing $c_0$

$$\begin{aligned}
2_3^* &\longmapsto k_2 \otimes Sq^2(k_1) + Sq^2(k_1) \otimes k_2 \\
3_4^* &\longmapsto k_3 \otimes Sq^4Sq^2(k_1) + Sq^4(k_2) \otimes Sq^4(k_2) + Sq^4Sq^2(k_1) \otimes k_3 \\
4_7^* &\longmapsto k_3 \otimes Sq^6(k_2) \\
5_7^* &\longmapsto k_3 \otimes k_3 \,,
\end{aligned}$$

sending the remaining $s_g^*$ to zero. Hence $(y \otimes y)D_1 \colon C_5 \to \mathbb{F}_2$ is dual to $5_7^*$, proving that $Sq^2(c_0) = 5_7 = h_0e_0$.

A chain homotopy $D_2 \colon C_{*-2} \to K_* \otimes K_*$ from $D_1$ to $\tau D_1$ is given by

$$4_7^* \longmapsto k_3 \otimes k_3 \,,$$

sending the remaining $s_g^*$ to zero. Hence $(y \otimes y)D_2 \colon C_4 \to \mathbb{F}_2$ is dual to $4_7^*$, proving that $Sq^1(c_0) = 4_7 = h_2\beta$.

In this case, $D_2 = \tau D_2$, so we can take $D_3 = 0$, confirming that $Sq^0(c_0) = 0$. □

A little more generally, there are Steenrod operations

$$Sq^i \colon \operatorname{Ext}_\Gamma^{s,t}(L, \mathbb{F}_2) \longrightarrow \operatorname{Ext}_\Gamma^{s+i,2t}(L, \mathbb{F}_2)$$

for any cocommutative $\Gamma$-module coalgebra $L$. Let $C_* \to L$ be a free $\Gamma$-module resolution, and let $\Delta \colon W_* \otimes C_* \to C_* \otimes C_*$ be a $\Sigma_2$-equivariant map of $\Gamma$-module complexes covering the coproduct $\psi \colon L \to L \otimes L$. For each cocycle $x \colon C_s \to \Sigma^t \mathbb{F}_2$ the composite

$$\begin{aligned}
C_{2s-i} \cong \mathbb{F}_2\{e_i\} \otimes C_{2s-i} \subset W_i \otimes C_{2s-i} \subset (W_* \otimes C_*)_{2s} \\
\xrightarrow{\Delta} (C_* \otimes C_*)_{2s} \xrightarrow{x \otimes x} \Sigma^t \mathbb{F}_2 \otimes \Sigma^t \mathbb{F}_2 \cong \Sigma^{2t} \mathbb{F}_2
\end{aligned}$$

$$0_0^* \longmapsto k_0 \otimes k_0$$

$$\begin{aligned}
1_0^* &\longmapsto k_1 \otimes k_0 + k_0 \otimes k_1 \\
1_1^* &\longmapsto k_1 \otimes Sq^1(k_0)
\end{aligned}$$

$$\begin{aligned}
2_0^* &\longmapsto k_1 \otimes k_1 \\
2_1^* &\longmapsto k_1 \otimes Sq^2(k_1) \\
2_2^* &\longmapsto k_2 \otimes k_0 + k_0 \otimes k_2 \\
2_3^* &\longmapsto k_2 \otimes Sq^2 Sq^1(k_0)
\end{aligned}$$

$$\begin{aligned}
3_1^* &\longmapsto k_2 \otimes k_1 + k_1 \otimes k_2 \\
3_2^* &\longmapsto k_3 \otimes k_0 + k_0 \otimes k_3 \\
3_4^* &\longmapsto Sq^4 Sq^2 Sq^1(k_0) \otimes k_3
\end{aligned}$$

$$4_4^* \longmapsto k_3 \otimes Sq^4 Sq^2(k_1) + Sq^4(k_2) \otimes Sq^4(k_2) + Sq^4 Sq^2(k_1) \otimes k_3$$

$$\begin{aligned}
5_6^* &\longmapsto k_3 \otimes Sq^4(k_2) + Sq^4(k_2) \otimes k_3 \\
5_7^* &\longmapsto k_3 \otimes Sq^6(k_2)
\end{aligned}$$

FIGURE 1.23. Chain map $D_0 \colon C_* \to K_* \otimes K_*$ covering $\mathbb{F}_2$

defines a cocycle $Sq^{s-i}(x) \colon C_{2s-i} \to \Sigma^{2t}\mathbb{F}_2$. This construction induces the Steenrod operation upon passage to cohomology classes. For later reference we record the following compatibility between the change-of-algebra isomorphism recalled in Lemma 2.1 and these Steenrod operations.

LEMMA 1.22. *Let $\Lambda \subset \Gamma$ be a pair of cocommutative Hopf algebras, and let $L$ be a cocommutative $\Lambda$-module coalgebra. Under the change-of-algebra isomorphisms*

$$\mathrm{Ext}_\Gamma^{s,t}(\Gamma \otimes_\Lambda L, \mathbb{F}_2) \cong \mathrm{Ext}_\Lambda^{s,t}(L, \mathbb{F}_2)$$

*the Steenrod operation $Sq^i \colon \mathrm{Ext}_\Gamma^{s,t}(\Gamma \otimes_\Lambda L, \mathbb{F}_2) \to \mathrm{Ext}_\Gamma^{s+i,2t}(\Gamma \otimes_\Lambda L, \mathbb{F}_2)$ corresponds to the Steenrod operation $Sq^i \colon \mathrm{Ext}_\Lambda^{s,t}(L, \mathbb{F}_2) \to \mathrm{Ext}_\Lambda^{s+i,2t}(L, \mathbb{F}_2)$.*

PROOF. Let $C_* \to L$ be a free $\Lambda$-module resolution. Then $F_* = \Gamma \otimes_\Lambda C_* \to \Gamma \otimes_\Lambda L$ is a free $\Gamma$-module resolution. Let $\Delta \colon W_* \otimes C_* \to C_* \otimes C_*$ be a $\Sigma_2$-equivariant map of $\Lambda$-module complexes that lifts the coproduct $\psi \colon L \to L \otimes L$. Then the composite

$$W_* \otimes \Gamma \otimes_\Lambda C_* \xrightarrow{1 \otimes \Delta} \Gamma \otimes_\Lambda (C_* \otimes C_*) \xrightarrow{\psi \otimes 1} (\Gamma \otimes_\Lambda C_*) \otimes (\Gamma \otimes_\Lambda C_*)$$

(with some twist isomorphisms suppressed) is a $\Sigma_2$-equivariant map $W_* \otimes F_* \to F_* \otimes F_*$ of $\Gamma$-module complexes that lifts the coproduct $\psi \colon \Gamma \otimes_\Lambda L \to (\Gamma \otimes_\Lambda L) \otimes (\Gamma \otimes_\Lambda L)$. A chase of definitions then shows that $Sq^{s-i}$ applied to the $\Gamma$-module extension of any $\Lambda$-linear cocycle $x\colon C_s \to \Sigma^t\mathbb{F}_2$ equals the $\Gamma$-module extension of $Sq^{s-i}(x)\colon C_{2s-i} \to \Sigma^{2t}\mathbb{F}_2$. □

## 1.4. The Adams $E_2$-term for $tmf/2$, $tmf/\eta$ and $tmf/\nu$

DEFINITION 1.23. Let

$$C2 = S/2 = S \cup_2 e^1$$
$$C\eta = S/\eta = S \cup_\eta e^2$$
$$C\nu = S/\nu = S \cup_\nu e^4$$
$$C\sigma = S/\sigma = S \cup_\sigma e^8$$

be the homotopy cofibers of the real Hopf map (degree two map) $2\colon S \to S$, the complex Hopf map $\eta\colon S^1 \to S$, the quaternionic Hopf map $\nu\colon S^3 \to S$ and the octonionic Hopf map $\sigma\colon S^7 \to S$. Let

$$tmf/2 = tmf \wedge C2$$
$$tmf/\eta = tmf \wedge C\eta$$
$$tmf/\nu = tmf \wedge C\nu\,.$$

We need not discuss the octonionic case, since $\sigma$ acts trivially on $tmf$ and $tmf/\sigma = tmf \wedge C\sigma \simeq tmf \vee \Sigma^8 tmf$. The defining homotopy cofiber sequences

$$S \xrightarrow{2} S \xrightarrow{i} C2 \xrightarrow{j} S^1$$
$$S^1 \xrightarrow{\eta} S \xrightarrow{i} C\eta \xrightarrow{j} S^2$$
$$S^3 \xrightarrow{\nu} S \xrightarrow{i} C\nu \xrightarrow{j} S^4$$

of spectra induce homotopy cofiber sequences

$$tmf \xrightarrow{2} tmf \xrightarrow{i} tmf/2 \xrightarrow{j} \Sigma tmf$$
$$\Sigma tmf \xrightarrow{\eta} tmf \xrightarrow{i} tmf/\eta \xrightarrow{j} \Sigma^2 tmf$$
$$\Sigma^3 tmf \xrightarrow{\nu} tmf \xrightarrow{i} tmf/\nu \xrightarrow{j} \Sigma^4 tmf$$

of $tmf$-modules. Let $M_1 = H^*(C2) = \mathbb{F}_2\{1, Sq^1\}$, $M_2 = H^*(C\eta) = \mathbb{F}_2\{1, Sq^2\}$ and $M_4 = H^*(C\nu) = \mathbb{F}_2\{1, Sq^4\}$.

REMARK 1.24. We follow [**52**, Def. 2.5], writing $M_i$ for a minimal $A(2)$-module with nontrivial action by $Sq^i$ from degree 0 to degree $i$.

LEMMA 1.25. *There are $A$-module isomorphisms*

$$H^*(tmf/2) \cong A/\!/A(2) \otimes M_1 \cong A \otimes_{A(2)} M_1$$
$$H^*(tmf/\eta) \cong A/\!/A(2) \otimes M_2 \cong A \otimes_{A(2)} M_2$$
$$H^*(tmf/\nu) \cong A/\!/A(2) \otimes M_4 \cong A \otimes_{A(2)} M_4\,.$$

PROOF. These are Künneth and untwisting isomorphisms, cf. Lemma 2.2. □

The Adams spectral sequences

$$E_2^{s,t}(tmf/2) = \mathrm{Ext}_A^{s,t}(H^*(tmf/2), \mathbb{F}_2) \Longrightarrow_s \pi_{t-s}(tmf/2)$$
$$E_2^{s,t}(tmf/\eta) = \mathrm{Ext}_A^{s,t}(H^*(tmf/\eta), \mathbb{F}_2) \Longrightarrow_s \pi_{t-s}(tmf/\eta)^\wedge_2$$

$$E_2^{s,t}(tmf/\nu) = \mathrm{Ext}_A^{s,t}(H^*(tmf/\nu), \mathbb{F}_2) \Longrightarrow_s \pi_{t-s}(tmf/\nu)_2^\wedge$$

for $tmf/2$, $tmf/\eta$ and $tmf/\nu$, respectively, are all strongly convergent module spectral sequences over the Adams spectral sequence for *tmf*. By change-of-algebras, the $E_2$-terms can be rewritten as

$$E_2(tmf/2) \cong \mathrm{Ext}_{A(2)}(M_1, \mathbb{F}_2)$$
$$E_2(tmf/\eta) \cong \mathrm{Ext}_{A(2)}(M_2, \mathbb{F}_2)$$
$$E_2(tmf/\nu) \cong \mathrm{Ext}_{A(2)}(M_4, \mathbb{F}_2)\,.$$

In each case the action by $E_2(tmf) \cong \mathrm{Ext}_{A(2)}(\mathbb{F}_2, \mathbb{F}_2)$ is induced by the tensor product of $A(2)$-modules, and agrees with the Yoneda product.

Using `ext` we can calculate a minimal free $A(2)$-module resolution $(D_*, \partial)$ of $M_1$, in a finite range.

$$\dots \xrightarrow{\partial} D_2 \xrightarrow{\partial} D_1 \xrightarrow{\partial} D_0 \xrightarrow{\epsilon} M_1 \to 0$$

The program will choose an $A(2)$-module basis $\{s_g^*\}_g$ for each $D_s$, and we let $s_g \in \mathrm{Hom}_{A(2)}(D_s, \mathbb{F}_2)$ be the dual cocycles, giving an $\mathbb{F}_2$-basis for $\mathrm{Ext}_{A(2)}^{s,*}(M_1, \mathbb{F}_2)$. The resulting charts for $0 \le t-s \le 96$ are shown in Figures 1.24 to 1.27. Similar calculations of minimal $A(2)$-module resolutions of $M_2$ and $M_4$ give $\mathbb{F}_2$-bases for $\mathrm{Ext}_{A(2)}^{s,t}(M_2, \mathbb{F}_2)$ and $\mathrm{Ext}_{A(2)}^{s,t}(M_4, \mathbb{F}_2)$, respectively, as shown for $0 \le t-s \le 96$ in Figures 1.28 to 1.31 and Figures 1.32 to 1.35.

REMARK 1.26. To make these calculations with `ext`, go to the directory `A2` and create a text file `tmfC2.def` with the following content.

```
2
0 1
0 1 1 1
```

This defines an $A(2)$-module with two generators, in degrees 0 and 1. There is a nontrivial action on the zeroth generator by $Sq^1$, with value a sum with one term, namely the first generator. Use `newmodule tmfC2 tmfC2.def` to create `tmfC2`. In this subdirectory run `dims 0 240` to calculate the minimal resolution for $0 \le s \le 40$ and $0 \le t \le 240$. Thereafter call on `report` to extract the results, and use

```
chart 0 16 0 24 Shape himults Ext-A2-M1-0-24.tex Ext-A2-M1
pdflatex himults Ext-A2-M1-0-24.tex
```

to obtain the Adams chart in Figure 1.24. The module definition files `tmfCeta.def` and `tmfCnu.def` for $M_2$ and $M_4$ should contain the lines

```
2
0 2
0 2 1 1
```

and

```
2
0 4
0 4 1 1
```

respectively.

DEFINITION 1.27. The nonzero homomorphisms $M_1 \to \mathbb{F}_2$ and $\Sigma\mathbb{F}_2 \to M_1$ induce $\mathrm{Ext}_{A(2)}(\mathbb{F}_2, \mathbb{F}_2)$-module homomorphisms

$$i\colon \mathrm{Ext}_{A(2)}(\mathbb{F}_2, \mathbb{F}_2) \longrightarrow \mathrm{Ext}_{A(2)}(M_1, \mathbb{F}_2)$$

$$j\colon \mathrm{Ext}_{A(2)}(M_1,\mathbb{F}_2) \longrightarrow \mathrm{Ext}_{A(2)}(\Sigma\mathbb{F}_2,\mathbb{F}_2) = \mathrm{Ext}_{A(2)}^{*,*-1}(\mathbb{F}_2,\mathbb{F}_2)\,.$$

Let

$$i(1), \widetilde{h_1}, \widetilde{h_2^2}, \widetilde{c_0}, \widetilde{h_0^2 e_0}, \widetilde{\gamma}, \widetilde{\beta^2}, \widetilde{d_0 e_0}, \widetilde{\delta'}, \widetilde{\beta g}, \widetilde{\alpha^2 e_0}$$

in $\mathrm{Ext}_{A(2)}(M_1,\mathbb{F}_2)$ be the classes represented by the cocycles

$$0_0, 1_1, 2_3, 3_2, 6_3, 5_8, 6_{10}, 8_7, 7_{10}, 7_{12}, 10_{12}\,,$$

respectively, as listed in Table 1.5 and illustrated in Figure 4.1. In each case the class is the only nonzero class in its $(t-s,s)$-bidegree. Each class denoted $\widetilde{x}$ maps to $x$ under $j$.

PROPOSITION 1.28. *In topological degrees $t-s \le 200$, $\mathrm{Ext}_{A(2)}(M_1,\mathbb{F}_2)$ is generated as an $\mathrm{Ext}_{A(2)}(\mathbb{F}_2,\mathbb{F}_2)$-module by the classes listed in Table 1.5.*

Table 1.5: $\mathrm{Ext}_{A(2)}(\mathbb{F}_2,\mathbb{F}_2)$-module generators for $\mathrm{Ext}_{A(2)}(M_1,\mathbb{F}_2)$ (for $t-s \le 200$)

| $t-s$ | $s$ | $g$ | $x$ |
|---|---|---|---|
| 0 | 0 | 0 | $i(1)$ |
| 2 | 1 | 1 | $\widetilde{h_1}$ |
| 7 | 2 | 3 | $\widetilde{h_2^2}$ |
| 9 | 3 | 2 | $\widetilde{c_0}$ |
| 18 | 6 | 3 | $\widetilde{h_0^2 e_0}$ |
| 26 | 5 | 8 | $\widetilde{\gamma}$ |
| 31 | 6 | 10 | $\widetilde{\beta^2}$ |
| 32 | 8 | 7 | $\widetilde{d_0 e_0}$ |
| 33 | 7 | 10 | $\widetilde{\delta'}$ |
| 36 | 7 | 12 | $\widetilde{\beta g}$ |
| 42 | 10 | 12 | $\widetilde{\alpha^2 e_0}$ |

SKETCH PROOF. We will make no formal use of this proposition, other than to introduce notation, and will therefore allow ourselves to assert that the claim can be verified by machine computation, as explained in Remarks 1.26 and 1.29. We will see later, in Corollary 4.3, that these classes generate all of $\mathrm{Ext}_{A(2)}(M_1,\mathbb{F}_2)$ as a module over $\mathrm{Ext}_{A(2)}(\mathbb{F}_2,\mathbb{F}_2)$. □

REMARK 1.29. To verify this calculation using `ext`, go to `A2` and use `cocycle tmfC2 0 0`, ..., `cocycle tmfC2 10 12` to create the cocycles `0_0`, ..., `10_12` in `A2/tmfC2`. Go to `A2/tmfC2` and run `dolifts 0 40 maps` to lift these cocycles to chain maps. Use `collect maps all` to obtain the text file `all`, with one row

```
s g (s' g' F2) s''_g''
```

for each summand $s_g \in \mathrm{Ext}_{A(2)}(M_1,\mathbb{F}_2)$ in the product of $s'_{g'} \in \mathrm{Ext}_{A(2)}(\mathbb{F}_2,\mathbb{F}_2)$ and $s''_{g''} \in \mathrm{Ext}_{A(2)}(M_1,\mathbb{F}_2)$.

DEFINITION 1.30. The nonzero homomorphisms $M_2 \to \mathbb{F}_2$ and $\Sigma^2\mathbb{F}_2 \to M_2$ induce $\mathrm{Ext}_{A(2)}(\mathbb{F}_2, \mathbb{F}_2)$-module homomorphisms

$$i\colon \mathrm{Ext}_{A(2)}(\mathbb{F}_2, \mathbb{F}_2) \longrightarrow \mathrm{Ext}_{A(2)}(M_2, \mathbb{F}_2)$$
$$j\colon \mathrm{Ext}_{A(2)}(M_2, \mathbb{F}_2) \longrightarrow \mathrm{Ext}_{A(2)}(\Sigma^2\mathbb{F}_2, \mathbb{F}_2) = \mathrm{Ext}^{*,*-2}_{A(2)}(\mathbb{F}_2, \mathbb{F}_2)\,.$$

Let

$$i(1), \widehat{h_0}, \widehat{h_2}, \widehat{h_1c_0}, \widehat{\alpha}, \widehat{\beta}, \widehat{d_0g}$$

in $\mathrm{Ext}_{A(2)}(M_2, \mathbb{F}_2)$ be the classes represented by the cocycles

$$0_0, 1_1, 1_3, 4_3, 3_5, 3_7, 8_{25}\,,$$

respectively, as listed in Table 1.6 and illustrated in Figure 4.2. In each case the class is the only nonzero class in its $(t-s,s)$-bidegree. Each class denoted $\widehat{x}$ maps to $x$ under $j$.

PROPOSITION 1.31. *In topological degrees* $t-s \le 200$, $\mathrm{Ext}_{A(2)}(M_2, \mathbb{F}_2)$ *is generated as an* $\mathrm{Ext}_{A(2)}(\mathbb{F}_2, \mathbb{F}_2)$*-module by the classes listed in Table* 1.6.

Table 1.6: $\mathrm{Ext}_{A(2)}(\mathbb{F}_2, \mathbb{F}_2)$-module generators for $\mathrm{Ext}_{A(2)}(M_2, \mathbb{F}_2)$ (for $t-s \le 200$)

| $t-s$ | $s$ | $g$ | $x$ |
|---|---|---|---|
| 0 | 0 | 0 | $i(1)$ |
| 2 | 1 | 1 | $\widehat{h_0}$ |
| 5 | 1 | 3 | $\widehat{h_2}$ |
| 11 | 4 | 3 | $\widehat{h_1c_0}$ |
| 14 | 3 | 5 | $\widehat{\alpha}$ |
| 17 | 3 | 7 | $\widehat{\beta}$ |
| 36 | 8 | 25 | $\widehat{d_0g}$ |

REMARK 1.32. We will see later, in Corollary 4.13, that these classes generate all of $\mathrm{Ext}_{A(2)}(M_2, \mathbb{F}_2)$ as a module over $\mathrm{Ext}_{A(2)}(\mathbb{F}_2, \mathbb{F}_2)$.

DEFINITION 1.33. The nonzero homomorphisms $M_4 \to \mathbb{F}_2$ and $\Sigma^4\mathbb{F}_2 \to M_4$ induce $\mathrm{Ext}_{A(2)}(\mathbb{F}_2, \mathbb{F}_2)$-module homomorphisms

$$i\colon \mathrm{Ext}_{A(2)}(\mathbb{F}_2, \mathbb{F}_2) \longrightarrow \mathrm{Ext}_{A(2)}(M_4, \mathbb{F}_2)$$
$$j\colon \mathrm{Ext}_{A(2)}(M_4, \mathbb{F}_2) \longrightarrow \mathrm{Ext}_{A(2)}(\Sigma^4\mathbb{F}_2, \mathbb{F}_2) = \mathrm{Ext}^{*,*-4}_{A(2)}(\mathbb{F}_2, \mathbb{F}_2)\,.$$

Let

$$i(1), \overline{h_0^3}, \overline{h_1}, \overline{h_0h_2}, \overline{h_2^2}, \overline{c_0}, \overline{h_0^2\alpha}, \overline{g}, \overline{h_0\alpha^2}, \overline{\gamma}, \overline{\alpha\beta}, \overline{\beta^2}, \overline{\delta}, \overline{\alpha^3}$$

in $\mathrm{Ext}_{A(2)}(M_4, \mathbb{F}_2)$ be the classes represented by the cocycles

$$0_0, 3_1, 1_2, 2_3, 2_4, 3_4, 5_7, 4_9, 7_{13}, 5_{13}, 6_{16}, 6_{17}, 7_{19}, 9_{24}\,,$$

respectively, as listed in Table 1.7 and illustrated in Figure 4.3. In most cases the class is the only nonzero class in its $(t-s,s)$-bidegree. The exceptions are $3_4 = \overline{c_0}$, which we prefer over $3_5$, and $7_{19} = \overline{\delta}$, which is the lift of $7_{11} = \delta$. Each class denoted $\overline{x}$ maps to $x$ under $j$.

PROPOSITION 1.34. *In topological degrees $t - s \leq 200$, $\mathrm{Ext}_{A(2)}(M_4, \mathbb{F}_2)$ is generated as an $\mathrm{Ext}_{A(2)}(\mathbb{F}_2, \mathbb{F}_2)$-module by the classes listed in Table 1.7.*

Table 1.7: $\mathrm{Ext}_{A(2)}(\mathbb{F}_2, \mathbb{F}_2)$-module generators for $\mathrm{Ext}_{A(2)}(M_4, \mathbb{F}_2)$ (for $t - s \leq 200$)

| $t-s$ | $s$ | $g$ | $x$ |
|---|---|---|---|
| 0 | 0 | 0 | $i(1)$ |
| 4 | 3 | 1 | $\overline{h_0^3}$ |
| 5 | 1 | 2 | $\overline{h_1}$ |
| 7 | 2 | 3 | $\overline{h_0 h_2}$ |
| 10 | 2 | 4 | $\overline{h_2^2}$ |
| 12 | 3 | 4 | $\overline{c_0}$ |
| 16 | 5 | 7 | $\overline{h_0^2 \alpha}$ |
| 24 | 4 | 9 | $\overline{g}$ |
| 28 | 7 | 13 | $\overline{h_0 \alpha^2}$ |
| 29 | 5 | 13 | $\overline{\gamma}$ |
| 31 | 6 | 16 | $\overline{\alpha\beta}$ |
| 34 | 6 | 17 | $\overline{\beta^2}$ |
| 36 | 7 | 19 | $\overline{\delta}$ |
| 40 | 9 | 24 | $\overline{\alpha^3}$ |

REMARK 1.35. We will see later, in Corollary 4.16, that these classes generate all of $\mathrm{Ext}_{A(2)}(M_4, \mathbb{F}_2)$ as a module over $\mathrm{Ext}_{A(2)}(\mathbb{F}_2, \mathbb{F}_2)$.

LEMMA 1.36. *The spectra $tmf/2$ and $tmf/\eta$ are not ring spectra (in the stable homotopy category).*

PROOF. If $tmf/2$ were a ring spectrum, then its Adams $E_2$-term would be a bigraded algebra over $\mathrm{Ext}_A(\mathbb{F}_2, \mathbb{F}_2)$, with unit $i(1)$. Since $h_0 \cdot i(1) = 0$, it would follow that $h_0 \cdot x = 0$ for all $x \in \mathrm{Ext}_{A(2)}(M_1, \mathbb{F}_2)$. This is not the case, e.g. for $x = 1_1 = \widetilde{h_1}$, as can be seen in Figure 1.24.

Likewise, if $tmf/\eta$ were a ring spectrum, then its Adams $E_2$-term would be a bigraded algebra. Since $h_1 \cdot i(1) = 0$, it would follow that $h_1 \cdot x = 0$ for all $x \in \mathrm{Ext}_{A(2)}(M_2, \mathbb{F}_2)$. This is not the case, e.g. for $x = 1_1 = \widehat{h_0}$, as can be seen in Figure 1.28. □

To discuss $tmf/\nu$, we let $A_* = \mathbb{F}_2[\xi_i \mid i \geq 1]$ denote the dual Steenrod algebra, reviewed in more detail in Section 3.1, and recall the following construction of sub Hopf algebras of the Steenrod algebra.

PROPOSITION 1.37 (Adams–Margolis). *For each profile function*

$$h\colon \{1,2,3,\dots\} \to \{0,1,2,\dots,\infty\}$$

*let*

$$B(h)_* = A_*/(\xi_i^{2^{h(i)}} \mid i \geq 1)$$

*be a quotient algebra of* $A_*$*, and let* $B(h) \subset A$ *be the dual coalgebra. Suppose that* $h$ *satisfies the condition*

- $h(i) \leq j + h(i+j)$ *or* $h(j) \leq h(i+j)$*, for all* $i, j \geq 1$*.*

*Then* $B(h)_*$ *is a quotient Hopf algebra of* $A_*$ *and* $B(h)$ *is a sub Hopf algebra of* $A$*. Conversely, all quotient Hopf algebras of* $A_*$ *and sub Hopf algebras of* $A$ *arise in this manner.*

PROOF. See [**12**, Prop. 2.3, Thm. 2.4] or [**112**, Thm. 15.6]. □

EXAMPLE 1.38. The Hopf algebra

$$A(n) = \langle Sq^1, Sq^2, \dots, Sq^{2^n} \rangle$$

corresponds to the function given by $h(i) = n+2-i$ for $1 \leq i \leq n+1$ and $h(i) = 0$ for $i \geq n+2$. It is the minimal sub Hopf algebra of $A$ that contains $Sq^{2^n}$.

The sub Hopf algebra $B(2,2,1)$ of $A(2)$, corresponding to the function $h(1) = h(2) = 2$, $h(3) = 1$ and $h(i) = 0$ for $i \geq 4$, is generated by $Sq^1$, $Sq^2$ and $Sq^{(0,2)}$.

LEMMA 1.39. *There is an isomorphism* $A(2)//B(2,2,1) \cong M_4$ *of* $A(2)$*-module coalgebras. Hence there is an isomorphism*

$$\mathrm{Ext}_{A(2)}(M_4, \mathbb{F}_2) \cong \mathrm{Ext}_{B(2,2,1)}(\mathbb{F}_2, \mathbb{F}_2)$$

*of bigraded algebras.*

PROOF. This follows by dualization from the $A(2)_*$-comodule algebra isomorphism

$$E(\xi_1^4) \cong A(2)_* \,\square_{B(2,2,1)_*}\, \mathbb{F}_2\,,$$

where □ denotes the cotensor product. See Section 2.2 for a more detailed review of that construction. □

LEMMA 1.40 (Oka). *The spectrum* $tmf/\nu$ *is not a ring spectrum.*

PROOF. Shichirô Oka proved [**139**, Lem. 1.2] that the primary obstruction to extending the module action $tmf \wedge tmf/\nu \to tmf/\nu$ over the unit map $i \wedge 1\colon tmf \wedge tmf/\nu \to tmf/\nu \wedge tmf/\nu$ is $2\nu$, which is nonzero in $\pi_3(tmf)$.

Alternatively, note that by the previous lemma the argument of Lemma 1.36 does not apply for $tmf/\nu$, since the Adams $E_2$-term for $tmf/\nu$ admits an algebra structure (over $\mathrm{Ext}_A(\mathbb{F}_2,\mathbb{F}_2)$ and over $\mathrm{Ext}_{A(2)}(\mathbb{F}_2,\mathbb{F}_2)$). However, we will see in Section 8.2 that the Adams $d_2$-differential does not satisfy the Leibniz rule. More specifically,

$$d_2(\overline{g} \cdot \overline{g}) = d_2(i(w_2)) = i(\alpha\beta g) = g^2 \cdot \overline{h_0 h_2} \neq 0$$

(where $\overline{g} \cdot \overline{g} = i(w_2)$ can be verified using `ext`) while

$$d_2(\overline{g}) \cdot \overline{g} + \overline{g} \cdot d_2(\overline{g}) = 0 \cdot \overline{g} + \overline{g} \cdot 0 = 0\,.$$

□

In view of Lemmas 1.36 and 1.40 it may be surprising that the $E_\infty$ ring spectrum $tmf_1(3) \simeq BP\langle 2\rangle$ can take the form $tmf \wedge \Phi$ for a finite cell spectrum $\Phi$. Here $H^*(\Phi)$ realizes $A(2)//E(2)$, after restricting the natural $A$-action to $A(2)$, so that $H^*(tmf \wedge \Phi) = A//A(2) \otimes A(2)//E(2) \cong A \otimes_{A(2)} A(2)//E(2) \cong A//E(2) = H^*(BP\langle 2\rangle)$. The $A(2)$-module $A(2)//E(2)$ is also known as "the double of $A(1)$", where $A(1) = \langle Sq^1, Sq^2\rangle \subset A$ is the subalgebra generated by $Sq^1$ and $Sq^2$. The latter has rank 8 and is concentrated in degrees $0 \le * \le 6$. After doubling all degrees, the resulting cyclic $A(2)$-module is concentrated in even degrees $0 \le * \le 12$, subject to the relations that the odd-degree generators $Q_0 = Sq^1$, $Q_1 = Sq^{(0,1)}$ and $Q_2 = Sq^{(0,0,1)}$ act trivially. In other words, it is isomorphic to $A(2)//E(2)$.

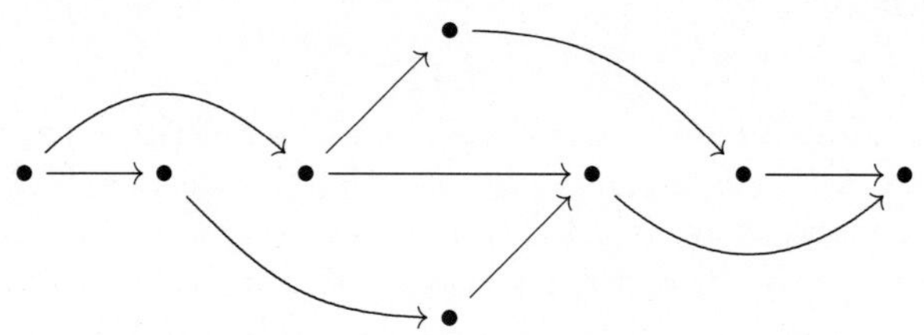

0 2 4 6 8 10 12

The figure above shows the generating actions by $Sq^2$ and $Sq^4$, and is identical to the picture for the generating actions by $Sq^1$ and $Sq^2$ in $A(1)$, except that the degrees have been doubled.

REMARK 1.41. It is elementary to check from the Adem relations that there are precisely four $A$-module structures on $A(2)//E(2)$ that extend the given $A(2)$-module structure, corresponding to the four possible pairs of values of $Sq^8$ acting on the generators 1 and $Sq^4$ in degrees 0 and 4, respectively. In each case, $Sq^8$ acts nontrivially on the generator $Sq^2$ in degree 2. Up to the evident degree shift, the two $A$-module structures where exactly one of $Sq^8(1)$ and $Sq^8(Sq^4)$ is nonzero are both self-dual, whereas the remaining two $A$-module structures are mutually dual. A direct cell-by-cell construction of 8-cell CW spectra realizing each of these four $A$-modules is possible, using `ext` to analyze the available attaching maps, and reveals that in each case some essential ambiguity remains in how the 10- and 12-cells are attached. We instead give the following less computational proof, which has the advantage of producing a self-dual model.

LEMMA 1.42 ([**76**, Lem. 6.1], [**114**, Def. 4.2]). *There exist finite CW spectra $\Phi = \Phi A(1)$ with cohomology $H^*(\Phi) \cong A(2)//E(2)$ realizing the double of $A(1)$. At least one such spectrum is Spanier–Whitehead self-dual, in the sense that there is a 2-adic equivalence $\Phi \simeq F(\Phi, S^{12})$, and of the form $C\gamma$, meaning that there is a homotopy cofiber sequence*

$$\Sigma^5 C\eta \wedge C\nu \xrightarrow{\gamma} C\eta \wedge C\nu \xrightarrow{i} \Phi \xrightarrow{j} \Sigma^6 C\eta \wedge C\nu$$

*(after implicit 2-completion) where $\gamma$ has Adams filtration 1.*

PROOF. Let $\tilde\nu\colon S^5 \to C\eta$ be the unique lift (up to homotopy) over $j\colon C\eta \to S^2$ of $\nu\colon S^5 \to S^2$. Its mapping cone $\Phi Q = C\eta \cup_{\tilde\nu} e^6$ is a finite CW spectrum, with cohomology $H^*(C\eta \cup_{\tilde\nu} e^6) = \mathbb{F}_2\{1, Sq^2, Sq^4Sq^2\} = M_{42}$, the minimal $A(2)$-module containing a generator in degree 0 with nontrivial action by $Sq^4Sq^2$. Its Spanier–Whitehead 6-dual $F(\Phi Q, S^6) = F(C\eta \cup_{\tilde{\nu}} e^6, S^6) \simeq C\nu \cup_{\tilde{\eta}} e^6$ is the mapping cone of the unique lift $\tilde{\eta}\colon S^5 \to C\nu$ over $j\colon C\nu \to S^4$ of $\eta\colon S^5 \to S^4$. Its cohomology $H^*(C\nu \cup_{\tilde{\eta}} e^6) = \mathbb{F}_2\{1, Sq^4, Sq^2Sq^4\} = M_{24}$ is minimal with nontrivial action by $Sq^2Sq^4$.

The evaluation map $e\colon F(\Phi Q, S^6) \wedge \Phi Q \to S^6$ induces

$$e^* : \Sigma^6\mathbb{F}_2 \longrightarrow M_{24} \otimes M_{42}$$

in cohomology, sending the generator to $Sq^2Sq^4 \otimes 1 + Sq^4 \otimes Sq^2 + 1 \otimes Sq^4Sq^2$. Its Spanier–Whitehead 12-dual $c = F(e, S^{12})$ is a coevaluation map $c\colon S^6 \to \Phi Q \wedge F(\Phi Q, S^6)$ inducing

$$c^* : M_{42} \otimes M_{24} \longrightarrow \Sigma^6\mathbb{F}_2$$

in cohomology, sending $Sq^4Sq^2 \otimes 1$, $Sq^2 \otimes Sq^4$ and $1 \otimes Sq^2Sq^4$ to the generator. The composite $e \circ \tau \circ c\colon S^6 \to S^6$, where $\tau$ is the twist equivalence, has degree 3, equal to the Euler characteristic of $\Phi Q$, hence is a 2-local equivalence. In particular, $c$ is 2-locally split injective. Direct calculation with the Cartan formula shows that the direct summand $\ker(c^*) \subset M_{42} \otimes M_{24}$ is isomorphic to $A(2)/\!/E(2)$ as an $A(2)$-module. Furthermore, $Sq^8$ acts nontrivially on the generator $1 \otimes 1$ in degree 0, but trivially on the generator $1 \otimes Sq^4$ in degree 4. Hence we can let $\Phi$ be the mapping cone of the coevaluation map $c$, and obtain a 2-locally split homotopy cofiber sequence as in the upper row of the following diagram.

$$\begin{array}{ccccc} S^6 & \xrightarrow{c} & \Phi Q \wedge F(\Phi Q, S^6) & \xrightarrow{d} & \Phi \\ & & \tau \big\updownarrow \simeq & & \\ F(\Phi, S^{12}) & \xrightarrow{f} & F(\Phi Q, S^6) \wedge \Phi Q & \xrightarrow{e} & S^6 \end{array}$$

Here the homotopy fiber map $f = F(d, S^{12})$ of $e$ is 12-dual to $d$, so the lower row is also a 2-locally split homotopy (co-)fiber sequence. The composite $d \circ \tau \circ f$ exhibits a 2-local equivalence $F(\Phi, S^{12}) \simeq \Phi$.

Let $i$ be the composite map

$$C\eta \wedge C\nu \longrightarrow (C\eta \cup_{\tilde{\nu}} e^6) \wedge (C\nu \cup_{\tilde{\eta}} e^6) \simeq \Phi Q \wedge F(\Phi Q, S^6) \xrightarrow{d} \Phi\,,$$

where the first map is the smash product of the two evident inclusions. The induced homomorphism

$$i^* : A(2)/\!/E(2) \longrightarrow M_2 \otimes M_4$$

is surjective, with kernel isomorphic to $\Sigma^6 M_2 \otimes M_4$ and generated by the commutator $[Sq^4, Sq^2] = Sq^{(0,2)}$. Hence the mapping cone of $i$ has mod 2 cohomology isomorphic to that of $\Sigma^6 C\eta \wedge C\nu$, as a left $A(2)$-module. This characterizes this spectrum up to 2-adic equivalence, and yields the stated homotopy cofiber sequence. The connecting map $\gamma$ must have Adams filtration exactly 1, since

$$0 \to \Sigma^6 M_2 \otimes M_4 \xrightarrow{j^*} A(2)/\!/E(2) \xrightarrow{i^*} M_2 \otimes M_4 \to 0$$

is short exact, but not split as an extension of $A(2)$-modules. □

REMARK 1.43. The finite CW spectra $\Phi Q$ and $F(\Phi Q, S^6)$ are doubles of $Q = C2 \cup_{\tilde{\eta}} e^3$ and $F(Q, S^3) = C\eta \cup_{\tilde{2}} e^3$, usually known as the question mark and inverted question mark complexes, respectively.

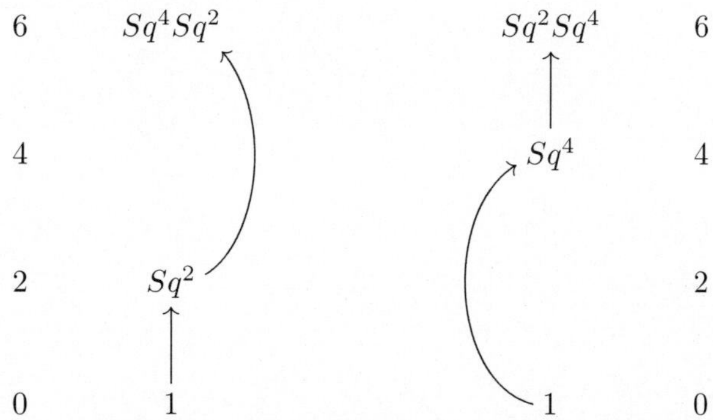


The proof of the lemma above is effectively a double of the construction given in [**73**, Cor. 1.7.7] of a spectrum realizing $A(1)$. A different proof can be given by doubling the construction given on pages 619–620 of [**51**, Thm. 1.4(i)].

PROPOSITION 1.44 ([**76**, Thm. 4.3]). *Let $\Phi$ be any finite CW spectrum realizing $A(2)//E(2)$. There is a 2-adic equivalence of tmf-modules*

$$tmf \wedge \Phi \simeq BP\langle 2\rangle$$

*extending the $E_\infty$ ring spectrum map $\iota'\colon tmf \to BP\langle 2\rangle$. Hence there is also a 2-adic equivalence of tmf-modules*

$$F(\Phi, tmf) \simeq \Sigma^{-12} BP\langle 2\rangle\,.$$

PROOF. Without loss of generality, we can build $\Phi$ from $S$ by attaching even-dimensional cells. Since $\pi_*(BP\langle 2\rangle)$ is trivial in odd degrees, there is no obstruction to extending the unit map $S \to BP\langle 2\rangle$ over $\Phi$. Any such extension $\Phi \to BP\langle 2\rangle$ then induces a $tmf$-module map

$$tmf \wedge \Phi \longrightarrow tmf \wedge BP\langle 2\rangle \longrightarrow BP\langle 2\rangle$$

that extends $\iota'$. The induced $A$-module homomorphism

$$A//E(2) \longrightarrow A//A(2) \otimes A(2)//E(2) \cong A//E(2)$$

is the identity in degree 0, hence is an isomorphism. Thus $tmf \wedge \Phi \to BP\langle 2\rangle$ is a 2-adic equivalence.

For the dual statement, we use that the Spanier–Whitehead dual $D\Phi = F(\Phi, S)$ is equivalent to a finite CW spectrum, with

$$H^*(D\Phi) \cong \mathrm{Hom}(H^*(\Phi), \mathbb{F}_2) \cong \Sigma^{-12} A(2)//E(2)$$

as an $A(2)$-module, so that the previous argument applies to $\Sigma^{12} D\Phi \simeq F(\Phi, S^{12})$. □

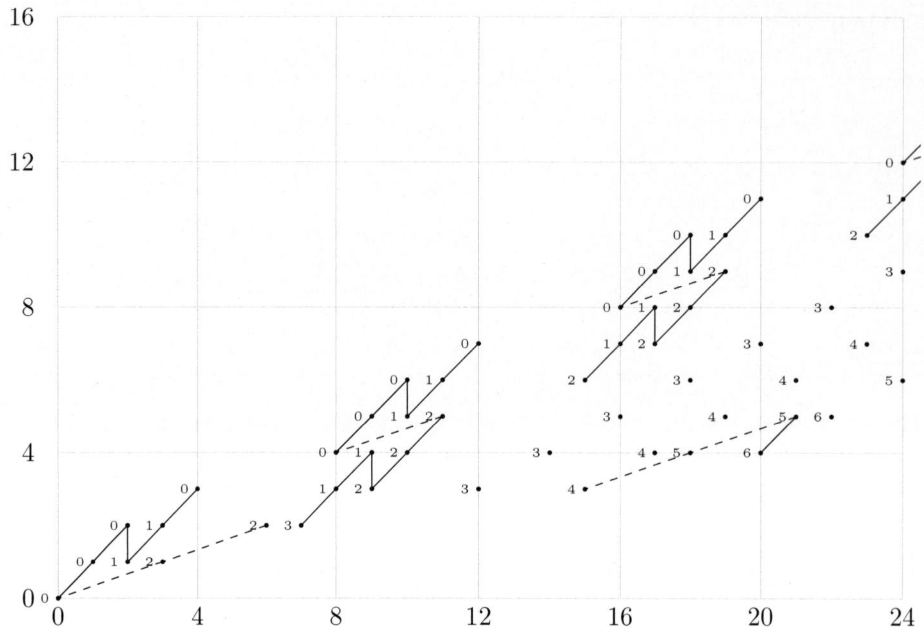

FIGURE 1.24. $\mathrm{Ext}^{s,t}_{A(2)}(M_1, \mathbb{F}_2)$ for $0 \le t-s \le 24$

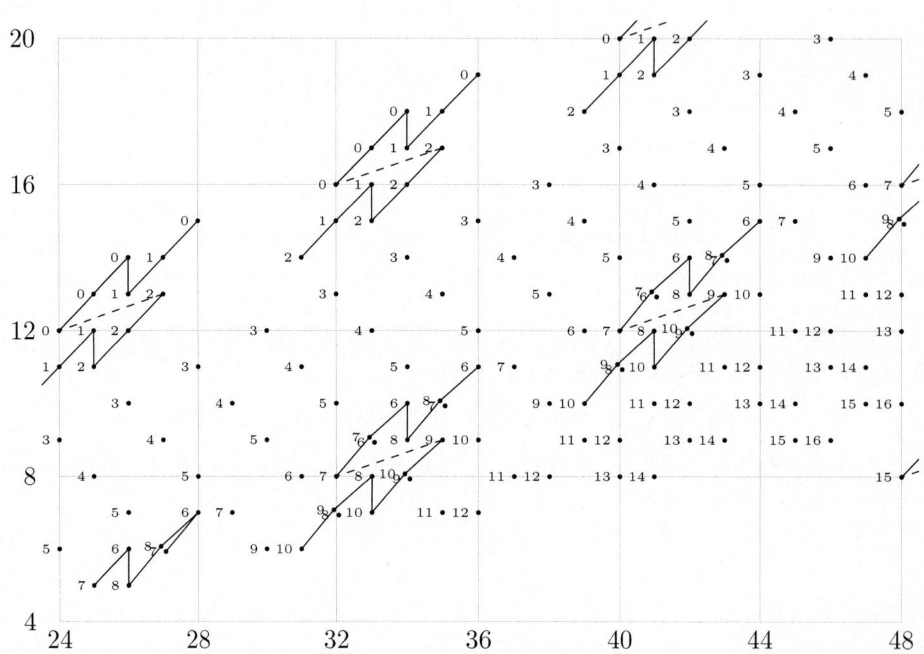

FIGURE 1.25. $\mathrm{Ext}^{s,t}_{A(2)}(M_1, \mathbb{F}_2)$ for $24 \le t-s \le 48$

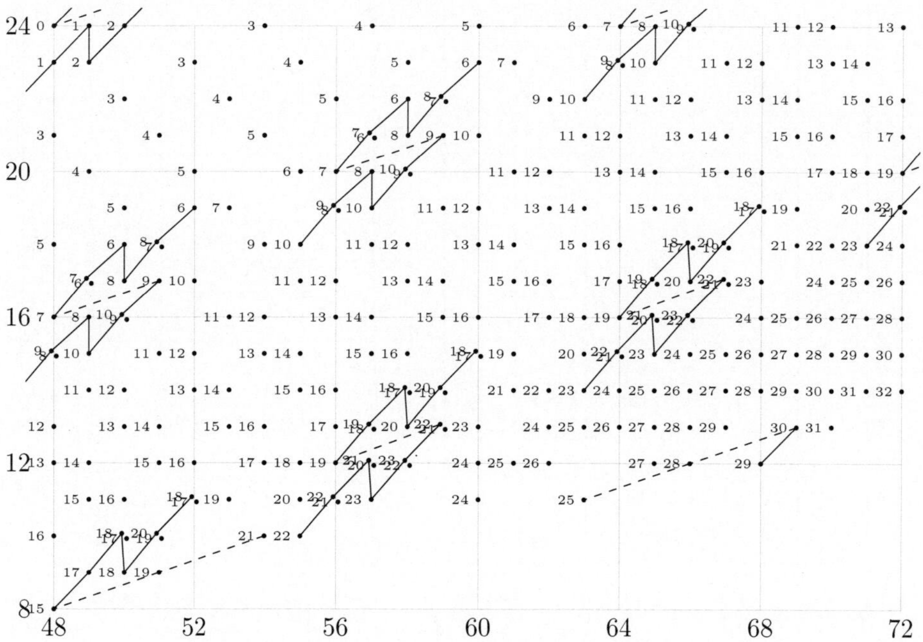

FIGURE 1.26. $\mathrm{Ext}_{A(2)}^{s,t}(M_1, \mathbb{F}_2)$ for $48 \leq t - s \leq 72$

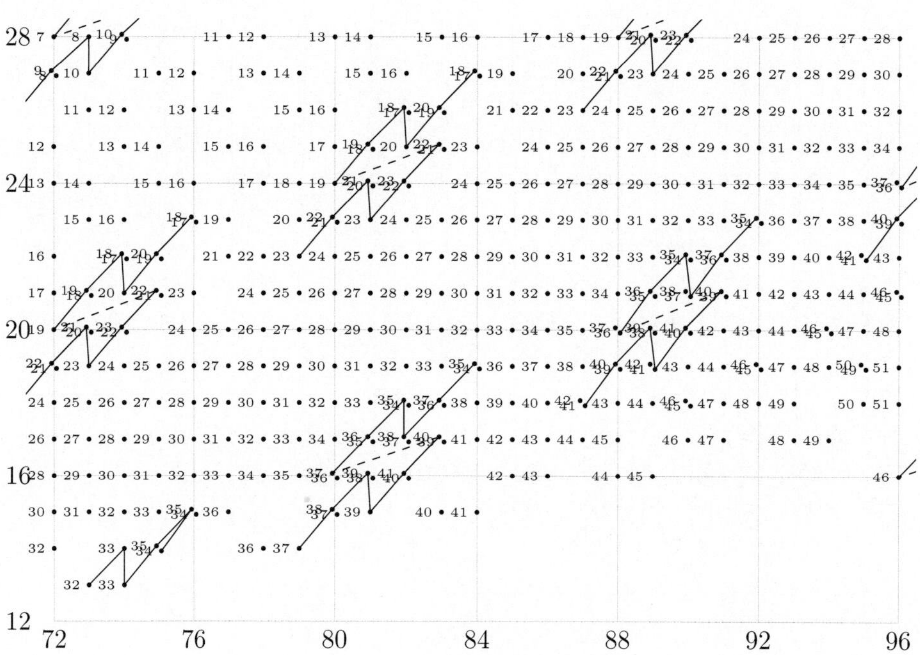

FIGURE 1.27. $\mathrm{Ext}_{A(2)}^{s,t}(M_1, \mathbb{F}_2)$ for $72 \leq t - s \leq 96$

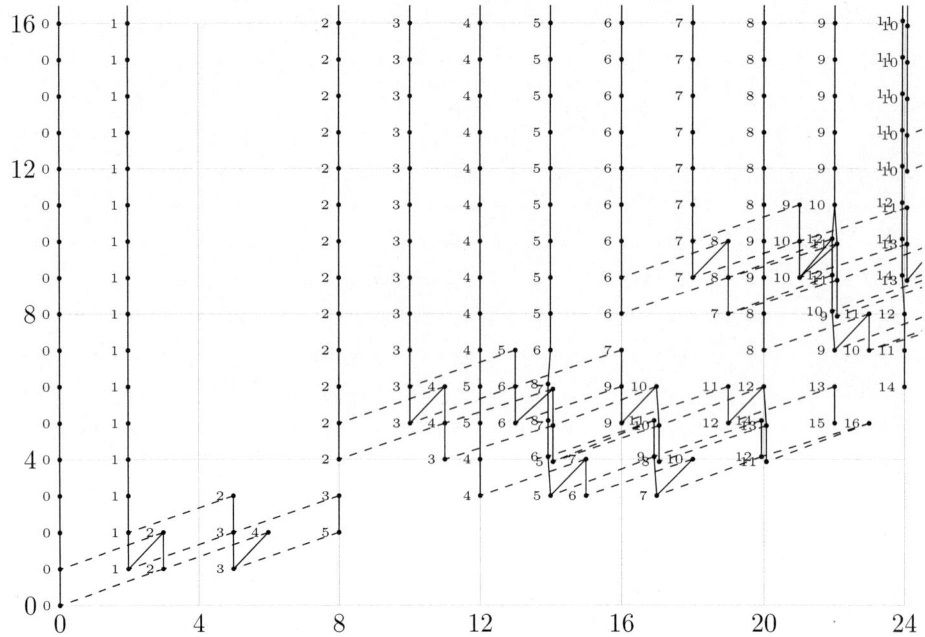

FIGURE 1.28. $\mathrm{Ext}_{A(2)}^{s,t}(M_2, \mathbb{F}_2)$ for $0 \leq t - s \leq 24$

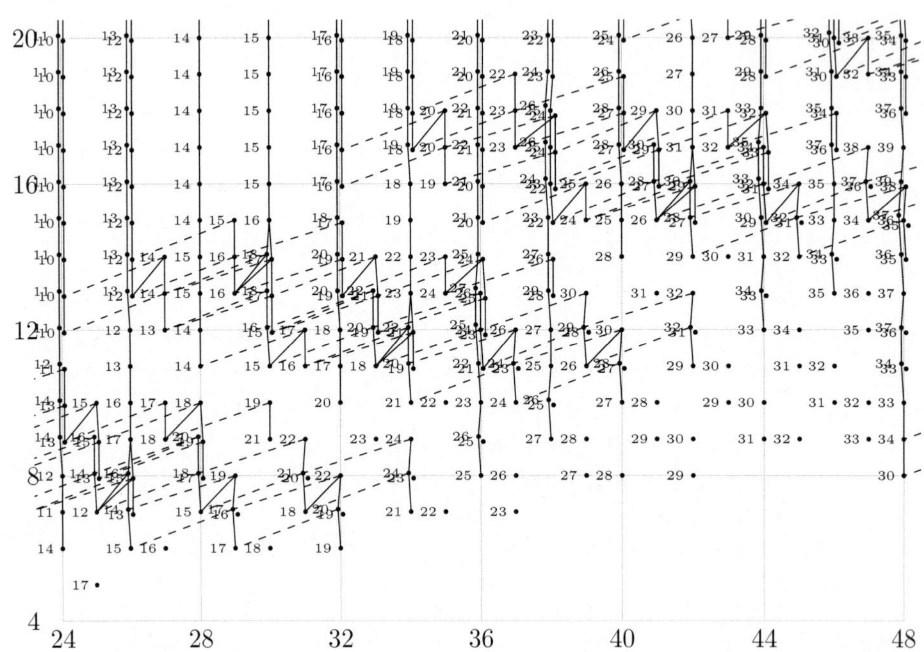

FIGURE 1.29. $\mathrm{Ext}_{A(2)}^{s,t}(M_2, \mathbb{F}_2)$ for $24 \leq t - s \leq 48$

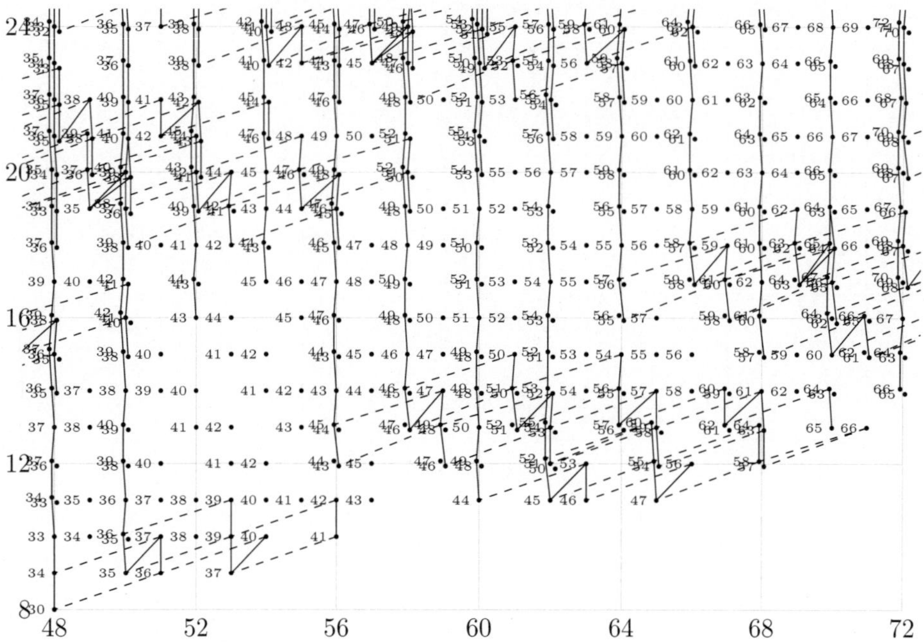

FIGURE 1.30. $\mathrm{Ext}_{A(2)}^{s,t}(M_2, \mathbb{F}_2)$ for $48 \le t-s \le 72$

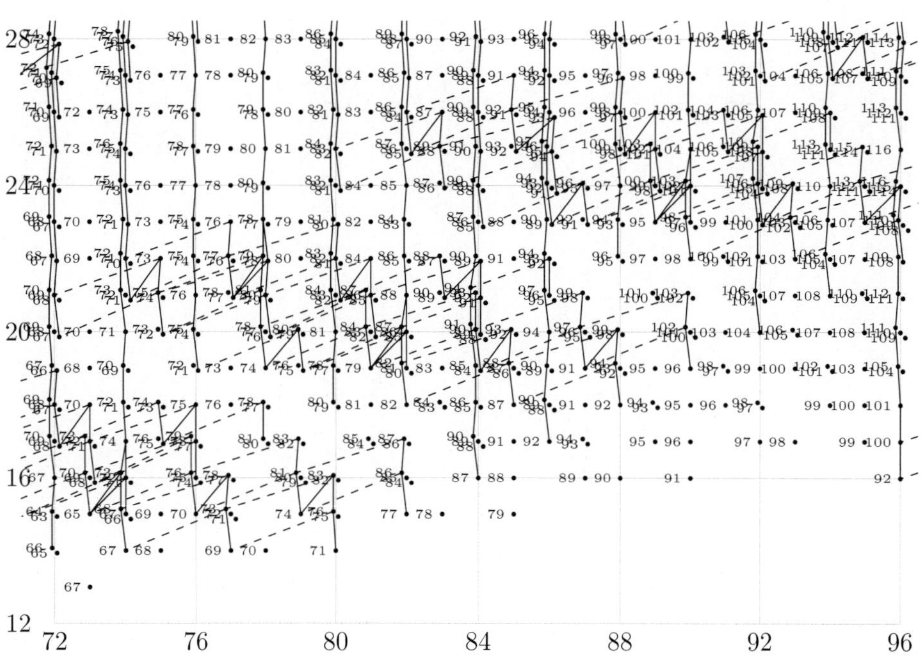

FIGURE 1.31. $\mathrm{Ext}_{A(2)}^{s,t}(M_2, \mathbb{F}_2)$ for $72 \le t-s \le 96$

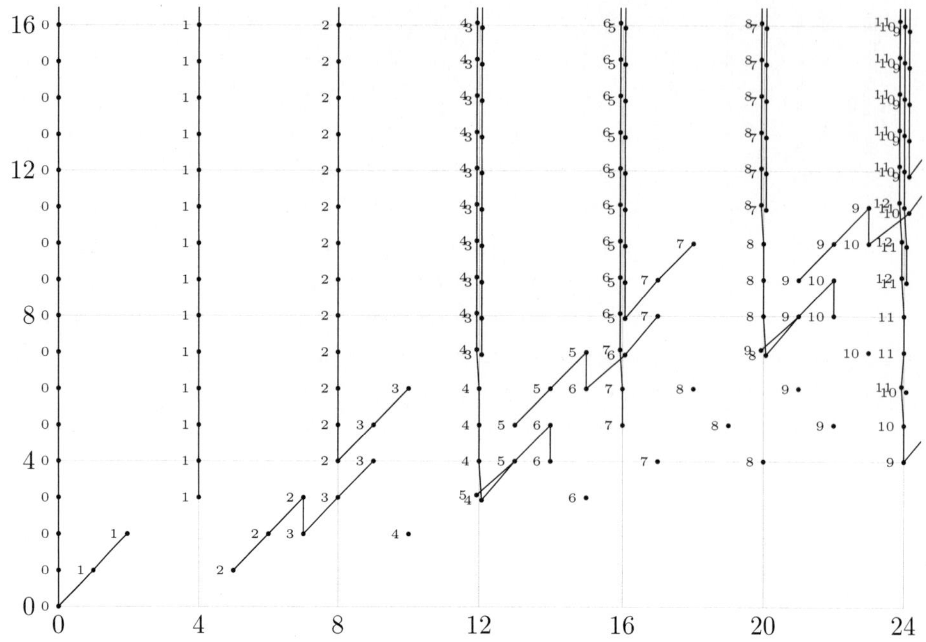

FIGURE 1.32. $\mathrm{Ext}^{s,t}_{A(2)}(M_4, \mathbb{F}_2)$ for $0 \le t-s \le 24$

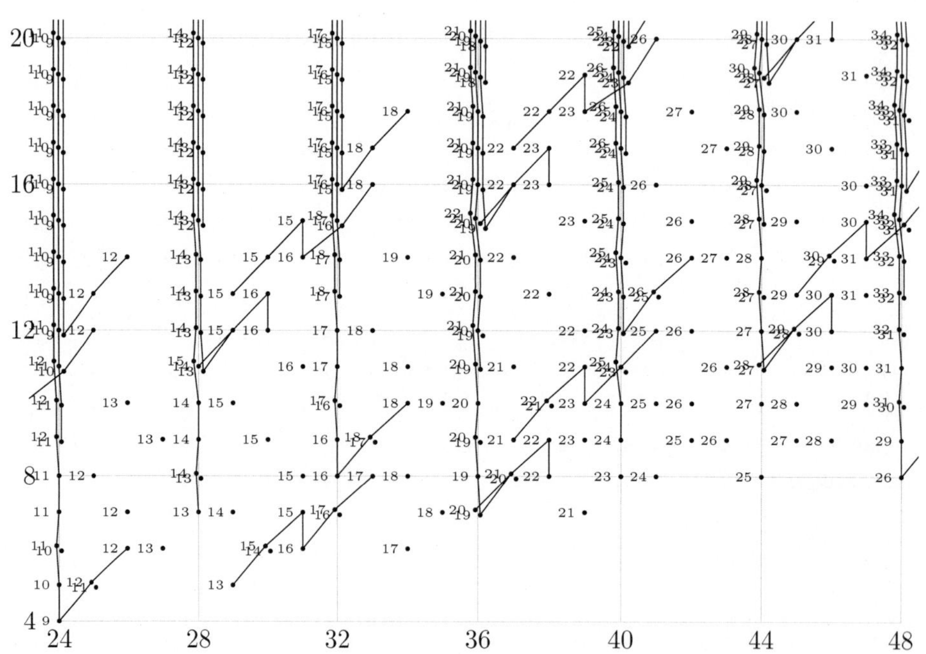

FIGURE 1.33. $\mathrm{Ext}^{s,t}_{A(2)}(M_4, \mathbb{F}_2)$ for $24 \le t-s \le 48$

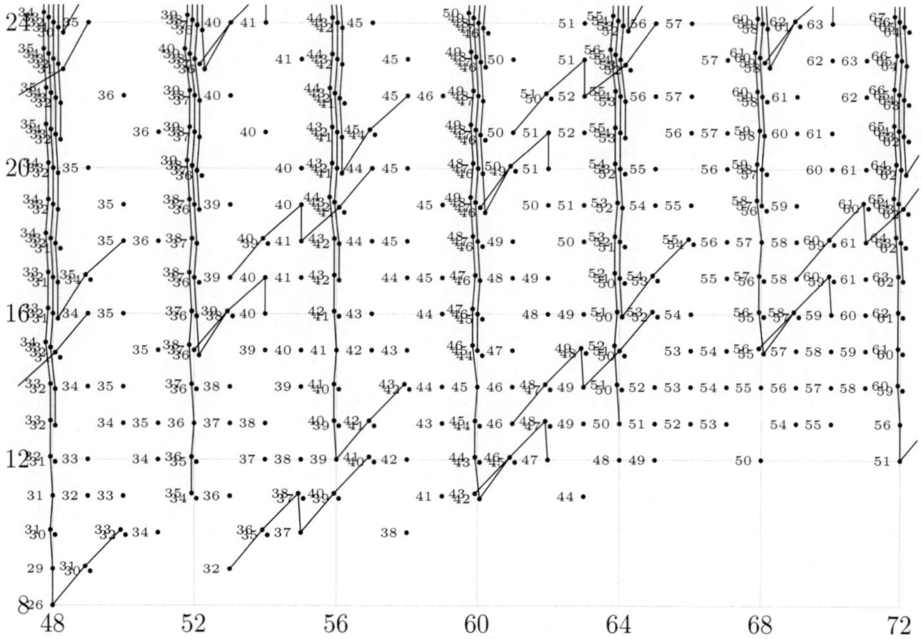

FIGURE 1.34. $\mathrm{Ext}^{s,t}_{A(2)}(M_4, \mathbb{F}_2)$ for $48 \le t - s \le 72$

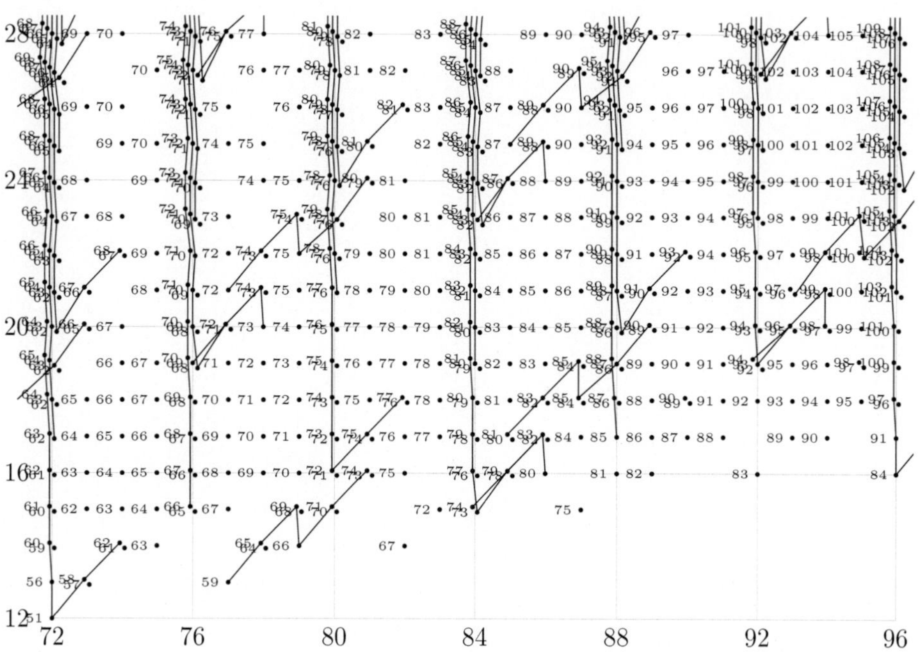

FIGURE 1.35. $\mathrm{Ext}^{s,t}_{A(2)}(M_4, \mathbb{F}_2)$ for $72 \le t - s \le 96$

CHAPTER 2

# The Davis–Mahowald spectral sequence

Davis and Mahowald [**52**] introduced a spectral sequence to calculate Ext for an $A(n)$-module $M$, in terms of Ext for a sequence of $A(n-1)$-modules $N_\sigma \otimes M$ indexed by weights $\sigma \geq 0$. The same spectral sequence was studied by Mahowald and Shick [**106**], who called it the Koszul spectral sequence. We generalize their work to the case of a pair $\Lambda \subset \Gamma$ of Hopf algebras, and clarify the origin of the multiplicative structure in the spectral sequence in the cocommutative case. Propositions 2.3 and 2.10 give additive forms of the spectral sequences calculating Ext in the categories of $\Gamma$-modules and $\Gamma_*$-comodules, respectively, while Theorems 2.25 and 2.24 give multiplicative forms of these spectral sequences.

## 2.1. Ext over a pair of Hopf algebras

Let $k$ be a field, and write $\otimes$ and Hom for $\otimes_k$ and $\mathrm{Hom}_k$, respectively. Let $\Gamma$ be a connected Hopf algebra over $k$, and let $\Lambda$ be a sub Hopf algebra of $\Gamma$. We follow the convention that a Hopf algebra comes equipped with a conjugation $\chi$ as part of the structure. By Milnor–Moore [**128**, Thm. 4.4], $\Gamma$ is free as a right $\Lambda$-module, i.e., it is isomorphic to a direct sum of suspensions of copies of $\Lambda$. Let $L$ be a left $\Lambda$-module, and give $\Gamma \otimes_\Lambda L$ the induced left $\Gamma$-module structure.

LEMMA 2.1. *There is a natural change-of-algebra isomorphism*

$$\mathrm{Ext}_\Gamma(\Gamma \otimes_\Lambda L, k) \cong \mathrm{Ext}_\Lambda(L, k)\,.$$

PROOF. Let $C_* \to L$ be a free $\Lambda$-module resolution of $L$. Then $\Gamma \otimes_\Lambda C_* \to \Gamma \otimes_\Lambda L$ is a free $\Gamma$-module resolution of $\Gamma \otimes_\Lambda L$. Hence the natural isomorphism $\mathrm{Hom}_\Gamma(\Gamma \otimes_\Lambda C_*, k) \cong \mathrm{Hom}_\Lambda(C_*, k)$ of cochain complexes induces the asserted change-of-algebra isomorphism upon passage to cohomology. □

Let $\Gamma/\!/\Lambda = \Gamma \otimes_\Lambda k$. Let $M$ be a left $\Gamma$-module, give $\Gamma/\!/\Lambda \otimes M$ the diagonal $\Gamma$-module structure, and give $\Gamma \otimes_\Lambda M$ the $\Gamma$-module structure induced up from the restricted $\Lambda$-module structure on $M$.

LEMMA 2.2 ([**14**, Cor. 3.5]). *There is a natural untwisting isomorphism of* $\Gamma$-*modules*

$$\zeta\colon \Gamma \otimes_\Lambda M \xrightarrow{\cong} \Gamma/\!/\Lambda \otimes M$$

*induced by the composite*

$$\Gamma \otimes M \xrightarrow{\psi\otimes 1} \Gamma \otimes \Gamma \otimes M \xrightarrow{\pi\otimes\lambda} \Gamma/\!/\Lambda \otimes M\,.$$

*Here* $\psi\colon \Gamma \to \Gamma \otimes \Gamma$ *denotes the coproduct,* $\pi\colon \Gamma \to \Gamma/\!/\Lambda$ *denotes the projection, and* $\lambda\colon \Gamma \otimes M \to M$ *denotes the left module action.*

PROOF. The homomorphism $\zeta$ is well defined because $\Lambda$ is a sub Hopf algebra of $\Gamma$. An inverse is induced by the composite

$$\Gamma \otimes M \xrightarrow{\psi\otimes 1} \Gamma \otimes \Gamma \otimes M \xrightarrow{1\otimes\chi\otimes 1} \Gamma \otimes \Gamma \otimes M \xrightarrow{1\otimes\lambda} \Gamma \otimes M \xrightarrow{\pi} \Gamma \otimes_\Lambda M\,,$$

where $\chi\colon \Gamma \to \Gamma$ is the conjugation. □

PROPOSITION 2.3. *Suppose that we have chosen a sequence of* $\Gamma$*-modules* $N_\sigma$*, for* $\sigma \geq 0$*, and an exact chain complex*

$$\dots \xrightarrow{\partial_3} \Gamma/\!/\Lambda \otimes N_2 \xrightarrow{\partial_2} \Gamma/\!/\Lambda \otimes N_1 \xrightarrow{\partial_1} \Gamma/\!/\Lambda \otimes N_0 \xrightarrow{\epsilon} k \to 0$$

*of* $\Gamma$*-modules with diagonal* $\Gamma$*-action. Then there is a strongly convergent trigraded spectral sequence*

$$E_1^{\sigma,s,t} = \operatorname{Ext}_\Lambda^{s-\sigma,t}(N_\sigma \otimes M, k) \Longrightarrow_\sigma \operatorname{Ext}_\Gamma^{s,t}(M, k)\,.$$

*The* $d_r$*-differentials have* $(\sigma, s, t)$*-tridegree* $(r, 1, 0)$ *and there are isomorphisms*

$$E_\infty^{\sigma,s,t} \cong F^\sigma \operatorname{Ext}^{s,t}(M)/F^{\sigma+1} \operatorname{Ext}^{s,t}(M)$$

*for all* $\sigma$*,* $s$ *and* $t$*, where* $\{F^\sigma \operatorname{Ext}^{s,t}(M)\}_\sigma$ *is a finite and exhaustive filtration of* $\operatorname{Ext}^{s,t}(M) = \operatorname{Ext}_\Gamma^{s,t}(M, k)$.

PROOF. For each $\sigma \geq 0$ we have a short exact sequence of $\Gamma$-modules

$$0 \to \operatorname{im}(\partial_{\sigma+1}) \otimes M \longrightarrow \Gamma/\!/\Lambda \otimes N_\sigma \otimes M \longrightarrow \operatorname{im}(\partial_\sigma) \otimes M \to 0\,,$$

where we interpret $\operatorname{im}(\partial_0)$ as $\operatorname{im}(\epsilon) = k$. These induce long exact sequences

$$\begin{aligned} \dots \xrightarrow{\delta} \operatorname{Ext}_\Gamma^{s,t}(\operatorname{im}(\partial_\sigma) \otimes M, k) &\longrightarrow \operatorname{Ext}_\Gamma^{s,t}(\Gamma/\!/\Lambda \otimes N_\sigma \otimes M, k) \\ &\longrightarrow \operatorname{Ext}_\Gamma^{s,t}(\operatorname{im}(\partial_{\sigma+1}) \otimes M, k) \xrightarrow{\delta} \operatorname{Ext}_\Gamma^{s+1,t}(\operatorname{im}(\partial_\sigma) \otimes M, k) \longrightarrow \dots \end{aligned}$$

for each $\sigma \geq 0$. Rewriting $\operatorname{Ext}_\Gamma^{s,t}(\Gamma/\!/\Lambda \otimes N_\sigma \otimes M, k)$ as $\operatorname{Ext}_\Lambda^{s,t}(N_\sigma \otimes M, k)$ by means of the isomorphisms of Lemmas 2.1 and 2.2, we can combine these into the following unrolled exact couple:

$$\begin{array}{ccccc} \dots \xrightarrow{\ \delta\ } & \operatorname{Ext}_\Gamma^{s-1,t}(\operatorname{im}(\partial_1) \otimes M, k) & \xrightarrow{\quad\delta\quad} & \operatorname{Ext}_\Gamma^{s,t}(M, k) \\ & \downarrow & \nwarrow & \downarrow{\scriptstyle \epsilon^*} \\ & \operatorname{Ext}_\Lambda^{s-1,t}(N_1 \otimes M, k) & & \operatorname{Ext}_\Lambda^{s,t}(N_0 \otimes M, k) \end{array} \tag{2.1}$$

Here

$$\begin{aligned} A^{\sigma,s,t} &= \operatorname{Ext}_\Gamma^{s-\sigma,t}(\operatorname{im}(\partial_\sigma) \otimes M, k) \\ E^{\sigma,s,t} &= \operatorname{Ext}_\Lambda^{s-\sigma,t}(N_\sigma \otimes M, k)\,. \end{aligned}$$

The homomorphisms $i = \delta\colon A^{\sigma+1,s,t} \to A^{\sigma,s,t}$ and $j\colon A^{\sigma,s,t} \to E^{\sigma,s,t}$ preserve the $(s,t)$-bigrading, whereas $k\colon E^{\sigma,s,t} \to A^{\sigma+1,s+1,t}$ has $(s,t)$-bidegree $(1,0)$ and $(t-s,s)$-bidegree $(-1,1)$. The resulting trigraded spectral sequence has $E_1$-term

$$E_1^{\sigma,s,t} = \operatorname{Ext}_\Lambda^{s-\sigma,t}(N_\sigma \otimes M, k)$$

and $d_r$-differentials

$$d_r\colon E_r^{\sigma,s,t} \longrightarrow E_r^{\sigma+r,s+1,t}$$

of $(s,t)$-bidegree $(1,0)$ and $(t-s,s)$-bidegree $(-1,1)$, for each $r \geq 1$. Neglecting the internal degree $t$, the spectral sequence can be considered as a first quadrant cohomological spectral sequence in the $(\sigma, s-\sigma)$-plane. In this grading the $d_r$-differential has the traditional bidegree $(r, 1-r)$.

It follows that the spectral sequence converges strongly to $A^{0,s,t} = \operatorname{Ext}_\Gamma^{s,t}(M,k)$, which is filtered by the images $F^\sigma \operatorname{Ext}^{s,t}(M) = \operatorname{im}(\delta^\sigma)$ of the iterated coboundary homomorphisms

$$\delta^\sigma \colon \operatorname{Ext}_\Gamma^{s-\sigma,t}(\operatorname{im}(\partial_\sigma) \otimes M, k) \longrightarrow \operatorname{Ext}_\Gamma^{s,t}(M,k) .$$

This is a finite filtration in each $(s,t)$-bidegree, since $F^\sigma \operatorname{Ext}^{s,t}(M) = 0$ for all $\sigma > s$. □

## 2.2. A dual formulation

To make use of the multiplicative structure present in our main examples, it will be convenient to pass from the categories of $\Gamma$- and $\Lambda$-modules to the dual categories of $\Gamma_*$- and $\Lambda_*$-comodules, respectively.

DEFINITION 2.4. Let $\Gamma_*$ be a connected coalgebra over $k$. Given a right $\Gamma_*$-comodule $M_*$ and a left $\Gamma_*$-comodule $N_*$, the cotensor product $M_* \square_{\Gamma_*} N_*$ is the graded $k$-vector space defined [**128**, Def. 2.2] as the equalizer of the two homomorphisms

$$M_* \otimes N_* \overset{\nu\otimes 1}{\underset{1\otimes\nu}{\rightrightarrows}} M_* \otimes \Gamma_* \otimes N_*$$

induced by the right $\Gamma_*$-coaction $\nu\colon M_* \to M_*\otimes\Gamma_*$ and the left $\Gamma_*$-coaction $\nu\colon N_* \to \Gamma_* \otimes N_*$, respectively. Given a second left $\Gamma_*$-comodule $L_*$, the graded $k$-vector space of $\Gamma_*$-comodule homomorphisms $\operatorname{Hom}_{\Gamma_*}(L_*, N_*)$ is the equalizer of the two homomorphisms

$$\operatorname{Hom}(L_*, N_*) \overset{\nu^*}{\underset{\nu_*}{\rightrightarrows}} \operatorname{Hom}(L_*, \Gamma_* \otimes N_*)$$

induced by the left $\Gamma_*$-coactions $\nu\colon L_* \to \Gamma_*\otimes L_*$ and $\nu\colon N_* \to \Gamma_*\otimes N_*$, respectively. When $M_* = k = L_*$, the two diagrams above specialize to the diagram

$$N_* \overset{\eta}{\underset{\nu}{\rightrightarrows}} \Gamma_* \otimes N_*$$

with equalizer the graded $k$-vector space of $\Gamma_*$-comodule primitives

$$P_{\Gamma_*}(N_*) = \{x \in N_* \mid \nu(x) = 1 \otimes x\}$$

in $N_*$. Hence there are identifications

$$k \square_{\Gamma_*} N_* \cong P_{\Gamma_*}(N_*) \cong \operatorname{Hom}_{\Gamma_*}(k, N_*) .$$

DEFINITION 2.5. The forgetful functor from left $\Gamma_*$-comodules to graded $k$-vector spaces has a right adjoint, mapping a vector space $V$ to the extended comodule $\Gamma_* \otimes V$, with coaction induced by the coproduct $\psi\colon \Gamma_* \to \Gamma_* \otimes \Gamma_*$. By definition [**57**, §3], a $\Gamma_*$-comodule is said to be injective if it is a direct summand of an extended $\Gamma_*$-comodule. Each left $\Gamma_*$-comodule $N_*$ admits an injective left $\Gamma_*$-comodule resolution

$$0 \to N_* \longrightarrow X_*^0 \overset{\delta}{\longrightarrow} X_*^1 \overset{\delta}{\longrightarrow} \dots ,$$

i.e., an exact complex where each $X_*^s$ is injective. By definition, $\operatorname{Cotor}_{\Gamma_*}^s(M_*, N_*)$ is the cohomology of the induced complex

$$\dots \overset{\delta_*}{\longrightarrow} M_* \square_{\Gamma_*} X_*^{s-1} \overset{\delta_*}{\longrightarrow} M_* \square_{\Gamma_*} X_*^s \overset{\delta_*}{\longrightarrow} M_* \square_{\Gamma_*} X_*^{s+1} \overset{\delta_*}{\longrightarrow} \dots$$

and $\mathrm{Ext}^s_{\Gamma_*}(L_*, N_*)$ is the cohomology of the induced complex

$$\dots \xrightarrow{\delta_*} \mathrm{Hom}_{\Gamma_*}(L_*, X^{s-1}_*) \xrightarrow{\delta_*} \mathrm{Hom}_{\Gamma_*}(L_*, X^s_*) \xrightarrow{\delta_*} \mathrm{Hom}_{\Gamma_*}(L_*, X^{s+1}_*) \xrightarrow{\delta_*} \dots .$$

In particular, there are canonical isomorphisms $\mathrm{Cotor}^0_{\Gamma_*}(L_*, N_*) \cong L_* \,\square_{\Gamma_*} N_*$, $\mathrm{Ext}^0_{\Gamma_*}(L_*, N_*) \cong \mathrm{Hom}_{\Gamma_*}(L_*, N_*)$, and

$$\mathrm{Cotor}^s_{\Gamma_*}(k, N_*) \cong \mathrm{Ext}^s_{\Gamma_*}(k, N_*)$$

for each $s \geq 0$.

The coalgebra $\Gamma_*$ gives rise to a dual algebra $\Gamma = \mathrm{Hom}(\Gamma_*, k)$, with multiplication $\phi$ given by the composite

$$\mathrm{Hom}(\Gamma_*, k) \otimes \mathrm{Hom}(\Gamma_*, k) \xrightarrow{\otimes} \mathrm{Hom}(\Gamma_* \otimes \Gamma_*, k) \xrightarrow{\psi^*} \mathrm{Hom}(\Gamma_*, k) .$$

If $\Gamma$ is bounded below (e.g., connected) and of finite type as a graded $k$-vector space, we can recover $\Gamma_*$ as the dual $\mathrm{Hom}(\Gamma, k)$, with coproduct given by the composite

$$\mathrm{Hom}(\Gamma, k) \xrightarrow{\phi^*} \mathrm{Hom}(\Gamma \otimes \Gamma, k) \xleftarrow{\cong} \mathrm{Hom}(\Gamma, k) \otimes \mathrm{Hom}(\Gamma, k) .$$

Similarly, each left $\Gamma_*$-comodule $N_*$ gives rise to a left $\Gamma$-module $N = \mathrm{Hom}(N_*, k)$, with action $\lambda$ given by the composite

$$\mathrm{Hom}(\Gamma_*, k) \otimes \mathrm{Hom}(N_*, k) \xrightarrow{\otimes} \mathrm{Hom}(\Gamma_* \otimes N_*, k) \xrightarrow{\nu^*} \mathrm{Hom}(N_*, k) .$$

If $\Gamma$ and $N$ are bounded below and of finite type, we can recover $N_*$ as the dual $\mathrm{Hom}(N, k)$, with coaction given by the composite

$$\mathrm{Hom}(N, k) \xrightarrow{\lambda^*} \mathrm{Hom}(\Gamma \otimes N, k) \xleftarrow{\cong} \mathrm{Hom}(\Gamma, k) \otimes \mathrm{Hom}(N, k) .$$

(Alternatively, if $\Gamma$ is finite, i.e., finite-dimensional as a $k$-vector space, it suffices to assume that $N$ is of finite type.)

LEMMA 2.6. *Let $L_*$ and $N_*$ be $\Gamma_*$-comodules, dual to $\Gamma$-modules $L$ and $N$, where $\Gamma$, $L$ and $N$ are bounded below and of finite type. Then there is a natural isomorphism*

$$D\colon \mathrm{Ext}^s_{\Gamma_*}(L_*, N_*) \cong \mathrm{Ext}^s_{\Gamma}(N, L)$$

*for each $s \geq 0$. (Alternatively, the same conclusion holds if $\Gamma$ is finite and $L$ and $N$ are of finite type.)*

PROOF. Each $\Gamma_*$-comodule homomorphism $f_*\colon L_* \to N_*$ gives rise to a $\Gamma$-module homomorphism $f = \mathrm{Hom}(f_*, k)\colon N \to L$, defining a duality homomorphism

$$D\colon \mathrm{Hom}_{\Gamma_*}(L_*, N_*) \longrightarrow \mathrm{Hom}_{\Gamma}(N, L) .$$

Conversely, if $\Gamma$, $L$ and $N$ are bounded below and of finite type (or if $\Gamma$ is finite and $L$ and $N$ are of finite type), then we can recover $f_*$ as $\mathrm{Hom}(f, k)$. Under these hypotheses, $D$ is an isomorphism.

If $\Gamma$ and $N$ are bounded below and of finite type, then there exists an injective $\Gamma_*$-comodule resolution

$$0 \to N_* \longrightarrow X^0_* \xrightarrow{\delta} X^1_* \xrightarrow{\delta} \dots$$

such that each $X_*^s$ is an extended $\Gamma_*$-comodule that is bounded below and of finite type, i.e., of the form $\Gamma_* \otimes V$ with $V$ bounded below and of finite type. (Alternatively, if $\Gamma$ is finite and $N$ is of finite type, then there exists such a resolution with each $X_*^s$ extended and of finite type.) The isomorphism

$$\mathrm{Hom}(\Gamma_* \otimes V, k) \xleftarrow{\cong} \mathrm{Hom}(\Gamma_*, k) \otimes \mathrm{Hom}(V, k)$$

shows that the dual $X_s = \mathrm{Hom}(X_*^s, k)$ is an extended, hence free, $\Gamma$-module. The duality isomorphisms

$$D\colon \mathrm{Hom}_{\Gamma_*}(L_*, X_*^s) \xrightarrow{\cong} \mathrm{Hom}_\Gamma(X_s, L)$$

show that the complexes with cohomology defining $\mathrm{Ext}_{\Gamma_*}^s(L_*, N_*)$ and $\mathrm{Ext}_\Gamma^s(N, L)$ are isomorphic. □

REMARK 2.7. This lemma shows that when considering Ext over one of the finite subalgebras of the Steenrod algebra, with coefficients in modules of finite type, we may pass freely between the module and comodule contexts. On the other hand, when considering Ext over the full Steenrod algebra, the bounded-below condition plays a significant role. There are interesting examples of $A$-modules of finite type that are not the dual of any $A_*$-comodule, such as the localization

$$L = H^*(P_{-\infty}^\infty; \mathbb{F}_2) = \mathbb{F}_2[x, x^{-1}]$$

of $H^*(P^\infty; \mathbb{F}_2) = \mathbb{F}_2[x]$, with the $A$-module action given by $Sq^i(x^j) = \binom{j}{i} x^{i+j}$. Here $Sq^n(x^{-n}) = 1$ for infinitely many values of $n$, so the dual of $\lambda\colon A \otimes L \to L$ does not factor through $A_* \otimes L_* \to \mathrm{Hom}(A \otimes L, \mathbb{F}_2)$. See [**10**, Part II] and [**94**].

We now specialize to the situation where $\Gamma_*$ is a connected Hopf algebra, and let $\Lambda_*$ be a quotient Hopf algebra of $\Gamma_*$. By Milnor–Moore [**128**, Thm. 4.7], $\Gamma_*$ is isomorphic as a right $\Lambda_*$-comodule to an extended $\Lambda_*$-comodule (of the form $V \otimes \Lambda_*$). Let $L_*$ be a left $\Lambda_*$-comodule, and let $\Gamma_* \,\square_{\Lambda_*}\, L_*$ be the coinduced left $\Gamma_*$-comodule. This is the equalizer of the two homomorphisms

$$\Gamma_* \otimes L_* \underset{1\otimes\nu}{\overset{\nu\otimes 1}{\rightrightarrows}} \Gamma_* \otimes \Lambda_* \otimes L_*$$

given by the right $\Lambda_*$-coaction on $\Gamma_*$ and the left $\Lambda_*$-coaction on $L_*$, respectively.

LEMMA 2.8. *There is a natural change-of-coalgebra isomorphism*

$$\mathrm{Ext}_{\Gamma_*}(k, \Gamma_* \,\square_{\Lambda_*}\, L_*) \cong \mathrm{Ext}_{\Lambda_*}(k, L_*)\,.$$

PROOF. Let

$$0 \to L_* \longrightarrow X_*^0 \longrightarrow X_*^1 \longrightarrow \dots$$

be an injective $\Lambda_*$-comodule resolution of $L_*$. Applying $\Gamma_* \,\square_{\Lambda_*} -$ gives an injective $\Gamma_*$-comodule resolution

$$0 \to \Gamma_* \,\square_{\Lambda_*}\, L_* \longrightarrow \Gamma_* \,\square_{\Lambda_*}\, X_*^0 \longrightarrow \Gamma_* \,\square_{\Lambda_*}\, X_*^1 \longrightarrow \dots\,.$$

Hence the natural isomorphism $\mathrm{Hom}_{\Lambda_*}(k, X_*^s) \cong \mathrm{Hom}_{\Gamma_*}(k, \Gamma_* \,\square_{\Lambda_*}\, X_*^s)$ induces the asserted change-of-coalgebra isomorphism upon passage to cohomology. □

We will allow ourselves to write $(\Gamma/\!/\Lambda)_*$ for $\Gamma_* \,\square_{\Lambda_*}\, k$. Let $M_*$ be a left $\Gamma_*$-comodule, give $(\Gamma/\!/\Lambda)_* \otimes M_*$ the diagonal $\Gamma_*$-comodule structure, and give $\Gamma_* \square_{\Lambda_*} M_*$ the $\Gamma_*$-comodule structure coinduced from the corestricted $\Lambda_*$-comodule structure on $M_*$.

LEMMA 2.9. *There is a natural twisting isomorphism of $\Gamma_*$-comodules*

$$\zeta_* \colon (\Gamma//\Lambda)_* \otimes M_* \xrightarrow{\cong} \Gamma_* \,\square_{\Lambda_*} M_*$$

*lifting the composite*

$$(\Gamma//\Lambda)_* \otimes M_* \xrightarrow{\iota\otimes\nu} \Gamma_* \otimes \Gamma_* \otimes M_* \xrightarrow{\phi\otimes 1} \Gamma_* \otimes M_* \,.$$

*Here $\iota \colon (\Gamma//\Lambda)_* \to \Gamma_*$ denotes the inclusion, $\nu \colon M_* \to \Gamma_* \otimes M_*$ denotes the comodule coaction, and $\phi \colon \Gamma_* \otimes \Gamma_* \to \Gamma_*$ denotes the product pairing.*

PROOF. The lift exists because $\Lambda_*$ is a quotient Hopf algebra of $\Gamma_*$. An inverse is obtained by factoring the composite

$$\Gamma_* \,\square_{\Lambda_*} M_* \xrightarrow{\iota} \Gamma_* \otimes M_* \xrightarrow{1\otimes\nu} \Gamma_* \otimes \Gamma_* \otimes M_* \xrightarrow{1\otimes\chi\otimes 1} \Gamma_* \otimes \Gamma_* \otimes M_* \xrightarrow{\phi\otimes 1} \Gamma_* \otimes M_* \,.$$

□

PROPOSITION 2.10. *Suppose we have chosen a sequence of $\Gamma_*$-comodules $R^\sigma$, for $\sigma \ge 0$, and an exact cochain complex*

$$0 \to k \xrightarrow{\eta} (\Gamma//\Lambda)_* \otimes R^0 \xrightarrow{\delta^0} (\Gamma//\Lambda)_* \otimes R^1 \xrightarrow{\delta^1} (\Gamma//\Lambda)_* \otimes R^2 \xrightarrow{\delta^2} \dots$$

*of $\Gamma_*$-comodules with diagonal $\Gamma_*$-coaction. Then there is a strongly convergent trigraded spectral sequence*

$$E_1^{\sigma,s,t} = \mathrm{Ext}_{\Lambda_*}^{s-\sigma,t}(k, R^\sigma \otimes M_*) \Longrightarrow_\sigma \mathrm{Ext}_{\Gamma_*}^{s,t}(k, M_*) \,.$$

*The $d_r$-differentials have $(\sigma, s, t)$-tridegree $(r, 1, 0)$, and there are isomorphisms*

$$E_\infty^{\sigma,s,t} \cong F^\sigma \,\mathrm{Ext}^{s,t}(M_*)/F^{\sigma+1}\,\mathrm{Ext}^{s,t}(M_*)$$

*for all $\sigma$, $s$ and $t$, where $\{F^\sigma \,\mathrm{Ext}^{s,t}(M_*)\}_\sigma$ is a finite and exhaustive filtration of $\mathrm{Ext}_{\Gamma_*}^{s,t}(k, M_*)$.*

PROOF. For each $\sigma \ge 0$ we have a short exact sequence of $\Gamma_*$-comodules

$$0 \to \ker(\delta^\sigma) \otimes M_* \longrightarrow (\Gamma//\Lambda)_* \otimes R^\sigma \otimes M_* \longrightarrow \ker(\delta^{\sigma+1}) \otimes M_* \to 0 \,.$$

Note that $k = \ker(\delta^0)$. These induce long exact sequences

$$\begin{aligned} \dots \xrightarrow{\delta} \mathrm{Ext}_{\Gamma_*}^{s,t}(k, \ker(\delta^\sigma) \otimes M_*) &\longrightarrow \mathrm{Ext}_{\Gamma_*}^{s,t}(k, (\Gamma//\Lambda)_* \otimes R^\sigma \otimes M_*) \\ &\longrightarrow \mathrm{Ext}_{\Gamma_*}^{s,t}(k, \ker(\delta^{\sigma+1}) \otimes M_*) \xrightarrow{\delta} \mathrm{Ext}_{\Gamma_*}^{s+1,t}(k, \ker(\delta^\sigma) \otimes M_*) \longrightarrow \dots \end{aligned}$$

for each $\sigma \ge 0$. Rewriting $\mathrm{Ext}_{\Gamma_*}^{s,t}(k, (\Gamma//\Lambda)_* \otimes R^\sigma \otimes M_*)$ as $\mathrm{Ext}_{\Lambda_*}^{s,t}(k, R^\sigma \otimes M_*)$, using Lemmas 2.8 and 2.9, we obtain the following exact couple:

$$\begin{array}{ccccc} \dots \xrightarrow{\delta} & \mathrm{Ext}_{\Gamma_*}^{s-1,t}(k, \ker(\delta^1) \otimes M_*) & \xrightarrow{\delta} & \mathrm{Ext}_{\Gamma_*}^{s,t}(k, M_*) \\ & \downarrow & & \downarrow \eta_* \\ & \mathrm{Ext}_{\Lambda_*}^{s-1,t}(k, R^1 \otimes M_*) & & \mathrm{Ext}_{\Lambda_*}^{s,t}(k, R^0 \otimes M_*) \end{array} \tag{2.2}$$

Here

$$\begin{aligned} A^{\sigma,s,t} &= \mathrm{Ext}_{\Gamma_*}^{s-\sigma,t}(k, \ker(\delta^\sigma) \otimes M_*) \\ E^{\sigma,s,t} &= \mathrm{Ext}_{\Lambda_*}^{s-\sigma,t}(k, R^\sigma \otimes M_*) \,. \end{aligned}$$

The remainder of the proof follows that of Proposition 2.3. □

REMARK 2.11. When $\Gamma$ is of finite type over $k$, and $M$ and each $N_\sigma$ is of finite type and bounded below, we can let $\Gamma_* = \mathrm{Hom}(\Gamma, k)$, $\Lambda_* = \mathrm{Hom}(\Lambda, k)$, $M_* = \mathrm{Hom}(M, k)$ and $R^\sigma = \mathrm{Hom}(N_\sigma, k)$. Then $(\Gamma//\Lambda)_* = \mathrm{Hom}(\Gamma//\Lambda, k)$, and the spectral sequences of Propositions 2.3 and 2.10 are canonically isomorphic. If $\Gamma$ is finite, i.e., finite-dimensional over $k$, then we can omit the hypothesis that $M$ and the $N_\sigma$ are bounded below.

## 2.3. A filtered cobar complex

We now give an alternative construction of the Davis–Mahowald spectral sequence, starting from a filtered cochain complex. For this purpose we will use the cobar construction [**1**, §2], [**145**, 1.6] to obtain functorial injective resolutions. The following discussion is dual to that of the two-sided bar construction and the associated bar complex [**56**, §2], [**119**, §9].

DEFINITION 2.12. Let $\Gamma_*$ be a connected coalgebra over $k$, with coproduct $\psi\colon \Gamma_* \to \Gamma_* \otimes \Gamma_*$ and counit $\epsilon\colon \Gamma_* \to k$, let $M_*$ be a right $\Gamma_*$-comodule with coaction $\nu\colon M_* \to M_* \otimes \Gamma_*$, and let $N_*$ be a left $\Gamma_*$-comodule with coaction $\nu\colon N_* \to \Gamma_* \otimes N_*$. The two-sided cobar construction $C^\bullet(M_*, \Gamma_*, N_*)$ is the cosimplicial graded $k$-vector space with

$$C^p(M_*, \Gamma_*, N_*) = M_* \otimes \Gamma_*^{\otimes p} \otimes N_*$$

in cosimplicial degree $p$, coface operators $d^i\colon C^{p-1}(M_*, \Gamma_*, N_*) \to C^p(M_*, \Gamma_*, N_*)$ given by

$$d^i = \begin{cases} \nu \otimes 1^{\otimes p} & \text{for } i = 0 \\ 1^{\otimes i} \otimes \psi \otimes 1^{\otimes p-i} & \text{for } 0 < i < p \\ 1^{\otimes p} \otimes \nu & \text{for } i = p\,, \end{cases}$$

and codegeneracy operators $s^j\colon C^{p+1}(M_*, \Gamma_*, N_*) \to C^p(M_*, \Gamma_*, N_*)$ given by

$$s^j = 1^{\otimes j+1} \otimes \epsilon \otimes 1^{\otimes p-j+1}$$

for $0 \le j \le p$. In these formulas, each tensor power of 1 refers to the identity map of a number of copies of $M_*$, $\Gamma_*$ or $N_*$. The cobar construction is coaugmented by the canonical map $\eta\colon M_* \,\Box_{\Gamma_*}\, N_* \to M_* \otimes N_* = C^0(M_*, \Gamma_*, N_*)$ from the cotensor product. In the special case when $M_* = \Gamma_*$, viewed as a $\Gamma_*$-$\Gamma_*$-bicomodule, the cobar construction $C^\bullet(\Gamma_*, \Gamma_*, N_*)$ is a cosimplicial left $\Gamma_*$-comodule. The underlying cosimplicial graded $k$-vector space admits a cosimplicial contraction to $\Gamma_* \,\Box_{\Gamma_*}\, N_* \cong N_*$.

The cobar complex $C^*_{\Gamma_*}(M_*, N_*)$ is the associated normalized cochain complex of graded $k$-vector spaces, given by

$$C^p_{\Gamma_*}(M_*, N_*) = M_* \otimes \bar{\Gamma}_*^{\otimes p} \otimes N_* = \bigcap_{j=0}^{p-1} \ker(s^j)$$

in degree $p \ge 0$, with coboundary

$$\delta = \sum_{i=0}^{p+1} (-1)^i d^i\colon C^p_{\Gamma_*}(M_*, N_*) \longrightarrow C^{p+1}_{\Gamma_*}(M_*, N_*)\,.$$

Here $\bar{\Gamma}_* = \ker(\epsilon)$ is the augmentation ideal of $\Gamma_*$. The cobar complex is coaugmented by $\eta\colon M_* \,\Box_{\Gamma_*}\, N_* \to M_* \otimes N_* = C^0_{\Gamma_*}(M_*, N_*)$. In the special case $M_* = \Gamma_*$, the cobar complex $C^*_{\Gamma_*}(\Gamma_*, N_*)$ admits a cochain contraction to $N_*$, and in each degree

$$C^p_{\Gamma_*}(\Gamma_*, N_*) = \Gamma_* \otimes \bar{\Gamma}_*^{\otimes p} \otimes N_*$$

is an extended $\Gamma_*$-comodule. Hence

$$0 \to N_* \xrightarrow{\eta} C^0_{\Gamma_*}(\Gamma_*, N_*) \xrightarrow{\delta} C^1_{\Gamma_*}(\Gamma_*, N_*) \xrightarrow{\delta} \dots$$

is an injective resolution of the left $\Gamma_*$-comodule $N_*$.

The isomorphisms of cochain complexes

$$k \,\Box_{\Gamma_*} C^*_{\Gamma_*}(\Gamma_*, N_*) \cong C^*_{\Gamma_*}(k, N_*) \cong \mathrm{Hom}_{\Gamma_*}(k, C^*_{\Gamma_*}(\Gamma_*, N_*))$$

induce isomorphisms of graded $k$-vector spaces

$$\mathrm{Cotor}^p_{\Gamma_*}(k, N_*) \cong H^p(C^*_{\Gamma_*}(k, N_*), \delta) \cong \mathrm{Ext}^p_{\Gamma_*}(k, N_*)$$

upon passage to cohomology, for each $p \geq 0$. In other words, $\mathrm{Ext}^*_{\Gamma_*}(k, N_*)$ can be calculated as the cohomology of the cobar complex with

$$C^p_{\Gamma_*}(k, N_*) = \bar{\Gamma}^{\otimes p} \otimes N_*$$

for $p \geq 0$ and

$$\delta([\gamma_1|\dots|\gamma_p]n) = [1|\gamma_1|\dots|\gamma_p]n + \sum_{i=1}^{p}(-1)^i[\gamma_1|\dots|\gamma_i'|\gamma_i''|\dots|\gamma_p]n + (-1)^{p+1}[\gamma_1|\dots|\gamma_p|\gamma']n'' .$$

Here $\psi(\gamma_i) = \sum \gamma_i' \otimes \gamma_i''$ and $\nu(n) = \sum \gamma' \otimes n''$, and these summations are implicit in the formula. More generally, for each right $\Gamma_*$-comodule $M_*$, the isomorphism $M_* \,\Box_{\Gamma_*} C^*_{\Gamma_*}(\Gamma_*, N_*) \cong C^*_{\Gamma_*}(M_*, N_*)$ induces an isomorphism $\mathrm{Cotor}^p_{\Gamma_*}(M_*, N_*) \cong H^p(C^*_{\Gamma_*}(M_*, N_*), \delta)$ for each $p \geq 0$.

REMARK 2.13. The cobar complex $C^*_{\Gamma_*}(k, N_*)$ would be denoted $\Omega(\Gamma_*, N_*)$ in the notation of [**123**, p. 75], and written as $\Omega(\Gamma_*) \otimes_\tau N_*$ in the notation of [**80**, §II.3]. In the special case $N_* = k$, it is the construction denoted $F(\Gamma_*)$ by Adams [**1**], up to signs. The cobar resolution $C^*_{\Gamma_*}(\Gamma_*, N_*)$ agrees with the cobar resolution of [**144**, Def. A1.2.11]. For right $\Gamma_*$-comodules, the cobar resolution $C^*_{\Gamma_*}(M_*, \Gamma_*)$ is isomorphic to, but not equal to, the canonical resolution $C(\Gamma_*, M_*)$ of [**45**, Def. IV.1.1], cf. [**119**, Prop. 10.3].

We again specialize to the situation where $\Gamma_*$ is a connected Hopf algebra and $\Lambda_*$ is a quotient Hopf algebra of $\Gamma_*$. With notation as in Proposition 2.10, the quasi-isomorphism

$$\eta\colon k \xrightarrow{\sim} (\Gamma//\Lambda)_* \otimes R^*$$

induces a quasi-isomorphism of cobar complexes

$$\eta\colon C^*_{\Gamma_*}(k, M_*) \xrightarrow{\sim} C^*_{\Gamma_*}(k, (\Gamma//\Lambda)_* \otimes R^* \otimes M_*)$$

The object on the right hand side is the bigraded total complex of a trigraded bicomplex, with $C^p_{\Gamma_*}(k, (\Gamma//\Lambda)_* \otimes R^\sigma \otimes M_*)$ contributing to cohomological degree $s = p+\sigma$.

DEFINITION 2.14. For each $\sigma \geq 0$ let $(\Gamma//\Lambda)_* \otimes R^{*\geq\sigma}$ be the subcomplex of $(\Gamma//\Lambda)_* \otimes R^*$ consisting of the terms $(\Gamma//\Lambda)_* \otimes R^\tau$ with $\tau \geq \sigma$, and let

$$F^\sigma C^*(M_*) = C^*_{\Gamma_*}(k, (\Gamma//\Lambda)_* \otimes R^{*\geq\sigma} \otimes M_*) .$$

We obtain a filtered cochain complex

$$\cdots \subset F^2C^*(M_*) \subset F^1C^*(M_*) \subset F^0C^*(M_*) = C^*_{\Gamma_*}(k, (\Gamma/\!/\Lambda)_* \otimes R^* \otimes M_*)$$

with filtration quotients

$$F^\sigma C^*(M_*)/F^{\sigma+1}C^*(M_*) = C^*_{\Gamma_*}(k, (\Gamma/\!/\Lambda)_* \otimes \Sigma^\sigma R^\sigma \otimes M_*)\,.$$

Here $\Sigma^\sigma R^\sigma$ refers to $R^\sigma$ located in cohomological degree $\sigma$. The associated exact couple

$$(2.3)\qquad \begin{array}{ccccc} \cdots & \xrightarrow{\ i\ } & H^*(F^1C^*(M_*)) & \xrightarrow{\ i\ } & H^*(F^0C^*(M_*)) \\ & \nwarrow{\scriptstyle k} & \downarrow{\scriptstyle j} & \nwarrow{\scriptstyle k} & \downarrow{\scriptstyle j} \\ & & H^*(F^1C^*(M_*)/F^2C^*(M_*)) & & H^*(F^0C^*(M_*)/F^1C^*(M_*)) \end{array}$$

gives rise to a trigraded spectral sequence with

$$\begin{aligned} E_1^{\sigma,s,t} &= H^{s,t}(F^\sigma C^*(M_*)/F^{\sigma+1}C^*(M_*)) = H^{s,t}(C^*_{\Gamma_*}(k, (\Gamma/\!/\Lambda)_* \otimes \Sigma^\sigma R^\sigma \otimes M_*)) \\ &= \operatorname{Ext}^{s-\sigma,t}_{\Gamma_*}(k, (\Gamma/\!/\Lambda)_* \otimes R^\sigma \otimes M_*) \cong \operatorname{Ext}^{s-\sigma,t}_{\Gamma_*}(k, \Gamma_* \,\square_{\Lambda_*} (R^\sigma \otimes M_*)) \\ &\cong \operatorname{Ext}^{s-\sigma,t}_{\Lambda_*}(k, R^\sigma \otimes M_*) \end{aligned}$$

and differentials

$$d_r \colon E_r^{\sigma,s,t} \longrightarrow E_r^{\sigma+r,s+1,t}$$

characterized by $d_r([x]) = [j(y)]$ where $k(x) = i^{r-1}(y)$. It converges strongly to

$$\begin{aligned} H^{s,t}(F^0C^*(M_*)) &= H^{s,t}(C^*_{\Gamma_*}(k, (\Gamma/\!/\Lambda)_* \otimes R^* \otimes M_*)) \\ &\cong H^{s,t}(C^*_{\Gamma_*}(k, M_*)) = \operatorname{Ext}^{s,t}_{\Gamma_*}(k, M_*)\,. \end{aligned}$$

Here $\operatorname{Ext}^{s,t}_{\Gamma_*}(k, M_*)$ is filtered by the images $F^\sigma \operatorname{Ext}^{s,t}(M_*) = \operatorname{im}(i^\sigma)$ of the homomorphisms

$$i^\sigma \colon H^{s,t}(F^\sigma C^*(M_*)) \longrightarrow H^{s,t}(F^0C^*(M_*))\,,$$

and there are isomorphisms

$$E_\infty^{\sigma,s,t} \cong F^\sigma \operatorname{Ext}^{s,t}(M_*)/F^{\sigma+1} \operatorname{Ext}^{s,t}(M_*)$$

for all $\sigma \ge 0$. This is a finite filtration, since $F^\sigma \operatorname{Ext}^{s,t}(M_*) = 0$ for $\sigma > s$.

DEFINITION 2.15. We call

$$E_1^{\sigma,s,t} = E_1^{\sigma,s,t}(M_*) = \operatorname{Ext}^{s-\sigma,t}_{\Lambda_*}(k, R^\sigma \otimes M_*) \Longrightarrow_\sigma \operatorname{Ext}^{s,t}_{\Gamma_*}(k, M_*)$$

the Davis–Mahowald spectral sequence for $\Gamma_* \to \Lambda_*$ with coefficients in $M_*$. It is strongly convergent, with $d_r$-differentials of $(\sigma, s, t)$-tridegree $(r, 1, 0)$.

We prove in Theorem 2.24 that this spectral sequence is monoidal, in the sense that pairings of $\Gamma_*$-comodules lead to pairings of Davis–Mahowald spectral sequences. The dual statement for pairings of $\Gamma$-modules appears in Theorem 2.25.

LEMMA 2.16. *For each $\sigma \ge 0$ there is a quasi-isomorphism*

$$f^\sigma \colon \Sigma^\sigma \ker(\delta^\sigma) \xrightarrow{\ \sim\ } (\Gamma/\!/\Lambda)_* \otimes R^{*\ge\sigma}$$

*of complexes of $\Gamma_*$-comodules. Here $\Sigma^\sigma \ker(\delta^\sigma)$ denotes the complex with $\ker(\delta^\sigma)$ concentrated in degree $\sigma$. The induced morphism of cobar complexes*

$$f^\sigma \colon C^*_{\Gamma_*}(k, \Sigma^\sigma \ker(\delta^\sigma) \otimes M_*) \xrightarrow{\ \sim\ } F^\sigma C^*(M_*)$$

*is also a quasi-isomorphism.*

PROOF. The cohomology of the truncated complex $(\Gamma//\Lambda)_* \otimes R^{*\geq\sigma}$ is $\ker(\delta^\sigma)$ concentrated in degree $\sigma$. □

LEMMA 2.17. *In the diagram*

$$\begin{array}{ccc}
\mathrm{Ext}^{s-\sigma-1}_{\Gamma_*}(k, \ker(\delta^{\sigma+1}) \otimes M_*) & \xrightarrow{f^{\sigma+1}} & H^s(F^{\sigma+1}C^*(M_*)) \\
\downarrow{\scriptstyle \delta} & & \downarrow{\scriptstyle i} \\
\mathrm{Ext}^{s-\sigma}_{\Gamma_*}(k, \ker(\delta^{\sigma}) \otimes M_*) & \xrightarrow{f^{\sigma}} & H^s(F^{\sigma}C^*(M_*)) \\
\downarrow & & \downarrow{\scriptstyle j} \\
\mathrm{Ext}^{s-\sigma}_{\Gamma_*}(k, (\Gamma//\Lambda)_* \otimes R^\sigma \otimes M_*) & = & H^s(F^{\sigma}C^*(M_*)/F^{\sigma+1}C^*(M_*)) \\
\downarrow & & \downarrow{\scriptstyle k} \\
\mathrm{Ext}^{s-\sigma}_{\Gamma_*}(k, \ker(\delta^{\sigma+1}) \otimes M_*) & \xrightarrow{f^{\sigma+1}} & H^{s+1}(F^{\sigma+1}C^*(M_*))\,,
\end{array}$$

*with exact columns, the upper square commutes up to sign and the middle and lower squares commute strictly, for each $\sigma \geq 0$.*

PROOF. For brevity, let $T^\sigma = (\Gamma//\Lambda)_* \otimes \Sigma^\sigma R^\sigma \otimes M_*$, $Z^\sigma = \Sigma^\sigma \ker(\delta^\sigma) \otimes M_*$, $C^p N_* = C^p_{\Gamma_*}(k, N_*)$ and $s = p + \sigma$. The homomorphisms in the upper square are derived from the following diagram.

$$\begin{array}{ccccc}
C^{p-1}Z^{\sigma+1} & & \xrightarrow{f^{\sigma+1}} & & (C^*T^{*\geq\sigma+1})^s \\
\uparrow{\scriptstyle 1\otimes\delta} & & & & \\
C^{p-1}T^\sigma & \xrightarrow{\delta\otimes 1} & C^pT^\sigma & & \downarrow{\scriptstyle i} \\
& & \uparrow & & \\
& & C^pZ^\sigma & \xrightarrow{f^\sigma} & (C^*T^{*\geq\sigma})^s
\end{array}$$

Start with a $(p-1)$-cocycle $x \in C^{p-1}Z^{\sigma+1}$, i.e., an element with $(\delta \otimes 1)(x) = 0$ in $C^pZ^{\sigma+1}$. By exactness of

$$0 \to Z^\sigma \longrightarrow T^\sigma \xrightarrow{\delta} Z^{\sigma+1} \to 0$$

we have $x = (1 \otimes \delta)(y)$ for some $y \in C^{p-1}T^\sigma$, and $(\delta \otimes 1)(y) = z$ for some $z \in C^pZ^\sigma$. The image $f^\sigma\delta([x])$ is then the class of $z$ viewed as an $s$-cocycle in $C^*T^{*\geq\sigma}$. On the other hand, $if^{\sigma+1}([x])$ is the class of $x$ viewed as an $s$-cocycle in $C^*T^{*\geq\sigma}$. We can also view $y$ as an $(s-1)$-cochain in $C^*T^{*\geq\sigma}$, with total coboundary $(\delta \otimes 1 + 1 \otimes \delta)(y) = z + x$. Hence $[x] = -[z]$ in cohomology, so $if^{\sigma+1} = -f^\sigma\delta$.

The middle square is derived from the following commutative square, hence commutes strictly.

$$\begin{array}{ccc}
C^pZ^\sigma & \xrightarrow{f^\sigma} & (C^*T^{*\geq\sigma})^s \\
\downarrow & & \downarrow{\scriptstyle j} \\
C^pT^\sigma & \xrightarrow{=} & (C^*T^\sigma)^s
\end{array}$$

The lower square is derived from the following diagram.

$$
\begin{array}{ccccc}
C^pT^\sigma & \xrightarrow{=} & (C^*T^\sigma)^s & & \\
 & & \big\uparrow{\scriptstyle j} & & \\
\big\downarrow{\scriptstyle 1\otimes\delta} & & (C^*T^{*\geq\sigma})^s & \xrightarrow{\delta\otimes 1+1\otimes\delta} & (C^*T^{*\geq\sigma})^{s+1} \\
 & & & & \big\uparrow{\scriptstyle i} \\
C^pZ^{\sigma+1} & & \xrightarrow{\quad f^{\sigma+1}\quad} & & (C^*T^{*\geq\sigma+1})^{s+1}
\end{array}
$$

Start with a $p$-cocycle $x \in C^pT^\sigma$, meaning that $(\delta \otimes 1)(x) = 0$ in $C^{p+1}T^\sigma$. Letting $y = x$ as an element in $C^pT^\sigma \subset (C^*T^{*\geq\sigma})^s$, we have $j(y) = x$ and $(\delta\otimes 1+1\otimes\delta)(y) = (1 \otimes \delta)(x) = z$ where $z \in C^pZ^{\sigma+1} \subset (C^*T^{*\geq\sigma+1})^{s+1}$. Hence both composites in the lower square map $[x]$ to $[z]$. □

PROPOSITION 2.18. *The Davis–Mahowald spectral sequence of Definition* 2.15 *agrees, up to signs in the differentials, with the spectral sequences in Propositions* 2.3 *and* 2.10.

PROOF. By Lemma 2.17, the exact couples (2.1) and (2.2) agree up to signs with the exact couple (2.3). □

## 2.4. Multiplicative structure

In the setting they studied, Davis and Mahowald verified through case-by-case calculation [**52**, pp. 322–325] that their spectral sequence is an algebra spectral sequence. With our modified construction, this is an instance of the standard algebra spectral sequence associated to a filtered differential graded (DG) algebra.

DEFINITION 2.19. Let $\Gamma_*$ be a connected coalgebra over $k$, and let $M_*$ and $N_*$ be right and left $\Gamma_*$-comodules, respectively. Likewise, let $\Gamma'_*$ be a connected coalgebra and let $M'_*$ and $N'_*$ be right and left $\Gamma'_*$-comodules. There is an Alexander–Whitney chain map

$$f\colon C^*_{\Gamma_*}(M_*, N_*) \otimes C^*_{\Gamma'_*}(M'_*, N'_*) \longrightarrow C^*_{\Gamma_*\otimes\Gamma'_*}(M_* \otimes M'_*, N_* \otimes N'_*)$$

given for $p, q \geq 0$ by the composite

$$
\begin{aligned}
C^p(M_*, \Gamma_*, N_*) \otimes C^q(M'_*, \Gamma'_*, N'_*) &\overset{\lambda_p\otimes\rho_q}{\longrightarrow} C^{p+q}(M_*, \Gamma_*, N_*) \otimes C^{p+q}(M'_*, \Gamma'_*, N'_*) \\
&\overset{\tau}{\cong} C^{p+q}(M_* \otimes M'_*, \Gamma_* \otimes \Gamma'_*, N_* \otimes N'_*)
\end{aligned}
$$

where $\lambda_p = d^{p+q} \cdots d^{p+1}$ and $\rho_q = d^0 \cdots d^0$ are the front $p$-coface and back $q$-coface operators, respectively, and the right hand isomorphism $\tau$ is given by a shuffle permutation. More explicitly,

$$m[\gamma_1|\dots|\gamma_p]n \otimes m'[\gamma'_1|\dots|\gamma'_q]n'$$

in $M_* \otimes \Gamma_*^{\otimes p} \otimes N_* \otimes M'_* \otimes {\Gamma'_*}^{\otimes q} \otimes N'_*$ maps by $f$ to

$$\pm m \otimes m'_0[\gamma_1 \otimes m'_1|\dots|\gamma_p \otimes m'_p|n_1 \otimes \gamma'_1|\dots|n_q \otimes \gamma'_q]n_{q+1} \otimes n' \tag{2.4}$$

in $M_* \otimes M'_* \otimes (\Gamma_* \otimes \Gamma'_*)^{\otimes p+q} \otimes N_* \otimes N'_*$. Here the sign is that induced by $\tau$, the iterated coactions $\nu^q\colon N_* \to \Gamma_*^{\otimes q} \otimes N_*$ and $\nu^p\colon M'_* \to M'_* \otimes {\Gamma'_*}^{\otimes p}$ are given on $n$

and $m'$ by $\nu^q(n) = \sum n_1 \otimes \cdots \otimes n_q \otimes n_{q+1}$ and $\nu^p(m') = \sum m'_0 \otimes m'_1 \otimes \cdots \otimes m'_p$, and the latter two summations are implicit in (2.4).

When $M_* = \Gamma_*$ and $M'_* = \Gamma'_*$, the Alexander–Whitney map is a chain equivalence between two injective $\Gamma_* \otimes \Gamma'_*$-comodule resolutions of $N_* \otimes N'_*$. Hence the Alexander–Whitney map

$$f\colon C^*_{\Gamma_*}(k, N_*) \otimes C^*_{\Gamma'_*}(k, N'_*) \longrightarrow C^*_{\Gamma_*\otimes\Gamma'_*}(k, N_* \otimes N'_*)$$

for $M_* = k$ and $M'_* = k$ induces the standard external pairing

$$\mathrm{Ext}^p_{\Gamma_*}(k, N_*) \otimes \mathrm{Ext}^q_{\Gamma'_*}(k, N'_*) \longrightarrow \mathrm{Ext}^{p+q}_{\Gamma_*\otimes\Gamma'_*}(k, N_* \otimes N'_*)$$

by passage to cohomology.

When $\Gamma_* = \Gamma'_*$ is a Hopf algebra, we can internalize the pairing above by composing $f$ with the chain map $C^*_{\Gamma_*\otimes\Gamma_*}(k, N_* \otimes N'_*) \to C^*_{\Gamma_*}(k, N_* \otimes N'_*)$ induced by the algebra multiplication $\phi\colon \Gamma_* \otimes \Gamma_* \to \Gamma_*$. The composite chain map

$$\phi f\colon C^*_{\Gamma_*}(k, N_*) \otimes C^*_{\Gamma_*}(k, N'_*) \longrightarrow C^*_{\Gamma_*}(k, N_* \otimes N'_*)$$

takes

$$[\gamma_1|\dots|\gamma_p]n \otimes [\gamma'_1|\dots|\gamma'_q]n'$$

in $\Gamma_*^{\otimes p} \otimes N_* \otimes \Gamma_*^{\otimes q} \otimes N'_*$ to

$$\pm[\gamma_1|\dots|\gamma_p|n_1\gamma'_1|\dots|n_q\gamma'_q]n_{q+1} \otimes n' \tag{2.5}$$

in $\Gamma_*^{\otimes p+q} \otimes N_* \otimes N'_*$. As before the sign is that induced by $\tau$, and the $q$-fold iterated coaction $\nu^q\colon N_* \to \Gamma_*^{\otimes q} \otimes N_*$ is given by $\nu^q(n) = \sum n_1 \otimes \cdots \otimes n_q \otimes n_{q+1}$, with $n_1, \dots, n_q \in \Gamma_*$ and $n_{q+1} \in N_*$. This defines the internal pairing

$$\mathrm{Ext}^p_{\Gamma_*}(k, N_*) \otimes \mathrm{Ext}^q_{\Gamma_*}(k, N'_*) \longrightarrow \mathrm{Ext}^{p+q}_{\Gamma_*}(k, N_* \otimes N'_*)\,.$$

Finally, if $N_* = N'_*$ is a $\Gamma_*$-comodule algebra, composition with the chain map

$$\mu\colon C^*_{\Gamma_*}(k, N_* \otimes N_*) \longrightarrow C^*_{\Gamma_*}(k, N_*)\,,$$

induced by the multiplication $\mu\colon N_* \otimes N_* \to N_*$, defines a product that makes $\mathrm{Ext}^*_{\Gamma_*}(k, N_*)$ a bigraded algebra.

Remark 2.20. Formula (2.4) is given in [**144**, A1.2.15], and the special case with $M_* = k = M'_*$ is given in [**123**, (1.3)]. If $N_* = k$, the product (2.5) simplifies to

$$[\gamma_1|\dots|\gamma_p] \otimes [\gamma'_1|\dots|\gamma'_q]n' \longmapsto [\gamma_1|\dots|\gamma_p|\gamma'_1|\dots|\gamma'_q]n'$$

so that the algebra structure in $\mathrm{Ext}^*_{\Gamma_*}(k, k)$ and the left module pairing

$$\mathrm{Ext}^*_{\Gamma_*}(k, k) \otimes \mathrm{Ext}^*_{\Gamma_*}(k, N'_*) \longrightarrow \mathrm{Ext}^*_{\Gamma_*}(k, N'_*)$$

are induced by juxtaposition, as in [**3**, p. 33] and [**126**, §3].

We return to the situation where $\Gamma_*$ is a connected Hopf algebra, and $\Lambda_*$ is a quotient Hopf algebra of $\Gamma_*$. The right coaction $\nu\colon \Gamma_* \to \Gamma_* \otimes \Lambda_*$ and the left coaction $\nu\colon k \to \Lambda_* \otimes k$ are both algebra homomorphisms, so the equalizer diagram

$$\Gamma_* \otimes k \overset{\nu\otimes 1}{\underset{1\otimes\nu}{\rightrightarrows}} \Gamma_* \otimes \Lambda_* \otimes k$$

defining $(\Gamma/\!/\Lambda)_* = \Gamma_* \,\square_{\Lambda_*}\, k$ exhibits $(\Gamma/\!/\Lambda)_*$ as a sub $\Gamma_*$-comodule algebra of $\Gamma_*$.

PROPOSITION 2.21. *Suppose that $R^*$ is a graded $\Gamma_*$-comodule algebra, and that there are differentials $\delta^\sigma\colon (\Gamma//\Lambda)_* \otimes R^\sigma \to (\Gamma//\Lambda)_* \otimes R^{\sigma+1}$ making $(\Gamma//\Lambda)_* \otimes R^*$, with the diagonal $\Gamma_*$-coaction, a differential graded $\Gamma_*$-comodule algebra. Suppose also that the unit map*

$$\eta\colon k \xrightarrow{\sim} (\Gamma//\Lambda)_* \otimes R^*$$

*is a quasi-isomorphism. Then there is a pairing of spectral sequences*

$$E_r^{\sigma,s,t}(M_*) \otimes E_r^{\sigma',s',t'}(M'_*) \longrightarrow E_r^{\sigma+\sigma',s+s',t+t'}(M_* \otimes M'_*)\,,$$

*converging to*

$$\mathrm{Ext}_{\Gamma_*}^{s,t}(k, M_*) \otimes \mathrm{Ext}_{\Gamma_*}^{s',t'}(k, M'_*) \longrightarrow \mathrm{Ext}_{\Gamma_*}^{s+s',t+t'}(k, M_* \otimes M'_*)\,.$$

*The pairing of $E_1$-terms*

$$\begin{aligned}\mathrm{Ext}_{\Gamma_*}^{s-\sigma,t}(k, (\Gamma//\Lambda)_* \otimes R^\sigma \otimes M_*) \otimes \mathrm{Ext}_{\Gamma_*}^{s'-\sigma',t'}(k, (\Gamma//\Lambda)_* \otimes R^{\sigma'} \otimes M'_*)& \\ \longrightarrow \mathrm{Ext}_{\Gamma_*}^{s-\sigma+s'-\sigma',t+t'}(k, (\Gamma//\Lambda)_* \otimes R^{\sigma+\sigma'} \otimes M_* \otimes M'_*)&\end{aligned}$$

*is induced by the pairing*

$$(\Gamma//\Lambda)_* \otimes R^\sigma \otimes (\Gamma//\Lambda)_* \otimes R^{\sigma'} \longrightarrow (\Gamma//\Lambda)_* \otimes R^{\sigma+\sigma'} \tag{2.6}$$

*obtained from the product on $(\Gamma//\Lambda)_*$ and the multiplication $R^\sigma \otimes R^{\sigma'} \to R^{\sigma+\sigma'}$. In particular, if $M_*$ is a $\Gamma_*$-comodule algebra then*

$$E_1^{\sigma,s,t}(M_*) = \mathrm{Ext}_{\Gamma_*}^{s-\sigma,t}(k, (\Gamma//\Lambda)_* \otimes R^\sigma \otimes M_*) \Longrightarrow_\sigma \mathrm{Ext}_{\Gamma_*}^{s,t}(k, M_*)$$

*is an algebra spectral sequence.*

PROOF. By assumption, the unit map $k \to R^0$, the multiplications $R^\sigma \otimes R^{\sigma'} \to R^{\sigma+\sigma'}$ and the differential $\delta^\sigma\colon (\Gamma//\Lambda)_* \otimes R^\sigma \to (\Gamma//\Lambda)_* \otimes R^{\sigma+1}$ are $\Gamma_*$-comodule homomorphisms, the differential satisfies $\delta\delta = 0$ and $\delta(x \cdot y) = \delta x \cdot y + (-1)^{|x|} x \cdot \delta y$ for $x$ in degree $|x| = t - \sigma$, and the cochain complex

$$0 \to k \xrightarrow{\eta} (\Gamma//\Lambda)_* \otimes R^0 \xrightarrow{\delta^0} (\Gamma//\Lambda)_* \otimes R^1 \xrightarrow{\delta^1} (\Gamma//\Lambda)_* \otimes R^2 \xrightarrow{\delta^2} \dots$$

is exact. Hence

$$C^*_{\Gamma_*}(k, (\Gamma//\Lambda)_* \otimes R^*)$$

is a differential graded algebra, and the unit map

$$\eta\colon C^*_{\Gamma_*}(k,k) \longrightarrow C^*_{\Gamma_*}(k, (\Gamma//\Lambda)_* \otimes R^*)$$

is a quasi-isomorphism. The $\Gamma_*$-comodule pairing

$$R^{*\ge\sigma} \otimes R^{*\ge\sigma'} \longrightarrow R^{*\ge\sigma+\sigma'}$$

induces a pairing of cochain complexes

$$F^\sigma C^*(M_*) \otimes F^{\sigma'} C^*(M'_*) \longrightarrow F^{\sigma+\sigma'} C^*(M_* \otimes M'_*)\,.$$

For varying $\sigma$ and $\sigma'$, these combine to a pairing of filtered cochain complexes. It follows, as in [**113**, §7, §8], that there is an induced pairing of the associated spectral sequences. □

LEMMA 2.22. *If $\Gamma_*$ is commutative, then the pairing (2.6) corresponds under the twisting isomorphisms $\zeta_*$ for $R^\sigma$, $R^{\sigma'}$ and $R^{\sigma+\sigma'}$ to the $\Gamma_*$-comodule pairing*

$$(\Gamma_* \,\square_{\Lambda_*}\, R^\sigma) \otimes (\Gamma_* \,\square_{\Lambda_*}\, R^{\sigma'}) \longrightarrow \Gamma_* \,\square_{\Lambda_*}\, R^{\sigma+\sigma'} \tag{2.7}$$

*induced by the product $\phi$ on $\Gamma_*$ and the pairing $\phi\colon R^\sigma \otimes R^{\sigma'} \to R^{\sigma+\sigma'}$.*

PROOF. When $\Gamma_*$ is commutative, the diagram

$$\begin{array}{ccccc}
\Gamma_* \otimes R^\sigma \otimes \Gamma_* \otimes R^{\sigma'} & \xrightarrow{1\otimes\tau\otimes 1} & \Gamma_* \otimes \Gamma_* \otimes R^\sigma \otimes R^{\sigma'} & \xrightarrow{\phi\otimes\phi} & \Gamma_* \otimes R^{\sigma+\sigma'} \\
\downarrow{\scriptstyle 1\otimes\nu\otimes 1\otimes\nu} & & & & \downarrow{\scriptstyle 1\otimes\nu} \\
\Gamma_* \otimes \Gamma_* \otimes R^\sigma \otimes \Gamma_* \otimes \Gamma_* \otimes R^{\sigma'} & & & & \Gamma_* \otimes \Gamma_* \otimes R^{\sigma+\sigma'} \\
\downarrow{\scriptstyle \phi\otimes 1\otimes\phi\otimes 1} & & & & \downarrow{\scriptstyle \phi\otimes 1} \\
\Gamma_* \otimes R^\sigma \otimes \Gamma_* \otimes R^{\sigma'} & \xrightarrow{1\otimes\tau\otimes 1} & \Gamma_* \otimes \Gamma_* \otimes R^\sigma \otimes R^{\sigma'} & \xrightarrow{\phi\otimes\phi} & \Gamma_* \otimes R^{\sigma+\sigma'}
\end{array}$$

commutes. Here $\tau$ denotes the symmetry isomorphism. Hence the induced square

$$\begin{array}{ccc}
(\Gamma/\!/\Lambda)_* \otimes R^\sigma \otimes (\Gamma/\!/\Lambda)_* \otimes R^{\sigma'} & \xrightarrow{(2.6)} & (\Gamma/\!/\Lambda)_* \otimes R^{\sigma+\sigma'} \\
{\scriptstyle \zeta_*\otimes\zeta_*}\downarrow{\scriptstyle \cong} & & {\scriptstyle \cong}\downarrow{\scriptstyle \zeta_*} \\
\Gamma_* \,\square_{\Lambda_*}\, R^\sigma \otimes \Gamma_* \,\square_{\Lambda_*}\, R^{\sigma'} & \xrightarrow{(2.7)} & \Gamma_* \,\square_{\Lambda_*}\, R^{\sigma+\sigma'}
\end{array}$$

and its generalization

$$\begin{array}{ccc}
(\Gamma/\!/\Lambda)_* \otimes R^\sigma \otimes M_* \otimes (\Gamma/\!/\Lambda)_* \otimes R^{\sigma'} \otimes M'_* & \longrightarrow & (\Gamma/\!/\Lambda)_* \otimes R^{\sigma+\sigma'} \otimes M_* \otimes M'_* \\
{\scriptstyle \zeta_*\otimes\zeta_*}\downarrow{\scriptstyle \cong} & & {\scriptstyle \cong}\downarrow{\scriptstyle \zeta_*} \\
\Gamma_* \,\square_{\Lambda_*}\, (R^\sigma \otimes M_*) \otimes \Gamma_* \,\square_{\Lambda_*}\, (R^{\sigma'} \otimes M'_*) & \longrightarrow & \Gamma_* \,\square_{\Lambda_*}\, (R^{\sigma+\sigma'} \otimes M_* \otimes M'_*)
\end{array}$$

commute. □

LEMMA 2.23. *Under the change-of-coalgebra isomorphisms, the pairing*

$$\mathrm{Ext}_{\Lambda_*}(k, R^\sigma \otimes M_*) \otimes \mathrm{Ext}_{\Lambda_*}(k, R^{\sigma'} \otimes M'_*) \longrightarrow \mathrm{Ext}_{\Lambda_*}(k, R^{\sigma+\sigma'} \otimes M_* \otimes M'_*)$$

*induced by* $R^\sigma \otimes R^{\sigma'} \to R^{\sigma+\sigma'}$ *corresponds to the pairing*

$$\mathrm{Ext}_{\Gamma_*}(k, \Gamma_* \,\square_{\Lambda_*}\, (R^\sigma \otimes M_*)) \otimes \mathrm{Ext}_{\Gamma_*}(k, \Gamma_* \,\square_{\Lambda_*}\, (R^{\sigma'} \otimes M'_*)) \\ \longrightarrow \mathrm{Ext}_{\Gamma_*}(k, \Gamma_* \,\square_{\Lambda_*}\, (R^{\sigma+\sigma'} \otimes M_* \otimes M'_*))$$

*that is induced by* (2.7) *and its generalization.*

PROOF. This follows from the adjunctions underlying the change-of-coalgebra isomorphisms. □

THEOREM 2.24. *Let* $\Gamma_*$ *be a connected, commutative Hopf algebra over a field* $k$, *and let* $\Lambda_*$ *be a quotient Hopf algebra of* $\Gamma_*$. *Suppose that* $R^*$ *is a graded* $\Gamma_*$*-comodule algebra, and that*

$$\eta\colon k \xrightarrow{\sim} ((\Gamma/\!/\Lambda)_* \otimes R^*, \delta)$$

*is a differential (cohomologically) graded* $\Gamma_*$*-comodule algebra resolution of* $k$, *where each term* $(\Gamma/\!/\Lambda)_* \otimes R^\sigma$ *has the diagonal* $\Gamma_*$*-comodule structure. Let* $M_*$ *and* $M'_*$ *be* $\Gamma_*$*-comodules. Then there is a pairing of trigraded spectral sequences*

$$E_r^{\sigma,s,t}(M_*) \otimes E_r^{\sigma',s',t'}(M'_*) \longrightarrow E_r^{\sigma+\sigma',s+s',t+t'}(M_* \otimes M'_*)$$

*converging to*

$$\mathrm{Ext}_{\Gamma_*}^{s,t}(k, M_*) \otimes \mathrm{Ext}_{\Gamma_*}^{s',t'}(k, M'_*) \longrightarrow \mathrm{Ext}_{\Gamma_*}^{s+s',t+t'}(k, M_* \otimes M'_*)\,.$$

*The pairing of $E_1$-terms*

$$\operatorname{Ext}_{\Lambda_*}^{s-\sigma,t}(k, R^\sigma \otimes M_*) \otimes \operatorname{Ext}_{\Lambda_*}^{s'-\sigma',t'}(k, R^{\sigma'} \otimes M'_*) \longrightarrow \operatorname{Ext}_{\Lambda_*}^{s-\sigma+s'-\sigma',t+t'}(k, R^{\sigma+\sigma'} \otimes M_* \otimes M'_*)$$

*is induced by the component $R^\sigma \otimes R^{\sigma'} \longrightarrow R^{\sigma+\sigma'}$ of the graded algebra structure on $R^*$. In particular, if $M_*$ is a $\Gamma_*$-comodule algebra then*

$$E_1^{\sigma,s,t}(M_*) = \operatorname{Ext}_{\Lambda_*}^{s-\sigma,t}(k, R^\sigma \otimes M_*) \Longrightarrow_\sigma \operatorname{Ext}_{\Gamma_*}^{s,t}(k, M_*)$$

*is an algebra spectral sequence.*

PROOF. Combine Proposition 2.21 with Lemmas 2.22 and 2.23. □

Before we give the dual statement, note that the arrows in the coequalizer diagram defining $\Gamma//\Lambda = \Gamma \otimes_\Lambda k$ are coalgebra homomorphisms, so that $\Gamma//\Lambda$ is a quotient $\Gamma$-module coalgebra of $\Gamma$.

THEOREM 2.25. *Let $\Gamma$ be a connected, cocommutative Hopf algebra over a field $k$, and let $\Lambda$ be a sub Hopf algebra of $\Gamma$. Suppose that $N_*$ is a graded $\Gamma$-module coalgebra, and that*

$$\epsilon\colon (\Gamma//\Lambda \otimes N_*, \partial) \xrightarrow{\sim} k$$

*is a differential (homologically) graded $\Gamma$-module coalgebra resolution of $k$, where each term $\Gamma//\Lambda \otimes N_\sigma$ has the diagonal $\Gamma$-module structure. Let $M$ and $M'$ be $\Gamma$-modules. Then there is a pairing of trigraded spectral sequences*

$$E_r^{\sigma,s,t}(M) \otimes E_r^{\sigma',s',t'}(M') \longrightarrow E_r^{\sigma+\sigma',s+s',t+t'}(M \otimes M')$$

*converging to*

$$\operatorname{Ext}_\Gamma^{s,t}(M, k) \otimes \operatorname{Ext}_\Gamma^{s',t'}(M', k) \longrightarrow \operatorname{Ext}_\Gamma^{s+s',t+t'}(M \otimes M', k)\,.$$

*The pairing of $E_1$-terms*

$$\operatorname{Ext}_\Lambda^{s-\sigma,t}(N_\sigma \otimes M, k) \otimes \operatorname{Ext}_\Lambda^{s'-\sigma',t'}(N_{\sigma'} \otimes M', k) \longrightarrow \operatorname{Ext}_\Lambda^{s-\sigma+s'-\sigma',t+t'}(N_{\sigma+\sigma'} \otimes M \otimes M', k)$$

*is induced by the component $N_{\sigma+\sigma'} \to N_\sigma \otimes N_{\sigma'}$ of the graded coalgebra structure on $N_*$. In particular, if $M$ is a $\Gamma$-module coalgebra then*

$$E_1^{\sigma,s,t}(M) = \operatorname{Ext}_\Lambda^{s-\sigma,t}(N_\sigma \otimes M, k) \Longrightarrow_\sigma \operatorname{Ext}_\Gamma^{s,t}(M, k)$$

*is an algebra spectral sequence.*

We omit the proof, which is similar to that in the $\Gamma_*$-comodule case, using the bar construction in place of the cobar construction. When $\Gamma$ is finite-dimensional over $k$, and each of $N_\sigma$, $M$ and $M'$ is of finite type, then Lemma 2.6 shows that the two statements are equivalent. If $\Gamma$ is just of finite type, then we must also assume that $N_\sigma$, $M$ and $M'$ are bounded below.

REMARK 2.26. When $(\Gamma//\Lambda)_* = E_* = E(e_1, \dots, e_n)$ is an exterior algebra on $n$ generators we can let $R^* = k[x_1, \dots, x_n]$ be a polynomial algebra on the same number of generators and equip

$$E_* \otimes R^* = E(e_1, \dots, e_n) \otimes k[x_1, \dots, x_n]$$

with the differential $d$ given by $d(e_i) = x_i$ for $1 \le i \le n$. From $d^2 = 0$ it follows that $d(x_i) = 0$. If we give $E_* \otimes R^*$ a homological grading, with each $e_i$ in degree 1 and each $x_i$ in degree 0, then the underlying exact chain complex

$$0 \to E_n \otimes R^* \xrightarrow{d} E_{n-1} \otimes R^* \xrightarrow{d} \dots \xrightarrow{d} E_1 \otimes R^* \xrightarrow{d} E_0 \otimes R^* \xrightarrow{\epsilon} k \to 0$$

is the Koszul resolution associated to the regular sequence $(x_1, \dots, x_n)$. If we instead give $E_* \otimes R^*$ a cohomological grading, with each $e_i$ in degree 0 and each $x_i$ in degree 1, then the underlying exact cochain complex

$$0 \to k \xrightarrow{\eta} E_* \otimes R^0 \xrightarrow{d} E_* \otimes R^1 \xrightarrow{d} E_* \otimes R^2 \xrightarrow{d} \dots$$

is a resolution of the sort considered by Davis–Mahowald. In this sense a Davis–Mahowald resolution can arise as a modified Koszul resolution, and justifies the name "Koszul spectral sequence" used in [**106**]. We use the name "Davis–Mahowald spectral sequence" to acknowledge the origin of its construction, and to allow for the more general case where $\eta \colon k \to (\Gamma//\Lambda)_* \otimes R^*$ is not necessarily a Koszul resolution.

## 2.5. The spectral sequence for $A(1)$

As a warm-up to the calculation in Chapter 3, we first consider a simpler case. Let $k = \mathbb{F}_2$, and consider the subalgebras $A(1) = \langle Sq^1, Sq^2 \rangle$ and $A(0) = E(Sq^1)$ of the mod 2 Steenrod algebra $A$, which are generated by $Sq^1$ and $Sq^2$, and by $Sq^1$, respectively. These are connected, cocommutative sub Hopf algebras of $A$, with dual Hopf algebras

$$A(1)_* = \mathbb{F}_2[\xi_1, \bar{\xi}_2]/(\xi_1^4, \bar{\xi}_2^2)$$

and

$$A(0)_* = \mathbb{F}_2[\xi_1]/(\xi_1^2) = E(\xi_1) .$$

The coproduct in $A(1)_*$ is given by

$$\psi(\xi_1) = 1 \otimes \xi_1 + \xi_1 \otimes 1$$
$$\psi(\bar{\xi}_2) = 1 \otimes \bar{\xi}_2 + \xi_1 \otimes \xi_1^2 + \bar{\xi}_2 \otimes 1$$

so that $(A(1)//A(0))_* = E(\xi_1^2, \bar{\xi}_2)$ as a sub $A(1)_*$-comodule algebra of $A(1)_*$. In this section, let $R^* = \mathbb{F}_2[x_2, x_3]$ be the graded $A(1)_*$-comodule algebra with $x_i$ in internal degree $i$ and cohomological degree 1, having coaction given by

$$\nu(x_2) = 1 \otimes x_2$$
$$\nu(x_3) = 1 \otimes x_3 + \xi_1 \otimes x_2 .$$

We equip

$$(A(1)//A(0))_* \otimes R^* = E(\xi_1^2, \bar{\xi}_2) \otimes \mathbb{F}_2[x_2, x_3]$$

with the diagonal $A(1)_*$-comodule algebra structure. It becomes a differential graded $A(1)_*$-comodule algebra with the differential $\delta$ given by

$$\delta(\xi_1^2) = x_2$$
$$\delta(\bar{\xi}_2) = x_3 ,$$

and the resulting cochain complex

$$0 \to \mathbb{F}_2 \xrightarrow{\eta} E(\xi_1^2, \bar{\xi}_2) \otimes R^0 \xrightarrow{\delta^0} E(\xi_1^2, \bar{\xi}_2) \otimes R^1 \xrightarrow{\delta^1} E(\xi_1^2, \bar{\xi}_2) \otimes R^2 \xrightarrow{\delta^2} \dots$$

is exact. Here

$$R^\sigma = \mathbb{F}_2\{x_2^i x_3^j \mid i + j = \sigma\}$$

is the $A(1)_*$-comodule of homogeneous polynomials in $\mathbb{F}_2[x_2, x_3]$ of degree $\sigma$.

The Davis–Mahowald spectral sequence for $A(1)_* \to A(0)_*$ with coefficients in $\mathbb{F}_2$ is thus the algebra spectral sequence

$$E_1^{\sigma,s,t} = \mathrm{Ext}_{A(0)_*}^{s-\sigma,t}(\mathbb{F}_2, R^\sigma) \Longrightarrow_\sigma \mathrm{Ext}_{A(1)_*}^{s,t}(\mathbb{F}_2, \mathbb{F}_2)\,.$$

Recall that for a $\Gamma_*$-comodule $M_*$ the group $\mathrm{Ext}_{\Gamma_*}^{0,*}(k, M_*)$ consists of the $\Gamma_*$-comodule primitives in $M_*$, i.e., the elements $x \in M_*$ with $\nu(x) = 1 \otimes x$. We note that $x_2$ and $x_3^2$ are $A(0)_*$-comodule primitives, and that $R^*$ is free as a module over $\mathbb{F}_2[x_3^2]$. We obtain an extension of graded $A(0)_*$-comodule algebras

$$\mathbb{F}_2[x_3^2] \longrightarrow R^* \longrightarrow \bar{R}^*$$

where, by definition,

$$\bar{R}^* = R^* \otimes_{\mathbb{F}_2[x_3^2]} \mathbb{F}_2 = R^*/(x_3^2) = \mathbb{F}_2[x_2, x_3]/(x_3^2)\,.$$

Here $\bar{R}^0 = \mathbb{F}_2\{1\} \cong \mathbb{F}_2$, and

$$\bar{R}^\sigma = \mathbb{F}_2\{x_2^\sigma, x_2^{\sigma-1}x_3\} \cong \Sigma^{2\sigma} A(0)_*$$

for $\sigma \geq 1$. Hence we obtain an extension of trigraded algebras

$$\mathbb{F}_2[x_3^2] \longrightarrow E_1^{*,*,*} \longrightarrow \bar{E}_1^{*,*,*}$$

where $E_1^{*,*,*}$ is free as a module over $\mathbb{F}_2[x_3^2]$. By abuse of notation,

$$\bar{E}_1^{\sigma,s,t} = \mathrm{Ext}_{A(0)_*}^{s-\sigma,t}(\mathbb{F}_2, \bar{R}^\sigma)$$

is given by

$$\bar{E}_1^{0,*,*} \cong \mathrm{Ext}_{A(0)_*}^{*,*}(\mathbb{F}_2, \mathbb{F}_2) = \mathbb{F}_2[h_0]$$

and

$$\bar{E}_1^{\sigma,*,*} \cong \mathrm{Ext}_{A(0)_*}^{*-\sigma,*}(\mathbb{F}_2, \Sigma^{2\sigma} A(0)_*) \cong \mathbb{F}_2\{x_2^\sigma\}$$

for $\sigma \geq 1$. Here $h_0 \in \bar{E}_1^{0,1,1} = \mathrm{Ext}_{A(0)_*}^{1,1}(\mathbb{F}_2, \mathbb{F}_2)$ corresponds to the coalgebra primitive $\xi_1$ dual to $Sq^1$. We write $x_2^\sigma$ for the class in $\bar{E}_1^{\sigma,\sigma,2\sigma}$ that corresponds to $x_2^\sigma \in \mathrm{Ext}_{A(0)_*}^{0,2\sigma}(\mathbb{F}_2, \bar{R}^\sigma)$. Thus,

$$E_1^{*,*,*} = \mathbb{F}_2[h_0, x_2, x_3^2]/(h_0 x_2)$$

with generators in $(\sigma, s, t)$-degrees $|h_0| = (0, 1, 1)$, $|x_2| = (1, 1, 2)$ and $|x_3^2| = (2, 2, 6)$. The algebra extension $E_1 \to \bar{E}_1$ splits, because $h_0 x_2$ lies in weight $\sigma = 1$, where $(x_3^2)$ is trivial.

LEMMA 2.27. *$d_1(h_0) = 0$, $d_1(x_2) = 0$ and $d_1(x_3^2) = x_2^3$.*

PROOF. The target groups of the first two differentials, $E_1^{1,2,1}$ and $E_1^{2,2,2}$, are both zero. The differential $d_1 \colon E_1^{2,2,6} \to E_1^{3,3,6}$ is the homomorphism

$$\delta_*^2 \colon \mathrm{Ext}_{A(1)_*}^{0,6}(\mathbb{F}_2, (A(1)/\!/A(0))_* \otimes R^2) \longrightarrow \mathrm{Ext}_{A(1)_*}^{0,6}(\mathbb{F}_2, (A(1)/\!/A(0))_* \otimes R^3)$$

induced by $\delta^2$. In internal degree 6 the only nonzero $A(1)_*$-comodule primitive in $(A(1)/\!/A(0))_* \otimes R^2$ is $\xi_1^2 x_2^2 + x_3^2$, which is mapped by $\delta^2$ to the nonzero $A(1)_*$-comodule primitive $\delta(\xi_1^2)x_2^2 + 0 = x_2^3$ in $(A(1)/\!/A(0))_* \otimes R^3$. Hence $d_1(x_3^2) = x_2^3$. □

LEMMA 2.28.

$$E_2^{*,*,*} = \mathbb{F}_2[h_0, x_2, h_0 x_3^2, x_3^4]/(h_0 x_2, x_2^3, x_2(h_0 x_3^2), (h_0 x_3^2)^2 - h_0^2(x_3^4))$$

*is equal to $E_\infty^{*,*,*}$.*

PROOF. See Figures 2.1 and 2.2. The differentials $d_r(x_2)$ lie in the groups $E_r^{1+r,2,2}$, which are trivial. □

PROPOSITION 2.29.

$$\mathrm{Ext}_{A(1)}(\mathbb{F}_2, \mathbb{F}_2) \cong \mathrm{Ext}_{A(1)_*}(\mathbb{F}_2, \mathbb{F}_2) \cong \mathbb{F}_2[h_0, h_1, v, w_1]/(h_0h_1, h_1^3, h_1v, v^2 - h_0^2w_1)$$

*with* $(t-s,s)$*-bigradings* $|h_0| = (0,1)$, $|h_1| = (1,1)$, $|v| = (4,3)$ *and* $|w_1| = (8,4)$.

PROOF. There are unique classes $h_0$, $h_1$, $v$ and $w_1$ in $\mathrm{Ext}_{A(1)_*}(\mathbb{F}_2, \mathbb{F}_2)$ that are detected by $h_0$, $x_2$, $h_0x_3^2$ and $x_3^4$ in $E_\infty^{*,*,*}$, respectively. Each multiplicative relation in $E_\infty^{\sigma,s,t}$ lifts unchanged to $\mathrm{Ext}_{A(1)_*}^{s,t}(\mathbb{F}_2, \mathbb{F}_2)$, since in each case there are no classes in $E_\infty^{*,s,t}$ of higher weight than $\sigma$. See Figure 2.3. □

## 2.6. Real, quaternionic and complex $K$-theory spectra

Having calculated $\mathrm{Ext}_{A(1)}(\mathbb{F}_2, \mathbb{F}_2)$, we round off this chapter with some examples of spectra with mod 2 cohomology induced up from various $A(1)$-modules, namely the connective topological $K$-theory spectra. To each permutative (or symmetric monoidal) topological category $(\mathcal{C}, \oplus)$ one can associate a $K$-theory spectrum $K(\mathcal{C})$ [**151**], [**120**], [**59**, Thm. 1.1]. When $(\mathcal{C}, \oplus, \otimes)$ is bipermutative (or symmetric bimonoidal), the $K$-theory spectrum becomes an $E_\infty$ ring spectrum [**121**], [**59**, Thm. 1.2]. Furthermore, if $\mathcal{D}$ is a suitably defined module category over $\mathcal{C}$, then $K(\mathcal{D})$ is a module spectrum over $K(\mathcal{C})$, see [**59**, §9].

EXAMPLE 2.30. The connective real $K$-theory spectrum $ko$ is the $K$-theory spectrum of a bipermutative topological category $\mathcal{GL}(\mathbb{R})$ [**121**, Ex. VI.5.4] equivalent to the symmetric bimonoidal topological category of finite dimensional real vector spaces, with respect to the usual direct sum and tensor product. It is an $E_\infty$ ring spectrum with mod 2 cohomology

$$H^*(ko) = A/A(Sq^1, Sq^2) = A \otimes_{A(1)} \mathbb{F}_2 = A/\!/A(1)$$

and mod 2 homology

$$H_*(ko) = A_* \,\square_{A(1)_*}\, \mathbb{F}_2 = \mathbb{F}_2[\bar\xi_1^4, \bar\xi_2^2, \bar\xi_i \mid i \geq 3]\,,$$

see [**163**, Thm. A] or Proposition 16.6 of [**9**, Part III]. The Adams spectral sequence

$$E_2^{s,t}(ko) = \mathrm{Ext}_{A_*}^{s,t}(\mathbb{F}_2, H_*(ko)) \Longrightarrow_s \pi_{t-s}(ko)_2^\wedge$$

is an algebra spectral sequence with $E_2$-term

$$\begin{aligned} ko^{*,*} &= \mathrm{Ext}_{A_*}^{*,*}(\mathbb{F}_2, A_* \,\square_{A(1)_*}\, \mathbb{F}_2) \cong \mathrm{Ext}_{A(1)_*}^{*,*}(\mathbb{F}_2, \mathbb{F}_2) \\ &= \mathbb{F}_2[h_0, h_1, v, w_1]/(h_0h_1, h_1^3, h_1v, v^2 - h_0^2w_1)\,. \end{aligned}$$

See Figure 2.3. The classes $h_0$ and $h_1$ in $(t-s,s)$-bidegrees $(0,1)$ and $(1,1)$ are dual to $Sq^1$ and $Sq^2$, respectively. The Adams spectral sequence collapses at the $E_2$-term, and converges to

$$\pi_*(ko)_2^\wedge = \mathbb{Z}_2[\eta, A, B]/(2\eta, \eta^3, \eta A, A^2 - 4B)\,,$$

where $\eta$, $A$ and $B$, in topological degrees 1, 4 and 8, are detected by $h_1$, $v$ and $w_1$, respectively. By real Bott periodicity, $\Sigma^8 ko$ is equivalent to the 7-connected cover *bstring* of real $K$-theory, and $\pi_*(ko) = \mathbb{Z}[\eta, A, B]/(2\eta, \eta^3, \eta A, A^2 - 4B)$, before 2-adic completion.

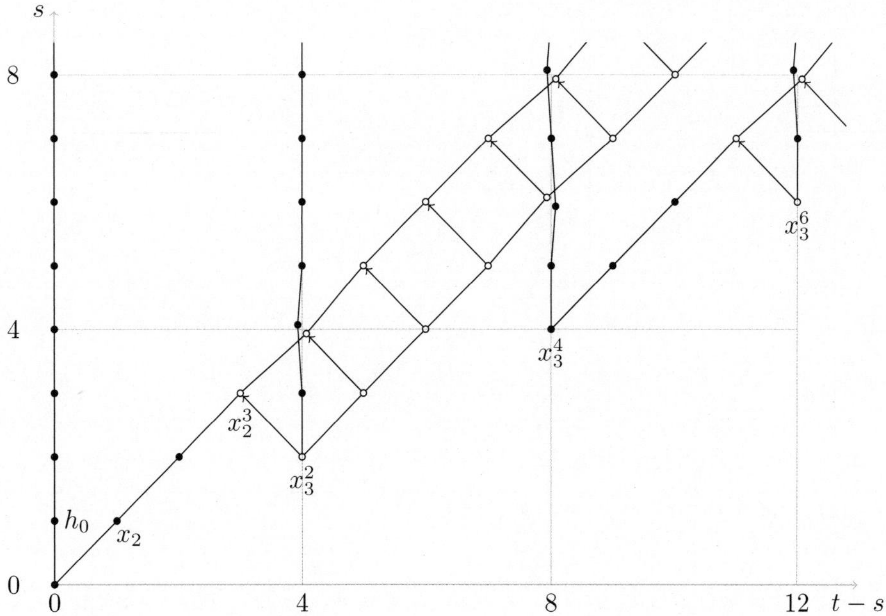


FIGURE 2.1. $(E_1, d_1)$-term of Davis–Mahowald spectral sequence for $A(1)$

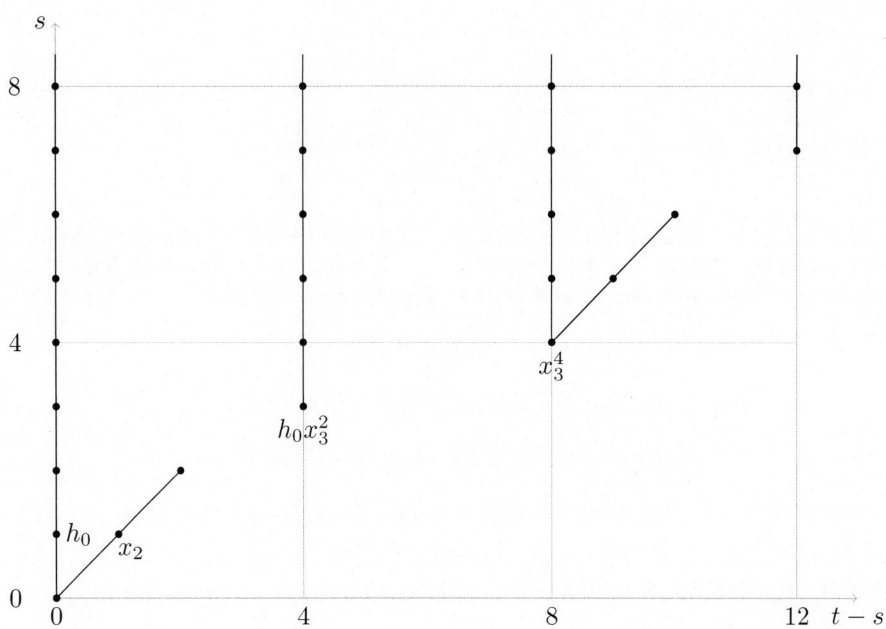


FIGURE 2.2. $E_2 = E_\infty$-term of Davis–Mahowald spectral sequence for $A(1)$

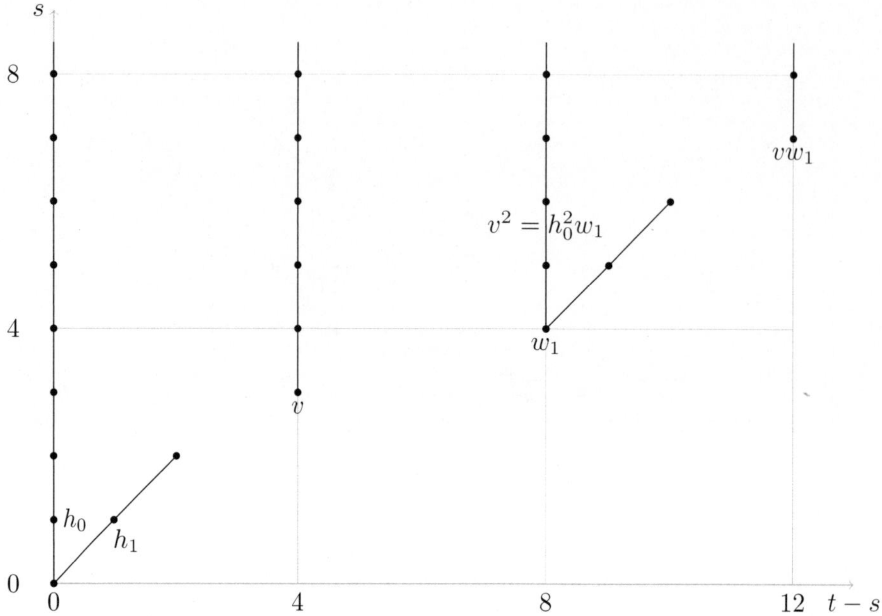


FIGURE 2.3. $E_2$-term $ko^{*,*} = \mathrm{Ext}^{*,*}_{A(1)}(\mathbb{F}_2, \mathbb{F}_2)$ of Adams spectral sequence for $ko$

EXAMPLE 2.31. There is a tower of $ko$-modules

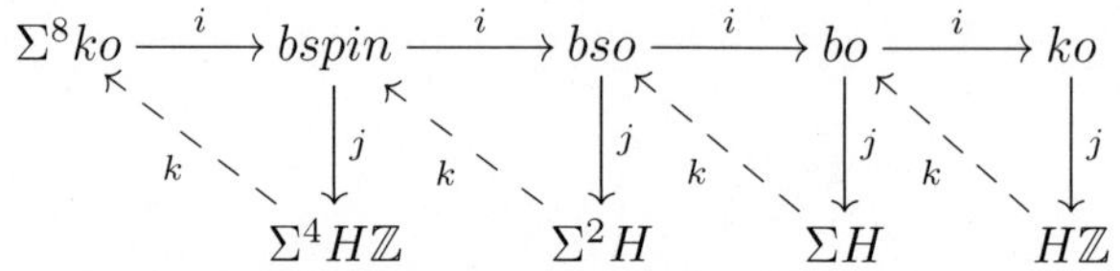


relating the 0-, 1- and 3-connected covers $bo$, $bso$ and $bspin$ of real $K$-theory. The dashed arrows represent maps of degree $-1$. The induced long exact sequences in cohomology break up into short exact sequences of $A$-modules

$$0 \to \Sigma H^*(bo) \xrightarrow{k^*} H^*(H\mathbb{Z}) \xrightarrow{j^*} H^*(ko) \to 0$$

$$0 \to \Sigma H^*(bso) \xrightarrow{k^*} \Sigma H^*(H) \xrightarrow{j^*} H^*(bo) \to 0$$

$$0 \to \Sigma H^*(bspin) \xrightarrow{k^*} \Sigma^2 H^*(H) \xrightarrow{j^*} H^*(bso) \to 0$$

$$0 \to \Sigma H^*(\Sigma^8 ko) \xrightarrow{k^*} \Sigma^4 H^*(H\mathbb{Z}) \xrightarrow{j^*} H^*(bspin) \to 0\,,$$

with $H^*(H) = A$ and $H^*(H\mathbb{Z}) = A/A(Sq^1) = A/\!/A(0)$. These are induced up along $A(1) \subset A$ from the following short exact sequences of $A(1)$-modules

$$0 \to \Sigma^2 A(1)/A(1)(Sq^2) \xrightarrow{Sq^2} A(1)/\!/A(0) \longrightarrow \mathbb{F}_2 \to 0$$

$$0 \to \Sigma^3 A(1)/A(1)(Sq^3) \xrightarrow{Sq^2} \Sigma A(1) \longrightarrow \Sigma A(1)/A(1)(Sq^2) \to 0$$

$$0 \to \Sigma^5 A(1)/A(1)(Sq^1, Sq^2Sq^3) \xrightarrow{Sq^3} \Sigma^2 A(1) \longrightarrow \Sigma^2 A(1)/A(1)(Sq^3) \to 0$$

$$0 \to \Sigma^9\mathbb{F}_2 \xrightarrow{Sq^2Sq^3} \Sigma^4 A(1)/\!/A(0) \longrightarrow \Sigma^4 A(1)/A(1)(Sq^1, Sq^2Sq^3) \to 0\,,$$

which can be spliced together as in the following diagram.

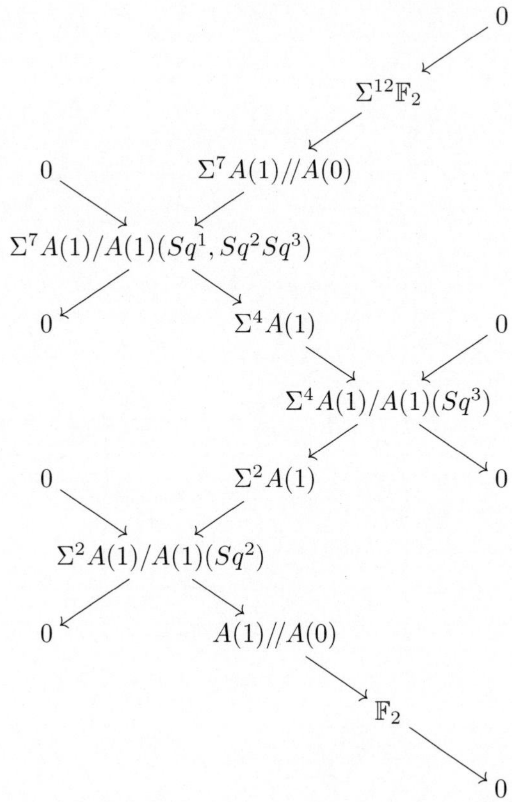


Hence

$$\begin{aligned} H^*(bo) &= \Sigma A/A(Sq^2) \\ H^*(bso) &= \Sigma^2 A/A(Sq^3) \\ H^*(bspin) &= \Sigma^4 A/A(Sq^1, Sq^2Sq^3)\,, \end{aligned}$$

as was proved by Stong [**163**, Thm. A]. The exactness of the underlying algebraic sequences of $A$-modules was established earlier by Toda in [**170**, Thm. I]. See also Figure 2.4, where the short and long solid arrows show the nonzero multiplications by $Sq^1$ and $Sq^2$, respectively, and the dotted arrows show the nonzero homomorphisms in the diagram above.

EXAMPLE 2.32. The connective quaternionic $K$-theory spectrum $ksp$ is the $K$-theory spectrum of a permutative topological category $\mathcal{GL}(\mathbb{H})$ [**121**, Ex. VI.5.4] equivalent to the symmetric monoidal topological category of finite-dimensional (right) quaternionic vector spaces, with respect to their usual direct sum. The tensor product of real and quaternionic vector spaces makes $ksp$ a $ko$-module spectrum. By real Bott periodicity, $ksp$ satisfies $\Sigma^4 ksp \simeq bspin$. Hence

$$H^*(ksp) = A/A(Sq^1, Sq^2Sq^3)$$

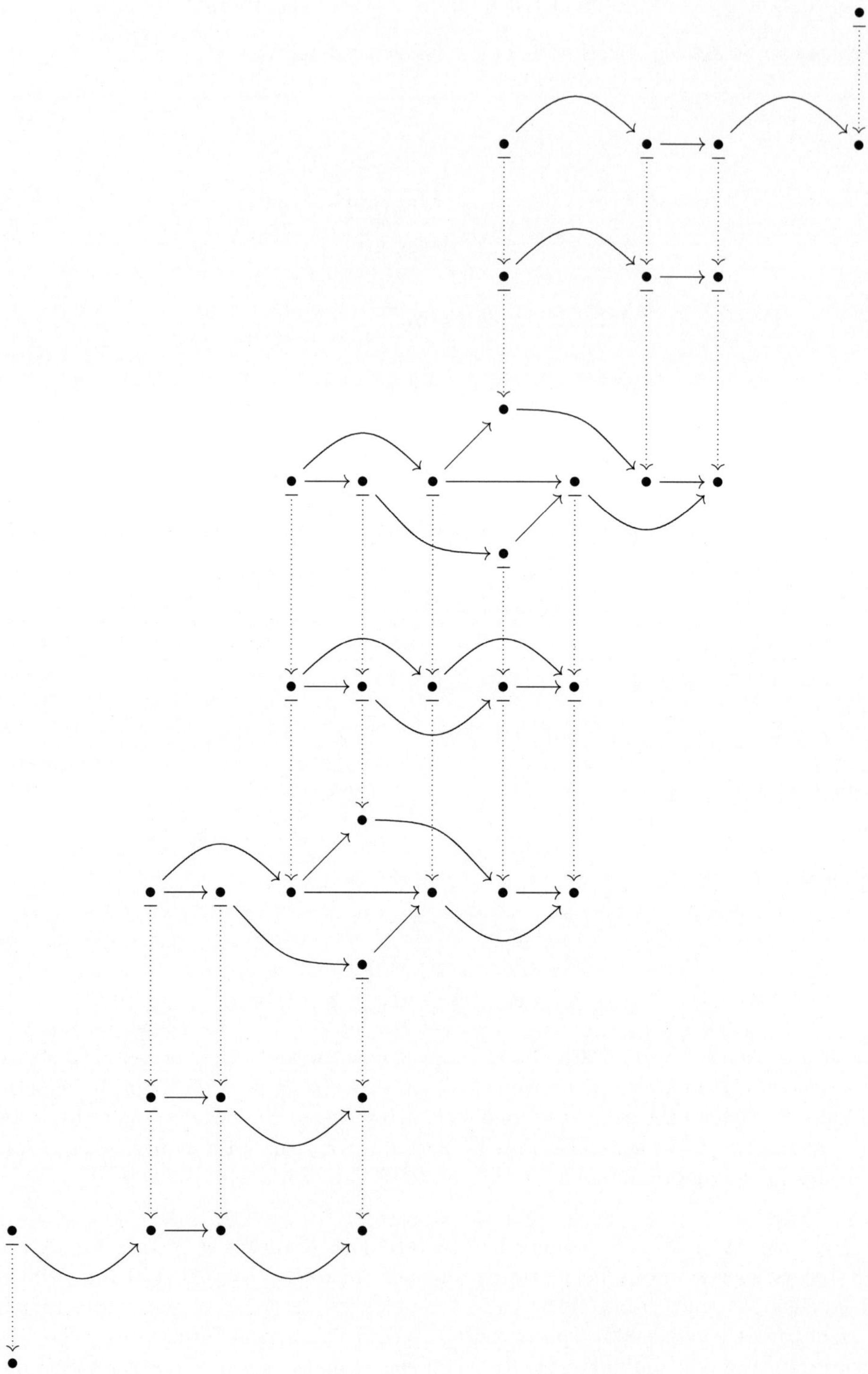

FIGURE 2.4. Spliced $A(1)$-extensions

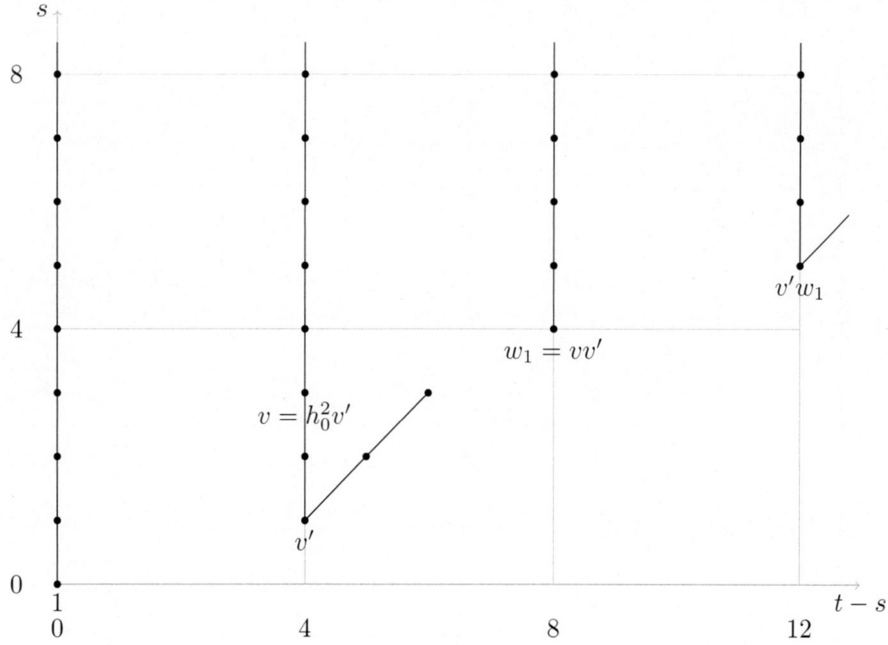


FIGURE 2.5. $E_2$-term $ksp^{*,*} = \mathrm{Ext}^{*,*}_{A(1)}(\mathbb{F}_2\{1, Sq^2, Sq^3\}, \mathbb{F}_2)$ of Adams spectral sequence for $ksp$

is induced up along $A(1) \subset A$ from

$$A(1)/A(1)(Sq^1, Sq^2Sq^3) = \mathbb{F}_2\{1, Sq^2, Sq^3\}.$$

Dually,

$$H_*(ksp) = A_* \,\square_{A(1)_*}\, \mathbb{F}_2\{1, \xi_1^2, \bar{\xi}_2\}.$$

The Adams spectral sequence

$$E_2^{s,t}(ksp) = \mathrm{Ext}^{s,t}_{A_*}(\mathbb{F}_2, H_*(ksp)) \Longrightarrow_s \pi_{t-s}(ksp)^\wedge_2$$

is a module spectral sequence over the Adams spectral sequence for $ko$, with $E_2$-term

$$\begin{aligned} ksp^{*,*} &= \mathrm{Ext}^{*,*}_{A_*}(\mathbb{F}_2, A_* \,\square_{A(1)_*}\, \mathbb{F}_2\{1, \xi_1^2, \bar{\xi}_2\}) \\ &\cong \mathrm{Ext}^{*,*}_{A(1)_*}(\mathbb{F}_2, \mathbb{F}_2\{1, \xi_1^2, \bar{\xi}_2\}) \\ &= \mathbb{F}_2[w_1]\{h_0^i, h_0^i v', h_1 v', h_1^2 v' \mid i \geq 0\} \\ &= ko^{*,*}\{1, v'\}/(h_1 \cdot 1, v \cdot 1 - h_0^2 \cdot v', v \cdot v' - w_1 \cdot 1). \end{aligned}$$

Here $v'$ has $(t-s, s)$-bidegree $|v'| = (4, 1)$, see Figure 2.5. (This can be verified using the Davis–Mahowald spectral sequence for $A(1)_* \to A(0)_*$ with coefficients in $\mathbb{F}_2\{1, \xi_1^2, \bar{\xi}_2\}$, which we leave as an exercise for the interested reader.) The Adams spectral sequence collapses at the $E_2$-term, and converges to

$$\pi_*(ksp)^\wedge_2 = \pi_*(ko)^\wedge_2\{1, A'\}/(\eta \cdot 1, A \cdot 1 - 4 \cdot A', A \cdot A' - B \cdot 1),$$

where 1 and $A'$ are detected by 1 and $v'$, respectively.

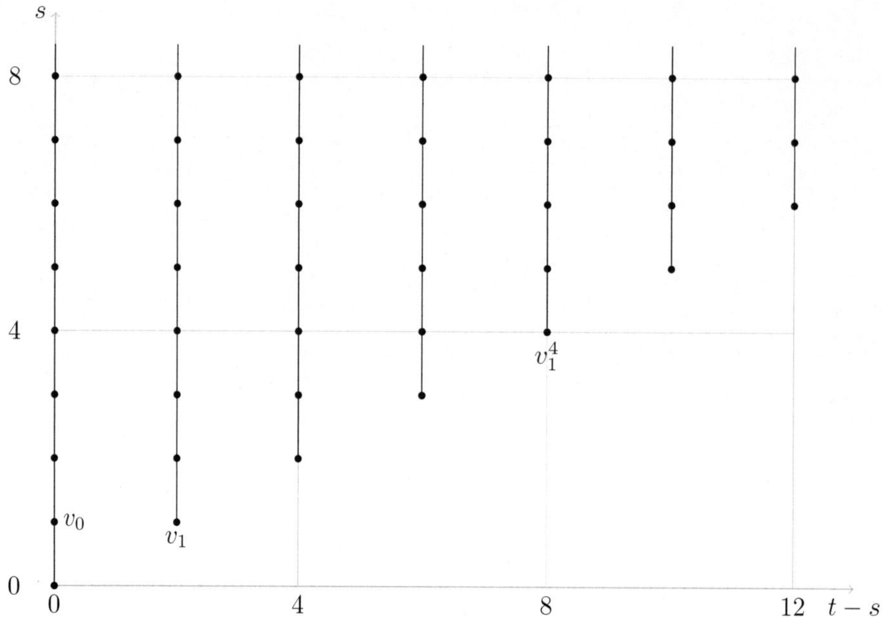


FIGURE 2.6. $E_2$-term $ku^{*,*} = \mathrm{Ext}^{*,*}_{E(1)}(\mathbb{F}_2, \mathbb{F}_2)$ of Adams spectral sequence for $ku$

EXAMPLE 2.33. The connective complex $K$-theory spectrum $ku$ is the $K$-theory spectrum of a bipermutative topological category $\mathcal{GL}(\mathbb{C})$ [**121**, Ex. VI.5.4] equivalent to the symmetric bimonoidal topological category of finite dimensional complex vector spaces, with respect to their usual direct sum and tensor product. It is an $E_\infty$ ring spectrum with mod 2 cohomology

$$H^*(ku) = A/A(Q_0, Q_1) = A \otimes_{E(1)} \mathbb{F}_2 = A//E(1)$$

and mod 2 homology

$$H_*(ku) = A_* \,\Box_{E(1)_*}\, \mathbb{F}_2 = \mathbb{F}_2[\bar\xi_1^2, \bar\xi_2^2, \bar\xi_i \mid i \geq 3]\,,$$

see [**4**, Lem. 4], [**163**, Thm. B] or Proposition 16.6 of [**9**, Part III]. Here $E(1) = E(Q_0, Q_1)$ denotes the sub Hopf algebra of $A(1) \subset A$ generated by the Milnor primitives $Q_0 = Sq^1$ and $Q_1 = [Sq^2, Sq^1]$. The dual Hopf algebra is $E(1)_* = E(\xi_1, \bar\xi_2)$, where $\bar\xi_2 = \xi_2 + \xi_1^3 \equiv \xi_2$. The Adams spectral sequence

$$E_2^{s,t}(ku) = \mathrm{Ext}^{s,t}_{A_*}(\mathbb{F}_2, H_*(ku)) \Longrightarrow_s \pi_{t-s}(ku)^\wedge_2$$

is an algebra spectral sequence with $E_2$-term

$$\begin{aligned} ku^{*,*} &= \mathrm{Ext}^{*,*}_{A_*}(\mathbb{F}_2, A_* \,\Box_{E(1)_*}\, \mathbb{F}_2) \\ &\cong \mathrm{Ext}^{*,*}_{E(1)_*}(\mathbb{F}_2, \mathbb{F}_2) = \mathbb{F}_2[v_0, v_1]\,. \end{aligned}$$

See Figure 2.6. The classes $v_0$ and $v_1$ in $(t-s, s)$-bidegrees $(0, 1)$ and $(2, 1)$ are dual to $Q_0$ and $Q_1$, respectively. The spectral sequence collapses at the $E_2$-term, and converges to

$$\pi_*(ku)^\wedge_2 = \mathbb{Z}_2[v_1]\,.$$

By complex Bott periodicity, $\Sigma^2 ku \simeq bu$ is equivalent to the 1-connected cover of complex $K$-theory. Hence $\pi_*(ku) = \mathbb{Z}[u]$ integrally, with $u$ in degree 2 mapping to $v_1$ under 2-completion. The homotopy cofiber sequence

$$bu \xrightarrow{i} ku \xrightarrow{j} H\mathbb{Z} \xrightarrow{k} \Sigma bu$$

induces a long exact sequence in cohomology, which breaks up into a short exact sequence of $A$-modules

$$0 \to \Sigma H^*(bu) \xrightarrow{k^*} H^*(H\mathbb{Z}) \xrightarrow{j^*} H^*(ku) \to 0\,,$$

induced up along $E(1) \subset A$ from the short exact sequence

$$0 \to \Sigma^3 \mathbb{F}_2 \longrightarrow E(1)//A(0) \longrightarrow \mathbb{F}_2 \to 0$$

of $E(1)$-modules.

CHAPTER 3

# Ext over $A(2)$

We use the Davis–Mahowald spectral sequence from Chapter 2 to calculate $\mathrm{Ext}_{A(2)}(\mathbb{F}_2, \mathbb{F}_2)$ as a free module over $\mathbb{F}_2[w_1, w_2]$, and then combine this result with the `ext`-calculations of Chapter 1 to verify the presentation given by Shimada–Iwai [**155**] of $\mathrm{Ext}_{A(2)}(\mathbb{F}_2, \mathbb{F}_2)$ as a bigraded commutative algebra with 13 generators and 54 relations. We also obtain a Gröbner basis for the ideal of relations, which allows for algorithmic computations in this algebra. Finally, we give an additive decomposition of $\mathrm{Ext}_{A(2)}(\mathbb{F}_2, \mathbb{F}_2)$ as a direct sum of cyclic $\mathbb{F}_2[g, w_1, w_2]$-modules.

## 3.1. The Davis–Mahowald $E_1$-term for $A(2)$

The mod 2 Steenrod algebra $A$ is the connected $\mathbb{F}_2$-algebra generated by the Steenrod squaring operations $Sq^i$ for $i \geq 1$, subject to the Adem relations, and graded by $|Sq^i| = i$. It becomes a cocommutative Hopf algebra when equipped with the coproduct $\psi(Sq^k) = \sum_{i+j=k} Sq^i \otimes Sq^j$, where $Sq^0$ is interpreted as 1.

The dual Steenrod algebra $A_* = \mathrm{Hom}(A, \mathbb{F}_2)$ is the connected commutative Hopf algebra $A_* = \mathbb{F}_2[\xi_i \mid i \geq 1]$ with $|\xi_i| = 2^i - 1$ and coproduct $\psi(\xi_k) = \sum_{i+j=k} \xi_i^{2^j} \otimes \xi_j$, where $\xi_0$ is interpreted as 1. The canonical conjugation $\chi\colon A_* \to A_*$ satisfies $\sum_{i+j=k} \xi_i^{2^j} \chi(\xi_j) = 0$ for each $k \geq 1$. We let $\bar{\xi}_i = \chi(\xi_i)$ denote the conjugate generators of $A_*$. This leads to the alternative presentation

$$A_* = \mathbb{F}_2[\bar{\xi}_i \mid i \geq 1]$$

of the dual Steenrod algebra, with $|\bar{\xi}_i| = 2^i - 1$ and

$$\psi(\bar{\xi}_k) = \sum_{i+j=k} \bar{\xi}_i \otimes \bar{\xi}_j^{2^i} .$$

Again $\bar{\xi}_0$ is interpreted as 1. We will write $\xi_1$ in place of $\bar{\xi}_1 = -\xi_1$, since we are working over $\mathbb{F}_2$.

Consider the subalgebras $A(2) = \langle Sq^1, Sq^2, Sq^4 \rangle$ and $A(1) = \langle Sq^1, Sq^2 \rangle$ of the Steenrod algebra, generated by $Sq^1$, $Sq^2$ and $Sq^4$, and by $Sq^1$ and $Sq^2$, respectively. These are connected, cocommutative sub Hopf algebras of $A$, with dual Hopf algebras

$$A(2)_* = \mathbb{F}_2[\xi_1, \bar{\xi}_2, \bar{\xi}_3]/(\xi_1^8, \bar{\xi}_2^4, \bar{\xi}_3^2)$$

and

$$A(1)_* = \mathbb{F}_2[\xi_1, \bar{\xi}_2]/(\xi_1^4, \bar{\xi}_2^2) .$$

The coproduct in $A(2)_*$ is given by

$$\begin{aligned}
\psi(\xi_1) &= 1 \otimes \xi_1 + \xi_1 \otimes 1 \\
\psi(\bar{\xi}_2) &= 1 \otimes \bar{\xi}_2 + \xi_1 \otimes \xi_1^2 + \bar{\xi}_2 \otimes 1 \\
\psi(\bar{\xi}_3) &= 1 \otimes \bar{\xi}_3 + \xi_1 \otimes \bar{\xi}_2^2 + \bar{\xi}_2 \otimes \xi_1^4 + \bar{\xi}_3 \otimes 1 ,
\end{aligned}$$

so that

$$(A(2)//A(1))_* = A(2)_* \square_{A(1)_*} \mathbb{F}_2 = E(\xi_1^4, \bar{\xi}_2^2, \bar{\xi}_3)$$

as a sub $A(2)_*$-comodule algebra of $A(2)_*$.

DEFINITION 3.1. In this chapter, let $R^* = \mathbb{F}_2[x_4, x_6, x_7]$ be the graded $A(2)_*$-comodule algebra with coaction given by

$$\begin{aligned}\nu(x_4) &= 1 \otimes x_4 \\ \nu(x_6) &= 1 \otimes x_6 + \xi_1^2 \otimes x_4 \\ \nu(x_7) &= 1 \otimes x_7 + \xi_1 \otimes x_6 + \bar{\xi}_2 \otimes x_4 .\end{aligned}$$

We assign internal degree $i$ and cohomological degree 1 to $x_i$, for $i = 4$, 6 and 7, and give

$$(A(2)//A(1))_* \otimes R^* = E(\xi_1^4, \bar{\xi}_2^2, \bar{\xi}_3) \otimes \mathbb{F}_2[x_4, x_6, x_7]$$

the diagonal $A(2)_*$-comodule structure. It becomes a differential graded $A(2)_*$-comodule algebra with the differential $\delta$ given by

$$\begin{aligned}\delta(\xi_1^4) &= x_4 \\ \delta(\bar{\xi}_2^2) &= x_6 \\ \delta(\bar{\xi}_3) &= x_7 .\end{aligned}$$

It follows that $\delta(x_4) = 0$, $\delta(x_6) = 0$ and $\delta(x_7) = 0$. The underlying cochain complex

$$0 \to \mathbb{F}_2 \xrightarrow{\eta} E(\xi_1^4, \bar{\xi}_2^2, \bar{\xi}_3) \otimes R^0 \xrightarrow{\delta^0} E(\xi_1^4, \bar{\xi}_2^2, \bar{\xi}_3) \otimes R^1 \xrightarrow{\delta^1} E(\xi_1^4, \bar{\xi}_2^2, \bar{\xi}_3) \otimes R^2 \longrightarrow \dots$$

is exact. Here

$$R^\sigma = \mathbb{F}_2\{x_4^i x_6^j x_7^k \mid i + j + k = \sigma\}$$

is the $A(2)_*$-comodule of homogeneous polynomials in $\mathbb{F}_2[x_4, x_6, x_7]$ of (cohomological) degree $\sigma$.

The Davis–Mahowald spectral sequence for $\pi \colon A(2)_* \to A(1)_*$ with coefficients in $\mathbb{F}_2$ is an algebra spectral sequence

$$E_1^{\sigma,s,t} = \mathrm{Ext}_{A(1)_*}^{s-\sigma,t}(\mathbb{F}_2, R^\sigma) \Longrightarrow_\sigma \mathrm{Ext}_{A(2)_*}^{s,t}(\mathbb{F}_2, \mathbb{F}_2) \tag{3.1}$$

converging strongly to the $E_2$-term $\mathrm{Ext}_{A(2)_*}^{s,t}(\mathbb{F}_2, \mathbb{F}_2) = \mathrm{Ext}_{A(2)}^{s,t}(\mathbb{F}_2, \mathbb{F}_2)$ of the mod 2 Adams spectral sequence for *tmf*. The $A(1)_*$-coaction on $R^*$ is given by the composite

$$R^* \xrightarrow{\nu} A(2)_* \otimes R^* \xrightarrow{\pi \otimes 1} A(1)_* \otimes R^* .$$

Note that $x_4$, $x_6^2$, $x_6^3 + x_4 x_7^2$ and $x_7^4$ are $A(1)_*$-comodule primitive, and that $R^*$ is free as a module over $\mathbb{F}_2[x_7^4]$. We obtain an extension of graded $A(1)_*$-comodule algebras

$$\mathbb{F}_2[x_7^4] \longrightarrow R^* \longrightarrow \bar{R}^*$$

where, by definition,

$$\bar{R}^* = R^* \otimes_{\mathbb{F}_2[x_7^4]} \mathbb{F}_2 = R^*/(x_7^4) = \mathbb{F}_2[x_4, x_6, x_7]/(x_7^4) .$$

Thus

$$\bar{R}^\sigma = \mathbb{F}_2\{x_4^i x_6^j x_7^k \mid i + j + k = \sigma,\ 0 \le k \le 3\} .$$

Applying $\mathrm{Ext}_{A(1)_*}(\mathbb{F}_2, -)$ yields an extension of trigraded algebras

$$\mathbb{F}_2[x_7^4] \longrightarrow E_1^{*,*,*} \longrightarrow \bar{E}_1^{*,*,*} .$$

Here $E_1^{*,*,*}$ is free as a module over $\mathbb{F}_2[x_7^4]$, and

$$\bar{E}_1^{\sigma,s,t} = \mathrm{Ext}_{A(1)_*}^{s-\sigma,t}(\mathbb{F}_2, \bar{R}^\sigma)\,.$$

In the following sections we shall express $\mathrm{Ext}_{A(1)_*}(\mathbb{F}_2, \bar{R}^\sigma)$ by means of the Adams $E_2$-terms for spectra $ko$, $ksp$ and $ku\langle\sigma\rangle$, cf. Proposition 3.26. Thereafter we shall use these expressions to calculate the Davis–Mahowald $d_1$-differentials, leading to the description of the $E_2$-term given in Proposition 3.33. This turns out to also be the $E_\infty$-term of this Davis–Mahowald spectral sequence.

DEFINITION 3.2. Let $S\colon \bar{R}^* \to R^*$ be the section to $R^* \to \bar{R}^*$ given by

$$S(x_4^i x_6^j x_7^k) = x_4^i x_6^j x_7^k$$

for $0 \le k \le 3$. It is an $\mathbb{F}_2[x_4, x_6^2]$-linear $A(1)_*$-comodule homomorphism.

Using $S$ and multiplication by powers of $x_7^4$ we obtain finite $\mathbb{F}_2[x_4, x_6^2]$-linear sum decompositions

$$R^\sigma \cong \bar{R}^\sigma \oplus \bar{R}^{\sigma-4}\{x_7^4\} \oplus \bar{R}^{\sigma-8}\{x_7^8\} \oplus \dots$$

of $A(1)_*$-comodules. Applying $\mathrm{Ext}_{A(1)_*}(\mathbb{F}_2, -)$ we obtain finite $\mathbb{F}_2[x_4, x_6^2]$-linear sum decompositions

$$E_1^{\sigma,*,*} \cong \bar{E}_1^{\sigma,*,*} \oplus \bar{E}_1^{\sigma-4,*,*}\{x_7^4\} \oplus \bar{E}_1^{\sigma-8,*,*}\{x_7^8\} \oplus \dots$$

of $ko^{*,*}$-modules.

EXAMPLE 3.3. For $0 \le \sigma \le 3$ the $A(1)_*$-modules $R^\sigma = \bar{R}^\sigma$ can be depicted as follows, with a short line connecting $x$ and $y$ when $\nu(x)$ contains $\xi_1 \otimes y$, and a longer curve connecting $x$ and $z$ when $\nu(x)$ contains $\xi_1^2 \otimes z$. These correspond to nontrivial operations $Sq^1$ and $Sq^2$, respectively, in the dual $A(1)$-modules $N_\sigma$.

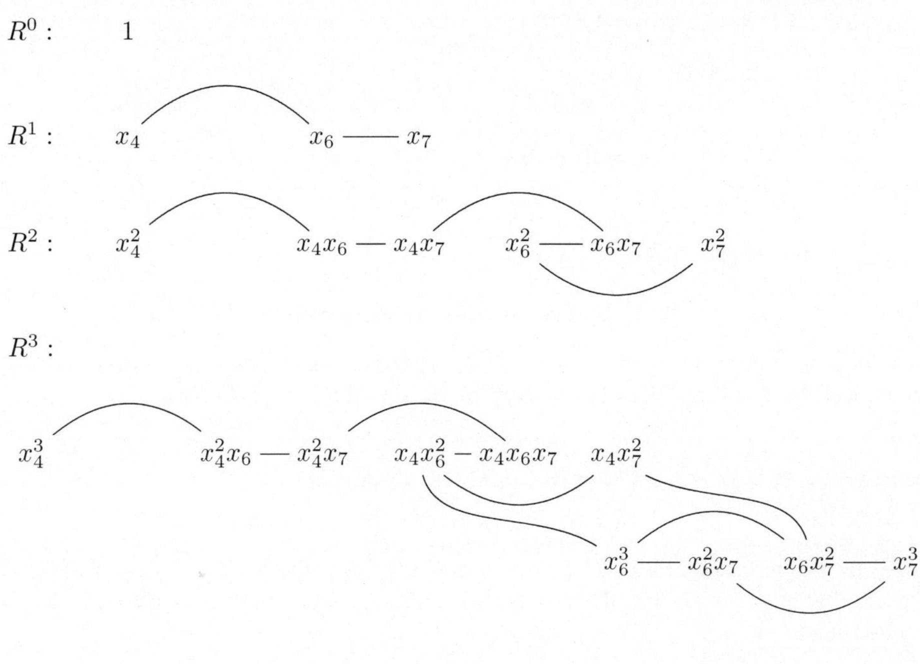


LEMMA 3.4. *$R^0 = \mathbb{F}_2$ is dual to $N_0 = \mathbb{F}_2$, and $R^1 = \mathbb{F}_2\{x_4, x_6, x_7\}$ is dual to $N_1 \cong \Sigma^4 A(1)/A(1)(Sq^1, Sq^2Sq^3)$.*

PROOF. The $A(1)_*$-comodule

is dual to $A(1)/A(1)(Sq^1, Sq^2Sq^3)$. $\square$

LEMMA 3.5. *For each $\sigma \geq 3$ there is a short exact sequence of $A(1)_*$-comodules*

$$0 \to \Sigma^4 \bar{R}^{\sigma-1} \xrightarrow{x_4} \bar{R}^\sigma \longrightarrow \operatorname{cok}(x_4) \to 0\,,$$

*where* $\operatorname{cok}(x_4) = \mathbb{F}_2\{x_6^\sigma, x_6^{\sigma-1}x_7, x_6^{\sigma-2}x_7^2, x_6^{\sigma-3}x_7^3\}$ *is dual to* $\Sigma^{6\sigma}A(1)//E(Q_1)$.

PROOF. The $A(1)_*$-comodule

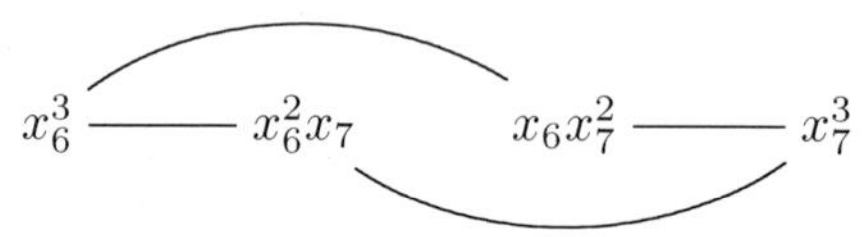


is dual to $\Sigma^{18}A(1)//E(Q_1)$, where $Q_1 = [Sq^2, Sq^1]$ is the Milnor primitive. $\square$

LEMMA 3.6. *For each $\sigma \geq 4$ there is a short exact sequence of $A(1)_*$-comodules*

$$0 \to \Sigma^{12}\bar{R}^{\sigma-2} \xrightarrow{x_6^2} \bar{R}^\sigma \longrightarrow \operatorname{cok}(x_6^2) \to 0\,,$$

*where* $\operatorname{cok}(x_6^2) = \mathbb{F}_2\{x_4^i x_6^j x_7^k \mid i+j+k = \sigma,\ 0 \leq j \leq 1,\ 0 \leq k \leq 3\}$ *is dual to the direct sum* $\Sigma^{4\sigma}A(1)//A(0) \oplus \Sigma^{4\sigma+6}A(1)//A(0)$.

PROOF. The $A(1)_*$-comodule

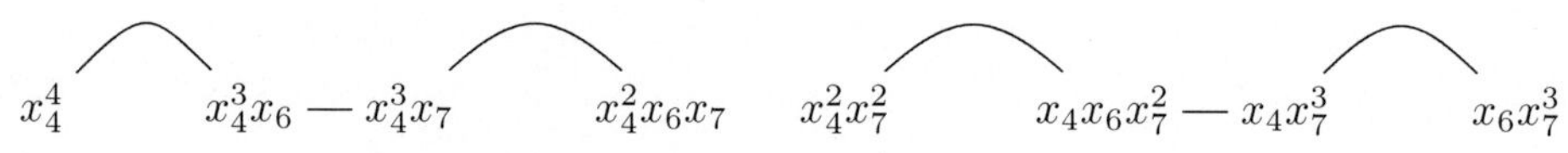


is dual to $\Sigma^{16}A(1)//A(0) \oplus \Sigma^{22}A(1)//A(0)$. $\square$

### 3.2. Syzygies and Adams covers

We continue to write $E(1) = E(Q_0, Q_1)$ for the sub Hopf algebra of $A(1) \subset A$ generated by $Q_0 = Sq^1$ and $Q_1 = [Sq^2, Sq^1]$. The dual Hopf algebra is

$$E(1)_* = E(\xi_1, \bar{\xi}_2)\,,$$

where $\bar{\xi}_2 = \xi_2 + \xi_1^3 \equiv \xi_2$. There is a minimal resolution

$$\eta\colon \mathbb{F}_2 \xrightarrow{\sim} E(1)_* \otimes \mathbb{F}_2[v_0, v_1]$$

of $\mathbb{F}_2$ by a differential graded $E(1)_*$-comodule algebra, where $\delta(\xi_1) = v_0$ and $\delta(\bar{\xi}_2) = v_1$, so that $\delta(v_0) = 0$ and $\delta(v_1) = 0$. The underlying cochain complex of $E(1)_*$-comodules

$$0 \to \mathbb{F}_2 \xrightarrow{\eta} E(1)_*\{1\} \xrightarrow{\delta^0} E(1)_*\{v_0, v_1\} \xrightarrow{\delta^1} E(1)_*\{v_0^2, v_0v_1, v_1^2\} \xrightarrow{\delta^2} \dots \tag{3.2}$$

is exact.

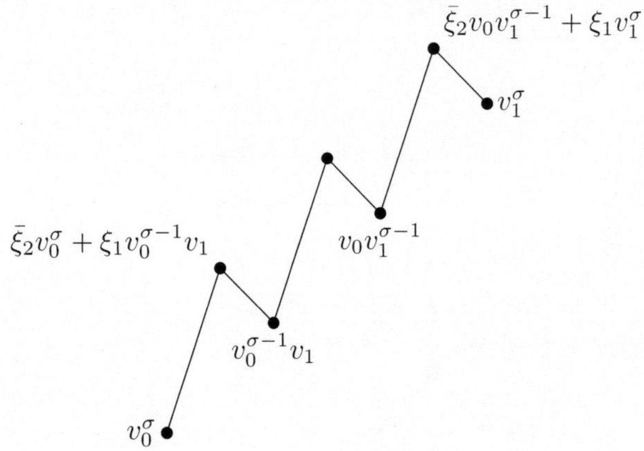


FIGURE 3.1. The syzygy $\Omega^{\sigma}_{E(1)_*}(\mathbb{F}_2)$ for $\sigma = 3$

DEFINITION 3.7. We write

$$\begin{aligned}\Omega^{\sigma}_{E(1)_*}(\mathbb{F}_2) &= \ker(\delta^{\sigma}) \\ &= \mathbb{F}_2\{v_0^{\sigma}\,,\, \bar{\xi}_2 v_0^{\sigma} + \xi_1 v_0^{\sigma-1} v_1\,,\, v_0^{\sigma-1} v_1\,,\, \dots\,,\, v_0 v_1^{\sigma-1}\,,\, \bar{\xi}_2 v_0 v_1^{\sigma-1} + \xi_1 v_1^{\sigma}\,,\, v_1^{\sigma}\}\,,\end{aligned}$$

to denote the $\sigma$-th $E(1)_*$-comodule syzygy of $\mathbb{F}_2$.

EXAMPLE 3.8. The syzygy $\Omega^3_{E(1)_*}(\mathbb{F}_2)$ is illustrated in Figure 3.1. A short line connects $x$ and $y$ when $\nu(x)$ contains $\xi_1 \otimes y$, and a long line connects $x$ and $z$ when $\nu(x)$ contains $\bar{\xi}_2 \otimes z$. These correspond to nontrivial operations $Q_0$ and $Q_1$, respectively, in the dual $E(1)$-module $\Omega^3_{E(1)}(\mathbb{F}_2)$.

Applying $A(1)_* \,\square_{E(1)_*} (-)$ to (3.2) we obtain an exact cochain complex of $A(1)_*$-comodules

$$0 \to E(\xi_1^2) \xrightarrow{1\otimes\eta} A(1)_*\{1\} \xrightarrow{1\otimes\delta^0} A(1)_*\{v_0, v_1\} \xrightarrow{1\otimes\delta^1} A(1)_*\{v_0^2, v_0v_1, v_1^2\} \xrightarrow{1\otimes\delta^2} \dots .$$

Here

$$A(1)_* \,\square_{E(1)_*} \mathbb{F}_2 = (A(1)/\!/E(1))_* = E(\xi_1^2)$$

and

$$\begin{aligned}\Omega^{\sigma}_{A(1)_*}(E(\xi_1^2)) &= \ker(1\otimes\delta^{\sigma}) = A(1)_* \,\square_{E(1)_*} \Omega^{\sigma}_{E(1)_*}(\mathbb{F}_2) \\ &= \mathbb{F}_2\{v_0^{\sigma}\,,\, \xi_1^2 v_0^{\sigma}\,,\, \bar{\xi}_2 v_0^{\sigma} + \xi_1 v_0^{\sigma-1} v_1\,,\, \xi_1^2(\bar{\xi}_2 v_0^{\sigma} + \xi_1 v_0^{\sigma-1} v_1)\,,\, v_0^{\sigma-1} v_1\,,\, \xi_1^2 v_0^{\sigma-1} v_1\,, \\ &\qquad \dots\,,\, v_0 v_1^{\sigma-1}\,,\, \xi_1^2 v_0 v_1^{\sigma-1}\,,\, \bar{\xi}_2 v_0 v_1^{\sigma-1} + \xi_1 v_1^{\sigma}\,,\, \xi_1^2(\bar{\xi}_2 v_0 v_1^{\sigma-1} + \xi_1 v_1^{\sigma})\,,\, v_1^{\sigma}\,,\, \xi_1^2 v_1^{\sigma}\}\end{aligned}$$

is the $\sigma$-th $A(1)_*$-comodule syzygy of $E(\xi_1^2)$.

EXAMPLE 3.9. The syzygy $\Omega^3_{A(1)_*}(E(\xi_1^2))$ is illustrated in Figure 3.2. A short line connects $x$ and $y$ when $\nu(x)$ contains $\xi_1 \otimes y$, and a vertical line connects $x$ and $z$ when $\nu(x)$ contains $\xi_1^2 \otimes z$. These correspond to nontrivial operations $Sq^1$ and $Sq^2$, respectively, in the dual $A(1)$-module $\Omega^3_{A(1)}(\mathbb{F}_2\{1, Sq^2\})$.

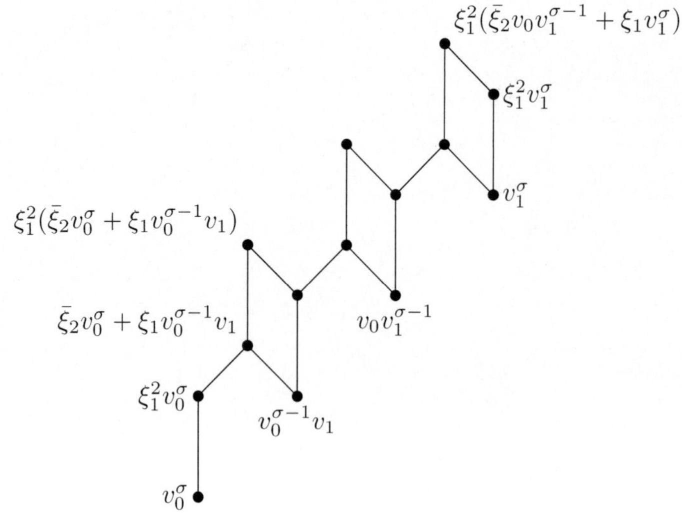


FIGURE 3.2. The syzygy $\Omega^\sigma_{A(1)_*}(E(\xi_1^2))$ for $\sigma = 3$

LEMMA 3.10. *For each $\sigma \geq 1$ there is a short exact sequence of $A(1)_*$-comodules*

$$0 \to \Sigma\Omega^{\sigma-1}_{A(1)_*}(E(\xi_1^2)) \xrightarrow{v_0} \Omega^\sigma_{A(1)_*}(E(\xi_1^2)) \longrightarrow \operatorname{cok}(v_0) \to 0\,,$$

*where* $\operatorname{cok}(v_0) = \mathbb{F}_2\{v_1^\sigma\,,\, \bar{\xi}_2 v_0 v_1^{\sigma-1} + \xi_1 v_1^\sigma\,,\, \xi_1^2 v_1^\sigma\,,\, \xi_1^2(\bar{\xi}_2 v_0 v_1^{\sigma-1} + \xi_1 v_1^\sigma)\}$ *is dual to* $\Sigma^{3\sigma} A(1)//E(Q_1)$.

PROOF. The $A(1)_*$-comodule

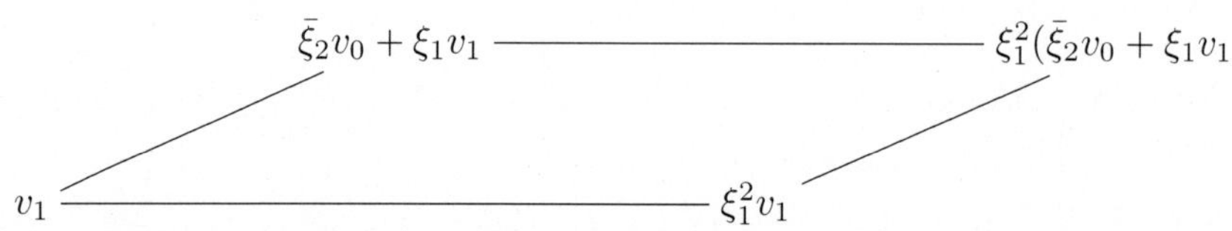


is dual to $\Sigma^3 A(1)//E(Q_1)$. □

REMARK 3.11. In the next section we shall see that $\Omega^\sigma_{A(1)_*}(E(\xi_1^2))$ is closely related to the $A(1)_*$-comodule $\bar{R}^\sigma$ from the previous section.

Recall the connective complex $K$-theory spectrum $ku$ from Section 2.6. There is a minimal Adams resolution (= Adams tower)

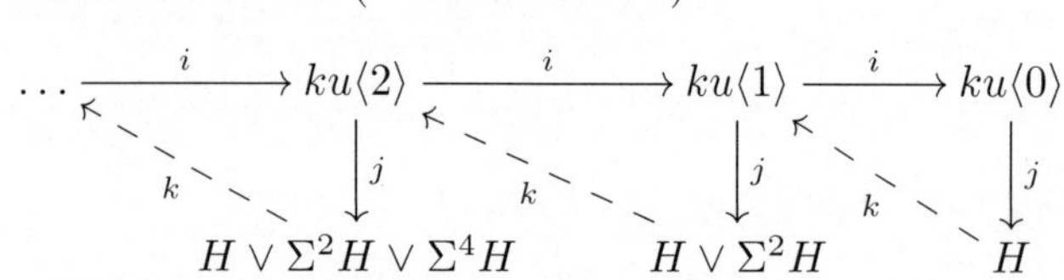


with $ku\langle 0\rangle = ku$, and there are short exact sequences

$$0 \to \pi_* ku\langle \sigma+1\rangle \xrightarrow{i} \pi_* ku\langle\sigma\rangle \xrightarrow{j} \mathbb{Z}/2\{v_0^\sigma, v_0^{\sigma-1}v_1, \dots, v_0 v_1^{\sigma-1}, v_1^\sigma\} \to 0$$

for each $\sigma \geq 0$.

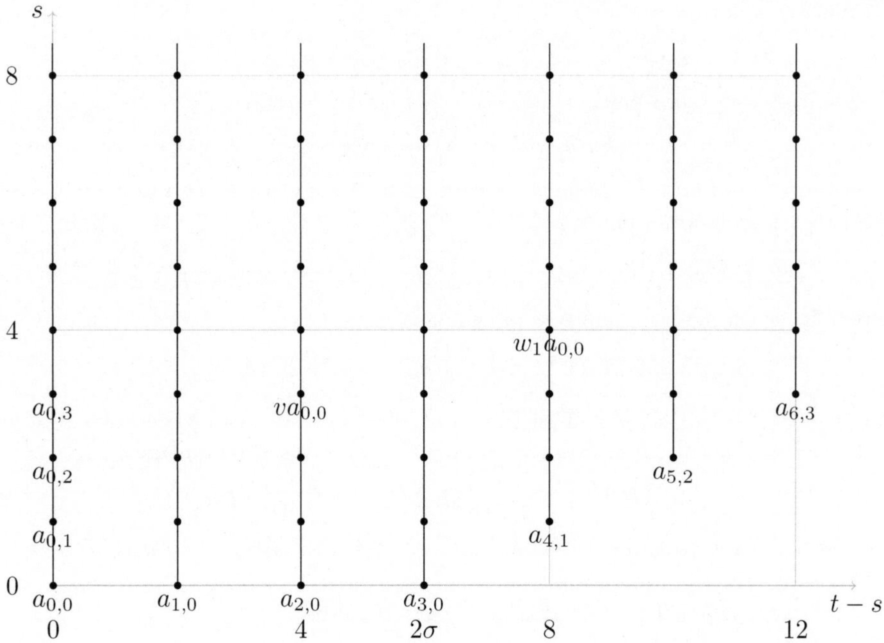


FIGURE 3.3. $E_2$-term $ku\langle\sigma\rangle^{*,*}$ of Adams spectral sequence for $ku\langle\sigma\rangle$ for $\sigma = 3$

DEFINITION 3.12. We call $ku\langle\sigma\rangle$ the $\sigma$-th Adams cover of $ku$.

In homology, the associated exact complex of $A_*$-comodules

$$0 \to H_*(ku) \xrightarrow{j_*} A_*\{1\} \xrightarrow{(jk)_*} A_*\{v_0, v_1\} \xrightarrow{(jk)_*} A_*\{v_0^2, v_0v_1, v_1^2\} \xrightarrow{(jk)_*} \dots$$

equals that obtained by applying $A_* \square_{E(1)_*} (-)$ to (3.2). Hence

$$\Sigma^\sigma H_*(ku\langle\sigma\rangle) = A_* \square_{E(1)_*} \Omega^\sigma_{E(1)_*}(\mathbb{F}_2)$$

is the $\sigma$-th $A_*$-comodule syzygy of $H_*(ku)$. The Adams spectral sequence

$$E_2^{s,t}(ku\langle\sigma\rangle) = \mathrm{Ext}^{s,t}_{A_*}(\mathbb{F}_2, H_*(ku\langle\sigma\rangle)) \Longrightarrow_s \pi_{t-s}(ku\langle\sigma\rangle)^\wedge_2$$

has $E_2$-term

$$\begin{aligned} ku\langle\sigma\rangle^{s,t} &= \mathrm{Ext}^{s,t+\sigma}_{A_*}(\mathbb{F}_2, A_* \square_{E(1)_*} \Omega^\sigma_{E(1)_*}(\mathbb{F}_2)) \\ &\cong \mathrm{Ext}^{s,t+\sigma}_{E(1)_*}(\mathbb{F}_2, \Omega^\sigma_{E(1)_*}(\mathbb{F}_2)) \\ &\cong \mathrm{Ext}^{s+\sigma,t+\sigma}_{E(1)_*}(\mathbb{F}_2, \mathbb{F}_2) = E_2^{s+\sigma,t+\sigma}(ku) \end{aligned}$$

for $s \geq 0$, hence appears as illustrated in Figure 3.3.

DEFINITION 3.13. For $s \geq 0$ and $0 \leq k \leq s + \sigma$ let

$$a_{k,s} \in E_2^{s,s+2k}(ku\langle\sigma\rangle)$$

be the generator in $(t-s, s)$-bidegree $(2k, s)$, corresponding to

$$v_0^{s+\sigma-k} v_1^k \in E_2^{s+\sigma,s+2k+\sigma}(ku)\,.$$

With this notation,

$$ku\langle\sigma\rangle^{*,*} = \mathbb{F}_2\{a_{k,s} \mid 0 \le k \le s+\sigma,\, s \ge 0\}\,.$$

The ring spectrum pairing $ku \wedge ku \to ku$ lifts to a pairing

$$ku\langle\sigma\rangle \wedge ku\langle\sigma'\rangle \longrightarrow ku\langle\sigma+\sigma'\rangle$$

for each $\sigma, \sigma' \ge 0$, and the induced pairing in homology equals (up to some suspensions) the $A_*$-comodule pairing

$$A_* \,\square_{E(1)_*}\, \Omega^{\sigma}_{E(1)_*}(\mathbb{F}_2) \otimes A_* \,\square_{E(1)_*}\, \Omega^{\sigma'}_{E(1)_*}(\mathbb{F}_2) \longrightarrow A_* \,\square_{E(1)_*}\, \Omega^{\sigma+\sigma'}_{E(1)_*}(\mathbb{F}_2)$$

derived from the $E(1)_*$-comodule pairing

$$\Omega^{\sigma}_{E(1)_*}(\mathbb{F}_2) \otimes \Omega^{\sigma'}_{E(1)_*}(\mathbb{F}_2) \longrightarrow \Omega^{\sigma+\sigma'}_{E(1)_*}(\mathbb{F}_2)$$

obtained by restricting the multiplication on $E(1)_* \otimes \mathbb{F}_2[v_0, v_1]$ to $\ker(\delta^\sigma)$ and $\ker(\delta^{\sigma'})$. Equivalently, it is derived from the $A(1)_*$-comodule pairing

$$\Omega^{\sigma}_{A(1)_*}(E(\xi_1^2)) \otimes \Omega^{\sigma'}_{A(1)_*}(E(\xi_1^2)) \longrightarrow \Omega^{\sigma+\sigma'}_{A(1)_*}(E(\xi_1^2)) \tag{3.3}$$

obtained by restricting the multiplication on $A(1)_* \otimes \mathbb{F}_2[v_0, v_1]$ to $\ker(1 \otimes \delta^\sigma)$ and $\ker(1 \otimes \delta^{\sigma'})$.

The induced pairing of Adams spectral sequences

$$E_r(ku\langle\sigma\rangle) \otimes E_r(ku\langle\sigma'\rangle) \longrightarrow E_r(ku\langle\sigma+\sigma'\rangle)$$

converges to the pairing

$$\pi_*(ku\langle\sigma\rangle)^\wedge_2 \otimes \pi_*(ku\langle\sigma'\rangle)^\wedge_2 \longrightarrow \pi_*(ku\langle\sigma+\sigma'\rangle)^\wedge_2$$

given by restriction of the product in $\pi_*(ku)^\wedge_2 = \mathbb{Z}_2[v_1]$. At the level of $E_2$-terms,

$$ku\langle\sigma\rangle^{*,*} \otimes ku\langle\sigma'\rangle^{*,*} \longrightarrow ku\langle\sigma+\sigma'\rangle^{*,*}$$

is given by the pairing

$$\begin{aligned}\mathrm{Ext}^{*,*+\sigma}_{E(1)_*}(\mathbb{F}_2, \Omega^{\sigma}_{E(1)_*}(\mathbb{F}_2)) \otimes \mathrm{Ext}^{*,*+\sigma'}_{E(1)_*}(\mathbb{F}_2, \Omega^{\sigma'}_{E(1)_*}(\mathbb{F}_2)) \\ \longrightarrow \mathrm{Ext}^{*,*+\sigma+\sigma'}_{E(1)_*}(\mathbb{F}_2, \Omega^{\sigma+\sigma'}_{E(1)_*}(\mathbb{F}_2))\,.\end{aligned}$$

Equivalently, it is given by the pairing

$$\begin{aligned}\mathrm{Ext}^{*,*+\sigma}_{A(1)_*}(\mathbb{F}_2, \Omega^{\sigma}_{A(1)_*}(E(\xi_1^2))) \otimes \mathrm{Ext}^{*,*+\sigma'}_{A(1)_*}(\mathbb{F}_2, \Omega^{\sigma'}_{A(1)_*}(E(\xi_1^2))) \\ \longrightarrow \mathrm{Ext}^{*,*+\sigma+\sigma'}_{A(1)_*}(\mathbb{F}_2, \Omega^{\sigma+\sigma'}_{A(1)_*}(E(\xi_1^2)))\,.\end{aligned} \tag{3.4}$$

LEMMA 3.14. *The pairing (3.4) is given by*

$$a_{k,s} \otimes a_{k',s'} \longmapsto a_{k+k',s+s'}$$

*whenever these classes are defined, i.e., for* $0 \le k \le s+\sigma$, $s \ge 0$, $0 \le k' \le s'+\sigma'$ *and* $s' \ge 0$. *In particular, the* $ko^{*,*}$*-module structure on* $ku\langle\sigma\rangle^{*,*}$ *is given by*

$$\begin{aligned} h_0 \cdot a_{k,s} &= a_{k,s+1} \\ h_1 \cdot a_{k,s} &= 0 \\ v \cdot a_{k,s} &= a_{k+2,s+3} \\ w_1 \cdot a_{k,s} &= a_{k+4,s+4}\,. \end{aligned}$$

PROOF. $v_0^{s+\sigma-k} v_1^k \cdot v_0^{s'+\sigma'-k'} v_1^{k'} = v_0^{s+s'-\sigma-\sigma'+k+k'} v_1^{k+k'}$. □

Recall the discussion of topological $K$-theory spectra from Section 2.6.

EXAMPLE 3.15. The induction functor $\mathcal{GL}(\mathbb{R}) \to \mathcal{GL}(\mathbb{C})$ from real to complex vector spaces respects the direct sum and tensor product pairings. Hence it induces a complexification map $c\colon ko \to ku$ of $E_\infty$ ring spectra. By real Bott periodicity it appears in a homotopy cofiber sequence

$$\Sigma ko \xrightarrow{\eta} ko \xrightarrow{c} ku \longrightarrow \Sigma^2 ko$$

of $ko$-modules. It induces the surjection $c^*\colon A/\!/E(1) \to A/\!/A(1)$ in cohomology, and the injection

$$c_*\colon \mathbb{F}_2[\bar\xi_1^4, \bar\xi_2^2, \bar\xi_i \mid i \geq 3] \longrightarrow \mathbb{F}_2[\bar\xi_1^2, \bar\xi_2^2, \bar\xi_i \mid i \geq 3]$$

in homology. The induced algebra homomorphism of Adams $E_2$-terms $c\colon ko^{*,*} \to ku^{*,*}$ is given by

$$\begin{aligned} h_0 &\longmapsto v_0 \\ h_1 &\longmapsto 0 \\ v &\longmapsto v_0 v_1^2 \\ w_1 &\longmapsto v_1^4 \,, \end{aligned}$$

as can be deduced from the associated morphism

$$\mathrm{Ext}_{A(0)_*}(\mathbb{F}_2, \mathbb{F}_2[x_2, x_3]) \longrightarrow \mathrm{Ext}_{A(0)_*}(\mathbb{F}_2, \mathbb{F}_2[v_1])$$

of Davis–Mahowald spectral sequences, mapping $x_2 \mapsto 0$ and $x_3 \mapsto v_1$. The induced ring homomorphism $\pi_*(c)\colon \pi_*(ko) \to \pi_*(ku)$ is given by $\eta \mapsto 0$, $A \mapsto 2v_1^2$ and $B \mapsto v_1^4$.

EXAMPLE 3.16. The restriction functor $\mathcal{GL}(\mathbb{H}) \to \mathcal{GL}(\mathbb{C})$ from quaternionic to complex vector spaces respects the direct sum pairings, as well as the tensor product with real vector spaces. Hence it induces a $ko$-module map $ksp \to ku$. It admits a unique lift $c'\colon ksp \to ku\langle 1\rangle$, reflecting the fact that quaternionic vector spaces have even-dimensional underlying complex vector spaces. By real Bott periodicity it is part of a homotopy cofiber sequence

$$\Sigma ksp \xrightarrow{\eta} ksp \xrightarrow{c'} ku\langle 1\rangle \longrightarrow \Sigma^2 ksp$$

of $ko$-modules. It induces a surjection $c'^*$ in cohomology, and an injection

$$c'_*\colon A_* \,\Box_{A(1)_*}\, \mathbb{F}_2\{1, \bar\xi_1^2, \bar\xi_2\} \longrightarrow A_* \,\Box_{A(1)_*}\, \Sigma^{-1}\Omega^1_{A(1)_*}(E(\bar\xi_1^2))$$

in homology. The induced $ko^{*,*}$-module homomorphism $c'\colon ksp^{*,*} \to ku\langle 1\rangle^{*,*}$ is given by

$$\begin{aligned} 1 &\longmapsto a_{0,0} \\ v' &\longmapsto a_{2,1} \,, \end{aligned}$$

since $h_0^2 \cdot v' = v \cdot 1$ maps to $a_{2,3}$ and $h_0^2 x = a_{2,3}$ only for $x = a_{2,1}$.

## 3.3. A comparison of $A(1)_*$-comodule algebras

DEFINITION 3.17. Consider

$$\bigoplus_{\sigma\geq 0} \Sigma^{3\sigma}\Omega^{\sigma}_{A(1)_*}(E(\xi_1^2))$$

as a graded $A(1)_*$-comodule algebra, with the multiplication given by the pairings (3.3) for $\sigma, \sigma' \geq 0$. Let

$$\phi\colon \bar{R}^* \longrightarrow \bigoplus_{\sigma\geq 0} \Sigma^{3\sigma}\Omega^{\sigma}_{A(1)_*}(E(\xi_1^2))$$

be the algebra homomorphism determined by

$$\begin{aligned}
\phi(x_4) &= \Sigma^3 v_0 \\
\phi(x_6) &= \Sigma^3(\xi_1^2 v_0 + v_1) \\
\phi(x_7) &= \Sigma^3(\bar{\xi}_2 v_0 + \xi_1 v_1)\,.
\end{aligned}$$

Let

$$\phi^\sigma : \bar{R}^\sigma \longrightarrow \Sigma^{3\sigma}\Omega^{\sigma}_{A(1)_*}(E(\xi_1^2))$$

be the restriction of $\phi$ to degree $\sigma$, and let

$$\psi^\sigma : \Sigma^{3\sigma}\Omega^{\sigma}_{A(1)_*}(E(\xi_1^2)) \longrightarrow \operatorname{cok}(\phi^\sigma)$$

be the projection onto its cokernel.

LEMMA 3.18. *$\phi$ is a well-defined $A(1)_*$-comodule algebra homomorphism.*

PROOF. $\phi$ is well defined, because

$$\phi(x_7^4) = \Sigma^{12}(\bar{\xi}_2 v_0 + \xi_1 v_1)^4 = 0$$

in $\Sigma^{12}\Omega^4_{A(1)_*}(E(\xi_1^2)) \subset \Sigma^{12}A(1)_*\{v_0^4, \dots, v_1^4\}$. To check that $\phi$ respects the $A(1)_*$-coactions, recall Definition 3.1 and note that

$$\begin{aligned}
\nu(v_0) &= 1 \otimes v_0 \\
\nu(\xi_1^2 v_0 + v_1) &= 1 \otimes (\xi_1^2 v_0 + v_1) + \xi_1^2 \otimes v_0 \\
\nu(\bar{\xi}_2 v_0 + \xi_1 v_1) &= 1 \otimes (\bar{\xi}_2 v_0 + \xi_1 v_1) + \xi_1 \otimes (\xi_1^2 v_0 + v_1) + \bar{\xi}_2 \otimes v_0\,.
\end{aligned}$$

□

LEMMA 3.19. *$\phi^0 : \bar{R}^0 \to E(\xi_1^2)$ is the inclusion $\mathbb{F}_2\{1\} \to \mathbb{F}_2\{1, \xi_1^2\}$. The induced map*

$$\phi^0_* \colon \operatorname{Ext}_{A(1)_*}(\mathbb{F}_2, \mathbb{F}_2) \longrightarrow \operatorname{Ext}_{A(1)_*}(\mathbb{F}_2, E(\xi_1^2)) \cong \operatorname{Ext}_{E(1)_*}(\mathbb{F}_2, \mathbb{F}_2)$$

*is the algebra homomorphism*

$$c\colon ko^{*,*} \longrightarrow ku^{*,*} = \mathbb{F}_2\{a_{k,s} \mid 0 \leq k \leq s\} = \mathbb{F}_2[v_0, v_1]$$

*given by*

$$\begin{aligned}
h_0 &\longmapsto a_{0,1} = v_0 \\
h_1 &\longmapsto 0 \\
v &\longmapsto a_{2,3} = v_0 v_1^2 \\
w_1 &\longmapsto a_{4,4} = v_1^4
\end{aligned}$$

PROOF. See Example 3.15. □

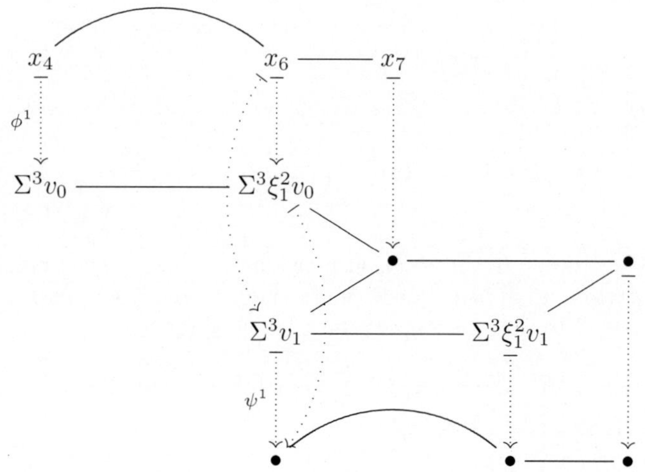


FIGURE 3.4. $\phi^1\colon \bar{R}^1 \to \Sigma^3\Omega^1_{A(1)_*}(E(\xi_1^2))$ and its cokernel $\psi^1$

LEMMA 3.20. *$\phi^1\colon \bar{R}^1 \to \Sigma^3\Omega^1_{A(1)_*}(E(\xi_1^2))$ is the monomorphism*

$$\begin{aligned} x_4 &\longmapsto \Sigma^3 v_0 \\ x_6 &\longmapsto \Sigma^3(\xi_1^2 v_0 + v_1) \\ x_7 &\longmapsto \Sigma^3(\bar{\xi}_2 v_0 + \xi_1 v_1)\,. \end{aligned}$$

*The induced map*

$$\phi^1_*\colon \operatorname{Ext}_{A(1)_*}(\mathbb{F}_2, \bar{R}^1) \longrightarrow \operatorname{Ext}_{A(1)_*}(\mathbb{F}_2, \Sigma^3\Omega^1_{A(1)_*}(E(\xi_1^2)))$$

*is the $ko^{*,*}$-module homomorphism*

$$\Sigma^4 c'\colon \Sigma^4 ksp^{*,*} \longrightarrow \Sigma^4 ku\langle 1\rangle^{*,*} = \Sigma^4\mathbb{F}_2\{a_{k,s} \mid 0 \le k \le s+1,\ s \ge 0\}$$

*given by*

$$\begin{aligned} \Sigma^4 1 &\longmapsto \Sigma^4 a_{0,0} \\ \Sigma^4 v' &\longmapsto \Sigma^4 a_{2,1} \end{aligned}$$

PROOF. See Example 3.16 and Figure 3.4. □

PROPOSITION 3.21. *For each $\sigma \ge 2$ there is a short exact sequence of $A(1)_*$-comodules*

$$0 \to \bar{R}^\sigma \xrightarrow{\phi^\sigma} \Sigma^{3\sigma}\Omega^\sigma_{A(1)_*}(E(\xi_1^2)) \xrightarrow{\psi^\sigma} \Sigma^{4\sigma+2}(A(1)/\!/A(0))_* \to 0\,,$$

*with $\psi^\sigma(\Sigma^{3\sigma} v_0^{\sigma-1} v_1) \ne 0$.*

PROOF. $\phi^2\colon \bar{R}^2 \to \Sigma^6\Omega^2_{A(1)_*}(E(\xi_1^2))$ is the monomorphism

$$\begin{aligned} x_4^2 &\longmapsto \Sigma^6 v_0^2 \\ x_4x_6 &\longmapsto \Sigma^6(\xi_1^2 v_0^2 + v_0 v_1) \\ x_4x_7 &\longmapsto \Sigma^6(\bar{\xi}_2 v_0^2 + \xi_1 v_0 v_1) \end{aligned}$$

$$\begin{aligned} x_6^2 &\longmapsto \Sigma^6 v_1^2 \\ x_6x_7 &\longmapsto \Sigma^6(\xi_1^2(\bar{\xi}_2v_0^2+\xi_1v_0v_1)+(\bar{\xi}_2v_0v_1+\xi_1v_1^2)) \\ x_7^2 &\longmapsto \Sigma^6\xi_1^2v_1^2 \end{aligned}$$

with cokernel

$$\Sigma^6\mathbb{F}_2\{\xi_1^2v_0^2 \equiv v_0v_1\,,\ \xi_1^2v_0v_1\,,\ \xi_1^2(\bar{\xi}_2v_0^2+\xi_1v_0v_1) \equiv (\bar{\xi}_2v_0v_1+\xi_1v_1^2)\,,\ \xi_1^2(\bar{\xi}_2v_0v_1+\xi_1v_1^2)\} \cong \Sigma^{10}(A(1)//A(0))_*\,.$$

See Figure 3.5, where the internal suspensions $\Sigma^6$ have been omitted from the notation. In particular, $\Sigma^6v_0v_1$ maps nontrivially to the cokernel of $\phi^2$.

For each $\sigma \geq 3$ we have a map of short exact sequences

$$\begin{array}{ccccccccc} 0 & \longrightarrow & \Sigma^4\bar{R}^{\sigma-1} & \xrightarrow{x_4} & \bar{R}^{\sigma} & \longrightarrow & \mathrm{cok}(x_4) & \longrightarrow & 0 \\ & & \downarrow{\scriptstyle \Sigma^4\phi^{\sigma-1}} & & \downarrow{\scriptstyle \phi^{\sigma}} & & \downarrow{\scriptstyle \cong} & & \\ 0 & \longrightarrow & \Sigma^{3\sigma+1}\Omega^{\sigma-1}_{A(1)_*}(E(\xi_1^2)) & \xrightarrow{v_0} & \Sigma^{3\sigma}\Omega^{\sigma}_{A(1)_*}(E(\xi_1^2)) & \longrightarrow & \mathrm{cok}(v_0) & \longrightarrow & 0 \end{array}$$

of $A(1)_*$-comodules, where

$$\mathrm{cok}(x_4) = \mathbb{F}_2\{x_6^\sigma\,,\ x_6^{\sigma-1}x_7\,,\ x_6^{\sigma-2}x_7^2\,,\ x_6^{\sigma-3}x_7^3\}$$

maps isomorphically to

$$\mathrm{cok}(v_0) = \Sigma^{3\sigma}\mathbb{F}_2\{v_1^\sigma\,,\ \bar{\xi}_2v_0v_1^{\sigma-1}+\xi_1v_1^\sigma\,,\ \xi_1^2v_1^\sigma\,,\ \xi_1^2(\bar{\xi}_2v_0v_1^{\sigma-1}+\xi_1v_1^\sigma)\}\,.$$

This follows from

$$\phi^\sigma(x_6^\sigma) = \Sigma^{3\sigma}(\xi_1^2v_0+v_1)^\sigma \equiv \Sigma^{3\sigma}v_1^\sigma \mod \mathrm{im}(v_0)$$

and Lemmas 3.5 and 3.10. The claims of the proposition now follow for all $\sigma \geq 2$, by induction on $\sigma$ and the snake lemma. □

LEMMA 3.22. *For each $\sigma \geq 2$ the induced map*

$$\psi_*^\sigma\colon \mathrm{Ext}_{A(1)_*}(\mathbb{F}_2, \Sigma^{3\sigma}\Omega^\sigma_{A(1)_*}(E(\xi_1^2))) \longrightarrow \mathrm{Ext}_{A(1)_*}(\mathbb{F}_2, \Sigma^{4\sigma+2}(A(1)//A(0))_*)$$

*is the $ko^{*,*}$-module epimorphism*

$$\Sigma^{4\sigma}ku\langle\sigma\rangle^{*,*} = \Sigma^{4\sigma}\mathbb{F}_2\{a_{k,s} \mid 0 \leq k \leq s+\sigma,\ s \geq 0\} \longrightarrow \Sigma^{4\sigma+2}\mathbb{F}_2[h_0]$$

*given by*

$$\Sigma^{4\sigma}a_{1,s} \longmapsto \Sigma^{4\sigma+2}h_0^s$$

*and $\Sigma^{4\sigma}a_{k,s} \mapsto 0$ for $k \neq 1$.*

PROOF. The class $\Sigma^{4\sigma}a_{1,0}$ is represented by the $A(1)_*$-comodule primitive $\Sigma^{3\sigma}v_0^{\sigma-1}v_1$, which maps nontrivially under $\psi^\sigma$. Hence $\psi_*^\sigma$ maps $\Sigma^{4\sigma}a_{1,0}$ to $\Sigma^{4\sigma+2}1$. By $ko^{*,*}$-linearity it follows that $\psi_*^\sigma$ maps $\Sigma^{4\sigma}a_{1,s}$ to $\Sigma^{4\sigma+2}h_0^s$ for each $s \geq 0$. □

DEFINITION 3.23. For each $\sigma \geq 2$, let

$$G\langle\sigma\rangle^{*,*} = \mathbb{F}_2\{a_{k,s} \mid 0 \leq k \leq s+\sigma,\ k \neq 1,\ s \geq 0\}$$

be the $ko^{*,*}$-submodule $\Sigma^{-4\sigma}\ker(\psi_*^\sigma)$ of $ku\langle\sigma\rangle^{*,*}$. See Figure 3.6.

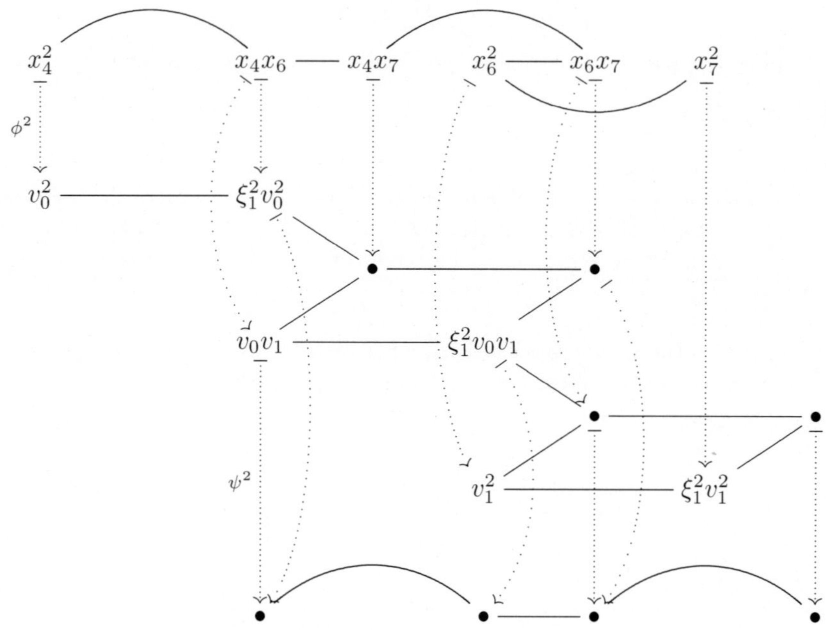


FIGURE 3.5. $\phi^2\colon \bar{R}^2 \to \Sigma^6\Omega^2_{A(1)_*}(E(\xi_1^2))$ and its cokernel $\psi^2$

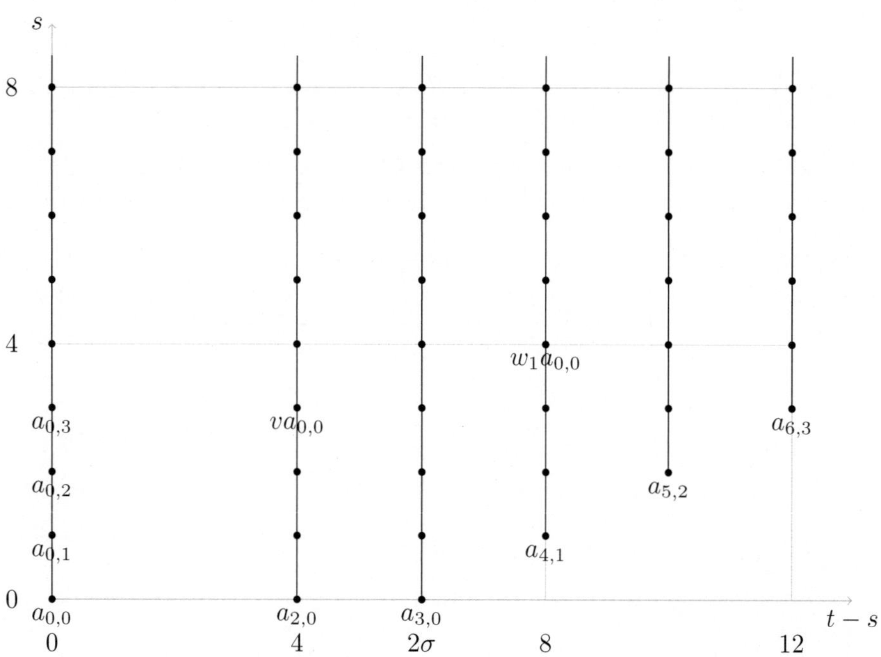


FIGURE 3.6. The Adams chart $G\langle\sigma\rangle^{*,*}$ for $\sigma = 3$

LEMMA 3.24. *For each $\sigma \geq 2$ there is an isomorphism*

$$\mathrm{Ext}_{A(1)_*}(\mathbb{F}_2, \bar{R}^\sigma) \cong \Sigma^{4\sigma} G\langle\sigma\rangle^{*,*}$$

*of $ko^{*,*}$-modules, which identifies $\phi_*^\sigma$ with the inclusion $\Sigma^{4\sigma} G\langle\sigma\rangle^{*,*} \subset \Sigma^{4\sigma} ku\langle\sigma\rangle^{*,*}$.*

PROOF. Clear. □

REMARK 3.25. For each $\sigma \geq 1$ there is a $ko$-module map $\psi^\sigma \colon ku\langle\sigma\rangle \to \Sigma^2 H\mathbb{Z}$ such that $\pi_2(\psi^\sigma)$ is an isomorphism. This follows by a comparison of Postnikov sections, since the $ko$-module $k$-invariant $k^3 \in H^3_{ko}(H\mathbb{Z}; \mathbb{Z}) \cong \mathbb{Z}/2$ of $ku$ has order 2. For $\sigma \geq 2$ we can define $G\langle\sigma\rangle$ to be the homotopy fiber of $\psi^\sigma$, so that

$$G\langle\sigma\rangle^{*,*} = E_2(G\langle\sigma\rangle)$$

is the $E_2$-term of the Adams spectral sequence for this spectrum. We set $G\langle 0\rangle = ko$ and $G\langle 1\rangle = ksp$, so that $G\langle 0\rangle^{*,*} = ko^{*,*}$ and $G\langle 1\rangle^{*,*} = ksp^{*,*}$.

PROPOSITION 3.26.

$$\mathrm{Ext}_{A(1)_*}(\mathbb{F}_2, \bar{R}^\sigma) \cong \begin{cases} ko^{*,*} & \text{for } \sigma = 0, \\ \Sigma^4 ksp^{*,*} & \text{for } \sigma = 1, \\ \Sigma^{4\sigma} G\langle\sigma\rangle^{*,*} & \text{for } \sigma \geq 2. \end{cases}$$

*The pairing*

$$\mathrm{Ext}_{A(1)_*}(\mathbb{F}_2, \bar{R}^\sigma) \otimes \mathrm{Ext}_{A(1)_*}(\mathbb{F}_2, \bar{R}^{\sigma'}) \longrightarrow \mathrm{Ext}_{A(1)_*}(\mathbb{F}_2, \bar{R}^{\sigma+\sigma'})$$

*is given by the $ko^{*,*}$-module structure if $\sigma = 0$ or $\sigma' = 0$. Otherwise $\sigma + \sigma' \geq 2$, and the pairing is given by the formula*

$$a_{k,s} \cdot a_{k',s'} = a_{k+k',s+s'}$$

*for $a_{k,s} \in G\langle\sigma\rangle^{*,*}$ and $a_{k',s'} \in G\langle\sigma'\rangle^{*,*}$. Here, if $\sigma = 1$ or $\sigma' = 1$, classes in $ksp^{*,*}$ are implicitly replaced by their images under $c'$ in $ku\langle 1\rangle^{*,*}$.*

PROOF. The additive claim summarizes Lemmas 3.19, 3.20 and 3.24. The $ko^{*,*}$-module claim for $\sigma = 0$ or $\sigma' = 0$ is also clear. It remains to consider the case $\sigma, \sigma' \geq 1$.

Applying $\mathrm{Ext}_{A(1)_*}(\mathbb{F}_2, -)$ to the commutative square

$$\begin{array}{ccc} \bar{R}^\sigma \otimes \bar{R}^{\sigma'} & \longrightarrow & \bar{R}^{\sigma+\sigma'} \\ \downarrow{\scriptstyle \phi^\sigma \otimes \phi^{\sigma'}} & & \downarrow{\scriptstyle \phi^{\sigma+\sigma'}} \\ \Sigma^{3\sigma}\Omega^\sigma_{A(1)_*}(E(\xi_1^2)) \otimes \Sigma^{3\sigma'}\Omega^{\sigma'}_{A(1)_*}(E(\xi_1^2)) & \xrightarrow{(3.3)} & \Sigma^{3(\sigma+\sigma')}\Omega^{\sigma+\sigma'}_{A(1)_*}(E(\xi_1^2)) \end{array}$$

of $A(1)_*$-comodules yields a commutative diagram

$$\begin{array}{ccc} \mathrm{Ext}_{A(1)_*}(\mathbb{F}_2, \bar{R}^\sigma) \otimes \mathrm{Ext}_{A(1)_*}(\mathbb{F}_2, \bar{R}^{\sigma'}) & \longrightarrow & \mathrm{Ext}_{A(1)_*}(\mathbb{F}_2, \bar{R}^{\sigma+\sigma'}) \\ \downarrow{\scriptstyle \phi_*^\sigma \otimes \phi_*^{\sigma'}} & & \downarrow{\scriptstyle \phi_*^{\sigma+\sigma'}} \\ \Sigma^{4\sigma} ku\langle\sigma\rangle^{*,*} \otimes \Sigma^{4\sigma'} ku\langle\sigma'\rangle^{*,*} & \longrightarrow & \Sigma^{4(\sigma+\sigma')} ku\langle\sigma+\sigma'\rangle^{*,*}. \end{array}$$

For $\sigma + \sigma' \geq 2$ the right hand vertical map is injective, so to verify the asserted product formula in $\mathrm{Ext}_{A(1)_*}(\mathbb{F}_2, \bar{R}^{\sigma+\sigma'})$ it suffices to verify it in the lower right hand corner. Here the formula follows from Lemma 3.14, under the assumption that classes in $\mathrm{Ext}_{A(1)_*}(\mathbb{F}_2, \bar{R}^1) = \Sigma^4 ksp^{*,*}$ are replaced by their images in $\Sigma^4 ku\langle 1\rangle^{*,*}$, i.e., that 1 is read as $a_{0,0}$ and $v'$ is read as $a_{2,1}$. □

REMARK 3.27. Note that each $G\langle\sigma\rangle^{*,*}$ is free as an $\mathbb{F}_2[w_1]$-module, and that $G\langle\sigma\rangle^{*,*}$ for $\sigma \geq 2$ is torsion-free as an $\mathbb{F}_2[h_0, w_1]$-module.

## 3.4. The $d_1$-differential for $A(2)$

In the extension $\mathbb{F}_2[x_7^4] \longrightarrow E_1^{*,*,*} \longrightarrow \bar{E}_1^{*,*,*}$ we have $\bar{E}_1^{\sigma,*,*} = \mathrm{Ext}^{*-\sigma,*}_{A(1)_*}(\mathbb{F}_2, \bar{R}^\sigma)$ and $\mathrm{Ext}_{A(1)_*}(\mathbb{F}_2, \bar{R}^\sigma) \cong \Sigma^{4\sigma} G\langle\sigma\rangle^{*,*}$. For $a \in G\langle\sigma\rangle^{s,t}$ we write

$$a x_4^\sigma \in \bar{E}_1^{\sigma, s+\sigma, t+4\sigma}$$

for the class that corresponds to $\Sigma^{4\sigma} a \in \Sigma^{4\sigma} G\langle\sigma\rangle^{*,*}$. In other words, we identify

$$\bar{E}_1^{\sigma,*,*} \cong G\langle\sigma\rangle^{*,*}\{x_4^\sigma\},$$

where $x_4^\sigma \in \bar{E}_1^{\sigma,\sigma,4\sigma}$ has $(t-s,s)$-bidegree $(3\sigma,\sigma)$. Using the splitting induced by $S\colon \bar{R}^* \to R^*$ from Definition 3.2, we can write

$$\begin{aligned} E_1^{\sigma,*,*} &\cong \bar{E}_1^{\sigma,*,*} \oplus \bar{E}_1^{\sigma-4,*,*}\{x_7^4\} \oplus \bar{E}_1^{\sigma-8,*,*}\{x_7^8\} \oplus \dots \\ &\cong G\langle\sigma\rangle\{x_4^\sigma\} \oplus G\langle\sigma-4\rangle\{x_4^{\sigma-4}x_7^4\} \oplus G\langle\sigma-8\rangle\{x_4^{\sigma-8}x_7^8\} \oplus \dots . \end{aligned} \tag{3.5}$$

LEMMA 3.28. *(0) $d_1^0\colon E_1^{0,*,*} \to E_1^{1,*+1,*}$ is the derivation*

$$d_1^0\colon G\langle 0\rangle^{*,*} = ko^{*,*} \longrightarrow G\langle 1\rangle^{*,*}\{x_4\} = ksp^{*,*}\{x_4\}$$

*given by*

$$\begin{aligned} h_0 &\longmapsto 0 \\ h_1 &\longmapsto 0 \\ v &\longmapsto h_0^3 x_4 \\ w_1 &\longmapsto 0 . \end{aligned}$$

*Hence each $d_1^\sigma$ is $\mathbb{F}_2[h_0, h_1, w_1]/(h_0h_1, h_1^3)$-linear.*

*(1) $d_1^1\colon E_1^{1,*,*} \to E_1^{2,*+1,*}$ is the homomorphism*

$$d_1^1\colon G\langle 1\rangle^{*,*} = ksp^{*,*}\{x_4\} \longrightarrow G\langle 2\rangle^{*,*}\{x_4^2\}$$

*given by*

$$\begin{aligned} x_4 &\longmapsto 0 \\ v' x_4 &\longmapsto h_0 x_4^2 . \end{aligned}$$

*(2) $d_1^2\colon E_1^{2,*,*} \to E_1^{3,*+1,*}$ is the homomorphism*

$$d_1^2\colon G\langle 2\rangle^{*,*}\{x_4^2\} \longrightarrow G\langle 3\rangle^{*,*}\{x_4^3\}$$

*given by*

$$\begin{aligned} x_4^2 &\longmapsto 0 \\ a_{2,0}x_4^2 &\longmapsto x_4^3 \\ a_{3,1}x_4^2 &\longmapsto 0 \\ a_{4,2}x_4^2 &\longmapsto 0 \\ a_{5,3}x_4^2 &\longmapsto a_{3,3}x_4^3 . \end{aligned}$$

*(3)* $d_1^3\colon E_1^{3,*,*} \to E_1^{4,*+1,*}$ *is the homomorphism*

$$d_1^3\colon G\langle 3\rangle^{*,*}\{x_4^3\} \longrightarrow G\langle 4\rangle^{*,*}\{x_4^4\} \oplus G\langle 0\rangle^{*,*}\{x_7^4\}$$

*given by*

$$\begin{aligned} x_4^3 &\longmapsto (0,0) \\ a_{2,0}x_4^3 &\longmapsto (x_4^4,0) \\ a_{3,0}x_4^3 &\longmapsto (0,0) \\ a_{4,1}x_4^3 &\longmapsto (0,0) \\ a_{5,2}x_4^3 &\longmapsto (a_{3,2}x_4^4,0) \\ a_{6,3}x_4^3 &\longmapsto (a_{4,3}x_4^4,0)\,. \end{aligned}$$

*(4)* $d_1^4\colon E_1^{4,*,*} \to E_1^{5,*+1,*}$ *is a homomorphism*

$$d_1^4\colon G\langle 4\rangle^{*,*}\{x_4^4\} \oplus G\langle 0\rangle^{*,*}\{x_7^4\} \longrightarrow G\langle 5\rangle^{*,*}\{x_4^5\} \oplus G\langle 1\rangle^{*,*}\{x_4x_7^4\}$$

*satisfying*

$$(0,x_7^4) \longmapsto (a_{4,0}x_4^5,0)\,.$$

PROOF. The classes $h_0 \in E_1^{0,1,1}$ and $h_1 \in E_1^{0,1,2}$ are infinite cycles, meaning that $d_r^0(h_0) = 0$ and $d_r^0(h_1) = 0$ for all $r \geq 1$, because the target groups $E_1^{\sigma,2,1}$ and $E_1^{\sigma,2,2}$ are trivial for all $\sigma \geq 1$. Similarly, the class $x_4 \in E_1^{1,1,4}$ is an infinite cycle because $E_1^{\sigma,2,4} = 0$ for all $\sigma \geq 2$.

Under the twisting isomorphism

$$E_1^{\sigma,\sigma,12} = \mathrm{Ext}_{A(2)_*}^{0,12}(\mathbb{F}_2, (A(2)/\!/A(1))_* \otimes R^\sigma) \cong \mathrm{Ext}_{A(1)_*}^{0,12}(\mathbb{F}_2, R^\sigma)\,,$$

the $A(1)_*$-comodule primitive $x_6^2 \in R^2$ corresponds to the $A(2)_*$-comodule primitive $x_6^2 + \xi_1^4 x_4^2 \in (A(2)/\!/A(1))_* \otimes R^2$, and the $A(2)_*$-comodule primitive

$$\delta(x_6^2 + \xi_1^4 x_4^2) = x_4^3$$

in $(A(2)/\!/A(1))_* \otimes R^3$ corresponds to the $A(1)_*$-comodule primitive $x_4^3$ in $R^3$. Under the isomorphism $E_1^{2,*,*} \cong G\langle 2\rangle^{*,*}\{x_4^2\}$, the class $x_6^2$ corresponds to $a_{2,0}x_4^2$. Hence $d_1^2(a_{2,0}x_4^2) = x_4^3$.

It follows by $x_4$- and $h_0$-linearity that $d_1^0(v) = h_0^3 x_4$, since

$$d_1^0(v)\cdot x_4^2 = d_1^2(v\cdot x_4^2) = d_1^2(h_0^3\cdot a_{2,0}x_4^2) = h_0^3\cdot d_1^2(a_{2,0}x_4^2) = h_0^3\cdot x_4^3\,.$$

Likewise, $d_1^1(v'x_4) = h_0 x_4^2$, and $d_1^\sigma(a_{2,0}x_4^\sigma) = x_4^{\sigma+1}$ for all $\sigma \geq 2$. This completes the proof of (1).

The class $w_1 \in E_1^{0,4,12}$ is an infinite cycle. First, $d_1^0(w_1)$ lies in $E_1^{1,5,12} = ksp^{4,4}\{x_4\} = \mathbb{F}_2\{h_0^3 v' x_4\}$ and $d_1^1(h_0^3 v' x_4) = h_0^4 x_4^2 \neq 0$, so we cannot have $d_1^0(w_1) \neq 0$ because $d_1^1 \circ d_1^0 = 0$. Next, $E_1^{\sigma,5,12} = 0$ for all $\sigma \geq 2$, so $d_r^0(w_1) = 0$ for all $r \geq 2$. This completes the proof of (0). It follows by $w_1$- and $h_0$-linearity that $d_1^2(a_{4,2}x_4^2) = 0$, $d_1^3(a_{4,1}x_4^3) = 0$ and $d_1^3(a_{6,3}x_4^3) = a_{4,3}x_4^4$.

The class $a_{3,1}x_4^2 \in E_1^{2,3,15}$ is an infinite cycle, because $E_1^{\sigma,4,15} = 0$ for all $\sigma \geq 3$. It follows by $x_4$- and $h_0$-linearity that $d_1^3(a_{3,0}x_4^3) = 0$. Multiplying by $v \in ko^{*,*}$, the Leibniz rule gives

$$d_1^2(a_{5,4}x_4^2) = d_1^2(v\cdot a_{3,1}x_4^2) = h_0^3 x_4\cdot a_{3,1}x_4^2 + v\cdot 0 = a_{3,4}x_4^3\,.$$

By $h_0$- and $x_4$-linearity it follows that $d_1^2(a_{5,3}x_4^2) = a_{3,3}x_4^3$ and $d_1^3(a_{5,2}x_4^3) = a_{3,2}x_4^4$. This completes the proof of (2) and (3).

In

$$E_1^{\sigma,\sigma,28} = \mathrm{Ext}^{0,28}_{A(2)_*}(\mathbb{F}_2, (A(2)//A(1))_* \otimes R^\sigma) \cong \mathrm{Ext}^{0,28}_{A(1)_*}(\mathbb{F}_2, R^\sigma) ,$$

for $\sigma \in \{4,5\}$, the $A(1)_*$-comodule primitive $x_7^4 \in R^4$ corresponds to the $A(2)_*$-comodule primitive $x_7^4 + \xi_1^4 x_6^4 \in (A(2)//A(1))_* \otimes R^4$, and the $A(2)_*$-comodule primitive

$$\delta(x_7^4 + \xi_1^4 x_6^4) = x_4 x_6^4$$

in $(A(2)//A(1))_* \otimes R^5$ corresponds to the $A(1)_*$-comodule primitive $x_4x_6^4$ in $R^5$. Under $E_1^{5,*,*} \cong G\langle 5\rangle^{*,*}\{x_4^5\} \oplus G\langle 1\rangle^{*,*}\{x_4x_7^4\}$ the class $x_4x_6^4$ corresponds to $(a_{4,0}x_4^5, 0)$. This completes the proof of (4). □

Lemma 3.29. *In terms of the splitting (3.5), the differential $d_1^\sigma\colon E_1^{\sigma,*,*} \to E_1^{\sigma+1,*,*}$ maps $\bar{E}_1^{\sigma,*,*}$ into $\bar{E}_1^{\sigma+1,*,*}$, and it maps $\bar{E}_1^{\sigma-4,*,*}\{x_7^4\}$ into $\bar{E}_1^{\sigma+1,*,*} \oplus \bar{E}_1^{\sigma-3,*,*}\{x_7^4\}$.*

Proof. By Lemma 3.5, multiplication by $x_4$ induces a short exact sequence

$$0 \to G\langle \sigma - 1\rangle^{*,*} \longrightarrow G\langle\sigma\rangle^{*,*} \longrightarrow \Sigma^{2\sigma}\mathbb{F}_2[v_1] \to 0$$

for each $\sigma \geq 3$. By Lemma 3.6, multiplication by $x_6^2$ induces a short exact sequence

$$0 \to \Sigma^4 G\langle \sigma - 2\rangle^{*,*} \longrightarrow G\langle\sigma\rangle^{*,*} \longrightarrow \mathbb{F}_2[h_0]\{a_{0,0}, a_{3,0}\} \to 0$$

for each $\sigma \geq 4$. Hence, for $\sigma \geq 4$ the images of $x_4\colon \bar{E}_1^{\sigma-1,*,*} \to \bar{E}_1^{\sigma,*,*}$ and $x_6^2\colon \bar{E}_1^{\sigma-2,*,*} \to \bar{E}_1^{\sigma,*,*}$ span $\bar{E}_1^{\sigma,*,*}$. By Lemma 3.28(1,2) and the Leibniz rule, $d_1^\sigma(ax_4) = d_1^{\sigma-1}(a)x_4$ and $d_1^\sigma(bx_6^2) = d_1^{\sigma-2}(b)x_6^2 + bx_4^3$ in $E_1^{*,*,*}$. Thus, if $d_1(a)$ and $d_1(b)$ lie in the image of $S\colon \bar{E}_1^{*,*,*} \to E_1^{*,*,*}$ then so do $d_1(ax_4)$ and $d_1(bx_6^2)$. The first claim of the lemma therefore follows by induction on $\sigma$.

By Lemma 3.28(4), $d_1^4(x_7^4) = a_{4,0}x_4^5$ is contained in the summand $\bar{E}_1^{5,*,*}$ of $E_1^{5,*,*}$. Hence $d_1^\sigma(cx_7^4) = d_1^{\sigma-4}(c)x_7^4 + ca_{4,0}x_4^5$ in $E_1^{*,*,*}$. Thus, if $c$ lies in the summand $\bar{E}_1^{*,*,*}$ then $d_1^\sigma(cx_7^4)$ lies in the direct sum $\bar{E}_1^{*,*,*} \oplus \bar{E}_1^{*,*,*}\{x_7^4\}$, as asserted. □

Schematically, the Davis–Mahowald $(E_1, d_1)$-term appears as in Figure 3.9, repeating $x_7^8$-periodically. The colors red, green, mustard and blue show classes of weight $\sigma \equiv 0, 1, 2, 3 \mod 4$, respectively. By the Leibniz rule, $x_7^8 = (x_7^4)^2$ is a $d_1$-cycle, so there is an extension of differential trigraded algebras

$$\mathbb{F}_2[x_7^8] \longrightarrow (E_1^{*,*,*}, d_1) \longrightarrow (\bar{\bar{E}}_1^{*,*,*}, d_1) ,$$

with $E_1^{*,*,*}$ free as a module over $\mathbb{F}_2[x_7^8]$, and with

$$\bar{\bar{E}}_1^{*,*,*} = E_1^{*,*,*} \otimes_{\mathbb{F}_2[x_7^8]} \mathbb{F}_2 = E_1^{*,*,*}/(x_7^8)$$

sitting in a short exact sequence of cochain complexes

$$0 \to (\bar{E}_1^{*,*,*}, d_1) \xrightarrow{S} (\bar{\bar{E}}_1^{*,*,*}, d_1) \longrightarrow (\bar{E}_1^{*-4,*,*}\{x_7^4\}, d_1) \to 0 .$$

It follows that there is an extension of trigraded algebras

$$\mathbb{F}_2[x_7^8] \longrightarrow E_2^{*,*,*} \longrightarrow \bar{\bar{E}}_2^{*,*,*} ,$$

with $E_2^{*,*,*}$ free as a module over $\mathbb{F}_2[x_7^8]$, and a long exact sequence

$$\dots \xrightarrow{\delta} \bar{E}_2^{\sigma,*,*} \xrightarrow{S} \bar{\bar{E}}_2^{\sigma,*,*} \longrightarrow \bar{E}_2^{\sigma-4,*,*}\{x_7^4\} \xrightarrow{\delta} \bar{E}_2^{\sigma+1,*,*} \xrightarrow{S} \dots . \tag{3.6}$$

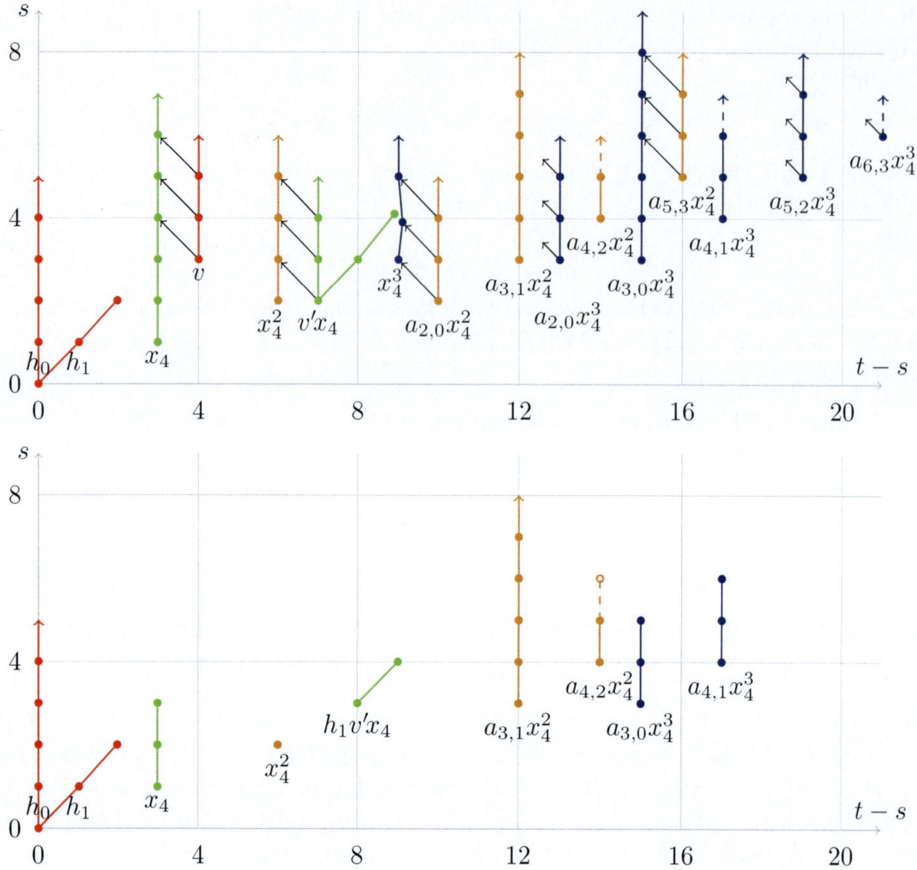


FIGURE 3.7. $(\bar{E}_1^{\sigma,*,*}, d_1^\sigma)$ and $\bar{E}_2^{\sigma,*,*}$ for $0 \le \sigma \le 3$

Here $\bar{E}_2^{*,*,*}$ is equal to the cohomology of $(\bar{E}_1^{*,*,*}, d_1)$, and $\bar{\bar{E}}_2^{*,*,*}$ is equal to the cohomology of $(\bar{\bar{E}}_1^{*,*,*}, d_1)$.

LEMMA 3.30. *For each $\sigma \ge 2$, $d_1^\sigma : \bar{E}_1^{\sigma,*,*} \to \bar{E}_1^{\sigma+1,*,*}$ is the homomorphism*

$$d_1^\sigma : G\langle\sigma\rangle^{*,*}\{x_4^\sigma\} \longrightarrow G\langle\sigma+1\rangle^{*,*}\{x_4^{\sigma+1}\}$$

*given by*

$$a_{k,s}x_4^\sigma \longmapsto \begin{cases} a_{k-2,s}x_4^{\sigma+1} & \text{for } k \equiv 2,5 \mod 4, \\ 0 & \text{otherwise.} \end{cases}$$

*Here $0 \le k \le s+\sigma$, $k \ne 1$ and $s \ge 0$, so that $a_{k,s}$ is defined.*

PROOF. We verified this in Lemma 3.28(2) for $\sigma = 2$ and $(k,s) = (0,0)$, $(2,0)$, $(3,1)$, $(4,2)$ and $(5,3)$. By $h_0$- and $w_1$-linearity the formula for $d_1^2$ holds for all $a_{k,s} \in G\langle 2\rangle^{*,*}$.

By $x_4$-linearity, the formula for $d_1^\sigma$ holds for the $a_{k,s} \in G\langle\sigma\rangle$ with $0 \le k \le s+2$, $k \ne 1$ and $s \ge 0$. By $h_0$-linearity, the formula also holds for the remaining $a_{k,s}$, with $s+2 < k \le s+\sigma$, since $G\langle\sigma+1\rangle$ is $h_0$-torsion free. □

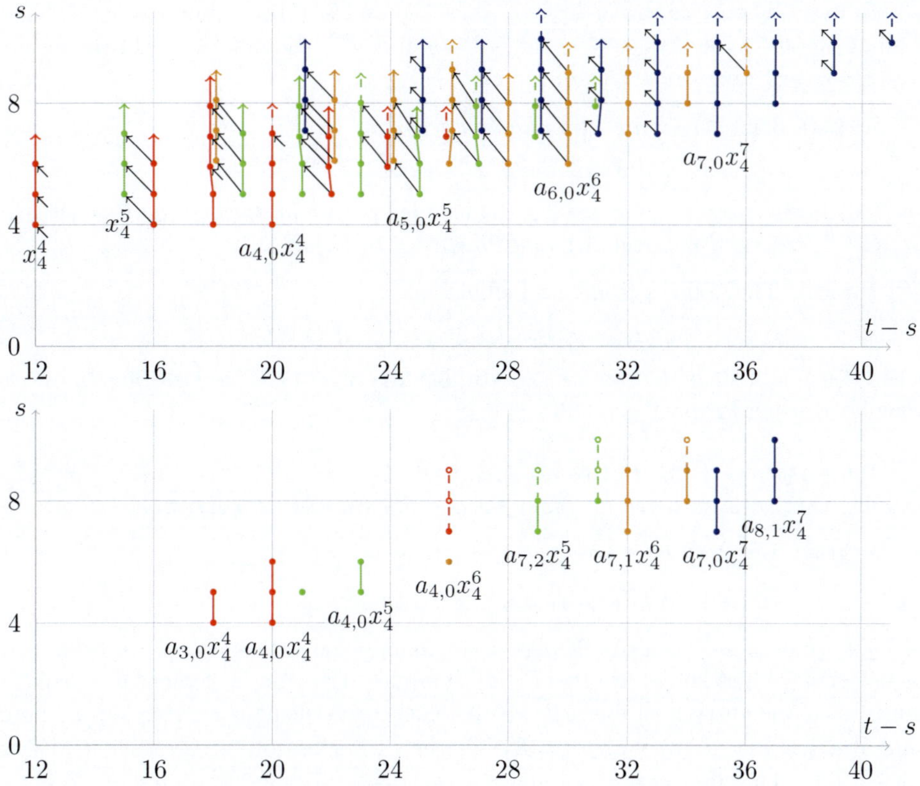


FIGURE 3.8. $(\bar{E}_1^{\sigma,*,*}, d_1^\sigma)$ and $\bar{E}_2^{\sigma,*,*}$ for $4 \le \sigma \le 7$

The complex

$$0 \to \bar{E}_1^{0,*,*} \xrightarrow{d_1^0} \bar{E}_1^{1,*,*} \xrightarrow{d_1^1} \bar{E}_1^{2,*,*} \xrightarrow{d_1^2} \bar{E}_1^{3,*,*} \xrightarrow{d_1^3} \dots$$

is illustrated in the upper part of Figure 3.7. Each term is free as a module over $\mathbb{F}_2[w_1]$, and only a basis for this module structure is shown. Dashed vertical arrows indicate $h_0$-multiplications taking $w_1$-divisible values. Its cohomology, $\bar{E}_2^{\sigma,*,*}$, is shown in the lower part of Figure 3.7. Again, each term is free over $\mathbb{F}_2[w_1]$, with a basis given by the filled circles. The open circle shows a $w_1$-multiple, and the dashed vertical line from $h_0 a_{4,2}x_4^2$ exhibits the relation $h_0^2 \cdot a_{4,2}x_4^2 = w_1 \cdot x_4^2$.

The complex

$$\dots \xrightarrow{d_1^3} \bar{E}_1^{4,*,*} \xrightarrow{d_1^4} \bar{E}_1^{5,*,*} \xrightarrow{d_1^5} \bar{E}_1^{6,*,*} \xrightarrow{d_1^6} \bar{E}_1^{7,*,*} \xrightarrow{d_1^7} \dots$$

and its cohomology, $\bar{E}_2^{\sigma,*,*}$ for $4 \le \sigma \le 7$, are shown in Figure 3.8. For larger $\sigma$, this pattern continues $(x_6^2)^2 = x_6^4$-periodically.

LEMMA 3.31. *Suppose $\sigma \ge 3$. Then $\bar{E}_2^{\sigma,*,*}$ is a free $\mathbb{F}_2[w_1]$-module with basis the six classes $a_{k,s}x_4^\sigma$ with $s + \sigma - 2 \le k \le s + \sigma$, $0 \le s \le 3$ and $k \equiv 0, 3 \mod 4$. Furthermore, multiplication by $x_6^4 = a_{4,0}x_4^4$ induces an isomorphism*

$$x_6^4 \colon \bar{E}_2^{\sigma,*,*} \xrightarrow{\cong} \bar{E}_2^{\sigma+4,*,*}$$

*of $(t-s, s)$-bidegree $(20, 4)$.*

PROOF. The $a_{k,s}x_4^\sigma$ with $k \le s+\sigma$, $s \ge 0$ and $k \equiv 0, 3 \mod 4$ are $d_1^\sigma$-cycles. Among these, those with $k \le s+\sigma-3$ are also $d_1^{\sigma-1}$-boundaries. Multiplication by $x_6^4 = a_{4,0}x_4^4$ takes $a_{k,s}x_4^\sigma$ to $a_{k+4,s}x_4^{\sigma+4}$. □

LEMMA 3.32. *The connecting homomorphism*

$$\delta\colon \bar{E}_2^{\sigma-4,*,*}\{x_7^4\} \longrightarrow \bar{E}_2^{\sigma+1,*,*}$$

*in* (3.6) *takes* $c \cdot x_7^4$ *to* $c \cdot a_{4,0}x_4^5$. *Its values for* $c$ *ranging through an* $\mathbb{F}_2[w_1]$*-basis for* $\bar{E}_2^{*,*,*}$ *are listed in Table* 3.1 *and illustrated in Figure* 3.10.

PROOF. This follows from the Leibniz rule

$$d_1(c \cdot x_7^4) = d_1(c) \cdot x_7^4 + c \cdot d_1(x_7^4)$$

when $d_1(c) = 0$, since $d_1(x_7^4) = a_{4,0}x_4^5$ by Lemma 3.28(4). The multiplications are calculated using Proposition 3.26. □

PROPOSITION 3.33. *The Davis–Mahowald* $E_2$*-term* $E_2^{*,*,*}$ *is a free* $\mathbb{F}_2[w_1, x_7^8]$*-module, with basis as listed in Table* 3.2 *and illustrated in Figure* 3.11.

PROOF. By (3.6) we have a short exact sequence

$$0 \to \mathrm{cok}(\delta) \xrightarrow{S} \bar{\bar{E}}_2^{*,*,*} \longrightarrow \ker(\delta) \to 0$$

of $\mathbb{F}_2[w_1]$-modules. No $w_1$-multiples occur among the values $\delta(cx_7^4)$ in Table 3.1, so both $\mathrm{cok}(\delta)$ and $\ker(\delta)$ are free $\mathbb{F}_2[w_1]$-modules. Each basis element $b$ for $\mathrm{cok}(\delta)$ appears as one entry in Table 3.2. To lift each basis element $cx_7^4$ for $\ker(\delta)$, note that if $d_1(cx_7^4) = d_1(a)$ with $a \in \bar{E}_1^{*,*,*}$, then the class of $-a + cx_7^4$ in $\bar{\bar{E}}_2^{*,*,*}$ is such a lift. This produces the remaining entries in Table 3.2, giving $\bar{\bar{E}}_2^{*,*,*}$ as a free $\mathbb{F}_2[w_1]$-module. It follows that $E_2^{*,*,*}$ is a free $\mathbb{F}_2[w_1, x_7^8]$ on the same list of generators. □

REMARK 3.34. The projection $E_1^{*,*,*} \to \bar{E}_1^{*,*,*}$ does not commute with $d_1$, and the section $S\colon \bar{E}_1^{*,*,*} \to E_1^{*,*,*}$ is not multiplicative. Hence the algebra structures in $E_2^{*,*,*}$ and $\bar{E}_2^{*,*,*}$ are not fully compatible. For example, in $E_2^{*,*,*}$ the square of $a_{3,0}x_4^3 = x_6^3 + x_4x_7^2$ is $(x_6^3 + x_4x_7^2)^2 = x_6^6 + x_4^2x_7^4 = a_{6,0}x_4^6 + x_4^2x_7^4$, while in $\bar{E}_2^{*,*,*}$ the square is $a_{6,0}x_4^6$.

PROPOSITION 3.35. *The Davis–Mahowald spectral sequence* (3.1) *collapses at the* $E_2$*-term, so* $E_2^{*,*,*} = E_\infty^{*,*,*}$ *is the associated graded of a multiplicative filtration of* $\mathrm{Ext}_{A(2)_*}(\mathbb{F}_2, \mathbb{F}_2) \cong \mathrm{Ext}_{A(2)}(\mathbb{F}_2, \mathbb{F}_2)$. *In particular,* $\mathrm{Ext}_{A(2)}(\mathbb{F}_2, \mathbb{F}_2)$ *is a free* $\mathbb{F}_2[w_1, w_2]$*-module, generated by classes that are detected by the generators listed in Table* 3.2, *where* $w_2$ *is a class that is detected by* $x_7^8$.

PROOF. The class $x_7^8 \in E_2^{8,8,56}$ is an infinite cycle, because $E_2^{\sigma,9,56} = 0$ for all $\sigma \ge 10$. For each $\mathbb{F}_2[w_1, x_7^8]$-module generator $c \in E_2^{\sigma,s,t}$ in Table 3.2, and each $r \ge 2$, the target group $E_2^{\sigma+r,s+1,t}$ is zero. Hence $d_r = 0$ for each $r \ge 2$. □

REMARK 3.36. Our conclusions agree with those of Davis and Mahowald [**52**, p. 325], except for one tiny typographical error: their class $h_2^5\alpha_{8,4}$ should have been $h_2^5\alpha_{8,3}$, and is the class we denote by $a_{8,3}x_4^5$.

TABLE 3.1. $\mathbb{F}_2[w_1]$-basis for $\bar{E}_2^{*,*,*}$ ($x_6^4$-periodic for $\sigma \geq 3$)

| $\sigma$ | $t-s$ | $s$ | $c$ | | $\delta(cx_7^4)$ |
|---|---|---|---|---|---|
| 0 | 0 | $i$ | $h_0^i$ | $i \in \{0,1\}$ | $a_{4,i}x_4^5$ |
| 0 | 0 | $i$ | $h_0^i$ | $i \geq 2$ | 0 |
| 0 | 1 | 1 | $h_1$ | | 0 |
| 0 | 2 | 2 | $h_1^2$ | | 0 |
| 1 | 3 | 1 | $x_4$ | | $a_{4,0}x_4^6$ |
| 1 | 3 | $1+i$ | $h_0^i x_4$ | $i \in \{1,2\}$ | 0 |
| 1 | 8 | 3 | $h_1 v' x_4$ | | 0 |
| 1 | 9 | 4 | $h_1^2 v' x_4$ | | 0 |
| 2 | 6 | 2 | $x_4^2$ | | 0 |
| 2 | 12 | $3+i$ | $a_{3,1+i}x_4^2$ | $i \in \{0,1\}$ | $a_{7,1+i}x_4^7$ |
| 2 | 12 | $3+i$ | $a_{3,1+i}x_4^2$ | $i \geq 2$ | 0 |
| 2 | 14 | $4+i$ | $a_{4,2+i}x_4^2$ | $i \in \{0,1\}$ | $a_{8,2+i}x_4^7$ |
| 3 | 15 | $3+i$ | $a_{3,i}x_4^3$ | $i \in \{0,1\}$ | $a_{7,i}x_4^8$ |
| 3 | 15 | 5 | $a_{3,2}x_4^3$ | | 0 |
| 3 | 17 | $4+i$ | $a_{4,1+i}x_4^3$ | $i \in \{0,1\}$ | $a_{8,1+i}x_4^8$ |
| 3 | 17 | 6 | $a_{4,3}x_4^3$ | | 0 |
| 4 | 18 | 4 | $a_{3,0}x_4^4$ | | $a_{7,0}x_4^9$ |
| 4 | 18 | 5 | $a_{3,1}x_4^4$ | | 0 |
| 4 | 20 | $4+i$ | $a_{4,i}x_4^4$ | $i \in \{0,1\}$ | $a_{8,i}x_4^9$ |
| 4 | 20 | 6 | $a_{4,2}x_4^4$ | | 0 |
| 4 | 26 | 7 | $a_{7,3}x_4^4$ | | $a_{11,3}x_4^9$ |
| 5 | 21 | 5 | $a_{3,0}x_4^5$ | | 0 |
| 5 | 23 | 5 | $a_{4,0}x_4^5$ | | $a_{8,0}x_4^{10}$ |
| 5 | 23 | 6 | $a_{4,1}x_4^5$ | | 0 |
| 5 | 29 | $7+i$ | $a_{7,2+i}x_4^5$ | $i \in \{0,1\}$ | $a_{11,2+i}x_4^{10}$ |
| 5 | 31 | 8 | $a_{8,3}x_4^5$ | | $a_{12,3}x_4^{10}$ |
| 6 | 26 | 6 | $a_{4,0}x_4^6$ | | 0 |
| 6 | 32 | $7+i$ | $a_{7,1+i}x_4^6$ | $i \in \{0,1\}$ | $a_{11,1+i}x_4^{11}$ |
| 6 | 32 | 9 | $a_{7,3}x_4^6$ | | 0 |
| 6 | 34 | $8+i$ | $a_{8,2+i}x_4^6$ | $i \in \{0,1\}$ | $a_{12,2+i}x_4^{11}$ |

Table 3.2: $\mathbb{F}_2[w_1, x_7^8]$-basis for $E_2^{*,*,*}$ ($x_6^4$-periodic for $\sigma \geq 7$)

| $\sigma$ | $t-s$ | $s$ | generator | | Ext |
|---|---|---|---|---|---|
| 0 | 0 | $i$ | $h_0^i$ | $i \geq 0$ | $h_0^i$ |
| 0 | 1 | 1 | $h_1$ | | $h_1$ |
| 0 | 2 | 2 | $h_1^2$ | | $h_1^2$ |
| 1 | 3 | $1+i$ | $h_0^i x_4$ | $i \in \{0,1,2\}$ | $h_0^i h_2$ |
| 1 | 8 | 3 | $h_1 v' x_4$ | | $c_0$ |
| 1 | 9 | 4 | $h_1^2 v' x_4$ | | $h_1 c_0$ |
| 2 | 6 | 2 | $x_4^2$ | | $h_2^2$ |
| 2 | 12 | $3+i$ | $a_{3,1+i} x_4^2$ | $i \geq 0$ | $h_0^i \alpha$ |
| 2 | 14 | $4+i$ | $a_{4,2+i} x_4^2$ | $i \in \{0,1\}$ | $h_0^i d_0$ |
| 3 | 15 | $3+i$ | $a_{3,i} x_4^3$ | $i \in \{0,1,2\}$ | $h_0^i \beta$ |
| 3 | 17 | $4+i$ | $a_{4,1+i} x_4^3$ | $i \in \{0,1,2\}$ | $h_0^i e_0$ |
| 4 | 18 | $4+i$ | $a_{3,i} x_4^4$ | $i \in \{0,1\}$ | $h_0^i h_2 \beta$ |
| 4 | 20 | $4+i$ | $a_{4,i} x_4^4$ | $i \in \{0,1,2\}$ | $h_0^i g$ |
| 4 | 24 | $6+i$ | $h_0^i(a_{6,2} x_4^4 + h_0^2 x_7^4)$ | $i \geq 0$ | $h_0^i \alpha^2$ |
| 4 | 25 | 5 | $h_1 x_7^4$ | | $\gamma$ |
| 4 | 26 | 6 | $h_1^2 x_7^4$ | | $h_1 \gamma$ |
| 4 | 26 | 7 | $a_{7,3} x_4^4$ | | $\alpha d_0$ |
| 5 | 21 | 5 | $a_{3,0} x_4^5$ | | $h_1 g$ |
| 5 | 27 | $6+i$ | $h_0^i(a_{6,1} x_4^5 + h_0 x_4 x_7^4)$ | $i \in \{0,1\}$ | $h_0^i \alpha\beta$ |
| 5 | 29 | 7 | $a_{7,2+i} x_4^5$ | $i \in \{0,1\}$ | $h_0^i \alpha e_0$ |
| 5 | 31 | 8 | $a_{8,3} x_4^5$ | | $d_0 e_0$ |
| 5 | 32 | 7 | $h_1 v' x_4 x_7^4$ | | $\delta$ |
| 5 | 33 | 8 | $h_1^2 v' x_4 x_7^4$ | | $h_1 \delta$ |
| 6 | 30 | 6 | $a_{6,0} x_4^6 + x_4^2 x_7^4$ | | $\beta^2$ |
| 6 | 32 | $7+i$ | $a_{7,1+i} x_4^6$ | $i \in \{0,1,2\}$ | $h_0^i \alpha g$ |
| 6 | 34 | $8+i$ | $a_{8,2+i} x_4^6$ | $i \in \{0,1\}$ | $h_0^i d_0 g$ |
| 6 | 36 | $9+i$ | $h_0^i(a_{9,3} x_4^6 + a_{3,3} x_4^2 x_7^4)$ | $i \geq 0$ | $h_0^i \alpha^3$ |
| 7 | 35 | 7 | $a_{7,0} x_4^7$ | | $\beta g$ |
| 7 | 37 | 8 | $a_{8,1} x_4^7$ | | $e_0 g$ |
| 7 | 39 | 9 | $a_{9,2} x_4^7 + a_{3,2} x_4^3 x_7^4$ | | $d_0 \gamma$ |
| 7 | 41 | 10 | $a_{10,3} x_4^7 + a_{4,3} x_4^3 x_7^4$ | | $\alpha^2 e_0$ |

Table 3.2: $\mathbb{F}_2[w_1, x_7^8]$-basis for $E_2^{*,*,*}$ ($x_6^4$-periodic for $\sigma \geq 7$) (cont.)

| $\sigma$ | $t-s$ | $s$ | generator | | Ext |
|---|---|---|---|---|---|
| 8 | 40 | 8 | $a_{8,0}x_4^8$ | | $g^2$ |
| 8 | 42 | 9 | $a_{9,1}x_4^8 + a_{3,1}x_4^4x_7^4$ | | $e_0\gamma$ |
| 8 | 44 | 10 | $a_{10,2}x_4^8 + a_{4,2}x_4^4x_7^4$ | | $\alpha^2 g$ |
| 8 | 46 | 11 | $a_{11,3}x_4^8$ | | $\alpha d_0 g$ |
| 9 | 45 | 9 | $a_{9,0}x_4^9 + a_{3,0}x_4^5x_7^4$ | | $\gamma g$ |
| 9 | 47 | 10 | $a_{10,1}x_4^9 + a_{4,1}x_4^5x_7^4$ | | $\alpha\beta g$ |
| 9 | 49 | 11 | $a_{11,2}x_4^9$ | | $\alpha e_0 g$ |
| 9 | 51 | 12 | $a_{12,3}x_4^9$ | | $d_0 e_0 g$ |
| 10 | 50 | 10 | $a_{10,0}x_4^{10} + a_{4,0}x_4^6x_7^4$ | | $\beta^2 g$ |
| 10 | 52 | 11 | $a_{11,1}x_4^{10}$ | | $\alpha g^2$ |
| 10 | 54 | 12 | $a_{12,2}x_4^{10}$ | | $d_0 g^2$ |
| 10 | 56 | 13 | $a_{13,3}x_4^{10} + a_{7,3}x_4^6x_7^4$ | | $\alpha^3 g$ |

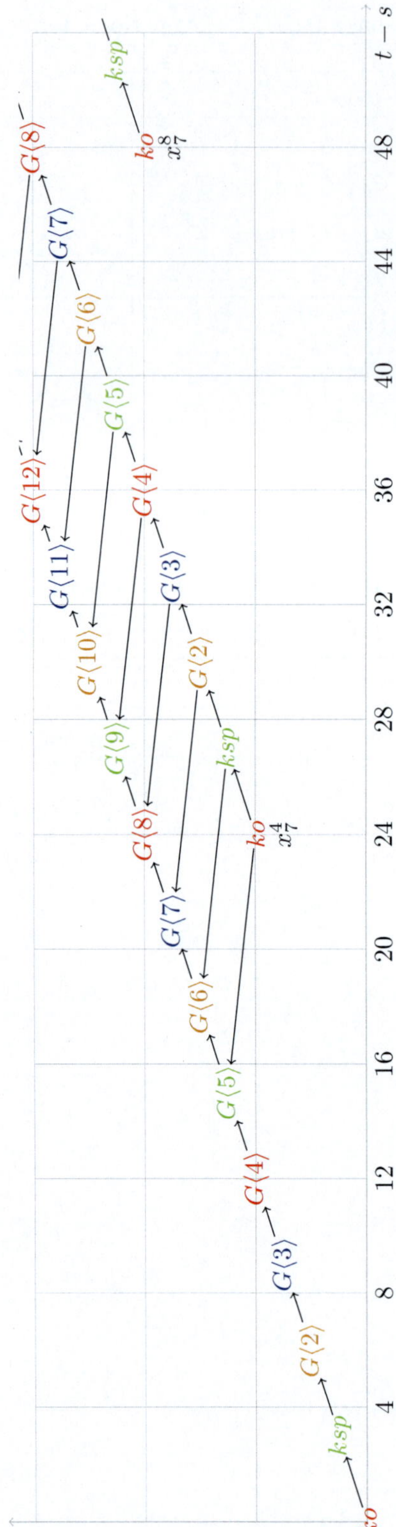

FIGURE 3.9. Schematic view of the Davis–Mahowald $(E_1, d_1)$-term

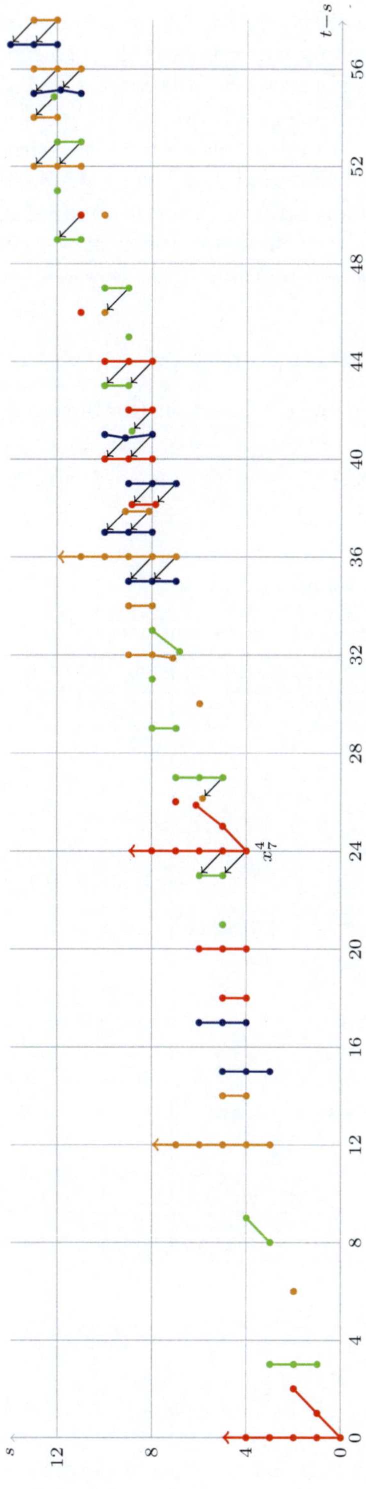

FIGURE 3.10. Connecting homomorphism $\delta \colon \bar{E}_2^{\sigma-4,*,*}\{x_7^4\} \to \bar{E}_2^{\sigma+1,*,*}$

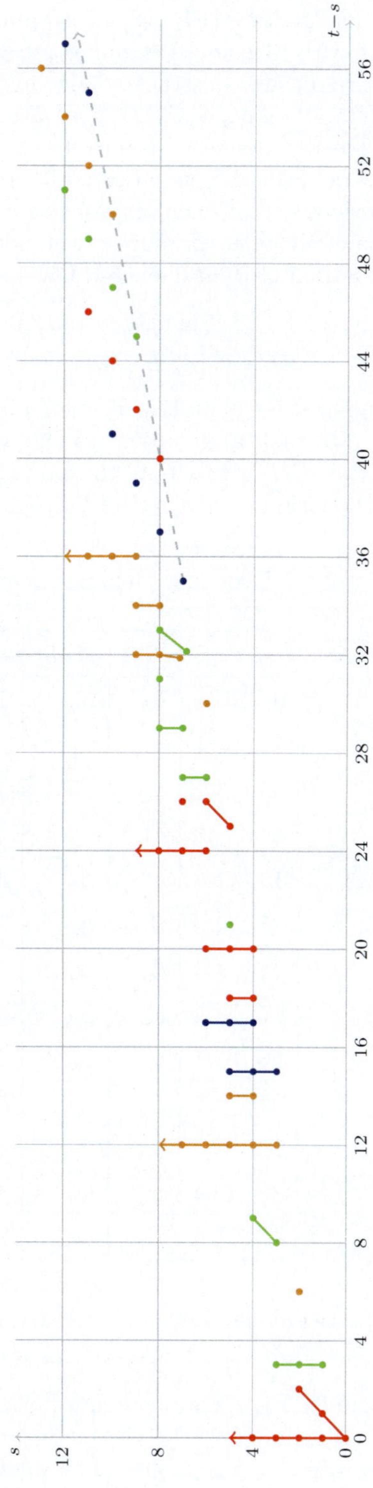

FIGURE 3.11. $\mathbb{F}_2[w_1, x_7^8]$-basis for $E_2^{*,*,*} = E_\infty^{*,*,*} \Longrightarrow \mathrm{Ext}_{A(2)}^{*,*}(\mathbb{F}_2, \mathbb{F}_2)$

## 3.5. The Shimada–Iwai presentation

Shimada and Iwai [**155**, §8] gave a presentation of $\mathrm{Ext}_{A(2)}(\mathbb{F}_2, \mathbb{F}_2)$ as a bigraded $\mathbb{F}_2$-algebra with 13 generators and 54 relations, which we will denote by $SI$. To confirm their result, we construct an algebra homomorphism $\phi\colon SI \to \mathrm{Ext}_{A(2)}(\mathbb{F}_2, \mathbb{F}_2)$ by specifying the images of the generators and then using `ext` to verify the relations. Thereafter we use Gröbner basis methods to find a basis for $SI$ as a free $\mathbb{F}_2[w_1, w_2]$-module. A comparison with the Davis–Mahowald $E_\infty$-term calculated in the previous section then proves that $\phi$ is an isomorphism of $\mathbb{F}_2[w_1, w_2]$-modules, hence also of algebras. In place of the notation used by Shimada and Iwai we will use the notation of Henriques [**54**, Ch. 13], as reviewed in Table 1.3.

Definition 3.37 (Shimada–Iwai). Let

$$SI = \mathbb{F}_2[h_0, h_1, h_2, c_0, \alpha, \beta, d_0, e_0, \gamma, \delta, g, w_1, w_2]/(\sim)$$

be the bigraded commutative $\mathbb{F}_2$-algebra generated by 13 classes in the bidegrees listed in Table 3.3, and subject to the 54 relations listed in Table 3.4. In other words, $SI = P/I$ where $P$ is the polynomial algebra $\mathbb{F}_2[h_0, h_1, h_2, \dots, g, w_1, w_2]$ and $I$ is the ideal $(h_0h_1, h_0^2h_2 + h_1^3, h_1h_2, \dots, \delta g, \gamma\delta + h_1c_0w_2, \delta^2) \subset P$.

Table 3.3: Generators of $SI \cong \mathrm{Ext}_{A(2)}(\mathbb{F}_2, \mathbb{F}_2)$

| $t-s$ | $s$ | [**54**] | [**155**] | `ext` | $E_\infty$ |
|---|---|---|---|---|---|
| 0 | 1 | $h_0$ | $h_0$ | $1_0$ | $h_0$ |
| 1 | 1 | $h_1$ | $h_1$ | $1_1$ | $h_1$ |
| 3 | 1 | $h_2$ | $h_2$ | $1_2$ | $x_4$ |
| 8 | 3 | $c_0$ | $\alpha_1$ | $3_2$ | $h_1v'x_4$ |
| 12 | 3 | $\alpha$ | $\alpha_2$ | $3_3$ | $a_{3,1}x_4^2$ |
| 15 | 3 | $\beta$ | $\alpha_3$ | $3_4$ | $a_{3,0}x_4^3 = x_6^3 + x_4x_7^2$ |
| 14 | 4 | $d_0$ | $\alpha_4$ | $4_4$ | $a_{4,2}x_4^2$ |
| 17 | 4 | $e_0$ | $\alpha_5$ | $4_6$ | $a_{4,1}x_4^3$ |
| 25 | 5 | $\gamma$ | $\alpha_6$ | $5_{11}$ | $h_1x_7^4$ |
| 32 | 7 | $\delta$ | $\alpha_7$ | $7_{11}$ | $h_1v'x_4x_7^4$ |
| 20 | 4 | $g$ | $\omega_1$ | $4_8$ | $a_{4,0}x_4^4 = x_6^4$ |
| 8 | 4 | $w_1$ | $\omega_0$ | $4_1$ | $w_1$ |
| 48 | 8 | $w_2$ | $\alpha_0$ | $8_{19}$ | $x_7^8$ |

Definition 3.38. Let

$$\phi\colon SI \longrightarrow \mathrm{Ext}_{A(2)}(\mathbb{F}_2, \mathbb{F}_2)$$

be the bigraded $\mathbb{F}_2$-algebra homomorphism given by sending each algebra generator $x = h_0, \dots, w_2$ in $SI$ to the class $\phi(x)$ in $\mathrm{Ext}_{A(2)}(\mathbb{F}_2, \mathbb{F}_2)$ represented by the `ext`-cocycle $s_g = 1_0, \dots, 8_{19}$, as given in Table 3.3. We usually omit $\phi$ from the notation, writing $h_0$ in place of $\phi(h_0)$, etc.

TABLE 3.4. Relations in $SI \cong \mathrm{Ext}_{A(2)}(\mathbb{F}_2, \mathbb{F}_2)$

| $t-s$ | $s$ | relation | |
|---|---|---|---|
| 1 | 2 | $h_0h_1 = 0$ | $z$ |
| 3 | 3 | $h_0^2h_2 = h_1^3$ | $h_i$ |
| 4 | 2 | $h_1h_2 = 0$ | $z$ |
| 6 | 3 | $h_0h_2^2 = 0$ | $z$ |
| 8 | 4 | $h_0c_0 = 0$ | $h_i$ |
| 9 | 3 | $h_2^3 = 0$ | $z$ |
| 10 | 5 | $h_1^2c_0 = 0$ | $z$ |
| 11 | 4 | $h_2c_0 = 0$ | $z$ |
| 13 | 4 | $h_1\alpha = 0$ | $z$ |
| 14 | 6 | $h_0^2d_0 = h_2^2w_1$ | $h_i$ |
| 15 | 4 | $h_0\beta = h_2\alpha$ | $h_i$ |
| 15 | 5 | $h_1d_0 = h_0h_2\alpha$ | $h_i$ |
| 16 | 4 | $h_1\beta = 0$ | $z$ |
| 16 | 6 | $c_0^2 = 0$ | $z$ |
| 17 | 5 | $h_0e_0 = h_2d_0$ | $h_i$ |
| 18 | 5 | $h_1e_0 = h_2^2\alpha$ | $h_i$ |
| 20 | 5 | $h_2e_0 = h_0g$ | $h_i$ |
| 20 | 6 | $c_0\alpha = h_0^2g$ | |
| 21 | 5 | $h_2^2\beta = h_1g$ | $h_i$ |
| 22 | 7 | $c_0d_0 = 0$ | $z$ |
| 23 | 5 | $h_2g = 0$ | $z$ |
| 23 | 6 | $c_0\beta = 0$ | $z$ |
| 25 | 6 | $h_0\gamma = 0$ | $z$ |
| 25 | 7 | $c_0e_0 = 0$ | $z$ |
| 26 | 8 | $h_0\alpha d_0 = h_2\beta w_1$ | |
| 27 | 7 | $h_2\alpha^2 = h_1^2\gamma$ | |
| 28 | 6 | $h_2\gamma = 0$ | $z$ |

| $t-s$ | $s$ | relation | |
|---|---|---|---|
| 28 | 7 | $c_0g = 0$ | $z$ |
| 28 | 8 | $d_0^2 = gw_1$ | |
| 29 | 7 | $\beta d_0 = \alpha e_0$ | |
| 30 | 7 | $h_2\alpha\beta = 0$ | $z$ |
| 32 | 7 | $\beta e_0 = \alpha g$ | |
| 32 | 8 | $h_0\delta = h_0\alpha g$ | |
| 33 | 7 | $h_2\beta^2 = 0$ | $z$ |
| 33 | 8 | $c_0\gamma = h_1\delta$ | |
| 34 | 8 | $e_0^2 = d_0g$ | |
| 34 | 9 | $h_1^2\delta = h_0d_0g$ | |
| 35 | 8 | $h_2\delta = 0$ | $z$ |
| 37 | 8 | $\alpha\gamma = e_0g$ | |
| 38 | 10 | $\alpha^2d_0 = \beta^2w_1$ | |
| 39 | 9 | $\alpha^2\beta = d_0\gamma$ | |
| 40 | 8 | $\beta\gamma = g^2$ | |
| 40 | 10 | $c_0\delta = 0$ | $z$ |
| 42 | 9 | $\alpha\beta^2 = e_0\gamma$ | |
| 44 | 10 | $\alpha\delta = 0$ | |
| 45 | 9 | $\beta^3 = \gamma g$ | |
| 46 | 11 | $d_0\delta = 0$ | |
| 47 | 10 | $\beta\delta = 0$ | |
| 48 | 12 | $\alpha^4 = h_0^4w_2 + g^2w_1$ | |
| 49 | 11 | $e_0\delta = 0$ | |
| 50 | 10 | $\gamma^2 = h_1^2w_2 + \beta^2g$ | |
| 52 | 11 | $\delta g = 0$ | |
| 57 | 12 | $\gamma\delta = h_1c_0w_2$ | |
| 64 | 14 | $\delta^2 = 0$ | |

In particular, $\delta$ in $(t-s,s)$-bidegree $(32,7)$ is sent to the class $\delta$ of $7_{11}$, with $h_0\delta = 8_{14} \neq 0$ and $h_1\delta = 8_{15} \neq 0$. In the remaining cases the cocycle $s_g$ is the only nonzero class in its bidegree.

LEMMA 3.39. *$\phi$ is well-defined.*

PROOF. The relations labeled "$z$" in Table 3.4 take place in bidegrees where $\mathrm{Ext}_{A(2)}(\mathbb{F}_2, \mathbb{F}_2)$ is zero. The relations labeled "$h_i$" are evident from the $h_0$-, $h_1$- and $h_2$-multiplications shown in Figures 1.19 and 1.20. To verify the remaining relations we use `ext` to calculate products in $\mathrm{Ext}_{A(2)}(\mathbb{F}_2, \mathbb{F}_2)$, as explained in Remarks 1.6 and 1.13. For instance, this gives $\gamma^2 = 10_{20} + 10_{21}$, $\beta^2 g = 10_{20}$ and $h_1^2 w_2 = 10_{21}$, confirming the relation $\gamma^2 = \beta^2 g + h_1^2 w_2$ in $(t-s, s)$-bidegree $(50, 10)$. □

REMARK 3.40. We use the method of Gröbner bases to make $I \subset P$ and $SI$ computationally accessible. We order the 13 algebra generators as in Table 3.3

$$h_0 > h_1 > h_2 > c_0 > \alpha > \beta > d_0 > e_0 > \gamma > \delta > g > w_1 > w_2 \,, \tag{3.7}$$

and write monomials in these generators in the format

$$m = h_0^{n_1} h_1^{n_2} h_2^{n_3} \cdots g^{n_{11}} w_1^{n_{12}} w_2^{n_{13}}$$

with $n_1, n_2, n_3, \dots, n_{11}, n_{12}, n_{13} \geq 0$. Since we are working over $\mathbb{F}_2$, where 1 is the only nonzero coefficient, there is no need to distinguish between monomials and terms.

Polynomials, which are sums of monomials, are written in reverse lexicographic order. This means that the terms with $n_{13} = 0$ (not containing $w_2$) are followed by the terms with $n_{13} = 1$ (containing a single copy of $w_2$), etc. Ties are broken by considering $n_{12}$ (the number of copies of $w_1$), and so on. For instance, the sum of $h_0^2 h_2$ and $h_1^3$ is written as $h_1^3 + h_0^2 h_2$, with $h_1^3$ preceding $h_0^2 h_2$, because neither term contains $c_0, \dots, w_2$, and $h_1^3$ contains fewer copies of $h_2$ ($n_3 = 0$) than $h_0^2 h_2$ does ($n_3 = 1$). The first monomial in a nonempty sum of terms is called the leading term.

Computer algebra systems like `MAGMA` and `sage` can effectively calculate a reduced Gröbner basis for a given ideal in a finitely generated polynomial ring, such as $I$ in $P$. The Gröbner basis is a generating set $B$ for the ideal $I$. Each element $b \in B$ is a sum of terms $\ell + r$, with $\ell$ the leading term and $r$ the (possibly empty) sum of the remaining terms. Then $\ell \equiv -r \mod I$, and more generally $m\ell \equiv -mr \mod I$ for any monomial $m$. A monomial in $P$ that is divisible by the leading term $\ell$ of an element $b \in B$, i.e., that is a product $m\ell$, is thus equivalent modulo $I$ to the product $-mr$. A monomial is irreducible if it is not divisible by the leading term of any element $b \in B$.

For a Gröbner basis $B$, the set of irreducible monomials $\{m_1, m_2, \dots\}$ in $P$ projects to give a vector space basis $\{m_1+I, m_2+I, \dots\}$ for $P/I$. Each polynomial $p$ in $P$ is equivalent modulo $I$ to a unique sum of irreducible monomials, which can be found by repeatedly replacing each reducible monomial $m\ell$ in $p$ with the sum $-mr$, which consists of monomials later than $m\ell$ in the reverse lexicographic term order. Eventually this process stops, and the resulting sum of irreducible monomials is called the normal form of $p$.

PROPOSITION 3.41. *The reduced Gröbner basis for the ideal $I \subset P$ generated by the Shimada–Iwai relations in Table* 3.4*, with respect to the ordering* (3.7) *of the algebra generators and the graded reverse lexicographic ordering of monomials, is given by the list of* 77 *polynomials in Table* 3.5.

PROOF. This is best verified by a computer algebra system. □

TABLE 3.5. Gröbner basis for the Shimada–Iwai relations

| $t-s$ | $s$ | basis element | $t-s$ | $s$ | basis element |
|---|---|---|---|---|---|
| 1 | 2 | $h_0h_1$ | 32 | 7 | $\beta e_0 + \alpha g$ |
| 3 | 3 | $h_1^3 + h_0^2h_2$ | 32 | 8 | $h_0\delta + h_0\alpha g$ |
| 3 | 4 | $h_0^3h_2$ | 32 | 9 | $h_1d_0e_0 + h_0^2\alpha g$ |
| 4 | 2 | $h_1h_2$ | 33 | 7 | $h_2\beta^2$ |
| 6 | 3 | $h_0h_2^2$ | 33 | 8 | $c_0\gamma + h_1\delta$ |
| 8 | 4 | $h_0c_0$ | 34 | 8 | $e_0^2 + d_0g$ |
| 9 | 3 | $h_2^3$ | 34 | 9 | $h_1^2\delta + h_0d_0g$ |
| 10 | 5 | $h_1^2c_0$ | 35 | 8 | $h_0\beta g$ |
| 11 | 4 | $h_2c_0$ | 35 | 8 | $h_2\delta$ |
| 13 | 4 | $h_1\alpha$ | 35 | 9 | $h_1d_0g$ |
| 14 | 6 | $h_0^2d_0 + h_2^2w_1$ | 37 | 8 | $\alpha\gamma + e_0g$ |
| 15 | 4 | $h_2\alpha + h_0\beta$ | 37 | 9 | $h_0e_0g$ |
| 15 | 5 | $h_0^2\beta + h_1d_0$ | 38 | 9 | $h_1e_0g$ |
| 16 | 4 | $h_1\beta$ | 38 | 10 | $\alpha^2d_0 + \beta^2w_1$ |
| 16 | 6 | $c_0^2$ | 39 | 9 | $\alpha^2\beta + d_0\gamma$ |
| 16 | 6 | $h_1^2d_0$ | 40 | 8 | $\beta\gamma + g^2$ |
| 17 | 5 | $h_2d_0 + h_0e_0$ | 40 | 9 | $h_0g^2$ |
| 17 | 7 | $h_0^3e_0$ | 40 | 10 | $c_0\delta$ |
| 18 | 5 | $h_0h_2\beta + h_1e_0$ | 40 | 10 | $h_1d_0\gamma$ |
| 19 | 6 | $h_1^2e_0$ | 41 | 9 | $h_1g^2$ |
| 20 | 5 | $h_2e_0 + h_0g$ | 41 | 11 | $h_0\alpha^2e_0$ |
| 20 | 6 | $c_0\alpha + h_0^2g$ | 42 | 9 | $\alpha\beta^2 + e_0\gamma$ |
| 20 | 7 | $h_0^3g$ | 43 | 10 | $h_1e_0\gamma$ |
| 21 | 5 | $h_2^2\beta + h_1g$ | 43 | 11 | $\alpha d_0e_0 + \beta gw_1$ |
| 22 | 6 | $h_1^2g$ | 44 | 10 | $\alpha\delta$ |
| 22 | 7 | $c_0d_0$ | 44 | 11 | $h_0\alpha^2g$ |
| 23 | 5 | $h_2g$ | 45 | 9 | $\beta^3 + \gamma g$ |
| 23 | 6 | $c_0\beta$ | 46 | 10 | $h_1\gamma g$ |
| 25 | 6 | $h_0\gamma$ | 46 | 11 | $d_0\delta$ |
| 25 | 7 | $c_0e_0$ | 47 | 10 | $\beta\delta$ |
| 26 | 8 | $h_0\alpha d_0 + h_2\beta w_1$ | 48 | 12 | $\alpha^4 + g^2w_1 + h_0^4w_2$ |
| 27 | 7 | $h_0\alpha\beta + h_1^2\gamma$ | 49 | 11 | $e_0\delta$ |
| 28 | 6 | $h_2\gamma$ | 50 | 10 | $\gamma^2 + \beta^2g + h_1^2w_2$ |
| 28 | 7 | $c_0g$ | 52 | 11 | $\delta g$ |
| 28 | 8 | $d_0^2 + gw_1$ | 53 | 13 | $\alpha^3e_0 + \gamma gw_1$ |
| 29 | 7 | $\beta d_0 + \alpha e_0$ | 56 | 13 | $d_0e_0\gamma + \alpha^3g$ |
| 29 | 9 | $h_0^2\alpha e_0 + h_1gw_1$ | 57 | 12 | $\gamma\delta + h_1c_0w_2$ |
| 30 | 7 | $h_0\beta^2$ | 64 | 14 | $\delta^2$ |
| 31 | 9 | $h_0d_0e_0$ | | | |

PROPOSITION 3.42 (Shimada–Iwai). *$SI$ is free as a module over $\mathbb{F}_2[w_1, w_2]$.*

PROOF. The ordering (3.7) is chosen so that the normal form of polynomials will emphasize terms containing $w_1$ or $w_2$. More precisely, no leading term $\ell$ in Table 3.5 contains $w_1$ or $w_2$. Hence a monomial of the form $mw_1^{n_{12}}w_2^{n_{13}}$ is irreducible if and only if $m$ is irreducible. As $m \in P$ ranges over the irreducible monomials that do not contain $w_1$ or $w_2$, the products $mw_1^{n_{12}}w_2^{n_{13}}$ with $n_{12}, n_{13} \geq 0$ range over all the irreducible monomials, so the cosets $mw_1^{n_{12}}w_2^{n_{13}} + I$ give an $\mathbb{F}_2$-basis for $P/I$. It follows that the cosets $m+I$ give an $\mathbb{F}_2[w_1, w_2]$-basis for $P/I = SI$. □

DEFINITION 3.43. Let $R_0 = \mathbb{F}_2[g, w_1, w_2]$.

REMARK 3.44. The algebra presentation of $SI \cong \operatorname{Ext}_{A(2)}(\mathbb{F}_2, \mathbb{F}_2)$ given in Tables 3.3, 3.4 and 3.5 is precise, but complex. In view of the previous proposition, the module structure over $\mathbb{F}_2[w_1, w_2] \subset SI$ is far simpler. However, $SI$ is infinitely generated over $\mathbb{F}_2[w_1, w_2]$, due to infinite $h_0$- and $g$-towers. For the purposes of the Adams spectral sequence calculations that follow, it will be convenient to view $SI$ as a module over the intermediate algebra $R_0$, as just defined. The $R_0$-module structure of $SI$ is still quite simple, as shown in the following proposition. It is not finitely generated, but this is only due to the presence of $h_0$-towers, which will turn out to remain manageable in our calculations.

PROPOSITION 3.45. *$SI$ is a direct sum of cyclic modules over $R_0 = \mathbb{F}_2[g, w_1, w_2]$, as listed in Table 3.6. Here $\operatorname{Ann}(x)$ denotes the annihilator ideal of $x$, so that*

$$SI \cong \bigoplus_x \langle x \rangle \cong \bigoplus_x \frac{\mathbb{F}_2[g, w_1, w_2]}{\operatorname{Ann}(x)}\{x\}\,.$$

PROOF. Each irreducible monomial in $P$ can be written in the form

$$q = mg^{n_{11}}w_1^{n_{12}}w_2^{n_{13}}\,,$$

where $m$ is an irreducible monomial that does not contain $g$, $w_1$ or $w_2$. However, not all of these products $q$ are irreducible. The elements in Table 3.5 with leading term containing $g$ are

$$h_0^3 g, h_1^2 g, h_2 g, c_0 g, h_0\beta g, h_1 d_0 g, h_0 e_0 g, h_1 e_0 g, h_0\alpha^2 g, h_1\gamma g, \delta g, h_0 g^2, h_1 g^2\,.$$

Hence a product $q$ is reducible precisely if $n_{11} \geq 1$ and $m$ is divisible by one of the coefficients $h_0^3, h_1^2, \dots, h_1\gamma$ or $\delta$, or if $n_{11} \geq 2$ and $m$ is divisible by $h_0$ or $h_1$. In these cases the monomial $q$ represents 0 in $P/I$.

Thus, as $m \in P$ ranges over the irreducible monomials that do not contain $g$, $w_1$ or $w_2$, the images $x = m + I$ generate $P/I = SI$ as a direct sum of cyclic $R_0 = \mathbb{F}_2[g, w_1, w_2]$-modules. If $m$ is divisible by $h_0^3, h_1^2, \dots, h_1\gamma$ or $\delta$, then the annihilator ideal of $x$ is $\operatorname{Ann}(x) = (g)$. Otherwise, if $m$ is divisible by $h_0$ or $h_1$, then $\operatorname{Ann}(x) = (g^2)$. In the remaining cases, $\operatorname{Ann}(x) = (0)$, so $x$ generates a free summand.

The generators $x$ of the cyclic summands in $SI$ project to an $\mathbb{F}_2$-basis for $SI/(g, w_1, w_2)$. We filter this algebra by the powers of its maximal ideal

$$\mathfrak{m} = (h_0, h_1, h_2, c_0, \alpha, d_0, \beta, e_0, \gamma, \delta)\,,$$

which are

$$\mathfrak{m}^2 = (h_0^2, h_1^2, h_0h_2, h_2^2, h_1c_0, h_0\alpha, h_0d_0, h_0\beta, h_1d_0, h_0e_0, h_2\beta, h_1e_0, \\ \alpha^2, h_1\gamma, \alpha d_0, \alpha\beta, \alpha e_0, \beta^2, d_0e_0, h_1\delta, d_0\gamma, e_0\gamma)\,,$$

$$\mathfrak{m}^3 = (h_0^3, h_0^2h_2, h_0^2\alpha, h_1d_0, h_0^2e_0, h_1e_0, h_0\alpha^2, h_1^2\gamma, h_0\alpha e_0, \alpha^3, d_0\gamma, \alpha^2e_0, e_0\gamma)$$

and

$$\mathfrak{m}^i = (h_0^i, h_0^{i-1}\alpha, h_0^{i-2}\alpha^2, h_0^{i-3}\alpha^3)$$

for each $i \geq 4$. Here the generators of $\mathfrak{m}^2$ are the nonzero normal forms of the products of pairs of generators of $\mathfrak{m}$, etc. Furthermore,

$$\begin{aligned}
\mathfrak{m}/\mathfrak{m}^2 &= \mathbb{F}_2\{h_0, h_1, h_2, c_0, \alpha, d_0, \beta, e_0, \gamma, \delta\}, \\
\mathfrak{m}^2/\mathfrak{m}^3 &= \mathbb{F}_2\{h_0^2, h_1^2, h_0h_2, h_2^2, h_1c_0, h_0\alpha, h_0d_0, h_0\beta, h_0e_0, h_2\beta, \\
&\qquad \alpha^2, h_1\gamma, \alpha d_0, \alpha\beta, \alpha e_0, \beta^2, d_0e_0, h_1\delta\}, \\
\mathfrak{m}^3/\mathfrak{m}^4 &= \mathbb{F}_2\{h_0^3, h_0^2h_2, h_0^2\alpha, h_1d_0, h_0^2e_0, h_1e_0, h_0\alpha^2, h_1^2\gamma, h_0\alpha e_0, \\
&\qquad \alpha^3, d_0\gamma, \alpha^2e_0, e_0\gamma\}
\end{aligned}$$

and

$$\mathfrak{m}^i/\mathfrak{m}^{i+1} = \mathbb{F}_2\{h_0^i, h_0^{i-1}\alpha, h_0^{i-2}\alpha^2, h_0^{i-3}\alpha^3\}$$

for each $i \geq 4$. Letting $m$ range over these $\mathbb{F}_2$-bases for $\mathfrak{m}^i/\mathfrak{m}^{i+1}$ for $i \geq 0$, the corresponding classes $x = m + I$ give the module generators of $SI$ over $R_0 = \mathbb{F}_2[g, w_1, w_2]$, as listed in Table 3.6 and illustrated in Figures 3.12 and 3.13. □

Table 3.6: $R_0 = \mathbb{F}_2[g, w_1, w_2]$-module generators of $SI \cong \mathrm{Ext}_{A(2)}(\mathbb{F}_2, \mathbb{F}_2)$, with $i \geq 0$ in each $h_0$-tower

| $t-s$ | $s$ | $g$ | $x$ | $\mathrm{Ann}(x)$ |
|---|---|---|---|---|
| 0 | 0 | 0 | $1$ | $(0)$ |
| 0 | 1 | 0 | $h_0$ | $(g^2)$ |
| 0 | 2 | 0 | $h_0^2$ | $(g^2)$ |
| 0 | 3 | 0 | $h_0^3$ | $(g)$ |
| 0 | $4+i$ | 0 | $h_0^{4+i}$ | $(g)$ |
| 1 | 1 | 1 | $h_1$ | $(g^2)$ |
| 2 | 2 | 1 | $h_1^2$ | $(g)$ |
| 3 | 1 | 2 | $h_2$ | $(g)$ |
| 3 | 2 | 2 | $h_0h_2$ | $(g)$ |
| 3 | 3 | 1 | $h_0^2h_2$ | $(g)$ |
| 6 | 2 | 3 | $h_2^2$ | $(g)$ |
| 8 | 3 | 2 | $c_0$ | $(g)$ |
| 9 | 4 | 2 | $h_1c_0$ | $(g)$ |
| 12 | 3 | 3 | $\alpha$ | $(0)$ |
| 12 | 4 | 3 | $h_0\alpha$ | $(g^2)$ |
| 12 | 5 | 4 | $h_0^2\alpha$ | $(g^2)$ |
| 12 | $6+i$ | 4 | $h_0^{3+i}\alpha$ | $(g)$ |

Table 3.6: $R_0 = \mathbb{F}_2[g, w_1, w_2]$-module generators of $SI \cong \mathrm{Ext}_{A(2)}(\mathbb{F}_2, \mathbb{F}_2)$, with $i \geq 0$ in each $h_0$-tower (cont.)

| $t-s$ | $s$ | $g$ | $x$ | $\mathrm{Ann}(x)$ |
|---|---|---|---|---|
| 14 | 4 | 4 | $d_0$ | $(0)$ |
| 14 | 5 | 5 | $h_0d_0$ | $(g^2)$ |
| 15 | 3 | 4 | $\beta$ | $(0)$ |
| 15 | 4 | 5 | $h_0\beta$ | $(g)$ |
| 15 | 5 | 6 | $h_1d_0$ | $(g)$ |
| 17 | 4 | 6 | $e_0$ | $(0)$ |
| 17 | 5 | 7 | $h_0e_0$ | $(g)$ |
| 17 | 6 | 6 | $h_0^2e_0$ | $(g)$ |
| 18 | 4 | 7 | $h_2\beta$ | $(g)$ |
| 18 | 5 | 8 | $h_1e_0$ | $(g)$ |
| 24 | 6 | 8 | $\alpha^2$ | $(0)$ |
| 24 | 7 | 7 | $h_0\alpha^2$ | $(g)$ |
| 24 | $8+i$ | 8 | $h_0^{2+i}\alpha^2$ | $(g)$ |
| 25 | 5 | 11 | $\gamma$ | $(0)$ |
| 26 | 6 | 9 | $h_1\gamma$ | $(g)$ |
| 26 | 7 | 8 | $\alpha d_0$ | $(0)$ |
| 27 | 6 | 10 | $\alpha\beta$ | $(0)$ |
| 27 | 7 | 9 | $h_1^2\gamma$ | $(g)$ |
| 29 | 7 | 10 | $\alpha e_0$ | $(0)$ |
| 29 | 8 | 12 | $h_0\alpha e_0$ | $(g)$ |
| 30 | 6 | 11 | $\beta^2$ | $(0)$ |
| 31 | 8 | 13 | $d_0e_0$ | $(0)$ |
| 32 | 7 | 11 | $\delta$ | $(g)$ |
| 33 | 8 | 15 | $h_1\delta$ | $(g)$ |
| 36 | 9 | 17 | $\alpha^3$ | $(0)$ |
| 36 | $10+i$ | 14 | $h_0^{1+i}\alpha^3$ | $(g)$ |
| 39 | 9 | 18 | $d_0\gamma$ | $(0)$ |
| 41 | 10 | 16 | $\alpha^2e_0$ | $(0)$ |
| 42 | 9 | 19 | $e_0\gamma$ | $(0)$ |

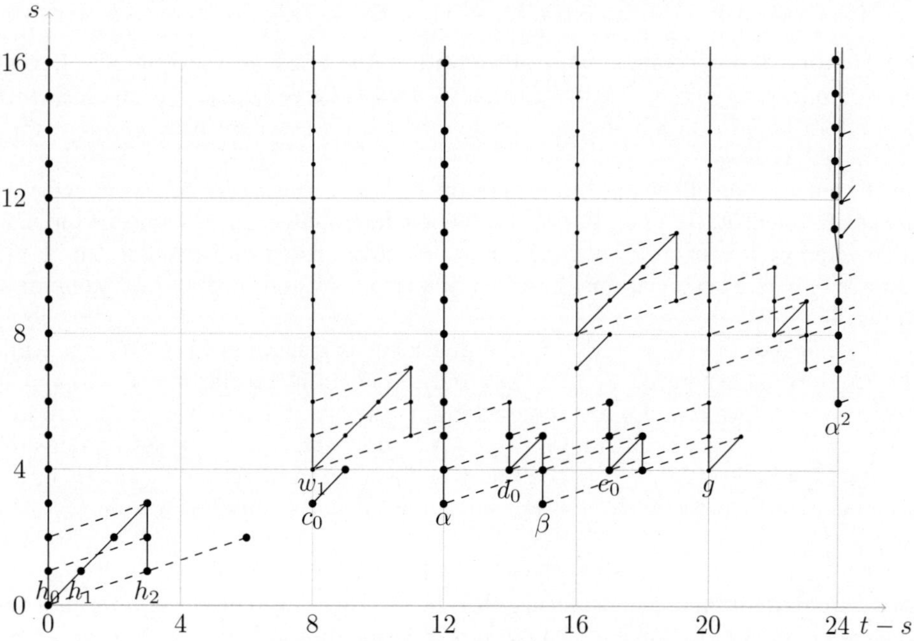


FIGURE 3.12. $R_0 = \mathbb{F}_2[g, w_1, w_2]$-module generators, indicated by •, of $\mathrm{Ext}_{A(2)}(\mathbb{F}_2, \mathbb{F}_2)$ for $0 \le t - s \le 24$

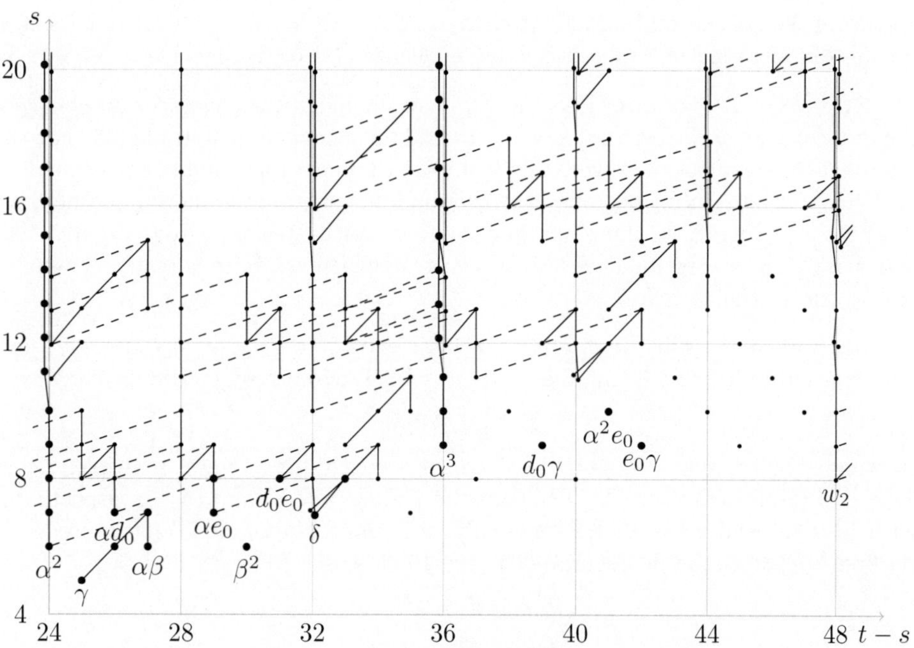


FIGURE 3.13. $R_0 = \mathbb{F}_2[g, w_1, w_2]$-module generators, indicated by •, of $\mathrm{Ext}_{A(2)}(\mathbb{F}_2, \mathbb{F}_2)$ for $24 \le t - s \le 48$

THEOREM 3.46 (Shimada–Iwai). *$\phi\colon SI \to \mathrm{Ext}_{A(2)}(\mathbb{F}_2, \mathbb{F}_2)$ is an isomorphism.*

PROOF. The decomposition in Proposition 3.45 of $SI$ as a direct sum of cyclic $R_0 = \mathbb{F}_2[g, w_1, w_2]$-modules splits further as a sum of free $\mathbb{F}_2[w_1, w_2]$-modules, with generators $\{x\}$, $\{x, xg\}$ or $\{x, xg, xg^2, \dots\}$ in the cases where $\mathrm{Ann}(x) = (g)$, $(g^2)$ or $(0)$, respectively.

When $x$ is one of the algebra generators $h_0, h_1, \dots, w_1, w_2$ of $SI$, its image $\phi(x)$ in the abutment $\mathrm{Ext}_{A(2)}(\mathbb{F}_2, \mathbb{F}_2)$ of the Davis–Mahowald spectral sequence for $A(2)$ is detected by a nonzero class in the $E_2 = E_\infty$-term listed in Table 3.2. For $x \neq \delta$ there is only one such class in the given bidegree, as listed in the "$E_\infty$"-column of Table 3.3.

In bidegree $(t-s, s) = (32, 7)$ the abutment is generated by $\delta$ and $\alpha g$, while the $E_\infty$-term is generated by $h_1 v' x_4 x_7^4$ and $a_{7,1} x_4^6$, in filtrations $\sigma = 5$ and $\sigma = 6$, respectively. Since $\alpha$ and $g$ are detected in filtrations $\sigma = 2$ and $\sigma = 4$, the product $\alpha g$ must be detected in filtration $\sigma \geq 6$. Alternatively, $h_1\delta \neq 0$ must be detected by $h_1^2 v' x_4 x_7^4$ in filtration $\sigma = 5$, so $\delta$ must be detected in filtration $\sigma \leq 5$. By either argument, $\alpha g$ is detected by $a_{7,1} x_4^6$, and $\delta$ and $\delta' = \delta + \alpha g$ are both detected by $h_1 v' x_4 x_7^4$.

For each $\mathbb{F}_2[w_1, w_2]$-module generator $x$ in $SI$ we can now use the multiplicative structure to determine the detecting class in the Davis–Mahowald $E_\infty$-term of the image $\phi(x)$ in $\mathrm{Ext}_{A(2)}(\mathbb{F}_2, \mathbb{F}_2)$. The results are listed in the "Ext"- and "generator"-columns of Table 3.2, and show that $\phi$ induces a bijection between the $\mathbb{F}_2[w_1, w_2]$-module generators of $SI$ and the $\mathbb{F}_2[w_1, x_7^8]$-module generators of $E_\infty^{*,*,*}$. A few cases require special attention: Each product $h_1 g$, $\alpha^2$, $\gamma g$, $\alpha\beta$, $\beta^2$, $\alpha^3$, $d_0\gamma$, $\alpha^2 e_0$ and $e_0\gamma$ is the unique nonzero class in its bidegree, as calculated by `ext`, and this determines its detecting class in the Davis–Mahowald $E_\infty$-term. The products $h_1 d_0$, $h_1 e_0$ and $h_1^2\gamma$ appear in the non-normal forms $h_0^2\beta$, $h_0 h_2 \beta$ and $h_0\alpha\beta$, respectively. It follows that $\phi$ is an isomorphism of $\mathbb{F}_2[w_1, w_2]$-modules, hence also of $\mathbb{F}_2$-algebras. □

REMARK 3.47. For later reference, we have included the generator number $g$ of the `ext`-cocycle $s_g$ corresponding to each module generator $x$ in Table 3.6. For the infinite $h_0$-towers, parameterized by $i \geq 0$, only the generator number corresponding to $i = 0$ is given. In all but one case the module generator is the unique nonzero class in its bidegree, so the generator number can be read off from Figures 1.11 and 1.12. The exceptional case is that of $\delta$, which we have already chosen to correspond to the cocycle $7_{11}$.

REMARK 3.48. The direct sum of the 16 free $R_0 = \mathbb{F}_2[g, w_1, w_2]$-module summands listed in Table 3.6 contains a Mahowald–Tangora wedge [**108**] of the form

$$\mathbb{F}_2[v_1, w]\{\beta g\},$$

starting in bidegree $(t-s, s) = (35, 7)$, together with its $w_2$-power multiples. Here $v_1$ and $w$ are formal symbols of bidegree $(t-s, s) = (2, 1)$ and $(5, 1)$, respectively, with $v_1^4 = w_1$ and $w^4 = g$. Less formally, the (first) Mahowald–Tangora wedge is the free $\mathbb{F}_2[g, w_1]$-module generated by the 16 classes

$$\beta g, e_0 g, d_0\gamma, \alpha^2 e_0,$$
$$g^2, e_0\gamma, \alpha^2 g, \alpha d_0 g,$$
$$\gamma g, \alpha\beta g, \alpha e_0 g, d_0 e_0 g,$$
$$\beta^2 g, \alpha g^2, d_0 g^2, \alpha^3 g.$$

See Figure 3.14.

Our discussion of the Adams spectral sequence for *tmf* continues in Chapter 5, where we determine the differential pattern that leads from $E_2(tmf) = \mathrm{Ext}_{A(2)}(\mathbb{F}_2, \mathbb{F}_2)$ to $E_\infty(tmf)$.

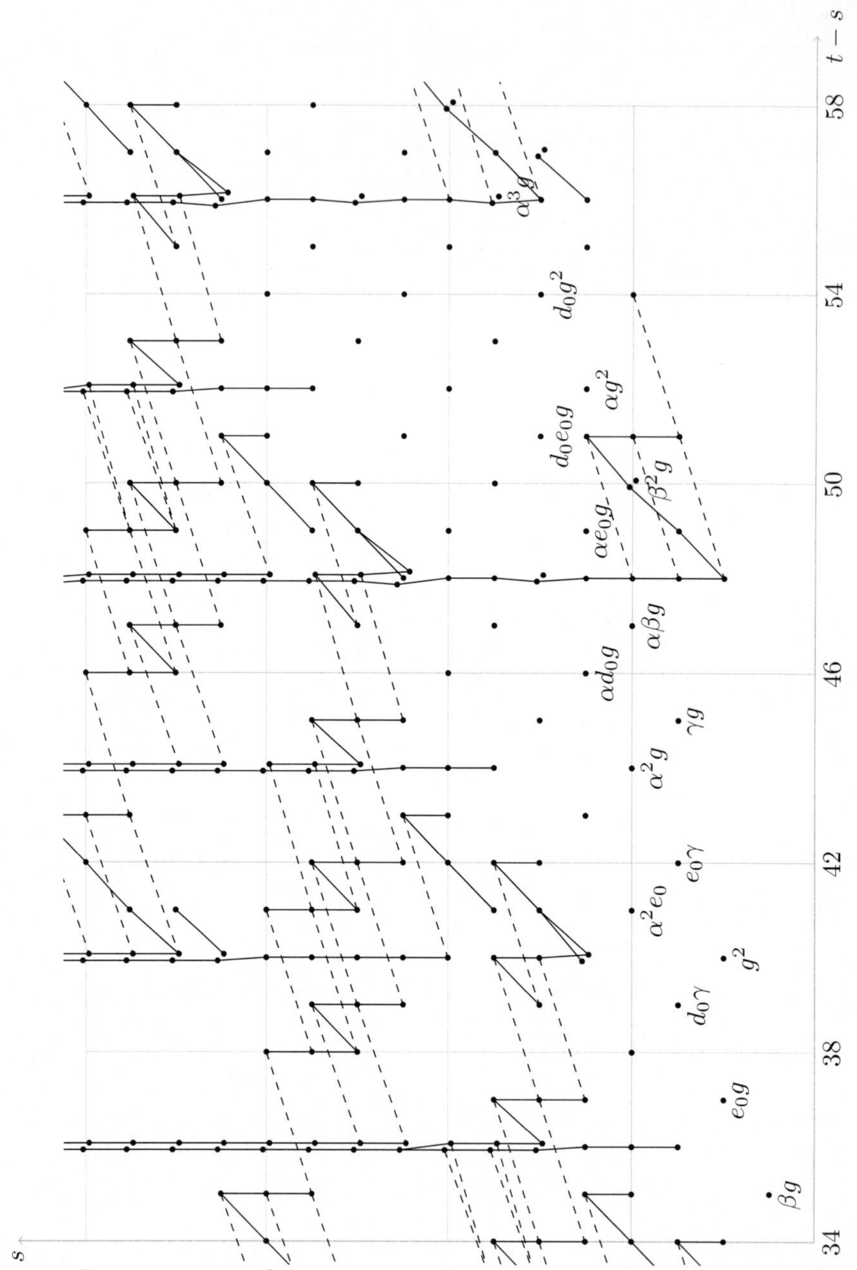


FIGURE 3.14. $\mathbb{F}_2[g, w_1]$-module basis for the (first) Mahowald–Tangora wedge in $\mathrm{Ext}_{A(2)}(\mathbb{F}_2, \mathbb{F}_2)$

CHAPTER 4

# Ext with coefficients

We use long exact sequences of Ext-groups to determine $\mathrm{Ext}_{A(2)}(M, \mathbb{F}_2)$ as an $R_0 = \mathbb{F}_2[g, w_1, w_2]$-module, for $M$ equal to $M_1$, $M_2$ and $M_4$. In each case we also determine a minimal generating set for $\mathrm{Ext}_{A(2)}(M, \mathbb{F}_2)$ as a module over $\mathrm{Ext}_{A(2)}(\mathbb{F}_2, \mathbb{F}_2)$.

## 4.1. Coefficients in $M_1$

Recall our notations from Section 1.4. The short exact sequence of $A(2)$-modules

$$0 \to \Sigma\mathbb{F}_2 \longrightarrow M_1 \longrightarrow \mathbb{F}_2 \to 0$$

represents $h_0$ in $\mathrm{Ext}^{1,1}_{A(2)}(\mathbb{F}_2, \mathbb{F}_2)$. In the induced long exact sequence

$$\dots \xrightarrow{\delta} \mathrm{Ext}^{*,*}_{A(2)}(\mathbb{F}_2, \mathbb{F}_2) \xrightarrow{i} \mathrm{Ext}^{*,*}_{A(2)}(M_1, \mathbb{F}_2)$$
$$\xrightarrow{j} \mathrm{Ext}^{*,*}_{A(2)}(\Sigma\mathbb{F}_2, \mathbb{F}_2) \xrightarrow{\delta} \mathrm{Ext}^{*+1,*}_{A(2)}(\mathbb{F}_2, \mathbb{F}_2) \longrightarrow \dots$$

the connecting homomorphism $\delta$ is therefore given by multiplication by $h_0$. Hence the long exact sequence breaks up into short exact sequences

$$0 \to \mathrm{cok}(h_0)^{s,t} \xrightarrow{i} \mathrm{Ext}^{s,t}_{A(2)}(M_1, \mathbb{F}_2) \xrightarrow{j} \ker(h_0)^{s,t-1} \to 0\,,$$

where $\mathrm{Ext}^{s,t}_{A(2)}(\mathbb{F}_2, \mathbb{F}_2)/\operatorname{im}(h_0) = \mathrm{cok}(h_0)^{s,t}$ and $\ker(h_0)^{s,t} \subset \mathrm{Ext}^{s,t}_{A(2)}(\mathbb{F}_2, \mathbb{F}_2)$.

LEMMA 4.1. *The kernel and cokernel of $h_0$ are both direct sums of cyclic $R_0$-modules, with generators and annihilator ideals as listed in Table* 4.1.

PROOF. For each class $x$ listed in Table 3.6, spanning a cyclic $R_0$-module summand $\langle x\rangle$ of $SI \cong \mathrm{Ext}_{A(2)}(\mathbb{F}_2, \mathbb{F}_2)$, we express $h_0 x$ as an element in a summand $\langle y\rangle$ and record the kernel and cokernel of the $R_0$-module homomorphism $h_0 \colon \langle x\rangle \to \langle y\rangle$. In most cases $h_0 x = 0$ or $y = h_0 x$. The less obvious cases are

$$\begin{aligned} h_0 \cdot h_0 d_0 &= w_1 \cdot h_2^2 \\ h_0 \cdot h_0\beta &= h_1 d_0 \\ h_0 \cdot h_2\beta &= h_1 e_0 \\ h_0 \cdot \alpha d_0 &= w_1 \cdot h_2\beta \\ h_0 \cdot \alpha\beta &= h_1^2\gamma \\ h_0 \cdot h_0\alpha e_0 &= g w_1 \cdot h_1 \\ h_0 \cdot \delta &= g \cdot h_0\alpha\,, \end{aligned}$$

which are clear from Table 3.5 and visible in Figures 3.12 and 3.13. Only in the last case is there some interaction between several cyclic summands, with $h_0 \colon R_0 \oplus R_0/(g) \cong \langle\alpha\rangle \oplus \langle\delta\rangle \longrightarrow \langle h_0\alpha\rangle \cong R_0/(g^2)$. Its kernel is $\langle \delta + \alpha g\rangle = \langle\delta'\rangle \cong R_0$, while the cokernel is zero. In Table 4.1 the $\ker(h_0)$-entries for $x = \alpha$ and $x = \delta$ have therefore been combined, and appear together with the latter generator. □

Table 4.1: Direct sum decompositions of the kernel and cokernel of multiplication by $h_0$, with $i \geq 0$ in each $h_0$-tower

| $t-s$ | $s$ | $g$ | $\ker(h_0)$ | $x$ | $h_0x$ | $\mathrm{cok}(h_0)$ |
|---|---|---|---|---|---|---|
| 0 | 0 | 0 | $\langle g^2\rangle = R_0$ | 1 | $h_0$ | $\langle 1\rangle = R_0$ |
| 0 | 1 | 0 | 0 | $h_0$ | $h_0^2$ | 0 |
| 0 | 2 | 0 | $\langle h_0^2 g\rangle = R_0/(g)$ | $h_0^2$ | $h_0^3$ | 0 |
| 0 | $3+i$ | 0 | 0 | $h_0^{3+i}$ | $h_0^{4+i}$ | 0 |
| 1 | 1 | 1 | $\langle h_1\rangle = R_0/(g^2)$ | $h_1$ | 0 | $\langle h_1\rangle = R_0/(g^2, gw_1)$ |
| 2 | 2 | 1 | $\langle h_1^2\rangle = R_0/(g)$ | $h_1^2$ | 0 | $\langle h_1^2\rangle = R_0/(g)$ |
| 3 | 1 | 2 | 0 | $h_2$ | $h_0h_2$ | $\langle h_2\rangle = R_0/(g)$ |
| 3 | 2 | 2 | 0 | $h_0h_2$ | $h_0^2h_2$ | 0 |
| 3 | 3 | 1 | $\langle h_0^2h_2\rangle = R_0/(g)$ | $h_0^2h_2$ | 0 | 0 |
| 6 | 2 | 3 | $\langle h_2^2\rangle = R_0/(g)$ | $h_2^2$ | 0 | $\langle h_2^2\rangle = R_0/(g, w_1)$ |
| 8 | 3 | 2 | $\langle c_0\rangle = R_0/(g)$ | $c_0$ | 0 | $\langle c_0\rangle = R_0/(g)$ |
| 9 | 4 | 2 | $\langle h_1c_0\rangle = R_0/(g)$ | $h_1c_0$ | 0 | $\langle h_1c_0\rangle = R_0/(g)$ |
| 12 | 3 | 3 | – (cf. $x = \delta$) | $\alpha$ | $h_0\alpha$ | $\langle \alpha\rangle = R_0$ |
| 12 | 4 | 3 | 0 | $h_0\alpha$ | $h_0^2\alpha$ | 0 |
| 12 | 5 | 4 | $\langle h_0^2\alpha g\rangle = R_0/(g)$ | $h_0^2\alpha$ | $h_0^3\alpha$ | 0 |
| 12 | $6+i$ | 4 | 0 | $h_0^{3+i}\alpha$ | $h_0^{4+i}\alpha$ | 0 |
| 14 | 4 | 4 | $\langle d_0g^2\rangle = R_0$ | $d_0$ | $h_0d_0$ | $\langle d_0\rangle = R_0$ |
| 14 | 5 | 5 | $\langle h_0d_0g\rangle = R_0/(g)$ | $h_0d_0$ | $w_1 \cdot h_2^2$ | 0 |
| 15 | 3 | 4 | $\langle \beta g\rangle = R_0$ | $\beta$ | $h_0\beta$ | $\langle \beta\rangle = R_0$ |
| 15 | 4 | 5 | 0 | $h_0\beta$ | $h_1d_0$ | 0 |
| 15 | 5 | 6 | $\langle h_1d_0\rangle = R_0/(g)$ | $h_1d_0$ | 0 | 0 |
| 17 | 4 | 6 | $\langle e_0g\rangle = R_0$ | $e_0$ | $h_0e_0$ | $\langle e_0\rangle = R_0$ |
| 17 | 5 | 7 | 0 | $h_0e_0$ | $h_0^2e_0$ | 0 |
| 17 | 6 | 6 | $\langle h_0^2e_0\rangle = R_0/(g)$ | $h_0^2e_0$ | 0 | 0 |
| 18 | 4 | 7 | 0 | $h_2\beta$ | $h_1e_0$ | $\langle h_2\beta\rangle = R_0/(g, w_1)$ |
| 18 | 5 | 8 | $\langle h_1e_0\rangle = R_0/(g)$ | $h_1e_0$ | 0 | 0 |
| 24 | 6 | 8 | $\langle \alpha^2 g\rangle = R_0$ | $\alpha^2$ | $h_0\alpha^2$ | $\langle \alpha^2\rangle = R_0$ |
| 24 | $7+i$ | 7 | 0 | $h_0^{1+i}\alpha^2$ | $h_0^{2+i}\alpha^2$ | 0 |
| 25 | 5 | 11 | $\langle \gamma\rangle = R_0$ | $\gamma$ | 0 | $\langle \gamma\rangle = R_0$ |

Table 4.1: Direct sum decompositions of the kernel and cokernel of multiplication by $h_0$, with $i \geq 0$ in each $h_0$-tower (cont.)

| $t-s$ | $s$ | $g$ | $\ker(h_0)$ | $x$ | $h_0x$ | $\mathrm{cok}(h_0)$ |
|---|---|---|---|---|---|---|
| 26 | 6 | 9 | $\langle h_1\gamma\rangle = R_0/(g)$ | $h_1\gamma$ | 0 | $\langle h_1\gamma\rangle = R_0/(g)$ |
| 26 | 7 | 8 | $\langle \alpha d_0 g\rangle = R_0$ | $\alpha d_0$ | $w_1 \cdot h_2\beta$ | $\langle \alpha d_0\rangle = R_0$ |
| 27 | 6 | 10 | $\langle \alpha\beta g\rangle = R_0$ | $\alpha\beta$ | $h_1^2\gamma$ | $\langle \alpha\beta\rangle = R_0$ |
| 27 | 7 | 9 | $\langle h_1^2\gamma\rangle = R_0/(g)$ | $h_1^2\gamma$ | 0 | 0 |
| 29 | 7 | 10 | $\langle \alpha e_0 g\rangle = R_0$ | $\alpha e_0$ | $h_0\alpha e_0$ | $\langle \alpha e_0\rangle = R_0$ |
| 29 | 8 | 12 | 0 | $h_0\alpha e_0$ | $gw_1 \cdot h_1$ | 0 |
| 30 | 6 | 11 | $\langle \beta^2\rangle = R_0$ | $\beta^2$ | 0 | $\langle \beta^2\rangle = R_0$ |
| 31 | 8 | 13 | $\langle d_0e_0\rangle = R_0$ | $d_0e_0$ | 0 | $\langle d_0e_0\rangle = R_0$ |
| 32 | 7 | 11 | $\langle \delta'\rangle = R_0$ | $\delta$ | $g \cdot h_0\alpha$ | $\langle \delta\rangle = R_0/(g)$ |
| 33 | 8 | 15 | $\langle h_1\delta\rangle = R_0/(g)$ | $h_1\delta$ | 0 | $\langle h_1\delta\rangle = R_0/(g)$ |
| 36 | 9 | 17 | $\langle \alpha^3 g\rangle = R_0$ | $\alpha^3$ | $h_0\alpha^3$ | $\langle \alpha^3\rangle = R_0$ |
| 36 | $10+i$ | 14 | 0 | $h_0^{1+i}\alpha^3$ | $h_0^{2+i}\alpha^3$ | 0 |
| 39 | 9 | 18 | $\langle d_0\gamma\rangle = R_0$ | $d_0\gamma$ | 0 | $\langle d_0\gamma\rangle = R_0$ |
| 41 | 10 | 16 | $\langle \alpha^2 e_0\rangle = R_0$ | $\alpha^2 e_0$ | 0 | $\langle \alpha^2 e_0\rangle = R_0$ |
| 42 | 9 | 19 | $\langle e_0\gamma\rangle = R_0$ | $e_0\gamma$ | 0 | $\langle e_0\gamma\rangle = R_0$ |

PROPOSITION 4.2. *$\mathrm{Ext}_{A(2)}(M_1, \mathbb{F}_2)$ is a direct sum of cyclic $R_0$-modules, together with one non-cyclic $R_0$-module, with generators and annihilator ideals as listed in Table* 4.2.

PROOF. We use `ext` as discussed in Remark 1.29 to determine the $R_0$-module extensions of summands in $\ker(h_0)$ by summands in $\mathrm{cok}(h_0)$. Each summand in $\ker(h_0)$ has a generator of the form $y = xg^n$, and we choose a lift $\widetilde{y}$ in $\mathrm{Ext}_{A(2)}(M_1, \mathbb{F}_2)$ with $j(\widetilde{y}) = y$. In most cases the lift is unique, but for $xg^n = h_1\gamma$ we prefer $6_8 = h_1\widetilde{\gamma}$ over $6_7$, for $xg^n = h_0^2\alpha g$ we prefer $9_7 = h_1\widetilde{d_0e_0}$ over $9_6$, for $xg^n = h_1\delta$ we prefer $8_{10} = h_1\widetilde{\delta'}$ over $8_9 + 8_{10}$, and for $xg^n = \alpha^3 g$ we prefer $13_{18} = d_0e_0\widetilde{\gamma}$ over $13_{18} + 13_{19}$. The first three choices, each with $g \cdot \widetilde{y} = 0$, are forced by our aim to split $\mathrm{Ext}_{A(2)}(M_1, \mathbb{F}_2)$ into indecomposable $R_0$-modules. The fourth choice will turn out to be more convenient when we get to $E_4(tmf/2)$.

We then use `ext` to write $\widetilde{y}$ as the product of a class in $\mathrm{Ext}_{A(2)}(\mathbb{F}_2, \mathbb{F}_2)$ and one of the module generators from Table 1.5. When given a choice, we prefer factorizations that (in hindsight will turn out to) last as long as possible in the Adams spectral sequence for $tmf/2$, and we emphasize $h_i$-multiplications and other products with coefficients in low topological degree. In most cases the given presentation of $\widetilde{y}$ is evidently a lift of $y$. The less obvious cases are $j(h_1^2\widetilde{h_1}) = h_0^2h_2$, $j(d_0\widetilde{h_2^2}) = h_0^2g$, $j(h_1\widetilde{d_0e_0}) = h_0^2\alpha g$, $j(\widetilde{h_1^2\delta'}) = h_0d_0g$, $j(\alpha\widetilde{\gamma}) = e_0g$, $j(\beta\widetilde{\gamma}) = g^2$, $j(d_0\widetilde{\beta^2}) = \alpha^2 g$, $j(d_0\widetilde{\delta'}) = \alpha d_0 g$, $j(e_0\widetilde{\beta^2}) = \alpha\beta g$, $j(d_0\widetilde{\beta g}) = \alpha e_0 g$, $j(\alpha^2\widetilde{\beta^2}) = d_0g^2$ and $j(d_0e_0\widetilde{\gamma}) = \alpha^3 g$, all of which follow from the relations in Table 3.5.

If $\langle y\rangle = R_0$ then $\langle\widetilde{y}\rangle = R_0$. Otherwise, if $\langle y\rangle = R_0/(g^m)$ we use `ext` to calculate $g^m\cdot\widetilde{y}$. If the answer is 0, then $\langle\widetilde{y}\rangle = R_0/(g^m)$, but if $g^m\cdot\widetilde{y} = i(z) \neq 0$ then $\langle\widetilde{y}\rangle$ is an extension of $R_0/(g^m)$ by the summand containing $z$. This happens in the following seven cases.

$$\begin{aligned}
g\cdot \widetilde{d_0h_2^2} &= i(\alpha^2 e_0)\\
g^2\cdot \widetilde{h_1} &= i(e_0\gamma)\\
g\cdot \widetilde{h_2^2} &= i(\alpha\beta)\\
g\cdot \widetilde{c_0} &= i(\alpha e_0)\\
g\cdot \widetilde{d_0h_1} &= i(\alpha^3)\\
g\cdot \widetilde{h_0^2e_0} &= w_1\cdot i(\beta^2)\\
g\cdot \widetilde{e_0h_1} &= i(d_0\gamma)
\end{aligned}$$

In most instances $z$ generates that summand, and $\langle\widetilde{y}\rangle$ is cyclic, but in the case of $xg^n = h_0^2e_0$ with $\widetilde{y} = \widetilde{h_0^2e_0}$ we have $g\cdot\widetilde{y} = i(z) = w_1\cdot i(\beta^2)$, resulting in a non-cyclic $R_0$-module summand in $\mathrm{Ext}_{A(2)}(M_1,\mathbb{F}_2)$ generated by $\widetilde{h_0^2e_0}$ and $i(\beta^2)$. See Table 4.3.

This accounts for the summands in $\ker(h_0)$ and the seven summands in $\mathrm{cok}(h_0)$ that appear in the $R_0$-module extensions listed above. Each of the remaining summands $\langle z\rangle$ in $\mathrm{cok}(h_0)$ contributes a new summand $\langle i(z)\rangle$ in $\mathrm{Ext}_{A(2)}(M_1,\mathbb{F}_2)$. Gathering these together, and renaming $\widetilde{y}$ or $i(z)$ as $x$, leads to Table 4.2. □

Table 4.2: $R_0$-module generators of $\mathrm{Ext}_{A(2)}(M_1,\mathbb{F}_2)$

| $t-s$ | $s$ | $g$ | $x$ | $\mathrm{Ann}(x)$ | $j(x)$ |
|---|---|---|---|---|---|
| 0 | 0 | 0 | $i(1)$ | $(0)$ | 0 |
| 1 | 1 | 0 | $i(h_1)$ | $(g^2, gw_1)$ | 0 |
| 2 | 1 | 1 | $\widetilde{h_1}$ | $(0)$ | $h_1$ |
| 2 | 2 | 0 | $i(h_1^2)$ | $(g)$ | 0 |
| 3 | 1 | 2 | $i(h_2)$ | $(g)$ | 0 |
| 3 | 2 | 1 | $h_1\widetilde{h_1}$ | $(g)$ | $h_1^2$ |
| 4 | 3 | 0 | $h_1^2\widetilde{h_1}$ | $(g)$ | $h_0^2h_2$ |
| 6 | 2 | 2 | $i(h_2^2)$ | $(g, w_1)$ | 0 |
| 7 | 2 | 3 | $\widetilde{h_2^2}$ | $(0)$ | $h_2^2$ |
| 8 | 3 | 1 | $i(c_0)$ | $(g)$ | 0 |
| 9 | 3 | 2 | $\widetilde{c_0}$ | $(0)$ | $c_0$ |
| 9 | 4 | 1 | $i(h_1c_0)$ | $(g)$ | 0 |
| 10 | 4 | 2 | $h_1\widetilde{c_0}$ | $(g)$ | $h_1c_0$ |
| 12 | 3 | 3 | $i(\alpha)$ | $(0)$ | 0 |

Table 4.2: $R_0$-module generators of $\mathrm{Ext}_{A(2)}(M_1, \mathbb{F}_2)$ (cont.)

| $t-s$ | $s$ | $g$ | $x$ | $\mathrm{Ann}(x)$ | $j(x)$ |
|---|---|---|---|---|---|
| 14 | 4 | 3 | $i(d_0)$ | $(0)$ | $0$ |
| 15 | 3 | 4 | $i(\beta)$ | $(0)$ | $0$ |
| 16 | 5 | 3 | $d_0\widetilde{h_1}$ | $(0)$ | $h_1d_0$ |
| 17 | 4 | 4 | $i(e_0)$ | $(0)$ | $0$ |
| 18 | 4 | 5 | $i(h_2\beta)$ | $(g, w_1)$ | $0$ |
| 18 | 6 | 3 | $\widetilde{h_0^2e_0}$ | $-$ | $h_0^2e_0$ |
| 19 | 5 | 4 | $e_0\widetilde{h_1}$ | $(0)$ | $h_1e_0$ |
| 21 | 6 | 4 | $d_0\widetilde{h_2^2}$ | $(0)$ | $h_0^2g$ |
| 24 | 6 | 5 | $i(\alpha^2)$ | $(0)$ | $0$ |
| 25 | 5 | 7 | $i(\gamma)$ | $(0)$ | $0$ |
| 26 | 5 | 8 | $\widetilde{\gamma}$ | $(0)$ | $\gamma$ |
| 26 | 6 | 6 | $i(h_1\gamma)$ | $(g)$ | $0$ |
| 26 | 7 | 5 | $i(\alpha d_0)$ | $(0)$ | $0$ |
| 27 | 6 | 8 | $h_1\widetilde{\gamma}$ | $(g)$ | $h_1\gamma$ |
| 28 | 7 | 6 | $h_1^2\widetilde{\gamma}$ | $(g)$ | $h_1^2\gamma$ |
| 30 | 6 | 9 | $i(\beta^2)$ | $-$ | $0$ |
| 31 | 6 | 10 | $\widetilde{\beta^2}$ | $(0)$ | $\beta^2$ |
| 31 | 8 | 6 | $i(d_0e_0)$ | $(0)$ | $0$ |
| 32 | 7 | 9 | $i(\delta)$ | $(g)$ | $0$ |
| 32 | 8 | 7 | $\widetilde{d_0e_0}$ | $(0)$ | $d_0e_0$ |
| 33 | 7 | 10 | $\widetilde{\delta'}$ | $(0)$ | $\delta'$ |
| 33 | 8 | 8 | $i(h_1\delta)$ | $(g)$ | $0$ |
| 33 | 9 | 7 | $h_1\widetilde{d_0e_0}$ | $(g)$ | $h_0^2\alpha g$ |
| 34 | 8 | 10 | $h_1\widetilde{\delta'}$ | $(g)$ | $h_1\delta$ |
| 35 | 9 | 9 | $h_1^2\widetilde{\delta'}$ | $(g)$ | $h_0d_0g$ |
| 36 | 7 | 12 | $\widetilde{\beta g}$ | $(0)$ | $\beta g$ |
| 38 | 8 | 12 | $\alpha\widetilde{\gamma}$ | $(0)$ | $e_0g$ |
| 40 | 9 | 12 | $d_0\widetilde{\gamma}$ | $(0)$ | $d_0\gamma$ |
| 41 | 8 | 14 | $\beta\widetilde{\gamma}$ | $(0)$ | $g^2$ |
| 42 | 10 | 12 | $\widetilde{\alpha^2e_0}$ | $(0)$ | $\alpha^2e_0$ |
| 43 | 9 | 14 | $e_0\widetilde{\gamma}$ | $(0)$ | $e_0\gamma$ |
| 45 | 10 | 14 | $d_0\widetilde{\beta^2}$ | $(0)$ | $\alpha^2g$ |

Table 4.2: $R_0$-module generators of $\mathrm{Ext}_{A(2)}(M_1, \mathbb{F}_2)$ (cont.)

| $t-s$ | $s$ | $g$ | $x$ | $\mathrm{Ann}(x)$ | $j(x)$ |
|---|---|---|---|---|---|
| 47 | 11 | 14 | $d_0\widetilde{\delta'}$ | $(0)$ | $\alpha d_0 g$ |
| 48 | 10 | 16 | $e_0\widetilde{\beta^2}$ | $(0)$ | $\alpha\beta g$ |
| 50 | 11 | 16 | $d_0\widetilde{\beta g}$ | $(0)$ | $\alpha e_0 g$ |
| 55 | 12 | 18 | $\alpha^2\widetilde{\beta^2}$ | $(0)$ | $d_0 g^2$ |
| 57 | 13 | 18 | $d_0 e_0\widetilde{\gamma}$ | $(0)$ | $\alpha^3 g$ |

Table 4.3: The non-cyclic $R_0$-module summand in $\mathrm{Ext}_{A(2)}(M_1, \mathbb{F}_2)$

$$\langle \widetilde{h_0^2 e_0}, i(\beta^2)\rangle \cong \frac{\Sigma^{6,24}R_0 \oplus \Sigma^{6,36}R_0}{\langle (g, w_1)\rangle}$$

COROLLARY 4.3. *$\mathrm{Ext}_{A(2)}(M_1, \mathbb{F}_2)$ is generated as an $\mathrm{Ext}_{A(2)}(\mathbb{F}_2, \mathbb{F}_2)$-module by the classes*

$$i(1), \widetilde{h_1}, \widetilde{h_2^2}, \widetilde{c_0}, \widetilde{h_0^2 e_0}, \widetilde{\gamma}, \widetilde{\beta^2}, \widetilde{d_0 e_0}, \widetilde{\delta'}, \widetilde{\beta g}, \widetilde{\alpha^2 e_0}$$

*listed in Table* 1.5 *and shown in Figure* 4.1.

PROOF. Each $R_0$-module generator $x$ in Table 4.2 is an $\mathrm{Ext}_{A(2)}(\mathbb{F}_2, \mathbb{F}_2)$-multiple of one of these eleven classes. □

## 4.2. Adams periodicity

As an application of our calculation in Section 4.1, we establish an improved form of the Adams periodicity theorem from [**7**], originally due to Peter May (ca. 1968, unpublished). Our statement of Theorem 4.9 implies the formulation quoted in [**144**, Thm. 3.4.6(a)].

Define functions $F(s)$ and $G(s)$ as follows.

| $s$ | $\le -5$ | $-4$ | $-3$ | $-2$ | $-1$ | $0$ | $1$ | $2$ | $3$ | $4$ | $5$ | $\ge 6$ |
|---|---|---|---|---|---|---|---|---|---|---|---|---|
| $F(s)$ | $+\infty$ | $-8$ | $-7$ | $-6$ | $-4$ | $1$ | $8$ | $6$ | $18$ | $18$ | $21$ | $5s+3$ |
| $G(s)$ | $+\infty$ | $-8$ | $-7$ | $-6$ | $-4$ | $1$ | $8$ | $10$ | $18$ | $23$ | $25$ | $5s+3$ |

PROPOSITION 4.4. *Let $M_1 = H^*(S/2)$. Then*

$$w_1 \colon \mathrm{Ext}_{A(2)}^{s,t}(M_1, \mathbb{F}_2) \longrightarrow \mathrm{Ext}_{A(2)}^{s+4,t+12}(M_1, \mathbb{F}_2)$$

*is an isomorphism for $t-s < F(s)$, and is surjective for $t-s < G(s)$.*

PROOF. This follows by inspection from the $w_1$-action on $\mathrm{Ext}_{A(2)}(M_1, \mathbb{F}_2)$ given in Table 4.2. The classes $d_0\widetilde{h_1}$, $d_0\widetilde{h_2^2}$, $i(\alpha d_0)$, $i(d_0 e_0)$ and their $g$-power multiplies are not $w_1$-multiples, and lead to the bound $t-s < 5s+3$ for $s \ge 6$. □

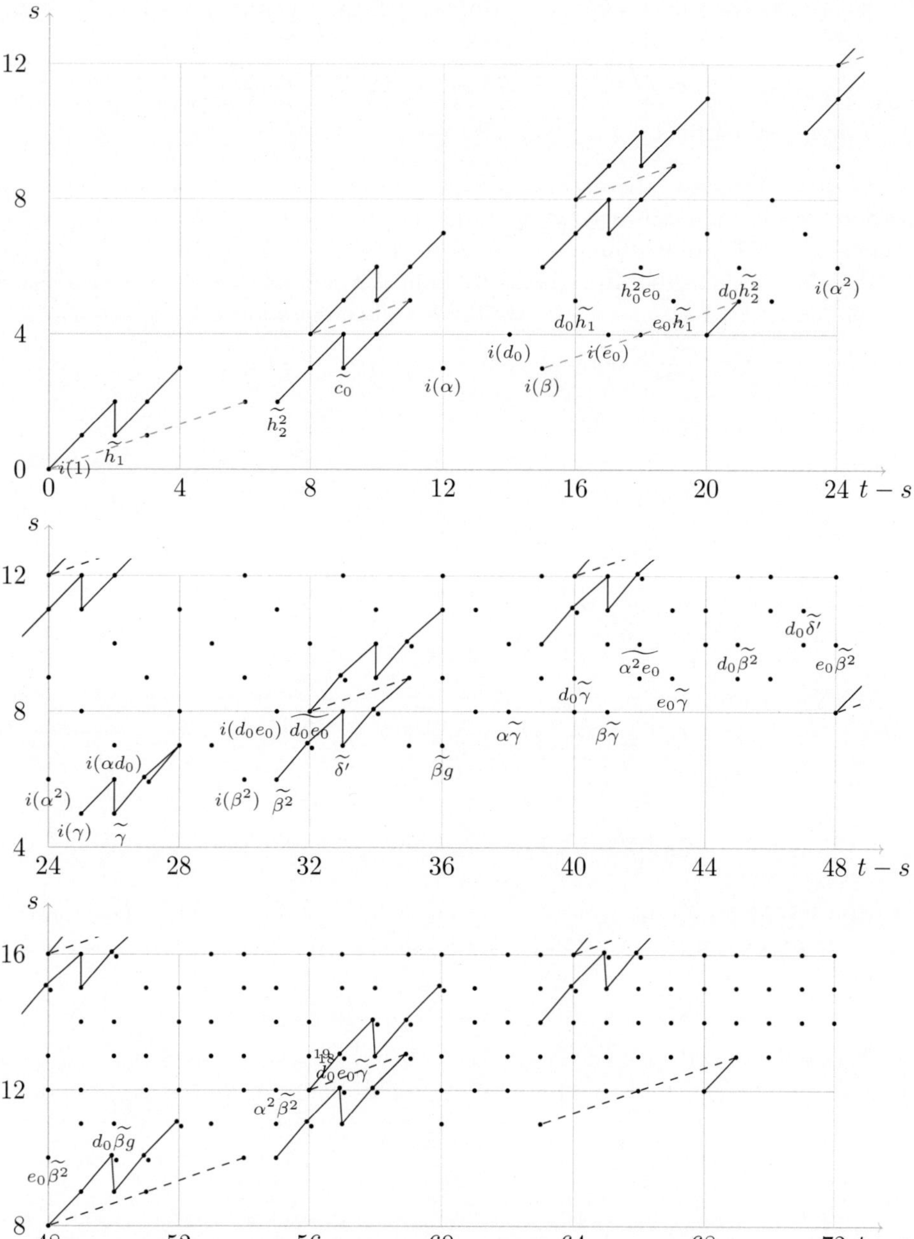


FIGURE 4.1. $R_0$-module generators of $\mathrm{Ext}_{A(2)}(M_1, \mathbb{F}_2)$. Note that $d_0e_0\widetilde{\gamma} = 13_{18}$.

Define functions $H(s)$ and $I(s)$ as follows.

| $s$ | $\leq -4$ | $-3$ | $-2$ | $-1$ | $0$ | $1$ | $2$ | $3$ | $4$ | $5$ | $6$ | $\geq 7$ |
|---|---|---|---|---|---|---|---|---|---|---|---|---|
| $H(s)$ | $-s-12$ | $-8$ | $-7$ | $-6$ | $-4$ | $1$ | $6$ | $10$ | $18$ | $21$ | $25$ | $5s-2$ |
| $I(s)$ | $-s-12$ | $-7$ | $-6$ | $-4$ | $1$ | $7$ | $10$ | $18$ | $22$ | $25$ | $33$ | $5s+3$ |

PROPOSITION 4.5. *Let $L$ be an $A(2)$-module that is $A(0)$-free and connective. Then*

$$w_1 \colon \operatorname{Ext}_{A(2)}^{s,t}(L, \mathbb{F}_2) \longrightarrow \operatorname{Ext}_{A(2)}^{s+4,t+12}(L, \mathbb{F}_2)$$

*is an isomorphism for $t - s < H(s)$, and is surjective for $t - s < I(s)$.*

PROOF. The case where $L$ is a direct sum of copies of $M_1$ follows from the previous proposition, since $H(s) \le F(s)$ and $I(s) \le G(s)$. As in the proof of Lemma 2.3 in [**7**], we can suppose that $L' \to L \to L''$ is an extension of $A(0)$-free $A(2)$-modules, where the result holds by induction for $L''$, and $L'$ is a direct sum of copies of $\Sigma^\nu M_1$ for some $\nu \ge 1$. By the Five Lemma applied to

$$\begin{array}{ccc}
\operatorname{Ext}_{A(2)}^{s-1,t}(L', \mathbb{F}_2) & \xrightarrow{w_1} & \operatorname{Ext}_{A(2)}^{s+3,t+12}(L', \mathbb{F}_2) \\
\downarrow \delta & & \downarrow \delta \\
\operatorname{Ext}_{A(2)}^{s,t}(L'', \mathbb{F}_2) & \xrightarrow{w_1} & \operatorname{Ext}_{A(2)}^{s+4,t+12}(L'', \mathbb{F}_2) \\
\downarrow & & \downarrow \\
\operatorname{Ext}_{A(2)}^{s,t}(L, \mathbb{F}_2) & \xrightarrow{w_1} & \operatorname{Ext}_{A(2)}^{s+4,t+12}(L, \mathbb{F}_2) \\
\downarrow & & \downarrow \\
\operatorname{Ext}_{A(2)}^{s,t}(L', \mathbb{F}_2) & \xrightarrow{w_1} & \operatorname{Ext}_{A(2)}^{s+4,t+12}(L', \mathbb{F}_2) \\
\downarrow \delta & & \downarrow \delta \\
\operatorname{Ext}_{A(2)}^{s+1,t}(L'', \mathbb{F}_2) & \xrightarrow{w_1} & \operatorname{Ext}_{A(2)}^{s+5,t+12}(L'', \mathbb{F}_2)
\end{array}$$

we deduce that $w_1$ of the proposition is an isomorphism if $t-(s-1) < G(s-1)+\nu$, $t-s < H(s)$, $t-s < F(s)+\nu$ and $t-(s+1) < H(s+1)$. This holds for $t-s < H(s)$ because

$$H(s) \le \min\{F(s), G(s-1), H(s+1)+1\}\,.$$

Furthermore, we deduce that $w_1$ of the proposition is surjective if $t - s < I(s)$, $t - s < G(s) + \nu$ and $t - (s+1) < H(s+1)$. This holds for $t - s < I(s)$ because

$$I(s) \le \min\{G(s), H(s+1)+1\}\,.$$

□

Define functions $J(s)$ and $K(s)$ as follows, where $i < 0$.

| $s$ | $4i$ | $4i+1$ | $4i+2$ | $4i+3$ | $0$ | $1$ | $2$ | $3$ | $4$ | $5$ | $6$ | $\ge 7$ |
|---|---|---|---|---|---|---|---|---|---|---|---|---|
| $J(s)$ | $8i-4$ | $8i$ | $8i+1$ | $8i+2$ | $-4$ | $1$ | $6$ | $10$ | $18$ | $21$ | $25$ | $5s-2$ |
| $K(s)$ | $8i$ | $8i+1$ | $8i+2$ | $8i+4$ | $1$ | $7$ | $10$ | $18$ | $22$ | $25$ | $33$ | $5s+3$ |

COROLLARY 4.6. *Let $m \ge 1$. Then*

$$w_1^m \colon \operatorname{Ext}_{A(2)}^{s,t}(L, \mathbb{F}_2) \longrightarrow \operatorname{Ext}_{A(2)}^{s+4m,t+12m}(L, \mathbb{F}_2)$$

*is an isomorphism for $t - s < J(s)$, and is surjective for $t - s < K(s)$.*

PROOF. If $t - s < J(s)$ then $(t + 12k) - (s + 4k) < H(s + 4k)$ for all $k \geq 0$, so $w_1^m$ is the composite of $m$ isomorphisms, by the previous proposition. Likewise, if $t - s < K(s)$ then $(t + 12k) - (s + 4k) < I(s + 4k)$ for all $k \geq 0$, so $w_1^m$ is the composite of $m$ surjections. □

For $n \geq 2$ let $\varpi_n \in \mathrm{Ext}_{A(n)}^{2^n, 3\cdot 2^n}(\mathbb{F}_2, \mathbb{F}_2)$ be the Adams periodicity element from [**7**, §4], which restricts to $w_1^{2^{n-2}}$ in $\mathrm{Ext}_{A(2)}^{2^n, 3\cdot 2^n}(\mathbb{F}_2, \mathbb{F}_2)$. Define functions $L(s)$, $M(s)$ and $N(s)$ as follows.

| $s$ | $4i$ | $4i{+}1$ | $4i{+}2$ | $4i{+}3$ | 0 | 1 | 2 | 3 | 4 | 5 | 6 | 7 | $\geq 8$ |
|---|---|---|---|---|---|---|---|---|---|---|---|---|---|
| $L(s)$ | $8i{-}5$ | $8i{-}1$ | $8i$ | $8i{+}1$ | $-5$ | 0 | 6 | 9 | 16 | 21 | 24 | 31 | $5s{-}3$ |
| $M(s)$ | $8i{-}5$ | $8i{-}1$ | $8i$ | $8i{+}1$ | 0 | 3 | 6 | 9 | 16 | 21 | 24 | 31 | $5s{-}3$ |
| $N(s)$ | $8i$ | $8i{+}1$ | $8i{+}2$ | $8i{+}4$ | 1 | 7 | 10 | 17 | 22 | 25 | 32 | 38 | $5s{+}3$ |

PROPOSITION 4.7. *Let $n \geq 2$, and let $L$ be an $A(n)$-module that is $A(0)$-free and connective. Then*

$$\varpi_n \colon \mathrm{Ext}_{A(n)}^{s,t}(L, \mathbb{F}_2) \longrightarrow \mathrm{Ext}_{A(n)}^{s+2^n, t+3\cdot 2^n}(L, \mathbb{F}_2)$$

*is an isomorphism for $t - s < M(s)$, and is surjective for $t - s < N(s)$.*

PROOF. We first prove the claim with $L(s)$ in place of $M(s)$. Consider the extension $\Sigma^8 K \to A(n) \otimes_{A(2)} L \to L$ of $A(n)$-modules. Here $K$ is $A(0)$-free and connective. By induction on $t$ we may assume that the proposition applies to $K$. By the Five Lemma applied to

$$\begin{array}{ccc}
\mathrm{Ext}_{A(2)}^{s-1,t}(L, \mathbb{F}_2) & \xrightarrow{w_1^{2^{n-2}}} & \mathrm{Ext}_{A(2)}^{s+2^n-1, t+3\cdot 2^n}(L, \mathbb{F}_2) \\
\downarrow & & \downarrow \\
\mathrm{Ext}_{A(n)}^{s-1,t}(\Sigma^8 K, \mathbb{F}_2) & \xrightarrow{\varpi_n} & \mathrm{Ext}_{A(n)}^{s+2^n-1, t+3\cdot 2^n}(\Sigma^8 K, \mathbb{F}_2) \\
\downarrow{\scriptstyle\delta} & & \downarrow{\scriptstyle\delta} \\
\mathrm{Ext}_{A(n)}^{s,t}(L, \mathbb{F}_2) & \xrightarrow{\varpi_n} & \mathrm{Ext}_{A(n)}^{s+2^n, t+3\cdot 2^n}(L, \mathbb{F}_2) \\
\downarrow & & \downarrow \\
\mathrm{Ext}_{A(2)}^{s,t}(L, \mathbb{F}_2) & \xrightarrow{w_1^{2^{n-2}}} & \mathrm{Ext}_{A(2)}^{s+2^n, t+3\cdot 2^n}(L, \mathbb{F}_2) \\
\downarrow & & \downarrow \\
\mathrm{Ext}_{A(n)}^{s,t}(\Sigma^8 K, \mathbb{F}_2) & \xrightarrow{\varpi_n} & \mathrm{Ext}_{A(n)}^{s+2^n, t+3\cdot 2^n}(\Sigma^8 K, \mathbb{F}_2)
\end{array}$$

we deduce that $\varpi_n$ of the proposition is an isomorphism if $t - (s-1) < K(s-1)$, $t - (s-1) < L(s-1) + 8$, $t - s < J(s)$ and $t - s < L(s) + 8$. This holds for $t - s < L(s)$ because

$$L(s) \leq \min\{J(s), K(s-1) - 1, L(s-1) + 7\} .$$

Furthermore, we deduce that $\varpi_n$ of the proposition is surjective if $t - (s-1) < N(s-1) + 8$, $t - s < K(s)$ and $t - s < L(s) + 8$. This holds for $t - s < N(s)$ because

$$N(s) \leq \min\{K(s), L(s) + 8, N(s-1) + 7\} .$$

To finish the proof, we appeal to [**7**, Thm. 5.3], showing that $\varpi_n$ is an isomorphism for $s \geq 0$ and $t - s < 3s$. This lets us improve $L(s)$ to $M(s)$, as shown, for $s = 0$ and $s = 1$. □

Let $L$ be an $A$-module that is $A(0)$-free and connective. Adapting [**7**, §2], we let

$$P(4k) = 8k, \quad P(4k+1) = 8k+1, \quad P(4k+2) = 8k+2, \quad P(4k+3) = 8k+4$$

for all integers $k$. By the Adams vanishing theorem [**7**, Thm. 2.1], $\mathrm{Ext}_A^{s,t}(L, \mathbb{F}_2) = 0$ for $t - s < P(s)$. For $n \geq 2$ the Massey product $\pi_n(x) = \langle h_{n+1}, h_0^{2^n}, x\rangle$ defines a homomorphism

$$\pi_n \colon \ker(h_0^{2^n}) \longrightarrow \frac{\mathrm{Ext}_A^{s+2^n, t+3\cdot 2^n}(L, \mathbb{F}_2)}{h_{n+1}\,\mathrm{Ext}_A^{s+2^n-1, t+2^n}(L, \mathbb{F}_2)}$$

from $\ker(h_0^{2^n}) \subset \mathrm{Ext}_A^{s,t}(L, \mathbb{F}_2)$. By the Adams approximation theorem [**7**, Thm. 3.1] the restriction homomorphism

$$\mathrm{Ext}_A^{s,t}(L, \mathbb{F}_2) \longrightarrow \mathrm{Ext}_{A(n)}^{s,t}(L, \mathbb{F}_2)$$

is an isomorphism for $(s, t)$ such that

$$t - s < 2^{n+1} - 1 + P(s-1)\,.$$

For $x \in \mathrm{Ext}_A^{s,t}(L, \mathbb{F}_2)$ the product $h_0^{2^n} x$ lies in bidegree $(s + 2^n, t + 2^n)$, and the inequality above implies that

$$t - s < 2^{n+1} - 1 + P(s-1) = P(s + 2^n - 1) - 1 \leq P(s + 2^n)\,.$$

Hence $h_0^{2^n} x = 0$ by the vanishing theorem, so that $\ker(h_0^{2^n}) = \mathrm{Ext}_A^{s,t}(L, \mathbb{F}_2)$. By the same theorem, $\mathrm{Ext}_A^{s+2^n-1, t+2^n}(L, \mathbb{F}_2) = 0$. We therefore have a commutative square

$$\begin{array}{ccc} \mathrm{Ext}_A^{s,t}(L, \mathbb{F}_2) & \xrightarrow{\pi_n} & \mathrm{Ext}_A^{s+2^n, t+3\cdot 2^n}(L, \mathbb{F}_2) \\ {\cong}\downarrow & & \downarrow{\cong} \\ \mathrm{Ext}_{A(n)}^{s,t}(L, \mathbb{F}_2) & \xrightarrow{\varpi_n} & \mathrm{Ext}_{A(n)}^{s+2^n, t+3\cdot 2^n}(L, \mathbb{F}_2) \end{array}$$

with vertical isomorphisms, for these $(s, t)$. This proves the following theorem.

THEOREM 4.8 (Adams [**7**, Thm. 5.4], May). *Let $L$ be an $A$-module that is $A(0)$-free and connective. Let $n \geq 2$, and assume that $t - s < 2^{n+1} - 1 + P(s-1)$. Then*

$$\pi_n \colon \mathrm{Ext}_A^{s,t}(L, \mathbb{F}_2) \longrightarrow \mathrm{Ext}_A^{s+2^n, t+3\cdot 2^n}(L, \mathbb{F}_2)$$

*is an isomorphism for $t - s < M(s)$, and is surjective for $t - s < N(s)$.* □

Define functions $Q(s) = M(s-1) + 1$ and $R(s) = N(s-1) + 1$, as in the following table.

| $s$ | 1 | 2 | 3 | 4 | 5 | 6 | 7 | 8 | $\geq 9$ |
|---|---|---|---|---|---|---|---|---|---|
| $Q(s)$ | 1 | 4 | 7 | 10 | 17 | 22 | 25 | 32 | $5s - 7$ |
| $R(s)$ | 2 | 8 | 11 | 18 | 23 | 26 | 33 | 39 | $5s - 1$ |

THEOREM 4.9 (Adams [7, Cor. 5.5], May). *Let $n \geq 2$, and consider $(s,t)$ satisfying $0 < t-s < 2^{n+1} + P(s-2)$. The operator*

$$\pi_n \colon \mathrm{Ext}_A^{s,t}(\mathbb{F}_2, \mathbb{F}_2) \longrightarrow \mathrm{Ext}_A^{s+2^n, t+3\cdot 2^n}(\mathbb{F}_2, \mathbb{F}_2)$$

*is an isomorphism for $t-s < Q(s)$, and is surjective for $t-s < R(s)$.*

PROOF. The homotopy cofiber sequence $S \to H\mathbb{Z} \to H\mathbb{Z}/S$ induces an extension $\Sigma^2 L \to A//A(0) \to \mathbb{F}_2$ in cohomology, with $L$ an $A$-module that is $A(0)$-free and connective. The connecting homomorphisms $\delta$ in the commutative diagram

$$\begin{array}{ccc} \mathrm{Ext}_A^{s-1,t}(\Sigma^2 L, \mathbb{F}_2) & \xrightarrow{\pi_n} & \mathrm{Ext}_A^{s+2^n-1, t+3\cdot 2^n}(\Sigma^2 L, \mathbb{F}_2) \\ \delta \downarrow & & \downarrow \delta \\ \mathrm{Ext}_A^{s,t}(\mathbb{F}_2, \mathbb{F}_2) & \xrightarrow{\pi_n} & \mathrm{Ext}_A^{s+2^n, t+3\cdot 2^n}(\mathbb{F}_2, \mathbb{F}_2) \end{array}$$

are isomorphisms for $t-s > 0$. □

## 4.3. Coefficients in $M_2$

The short exact sequence of $A(2)$-modules

$$0 \to \Sigma^2 \mathbb{F}_2 \longrightarrow M_2 \longrightarrow \mathbb{F}_2 \to 0$$

represents $h_1$ in $\mathrm{Ext}_{A(2)}^{1,2}(\mathbb{F}_2, \mathbb{F}_2)$. It follows that in the induced long exact sequence

$$\dots \xrightarrow{\delta} \mathrm{Ext}_{A(2)}^{*,*}(\mathbb{F}_2, \mathbb{F}_2) \xrightarrow{i} \mathrm{Ext}_{A(2)}^{*,*}(M_2, \mathbb{F}_2)$$
$$\xrightarrow{j} \mathrm{Ext}_{A(2)}^{*,*}(\Sigma^2 \mathbb{F}_2, \mathbb{F}_2) \xrightarrow{\delta} \mathrm{Ext}_{A(2)}^{*+1,*}(\mathbb{F}_2, \mathbb{F}_2) \longrightarrow \dots$$

the connecting homomorphism $\delta$ is given by multiplication by $h_1$. The long exact sequence therefore breaks up into short exact sequences

$$0 \to \mathrm{cok}(h_1)^{s,t} \xrightarrow{i} \mathrm{Ext}_{A(2)}^{s,t}(M_2, \mathbb{F}_2) \xrightarrow{j} \ker(h_1)^{s,t-2} \to 0\,,$$

where $\mathrm{Ext}_{A(2)}^{s,t}(\mathbb{F}_2, \mathbb{F}_2)/\operatorname{im}(h_1) = \mathrm{cok}(h_1)^{s,t}$ and $\ker(h_1)^{s,t} \subset \mathrm{Ext}_{A(2)}^{s,t}(\mathbb{F}_2, \mathbb{F}_2)$.

LEMMA 4.10. *The kernel and cokernel of $h_1$ are both direct sums of cyclic $R_0$-modules, with generators and annihilator ideals as listed in Table 4.4.*

PROOF. For each class $x$ listed in Table 3.6, spanning a cyclic $R_0$-module summand $\langle x \rangle$ of $SI \cong \mathrm{Ext}_{A(2)}(\mathbb{F}_2, \mathbb{F}_2)$, we are able to express $h_1 x$ as an element in a summand $\langle y \rangle$. We record the kernel and cokernel of the $R_0$-module homomorphism $h_1 \colon \langle x \rangle \to \langle y \rangle$ in Table 4.4. These $h_1$-multiplications are visible in Figures 3.12 and 3.13. In most cases $h_1 x = 0$ or $y = h_1 x$. The less obvious cases are

$$h_1 \cdot d_0 e_0 = g \cdot h_0^2 \alpha$$
$$h_1 \cdot h_1 \delta = g \cdot h_0 d_0\,,$$

which are clear from Table 3.5. □

Table 4.4: Direct sum decompositions of the kernel and cokernel of multiplication by $h_1$, with $i \geq 0$ in each $h_0$-tower

| $t-s$ | $s$ | $g$ | $\ker(h_1)$ | $x$ | $h_1 x$ | $\mathrm{cok}(h_1)$ |
|---|---|---|---|---|---|---|
| 0 | 0 | 0 | $\langle g^2 \rangle = R_0$ | 1 | $h_1$ | $\langle 1 \rangle = R_0$ |
| 0 | 1 | 0 | $\langle h_0 \rangle = R_0/(g^2)$ | $h_0$ | 0 | $\langle h_0 \rangle = R_0/(g^2)$ |
| 0 | 2 | 0 | $\langle h_0^2 \rangle = R_0/(g^2)$ | $h_0^2$ | 0 | $\langle h_0^2 \rangle = R_0/(g^2)$ |
| 0 | $3+i$ | 0 | $\langle h_0^{3+i} \rangle = R_0/(g)$ | $h_0^{3+i}$ | 0 | $\langle h_0^{3+i} \rangle = R_0/(g)$ |
| 1 | 1 | 1 | $\langle h_1 g \rangle = R_0/(g)$ | $h_1$ | $h_1^2$ | 0 |
| 2 | 2 | 1 | 0 | $h_1^2$ | $h_0^2 h_2$ | 0 |
| 3 | 1 | 2 | $\langle h_2 \rangle = R_0/(g)$ | $h_2$ | 0 | $\langle h_2 \rangle = R_0/(g)$ |
| 3 | 2 | 2 | $\langle h_0 h_2 \rangle = R_0/(g)$ | $h_0 h_2$ | 0 | $\langle h_0 h_2 \rangle = R_0/(g)$ |
| 3 | 3 | 1 | $\langle h_0^2 h_2 \rangle = R_0/(g)$ | $h_0^2 h_2$ | 0 | 0 |
| 6 | 2 | 3 | $\langle h_2^2 \rangle = R_0/(g)$ | $h_2^2$ | 0 | $\langle h_2^2 \rangle = R_0/(g)$ |
| 8 | 3 | 2 | 0 | $c_0$ | $h_1 c_0$ | $\langle c_0 \rangle = R_0/(g)$ |
| 9 | 4 | 2 | $\langle h_1 c_0 \rangle = R_0/(g)$ | $h_1 c_0$ | 0 | 0 |
| 12 | 3 | 3 | $\langle \alpha \rangle = R_0$ | $\alpha$ | 0 | $\langle \alpha \rangle = R_0$ |
| 12 | 4 | 3 | $\langle h_0 \alpha \rangle = R_0/(g^2)$ | $h_0 \alpha$ | 0 | $\langle h_0 \alpha \rangle = R_0/(g^2)$ |
| 12 | 5 | 4 | $\langle h_0^2 \alpha \rangle = R_0/(g^2)$ | $h_0^2 \alpha$ | 0 | $\langle h_0^2 \alpha \rangle = R_0/(g)$ |
| 12 | $6+i$ | 4 | $\langle h_0^{3+i} \alpha \rangle = R_0/(g)$ | $h_0^{3+i} \alpha$ | 0 | $\langle h_0^{3+i} \alpha \rangle = R_0/(g)$ |
| 14 | 4 | 4 | $\langle d_0 g \rangle = R_0$ | $d_0$ | $h_1 d_0$ | $\langle d_0 \rangle = R_0$ |
| 14 | 5 | 5 | $\langle h_0 d_0 \rangle = R_0/(g^2)$ | $h_0 d_0$ | 0 | $\langle h_0 d_0 \rangle = R_0/(g)$ |
| 15 | 3 | 4 | $\langle \beta \rangle = R_0$ | $\beta$ | 0 | $\langle \beta \rangle = R_0$ |
| 15 | 4 | 5 | $\langle h_0 \beta \rangle = R_0/(g)$ | $h_0 \beta$ | 0 | $\langle h_0 \beta \rangle = R_0/(g)$ |
| 15 | 5 | 6 | $\langle h_1 d_0 \rangle = R_0/(g)$ | $h_1 d_0$ | 0 | 0 |
| 17 | 4 | 6 | $\langle e_0 g \rangle = R_0$ | $e_0$ | $h_1 e_0$ | $\langle e_0 \rangle = R_0$ |
| 17 | 5 | 7 | $\langle h_0 e_0 \rangle = R_0/(g)$ | $h_0 e_0$ | 0 | $\langle h_0 e_0 \rangle = R_0/(g)$ |
| 17 | 6 | 6 | $\langle h_0^2 e_0 \rangle = R_0/(g)$ | $h_0^2 e_0$ | 0 | $\langle h_0^2 e_0 \rangle = R_0/(g)$ |
| 18 | 4 | 7 | $\langle h_2 \beta \rangle = R_0/(g)$ | $h_2 \beta$ | 0 | $\langle h_2 \beta \rangle = R_0/(g)$ |
| 18 | 5 | 8 | $\langle h_1 e_0 \rangle = R_0/(g)$ | $h_1 e_0$ | 0 | 0 |
| 24 | 6 | 8 | $\langle \alpha^2 \rangle = R_0$ | $\alpha^2$ | 0 | $\langle \alpha^2 \rangle = R_0$ |
| 24 | $7+i$ | 7 | $\langle h_0^{1+i} \alpha^2 \rangle = R_0/(g)$ | $h_0^{1+i} \alpha^2$ | 0 | $\langle h_0^{1+i} \alpha^2 \rangle = R_0/(g)$ |
| 25 | 5 | 11 | $\langle \gamma g \rangle = R_0$ | $\gamma$ | $h_1 \gamma$ | $\langle \gamma \rangle = R_0$ |
| 26 | 6 | 9 | 0 | $h_1 \gamma$ | $h_1^2 \gamma$ | 0 |
| 26 | 7 | 8 | $\langle \alpha d_0 \rangle = R_0$ | $\alpha d_0$ | 0 | $\langle \alpha d_0 \rangle = R_0$ |

Table 4.4: Direct sum decompositions of the kernel and cokernel of multiplication by $h_1$, with $i \geq 0$ in each $h_0$-tower (cont.)

| $t-s$ | $s$ | $g$ | $\ker(h_1)$ | $x$ | $h_1x$ | $\mathrm{cok}(h_1)$ |
|---|---|---|---|---|---|---|
| 27 | 6 | 10 | $\langle \alpha\beta \rangle = R_0$ | $\alpha\beta$ | 0 | $\langle \alpha\beta \rangle = R_0$ |
| 27 | 7 | 9 | $\langle h_1^2\gamma \rangle = R_0/(g)$ | $h_1^2\gamma$ | 0 | 0 |
| 29 | 7 | 10 | $\langle \alpha e_0 \rangle = R_0$ | $\alpha e_0$ | 0 | $\langle \alpha e_0 \rangle = R_0$ |
| 29 | 8 | 12 | $\langle h_0\alpha e_0 \rangle = R_0/(g)$ | $h_0\alpha e_0$ | 0 | $\langle h_0\alpha e_0 \rangle = R_0/(g)$ |
| 30 | 6 | 11 | $\langle \beta^2 \rangle = R_0$ | $\beta^2$ | 0 | $\langle \beta^2 \rangle = R_0$ |
| 31 | 8 | 13 | $\langle d_0e_0g \rangle = R_0$ | $d_0e_0$ | $h_0^2\alpha g$ | $\langle d_0e_0 \rangle = R_0$ |
| 32 | 7 | 11 | 0 | $\delta$ | $h_1\delta$ | $\langle \delta \rangle = R_0/(g)$ |
| 33 | 8 | 15 | 0 | $h_1\delta$ | $h_0d_0g$ | 0 |
| 36 | 9 | 17 | $\langle \alpha^3 \rangle = R_0$ | $\alpha^3$ | 0 | $\langle \alpha^3 \rangle = R_0$ |
| 36 | $10+i$ | 14 | $\langle h_0^{1+i}\alpha^3 \rangle = R_0/(g)$ | $h_0^{1+i}\alpha^3$ | 0 | $\langle h_0^{1+i}\alpha^3 \rangle = R_0/(g)$ |
| 39 | 9 | 18 | $\langle d_0\gamma \rangle = R_0$ | $d_0\gamma$ | 0 | $\langle d_0\gamma \rangle = R_0$ |
| 41 | 10 | 16 | $\langle \alpha^2e_0 \rangle = R_0$ | $\alpha^2e_0$ | 0 | $\langle \alpha^2e_0 \rangle = R_0$ |
| 42 | 9 | 19 | $\langle e_0\gamma \rangle = R_0$ | $e_0\gamma$ | 0 | $\langle e_0\gamma \rangle = R_0$ |

PROPOSITION 4.11. $\mathrm{Ext}_{A(2)}(M_2, \mathbb{F}_2)$ *is a direct sum of cyclic $R_0$-modules, with generators and annihilator ideals as listed in Table* 4.5.

PROOF. We use `ext` to determine the $R_0$-module extensions of summands in $\ker(h_1)$ by summands in $\mathrm{cok}(h_1)$. Each summand in $\ker(h_1)$ has a generator of the form $y = xg^n$, and we choose a lift $\widehat{y}$ in $\mathrm{Ext}_{A(2)}(M_2, \mathbb{F}_2)$ with $j(\widehat{y}) = y$. In most cases the lift is unique, but when given a choice we prefer classes that emphasize the $h_i$-multiplications.

We then use `ext` to write $\widehat{y}$ as the product of a class in $\mathrm{Ext}_{A(2)}(\mathbb{F}_2, \mathbb{F}_2)$ and one of the module generators from Table 1.6. When given a choice, we prefer factorizations that last as long as possible in the Adams spectral sequence for $tmf/\eta$, and we emphasize $h_i$-multiplications and other products with coefficients in low topological degree. In most cases the given presentation of $\widehat{y}$ is evidently a lift of $y$. The less obvious cases are $j(h_0^2\widehat{\beta}) = h_1d_0$, $j(d_0\widehat{h_2}) = h_0e_0$, $j(h_0d_0\widehat{h_2}) = h_0^2e_0$, $j(h_0h_2\widehat{\beta}) = h_1e_0$, $j(h_2^2\widehat{\beta}) = h_1g$, $j(h_0\alpha\widehat{\beta}) = h_1^2\gamma$, $j(d_0\widehat{\beta}) = \alpha e_0$, $j(h_0d_0\widehat{\beta}) = h_0\alpha e_0$, $j(\gamma\widehat{\alpha}) = e_0g$, $j(\alpha^2\widehat{\beta}) = d_0\gamma$, $j(\gamma\widehat{\beta}) = g^2$, $j(\alpha d_0\widehat{\beta}) = \alpha^2e_0$, $j(\alpha\beta\widehat{\beta}) = e_0\gamma$, $j(\beta^2\widehat{\beta}) = \gamma g$ and $j(d_0\gamma\widehat{\alpha}) = d_0e_0g$, all of which follow from the relations in Table 3.5.

If $\langle y \rangle = R_0$ then $\langle \widehat{y} \rangle = R_0$. Otherwise, if $\langle y \rangle = R_0/(g^m)$ we use `ext` to calculate $g^m \cdot \widehat{y}$. If the answer is 0, then $\langle \widehat{y} \rangle = R_0/(g^m)$, but if $g^m \cdot \widehat{y} = i(z) \neq 0$ then $\langle \widehat{y} \rangle$ is an extension of $R_0/(g^m)$ by the summand containing $z$. This happens in the following five cases.

$$g \cdot \widehat{h_2} = 5_{17} = i(\gamma)$$

$$g \cdot \widehat{h_1c_0} = 8_{20} = i(d_0e_0)$$

$$g \cdot d_0\widehat{h_2} = 9_{28} = i(d_0\gamma)$$

$$g^2 \cdot \widehat{h_0} = 9_{30} = i(e_0\gamma)$$
$$g^2 \cdot d_0\widehat{h_0} = 13_{44} = g \cdot i(\alpha^3)$$

In the first four cases $z$ generates a direct summand, and $\langle \widehat{y} \rangle$ is cyclic. In the final case, corresponding to $\widehat{y} = d_0\widehat{h_0}$, we make a change of basis, replacing the generator $i(\alpha^3)$ with

$$h_0\widehat{d_0 g} = 9_{26} = i(\alpha^3) + g \cdot d_0\widehat{h_0}\,.$$

This yields the splitting

$$\langle d_0\widehat{h_0}, i(\alpha^3)\rangle = \langle d_0\widehat{h_0}\rangle \oplus \langle h_0\widehat{d_0 g}\rangle \cong R_0 \oplus R_0/(g)\,.$$

It then makes sense to rewrite the $h_0$-tower $i(h_0^{1+i}\alpha^3)$ in the form $h_0^{2+i}\widehat{d_0 g}$. Each of the remaining summands $\langle z\rangle$ in $\mathrm{cok}(h_1)$ contributes a new summand $\langle i(z)\rangle$ in $\mathrm{Ext}_{A(2)}(M_2, \mathbb{F}_2)$. Gathering these together, and writing $x$ in place of $\widehat{y}$ or $i(z)$, leads to Table 4.5. □

REMARK 4.12. In our tables, we use $i$ to denote the running index in $h_0$-towers, as well as the inclusion of a bottom cell. This leads to notation such as $i(h_0^{3+i})$, where $i$ has both meanings. Given this warning, we hope the reader will not be confused. The $s$-, $g$- and $j(x)$-entries for $h_0$-towers refer to the $i = 0$ case. The notation "$g_1 + g_2$" in the $g$-column means that $x$ is represented by the cocycle $s_{g_1} + s_{g_2}$.

Table 4.5: $R_0$-module generators of $\mathrm{Ext}_{A(2)}(M_2, \mathbb{F}_2)$, with $i \geq 0$ in each $h_0$-tower

| $t-s$ | $s$ | $g$ | $x$ | $\mathrm{Ann}(x)$ | $j(x)$ |
|---|---|---|---|---|---|
| 0 | 0 | 0 | $i(1)$ | $(0)$ | 0 |
| 0 | 1 | 0 | $i(h_0)$ | $(g^2)$ | 0 |
| 0 | 2 | 0 | $i(h_0^2)$ | $(g^2)$ | 0 |
| 0 | $3+i$ | 0 | $i(h_0^{3+i})$ | $(g)$ | 0 |
| 2 | 1 | 1 | $\widehat{h_0}$ | $(0)$ | $h_0$ |
| 2 | 2 | 1 | $h_0\widehat{h_0}$ | $(g^2)$ | $h_0^2$ |
| 2 | $3+i$ | 1 | $h_0^{2+i}\widehat{h_0}$ | $(g)$ | $h_0^{3+i}$ |
| 3 | 1 | 2 | $i(h_2)$ | $(g)$ | 0 |
| 3 | 2 | 2 | $i(h_0h_2)$ | $(g)$ | 0 |
| 5 | 1 | 3 | $\widehat{h_2}$ | $(0)$ | $h_2$ |
| 5 | 2 | 3 | $h_0\widehat{h_2}$ | $(g)$ | $h_0h_2$ |
| 5 | 3 | 2 | $h_0^2\widehat{h_2}$ | $(g)$ | $h_0^2h_2$ |
| 6 | 2 | 4 | $i(h_2^2)$ | $(g)$ | 0 |
| 8 | 2 | 5 | $h_2\widehat{h_2}$ | $(g)$ | $h_2^2$ |
| 8 | 3 | 3 | $i(c_0)$ | $(g)$ | 0 |
| 11 | 4 | 3 | $\widehat{h_1 c_0}$ | $(0)$ | $h_1c_0$ |

Table 4.5: $R_0$-module generators of $\mathrm{Ext}_{A(2)}(M_2, \mathbb{F}_2)$, with $i \geq 0$ in each $h_0$-tower (cont.)

| $t-s$ | $s$ | $g$ | $x$ | $\mathrm{Ann}(x)$ | $j(x)$ |
|---|---|---|---|---|---|
| 12 | 3 | 4 | $i(\alpha)$ | $(0)$ | 0 |
| 12 | 4 | 4 | $i(h_0\alpha)$ | $(g^2)$ | 0 |
| 12 | $5+i$ | 5 | $i(h_0^{2+i}\alpha)$ | $(g)$ | 0 |
| 14 | 3 | 5 | $\widehat{\alpha}$ | $(0)$ | $\alpha$ |
| 14 | 4 | 5 | $i(d_0)$ | $(0)$ | 0 |
| 14 | 4 | 6 | $h_0\widehat{\alpha}$ | $(g^2)$ | $h_0\alpha$ |
| 14 | 5 | 7 | $i(h_0 d_0)$ | $(g)$ | 0 |
| 14 | 5 | 8 | $h_0^2\widehat{\alpha}$ | $(g^2)$ | $h_0^2\alpha$ |
| 14 | $6+i$ | 8 | $h_0^{3+i}\widehat{\alpha}$ | $(g)$ | $h_0^{3+i}\alpha$ |
| 15 | 3 | 6 | $i(\beta)$ | $(0)$ | 0 |
| 15 | 4 | 7 | $i(h_0\beta)$ | $(g)$ | 0 |
| 16 | 5 | 9 | $d_0\widehat{h_0}$ | $(0)$ | $h_0 d_0$ |
| 17 | 3 | 7 | $\widehat{\beta}$ | $(0)$ | $\beta$ |
| 17 | 4 | $8+9$ | $i(e_0)$ | $(0)$ | 0 |
| 17 | 4 | 9 | $h_0\widehat{\beta}$ | $(g)$ | $h_0\beta$ |
| 17 | 5 | $10+11$ | $i(h_0 e_0)$ | $(g)$ | 0 |
| 17 | 5 | 11 | $h_0^2\widehat{\beta}$ | $(g)$ | $h_1 d_0$ |
| 17 | 6 | 10 | $i(h_0^2 e_0)$ | $(g)$ | 0 |
| 18 | 4 | 10 | $i(h_2\beta)$ | $(g)$ | 0 |
| 19 | 5 | 12 | $d_0\widehat{h_2}$ | $(0)$ | $h_0 e_0$ |
| 19 | 6 | 11 | $h_0 d_0\widehat{h_2}$ | $(g)$ | $h_0^2 e_0$ |
| 20 | 4 | 12 | $h_2\widehat{\beta}$ | $(g)$ | $h_2\beta$ |
| 20 | 5 | 14 | $h_0 h_2\widehat{\beta}$ | $(g)$ | $h_1 e_0$ |
| 23 | 5 | 16 | $h_2^2\widehat{\beta}$ | $(g)$ | $h_1 g$ |
| 24 | 6 | 14 | $i(\alpha^2)$ | $(0)$ | 0 |
| 24 | $7+i$ | 11 | $i(h_0^{1+i}\alpha^2)$ | $(g)$ | 0 |
| 26 | 6 | 15 | $\alpha\widehat{\alpha}$ | $(0)$ | $\alpha^2$ |
| 26 | 7 | $13+14$ | $i(\alpha d_0)$ | $(0)$ | 0 |
| 26 | $7+i$ | 14 | $h_0^{1+i}\alpha\widehat{\alpha}$ | $(g)$ | $h_0^{1+i}\alpha^2$ |
| 27 | 6 | 16 | $i(\alpha\beta)$ | $(0)$ | 0 |
| 28 | 7 | 15 | $d_0\widehat{\alpha}$ | $(0)$ | $\alpha d_0$ |

Table 4.5: $R_0$-module generators of $\mathrm{Ext}_{A(2)}(M_2, \mathbb{F}_2)$, with $i \geq 0$ in each $h_0$-tower (cont.)

| $t-s$ | $s$ | $g$ | $x$ | $\mathrm{Ann}(x)$ | $j(x)$ |
|---|---|---|---|---|---|
| 29 | 6 | 17 | $\alpha\widehat{\beta}$ | $(0)$ | $\alpha\beta$ |
| 29 | 7 | $16+17$ | $i(\alpha e_0)$ | $(0)$ | $0$ |
| 29 | 7 | 17 | $h_0\alpha\widehat{\beta}$ | $(g)$ | $h_1^2\gamma$ |
| 29 | 8 | 19 | $i(h_0\alpha e_0)$ | $(g)$ | $0$ |
| 30 | 6 | 18 | $i(\beta^2)$ | $(0)$ | $0$ |
| 31 | 7 | 18 | $d_0\widehat{\beta}$ | $(0)$ | $\alpha e_0$ |
| 31 | 8 | 21 | $h_0d_0\widehat{\beta}$ | $(g)$ | $h_0\alpha e_0$ |
| 32 | 6 | 19 | $\beta\widehat{\beta}$ | $(0)$ | $\beta^2$ |
| 32 | 7 | 20 | $i(\delta)$ | $(g)$ | $0$ |
| 36 | 8 | 25 | $\widehat{d_0g}$ | $(0)$ | $d_0g$ |
| 36 | 9 | 26 | $h_0\widehat{d_0g}$ | $(g)$ | $h_0d_0g$ |
| 36 | $10+i$ | 23 | $h_0^{2+i}\widehat{d_0g}$ | $(g)$ | $0$ |
| 38 | 9 | 27 | $\alpha^2\widehat{\alpha}$ | $(0)$ | $\alpha^3$ |
| 38 | $10+i$ | 26 | $h_0^{1+i}\alpha^2\widehat{\alpha}$ | $(g)$ | $h_0^{1+i}\alpha^3$ |
| 39 | 8 | 27 | $\gamma\widehat{\alpha}$ | $(0)$ | $e_0g$ |
| 41 | 9 | 29 | $\alpha^2\widehat{\beta}$ | $(0)$ | $d_0\gamma$ |
| 41 | 10 | 28 | $i(\alpha^2e_0)$ | $(0)$ | $0$ |
| 42 | 8 | 29 | $\gamma\widehat{\beta}$ | $(0)$ | $g^2$ |
| 43 | 10 | 29 | $\alpha d_0\widehat{\beta}$ | $(0)$ | $\alpha^2e_0$ |
| 44 | 9 | 31 | $\alpha\beta\widehat{\beta}$ | $(0)$ | $e_0\gamma$ |
| 47 | 9 | 33 | $\beta^2\widehat{\beta}$ | $(0)$ | $\gamma g$ |
| 53 | 12 | 41 | $d_0\gamma\widehat{\alpha}$ | $(0)$ | $d_0e_0g$ |

COROLLARY 4.13. *$\mathrm{Ext}_{A(2)}(M_2, \mathbb{F}_2)$ is generated as an $\mathrm{Ext}_{A(2)}(\mathbb{F}_2, \mathbb{F}_2)$-module by the classes*

$$i(1), \widehat{h_0}, \widehat{h_2}, \widehat{h_1c_0}, \widehat{\alpha}, \widehat{\beta}, \widehat{d_0g}$$

*listed in Table* 1.6 *and shown in Figure* 4.2.

PROOF. Each $R_0$-module generator $x$ in Table 4.5 is an $\mathrm{Ext}_{A(2)}(\mathbb{F}_2, \mathbb{F}_2)$-multiple of one of these seven classes. □

## 4.4. Coefficients in $M_4$

The short exact sequence of $A(2)$-modules

$$0 \to \Sigma^4\mathbb{F}_2 \longrightarrow M_4 \longrightarrow \mathbb{F}_2 \to 0$$

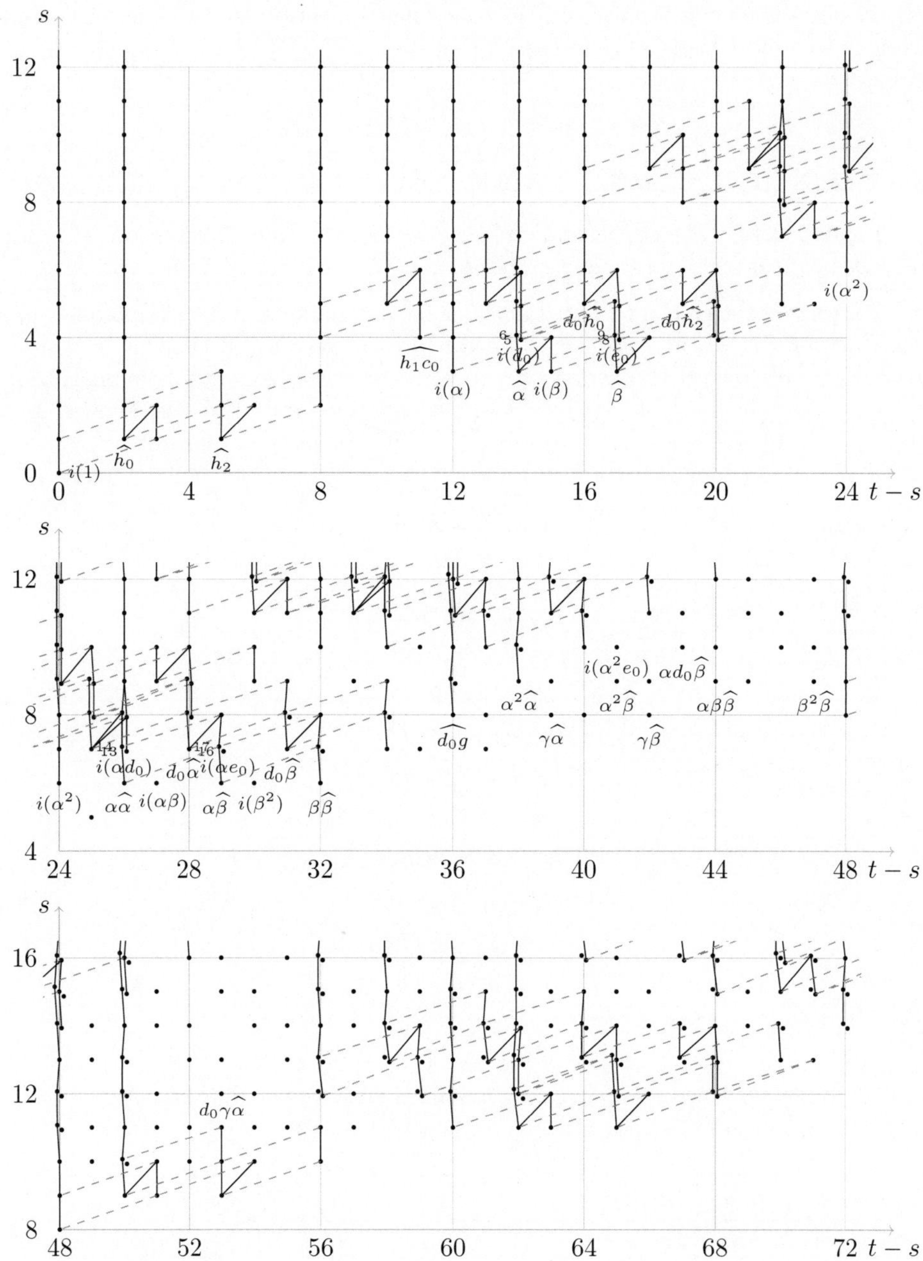


FIGURE 4.2. $R_0$-module generators of $\operatorname{Ext}_{A(2)}(M_2, \mathbb{F}_2)$. Note that $i(d_0) = 4_5$, $i(e_0) = 4_8 + 4_9$, $i(\alpha d_0) = 7_{13} + 7_{14}$ and $i(\alpha e_0) = 7_{16} + 7_{17}$.

represents $h_2$ in $\operatorname{Ext}^{1,4}_{A(2)}(\mathbb{F}_2, \mathbb{F}_2)$. Hence, in the induced long exact sequence

$$\dots \xrightarrow{\delta} \operatorname{Ext}^{*,*}_{A(2)}(\mathbb{F}_2, \mathbb{F}_2) \xrightarrow{i} \operatorname{Ext}^{*,*}_{A(2)}(M_4, \mathbb{F}_2) \xrightarrow{j} \operatorname{Ext}^{*,*}_{A(2)}(\Sigma^4\mathbb{F}_2, \mathbb{F}_2) \xrightarrow{\delta} \operatorname{Ext}^{*+1,*}_{A(2)}(\mathbb{F}_2, \mathbb{F}_2) \longrightarrow \dots$$

the connecting homomorphism $\delta$ is given by multiplication by $h_2$. The long exact sequence therefore leads to a short exact sequence of $\mathrm{Ext}_{A(2)}(\mathbb{F}_2,\mathbb{F}_2)$-modules, given in bidegree $(s,t)$ by

$$0 \to \mathrm{cok}(h_2)^{s,t} \xrightarrow{i} \mathrm{Ext}^{s,t}_{A(2)}(M_4,\mathbb{F}_2) \xrightarrow{j} \ker(h_2)^{s,t-4} \to 0\,,$$

where $\mathrm{Ext}^{s,t}_{A(2)}(\mathbb{F}_2,\mathbb{F}_2)/\mathrm{im}(h_2) = \mathrm{cok}(h_2)^{s,t}$ and $\ker(h_2)^{s,t} \subset \mathrm{Ext}^{s,t}_{A(2)}(\mathbb{F}_2,\mathbb{F}_2)$.

LEMMA 4.14. *The kernel and cokernel of $h_2$ are both direct sums of cyclic $R_0$-modules, with generators and annihilator ideals as listed in Table 4.6.*

PROOF. For each class $x$ listed in Table 3.6, spanning a cyclic $R_0$-module summand $\langle x\rangle$ of $SI \cong \mathrm{Ext}_{A(2)}(\mathbb{F}_2,\mathbb{F}_2)$, we can present $h_2 x$ as an element in a summand $\langle y\rangle$. We record the kernel and cokernel of the $R_0$-module homomorphism $h_2\colon \langle x\rangle \to \langle y\rangle$ in Table 4.6. These $h_2$-multiplications are visible in Figures 3.12 and 3.13. In most cases $h_2 x = 0$ or $y = h_2 x$. The less obvious cases are

$$\begin{aligned}
h_2 \cdot \alpha &= h_0\beta\\
h_2 \cdot h_0\alpha &= h_1 d_0\\
h_2 \cdot d_0 &= h_0 e_0\\
h_2 \cdot h_0 d_0 &= h_0^2 e_0\\
h_2 \cdot h_0\beta &= h_1 e_0\\
h_2 \cdot e_0 &= g \cdot h_0\\
h_2 \cdot h_0 e_0 &= g \cdot h_0^2\\
h_2 \cdot h_2\beta &= g \cdot h_1\\
h_2 \cdot \alpha^2 &= h_1^2\gamma\\
h_2 \cdot \alpha d_0 &= h_0\alpha e_0\\
h_2 \cdot \alpha e_0 &= g \cdot h_0\alpha\\
h_2 \cdot h_0\alpha e_0 &= g \cdot h_0^2\alpha\\
h_2 \cdot d_0 e_0 &= g \cdot h_0 d_0\,,
\end{aligned}$$

which are clear from Table 3.5. □

Table 4.6: Direct sum decompositions of the kernel and cokernel of multiplication by $h_2$, with $i \ge 0$ in each $h_0$-tower

| $t-s$ | $s$ | $g$ | $\ker(h_2)$ | $x$ | $h_2x$ | $\mathrm{cok}(h_2)$ |
|---|---|---|---|---|---|---|
| 0 | 0 | 0 | $\langle g\rangle = R_0$ | 1 | $h_2$ | $\langle 1\rangle = R_0$ |
| 0 | 1 | 0 | $\langle h_0 g\rangle = R_0/(g)$ | $h_0$ | $h_0h_2$ | $\langle h_0\rangle = R_0/(g)$ |
| 0 | 2 | 0 | $\langle h_0^2 g\rangle = R_0/(g)$ | $h_0^2$ | $h_0^2h_2$ | $\langle h_0^2\rangle = R_0/(g)$ |
| 0 | $3+i$ | 0 | $\langle h_0^{3+i}\rangle = R_0/(g)$ | $h_0^{3+i}$ | 0 | $\langle h_0^{3+i}\rangle = R_0/(g)$ |
| 1 | 1 | 1 | $\langle h_1\rangle = R_0/(g^2)$ | $h_1$ | 0 | $\langle h_1\rangle = R_0/(g)$ |
| 2 | 2 | 1 | $\langle h_1^2\rangle = R_0/(g)$ | $h_1^2$ | 0 | $\langle h_1^2\rangle = R_0/(g)$ |
| 3 | 1 | 2 | 0 | $h_2$ | $h_2^2$ | 0 |

Table 4.6: Direct sum decompositions of the kernel and cokernel of multiplication by $h_2$, with $i \geq 0$ in each $h_0$-tower (cont.)

| $t-s$ | $s$ | $g$ | $\ker(h_2)$ | $x$ | $h_2x$ | $\mathrm{cok}(h_2)$ |
|---|---|---|---|---|---|---|
| 3 | 2 | 2 | $\langle h_0h_2\rangle = R_0/(g)$ | $h_0h_2$ | 0 | 0 |
| 3 | 3 | 1 | $\langle h_0^2h_2\rangle = R_0/(g)$ | $h_0^2h_2$ | 0 | 0 |
| 6 | 2 | 3 | $\langle h_2^2\rangle = R_0/(g)$ | $h_2^2$ | 0 | 0 |
| 8 | 3 | 2 | $\langle c_0\rangle = R_0/(g)$ | $c_0$ | 0 | $\langle c_0\rangle = R_0/(g)$ |
| 9 | 4 | 2 | $\langle h_1c_0\rangle = R_0/(g)$ | $h_1c_0$ | 0 | $\langle h_1c_0\rangle = R_0/(g)$ |
| 12 | 3 | 3 | $\langle \alpha g\rangle = R_0$ | $\alpha$ | $h_0\beta$ | $\langle \alpha\rangle = R_0$ |
| 12 | 4 | 3 | $\langle h_0\alpha g\rangle = R_0/(g)$ | $h_0\alpha$ | $h_1d_0$ | $\langle h_0\alpha\rangle = R_0/(g)$ |
| 12 | 5 | 4 | $\langle h_0^2\alpha\rangle = R_0/(g^2)$ | $h_0^2\alpha$ | 0 | $\langle h_0^2\alpha\rangle = R_0/(g)$ |
| 12 | $6+i$ | 4 | $\langle h_0^{3+i}\alpha\rangle = R_0/(g)$ | $h_0^{3+i}\alpha$ | 0 | $\langle h_0^{3+i}\alpha\rangle = R_0/(g)$ |
| 14 | 4 | 4 | $\langle d_0g\rangle = R_0$ | $d_0$ | $h_0e_0$ | $\langle d_0\rangle = R_0$ |
| 14 | 5 | 5 | $\langle h_0d_0g\rangle = R_0/(g)$ | $h_0d_0$ | $h_0^2e_0$ | $\langle h_0d_0\rangle = R_0/(g)$ |
| 15 | 3 | 4 | $\langle \beta g\rangle = R_0$ | $\beta$ | $h_2\beta$ | $\langle \beta\rangle = R_0$ |
| 15 | 4 | 5 | 0 | $h_0\beta$ | $h_1e_0$ | 0 |
| 15 | 5 | 6 | $\langle h_1d_0\rangle = R_0/(g)$ | $h_1d_0$ | 0 | 0 |
| 17 | 4 | 6 | $\langle e_0g\rangle = R_0$ | $e_0$ | $g\cdot h_0$ | $\langle e_0\rangle = R_0$ |
| 17 | 5 | 7 | 0 | $h_0e_0$ | $g\cdot h_0^2$ | 0 |
| 17 | 6 | 6 | $\langle h_0^2e_0\rangle = R_0/(g)$ | $h_0^2e_0$ | 0 | 0 |
| 18 | 4 | 7 | 0 | $h_2\beta$ | $g\cdot h_1$ | 0 |
| 18 | 5 | 8 | $\langle h_1e_0\rangle = R_0/(g)$ | $h_1e_0$ | 0 | 0 |
| 24 | 6 | 8 | $\langle \alpha^2g\rangle = R_0$ | $\alpha^2$ | $h_1^2\gamma$ | $\langle \alpha^2\rangle = R_0$ |
| 24 | $7+i$ | 7 | $\langle h_0^{1+i}\alpha^2\rangle = R_0/(g)$ | $h_0^{1+i}\alpha^2$ | 0 | $\langle h_0^{1+i}\alpha^2\rangle = R_0/(g)$ |
| 25 | 5 | 11 | $\langle \gamma\rangle = R_0$ | $\gamma$ | 0 | $\langle \gamma\rangle = R_0$ |
| 26 | 6 | 9 | $\langle h_1\gamma\rangle = R_0/(g)$ | $h_1\gamma$ | 0 | $\langle h_1\gamma\rangle = R_0/(g)$ |
| 26 | 7 | 8 | $\langle \alpha d_0g\rangle = R_0$ | $\alpha d_0$ | $h_0\alpha e_0$ | $\langle \alpha d_0\rangle = R_0$ |
| 27 | 6 | 10 | $\langle \alpha\beta\rangle = R_0$ | $\alpha\beta$ | 0 | $\langle \alpha\beta\rangle = R_0$ |
| 27 | 7 | 9 | $\langle h_1^2\gamma\rangle = R_0/(g)$ | $h_1^2\gamma$ | 0 | 0 |
| 29 | 7 | 10 | $\langle \alpha e_0g\rangle = R_0$ | $\alpha e_0$ | $g\cdot h_0\alpha$ | $\langle \alpha e_0\rangle = R_0$ |
| 29 | 8 | 12 | 0 | $h_0\alpha e_0$ | $g\cdot h_0^2\alpha$ | 0 |
| 30 | 6 | 11 | $\langle \beta^2\rangle = R_0$ | $\beta^2$ | 0 | $\langle \beta^2\rangle = R_0$ |
| 31 | 8 | 13 | $\langle d_0e_0g\rangle = R_0$ | $d_0e_0$ | $g\cdot h_0d_0$ | $\langle d_0e_0\rangle = R_0$ |
| 32 | 7 | 11 | $\langle \delta\rangle = R_0/(g)$ | $\delta$ | 0 | $\langle \delta\rangle = R_0/(g)$ |

Table 4.6: Direct sum decompositions of the kernel and cokernel of multiplication by $h_2$, with $i \geq 0$ in each $h_0$-tower (cont.)

| $t-s$ | $s$ | $g$ | $\ker(h_2)$ | $x$ | $h_2x$ | $\mathrm{cok}(h_2)$ |
|---|---|---|---|---|---|---|
| 33 | 8 | 15 | $\langle h_1\delta\rangle = R_0/(g)$ | $h_1\delta$ | 0 | $\langle h_1\delta\rangle = R_0/(g)$ |
| 36 | 9 | 17 | $\langle \alpha^3\rangle = R_0$ | $\alpha^3$ | 0 | $\langle \alpha^3\rangle = R_0$ |
| 36 | $10+i$ | 14 | $\langle h_0^{1+i}\alpha^3\rangle = R_0/(g)$ | $h_0^{1+i}\alpha^3$ | 0 | $\langle h_0^{1+i}\alpha^3\rangle = R_0/(g)$ |
| 39 | 9 | 18 | $\langle d_0\gamma\rangle = R_0$ | $d_0\gamma$ | 0 | $\langle d_0\gamma\rangle = R_0$ |
| 41 | 10 | 16 | $\langle \alpha^2 e_0\rangle = R_0$ | $\alpha^2 e_0$ | 0 | $\langle \alpha^2 e_0\rangle = R_0$ |
| 42 | 9 | 19 | $\langle e_0\gamma\rangle = R_0$ | $e_0\gamma$ | 0 | $\langle e_0\gamma\rangle = R_0$ |

PROPOSITION 4.15. *$\mathrm{Ext}_{A(2)}(M_4, \mathbb{F}_2)$ is a direct sum of cyclic $R_0$-modules, with generators and annihilator ideals as listed in Table* 4.7.

PROOF. We use `ext` to determine the $R_0$-module extensions of summands in $\ker(h_2)$ by summands in $\mathrm{cok}(h_2)$. Each summand in $\ker(h_2)$ has a generator of the form $y = xg^n$, and we choose a lift $\overline{y}$ in $\mathrm{Ext}_{A(2)}(M_4, \mathbb{F}_2)$ with $j(\overline{y}) = y$. In most cases the lift is unique, but for $xg^n = c_0$ we have already chosen $3_4 = \overline{c_0}$ as the lift of $c_0$, for $xg^n = h_0^2 g$ we prefer $6_{11} = h_0^2\overline{g}$ over $6_{12}$, for $xg^n = h_1\gamma$ we prefer $6_{15} = h_1\overline{\gamma}$ over $6_{14} + 6_{15}$, for $xg^n = h_1\delta$ we prefer $8_{21} = h_1\overline{\delta}$ over $8_{20} + 8_{21}$, and for $xg^n = \alpha^2 g$ we prefer $10_{30} + 10_{31} = \alpha^2\overline{g}$ over $10_{30}$.

We then use `ext` to write $\overline{y}$ as the product of a class in $\mathrm{Ext}_{A(2)}(\mathbb{F}_2, \mathbb{F}_2)$ and one of the module generators from Table 1.7. When given a choice, we prefer factorizations that last as long as possible in the Adams spectral sequence for $tmf/\nu$, and we emphasize $h_i$-multiplications and other products with coefficients in low topological degree. In most cases the given presentation of $\overline{y}$ is evidently a lift of $y$. The less obvious cases are $j(d_0\overline{h_0h_2}) = h_0^2 e_0$, $j(h_0\overline{\alpha\beta}) = h_1^2\gamma$, $j(h_0\overline{\delta}) = h_0\alpha g$, $j(d_0\overline{\alpha\beta}) = \alpha^2 e_0$ and $j(\alpha^2\overline{\gamma}) = \alpha e_0 g$, all of which follow from the relations in Table 3.5.

If $\langle y\rangle = R_0$ then $\langle\overline{y}\rangle = R_0$. Otherwise, if $\langle y\rangle = R_0/(g^m)$ we use `ext` to calculate $g^m\cdot\overline{y}$. If the answer is 0, then $\langle\overline{y}\rangle = R_0/(g^m)$, but if $g^m\cdot\overline{y} = i(z) \neq 0$ then $\langle\overline{y}\rangle$ is an extension of $R_0/(g^m)$ by the summand containing $z$. This happens in the following seven cases.

$$g^2\cdot\overline{h_1} = g\cdot i(\gamma)$$
$$g\cdot\overline{h_0h_2} = i(\alpha\beta)$$
$$g\cdot\overline{h_2^2} = i(\beta^2)$$
$$g^2\cdot\overline{h_0^2\alpha} = g\cdot i(\alpha^3)$$
$$g\cdot d_0\overline{h_1} = i(d_0\gamma)$$
$$g\cdot d_0\overline{h_0h_2} = i(\alpha^2 e_0)$$
$$g\cdot e_0\overline{h_1} = i(e_0\gamma)$$

In most instances $z$ generates that summand, and $\langle\overline{y}\rangle$ is cyclic. In two exceptional cases, corresponding to $\overline{y} = \overline{h_1}$ and $\overline{y} = \overline{h_0^2\alpha}$, we make a change of basis, replacing the generators $i(\gamma)$ and $i(\alpha^3)$ with

$$h_1\overline{g} = i(\gamma) + g \cdot \overline{h_1}$$

and

$$h_0^2\overline{\delta} = i(\alpha^3) + g \cdot \overline{h_0^2\alpha}\,,$$

respectively. This yields the splittings

$$\langle \overline{h_1}, i(\gamma)\rangle = \langle \overline{h_1}\rangle \oplus \langle h_1\overline{g}\rangle \cong R_0 \oplus R_0/(g)$$

and

$$\langle \overline{h_0^2\alpha}, i(\alpha^3)\rangle = \langle \overline{h_0^2\alpha}\rangle \oplus \langle h_0^2\overline{\delta}\rangle \cong R_0 \oplus R_0/(g)\,.$$

Each of the remaining summands $\langle z\rangle$ in $\mathrm{cok}(h_2)$ contributes a new summand $\langle i(z)\rangle$ in $\mathrm{Ext}_{A(2)}(M_4, \mathbb{F}_2)$. Gathering these together, and writing $x$ in place of $\overline{y}$ or $i(z)$, leads to Table 4.7. □

Table 4.7: $R_0$-module generators of $\mathrm{Ext}_{A(2)}(M_4, \mathbb{F}_2)$, with $i \geq 0$ in each $h_0$-tower

| $t-s$ | $s$ | $g$ | $x$ | $\mathrm{Ann}(x)$ | $j(x)$ |
|---|---|---|---|---|---|
| 0 | 0 | 0 | $i(1)$ | $(0)$ | 0 |
| 0 | $1+i$ | 0 | $i(h_0^{1+i})$ | $(g)$ | 0 |
| 1 | 1 | 1 | $i(h_1)$ | $(g)$ | 0 |
| 2 | 2 | 1 | $i(h_1^2)$ | $(g)$ | 0 |
| 4 | $3+i$ | 1 | $h_0^i\overline{h_0^3}$ | $(g)$ | $h_0^{3+i}$ |
| 5 | 1 | 2 | $\overline{h_1}$ | $(0)$ | $h_1$ |
| 6 | 2 | 2 | $h_1\overline{h_1}$ | $(g)$ | $h_1^2$ |
| 7 | 2 | 3 | $\overline{h_0h_2}$ | $(0)$ | $h_0h_2$ |
| 7 | 3 | 2 | $h_0\overline{h_0h_2}$ | $(g)$ | $h_0^2h_2$ |
| 8 | 3 | 3 | $i(c_0)$ | $(g)$ | 0 |
| 9 | 4 | 3 | $i(h_1c_0)$ | $(g)$ | 0 |
| 10 | 2 | 4 | $\overline{h_2^2}$ | $(0)$ | $h_2^2$ |
| 12 | 3 | 4 | $\overline{c_0}$ | $(g)$ | $c_0$ |
| 12 | 3 | $4+5$ | $i(\alpha)$ | $(0)$ | 0 |
| 12 | $4+i$ | 4 | $i(h_0^{1+i}\alpha)$ | $(g)$ | 0 |
| 13 | 4 | 5 | $h_1\overline{c_0}$ | $(g)$ | $h_1c_0$ |
| 14 | 4 | 6 | $i(d_0)$ | $(0)$ | 0 |
| 14 | 5 | 6 | $i(h_0d_0)$ | $(g)$ | 0 |
| 15 | 3 | 6 | $i(\beta)$ | $(0)$ | 0 |
| 16 | 5 | 7 | $\overline{h_0^2\alpha}$ | $(0)$ | $h_0^2\alpha$ |
| 16 | $6+i$ | 7 | $h_0^{1+i}\overline{h_0^2\alpha}$ | $(g)$ | $h_0^{3+i}\alpha$ |

Table 4.7: $R_0$-module generators of $\mathrm{Ext}_{A(2)}(M_4, \mathbb{F}_2)$, with $i \geq 0$ in each $h_0$-tower (cont.)

| $t-s$ | $s$ | $g$ | $x$ | $\mathrm{Ann}(x)$ | $j(x)$ |
|---|---|---|---|---|---|
| 17 | 4 | 7 | $i(e_0)$ | $(0)$ | $0$ |
| 19 | 5 | 8 | $d_0\overline{h_1}$ | $(0)$ | $h_1 d_0$ |
| 21 | 6 | 9 | $d_0\overline{h_0 h_2}$ | $(0)$ | $h_0^2 e_0$ |
| 22 | 5 | 9 | $e_0\overline{h_1}$ | $(0)$ | $h_1 e_0$ |
| 24 | 4 | 9 | $\overline{g}$ | $(0)$ | $g$ |
| 24 | 5 | 10 | $h_0\overline{g}$ | $(g)$ | $h_0 g$ |
| 24 | 6 | $10+11$ | $i(\alpha^2)$ | $(0)$ | $0$ |
| 24 | 6 | 11 | $h_0^2\overline{g}$ | $(g)$ | $h_0^2 g$ |
| 24 | $7+i$ | 11 | $i(h_0^{1+i}\alpha^2)$ | $(g)$ | $0$ |
| 25 | 5 | 12 | $h_1\overline{g}$ | $(g)$ | $0$ |
| 26 | 6 | 12 | $i(h_1\gamma)$ | $(g)$ | $0$ |
| 26 | 7 | 12 | $i(\alpha d_0)$ | $(0)$ | $0$ |
| 28 | $7+i$ | 13 | $h_0^i\overline{h_0\alpha^2}$ | $(g)$ | $h_0^{1+i}\alpha^2$ |
| 29 | 5 | 13 | $\overline{\gamma}$ | $(0)$ | $\gamma$ |
| 29 | 7 | 14 | $i(\alpha e_0)$ | $(0)$ | $0$ |
| 30 | 6 | 15 | $h_1\overline{\gamma}$ | $(g)$ | $h_1\gamma$ |
| 31 | 6 | 16 | $\overline{\alpha\beta}$ | $(0)$ | $\alpha\beta$ |
| 31 | 7 | 15 | $h_0\overline{\alpha\beta}$ | $(g)$ | $h_1^2\gamma$ |
| 31 | 8 | 15 | $i(d_0 e_0)$ | $(0)$ | $0$ |
| 32 | 7 | 17 | $i(\delta)$ | $(g)$ | $0$ |
| 33 | 8 | 17 | $i(h_1\delta)$ | $(g)$ | $0$ |
| 34 | 6 | 17 | $\overline{\beta^2}$ | $(0)$ | $\beta^2$ |
| 36 | 7 | 19 | $\overline{\delta}$ | $(g)$ | $\delta$ |
| 36 | 7 | $19+20$ | $\alpha\overline{g}$ | $(0)$ | $\alpha g$ |
| 36 | 8 | 19 | $h_0\overline{\delta}$ | $(g)$ | $h_0\alpha g$ |
| 36 | 9 | 20 | $h_0^2\overline{\delta}$ | $(g)$ | $0$ |
| 36 | $10+i$ | 20 | $i(h_0^{1+i}\alpha^3)$ | $(g)$ | $0$ |
| 37 | 8 | 21 | $h_1\overline{\delta}$ | $(g)$ | $h_1\delta$ |
| 38 | 8 | 22 | $d_0\overline{g}$ | $(0)$ | $d_0 g$ |
| 38 | 9 | 22 | $h_0 d_0\overline{g}$ | $(g)$ | $h_0 d_0 g$ |
| 39 | 7 | 21 | $\beta\overline{g}$ | $(0)$ | $\beta g$ |

Table 4.7: $R_0$-module generators of $\mathrm{Ext}_{A(2)}(M_4, \mathbb{F}_2)$, with $i \geq 0$ in each $h_0$-tower (cont.)

| $t-s$ | $s$ | $g$ | $x$ | $\mathrm{Ann}(x)$ | $j(x)$ |
|---|---|---|---|---|---|
| 40 | 9 | 24 | $\overline{\alpha^3}$ | $(0)$ | $\alpha^3$ |
| 40 | $10+i$ | 24 | $h_0^{1+i}\overline{\alpha^3}$ | $(g)$ | $h_0^{1+i}\alpha^3$ |
| 41 | 8 | 24 | $e_0\overline{g}$ | $(0)$ | $e_0 g$ |
| 43 | 9 | 26 | $d_0\overline{\gamma}$ | $(0)$ | $d_0\gamma$ |
| 45 | 10 | 28 | $d_0\overline{\alpha\beta}$ | $(0)$ | $\alpha^2 e_0$ |
| 46 | 9 | 28 | $e_0\overline{\gamma}$ | $(0)$ | $e_0\gamma$ |
| 48 | 10 | $30+31$ | $\alpha^2\overline{g}$ | $(0)$ | $\alpha^2 g$ |
| 50 | 11 | 33 | $\alpha d_0\overline{g}$ | $(0)$ | $\alpha d_0 g$ |
| 53 | 11 | 36 | $\alpha^2\overline{\gamma}$ | $(0)$ | $\alpha e_0 g$ |
| 55 | 12 | 38 | $d_0 e_0\overline{g}$ | $(0)$ | $d_0 e_0 g$ |

COROLLARY 4.16. *$\mathrm{Ext}_{A(2)}(M_4, \mathbb{F}_2) \cong \mathrm{Ext}_{B(2,2,1)}(\mathbb{F}_2, \mathbb{F}_2)$ is generated as an $\mathrm{Ext}_{A(2)}(\mathbb{F}_2, \mathbb{F}_2)$-module by the classes*

$$i(1), \overline{h_0^3}, \overline{h_1}, \overline{h_0 h_2}, \overline{h_2^2}, \overline{c_0}, \overline{h_0^2\alpha}, \overline{g}, \overline{h_0\alpha^2}, \overline{\gamma}, \overline{\alpha\beta}, \overline{\beta^2}, \overline{\delta}, \overline{\alpha^3}$$

*listed in Table* 1.7 *and shown in Figure* 4.3.

PROOF. Each $R_0$-module generator $x$ in Table 4.7 is an $\mathrm{Ext}_{A(2)}(\mathbb{F}_2, \mathbb{F}_2)$-multiple of one of these 14 classes. □

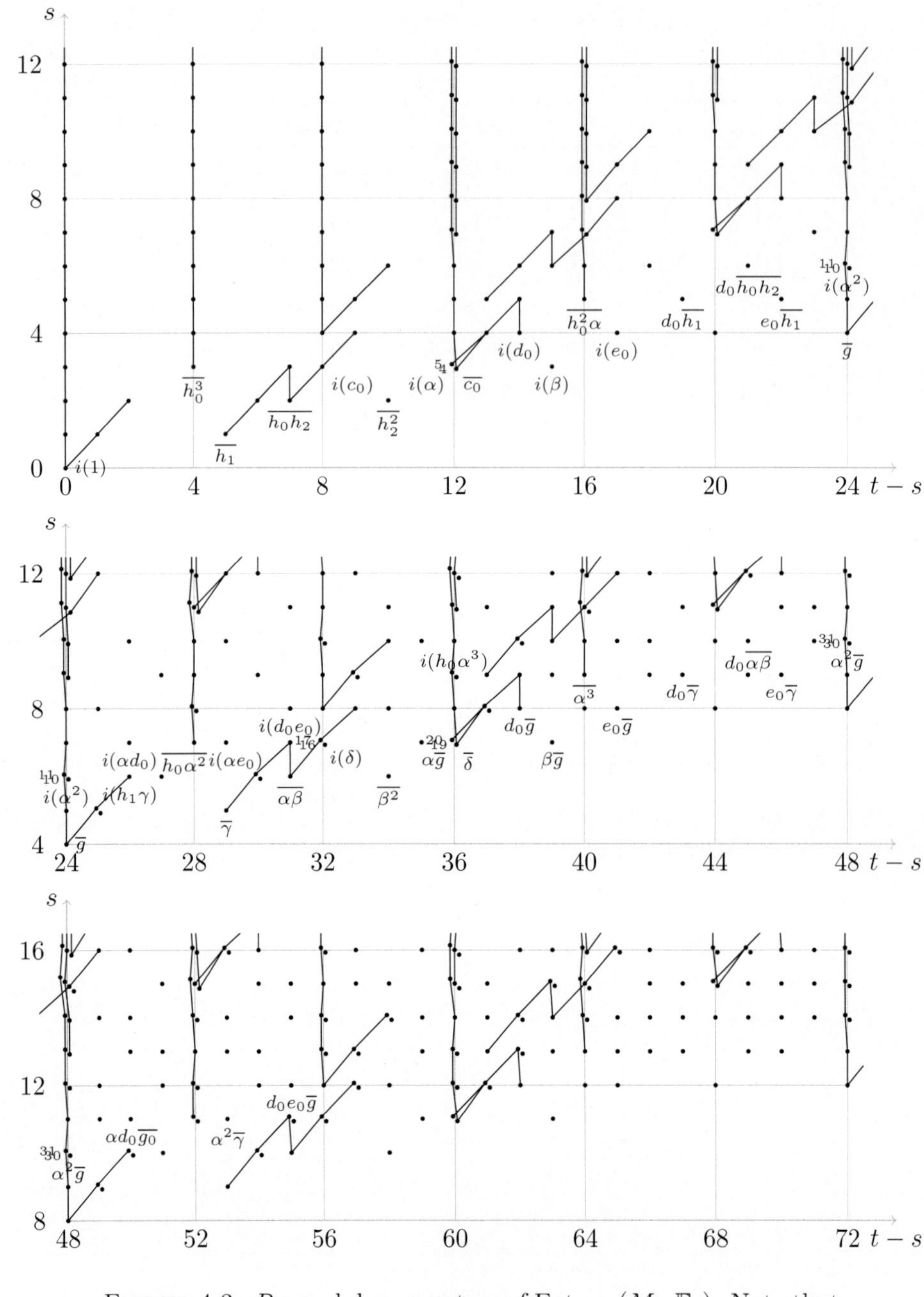

FIGURE 4.3. $R_0$-module generators of $\mathrm{Ext}_{A(2)}(M_4, \mathbb{F}_2)$. Note that $\overline{c_0} = 3_4$, $i(\alpha) = 3_4 + 3_5$, $i(\alpha^2) = 6_{10} + 6_{11}$ $i(\delta) = 7_{17}$, $\overline{\delta} = 7_{19}$, $\alpha\overline{g} = 7_{19} + 7_{20}$ and $\alpha^2\overline{g} = 10_{30} + 10_{31}$.

# Part 2

# The Adams differentials

CHAPTER 5

# The Adams spectral sequence for *tmf*

We calculate the $d_r$-differentials in the Adams spectral sequence for the topological modular forms spectrum. These are nontrivial for $r \in \{2, 3, 4\}$, and zero for $r \geq 5$, so the spectral sequence collapses at the $E_5$-term. The $E_\infty$ (or $H_\infty$) ring structure on *tmf* suffices to determine most of these differentials, due to their interaction with the Steenrod operations in $\mathrm{Ext}_{A(2)}(\mathbb{F}_2, \mathbb{F}_2)$. Two further differentials are determined by naturality with respect to the unit map $\iota\colon S \to tmf$. The resulting $E_\infty$-term is the associated graded of a complete Hausdorff filtration of $\pi_*(tmf)^\wedge_2$.

## 5.1. The $E_2$-term for *tmf*

The initial term

$$E_2 = E_2(tmf) \cong \mathrm{Ext}_{A(2)}(\mathbb{F}_2, \mathbb{F}_2)$$

of the mod 2 Adams spectral sequence for the topological modular forms spectrum *tmf* was calculated in Part I. The groups $E_2^{s,t}$ for $0 \leq t - s \leq 192$ are displayed in Figures 1.11 to 1.18. As a bigraded commutative algebra, the $E_2$-term is generated by the 13 classes

$$h_0, h_1, h_2, c_0, \alpha, \beta, d_0, e_0, \gamma, \delta, g, w_1, w_2$$

listed in Tables 1.3 and 3.3. These are subject to the ideal of relations generated by the 54 relations listed in Table 3.4. A Gröbner basis for this ideal is given by the 77 relations listed in Table 3.5.

The $E_2$-term is free as a module over $\mathbb{F}_2[w_1, w_2]$, and is finitely generated as a module over $\mathbb{F}_2[h_0, g, w_1, w_2]$, but we choose to primarily keep track of its module structure over the intermediate algebra $R_0 = \mathbb{F}_2[g, w_1, w_2]$. The classes $g$, $w_1$ and $w_2^4$ will be seen to be infinite cycles in the Adams spectral sequence for *tmf*, meaning that they are $d_r$-cycles for all $r \geq 2$, but there are nonzero differentials $d_2(w_2) = \alpha\beta g$ and $d_3(w_2^2) = \beta g^4$. We will therefore consider the $E_3$-term as a module over $R_1 = \mathbb{F}_2[g, w_1, w_2^2]$, and regard the $E_r$-terms for $r \geq 4$ as modules over $R_2 = \mathbb{F}_2[g, w_1, w_2^4]$.

Definition 5.1. For $i \in \{0, 1, 2\}$ let $R_i = \mathbb{F}_2[g, w_1, w_2^{2^i}]$. Then $R_0 \cong R_1\{1, w_2\}$ and $R_1 \cong R_2\{1, w_2^2\}$ as $R_1$- and $R_2$-modules, respectively.

A presentation of the $E_2$-term as a direct sum of cyclic $R_0$-modules is given in Table 5.1, most of which is obtained from Table 3.6 by combining a few of the rows. By Proposition 3.45 we have an isomorphism

$$E_2 = E_2(tmf) \cong \bigoplus_x \frac{R_0}{\mathrm{Ann}(x)}\{x\}$$

of $R_0$-modules, where $x$ ranges over the generators listed in Table 5.1 and $\mathrm{Ann}(x) \subset R_0$ denotes the annihilator ideal of $x$. The $R_0$-module generators are indicated by large dots (•) in Figures 3.12 and 3.13. The four $h_0$-towers, in topological degrees $t-s \in \{0, 12, 24, 36\}$, continue indefinitely. When enumerating the infinitely repeating parts of such $h_0$-towers we will always use an index $i$ that runs over the non-negative integers. In other words, we systematically let $i \geq 0$ in these tables. The columns $t-s$ and $s$ give the topological degree and Adams filtration of the generator $x$, respectively. The column $g$ gives the generator number in the minimal $A(2)$-module resolution calculated by `ext`, see Definition 1.8, so that $x$ corresponds to the cocycle denoted $s_g$. In the case of an $h_0$-tower of the form $\{h_0^i x\}$ with $i \geq 0$, the generator number $g$ is given for the element corresponding to $i = 0$.

Table 5.1: $R_0$-module generators of $E_2(tmf)$, with $i \geq 0$ in each $h_0$-tower

| $t-s$ | $s$ | $g$ | $x$ | $\mathrm{Ann}(x)$ | $d_2(x)$ | $d_2(xw_2)$ |
|---|---|---|---|---|---|---|
| 0 | 0 | 0 | $1$ | $(0)$ | 0 | $\alpha\beta g$ |
| 0 | 1 | 0 | $h_0$ | $(g^2)$ | 0 | 0 |
| 0 | 2 | 0 | $h_0^2$ | $(g^2)$ | 0 | 0 |
| 0 | $3+i$ | 0 | $h_0^{3+i}$ | $(g)$ | 0 | 0 |
| 1 | 1 | 1 | $h_1$ | $(g^2)$ | 0 | 0 |
| 2 | 2 | 1 | $h_1^2$ | $(g)$ | 0 | 0 |
| 3 | 1 | 2 | $h_2$ | $(g)$ | 0 | 0 |
| 3 | 2 | 2 | $h_0h_2$ | $(g)$ | 0 | 0 |
| 3 | 3 | 1 | $h_0^2h_2$ | $(g)$ | 0 | 0 |
| 6 | 2 | 3 | $h_2^2$ | $(g)$ | 0 | 0 |
| 8 | 3 | 2 | $c_0$ | $(g)$ | 0 | 0 |
| 9 | 4 | 2 | $h_1c_0$ | $(g)$ | 0 | 0 |
| 12 | 3 | 3 | $\alpha$ | $(0)$ | $h_2w_1$ | $d_0\gamma g + h_2w_1w_2$ |
| 12 | 4 | 3 | $h_0\alpha$ | $(g^2)$ | $h_0h_2w_1$ | $h_0h_2w_1w_2$ |
| 12 | 5 | 4 | $h_0^2\alpha$ | $(g^2)$ | $h_0^2h_2w_1$ | $h_0^2h_2w_1w_2$ |
| 12 | $6+i$ | 4 | $h_0^{3+i}\alpha$ | $(g)$ | 0 | 0 |
| 14 | 4 | 4 | $d_0$ | $(0)$ | 0 | $\alpha^2e_0g$ |
| 14 | 5 | 5 | $h_0d_0$ | $(g^2)$ | 0 | 0 |
| 15 | 3 | 4 | $\beta$ | $(0)$ | $h_0d_0$ | $e_0\gamma g + h_0d_0w_2$ |
| 15 | 4 | 5 | $h_0\beta$ | $(g)$ | $h_2^2w_1$ | $h_2^2w_1w_2$ |
| 15 | 5 | 6 | $h_1d_0$ | $(g)$ | 0 | 0 |
| 17 | 4 | 6 | $e_0$ | $(0)$ | 0 | $\alpha^2g^2$ |
| 17 | 5 | 7 | $h_0e_0$ | $(g)$ | 0 | 0 |
| 17 | 6 | 6 | $h_0^2e_0$ | $(g)$ | 0 | 0 |

Table 5.1: $R_0$-module generators of $E_2(tmf)$, with $i \geq 0$ in each $h_0$-tower (cont.)

| $t-s$ | $s$ | $g$ | $x$ | $\mathrm{Ann}(x)$ | $d_2(x)$ | $d_2(xw_2)$ |
|---|---|---|---|---|---|---|
| 18 | 4 | 7 | $h_2\beta$ | $(g)$ | $h_0^2e_0$ | $h_0^2e_0w_2$ |
| 18 | 5 | 8 | $h_1e_0$ | $(g)$ | 0 | 0 |
| 24 | 6 | 8 | $\alpha^2$ | $(0)$ | 0 | $d_0e_0g^2$ |
| 24 | $7+i$ | 7 | $h_0^{1+i}\alpha^2$ | $(g)$ | 0 | 0 |
| 25 | 5 | 11 | $\gamma$ | $(0)$ | 0 | $\alpha g^3$ |
| 26 | 6 | 9 | $h_1\gamma$ | $(g)$ | 0 | 0 |
| 26 | 7 | 8 | $\alpha d_0$ | $(0)$ | $h_0e_0w_1$ | $\gamma g^2w_1 + h_0e_0w_1w_2$ |
| 27 | 6 | 10 | $\alpha\beta$ | $(0)$ | 0 | $d_0g^3$ |
| 27 | 7 | 9 | $h_1^2\gamma$ | $(g)$ | 0 | 0 |
| 29 | 7 | 10 | $\alpha e_0$ | $(0)$ | $h_0gw_1$ | $\alpha^3g^2 + h_0gw_1w_2$ |
| 29 | 8 | 12 | $h_0\alpha e_0$ | $(g)$ | $h_0^2gw_1$ | $h_0^2gw_1w_2$ |
| 30 | 6 | 11 | $\beta^2$ | $(0)$ | 0 | $e_0g^3$ |
| 31 | 8 | 13 | $d_0e_0$ | $(0)$ | 0 | $\beta^2g^2w_1$ |
| 32 | 7 | 11 | $\delta$ | $(g)$ | 0 | 0 |
| 33 | 8 | 15 | $h_1\delta$ | $(g)$ | 0 | 0 |
| 36 | 9 | 17 | $\alpha^3$ | $(0)$ | $h_1^2\gamma w_1$ | $\beta g^3w_1 + h_1^2\gamma w_1w_2$ |
| 36 | $10+i$ | 14 | $h_0^{1+i}\alpha^3$ | $(g)$ | 0 | 0 |
| 39 | 9 | 18 | $d_0\gamma$ | $(0)$ | 0 | $\alpha d_0g^3$ |
| 41 | 10 | 16 | $\alpha^2e_0$ | $(0)$ | 0 | $g^4w_1$ |
| 42 | 9 | 19 | $e_0\gamma$ | $(0)$ | 0 | $\alpha e_0g^3$ |

## 5.2. The $d_2$-differentials for *tmf*

The main purpose of this section is to determine the $d_2$-differentials in the Adams spectral sequence for *tmf*. We will see that $g$, $w_1$ and $w_2^2$ are $d_2$-cycles, so that the $d_2$-differential is $R_1$-linear. Hence it suffices to determine $d_2(x)$ and $d_2(xw_2)$ as $x$ ranges through a set of $R_0$-module generators for the $E_2$-term, since the classes $x$ and $xw_2$ will then range through a set of $R_1$-module generators for the same $E_2$-term. We first determine $d_2$ on the 13 algebra generators of $E_2$. The values of $d_2$ on the remaining $R_1$-module generators will then follow by the Leibniz rule

$$d_r(xy) = d_r(x)y + xd_r(y)$$

(for $r = 2$), which holds because the Adams spectral sequence for *tmf* is an algebra spectral sequence.

Inspection of the $E_2$-term quickly shows that ten of the algebra generators are $d_2$-cycles. The three remaining generators are $\alpha$, $\beta$ and $w_2$. The $d_2$-differentials on $\alpha$ and $\beta$ follow from the known interaction between differentials and Steenrod operations in the $E_2$-term. To determine $d_2(w_2)$ we will rely on some external input, given by a comparison of the Adams spectral sequences for $S$ and $tmf$. The first two hidden $\eta$-multiplications in the Adams spectral sequence for $S$ (showing that $\eta\rho$ is detected by $Pc_0$ and $\eta^2\bar{\kappa}$ is detected by $Pd_0$) lead to two key differentials in the Adams spectral sequence for $tmf$ (namely, $d_3(e_0) = c_0 w_1$ and $d_4(e_0 g) = g w_1^2$), and the value of $d_2(w_2)$ follows from this.

First we have some easy vanishing results.

LEMMA 5.2.

(1) $h_0$, $h_1$, $h_2$, $c_0$, $w_1$ *and* $d_0$ *are infinite cycles.*
(2) $\alpha$, $\beta$ *and* $w_2$ *may support nonzero* $d_2$*-differentials.*
(3) $e_0$ *survives to* $E_3$.
(4) $g$ *survives (at least) to* $E_5$.
(5) $\gamma$ *survives (at least) to* $E_6$.
(6) $\delta$ *survives (at least) to* $E_4$.

PROOF. This follows by inspection of Figures 3.12 and 3.13. There are no nonzero targets for $d_r$-differentials for $r \geq 2$ on $h_0$, $h_2$, $c_0$, $w_1$ and $d_0$. By $h_0$-linearity and induction $d_r(h_1) = 0$ for each $r \geq 2$, since $h_0 h_1 = 0$ and $h_0^{r+2} \neq 0$ at the $E_r$-term. The target groups for $d_2(e_0)$, $d_r(g)$ for $r \in \{2,3,4\}$, and $d_r(\delta)$ for $r \in \{2,3\}$, are all trivial. Finally, multiplication by $h_0$ acts injectively on the target groups of $d_r(\gamma)$ for $r \in \{2,3,4,5\}$, and $h_0\gamma = 0$, so $d_r(\gamma) = 0$ for these values of $r$. □

Next, we use the Steenrod operations in

$$E_2(tmf) = \mathrm{Ext}_A(H^*(tmf), \mathbb{F}_2) \cong \mathrm{Ext}_{A(2)}(\mathbb{F}_2, \mathbb{F}_2)\,.$$

By Lemma 1.22 applied to $A(2) \subset A$, these are unambiguously defined. The operations

$$Sq^i \colon \mathrm{Ext}^{s,t}_{A(2)}(\mathbb{F}_2, \mathbb{F}_2) \longrightarrow \mathrm{Ext}^{s+i,2t}_{A(2)}(\mathbb{F}_2, \mathbb{F}_2) \tag{5.1}$$

were calculated on the algebra generators of the $E_2$-term in Theorem 1.20, and can be evaluated on the remaining classes by means of the Cartan formula

$$Sq^k(xy) = \sum_{i+j=k} Sq^i(x) Sq^j(y)\,.$$

By its construction as the connective cover of the global sections of a sheaf of $E_\infty$ ring spectra, see Section 0.1, the topological modular forms spectrum is an $E_\infty$ ring spectrum. Hence it is also an $H_\infty$ ring spectrum in the sense of [**45**, §I.3]. This implies a number of relations between the differentials $d_r$ in its Adams spectral sequence and the Steenrod operations $Sq^i$ in its $E_2$-term. These results are due, in increasing generality, to Daniel Kahn [**85**], James Milgram [**122**], Jukka Mäkinen [**109**] and the first author [**37**]. We now recall the first author's theorems from [**45**, §VI.1], translated to the cohomological indexing of the Steenrod squaring operations used in Section 1.3 and equation (5.1). (In [**45**] a different indexing convention was used, under which $Sq^j$ denotes the operation that increases the topological degree $t-s$ by $j$.) The theory will be more fully reviewed in Chapter 11, where we study the Adams spectral sequence for $S$.

DEFINITION 5.3 ([**45**, Def. V.2.15]). For $n \geq 0$, let $v = v(n)$ denote the "vector field number", i.e., the maximal number $v$ such that the attaching map of the $n$-cell in the real projective $n$-space $P^n$ factors up to homotopy as

$$S^{n-1} \xrightarrow{\alpha} P^{n-v} \subset P^{n-1} .$$

Let $a = a(n) \in \pi_{v-1}(S)$ denote the top component

$$S^{n-1} \xrightarrow{\alpha} P^{n-v} \longrightarrow S^{n-v}$$

of a maximal compression. Let $\bar{a} \in E_\infty^{f,f+v-1}(S)$ be the infinite cycle that detects $a$ in the mod 2 Adams spectral sequence for $S$. Here $f$ is the Adams filtration of $a$.

Adams' solution of the vector-field problem for spheres [**5**] leads to the following formulas.

PROPOSITION 5.4 ([**45**, Prop. V.2.16 and V.2.17]). *Let the* 2*-adic valuation of* $n+1$ *be* $4q+r$*, with* $0 \leq r \leq 3$*. Then* $v = v(n) = 8q + 2^r$*.*

*If $n$ is even, then $v = 1$, $a = 2$ and $\bar{a} = h_0$. If $n$ is odd, then $v \geq 2$ and $a$ generates the image of the $J$-homomorphism in $\pi_{v-1}(S)^\wedge_2$. In particular, if $n \equiv 1$* mod 4 *then $v = 2$, $a = \eta$ and $\bar{a} = h_1$. If $n \equiv 3$* mod 8 *then $v = 4$, $a \equiv \nu$* mod $2\nu$ *and $\bar{a} = h_2$.*

DEFINITION 5.5. Let $A \in E_2^{s,t}$, $B_1 \in E_2^{s+r_1,t+r_1-1}$ and $B_2 \in E_2^{s+r_2,t+r_2-1}$ be classes in a spectral sequence with differentials $d_r \colon E_r^{s,t} \to E_r^{s+r,t+r-1}$. The notation

$$d_*(A) = B_1 \dotplus B_2$$

means that $d_r(A) = 0$ for $2 \leq r < \min\{r_1, r_2\}$, while

$$\begin{cases} d_{r_1}(A) = B_1 & \text{if } r_1 < r_2, \\ d_r(A) = B_1 + B_2 & \text{if } r_1 = r = r_2, \\ d_{r_2}(A) = B_2 & \text{if } r_1 > r_2. \end{cases}$$

THEOREM 5.6 ([**45**, Thm. VI.1.1 and VI.1.2]). *Let $E_r(Y)$ be the mod 2 Adams spectral sequence for an $H_\infty$ ring spectrum $Y$, and let $x \in E_2^{s,t}(Y)$ be an element that survives to the $E_r$-term, where $r \geq 2$. Let $0 \leq i \leq s$, and let $v = v(t-i)$, $a = a(t-i)$ and $\bar{a}$ be as just defined. Then*

$$d_*(Sq^i(x)) = Sq^{i+r-1}(d_r(x)) \dotplus \begin{cases} 0 & \textit{if } v > s-i+1, \\ \bar{a}\, x\, d_r(x) & \textit{if } v = s-i+1, \\ \bar{a}\, Sq^{i+v}(x) & \textit{if } v \leq \min\{s-i, 10\}. \end{cases}$$

REMARK 5.7. If $r_1 < r_2$ and $B_1 = 0$, then $B_1 \dotplus B_2$ denotes the zero element in filtration $s+r_1$. In this case the theorem does not give information about $d_r(Sq^i(x))$ for $r > r_1$. Similar remarks apply if $r_1 > r_2$ and $B_2 = 0$. However, in the (first) case $v > s-i+1$ of the theorem the summand $B_2 = 0$ should be interpreted as lying in arbitrarily high Adams filtration $s + r_2$, so that

$$d_{2r-1}(Sq^i(x)) = Sq^{i+r-1}(d_r(x)) .$$

PROPOSITION 5.8.

(1) $d_2(\alpha) = h_2 w_1$ *and* $d_2(\beta) = h_0 d_0$.
(2) $d_3(\alpha^2) = h_1 d_0 w_1$.
(3) $d_3(\beta^2) = h_1 g w_1$.

(4) $d_3(w_2^2) = Sq^9(d_2(w_2))$.

PROOF. We apply Theorem 5.6 for classes $x \in E_2^{s,t} = E_2^{s,t}(tmf)$ with $r = 2$.

(1) For $x = c_0 \in E_2^{3,11}$ and $i = 1$ we get $v = 1$, $s - i + 1 = 3$ and

$$d_*(Sq^1(c_0)) = Sq^2(d_2(c_0)) \dotplus h_0 Sq^2(c_0) = h_0 Sq^2(c_0)\,,$$

so that $d_2(h_2\beta) = h_0^2 e_0$ by Proposition 1.21. Here $d_2(h_2\beta) = h_2 \cdot d_2(\beta)$, $h_0^2 e_0 = h_2 \cdot h_0 d_0$ and $h_2$-multiplication acts injectively on the group $E_2^{5,19}$ containing $d_2(\beta)$, so $d_2(\beta) = h_0 d_0$. By $h_0$- and $h_2$-linearity $h_2 \cdot d_2(\alpha) = h_0 \cdot d_2(\beta) = h_0^2 d_0 = h_2 \cdot h_2 w_1$. Multiplication by $h_2$ acts injectively on the group $E_2^{5,16}$ containing $d_2(\alpha)$, so $d_2(\alpha) = h_2 w_1$.

(2) For $x = \alpha \in E_2^{3,15}$ and $i = 3$ we get $v = 1$, $s - i + 1 = 1$ and

$$d_*(Sq^3(\alpha)) = Sq^4(d_2(\alpha)) \dotplus h_0 \alpha d_2(\alpha) = Sq^4(d_2(\alpha)) + h_0 \alpha d_2(\alpha)\,.$$

Hence $d_3(\alpha^2) = Sq^4(h_2 w_1) + h_0 \alpha h_2 w_1 = 0 + h_0 h_2 \alpha w_1 = h_1 d_0 w_1$, by Theorem 1.20 and case (1).

(3) For $x = \beta \in E_2^{3,18}$ and $i = 3$ we get $v = 9$, $s - i + 1 = 1$ and

$$d_*(Sq^3(\beta)) = Sq^4(d_2(\beta)) \dotplus 0 = Sq^4(d_2(\beta))\,.$$

Hence $d_3(\beta^2) = Sq^4(h_0 d_0) = h_1 g w_1$, by Theorem 1.20 and case (1).

(4) For $x = w_2 \in E_2^{8,56}$ and $i = 8$ we get $v = 1$, $s - i + 1 = 1$ and

$$d_*(Sq^8(w_2)) = Sq^9(d_2(w_2)) \dotplus h_0 w_2 d_2(w_2) = Sq^9(d_2(w_2)) + h_0 w_2 d_2(w_2)\,.$$

Hence $d_3(w_2^2) = Sq^9(d_2(w_2)) + h_0 w_2 d_2(w_2)$. Here $d_2(w_2) \in E_2^{10,57} = \mathbb{F}_2\{\alpha\beta g\}$, and $h_0 \cdot \alpha\beta g = 0$, so $d_3(w_2^2) = Sq^9(d_2(w_2))$. □

REMARK 5.9. Once we show that $d_2(w_2) = \alpha\beta g$, we can deduce that $d_3(w_2^2) = Sq^9(\alpha\beta g) = \gamma\beta^2 g^2 = \beta g^4$, using Theorem 1.20 and Table 3.5. In order to show that $d_2(w_2)$ is nonzero, we first use naturality with respect to $\iota\colon S \to tmf$ to determine the differentials $d_3(e_0)$ and $d_4(e_0 g)$, and then make use of the relation $\gamma^2 = \beta^2 g + h_1^2 w_2$ in $\mathrm{Ext}_{A(2)}(\mathbb{F}_2, \mathbb{F}_2)$.

THEOREM 5.10. $d_3(e_0) = c_0 w_1$.

PROOF. This is a consequence of the first hidden $\eta$-multiplication in the Adams spectral sequence for $S$, from $h_0^3 h_4$ detecting $\rho \in \pi_{15}(S)$ to $Pc_0$ detecting $\eta\rho \in \pi_{16}(S)$. See Figure 1.9 and case (16) of Theorem 11.61.

Let $\kappa \in \pi_{14}(S)$ be detected by $d_0$, so that $\eta\kappa$ is detected by $h_1 d_0$. The differentials $d_2(\beta) = h_0 d_0$ and $d_2(h_0\beta) = h_0^2 d_0 = h_2^2 w_1$ in $E_2(tmf)$ show that $\pi_{15}(tmf) = \mathbb{Z}/2\{\iota(\eta\kappa)\}$. See Figure 1.19. Hence $\iota(\rho)$ is 0 or $\iota(\eta\kappa)$. In either case $\iota(\eta\rho)$ is 0, since $\eta^2\kappa = 0$ in $\pi_{16}(S)$. Again, see case (16) of Theorem 11.61.

By Proposition 1.14, $\iota(Pc_0) = c_0 w_1$. It follows that $c_0 w_1$ is an infinite cycle that detects zero in $\pi_{16}(tmf)$, i.e., it is a $d_r$-boundary for some $r \geq 2$. Here $d_2(h_0 e_0) = h_0 d_2(e_0) = 0$ in $E_2(tmf)$, so the only remaining possibility is $d_3(e_0) = c_0 w_1$. □

COROLLARY 5.11. *$\alpha\beta$ survives (at least) to $E_8$.*

PROOF. See Figure 3.13. By the Leibniz rule, $d_2(\alpha\beta) = h_2 w_1 \cdot \beta + \alpha \cdot h_0 d_0 = 0$. To see that $d_3(\alpha\beta) = 0$, note that in its bidegree the $E_2$-term is $\mathbb{F}_2\{h_1 e_0 w_1\}$. By the theorem above, $d_3(h_1 e_0 w_1) = h_1 c_0 w_1^2 \neq 0$, since $h_1 c_0 w_1^2$ cannot be a $d_2$-boundary. Hence $d_3 \circ d_3 = 0$ implies $d_3(\alpha\beta) \neq h_1 e_0 w_1$. The differentials $d_r(\alpha\beta)$ for $4 \leq r \leq 7$ land in trivial groups, hence are zero. □

THEOREM 5.12. $d_4(e_0 g) = gw_1^2$.

PROOF. This is a consequence of the second hidden $\eta$-multiplication in the Adams spectral sequence for $S$, from $h_1 g$ detecting $\eta\bar{\kappa}$ to $Pd_0$ detecting $\eta^2\bar{\kappa}$. See Figure 1.9 and case (22) of Theorem 11.61.

Let $\bar{\kappa} \in \pi_{20}(S)$ be detected by $g$. Then $\kappa \cdot \eta^2\bar{\kappa} \in \pi_{36}(S)$ is detected by $d_0 \cdot Pd_0$ in $E_2(S)$, which must be a boundary because $\eta^2\kappa = 0$. Likewise, the image $\iota(d_0 \cdot Pd_0) = d_0 \cdot d_0 w_1 = gw_1^2$ in $E_2(tmf)$ must be a boundary, and in this spectral sequence the only possible source of such a differential is $e_0 g$, with $d_4(e_0 g) = gw_1^2$. See Figure 3.14. □

COROLLARY 5.13. *$d_4(d_0 e_0) = d_0 w_1^2$ and $d_4(\beta^2 g) = \alpha^2 e_0 w_1$ are nonzero.*

PROOF. See Figures 3.13 and 3.14. We deduce $d_4(d_0 e_0) = d_0 w_1^2$ by $w_1$- and $d_0$-linearity from $d_4(e_0 g) = gw_1^2$ and the relation $d_0^2 = gw_1$. First, $d_4(e_0 g w_1) = gw_1^3$ remains nonzero at $E_4$ because it cannot be a $d_2$- or $d_3$-boundary. Hence $d_0 \cdot d_4(d_0 e_0) = d_4(e_0 g w_1)$ is nonzero, which implies that $d_4(d_0 e_0)$ is nonzero. The only possible nonzero value is $d_0 w_1^2$.

Similarly, $d_4(\beta^2 g) = \alpha^2 e_0 w_1$ follows from $d_4(d_0 e_0) = d_0 w_1^2$ by $w_1$- and $\alpha\beta$-linearity at $E_4$ and the relations $\beta d_0 = \alpha e_0$ and $\alpha d_0 e_0 = \beta g w_1$. Here $d_4(\alpha\beta \cdot d_0 e_0) = \alpha\beta \cdot d_0 w_1^2 = \alpha^2 e_0 w_1^2$ in bidegree $(t-s, s) = (57, 18)$ remains nonzero at $E_4$ because there is no source for a $d_2$- or $d_3$-differential that could hit it. Hence $d_4(\beta^2 g) \cdot w_1 = d_4(\alpha\beta \cdot d_0 e_0)$ is nonzero, which implies that $d_4(\beta^2 g)$ is nonzero. The only possible value is $\alpha^2 e_0 w_1$. □

PROPOSITION 5.14. *$d_2(w_2) = \alpha\beta g$, $d_3(h_1 w_2) = g^2 w_1$ and $d_4(h_1^2 w_2) = \alpha^2 e_0 w_1$ are nonzero.*

PROOF. We use the relation $\gamma^2 = \beta^2 g + h_1^2 w_2$ in bidegree $(t-s, s) = (50, 10)$, see Figure 3.14. From Lemma 5.2 and Corollary 5.13 we deduce that $d_4(\gamma^2)$ is zero and $d_4(h_1^2 w_2) = \alpha^2 e_0 w_1$ is nonzero.

If $d_2(w_2)$ were zero, then $d_3(w_2) = 0$ and $d_4(w_2) = 0$ because these lie in trivial groups, so $d_4(h_1^2 w_2) = h_1^2 \cdot d_4(w_2)$ would be zero. This contradiction show that $d_2(w_2)$ is nonzero, and $\alpha\beta g$ is the only possible value.

It follows that $d_2(h_1 w_2) = h_1 \cdot \alpha\beta g = 0$. If $d_3(h_1 w_2)$ were zero, then $d_4(h_1 w_2)$ is defined and lies in bidegree $(t-s, s) = (48, 13)$. Multiplication by $h_1$ acts trivially on this bidegree, already at $E_2$, so $d_4(h_1^2 w_2) = h_1 \cdot d_4(h_1 w_2) = 0$. This is again a contradiction, so $d_3(h_1 w_2)$ is nonzero. Since $h_0 \cdot h_1 w_2 = 0$ we must have $h_0 \cdot d_3(h_1 w_2) = 0$ at $E_3$, and $g^2 w_1$ is therefore the only possible value. Alternatives involving $h_0^4 w_2$ are excluded because $d_2(\alpha e_0 g)$ must be $h_0$-torsion, hence is zero, so that $h_0^5 w_2$ remains nonzero at $E_3$. □

THEOREM 5.15. *The $d_2$-differential in $E_2(tmf)$ is $R_1$-linear. Table* 5.1 *gives its values on a list of $R_1$-module generators.*

PROOF. Lemma 5.2, Proposition 5.8 and Proposition 5.14 give the values of $d_2$ on the algebra generators of $E_2(tmf)$. In particular, $g$, $w_1$ and $w_2^2$ are $d_2$-cycles, which gives $R_1$-linearity. The $d_2$-differentials on the $R_0$-module generators $x$ of $E_2(tmf)$ can then be calculated with the Leibniz rule, using the relations in Table 3.5 to express them in normal form:

- $d_2(h_0\beta) = h_0 \cdot h_0 d_0 = h_2^2 w_1$
- $d_2(h_2\beta) = h_2 \cdot h_0 d_0 = h_0^2 e_0$

- $d_2(\alpha d_0) = h_2 w_1 \cdot d_0 = h_0 e_0 w_1$
- $d_2(\alpha\beta) = h_2 w_1 \cdot \beta + \alpha \cdot h_0 d_0 = 0$
- $d_2(\alpha e_0) = h_2 w_1 \cdot e_0 = h_0 g w_1$
- $d_2(\alpha^3) = h_2 w_1 \cdot \alpha^2 = h_0 \alpha\beta w_1 = h_1^2 \gamma w_1$.

The other cases are easier. The $d_2$-differentials on the remaining $R_1$-module generators $xw_2$ are also calculated with the Leibniz rule, in the form

$$d_2(xw_2) = d_2(x)w_2 + xd_2(w) = w_2 \cdot d_2(x) + \alpha\beta g \cdot x\,.$$

The first summand, $w_2 \cdot d_2(x)$, can be written down directly. The second summand, $\alpha\beta g \cdot x$, vanishes when $g \in \operatorname{Ann}(x) \subset R_0$. In the other cases, we calculate as follows:

- $\alpha\beta g \cdot 1 = \alpha\beta g$
- $\alpha\beta g \cdot \alpha = d_0 \gamma g$
- $\alpha\beta g \cdot d_0 = \alpha^2 e_0 g$
- $\alpha\beta g \cdot \beta = e_0 \gamma g$
- $\alpha\beta g \cdot e_0 = \alpha^2 g^2$
- $\alpha\beta g \cdot \alpha^2 = \alpha d_0 \gamma g = d_0 e_0 g^2$
- $\alpha\beta g \cdot \gamma = \alpha g^3$
- $\alpha\beta g \cdot \alpha d_0 = d_0^2 \gamma g = \gamma g^2 w_1$
- $\alpha\beta g \cdot \alpha\beta = \beta d_0 \gamma g = d_0 g^3$
- $\alpha\beta g \cdot \alpha e_0 = \alpha^3 g^2$
- $\alpha\beta g \cdot \beta^2 = e_0 \beta\gamma g = e_0 g^3$
- $\alpha\beta g \cdot d_0 e_0 = \beta^2 g^2 w_1$
- $\alpha\beta g \cdot \alpha^3 = \alpha^2 d_0 \gamma g = \beta^2 \gamma g w_1 = \beta g^3 w_1$
- $\alpha\beta g \cdot d_0 \gamma = \alpha^2 e_0 \gamma g = \alpha e_0^2 g^2 = \alpha d_0 g^3$
- $\alpha\beta g \cdot \alpha^2 e_0 = \alpha^3 \beta e_0 g = \beta\gamma g^2 w_1 = g^4 w_1$
- $\alpha\beta g \cdot e_0 \gamma = \alpha e_0 g^3$. □

REMARK 5.16. To use `ext` to assist in the calculation of the products $\alpha\beta g \cdot x$, use `cocycle tmf 10 18` and `dolifts` to lift the cocycle $10_{18}$ corresponding to $d_2(w_2) = \alpha\beta g$. The nonzero products $\alpha\beta g \cdot x$ can then be read off from the output of `collect`.

## 5.3. The $d_3$-differentials for *tmf*

Given Theorem 5.15, it is elementary to calculate $E_3(tmf)$ as an $R_1$-module. The details are given in Appendix A.1, and the results are recorded in Table 5.2. The $(t-s, s)$-bidegree of each generator $x$ is shown as before. Some generators correspond to a sum $s_g + s_{g'}$ of two `ext`-cocycles, which is indicated by a formal sum $g + g'$ in the $g$-column. For example, $\alpha g = 7_{11} + 7_{12}$. In most cases, $x$ generates a cyclic summand

$$\langle x \rangle \cong \Sigma^{s,t} R_1 / \operatorname{Ann}(x)$$

of $E_3(tmf)$, where $\operatorname{Ann}(x) \subset R_1$ is the annihilator ideal of $x$. The remaining cases are indicated by a dash (–) in the $\operatorname{Ann}(x)$-column, and the non-cyclic summand that contains $x$ is displayed in Table 5.3.

Table 5.2: $R_1$-module generators of $E_3(tmf)$, with $i \geq 0$ in each $h_0$-tower

| $t-s$ | $s$ | $g$ | $x$ | $\mathrm{Ann}(x)$ | $d_3(x)$ | $d_3(xw_2^2)$ |
|---|---|---|---|---|---|---|
| 0 | 0 | 0 | $1$ | $(g^4w_1)$ | 0 | $g^2 \cdot \beta g^2$ |
| 0 | 1 | 0 | $h_0$ | $(g^2, gw_1)$ | 0 | 0 |
| 0 | 2 | 0 | $h_0^2$ | $(g^2, gw_1)$ | 0 | 0 |
| 0 | $3+i$ | 0 | $h_0^{3+i}$ | $(g)$ | 0 | 0 |
| 1 | 1 | 1 | $h_1$ | $(g^2)$ | 0 | 0 |
| 2 | 2 | 1 | $h_1^2$ | $(g)$ | 0 | 0 |
| 3 | 1 | 2 | $h_2$ | $(g, w_1)$ | 0 | 0 |
| 3 | 2 | 2 | $h_0h_2$ | $(g, w_1)$ | 0 | 0 |
| 3 | 3 | 1 | $h_0^2h_2$ | $(g, w_1)$ | 0 | 0 |
| 6 | 2 | 3 | $h_2^2$ | $(g, w_1)$ | 0 | 0 |
| 8 | 3 | 2 | $c_0$ | $(g)$ | 0 | 0 |
| 9 | 4 | 2 | $h_1c_0$ | $(g)$ | 0 | 0 |
| 12 | $6+i$ | 4 | $h_0^{3+i}\alpha$ | $(g)$ | 0 | 0 |
| 14 | 4 | 4 | $d_0$ | $(g^3)$ | 0 | 0 |
| 15 | 5 | 6 | $h_1d_0$ | $(g)$ | 0 | 0 |
| 17 | 4 | 6 | $e_0$ | $(g^3)$ | $w_1 \cdot c_0$ | $w_1 \cdot c_0w_2^2$ |
| 17 | 5 | 7 | $h_0e_0$ | $(g, w_1)$ | 0 | 0 |
| 18 | 5 | 8 | $h_1e_0$ | $(g)$ | $w_1 \cdot h_1c_0$ | $w_1 \cdot h_1c_0w_2^2$ |
| 24 | 6 | 8 | $\alpha^2$ | $(g^2)$ | $w_1 \cdot h_1d_0$ | $w_1 \cdot h_1d_0w_2^2$ |
| 24 | $7+i$ | 7 | $h_0^{1+i}\alpha^2$ | $(g)$ | 0 | 0 |
| 25 | 5 | 11 | $\gamma$ | $-$ | 0 | $g^6 \cdot 1$ |
| 26 | 6 | 9 | $h_1\gamma$ | $(g)$ | 0 | 0 |
| 27 | 6 | 10 | $\alpha\beta$ | $(g)$ | 0 | 0 |
| 27 | 7 | 9 | $h_1^2\gamma$ | $(g, w_1)$ | 0 | 0 |
| 30 | 6 | 11 | $\beta^2$ | $(g^2w_1)$ | $gw_1 \cdot h_1$ | $g^5 \cdot \gamma$ $+ gw_1 \cdot h_1w_2^2$ |
| 31 | 8 | 13 | $d_0e_0$ | $(g^2)$ | 0 | 0 |
| 32 | 7 | 11 | $\delta$ | $(g)$ | 0 | 0 |
| 32 | 7 | $11+12$ | $\alpha g$ | $(g^2)$ | 0 | 0 |
| 32 | 8 | 14 | $h_0\alpha g$ | $(g)$ | 0 | 0 |
| 32 | 9 | 14 | $h_0^2\alpha g$ | $(g)$ | 0 | 0 |

Table 5.2: $R_1$-module generators of $E_3(tmf)$, with $i \geq 0$ in each $h_0$-tower (cont.)

| $t-s$ | $s$ | $g$ | $x$ | $\mathrm{Ann}(x)$ | $d_3(x)$ | $d_3(xw_2^2)$ |
|---|---|---|---|---|---|---|
| 33 | 8 | 15 | $h_1\delta$ | $(g)$ | 0 | 0 |
| 36 | $10+i$ | 14 | $h_0^{1+i}\alpha^3$ | $(g)$ | 0 | 0 |
| 39 | 9 | 18 | $d_0\gamma$ | – | 0 | 0 |
| 41 | 10 | 16 | $\alpha^2 e_0$ | $(g)$ | 0 | 0 |
| 42 | 9 | 19 | $e_0\gamma$ | $(g^3)$ | $w_1 \cdot h_1\delta$ | $w_1 \cdot h_1\delta w_2^2$ |
| 46 | 11 | 18 | $\alpha d_0 g$ | $(g^2)$ | 0 | 0 |
| 48 | 9 | 21 | $h_0 w_2$ | $(g^2)$ | 0 | 0 |
| 48 | 10 | 19 | $h_0^2 w_2$ | $(g^2, gw_1)$ | 0 | 0 |
| 48 | $11+i$ | 19 | $h_0^{3+i} w_2$ | $(g)$ | 0 | 0 |
| 49 | 9 | 22 | $h_1 w_2$ | $(g^2)$ | $g^2 w_1 \cdot 1$ | $g^2 w_1 \cdot w_2^2$ |
| 49 | 11 | 20 | $\alpha e_0 g$ | $(g^2)$ | 0 | 0 |
| 50 | 10 | 21 | $h_1^2 w_2$ | $(g)$ | 0 | 0 |
| 51 | 9 | 23 | $h_2 w_2$ | – | 0 | 0 |
| 51 | 10 | 22 | $h_0 h_2 w_2$ | $(g, w_1)$ | 0 | 0 |
| 51 | 11 | 21 | $h_0^2 h_2 w_2$ | $(g, w_1)$ | 0 | 0 |
| 54 | 10 | 23 | $h_2^2 w_2$ | $(g, w_1)$ | 0 | 0 |
| 55 | 11 | 23 | $\beta g^2$ | – | 0 | $g^6 \cdot \beta^2$ |
| 56 | 11 | 24 | $c_0 w_2$ | $(g)$ | 0 | 0 |
| 56 | 13 | $26+27$ | $\alpha^3 g$ $+ h_0 w_1 w_2$ | $(g)$ | 0 | 0 |
| 57 | 12 | 28 | $h_1 c_0 w_2$ | $(g)$ | 0 | 0 |
| 60 | $14+i$ | 28 | $h_0^{3+i}\alpha w_2$ | $(g)$ | 0 | 0 |
| 63 | 13 | 34 | $h_1 d_0 w_2$ | $(g)$ | $g^2 w_1 \cdot d_0$ | $g^2 w_1 \cdot d_0 w_2^2$ |
| 65 | 13 | 36 | $h_0 e_0 w_2$ | – | 0 | 0 |
| 66 | 13 | 37 | $h_1 e_0 w_2$ | $(g)$ | $g^2 w_1 \cdot e_0$ $+ w_1 \cdot h_1 c_0 w_2$ | $g^2 w_1 \cdot e_0 w_2^2$ $+ w_1 \cdot h_1 c_0 w_2^3$ |
| 72 | $15+i$ | 36 | $h_0^{1+i}\alpha^2 w_2$ | $(g)$ | 0 | 0 |
| 74 | 14 | 37 | $h_1\gamma w_2$ | $(g)$ | $g^2 w_1 \cdot \gamma$ | $g^2 w_1 \cdot \gamma w_2^2$ |
| 75 | 15 | 39 | $h_1^2\gamma w_2$ | – | 0 | 0 |
| 80 | 15 | 41 | $\delta w_2$ | $(g)$ | 0 | 0 |
| 80 | 16 | 49 | $h_0\alpha g w_2$ | $(g)$ | 0 | 0 |

Table 5.2: $R_1$-module generators of $E_3(tmf)$, with $i \geq 0$ in each $h_0$-tower (cont.)

| $t-s$ | $s$ | $g$ | $x$ | $\mathrm{Ann}(x)$ | $d_3(x)$ | $d_3(xw_2^2)$ |
|---|---|---|---|---|---|---|
| 80 | 17 | 49 | $h_0^2\alpha g w_2$ | $(g)$ | 0 | 0 |
| 81 | 16 | 50 | $h_1\delta w_2$ | $(g)$ | 0 | 0 |
| 84 | $18+i$ | 48 | $h_0^{1+i}\alpha^3 w_2$ | $(g)$ | 0 | 0 |

Table 5.3: The non-cyclic $R_1$-module summands in $E_3(tmf)$

| $\langle x_1, x_2\rangle$ |
|---|
| $\langle \gamma, h_0e_0w_2\rangle \cong \dfrac{\Sigma^{5,30}R_1 \oplus \Sigma^{13,78}R_1}{\langle (g^2w_1, w_1), (0,g)\rangle}$ |
| $\langle d_0\gamma, h_2w_2\rangle \cong \dfrac{\Sigma^{9,48}R_1 \oplus \Sigma^{9,60}R_1}{\langle (g, w_1), (0,g)\rangle}$ |
| $\langle \beta g^2, h_1^2\gamma w_2\rangle \cong \dfrac{\Sigma^{11,66}R_1 \oplus \Sigma^{15,90}R_1}{\langle (gw_1, w_1), (0,g)\rangle}$ |

In this section we determine the $d_3$-differentials in $E_3(tmf)$. Since $g$, $w_1$ and $w_2^4$ are $d_3$-cycles, we know that this differential is $R_2$-linear. When $x$ ranges through a set of $R_1$-module generators for the $E_3$-term, the classes $x$ and $xw_2^2$ will range through a set of $R_2$-module generators for the same $E_3$-term, so it will suffice to determine $d_3(x)$ and $d_3(xw_2^2)$ for the generators $x$ in Table 5.2. To do this, we first determine $d_3$ on a set of algebra generators for $E_3(tmf)$, and then use the Leibniz rule.

PROPOSITION 5.17. *A set of* 24 *algebra generators for* $E_3(tmf)$ *is listed in Table* 5.4.

PROOF. The remaining $R_1$-module generators in Table 5.2 can be expressed as polynomials in these elements. This is evident from their (Gröbner) normal forms at the $E_2$-term in almost all cases, and follows from the factorizations

$$\beta g^2 = \beta^2 \cdot \gamma$$

$$\alpha^3 g + h_0 w_1 w_2 = \alpha^2 \cdot \alpha g + w_1 \cdot h_0 w_2$$

in the two remaining cases. $\square$

Table 5.4: Algebra generators of $E_3(tmf)$

| $t-s$ | $s$ | $g$ | $x$ | $d_3(x)$ |
|---|---|---|---|---|
| 0 | 1 | 0 | $h_0$ | 0 |
| 1 | 1 | 1 | $h_1$ | 0 |
| 3 | 1 | 2 | $h_2$ | 0 |
| 8 | 3 | 2 | $c_0$ | 0 |
| 8 | 4 | 1 | $w_1$ | 0 |
| 12 | 6 | 4 | $h_0^3\alpha$ | 0 |
| 14 | 4 | 4 | $d_0$ | 0 |
| 17 | 4 | 6 | $e_0$ | $c_0w_1$ |
| 20 | 4 | 8 | $g$ | 0 |
| 24 | 6 | 8 | $\alpha^2$ | $h_1d_0w_1$ |
| 25 | 5 | 11 | $\gamma$ | 0 |
| 27 | 6 | 10 | $\alpha\beta$ | 0 |
| 30 | 6 | 11 | $\beta^2$ | $h_1gw_1$ |
| 32 | 7 | 11 | $\delta$ | 0 |
| 32 | 7 | $11+12$ | $\alpha g$ | 0 |
| 36 | 10 | 14 | $h_0\alpha^3$ | 0 |
| 48 | 9 | 21 | $h_0w_2$ | 0 |
| 49 | 9 | 22 | $h_1w_2$ | $g^2w_1$ |
| 51 | 9 | 23 | $h_2w_2$ | 0 |
| 56 | 11 | 24 | $c_0w_2$ | 0 |
| 60 | 14 | 28 | $h_0^3\alpha w_2$ | 0 |
| 80 | 15 | 41 | $\delta w_2$ | 0 |
| 84 | 18 | 48 | $h_0\alpha^3w_2$ | 0 |
| 96 | 16 | 54 | $w_2^2$ | $\beta g^4$ |

THEOREM 5.18. *The $d_3$-differential in $E_3(tmf)$ is $R_2$-linear. Its values on a set of algebra generators are as listed in Table* 5.4, *and its values on a set of $R_2$-module generators are as listed in Table* 5.2.

PROOF. Lemma 5.2 (on $h_0$, $h_1$, $h_2$, $c_0$, $w_1$, $d_0$, $g$, $\gamma$ and $\delta$), Proposition 5.8 (on $\alpha^2$ and $\beta^2$), Remark 5.9 (on $w_2^2$), Theorem 5.10 (on $e_0$), Corollary 5.11 (on $\alpha\beta$) and Proposition 5.14 (on $h_1w_2$) have already given us the values of $d_3$ on many of the algebra generators of $E_3(tmf)$.

The $d_3$-differentials on $h_0^3\alpha$, $\alpha g$, $h_0\alpha^3$, $h_0w_2$, $h_2w_2$, $h_0^3\alpha w_2$, $\delta w_2$ and $h_0\alpha^3w_2$ all vanish because the target groups are trivial, already at $E_2$, as can be seen from Figures 1.11 to 1.14.

Only $c_0w_2$ remains. In the bidegree $(t-s,s)=(55,14)$ of $d_3(c_0w_2)$ the $E_2$-term is $\mathbb{F}_2\{\alpha\beta gw_1\}$, but $d_2(w_1w_2)=\alpha\beta gw_1$, so the $E_3$-term is trivial in this bidegree. Hence $d_3(c_0w_2)=0$.

This verifies the formulas for $d_3(x)$ with $x$ one of the algebra generators in Table 5.4. We use the Leibniz rule to evaluate $d_3(x)$ for the decomposable $R_1$-module generators $x$ in Table 5.2:

- $d_3(d_0\cdot e_0)=d_0\cdot c_0w_1=0$
- $d_3(d_0\cdot\gamma)=0$
- $d_3(\alpha^2\cdot e_0)=h_1d_0w_1\cdot e_0+\alpha^2\cdot c_0w_1=h_0^2\alpha gw_1+h_0^2\alpha gw_1=0$
- $d_3(e_0\cdot\gamma)=c_0w_1\cdot\gamma=h_1\delta w_1$
- $d_3(d_0\cdot\alpha g)=0$
- $d_3(e_0\cdot\alpha g)=c_0w_1\cdot\alpha g=0$
- $d_3(\beta g^2)=d_3(\beta^2\cdot\gamma)=h_1gw_1\cdot\gamma=0$
- $d_3(\alpha^2\cdot\alpha g+w_1\cdot h_0w_2)=h_1d_0w_1\cdot\alpha g+0=0$
- $d_3(d_0\cdot h_1w_2)=d_0\cdot g^2w_1$
- $d_3(e_0\cdot h_0w_2)=c_0w_1\cdot h_0w_2=0$
- $d_3(e_0\cdot h_1w_2)=c_0w_1\cdot h_1w_2+e_0\cdot g^2w_1=h_1c_0w_1w_2+e_0g^2w_1$
- $d_3(\alpha^2\cdot h_0w_2)=h_1d_0w_1\cdot h_0w_2=0$
- $d_3(\gamma\cdot h_1w_2)=\gamma\cdot g^2w_1$.

The remaining cases follow by $h_0$-, $h_1$- and $h_2$-linearity, keeping in mind that $h_0\delta=h_0\alpha g$. For the $R_2$-module generators of the form $xw_2^2$, we use the Leibniz rule in the form

$$d_3(xw_2^2)=d_3(x)w_2^2+xd_3(w_2^2)=w_2^2\cdot d_3(x)+\beta g^4\cdot x\,.$$

The first summand is easy to write down in terms of our $R_2$-module generators. The second summand vanishes whenever $g^4\in\operatorname{Ann}(x)\subset R_1$. In the four other cases we can calculate $\beta g^4\cdot x$ using the known relations in $E_2(tmf)$ from Table 3.5, as follows:

- $\beta g^4\cdot 1=g^2\cdot\beta g^2$
- $\beta g^4\cdot\gamma=g^6\cdot 1$
- $\beta g^4\cdot\beta^2=g^5\cdot\gamma$
- $\beta g^4\cdot\beta g^2=g^6\cdot\beta^2$. □

REMARK 5.19. We can use `ext` to aid in the calculation of the products $\beta g^4\cdot x$ by using `cocycle tmf 19 56` and `dolifts` to calculate all products with the cocycle $19_{56}$ (which corresponds to $d_3(w_2^2)=\beta g^4$). The nonzero products $\beta g^4\cdot x$ can then be read off from the output of `collect`.

## 5.4. The $d_4$-differentials for *tmf*

Given Theorem 5.18, it is elementary to calculate $E_4(tmf)$ as an $R_2$-module. The details are given in Appendix A.2, and the results are recorded in Table 5.5. The Adams bidegree $(t-s,s)$ and `ext`-index $g$ of each $R_2$-module generator $x$ is shown as before. In most cases, $x$ generates a cyclic summand

$$\langle x\rangle\cong\Sigma^{s,t}R_2/\operatorname{Ann}(x)$$

of $E_4(tmf)$, where now $\operatorname{Ann}(x)\subset R_2$. The remaining cases are indicated by a dash $(-)$ in the $\operatorname{Ann}(x)$-column, and the non-cyclic summand that contains $x$ is displayed in Table 5.6.

Table 5.5: $R_2$-module generators of $E_4(tmf)$, with $i \geq 0$ in each $h_0$-tower

| $t-s$ | $s$ | $g$ | $x$ | $\mathrm{Ann}(x)$ | $d_4(x)$ |
|---|---|---|---|---|---|
| 0 | 0 | 0 | $1$ | $(g^6, g^2w_1)$ | 0 |
| 0 | 1 | 0 | $h_0$ | $(g^2, gw_1)$ | 0 |
| 0 | 2 | 0 | $h_0^2$ | $(g^2, gw_1)$ | 0 |
| 0 | $3+i$ | 0 | $h_0^{3+i}$ | $(g)$ | 0 |
| 1 | 1 | 1 | $h_1$ | $(g^2, gw_1)$ | 0 |
| 2 | 2 | 1 | $h_1^2$ | $(g)$ | 0 |
| 3 | 1 | 2 | $h_2$ | $(g, w_1)$ | 0 |
| 3 | 2 | 2 | $h_0h_2$ | $(g, w_1)$ | 0 |
| 3 | 3 | 1 | $h_0^2h_2$ | $(g, w_1)$ | 0 |
| 6 | 2 | 3 | $h_2^2$ | $(g, w_1)$ | 0 |
| 8 | 3 | 2 | $c_0$ | $(g, w_1)$ | 0 |
| 9 | 4 | 2 | $h_1c_0$ | $(g, w_1)$ | 0 |
| 12 | $6+i$ | 4 | $h_0^{3+i}\alpha$ | $(g)$ | 0 |
| 14 | 4 | 4 | $d_0$ | $(g^3, g^2w_1)$ | 0 |
| 15 | 5 | 6 | $h_1d_0$ | $(g, w_1)$ | 0 |
| 17 | 5 | 7 | $h_0e_0$ | $(g, w_1)$ | 0 |
| 24 | $7+i$ | 7 | $h_0^{1+i}\alpha^2$ | $(g)$ | 0 |
| 25 | 5 | 11 | $\gamma$ | – | 0 |
| 26 | 6 | 9 | $h_1\gamma$ | $(g)$ | 0 |
| 27 | 6 | 10 | $\alpha\beta$ | $(g)$ | 0 |
| 27 | 7 | 9 | $h_1^2\gamma$ | $(g, w_1)$ | 0 |
| 31 | 8 | 13 | $d_0e_0$ | $(g^2)$ | $w_1^2 \cdot d_0$ |
| 32 | 7 | 11 | $\delta$ | $(g)$ | 0 |
| 32 | 7 | $11+12$ | $\alpha g$ | $(g^2)$ | 0 |
| 32 | 8 | 14 | $h_0\alpha g$ | $(g)$ | 0 |
| 32 | 9 | 14 | $h_0^2\alpha g$ | $(g)$ | 0 |
| 33 | 8 | 15 | $h_1\delta$ | $(g, w_1)$ | 0 |
| 36 | $10+i$ | 14 | $h_0^{1+i}\alpha^3$ | $(g)$ | 0 |
| 37 | 8 | 17 | $e_0g$ | $(g^2)$ | $gw_1^2 \cdot 1$ |
| 39 | 9 | 18 | $d_0\gamma$ | – | 0 |
| 41 | 10 | 16 | $\alpha^2e_0$ | $(g)$ | 0 |

Table 5.5: $R_2$-module generators of $E_4(tmf)$, with $i \geq 0$ in each $h_0$-tower (cont.)

| $t-s$ | $s$ | $g$ | $x$ | $\mathrm{Ann}(x)$ | $d_4(x)$ |
|---|---|---|---|---|---|
| 44 | 10 | 17 | $\alpha^2 g$ | $(g)$ | $w_1^2 \cdot \alpha\beta$ |
| 46 | 11 | 18 | $\alpha d_0 g$ | $(g^2)$ | 0 |
| 48 | 9 | 21 | $h_0 w_2$ | $(g^2)$ | $w_1 \cdot d_0\gamma$ |
| 48 | 10 | 19 | $h_0^2 w_2$ | $(g^2, gw_1)$ | 0 |
| 48 | $11+i$ | 19 | $h_0^{3+i} w_2$ | $(g)$ | 0 |
| 49 | 11 | 20 | $\alpha e_0 g$ | $(g^2)$ | $w_1^2 \cdot \delta'$ |
| 50 | 10 | 20 | $\beta^2 g$ | $(g^5, gw_1)$ | $w_1 \cdot \alpha^2 e_0$ |
| 50 | 10 | 21 | $h_1^2 w_2$ | $(g)$ | $w_1 \cdot \alpha^2 e_0$ |
| 51 | 9 | 23 | $h_2 w_2$ | $-$ | 0 |
| 51 | 10 | 22 | $h_0 h_2 w_2$ | $(g, w_1)$ | 0 |
| 51 | 11 | 21 | $h_0^2 h_2 w_2$ | $(g, w_1)$ | 0 |
| 54 | 10 | 23 | $h_2^2 w_2$ | $(g, w_1)$ | 0 |
| 55 | 11 | 23 | $\beta g^2$ | $(g^2)$ | $w_1 \cdot \alpha d_0 g$ |
| 56 | 11 | 24 | $c_0 w_2$ | $(g)$ | 0 |
| 56 | 13 | $26+27$ | $\alpha^3 g + h_0 w_1 w_2$ | $(g)$ | 0 |
| 57 | 12 | $27+28$ | $\gamma\delta'$ | $(g, w_1)$ | 0 |
| 60 | $14+i$ | 28 | $h_0^{3+i} \alpha w_2$ | $(g)$ | 0 |
| 62 | 13 | 32 | $e_0 \gamma g$ | $(g^2)$ | $g w_1^2 \cdot \gamma$ |
| 65 | 13 | 36 | $h_0 e_0 w_2$ | $(g, w_1)$ | 0 |
| 72 | $15+i$ | 36 | $h_0^{1+i} \alpha^2 w_2$ | $(g)$ | 0 |
| 75 | 15 | $38+39$ | $\gamma^3$ | $(g, w_1)$ | 0 |
| 80 | 15 | 41 | $\delta w_2$ | $(g)$ | 0 |
| 80 | 16 | 49 | $h_0 \alpha g w_2$ | $(g)$ | 0 |
| 80 | 17 | 49 | $h_0^2 \alpha g w_2$ | $(g)$ | 0 |
| 81 | 16 | 50 | $h_1 \delta w_2$ | $(g)$ | 0 |
| 84 | $18+i$ | 48 | $h_0^{1+i} \alpha^3 w_2$ | $(g)$ | 0 |
| 96 | 17 | 58 | $h_0 w_2^2$ | $(g^2, gw_1)$ | 0 |
| 96 | 18 | 55 | $h_0^2 w_2^2$ | $(g^2, gw_1)$ | 0 |
| 96 | $19+i$ | 57 | $h_0^{3+i} w_2^2$ | $(g)$ | 0 |
| 97 | 17 | 59 | $h_1 w_2^2$ | $-$ | 0 |
| 98 | 18 | 57 | $h_1^2 w_2^2$ | $(g)$ | 0 |

Table 5.5: $R_2$-module generators of $E_4(tmf)$, with $i \geq 0$ in each $h_0$-tower (cont.)

| $t-s$ | $s$ | $g$ | $x$ | $\mathrm{Ann}(x)$ | $d_4(x)$ |
|---|---|---|---|---|---|
| 99 | 17 | 60 | $h_2 w_2^2$ | $(g, w_1)$ | 0 |
| 99 | 18 | 58 | $h_0 h_2 w_2^2$ | $(g, w_1)$ | 0 |
| 99 | 19 | 59 | $h_0^2 h_2 w_2^2$ | $(g, w_1)$ | 0 |
| 102 | 18 | 59 | $h_2^2 w_2^2$ | $(g, w_1)$ | 0 |
| 104 | 19 | 62 | $c_0 w_2^2$ | $(g, w_1)$ | 0 |
| 104 | 20 | 69 | $w_1 w_2^2$ | $(g^2)$ | 0 |
| 105 | 20 | 71 | $h_1 c_0 w_2^2$ | $(g, w_1)$ | 0 |
| 108 | $22+i$ | 71 | $h_0^{3+i} \alpha w_2^2$ | $(g)$ | 0 |
| 110 | 20 | 74 | $d_0 w_2^2$ | $(g^3, g^2 w_1)$ | 0 |
| 111 | 21 | 79 | $h_1 d_0 w_2^2$ | $(g, w_1)$ | 0 |
| 113 | 21 | 81 | $h_0 e_0 w_2^2$ | $(g, w_1)$ | 0 |
| 120 | $23+i$ | 82 | $h_0^{1+i} \alpha^2 w_2^2$ | $(g)$ | 0 |
| 122 | 22 | 81 | $h_1 \gamma w_2^2$ | $(g)$ | 0 |
| 123 | 22 | 82 | $\alpha \beta w_2^2$ | $(g)$ | 0 |
| 123 | 23 | 85 | $h_1^2 \gamma w_2^2$ | $(g, w_1)$ | 0 |
| 127 | 24 | 98 | $d_0 e_0 w_2^2$ | $(g^2)$ | $w_1^2 \cdot d_0 w_2^2$ |
| 128 | 23 | 87 | $\delta w_2^2$ | $(g)$ | 0 |
| 128 | 23 | $87+88$ | $\alpha g w_2^2$ | $(g^2)$ | 0 |
| 128 | 24 | 100 | $h_0 \alpha g w_2^2$ | $(g)$ | 0 |
| 128 | 25 | 102 | $h_0^2 \alpha g w_2^2$ | $(g)$ | 0 |
| 129 | 24 | 101 | $h_1 \delta w_2^2$ | $(g, w_1)$ | 0 |
| 129 | 25 | 103 | $\gamma w_1 w_2^2$ | $(g^2)$ | 0 |
| 132 | $26+i$ | 100 | $h_0^{1+i} \alpha^3 w_2^2$ | $(g)$ | 0 |
| 133 | 24 | 103 | $e_0 g w_2^2$ | $(g^2)$ | $g w_1 \cdot w_1 w_2^2$ |
| 135 | 25 | 108 | $d_0 \gamma w_2^2$ | – | 0 |
| 137 | 26 | 103 | $\alpha^2 e_0 w_2^2$ | $(g)$ | 0 |
| 140 | 26 | 105 | $\alpha^2 g w_2^2$ | $(g)$ | $w_1^2 \cdot \alpha \beta w_2^2$ |
| 142 | 27 | 109 | $\alpha d_0 g w_2^2$ | $(g^2)$ | 0 |
| 144 | 25 | 111 | $h_0 w_2^3$ | $(g^2)$ | $w_1 \cdot d_0 \gamma w_2^2$ |
| 144 | 26 | 107 | $h_0^2 w_2^3$ | $(g^2, g w_1)$ | 0 |
| 144 | $27+i$ | 111 | $h_0^{3+i} w_2^3$ | $(g)$ | 0 |

Table 5.5: $R_2$-module generators of $E_4(tmf)$, with $i \geq 0$ in each $h_0$-tower (cont.)

| $t-s$ | $s$ | $g$ | $x$ | $\mathrm{Ann}(x)$ | $d_4(x)$ |
|---|---|---|---|---|---|
| 145 | 27 | 112 | $\alpha e_0 g w_2^2$ | $(g^2)$ | $w_1^2 \cdot \delta' w_2^2$ |
| 146 | 26 | 109 | $h_1^2 w_2^3$ | $(g)$ | $w_1 \cdot \alpha^2 e_0 w_2^2$ |
| 147 | 25 | 113 | $h_2 w_2^3$ | $-$ | 0 |
| 147 | 26 | 110 | $h_0 h_2 w_2^3$ | $(g, w_1)$ | 0 |
| 147 | 27 | 113 | $h_0^2 h_2 w_2^3$ | $(g, w_1)$ | 0 |
| 150 | 26 | 111 | $h_2^2 w_2^3$ | $(g, w_1)$ | 0 |
| 152 | 27 | 116 | $c_0 w_2^3$ | $(g)$ | 0 |
| 152 | 29 | $131 + 132$ | $\alpha^3 g w_2^2 + h_0 w_1 w_2^3$ | $(g)$ | 0 |
| 153 | 28 | $129 + 130$ | $\gamma \delta' w_2^2$ | $(g, w_1)$ | 0 |
| 154 | 30 | 127 | $\beta^2 g w_1 w_2^2$ | $(g)$ | $w_1^2 \cdot \alpha^2 e_0 w_2^2$ |
| 156 | $30 + i$ | 131 | $h_0^{3+i} \alpha w_2^3$ | $(g)$ | 0 |
| 158 | 29 | 138 | $e_0 \gamma g w_2^2$ | $(g^2)$ | $g w_1 \cdot \gamma w_1 w_2^2$ |
| 159 | 31 | 135 | $\beta g^2 w_1 w_2^2$ | $-$ | $w_1^2 \cdot \alpha d_0 g w_2^2$ |
| 161 | 29 | 142 | $h_0 e_0 w_2^3$ | $(g, w_1)$ | 0 |
| 168 | $31 + i$ | 144 | $h_0^{1+i} \alpha^2 w_2^3$ | $(g)$ | 0 |
| 171 | 31 | 147 | $h_1^2 \gamma w_2^3$ | $-$ | $g w_1 \cdot \alpha d_0 g w_2^2$ |
| 176 | 31 | 149 | $\delta w_2^3$ | $(g)$ | 0 |
| 176 | 32 | 167 | $h_0 \alpha g w_2^3$ | $(g)$ | 0 |
| 176 | 33 | 171 | $h_0^2 \alpha g w_2^3$ | $(g)$ | 0 |
| 177 | 32 | 168 | $h_1 \delta w_2^3$ | $(g)$ | 0 |
| 180 | $34 + i$ | 168 | $h_0^{1+i} \alpha^3 w_2^3$ | $(g)$ | 0 |

Table 5.6: The non-cyclic $R_2$-module summands in $E_4(tmf)$

| $\langle x_1, x_2\rangle$ |
|---|
| $\langle\gamma, h_1w_2^2\rangle \cong \dfrac{\Sigma^{5,30}R_2 \oplus \Sigma^{17,114}R_2}{\langle(g^2w_1,0),(g^5,gw_1),(0,g^2)\rangle}$ |
| $\langle d_0\gamma, h_2w_2\rangle \cong \dfrac{\Sigma^{9,48}R_2 \oplus \Sigma^{9,60}R_2}{\langle(g,w_1),(0,g)\rangle}$ |
| $\langle d_0\gamma w_2^2, h_2w_2^3\rangle \cong \dfrac{\Sigma^{25,160}R_2 \oplus \Sigma^{25,172}R_2}{\langle(g,w_1),(0,g)\rangle}$ |
| $\langle\beta g^2w_1w_2^2, h_1^2\gamma w_2^3\rangle \cong \dfrac{\Sigma^{31,190}R_2 \oplus \Sigma^{31,202}R_2}{\langle(g,w_1),(0,g)\rangle}$ |

PROPOSITION 5.20. *A set of* 52 *algebra generators for* $E_4(tmf)$ *is listed in Table* 5.7.

PROOF. The remaining $R_2$-module generators in Table 5.5 can be expressed as polynomials in these elements. This is evident from their normal forms at $E_2$ in the great majority of cases. The factorizations

$$\begin{aligned} h_0e_0 &= h_2 \cdot d_0 \\ \alpha^2e_0 &= d_0 \cdot \alpha\beta \\ \beta^2g &= \gamma\cdot\gamma + 1\cdot h_1^2w_2 \\ \alpha^3g + h_0w_1w_2 &= \gamma\cdot d_0e_0 + w_1\cdot h_0w_2 \\ h_0e_0w_2 &= d_0\cdot h_2w_2 \\ h_0e_0w_2^2 &= h_2\cdot d_0w_2^2 \\ \alpha^2e_0w_2^2 &= d_0\cdot\alpha\beta w_2^2 \\ \alpha^3gw_2^2 + h_0w_1w_2^3 &= \gamma\cdot d_0e_0w_2^2 + w_1\cdot h_0w_2^3 \\ h_0e_0w_2^3 &= d_0\cdot h_2w_2^3 \end{aligned}$$

(all valid at the $E_2$-term), account for the remaining module generators. □

Table 5.7: Algebra generators of $E_4(tmf)$

| $t-s$ | $s$ | $g$ | $x$ | $d_4(x)$ |
|---|---|---|---|---|
| 0 | 1 | 0 | $h_0$ | 0 |
| 1 | 1 | 1 | $h_1$ | 0 |
| 3 | 1 | 2 | $h_2$ | 0 |

Table 5.7: Algebra generators of $E_4(tmf)$ (cont.)

| $t-s$ | $s$ | $g$ | $x$ | $d_4(x)$ |
|---|---|---|---|---|
| 8 | 3 | 2 | $c_0$ | 0 |
| 8 | 4 | 1 | $w_1$ | 0 |
| 12 | 6 | 4 | $h_0^3\alpha$ | 0 |
| 14 | 4 | 4 | $d_0$ | 0 |
| 20 | 4 | 8 | $g$ | 0 |
| 24 | 7 | 7 | $h_0\alpha^2$ | 0 |
| 25 | 5 | 11 | $\gamma$ | 0 |
| 27 | 6 | 10 | $\alpha\beta$ | 0 |
| 31 | 8 | 13 | $d_0e_0$ | $d_0w_1^2$ |
| 32 | 7 | 11 | $\delta$ | 0 |
| 32 | 7 | $11+12$ | $\alpha g$ | 0 |
| 36 | 10 | 14 | $h_0\alpha^3$ | 0 |
| 37 | 8 | 17 | $e_0g$ | $gw_1^2$ |
| 44 | 10 | 17 | $\alpha^2 g$ | $\alpha\beta w_1^2$ |
| 48 | 9 | 21 | $h_0w_2$ | $d_0\gamma w_1$ |
| 49 | 11 | 20 | $\alpha e_0 g$ | $\delta' w_1^2$ |
| 50 | 10 | 21 | $h_1^2w_2$ | $\alpha^2 e_0 w_1$ |
| 51 | 9 | 23 | $h_2w_2$ | 0 |
| 55 | 11 | 23 | $\beta g^2$ | $\alpha d_0 g w_1$ |
| 56 | 11 | 24 | $c_0w_2$ | 0 |
| 60 | 14 | 28 | $h_0^3\alpha w_2$ | 0 |
| 72 | 15 | 36 | $h_0\alpha^2 w_2$ | 0 |
| 80 | 15 | 41 | $\delta w_2$ | 0 |
| 84 | 18 | 48 | $h_0\alpha^3 w_2$ | 0 |
| 96 | 17 | 58 | $h_0w_2^2$ | 0 |
| 97 | 17 | 59 | $h_1w_2^2$ | 0 |
| 99 | 17 | 60 | $h_2w_2^2$ | 0 |
| 104 | 19 | 62 | $c_0w_2^2$ | 0 |
| 104 | 20 | 69 | $w_1w_2^2$ | 0 |
| 108 | 22 | 71 | $h_0^3\alpha w_2^2$ | 0 |
| 110 | 20 | 74 | $d_0w_2^2$ | 0 |
| 120 | 23 | 82 | $h_0\alpha^2 w_2^2$ | 0 |

Table 5.7: Algebra generators of $E_4(tmf)$ (cont.)

| $t-s$ | $s$ | $g$ | $x$ | $d_4(x)$ |
|---|---|---|---|---|
| 123 | 22 | 82 | $\alpha\beta w_2^2$ | 0 |
| 127 | 24 | 98 | $d_0e_0w_2^2$ | $d_0w_1^2w_2^2$ |
| 128 | 23 | 87 | $\delta w_2^2$ | 0 |
| 128 | 23 | $87+88$ | $\alpha g w_2^2$ | 0 |
| 132 | 26 | 100 | $h_0\alpha^3w_2^2$ | 0 |
| 133 | 24 | 103 | $e_0gw_2^2$ | $gw_1^2w_2^2$ |
| 140 | 26 | 105 | $\alpha^2gw_2^2$ | $\alpha\beta w_1^2w_2^2$ |
| 144 | 25 | 111 | $h_0w_2^3$ | $d_0\gamma w_1w_2^2$ |
| 145 | 27 | 112 | $\alpha e_0gw_2^2$ | $\delta' w_1^2w_2^2$ |
| 146 | 26 | 109 | $h_1^2w_2^3$ | $\alpha^2e_0w_1w_2^2$ |
| 147 | 25 | 113 | $h_2w_2^3$ | 0 |
| 152 | 27 | 116 | $c_0w_2^3$ | 0 |
| 156 | 30 | 131 | $h_0^3\alpha w_2^3$ | 0 |
| 168 | 31 | 144 | $h_0\alpha^2w_2^3$ | 0 |
| 176 | 31 | 149 | $\delta w_2^3$ | 0 |
| 180 | 34 | 168 | $h_0\alpha^3w_2^3$ | 0 |
| 192 | 32 | 172 | $w_2^4$ | 0 |

PROPOSITION 5.21. *The following classes are $d_4$-cycles:*

(1) $h_0$, $h_1$, $h_2$, $c_0$, $w_1$, $d_0$, $g$ *and* $\gamma$.
(2) $h_0^3\alpha$, $h_0\alpha^2$, $\alpha\beta$, $h_0\alpha^3$, $c_0w_2$, $h_0^3\alpha w_2$, $w_1w_2^2$, $h_0^3\alpha w_2^2$, $d_0w_2^2$, $\alpha\beta w_2^2$ *and* $w_2^4$.
(3) $\delta$, $\alpha g$, $\delta w_2$, $h_0\alpha^3w_2$, $h_0w_2^2$, $h_2w_2^2$, $c_0w_2^2$, $h_0\alpha^2w_2^2$, $\delta w_2^2$, $\alpha g w_2^2$, $h_0\alpha^3w_2^2$, $c_0w_2^3$, $h_0^3\alpha w_2^3$ *and* $h_0\alpha^3w_2^3$.
(4) $h_2w_2$, $h_2w_2^3$ *and* $\delta w_2^3$.

PROOF. (1) We proved that $d_4(x)=0$ for $x=h_0$, $h_1$, $h_2$, $c_0$, $w_1$, $d_0$, $g$ and $\gamma$ in Lemma 5.2.

(2) By inspection of Figures 1.11 to 1.18 we see that $d_4(x)=0$ for $x=h_0^3\alpha$, $h_0\alpha^2$, $\alpha\beta$, $h_0\alpha^3$, $c_0w_2$, $h_0^3\alpha w_2$, $w_1w_2^2$, $h_0^3\alpha w_2^2$, $d_0w_2^2$, $\alpha\beta w_2^2$ and $w_2^4$, because the target groups are trivial at the $E_2$-term.

(3) We can read off from Table 5.1 that $d_4(x)=0$ for $x=\delta$, $\alpha g$, $\delta w_2$, $h_0\alpha^3w_2$, $h_0w_2^2$, $h_2w_2^2$, $c_0w_2^2$, $h_0\alpha^2w_2^2$, $\delta w_2^2$, $\alpha g w_2^2$, $h_0\alpha^3w_2^2$, $c_0w_2^3$, $h_0^3\alpha w_2^3$ and $h_0\alpha^3w_2^3$, because the target groups become trivial at the $E_3$-term:

- For $x=\delta$, and for $x=\alpha g$, the $E_2$-term in the bidegree of $d_4(x)$ is $\mathbb{F}_2\{w_1^2\cdot\beta\}$, and $d_2(w_1^2\cdot\beta)=w_1^2\cdot h_0d_0\neq 0$.
- For $x=\delta w_2$ the target is $\mathbb{F}_2\{w_1^2\cdot\beta w_2\}$ at $E_2$, and $d_2(w_1^2\cdot\beta w_2)=gw_1^2\cdot e_0\gamma+w_1^2w_2\cdot h_0d_0\neq 0$.

- For $x = h_0\alpha^3 w_2$ the target is $\mathbb{F}_2\{g^2w_1^2 \cdot \alpha\beta\}$ at $E_2$, and $d_2(gw_1^2 \cdot w_2) = g^2w_1^2 \cdot \alpha\beta$.
- For $x = h_0 w_2^2$ the target is $\mathbb{F}_2\{w_1 \cdot d_0\gamma w_2\}$ at $E_2$, and $d_2(w_1 \cdot d_0\gamma w_2) = g^3 w_1 \cdot \alpha d_0 \neq 0$.
- For $x = h_2 w_2^2$ the target is $\mathbb{F}_2\{w_1 \cdot e_0\gamma w_2\}$ at $E_2$, and $d_2(w_1 \cdot e_0\gamma w_2) = g^3 w_1 \cdot \alpha e_0 \neq 0$.
- For $x = c_0 w_2^2$ the target is $\mathbb{F}_2\{g^4 w_1 \cdot \beta\}$ at $E_2$, and $d_2(g \cdot \alpha^3 w_2) = g^4 w_1 \cdot \beta + g w_1 w_2 \cdot h_1^2\gamma = g^4 w_1 \cdot \beta$.
- For $x = h_0\alpha^2 w_2^2$ the target is $\mathbb{F}_2\{g^2w_1^2 \cdot \beta w_2\}$ at $E_2$, and $d_2(g^2w_1^2 \cdot \beta w_2) = g^3w_1^2 \cdot e_0\gamma + g^2w_1^2w_2 \cdot h_0 d_0 = g^3 w_1^2 \cdot e_0\gamma \neq 0$.
- For $x = \delta w_2^2$, and for $x = \alpha g w_2^2$, the target is $\mathbb{F}_2\{w_1^2w_2^2 \cdot \beta\}$ at $E_2$, and $d_2(w_1^2w_2^2 \cdot \beta) = w_1^2w_2^2 \cdot h_0 d_0 \neq 0$.
- For $x = h_0\alpha^3 w_2^2$ the target is $\mathbb{F}_2\{g^2w_1^2 \cdot \alpha\beta w_2\}$ at $E_2$, and $d_2(g^2w_1^2 \cdot \alpha\beta w_2) = g^5 w_1^2 \cdot d_0 \neq 0$.
- For $x = c_0 w_2^3$ the target is $\mathbb{F}_2\{g^4 w_1 \cdot \beta w_2\}$ at $E_2$, and $d_2(g^4 w_1 \cdot \beta w_2) = g^5 w_1 \cdot e_0\gamma + g^4 w_1 w_2 \cdot h_0 d_0 = g^5 w_1 \cdot e_0\gamma \neq 0$.
- For $x = h_0^3 \alpha w_2^3$ the target is $\mathbb{F}_2\{g^6 w_1 \cdot \alpha\beta\}$ at $E_2$, and $d_2(g^5 w_1 \cdot w_2) = g^6 w_1 \cdot \alpha\beta$.
- For $x = h_0\alpha^3 w_2^3$ the target is $\mathbb{F}_2\{g^2w_1^2w_2^2 \cdot \alpha\beta\}$ at $E_2$, and $d_2(gw_1^2w_2^2 \cdot w_2) = g^2w_1^2w_2^2 \cdot \alpha\beta$.

(4) Similarly, we see from Table 5.2 that $d_4(x) = 0$ for $x = h_2 w_2$, $h_2 w_2^3$ and $\delta w_2^3$, because the target groups become trivial at the $E_4$-term:

- For $x = h_2 w_2$ the $E_2$-term in the bidegree of $d_4(x)$ is $\mathbb{F}_2\{w_1 \cdot e_0\gamma\}$, and $d_3(w_1 \cdot e_0\gamma) = w_1^2 \cdot h_1\delta \neq 0$.
- For $x = h_2 w_2^3$ the target is $\mathbb{F}_2\{w_1 w_2^2 \cdot e_0\gamma\}$ at $E_2$ and $E_3$, and $d_3(w_1 \cdot e_0\gamma w_2^2) = w_1^2w_2^2 \cdot h_1\delta \neq 0$.
- For $x = \delta w_2^3$ the target is $\mathbb{F}_2\{g^8 \cdot \beta, w_1^2w_2^2 \cdot \beta w_2\}$ at $E_2$. Here $d_2(g^8 \cdot \beta) = g^8 \cdot h_0 d_0 = 0$ and $d_2(w_1^2w_2^2 \cdot \beta w_2) = gw_1^2w_2^2 \cdot e_0\gamma + w_1^2w_2^3 \cdot h_0 d_0 \neq 0$. Hence the target at $E_3$ is $\mathbb{F}_2\{g^6 \cdot \beta g^2\}$, and $d_3(g^4 \cdot w_2^2) = g^6 \cdot \beta g^2$. □

PROPOSITION 5.22.

(1) $d_4(d_0 e_0) = d_0 w_1^2$.
(2) $d_4(e_0 g) = g w_1^2$.
(3) $d_4(h_1^2 w_2) = \alpha^2 e_0 w_1$.
(4) $d_4(\alpha^2 g) = \alpha\beta w_1^2$.
(5) $d_4(h_0 w_2) = d_0\gamma w_1$.
(6) $d_4(\alpha e_0 g) = (\delta + \alpha g) w_1^2 = \delta' w_1^2$.
(7) $d_4(\beta g^2) = \alpha d_0 g w_1$.
(8) $d_4(h_0\alpha^2 w_2) = 0$.
(9) $d_4(h_0\alpha^2 w_2^3) = 0$.
(10) $d_4(h_1 w_2^2) = 0$.
(11) $d_4(d_0 e_0 w_2^2) = d_0 w_1^2 w_2^2$.
(12) $d_4(e_0 g w_2^2) = g w_1^2 w_2^2$.
(13) $d_4(\alpha^2 g w_2^2) = \alpha\beta w_1^2 w_2^2$.
(14) $d_4(h_0 w_2^3) = d_0\gamma w_1 w_2^2$.
(15) $d_4(\alpha e_0 g w_2^2) = (\delta + \alpha g) w_1^2 w_2^2 = \delta' w_1^2 w_2^2$.
(16) $d_4(h_1^2 w_2^3) = \alpha^2 e_0 w_1 w_2^2$.

PROOF. The differentials on $x = d_0e_0$, $e_0g$ and $h_1^2w_2$ have already been identified. For $x = \alpha^2 g$, $h_0w_2$, $\alpha e_0 g$ and $\beta g^2$ we use multiplicative relations in the Adams spectral sequence to determine $d_4(x)$. For $x = h_0\alpha^2 w_2$ and $h_0\alpha^2 w_2^3$ we use $d_4 \circ d_4 = 0$ to show that $d_4(x) = 0$. Finally, for the remaining classes $x = h_1w_2^2$, $d_0e_0w_2^2$, $e_0gw_2^2$, $\alpha^2 gw_2^2$, $h_0w_2^3$, $\alpha e_0gw_2^2$ and $h_1^2w_2^3$ we use $w_1$- and $w_1w_2^2$-linearity to determine $d_4(x)$. In many cases we (implicitly) refer to Table 5.5 to determine whether a class is nonzero at $E_4$.

(1)–(3) We know that $d_4(e_0g) = gw_1^2$ by Theorem 5.12, $d_4(d_0e_0) = d_0w_1^2$ by Corollary 5.13, and $d_4(h_1^2w_2) = \alpha^2 e_0w_1$ by Proposition 5.14.

(4) From $d_4(w_1) = 0$, $d_4(d_0) = 0$, $d_4(\beta^2 g) = \alpha^2 e_0 w_1$ (by Corollary 5.13) and the relation

$$d_0 \cdot \alpha^2 g = w_1 \cdot \beta^2 g$$

we deduce that $d_0 \cdot d_4(\alpha^2 g) = w_1^2 \cdot \alpha^2 e_0$. A glance at Table 5.5 shows that this is nonzero at $E_4$, because $w_1^2$ is not in the annihilator ideal of $\alpha^2 e_0$. Hence $d_4(\alpha^2 g) \neq 0$, and $\alpha\beta w_1^2$ is the only possible value.

(5) From $d_4(\gamma) = 0$, $d_4(d_0e_0) = d_0w_1^2$ and the relation

$$\gamma \cdot d_0e_0 = \alpha^3 g$$

we deduce that $d_4(\alpha^3 g) = d_0\gamma w_1^2$. From the differential

$$d_2(\alpha e_0 w_2) = \alpha^3 g^2 + h_0 g w_1 w_2$$

we know that $g \cdot \alpha^3 g = gw_1 \cdot h_0w_2$ at the $E_4$-term. Hence $gw_1 \cdot d_4(h_0w_2) = g \cdot d_4(\alpha^3 g) = gw_1^2 \cdot d_0\gamma$, which is nonzero at $E_4$ (by Table 5.6). Thus $d_4(h_0w_2) \neq 0$, and $d_0\gamma w_1$ is the only possible value.

(6) The ($E_2$- and) $E_4$-term in the bidegree of $d_4(\alpha e_0 g)$ is $\mathbb{F}_2\{\delta w_1^2, \alpha g w_1^2, h_0^7 w_2\}$. Multiplication by $h_0$ annihilates only the subgroup $\mathbb{F}_2\{\delta' w_1^2\}$, where $\delta' w_1^2 = \delta w_1^2 + \alpha g w_1^2$. From $d_4(g) = 0$, $d_4(\alpha g) = 0$, $d_4(e_0 g) = gw_1^2$ and the factorization

$$g \cdot \alpha e_0 g = \alpha g \cdot e_0 g$$

we deduce that $g \cdot d_4(\alpha e_0 g) = \alpha g \cdot d_4(e_0 g) = \alpha g^2 w_1^2 \neq 0$. Furthermore, $h_0 \cdot \alpha e_0 g = 0$. Hence $d_4(\alpha e_0 g)$ is nonzero and $h_0$-annihilated, leaving $\delta' w_1^2$ as the only possible value.

(7) From $d_4(g) = 0$, $d_4(\gamma) = 0$, $d_4(h_1^2 w_2) = \alpha^2 e_0 w_1$ and the relation

$$\gamma^3 = g \cdot \beta g^2 + \gamma \cdot h_1^2 w_2$$

we deduce that $g \cdot d_4(\beta g^2) = \gamma \cdot \alpha^2 e_0 w_1 = gw_1 \cdot \alpha d_0 g \neq 0$ at $E_4$. Hence $d_4(\beta g^2)$ is nonzero, and $\alpha d_0 g w_1$ is the only possible value.

(8) The ($E_2$- and) $E_4$-term in the bidegree of $d_4(h_0\alpha^2 w_2)$ is $\mathbb{F}_2\{\beta g^2 w_1^2\}$, and

$$d_4(\beta g^2 w_1^2) = w_1^3 \cdot \alpha d_0 g \neq 0$$

by the previous case. We cannot have $d_4(h_0\alpha^2 w_2) = \beta g^2 w_1^2$, because $d_4 \circ d_4 = 0$. Hence $d_4(h_0\alpha^2 w_2) = 0$.

(9) The ($E_2$- and) $E_4$-term in the bidegree of $d_4(h_0\alpha^2 w_2^3)$ is $\mathbb{F}_2\{\beta g^2 w_1^2 w_2^2\}$, and

$$d_4(\beta g^2 w_1^2 w_2^2) = w_1^3 w_2^2 \cdot \alpha d_0 g \neq 0\,.$$

Hence $d_4 \circ d_4 = 0$ implies $d_4(h_0\alpha^2 w_2^3) \neq \beta g^2 w_1^2 w_2^2$, leaving 0 as the only possible value.

(10) The $E_2$-term in the bidegree of $d_4(h_1 w_2^2)$ is $\mathbb{F}_2\{\alpha^3 g^3, h_0^5 w_2^2\}$, and

$$d_2(g \cdot \alpha e_0 w_2) = g \cdot (\alpha^3 g^2 + h_0 g w_1 w_2) = \alpha^3 g^3\,,$$

so the target $E_4$-term is $\mathbb{F}_2\{h_0^5 w_2^2\}$. We have $w_1 \cdot d_4(h_1 w_2^2) = d_4(w_1 \cdot h_1 w_2^2) = d_4(w_1 w_2^2 \cdot h_1) = w_1 w_2^2 \cdot d_4(h_1) = 0$, since $w_1$ and $w_1 w_2^2$ are $d_4$-cycles. Furthermore, $w_1 \cdot h_0^5 w_2^2 \neq 0$ at $E_4$, so $d_4(h_1 w_2^2) = 0$.

(11) From $d_4(d_0 e_0) = d_0 w_1^2$ we deduce that $w_1 \cdot d_4(d_0 e_0 w_2^2) = w_1 w_2^2 \cdot d_4(d_0 e_0) = w_1^3 \cdot d_0 w_2^2 \neq 0$ at $E_4$. It follows that $d_4(d_0 e_0 w_2^2)$ is nonzero, and $d_0 w_1^2 w_2^2$ is the only possible value.

(12) The $E_2$-term in the bidegree of $d_4(e_0 g w_2^2)$ is $\mathbb{F}_2\{h_0^3 \alpha^3 w_2^2, g w_1^2 w_2^2\}$, which equals the $E_4$-term in this bidegree. From $h_0 \cdot h_0^3 \alpha^3 w_2^2 \neq 0$ and

$$h_0 \cdot g w_1^2 w_2^2 = d_2(w_1 w_2^2 \cdot \alpha e_0)$$

we see that multiplication by $h_0$ annihilates only the subgroup $\mathbb{F}_2\{g w_1^2 w_2^2\}$ of the $E_4$-term. From $d_4(e_0 g) = g w_1^2$ we deduce that $w_1 \cdot d_4(e_0 g w_2^2) = w_1 w_2^2 \cdot d_4(e_0 g) = g w_1^2 \cdot w_1 w_2^2 \neq 0$ at $E_4$. Furthermore, $h_0 \cdot e_0 g w_2^2 = 0$. Hence $d_4(e_0 g w_2^2)$ is nonzero and $h_0$-annihilated, and $g w_1^2 w_2^2$ is the only possible value.

(13) From $d_4(\alpha^2 g) = \alpha\beta w_1^2$ we deduce that $w_1 \cdot d_4(\alpha^2 g w_2^2) = w_1 w_2^2 \cdot d_4(\alpha^2 g) = w_1^3 \cdot \alpha\beta w_2^2 \neq 0$ at $E_4$. It follows that $d_4(\alpha^2 g w_2^2)$ is nonzero, and $\alpha\beta w_1^2 w_2^2$ is the only possible value.

(14) From $d_4(h_0 w_2) = d_0 \gamma w_1$ we deduce that $w_1 \cdot d_4(h_0 w_2^3) = w_1 w_2^2 \cdot d_4(h_0 w_2) = w_1^2 \cdot d_0 \gamma w_2^2 \neq 0$ at $E_4$. It follows that $d_4(h_0 w_2^3)$ is nonzero, and $d_0 \gamma w_1 w_2^2$ is the only possible value.

(15) The $E_2$-term in the bidegree of $d_4(\alpha e_0 g w_2^2)$ is $\mathbb{F}_2\{\delta w_1^2 w_2^2, \alpha g w_1^2 w_2^2, h_0^7 w_2^3\}$. At the $E_4$-term, multiplication by $h_0$ annihilates only the subgroup $\mathbb{F}_2\{\delta' w_1^2 w_2^2\}$. From $d_4(\alpha e_0 g) = \delta' w_1^2$ we deduce that $w_1 \cdot d_4(\alpha e_0 g w_2^2) = w_1 w_2^2 \cdot d_4(\alpha e_0 g) = \delta' w_1^3 w_2^2 \neq 0$ at $E_4$. Furthermore, $h_0 \cdot \alpha e_0 g w_2^2 = 0$. Hence $d_4(\alpha e_0 g w_2^2)$ is nonzero and $h_0$-annihilated, and $\delta' w_1^2 w_2^2$ is the only possible value.

(16) From $d_4(h_1^2 w_2) = \alpha^2 e_0 w_1$ we deduce that $w_1 \cdot d_4(h_1^2 w_2^3) = w_1 w_2^2 \cdot d_4(h_1^2 w_2) = w_1^2 \cdot \alpha^2 e_0 w_2^2 \neq 0$ at $E_4$. It follows that $d_4(h_1^2 w_2^3)$ is nonzero, and $\alpha^2 e_0 w_1 w_2^2$ is the only possible value. □

THEOREM 5.23. *The $d_4$-differential in $E_4(tmf)$ is $R_2$-linear. Its values on a set of algebra generators are as listed in Table* 5.7, *and its values on a set of $R_2$-module generators are as listed in Table* 5.5.

PROOF. The first two claims follow from Propositions 5.21 and 5.22. As before, the Leibniz rule lets us calculate $d_4(x)$ for the remaining $R_2$-module generators $x$, using the factorizations given in the proof of Proposition 5.20:

- $d_4(h_0 e_0) = d_4(h_2 \cdot d_0) = 0$
- $d_4(d_0 \cdot \gamma) = 0$
- $d_4(\alpha^2 e_0) = d_4(d_0 \cdot \alpha\beta) = 0$
- $d_4(d_0 \cdot \alpha g) = 0$
- $d_4(\beta^2 g) = d_4(\gamma \cdot \gamma + h_1^2 w_2) = 0 + \alpha^2 e_0 w_1$
- $d_4(\alpha^3 g + h_0 w_1 w_2) = d_4(\gamma \cdot d_0 e_0 + w_1 \cdot h_0 w_2) = \gamma \cdot d_0 w_1^2 + w_1 \cdot d_0 \gamma w_1 = 0$
- $d_4(\gamma\delta') = d_4(\gamma \cdot (\delta + \alpha g)) = 0$
- $d_4(\gamma \cdot e_0 g) = \gamma \cdot g w_1^2$
- $d_4(h_0 e_0 w_2) = d_4(d_0 \cdot h_2 w_2) = 0$
- $d_4(\gamma^3) = 0 \cdot \gamma^2 = 0$
- $d_4(h_0 e_0 w_2^2) = d_4(h_2 \cdot d_0 w_2^2) = 0$
- $d_4(\gamma \cdot h_1 w_2^2) = 0$
- $d_4(\gamma \cdot w_1 w_2^2) = 0$
- $d_4(\gamma \cdot d_0 w_2^2) = 0$

- $d_4(\alpha^2 e_0 w_2^2) = d_4(d_0 \cdot \alpha\beta w_2^2) = 0$
- $d_4(d_0 \cdot \alpha g w_2^2) = 0$
- $d_4(\alpha^3 g w_2^2 + h_0 w_1 w_2^3) = d_4(\gamma \cdot d_0 e_0 w_2^2 + w_1 \cdot h_0 w_2^3) = \gamma \cdot d_0 w_1^2 w_2^2 + w_1 \cdot d_0 \gamma w_1 w_2^2 = 0$
- $d_4(\gamma\delta' w_2^2) = d_4(\gamma \cdot (\delta w_2^2 + \alpha g w_2^2)) = 0$
- $d_4(\beta^2 g \cdot w_1 w_2^2) = \alpha^2 e_0 w_1 \cdot w_1 w_2^2$
- $d_4(\gamma \cdot e_0 g w_2^2) = \gamma \cdot g w_1^2 w_2^2$
- $d_4(\beta g^2 \cdot w_1 w_2^2) = \alpha d_0 g w_1 \cdot w_1 w_2^2$
- $d_4(h_0 e_0 w_2^3) = d_4(d_0 \cdot h_2 w_2^3) = 0$
- $d_4(\gamma \cdot h_1^2 w_2^3) = \gamma \cdot \alpha^2 e_0 w_1 w_2^2 = \alpha e_0^2 g w_1 w_2^2 = \alpha d_0 g^2 w_1 w_2^2$.

The other cases follow easily by $h_0$-, $h_1$- and $h_2$-linearity. □

## 5.5. The $E_\infty$-term for $tmf$

Given Theorem 5.23, it is elementary to calculate $E_5(tmf)$ as an $R_2$-module. The details are given in Appendix A.3, and with two minor modifications, explained in Remark 5.24, the results are recorded in Table 5.8. The non-cyclic summands are displayed in Table 5.9. We note that the $E_5$-term is free over $\mathbb{F}_2[w_2^4]$ and finitely generated over $\mathbb{F}_2[h_0, w_1, w_2^4]$. The class $g$ is nilpotent, with $g^6 = 0$.

REMARK 5.24. We rewrite the direct sum

$$\langle e_0 g^2 \rangle \oplus \langle \gamma\delta' \rangle \cong R_2/(g) \oplus R_2/(g, w_1)$$

as

$$\langle \gamma\delta' \rangle \oplus \langle h_1 c_0 w_2 \rangle \cong R_2/(g, w_1) \oplus R_2/(g)\,.$$

This makes the $h_1$-multiplication from $(t-s, s) = (56, 11)$ easier to display, since $h_1 \cdot c_0 w_1 = h_1 c_0 w_1$. Likewise, we rewrite the direct sum

$$\langle e_0 g^2 w_2^2 \rangle \oplus \langle \gamma\delta' w_2^2 \rangle \cong R_2/(g) \oplus R_2/(g, w_1)$$

as

$$\langle \gamma\delta' w_2^2 \rangle \oplus \langle h_1 c_0 w_2^3 \rangle \cong R_2/(g, w_1) \oplus R_2/(g)\,.$$

Table 5.8: $R_2$-module generators of $E_5(tmf)$, with $i \geq 0$ in each $h_0$-tower

| $t-s$ | $s$ | $g$ | $x$ | $\mathrm{Ann}(x)$ | dec. |
|---|---|---|---|---|---|
| 0 | 0 | 0 | $1$ | $(g^6, g^2 w_1, g w_1^2)$ | 1 |
| 0 | 1 | 0 | $h_0$ | $(g^2, g w_1)$ | **gen.** |
| 0 | 2 | 0 | $h_0^2$ | $(g^2, g w_1)$ | $h_0 \cdot h_0$ |
| 0 | $3+i$ | 0 | $h_0^{3+i}$ | $(g)$ | $h_0^{2+i} \cdot h_0$ |
| 1 | 1 | 1 | $h_1$ | $(g^2, g w_1)$ | **gen.** |
| 2 | 2 | 1 | $h_1^2$ | $(g)$ | $h_1 \cdot h_1$ |
| 3 | 1 | 2 | $h_2$ | $(g, w_1)$ | **gen.** |
| 3 | 2 | 2 | $h_0 h_2$ | $(g, w_1)$ | $h_0 \cdot h_2$ |
| 3 | 3 | 1 | $h_0^2 h_2$ | $(g, w_1)$ | $h_0^2 \cdot h_2$ |

Table 5.8: $R_2$-module generators of $E_5(tmf)$, with $i \geq 0$ in each $h_0$-tower (cont.)

| $t-s$ | $s$ | $g$ | $x$ | $\mathrm{Ann}(x)$ | dec. |
|---|---|---|---|---|---|
| 6 | 2 | 3 | $h_2^2$ | $(g, w_1)$ | $h_2 \cdot h_2$ |
| 8 | 3 | 2 | $c_0$ | $(g, w_1)$ | **gen.** |
| 9 | 4 | 2 | $h_1 c_0$ | $(g, w_1)$ | $h_1 \cdot c_0$ |
| 12 | $6+i$ | 4 | $h_0^{3+i}\alpha$ | $(g)$ | $h_0^i \cdot$ **gen.** |
| 14 | 4 | 4 | $d_0$ | $(g^3, g^2 w_1, w_1^2)$ | **gen.** |
| 15 | 5 | 6 | $h_1 d_0$ | $(g, w_1)$ | $h_1 \cdot d_0$ |
| 17 | 5 | 7 | $h_0 e_0$ | $(g, w_1)$ | $h_2 \cdot d_0$ |
| 24 | $7+i$ | 7 | $h_0^{1+i}\alpha^2$ | $(g)$ | $h_0^i \cdot$ **gen.** |
| 25 | 5 | 11 | $\gamma$ | – | **gen.** |
| 26 | 6 | 9 | $h_1\gamma$ | $(g)$ | $h_1 \cdot \gamma$ |
| 27 | 6 | 10 | $\alpha\beta$ | $(g, w_1^2)$ | **gen.** |
| 27 | 7 | 9 | $h_1^2\gamma$ | $(g, w_1)$ | $h_1^2 \cdot \gamma$ |
| 32 | 7 | 11 | $\delta$ | $(g)$ | **gen.** |
| 32 | 7 | 12 | $\delta'$ | $(g^2, w_1^2)$ | **gen.** |
| 32 | 8 | 14 | $h_0 \alpha g$ | $(g)$ | $h_0 \cdot \delta$ |
| 32 | 9 | 14 | $h_0^2 \alpha g$ | $(g)$ | $h_0^2 \cdot \delta$ |
| 33 | 8 | 15 | $h_1 \delta$ | $(g, w_1)$ | $h_1 \cdot \delta'$ |
| 36 | $10+i$ | 14 | $h_0^{1+i}\alpha^3$ | $(g)$ | $h_0^i \cdot$ **gen.** |
| 39 | 9 | 18 | $d_0\gamma$ | – | $d_0 \cdot \gamma$ |
| 41 | 10 | 16 | $\alpha^2 e_0$ | $(g, w_1)$ | $d_0 \cdot \alpha\beta$ |
| 46 | 11 | 18 | $\alpha d_0 g$ | $(g^2, w_1)$ | $d_0 \cdot \delta'$ |
| 48 | 10 | 19 | $h_0^2 w_2$ | $(g^2, g w_1)$ | **gen.** |
| 48 | $11+i$ | 19 | $h_0^{3+i} w_2$ | $(g)$ | $h_0^{1+i} \cdot h_0^2 w_2$ |
| 50 | 10 | $20+21$ | $\gamma^2$ | $(g^5, g w_1)$ | $\gamma \cdot \gamma$ |
| 51 | 9 | 23 | $h_2 w_2$ | – | **gen.** |
| 51 | 10 | 22 | $h_0 h_2 w_2$ | $(g, w_1)$ | $h_0 \cdot h_2 w_2$ |
| 51 | 11 | 21 | $h_0^2 h_2 w_2$ | $(g, w_1)$ | $h_0^2 \cdot h_2 w_2$ |
| 54 | 10 | 23 | $h_2^2 w_2$ | $(g, w_1)$ | $h_2 \cdot h_2 w_2$ |
| 56 | 11 | 24 | $c_0 w_2$ | $(g)$ | **gen.** |
| 56 | 13 | $26+27$ | $\alpha^3 g + h_0 w_1 w_2$ | $(g)$ | **gen.** |
| 57 | 12 | $27+28$ | $\gamma\delta'$ | $(g, w_1)$ | $\gamma \cdot \delta'$ |

Table 5.8: $R_2$-module generators of $E_5(tmf)$, with $i \geq 0$ in each $h_0$-tower (cont.)

| $t-s$ | $s$ | $g$ | $x$ | $\mathrm{Ann}(x)$ | dec. |
|---|---|---|---|---|---|
| 57 | 12 | 28 | $h_1 c_0 w_2$ | $(g)$ | $\gamma \cdot \delta$ |
| 60 | $14+i$ | 28 | $h_0^{3+i} \alpha w_2$ | $(g)$ | $h_0^i \cdot \mathbf{gen.}$ |
| 65 | 13 | 36 | $h_0 e_0 w_2$ | $(g, w_1)$ | $d_0 \cdot h_2 w_2$ |
| 72 | $15+i$ | 36 | $h_0^{1+i} \alpha^2 w_2$ | $(g)$ | $h_0^i \cdot \mathbf{gen.}$ |
| 75 | 15 | $38+39$ | $\gamma^3$ | $(g, w_1)$ | $\gamma^2 \cdot \gamma$ |
| 80 | 15 | 41 | $\delta w_2$ | $(g)$ | **gen.** |
| 80 | 16 | 49 | $h_0 \alpha g w_2$ | $(g)$ | $h_0 \cdot \delta w_2$ |
| 80 | 17 | 49 | $h_0^2 \alpha g w_2$ | $(g)$ | $h_0^2 \cdot \delta w_2$ |
| 81 | 16 | 50 | $h_1 \delta w_2$ | $(g)$ | $h_1 \cdot \delta w_2$ |
| 82 | 17 | 51 | $e_0 \gamma g^2$ | $(g)$ | $\gamma^2 \cdot (\delta + \delta')$ |
| 84 | $18+i$ | 48 | $h_0^{1+i} \alpha^3 w_2$ | $(g)$ | $h_0^i \cdot \mathbf{gen.}$ |
| 96 | 17 | 58 | $h_0 w_2^2$ | $(g^2, g w_1)$ | **gen.** |
| 96 | 18 | 55 | $h_0^2 w_2^2$ | $(g^2, g w_1)$ | $h_0 \cdot h_0 w_2^2$ |
| 96 | $19+i$ | 57 | $h_0^{3+i} w_2^2$ | $(g)$ | $h_0^{2+i} \cdot h_0 w_2^2$ |
| 97 | 17 | 59 | $h_1 w_2^2$ | $-$ | **gen.** |
| 98 | 18 | 57 | $h_1^2 w_2^2$ | $(g)$ | $h_1 \cdot h_1 w_2^2$ |
| 99 | 17 | 60 | $h_2 w_2^2$ | $(g, w_1)$ | **gen.** |
| 99 | 18 | 58 | $h_0 h_2 w_2^2$ | $(g, w_1)$ | $h_0 \cdot h_2 w_2^2$ |
| 99 | 19 | 59 | $h_0^2 h_2 w_2^2$ | $(g, w_1)$ | $h_0^2 \cdot h_2 w_2^2$ |
| 102 | 18 | 59 | $h_2^2 w_2^2$ | $(g, w_1)$ | $h_2 \cdot h_2 w_2^2$ |
| 104 | 19 | 62 | $c_0 w_2^2$ | $(g, w_1)$ | **gen.** |
| 104 | 20 | 69 | $w_1 w_2^2$ | $(g^2, g w_1)$ | **gen.** |
| 105 | 20 | 71 | $h_1 c_0 w_2^2$ | $(g, w_1)$ | $h_1 \cdot c_0 w_2^2$ |
| 108 | $22+i$ | 71 | $h_0^{3+i} \alpha w_2^2$ | $(g)$ | $h_0^i \cdot \mathbf{gen.}$ |
| 110 | 20 | 74 | $d_0 w_2^2$ | $(g^3, g^2 w_1, w_1^2)$ | **gen.** |
| 111 | 21 | 79 | $h_1 d_0 w_2^2$ | $(g, w_1)$ | $h_1 \cdot d_0 w_2^2$ |
| 113 | 21 | 81 | $h_0 e_0 w_2^2$ | $(g, w_1)$ | $h_2 \cdot d_0 w_2^2$ |
| 120 | $23+i$ | 82 | $h_0^{1+i} \alpha^2 w_2^2$ | $(g)$ | $h_0^i \cdot \mathbf{gen.}$ |
| 122 | 22 | 81 | $h_1 \gamma w_2^2$ | $(g)$ | $\gamma \cdot h_1 w_2^2$ |
| 123 | 22 | 82 | $\alpha \beta w_2^2$ | $(g, w_1^2)$ | **gen.** |
| 123 | 23 | 85 | $h_1^2 \gamma w_2^2$ | $(g, w_1)$ | $h_1 \gamma \cdot h_1 w_2^2$ |

Table 5.8: $R_2$-module generators of $E_5(tmf)$, with $i \geq 0$ in each $h_0$-tower (cont.)

| $t-s$ | $s$ | $g$ | $x$ | $\mathrm{Ann}(x)$ | dec. |
|---|---|---|---|---|---|
| 128 | 23 | 87 | $\delta w_2^2$ | $(g)$ | **gen.** |
| 128 | 23 | 88 | $\delta' w_2^2$ | $(g^2, w_1^2)$ | **gen.** |
| 128 | 24 | 100 | $h_0\alpha g w_2^2$ | $(g)$ | $h_0 \cdot \delta w_2^2$ |
| 128 | 25 | 102 | $h_0^2\alpha g w_2^2$ | $(g)$ | $h_0^2 \cdot \delta w_2^2$ |
| 129 | 24 | 101 | $h_1\delta w_2^2$ | $(g, w_1)$ | $h_1 \cdot \delta' w_2^2$ |
| 129 | 25 | 103 | $\gamma w_1 w_2^2$ | $(g^2, gw_1)$ | $\gamma \cdot w_1 w_2^2$ |
| 132 | $26+i$ | 100 | $h_0^{1+i}\alpha^3 w_2^2$ | $(g)$ | $h_0^i \cdot$ **gen.** |
| 135 | 25 | 108 | $d_0\gamma w_2^2$ | $-$ | $\gamma \cdot d_0 w_2^2$ |
| 137 | 26 | 103 | $\alpha^2 e_0 w_2^2$ | $(g, w_1)$ | $d_0 \cdot \alpha\beta w_2^2$ |
| 142 | 27 | 109 | $\alpha d_0 g w_2^2$ | $(g^2, gw_1, w_1^2)$ | $d_0 \cdot \delta' w_2^2$ |
| 144 | 26 | 107 | $h_0^2 w_2^3$ | $(g^2, gw_1)$ | **gen.** |
| 144 | $27+i$ | 111 | $h_0^{3+i} w_2^3$ | $(g)$ | $h_0^{1+i} \cdot h_0^2 w_2^3$ |
| 147 | 25 | 113 | $h_2 w_2^3$ | $-$ | **gen.** |
| 147 | 26 | 110 | $h_0 h_2 w_2^3$ | $(g, w_1)$ | $h_0 \cdot h_2 w_2^3$ |
| 147 | 27 | 113 | $h_0^2 h_2 w_2^3$ | $(g, w_1)$ | $h_0^2 \cdot h_2 w_2^3$ |
| 150 | 26 | 111 | $h_2^2 w_2^3$ | $(g, w_1)$ | $h_2 \cdot h_2 w_2^3$ |
| 152 | 27 | 116 | $c_0 w_2^3$ | $(g)$ | **gen.** |
| 152 | 29 | $131+132$ | $\alpha^3 g w_2^2 + h_0 w_1 w_2^3$ | $(g)$ | **gen.** |
| 153 | 28 | $129+130$ | $\gamma\delta' w_2^2$ | $(g, w_1)$ | $\gamma \cdot \delta' w_2^2$ |
| 153 | 28 | 130 | $h_1 c_0 w_2^3$ | $(g)$ | $\gamma \cdot \delta w_2^2$ |
| 154 | 30 | $127+128$ | $\gamma^2 w_1 w_2^2$ | $(g)$ | $\gamma^2 \cdot w_1 w_2^2$ |
| 156 | $30+i$ | 131 | $h_0^{3+i}\alpha w_2^3$ | $(g)$ | $h_0^i \cdot$ **gen.** |
| 161 | 29 | 142 | $h_0 e_0 w_2^3$ | $(g, w_1)$ | $d_0 \cdot h_2 w_2^3$ |
| 168 | $31+i$ | 144 | $h_0^{1+i}\alpha^2 w_2^3$ | $(g)$ | $h_0^i \cdot$ **gen.** |
| 176 | 31 | 149 | $\delta w_2^3$ | $(g)$ | **gen.** |
| 176 | 32 | 167 | $h_0\alpha g w_2^3$ | $(g)$ | $h_0 \cdot \delta w_2^3$ |
| 176 | 33 | 171 | $h_0^2\alpha g w_2^3$ | $(g)$ | $h_0^2 \cdot \delta w_2^3$ |
| 177 | 32 | 168 | $h_1\delta w_2^3$ | $(g)$ | $h_1 \cdot \delta w_2^3$ |
| 178 | 33 | 173 | $e_0\gamma g^2 w_2^2$ | $(g)$ | $\gamma^2 \cdot (\delta + \delta') w_2^2$ |
| 180 | $34+i$ | 168 | $h_0^{1+i}\alpha^3 w_2^3$ | $(g)$ | $h_0^i \cdot$ **gen.** |

Table 5.9: The non-cyclic $R_2$-module summands in $E_5(tmf)$

| $\langle x_1, x_2 \rangle$ |
|---|
| $\langle \gamma, h_1 w_2^2 \rangle \cong \dfrac{\Sigma^{5,30} R_2 \oplus \Sigma^{17,114} R_2}{\langle (g^2 w_1, 0), (g w_1^2, 0), (g^5, g w_1), (0, g^2) \rangle}$ |
| $\langle d_0 \gamma, h_2 w_2 \rangle \cong \dfrac{\Sigma^{9,48} R_2 \oplus \Sigma^{9,60} R_2}{\langle (w_1, 0), (g, w_1), (0, g) \rangle}$ |
| $\langle d_0 \gamma w_2^2, h_2 w_2^3 \rangle \cong \dfrac{\Sigma^{25,160} R_2 \oplus \Sigma^{25,172} R_2}{\langle (w_1, 0), (g, w_1), (0, g) \rangle}$ |

PROPOSITION 5.25. *A set of* 43 *algebra generators for* $E_5(tmf)$ *is listed in Table* 5.10.

PROOF. The dec.-column in Table 5.8 shows how each $R_2$-module generator can be decomposed as a polynomial in the listed algebra generators, using only relations that hold in the $E_2$-term. The algebra generators themselves are indicated by "**gen.**". For typographic reasons, $\gamma \cdot \delta + \gamma \cdot \delta'$ is abbreviated to $\gamma \cdot (\delta + \delta')$, etc. □

PROPOSITION 5.26. *Charts showing* $E_5(tmf)$ *for* $0 \le t - s \le 192$ *are given in Figures* 5.1 *to* 5.8. *All nonzero* $h_0$-, $h_1$- *and* $h_2$-*multiplications are displayed. The red dots indicate* $w_1$-*power torsion classes, and black dots indicate* $w_1$-*periodic classes. All* $\mathbb{F}_2[w_1]$-*module generators are labeled, except those that are also* $h_0$-, $h_1$- *or* $h_2$-*multiples.*

PROOF. The $R_2$-module structure of $E_5(tmf)$ is given by Table 5.8. We emphasize the algebra structure at the $E_5$-term by factorizing some of the module generators, as follows:

$$\begin{aligned} h_0 e_0 &= h_2 \cdot d_0 \\ h_0 \alpha g &= h_0 \cdot \delta \\ \alpha^2 e_0 &= \alpha\beta \cdot d_0 \\ \alpha d_0 g &= d_0 \cdot \delta' \\ \alpha d_0 g^2 &= d_0 \cdot \delta' \cdot g\,. \end{aligned}$$

Similar factorizations apply for $w_2$-, $w_2^2$- or $w_2^3$-multiples of some of these generators. These relations are all valid already at the $E_2$-term.

Most of the $h_0$-, $h_1$- and $h_2$-multiplications are evident from the normal form of the generators. The less obvious cases are

$$\begin{aligned} h_1 \cdot h_1^2 &= h_0 \cdot h_0 h_2 \\ h_2 \cdot h_2 d_0 &= h_0 \cdot h_0 g \\ h_0 \cdot \alpha\beta &= h_1^2 \gamma \\ h_1 \cdot \delta' &= h_1 \delta \end{aligned}$$

TABLE 5.10. Algebra generators of $E_5(tmf) = E_\infty(tmf)$

| $t-s$ | $s$ | $g$ | $x$ |
|---|---|---|---|
| 0 | 1 | 0 | $h_0$ |
| 1 | 1 | 1 | $h_1$ |
| 3 | 1 | 2 | $h_2$ |
| 8 | 3 | 2 | $c_0$ |
| 8 | 4 | 1 | $w_1$ |
| 12 | 6 | 4 | $h_0^3\alpha$ |
| 14 | 4 | 4 | $d_0$ |
| 20 | 4 | 8 | $g$ |
| 24 | 7 | 7 | $h_0\alpha^2$ |
| 25 | 5 | 11 | $\gamma$ |
| 27 | 6 | 10 | $\alpha\beta$ |
| 32 | 7 | 11 | $\delta$ |
| 32 | 7 | 12 | $\delta'$ |
| 36 | 10 | 14 | $h_0\alpha^3$ |
| 48 | 10 | 19 | $h_0^2w_2$ |
| 51 | 9 | 23 | $h_2w_2$ |
| 56 | 11 | 24 | $c_0w_2$ |
| 56 | 13 | $26+27$ | $\alpha^3g + h_0w_1w_2$ |
| 60 | 14 | 28 | $h_0^3\alpha w_2$ |
| 72 | 15 | 36 | $h_0\alpha^2w_2$ |
| 80 | 15 | 41 | $\delta w_2$ |
| 84 | 18 | 48 | $h_0\alpha^3w_2$ |
| 96 | 17 | 58 | $h_0w_2^2$ |
| 97 | 17 | 59 | $h_1w_2^2$ |
| 99 | 17 | 60 | $h_2w_2^2$ |
| 104 | 19 | 62 | $c_0w_2^2$ |
| 104 | 20 | 69 | $w_1w_2^2$ |
| 108 | 22 | 71 | $h_0^3\alpha w_2^2$ |
| 110 | 20 | 74 | $d_0w_2^2$ |
| 120 | 23 | 82 | $h_0\alpha^2w_2^2$ |
| 123 | 22 | 82 | $\alpha\beta w_2^2$ |
| 128 | 23 | 87 | $\delta w_2^2$ |
| 128 | 23 | 88 | $\delta' w_2^2$ |
| 132 | 26 | 100 | $h_0\alpha^3w_2^2$ |
| 144 | 26 | 107 | $h_0^2w_2^3$ |
| 147 | 25 | 113 | $h_2w_2^3$ |
| 152 | 27 | 116 | $c_0w_2^3$ |
| 152 | 29 | $131+132$ | $\alpha^3gw_2^2$ $+ h_0w_1w_2^3$ |
| 156 | 30 | 131 | $h_0^3\alpha w_2^3$ |
| 168 | 31 | 144 | $h_0\alpha^2w_2^3$ |
| 176 | 31 | 149 | $\delta w_2^3$ |
| 180 | 34 | 168 | $h_0\alpha^3w_2^3$ |
| 192 | 32 | 172 | $w_2^4$ |

$$h_1 \cdot \gamma^2 = h_0^2h_2w_2$$
$$h_0 \cdot (\alpha^3g + h_0w_1w_2) = w_1 \cdot h_0^2w_2\,,$$

together with some $w_2$-power multiples of these. Again, these relations are valid at the $E_2$-term. Note also the identites

$$g \cdot d_0\gamma = w_1 \cdot h_2w_2$$
$$h_1 \cdot h_1\delta w_2 = e_0\gamma g^2\,,$$

together with their $w_2^2$-multiples, which are valid starting at the $E_3$-term, and the relation

$$g^5 \cdot \gamma = gw_1 \cdot h_1w_2^2$$

which is valid from the $E_4$-term and onward.

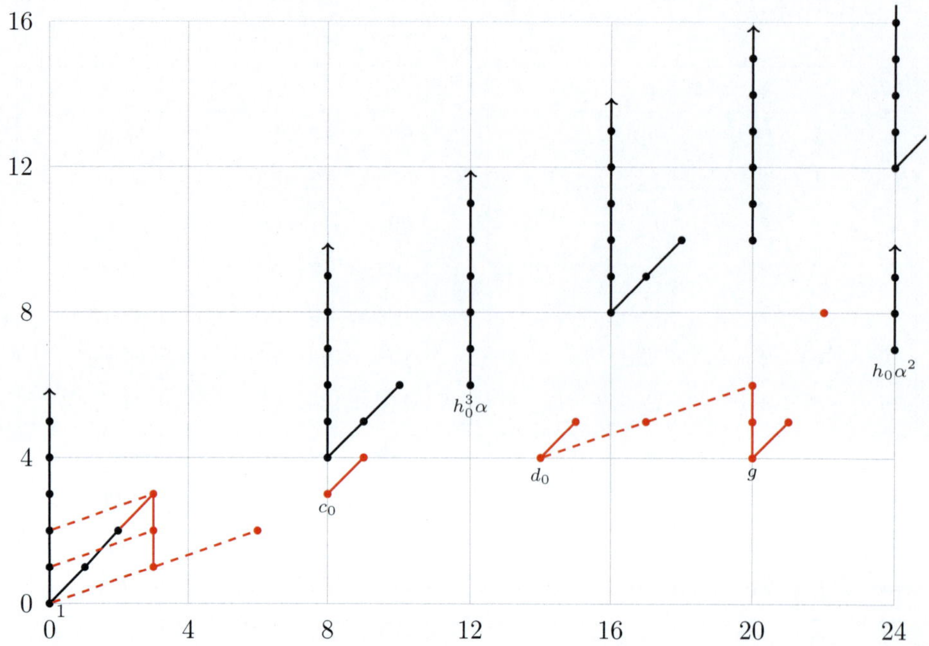


FIGURE 5.1. $E_5(tmf) = E_\infty(tmf)$ for $0 \le t - s \le 24$

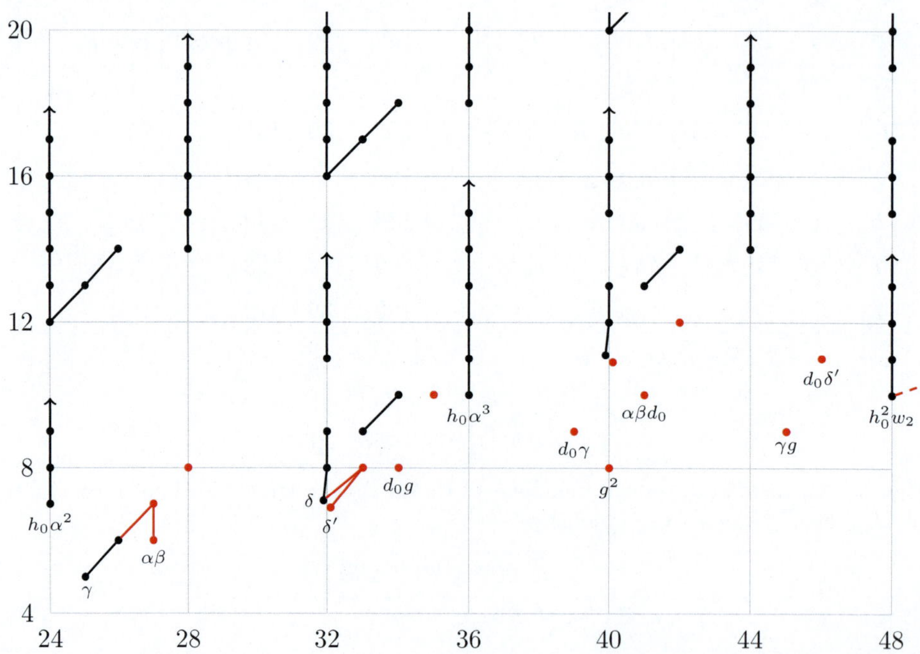


FIGURE 5.2. $E_5(tmf) = E_\infty(tmf)$ for $24 \le t - s \le 48$

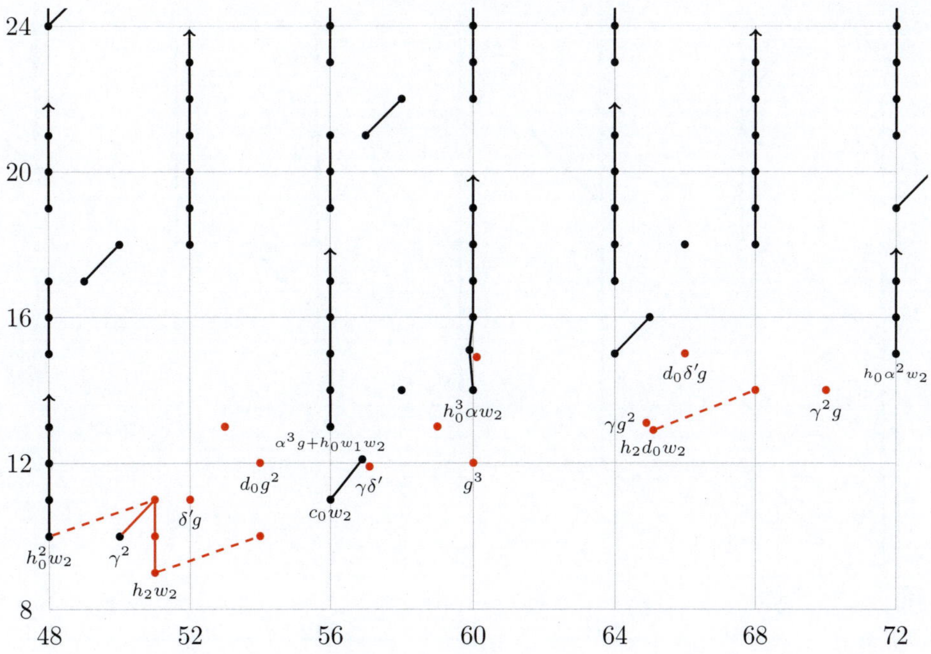


FIGURE 5.3. $E_5(tmf) = E_\infty(tmf)$ for $48 \leq t - s \leq 72$

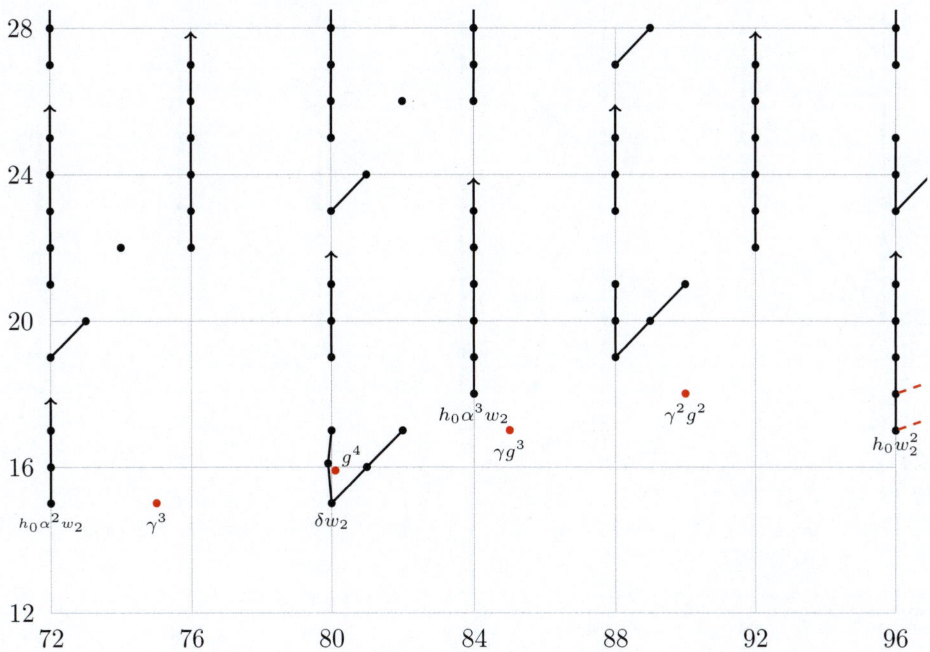


FIGURE 5.4. $E_5(tmf) = E_\infty(tmf)$ for $72 \leq t - s \leq 96$

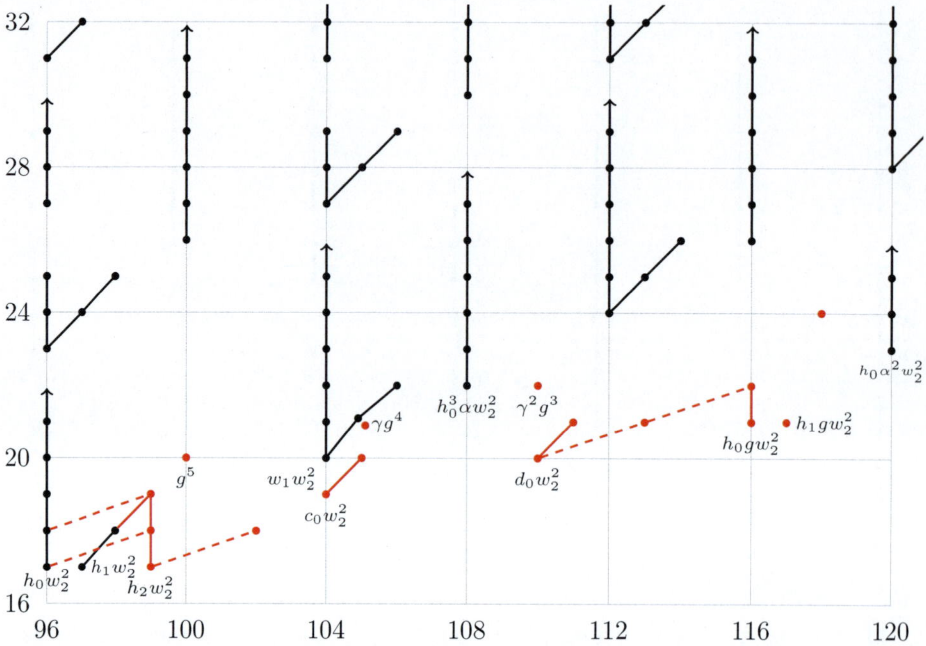


FIGURE 5.5. $E_5(tmf) = E_\infty(tmf)$ for $96 \leq t - s \leq 120$

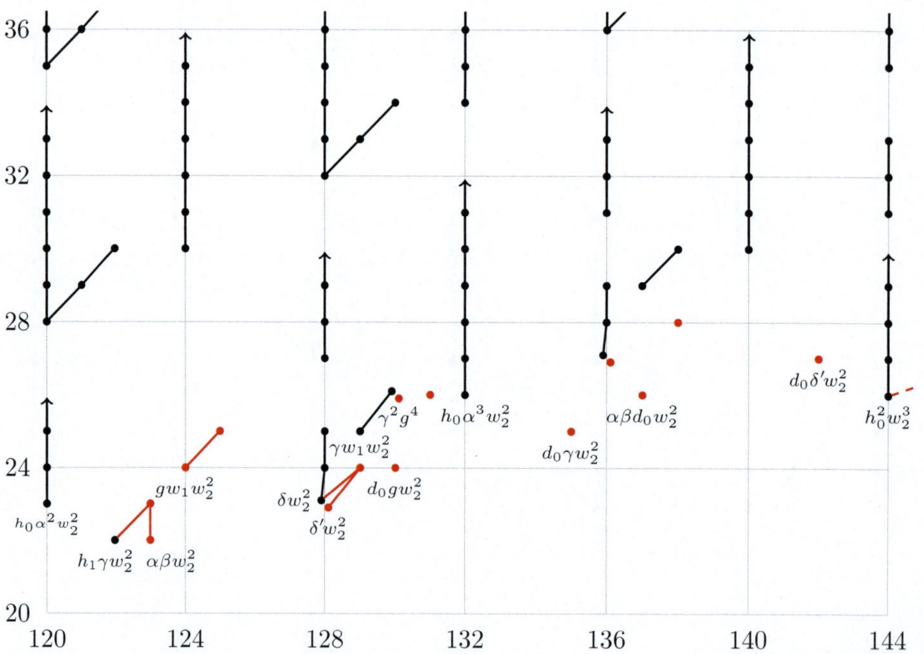


FIGURE 5.6. $E_5(tmf) = E_\infty(tmf)$ for $120 \leq t - s \leq 144$

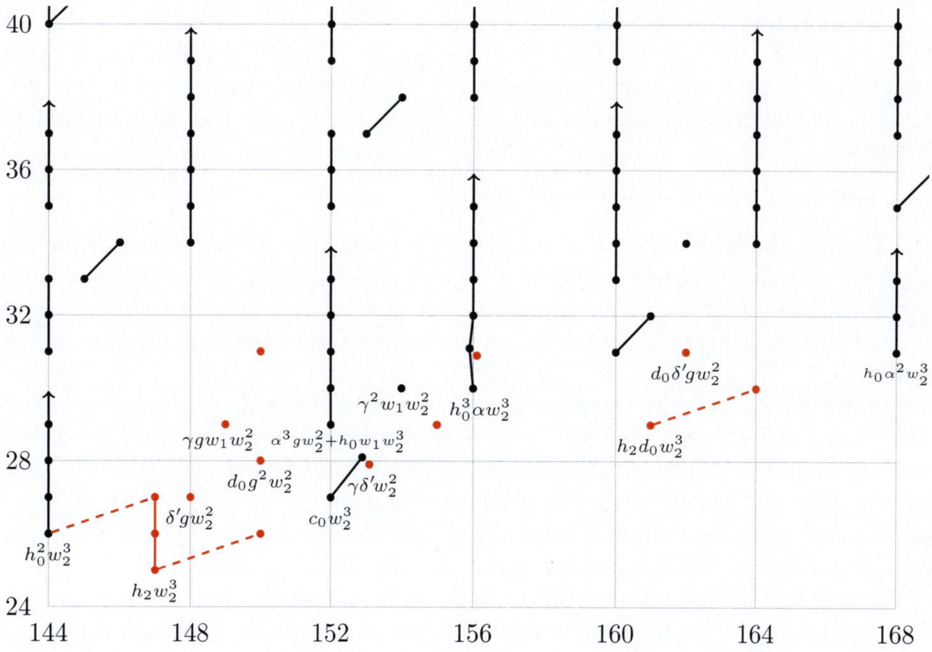


FIGURE 5.7. $E_5(tmf) = E_\infty(tmf)$ for $144 \leq t - s \leq 168$

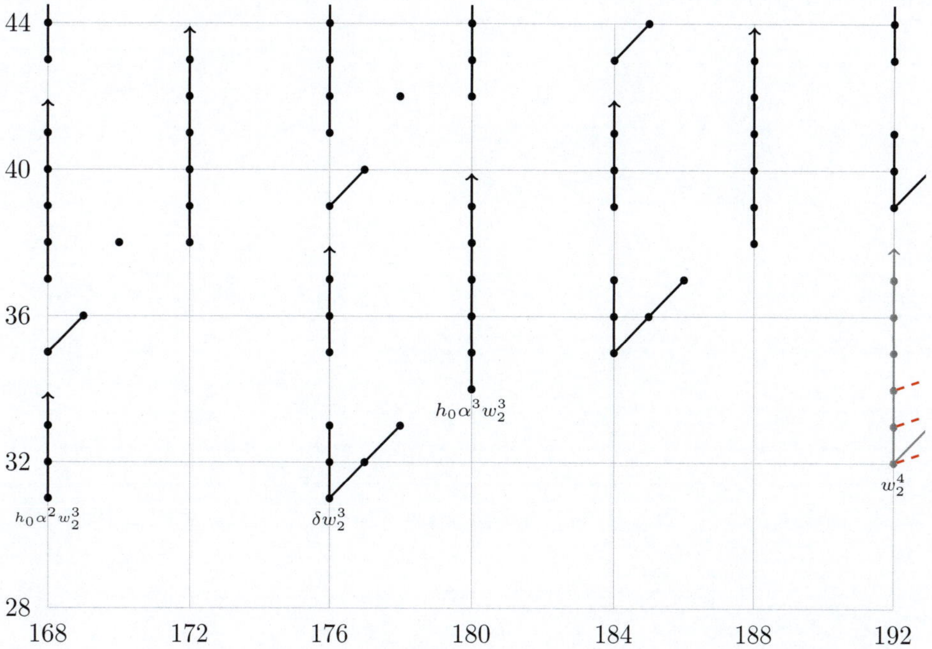


FIGURE 5.8. $E_5(tmf) = E_\infty(tmf)$ for $168 \leq t - s \leq 192$

By recording the $\mathbb{F}_2[w_1]$-module structure implicit in the $R_2$-module structure of the $E_5$-term, and accounting for the $h_0$-, $h_1$- and $h_2$-multiplications known from the $E_2$-term, we deduce that $E_5(tmf)$ is a free $\mathbb{F}_2[w_2^4]$-module, with $\mathbb{F}_2[w_1, w_2^4]$-module generators concentrated in the range $0 \le t-s \le 180$, as indicated in Figures 5.1 to 5.8. □

THEOREM 5.27. *$E_5(tmf) = E_\infty(tmf)$.*

PROOF. To prove that the Adams spectral sequence for *tmf* collapses at the $E_5$-term, we show that each algebra generator $x$ listed in Table 5.10 is an infinite cycle, i.e., that $d_r(x) = 0$ for each $r \ge 5$. For most of these algebra generators all possible target groups are trivial, as can be seen by inspection of Figures 5.1 to 5.8 and 0.8.

The remaining eight cases are $x = h_1, \gamma, \alpha\beta, h_2w_2, h_1w_2^2, h_2w_2^2, \alpha\beta w_2^2$ and $h_2w_2^3$. All differentials on $h_1$ and $\gamma$ vanish by $h_0$-linearity. All differentials on $h_2w_2^2$ vanish by $w_1$-linearity, since $w_1 \cdot d_r(h_2w_2^2) = d_r(w_1 \cdot h_2w_2^2) = 0$. This can only happen if $d_r(h_2w_2^2) = 0$, because $w_1$ acts injectively on $E_5^{s,t}(tmf)$ in all bidegrees with $t-s = 98$, and no intermediate differentials can change this. Similarly, all differentials on $\alpha\beta$, $h_2w_2$, $\alpha\beta w_2^2$ and $h_2w_2^3$ vanish by $w_1^2$-linearity.

Finally, all differentials vanish on $h_1w_2^2$ by $w_1$-linearity, since $w_1 \cdot d_r(h_1w_2^2) = h_1 \cdot d_r(w_1w_2^2) = 0$. Again, this can only happen if $d_r(h_1w_2^2) = 0$, because $w_1$ acts injectively on $E_5^{s,t}(tmf)$ for $t-s = 96$ and no earlier differentials can intervene. □

Our discussion of the Adams spectral sequence for *tmf* continues in Chapter 9, where we determine the additive and multiplicative extensions involved in the passage from $E_\infty(tmf)$ to $\pi_*(tmf)$.

CHAPTER 6

# The Adams spectral sequence for *tmf*/2

We calculate the $d_r$-differentials in the Adams spectral sequence for $tmf/2 = tmf \wedge C2$. These are nontrivial for $r \in \{2, 3, 4\}$, and zero for $r \geq 5$, so the spectral sequence collapses at the $E_5$-term. The module structure over the Adams spectral sequence for *tmf* suffices to determine all of these differentials. The resulting $E_\infty$-term is the associated graded of a degreewise finite length filtration of $\pi_*(tmf/2)$.

## 6.1. The $E_2$-term for *tmf*/2

The initial term

$$E_2 = E_2(tmf/2) \cong \mathrm{Ext}_{A(2)}(M_1, \mathbb{F}_2)$$

of the mod 2 Adams spectral sequence for *tmf*/2 was calculated in Part I. The groups $E_2^{s,t}$ for $0 \leq t-s \leq 96$ are displayed in Figures 1.24 to 1.27. By Corollary 4.3 the $E_2$-term for *tmf*/2 is generated as a module over $E_2(tmf) = \mathrm{Ext}_{A(2)}(\mathbb{F}_2, \mathbb{F}_2)$ by the eleven classes listed in Table 6.1. As a module over $R_0 = \mathbb{F}_2[g, w_1, w_2]$, the $E_2$-term for *tmf*/2 is presented in Tables 6.2 and 6.3 as a direct sum of cyclic modules, together with one non-cyclic module, and illustrated in Figure 4.1. Most entries in these tables are reproduced from Tables 4.2 and 4.3, but the information about $d_2$-differentials will be obtained in the next section. We note that the $E_2$-term is free over $\mathbb{F}_2[w_2]$, but not over $\mathbb{F}_2[w_1, w_2]$, and is finitely generated over $R_0$. Following the strategy of Chapter 5 we will keep track of $R_0$-module structure on the $E_2$-term, $R_1$-module structure on the $E_3$-term, and $R_2$-module structure on the $E_4$- and $E_5 = E_\infty$-terms of the Adams spectral sequence for *tmf*/2. Here $R_1 = \mathbb{F}_2[g, w_1, w_2^2]$ and $R_2 = \mathbb{F}_2[g, w_1, w_2^4]$, as introduced in Definition 5.1.

TABLE 6.1. $E_2(tmf)$-module generators of $E_2(tmf/2)$

| $t-s$ | $s$ | $g$ | $x$ | $d_2(x)$ |
|---|---|---|---|---|
| 0 | 0 | 0 | $i(1)$ | 0 |
| 2 | 1 | 1 | $\widetilde{h_1}$ | 0 |
| 7 | 2 | 3 | $\widetilde{h_2^2}$ | 0 |
| 9 | 3 | 2 | $\widetilde{c_0}$ | 0 |
| 18 | 6 | 3 | $\widetilde{h_0^2 e_0}$ | $i(h_1 c_0 w_1)$ |
| 26 | 5 | 8 | $\widetilde{\gamma}$ | 0 |

| $t-s$ | $s$ | $g$ | $x$ | $d_2(x)$ |
|---|---|---|---|---|
| 31 | 6 | 10 | $\widetilde{\beta^2}$ | 0 |
| 32 | 8 | 7 | $\widetilde{d_0 e_0}$ | 0 |
| 33 | 7 | 10 | $\widetilde{\delta'}$ | 0 |
| 36 | 7 | 12 | $\widetilde{\beta g}$ | $h_1^2 \widetilde{\delta'}$ |
| 42 | 10 | 12 | $\widetilde{\alpha^2 e_0}$ | $i(h_1 \delta w_1)$ |

Table 6.2: $R_0$-module generators of $E_2(tmf/2)$

| $t-s$ | $s$ | $g$ | $x$ | $\mathrm{Ann}(x)$ | $d_2(x)$ | $d_2(xw_2)$ |
|---|---|---|---|---|---|---|
| 0 | 0 | 0 | $i(1)$ | $(0)$ | 0 | $g^2 \cdot \widetilde{h_2^2}$ |
| 1 | 1 | 0 | $i(h_1)$ | $(g^2, gw_1)$ | 0 | 0 |
| 2 | 1 | 1 | $\widetilde{h_1}$ | $(0)$ | 0 | $g^2 \cdot \widetilde{c_0}$ |
| 2 | 2 | 0 | $i(h_1^2)$ | $(g)$ | 0 | 0 |
| 3 | 1 | 2 | $i(h_2)$ | $(g)$ | 0 | 0 |
| 3 | 2 | 1 | $h_1\widetilde{h_1}$ | $(g)$ | 0 | 0 |
| 4 | 3 | 0 | $h_1^2\widetilde{h_1}$ | $(g)$ | 0 | 0 |
| 6 | 2 | 2 | $i(h_2^2)$ | $(g, w_1)$ | 0 | 0 |
| 7 | 2 | 3 | $\widetilde{h_2^2}$ | $(0)$ | 0 | $g^2 \cdot i(d_0)$ |
| 8 | 3 | 1 | $i(c_0)$ | $(g)$ | 0 | 0 |
| 9 | 3 | 2 | $\widetilde{c_0}$ | $(0)$ | 0 | $g^2 \cdot d_0\widetilde{h_1}$ |
| 9 | 4 | 1 | $i(h_1c_0)$ | $(g)$ | 0 | 0 |
| 10 | 4 | 2 | $h_1\widetilde{c_0}$ | $(g)$ | 0 | 0 |
| 12 | 3 | 3 | $i(\alpha)$ | $(0)$ | $w_1 \cdot i(h_2)$ | $g^2 \cdot e_0\widetilde{h_1} + w_1 \cdot i(h_2w_2)$ |
| 14 | 4 | 3 | $i(d_0)$ | $(0)$ | 0 | $g^2 \cdot d_0\widetilde{h_2^2}$ |
| 15 | 3 | 4 | $i(\beta)$ | $(0)$ | 0 | $g^3 \cdot \widetilde{h_1}$ |
| 16 | 5 | 3 | $d_0\widetilde{h_1}$ | $(0)$ | 0 | $g^2w_1 \cdot i(\beta)$ |
| 17 | 4 | 4 | $i(e_0)$ | $(0)$ | 0 | $g^2 \cdot i(\alpha^2)$ |
| 18 | 4 | 5 | $i(h_2\beta)$ | $(g, w_1)$ | 0 | 0 |
| 18 | 6 | 3 | $\widetilde{h_0^2e_0}$ | $-$ | $w_1 \cdot i(h_1c_0)$ | $g^2w_1 \cdot i(e_0) + w_1 \cdot i(h_1c_0w_2)$ |
| 19 | 5 | 4 | $e_0\widetilde{h_1}$ | $(0)$ | 0 | $g^2 \cdot i(\alpha d_0)$ |
| 21 | 6 | 4 | $d_0\widetilde{h_2^2}$ | $(0)$ | 0 | $g^3w_1 \cdot i(1)$ |
| 24 | 6 | 5 | $i(\alpha^2)$ | $(0)$ | 0 | $g^2 \cdot i(d_0e_0)$ |
| 25 | 5 | 7 | $i(\gamma)$ | $(0)$ | 0 | $g^3 \cdot i(\alpha)$ |
| 26 | 5 | 8 | $\widetilde{\gamma}$ | $(0)$ | 0 | $g^2 \cdot \widetilde{\delta'}$ |
| 26 | 6 | 6 | $i(h_1\gamma)$ | $(g)$ | 0 | 0 |
| 26 | 7 | 5 | $i(\alpha d_0)$ | $(0)$ | 0 | $g^2w_1 \cdot i(\gamma)$ |
| 27 | 6 | 8 | $h_1\widetilde{\gamma}$ | $(g)$ | 0 | 0 |
| 28 | 7 | 6 | $h_1^2\widetilde{\gamma}$ | $(g)$ | 0 | 0 |
| 30 | 6 | 9 | $i(\beta^2)$ | $-$ | 0 | $g^3 \cdot i(e_0)$ |
| 31 | 6 | 10 | $\widetilde{\beta^2}$ | $(0)$ | 0 | $g^2 \cdot \alpha\widetilde{\gamma}$ |
| 31 | 8 | 6 | $i(d_0e_0)$ | $(0)$ | 0 | $g^2w_1 \cdot i(\beta^2)$ |

Table 6.2: $R_0$-module generators of $E_2(tmf/2)$ (cont.)

| $t-s$ | $s$ | $g$ | $x$ | $\mathrm{Ann}(x)$ | $d_2(x)$ | $d_2(xw_2)$ |
|---|---|---|---|---|---|---|
| 32 | 7 | 9 | $i(\delta)$ | $(g)$ | 0 | 0 |
| 32 | 8 | 7 | $\widetilde{d_0e_0}$ | $(0)$ | 0 | $g^2w_1\cdot\widetilde{\beta^2}$ |
| 33 | 7 | 10 | $\widetilde{\delta'}$ | $(0)$ | 0 | $g^2\cdot d_0\widetilde{\gamma}$ |
| 33 | 8 | 8 | $i(h_1\delta)$ | $(g)$ | 0 | 0 |
| 33 | 9 | 7 | $h_1\widetilde{d_0e_0}$ | $(g)$ | 0 | 0 |
| 34 | 8 | 10 | $h_1\widetilde{\delta'}$ | $(g)$ | 0 | 0 |
| 35 | 9 | 9 | $h_1^2\widetilde{\delta'}$ | $(g)$ | 0 | 0 |
| 36 | 7 | 12 | $\widetilde{\beta g}$ | $(0)$ | $h_1^2\widetilde{\delta'}$ | $g^2\cdot e_0\widetilde{\gamma}+h_1^2w_2\widetilde{\delta'}$ |
| 38 | 8 | 12 | $\alpha\widetilde{\gamma}$ | $(0)$ | 0 | $g^2\cdot d_0\widetilde{\beta^2}$ |
| 40 | 9 | 12 | $d_0\widetilde{\gamma}$ | $(0)$ | 0 | $g^2\cdot d_0\widetilde{\delta'}$ |
| 41 | 8 | 14 | $\beta\widetilde{\gamma}$ | $(0)$ | 0 | $g^2\cdot e_0\widetilde{\beta^2}$ |
| 42 | 10 | 12 | $\widetilde{\alpha^2e_0}$ | $(0)$ | $w_1\cdot i(h_1\delta)$ | $g^2w_1\cdot\beta\widetilde{\gamma}+w_1\cdot i(h_1\delta w_2)$ |
| 43 | 9 | 14 | $e_0\widetilde{\gamma}$ | $(0)$ | 0 | $g^2\cdot d_0\widetilde{\beta g}$ |
| 45 | 10 | 14 | $d_0\widetilde{\beta^2}$ | $(0)$ | 0 | $g^3\cdot\widetilde{d_0e_0}$ |
| 47 | 11 | 14 | $d_0\widetilde{\delta'}$ | $(0)$ | 0 | $g^3w_1\cdot\widetilde{\gamma}$ |
| 48 | 10 | 16 | $e_0\widetilde{\beta^2}$ | $(0)$ | 0 | $g^2\cdot\alpha^2\widetilde{\beta^2}$ |
| 50 | 11 | 16 | $d_0\widetilde{\beta g}$ | $(0)$ | 0 | $g^2\cdot d_0e_0\widetilde{\gamma}$ |
| 55 | 12 | 18 | $\alpha^2\widetilde{\beta^2}$ | $(0)$ | 0 | $g^3\cdot\widetilde{\alpha^2e_0}$ |
| 57 | 13 | 18 | $d_0e_0\widetilde{\gamma}$ | $(0)$ | 0 | $g^3w_1\cdot\widetilde{\beta g}$ |

Table 6.3: The non-cyclic $R_0$-module summand in $E_2(tmf/2)$

| $\langle x_1, x_2\rangle$ |
|---|
| $\langle\widetilde{h_0^2e_0}, i(\beta^2)\rangle \cong \dfrac{\Sigma^{6,24}R_0\oplus\Sigma^{6,36}R_0}{\langle(g,w_1)\rangle}$ |

## 6.2. The $d_2$-differentials for $tmf/2$

To determine the $d_2$-differentials for $tmf/2$, we use the following preliminary estimate. See Figures 1.25, 1.26 and 4.1.

LEMMA 6.1. *If* $d_3(\widetilde{d_0e_0}) = i(\beta w_1^2)$ *then* $d_3(\beta\widetilde{\gamma}) = i((\delta+\alpha g)w_1) = i(\delta' w_1)$. *Otherwise,* $d_3(\widetilde{d_0e_0}) = 0$ *and* $d_3(\beta\widetilde{\gamma}) = i(\delta w_1)$.

PROOF. From $w_1 \cdot \beta\widetilde{\gamma} = 12_{14} = e_0 \cdot \widetilde{d_0e_0}$ with $d_3(e_0) = c_0w_1$ and $d_3(w_1) = 0$ we get $w_1 \cdot d_3(\beta\widetilde{\gamma}) = c_0w_1 \cdot \widetilde{d_0e_0} + e_0 \cdot d_3(\widetilde{d_0e_0})$. Here $c_0w_1 \cdot \widetilde{d_0e_0} = 15_9 = i(\delta w_1^2)$. Note that $E_2 = E_3$ in the bidegree generated by $15_8$ and $15_9$.

If $d_3(\widetilde{d_0e_0}) = i(\beta w_1^2)$ then $e_0 \cdot d_3(\widetilde{d_0e_0}) = i(\beta e_0 w_1^2) = i(\alpha g w_1^2) = 15_8$, and $w_1 \cdot d_3(\beta\widetilde{\gamma}) = i(\delta w_1^2) + i(\alpha g w_1^2) = 15_9 + 15_8$. This implies $d_3(\beta\widetilde{\gamma}) = i(\delta w_1) + i(\alpha g w_1) = 11_9 + 11_8$.

Otherwise, $d_3(\widetilde{d_0e_0}) = 0$, so $e_0 \cdot d_3(\widetilde{d_0e_0}) = 0$ and $w_1 \cdot d_3(\beta\widetilde{\gamma}) = i(\delta w_1^2) = 15_9$. This implies $d_3(\beta\widetilde{\gamma}) = i(\delta w_1) = 11_9$. □

THEOREM 6.2. *The $d_2$-differential in $E_2(tmf/2)$ is $R_1$-linear. Its values on a set of $E_2(tmf)$-module generators are listed in Table* 6.1, *and its values on a set of $R_1$-module generators are listed in Table* 6.2.

PROOF. The classes $g$, $w_1$ and $w_2^2$ are $d_2$-cycles in $E_2(tmf)$, so the Leibniz rule implies that multiplication by each of these elements commutes with the $d_2$-differential in $E_2(tmf/2)$. Hence $d_2$ is $R_1$-linear.

The $d_2$-differentials on the $E_2(tmf)$-module generators $i(1)$, $\widetilde{h_1}$, $\widetilde{h_2^2}$, $\widetilde{c_0}$, $\widetilde{\gamma}$, $\widetilde{\beta^2}$, $\widetilde{d_0e_0}$ and $\widetilde{\delta'}$ are zero because the target groups are trivial.

The $d_3$-differential $d_3(e_0) = c_0w_1$ in $E_3(tmf)$ (see Table 5.2) implies $d_3(i(e_0)) = i(c_0w_1)$ in $E_3 = E_3(tmf/2)$, by naturality with respect to $i$. Here $i(h_1e_0) = 0$, so $i(h_1c_0w_1) = 0$ at $E_3$. Since $i(h_1c_0w_1) \neq 0$ at $E_2$, we must have $d_2(x) = i(h_1c_0w_1)$ for some nonzero $x$. The only possibility is $x = \widetilde{h_0^2e_0}$.

The $d_2$-boundary $d_2(\widetilde{\beta g})$ maps by $j$ to $d_2(\beta g) = h_0d_0g \neq 0$ in $E_2(tmf)$, using Table 5.1. Hence $d_2(\widetilde{\beta g})$ is nonzero, and $h_1^2\widetilde{\delta'}$ is the only possible value.

To determine $d_2(\widetilde{\alpha^2e_0})$ we use Lemma 6.1, showing that $d_3(\beta\widetilde{\gamma}) = i(\delta'w_1)$ or $i(\delta w_1)$. From $h_1 \cdot \beta\widetilde{\gamma} = 0$ and $h_1\delta' = h_1\delta$ we deduce that $0 = d_3(h_1 \cdot \beta\widetilde{\gamma}) = h_1 \cdot d_3(\beta\widetilde{\gamma}) = i(h_1\delta w_1)$ at $E_3$. But $i(h_1\delta w_1) \neq 0$ at $E_2$, so $i(h_1\delta w_1) = d_2(y)$ for some nonzero $y$. The only possibility is $y = \widetilde{\alpha^2e_0}$.

We use Table 5.1 and the Leibniz rule to calculate $d_2(x)$ for $x$ ranging through the $R_0$-module generators for $E_2(tmf/2)$ listed in Table 6.2. The less obvious cases are:

- $d_2(\alpha \cdot \widetilde{\gamma}) = h_2w_1 \cdot \widetilde{\gamma} = 0$
- $d_2(\beta \cdot \widetilde{\gamma}) = h_0d_0 \cdot \widetilde{\gamma} = 0$
- $d_2(d_0 \cdot \widetilde{\beta g}) = d_0 \cdot h_1^2\widetilde{\delta'} = 0$.

The vanishing of these products is readily seen in Figure 4.1.

To finish the proof we calculate $d_2(w_2 \cdot x) = d_2(w_2) \cdot x + w_2 \cdot d_2(x) = \alpha\beta g \cdot x + w_2 \cdot d_2(x)$, for the same generators $x$, so that $xw_2$ ranges through the remaining $R_1$-module generators for $E_2(tmf/2)$. This is easy when $g \in \mathrm{Ann}(x)$. In the remaining cases we use `ext` to calculate the product $\alpha\beta g \cdot x$ and to present it in terms of our $R_1$-module generators for $E_2(tmf/2)$:

- $\alpha\beta g \cdot i(1) = 10_{15} = g^2 \cdot \widetilde{h_2^2}$
- $\alpha\beta g \cdot i(h_1) = 0$
- $\alpha\beta g \cdot \widetilde{h_1} = 11_{15} = g^2 \cdot \widetilde{c_0}$
- $\alpha\beta g \cdot \widetilde{h_2^2} = 12_{17} = g^2 \cdot i(d_0)$
- $\alpha\beta g \cdot \widetilde{c_0} = 13_{17} = g^2 \cdot \widetilde{d_0h_1}$
- $\alpha\beta g \cdot i(\alpha) = 13_{21} = g^2 \cdot \widetilde{e_0h_1}$

- $\alpha\beta g \cdot i(d_0) = 14_{21} = g^2 \cdot d_0\widetilde{h_2^2}$
- $\alpha\beta g \cdot i(\beta) = 13_{24} = g^3 \cdot \widetilde{h_1}$
- $\alpha\beta g \cdot d_0\widetilde{h_1} = 15_{20} = g^2 w_1 \cdot i(\beta)$
- $\alpha\beta g \cdot i(e_0) = 14_{24} = g^2 \cdot i(\alpha^2)$
- $\alpha\beta g \cdot \widetilde{h_0^2 e_0} = 16_{20} = g^2 w_1 \cdot i(e_0)$
- $\alpha\beta g \cdot e_0\widetilde{h_1} = 15_{24} = g^2 \cdot i(\alpha d_0)$
- $\alpha\beta g \cdot d_0\widetilde{h_2^2} = 16_{24} = g^3 w_1 \cdot i(1)$
- $\alpha\beta g \cdot i(\alpha^2) = 16_{27} = g^2 \cdot i(d_0 e_0)$
- $\alpha\beta g \cdot i(\gamma) = 15_{30} = g^3 \cdot i(\alpha)$
- $\alpha\beta g \cdot \widetilde{\gamma} = 15_{31} = g^2 \cdot \widetilde{\delta}'$
- $\alpha\beta g \cdot i(\alpha d_0) = 17_{27} = g^2 w_1 \cdot i(\gamma)$
- $\alpha\beta g \cdot i(\beta^2) = 16_{33} = g^3 \cdot i(e_0)$
- $\alpha\beta g \cdot \widetilde{\beta^2} = 16_{34} = g^2 \cdot \alpha\widetilde{\gamma}$
- $\alpha\beta g \cdot i(d_0 e_0) = 18_{30} = g^2 w_1 \cdot i(\beta^2)$
- $\alpha\beta g \cdot \widetilde{d_0 e_0} = 18_{31} = g^2 w_1 \cdot \widetilde{\beta^2}$
- $\alpha\beta g \cdot \widetilde{\delta}' = 17_{34} = g^2 \cdot d_0\widetilde{\gamma}$
- $\alpha\beta g \cdot \widetilde{\beta g} = 17_{39} = g^2 \cdot e_0\widetilde{\gamma}$
- $\alpha\beta g \cdot \alpha\widetilde{\gamma} = 18_{39} = g^2 \cdot d_0\widetilde{\beta^2}$
- $\alpha\beta g \cdot d_0\widetilde{\gamma} = 19_{38} = g^2 \cdot d_0\widetilde{\delta}'$
- $\alpha\beta g \cdot \beta\widetilde{\gamma} = 18_{43} = g^2 \cdot e_0\widetilde{\beta^2}$
- $\alpha\beta g \cdot \widetilde{\alpha^2 e_0} = 20_{38} = g^2 w_1 \cdot \beta\widetilde{\gamma}$
- $\alpha\beta g \cdot e_0\widetilde{\gamma} = 19_{43} = g^2 \cdot d_0\widetilde{\beta g}$
- $\alpha\beta g \cdot d_0\widetilde{\beta^2} = 20_{43} = g^3 \cdot \widetilde{d_0 e_0}$
- $\alpha\beta g \cdot d_0\widetilde{\delta}' = 21_{43} = g^3 w_1 \cdot \widetilde{\gamma}$
- $\alpha\beta g \cdot e_0\widetilde{\beta^2} = 20_{47} = g^2 \cdot \alpha^2\widetilde{\beta^2}$
- $\alpha\beta g \cdot d_0\widetilde{\beta g} = 21_{47} = g^2 \cdot d_0 e_0\widetilde{\gamma}$
- $\alpha\beta g \cdot \alpha^2\widetilde{\beta^2} = 22_{51} = g^3 \cdot \widetilde{\alpha^2 e_0}$
- $\alpha\beta g \cdot d_0 e_0\widetilde{\gamma} = 23_{51} = g^3 w_1 \cdot \widetilde{\beta g}$. □

REMARK 6.3. To use `ext` to assist in calculating the products $\alpha\beta g \cdot x$ for $x \in E_2(tmf/2)$, use `cocycle tmfC2 0 0`, ..., `cocycle tmfC2 13 18`, `dolifts 0 40 maps` and `collect maps all`. The nonzero products with $\alpha\beta g = 10_{18}$ then appear as lines containing `(10 18 F2)` in the file `all`. If the product is a $g^2$-multiple, there will also appear a line containing `(8 18 F2)` in the same block, since $g^2 = 8_{18}$ in the minimal $A(2)$-module resolution for $\mathbb{F}_2$. Similarly, $g^2 w_1$-multiples appear with `(12 22 F2)`, $g^3$-multiples appear with `(12 29 F2)`, and $g^3 w_1$-multiples appear with `(16 35 F2)`.

## 6.3. The $d_3$-differentials for *tmf*/2

It is now an elementary matter to compute $E_3(tmf/2)$. This is done in Appendix B.1 and the results are recorded in Tables 6.4 and 6.5. In the process, we use the relations $i(e_0 g^2) = 12_{20} = \beta^2 g\widetilde{h_2^2}$, $i(h_1 c_0 w_2) = 12_{21} = h_1^2 w_2\widetilde{h_2^2}$ and $\gamma^2 = \beta^2 g + h_1^2 w_2$ to shorten the name of the generator in bidegree $(t-s, s) = (57, 12)$ from $i(h_1 c_0 w_2 + e_0 g^2)$ to $\gamma^2\widetilde{h_2^2}$. We also make the name change $g\widetilde{\beta g} = 11_{21} = \beta^2\widetilde{\gamma}$

in bidegree $(56, 11)$, as the latter decomposition is a more useful description of this element.

Table 6.4: $R_1$-module generators of $E_3(tmf/2)$

| $t-s$ | $s$ | $g$ | $x$ | $\mathrm{Ann}(x)$ | $d_3(x)$ | $d_3(xw_2^2)$ |
|---|---|---|---|---|---|---|
| 0 | 0 | 0 | $i(1)$ | $(g^3w_1)$ | 0 | $g^4 \cdot i(\beta)$ |
| 1 | 1 | 0 | $i(h_1)$ | $(g^2, gw_1)$ | 0 | 0 |
| 2 | 1 | 1 | $\widetilde{h_1}$ | $(g^3)$ | 0 | 0 |
| 2 | 2 | 0 | $i(h_1^2)$ | $(g)$ | 0 | 0 |
| 3 | 1 | 2 | $i(h_2)$ | $(g, w_1)$ | 0 | 0 |
| 3 | 2 | 1 | $h_1\widetilde{h_1}$ | $(g)$ | 0 | 0 |
| 4 | 3 | 0 | $h_1^2\widetilde{h_1}$ | $(g)$ | 0 | 0 |
| 6 | 2 | 2 | $i(h_2^2)$ | $(g, w_1)$ | 0 | 0 |
| 7 | 2 | 3 | $\widetilde{h_2^2}$ | $(g^2)$ | 0 | 0 |
| 8 | 3 | 1 | $i(c_0)$ | $(g)$ | 0 | 0 |
| 9 | 3 | 2 | $\widetilde{c_0}$ | $(g^2)$ | 0 | 0 |
| 9 | 4 | 1 | $i(h_1c_0)$ | $(g, w_1)$ | 0 | 0 |
| 10 | 4 | 2 | $h_1\widetilde{c_0}$ | $(g)$ | 0 | 0 |
| 14 | 4 | 3 | $i(d_0)$ | $(g^2)$ | 0 | 0 |
| 15 | 3 | 4 | $i(\beta)$ | $(g^2w_1)$ | 0 | $g^4 \cdot i(\beta^2)$ |
| 16 | 5 | 3 | $d_0\widetilde{h_1}$ | $(g^2)$ | 0 | 0 |
| 17 | 4 | 4 | $i(e_0)$ | $(g^3)$ | $w_1 \cdot i(c_0)$ | $w_1 \cdot i(c_0w_2^2)$ |
| 18 | 4 | 5 | $i(h_2\beta)$ | $(g, w_1)$ | 0 | 0 |
| 19 | 5 | 4 | $e_0\widetilde{h_1}$ | – | $w_1 \cdot h_1\widetilde{c_0}$ | $w_1 \cdot h_1w_2^2\widetilde{c_0}$ |
| 21 | 6 | 4 | $d_0\widetilde{h_2^2}$ | $(g^2)$ | 0 | 0 |
| 24 | 6 | 5 | $i(\alpha^2)$ | $(g^2)$ | 0 | 0 |
| 25 | 5 | 7 | $i(\gamma)$ | $(g^2w_1)$ | 0 | $g^6 \cdot i(1)$ |
| 26 | 5 | 8 | $\widetilde{\gamma}$ | $(g^3w_1)$ | 0 | $g^4 \cdot \beta\widetilde{\gamma}$ |
| 26 | 6 | 6 | $i(h_1\gamma)$ | $(g)$ | 0 | 0 |
| 26 | 7 | 5 | $i(\alpha d_0)$ | $(g^2)$ | 0 | 0 |
| 27 | 6 | 8 | $h_1\widetilde{\gamma}$ | $(g)$ | 0 | 0 |
| 28 | 7 | 6 | $h_1^2\widetilde{\gamma}$ | $(g)$ | 0 | 0 |
| 30 | 6 | 9 | $i(\beta^2)$ | $(g^2w_1)$ | 0 | $g^5 \cdot i(\gamma)$ |
| 31 | 6 | 10 | $\widetilde{\beta^2}$ | $(g^2w_1)$ | $gw_1 \cdot \widetilde{h_1}$ | $gw_1 \cdot w_2^2\widetilde{h_1} + g^5 \cdot \widetilde{\gamma}$ |
| 31 | 8 | 6 | $i(d_0e_0)$ | $(g^2)$ | 0 | 0 |

Table 6.4: $R_1$-module generators of $E_3(tmf/2)$ (cont.)

| $t-s$ | $s$ | $g$ | $x$ | $\mathrm{Ann}(x)$ | $d_3(x)$ | $d_3(xw_2^2)$ |
|---|---|---|---|---|---|---|
| 32 | 7 | 8 | $i(\alpha g)$ | $(g^2)$ | 0 | 0 |
| 32 | 7 | 9 | $i(\delta)$ | $(g)$ | 0 | 0 |
| 32 | 8 | 7 | $\widetilde{d_0 e_0}$ | $(g^3)$ | $w_1^2 \cdot i(\beta)$ | $w_1^2 \cdot i(\beta w_2^2)$ |
| 33 | 7 | 10 | $\widetilde{\delta'}$ | $(g^2)$ | 0 | 0 |
| 33 | 8 | 8 | $i(h_1\delta)$ | $(g, w_1)$ | 0 | 0 |
| 33 | 9 | 7 | $h_1\widetilde{d_0 e_0}$ | $(g)$ | 0 | 0 |
| 34 | 8 | 10 | $h_1\widetilde{\delta'}$ | $(g)$ | 0 | 0 |
| 38 | 8 | 12 | $\alpha\widetilde{\gamma}$ | $(g^2)$ | $gw_1 \cdot \widetilde{c_0}$ | $gw_1 \cdot w_2^2\widetilde{c_0}$ |
| 40 | 9 | 12 | $d_0\widetilde{\gamma}$ | $(g^2)$ | 0 | 0 |
| 41 | 8 | 14 | $\beta\widetilde{\gamma}$ | $-$ | $w_1 \cdot i(\delta')$ | $w_1 \cdot i(\delta' w_2^2) + g^4 \cdot \beta^2\widetilde{\gamma}$ |
| 43 | 9 | 14 | $e_0\widetilde{\gamma}$ | $(g^3)$ | $w_1 \cdot h_1\widetilde{\delta'}$ | $w_1 \cdot h_1 w_2^2\widetilde{\delta'}$ |
| 45 | 10 | 14 | $d_0\widetilde{\beta^2}$ | $(g^2)$ | $gw_1 \cdot d_0\widetilde{h_1}$ | $gw_1 \cdot d_0 w_2^2\widetilde{h_1}$ |
| 47 | 11 | 14 | $d_0\widetilde{\delta'}$ | $(g^2)$ | 0 | 0 |
| 48 | 10 | 16 | $e_0\widetilde{\beta^2}$ | $(g^2)$ | $gw_1 \cdot e_0\widetilde{h_1}$ | $gw_1 \cdot e_0 w_2^2\widetilde{h_1}$ |
| 49 | 9 | 17 | $i(h_1 w_2)$ | $(g^2, gw_1)$ | $g^2 w_1 \cdot i(1)$ | $g^2 w_1 \cdot i(w_2^2)$ |
| 50 | 10 | 18 | $i(h_1^2 w_2)$ | $(g)$ | 0 | 0 |
| 50 | 11 | 16 | $d_0\widetilde{\beta g}$ | $(g^2)$ | 0 | 0 |
| 51 | 9 | 19 | $i(h_2 w_2)$ | $-$ | 0 | 0 |
| 51 | 10 | 20 | $h_1 w_2\widetilde{h_1}$ | $(g)$ | $g^2 w_1 \cdot \widetilde{h_1}$ | $g^2 w_1 \cdot w_2^2\widetilde{h_1}$ |
| 52 | 11 | 18 | $h_1^2 w_2\widetilde{h_1}$ | $(g)$ | 0 | 0 |
| 54 | 10 | 21 | $i(h_2^2 w_2)$ | $(g, w_1)$ | 0 | 0 |
| 55 | 12 | 18 | $\alpha^2\widetilde{\beta^2}$ | $(g^2)$ | $gw_1 \cdot i(\alpha d_0)$ | $gw_1 \cdot i(\alpha d_0 w_2^2)$ |
| 56 | 11 | 21 | $\beta^2\widetilde{\gamma}$ | $(g^2 w_1)$ | 0 | $g^6 \cdot \widetilde{\beta^2}$ |
| 56 | 11 | 22 | $i(c_0 w_2)$ | $(g)$ | 0 | 0 |
| 57 | 12 | $20+21$ | $\gamma^2\widetilde{h_2^2}$ | $(g, w_1)$ | 0 | 0 |
| 57 | 13 | 18 | $d_0 e_0\widetilde{\gamma}$ | $(g^2)$ | 0 | 0 |
| 58 | 12 | 23 | $h_1 w_2\widetilde{c_0}$ | $(g)$ | 0 | 0 |
| 62 | 14 | 22 | $\widetilde{g\alpha^2 e_0}$ | $(g^2)$ | $gw_1^2 \cdot i(\gamma)$ | $gw_1^2 \cdot i(\gamma w_2^2)$ |
| 66 | 12 | 28 | $i(h_2\beta w_2)$ | $(g, w_1)$ | 0 | 0 |
| 74 | 14 | 33 | $i(h_1\gamma w_2)$ | $(g)$ | 0 | 0 |
| 75 | 14 | 35 | $h_1 w_2\widetilde{\gamma}$ | $(g)$ | $g^2 w_1 \cdot \widetilde{\gamma}$ | $g^2 w_1 \cdot w_2^2\widetilde{\gamma}$ |
| 76 | 15 | 35 | $h_1^2 w_2\widetilde{\gamma}$ | $(g)$ | 0 | 0 |

Table 6.4: $R_1$-module generators of $E_3(tmf/2)$ (cont.)

| $t-s$ | $s$ | $g$ | $x$ | $\mathrm{Ann}(x)$ | $d_3(x)$ | $d_3(xw_2^2)$ |
|---|---|---|---|---|---|---|
| 80 | 15 | 38 | $i(\delta w_2)$ | $(g)$ | 0 | 0 |
| 81 | 16 | 39 | $i(h_1\delta w_2)$ | – | 0 | 0 |
| 81 | 17 | 36 | $h_1w_2\widetilde{d_0e_0}$ | $(g)$ | $g^2w_1 \cdot \widetilde{d_0e_0}$ | $g^2w_1 \cdot \widetilde{w_2^2d_0e_0}$ |
| 82 | 16 | 41 | $h_1w_2\widetilde{\delta'}$ | $(g)$ | 0 | 0 |

Table 6.5: The non-cyclic $R_1$-module summands in $E_3(tmf/2)$

| $\langle x_1, x_2\rangle$ |
|---|
| $\langle e_0\widetilde{h_1}, i(h_2w_2)\rangle \cong \dfrac{\Sigma^{5,24}R_1 \oplus \Sigma^{9,60}R_1}{\langle (g^2, w_1), (0, g)\rangle}$ |
| $\langle \beta\widetilde{\gamma}, i(h_1\delta w_2)\rangle \cong \dfrac{\Sigma^{8,49}R_1 \oplus \Sigma^{16,97}R_1}{\langle (g^2w_1, w_1), (0, g)\rangle}$ |

PROPOSITION 6.4. *The eleven classes listed in Table* 6.6 *generate* $E_3(tmf/2)$ *as a module over* $E_3(tmf)$.

PROOF. Inspection of Tables 5.2 and 6.4 easily shows that most of the $R_1$-module generators of $E_3(tmf/2)$ are $E_3(tmf)$-multiples of the classes in Table 6.6. The less evident cases follow from the relations

$$i(\alpha d_0) = 7_5 = e_0 \cdot \widetilde{c_0}$$
$$\widetilde{d_0\beta g} = 11_{16} = e_0 \cdot \widetilde{\delta'}$$
$$\widetilde{g\alpha^2 e_0} = 14_{22} = d_0e_0 \cdot \widetilde{\beta^2}\,,$$

which we verify by calculating the relevant Yoneda products using `ext`. □

Table 6.6: $E_3(tmf)$-module generators of $E_3(tmf/2)$

| $t-s$ | $s$ | $g$ | $x$ | $d_3(x)$ |
|---|---|---|---|---|
| 0 | 0 | 0 | $i(1)$ | 0 |
| 2 | 1 | 1 | $\widetilde{h_1}$ | 0 |
| 7 | 2 | 3 | $\widetilde{h_2^2}$ | 0 |
| 9 | 3 | 2 | $\widetilde{c_0}$ | 0 |
| 15 | 3 | 4 | $i(\beta)$ | 0 |
| 26 | 5 | 8 | $\widetilde{\gamma}$ | 0 |

Table 6.6: $E_3(tmf)$-module generators for $E_3(tmf/2)$ (cont.)

| $t-s$ | $s$ | $g$ | $x$ | $d_3(x)$ |
|---|---|---|---|---|
| 31 | 6 | 10 | $\widetilde{\beta^2}$ | $gw_1\widetilde{h_1}$ |
| 32 | 8 | 7 | $\widetilde{d_0e_0}$ | $i(\beta w_1^2)$ |
| 33 | 7 | 10 | $\widetilde{\delta'}$ | 0 |
| 38 | 8 | 12 | $\alpha\widetilde{\gamma}$ | $gw_1\widetilde{c_0}$ |
| 41 | 8 | 14 | $\beta\widetilde{\gamma}$ | $i(\delta' w_1)$ |

PROPOSITION 6.5. *The $d_3$-differentials on the $E_3(tmf)$-module generators of $E_3(tmf/2)$ are as listed in Table* 6.6.

PROOF. The target groups of $d_3$ on $i(1)$, $\widetilde{h_1}$, $\widetilde{h_2^2}$, $\widetilde{c_0}$, $i(\beta)$ are trivial.

Since $d_3 \circ d_3 = 0$ and $d_3(i(e_0w_1)) = i(c_0w_1^2) \neq 0$, we cannot have $d_3(\widetilde{\gamma}) = i(e_0w_1)$. The only alternative is $d_3(\widetilde{\gamma}) = 0$.

The $d_3$-boundary $d_3(\widetilde{\beta^2})$ maps by $j$ to $d_3(\beta^2) = h_1gw_1 \neq 0$ in $E_3(tmf)$, so $d_3(\widetilde{\beta^2}) \neq 0$ and $gw_1\widetilde{h_1}$ is the only possible value.

To determine $d_3(\widetilde{d_0e_0})$ we use the relation $\beta^2 \cdot \widetilde{d_0e_0} = 14_{22} = d_0e_0\widetilde{\beta^2}$. We find that $d_3(\beta^2\widetilde{d_0e_0}) = h_1gw_1 \cdot \widetilde{d_0e_0} + \beta^2 \cdot d_3(\widetilde{d_0e_0}) = \beta^2 d_3(\widetilde{d_0e_0})$ since $h_1$ annihilates $E_2(tmf/2)$ in bidegree $(t-s, s) = (60, 16)$. This must equal $d_3(d_0e_0\widetilde{\beta^2}) = d_0e_0 \cdot gw_1\widetilde{h_1} = 17_{15} = gw_1^2 \cdot i(\gamma) \neq 0$. Therefore $d_3(\widetilde{d_0e_0})$ is nonzero, and $i(\beta w_1^2)$ is the only possible value.

Lemma 6.1 then shows that $d_3(\beta\widetilde{\gamma}) = i(\delta' w_1)$.

From the relation $e_0 \cdot \widetilde{\delta'} = 11_{16} = \alpha^2 \cdot \widetilde{\gamma}$ and the differentials $d_3(e_0) = c_0w_1$, $d_3(\alpha^2) = h_1d_0w_1$ and $d_3(\widetilde{\gamma}) = 0$ we obtain $c_0w_1 \cdot \widetilde{\delta'} + e_0 \cdot d_3(\widetilde{\delta'}) = h_1d_0w_1 \cdot \widetilde{\gamma}$. Here $c_0w_1 \cdot \widetilde{\delta'} = 0$ and $h_1d_0w_1 \cdot \widetilde{\gamma} = 0$, so $e_0 \cdot d_3(\widetilde{\delta'}) = 0$. On the other hand, $e_0 \cdot i(\alpha^2w_1) = i(\alpha^2e_0w_1) = 14_{11} \neq 0$ cannot be a $d_2$-boundary, hence remains nonzero at $E_3(tmf/2)$. Thus $d_3(\widetilde{\delta'}) \neq i(\alpha^2w_1)$, and 0 is the only possible value.

From the relation $e_0 \cdot \alpha\widetilde{\gamma} = 12_{18} = d_0 \cdot \beta\widetilde{\gamma}$ and the differentials $d_3(e_0) = c_0w_1$, $d_3(d_0) = 0$ and $d_3(\beta\widetilde{\gamma}) = i(\delta' w_1)$ we deduce that $c_0w_1 \cdot \alpha\widetilde{\gamma} + e_0 \cdot d_3(\alpha\widetilde{\gamma}) = d_0 \cdot i(\delta' w_1) = i(\alpha d_0 g w_1) = 15_{13}$. Here $c_0w_1 \cdot \alpha\widetilde{\gamma} = 0$, so $e_0 \cdot d_3(\alpha\widetilde{\gamma}) = 15_{13} \neq 0$ at $E_2(tmf/2)$. This class is not a $d_2$-boundary, hence remains nonzero at $E_3(tmf/2)$, so $d_3(\alpha\widetilde{\gamma}) \neq 0$. The only possible value is $gw_1\widetilde{c_0}$. □

THEOREM 6.6. *The $d_3$-differential in $E_3(tmf/2)$ is $R_2$-linear. Its values on a set of $R_2$-module generators are listed in Table* 6.4.

PROOF. The classes $g$, $w_1$ and $w_2^4$ are $d_3$-cycles in $E_3(tmf)$, so the Leibniz rule implies that multiplication by each of these elements commutes with the $d_3$-differential in $E_3(tmf/2)$. Hence $d_3$ is $R_2$-linear.

The $d_3$-differential on the $R_1$-module generators $x$ in Table 6.4 is given by the Leibniz rule applied to the (implicit and explicit) factorizations in the proof of Proposition 6.4, and the $d_3$-differentials from Tables 5.2 and 6.6. The less obvious cases are:

- $d_3(e_0 \cdot \widetilde{h_1}) = c_0w_1 \cdot \widetilde{h_1} = 8_2 = w_1 \cdot h_1\widetilde{c_0}$
- $d_3(i(\alpha^2)) = i(h_1d_0w_1) = 0$

- $d_3(i(\alpha d_0)) = d_3(e_0 \cdot \widetilde{c_0}) = c_0 w_1 \cdot \widetilde{c_0} = 0$
- $d_3(i(\beta^2)) = i(h_1 g w_1) = 0$
- $d_3(h_1 \cdot \widetilde{d_0 e_0}) = h_1 \cdot i(\beta w_1^2) = 0$
- $d_3(e_0 \cdot \widetilde{\gamma}) = c_0 w_1 \cdot \widetilde{\gamma} = 12_{10} = h_1 w_1 \widetilde{\delta'}$
- $d_3(e_0 \cdot \widetilde{\beta^2}) = c_0 w_1 \cdot \widetilde{\beta^2} + e_0 \cdot g w_1 \widetilde{h_1} = 0 + e_0 g w_1 \widetilde{h_1}$
- $d_3(\widetilde{d_0 \beta g}) = d_3(e_0 \cdot \widetilde{\delta'}) = c_0 w_1 \cdot \widetilde{\delta'} = 0$
- $d_3(\alpha^2 \cdot \widetilde{\beta^2}) = h_1 d_0 w_1 \cdot \widetilde{\beta^2} + \alpha^2 \cdot g w_1 \widetilde{h_1} = 0 + 15_{13} = g w_1 \cdot i(\alpha d_0)$
- $d_3(\beta^2 \cdot \widetilde{\gamma}) = h_1 g w_1 \cdot \widetilde{\gamma} = 0$
- $d_3(h_1 w_2 \cdot \widetilde{c_0}) = g^2 w_1 \cdot \widetilde{c_0} = 0$ at $E_3$
- $d_3(\widetilde{g \alpha^2 e_0}) = d_3(d_0 e_0 \cdot \widetilde{\beta^2}) = d_0 e_0 \cdot g w_1 \widetilde{h_1} = 17_{15} = g w_1^2 \cdot i(\gamma)$
- $d_3(i(h_1 \gamma w_2)) = g^2 w_1 \cdot i(\gamma) = 0$ at $E_3$
- $d_3(h_1 w_2 \cdot \widetilde{d_0 e_0}) = g^2 w_1 \cdot \widetilde{d_0 e_0} + h_1 w_2 \cdot i(\beta w_1^2) = g^2 w_1 \widetilde{d_0 e_0}$
- $d_3(h_1 w_2 \cdot \widetilde{\delta'}) = g^2 w_1 \cdot \widetilde{\delta'} = 0$ at $E_3$.

It remains to determine $d_3(w_2^2 \cdot x) = d_3(w_2^2) \cdot x + w_2^2 \cdot d_3(x) = \beta g^4 \cdot x + w_2^2 \cdot d_3(x)$ for the same generators $x$. This is easy when $g^4 \in \mathrm{Ann}(x)$. In the other cases we use `ext` to calculate the product $\beta g^4 \cdot x$ and to present it in terms of our $R_2$-module generators for $E_3(tmf/2)$:

- $\beta g^4 \cdot i(\gamma) = 24_{73} = g^6 \cdot i(1)$
- $\beta g^4 \cdot i(\beta^2) = 25_{78} = g^5 \cdot i(\gamma)$
- $\beta g^4 \cdot \widetilde{\beta^2} = 25_{80} = g^5 \cdot \widetilde{\gamma}$
- $\beta g^4 \cdot \beta^2 \widetilde{\gamma} = 30_{110} = g^6 \cdot \widetilde{\beta^2}$. □

REMARK 6.7. To calculate the products $\beta g^4 \cdot x$ with `ext`, use `cocycle`, `dolifts` and `collect` as in Remark 6.3. The nonzero products with $\beta g^4 = 19_{56}$ then appear as lines containing `(19 56 F2)` in the file `all`. If the product is a $g^5$-multiple, there will also appear a line containing `(20 67 F2)` in the same block, since $g^5 = 20_{67}$ in the minimal $A(2)$-module resolution for $\mathbb{F}_2$. Similarly, $g^6$-multiples appear with `(24 90 F2)`.

## 6.4. The $d_4$-differentials for $tmf/2$

The calculation of $E_4(tmf/2)$ as the homology of $(E_3(tmf/2), d_3)$ is carried out in Appendix B.2 and the results are recorded in Tables 6.7 and 6.8.

At this stage of the calculation it is convenient to change generators in order to simplify the calculation of the next stage of the spectral sequence and to give more informative or convenient names for some of the elements. This happens in bidegrees $(58, 12)$ and $(154, 28)$, where we change basis and also change names of elements. In bidegrees $(51, 10)$ and $(81, 16)$ we simply change the names of the generators.

In bidegree $(t-s, s) = (58, 12)$, we will replace $\alpha g \widetilde{\gamma} = 12_{22}$ by $\delta' \widetilde{\gamma} = 12_{22} + 12_{23} = (\alpha g + \delta)\widetilde{\gamma}$ so that the differential $d_4(\alpha^2 g \widetilde{\beta^2}) = w_1^2 \cdot \delta' \widetilde{\gamma}$ is simply a map between cyclic summands. We treat its $w_2^2$-multiple in $(t-s, s) = (154, 28)$ similarly: we replace $\alpha g w_2^2 \widetilde{\gamma} = 28_{116}$ by $\delta' w_2^2 \widetilde{\gamma} = 28_{116} + 28_{117} = (\alpha g + \delta) w_2^2 \widetilde{\gamma}$.

It is then also convenient to change the names of $h_1 w_2 \widetilde{c_0} = 12_{23} = \delta \widetilde{\gamma}$ and $h_1 w_2^3 \widetilde{c_0} = 28_{117} = \delta w_2^2 \widetilde{\gamma}$.

As in Appendix B.2, in bidegree $(t-s,s)=(51,10)$ we use the relation $\gamma\widetilde{\gamma} = g\widetilde{\beta^2} + h_1 w_2 \widetilde{h_1}$ to replace the latter expression. In bidegree $(t-s,s)=(81,16)$ we use the relation $\gamma^2\widetilde{\beta^2} = \beta g^2\widetilde{\gamma} + i(h_1\delta w_2)$ to replace the latter sum.

Table 6.7: $R_2$-module generators of $E_4(tmf/2)$

| $t-s$ | $s$ | $g$ | $x$ | $\mathrm{Ann}(x)$ | $d_4(x)$ |
|---|---|---|---|---|---|
| 0 | 0 | 0 | $i(1)$ | $(g^6, g^2w_1)$ | 0 |
| 1 | 1 | 0 | $i(h_1)$ | $(g^2, gw_1)$ | 0 |
| 2 | 1 | 1 | $\widetilde{h_1}$ | $(g^3, gw_1)$ | 0 |
| 2 | 2 | 0 | $i(h_1^2)$ | $(g)$ | 0 |
| 3 | 1 | 2 | $i(h_2)$ | $(g, w_1)$ | 0 |
| 3 | 2 | 1 | $h_1\widetilde{h_1}$ | $(g)$ | 0 |
| 4 | 3 | 0 | $h_1^2\widetilde{h_1}$ | $(g)$ | 0 |
| 6 | 2 | 2 | $i(h_2^2)$ | $(g, w_1)$ | 0 |
| 7 | 2 | 3 | $\widetilde{h_2^2}$ | $(g^2)$ | 0 |
| 8 | 3 | 1 | $i(c_0)$ | $(g, w_1)$ | 0 |
| 9 | 3 | 2 | $\widetilde{c_0}$ | $(g^2, gw_1)$ | 0 |
| 9 | 4 | 1 | $i(h_1c_0)$ | $(g, w_1)$ | 0 |
| 10 | 4 | 2 | $h_1\widetilde{c_0}$ | $(g, w_1)$ | 0 |
| 14 | 4 | 3 | $i(d_0)$ | $(g^2)$ | 0 |
| 15 | 3 | 4 | $i(\beta)$ | $(g^4, g^2w_1, w_1^2)$ | 0 |
| 16 | 5 | 3 | $d_0\widetilde{h_1}$ | $(g^2, gw_1)$ | 0 |
| 18 | 4 | 5 | $i(h_2\beta)$ | $(g, w_1)$ | 0 |
| 21 | 6 | 4 | $d_0\widetilde{h_2^2}$ | $(g^2)$ | 0 |
| 24 | 6 | 5 | $i(\alpha^2)$ | $(g^2)$ | $w_1^2 \cdot \widetilde{h_2^2}$ |
| 25 | 5 | 7 | $i(\gamma)$ | $(g^5, g^2w_1, gw_1^2)$ | 0 |
| 26 | 5 | 8 | $\widetilde{\gamma}$ | $-$ | 0 |
| 26 | 6 | 6 | $i(h_1\gamma)$ | $(g)$ | 0 |
| 26 | 7 | 5 | $i(\alpha d_0)$ | $(g^2, gw_1)$ | $w_1^2 \cdot \widetilde{c_0}$ |
| 27 | 6 | 8 | $h_1\widetilde{\gamma}$ | $(g)$ | 0 |
| 28 | 7 | 6 | $h_1^2\widetilde{\gamma}$ | $(g)$ | 0 |
| 30 | 6 | 9 | $i(\beta^2)$ | $(g^4, g^2w_1)$ | $w_1 \cdot d_0\widetilde{h_2^2}$ |
| 31 | 8 | 6 | $i(d_0e_0)$ | $(g^2)$ | $w_1^2 \cdot i(d_0)$ |
| 32 | 7 | 8+9 | $i(\delta')$ | $(g^2, w_1)$ | 0 |
| 32 | 7 | 9 | $i(\delta)$ | $(g)$ | 0 |

Table 6.7: $R_2$-module generators of $E_4(tmf/2)$ (cont.)

| $t-s$ | $s$ | $g$ | $x$ | $\mathrm{Ann}(x)$ | $d_4(x)$ |
|---|---|---|---|---|---|
| 33 | 7 | 10 | $\widetilde{\delta'}$ | $(g^2)$ | 0 |
| 33 | 8 | 8 | $i(h_1\delta)$ | $(g, w_1)$ | 0 |
| 33 | 9 | 7 | $h_1\widetilde{d_0e_0}$ | $(g)$ | $w_1^2 \cdot d_0\widetilde{h_1}$ |
| 34 | 8 | 10 | $h_1\widetilde{\delta'}$ | $(g, w_1)$ | 0 |
| 37 | 8 | 11 | $i(e_0g)$ | $(g^2)$ | $gw_1^2 \cdot i(1)$ |
| 39 | 9 | 11 | $e_0g\widetilde{h_1}$ | – | 0 |
| 40 | 9 | 12 | $d_0\widetilde{\gamma}$ | $(g^2)$ | 0 |
| 47 | 11 | 14 | $d_0\widetilde{\delta'}$ | $(g^2)$ | 0 |
| 50 | 10 | 18 | $i(h_1^2w_2)$ | $(g)$ | $gw_1 \cdot d_0\widetilde{h_2^2}$ |
| 50 | 11 | 16 | $d_0\widetilde{\beta g}$ | $(g^2)$ | $w_1^2 \cdot \widetilde{\delta'}$ |
| 51 | 9 | 19 | $i(h_2w_2)$ | – | 0 |
| 51 | 10 | $19+20$ | $\gamma\widetilde{\gamma}$ | $(g^5, gw_1)$ | 0 |
| 52 | 11 | 18 | $h_1^2w_2\widetilde{h_1}$ | $(g)$ | 0 |
| 54 | 10 | 21 | $i(h_2^2w_2)$ | $(g, w_1)$ | 0 |
| 56 | 11 | 21 | $\beta^2\widetilde{\gamma}$ | – | $w_1 \cdot d_0\widetilde{\delta'}$ |
| 56 | 11 | 22 | $i(c_0w_2)$ | $(g)$ | 0 |
| 57 | 12 | $20+21$ | $\gamma^2\widetilde{h_2^2}$ | $(g, w_1)$ | 0 |
| 57 | 13 | 18 | $d_0e_0\widetilde{\gamma}$ | $(g^2)$ | $w_1^2 \cdot d_0\widetilde{\gamma}$ |
| 58 | 12 | $22+23$ | $\delta'\widetilde{\gamma}$ | $(g)$ | 0 |
| 58 | 12 | 23 | $\delta\widetilde{\gamma}$ | $(g)$ | 0 |
| 63 | 13 | 25 | $e_0g\widetilde{\gamma}$ | $(g^2)$ | $gw_1^2 \cdot \widetilde{\gamma}$ |
| 65 | 14 | 25 | $d_0g\widetilde{\beta^2}$ | $(g)$ | 0 |
| 66 | 12 | 28 | $i(h_2\beta w_2)$ | $(g, w_1)$ | 0 |
| 69 | 13 | 30 | $i(h_1gw_2)$ | $(g, w_1)$ | 0 |
| 72 | 16 | 28 | $g^2\widetilde{d_0e_0}$ | $(g, w_1)$ | 0 |
| 74 | 14 | 33 | $i(h_1\gamma w_2)$ | $(g)$ | $w_1 \cdot d_0g\widetilde{\beta^2}$ |
| 75 | 16 | 31 | $\alpha^2g\widetilde{\beta^2}$ | $(g)$ | $w_1^2 \cdot \delta'\widetilde{\gamma}$ |
| 76 | 15 | 35 | $h_1^2w_2\widetilde{\gamma}$ | $(g)$ | $gw_1 \cdot d_0\widetilde{\delta'}$ |
| 80 | 15 | 38 | $i(\delta w_2)$ | $(g)$ | 0 |
| 81 | 16 | $38+39$ | $\gamma^2\widetilde{\beta^2}$ | $(g^2, w_1)$ | 0 |
| 81 | 16 | 39 | $i(h_1\delta w_2)$ | $(g)$ | 0 |
| 82 | 16 | 41 | $h_1w_2\widetilde{\delta'}$ | $(g)$ | 0 |

Table 6.7: $R_2$-module generators of $E_4(tmf/2)$ (cont.)

| $t-s$ | $s$ | $g$ | $x$ | $\mathrm{Ann}(x)$ | $d_4(x)$ |
|---|---|---|---|---|---|
| 82 | 18 | 34 | $g^2\widetilde{\alpha^2 e_0}$ | $(g)$ | $w_1^2 \cdot d_0 g\widetilde{\beta^2}$ |
| 97 | 17 | 50 | $i(h_1 w_2^2)$ | $(g^2, gw_1)$ | 0 |
| 98 | 17 | 51 | $w_2^2\widetilde{h_1}$ | – | 0 |
| 98 | 18 | 53 | $i(h_1^2 w_2^2)$ | $(g)$ | 0 |
| 99 | 17 | 52 | $i(h_2 w_2^2)$ | $(g, w_1)$ | 0 |
| 99 | 18 | 55 | $h_1 w_2^2\widetilde{h_1}$ | $(g)$ | 0 |
| 100 | 19 | 55 | $h_1^2 w_2^2\widetilde{h_1}$ | $(g)$ | 0 |
| 102 | 18 | 56 | $i(h_2^2 w_2^2)$ | $(g, w_1)$ | 0 |
| 103 | 18 | 57 | $w_2^2\widetilde{h_2^2}$ | $(g^2)$ | 0 |
| 104 | 19 | 59 | $i(c_0 w_2^2)$ | $(g, w_1)$ | 0 |
| 104 | 20 | 58 | $i(w_1 w_2^2)$ | $(g^2)$ | 0 |
| 105 | 19 | 60 | $w_2^2\widetilde{c_0}$ | $(g^2, gw_1)$ | 0 |
| 105 | 20 | 60 | $i(h_1 c_0 w_2^2)$ | $(g, w_1)$ | 0 |
| 106 | 20 | 62 | $h_1 w_2^2\widetilde{c_0}$ | $(g, w_1)$ | 0 |
| 110 | 20 | 65 | $i(d_0 w_2^2)$ | $(g^2)$ | 0 |
| 112 | 21 | 67 | $d_0 w_2^2\widetilde{h_1}$ | $(g^2, gw_1)$ | 0 |
| 114 | 20 | 67 | $i(h_2\beta w_2^2)$ | $(g, w_1)$ | 0 |
| 117 | 22 | 72 | $d_0 w_2^2\widetilde{h_2^2}$ | $(g^2)$ | 0 |
| 119 | 23 | 74 | $i(\beta w_1 w_2^2)$ | $(g^2, w_1)$ | 0 |
| 120 | 22 | 75 | $i(\alpha^2 w_2^2)$ | $(g^2)$ | $w_1^2 \cdot w_2^2\widetilde{h_2^2}$ |
| 122 | 22 | 76 | $i(h_1\gamma w_2^2)$ | $(g)$ | 0 |
| 122 | 23 | 77 | $i(\alpha d_0 w_2^2)$ | $(g^2, gw_1)$ | $w_1^2 \cdot w_2^2\widetilde{c_0}$ |
| 123 | 22 | 78 | $h_1 w_2^2\widetilde{\gamma}$ | $(g)$ | 0 |
| 124 | 23 | 80 | $h_1^2 w_2^2\widetilde{\gamma}$ | $(g)$ | 0 |
| 127 | 24 | 82 | $i(d_0 e_0 w_2^2)$ | $(g^2)$ | $w_1^2 \cdot i(d_0 w_2^2)$ |
| 128 | 23 | $82+83$ | $i(\delta' w_2^2)$ | – | 0 |
| 128 | 23 | 83 | $i(\delta w_2^2)$ | $(g)$ | 0 |
| 129 | 23 | 84 | $w_2^2\widetilde{\delta'}$ | $(g^2)$ | 0 |
| 129 | 24 | 86 | $i(h_1\delta w_2^2)$ | $(g, w_1)$ | 0 |
| 129 | 25 | $84+85$ | $i(\gamma w_1 w_2^2)$ | $(g^2, gw_1)$ | 0 |
| 129 | 25 | 85 | $h_1 w_2^2\widetilde{d_0 e_0}$ | $(g)$ | $w_1^2 \cdot d_0 w_2^2\widetilde{h_1}$ |
| 130 | 24 | 88 | $h_1 w_2^2\widetilde{\delta'}$ | $(g, w_1)$ | 0 |

Table 6.7: $R_2$-module generators of $E_4(tmf/2)$ (cont.)

| $t-s$ | $s$ | $g$ | $x$ | $\mathrm{Ann}(x)$ | $d_4(x)$ |
|---|---|---|---|---|---|
| 130 | 25 | 87 | $w_1 w_2^2\widetilde{\gamma}$ | $(g^2)$ | 0 |
| 133 | 24 | 89 | $i(e_0 g w_2^2)$ | $(g^2)$ | $gw_1 \cdot i(w_1 w_2^2)$ |
| 134 | 26 | 91 | $i(\beta^2 w_1 w_2^2)$ | $(g^2)$ | $w_1^2 \cdot d_0 w_2^2\widetilde{h_2^2}$ |
| 135 | 25 | 93 | $e_0 g w_2^2\widetilde{h_1}$ | $-$ | 0 |
| 136 | 25 | 94 | $d_0 w_2^2\widetilde{\gamma}$ | $(g^2)$ | 0 |
| 143 | 27 | 103 | $d_0 w_2^2\widetilde{\delta'}$ | $(g^2)$ | 0 |
| 146 | 26 | 104 | $i(h_1^2 w_2^3)$ | $(g)$ | $gw_1 \cdot d_0 w_2^2\widetilde{h_2^2}$ |
| 146 | 27 | 106 | $d_0 w_2^2\widetilde{\beta g}$ | $(g^2)$ | $w_1^2 \cdot w_2^2\widetilde{\delta'}$ |
| 147 | 25 | 101 | $i(h_2 w_2^3)$ | $-$ | 0 |
| 148 | 27 | 108 | $h_1^2 w_2^3\widetilde{h_1}$ | $(g)$ | 0 |
| 150 | 26 | 107 | $i(h_2^2 w_2^3)$ | $(g, w_1)$ | 0 |
| 152 | 27 | 112 | $i(c_0 w_2^3)$ | $(g)$ | 0 |
| 153 | 28 | $114+115$ | $\gamma^2 w_2^2\widetilde{h_2^2}$ | $(g, w_1)$ | 0 |
| 153 | 29 | 115 | $d_0 e_0 w_2^2\widetilde{\gamma}$ | $(g^2)$ | $w_1^2 \cdot d_0 w_2^2\widetilde{\gamma}$ |
| 154 | 28 | $116+117$ | $\delta' w_2^2\widetilde{\gamma}$ | $(g)$ | 0 |
| 154 | 28 | 117 | $\delta w_2^2\widetilde{\gamma}$ | $(g)$ | 0 |
| 155 | 30 | $118+119$ | $\gamma w_1 w_2^2\widetilde{\gamma}$ | $(g)$ | 0 |
| 159 | 29 | 123 | $e_0 g w_2^2\widetilde{\gamma}$ | $(g^2)$ | $gw_1 \cdot w_1 w_2^2\widetilde{\gamma}$ |
| 160 | 31 | 124 | $\beta^2 w_1 w_2^2\widetilde{\gamma}$ | $(g^2)$ | $w_1^2 \cdot d_0 w_2^2\widetilde{\delta'}$ |
| 161 | 30 | 127 | $d_0 g w_2^2\widetilde{\beta^2}$ | $(g)$ | 0 |
| 162 | 28 | 122 | $i(h_2 \beta w_2^3)$ | $(g, w_1)$ | 0 |
| 165 | 29 | 128 | $i(h_1 g w_2^3)$ | $(g, w_1)$ | 0 |
| 168 | 32 | 137 | $g^2 w_2^2\widetilde{d_0 e_0}$ | $(g, w_1)$ | 0 |
| 170 | 30 | 135 | $i(h_1 \gamma w_2^3)$ | $(g)$ | $w_1 \cdot d_0 g w_2^2\widetilde{\beta^2}$ |
| 171 | 32 | 141 | $\alpha^2 g w_2^2\widetilde{\beta^2}$ | $(g)$ | $w_1^2 \cdot \delta' w_2^2\widetilde{\gamma}$ |
| 172 | 31 | 141 | $h_1^2 w_2^3\widetilde{\gamma}$ | $(g)$ | $gw_1 \cdot d_0 w_2^2\widetilde{\delta'}$ |
| 176 | 31 | 144 | $i(\delta w_2^3)$ | $(g)$ | 0 |
| 177 | 32 | 149 | $i(h_1 \delta w_2^3)$ | $(g)$ | 0 |
| 178 | 32 | 151 | $h_1 w_2^3\widetilde{\delta'}$ | $(g)$ | 0 |
| 178 | 34 | 151 | $g^2 w_2^2\widetilde{\alpha^2 e_0}$ | $(g)$ | $w_1^2 \cdot d_0 g w_2^2\widetilde{\beta^2}$ |

Table 6.8: The non-cyclic $R_2$-module summands in $E_4(tmf/2)$

| $\langle x_1, x_2 \rangle$ |
|---|
| $\langle \widetilde{\gamma}, w_2^2\widetilde{h_1} \rangle \cong \dfrac{\Sigma^{5,31}R_2 \oplus \Sigma^{17,115}R_2}{\langle (g^2w_1, 0), (g^5, gw_1), (0, g^3), (0, g^2w_1) \rangle}$ |
| $\langle e_0g\widetilde{h_1}, i(h_2w_2) \rangle \cong \dfrac{\Sigma^{9,48}R_2 \oplus \Sigma^{9,60}R_2}{\langle (w_1, 0), (g, w_1), (0, g) \rangle}$ |
| $\langle \beta^2\widetilde{\gamma}, i(\delta' w_2^2) \rangle \cong \dfrac{\Sigma^{11,67}R_2 \oplus \Sigma^{23,151}R_2}{\langle (g^2w_1, 0), (g^4, w_1), (0, g^2) \rangle}$ |
| $\langle e_0gw_2^2\widetilde{h_1}, i(h_2w_2^3) \rangle \cong \dfrac{\Sigma^{25,160}R_2 \oplus \Sigma^{25,172}R_2}{\langle (w_1, 0), (g, w_1), (0, g) \rangle}$ |

PROPOSITION 6.8. *The* 19 *classes listed in Table* 6.9 *generate* $E_4(tmf/2)$ *as a module over* $E_4(tmf)$.

PROOF. Inspection of Tables 5.5 and 6.7 easily shows that most of the $R_2$-module generators of $E_4(tmf/2)$ are $E_4(tmf)$-multiples of the classes in Table 6.9. The less evident cases follow from the relations

$$
\begin{aligned}
h_1\widetilde{d_0e_0} &= w_1 \cdot i(\gamma) + d_0e_0 \cdot \widetilde{h_1} \\
d_0g\widetilde{\beta^2} &= \delta' \cdot \widetilde{\delta'} \\
i(h_1gw_2) &= h_2^2w_2 \cdot i(\beta) \\
g^2\widetilde{d_0e_0} &= \alpha d_0g \cdot \widetilde{\gamma} \\
i(h_1\gamma w_2) &= h_0w_2 \cdot \widetilde{\gamma} \\
\alpha^2g\widetilde{\beta^2} &= \alpha e_0g \cdot \widetilde{\gamma} \\
\gamma^2\widetilde{\beta^2} &= \beta g^2 \cdot \widetilde{\gamma} + i(h_1\delta w_2) \\
h_1w_2\widetilde{\delta'} &= c_0w_2 \cdot \widetilde{\gamma} \\
g^2\widetilde{\alpha^2e_0} &= \alpha^3g \cdot \widetilde{\gamma} \\
h_1w_2^2\widetilde{d_0e_0} &= i(\gamma w_1w_2^2) + d_0e_0w_2^2 \cdot \widetilde{h_1} \\
d_0gw_2^2\widetilde{\beta^2} &= d_0\gamma w_2^2 \cdot \widetilde{\gamma} \\
i(h_1gw_2^3) &= h_2^2w_2^3 \cdot i(\beta) \\
g^2w_2^2\widetilde{d_0e_0} &= d_0\gamma \cdot w_2^2\widetilde{\delta'} \\
i(h_1\gamma w_2^3) &= h_0w_2^3 \cdot \widetilde{\gamma} \\
\alpha^2gw_2^2\widetilde{\beta^2} &= \alpha e_0gw_2^2 \cdot \widetilde{\gamma}
\end{aligned}
$$

$$h_1 w_2^3 \widetilde{\delta'} = c_0 w_2^3 \cdot \widetilde{\gamma}$$
$$g^2 w_2^2 \widetilde{\alpha^2 e_0} = \alpha^3 g w_2^2 \cdot \widetilde{\gamma},$$

which we verify by calculating the relevant Yoneda products using `ext`. Note that $\alpha^3 g = (\alpha^3 g + h_0 w_1 w_2) + w_1 \cdot h_0 w_2$ lies in $E_4(tmf)$. □

Table 6.9: $E_4(tmf)$-module generators of $E_4(tmf/2)$

| $t-s$ | $s$ | $g$ | $x$ | $d_4(x)$ |
|---|---|---|---|---|
| 0 | 0 | 0 | $i(1)$ | 0 |
| 2 | 1 | 1 | $\widetilde{h_1}$ | 0 |
| 7 | 2 | 3 | $\widetilde{h_2^2}$ | 0 |
| 9 | 3 | 2 | $\widetilde{c_0}$ | 0 |
| 15 | 3 | 4 | $i(\beta)$ | 0 |
| 24 | 6 | 5 | $i(\alpha^2)$ | $w_1^2 \cdot \widetilde{h_2^2}$ |
| 26 | 5 | 8 | $\widetilde{\gamma}$ | 0 |
| 26 | 7 | 5 | $i(\alpha d_0)$ | $w_1^2 \cdot \widetilde{c_0}$ |
| 30 | 6 | 9 | $i(\beta^2)$ | $w_1 \cdot d_0\widetilde{h_2^2}$ |
| 33 | 7 | 10 | $\widetilde{\delta'}$ | 0 |
| 50 | 11 | 16 | $\widetilde{d_0\beta g}$ | $w_1^2 \cdot \widetilde{\delta'}$ |
| 56 | 11 | 21 | $\beta^2\widetilde{\gamma}$ | $w_1 \cdot d_0\widetilde{\delta'}$ |
| 98 | 17 | 51 | $\widetilde{w_2^2 h_1}$ | 0 |
| 103 | 18 | 57 | $\widetilde{w_2^2 h_2^2}$ | 0 |
| 105 | 19 | 60 | $w_2^2\widetilde{c_0}$ | 0 |
| 120 | 22 | 75 | $i(\alpha^2 w_2^2)$ | $w_1^2 \cdot \widetilde{w_2^2 h_2^2}$ |
| 122 | 23 | 77 | $i(\alpha d_0 w_2^2)$ | $w_1^2 \cdot w_2^2\widetilde{c_0}$ |
| 129 | 23 | 84 | $w_2^2\widetilde{\delta'}$ | 0 |
| 146 | 27 | 106 | $d_0\widetilde{w_2^2\beta g}$ | $w_1^2 \cdot w_2^2\widetilde{\delta'}$ |

PROPOSITION 6.9. *The $d_4$-differentials on the $E_4(tmf)$-module generators of $E_4(tmf/2)$ are as given in Table 6.9.*

PROOF. For $x \in \{i(1), \widetilde{h_1}, \widetilde{h_2^2}, \widetilde{c_0}, i(\beta), \widetilde{\gamma}, \widetilde{\delta'}\}$ the bidegree of $d_4(x)$ is zero at $E_2$.

For $x = \widetilde{w_2^2 h_1}$, $x = \widetilde{w_2^2 h_2^2}$ and $x = w_2^2\widetilde{c_0}$ the target of $d_4$ on $x$ is generated at $E_2$ by $21_{47} = g^2 d_0 e_0 \widetilde{\gamma} = d_2(w_2 \cdot \widetilde{d_0\beta g})$, $22_{51} = g^3 \widetilde{\alpha^2 e_0} = d_2(w_2 \cdot \widetilde{\alpha^2 \beta^2})$ and $23_{51} = g^3 w_1 \widetilde{\beta g} = d_2(w_2 \cdot d_0 e_0 \widetilde{\gamma})$, respectively. For $x = w_2^2\widetilde{\delta'}$, the target is generated at $E_2$ by $y = 27_{77} = i(\alpha g^3 w_1 w_2)$, and $d_2(y) = g^5 w_1 \cdot e_0 \widetilde{h_1} \neq 0$. Hence the target is zero at $E_3$ for $x \in \{\widetilde{w_2^2 h_1}, \widetilde{w_2^2 h_2^2}, w_2^2\widetilde{c_0}, w_2^2\widetilde{\delta'}\}$.

The remaining differentials are consequences of the differential $d_4(d_0e_0) = d_0w_1^2$ in $E_4(tmf)$, cf. Corollary 5.13, or of naturality with respect to $j\colon tmf/2 \to \Sigma tmf$, as we now show.

For $x = i(\alpha^2)$ the relation $d_0 \cdot x = 10_9 = d_0e_0 \cdot \widetilde{h_2^2}$, verified by `ext`, gives $d_0 \cdot d_4(x) = w_1^2 \cdot d_0\widetilde{h_2^2} \neq 0$ at $E_4$, so $d_4(x) \neq 0$ and $10_2 = w_1^2 \cdot \widetilde{h_2^2}$ is the only possible value.

For $x = i(\alpha d_0)$ the relation $\gamma \cdot x = 12_{15} = i(d_0e_0g)$, verified by `ext`, gives $\gamma \cdot d_4(x) = gw_1^2 \cdot i(d_0) \neq 0$ at $E_4$, so $d_4(x) \neq 0$ and $11_2 = w_1^2 \cdot \widetilde{c_0}$ is the only possible value.

For $x = i(\beta^2)$ the relation $w_1 \cdot x = 10_9 = d_0e_0 \cdot \widetilde{h_2^2}$, verified by `ext`, gives $w_1 \cdot d_4(x) = w_1^2 \cdot d_0\widetilde{h_2^2} \neq 0$ at $E_4$, so $d_4(x) \neq 0$ and $10_4 = w_1 \cdot d_0\widetilde{h_2^2}$ is the only possible value.

For $x = d_0\widetilde{\beta g}$ naturality with respect to $j$, and $j(x) = d_0 \cdot \beta g = \alpha e_0g$, show that $d_4(x)$ maps by $j$ to $d_4(\alpha e_0 g) = w_1^2 \cdot \delta' \neq 0$, so $d_4(x) = 15_{10} = w_1^2 \cdot \widetilde{\delta'}$ is the only possibility.

For $x = \beta^2\widetilde{\gamma}$ naturality with respect to $j$, and $j(x) = \beta^2\gamma = \beta g^2$, show that $d_4(x)$ maps to $d_4(\beta g^2) = w_1 \cdot \alpha d_0 g \neq 0$, and $d_4(x) = 15_{14} = w_1 \cdot d_0\widetilde{\delta'}$ is the only possibility.

For $x = i(\alpha^2w_2^2)$ the target bidegree of $d_4$ is generated at $E_2$ by $26_{65} = w_1^2 \cdot w_2^2\widetilde{h_2^2}$ and $26_{66} = g^4w_1\widetilde{\beta^2} = d_2(g^2 \cdot w_2\widetilde{d_0e_0})$. Hence $w_1^2 \cdot w_2^2\widetilde{h_2^2}$ is the only nonzero class at $E_3$ and $E_4$. The relation $d_0 \cdot x = d_0e_0 \cdot w_2^2\widetilde{h_2^2}$ gives $d_0 \cdot d_4(x) = w_1^2 \cdot d_0w_2^2\widetilde{h_2^2} \neq 0$, so $d_4(x) = 26_{65} = w_1^2 \cdot w_2^2\widetilde{h_2^2}$.

For $x = i(\alpha d_0 w_2^2)$ the target bidegree of $d_4$ is generated at $E_2$ by $27_{65} = w_1^2 \cdot w_2^2\widetilde{c_0}$ and $27_{66} = g^4w_1\widetilde{\delta'} = d_2(g^2w_1 \cdot w_2\widetilde{\gamma})$. Hence $w_1^2 \cdot w_2^2\widetilde{c_0}$ is the only nonzero class at $E_3$ and $E_4$. The relation $\gamma \cdot x = i(d_0e_0gw_2^2)$ gives $\gamma \cdot d_4(x) = gw_1^2 \cdot i(d_0w_2^2) \neq 0$, so $d_4(x) = 27_{65} = w_1^2 \cdot w_2^2\widetilde{c_0}$.

For $x = d_0w_2^2\widetilde{\beta g}$ the target bidegree of $d_4$ is generated at $E_2$ by $31_{94} = w_1^2 \cdot w_2^2\widetilde{\delta'}$ and $y = 31_{95} = g^4w_1 \cdot w_2\widetilde{c_0}$. Here $d_2(y) = g^6w_1 \cdot d_0\widetilde{h_1} \neq 0$. Hence $w_1^2 \cdot w_2^2\widetilde{\delta'}$ is the only nonzero class at $E_3$ and $E_4$. Naturality with respect to $j$, and $j(x) = d_0w_2^2 \cdot \beta g = \alpha e_0 g w_2^2$, show that $d_4(x)$ maps to $d_4(\alpha e_0 g w_2^2) = \delta' w_1^2 w_2^2 \neq 0$, which implies that $d_4(x) = 31_{94} = w_1^2 \cdot w_2^2\widetilde{\delta'}$. □

THEOREM 6.10. *The $d_4$-differential in $E_4(tmf/2)$ is $R_2$-linear. Its values on a set of $R_2$-module generators are listed in Table* 6.7.

PROOF. The classes $g$, $w_1$ and $w_2^4$ are $d_4$-cycles in $E_4(tmf)$, so multiplication by each of these commutes with the $d_4$-differential in $E_4(tmf/2)$.

The $d_4$-differential on the $E_4(tmf)$-module generators was computed in Proposition 6.9. The Leibniz rule then gives the value of $d_4$ on the remaining elements in terms of $E_4(tmf)$-multiples of the $R_2$-module generators. We use `ext` to rewrite these as $R_2$-multiples of the $R_2$-module generators, leaving out some straightforward cases:

- $d_4(h_1\widetilde{d_0e_0}) = d_4(w_1 \cdot i(\gamma) + d_0e_0 \cdot \widetilde{h_1}) = 0 + d_0w_1^2 \cdot \widetilde{h_1}$
- $d_4(e_0g\widetilde{h_1}) = gw_1^2 \cdot \widetilde{h_1} = 0$ at $E_4$
- $d_4(i(h_1^2w_2)) = i(\alpha^2e_0w_1) = 14_{11} = gw_1 \cdot d_0\widetilde{h_2^2}$
- $d_4(h_1^2w_2 \cdot \widetilde{h_1}) = \alpha^2e_0w_1 \cdot \widetilde{h_1} = 15_{11} = gw_1^2 \cdot i(\beta) = 0$ at $E_4$
- $d_4(d_0e_0 \cdot \widetilde{\gamma}) = d_0w_1^2 \cdot \widetilde{\gamma}$

- $d_4(e_0 g \cdot \widetilde{\gamma}) = g w_1^2 \cdot \widetilde{\gamma}$
- $d_4(d_0 g \widetilde{\beta^2}) = d_4(\delta' \cdot \widetilde{\delta'}) = 0$
- $d_4(h_2 w_2 \cdot i(\beta)) = 0$
- $d_4(h_1 g w_2) = d_4(h_2^2 w_2 \cdot i(\beta)) = 0$
- $d_4(g^2 \widetilde{d_0 e_0}) = d_4(\alpha d_0 g \cdot \widetilde{\gamma}) = 0$
- $d_4(i(h_1 \gamma w_2)) = d_4(h_0 w_2 \cdot \widetilde{\gamma}) = d_0 \gamma w_1 \cdot \widetilde{\gamma} = 18_{25} = w_1 \cdot d_0 g \widetilde{\beta^2}$
- $d_4(\alpha^2 g \widetilde{\beta^2}) = d_4(\alpha e_0 g \cdot \widetilde{\gamma}) = \delta' w_1^2 \cdot \widetilde{\gamma}$
- $d_4(h_1^2 w_2 \cdot \widetilde{\gamma}) = \alpha^2 e_0 w_1 \cdot \widetilde{\gamma} = 19_{25} = g w_1 \cdot d_0 \widetilde{\delta'}$
- $d_4(\gamma^2 \widetilde{\beta^2}) = d_4(\beta g^2 \cdot \widetilde{\gamma} + i(h_1 \delta w_2)) = \alpha d_0 g w_1 \cdot \widetilde{\gamma} + 0 = 20_{28} = w_1 \cdot g^2 \widetilde{d_0 e_0} = 0$ at $E_4$
- $d_4(h_1 w_2 \widetilde{\delta'}) = d_4(c_0 w_2 \cdot \widetilde{\gamma}) = 0$
- $d_4(g^2 \widetilde{\alpha^2 e_0}) = d_4(\alpha^3 g \cdot \widetilde{\gamma}) = d_0 \gamma w_1^2 \cdot \widetilde{\gamma} = 22_{25} = w_1^2 \cdot d_0 g \widetilde{\beta^2}$
- $d_4(h_1 w_2^2 \widetilde{d_0 e_0}) = d_4(i(\gamma w_1 w_2^2) + d_0 e_0 w_2^2 \cdot \widetilde{h_1}) = 0 + d_0 w_1^2 w_2^2 \cdot \widetilde{h_1}$
- $d_4(w_1 w_2^2 \cdot i(\beta^2)) = w_1 w_2^2 \cdot d_0 w_1 \widetilde{h_2^2}$
- $d_4(e_0 g w_2^2 \cdot \widetilde{h_1}) = g w_1^2 w_2^2 \cdot \widetilde{h_1} = 0$ at $E_4$
- $d_4(i(h_1^2 w_2^3)) = i(\alpha^2 e_0 w_1 w_2^2) = 30_{99} = g w_1 \cdot d_0 w_2^2 \widetilde{h_2^2}$
- $d_4(h_1^2 w_2^3 \cdot \widetilde{h_1}) = \alpha^2 e_0 w_1 w_2^2 \cdot \widetilde{h_1} = 31_{99} = g w_1 \cdot i(\beta w_1 w_2^2) = 0$ at $E_4$
- $d_4(w_1 w_2^2 \cdot \beta^2 \widetilde{\gamma}) = w_1 w_2^2 \cdot d_0 w_1 \widetilde{\delta'}$
- $d_4(d_0 g w_2^2 \widetilde{\beta^2}) = d_4(d_0 \gamma w_2^2 \cdot \widetilde{\gamma}) = 0$
- $d_4(h_2 w_2^3 \cdot i(\beta)) = 0$
- $d_4(i(h_1 g w_2^3)) = d_4(h_2^2 w_2^3 \cdot i(\beta)) = 0$
- $d_4(g^2 w_2^2 \widetilde{d_0 e_0}) = d_4(d_0 \gamma \cdot w_2^2 \widetilde{\delta'}) = 0$
- $d_4(i(h_1 \gamma w_2^3)) = d_4(h_0 w_2^3 \cdot \widetilde{\gamma}) = d_0 \gamma w_1 w_2^2 \cdot \widetilde{\gamma} = 34_{134} = w_1 \cdot d_0 g w_2^2 \widetilde{\beta^2}$
- $d_4(\alpha^2 g w_2^2 \widetilde{\beta^2}) = d_4(\alpha e_0 g w_2^2 \cdot \widetilde{\gamma}) = \delta' w_1^2 w_2^2 \cdot \widetilde{\gamma}$
- $d_4(h_1^2 w_2^3 \cdot \widetilde{\gamma}) = \alpha^2 e_0 w_1 w_2^2 \cdot \widetilde{\gamma} = 35_{134} = g w_1 \cdot d_0 w_2^2 \widetilde{\delta'}$
- $d_4(h_1 w_2^3 \widetilde{\delta'}) = d_4(c_0 w_2^3 \cdot \widetilde{\gamma}) = 0$
- $d_4(g^2 w_2^2 \widetilde{\alpha^2 e_0}) = d_4(\alpha^3 g w_2^2 \cdot \widetilde{\gamma}) = d_0 \gamma w_1^2 w_2^2 \cdot \widetilde{\gamma} = 38_{134} = w_1^2 \cdot d_0 g w_2^2 \widetilde{\beta^2}$. □

## 6.5. The $E_\infty$-term for $tmf/2$

The calculation of $E_5(tmf/2)$ as the homology of $(E_4(tmf/2), d_4)$ is carried out in Appendix B.3, and the result is displayed in Tables 6.10 and 6.11. In this section we show that the twelve classes in Table 6.12 generate $E_5(tmf/2)$ as a module over $E_5(tmf) = E_\infty(tmf)$, and use this to deduce that $E_5(tmf/2) = E_\infty(tmf/2)$.

Passing from $E_4$ to $E_5$, we introduced shorter monomial names for some $d_4$-cycles which were sums in $E_4(tmf/2)$. These were

$$\begin{aligned}
i(\gamma^2) &= 10_{17} + 10_{18} = g \cdot i(\beta^2) + i(h_1^2 w_2) \\
\gamma^2 \widetilde{\gamma} &= 15_{34} + 15_{35} = g \cdot \beta^2 \widetilde{\gamma} + h_1^2 w_2 \widetilde{\gamma} \\
\gamma^2 \widetilde{d_0 e_0} &= 18_{34} + 18_{35} = w_1 \cdot i(h_1 \gamma w_2) + g^2 \widetilde{\alpha^2 e_0} \\
i(\gamma^2 w_1 w_2^2) &= 30_{115} + 30_{116} = g \cdot i(\beta^2 w_1 w_2^2) + w_1 \cdot i(h_1^2 w_2^3) \\
\gamma^2 w_1 w_2^2 \widetilde{\gamma} &= 35_{153} + 35_{154} = g \cdot \beta^2 w_1 w_2^2 \widetilde{\gamma} + w_1 \cdot h_1^2 w_2^3 \widetilde{\gamma} .
\end{aligned}$$

In order to make the $h_1$-action on $i(c_0 w_2)$ more visible, we now make a change of basis, replacing $i(e_0 g^2) = 12_{20}$ by $i(e_0 g^2) + \gamma^2 \widetilde{h_2^2} = 12_{21} = i(h_1 c_0 w_2)$ and keeping $\gamma^2 \widetilde{h_2^2}$

as a generator. Similarly, we replace $i(e_0 g^2 w_2^2) = 28_{114}$ by $i(e_0 g^2 w_2^2) + \gamma^2 w_2^2 \widetilde{h_2^2} = 28_{115} = i(h_1 c_0 w_2^3)$, while keeping $\gamma^2 w_2^2 \widetilde{h_2^2}$.

Table 6.10: $R_2$-module generators of $E_5(tmf/2)$

| $t-s$ | $s$ | $g$ | $x$ | $\mathrm{Ann}(x)$ | dec. |
|---|---|---|---|---|---|
| 0 | 0 | 0 | $i(1)$ | $(g^6, g^2w_1, gw_1^2)$ | **gen.** |
| 1 | 1 | 0 | $i(h_1)$ | $(g^2, gw_1)$ | $h_1 \cdot i(1)$ |
| 2 | 1 | 1 | $\widetilde{h_1}$ | $(g^3, gw_1)$ | **gen.** |
| 2 | 2 | 0 | $i(h_1^2)$ | $(g)$ | $h_1^2 \cdot i(1) = h_0 \cdot \widetilde{h_1}$ |
| 3 | 1 | 2 | $i(h_2)$ | $(g, w_1)$ | $h_2 \cdot i(1)$ |
| 3 | 2 | 1 | $h_1\widetilde{h_1}$ | $(g)$ | $h_1 \cdot \widetilde{h_1}$ |
| 4 | 3 | 0 | $h_1^2\widetilde{h_1}$ | $(g)$ | $h_1^2 \cdot \widetilde{h_1}$ |
| 6 | 2 | 2 | $i(h_2^2)$ | $(g, w_1)$ | $h_2^2 \cdot i(1)$ |
| 7 | 2 | 3 | $\widetilde{h_2^2}$ | $(g^2, w_1^2)$ | **gen.** |
| 8 | 3 | 1 | $i(c_0)$ | $(g, w_1)$ | $h_1 \cdot \widetilde{h_2^2}$ |
| 9 | 3 | 2 | $\widetilde{c_0}$ | $(g^2, gw_1, w_1^2)$ | **gen.** |
| 9 | 4 | 1 | $i(h_1c_0)$ | $(g, w_1)$ | $h_1^2 \cdot \widetilde{h_2^2} = h_0 \cdot \widetilde{c_0}$ |
| 10 | 4 | 2 | $h_1\widetilde{c_0}$ | $(g, w_1)$ | $h_1 \cdot \widetilde{c_0}$ |
| 14 | 4 | 3 | $i(d_0)$ | $(g^2, w_1^2)$ | $d_0 \cdot i(1)$ |
| 15 | 3 | 4 | $i(\beta)$ | $(g^4, g^2w_1, w_1^2)$ | **gen.** |
| 16 | 5 | 3 | $d_0\widetilde{h_1}$ | $(g^2, gw_1, w_1^2)$ | $d_0 \cdot \widetilde{h_1}$ |
| 18 | 4 | 5 | $i(h_2\beta)$ | $(g, w_1)$ | $h_2 \cdot i(\beta)$ |
| 21 | 6 | 4 | $d_0\widetilde{h_2^2}$ | $(g^2, w_1)$ | $d_0 \cdot \widetilde{h_2^2}$ |
| 25 | 5 | 7 | $i(\gamma)$ | $(g^5, g^2w_1, gw_1^2)$ | $\gamma \cdot i(1)$ |
| 26 | 5 | 8 | $\widetilde{\gamma}$ | $-$ | **gen.** |
| 26 | 6 | 6 | $i(h_1\gamma)$ | $(g)$ | $h_1\gamma \cdot i(1) = h_0 \cdot \widetilde{\gamma}$ |
| 27 | 6 | 8 | $h_1\widetilde{\gamma}$ | $(g)$ | $h_1 \cdot \widetilde{\gamma}$ |
| 28 | 7 | 6 | $h_1^2\widetilde{\gamma}$ | $(g)$ | $h_1^2 \cdot \widetilde{\gamma}$ |
| 32 | 7 | $8+9$ | $i(\delta')$ | $(g^2, w_1)$ | $\delta' \cdot i(1)$ |
| 32 | 7 | 9 | $i(\delta)$ | $(g)$ | $\delta \cdot i(1)$ |
| 33 | 7 | 10 | $\widetilde{\delta'}$ | $(g^2, w_1^2)$ | **gen.** |
| 33 | 8 | 8 | $i(h_1\delta)$ | $(g, w_1)$ | $h_1\delta \cdot i(1) = h_0 \cdot \widetilde{\delta'}$ |
| 34 | 8 | 10 | $h_1\widetilde{\delta'}$ | $(g, w_1)$ | $h_1 \cdot \widetilde{\delta'}$ |
| 39 | 9 | 11 | $e_0g\widetilde{h_1}$ | $-$ | $d_0\gamma \cdot i(1)$ |
| 40 | 9 | 12 | $d_0\widetilde{\gamma}$ | $(g^2, w_1^2)$ | $d_0 \cdot \widetilde{\gamma}$ |

Table 6.10: $R_2$-module generators of $E_5(tmf/2)$ (cont.)

| $t-s$ | $s$ | $g$ | $x$ | $\mathrm{Ann}(x)$ | dec. |
|---|---|---|---|---|---|
| 46 | 11 | 13 | $i(\alpha d_0 g)$ | $(g, w_1)$ | $\alpha d_0 g \cdot i(1)$ |
| 47 | 11 | 14 | $d_0\widetilde{\delta'}$ | $(g^2, w_1)$ | $d_0 \cdot \widetilde{\delta'}$ |
| 50 | 10 | $17+18$ | $i(\gamma^2)$ | $(g^3, gw_1)$ | $\gamma^2 \cdot i(1)$ |
| 51 | 9 | 19 | $i(h_2 w_2)$ | – | $h_2 w_2 \cdot i(1)$ |
| 51 | 10 | $19+20$ | $\gamma\widetilde{\gamma}$ | $(g^5, gw_1)$ | $\gamma \cdot \widetilde{\gamma}$ |
| 52 | 11 | 18 | $h_1^2 w_2 \widetilde{h_1}$ | $(g)$ | $h_1\gamma \cdot \widetilde{\gamma}$ |
| 54 | 10 | 21 | $i(h_2^2 w_2)$ | $(g, w_1)$ | $h_2^2 w_2 \cdot i(1)$ |
| 56 | 11 | 22 | $i(c_0 w_2)$ | $(g)$ | $c_0 w_2 \cdot i(1)$ |
| 57 | 12 | 21 | $i(h_1 c_0 w_2)$ | $(g)$ | $h_1 c_0 w_2 \cdot i(1)$ |
| 57 | 12 | $20+21$ | $\gamma^2\widetilde{h_2^2}$ | $(g, w_1)$ | $\gamma\delta' \cdot i(1)$ |
| 58 | 12 | $22+23$ | $\delta'\widetilde{\gamma}$ | $(g, w_1^2)$ | $\delta' \cdot \widetilde{\gamma}$ |
| 58 | 12 | 23 | $\delta\widetilde{\gamma}$ | $(g)$ | $\delta \cdot \widetilde{\gamma}$ |
| 65 | 14 | 25 | $d_0 g\widetilde{\beta^2}$ | $(g, w_1)$ | $\delta' \cdot \widetilde{\delta'}$ |
| 66 | 12 | 28 | $i(h_2\beta w_2)$ | $(g, w_1)$ | $h_2 w_2 \cdot i(\beta)$ |
| 69 | 13 | 30 | $i(h_1 g w_2)$ | $(g, w_1)$ | $h_2^2 w_2 \cdot i(\beta)$ |
| 72 | 16 | 28 | $g^2\widetilde{d_0 e_0}$ | $(g, w_1)$ | $d_0\gamma \cdot \widetilde{\delta'}$ |
| 76 | 15 | $34+35$ | $\gamma^2\widetilde{\gamma}$ | – | $\gamma^2 \cdot \widetilde{\gamma}$ |
| 80 | 15 | 38 | $i(\delta w_2)$ | $(g)$ | $\delta w_2 \cdot i(1)$ |
| 81 | 16 | $38+39$ | $\gamma^2\widetilde{\beta^2}$ | $(g^2, w_1)$ | **gen.** |
| 81 | 16 | 39 | $i(h_1\delta w_2)$ | $(g)$ | $h_1\delta w_2 \cdot i(1)$ |
| 82 | 16 | 41 | $h_1 w_2\widetilde{\delta'}$ | $(g)$ | $\delta w_2 \cdot \widetilde{h_1}$ |
| 82 | 18 | $34+35$ | $\gamma^2\widetilde{d_0 e_0}$ | $(g)$ | $(\alpha^3 g + h_0 w_1 w_2) \cdot \widetilde{\gamma}$ |
| 83 | 17 | 39 | $e_0 g^2\widetilde{\gamma}$ | $(g)$ | $h_1\delta w_2 \cdot \widetilde{h_1}$ |
| 97 | 17 | 50 | $i(h_1 w_2^2)$ | $(g^2, gw_1)$ | $h_1 w_2^2 \cdot i(1)$ |
| 98 | 17 | 51 | $w_2^2\widetilde{h_1}$ | – | **gen.** |
| 98 | 18 | 53 | $i(h_1^2 w_2^2)$ | $(g)$ | $h_1^2 w_2^2 \cdot i(1) = h_0 \cdot w_2^2\widetilde{h_1}$ |
| 99 | 17 | 52 | $i(h_2 w_2^2)$ | $(g, w_1)$ | $h_2 w_2^2 \cdot i(1)$ |
| 99 | 18 | 55 | $h_1 w_2^2\widetilde{h_1}$ | $(g)$ | $h_1 \cdot w_2^2\widetilde{h_1}$ |
| 100 | 19 | 55 | $h_1^2 w_2^2\widetilde{h_1}$ | $(g)$ | $h_1^2 \cdot w_2^2\widetilde{h_1}$ |
| 102 | 18 | 56 | $i(h_2^2 w_2^2)$ | $(g, w_1)$ | $h_2^2 w_2^2 \cdot i(1)$ |
| 103 | 18 | 57 | $w_2^2\widetilde{h_2^2}$ | $(g^2, w_1^2)$ | **gen.** |
| 104 | 19 | 59 | $i(c_0 w_2^2)$ | $(g, w_1)$ | $h_1 \cdot w_2^2\widetilde{h_2^2}$ |

Table 6.10: $R_2$-module generators of $E_5(tmf/2)$ (cont.)

| $t-s$ | $s$ | $g$ | $x$ | $\mathrm{Ann}(x)$ | dec. |
|---|---|---|---|---|---|
| 104 | 20 | 58 | $i(w_1w_2^2)$ | $(g^2, gw_1)$ | $w_1w_2^2 \cdot i(1)$ |
| 105 | 19 | 60 | $w_2^2\widetilde{c_0}$ | $(g^2, gw_1, w_1^2)$ | **gen.** |
| 105 | 20 | 60 | $i(h_1c_0w_2^2)$ | $(g, w_1)$ | $h_1^2 \cdot w_2^2\widetilde{h_2^2} = h_0 \cdot w_2^2\widetilde{c_0}$ |
| 106 | 20 | 62 | $h_1w_2^2\widetilde{c_0}$ | $(g, w_1)$ | $h_1 \cdot w_2^2\widetilde{c_0}$ |
| 110 | 20 | 65 | $i(d_0w_2^2)$ | $(g^2, w_1^2)$ | $d_0w_2^2 \cdot i(1)$ |
| 112 | 21 | 67 | $d_0w_2^2\widetilde{h_1}$ | $(g^2, gw_1, w_1^2)$ | $d_0w_2^2 \cdot \widetilde{h_1}$ |
| 114 | 20 | 67 | $i(h_2\beta w_2^2)$ | $(g, w_1)$ | $h_2w_2^2 \cdot i(\beta)$ |
| 117 | 22 | 72 | $d_0w_2^2\widetilde{h_2^2}$ | $(g^2, gw_1, w_1^2)$ | $d_0w_2^2 \cdot \widetilde{h_2^2}$ |
| 119 | 23 | 74 | $i(\beta w_1w_2^2)$ | $(g^2, w_1)$ | $d_0w_2^2 \cdot \widetilde{c_0}$ |
| 122 | 22 | 76 | $i(h_1\gamma w_2^2)$ | $(g)$ | $h_1\gamma w_2^2 \cdot i(1)$ |
| 123 | 22 | 78 | $h_1w_2^2\widetilde{\gamma}$ | $(g)$ | $h_1w_2^2 \cdot \widetilde{\gamma}$ |
| 124 | 23 | 80 | $h_1^2w_2^2\widetilde{\gamma}$ | $(g)$ | $h_1\gamma w_2^2 \cdot \widetilde{h_1}$ |
| 128 | 23 | $82+83$ | $i(\delta' w_2^2)$ | – | $\delta' w_2^2 \cdot i(1)$ |
| 128 | 23 | 83 | $i(\delta w_2^2)$ | $(g)$ | $\delta w_2^2 \cdot i(1)$ |
| 129 | 23 | 84 | $w_2^2\widetilde{\delta'}$ | $(g^2, w_1^2)$ | **gen.** |
| 129 | 24 | 86 | $i(h_1\delta w_2^2)$ | $(g, w_1)$ | $h_1\delta w_2^2 \cdot i(1) = h_0 \cdot w_2^2\widetilde{\delta'}$ |
| 129 | 25 | $84+85$ | $i(\gamma w_1w_2^2)$ | $(g^2, gw_1)$ | $\gamma w_1w_2^2 \cdot i(1)$ |
| 130 | 24 | 88 | $h_1w_2^2\widetilde{\delta'}$ | $(g, w_1)$ | $h_1 \cdot w_2^2\widetilde{\delta'}$ |
| 130 | 25 | 87 | $w_1w_2^2\widetilde{\gamma}$ | $(g^2, gw_1)$ | $w_1w_2^2 \cdot \widetilde{\gamma}$ |
| 135 | 25 | 93 | $e_0gw_2^2\widetilde{h_1}$ | – | $d_0\gamma w_2^2 \cdot i(1)$ |
| 136 | 25 | 94 | $d_0w_2^2\widetilde{\gamma}$ | $(g^2, w_1^2)$ | $d_0w_2^2 \cdot \widetilde{\gamma}$ |
| 142 | 27 | 101 | $i(\alpha d_0gw_2^2)$ | $(g, w_1)$ | $\alpha d_0gw_2^2 \cdot i(1)$ |
| 143 | 27 | 103 | $d_0w_2^2\widetilde{\delta'}$ | $(g^2, gw_1, w_1^2)$ | $d_0w_2^2 \cdot \widetilde{\delta'}$ |
| 147 | 25 | 101 | $i(h_2w_2^3)$ | – | $h_2w_2^3 \cdot i(1)$ |
| 148 | 27 | 108 | $h_1^2w_2^3\widetilde{h_1}$ | $(g)$ | $h_1\gamma w_2^2 \cdot \widetilde{\gamma}$ |
| 150 | 26 | 107 | $i(h_2^2w_2^3)$ | $(g, w_1)$ | $h_2^2w_2^3 \cdot i(1)$ |
| 152 | 27 | 112 | $i(c_0w_2^3)$ | $(g)$ | $c_0w_2^3 \cdot i(1)$ |
| 153 | 28 | 115 | $i(h_1c_0w_2^3)$ | $(g)$ | $h_1c_0w_2^3 \cdot i(1)$ |
| 153 | 28 | $114+115$ | $\gamma^2w_2^2\widetilde{h_2^2}$ | $(g, w_1)$ | $\gamma\delta' w_2^2 \cdot i(1)$ |
| 154 | 28 | $116+117$ | $\delta' w_2^2\widetilde{\gamma}$ | $(g, w_1^2)$ | $\delta' w_2^2 \cdot \widetilde{\gamma}$ |
| 154 | 28 | 117 | $\delta w_2^2\widetilde{\gamma}$ | $(g)$ | $\delta w_2^2 \cdot \widetilde{\gamma}$ |
| 154 | 30 | $115+116$ | $i(\gamma^2w_1w_2^2)$ | $(g)$ | $\gamma^2w_1w_2^2 \cdot i(1)$ |

Table 6.10: $R_2$-module generators of $E_5(tmf/2)$ (cont.)

| $t-s$ | $s$ | $g$ | $x$ | $\mathrm{Ann}(x)$ | dec. |
|---|---|---|---|---|---|
| 155 | 30 | $118+119$ | $\gamma w_1 w_2^2 \widetilde{\gamma}$ | $(g)$ | $\gamma w_1 w_2^2 \cdot \widetilde{\gamma}$ |
| 161 | 30 | 127 | $d_0 g w_2^2 \widetilde{\beta^2}$ | $(g, w_1)$ | $d_0 \gamma w_2^2 \cdot \widetilde{\gamma}$ |
| 162 | 28 | 122 | $i(h_2 \beta w_2^3)$ | $(g, w_1)$ | $h_2 w_2^3 \cdot i(\beta)$ |
| 165 | 29 | 128 | $i(h_1 g w_2^3)$ | $(g, w_1)$ | $h_2^2 w_2^3 \cdot i(\beta)$ |
| 168 | 32 | 137 | $g^2 w_2^2 \widetilde{d_0 e_0}$ | $(g, w_1)$ | $\alpha d_0 g w_2^2 \cdot \widetilde{\gamma}$ |
| 176 | 31 | 144 | $i(\delta w_2^3)$ | $(g)$ | $\delta w_2^3 \cdot i(1)$ |
| 177 | 32 | 149 | $i(h_1 \delta w_2^3)$ | $(g)$ | $h_1 \delta w_2^3 \cdot i(1)$ |
| 178 | 32 | 151 | $h_1 w_2^3 \widetilde{\delta'}$ | $(g)$ | $\delta w_2^3 \cdot \widetilde{h_1}$ |
| 178 | 34 | $151+152$ | $\gamma^2 w_2^2 \widetilde{d_0 e_0}$ | $(g)$ | $(\alpha^3 g w_2^2 + h_0 w_1 w_2^3) \cdot \widetilde{\gamma}$ |
| 179 | 33 | 153 | $e_0 g^2 w_2^2 \widetilde{\gamma}$ | $(g)$ | $h_1 \delta w_2^3 \cdot \widetilde{h_1}$ |
| 180 | 35 | $153+154$ | $\gamma^2 w_1 w_2^2 \widetilde{\gamma}$ | $(g)$ | $\gamma^2 w_1 w_2^2 \cdot \widetilde{\gamma}$ |

Table 6.11: The non-cyclic $R_2$-module summands in $E_5(tmf/2)$

| $\langle x_1, x_2 \rangle$ |
|---|
| $\langle \widetilde{\gamma}, w_2^2 \widetilde{h_1} \rangle \cong \dfrac{\Sigma^{5,31} R_2 \oplus \Sigma^{17,115} R_2}{\langle (g^2 w_1, 0), (g w_1^2, 0), (g^5, g w_1), (0, g^3), (0, g^2 w_1) \rangle}$ |
| $\langle e_0 g \widetilde{h_1}, i(h_2 w_2) \rangle \cong \dfrac{\Sigma^{9,48} R_2 \oplus \Sigma^{9,60} R_2}{\langle (w_1, 0), (g, w_1), (0, g) \rangle}$ |
| $\langle \gamma^2 \widetilde{\gamma}, i(\delta' w_2^2) \rangle \cong \dfrac{\Sigma^{15,91} R_2 \oplus \Sigma^{23,151} R_2}{\langle (g w_1, 0), (g^3, w_1), (0, g^2) \rangle}$ |
| $\langle e_0 g w_2^2 \widetilde{h_1}, i(h_2 w_2^3) \rangle \cong \dfrac{\Sigma^{25,160} R_2 \oplus \Sigma^{25,172} R_2}{\langle (w_1, 0), (g, w_1), (0, g) \rangle}$ |

PROPOSITION 6.11. *The twelve classes in Table* 6.12 *generate* $E_5(tmf/2)$ *as an* $E_5(tmf)$*-module.*

PROOF. In view of Table 5.8, this is clear from the factorizations in the "dec."-column of Table 6.10, which can be verified using `ext`.

In bidegree $(t-s, s) = (83, 17)$, the differential $d_2(\widetilde{w_2 \beta g}) = e_0 g^2 \widetilde{\gamma} + h_1^2 w_2 \widetilde{\delta'}$ makes $e_0 g^2 \widetilde{\gamma} = 17_{39}$ equal to $h_1^2 w_2 \widetilde{\delta'} = 17_{40} = h_1 \delta w_2 \cdot \widetilde{h_1}$, starting at the $E_3$-term.

Likewise, in bidegree $(t-s,s) = (179,33)$ the generator $e_0 g^2 w_2^2 \widetilde{\gamma} = 33_{153}$ becomes equal to $h_1^2 w_2^3 \widetilde{\delta}' = 33_{154} = h_1 \delta w_2^3 \cdot \widetilde{h_1}$ at $E_3$. □

Table 6.12: $E_5(tmf)$-module generators of $E_5(tmf/2)$

| $t-s$ | $s$ | $g$ | $x$ |
|---|---|---|---|
| 0 | 0 | 0 | $i(1)$ |
| 2 | 1 | 1 | $\widetilde{h_1}$ |
| 7 | 2 | 3 | $\widetilde{h_2^2}$ |
| 9 | 3 | 2 | $\widetilde{c_0}$ |
| 15 | 3 | 4 | $i(\beta)$ |
| 26 | 5 | 8 | $\widetilde{\gamma}$ |
| 33 | 7 | 10 | $\widetilde{\delta}'$ |
| 81 | 16 | $38+39$ | $\gamma^2\widetilde{\beta^2}$ |
| 98 | 17 | 51 | $w_2^2\widetilde{h_1}$ |
| 103 | 18 | 57 | $w_2^2\widetilde{h_2^2}$ |
| 105 | 19 | 60 | $w_2^2\widetilde{c_0}$ |
| 129 | 23 | 84 | $w_2^2\widetilde{\delta}'$ |

PROPOSITION 6.12. *Charts showing $E_5(tmf/2)$ for $0 \le t-s \le 192$ are given in Figures 6.1 to 6.8. All nonzero $h_0$-, $h_1$- and $h_2$-multiplications are displayed. The red dots indicate $w_1$-power torsion classes, and black dots indicate $w_1$-periodic classes. All $R_2$-module generators are labeled, except those that are also $h_0$-, $h_1$- or $h_2$-multiples.*

PROOF. The $R_2$-module structure shown in these charts is made explicit in Tables 6.10 and 6.11. The $h_0$-, $h_1$- and $h_2$-multiplications follow by comparison with the $E_2$-term, shown for $0 \le t-s \le 96$ in Figures 1.24 to 1.27. In many cases the $h_i$-multiplications are also visible from the decompositions given in Table 6.10. □

THEOREM 6.13. *$E_5 = E_\infty$ in the Adams spectral sequence for $tmf/2$.*

PROOF. It will be useful to consult the charts of $E_5(tmf/2)$ in Figures 6.1 to 6.8. The generators $i(1)$, $\widetilde{h_1}$, $\widetilde{h_2^2}$, $\widetilde{c_0}$, $i(\beta)$ and $w_2^2\widetilde{h_2^2}$ are infinite cycles because all possible differentials on these classes land in trivial groups.

The generators $\widetilde{\delta}'$, $\gamma^2\widetilde{\beta^2}$, $w_2^2\widetilde{c_0}$ and $w_2^2\widetilde{\delta}'$ are all annihilated by $w_1^2$ at $E_5$, while their possible targets are all $w_1$-torsion free at $E_5$. Formally, we have $w_1^2 x = 0$, so, to rule out the possibility that $d_r(x) = y$, it suffices to show that $w_1^2 y \ne 0$ at $E_r$.

For $x = \widetilde{\delta}'$ we must rule out $d_9(x) = i(w_1^4)$. This is impossible because $i(w_1^6)$ could only have been hit by $i(\gamma w_1^3)$, which is an infinite cycle because $\gamma$ and $w_1$ are infinite cycles in the Adams spectral sequence for $tmf$.

For $x = \gamma^2\widetilde{\beta^2}$ we must rule out a $d_7$, a $d_{15}$ and a $d_{24}$. In this case, since $w_1 x = 0$, we can rule out $d_r(x) = y$ by showing that $w_1 y \ne 0$ at $E_r$. The possible sources for

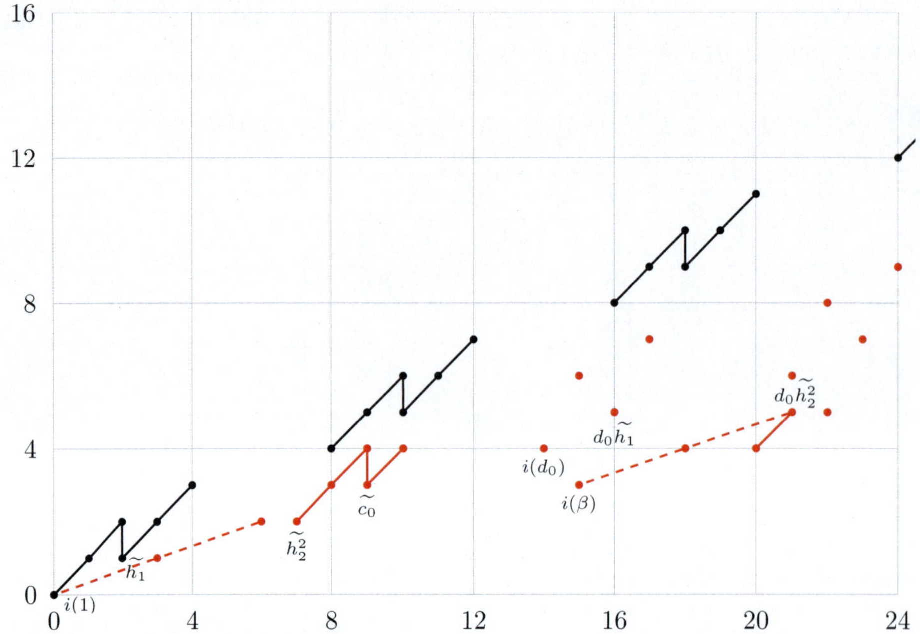


FIGURE 6.1. $E_5(tmf/2) = E_\infty(tmf/2)$ for $0 \leq t - s \leq 24$

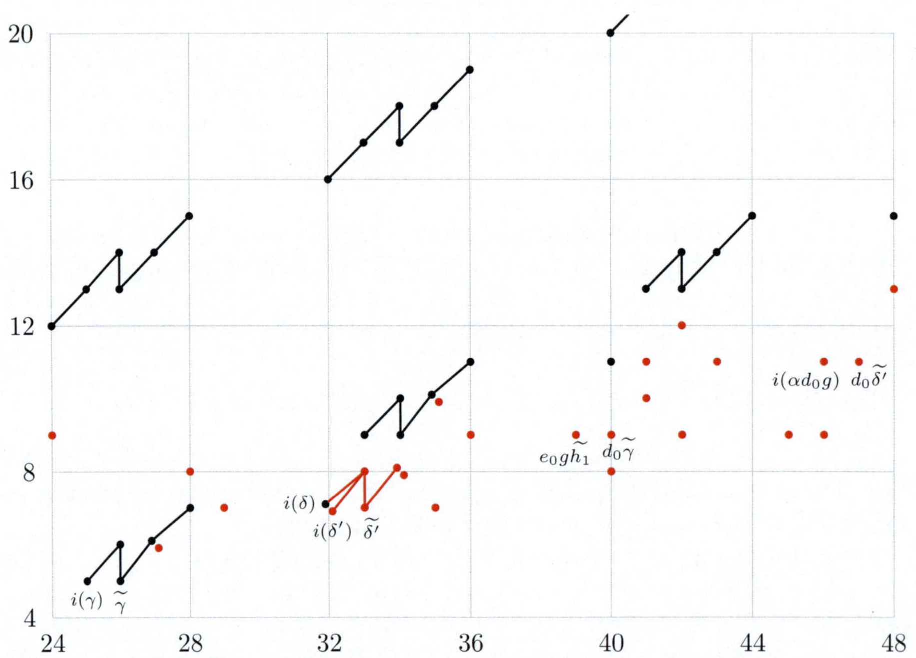


FIGURE 6.2. $E_5(tmf/2) = E_\infty(tmf/2)$ for $24 \leq t - s \leq 48$

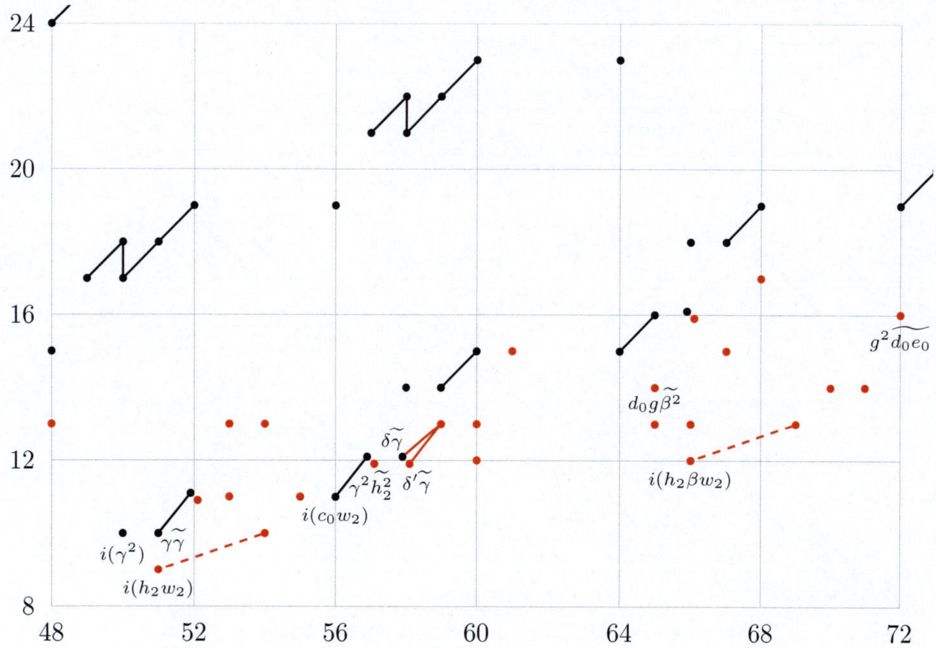


FIGURE 6.3. $E_5(tmf/2) = E_\infty(tmf/2)$ for $48 \leq t - s \leq 72$

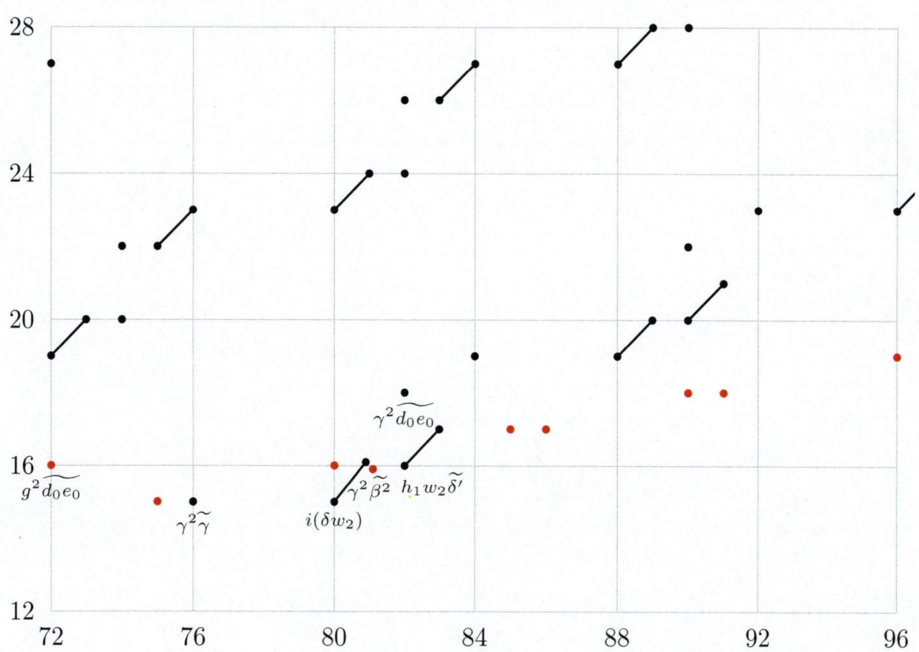


FIGURE 6.4. $E_5(tmf/2) = E_\infty(tmf/2)$ for $72 \leq t - s \leq 96$

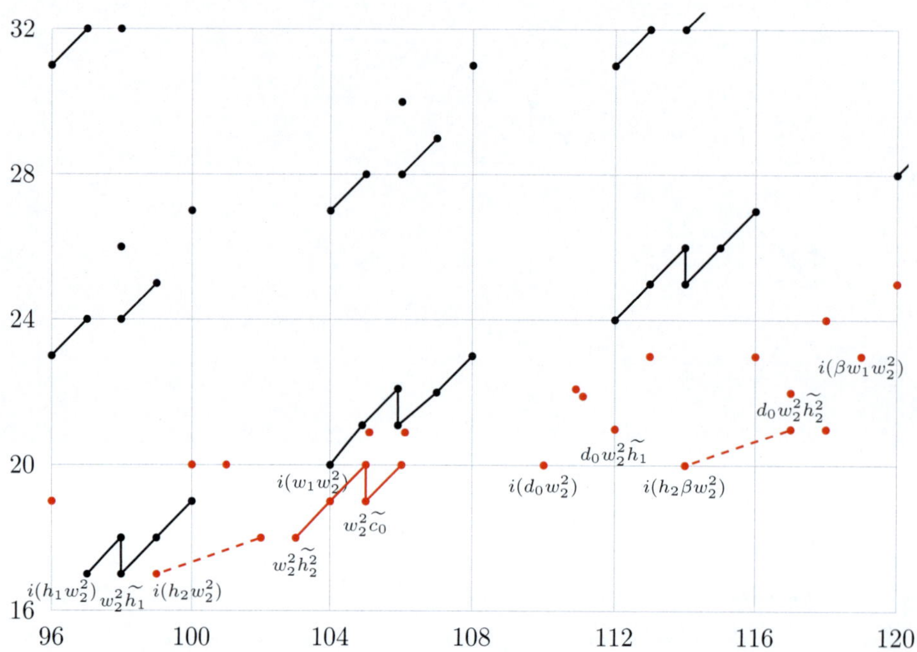


FIGURE 6.5. $E_5(tmf/2) = E_\infty(tmf/2)$ for $96 \leq t-s \leq 120$

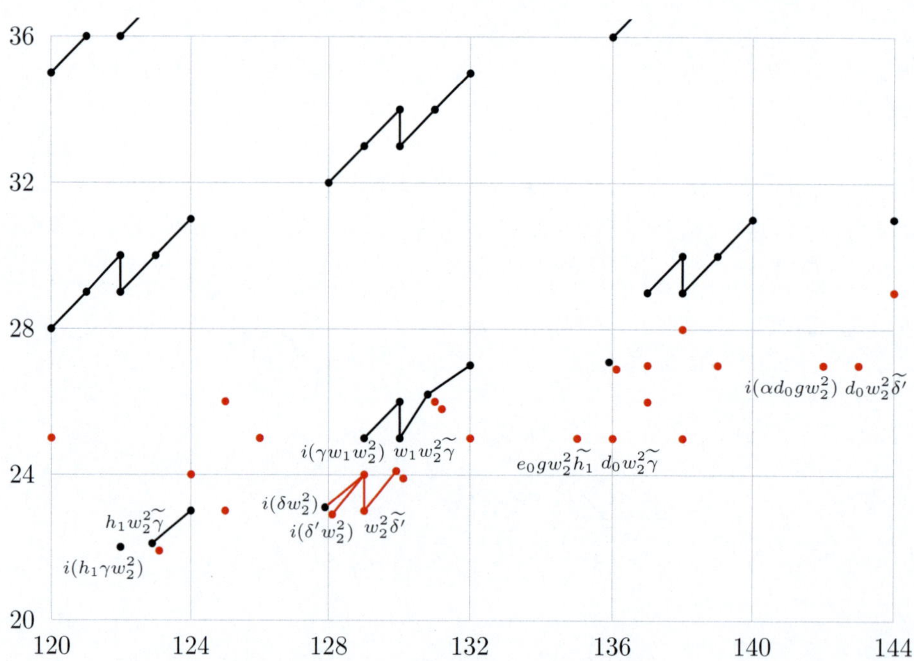


FIGURE 6.6. $E_5(tmf/2) = E_\infty(tmf/2)$ for $120 \leq t-s \leq 144$

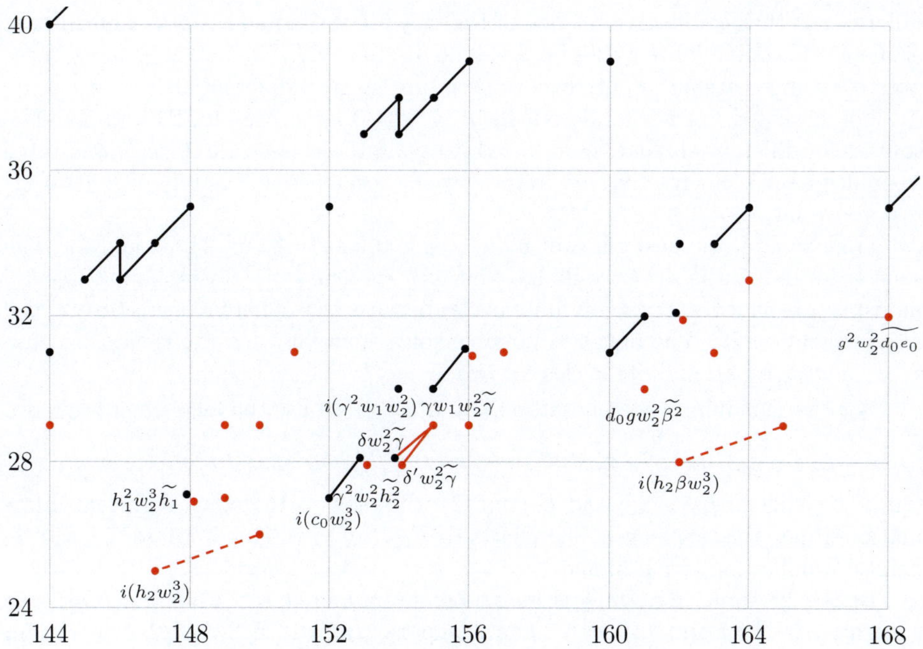


FIGURE 6.7. $E_5(tmf/2) = E_\infty(tmf/2)$ for $144 \leq t - s \leq 168$

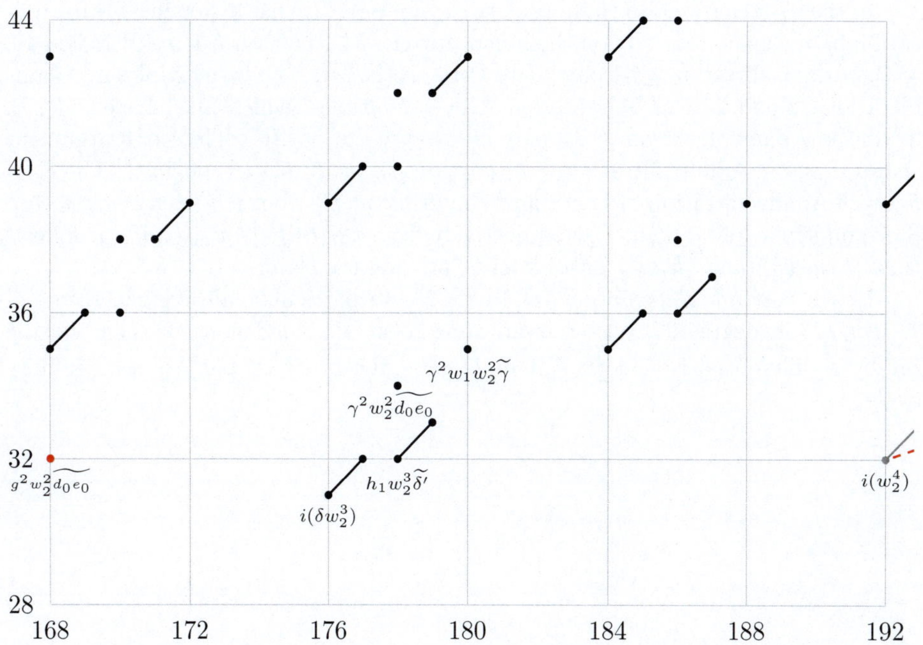


FIGURE 6.8. $E_5(tmf/2) = E_\infty(tmf/2)$ for $168 \leq t - s \leq 192$

differentials that could have hit one of the $w_1 y$ are $i(\gamma w_1^8)$ and two $h_1$-multiples of elements that are infinite cycles because the 87-stem of $E_5(tmf/2)$ is 0. Since these possible sources are all infinite cycles, there are no such differentials.

For $x = w_2^2\widetilde{c_0}$ we must rule out $d_r(x) = y \neq 0$ for $r = 8$, 16, 24 and 33. The sources for differentials that could have hit $w_1^2 y$ in these cases are $i(\gamma w_1^{12})$ and three $h_1$-multiples of elements that are infinite cycles because the 119-stem of $E_5(tmf/2)$ is 0 above filtration 23.

For $x = w_2^2\widetilde{\delta'}$ we must rule out $d_r(x) = y \neq 0$ for $r = 9$, 16, 24, 32 and 41. The sources for differentials that could have hit $w_1^2 y$ are $i(\gamma w_1^3 w_2^2)$, $i(\gamma w_1^{15})$ and three $h_1$-multiples of elements that are infinite cycles because the 143-stem of $E_5(tmf/2)$ is 0 above filtration 27. The first two possible sources are also infinite cycles, because $\gamma$, $w_1$ and $w_1 w_2^2$ are infinite cycles for $tmf$.

For the remaining two generators, $\widetilde{\gamma}$ and $\widetilde{w_2^2 h_1}$, we use the long exact sequence

$$\cdots \longrightarrow \pi_n(tmf) \xrightarrow{2} \pi_n(tmf) \xrightarrow{i} \pi_n(tmf/2) \xrightarrow{j} \pi_{n-1}(tmf) \longrightarrow \dots .$$

Our knowledge of $E_\infty(tmf)$ and $E_5(tmf/2)$ will allow us to deduce sufficient information about this sequence. The charts of $E_\infty(tmf)$ in Figures 5.1 to 5.5 will be helpful in following the argument.

In the 25-stem, $E_\infty(tmf)$ is generated by $\gamma$ and $h_1 w_1^3$, while $E_5(tmf/2)$ is generated by $i(\gamma)$ and $i(h_1 w_1^3)$. The group $\pi_{25}(tmf)$ is $\mathbb{Z}/2 \oplus \mathbb{Z}/2$ rather than $\mathbb{Z}/4$, because multiplication by $\eta$ acts nontrivially on the homotopy class $\{h_1 w_1^3\}$ detected by $h_1 w_1^3$, so this class cannot be a multiple of 2. Hence $\pi_{25}(tmf/2)$ has order (at least) 4, so both $i(\gamma)$ and $i(h_1 w_1^3)$ survive to $E_\infty(tmf/2)$. In particular, $d_8(\widetilde{\gamma}) \neq i(h_1 w_1^3)$ must be zero.

In the 97-stem, we also claim that $\pi_{97}(tmf) \cong (\mathbb{Z}/2)^5$ has exponent 2. Multiplication by $\eta$ acts nontrivially on the homotopy class $\{h_1 w_1^{12}\}$ in Adams filtration 49, so the Adams filtration $\geq 41$ part of $\pi_{97}(tmf)$ is $(\mathbb{Z}/2)^2$. The $h_1$-multiples in Adams filtration 32 and 24 can be represented by $\eta$-multiples, which must have order 2. Thus the Adams filtration $\geq 24$ part of $\pi_{97}(tmf)$ is $(\mathbb{Z}/2)^4$. The same argument shows that the Adams filtration $\geq 28$ part of $\pi_{105}(tmf)$ has exponent 2. The class $h_1 w_2^2$ in Adams filtration 17 is not an $h_1$-multiple at $E_\infty$, but $w_1 \cdot h_1 w_2^2 = h_1 \cdot w_1 w_2^2$. Since multiplication by $\{w_1\}$ acts injectively from $\pi_{97}(tmf)$ to $\pi_{105}(tmf)$, it follows that $\{h_1 w_1 w_2^2\}$ and $\{h_1 w_2^2\}$ have order 2, proving the claim.

Hence $\pi_{97}(tmf/2)$ has order (at least) $2^5$. In particular, all five generators of $E_5(tmf/2)$ in degree 97 must remain nonzero at $E_\infty$, and none of them can be hit by a differential from $\widetilde{w_2^2 h_1}$. This finishes the proof that $\widetilde{w_2^2 h_1}$ is an infinite cycle. □

CHAPTER 7

# The Adams spectral sequence for $tmf/\eta$

We calculate the $d_r$-differentials in the Adams spectral sequence for $tmf/\eta = tmf \wedge C\eta$. These are nontrivial for $r \in \{2, 3\}$, and zero for $r \geq 4$, so the spectral sequence collapses at the $E_4$-term. The module structure over the Adams spectral sequence for $tmf$ suffices to determine almost all of these differentials. There is one exceptional case, concerning $d_3(h_2^2\widehat{\beta})$, for which we also use the hidden $\eta$-extension to $d_0 w_1$ for $tmf$. The resulting $E_\infty$-term is the associated graded of a complete Hausdorff filtration of $\pi_*(tmf/\eta)^\wedge_2$.

## 7.1. The $E_2$-term for $tmf/\eta$

The initial term

$$E_2 = E_2(tmf/\eta) \cong \mathrm{Ext}_{A(2)}(M_2, \mathbb{F}_2)$$

of the mod 2 Adams spectral sequence for $tmf/\eta$ was calculated in Part I. The groups $E_2^{s,t}$ for $0 \leq t-s \leq 96$ are displayed in Figures 1.28 to 1.31. By Corollary 4.13 the $E_2$-term for $tmf/\eta$ is generated as a module over $E_2(tmf) = \mathrm{Ext}_{A(2)}(\mathbb{F}_2, \mathbb{F}_2)$ by the seven classes listed in Table 7.1. As a module over $R_0 = \mathbb{F}_2[g, w_1, w_2]$ the $E_2$-term for $tmf/\eta$ is presented as a direct sum of cyclic modules in Table 7.2, most of which is reproduced from Table 4.5. We note that the $E_2$-term is free over $\mathbb{F}_2[w_1, w_2]$, and finitely generated over $R_0[h_0] = \mathbb{F}_2[h_0, g, w_1, w_2]$.

We have made the following changes in our choice of $R_0$-module generators in order to simplify the description of $d_2$ and $E_3$.

(1) In bidegree $(t-s, s) = (26, 7)$, we replace the generator $i(\alpha d_0) = 7_{13} + 7_{14}$ by $\alpha^2\widehat{h_0} = 7_{13}$. We then also replace the tower $h_0^{1+i}\alpha\widehat{\alpha}$ by the element $h_0\alpha\widehat{\alpha} = 7_{14}$ and the tower $h_0^{1+i}\alpha^2\widehat{h_0}$. These substitutions make use of the relations
   - $i(\alpha d_0) = \alpha^2\widehat{h_0} + h_0\alpha\widehat{\alpha}$
   - $g \cdot i(\alpha d_0) = g \cdot \alpha^2\widehat{h_0}$
   - $h_0\alpha^2\widehat{h_0} = h_0^2\alpha\widehat{\alpha} + w_1 \cdot i(h_2\beta)$.

(2) In bidegree $(t-s, s) = (29, 7)$, we replace the generator $i(\alpha e_0) = 7_{16} + 7_{17}$ by $\alpha\beta\widehat{h_0} = 7_{16}$ keeping $h_0\alpha\widehat{\beta} = 7_{17}$. We then also write the generator in bidegree $(29, 8)$ as $h_0^2\alpha\widehat{\beta} = 8_{19}$ rather than as $i(h_0\alpha e_0)$. These substitutions make use of the relations
   - $i(\alpha e_0) = \alpha\beta\widehat{h_0} + h_0\alpha\widehat{\beta}$
   - $g \cdot i(\alpha e_0) = g \cdot \alpha\beta\widehat{h_0}$
   - $h_0^2\alpha\widehat{\beta} = i(h_0\alpha e_0)$.

We also use the notation $\delta' = \delta + \alpha g$ from Chapter 5 to shorten some formulas. Recall Definition 5.1: $R_i = \mathbb{F}_2[g, w_1, w_2^{2^i}]$. Following the strategy of Chapter 5 we will keep track of $R_0$-module structure on the $E_2$-term, $R_1$-module structure on the $E_3$-term, and $R_2$-module structure on the $E_4 = E_\infty$-terms of the Adams spectral sequence for $tmf/\eta$.

Table 7.1: $E_2(tmf)$-module generators of $E_2(tmf/\eta)$

| $t-s$ | $s$ | $g$ | $x$ | $d_2(x)$ |
|---|---|---|---|---|
| 0 | 0 | 0 | $i(1)$ | 0 |
| 2 | 1 | 1 | $\widehat{h_0}$ | 0 |
| 5 | 1 | 3 | $\widehat{h_2}$ | 0 |
| 11 | 4 | 3 | $\widehat{h_1 c_0}$ | 0 |
| 14 | 3 | 5 | $\widehat{\alpha}$ | $w_1\widehat{h_2}$ |
| 17 | 3 | 7 | $\widehat{\beta}$ | $d_0\widehat{h_0}$ |
| 36 | 8 | 25 | $\widehat{d_0 g}$ | $w_1 i(\alpha\beta)$ |

Table 7.2: $R_0$-module generators of $E_2(tmf/\eta)$, with $i \geq 0$ in each $h_0$-tower

| $t-s$ | $s$ | $g$ | $x$ | $\mathrm{Ann}(x)$ | $d_2(x)$ | $d_2(xw_2)$ |
|---|---|---|---|---|---|---|
| 0 | 0 | 0 | $i(1)$ | $(0)$ | 0 | $g \cdot i(\alpha\beta)$ |
| 0 | 1 | 0 | $i(h_0)$ | $(g^2)$ | 0 | 0 |
| 0 | 2 | 0 | $i(h_0^2)$ | $(g^2)$ | 0 | 0 |
| 0 | $3+i$ | 0 | $i(h_0^{3+i})$ | $(g)$ | 0 | 0 |
| 2 | 1 | 1 | $\widehat{h_0}$ | $(0)$ | 0 | $g \cdot \alpha\beta\widehat{h_0}$ |
| 2 | 2 | 1 | $h_0\widehat{h_0}$ | $(g^2)$ | 0 | 0 |
| 2 | $3+i$ | 1 | $h_0^{2+i}\widehat{h_0}$ | $(g)$ | 0 | 0 |
| 3 | 1 | 2 | $i(h_2)$ | $(g)$ | 0 | 0 |
| 3 | 2 | 2 | $i(h_0h_2)$ | $(g)$ | 0 | 0 |
| 5 | 1 | 3 | $\widehat{h_2}$ | $(0)$ | 0 | $g^2 \cdot i(\alpha)$ |
| 5 | 2 | 3 | $h_0\widehat{h_2}$ | $(g)$ | 0 | 0 |
| 5 | 3 | 2 | $h_0^2\widehat{h_2}$ | $(g)$ | 0 | 0 |
| 6 | 2 | 4 | $i(h_2^2)$ | $(g)$ | 0 | 0 |
| 8 | 2 | 5 | $h_2\widehat{h_2}$ | $(g)$ | 0 | 0 |
| 8 | 3 | 3 | $i(c_0)$ | $(g)$ | 0 | 0 |
| 11 | 4 | 3 | $\widehat{h_1c_0}$ | $(0)$ | 0 | $gw_1 \cdot i(\beta^2)$ |
| 12 | 3 | 4 | $i(\alpha)$ | $(0)$ | $w_1 \cdot i(h_2)$ | $w_1 \cdot i(h_2w_2)$ $+ g^2 \cdot d_0\widehat{h_2}$ |

Table 7.2: $R_0$-module generators of $E_2(tmf/\eta)$, with $i \geq 0$ in each $h_0$-tower (cont.)

| $t-s$ | $s$ | $g$ | $x$ | $\mathrm{Ann}(x)$ | $d_2(x)$ | $d_2(xw_2)$ |
|---|---|---|---|---|---|---|
| 12 | 4 | 4 | $i(h_0\alpha)$ | $(g^2)$ | $w_1 \cdot i(h_0h_2)$ | $w_1 \cdot i(h_0h_2w_2)$ |
| 12 | $5+i$ | 5 | $i(h_0^{2+i}\alpha)$ | $(g)$ | 0 | 0 |
| 14 | 3 | 5 | $\widehat{\alpha}$ | $(0)$ | $w_1 \cdot \widehat{h_2}$ | $w_1 \cdot w_2\widehat{h_2}$ $+ g \cdot \alpha^2\widehat{\beta}$ |
| 14 | 4 | 5 | $i(d_0)$ | $(0)$ | 0 | $g \cdot i(\alpha^2 e_0)$ |
| 14 | 4 | 6 | $h_0\widehat{\alpha}$ | $(g^2)$ | $w_1 \cdot h_0\widehat{h_2}$ | $w_1 \cdot h_0w_2\widehat{h_2}$ |
| 14 | 5 | 7 | $i(h_0d_0)$ | $(g)$ | 0 | 0 |
| 14 | 5 | 8 | $h_0^2\widehat{\alpha}$ | $(g^2)$ | $w_1 \cdot h_0^2\widehat{h_2}$ | $w_1 \cdot h_0^2w_2\widehat{h_2}$ |
| 14 | $6+i$ | 8 | $h_0^{3+i}\widehat{\alpha}$ | $(g)$ | 0 | 0 |
| 15 | 3 | 6 | $i(\beta)$ | $(0)$ | $i(h_0d_0)$ | $i(h_0d_0w_2)$ $+ g^3 \cdot \widehat{h_0}$ |
| 15 | 4 | 7 | $i(h_0\beta)$ | $(g)$ | $w_1 \cdot i(h_2^2)$ | $w_1 \cdot i(h_2^2w_2)$ |
| 16 | 5 | 9 | $d_0\widehat{h_0}$ | $(0)$ | 0 | $g^2w_1 \cdot i(\beta)$ |
| 17 | 3 | 7 | $\widehat{\beta}$ | $(0)$ | $d_0\widehat{h_0}$ | $d_0w_2\widehat{h_0}$ $+ g \cdot \alpha\beta\widehat{\beta}$ |
| 17 | 4 | $8+9$ | $i(e_0)$ | $(0)$ | 0 | $g^2 \cdot i(\alpha^2)$ |
| 17 | 4 | 9 | $h_0\widehat{\beta}$ | $(g)$ | $w_1 \cdot h_2\widehat{h_2}$ | $w_1 \cdot h_2w_2\widehat{h_2}$ |
| 17 | 5 | $10+11$ | $i(h_0e_0)$ | $(g)$ | 0 | 0 |
| 17 | 5 | 11 | $h_0^2\widehat{\beta}$ | $(g)$ | $w_1 \cdot i(c_0)$ | $w_1 \cdot i(c_0w_2)$ |
| 17 | 6 | 10 | $i(h_0^2e_0)$ | $(g)$ | 0 | 0 |
| 18 | 4 | 10 | $i(h_2\beta)$ | $(g)$ | $i(h_0^2e_0)$ | $i(h_0^2e_0w_2)$ |
| 19 | 5 | 12 | $d_0\widehat{h_2}$ | $(0)$ | 0 | $g^2 \cdot \alpha^2\widehat{h_0}$ |
| 19 | 6 | 11 | $h_0d_0\widehat{h_2}$ | $(g)$ | 0 | 0 |
| 20 | 4 | 12 | $h_2\widehat{\beta}$ | $(g)$ | $h_0d_0\widehat{h_2}$ | $h_0d_0w_2\widehat{h_2}$ |
| 20 | 5 | 14 | $h_0h_2\widehat{\beta}$ | $(g)$ | 0 | 0 |
| 23 | 5 | 16 | $h_2^2\widehat{\beta}$ | $(g)$ | 0 | 0 |
| 24 | 6 | 14 | $i(\alpha^2)$ | $(0)$ | 0 | $g^3 \cdot \widehat{h_1c_0}$ |
| 24 | $7+i$ | 11 | $i(h_0^{1+i}\alpha^2)$ | $(g)$ | 0 | 0 |
| 26 | 6 | 15 | $\alpha\widehat{\alpha}$ | $(0)$ | $w_1 \cdot i(e_0)$ | $w_1 \cdot i(e_0w_2)$ $+ g \cdot d_0\gamma\widehat{\alpha}$ |
| 26 | 7 | 13 | $\alpha^2\widehat{h_0}$ | $(0)$ | 0 | $g^3w_1 \cdot \widehat{h_2}$ |

Table 7.2: $R_0$-module generators of $E_2(tmf/\eta)$, with $i \geq 0$ in each $h_0$-tower (cont.)

| $t-s$ | $s$ | $g$ | $x$ | $\mathrm{Ann}(x)$ | $d_2(x)$ | $d_2(xw_2)$ |
|---|---|---|---|---|---|---|
| 26 | 7 | 14 | $h_0\alpha\widehat{\alpha}$ | $(g)$ | $w_1 \cdot i(h_0e_0)$ | $w_1 \cdot i(h_0e_0w_2)$ |
| 26 | $8+i$ | 15 | $h_0^{1+i}\alpha^2\widehat{h_0}$ | $(g)$ | 0 | 0 |
| 27 | 6 | 16 | $i(\alpha\beta)$ | $(0)$ | 0 | $g^3 \cdot i(d_0)$ |
| 28 | 7 | 15 | $d_0\widehat{\alpha}$ | $(0)$ | $w_1 \cdot d_0\widehat{h_2}$ | $w_1 \cdot d_0w_2\widehat{h_2}$ $+ gw_1 \cdot \beta^2\widehat{\beta}$ |
| 29 | 6 | 17 | $\alpha\widehat{\beta}$ | $(0)$ | $gw_1 \cdot i(1)$ | $gw_1 \cdot i(w_2)$ $+ g^2 \cdot \widehat{d_0g}$ |
| 29 | 7 | 16 | $\alpha\beta\widehat{h_0}$ | $(0)$ | 0 | $g^3 \cdot d_0\widehat{h_0}$ |
| 29 | 7 | 17 | $h_0\alpha\widehat{\beta}$ | $(g)$ | $gw_1 \cdot i(h_0)$ | $gw_1 \cdot i(h_0w_2)$ |
| 29 | 8 | 19 | $h_0^2\alpha\widehat{\beta}$ | $(g)$ | $gw_1 \cdot i(h_0^2)$ | $gw_1 \cdot i(h_0^2w_2)$ |
| 30 | 6 | 18 | $i(\beta^2)$ | $(0)$ | 0 | $g^3 \cdot i(e_0)$ |
| 31 | 7 | 18 | $d_0\widehat{\beta}$ | $(0)$ | $gw_1 \cdot \widehat{h_0}$ | $gw_1 \cdot w_2\widehat{h_0}$ $+ g^2 \cdot \alpha^2\widehat{\alpha}$ |
| 31 | 8 | 21 | $h_0d_0\widehat{\beta}$ | $(g)$ | $gw_1 \cdot h_0\widehat{h_0}$ | $gw_1 \cdot h_0w_2\widehat{h_0}$ |
| 32 | 6 | 19 | $\beta\widehat{\beta}$ | $(0)$ | $g \cdot \widehat{h_1c_0}$ | $g \cdot w_2\widehat{h_1c_0}$ $+ g^2 \cdot \gamma\widehat{\alpha}$ |
| 32 | 7 | 20 | $i(\delta)$ | $(g)$ | 0 | 0 |
| 36 | 8 | 25 | $\widehat{d_0g}$ | $(0)$ | $w_1 \cdot i(\alpha\beta)$ | $w_1 \cdot i(\alpha\beta w_2)$ $+ g^2 \cdot \alpha d_0\widehat{\beta}$ |
| 36 | $9+i$ | 26 | $h_0^{1+i}\widehat{d_0g}$ | $(g)$ | 0 | 0 |
| 38 | 9 | 27 | $\alpha^2\widehat{\alpha}$ | $(0)$ | $w_1 \cdot \alpha\beta\widehat{h_0}$ | $w_1 \cdot \alpha\beta w_2\widehat{h_0}$ $+ g^3w_1 \cdot \widehat{\beta}$ |
| 38 | $10+i$ | 26 | $h_0^{1+i}\alpha^2\widehat{\alpha}$ | $(g)$ | 0 | 0 |
| 39 | 8 | 27 | $\gamma\widehat{\alpha}$ | $(0)$ | $w_1 \cdot i(\beta^2)$ | $w_1 \cdot i(\beta^2w_2)$ $+ g^3 \cdot \alpha\widehat{\alpha}$ |
| 41 | 9 | 29 | $\alpha^2\widehat{\beta}$ | $(0)$ | $w_1 \cdot i(\delta')$ | $w_1 \cdot i(\delta'w_2)$ $+ g^3 \cdot d_0\widehat{\alpha}$ |
| 41 | 10 | 28 | $i(\alpha^2e_0)$ | $(0)$ | 0 | $g^4w_1 \cdot i(1)$ |
| 42 | 8 | 29 | $\gamma\widehat{\beta}$ | $(0)$ | $i(\alpha^2e_0)$ | $i(\alpha^2e_0w_2)$ $+ g^3 \cdot \alpha\widehat{\beta}$ |

Table 7.2: $R_0$-module generators of $E_2(tmf/\eta)$, with $i \geq 0$ in each $h_0$-tower (cont.)

| $t-s$ | $s$ | $g$ | $x$ | Ann$(x)$ | $d_2(x)$ | $d_2(xw_2)$ |
|---|---|---|---|---|---|---|
| 43 | 10 | 29 | $\alpha d_0\widehat{\beta}$ | (0) | $gw_1 \cdot i(d_0)$ | $gw_1 \cdot i(d_0w_2)$ $+ g^2w_1 \cdot \gamma\widehat{\beta}$ |
| 44 | 9 | 31 | $\alpha\beta\widehat{\beta}$ | (0) | $gw_1 \cdot i(\beta)$ | $gw_1 \cdot i(\beta w_2)$ $+ g^3 \cdot d_0\widehat{\beta}$ |
| 47 | 9 | 33 | $\beta^2\widehat{\beta}$ | (0) | $g \cdot \alpha^2\widehat{h_0}$ | $g \cdot \alpha^2 w_2\widehat{h_0}$ $+ g^4 \cdot \widehat{\alpha}$ |
| 53 | 12 | 41 | $d_0\gamma\widehat{\alpha}$ | (0) | $gw_1 \cdot i(\alpha^2)$ | $gw_1 \cdot i(\alpha^2 w_2)$ $+ g^3w_1 \cdot \beta\widehat{\beta}$ |

## 7.2. The $d_2$-differentials for $tmf/\eta$

THEOREM 7.1. *The $d_2$-differential in $E_2(tmf/\eta)$ is $R_1$-linear. Its values on a set of $E_2(tmf)$-module generators are listed in Table 7.1, and its values on a set of $R_1$-module generators are listed in Table 7.2.*

PROOF. The classes $g$, $w_1$ and $w_2^2$ are $d_2$-cycles in $E_2(tmf)$, so the Leibniz rule implies that multiplication by each of these elements commutes with the $d_2$-differential in $E_2(tmf/\eta)$.

Next, we determine $d_2$ on the module generators of $E_2(tmf/\eta)$ over $E_2(tmf)$. See Figures 1.28 and 1.29. The $d_2$-differentials on $i(1)$, $\widehat{h_0}$ and $\widehat{h_2}$ are zero because the target groups are trivial. The $d_2$-differential on $\widehat{h_1c_0}$ is zero by $h_0$-linearity. The map $j\colon C\eta \to S^2$ induces a morphism of Adams spectral sequences

$$E_r(tmf/\eta) \xrightarrow{j} E_r^{*,*-2}(tmf)\,.$$

By Proposition 5.8 (or Table 5.1) the classes $\alpha$ and $\beta$ both support nontrivial $d_2$-differentials. Hence their lifts $\widehat{\alpha}$ and $\widehat{\beta}$ must also support nonzero $d_2$-differentials, and the only possible values are $5_6 = w_1\widehat{h_2}$ and $5_9 = d_0\widehat{h_0}$, respectively.

The case of $d_2(\widehat{d_0g})$ remains. Here we use the relation $e_0 \cdot \widehat{d_0g} = 12_{41} = d_0\gamma \cdot \widehat{\alpha}$ and the Leibniz rule to calculate $e_0 \cdot d_2(\widehat{d_0g}) = d_0\gamma \cdot d_2(\widehat{\alpha}) = d_0\gamma \cdot w_1\widehat{h_2} = 14_{40} = gw_1 \cdot i(\alpha^2) \neq 0$. Hence $d_2(\widehat{d_0g}) \neq 0$, and the only possible value is $10_{22} = w_1 \cdot i(\alpha\beta)$.

Finally, we use Table 5.1 and the Leibniz rule to calculate $d_2$ for $x$ and $xw_2 = w_2 \cdot x$, with $x$ ranging through the list of $R_0$-module generators for $E_2(tmf/\eta)$. (These elements then range through a list of $R_1$-module generators for the same $E_2$-term.) In particular $d_2(w_2 \cdot x) = d_2(w_2) \cdot x + w_2 \cdot d_2(x) = \alpha\beta g \cdot x + w_2 \cdot d_2(x)$. In this finite range, the action of $E_2(tmf)$ on $E_2(tmf/\eta)$ is calculated using `ext`. □

REMARK 7.2. To use `ext` to assist in calculating the products $\alpha\beta g \cdot x$ for $x \in E_2(tmf/\eta)$, use `cocycle tmfCeta 0 0`, ..., `cocycle tmfCeta 12 41`, `dolifts 0 40 maps` and `collect maps all`. The nonzero products with $\alpha\beta g = 10_{18}$ then appear as lines containing `(10 18 F2)` in the file `all`. If the product is a $g$-multiple, there will also appear a line containing `(4 8 F2)` in the same block, since $g = 4_8$ in the minimal $A(2)$-module resolution for $\mathbb{F}_2$. Similarly, $gw_1$-multiples appear with `(8 11 F2)`, $g^2$-multiples appear with `(8 18 F2)`, and so on.

## 7.3. The $d_3$-differentials for $tmf/\eta$

It is now a simple matter to compute the $E_3$-term of the Adams spectral sequence for $tmf/\eta$, as a direct sum of $R_1 = \mathbb{F}_2[g, w_1, w_2^2]$-modules. This is carried out in Appendix C.1 and the results are recorded in Tables 7.3 and 7.4, where we also record the results of this section, calculating the $d_3$-differential. Among the new relations that appear at the $E_3$-term we emphasize

$$i(h_0 d_0 w_2) = g^3 \cdot \widehat{h_0}$$

in bidegree $(t-s, s) = (62, 13)$, which follows from the $d_2$-differential on $i(\beta w_2)$.

Table 7.3: $R_1$-module generators of $E_3(tmf/\eta)$, with $i \geq 0$ in each $h_0$-tower

| $t-s$ | $s$ | $g$ | $x$ | $\mathrm{Ann}(x)$ | $d_3(x)$ | $d_3(xw_2^2)$ |
|---|---|---|---|---|---|---|
| 0 | 0 | 0 | $i(1)$ | $(gw_1)$ | 0 | $g^3 \cdot i(\beta g)$ |
| 0 | 1 | 0 | $i(h_0)$ | $(g^2, gw_1)$ | 0 | 0 |
| 0 | 2 | 0 | $i(h_0^2)$ | $(g^2, gw_1)$ | 0 | 0 |
| 0 | $3+i$ | 0 | $i(h_0^{3+i})$ | $(g)$ | 0 | 0 |
| 2 | 1 | 1 | $\widehat{h_0}$ | $(g^4, gw_1)$ | 0 | 0 |
| 2 | 2 | 1 | $h_0\widehat{h_0}$ | $(g^2, gw_1)$ | 0 | 0 |
| 2 | $3+i$ | 1 | $h_0^{2+i}\widehat{h_0}$ | $(g)$ | 0 | 0 |
| 3 | 1 | 2 | $i(h_2)$ | $(g, w_1)$ | 0 | 0 |
| 3 | 2 | 2 | $i(h_0 h_2)$ | $(g, w_1)$ | 0 | 0 |
| 5 | 1 | 3 | $\widehat{h_2}$ | $(w_1)$ | 0 | $g^5 \cdot i(1)$ |
| 5 | 2 | 3 | $h_0\widehat{h_2}$ | $(g, w_1)$ | 0 | 0 |
| 5 | 3 | 2 | $h_0^2\widehat{h_2}$ | $(g, w_1)$ | 0 | 0 |
| 6 | 2 | 4 | $i(h_2^2)$ | $(g, w_1)$ | 0 | 0 |
| 8 | 2 | 5 | $h_2\widehat{h_2}$ | $(g, w_1)$ | 0 | 0 |
| 8 | 3 | 3 | $i(c_0)$ | $(g, w_1)$ | 0 | 0 |
| 11 | 4 | 3 | $\widehat{h_1 c_0}$ | $(g)$ | 0 | 0 |
| 12 | $5+i$ | 5 | $i(h_0^{2+i}\alpha)$ | $(g)$ | 0 | 0 |
| 14 | 4 | 5 | $i(d_0)$ | $(g^3, gw_1)$ | 0 | 0 |
| 14 | $6+i$ | 8 | $h_0^{3+i}\widehat{\alpha}$ | $(g)$ | 0 | 0 |
| 17 | 4 | $8+9$ | $i(e_0)$ | $(g^3, w_1)$ | 0 | 0 |
| 17 | 5 | $10+11$ | $i(h_0 e_0)$ | $(g, w_1)$ | 0 | 0 |
| 19 | 5 | 12 | $d_0\widehat{h_2}$ | – | 0 | 0 |

Table 7.3: $R_1$-module generators of $E_3(tmf/\eta)$, with $i \geq 0$ in each $h_0$-tower (cont.)

| $t-s$ | $s$ | $g$ | $x$ | $\mathrm{Ann}(x)$ | $d_3(x)$ | $d_3(xw_2^2)$ |
|---|---|---|---|---|---|---|
| 20 | 5 | 14 | $h_0h_2\widehat{\beta}$ | $(g)$ | $w_1 \cdot \widehat{h_1c_0}$ | $w_1 \cdot w_2^2\widehat{h_1c_0}$ |
| 23 | 5 | 16 | $h_2^2\widehat{\beta}$ | $(g)$ | $w_1 \cdot i(d_0)$ | $w_1 \cdot i(d_0w_2^2)$ |
| 24 | 6 | 14 | $i(\alpha^2)$ | $(g^2, gw_1)$ | 0 | 0 |
| 24 | $7+i$ | 11 | $i(h_0^{1+i}\alpha^2)$ | $(g)$ | 0 | 0 |
| 26 | $7+i$ | 13 | $h_0^i\alpha^2\widehat{h_0}$ | $(g)$ | 0 | 0 |
| 27 | 6 | 16 | $i(\alpha\beta)$ | $(g, w_1)$ | 0 | 0 |
| 29 | 7 | 16 | $\alpha\beta\widehat{h_0}$ | $(g, w_1)$ | 0 | 0 |
| 30 | 6 | 18 | $i(\beta^2)$ | $(w_1)$ | 0 | $g^6 \cdot \widehat{h_2}$ |
| 32 | 7 | $19+20$ | $i(\alpha g)$ | $(g)$ | 0 | 0 |
| 32 | 7 | 19 | $i(\delta')$ | $(g, w_1)$ | 0 | 0 |
| 32 | 8 | 22 | $i(h_0\alpha g)$ | $(g)$ | 0 | 0 |
| 34 | 8 | 24 | $h_0g\widehat{\alpha}$ | $(g)$ | 0 | 0 |
| 34 | 9 | 24 | $h_0^2g\widehat{\alpha}$ | $(g)$ | 0 | 0 |
| 35 | 7 | 22 | $i(\beta g)$ | $(w_1)$ | 0 | $g^5 \cdot i(\beta^2)$ |
| 36 | $9+i$ | 26 | $h_0^{1+i}\widehat{d_0g}$ | $(g)$ | 0 | 0 |
| 38 | $10+i$ | 26 | $h_0^{1+i}\alpha^2\widehat{\alpha}$ | $(g)$ | 0 | 0 |
| 48 | 9 | 34 | $i(h_0w_2)$ | $(g^2, gw_1)$ | 0 | 0 |
| 48 | 10 | 33 | $i(h_0^2w_2)$ | $(g^2, gw_1)$ | 0 | 0 |
| 48 | $11+i$ | 34 | $i(h_0^{3+i}w_2)$ | $(g)$ | 0 | 0 |
| 50 | 10 | 36 | $h_0w_2\widehat{h_0}$ | $(g^2, gw_1)$ | 0 | 0 |
| 50 | $11+i$ | 36 | $h_0^{2+i}w_2\widehat{h_0}$ | $(g)$ | 0 | 0 |
| 51 | 9 | 36 | $i(h_2w_2)$ | – | 0 | 0 |
| 51 | 10 | 37 | $i(h_0h_2w_2)$ | $(g, w_1)$ | 0 | 0 |
| 53 | 10 | 39 | $h_0w_2\widehat{h_2}$ | $(g, w_1)$ | 0 | 0 |
| 53 | 11 | 39 | $h_0^2w_2\widehat{h_2}$ | $(g, w_1)$ | 0 | 0 |
| 54 | 10 | 40 | $i(h_2^2w_2)$ | $(g, w_1)$ | 0 | 0 |
| 56 | 10 | 41 | $h_2w_2\widehat{h_2}$ | $(g, w_1)$ | 0 | 0 |
| 56 | 11 | 42 | $i(c_0w_2)$ | $(g, w_1)$ | 0 | 0 |
| 56 | 12 | $43+44$ | $g\widehat{d_0g} + i(w_1w_2)$ | $(g)$ | 0 | 0 |
| 58 | 13 | $46+47$ | $\alpha^2g\widehat{\alpha} + w_1w_2\widehat{h_0}$ | $(g)$ | 0 | 0 |
| 59 | 12 | $46+47$ | $\gamma g\widehat{\alpha} + w_2\widehat{h_1c_0}$ | $(g)$ | 0 | 0 |

Table 7.3: $R_1$-module generators of $E_3(tmf/\eta)$, with $i \geq 0$ in each $h_0$-tower (cont.)

| $t-s$ | $s$ | $g$ | $x$ | $\mathrm{Ann}(x)$ | $d_3(x)$ | $d_3(xw_2^2)$ |
|---|---|---|---|---|---|---|
| 60 | $13+i$ | 50 | $i(h_0^{2+i}\alpha w_2)$ | $(g)$ | 0 | 0 |
| 62 | 12 | $50+51$ | $\gamma g\widehat{\beta} + i(d_0 w_2)$ | $(gw_1)$ | 0 | $g^4 \cdot (g^3\widehat{\beta}$ $+ \alpha\beta w_2\widehat{h_0})$ |
| 62 | $14+i$ | 53 | $h_0^{3+i} w_2\widehat{\alpha}$ | $(g)$ | 0 | 0 |
| 65 | 13 | $59+60$ | $i(h_0 e_0 w_2)$ | $(g, w_1)$ | 0 | 0 |
| 67 | 13 | $61+62$ | $\beta^2 g\widehat{\beta} + d_0 w_2\widehat{h_2}$ | $(w_1)$ | 0 | $g^5 \cdot (\gamma g\widehat{\beta}$ $+ i(d_0 w_2))$ |
| 68 | 13 | 64 | $h_0 h_2 w_2\widehat{\beta}$ | $(g)$ | $w_1 \cdot (\gamma g\widehat{\alpha}$ $+ w_2\widehat{h_1 c_0})$ | $w_1 \cdot (\gamma g w_2^2\widehat{\alpha}$ $+ w_2^3\widehat{h_1 c_0})$ |
| 71 | 13 | 66 | $h_2^2 w_2\widehat{\beta}$ | $(g)$ | $w_1 \cdot (\gamma g\widehat{\beta}$ $+ i(d_0 w_2))$ | $w_1 \cdot (\gamma g w_2^2\widehat{\beta}$ $+ i(d_0 w_2^3))$ |
| 72 | 14 | $65+66$ | $\beta g^2\widehat{\beta} + i(\alpha^2 w_2)$ | $(gw_1)$ | 0 | $g^5 \cdot (\beta^2 g\widehat{\beta}$ $+ d_0 w_2\widehat{h_2})$ |
| 72 | $15+i$ | 64 | $i(h_0^{1+i}\alpha^2 w_2)$ | $(g)$ | 0 | 0 |
| 74 | 15 | $66+67$ | $g^3\widehat{\alpha} + \alpha^2 w_2\widehat{h_0}$ | $(g)$ | 0 | 0 |
| 74 | $16+i$ | 72 | $h_0^{1+i}\alpha^2 w_2\widehat{h_0}$ | $(g)$ | 0 | 0 |
| 77 | 15 | $71+72$ | $g^3\widehat{\beta} + \alpha\beta w_2\widehat{h_0}$ | $(w_1)$ | 0 | $g^5 \cdot (\beta g^2\widehat{\beta}$ $+ i(\alpha^2 w_2))$ |
| 80 | 15 | 76 | $i(\delta w_2)$ | $(g)$ | 0 | 0 |
| 80 | 16 | 83 | $i(h_0\alpha g w_2)$ | $(g)$ | 0 | 0 |
| 82 | 16 | 86 | $h_0 g w_2\widehat{\alpha}$ | $(g)$ | 0 | 0 |
| 82 | 17 | 87 | $h_0^2 g w_2\widehat{\alpha}$ | $(g)$ | 0 | 0 |
| 84 | $17+i$ | 90 | $h_0^{1+i} w_2\widehat{d_0 g}$ | $(g)$ | 0 | 0 |
| 86 | $18+i$ | 90 | $h_0^{1+i}\alpha^2 w_2\widehat{\alpha}$ | $(g)$ | 0 | 0 |

Table 7.4: The non-cyclic $R_1$-module summand in $E_3(tmf/\eta)$

$$\langle d_0\widehat{h_2}, i(h_2 w_2)\rangle \cong \frac{\Sigma^{5,24}R_1 \oplus \Sigma^{9,60}R_1}{\langle (w_1, 0), (g^2, w_1), (0, g)\rangle}$$

THEOREM 7.3. *The $d_3$-differential in $E_3(tmf/\eta)$ is $R_2$-linear. Its values on a set of $R_2$-module generators are listed in Table 7.3.*

PROOF. The $E_3$-term is so sparse that the only $R_1[h_0]$-module generators whose $d_3$ lies in a nonzero bidegree are:

(3,1): $i(h_2)$
(11,4): $\widehat{h_1c_0}$
(20,5): $h_0h_2\widehat{\beta}$
(23,5): $h_2^2\widehat{\beta}$
(27,6): $i(\alpha\beta)$
(51,9): $i(h_2w_2)$
(59,12): $\gamma g\widehat{\alpha} + w_2\widehat{h_1c_0}$
(65,13): $i(h_0e_0w_2)$
(68,13): $h_0h_2w_2\widehat{\beta}$
(71,13): $h_2^2w_2\widehat{\beta}$.

Those of the form $i(x)$, where $x \in E_3(tmf)$, are immediate by naturality. Eliminating these, we have only the following left to consider:

(11,4): $\widehat{h_1c_0}$
(20,5): $h_0h_2\widehat{\beta}$
(23,5): $h_2^2\widehat{\beta}$
(59,12): $\gamma g\widehat{\alpha} + w_2\widehat{h_1c_0}$
(68,13): $h_0h_2w_2\widehat{\beta}$
(71,13): $h_2^2w_2\widehat{\beta}$.

We deal with these individually.

Applying $j\colon E_3^{s,t}(tmf/\eta) \to E_3^{s,t-2}(tmf)$, we get $j(d_3(h_0h_2\widehat{\beta})) = d_3(h_0h_2\beta) = d_3(h_1e_0) = h_1c_0w_1$. The only lift is $d_3(h_0h_2\widehat{\beta}) = w_1 \cdot \widehat{h_1c_0}$. This then eliminates the only possibility of a nonzero differential on $\widehat{h_1c_0}$, which is $d_3(\widehat{h_1c_0}) = h_0^2w_1\widehat{h_0}$, since this would imply that $d_3(w_1\widehat{h_1c_0}) = h_0^2w_1^2\widehat{h_0} \neq 0$.

Similarly, naturality with respect to $j$ implies that $d_3(h_0h_2w_2\widehat{\beta}) = w_1 \cdot (\gamma g\widehat{\alpha} + w_2\widehat{h_1c_0})$ and $d_3(\gamma g\widehat{\alpha} + w_2\widehat{h_1c_0}) = 0$.

Again, by naturality with respect to $j$ we have that $d_3(h_2^2w_2\widehat{\beta})$ must map to $d_3(h_2^2\beta w_2) = d_3(h_1gw_2) = g^3w_1 = \beta\gamma gw_1 \neq 0$, and $d_3(h_2^2w_2\widehat{\beta}) = w_1 \cdot (\gamma g\widehat{\beta} + i(d_0w_2))$ is the only possibility.

Finally, by Theorem 11.71 due to Mimura and Mahowald–Tangora, we know that $\eta^2\bar{\kappa} \in \pi_{22}(S)$ is detected by $Pd_0$ in $E_\infty(S)$. Hence $\eta^2\bar{\kappa} \in \pi_{22}(tmf)$ is detected by $d_0w_1$ in $E_\infty(tmf)$, as a consequence of Proposition 1.14 due to Adams. This $\eta$-multiple must map to zero in $\pi_{22}(tmf/\eta)$, so $i(d_0w_1) = 8_9 + 8_{10}$ must be a boundary. The only possibility is that $d_3(h_2^2\widehat{\beta}) = w_1 \cdot i(d_0)$.

The $w_2^2$-multiples now follow by the Leibniz rule, $d_3(w_2^2 \cdot x) = d_3(w_2^2) \cdot x + w_2^2 \cdot d_3(x) = \beta g^4 \cdot x + w_2^2 \cdot d_3(x)$. The second summand is straightforward to write down. The first summand vanishes whenever $g^4 \in \mathrm{Ann}(x)$. In the remaining eight cases we use `ext` to calculate $\beta g^4 \cdot x$ and to express it in terms of our module generators, as follows:

- $d_3(w_2^2 \cdot i(1)) = \beta g^4 \cdot i(1) = 19_{103} = g^3 \cdot i(\beta g)$.
- $d_3(w_2^2 \cdot \widehat{h_2}) = \beta g^4 \cdot \widehat{h_2} = 20_{117} = g^5 \cdot i(1)$.

- $d_3(w_2^2 \cdot i(\beta^2)) = \beta g^4 \cdot i(\beta^2) = 25_{176} = g^6 \cdot \widehat{h_2}$.
- $d_3(w_2^2 \cdot i(\beta g)) = \beta g^4 \cdot i(\beta g) = 26_{180} = g^5 \cdot i(\beta^2)$.
- $d_3(w_2^2 \cdot (\gamma g\widehat{\beta} + i(d_0 w_2))) = \beta g^4 \cdot (\gamma g\widehat{\beta} + i(d_0 w_2)) = 31_{251} + 31_{252} = g^4 \cdot (g^3\widehat{\beta} + \alpha\beta w_2\widehat{h_0})$.
- $d_3(w_2^2 \cdot (\beta^2 g\widehat{\beta} + d_0 w_2\widehat{h_2})) = \beta g^4 \cdot (\beta^2 g\widehat{\beta} + d_0 w_2\widehat{h_2}) = 32_{271} + 32_{272} = g^5 \cdot (\gamma g\widehat{\beta} + i(d_0 w_2))$.
- $d_3(w_2^2 \cdot (\beta g^2\widehat{\beta} + i(\alpha^2 w_2))) = \beta g^4 \cdot (\beta g^2\widehat{\beta} + i(\alpha^2 w_2)) = 33_{289} + 33_{290} = g^5 \cdot (\beta^2 g\widehat{\beta} + d_0 w_2\widehat{h_2})$.
- $d_3(w_2^2 \cdot (g^3\widehat{\beta} + \alpha\beta w_2\widehat{h_0})) = \beta g^4 \cdot (g^3\widehat{\beta} + \alpha\beta w_2\widehat{h_0}) = 34_{300} + 34_{301} = g^5 \cdot (\beta g^2\widehat{\beta} + i(\alpha^2 w_2))$. □

REMARK 7.4. To calculate the products $\beta g^4 \cdot x$ with `ext`, use `cocycle`, `dolifts` and `collect` as in Remark 7.2. The nonzero products with $\beta g^4 = 19_{56}$ then appear as lines containing `(19 56 F2)` in the file `all`. If the product is a $g^3$-multiple, there will also appear a line containing `(12 29 F2)` in the same block, since $g^3 = 12_{29}$ in the minimal $A(2)$-module resolution for $\mathbb{F}_2$. Similarly, $g^4$-multiples appear with `(16 48 F2)`, and so on.

## 7.4. The $E_\infty$-term for $tmf/\eta$

It is now a simple matter to compute the $E_4$-term of the Adams spectral sequence for $tmf/\eta$, as a direct sum of $R_2 = \mathbb{F}_2[g, w_1, w_2^4]$-modules. This is carried out in Appendix C.2 and the results are recorded in Tables 7.5 and 7.6. We show in Theorem 7.6 that there are no further nonzero differentials, so that $E_4(tmf/\eta) = E_\infty(tmf/\eta)$.

We make one pair of basis changes. We replace $i(\alpha g) = 7_{19} + 7_{20}$ by $i(\delta) = 7_{20}$. This has the same $R_2$-annihilator and is consistent with the basis chosen for $tmf$. Similarly, we replace $i(\alpha g w_2^2) = 23_{163} + 23_{164}$ by $i(\delta w_2^2) = 23_{164}$. Note that already at $E_2$, $i(h_0\alpha g) = i(h_0\delta)$, so we also make this name change in degrees 32, 80, 128 and 176.

Table 7.5: $R_2$-module generators of $E_4(tmf/\eta) = E_\infty(tmf/\eta)$, with $i \geq 0$ in each $h_0$-tower

| $t-s$ | $s$ | $g$ | $x$ | $\mathrm{Ann}(x)$ | dec. |
|---|---|---|---|---|---|
| 0 | 0 | 0 | $i(1)$ | $(g^5, gw_1)$ | **gen.** |
| 0 | 1 | 0 | $i(h_0)$ | $(g^2, gw_1)$ | $h_0 \cdot i(1)$ |
| 0 | 2 | 0 | $i(h_0^2)$ | $(g^2, gw_1)$ | $h_0^2 \cdot i(1)$ |
| 0 | $3+i$ | 0 | $i(h_0^{3+i})$ | $(g)$ | $h_0^{3+i} \cdot i(1)$ |
| 2 | 1 | 1 | $\widehat{h_0}$ | $(g^4, gw_1)$ | **gen.** |
| 2 | 2 | 1 | $h_0\widehat{h_0}$ | $(g^2, gw_1)$ | $h_0 \cdot \widehat{h_0}$ |
| 2 | $3+i$ | 1 | $h_0^{2+i}\widehat{h_0}$ | $(g)$ | $h_0^{2+i} \cdot \widehat{h_0}$ |
| 3 | 1 | 2 | $i(h_2)$ | $(g, w_1)$ | $h_2 \cdot i(1)$ |

Table 7.5: $R_2$-module generators of $E_4(tmf/\eta) = E_\infty(tmf/\eta)$, with $i \geq 0$ in each $h_0$-tower (cont.)

| $t-s$ | $s$ | $g$ | $x$ | $\mathrm{Ann}(x)$ | dec. |
|---|---|---|---|---|---|
| 3 | 2 | 2 | $i(h_0h_2)$ | $(g, w_1)$ | $h_0h_2 \cdot i(1)$ |
| 5 | 1 | 3 | $\widehat{h_2}$ | $(g^6, w_1)$ | **gen.** |
| 5 | 2 | 3 | $h_0\widehat{h_2}$ | $(g, w_1)$ | $h_0 \cdot \widehat{h_2}$ |
| 5 | 3 | 2 | $h_0^2\widehat{h_2}$ | $(g, w_1)$ | $h_0^2 \cdot \widehat{h_2}$ |
| 6 | 2 | 4 | $i(h_2^2)$ | $(g, w_1)$ | $h_2^2 \cdot i(1)$ |
| 8 | 2 | 5 | $h_2\widehat{h_2}$ | $(g, w_1)$ | $h_2 \cdot \widehat{h_2}$ |
| 8 | 3 | 3 | $i(c_0)$ | $(g, w_1)$ | $c_0 \cdot i(1)$ |
| 11 | 4 | 3 | $\widehat{h_1c_0}$ | $(g, w_1)$ | **gen.** |
| 12 | $5+i$ | 5 | $i(h_0^{2+i}\alpha)$ | $(g)$ | $h_0^i \cdot$ **gen.** |
| 14 | 4 | 5 | $i(d_0)$ | $(g^3, w_1)$ | $d_0 \cdot i(1)$ |
| 14 | $6+i$ | 8 | $h_0^{3+i}\widehat{\alpha}$ | $(g)$ | $h_0^i \cdot$ **gen.** |
| 17 | 4 | $8+9$ | $i(e_0)$ | $(g^3, w_1)$ | **gen.** |
| 17 | 5 | $10+11$ | $i(h_0e_0)$ | $(g, w_1)$ | $h_0 \cdot i(e_0)$ |
| 19 | 5 | 12 | $d_0\widehat{h_2}$ | – | $d_0 \cdot \widehat{h_2}$ |
| 24 | 6 | 14 | $i(\alpha^2)$ | $(g^2, gw_1)$ | **gen.** |
| 24 | $7+i$ | 11 | $i(h_0^{1+i}\alpha^2)$ | $(g)$ | $h_0^{1+i} \cdot i(\alpha^2)$ |
| 26 | $7+i$ | 13 | $h_0^i\alpha^2\widehat{h_0}$ | $(g)$ | $h_0^i \cdot$ **gen.** |
| 27 | 6 | 16 | $i(\alpha\beta)$ | $(g, w_1)$ | $\alpha\beta \cdot i(1)$ |
| 29 | 7 | 16 | $\alpha\beta\widehat{h_0}$ | $(g, w_1)$ | $\alpha\beta \cdot \widehat{h_0}$ |
| 30 | 6 | 18 | $i(\beta^2)$ | $(g^5, w_1)$ | $\gamma \cdot \widehat{h_2}$ |
| 32 | 7 | 19 | $i(\delta')$ | $(g, w_1)$ | $\delta' \cdot i(1)$ |
| 32 | 7 | 20 | $i(\delta)$ | $(g)$ | $\delta \cdot i(1)$ |
| 32 | 8 | 22 | $i(h_0\delta)$ | $(g)$ | $h_0\delta \cdot i(1)$ |
| 34 | 8 | 24 | $h_0g\widehat{\alpha}$ | $(g)$ | $\delta \cdot \widehat{h_0}$ |
| 34 | 9 | 24 | $h_0^2g\widehat{\alpha}$ | $(g)$ | $h_0\delta \cdot \widehat{h_0}$ |
| 35 | 7 | 22 | $i(\beta g)$ | $(g^3, w_1)$ | **gen.** |
| 36 | $9+i$ | 26 | $h_0^{1+i}\widehat{d_0g}$ | $(g)$ | $h_0^i \cdot$ **gen.** |
| 38 | $10+i$ | 26 | $h_0^{1+i}\alpha^2\widehat{\alpha}$ | $(g)$ | $h_0^i \cdot$ **gen.** |
| 48 | 9 | 34 | $i(h_0w_2)$ | $(g^2, gw_1)$ | **gen.** |
| 48 | 10 | 33 | $i(h_0^2w_2)$ | $(g^2, gw_1)$ | $h_0 \cdot i(h_0w_2)$ |
| 48 | $11+i$ | 34 | $i(h_0^{3+i}w_2)$ | $(g)$ | $h_0^{2+i} \cdot i(h_0w_2)$ |

Table 7.5: $R_2$-module generators of $E_4(tmf/\eta) = E_\infty(tmf/\eta)$, with $i \geq 0$ in each $h_0$-tower (cont.)

| $t-s$ | $s$ | $g$ | $x$ | Ann$(x)$ | dec. |
|---|---|---|---|---|---|
| 50 | 10 | 36 | $h_0 w_2 \widehat{h_0}$ | $(g^2, gw_1)$ | **gen.** |
| 50 | $11+i$ | 36 | $h_0^{2+i} w_2 \widehat{h_0}$ | $(g)$ | $h_0^{1+i} \cdot h_0 w_2 \widehat{h_0}$ |
| 51 | 9 | 36 | $i(h_2 w_2)$ | $-$ | $h_2 w_2 \cdot i(1)$ |
| 51 | 10 | 37 | $i(h_0 h_2 w_2)$ | $(g, w_1)$ | $h_0 h_2 w_2 \cdot i(1)$ |
| 53 | 10 | 39 | $h_0 w_2 \widehat{h_2}$ | $(g, w_1)$ | $h_2 w_2 \cdot \widehat{h_0}$ |
| 53 | 11 | 39 | $h_0^2 w_2 \widehat{h_2}$ | $(g, w_1)$ | $h_0 h_2 w_2 \cdot \widehat{h_0}$ |
| 54 | 10 | 40 | $i(h_2^2 w_2)$ | $(g, w_1)$ | $h_2^2 w_2 \cdot i(1)$ |
| 56 | 10 | 41 | $h_2 w_2 \widehat{h_2}$ | $(g, w_1)$ | $h_2 w_2 \cdot \widehat{h_2}$ |
| 56 | 11 | 42 | $i(c_0 w_2)$ | $(g, w_1)$ | $c_0 w_2 \cdot i(1)$ |
| 56 | 12 | $43+44$ | $g\widehat{d_0 g} + i(w_1 w_2)$ | $(g)$ | **gen.** |
| 58 | 13 | $46+47$ | $\alpha^2 g \widehat{\alpha} + w_1 w_2 \widehat{h_0}$ | $(g)$ | **gen.** |
| 59 | 12 | $46+47$ | $\gamma g \widehat{\alpha} + w_2 \widehat{h_1 c_0}$ | $(g, w_1)$ | **gen.** |
| 60 | $13+i$ | 50 | $i(h_0^{2+i} \alpha w_2)$ | $(g)$ | $h_0^i \cdot$ **gen.** |
| 62 | 12 | $50+51$ | $\gamma g \widehat{\beta} + i(d_0 w_2)$ | $(g^5, w_1)$ | **gen.** |
| 62 | $14+i$ | 53 | $h_0^{3+i} w_2 \widehat{\alpha}$ | $(g)$ | $h_0^i \cdot$ **gen.** |
| 65 | 13 | $59+60$ | $i(h_0 e_0 w_2)$ | $(g, w_1)$ | $h_2 \cdot (\gamma g \widehat{\beta}$ $+ i(d_0 w_2))$ |
| 67 | 13 | $61+62$ | $\beta^2 g \widehat{\beta} + d_0 w_2 \widehat{h_2}$ | $(g^5, w_1)$ | **gen.** |
| 72 | 14 | $65+66$ | $\beta g^2 \widehat{\beta} + i(\alpha^2 w_2)$ | $(g^5, gw_1)$ | **gen.** |
| 72 | $15+i$ | 64 | $i(h_0^{1+i} \alpha^2 w_2)$ | $(g)$ | $h_0^{1+i} \cdot (\beta g^2 \widehat{\beta}$ $+ i(\alpha^2 w_2))$ |
| 74 | 15 | $66+67$ | $g^3 \widehat{\alpha} + \alpha^2 w_2 \widehat{h_0}$ | $(g)$ | **gen.** |
| 74 | $16+i$ | 72 | $h_0^{1+i} \alpha^2 w_2 \widehat{h_0}$ | $(g)$ | $h_0^{1+i} \cdot (g^3 \widehat{\alpha}$ $+ \alpha^2 w_2 \widehat{h_0})$ |
| 77 | 15 | $71+72$ | $g^3 \widehat{\beta} + \alpha\beta w_2 \widehat{h_0}$ | $(g^4, w_1)$ | **gen.** |
| 80 | 15 | 76 | $i(\delta w_2)$ | $(g)$ | $\delta w_2 \cdot i(1)$ |
| 80 | 16 | 83 | $i(h_0 \delta w_2)$ | $(g)$ | $h_0 \delta w_2 \cdot i(1)$ |
| 82 | 16 | 86 | $h_0 g w_2 \widehat{\alpha}$ | $(g)$ | $\delta w_2 \cdot \widehat{h_0}$ |
| 82 | 17 | 87 | $h_0^2 g w_2 \widehat{\alpha}$ | $(g)$ | $h_0 \delta w_2 \cdot \widehat{h_0}$ |
| 84 | $17+i$ | 90 | $h_0^{1+i} w_2 \widehat{d_0 g}$ | $(g)$ | $h_0^i \cdot$ **gen.** |
| 86 | $18+i$ | 90 | $h_0^{1+i} \alpha^2 w_2 \widehat{\alpha}$ | $(g)$ | $h_0^i \cdot$ **gen.** |

Table 7.5: $R_2$-module generators of $E_4(tmf/\eta) = E_\infty(tmf/\eta)$, with $i \geq 0$ in each $h_0$-tower (cont.)

| $t-s$ | $s$ | $g$ | $x$ | Ann$(x)$ | dec. |
|---|---|---|---|---|---|
| 96 | 17 | 100 | $i(h_0 w_2^2)$ | $(g^2, gw_1)$ | $h_0 w_2^2 \cdot i(1)$ |
| 96 | 18 | 101 | $i(h_0^2 w_2^2)$ | $(g^2, gw_1)$ | $h_0^2 w_2^2 \cdot i(1)$ |
| 96 | $19+i$ | 105 | $i(h_0^{3+i} w_2^2)$ | $(g)$ | $h_0^{3+i} w_2^2 \cdot i(1)$ |
| 98 | 17 | 101 | $w_2^2 \widehat{h_0}$ | $(g^4, gw_1)$ | **gen.** |
| 98 | 18 | 104 | $h_0 w_2^2 \widehat{h_0}$ | $(g^2, gw_1)$ | $h_0 \cdot w_2^2 \widehat{h_0}$ |
| 98 | $19+i$ | 108 | $h_0^{2+i} w_2^2 \widehat{h_0}$ | $(g)$ | $h_0^{2+i} \cdot w_2^2 \widehat{h_0}$ |
| 99 | 17 | 102 | $i(h_2 w_2^2)$ | $(g, w_1)$ | $h_2 w_2^2 \cdot i(1)$ |
| 99 | 18 | 105 | $i(h_0 h_2 w_2^2)$ | $(g, w_1)$ | $h_0 h_2 w_2^2 \cdot i(1)$ |
| 101 | 18 | 107 | $h_0 w_2^2 \widehat{h_2}$ | $(g, w_1)$ | $h_0 w_2^2 \cdot \widehat{h_2}$ |
| 101 | 19 | 111 | $h_0^2 w_2^2 \widehat{h_2}$ | $(g, w_1)$ | $h_0^2 w_2^2 \cdot \widehat{h_2}$ |
| 102 | 18 | 108 | $i(h_2^2 w_2^2)$ | $(g, w_1)$ | $h_2^2 w_2^2 \cdot i(1)$ |
| 104 | 18 | 109 | $h_2 w_2^2 \widehat{h_2}$ | $(g, w_1)$ | $h_2 w_2^2 \cdot \widehat{h_2}$ |
| 104 | 19 | 114 | $i(c_0 w_2^2)$ | $(g, w_1)$ | $c_0 w_2^2 \cdot i(1)$ |
| 104 | 20 | 121 | $i(w_1 w_2^2)$ | $(g)$ | $w_1 w_2^2 \cdot i(1)$ |
| 107 | 20 | 124 | $w_2^2 \widehat{h_1 c_0}$ | $(g, w_1)$ | **gen.** |
| 108 | $21+i$ | 132 | $i(h_0^{2+i} \alpha w_2^2)$ | $(g)$ | $h_0^i \cdot$ **gen.** |
| 110 | 20 | 129 | $i(d_0 w_2^2)$ | $(g^3, w_1)$ | $d_0 w_2^2 \cdot i(1)$ |
| 110 | $22+i$ | 136 | $h_0^{3+i} w_2^2 \widehat{\alpha}$ | $(g)$ | $h_0^i \cdot$ **gen.** |
| 113 | 20 | $132+133$ | $i(e_0 w_2^2)$ | $(g^3, w_1)$ | **gen.** |
| 113 | 21 | $141+142$ | $i(h_0 e_0 w_2^2)$ | $(g, w_1)$ | $h_0 \cdot i(e_0 w_2^2)$ |
| 115 | 21 | 144 | $d_0 w_2^2 \widehat{h_2}$ | $-$ | $d_0 w_2^2 \cdot \widehat{h_2}$ |
| 120 | 22 | 150 | $i(\alpha^2 w_2^2)$ | $(g^2, gw_1)$ | **gen.** |
| 120 | $23+i$ | 152 | $i(h_0^{1+i} \alpha^2 w_2^2)$ | $(g)$ | $h_0^{1+i} \cdot i(\alpha^2 w_2^2)$ |
| 122 | $23+i$ | 155 | $h_0^i \alpha^2 w_2^2 \widehat{h_0}$ | $(g)$ | $h_0^i \cdot$ **gen.** |
| 123 | 22 | 152 | $i(\alpha\beta w_2^2)$ | $(g, w_1)$ | $\alpha\beta w_2^2 \cdot i(1)$ |
| 125 | 23 | 160 | $\alpha\beta w_2^2 \widehat{h_0}$ | $(g, w_1)$ | $\alpha\beta w_2^2 \cdot \widehat{h_0}$ |
| 128 | 23 | 163 | $i(\delta' w_2^2)$ | $(g, w_1)$ | $\delta' w_2^2 \cdot i(1)$ |
| 128 | 23 | 164 | $i(\delta w_2^2)$ | $(g)$ | $\delta w_2^2 \cdot i(1)$ |
| 128 | 24 | 177 | $i(h_0 \delta w_2^2)$ | $(g)$ | $h_0 \delta w_2^2 \cdot i(1)$ |
| 130 | 24 | 180 | $h_0 g w_2^2 \widehat{\alpha}$ | $(g)$ | $\delta w_2^2 \cdot \widehat{h_0}$ |
| 130 | 25 | 185 | $h_0^2 g w_2^2 \widehat{\alpha}$ | $(g)$ | $h_0 \delta w_2^2 \cdot \widehat{h_0}$ |

Table 7.5: $R_2$-module generators of $E_4(tmf/\eta) = E_\infty(tmf/\eta)$, with $i \geq 0$ in each $h_0$-tower (cont.)

| $t-s$ | $s$ | $g$ | $x$ | Ann$(x)$ | dec. |
|---|---|---|---|---|---|
| 132 | $25+i$ | 188 | $h_0^{1+i} w_2^2 \widehat{d_0 g}$ | $(g)$ | $h_0^i \cdot$ **gen.** |
| 134 | $26+i$ | 190 | $h_0^{1+i} \alpha^2 w_2^2 \widehat{\alpha}$ | $(g)$ | $h_0^i \cdot$ **gen.** |
| 144 | 25 | 198 | $i(h_0 w_2^3)$ | $(g^2, gw_1)$ | **gen.** |
| 144 | 26 | 201 | $i(h_0^2 w_2^3)$ | $(g^2, gw_1)$ | $h_0 \cdot i(h_0 w_2^3)$ |
| 144 | $27+i$ | 209 | $i(h_0^{3+i} w_2^3)$ | $(g)$ | $h_0^{2+i} \cdot i(h_0 w_2^3)$ |
| 146 | 26 | 204 | $h_0 w_2^3 \widehat{h_0}$ | $(g^2, gw_1)$ | **gen.** |
| 146 | $27+i$ | 212 | $h_0^{2+i} w_2^3 \widehat{h_0}$ | $(g)$ | $h_0^{1+i} \cdot h_0 w_2^3 \widehat{h_0}$ |
| 147 | 25 | 200 | $i(h_2 w_2^3)$ | $-$ | $h_2 w_2^3 \cdot i(1)$ |
| 147 | 26 | 205 | $i(h_0 h_2 w_2^3)$ | $(g, w_1)$ | $h_0 h_2 w_2^3 \cdot i(1)$ |
| 149 | 26 | 207 | $h_0 w_2^3 \widehat{h_2}$ | $(g, w_1)$ | $h_2 w_2^3 \cdot \widehat{h_0}$ |
| 149 | 27 | 215 | $h_0^2 w_2^3 \widehat{h_2}$ | $(g, w_1)$ | $h_0 h_2 w_2^3 \cdot \widehat{h_0}$ |
| 150 | 26 | 208 | $i(h_2^2 w_2^3)$ | $(g, w_1)$ | $h_2^2 w_2^3 \cdot i(1)$ |
| 152 | 26 | 209 | $h_2 w_2^3 \widehat{h_2}$ | $(g, w_1)$ | $h_2 w_2^3 \cdot \widehat{h_2}$ |
| 152 | 27 | 218 | $i(c_0 w_2^3)$ | $(g, w_1)$ | $c_0 w_2^3 \cdot i(1)$ |
| 152 | 28 | $231+232$ | $g w_2^2 \widehat{d_0 g} + i(w_1 w_2^3)$ | $(g)$ | **gen.** |
| 154 | 29 | $241+242$ | $\alpha^2 g w_2^2 \widehat{\alpha} + w_1 w_2^3 \widehat{h_0}$ | $(g)$ | **gen.** |
| 155 | 28 | $234+235$ | $\gamma g w_2^2 \widehat{\alpha} + w_2^3 \widehat{h_1 c_0}$ | $(g, w_1)$ | **gen.** |
| 156 | $29+i$ | 246 | $i(h_0^{2+i} \alpha w_2^3)$ | $(g)$ | $h_0^i \cdot$ **gen.** |
| 158 | $30+i$ | 252 | $h_0^{3+i} w_2^3 \widehat{\alpha}$ | $(g)$ | $h_0^i \cdot$ **gen.** |
| 161 | 29 | $255+256$ | $i(h_0 e_0 w_2^3)$ | $(g, w_1)$ | $h_2 w_2^2 \cdot (\gamma g \widehat{\beta} + i(d_0 w_2))$ |
| 168 | $31+i$ | 272 | $i(h_0^{1+i} \alpha^2 w_2^3)$ | $(g)$ | $h_0^{1+i} w_2^2 \cdot (\beta g^2 \widehat{\beta} + i(\alpha^2 w_2))$ |
| 170 | 31 | $274+275$ | $g^3 w_2^2 \widehat{\alpha} + \alpha^2 w_2^3 \widehat{h_0}$ | $(g)$ | **gen.** |
| 170 | $32+i$ | 292 | $h_0^{1+i} \alpha^2 w_2^3 \widehat{h_0}$ | $(g)$ | $h_0^{1+i} \cdot (g^3 w_2^2 \widehat{\alpha} + \alpha^2 w_2^3 \widehat{h_0})$ |
| 176 | 31 | 284 | $i(\delta w_2^3)$ | $(g)$ | $\delta w_2^3 \cdot i(1)$ |
| 176 | 32 | 303 | $i(h_0 \delta w_2^3)$ | $(g)$ | $h_0 \delta w_2^3 \cdot i(1)$ |
| 176 | 34 | $308+309$ | $\beta g^2 w_1 w_2^2 \widehat{\beta} + i(\alpha^2 w_1 w_2^3)$ | $(g)$ | $w_1 w_2^2 \cdot (\beta g^2 \widehat{\beta} + i(\alpha^2 w_2))$ |
| 178 | 32 | 306 | $h_0 g w_2^3 \widehat{\alpha}$ | $(g)$ | $\delta w_2^3 \cdot \widehat{h_0}$ |

Table 7.5: $R_2$-module generators of $E_4(tmf/\eta) = E_\infty(tmf/\eta)$, with $i \geq 0$ in each $h_0$-tower (cont.)

| $t-s$ | $s$ | $g$ | $x$ | $\mathrm{Ann}(x)$ | dec. |
|---|---|---|---|---|---|
| 178 | 33 | 315 | $h_0^2 g w_2^3 \widehat{\alpha}$ | $(g)$ | $h_0 \delta w_2^3 \cdot \widehat{h_0}$ |
| 180 | $33+i$ | 318 | $h_0^{1+i} w_2^3 \widehat{d_0 g}$ | $(g)$ | $h_0^i \cdot$ **gen.** |
| 182 | $34+i$ | 322 | $h_0^{1+i} \alpha^2 w_2^3 \widehat{\alpha}$ | $(g)$ | $h_0^i \cdot$ **gen.** |

Table 7.6: The non-cyclic $R_2$-module summands in $E_4(tmf/\eta)$

$$\langle d_0 \widehat{h_2}, i(h_2 w_2) \rangle \cong \frac{\Sigma^{5,24} R_1 \oplus \Sigma^{9,60} R_1}{\langle (w_1, 0), (g^2, w_1), (0, g) \rangle}$$

$$\langle d_0 w_2^2 \widehat{h_2}, i(h_2 w_2^3) \rangle \cong \frac{\Sigma^{21,136} R_1 \oplus \Sigma^{25,172} R_1}{\langle (w_1, 0), (g^2, w_1), (0, g) \rangle}$$

PROPOSITION 7.5. *Charts showing $E_4(tmf/\eta)$ for $0 \leq t-s \leq 192$ are given in Figures 7.1 to 7.8. All nonzero $h_0$-, $h_1$- and $h_2$-multiplications are displayed. The red dots indicate $w_1$-power torsion classes, and black dots indicate $w_1$-periodic classes. All $R_2$-module generators are labeled, except those that are also $h_0$-, $h_1$- or $h_2$-multiples.*

PROOF. The $R_2$-module structure shown in these charts is made explicit in Tables 7.5 and 7.6. The $h_0$-, $h_1$- and $h_2$-multiplications mostly follow by comparison with the $E_2$-term, shown for $0 \leq t-s \leq 96$ in Figures 1.28 to 1.31. This also shows that

$$h_2 \cdot (\gamma g \widehat{\alpha} + w_2 \widehat{h_1 c_0}) = h_0 \cdot (\gamma g \widehat{\beta} + i(d_0 w_2)) = i(h_0 d_0 w_2)\,,$$

which we have noted becomes equal to $g^3 \cdot \widehat{h_0}$ at the $E_3$-term. Similarly $h_2 \cdot (\gamma g w_2^2 \widehat{\alpha} + w_2^3 \widehat{h_1 c_0}) = i(h_0 d_0 w_2^3)$ becomes equal to $g^3 \cdot w_2^2 \widehat{h_0}$ in $E_3(tmf/\eta)$. □

THEOREM 7.6. *$E_4(tmf/\eta) = E_\infty(tmf/\eta)$.*

PROOF. It suffices to verify that $d_r(x) = 0$ for each $R_2$-module generator $x$ in Table 7.5, for each $r \geq 4$. In most cases this is clear because all target groups are trivial. In the remaining cases, $x$ is ($w_1$- or) $w_1^2$-torsion, so if $d_r(x) = y$ then $w_1^2 y = 0$ at the $E_r$-term. Moreover, in each of these cases the $E_4$-term is trivial in and above the bidegree of $w_1^2 x$, so none of the differentials $d_4, \ldots, d_r$ can hit $w_1^2 y$. Hence $w_1^2 y = 0$ in $E_4(tmf/\eta)$. Furthermore, $w_1^2$ acts injectively on the $E_4$-term in the bidegree containing $d_r(x)$, and this implies that $y = 0$. □

We have also determined a set of $E_\infty(tmf)$-module generators for $E_\infty(tmf/\eta)$, and expressed the remaining $R_2$-module generators in terms of this module structure. The results are listed in the following proposition, and in the dec.-column of Table 7.5.

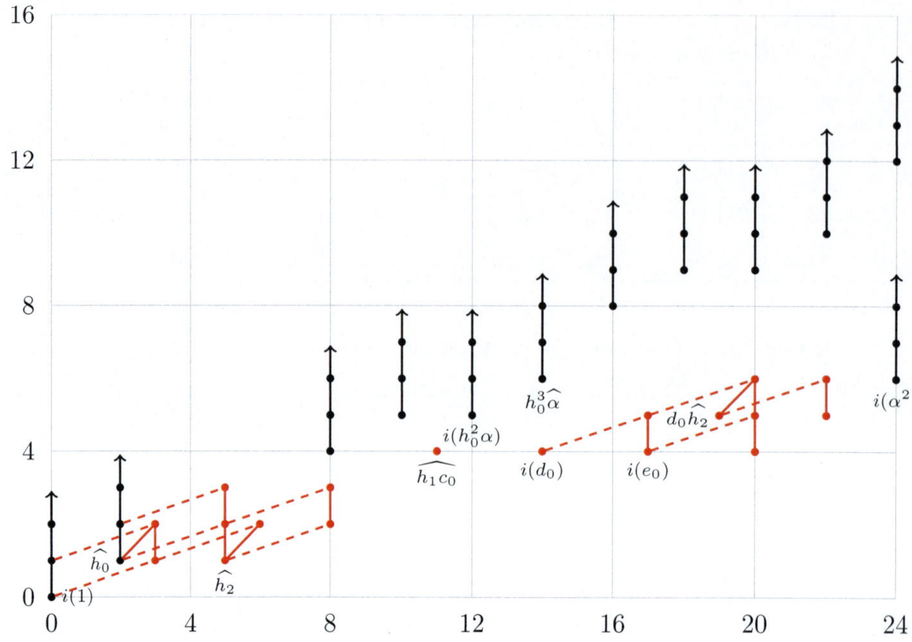


FIGURE 7.1. $E_4(tmf/\eta) = E_\infty(tmf/\eta)$ for $0 \leq t - s \leq 24$

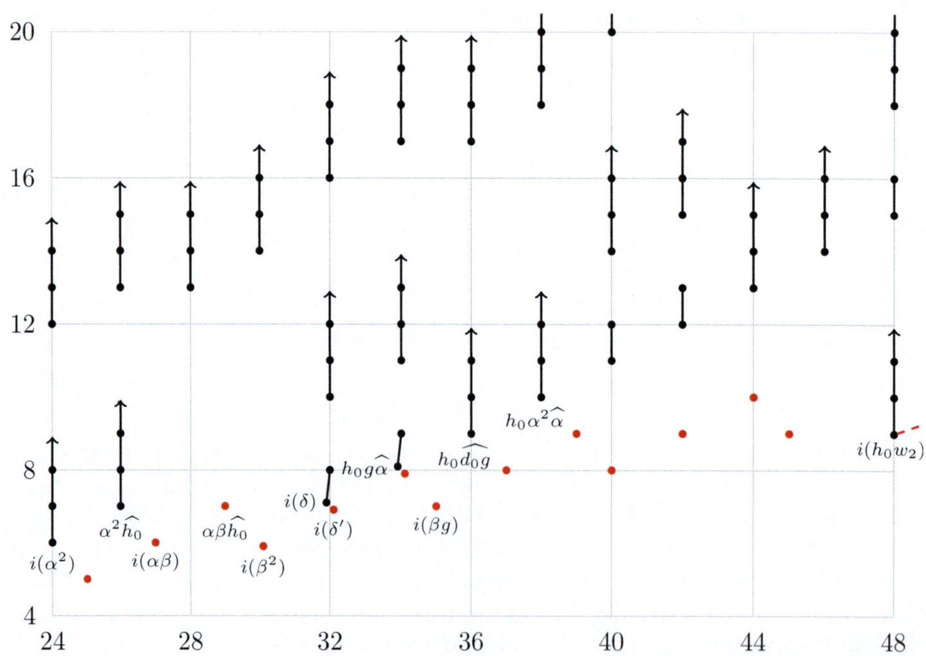


FIGURE 7.2. $E_4(tmf/\eta) = E_\infty(tmf/\eta)$ for $24 \leq t - s \leq 48$

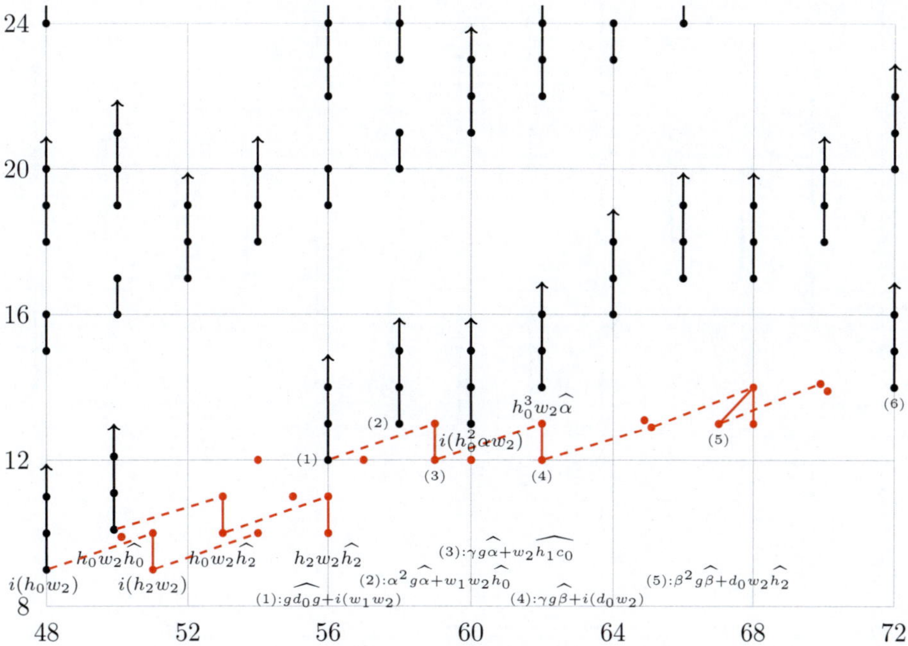

FIGURE 7.3. $E_4(tmf/\eta) = E_\infty(tmf/\eta)$ for $48 \le t - s \le 72$

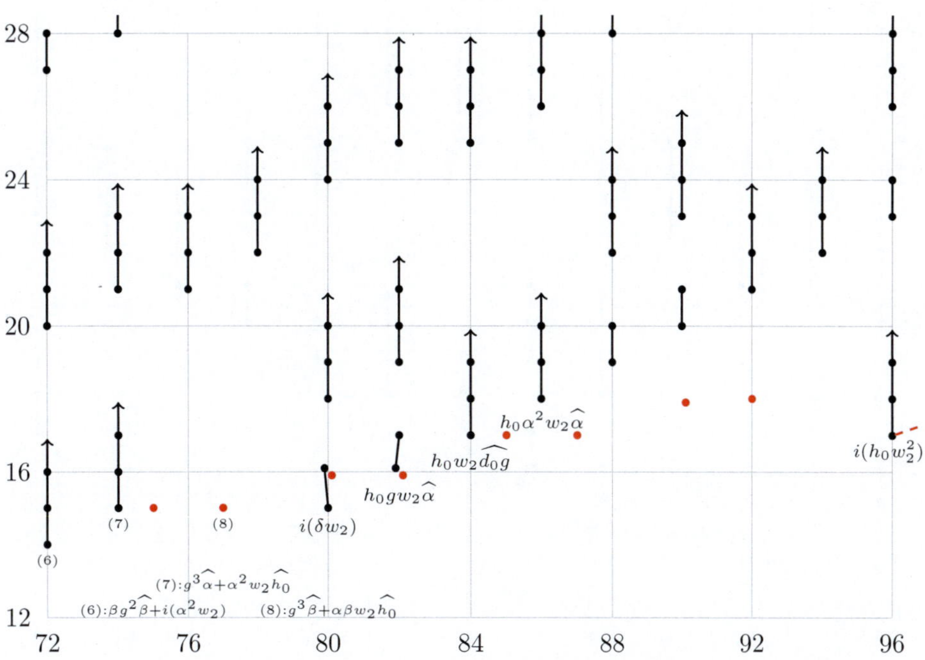

FIGURE 7.4. $E_4(tmf/\eta) = E_\infty(tmf/\eta)$ for $72 \le t - s \le 96$

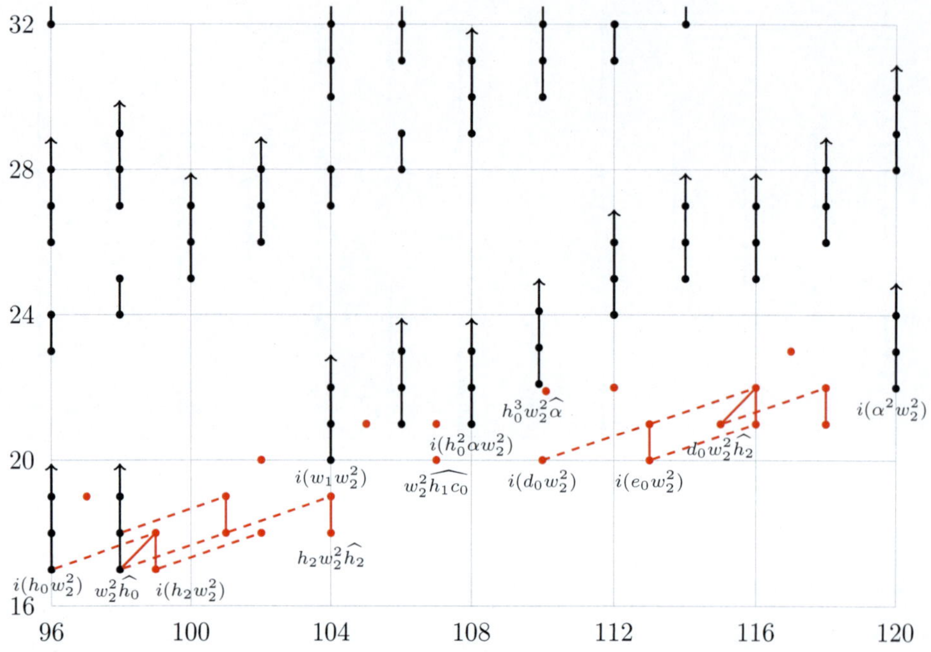


FIGURE 7.5. $E_4(tmf/\eta) = E_\infty(tmf/\eta)$ for $96 \leq t-s \leq 120$

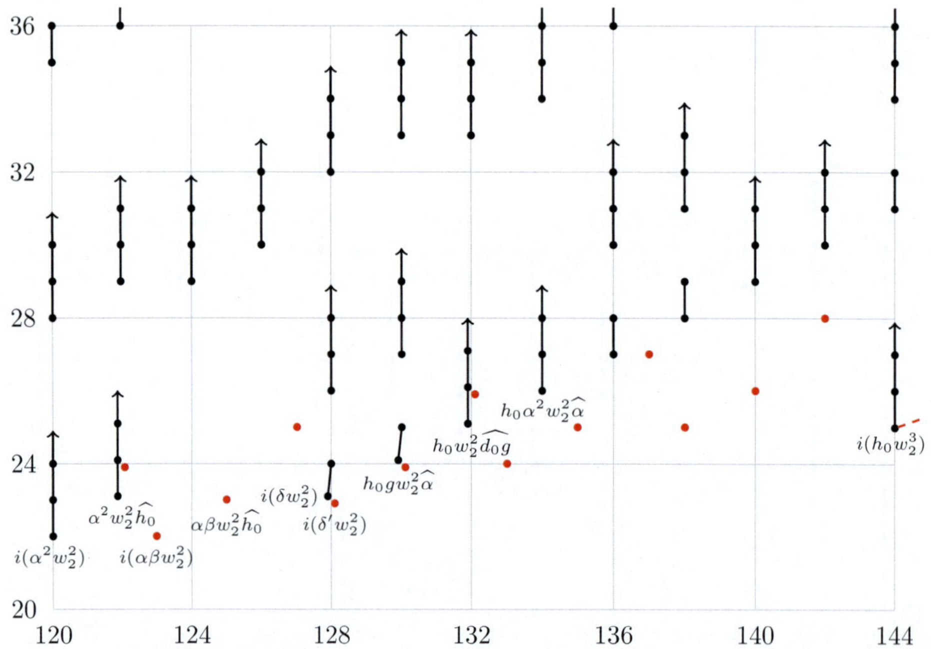


FIGURE 7.6. $E_4(tmf/\eta) = E_\infty(tmf/\eta)$ for $120 \leq t-s \leq 144$

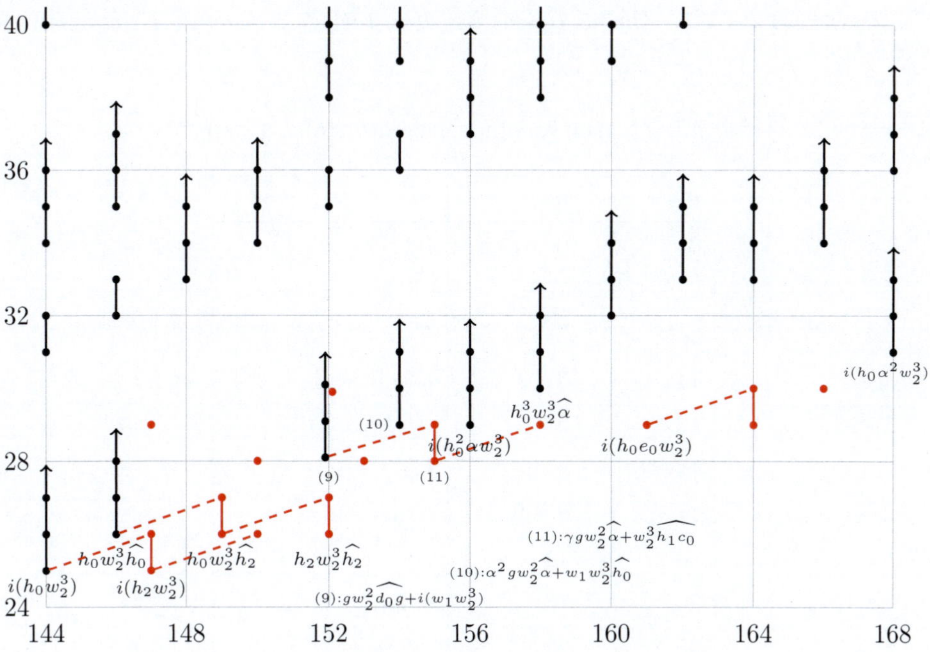


FIGURE 7.7. $E_4(tmf/\eta) = E_\infty(tmf/\eta)$ for $144 \le t-s \le 168$

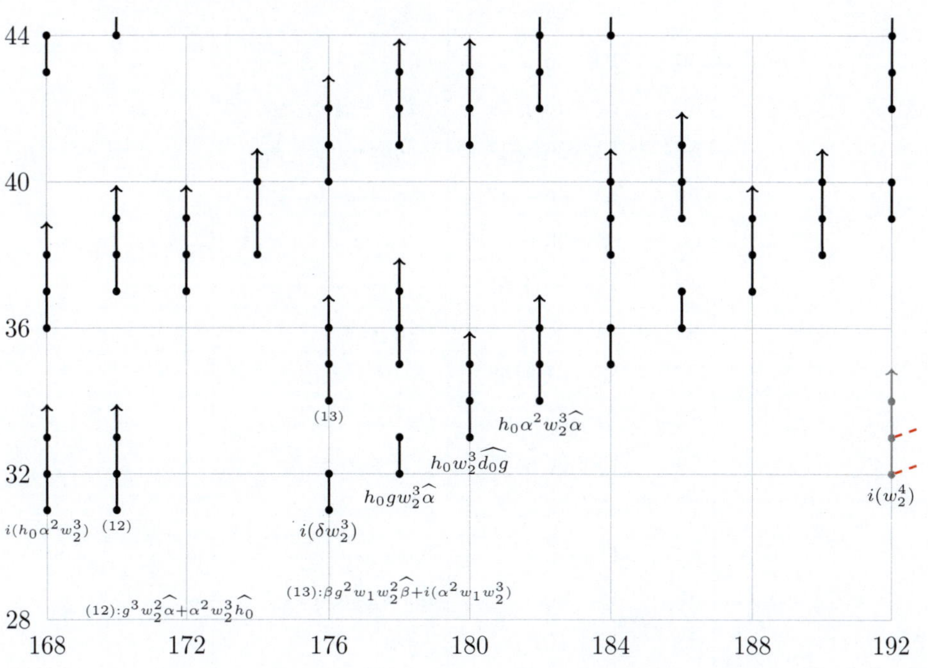


FIGURE 7.8. $E_4(tmf/\eta) = E_\infty(tmf/\eta)$ for $168 \le t-s \le 192$

PROPOSITION 7.7. *The* 45 *classes listed in Table* 7.7 *generate* $E_\infty(tmf/\eta)$ *as a module over* $E_\infty(tmf)$.

Table 7.7: $E_\infty(tmf)$-module generators of $E_\infty(tmf/\eta)$

| $t-s$ | $s$ | $g$ | $x$ |
|---|---|---|---|
| 0 | 0 | 0 | $i(1)$ |
| 2 | 1 | 1 | $\widehat{h_0}$ |
| 5 | 1 | 3 | $\widehat{h_2}$ |
| 11 | 4 | 3 | $\widehat{h_1 c_0}$ |
| 12 | 5 | 5 | $i(h_0^2\alpha)$ |
| 14 | 6 | 8 | $h_0^3\widehat{\alpha}$ |
| 17 | 4 | $8+9$ | $i(e_0)$ |
| 24 | 6 | 14 | $i(\alpha^2)$ |
| 26 | 7 | 13 | $\alpha^2\widehat{h_0}$ |
| 35 | 7 | 22 | $i(\beta g)$ |
| 36 | 9 | 26 | $h_0\widehat{d_0 g}$ |
| 38 | 10 | 26 | $h_0\alpha^2\widehat{\alpha}$ |
| 48 | 9 | 34 | $i(h_0 w_2)$ |
| 50 | 10 | 36 | $h_0 w_2\widehat{h_0}$ |
| 56 | 12 | $43+44$ | $g\widehat{d_0 g} + i(w_1 w_2)$ |
| 58 | 13 | $46+47$ | $\alpha^2 g\widehat{\alpha} + w_1 w_2\widehat{h_0}$ |
| 59 | 12 | $46+47$ | $\gamma g\widehat{\alpha} + w_2\widehat{h_1 c_0}$ |
| 60 | 13 | 50 | $i(h_0^2\alpha w_2)$ |
| 62 | 12 | $50+51$ | $\gamma g\widehat{\beta} + i(d_0 w_2)$ |
| 62 | 14 | 53 | $h_0^3 w_2\widehat{\alpha}$ |
| 67 | 13 | $61+62$ | $\beta^2 g\widehat{\beta} + d_0 w_2\widehat{h_2}$ |
| 72 | 14 | $65+66$ | $\beta g^2\widehat{\beta} + i(\alpha^2 w_2)$ |
| 74 | 15 | $66+67$ | $g^3\widehat{\alpha} + \alpha^2 w_2\widehat{h_0}$ |
| 77 | 15 | $71+72$ | $g^3\widehat{\beta} + \alpha\beta w_2\widehat{h_0}$ |
| 84 | 17 | 90 | $h_0 w_2\widehat{d_0 g}$ |
| 86 | 18 | 90 | $h_0\alpha^2 w_2\widehat{\alpha}$ |
| 98 | 17 | 101 | $w_2^2\widehat{h_0}$ |
| 107 | 20 | 124 | $w_2^2\widehat{h_1 c_0}$ |
| 108 | 21 | 132 | $i(h_0^2\alpha w_2^2)$ |
| 110 | 22 | 136 | $h_0^3 w_2^2\widehat{\alpha}$ |

Table 7.7: $E_\infty(tmf)$-module generators of $E_\infty(tmf/\eta)$ (cont.)

| $t-s$ | $s$ | $g$ | $x$ |
|---|---|---|---|
| 113 | 20 | $132+133$ | $i(e_0w_2^2)$ |
| 120 | 22 | 150 | $i(\alpha^2w_2^2)$ |
| 122 | 23 | 155 | $\alpha^2w_2^2\widehat{h_0}$ |
| 132 | 25 | 188 | $h_0w_2^2\widehat{d_0g}$ |
| 134 | 26 | 190 | $h_0\alpha^2w_2^2\widehat{\alpha}$ |
| 144 | 25 | 198 | $i(h_0w_2^3)$ |
| 146 | 26 | 204 | $h_0w_2^3\widehat{h_0}$ |
| 152 | 28 | $231+232$ | $gw_2^2\widehat{d_0g}+i(w_1w_2^3)$ |
| 154 | 29 | $241+242$ | $\alpha^2gw_2^2\widehat{\alpha}+w_1w_2^3\widehat{h_0}$ |
| 155 | 28 | $234+235$ | $\gamma gw_2^2\widehat{\alpha}+w_2^3\widehat{h_1c_0}$ |
| 156 | 29 | 246 | $i(h_0^2\alpha w_2^3)$ |
| 158 | 30 | 252 | $h_0^3w_2^3\widehat{\alpha}$ |
| 170 | 31 | $274+275$ | $g^3w_2^2\widehat{\alpha}+\alpha^2w_2^3\widehat{h_0}$ |
| 180 | 33 | 318 | $h_0w_2^3\widehat{d_0g}$ |
| 182 | 34 | 322 | $h_0\alpha^2w_2^3\widehat{\alpha}$ |

CHAPTER 8

# The Adams spectral sequence for $tmf/\nu$

We calculate the $d_r$-differentials in the Adams spectral sequence for $tmf/\nu = tmf \wedge C\nu$. These are nontrivial for $r \in \{2, 3, 4\}$, and zero for $r \geq 5$, so the spectral sequence collapses at the $E_5$-term. The module structure over the Adams spectral sequence for $tmf$ suffices to determine almost all of these differentials. There is one exceptional case, concerning $d_2(\overline{\beta^2})$, which we settle by means of an external smash product pairing. The resulting $E_\infty$-term is the associated graded of a complete Hausdorff filtration of $\pi_*(tmf/\nu)^\wedge_2$.

## 8.1. The $E_2$-term for $tmf/\nu$

The initial term

$$E_2 = E_2(tmf/\nu) \cong \mathrm{Ext}_{A(2)}(M_4, \mathbb{F}_2)$$

of the mod 2 Adams spectral sequence for $tmf/\nu$ was calculated in Part I. The groups $E_2^{s,t}$ for $0 \leq t-s \leq 96$ are displayed in Figures 1.32 to 1.35. By Corollary 4.16 the $E_2$-term for $tmf/\nu$ is generated as a module over $E_2(tmf) = \mathrm{Ext}_{A(2)}(\mathbb{F}_2, \mathbb{F}_2)$ by the 14 classes listed in Table 8.1. As a module over $R_0 = \mathbb{F}_2[g, w_1, w_2]$ the $E_2$-term for $tmf/\nu$ is presented as a direct sum of cyclic modules in Table 8.2, most of which is reproduced from Table 4.7 and illustrated in Figure 4.3. We note that the $E_2$-term is free over $\mathbb{F}_2[w_1, w_2]$, and finitely generated over $R_0[h_0] = \mathbb{F}_2[h_0, g, w_1, w_2]$. Recall Definition 5.1. Following the strategy of Chapter 5 we will keep track of $R_0$-module structure on the $E_2$-term, $R_1$-module structure on the $E_3$-term, and $R_2$-module structure on the $E_4$- and $E_5 = E_\infty$-terms of the Adams spectral sequence for $tmf/\nu$.

Table 8.1: $E_2(tmf)$-module generators of $E_2(tmf/\nu)$

| $t-s$ | $s$ | $g$ | $x$ | $d_2(x)$ |
|---|---|---|---|---|
| 0 | 0 | 0 | $i(1)$ | 0 |
| 4 | 3 | 1 | $\overline{h_0^3}$ | 0 |
| 5 | 1 | 2 | $\overline{h_1}$ | 0 |
| 7 | 2 | 3 | $\overline{h_0h_2}$ | 0 |
| 10 | 2 | 4 | $\overline{h_2^2}$ | $i(h_1c_0)$ |
| 12 | 3 | 4 | $\overline{c_0}$ | 0 |
| 16 | 5 | 7 | $\overline{h_0^2\alpha}$ | $h_0w_1\overline{h_0h_2}$ |
| 24 | 4 | 9 | $\overline{g}$ | 0 |

Table 8.1: $E_2(tmf)$-module generators of $E_2(tmf/\nu)$ (cont.)

| $t-s$ | $s$ | $g$ | $x$ | $d_2(x)$ |
|---|---|---|---|---|
| 28 | 7 | 13 | $\overline{h_0\alpha^2}$ | 0 |
| 29 | 5 | 13 | $\overline{\gamma}$ | 0 |
| 31 | 6 | 16 | $\overline{\alpha\beta}$ | 0 |
| 34 | 6 | 17 | $\overline{\beta^2}$ | $i(h_1\delta)$ |
| 36 | 7 | 19 | $\overline{\delta}$ | 0 |
| 40 | 9 | 24 | $\overline{\alpha^3}$ | $h_0w_1\overline{\alpha\beta}$ |

Table 8.2: $R_0$-module generators of $E_2(tmf/\nu)$, with $i \geq 0$ in each $h_0$-tower

| $t-s$ | $s$ | $g$ | $x$ | $\mathrm{Ann}(x)$ | $d_2(x)$ | $d_2(xw_2)$ |
|---|---|---|---|---|---|---|
| 0 | 0 | 0 | $i(1)$ | $(0)$ | 0 | $g^2 \cdot \overline{h_0h_2}$ |
| 0 | $1+i$ | 0 | $i(h_0^{1+i})$ | $(g)$ | 0 | 0 |
| 1 | 1 | 1 | $i(h_1)$ | $(g)$ | 0 | 0 |
| 2 | 2 | 1 | $i(h_1^2)$ | $(g)$ | 0 | 0 |
| 4 | $3+i$ | 1 | $h_0^i\overline{h_0^3}$ | $(g)$ | 0 | 0 |
| 5 | 1 | 2 | $\overline{h_1}$ | $(0)$ | 0 | $g^2 \cdot i(\alpha)$ |
| 6 | 2 | 2 | $h_1\overline{h_1}$ | $(g)$ | 0 | 0 |
| 7 | 2 | 3 | $\overline{h_0h_2}$ | $(0)$ | 0 | $g^2 \cdot i(d_0)$ |
| 7 | 3 | 2 | $h_0\overline{h_0h_2}$ | $(g)$ | 0 | 0 |
| 8 | 3 | 3 | $i(c_0)$ | $(g)$ | 0 | 0 |
| 9 | 4 | 3 | $i(h_1c_0)$ | $(g)$ | 0 | 0 |
| 10 | 2 | 4 | $\overline{h_2^2}$ | $(0)$ | $i(h_1c_0)$ | $g^2 \cdot i(e_0)$ $+ i(h_1c_0w_2)$ |
| 12 | 3 | 4 | $\overline{c_0}$ | $(g)$ | 0 | 0 |
| 12 | 3 | $4+5$ | $i(\alpha)$ | $(0)$ | 0 | $g^2 \cdot d_0\overline{h_1}$ |
| 12 | $4+i$ | 4 | $i(h_0^{1+i}\alpha)$ | $(g)$ | 0 | 0 |
| 13 | 4 | 5 | $h_1\overline{c_0}$ | $(g)$ | 0 | 0 |
| 14 | 4 | 6 | $i(d_0)$ | $(0)$ | 0 | $g^2 \cdot d_0\overline{h_0h_2}$ |
| 14 | 5 | 6 | $i(h_0d_0)$ | $(g)$ | 0 | 0 |
| 15 | 3 | 6 | $i(\beta)$ | $(0)$ | $i(h_0d_0)$ | $g^2 \cdot e_0\overline{h_1}$ $+ i(h_0d_0w_2)$ |

Table 8.2: $R_0$-module generators of $E_2(tmf/\nu)$, with $i \geq 0$ in each $h_0$-tower (cont.)

| $t-s$ | $s$ | $g$ | $x$ | $\mathrm{Ann}(x)$ | $d_2(x)$ | $d_2(xw_2)$ |
|---|---|---|---|---|---|---|
| 16 | 5 | 7 | $\overline{h_0^2\alpha}$ | $(0)$ | $w_1 \cdot h_0\overline{h_0h_2}$ | $g^2w_1 \cdot i(\beta)$ $+ w_1 \cdot h_0w_2\overline{h_0h_2}$ |
| 16 | $6+i$ | 7 | $h_0^{1+i}\overline{h_0^2\alpha}$ | $(g)$ | 0 | 0 |
| 17 | 4 | 7 | $i(e_0)$ | $(0)$ | 0 | $g^2 \cdot i(\alpha^2)$ |
| 19 | 5 | 8 | $d_0\overline{h_1}$ | $(0)$ | 0 | $g^2 \cdot i(\alpha d_0)$ |
| 21 | 6 | 9 | $d_0\overline{h_0h_2}$ | $(0)$ | 0 | $g^3w_1 \cdot i(1)$ |
| 22 | 5 | 9 | $e_0\overline{h_1}$ | $(0)$ | 0 | $g^2 \cdot i(\alpha e_0)$ |
| 24 | 4 | 9 | $\overline{g}$ | $(0)$ | 0 | $g^2 \cdot \overline{\alpha\beta}$ |
| 24 | 5 | 10 | $h_0\overline{g}$ | $(g)$ | 0 | 0 |
| 24 | 6 | $10+11$ | $i(\alpha^2)$ | $(0)$ | 0 | $g^2 \cdot i(d_0e_0)$ |
| 24 | 6 | 11 | $h_0^2\overline{g}$ | $(g)$ | 0 | 0 |
| 24 | $7+i$ | 11 | $i(h_0^{1+i}\alpha^2)$ | $(g)$ | 0 | 0 |
| 25 | 5 | 12 | $h_1\overline{g}$ | $(g)$ | 0 | 0 |
| 26 | 6 | 12 | $i(h_1\gamma)$ | $(g)$ | 0 | 0 |
| 26 | 7 | 12 | $i(\alpha d_0)$ | $(0)$ | 0 | $g^3w_1 \cdot \overline{h_1}$ |
| 28 | $7+i$ | 13 | $h_0^i\overline{h_0\alpha^2}$ | $(g)$ | 0 | 0 |
| 29 | 5 | 13 | $\overline{\gamma}$ | $(0)$ | 0 | $g^2 \cdot \alpha\overline{g}$ |
| 29 | 7 | 14 | $i(\alpha e_0)$ | $(0)$ | 0 | $g^3 \cdot \overline{h_0^2\alpha}$ |
| 30 | 6 | 15 | $h_1\overline{\gamma}$ | $(g)$ | 0 | 0 |
| 31 | 6 | 16 | $\overline{\alpha\beta}$ | $(0)$ | 0 | $g^2 \cdot d_0\overline{g}$ |
| 31 | 7 | 15 | $h_0\overline{\alpha\beta}$ | $(g)$ | 0 | 0 |
| 31 | 8 | 15 | $i(d_0e_0)$ | $(0)$ | 0 | $g^3w_1 \cdot \overline{h_2^2}$ |
| 32 | 7 | 17 | $i(\delta)$ | $(g)$ | 0 | 0 |
| 33 | 8 | 17 | $i(h_1\delta)$ | $(g)$ | 0 | 0 |
| 34 | 6 | 17 | $\overline{\beta^2}$ | $(0)$ | $i(h_1\delta)$ | $g^2 \cdot e_0\overline{g}$ $+ i(h_1\delta w_2)$ |
| 36 | 7 | 19 | $\overline{\delta}$ | $(g)$ | 0 | 0 |
| 36 | 7 | $19+20$ | $\alpha\overline{g}$ | $(0)$ | 0 | $g^2 \cdot d_0\overline{\gamma}$ |
| 36 | 8 | 19 | $h_0\overline{\delta}$ | $(g)$ | 0 | 0 |
| 36 | 9 | 20 | $h_0^2\overline{\delta}$ | $(g)$ | 0 | 0 |
| 36 | $10+i$ | 20 | $i(h_0^{1+i}\alpha^3)$ | $(g)$ | 0 | 0 |

Table 8.2: $R_0$-module generators of $E_2(tmf/\nu)$, with $i \geq 0$ in each $h_0$-tower (cont.)

| $t-s$ | $s$ | $g$ | $x$ | $\mathrm{Ann}(x)$ | $d_2(x)$ | $d_2(xw_2)$ |
|---|---|---|---|---|---|---|
| 37 | 8 | 21 | $h_1\overline{\delta}$ | $(g)$ | 0 | 0 |
| 38 | 8 | 22 | $d_0\overline{g}$ | $(0)$ | 0 | $g^2 \cdot d_0\overline{\alpha\beta}$ |
| 38 | 9 | 22 | $h_0d_0\overline{g}$ | $(g)$ | 0 | 0 |
| 39 | 7 | 21 | $\beta\overline{g}$ | $(0)$ | $h_0d_0\overline{g}$ | $g^2 \cdot e_0\overline{\gamma}$ $+ h_0d_0w_2\overline{g}$ |
| 40 | 9 | 24 | $\overline{\alpha^3}$ | $(0)$ | $w_1 \cdot h_0\overline{\alpha\beta}$ | $g^2w_1 \cdot \beta\overline{g}$ $+ w_1 \cdot h_0w_2\overline{\alpha\beta}$ |
| 40 | $10+i$ | 24 | $h_0^{1+i}\overline{\alpha^3}$ | $(g)$ | 0 | 0 |
| 41 | 8 | 24 | $e_0\overline{g}$ | $(0)$ | 0 | $g^2 \cdot \alpha^2\overline{g}$ |
| 43 | 9 | 26 | $d_0\overline{\gamma}$ | $(0)$ | 0 | $g^2 \cdot \alpha d_0\overline{g}$ |
| 45 | 10 | 28 | $d_0\overline{\alpha\beta}$ | $(0)$ | 0 | $g^3w_1 \cdot \overline{g}$ |
| 46 | 9 | 28 | $e_0\overline{\gamma}$ | $(0)$ | 0 | $g^2 \cdot \alpha^2\overline{\gamma}$ |
| 48 | 10 | $30+31$ | $\alpha^2\overline{g}$ | $(0)$ | 0 | $g^2 \cdot d_0e_0\overline{g}$ |
| 50 | 11 | 33 | $\alpha d_0\overline{g}$ | $(0)$ | 0 | $g^3w_1 \cdot \overline{\gamma}$ |
| 53 | 11 | 36 | $\alpha^2\overline{\gamma}$ | $(0)$ | 0 | $g^3 \cdot \overline{\alpha^3}$ |
| 55 | 12 | 38 | $d_0e_0\overline{g}$ | $(0)$ | 0 | $g^3w_1 \cdot \overline{\beta^2}$ |

## 8.2. The $d_2$-differentials for $tmf/\nu$

THEOREM 8.1. *The $d_2$-differential in $E_2(tmf/\nu)$ is $R_1$-linear. Its values on a set of $E_2(tmf)$-module generators are listed in Table* 8.1, *and its values on a set of $R_1$-module generators are listed in Table* 8.2.

PROOF. The classes $g$, $w_1$ and $w_2^2$ are $d_2$-cycles in $E_2(tmf)$, so the Leibniz rule implies that multiplication by each of these elements commutes with the $d_2$-differential in $E_2(tmf/\nu)$.

Next, we determine $d_2$ on the module generators of $E_2(tmf/\nu)$ over $E_2(tmf)$. See Figures 1.32 and 1.33. The $d_2$-differentials on $i(1)$, $\overline{h_0^3}$, $\overline{h_0h_2}$, $\overline{c_0}$, $\overline{g}$, $\overline{\alpha\beta}$ and $\overline{\delta}$ are zero because the target groups are trivial. The $d_2$-differentials on $\overline{h_1}$ and $\overline{\gamma}$ are zero by $h_0$-linearity. The cofiber sequence

$$S \xrightarrow{i} C\nu \xrightarrow{j} S^4$$

induces maps of Adams spectral sequences

$$E_r(tmf) \xrightarrow{i} E_r(tmf/\nu) \xrightarrow{j} E_r^{*,*-4}(tmf)\,.$$

By Theorem 5.10 (or Table 5.2) the class $h_1c_0w_1$ is a $d_3$-boundary in the Adams spectral sequence for $tmf$, so its image $i(h_1c_0w_1)$ must be a $d_2$- or $d_3$-boundary in the Adams spectral sequence for $tmf/\nu$. For bidegree reasons, the only possibility is $d_2(w_1\overline{h_2^2}) = i(h_1c_0w_1)$. It follows that $d_2(\overline{h_2^2}) = i(h_1c_0)$, by injectivity of the $w_1$-multiplication from bidegree $(t-s,s) = (9,4)$.

By Proposition 5.8 (or Table 5.1) the classes $h_0^2\alpha$ and $\alpha^3$ both support nontrivial $d_2$-differentials. Hence their lifts $\overline{h_0^2\alpha}$ and $\overline{\alpha^3}$ must also support nonzero $d_2$-differentials, and the only possible values are $h_0w_1\overline{h_0h_2}$ and $h_0w_1\overline{\alpha\beta}$, respectively. The value of $d_2(\overline{h_0\alpha^2})$ is either 0 or $d_0w_1\overline{h_1}$. It maps under $j$ to $d_2(h_0\alpha^2)$ in the Adams spectral sequence for $tmf$, which is zero by the Leibniz rule (or Table 5.1). However, $j$ maps $d_0w_1\overline{h_1}$ to $h_1d_0w_1$, which is nonzero in $E_2(tmf)$. Hence $d_2$ vanishes on $\overline{h_0\alpha^2}$.

Only the case of $d_2(\overline{\beta^2})$ remains. The cofiber sequence

$$S \wedge C\nu \xrightarrow{i\wedge 1} C\nu \wedge C\nu \xrightarrow{j\wedge 1} S^4 \wedge C\nu$$

induces a long exact sequence

$$\dots \xrightarrow{\delta} E_2(tmf \wedge C\nu) \xrightarrow{i_*} E_2(tmf \wedge C\nu \wedge C\nu) \xrightarrow{j_*} E_2(tmf \wedge S^4 \wedge C\nu) \xrightarrow{\delta} \dots$$

of Adams $E_2$-terms. The connecting homomorphism $\delta$ induces multiplication by $h_2$ on $E_2(tmf \wedge C\nu) = \mathrm{Ext}_{A(2)}(M_4, \mathbb{F}_2)$. We know from Lemma 1.39 that this $E_2$-term is a graded algebra over $\mathrm{Ext}_{A(2)}(\mathbb{F}_2, \mathbb{F}_2)$, with $h_2 \cdot i(1) = 0$, so the $h_2$-multiplication is zero and the homomorphism $i_*$ is injective. Alternatively, we can see directly from Figures 1.32 and 1.33 that $h_2$-multiplication is zero, in the range of bidegrees shown.

The smash product of $tmf$-modules induces an external pairing

$$\wedge\colon E_2(tmf \wedge C\nu) \otimes E_2(tmf \wedge C\nu) \longrightarrow E_2(tmf \wedge C\nu \wedge C\nu)$$

of Adams spectral sequences, taking $\overline{h_2^2} \otimes \overline{g}$ to $\overline{h_2^2} \wedge \overline{g}$ in $(t-s,s) = (34,6)$, with

$$d_2(\overline{h_2^2} \wedge \overline{g}) = d_2(\overline{h_2^2}) \wedge \overline{g} + \overline{h_2^2} \wedge d_2(\overline{g}) = i(h_1c_0) \wedge \overline{g} + \overline{h_2^2} \wedge 0 = i_*(h_1c_0\overline{g})$$

in bidegree $(t-s,s) = (33,8)$. An `ext`-calculation shows that $h_1c_0\overline{g} = 8_{17} = i(h_1\delta)$, which is nonzero in $E_2(tmf \wedge C\nu)$. Hence $i_*(h_1c_0\overline{g}) \neq 0$ and $\overline{h_2^2} \wedge \overline{g} \neq 0$ in $E_2(tmf \wedge C\nu \wedge C\nu)$. Note that $j_*(\overline{h_2^2} \wedge \overline{g}) = \Sigma^4 h_2^2\overline{g} = 0$ in $E_2(tmf \wedge S^4 \wedge C\nu)$, by the vanishing of the $h_2$-multiplication. It follows that $\overline{h_2^2} \wedge \overline{g} = i_*(x)$ for a nonzero class $x$ in $(t-s,s) = (34,6)$, and the only possibility is $x = \overline{\beta^2}$. Thus $i_*d_2(\overline{\beta^2}) = d_2(i_*(\overline{\beta^2})) = d_2(\overline{h_2^2} \wedge \overline{g}) = i_*(h_1c_0\overline{g})$. The injectivity of $i_*$ then implies that $d_2(\overline{\beta^2}) = h_1c_0\overline{g} = i(h_1\delta)$.

Finally, we use Table 5.1 and the Leibniz rule to calculate $d_2$ for $x$ and $xw_2 = w_2 \cdot x$, with $x$ ranging through the list of $R_0$-module generators for $E_2(tmf/\nu)$. These elements then range through a list of $R_1$-module generators for the same $E_2$-term. In particular $d_2(w_2 \cdot x) = d_2(w_2) \cdot x + w_2 \cdot d_2(x)$, with $d_2(w_2) = \alpha\beta g = 10_{18}$ as in Table 5.1. In this finite range, the action of $E_2(tmf)$ on $E_2(tmf/\nu)$ is calculated using `ext`. □

REMARK 8.2. To use `ext` to assist in calculating the products $\alpha\beta g \cdot x$ for $x \in E_2(tmf/\nu)$, use `cocycle tmfCnu 0 0`, …, `cocycle tmfCnu 12 38`, `dolifts 0 40 maps` and `collect maps all`. The nonzero products with $\alpha\beta g = 10_{18}$ then appear as lines containing `(10 18 F2)` in the file `all`. If the product is a $g^2$-multiple, there will also appear a line containing `(8 18 F2)` in the same block, since

$g^2 = 8_{18}$ in the minimal $A(2)$-module resolution for $\mathbb{F}_2$. Similarly, $g^2 w_1$-multiples appear with (12 22 F2), $g^3$-multiples appear with (12 29 F2), etc.

## 8.3. The $d_3$-differentials for $tmf/\nu$

It is now a simple matter to compute the $E_3$-term of the Adams spectral sequence for $tmf/\nu$, as a direct sum of $R_1 = \mathbb{F}_2[g, w_1, w_2^2]$-modules. This is carried out in Appendix D.1 and the results are recorded in Tables 8.3 and 8.4, where we also record the results of this section, calculating the $d_3$-differential.

Table 8.3: $R_1$-module generators of $E_3(tmf/\nu)$, with $i \geq 0$ in each $h_0$-tower

| $t-s$ | $s$ | $g$ | $x$ | $\mathrm{Ann}(x)$ | $d_3(x)$ | $d_3(xw_2^2)$ |
|---|---|---|---|---|---|---|
| 0 | 0 | 0 | $i(1)$ | $(g^3w_1)$ | 0 | $g^3 \cdot i(\beta g)$ |
| 0 | $1+i$ | 0 | $i(h_0^{1+i})$ | $(g)$ | 0 | 0 |
| 1 | 1 | 1 | $i(h_1)$ | $(g)$ | 0 | 0 |
| 2 | 2 | 1 | $i(h_1^2)$ | $(g)$ | 0 | 0 |
| 4 | $3+i$ | 1 | $h_0^i\overline{h_0^3}$ | $(g)$ | 0 | 0 |
| 5 | 1 | 2 | $\overline{h_1}$ | $(g^3w_1)$ | 0 | $g^5 \cdot i(1)$ |
| 6 | 2 | 2 | $h_1\overline{h_1}$ | $(g)$ | 0 | 0 |
| 7 | 2 | 3 | $\overline{h_0h_2}$ | $(g^2)$ | 0 | 0 |
| 7 | 3 | 2 | $h_0\overline{h_0h_2}$ | $(g, w_1)$ | 0 | 0 |
| 8 | 3 | 3 | $i(c_0)$ | $(g)$ | 0 | 0 |
| 12 | 3 | 4 | $\overline{c_0}$ | $(g)$ | 0 | 0 |
| 12 | 3 | $4+5$ | $i(\alpha)$ | $(g^2)$ | 0 | 0 |
| 12 | $4+i$ | 4 | $i(h_0^{1+i}\alpha)$ | $(g)$ | 0 | 0 |
| 13 | 4 | 5 | $h_1\overline{c_0}$ | $(g)$ | 0 | 0 |
| 14 | 4 | 6 | $i(d_0)$ | $(g^2)$ | 0 | 0 |
| 16 | $6+i$ | 7 | $h_0^{1+i}\overline{h_0^2\alpha}$ | $(g)$ | 0 | 0 |
| 17 | 4 | 7 | $i(e_0)$ | $(g^3)$ | $w_1 \cdot i(c_0)$ | $w_1 \cdot i(c_0w_2^2)$ |
| 19 | 5 | 8 | $d_0\overline{h_1}$ | $(g^2)$ | 0 | 0 |
| 21 | 6 | 9 | $d_0\overline{h_0h_2}$ | $(g^2)$ | 0 | 0 |
| 22 | 5 | 9 | $e_0\overline{h_1}$ | $(g^3)$ | $w_1 \cdot h_1\overline{c_0}$ | $w_1 \cdot h_1 w_2^2\overline{c_0}$ |
| 24 | 4 | 9 | $\overline{g}$ | $(g^3w_1)$ | 0 | $g^3 \cdot \beta g\overline{g}$ |
| 24 | 5 | 10 | $h_0\overline{g}$ | $(g)$ | 0 | 0 |
| 24 | 6 | $10+11$ | $i(\alpha^2)$ | $(g^2)$ | 0 | 0 |
| 24 | 6 | 11 | $h_0^2\overline{g}$ | $(g)$ | 0 | 0 |
| 24 | $7+i$ | 11 | $i(h_0^{1+i}\alpha^2)$ | $(g)$ | 0 | 0 |

Table 8.3: $R_1$-module generators of $E_3(tmf/\nu)$, with $i \geq 0$ in each $h_0$-tower (cont.)

| $t-s$ | $s$ | $g$ | $x$ | $\mathrm{Ann}(x)$ | $d_3(x)$ | $d_3(xw_2^2)$ |
|---|---|---|---|---|---|---|
| 25 | 5 | 12 | $h_1\overline{g}$ | $(g)$ | 0 | 0 |
| 26 | 6 | 12 | $i(h_1\gamma)$ | $(g)$ | 0 | 0 |
| 26 | 7 | 12 | $i(\alpha d_0)$ | $(g^2)$ | 0 | 0 |
| 28 | $7+i$ | 13 | $h_0^i\overline{h_0\alpha^2}$ | $(g)$ | 0 | 0 |
| 29 | 5 | 13 | $\overline{\gamma}$ | $(g^3w_1)$ | $gw_1 \cdot i(1)$ | $g^5 \cdot \overline{g}$ $+ gw_1 \cdot i(w_2^2)$ |
| 29 | 7 | 14 | $i(\alpha e_0)$ | $(g^2)$ | 0 | 0 |
| 30 | 6 | 14 | $g\overline{h_2^2}$ | $(g^2w_1)$ | 0 | $g^6 \cdot \overline{h_1}$ |
| 30 | 6 | 15 | $h_1\overline{\gamma}$ | $(g)$ | 0 | 0 |
| 31 | 6 | 16 | $\overline{\alpha\beta}$ | $(g^2)$ | 0 | 0 |
| 31 | 7 | 15 | $h_0\overline{\alpha\beta}$ | $(g, w_1)$ | 0 | 0 |
| 31 | 8 | 15 | $i(d_0e_0)$ | $(g^2)$ | 0 | 0 |
| 32 | 7 | 17 | $i(\delta)$ | $(g)$ | 0 | 0 |
| 35 | 7 | 18 | $i(\beta g)$ | – | 0 | $g^5 \cdot g\overline{h_2^2}$ |
| 36 | 7 | 19 | $\overline{\delta}$ | $(g)$ | 0 | 0 |
| 36 | 7 | $19+20$ | $\alpha\overline{g}$ | $(g^2)$ | $gw_1 \cdot \overline{h_0h_2}$ | $gw_1 \cdot w_2^2\overline{h_0h_2}$ |
| 36 | 8 | 19 | $h_0\overline{\delta}$ | $(g)$ | 0 | 0 |
| 36 | 9 | 19 | $g\overline{h_0^2\alpha}$ | $(g^2)$ | 0 | 0 |
| 36 | 9 | 20 | $h_0^2\overline{\delta}$ | $(g)$ | 0 | 0 |
| 36 | $10+i$ | 20 | $i(h_0^{1+i}\alpha^3)$ | $(g)$ | 0 | 0 |
| 37 | 8 | 21 | $h_1\overline{\delta}$ | $(g)$ | 0 | 0 |
| 38 | 8 | 22 | $d_0\overline{g}$ | $(g^2)$ | 0 | 0 |
| 40 | $10+i$ | 24 | $h_0^{1+i}\overline{\alpha^3}$ | $(g)$ | 0 | 0 |
| 41 | 8 | 24 | $e_0\overline{g}$ | $(g^3)$ | $w_1 \cdot i(\delta)$ | $w_1 \cdot i(\delta w_2^2)$ |
| 43 | 9 | 26 | $d_0\overline{\gamma}$ | $(g^2)$ | $gw_1 \cdot i(d_0)$ | $gw_1 \cdot i(d_0w_2^2)$ |
| 45 | 10 | 28 | $d_0\overline{\alpha\beta}$ | $(g^2)$ | 0 | 0 |
| 46 | 9 | 28 | $e_0\overline{\gamma}$ | $(g^3)$ | $gw_1 \cdot i(e_0)$ $+ w_1 \cdot h_1\overline{\delta}$ | $gw_1 \cdot i(e_0w_2^2)$ $+ w_1 \cdot h_1w_2^2\overline{\delta}$ |
| 48 | $9+i$ | 29 | $i(h_0^{1+i}w_2)$ | $(g)$ | 0 | 0 |
| 48 | 10 | $30+31$ | $\alpha^2\overline{g}$ | $(g^2)$ | 0 | 0 |
| 49 | 9 | 31 | $i(h_1w_2)$ | $(g)$ | $g^2w_1 \cdot i(1)$ | $g^2w_1 \cdot i(w_2^2)$ |

Table 8.3: $R_1$-module generators of $E_3(tmf/\nu)$, with $i \geq 0$ in each $h_0$-tower (cont.)

| $t-s$ | $s$ | $g$ | $x$ | $\mathrm{Ann}(x)$ | $d_3(x)$ | $d_3(xw_2^2)$ |
|---|---|---|---|---|---|---|
| 50 | 10 | 33 | $i(h_1^2 w_2)$ | $(g)$ | 0 | 0 |
| 50 | 11 | 33 | $\alpha d_0 \overline{g}$ | $(g^2)$ | $gw_1 \cdot d_0 \overline{h_0 h_2}$ | $gw_1 \cdot d_0 w_2^2 \overline{h_0 h_2}$ |
| 52 | $11+i$ | 35 | $h_0^i w_2 \overline{h_0^3}$ | $(g)$ | 0 | 0 |
| 53 | 11 | 36 | $\alpha^2 \overline{\gamma}$ | $(g^2)$ | $gw_1 \cdot i(\alpha^2)$ | $gw_1 \cdot i(\alpha^2 w_2^2)$ |
| 54 | 10 | 35 | $g\overline{\beta^2}$ | $(g^2 w_1)$ | 0 | $g^6 \cdot \overline{\gamma}$ |
| 54 | 10 | 36 | $h_1 w_2 \overline{h_1}$ | $(g)$ | $g^2 w_1 \cdot \overline{h_1}$ | $g^2 w_1 \cdot w_2^2 \overline{h_1}$ |
| 55 | 11 | 38 | $h_0 w_2 \overline{h_0 h_2}$ | — | 0 | 0 |
| 55 | 12 | 38 | $d_0 e_0 \overline{g}$ | $(g^2)$ | 0 | 0 |
| 56 | 11 | 40 | $i(c_0 w_2)$ | $(g)$ | 0 | 0 |
| 59 | 11 | 41 | $\beta g \overline{g}$ | — | $gw_1 \cdot g\overline{h_2^2}$ | $g^5 \cdot g\overline{\beta^2}$ $+ gw_1 \cdot gw_2^2 \overline{h_2^2}$ |
| 60 | 11 | 42 | $w_2 \overline{c_0}$ | $(g)$ | 0 | 0 |
| 60 | $12+i$ | 44 | $i(h_0^{1+i} \alpha w_2)$ | $(g)$ | 0 | 0 |
| 60 | 13 | 44 | $g\overline{\alpha^3}$ | $(g^2)$ | $gw_1 \cdot i(d_0 e_0)$ | $gw_1 \cdot i(d_0 e_0 w_2^2)$ |
| 61 | 12 | 46 | $h_1 w_2 \overline{c_0}$ | $(g)$ | 0 | 0 |
| 64 | $14+i$ | 51 | $h_0^{1+i} w_2 \overline{h_0^2 \alpha}$ | $(g)$ | 0 | 0 |
| 72 | 13 | 56 | $h_0 w_2 \overline{g}$ | $(g)$ | 0 | 0 |
| 72 | 14 | 60 | $h_0^2 w_2 \overline{g}$ | $(g)$ | 0 | 0 |
| 72 | $15+i$ | 61 | $i(h_0^{1+i} \alpha^2 w_2)$ | $(g)$ | 0 | 0 |
| 73 | 13 | 58 | $h_1 w_2 \overline{g}$ | $(g)$ | $g^2 w_1 \cdot \overline{g}$ | $g^2 w_1 \cdot w_2^2 \overline{g}$ |
| 74 | 14 | 62 | $i(h_1 \gamma w_2)$ | $(g)$ | 0 | 0 |
| 76 | $15+i$ | 66 | $h_0^i w_2 \overline{h_0 \alpha^2}$ | $(g)$ | 0 | 0 |
| 78 | 14 | 65 | $h_1 w_2 \overline{\gamma}$ | $(g)$ | $g^2 w_1 \cdot \overline{\gamma}$ | $g^2 w_1 \cdot w_2^2 \overline{\gamma}$ |
| 79 | 15 | 69 | $h_0 w_2 \overline{\alpha\beta}$ | — | 0 | 0 |
| 80 | 15 | 71 | $i(\delta w_2)$ | $(g)$ | 0 | 0 |
| 84 | 15 | 73 | $w_2 \overline{\delta}$ | $(g)$ | 0 | 0 |
| 84 | 16 | 77 | $h_0 w_2 \overline{\delta}$ | $(g)$ | 0 | 0 |
| 84 | 17 | 79 | $h_0^2 w_2 \overline{\delta}$ | $(g)$ | 0 | 0 |
| 84 | $18+i$ | 80 | $i(h_0^{1+i} \alpha^3 w_2)$ | $(g)$ | 0 | 0 |
| 85 | 16 | 79 | $h_1 w_2 \overline{\delta}$ | $(g)$ | 0 | 0 |
| 88 | $18+i$ | 87 | $h_0^{1+i} w_2 \overline{\alpha^3}$ | $(g)$ | 0 | 0 |

Table 8.4: The non-cyclic $R_1$-module summands in $E_3(tmf/\nu)$

| $\langle x_1, x_2 \rangle$ |
|---|
| $\langle i(\beta g), h_0 w_2 \overline{h_0 h_2} \rangle \cong \dfrac{\Sigma^{7,42} R_1 \oplus \Sigma^{11,66} R_1}{\langle (g w_1, w_1), (0, g) \rangle}$ |
| $\langle \beta g \overline{g}, h_0 w_2 \overline{\alpha\beta} \rangle \cong \dfrac{\Sigma^{11,70} R_1 \oplus \Sigma^{15,94} R_1}{\langle (g w_1, w_1), (0, g) \rangle}$ |

PROPOSITION 8.3. *The* 20 *classes listed in Table* 8.5 *generate* $E_3(tmf/\nu)$ *as a module over* $E_3(tmf)$.

Table 8.5: $E_3(tmf)$-module generators of $E_3(tmf/\nu)$

| $t-s$ | $s$ | $g$ | $x$ | $d_3(x)$ |
|---|---|---|---|---|
| 0 | 0 | 0 | $i(1)$ | 0 |
| 4 | 3 | 1 | $\overline{h_0^3}$ | 0 |
| 5 | 1 | 2 | $\overline{h_1}$ | 0 |
| 7 | 2 | 3 | $\overline{h_0 h_2}$ | 0 |
| 12 | 3 | 4 | $\overline{c_0}$ | 0 |
| 12 | 3 | $4+5$ | $i(\alpha)$ | 0 |
| 16 | 6 | 7 | $h_0 \overline{h_0^2 \alpha}$ | 0 |
| 24 | 4 | 9 | $\overline{g}$ | 0 |
| 28 | 7 | 13 | $\overline{h_0 \alpha^2}$ | 0 |
| 29 | 5 | 13 | $\overline{\gamma}$ | $i(g w_1)$ |
| 31 | 6 | 16 | $\overline{\alpha\beta}$ | 0 |
| 36 | 7 | 19 | $\overline{\delta}$ | 0 |
| 36 | 7 | $19+20$ | $\alpha \overline{g}$ | $g w_1 \overline{h_0 h_2}$ |
| 40 | 10 | 24 | $h_0 \overline{\alpha^3}$ | 0 |
| 52 | 11 | 35 | $w_2 \overline{h_0^3}$ | 0 |
| 60 | 11 | 42 | $w_2 \overline{c_0}$ | 0 |
| 64 | 14 | 51 | $h_0 w_2 \overline{h_0^2 \alpha}$ | 0 |
| 76 | 15 | 66 | $w_2 \overline{h_0 \alpha^2}$ | 0 |
| 84 | 15 | 73 | $w_2 \overline{\delta}$ | 0 |
| 88 | 18 | 87 | $h_0 w_2 \overline{\alpha^3}$ | 0 |

PROOF. Inspection of Tables 5.2 and 8.3 easily shows that most of the $R_1$-module generators of $E_3(tmf/\nu)$ are $E_3(tmf)$-multiples of the classes in Table 8.5. The less evident cases follow from the relations

$$\begin{aligned} \overline{gh_2^2} &= \beta^2 \cdot i(1) \\ i(\beta g) &= \beta^2 \cdot \overline{h_1} \\ \overline{gh_0^2\alpha} &= d_0e_0 \cdot \overline{h_1} \\ g\overline{\beta^2} &= \beta^2 \cdot \overline{g} \\ \beta g\overline{g} &= \beta^2 \cdot \overline{\gamma} \\ g\overline{\alpha^3} &= d_0e_0 \cdot \overline{\gamma} \\ i(h_1\gamma w_2) &= h_1^2 w_2 \cdot \overline{g}\,, \end{aligned}$$

which we verify by calculating the relevant Yoneda products using `ext`. □

PROPOSITION 8.4. *The $d_3$-differentials on the $E_3(tmf)$-module generators for $E_3(tmf/\nu)$ are as listed in Table* 8.5.

PROOF. To determine $d_3$ on the $E_3(tmf)$-module generators of $E_3(tmf/\nu)$ we refer to Figures 1.32 to 1.35, keeping in mind that the $E_3$-term is a subquotient of the $E_2$-term shown in these charts. The $d_3$-differentials on $i(1)$, $\overline{h_0^3}$, $\overline{h_0h_2}$, $\overline{c_0}$, $i(\alpha)$, $h_0\overline{h_0^2\alpha}$, $\overline{h_0\alpha^2}$, $h_0\overline{\alpha^3}$, $w_2\overline{h_0^3}$ and $h_0w_2\overline{h_0^2\alpha}$ are zero because the target groups are trivial, already at the $E_2$-term.

The $d_3$-differentials on $\overline{g}$, $w_2\overline{h_0\alpha^2}$, $w_2\overline{\delta}$ and $h_0w_2\overline{\alpha^3}$ are zero because the target groups are trivial at the $E_3$-term:

- The bidegree $(t-s, s) = (23, 7)$ of $d_3(\overline{g})$ is generated at $E_2$ by $w_1 \cdot i(\beta)$, and $d_2(w_1 \cdot i(\beta)) = w_1 \cdot i(h_0d_0) \neq 0$.
- The bidegree $(75, 18)$ of $d_3(w_2\overline{h_0\alpha^2})$ is generated at $E_2$ by $g^3w_1 \cdot \overline{h_0h_2} = d_2(gw_1 \cdot i(w_2))$.
- The bidegree $(83, 18)$ of $d_3(w_2\overline{\delta})$ is generated at $E_2$ by $gw_1 \cdot w_2\overline{h_0h_2}$, and $d_2(gw_1 \cdot w_2\overline{h_0h_2}) = g^3w_1 \cdot i(d_0) \neq 0$.
- The bidegree $(87, 21)$ of $d_3(h_0w_2\overline{\alpha^3})$ is generated at $E_2$ by $g^3w_1 \cdot d_0\overline{h_1} = d_2(gw_1 \cdot i(\alpha w_2))$.

The $d_3$-differential on $\overline{h_1}$ is zero by $h_0$-linearity.

We show that $d_3(\overline{\gamma}) = i(gw_1)$ as a consequence of the relations $g\overline{\gamma} = 9_{30} = \gamma\overline{g} + i(h_1w_2)$ and $h_0\overline{\gamma} = 0$. We know from Table 5.2 that $d_3(g) = 0$, $d_3(\gamma) = 0$ and $d_3(h_1w_2) = g^2w_1$, and we have just seen that $d_3(\overline{g}) = 0$. Hence $gd_3(\overline{\gamma}) = i(g^2w_1) \neq 0$ in $E_3(tmf/\nu)$. It follows that $d_3(\overline{\gamma})$ is nonzero and annihilated by $h_0$, and the only possible value is $i(gw_1)$.

Furthermore, we show that $d_3(\alpha\overline{g}) = gw_1\overline{h_0h_2}$ as a consequence of the relation $e_0 \cdot \alpha\overline{g} = 11_{36} = \alpha^2 \cdot \overline{\gamma}$. We know from Table 5.2 that $d_3(e_0) = c_0w_1$ and $d_3(\alpha^2) = h_1d_0w_1$, and we have just seen that $d_3(\overline{\gamma}) = i(gw_1)$. Hence

$$d_3(e_0 \cdot \alpha\overline{g}) = c_0w_1 \cdot \alpha\overline{g} + e_0 \cdot d_3(\alpha\overline{g}) = e_0 \cdot d_3(\alpha\overline{g})$$

is equal to

$$d_3(\alpha^2 \cdot \overline{\gamma}) = h_1d_0w_1 \cdot \overline{\gamma} + \alpha^2 \cdot i(gw_1) = gw_1 \cdot i(\alpha^2) \neq 0\,,$$

where $c_0w_1 \cdot \alpha\overline{g} = 0$ and $h_1d_0w_1 \cdot \overline{\gamma} = 0$ can be verified with `ext`. Hence $d_3(\alpha\overline{g})$ is nonzero, and $gw_1\overline{h_0h_2}$ is the only possible value.

For $\overline{\alpha\beta}$ we use naturality with respect to $j: C\nu \to S^4$. We know that $d_3(\overline{\alpha\beta}) \in \{0, e_0 w_1 \overline{h_1}\}$ maps by $j$ to $d_3(\alpha\beta) = 0$ in $E_3(tmf)$. However, $j(e_0 w_1 \overline{h_1}) = w_1 \cdot h_1 e_0 \neq 0$. Hence $d_3(\overline{\alpha\beta}) = 0$.

For $\overline{\delta}$ and $w_2\overline{c_0}$ we use $d_0$-linearity of $d_3$, which follows from $d_3(d_0) = 0$. We know that $d_3(\overline{\delta}) \in \{0, gw_1\overline{h_0h_2}\}$, and $d_0 \cdot gw_1\overline{h_0h_2} = gw_1 \cdot d_0\overline{h_0h_2} \neq 0$. Since $d_0 \cdot \overline{\delta} = 0$ we must have $d_0 \cdot d_3(\overline{\delta}) = 0$. Hence $d_3(\overline{\delta}) = 0$.

Finally, $d_3(w_2\overline{c_0}) \in \{0, gw_1\overline{\alpha\beta}\}$, where $d_0 \cdot gw_1\overline{\alpha\beta} = gw_1 \cdot d_0\overline{\alpha\beta} \neq 0$. From $d_0 \cdot w_2\overline{c_0} = 0$ we deduce $d_0 \cdot d_3(w_2\overline{c_0}) = 0$. Hence $d_3(w_2\overline{c_0}) = 0$. □

THEOREM 8.5. *The $d_3$-differential in $E_3(tmf/\nu)$ is $R_2$-linear. Its values on a set of $R_2$-module generators are listed in Table 8.3.*

PROOF. The classes $g$, $w_1$ and $w_2^4$ are $d_3$-cycles in $E_3(tmf)$, so multiplication by each of these commutes with the $d_3$-differential in $E_3(tmf/\nu)$.

The $d_3$-differential on the $R_1$-module generators $x$ in Table 8.3 is given by the Leibniz rule applied to the (implicit and explicit) factorizations in the proof of Proposition 8.3, and the $d_3$-differentials from Tables 5.2 and 8.5. In several of the following cases we use `ext` to rewrite the output of the Leibniz rule in terms of the $R_1$-module presentation of $E_3(tmf/\nu)$.

- $d_3(e_0 \cdot \overline{h_1}) = c_0w_1 \cdot \overline{h_1} = 8_9 = w_1 \cdot h_1\overline{c_0}$
- $d_3(i(\alpha^2)) = i(h_1d_0w_1) = 0$
- $d_3(e_0 \cdot i(\alpha)) = c_0w_1 \cdot i(\alpha) = 0$
- $d_3(g\overline{h_2^2}) = d_3(\beta^2 \cdot i(1)) = h_1gw_1 \cdot i(1) = 0$
- $d_3(h_1\overline{\gamma}) = h_1 \cdot i(gw_1) = 0$
- $d_3(i(\beta g)) = d_3(\beta^2 \cdot \overline{h_1}) = h_1gw_1 \cdot \overline{h_1} = 0$
- $d_3(g\overline{h_0^2\alpha}) = d_3(d_0e_0 \cdot \overline{h_1}) = 0$
- $d_3(e_0 \cdot \overline{g}) = c_0w_1 \cdot \overline{g} = 11_{24} = w_1 \cdot i(\delta)$
- $d_3(d_0 \cdot \overline{\gamma}) = d_0 \cdot i(gw_1) = gw_1 \cdot i(d_0)$
- $d_3(e_0 \cdot \overline{\gamma}) = e_0 \cdot i(gw_1) + c_0w_1 \cdot \overline{\gamma} = 12_{28} + 12_{29} = gw_1 \cdot i(e_0) + w_1 \cdot h_1\overline{\delta}$
- $d_3(\alpha^2 \cdot \overline{g}) = h_1d_0w_1 \cdot \overline{g} = 0$
- $d_3(d_0 \cdot \alpha\overline{g}) = d_0 \cdot gw_1\overline{h_0h_2} = gw_1 \cdot d_0\overline{h_0h_2}$
- $d_3(\alpha^2 \cdot \overline{\gamma}) = h_1d_0w_1 \cdot \overline{\gamma} + \alpha^2 \cdot i(gw_1) = 14_{36} = gw_1 \cdot i(\alpha^2)$
- $d_3(g\overline{\beta^2}) = d_3(\beta^2 \cdot \overline{g}) = h_1gw_1 \cdot \overline{g} = 0$
- $d_3(\beta g\overline{g}) = d_3(\beta^2 \cdot \overline{\gamma}) = h_1gw_1 \cdot \overline{\gamma} + \beta^2 \cdot i(gw_1) = 14_{42} = gw_1 \cdot g\overline{h_2^2}$
- $d_3(h_0w_2 \cdot i(\alpha)) = 0$
- $d_3(g\overline{\alpha^3}) = d_3(d_0e_0 \cdot \overline{\gamma}) = d_0e_0 \cdot i(gw_1) = gw_1 \cdot i(d_0e_0)$
- $d_3(h_1w_2 \cdot \overline{c_0}) = g^2w_1 \cdot \overline{c_0} = 0$
- $d_3(i(h_1\gamma w_2)) = d_3(h_1^2w_2 \cdot \overline{g}) = 0$
- $d_3(h_1w_2 \cdot \overline{\gamma}) = g^2w_1 \cdot \overline{\gamma} + h_1w_2 \cdot i(gw_1) = 17_{69} = g^2w_1 \cdot \overline{\gamma}$.

It remains to determine $d_3(w_2^2 \cdot x) = d_3(w_2^2) \cdot x + w_2^2 \cdot d_3(x) = \beta g^4 \cdot x + w_2^2 \cdot d_3(x)$ for the same generators $x$. If $g^4 \in \mathrm{Ann}(x)$ this takes no effort. Otherwise we use `ext` to calculate:

- $\beta g^4 \cdot i(1) = g^3 \cdot i(\beta g)$
- $\beta g^4 \cdot \overline{h_1} = 20_{106} = g^5 \cdot i(1)$
- $\beta g^4 \cdot \overline{g} = 23_{144} = g^3 \cdot \beta g\overline{g}$
- $\beta g^4 \cdot \overline{\gamma} = 24_{155} = g^5 \cdot \overline{g}$
- $\beta g^4 \cdot g\overline{h_2^2} = 25_{160} = g^6 \cdot \overline{h_1}$
- $\beta g^4 \cdot i(\beta g) = 26_{171} = g^5 \cdot g\overline{h_2^2}$

- $\beta g^4 \cdot \overline{g\beta^2} = 29_{219} = g^6 \cdot \overline{\gamma}$
- $\beta g^4 \cdot \beta g\overline{g} = 30_{232} = g^5 \cdot \overline{g\beta^2}$. □

REMARK 8.6. To calculate the products $\beta g^4 \cdot x$ with `ext`, use `cocycle`, `dolifts` and `collect` as in Remark 8.2. The nonzero products with $\beta g^4 = 19_{56}$ then appear as lines containing `(19 56 F2)` in the file `all`. If the product is a $g^3$-multiple, there will also appear a line containing `(12 29 F2)` in the same block, since $g^3 = 12_{29}$ in the minimal $A(2)$-module resolution for $\mathbb{F}_2$. Similarly, $g^5$-multiples appear with `(20 67 F2)`, and $g^6$-multiples appear with `(24 90 F2)`.

## 8.4. The $d_4$-differentials for $tmf/\nu$

It is now an elementary matter to compute the $E_4$-term of the Adams spectral sequence for $tmf/\nu$, as a direct sum of $R_2 = \mathbb{F}_2[g, w_1, w_2^4]$-modules. This is carried out in Appendix D.2 and the results are recorded in Tables 8.6 and 8.7. In this section we determine the $d_4$-differentials on these $R_2$-module generators, and the results are also recorded in these tables.

Table 8.6: $R_2$-module generators of $E_4(tmf/\nu)$, with $i \geq 0$ in each $h_0$-tower

| $t-s$ | $s$ | $g$ | $x$ | $\mathrm{Ann}(x)$ | $d_4(x)$ |
|---|---|---|---|---|---|
| 0 | 0 | 0 | $i(1)$ | $(g^5, gw_1)$ | 0 |
| 0 | $1+i$ | 0 | $i(h_0^{1+i})$ | $(g)$ | 0 |
| 1 | 1 | 1 | $i(h_1)$ | $(g)$ | 0 |
| 2 | 2 | 1 | $i(h_1^2)$ | $(g)$ | 0 |
| 4 | $3+i$ | 1 | $h_0^i\overline{h_0^3}$ | $(g)$ | 0 |
| 5 | 1 | 2 | $\overline{h_1}$ | $(g^6, g^2w_1)$ | 0 |
| 6 | 2 | 2 | $h_1\overline{h_1}$ | $(g)$ | 0 |
| 7 | 2 | 3 | $\overline{h_0h_2}$ | $(g^2, gw_1)$ | 0 |
| 7 | 3 | 2 | $h_0\overline{h_0h_2}$ | $(g, w_1)$ | 0 |
| 8 | 3 | 3 | $i(c_0)$ | $(g, w_1)$ | 0 |
| 12 | 3 | 4 | $\overline{c_0}$ | $(g)$ | 0 |
| 12 | 3 | $4+5$ | $i(\alpha)$ | $(g^2)$ | 0 |
| 12 | $4+i$ | 4 | $i(h_0^{1+i}\alpha)$ | $(g)$ | 0 |
| 13 | 4 | 5 | $h_1\overline{c_0}$ | $(g, w_1)$ | 0 |
| 14 | 4 | 6 | $i(d_0)$ | $(g^2, gw_1)$ | 0 |
| 16 | $6+i$ | 7 | $h_0^{1+i}\overline{h_0^2\alpha}$ | $(g)$ | 0 |
| 19 | 5 | 8 | $d_0\overline{h_1}$ | $(g^2)$ | 0 |
| 21 | 6 | 9 | $d_0\overline{h_0h_2}$ | $(g^2, gw_1)$ | 0 |
| 24 | 4 | 9 | $\overline{g}$ | $(g^6, g^2w_1)$ | 0 |
| 24 | 5 | 10 | $h_0\overline{g}$ | $(g)$ | 0 |

Table 8.6: $R_2$-module generators of $E_4(tmf/\nu)$, with $i \geq 0$ in each $h_0$-tower (cont.)

| $t-s$ | $s$ | $g$ | $x$ | $\mathrm{Ann}(x)$ | $d_4(x)$ |
|---|---|---|---|---|---|
| 24 | 6 | $10+11$ | $i(\alpha^2)$ | $(g^2, gw_1)$ | $w_1^2 \cdot \overline{h_0h_2}$ |
| 24 | 6 | 11 | $h_0^2\overline{g}$ | $(g)$ | 0 |
| 24 | $7+i$ | 11 | $i(h_0^{1+i}\alpha^2)$ | $(g)$ | 0 |
| 25 | 5 | 12 | $h_1\overline{g}$ | $(g)$ | 0 |
| 26 | 6 | 12 | $i(h_1\gamma)$ | $(g)$ | 0 |
| 26 | 7 | 12 | $i(\alpha d_0)$ | $(g^2)$ | 0 |
| 28 | $7+i$ | 13 | $h_0^i\overline{h_0\alpha^2}$ | $(g)$ | 0 |
| 29 | 7 | 14 | $i(\alpha e_0)$ | $(g^2)$ | $w_1^2 \cdot (\overline{c_0} + i(\alpha))$ |
| 30 | 6 | 14 | $g\overline{h_2^2}$ | $(g^5, gw_1)$ | $w_1 \cdot d_0\overline{h_0h_2}$ |
| 30 | 6 | 15 | $h_1\overline{\gamma}$ | $(g)$ | $w_1 \cdot d_0\overline{h_0h_2}$ |
| 31 | 6 | 16 | $\overline{\alpha\beta}$ | $(g^2)$ | 0 |
| 31 | 7 | 15 | $h_0\overline{\alpha\beta}$ | $(g, w_1)$ | 0 |
| 31 | 8 | 15 | $i(d_0e_0)$ | $(g^2, gw_1)$ | $w_1^2 \cdot i(d_0)$ |
| 32 | 7 | 17 | $i(\delta)$ | $(g, w_1)$ | 0 |
| 35 | 7 | 18 | $i(\beta g)$ | – | $w_1 \cdot i(\alpha d_0)$ |
| 36 | 7 | 19 | $\overline{\delta}$ | $(g)$ | 0 |
| 36 | 8 | 19 | $h_0\overline{\delta}$ | $(g)$ | 0 |
| 36 | 9 | 19 | $g\overline{h_0^2\alpha}$ | $(g^2)$ | $w_1^2 \cdot d_0\overline{h_1}$ |
| 36 | 9 | 20 | $h_0^2\overline{\delta}$ | $(g)$ | 0 |
| 36 | $10+i$ | 20 | $i(h_0^{1+i}\alpha^3)$ | $(g)$ | 0 |
| 37 | 8 | $20+21$ | $\delta'\overline{h_1}$ | $(g^2, w_1)$ | 0 |
| 37 | 8 | 21 | $h_1\overline{\delta}$ | $(g)$ | 0 |
| 38 | 8 | 22 | $d_0\overline{g}$ | $(g^2)$ | 0 |
| 40 | $10+i$ | 24 | $h_0^{1+i}\overline{\alpha^3}$ | $(g)$ | 0 |
| 42 | 9 | 25 | $e_0g\overline{h_1}$ | $(g^2)$ | $gw_1^2 \cdot \overline{h_1}$ |
| 45 | 10 | 28 | $d_0\overline{\alpha\beta}$ | $(g^2)$ | 0 |
| 48 | 9 | 29 | $i(h_0w_2)$ | $(g)$ | $gw_1 \cdot d_0\overline{h_1}$ |
| 48 | 10 | $30+31$ | $\alpha^2\overline{g}$ | $(g^2)$ | $w_1^2 \cdot \overline{\alpha\beta}$ |
| 48 | $10+i$ | 31 | $i(h_0^{2+i}w_2)$ | $(g)$ | 0 |
| 49 | 9 | $30+31$ | $\gamma\overline{g}$ | $(g^5, gw_1)$ | 0 |
| 50 | 10 | 33 | $i(h_1^2w_2)$ | $(g)$ | 0 |

Table 8.6: $R_2$-module generators of $E_4(tmf/\nu)$, with $i \geq 0$ in each $h_0$-tower (cont.)

| $t-s$ | $s$ | $g$ | $x$ | $\mathrm{Ann}(x)$ | $d_4(x)$ |
|---|---|---|---|---|---|
| 52 | $11+i$ | 35 | $h_0^i w_2 \overline{h_0^3}$ | $(g)$ | 0 |
| 54 | 10 | 35 | $g\overline{\beta^2}$ | $(g^6, g^2 w_1)$ | $w_1 \cdot d_0\overline{\alpha\beta}$ |
| 55 | 11 | 38 | $h_0 w_2 \overline{h_0 h_2}$ | $-$ | $g w_1 \cdot i(\alpha d_0)$ |
| 55 | 12 | 38 | $d_0 e_0 \overline{g}$ | $(g^2)$ | $w_1^2 \cdot d_0\overline{g}$ |
| 56 | 11 | 39 | $\alpha g \overline{g}$ | $(g)$ | 0 |
| 56 | 11 | 40 | $i(c_0 w_2)$ | $(g)$ | 0 |
| 60 | 11 | 42 | $w_2 \overline{c_0}$ | $(g)$ | 0 |
| 60 | $12+i$ | 44 | $i(h_0^{1+i} \alpha w_2)$ | $(g)$ | 0 |
| 61 | 12 | 45 | $e_0 g \overline{g}$ | $(g^2)$ | $g w_1^2 \cdot \overline{g}$ |
| 61 | 12 | 46 | $h_1 w_2 \overline{c_0}$ | $(g)$ | 0 |
| 63 | 13 | 49 | $d_0 g \overline{\gamma}$ | $(g)$ | 0 |
| 64 | $14+i$ | 51 | $h_0^{1+i} w_2 \overline{h_0^2 \alpha}$ | $(g)$ | 0 |
| 70 | 15 | 58 | $\alpha d_0 g \overline{g}$ | $(g)$ | 0 |
| 72 | 13 | 56 | $h_0 w_2 \overline{g}$ | $(g)$ | $w_1 \cdot d_0 g \overline{\gamma}$ |
| 72 | 14 | 60 | $h_0^2 w_2 \overline{g}$ | $(g)$ | 0 |
| 72 | $15+i$ | 61 | $i(h_0^{1+i} \alpha^2 w_2)$ | $(g)$ | 0 |
| 73 | 15 | 62 | $\alpha^2 g \overline{\gamma}$ | $(g)$ | $w_1^2 \cdot \alpha g \overline{g}$ $+ w_1^2 \cdot i(c_0 w_2)$ |
| 74 | 14 | 62 | $i(h_1 \gamma w_2)$ | $(g)$ | $g w_1 \cdot d_0 \overline{\alpha\beta}$ |
| 76 | $15+i$ | 66 | $h_0^i w_2 \overline{h_0 \alpha^2}$ | $(g)$ | 0 |
| 79 | 15 | $68+69$ | $\gamma^2 \overline{\gamma}$ | $(g^2, w_1)$ | 0 |
| 79 | 15 | 69 | $h_0 w_2 \overline{\alpha\beta}$ | $(g)$ | $w_1 \cdot \alpha d_0 g \overline{g}$ |
| 80 | 15 | 71 | $i(\delta w_2)$ | $(g)$ | 0 |
| 80 | 17 | 72 | $g^2 \overline{\alpha^3}$ | $(g)$ | $w_1^2 \cdot d_0 g \overline{\gamma}$ |
| 84 | 15 | 73 | $w_2 \overline{\delta}$ | $(g)$ | 0 |
| 84 | 16 | 77 | $h_0 w_2 \overline{\delta}$ | $(g)$ | 0 |
| 84 | 17 | 79 | $h_0^2 w_2 \overline{\delta}$ | $(g)$ | 0 |
| 84 | $18+i$ | 80 | $i(h_0^{1+i} \alpha^3 w_2)$ | $(g)$ | 0 |
| 85 | 16 | 79 | $h_1 w_2 \overline{\delta}$ | $(g)$ | 0 |
| 86 | 17 | 82 | $e_0 g^2 \overline{\gamma}$ | $(g)$ | 0 |
| 88 | $18+i$ | 87 | $h_0^{1+i} w_2 \overline{\alpha^3}$ | $(g)$ | 0 |

Table 8.6: $R_2$-module generators of $E_4(tmf/\nu)$, with $i \geq 0$ in each $h_0$-tower (cont.)

| $t-s$ | $s$ | $g$ | $x$ | $\mathrm{Ann}(x)$ | $d_4(x)$ |
|---|---|---|---|---|---|
| 96 | $17+i$ | 91 | $i(h_0^{1+i}w_2^2)$ | $(g)$ | 0 |
| 97 | 17 | 93 | $i(h_1w_2^2)$ | $(g)$ | 0 |
| 98 | 18 | 99 | $i(h_1^2w_2^2)$ | $(g)$ | 0 |
| 100 | $19+i$ | 105 | $h_0^iw_2^2\overline{h_0^3}$ | $(g)$ | 0 |
| 102 | 18 | 102 | $h_1w_2^2\overline{h_1}$ | $(g)$ | 0 |
| 103 | 18 | 103 | $w_2^2\overline{h_0h_2}$ | $(g^2, gw_1)$ | 0 |
| 103 | 19 | 108 | $h_0w_2^2\overline{h_0h_2}$ | $(g, w_1)$ | 0 |
| 104 | 19 | 110 | $i(c_0w_2^2)$ | $(g, w_1)$ | 0 |
| 104 | 20 | $112+113$ | $g^4\overline{g}+i(w_1w_2^2)$ | $(g)$ | 0 |
| 108 | 19 | 112 | $w_2^2\overline{c_0}$ | $(g)$ | 0 |
| 108 | 19 | $112+113$ | $i(\alpha w_2^2)$ | $(g^2)$ | 0 |
| 108 | $20+i$ | 118 | $i(h_0^{1+i}\alpha w_2^2)$ | $(g)$ | 0 |
| 109 | 20 | 120 | $h_1w_2^2\overline{c_0}$ | $(g, w_1)$ | 0 |
| 109 | 21 | 124 | $w_1w_2^2\overline{h_1}$ | $(g^2)$ | 0 |
| 110 | 20 | 121 | $i(d_0w_2^2)$ | $(g^2, gw_1)$ | 0 |
| 112 | $22+i$ | 132 | $h_0^{1+i}w_2^2\overline{h_0^2\alpha}$ | $(g)$ | 0 |
| 115 | 21 | 131 | $d_0w_2^2\overline{h_1}$ | $(g^2)$ | 0 |
| 117 | 22 | 138 | $d_0w_2^2\overline{h_0h_2}$ | $(g^2, gw_1)$ | 0 |
| 120 | 21 | 134 | $h_0w_2^2\overline{g}$ | $(g)$ | 0 |
| 120 | 22 | $141+142$ | $i(\alpha^2w_2^2)$ | $(g^2, gw_1)$ | $w_1^2 \cdot w_2^2\overline{h_0h_2}$ |
| 120 | 22 | 142 | $h_0^2w_2^2\overline{g}$ | $(g)$ | 0 |
| 120 | $23+i$ | 147 | $i(h_0^{1+i}\alpha^2w_2^2)$ | $(g)$ | 0 |
| 121 | 21 | 136 | $h_1w_2^2\overline{g}$ | $(g)$ | 0 |
| 122 | 22 | 144 | $i(h_1\gamma w_2^2)$ | $(g)$ | 0 |
| 122 | 23 | 149 | $i(\alpha d_0w_2^2)$ | $(g^2)$ | 0 |
| 124 | $23+i$ | 152 | $h_0^iw_2^2\overline{h_0\alpha^2}$ | $(g)$ | 0 |
| 125 | 23 | 153 | $i(\alpha e_0w_2^2)$ | $(g^2)$ | $w_1^2 \cdot w_2^2(\overline{c_0}+i(\alpha))$ |
| 126 | 22 | 147 | $h_1w_2^2\overline{\gamma}$ | $(g)$ | $w_1 \cdot d_0w_2^2\overline{h_0h_2}$ |
| 127 | 22 | 148 | $w_2^2\overline{\alpha\beta}$ | $(g^2)$ | 0 |
| 127 | 23 | 155 | $h_0w_2^2\overline{\alpha\beta}$ | $(g, w_1)$ | 0 |
| 127 | 24 | 160 | $i(d_0e_0w_2^2)$ | $(g^2, gw_1)$ | $w_1^2 \cdot i(d_0w_2^2)$ |

Table 8.6: $R_2$-module generators of $E_4(tmf/\nu)$, with $i \geq 0$ in each $h_0$-tower (cont.)

| $t-s$ | $s$ | $g$ | $x$ | $\mathrm{Ann}(x)$ | $d_4(x)$ |
|---|---|---|---|---|---|
| 128 | 23 | 157 | $i(\delta w_2^2)$ | $(g, w_1)$ | 0 |
| 128 | 24 | 162 | $w_1 w_2^2 \overline{g}$ | $(g^2)$ | 0 |
| 132 | 23 | 159 | $w_2^2 \overline{\delta}$ | $(g)$ | 0 |
| 132 | 24 | 167 | $h_0 w_2^2 \overline{\delta}$ | $(g)$ | 0 |
| 132 | 25 | 172 | $g w_2^2 \overline{h_0^2 \alpha}$ | $(g^2)$ | $w_1^2 \cdot d_0 w_2^2 \overline{h_1}$ |
| 132 | 25 | 173 | $h_0^2 w_2^2 \overline{\delta}$ | $(g)$ | 0 |
| 132 | $26+i$ | 177 | $i(h_0^{1+i} \alpha^3 w_2^2)$ | $(g)$ | 0 |
| 133 | 24 | $168+169$ | $\delta' w_2^2 \overline{h_1}$ | $(g^2, w_1)$ | 0 |
| 133 | 24 | 169 | $h_1 w_2^2 \overline{\delta}$ | $(g)$ | 0 |
| 134 | 24 | 170 | $d_0 w_2^2 \overline{g}$ | $(g^2)$ | 0 |
| 134 | 26 | $179+180$ | $g^5 \overline{\beta^2} + g w_1 w_2^2 \overline{h_2^2}$ | $(g)$ | $w_1^2 \cdot d_0 w_2^2 \overline{h_0 h_2}$ |
| 136 | $26+i$ | 185 | $h_0^{1+i} w_2^2 \overline{\alpha^3}$ | $(g)$ | 0 |
| 138 | 25 | 181 | $e_0 g w_2^2 \overline{h_1}$ | $(g^2)$ | $g w_1 \cdot w_1 w_2^2 \overline{h_1}$ |
| 139 | 27 | 193 | $i(\beta g w_1 w_2^2)$ | $-$ | $w_1^2 \cdot i(\alpha d_0 w_2^2)$ |
| 141 | 26 | 191 | $d_0 w_2^2 \overline{\alpha\beta}$ | $(g^2)$ | 0 |
| 144 | 25 | 185 | $i(h_0 w_2^3)$ | $(g)$ | $g w_1 \cdot d_0 w_2^2 \overline{h_1}$ |
| 144 | 26 | $194+195$ | $\alpha^2 w_2^2 \overline{g}$ | $(g^2)$ | $w_1^2 \cdot w_2^2 \overline{\alpha\beta}$ |
| 144 | $26+i$ | 195 | $i(h_0^{2+i} w_2^3)$ | $(g)$ | 0 |
| 146 | 26 | 197 | $i(h_1^2 w_2^3)$ | $(g)$ | 0 |
| 148 | $27+i$ | 207 | $h_0^i w_2^3 \overline{h_0^3}$ | $(g)$ | 0 |
| 151 | 27 | 210 | $h_0 w_2^3 \overline{h_0 h_2}$ | $-$ | $g w_1 \cdot i(\alpha d_0 w_2^2)$ |
| 151 | 28 | 217 | $d_0 e_0 w_2^2 \overline{g}$ | $(g^2)$ | $w_1^2 \cdot d_0 w_2^2 \overline{g}$ |
| 152 | 27 | 211 | $\alpha g w_2^2 \overline{g}$ | $(g)$ | 0 |
| 152 | 27 | 212 | $i(c_0 w_2^3)$ | $(g)$ | 0 |
| 153 | 29 | $227+228$ | $\gamma w_1 w_2^2 \overline{g}$ | $(g)$ | 0 |
| 156 | 27 | 214 | $w_2^3 \overline{c_0}$ | $(g)$ | 0 |
| 156 | $28+i$ | 224 | $i(h_0^{1+i} \alpha w_2^3)$ | $(g)$ | 0 |
| 157 | 28 | 225 | $e_0 g w_2^2 \overline{g}$ | $(g^2)$ | $g w_1 \cdot w_1 w_2^2 \overline{g}$ |
| 157 | 28 | 226 | $h_1 w_2^3 \overline{c_0}$ | $(g)$ | 0 |
| 158 | 30 | 241 | $g w_1 w_2^2 \overline{\beta^2}$ | $(g^2)$ | $w_1^2 \cdot d_0 w_2^2 \overline{\alpha\beta}$ |
| 159 | 29 | 237 | $d_0 g w_2^2 \overline{\gamma}$ | $(g)$ | 0 |

Table 8.6: $R_2$-module generators of $E_4(tmf/\nu)$, with $i \geq 0$ in each $h_0$-tower (cont.)

| $t-s$ | $s$ | $g$ | $x$ | $\mathrm{Ann}(x)$ | $d_4(x)$ |
|---|---|---|---|---|---|
| 160 | $30+i$ | 246 | $h_0^{1+i} w_2^3 \overline{h_0^2 \alpha}$ | $(g)$ | 0 |
| 166 | 31 | 261 | $\alpha d_0 g w_2^2 \overline{g}$ | $(g)$ | 0 |
| 168 | 29 | 244 | $h_0 w_2^3 \overline{g}$ | $(g)$ | $w_1 \cdot d_0 g w_2^2 \overline{\gamma}$ |
| 168 | 30 | 256 | $h_0^2 w_2^3 \overline{g}$ | $(g)$ | 0 |
| 168 | $31+i$ | 265 | $i(h_0^{1+i} \alpha^2 w_2^3)$ | $(g)$ | 0 |
| 169 | 31 | 266 | $\alpha^2 g w_2^2 \overline{\gamma}$ | $(g)$ | $w_1^2 \cdot \alpha g w_2^2 \overline{g}$ $+ w_1^2 \cdot i(c_0 w_2^3)$ |
| 170 | 30 | 258 | $i(h_1 \gamma w_2^3)$ | $(g)$ | $g w_1 \cdot d_0 w_2^2 \overline{\alpha\beta}$ |
| 172 | $31+i$ | 270 | $h_0^i w_2^3 \overline{h_0 \alpha^2}$ | $(g)$ | 0 |
| 175 | 31 | 273 | $h_0 w_2^3 \overline{\alpha\beta}$ | $(g)$ | $w_1 \cdot \alpha d_0 g w_2^2 \overline{g}$ |
| 176 | 31 | 275 | $i(\delta w_2^3)$ | $(g)$ | 0 |
| 176 | 33 | 291 | $g^2 w_2^2 \overline{\alpha^3}$ | $(g)$ | $w_1^2 \cdot d_0 g w_2^2 \overline{\gamma}$ |
| 180 | 31 | 277 | $w_2^3 \overline{\delta}$ | $(g)$ | 0 |
| 180 | 32 | 289 | $h_0 w_2^3 \overline{\delta}$ | $(g)$ | 0 |
| 180 | 33 | 299 | $h_0^2 w_2^3 \overline{\delta}$ | $(g)$ | 0 |
| 180 | $34+i$ | 307 | $i(h_0^{1+i} \alpha^3 w_2^3)$ | $(g)$ | 0 |
| 181 | 32 | 291 | $h_1 w_2^3 \overline{\delta}$ | $(g)$ | 0 |
| 182 | 33 | 302 | $e_0 g^2 w_2^2 \overline{\gamma}$ | $(g)$ | 0 |
| 184 | $34+i$ | 315 | $h_0^{1+i} w_2^3 \overline{\alpha^3}$ | $(g)$ | 0 |

Table 8.7: The non-cyclic $R_2$-module summands in $E_4(tmf/\nu)$

| $\langle x_1, x_2 \rangle$ |
|---|
| $\langle i(\beta g), h_0 w_2 \overline{h_0 h_2} \rangle \cong \dfrac{\Sigma^{7,42} R_2 \oplus \Sigma^{11,66} R_2}{\langle (g^3, 0), (g w_1, w_1), (0, g) \rangle}$ |
| $\langle i(\beta g w_1 w_2^2), h_0 w_2^3 \overline{h_0 h_2} \rangle \cong \dfrac{\Sigma^{27,166} R_2 \oplus \Sigma^{27,178} R_2}{\langle (g, w_1), (0, g) \rangle}$ |

PROPOSITION 8.7. *The 43 classes listed in Table 8.8 generate $E_4(tmf/\nu)$ as a module over $E_4(tmf)$.*

Table 8.8: $E_4(tmf)$-module generators of $E_4(tmf/\nu)$

| $t-s$ | $s$ | $g$ | $x$ | $d_4(x)$ | reason |
|---|---|---|---|---|---|
| 0 | 0 | 0 | $i(1)$ | 0 | $E_2 = 0$ |
| 4 | 3 | 1 | $\overline{h_0^3}$ | 0 | $E_2 = 0$ |
| 5 | 1 | 2 | $\overline{h_1}$ | 0 | $h_0$-lin. |
| 7 | 2 | 3 | $\overline{h_0 h_2}$ | 0 | $E_2 = 0$ |
| 12 | 3 | 4 | $\overline{c_0}$ | 0 | $E_2 = 0$ |
| 12 | 3 | $4+5$ | $i(\alpha)$ | 0 | $E_2 = 0$ |
| 16 | 6 | 7 | $h_0\overline{h_0^2\alpha}$ | 0 | $E_2 = 0$ |
| 24 | 4 | 9 | $\overline{g}$ | 0 | $E_2 = 0$ |
| 24 | 6 | $10+11$ | $i(\alpha^2)$ | $w_1^2 \cdot \overline{h_0 h_2}$ | (1) |
| 28 | 7 | 13 | $\overline{h_0\alpha^2}$ | 0 | $E_2 = 0$ |
| 29 | 7 | 14 | $i(\alpha e_0)$ | $w_1^2 \cdot (\overline{c_0} + i(\alpha))$ | (2) |
| 30 | 6 | 14 | $g\overline{h_2^2}$ | $w_1 \cdot d_0\overline{h_0 h_2}$ | (3) |
| 31 | 6 | 16 | $\overline{\alpha\beta}$ | 0 | $E_2 = 0$ |
| 35 | 7 | 18 | $i(\beta g)$ | $w_1 \cdot i(\alpha d_0)$ | (4) |
| 36 | 7 | 19 | $\overline{\delta}$ | 0 | $E_2 = 0$ |
| 40 | 10 | 24 | $h_0\overline{\alpha^3}$ | 0 | $E_2 = 0$ |
| 48 | 10 | $30+31$ | $\alpha^2\overline{g}$ | $w_1^2 \cdot \overline{\alpha\beta}$ | $j$-nat. |
| 52 | 11 | 35 | $w_2\overline{h_0^3}$ | 0 | (5) |
| 54 | 10 | 35 | $g\overline{\beta^2}$ | $w_1 \cdot d_0\overline{\alpha\beta}$ | $j$-nat. |
| 60 | 11 | 42 | $w_2\overline{c_0}$ | 0 | $E_2 = 0$ |
| 64 | 14 | 51 | $h_0 w_2\overline{h_0^2\alpha}$ | 0 | $E_3 = 0$ |
| 76 | 15 | 66 | $w_2\overline{h_0\alpha^2}$ | 0 | $E_4 = 0$ |
| 84 | 15 | 73 | $w_2\overline{\delta}$ | 0 | $E_3 = 0$ |
| 88 | 18 | 87 | $h_0 w_2\overline{\alpha^3}$ | 0 | $E_3 = 0$ |
| 100 | 19 | 105 | $w_2^2\overline{h_0^3}$ | 0 | $E_3 = 0$ |
| 103 | 18 | 103 | $w_2^2\overline{h_0 h_2}$ | 0 | $E_3 = 0$ |
| 108 | 19 | 112 | $w_2^2\overline{c_0}$ | 0 | $E_3 = 0$ |
| 108 | 19 | $112+113$ | $i(\alpha w_2^2)$ | 0 | $E_3 = 0$ |
| 112 | 22 | 132 | $h_0 w_2^2\overline{h_0^2\alpha}$ | 0 | $E_3 = 0$ |
| 120 | 22 | $141+142$ | $i(\alpha^2 w_2^2)$ | $w_1^2 \cdot w_2^2\overline{h_0 h_2}$ | (6) |
| 124 | 23 | 152 | $w_2^2\overline{h_0\alpha^2}$ | 0 | $E_3 = 0$ |
| 125 | 23 | 153 | $i(\alpha e_0 w_2^2)$ | $w_1^2 \cdot w_2^2(\overline{c_0} + i(\alpha))$ | (7) |

Table 8.8: $E_4(tmf)$-module generators of $E_4(tmf/\nu)$ (cont.)

| $t-s$ | $s$ | $g$ | $x$ | $d_4(x)$ | reason |
|---|---|---|---|---|---|
| 126 | 22 | 147 | $h_1 w_2^2 \overline{\gamma}$ | $w_1 \cdot d_0 w_2^2 \overline{h_0 h_2}$ | (8) |
| 127 | 22 | 148 | $w_2^2 \overline{\alpha\beta}$ | 0 | $E_3 = 0$ |
| 132 | 23 | 159 | $w_2^2 \overline{\delta}$ | 0 | $E_3 = 0$ |
| 136 | 26 | 185 | $h_0 w_2^2 \overline{\alpha^3}$ | 0 | $E_3 = 0$ |
| 144 | 26 | $194 + 195$ | $\alpha^2 w_2^2 \overline{g}$ | $w_1^2 \cdot w_2^2 \overline{\alpha\beta}$ | $j$-nat. |
| 148 | 27 | 207 | $w_2^3 \overline{h_0^3}$ | 0 | (9) |
| 156 | 27 | 214 | $w_2^3 \overline{c_0}$ | 0 | $E_4 = 0$ |
| 160 | 30 | 246 | $h_0 w_2^3 \overline{h_0^2 \alpha}$ | 0 | $E_3 = 0$ |
| 172 | 31 | 270 | $w_2^3 \overline{h_0 \alpha^2}$ | 0 | $E_4 = 0$ |
| 180 | 31 | 277 | $w_2^3 \overline{\delta}$ | 0 | $E_4 = 0$ |
| 184 | 34 | 315 | $h_0 w_2^3 \overline{\alpha^3}$ | 0 | $E_3 = 0$ |

PROOF. Most of the factorizations are visible by comparing Table 8.6 with Tables 5.5 and 8.8. The non-obvious factorizations are

$$
\begin{aligned}
h_1 \overline{\gamma} &= 6_{15} = \gamma \cdot \overline{h_1} + g\overline{h_2^2} \\
g\overline{h_0^2 \alpha} &= 9_{19} = d_0 e_0 \cdot \overline{h_1} \\
i(h_0 \alpha w_2) &= 12_{44} = h_0 \cdot w_2 \overline{c_0} \\
d_0 g \overline{\gamma} &= 13_{49} = d_0 \gamma \cdot \overline{g} \\
\alpha^2 g \overline{\gamma} &= 15_{62} = \alpha e_0 g \cdot \overline{g} \\
i(h_1 \gamma w_2) &= 14_{62} = h_1^2 w_2 \cdot \overline{g} \\
\gamma^2 \overline{\gamma} &= 15_{68} + 15_{69} = \beta g^2 \cdot \overline{g} + h_0 w_2 \cdot \overline{\alpha\beta} \\
g^2 \overline{\alpha^3} &= 17_{72} = \alpha e_0 g \cdot \overline{\alpha\beta} \\
e_0 g^2 \overline{\gamma} &= 17_{82} = e_0 \gamma g \cdot \overline{g} \\
g w_2^2 \overline{h_0^2 \alpha} &= 25_{172} = d_0 e_0 w_2^2 \cdot \overline{h_1} \\
g^5 \overline{\beta^2} + g w_1 w_2^2 \overline{h_2^2} &= 26_{179} + 26_{180} = g^4 \cdot g\overline{\beta^2} + d_0 \cdot i(\alpha^2 w_2^2) \\
i(h_0 \alpha w_2^3) &= 28_{224} = h_0 \cdot w_2^3 \overline{c_0} \\
d_0 g w_2^2 \overline{\gamma} &= 29_{237} = d_0 \gamma w_2^2 \cdot \overline{g} \\
\alpha^2 g w_2^2 \overline{\gamma} &= 31_{266} = \alpha e_0 g w_2^2 \cdot \overline{g} \\
i(h_1 \gamma w_2^3) &= 30_{258} = h_1^2 w_2^3 \cdot \overline{g} \\
g^2 w_2^2 \overline{\alpha^3} &= 33_{291} = \alpha e_0 g w_2^2 \cdot \overline{\alpha\beta} \\
e_0 g^2 w_2^2 \overline{\gamma} &= 33_{302} = e_0 \gamma g w_2^2 \cdot \overline{g} \,,
\end{aligned}
$$

which we verify using `ext`. □

PROPOSITION 8.8. *The $d_4$-differentials on the $E_4(tmf)$-module generators for $E_4(tmf/\nu)$ are as listed in Table* 8.8.

PROOF. Many of the differentials vanish because the target bidegree is or becomes zero at the $E_2$-, $E_3$- or $E_4$-term. These are indicated by "$E_2 = 0$", "$E_3 = 0$" or "$E_4 = 0$", respectively.

First, $d_4(x) = 0$ for $x = i(1)$, $\overline{h_0^3}$, $\overline{h_0h_2}$, $\overline{c_0}$, $i(\alpha)$, $h_0\overline{h_0^2\alpha}$, $\overline{g}$, $\overline{h_0\alpha^2}$, $\overline{\alpha\beta}$, $\overline{\delta}$, $h_0\overline{\alpha^3}$ and $w_2\overline{c_0}$, because in each case the target group is zero at $E_2 = E_2(tmf/\nu)$. This is clear from Figures 1.32 to 1.34. Next, $d_4(x) = 0$ for $x = h_0w_2\overline{h_0^2\alpha}$, $w_2\overline{\delta}$, $h_0w_2\overline{\alpha^3}$, $w_2^2\overline{h_0^3}$, $w_2^2\overline{h_0h_2}$, $w_2^2\overline{c_0}$, $i(\alpha w_2^2)$, $h_0w_2^2\overline{h_0^2\alpha}$, $w_2^2\overline{h_0\alpha^2}$, $w_2^2\overline{\alpha\beta}$, $w_2^2\overline{\delta}$, $h_0w_2^2\overline{\alpha^3}$, $h_0w_2^3\overline{h_0^2\alpha}$ and $h_0w_2^3\overline{\alpha^3}$, because in each case the target group is zero at $E_3$.

$h_0w_2\overline{h_0^2\alpha}$: The target is generated by $18_{50} = g^2w_1^2\overline{h_0h_2} = d_2(w_1^2 \cdot i(w_2))$.

$w_2\overline{\delta}$: The target is generated by $19_{78} = i(\beta g^3w_1) = d_2(g \cdot w_2\overline{h_0^2\alpha})$, using $g \cdot h_0\overline{h_0h_2} = 0$.

$h_0w_2\overline{\alpha^3}$: The target is generated by $22_{87} = g^2w_1^2\overline{\alpha\beta} = d_2(w_1^2 \cdot w_2\overline{g})$.

$w_2^2\overline{h_0^3}$: The target bidegree is generated at $E_2$ by $23_{108} = gw_1^2i(\beta w_2)$. Here $d_2(gw_1^2 \cdot i(\beta w_2)) = gw_1^2 \cdot (g^2 \cdot e_0\overline{h_1} + i(h_0d_0w_2)) = g^3w_1^2 \cdot e_0\overline{h_1} \neq 0$ at $E_2$, using $g \cdot i(h_0d_0) = 0$.

$w_2^2\overline{h_0h_2}$: The target is generated by $22_{113} = g^3w_1\overline{\beta^2} = d_2(d_0e_0w_2\overline{g})$.

$w_2^2\overline{c_0}$: The target is generated by $23_{122} = \beta g^3w_1\overline{g} = d_2(g \cdot w_2\overline{\alpha^3})$, using $g \cdot h_0\overline{\alpha\beta} = 0$.

$i(\alpha w_2^2)$: The target is generated by $23_{122} = \beta g^3w_1\overline{g} = d_2(g \cdot w_2\overline{\alpha^3})$.

$h_0w_2^2\overline{h_0^2\alpha}$: The target bidegree is generated at $E_2$ by $26_{133} = g^2w_1^2w_2\overline{h_0h_2}$, and $d_2(g^2w_1^2 \cdot w_2\overline{h_0h_2}) = g^2w_1^2 \cdot g^2 \cdot i(d_0) = g^4w_1^2 \cdot i(d_0) \neq 0$ at $E_2$.

$w_2^2\overline{h_0\alpha^2}$: The target bidegree is generated at $E_2$ by $27_{158} = g^5w_1i(\beta)$ and $27_{159} = gw_1^2 \cdot \beta w_2\overline{g}$. Here $g^5w_1i(\beta) = d_2(g^3 \cdot w_2\overline{h_0^2\alpha})$, using $g \cdot h_0\overline{h_0h_2} = 0$, and $d_2(gw_1^2 \cdot \beta w_2\overline{g}) = gw_1^2 \cdot g^2 \cdot e_0\overline{\gamma} = g^3w_1^2 \cdot e_0\overline{\gamma} \neq 0$ at $E_2$, using $g \cdot h_0d_0\overline{g} = 0$.

$w_2^2\overline{\alpha\beta}$: The target bidegree is generated at $E_2$ by $26_{164} = g^3w_1 \cdot w_2\overline{h_2^2}$, and $d_2(g^3w_1 \cdot w_2\overline{h_2^2}) = g^3w_1 \cdot (g^2 \cdot i(e_0) + i(h_1c_0w_2)) = g^5w_1 \cdot i(e_0) \neq 0$, using $g \cdot i(h_1c_0) = 0$.

$w_2^2\overline{\delta}$: The target bidegree is generated at $E_2$ by $27_{175} = g^3w_1 \cdot i(\beta w_2)$, and $d_2(g^3w_1 \cdot i(\beta w_2)) = g^3w_1 \cdot (g^2 \cdot e_0\overline{h_1} + i(h_0d_0w_2)) = g^5w_1 \cdot e_0\overline{h_1} \neq 0$, using $g \cdot i(h_0d_0) = 0$.

$h_0w_2^2\overline{\alpha^3}$: The target bidegree is generated at $E_2$ by $30_{188} = g^6w_1\overline{h_0h_2}$ and $30_{189} = g^2w_1^2 \cdot w_2\overline{\alpha\beta}$. Here $g^6w_1\overline{h_0h_2} = d_2(g^4w_1 \cdot i(w_2))$, and $d_2(g^2w_1^2 \cdot w_2\overline{\alpha\beta}) = g^2w_1^2 \cdot g^2 \cdot d_0\overline{g} = g^4w_1^2 \cdot d_0\overline{g} \neq 0$ at $E_2$.

$h_0w_2^3\overline{h_0^2\alpha}$: The target is generated by $34_{253} = g^6w_1\overline{\alpha\beta} = d_2(g^4w_1 \cdot \overline{g})$ and $34_{254} = g^2w_1^2w_2^2\overline{h_0h_2} = d_2(w_1^2w_2^2 \cdot w_2)$.

$h_0w_2^3\overline{\alpha^3}$: The target bidegree is generated at $E_2$ by $38_{327} = g^6w_1 \cdot w_2\overline{h_0h_2}$ and $38_{328} = g^2w_1^2w_2^2\overline{\alpha\beta}$. Here $d_2(g^6w_1 \cdot w_2\overline{h_0h_2}) = g^6w_1 \cdot g^2 \cdot i(d_0) = g^8w_1 \cdot i(d_0) \neq 0$ at $E_2$. On the other hand, $g^2w_1^2w_2^2\overline{\alpha\beta} = d_2(w_1^2w_2^2 \cdot w_2\overline{g})$.

Finally, $d_4(x) = 0$ for $x = w_2\overline{h_0\alpha^2}$, $w_2^3\overline{c_0}$, $w_2^3\overline{h_0\alpha^2}$ and $w_2^3\overline{\delta}$ because in each case the target group is zero at $E_4$.

$w_2\overline{h_0\alpha^2}$: The target bidegree is generated at $E_2$ by $19_{67} = w_1^2 \cdot \beta g\overline{g}$, and $d_3(w_1^2 \cdot \beta g\overline{g}) = gw_1^3 \cdot g\overline{h_2^2} \neq 0$ at $E_3$.

$w_2^3\overline{c_0}$: The target bidegree is generated at $E_2$ by $31_{237} = g^6 \cdot i(\beta g)$ and $31_{238} = g^3 w_1 \cdot \beta w_2 \overline{g}$. Here $d_2(g^3 w_1 \cdot \beta w_2 \overline{g}) = g^3 w_1 \cdot (g^2 \cdot e_0 \overline{\gamma} + h_0 d_0 w_2 \overline{g}) = g^5 w_1 \cdot e_0 \overline{\gamma} \neq 0$ at $E_2$, using $g \cdot h_0 d_0 \overline{g} = 0$. Furthermore, $d_3(g^3 \cdot i(w_2^2)) = g^6 \cdot i(\beta g)$.

$w_2^3\overline{h_0\alpha^2}$: The target bidegree is generated at $E_2$ by $35_{287} = g^5 w_1 \cdot i(\beta w_2)$ and $35_{288} = w_1^2 \cdot \beta g w_2^2 \overline{g}$. Here $d_2(g^5 w_1 \cdot i(\beta w_2)) = g^5 w_1 \cdot (g^2 \cdot e_0 \overline{h_1} + i(h_0 d_0 w_2)) = g^7 w_1 \cdot e_0 \overline{h_1} \neq 0$ at $E_2$, using $g \cdot i(h_0 d_0) = 0$. Furthermore, $d_3(w_1^2 \cdot \beta g w_2^2 \overline{g}) = w_1^2 \cdot (g^5 \cdot g\overline{\beta^2} + g w_1 \cdot g w_2^2 \overline{h_2^2}) = g w_1^3 w_2^2 \cdot g\overline{h_2^2} \neq 0$ at $E_3$, using $g^2 w_1 \cdot g\overline{\beta^2} = 0$.

$w_2^3\overline{\delta}$: The target bidegree is generated at $E_2$ by $35_{308} = g^6 \cdot \beta g \overline{g}$ and $35_{309} = g^3 w_1 w_2^2 \cdot i(\beta)$. Here $d_2(g w_2^2 \cdot w_2 \overline{h_0^2 \alpha}) = g w_2^2 \cdot (g^2 w_1 \cdot i(\beta) + w_1 \cdot h_0 w_2 \overline{h_0 h_2}) = g^3 w_1 w_2^2 \cdot i(\beta)$, using $g \cdot h_0 \overline{h_0 h_2} = 0$. Finally, $d_3(g^3 \cdot w_2^2 \overline{g}) = g^6 \cdot \beta g \overline{g}$.

The reason "$h_0$-lin." refers to $h_0$-linearity, showing that $d_4(x) = 0$ for $x = \overline{h_1}$.

$\overline{h_1}$: We have $h_0 \cdot d_4(\overline{h_1}) = d_4(h_0 \cdot \overline{h_1}) = 0$, since $h_0 \cdot \overline{h_1} = 0$. This implies $d_4(\overline{h_1}) = 0$, because $h_0$ acts injectively on the target group at $E_4$.

The reason "$j$-nat." refers to naturality with respect to $j \colon C\nu \to S^4$, showing that $d_4(x)$ is nonzero for $x = \alpha^2\overline{g}$, $g\overline{\beta^2}$ and $\alpha^2 w_2^2 \overline{g}$. Recall Table 5.5.

$\alpha^2\overline{g}$: From $d_4(j(\alpha^2\overline{g})) = d_4(\alpha^2 g) = w_1^2 \cdot \alpha\beta \neq 0$ in $E_4(tmf)$ we deduce that $d_4(\alpha^2\overline{g}) \neq 0$ in $E_4(tmf/\nu)$, and the only possible target is $14_{31} = w_1^2 \cdot \overline{\alpha\beta}$.

$g\overline{\beta^2}$: From $d_4(j(g\overline{\beta^2})) = d_4(\beta^2 g) = w_1 \cdot \alpha^2 e_0 \neq 0$ in $E_4(tmf)$ we deduce that $d_4(g\overline{\beta^2}) \neq 0$ in $E_4(tmf/\nu)$, and the only possible target is $14_{38} = w_1 \cdot d_0 \overline{\alpha\beta}$.

$\alpha^2 w_2^2 \overline{g}$: From $d_4(j(\alpha^2 w_2^2 \overline{g})) = d_4(\alpha^2 g w_2^2) = w_1^2 \cdot \alpha\beta w_2^2 \neq 0$ in $E_4(tmf)$ we deduce that $d_4(\alpha^2 w_2^2 \overline{g}) \neq 0$ in $E_4(tmf/\nu)$. The target bidegree of the latter $d_4$ is generated at $E_2$ by $30_{207} = w_1^2 \cdot w_2^2 \overline{\alpha\beta}$ and $30_{208} = g^4 w_1 \cdot w_2 \overline{h_0 h_2}$. Here $w_1^2 \cdot w_2^2 \overline{\alpha\beta}$ survives to $E_4$, while $d_2(g^4 w_1 \cdot w_2 \overline{h_0 h_2}) = g^6 w_1 \cdot i(d_0) \neq 0$. Hence $d_4(\alpha^2 w_2^2 \overline{g}) = w_1^2 \cdot w_2^2 \overline{\alpha\beta}$.

The $d_4$-differentials on the remaining module generators, $x = i(\alpha^2)$, $i(\alpha e_0)$, $g\overline{h_2^2}$, $i(\beta g)$, $w_2 \overline{h_0^3}$, $i(\alpha^2 w_2^2)$, $i(\alpha e_0 w_2^2)$, $h_1 w_2^2 \overline{\gamma}$ and $w_2^3 \overline{h_0^3}$, can be found by the following arguments, numbered (1) to (9), respectively.

(1) From the relation $\alpha\gamma = e_0 g$ in $E_2(tmf)$ and the differential $d^4(\alpha e_0 g) = w_1^2 \cdot \delta'$ in $E_4(tmf)$ we deduce that $\gamma \cdot d_4(i(\alpha^2)) = i(d_4(\alpha^2\gamma)) = i(d_4(\alpha e_0 g)) = w_1^2 \cdot i(\delta') = g w_1^2 \cdot i(\alpha) \neq 0$ in $E_4(tmf/\nu)$. Hence $d_4(i(\alpha^2)) \neq 0$, and $10_{10} = w_1^2 \cdot \overline{h_0 h_2}$ is the only possible value.

(2) In case (1) we saw that $g \cdot d_4(i(\alpha e_0)) = i(d_4(\alpha e_0 g)) \neq 0$ in $E_4(tmf/\nu)$, so that $d_4(i(\alpha e_0)) \neq 0$. By $h_0$-linearity $11_{14} = w_1^2 \cdot (\overline{c_0} + i(\alpha))$ is the only possible value, since $h_0 \cdot i(\alpha e_0) = 0$.

(3) From the relation $\gamma \cdot h_1 \overline{\gamma} = 11_{38} = h_0 w_2 \cdot \overline{h_0 h_2}$, present at $E_2(tmf/\nu)$, and the differential $d_4(h_0 w_2) = d_0 \gamma w_1$ in $E_4(tmf)$, we deduce that $\gamma \cdot d_4(h_1\overline{\gamma}) = d_0 \gamma w_1 \cdot \overline{h_0 h_2} = 15_{39} = g w_1 \cdot i(\alpha d_0) \neq 0$ in $E_4(tmf/\nu)$. Hence $d_4(h_1\overline{\gamma}) \neq 0$, and $10_{15} = w_1 \cdot d_0 \overline{h_0 h_2}$ is the only possible value. Since $d_4(\gamma \cdot \overline{h_1}) = 0$, and $g\overline{h_2^2} = 6_{14} = h_1\overline{\gamma} + \gamma\overline{h_1}$, it follows that $d_4(g\overline{h_2^2}) = w_1 \cdot d_0 \overline{h_0 h_2}$.

(4) From the relation $g w_1 \cdot i(\beta g) = w_1 \cdot h_0 w_2 \cdot \overline{h_0 h_2}$ in $E_4(tmf/\nu)$, arising at $E_3$ from $d_2(w_2 \overline{h_0^2 \alpha})$, and the differential $d_4(h_0 w_2) = d_0 \gamma w_1$ in $E_4(tmf)$, we deduce as in case (3) that $g w_1 \cdot d_4(i(\beta g)) = w_1 \cdot d_0 \gamma w_1 \cdot \overline{h_0 h_2} = 19_{50} = g w_1^2 \cdot i(\alpha d_0) \neq 0$ in $E_4(tmf/\nu)$. Hence $d_4(i(\beta g)) \neq 0$, and $11_{18} = w_1 \cdot i(\alpha d_0)$ is the only possible value.

(5) The target group of $d_4$ on $w_2\overline{h_0^3}$ is generated by $15_{35} = w_1^2 \cdot i(\beta g)$, and $d_4(w_1^2 \cdot i(\beta g)) = w_1^3 \cdot i(\alpha d_0) \neq 0$ by case (4). Hence $d_4(w_2\overline{h_0^3}) = 0$, since $d_4^2 = 0$.

(6) From the relation $\alpha\gamma = e_0 g$ in $E_2(tmf)$ and the differential $d_4(\alpha e_0 g w_2^2) = w_1^2 \cdot \delta' w_2^2$ in $E_4(tmf)$ we deduce that $\gamma \cdot d_4(i(\alpha^2 w_2^2)) = i(d_4(\alpha^2\gamma w_2^2)) = i(d_4(\alpha e_0 g w_2^2)) = w_1^2 \cdot i(\delta' w_2^2) = g w_1^2 \cdot i(\alpha w_2^2) \neq 0$ in $E_4(tmf/\nu)$. Hence $d_4(i(\alpha^2 w_2^2)) \neq 0$.

The target group is generated at $E_2$ by $26_{149} = w_1^2 w_2^2 \overline{h_0 h_2}$ and $26_{150} = g^4 w_1 \overline{\alpha\beta}$. Here $w_1^2 \cdot w_2^2 \overline{h_0 h_2} \neq 0$ in $E_4(tmf/\nu)$, whereas $g^4 w_1 \overline{\alpha\beta} = d_2(g^2 w_1 \cdot w_2 \overline{g})$ is 0 at $E_3$ and $E_4$. Hence $d_4(i(\alpha^2 w_2^2)) = w_1^2 \cdot w_2^2 \overline{h_0 h_2}$.

(7) In case (6) we saw that $g \cdot d_4(i(\alpha e_0 w_2^2)) = i(d_4(\alpha e_0 g w_2^2)) \neq 0$ in $E_4(tmf/\nu)$, so that $d_4(i(\alpha e_0 w_2^2)) \neq 0$.

By $h_0$-linearity the possible targets are spanned at $E_2$ by $27_{160} = g^4 w_1 \cdot \alpha\overline{g}$ and $27_{162} = w_1^2 \cdot w_2^2(\overline{c_0} + i(\alpha))$. Here $g^4 w_1 \cdot \alpha\overline{g} = d_2(g^2 w_1 \cdot w_2\overline{\gamma})$ is 0 at $E_3$ and $E_4$, so the only possible value is $d_4(i(\alpha e_0 w_2^2)) = w_1^2 \cdot w_2^2(\overline{c_0} + i(\alpha))$.

(8) From the relation $\gamma \cdot h_1 w_2^2 \overline{\gamma} = 27_{210} = h_0 w_2^3 \cdot \overline{h_0 h_2}$, present at $E_2(tmf/\nu)$, and the differential $d_4(h_0 w_2^3) = d_0 \gamma w_1 w_2^2$ in $E_4(tmf)$, we deduce that $\gamma \cdot d_4(h_1 w_2^2 \overline{\gamma}) = d_0\gamma w_1 w_2^2 \cdot \overline{h_0 h_2} = 31_{227} = g w_1 \cdot i(\alpha d_0 w_2^2) \neq 0$ in $E_4(tmf/\nu)$. Hence $d_4(h_1 w_2^2 \overline{\gamma}) \neq 0$.

The target group is generated at $E_2$ by $26_{162} = g^4 \cdot d_0 \overline{\alpha\beta}$ and $26_{163} = w_1 w_2^2 \cdot d_0 \overline{h_0 h_2}$. Here $g^4 \cdot d_0 \overline{\alpha\beta} = d_2(g^2 \cdot d_0 w_2 \overline{g})$ is 0 at $E_3$ and $E_4$. Hence $d_4(h_1 w_2^2 \overline{\gamma}) = w_1 \cdot d_0 w_2^2 \overline{h_0 h_2}$.

(9) The target group of $d_4$ on $w_2^3 \overline{h_0^3}$ is generated at $E_2$ by $31_{218} = g^5 w_1 \cdot \beta\overline{g}$ and $31_{219} = g w_1^2 w_2^2 \cdot i(\beta)$. Here $g^5 w_1 \cdot \beta\overline{g} = d_2(g^3 \cdot w_2 \overline{\alpha^3})$ is 0 at $E_3$ and $E_4$, using $g \cdot h_0 \overline{\alpha\beta} = 0$. Furthermore, using the relation $\beta g w_1 = \alpha d_0 e_0$ and case (7) we calculate that $d_4(w_1 \cdot i(\beta g w_1 w_2^2)) = d_4(d_0 w_1 \cdot i(\alpha e_0 w_2^2)) = d_0 w_1^3 \cdot w_2^2(\overline{c_0} + i(\alpha)) = 35_{223} = w_1^3 \cdot i(\alpha d_0 w_2^2) \neq 0$ at $E_4$. Hence $d_4(w_2^3 \overline{h_0^3}) = 0$, since $d_4^2 = 0$. □

THEOREM 8.9. *The $d_4$-differential in $E_4(tmf/\nu)$ is $R_2$-linear. Its values on a set of $R_2$-module generators are listed in Table* 8.6.

PROOF. The classes $g$, $w_1$ and $w_2^4$ are $d_4$-cycles in $E_4(tmf)$, so multiplication by each of these commutes with the $d_4$-differential in $E_4(tmf/\nu)$.

The $d_4$-differential on the $R_2$-module generators $x$ in Table 8.6 is given by the Leibniz rule applied to the (implicit and explicit) factorizations in the proof of Proposition 8.7, and the $d_4$-differentials from Tables 5.5 and 8.8. In the following cases we use `ext` to rewrite the output of the Leibniz rule in terms of the $R_2$-module presentation of $E_4(tmf/\nu)$.

- $d_4(i(h_0 w_2)) = i(d_0 \gamma w_1) = 13_{31} = g w_1 \cdot d_0 \overline{h_1}$
- $d_4(i(h_1^2 w_2)) = i(\alpha^2 e_0 w_1) = 14_{34} = g w_1 \cdot d_0 \overline{h_0 h_2} = 0$ at $E_4$
- $d_4(h_0 w_2 \cdot \overline{h_0 h_2}) = d_0 \gamma w_1 \cdot \overline{h_0 h_2} = 15_{39} = g w_1 \cdot i(\alpha d_0)$
- $d_4(h_0 w_2 \cdot \overline{g}) = d_0 \gamma w_1 \cdot \overline{g} = 17_{61} = w_1 \cdot d_0 g \overline{\gamma}$
- $d_4(\alpha^2 g \overline{\gamma}) = d_4(\alpha e_0 g \cdot \overline{g}) = \delta' w_1^2 \cdot \overline{g} = 19_{62} + 19_{63} = w_1^2 \cdot \alpha g \overline{g} + w_1^2 \cdot i(c_0 w_2)$
- $d_4(i(h_1 \gamma w_2)) = d_4(h_1^2 w_2 \cdot \overline{g}) = \alpha^2 e_0 w_1 \cdot \overline{g} = 18_{65} = g w_1 \cdot d_0 \overline{\alpha\beta}$
- $d_4(\gamma^2 \overline{\gamma}) = d_4(\beta g^2 \cdot \overline{g} + h_0 w_2 \cdot \overline{\alpha\beta}) = \alpha d_0 g w_1 \cdot \overline{g} + d_0 \gamma w_1 \cdot \overline{\alpha\beta} = 19_{72} + 19_{72} = 0$
- $d_4(h_0 w_2 \cdot \overline{\alpha\beta}) = d_0 \gamma w_1 \cdot \overline{\alpha\beta} = 19_{72} = w_1 \cdot \alpha d_0 g \overline{g}$

- $d_4(g^2\overline{\alpha^3}) = d_4(\alpha e_0 g \cdot \overline{\alpha\beta}) = \delta' w_1^2 \cdot \overline{\alpha\beta} = 21_{75} = w_1^2 \cdot d_0 g\overline{\gamma}$
- $d_4(e_0 g^2\overline{\gamma}) = d_4(e_0\gamma g \cdot \overline{g}) = \gamma g w_1^2 \cdot \overline{g} = 21_{84} = g w_1^2 \cdot \gamma\overline{g} = 0$ at $E_4$
- $d_4(g^5\overline{\beta^2} + g w_1 w_2^2 \overline{h_2^2}) = d_4(g^4 \cdot g\overline{\beta^2} + d_0 \cdot i(\alpha^2 w_2^2)) = g^4 \cdot d_0 w_1 \overline{\alpha\beta} + d_0 \cdot w_1^2 w_2^2 \overline{h_0 h_2} = w_1^2 \cdot d_0 w_2^2 \overline{h_0 h_2}$ at $E_4$
- $d_4(i(h_0 w_2^3)) = i(w_1 \cdot d_0 \gamma w_2^2) = 29_{207} = g w_1 \cdot d_0 w_2^2 \overline{h_1}$
- $d_4(i(h_1^2 w_2^3)) = i(w_1 \cdot \alpha^2 e_0 w_2^2) = 30_{214} = g w_1 \cdot d_0 w_2^2 \overline{h_0 h_2} = 0$ at $E_4$
- $d_4(h_0 w_2^3 \cdot \overline{h_0 h_2}) = d_0 \gamma w_1 w_2^2 \cdot \overline{h_0 h_2} = 31_{227} = g w_1 \cdot i(\alpha d_0 w_2^2)$
- $d_4(h_0 w_2^3 \cdot \overline{g}) = d_0 \gamma w_1 w_2^2 \cdot \overline{g} = 33_{274} = w_1 \cdot d_0 g w_2^2 \overline{\gamma}$
- $d_4(\alpha^2 g w_2^2 \overline{\gamma}) = d_4(\alpha e_0 g w_2^2 \cdot \overline{g}) = \delta' w_1^2 w_2^2 \cdot \overline{g} = 35_{279} + 35_{280} = w_1^2 \cdot \alpha g w_2^2 \overline{g} + w_1^2 \cdot i(c_0 w_2^3)$
- $d_4(i(h_1 \gamma w_2^3)) = d_4(h_1^2 w_2^3 \cdot \overline{g}) = \alpha^2 e_0 w_1 w_2^2 \cdot \overline{g} = 34_{282} = g w_1 \cdot d_0 w_2^2 \overline{\alpha\beta}$
- $d_4(h_0 w_2^3 \cdot \overline{\alpha\beta}) = d_0 \gamma w_1 w_2^2 \cdot \overline{\alpha\beta} = 35_{297} = w_1 \cdot \alpha d_0 g w_2^2 \overline{g}$
- $d_4(g^2 w_2^2 \overline{\alpha^3}) = d_4(\alpha e_0 g w_2^2 \cdot \overline{\alpha\beta}) = \delta' w_1^2 w_2^2 \cdot \overline{\alpha\beta} = 37_{304} = w_1^2 \cdot d_0 g w_2^2 \overline{\gamma}$
- $d_4(e_0 g^2 w_2^2 \overline{\gamma}) = d_4(e_0 \gamma g w_2^2 \cdot \overline{g}) = \gamma g w_1^2 w_2^2 \cdot \overline{g} = g w_1 \cdot \gamma w_1 w_2^2 \overline{g} = 0$ at $E_4$. $\square$

## 8.5. The $E_\infty$-term for $tmf/\nu$

It is now a routine matter to compute $E_5$ of the Adams spectral sequence for $tmf/\nu$. This is carried out in Appendix D.3. The result is a direct sum of cyclic $R_2$-modules, and is recorded in Table 8.9.

Table 8.9: $R_2$-module generators of $E_5(tmf/\nu)$, with $i \geq 0$ in each $h_0$-tower

| $t-s$ | $s$ | $g$ | $x$ | $\mathrm{Ann}(x)$ | dec. |
|---|---|---|---|---|---|
| 0 | 0 | 0 | $i(1)$ | $(g^5, gw_1)$ | **gen.** |
| 0 | $1+i$ | 0 | $i(h_0^{1+i})$ | $(g)$ | $h_0^{1+i} \cdot i(1)$ |
| 1 | 1 | 1 | $i(h_1)$ | $(g)$ | $h_1 \cdot i(1)$ |
| 2 | 2 | 1 | $i(h_1^2)$ | $(g)$ | $h_1^2 \cdot i(1)$ |
| 4 | $3+i$ | 1 | $h_0^i \overline{h_0^3}$ | $(g)$ | $h_0^i \cdot$ **gen.** |
| 5 | 1 | 2 | $\overline{h_1}$ | $(g^6, g^2 w_1, g w_1^2)$ | **gen.** |
| 6 | 2 | 2 | $h_1 \overline{h_1}$ | $(g)$ | $h_1 \cdot \overline{h_1}$ |
| 7 | 2 | 3 | $\overline{h_0 h_2}$ | $(g^2, g w_1, w_1^2)$ | **gen.** |
| 7 | 3 | 2 | $h_0 \overline{h_0 h_2}$ | $(g, w_1)$ | $h_0 \cdot \overline{h_0 h_2}$ |
| 8 | 3 | 3 | $i(c_0)$ | $(g, w_1)$ | $h_1 \cdot \overline{h_0 h_2}$ |
| 12 | 3 | 4 | $\overline{c_0}$ | $(g)$ | **gen.** |
| 12 | 3 | 5 | $\overline{c_0} + i(\alpha)$ | $(g^2, w_1^2)$ | **gen.** |
| 12 | $4+i$ | 4 | $i(h_0^{1+i}\alpha)$ | $(g)$ | $h_0^{1+i} \cdot \overline{c_0}$ |
| 13 | 4 | 5 | $h_1 \overline{c_0}$ | $(g, w_1)$ | $h_1 \cdot \overline{c_0}$ |
| 14 | 4 | 6 | $i(d_0)$ | $(g^2, g w_1, w_1^2)$ | $d_0 \cdot i(1)$ |
| 16 | $6+i$ | 7 | $h_0^{1+i} \overline{h_0^2 \alpha}$ | $(g)$ | $h_0^i \cdot$ **gen.** |
| 19 | 5 | 8 | $d_0 \overline{h_1}$ | $(g^2, g w_1, w_1^2)$ | $d_0 \cdot \overline{h_1}$ |

Table 8.9: $R_2$-module generators of $E_5(tmf/\nu)$, with $i \geq 0$ in each $h_0$-tower (cont.)

| $t-s$ | $s$ | $g$ | $x$ | Ann$(x)$ | dec. |
|---|---|---|---|---|---|
| 21 | 6 | 9 | $d_0\overline{h_0h_2}$ | $(g^2, w_1)$ | $d_0 \cdot \overline{h_0h_2}$ |
| 24 | 4 | 9 | $\overline{g}$ | $(g^6, g^2w_1, gw_1^2)$ | **gen.** |
| 24 | 5 | 10 | $h_0\overline{g}$ | $(g)$ | $h_0 \cdot \overline{g}$ |
| 24 | 6 | 11 | $h_0^2\overline{g}$ | $(g)$ | $h_0^2 \cdot \overline{g}$ |
| 24 | $7+i$ | 11 | $i(h_0^{1+i}\alpha^2)$ | $(g)$ | $h_0^{3+i} \cdot \overline{g}$ |
| 25 | 5 | 12 | $h_1\overline{g}$ | $(g)$ | $h_1 \cdot \overline{g}$ |
| 26 | 6 | 12 | $i(h_1\gamma)$ | $(g)$ | $h_1^2 \cdot \overline{g}$ |
| 26 | 7 | 12 | $i(\alpha d_0)$ | $(g^2, w_1)$ | $d_0 \cdot (\overline{c_0} + i(\alpha))$ |
| 28 | $7+i$ | 13 | $h_0^i\overline{h_0\alpha^2}$ | $(g)$ | $h_0^i \cdot$ **gen.** |
| 30 | 6 | $14+15$ | $\gamma\overline{h_1}$ | $(g^5, gw_1)$ | $\gamma \cdot \overline{h_1}$ |
| 31 | 6 | 16 | $\overline{\alpha\beta}$ | $(g^2, w_1^2)$ | **gen.** |
| 31 | 7 | 15 | $h_0\overline{\alpha\beta}$ | $(g, w_1)$ | $h_0 \cdot \overline{\alpha\beta}$ |
| 32 | 7 | 17 | $i(\delta)$ | $(g, w_1)$ | $h_1 \cdot \overline{\alpha\beta}$ |
| 36 | 7 | 19 | $\overline{\delta}$ | $(g)$ | **gen.** |
| 36 | 8 | 19 | $h_0\overline{\delta}$ | $(g)$ | $h_0 \cdot \overline{\delta}$ |
| 36 | 9 | 20 | $h_0^2\overline{\delta}$ | $(g)$ | $h_0^2 \cdot \overline{\delta}$ |
| 36 | $10+i$ | 20 | $i(h_0^{1+i}\alpha^3)$ | $(g)$ | $h_0^{3+i} \cdot \overline{\delta}$ |
| 37 | 8 | $20+21$ | $\delta'\overline{h_1}$ | $(g^2, w_1)$ | $\delta' \cdot \overline{h_1}$ |
| 37 | 8 | 21 | $h_1\overline{\delta}$ | $(g)$ | $h_1 \cdot \overline{\delta}$ |
| 38 | 8 | 22 | $d_0\overline{g}$ | $(g^2, w_1^2)$ | $d_0 \cdot \overline{g}$ |
| 40 | $10+i$ | 24 | $h_0^{1+i}\overline{\alpha^3}$ | $(g)$ | $h_0^i \cdot$ **gen.** |
| 44 | 10 | 27 | $i(\alpha^2 g)$ | $(g, w_1)$ | $d_0\gamma \cdot \overline{h_1}$ |
| 45 | 10 | 28 | $d_0\overline{\alpha\beta}$ | $(g^2, w_1)$ | $d_0 \cdot \overline{\alpha\beta}$ |
| 48 | $10+i$ | 31 | $i(h_0^{2+i}w_2)$ | $(g)$ | $h_0^{2+i}w_2 \cdot i(1)$ |
| 49 | 9 | $30+31$ | $\gamma\overline{g}$ | $(g^5, gw_1)$ | $\gamma \cdot \overline{g}$ |
| 50 | 10 | 33 | $i(h_1^2w_2)$ | $(g)$ | $h_1\gamma \cdot \overline{g}$ |
| 51 | 12 | 34 | $i(d_0e_0g)$ | $(g, w_1)$ | $\alpha d_0 g \cdot \overline{h_1}$ |
| 52 | $11+i$ | 35 | $h_0^iw_2\overline{h_0^3}$ | $(g)$ | $h_0^i \cdot$ **gen.** |
| 55 | 11 | $37+38$ | $\gamma^2\overline{h_1}$ | $(g^2, w_1)$ | $\gamma^2 \cdot \overline{h_1}$ |
| 56 | 11 | $39+40$ | $\delta'\overline{g}$ | $(g, w_1^2)$ | $\delta' \cdot \overline{g}$ |
| 56 | 11 | 40 | $i(c_0w_2)$ | $(g)$ | $c_0w_2 \cdot i(1)$ |

Table 8.9: $R_2$-module generators of $E_5(tmf/\nu)$, with $i \geq 0$ in each $h_0$-tower (cont.)

| $t-s$ | $s$ | $g$ | $x$ | $\mathrm{Ann}(x)$ | dec. |
|---|---|---|---|---|---|
| 56 | 13 | $39+40$ | $i(\alpha^3 g + h_0 w_1 w_2)$ | $(g)$ | $(\alpha^3 g + h_0 w_1 w_2) \cdot i(1)$ |
| 60 | 11 | 42 | $w_2 \overline{c_0}$ | $(g)$ | **gen.** |
| 60 | $12+i$ | 44 | $i(h_0^{1+i} \alpha w_2)$ | $(g)$ | $h_0^{1+i} \cdot w_2 \overline{c_0}$ |
| 61 | 12 | 46 | $h_1 w_2 \overline{c_0}$ | $(g)$ | $h_1 \cdot w_2 \overline{c_0}$ |
| 62 | 13 | 47 | $e_0 g^2 \overline{h_1}$ | $(g)$ | $h_1^2 \cdot w_2 \overline{c_0}$ |
| 63 | 13 | 49 | $d_0 g \overline{\gamma}$ | $(g, w_1)$ | $d_0 \gamma \cdot \overline{g}$ |
| 64 | $14+i$ | 51 | $h_0^{1+i} w_2 \overline{h_0^2 \alpha}$ | $(g)$ | $h_0^i \cdot$ **gen.** |
| 70 | 15 | 58 | $\alpha d_0 g \overline{g}$ | $(g, w_1)$ | $\alpha d_0 g \cdot \overline{g}$ |
| 72 | 14 | 60 | $h_0^2 w_2 \overline{g}$ | $(g)$ | $h_0^2 w_2 \cdot \overline{g}$ |
| 72 | $15+i$ | 61 | $i(h_0^{1+i} \alpha^2 w_2)$ | $(g)$ | $h_0^{3+i} w_2 \cdot \overline{g}$ |
| 74 | 14 | $61+62$ | $\gamma^2 \overline{g}$ | $(g^5, g w_1)$ | $\gamma^2 \cdot \overline{g}$ |
| 76 | $15+i$ | 66 | $h_0^i w_2 \overline{h_0 \alpha^2}$ | $(g)$ | $h_0^i \cdot$ **gen.** |
| 79 | 15 | $68+69$ | $\gamma^2 \overline{\gamma}$ | $(g^2, w_1)$ | **gen.** |
| 80 | 15 | 71 | $i(\delta w_2)$ | $(g)$ | $\delta w_2 \cdot i(1)$ |
| 80 | 17 | $72+73$ | $(\alpha^3 g + h_0 w_1 w_2) \overline{g}$ | $(g)$ | $(\alpha^3 g + h_0 w_1 w_2) \cdot \overline{g}$ |
| 81 | 16 | 73 | $e_0 g^2 \overline{g}$ | $(g)$ | $h_1 \cdot i(\delta w_2)$ |
| 84 | 15 | 73 | $w_2 \overline{\delta}$ | $(g)$ | **gen.** |
| 84 | 16 | 77 | $h_0 w_2 \overline{\delta}$ | $(g)$ | $h_0 \cdot w_2 \overline{\delta}$ |
| 84 | 17 | 79 | $h_0^2 w_2 \overline{\delta}$ | $(g)$ | $h_0^2 \cdot w_2 \overline{\delta}$ |
| 84 | $18+i$ | 80 | $i(h_0^{1+i} \alpha^3 w_2)$ | $(g)$ | $h_0^{3+i} \cdot w_2 \overline{\delta}$ |
| 85 | 16 | 79 | $h_1 w_2 \overline{\delta}$ | $(g)$ | $h_1 \cdot w_2 \overline{\delta}$ |
| 86 | 17 | 82 | $e_0 g^2 \overline{\gamma}$ | $(g)$ | $h_1^2 \cdot w_2 \overline{\delta}$ |
| 88 | $18+i$ | 87 | $h_0^{1+i} w_2 \overline{\alpha^3}$ | $(g)$ | $h_0^i \cdot$ **gen.** |
| 96 | $17+i$ | 91 | $i(h_0^{1+i} w_2^2)$ | $(g)$ | $h_0^{1+i} w_2^2 \cdot i(1)$ |
| 97 | 17 | 93 | $i(h_1 w_2^2)$ | $(g)$ | $h_1 w_2^2 \cdot i(1)$ |
| 98 | 18 | 99 | $i(h_1^2 w_2^2)$ | $(g)$ | $h_1^2 w_2^2 \cdot i(1)$ |
| 100 | $19+i$ | 105 | $h_0^i w_2^2 \overline{h_0^3}$ | $(g)$ | $h_0^i \cdot$ **gen.** |
| 102 | 18 | 102 | $h_1 w_2^2 \overline{h_1}$ | $(g)$ | $h_1 w_2^2 \cdot \overline{h_1}$ |
| 103 | 18 | 103 | $w_2^2 \overline{h_0 h_2}$ | $(g^2, g w_1, w_1^2)$ | **gen.** |

Table 8.9: $R_2$-module generators of $E_5(tmf/\nu)$, with $i \geq 0$ in each $h_0$-tower (cont.)

| $t-s$ | $s$ | $g$ | $x$ | $\mathrm{Ann}(x)$ | dec. |
|---|---|---|---|---|---|
| 103 | 19 | 108 | $h_0 w_2^2 \overline{h_0 h_2}$ | $(g, w_1)$ | $h_0 \cdot w_2^2 \overline{h_0 h_2}$ |
| 104 | 19 | 110 | $i(c_0 w_2^2)$ | $(g, w_1)$ | $h_1 \cdot w_2^2 \overline{h_0 h_2}$ |
| 104 | 20 | $112 + 113$ | $g^4 \overline{g} + i(w_1 w_2^2)$ | $(g)$ | $g^4 \cdot \overline{g} + w_1 w_2^2 \cdot i(1)$ |
| 108 | 19 | 112 | $w_2^2 \overline{c_0}$ | $(g)$ | **gen.** |
| 108 | 19 | 113 | $w_2^2 \overline{c_0} + i(\alpha w_2^2)$ | $(g^2, w_1^2)$ | **gen.** |
| 108 | $20 + i$ | 118 | $i(h_0^{1+i} \alpha w_2^2)$ | $(g)$ | $h_0^{1+i} \cdot w_2^2 \overline{c_0}$ |
| 109 | 20 | 120 | $h_1 w_2^2 \overline{c_0}$ | $(g, w_1)$ | $h_1 \cdot w_2^2 \overline{c_0}$ |
| 109 | 21 | 124 | $w_1 w_2^2 \overline{h_1}$ | $(g^2, g w_1)$ | $w_1 w_2^2 \cdot \overline{h_1}$ |
| 110 | 20 | 121 | $i(d_0 w_2^2)$ | $(g^2, g w_1, w_1^2)$ | $d_0 w_2^2 \cdot i(1)$ |
| 112 | $22 + i$ | 132 | $h_0^{1+i} w_2^2 \overline{h_0^2 \alpha}$ | $(g)$ | $h_0^i \cdot$ **gen.** |
| 115 | 21 | 131 | $d_0 w_2^2 \overline{h_1}$ | $(g^2, g w_1, w_1^2)$ | $d_0 w_2^2 \cdot \overline{h_1}$ |
| 117 | 22 | 138 | $d_0 w_2^2 \overline{h_0 h_2}$ | $(g^2, w_1)$ | $d_0 w_2^2 \cdot \overline{h_0 h_2}$ |
| 120 | 21 | 134 | $h_0 w_2^2 \overline{g}$ | $(g)$ | $h_0 w_2^2 \cdot \overline{g}$ |
| 120 | 22 | 142 | $h_0^2 w_2^2 \overline{g}$ | $(g)$ | $h_0^2 w_2^2 \cdot \overline{g}$ |
| 120 | $23 + i$ | 147 | $i(h_0^{1+i} \alpha^2 w_2^2)$ | $(g)$ | $h_0^{3+i} w_2^2 \cdot \overline{g}$ |
| 121 | 21 | 136 | $h_1 w_2^2 \overline{g}$ | $(g)$ | $h_1 w_2^2 \cdot \overline{g}$ |
| 122 | 22 | 144 | $i(h_1 \gamma w_2^2)$ | $(g)$ | $h_1^2 w_2^2 \cdot \overline{g}$ |
| 122 | 23 | 149 | $i(\alpha d_0 w_2^2)$ | $(g^2, g w_1, w_1^2)$ | $d_0 w_2^2 \cdot (\overline{c_0} + i(\alpha))$ |
| 124 | $23 + i$ | 152 | $h_0^i w_2^2 \overline{h_0 \alpha^2}$ | $(g)$ | $h_0^i \cdot$ **gen.** |
| 127 | 22 | 148 | $w_2^2 \overline{\alpha \beta}$ | $(g^2, w_1^2)$ | **gen.** |
| 127 | 23 | 155 | $h_0 w_2^2 \overline{\alpha \beta}$ | $(g, w_1)$ | $h_0 \cdot w_2^2 \overline{\alpha \beta}$ |
| 128 | 23 | 157 | $i(\delta w_2^2)$ | $(g, w_1)$ | $h_1 \cdot w_2^2 \overline{\alpha \beta}$ |
| 128 | 24 | 162 | $w_1 w_2^2 \overline{g}$ | $(g^2, g w_1)$ | $w_1 w_2^2 \cdot \overline{g}$ |
| 132 | 23 | 159 | $w_2^2 \overline{\delta}$ | $(g)$ | **gen.** |
| 132 | 24 | 167 | $h_0 w_2^2 \overline{\delta}$ | $(g)$ | $h_0 \cdot w_2^2 \overline{\delta}$ |
| 132 | 25 | 173 | $h_0^2 w_2^2 \overline{\delta}$ | $(g)$ | $h_0^2 \cdot w_2^2 \overline{\delta}$ |
| 132 | $26 + i$ | 177 | $i(h_0^{1+i} \alpha^3 w_2^2)$ | $(g)$ | $h_0^{3+i} \cdot w_2^2 \overline{\delta}$ |
| 133 | 24 | $168 + 169$ | $\delta' w_2^2 \overline{h_1}$ | $(g^2, w_1)$ | $\delta' w_2^2 \cdot \overline{h_1}$ |
| 133 | 24 | 169 | $h_1 w_2^2 \overline{\delta}$ | $(g)$ | $h_1 \cdot w_2^2 \overline{\delta}$ |
| 134 | 24 | 170 | $d_0 w_2^2 \overline{g}$ | $(g^2, w_1^2)$ | $d_0 w_2^2 \cdot \overline{g}$ |

Table 8.9: $R_2$-module generators of $E_5(tmf/\nu)$, with $i \geq 0$ in each $h_0$-tower (cont.)

| $t-s$ | $s$ | $g$ | $x$ | $\mathrm{Ann}(x)$ | dec. |
|---|---|---|---|---|---|
| 134 | 26 | $179+180$ $+181$ | $\gamma^2 g^3\overline{g}$ $+\gamma w_1 w_2^2\overline{h_1}$ | $(g)$ | $\gamma^2 g^3\cdot\overline{g}$ $+\gamma w_1 w_2^2\cdot\overline{h_1}$ |
| 136 | $26+i$ | 185 | $h_0^{1+i}w_2^2\overline{\alpha^3}$ | $(g)$ | $h_0^i\cdot$ **gen.** |
| 140 | 26 | 190 | $i(\alpha^2 g w_2^2)$ | $(g, w_1)$ | $d_0\gamma w_2^2\cdot\overline{h_1}$ |
| 141 | 26 | 191 | $d_0 w_2^2\overline{\alpha\beta}$ | $(g^2, g w_1, w_1^2)$ | $d_0 w_2^2\cdot\overline{\alpha\beta}$ |
| 144 | $26+i$ | 195 | $i(h_0^{2+i}w_2^3)$ | $(g)$ | $h_0^{2+i}w_2^3\cdot i(1)$ |
| 146 | 26 | 197 | $i(h_1^2 w_2^3)$ | $(g)$ | $h_1\gamma w_2^2\cdot\overline{g}$ |
| 147 | 28 | 211 | $i(d_0 e_0 g w_2^2)$ | $(g, w_1)$ | $\alpha d_0 g w_2^2\cdot\overline{h_1}$ |
| 148 | $27+i$ | 207 | $h_0^i w_2^3\overline{h_0^3}$ | $(g)$ | $h_0^i\cdot$ **gen.** |
| 152 | 27 | $211+212$ | $\delta' w_2^2\overline{g}$ | $(g, w_1^2)$ | $\delta' w_2^2\cdot\overline{g}$ |
| 152 | 27 | 212 | $i(c_0 w_2^3)$ | $(g)$ | $c_0 w_2^3\cdot i(1)$ |
| 152 | 29 | $224+225$ | $i(\alpha^3 g w_2^2$ $+h_0 w_1 w_2^3)$ | $(g)$ | $(\alpha^3 g w_2^2$ $+h_0 w_1 w_2^3)\cdot i(1)$ |
| 153 | 29 | $227+228$ | $\gamma w_1 w_2^2\overline{g}$ | $(g)$ | $\gamma w_1 w_2^2\cdot\overline{g}$ |
| 156 | 27 | 214 | $w_2^3\overline{c_0}$ | $(g)$ | **gen.** |
| 156 | $28+i$ | 224 | $i(h_0^{1+i}\alpha w_2^3)$ | $(g)$ | $h_0^{1+i}\cdot w_2^3\overline{c_0}$ |
| 157 | 28 | 226 | $h_1 w_2^3\overline{c_0}$ | $(g)$ | $h_1\cdot w_2^3\overline{c_0}$ |
| 158 | 29 | 235 | $e_0 g^2 w_2^2\overline{h_1}$ | $(g)$ | $h_1^2\cdot w_2^3\overline{c_0}$ |
| 159 | 29 | 237 | $d_0 g w_2^2\overline{\gamma}$ | $(g, w_1)$ | $d_0\gamma w_2^2\cdot\overline{g}$ |
| 160 | $30+i$ | 246 | $h_0^{1+i}w_2^3\overline{h_0^2\alpha}$ | $(g)$ | $h_0^i\cdot$ **gen.** |
| 166 | 31 | 261 | $\alpha d_0 g w_2^2\overline{g}$ | $(g, w_1)$ | $\alpha d_0 g w_2^2\cdot\overline{g}$ |
| 168 | 30 | 256 | $h_0^2 w_2^3\overline{g}$ | $(g)$ | $h_0^2 w_2^3\cdot\overline{g}$ |
| 168 | $31+i$ | 265 | $i(h_0^{1+i}\alpha^2 w_2^3)$ | $(g)$ | $h_0^{3+i}w_2^3\cdot\overline{g}$ |
| 172 | $31+i$ | 270 | $h_0^i w_2^3\overline{h_0\alpha^2}$ | $(g)$ | $h_0^i\cdot$ **gen.** |
| 176 | 31 | 275 | $i(\delta w_2^3)$ | $(g)$ | $\delta w_2^3\cdot i(1)$ |
| 176 | 33 | $291+292$ | $(\alpha^3 g w_2^2$ $+h_0 w_1 w_2^3)\overline{g}$ | $(g)$ | $(\alpha^3 g w_2^2$ $+h_0 w_1 w_2^3)\cdot\overline{g}$ |
| 177 | 32 | 285 | $e_0 g^2 w_2^2\overline{g}$ | $(g)$ | $h_1\cdot i(\delta w_2^3)$ |
| 178 | 34 | $302+303$ | $\gamma^2 w_1 w_2^2\overline{g}$ | $(g)$ | $\gamma^2 w_1 w_2^2\cdot\overline{g}$ |
| 180 | 31 | 277 | $w_2^3\overline{\delta}$ | $(g)$ | **gen.** |
| 180 | 32 | 289 | $h_0 w_2^3\overline{\delta}$ | $(g)$ | $h_0\cdot w_2^3\overline{\delta}$ |

Table 8.9: $R_2$-module generators of $E_5(tmf/\nu)$, with $i \geq 0$ in each $h_0$-tower (cont.)

| $t-s$ | $s$ | $g$ | $x$ | $\mathrm{Ann}(x)$ | dec. |
|---|---|---|---|---|---|
| 180 | 33 | 299 | $h_0^2 w_2^3 \overline{\delta}$ | $(g)$ | $h_0^2 \cdot w_2^3 \overline{\delta}$ |
| 180 | $34+i$ | 307 | $i(h_0^{1+i} \alpha^3 w_2^3)$ | $(g)$ | $h_0^{3+i} \cdot w_2^3 \overline{\delta}$ |
| 181 | 32 | 291 | $h_1 w_2^3 \overline{\delta}$ | $(g)$ | $h_1 \cdot w_2^3 \overline{\delta}$ |
| 182 | 33 | 302 | $e_0 g^2 w_2^2 \overline{\gamma}$ | $(g)$ | $h_1^2 \cdot w_2^3 \overline{\delta}$ |
| 184 | $34+i$ | 315 | $h_0^{1+i} w_2^3 \overline{\alpha^3}$ | $(g)$ | $h_0^i \cdot \mathbf{gen.}$ |

Next, we determine a set of $E_5(tmf)$-module generators for $E_5(tmf/\nu)$, and express the remaining $R_2$-module generators in terms of this module structure. The results are listed in the following proposition, and in the dec.-column of Table 8.9.

PROPOSITION 8.10. *The* 34 *classes listed in Table* 8.10 *generate* $E_5(tmf/\nu)$ *as a module over* $E_5(tmf)$.

TABLE 8.10. $E_5(tmf)$-module generators of $E_5(tmf/\nu)$

| $t-s$ | $s$ | $g$ | $x$ |
|---|---|---|---|
| 0 | 0 | 0 | $i(1)$ |
| 4 | 3 | 1 | $\overline{h_0^3}$ |
| 5 | 1 | 2 | $\overline{h_1}$ |
| 7 | 2 | 3 | $\overline{h_0 h_2}$ |
| 12 | 3 | 4 | $\overline{c_0}$ |
| 12 | 3 | 5 | $\overline{c_0} + i(\alpha)$ |
| 16 | 6 | 7 | $h_0 \overline{h_0^2 \alpha}$ |
| 24 | 4 | 9 | $\overline{g}$ |
| 28 | 7 | 13 | $\overline{h_0 \alpha^2}$ |
| 31 | 6 | 16 | $\overline{\alpha\beta}$ |
| 36 | 7 | 19 | $\overline{\delta}$ |
| 40 | 10 | 24 | $h_0 \overline{\alpha^3}$ |
| 52 | 11 | 35 | $w_2 \overline{h_0^3}$ |
| 60 | 11 | 42 | $w_2 \overline{c_0}$ |
| 64 | 14 | 51 | $h_0 w_2 \overline{h_0^2 \alpha}$ |
| 76 | 15 | 66 | $w_2 \overline{h_0 \alpha^2}$ |
| 79 | 15 | $68+69$ | $\gamma^2 \overline{\gamma}$ |

| $t-s$ | $s$ | $g$ | $x$ |
|---|---|---|---|
| 84 | 15 | 73 | $w_2 \overline{\delta}$ |
| 88 | 18 | 87 | $h_0 w_2 \overline{\alpha^3}$ |
| 100 | 19 | 105 | $w_2^2 \overline{h_0^3}$ |
| 103 | 18 | 103 | $w_2^2 \overline{h_0 h_2}$ |
| 108 | 19 | 112 | $w_2^2 \overline{c_0}$ |
| 108 | 19 | 113 | $w_2^2 \overline{c_0} + i(\alpha w_2^2)$ |
| 112 | 22 | 132 | $h_0 w_2^2 \overline{h_0^2 \alpha}$ |
| 124 | 23 | 152 | $w_2^2 \overline{h_0 \alpha^2}$ |
| 127 | 22 | 148 | $w_2^2 \overline{\alpha\beta}$ |
| 132 | 23 | 159 | $w_2^2 \overline{\delta}$ |
| 136 | 26 | 185 | $h_0 w_2^2 \overline{\alpha^3}$ |
| 148 | 27 | 207 | $w_2^3 \overline{h_0^3}$ |
| 156 | 27 | 214 | $w_2^3 \overline{c_0}$ |
| 160 | 30 | 246 | $h_0 w_2^3 \overline{h_0^2 \alpha}$ |
| 172 | 31 | 270 | $w_2^3 \overline{h_0 \alpha^2}$ |
| 180 | 31 | 277 | $w_2^3 \overline{\delta}$ |
| 184 | 34 | 315 | $h_0 w_2^3 \overline{\alpha^3}$ |

PROOF. Most of the factorizations are evident from Tables 5.8 and 8.9. Using ext, we find the following less obvious factorizations, which are valid at $E_2$:

$$\begin{aligned}
i(c_0) &= 3_3 = h_1 \cdot \overline{h_0h_2} \\
i(h_0\alpha) &= 4_4 = h_0 \cdot \overline{c_0} \\
i(h_0\alpha^2) &= 7_{11} = h_0^3 \cdot \overline{g} \\
i(h_1\gamma) &= 6_{12} = h_1^2 \cdot \overline{g} \\
i(\alpha d_0) &= 7_{12} = d_0 \cdot (\overline{c_0} + i(\alpha)) \\
i(\delta) &= 7_{17} = h_1 \cdot \overline{\alpha\beta} \\
i(h_0\alpha^3) &= 10_{20} = h_0^3 \cdot \overline{\delta} \\
i(\alpha^2 g) &= 10_{27} = d_0\gamma \cdot \overline{h_1} \\
i(h_1^2 w_2) &= 10_{33} = h_1\gamma \cdot \overline{g} \\
i(d_0e_0g) &= 12_{34} = \alpha d_0 g \cdot \overline{h_1} \\
i(h_0\alpha w_2) &= 12_{44} = h_0 \cdot w_2\overline{c_0} \\
i(h_0d_0w_2) &= 13_{48} = h_1^2 \cdot w_2\overline{c_0} \\
d_0g\overline{\gamma} &= 13_{49} = d_0\gamma \cdot \overline{g} \\
i(h_0\alpha^2 w_2) &= 15_{61} = h_0^3 w_2 \cdot \overline{g} \\
i(h_0\alpha^3 w_2) &= 18_{80} = h_0^3 \cdot w_2\overline{\delta} \\
h_0d_0w_2\overline{g} &= 17_{83} = h_1^2 \cdot w_2\overline{\delta}\,.
\end{aligned}$$

Finally, we use the identities

$$\begin{aligned}
e_0g^2\overline{h_1} &= i(h_0d_0w_2) + d_2(i(\beta w_2)) \\
e_0g^2\overline{g} &= h_1 \cdot i(\delta w_2) + d_2(w_2\overline{\beta^2}) \\
e_0g^2\overline{\gamma} &= h_0d_0w_2\overline{g} + d_2(\beta w_2\overline{g})
\end{aligned}$$

and their $w_2^2$-multiples, which can be read off from Table 8.2. □

PROPOSITION 8.11. *Charts showing $E_5(tmf/\nu)$ for $0 \le t-s \le 192$ are given in Figures 8.1 to 8.8. All nonzero $h_0$- and $h_1$-multiplications are displayed. All $h_2$-multiplications are zero. The red dots indicate $w_1$-power torsion classes, and black dots indicate $w_1$-periodic classes. All $R_2$-module generators are labeled, except those that are also $h_0$-, $h_1$- or $h_2$-multiples.*

PROOF. Table 8.9 exhibits the $E_5$-term as a direct sum of cyclic $R_2$-modules. Most of the nontrivial $h_0$- and $h_1$-multiplications are evident from the $x$- and dec.-columns of that table. The remaining nonzero $h_0$- and $h_1$-multiplications are found by inspection of the $E_2$-term, as calculated by ext and displayed in Figures 1.32 to 1.35. Those valid at the $E_2$-term are:

$$\begin{aligned}
h_1 \cdot h_1\overline{h_1} &= h_0\overline{h_0h_2} \\
h_1 \cdot (\overline{c_0} + i(\alpha)) &= h_1\overline{c_0} \\
h_1 \cdot \gamma\overline{h_1} &= h_0\overline{\alpha\beta} \\
h_0 \cdot i(\alpha^3 g + h_0w_1w_2) &= w_1 \cdot i(h_0^2 w_2)
\end{aligned}$$

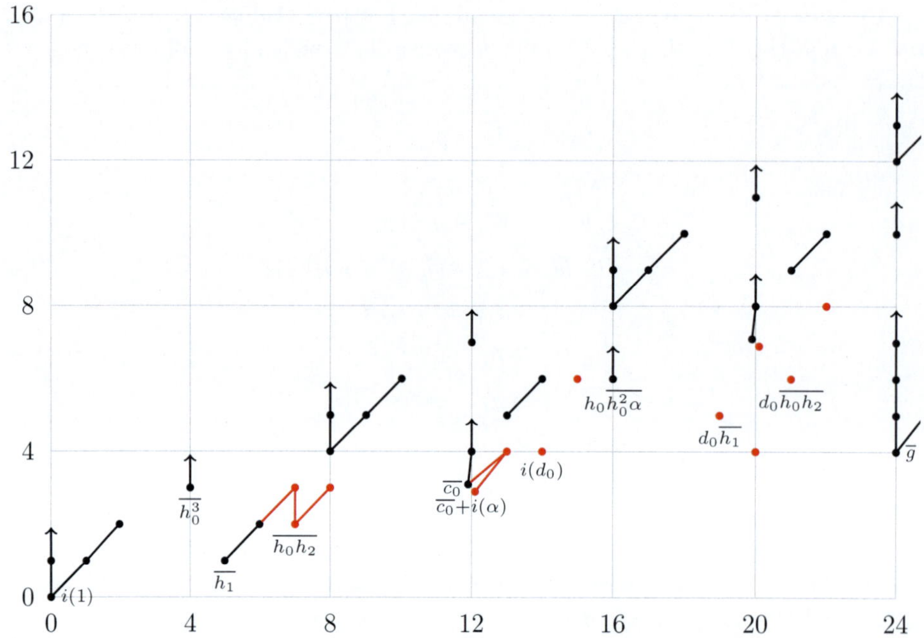


FIGURE 8.1. $E_5(tmf/\nu) = E_\infty(tmf/\nu)$ for $0 \le t - s \le 24$

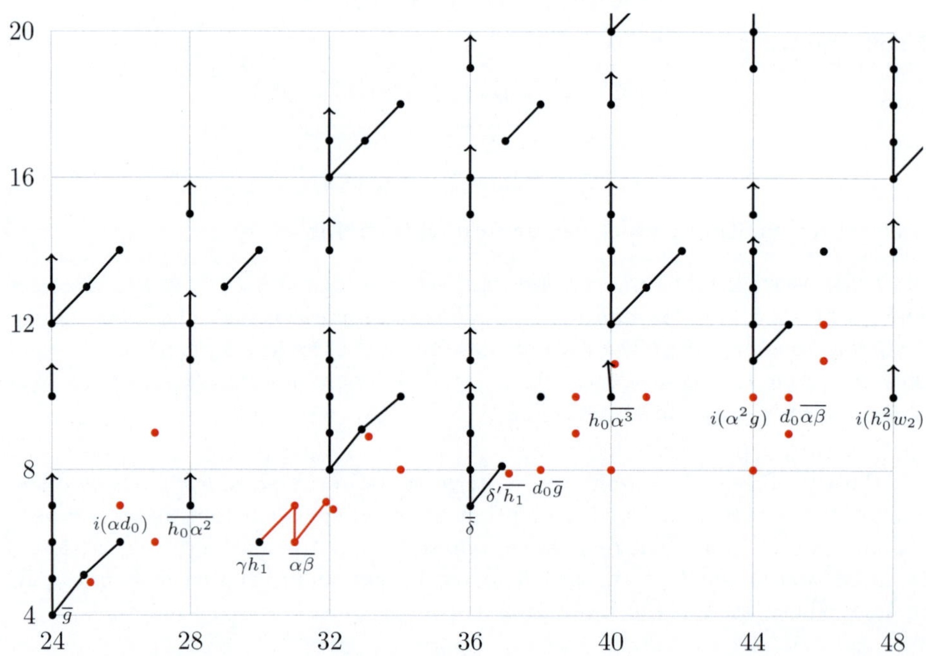


FIGURE 8.2. $E_5(tmf/\nu) = E_\infty(tmf/\nu)$ for $24 \le t - s \le 48$

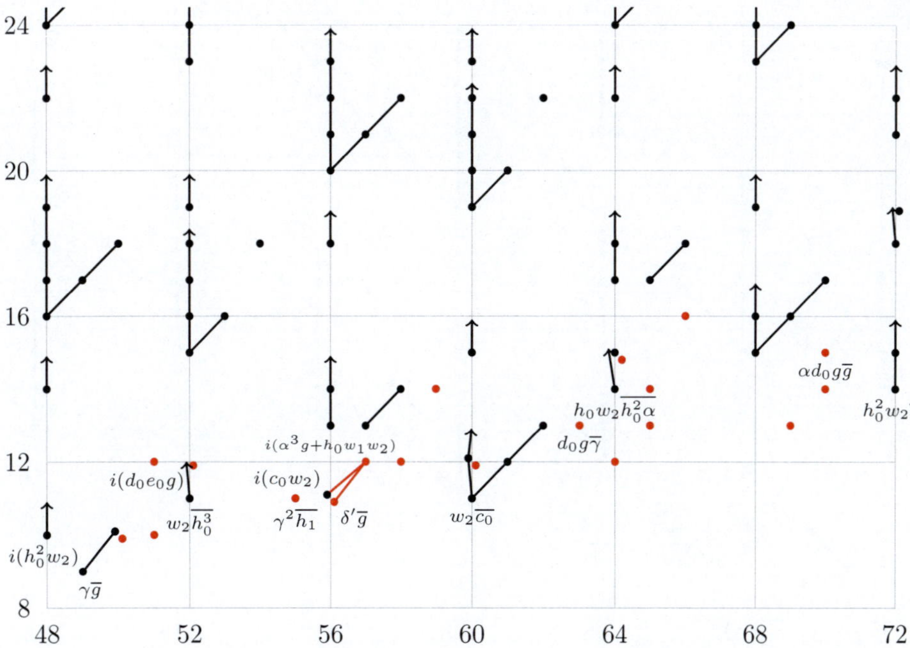


FIGURE 8.3. $E_5(tmf/\nu) = E_\infty(tmf/\nu)$ for $48 \leq t-s \leq 72$

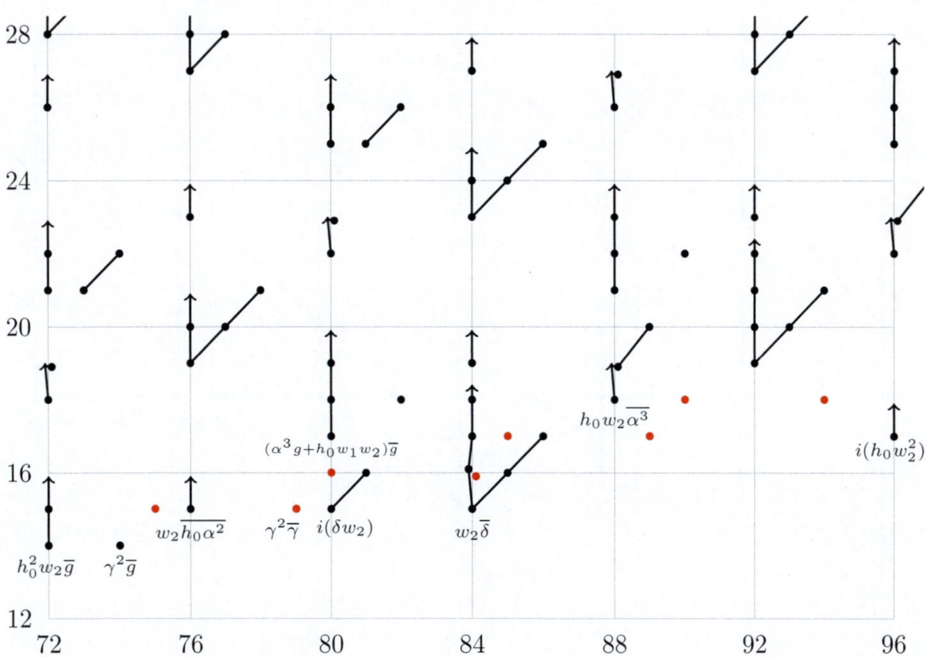


FIGURE 8.4. $E_5(tmf/\nu) = E_\infty(tmf/\nu)$ for $72 \leq t-s \leq 96$

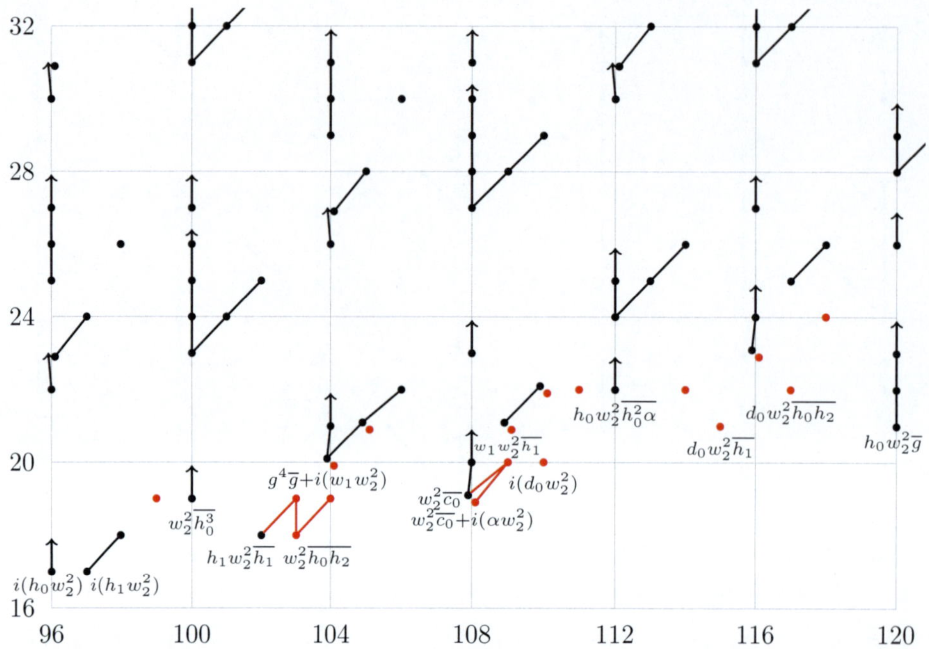


FIGURE 8.5. $E_5(tmf/\nu) = E_\infty(tmf/\nu)$ for $96 \le t-s \le 120$

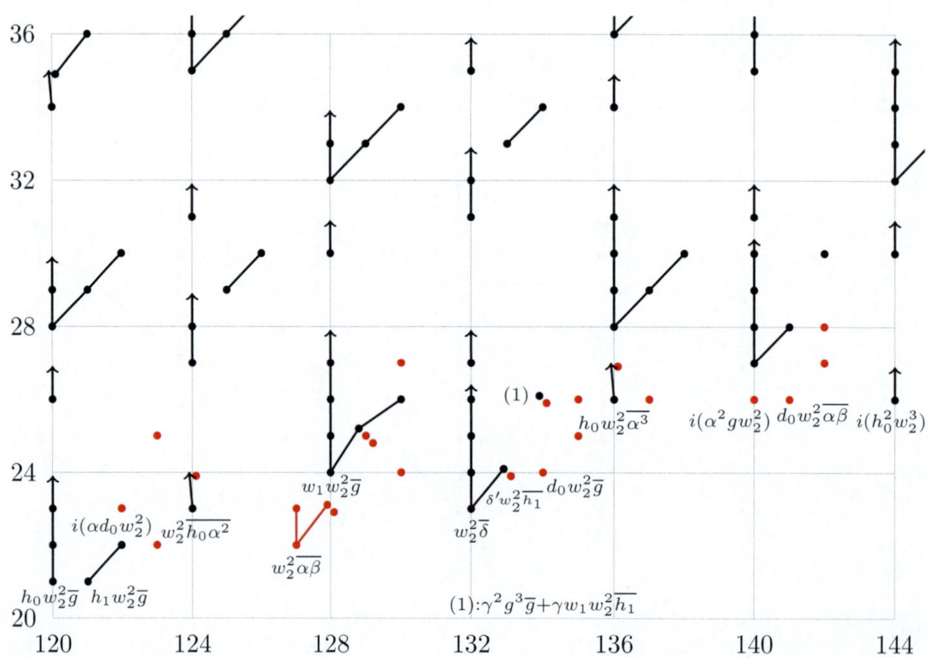


FIGURE 8.6. $E_5(tmf/\nu) = E_\infty(tmf/\nu)$ for $120 \le t-s \le 144$

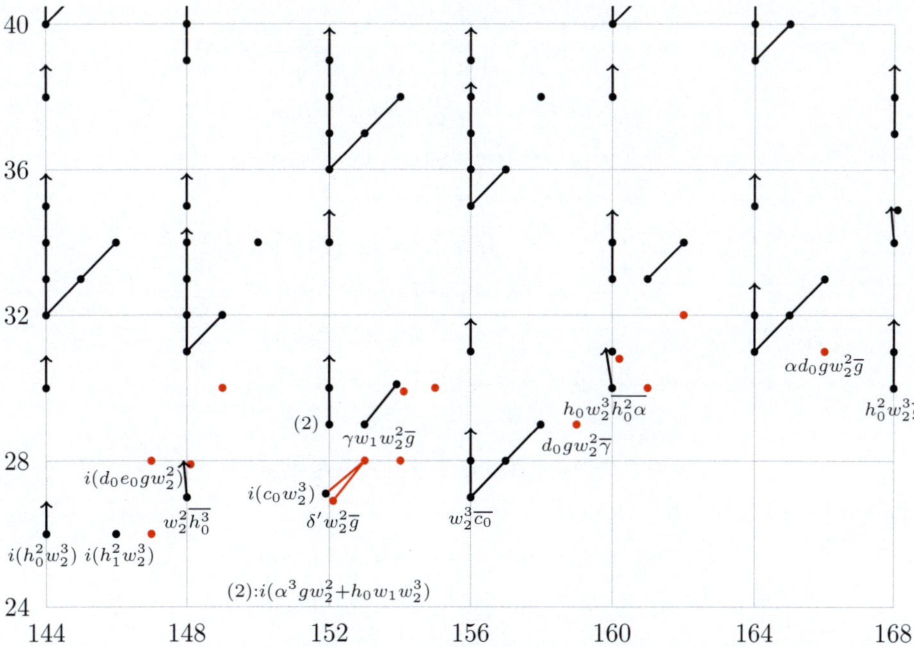


FIGURE 8.7. $E_5(tmf/\nu) = E_\infty(tmf/\nu)$ for $144 \le t - s \le 168$

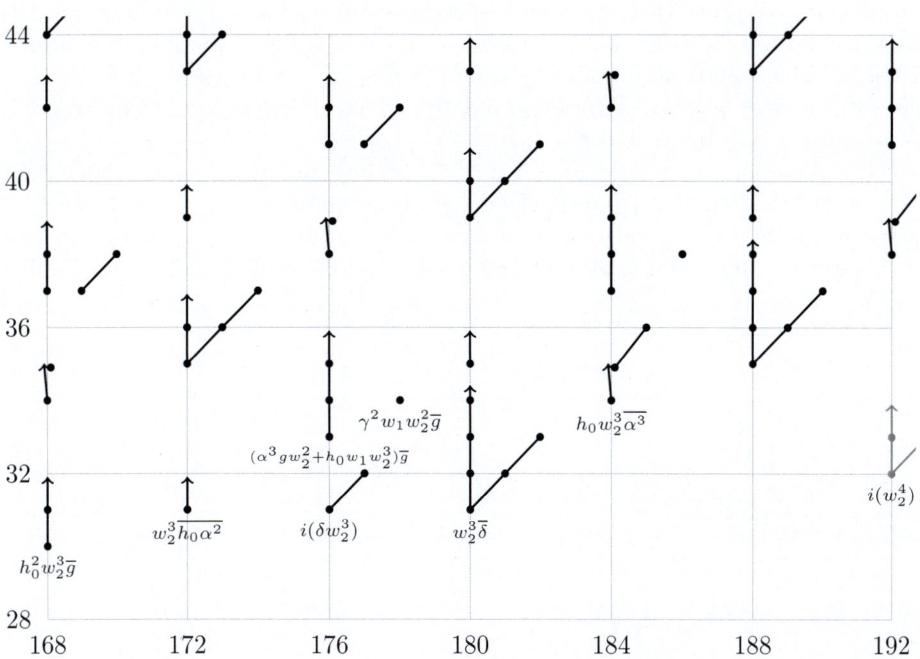


FIGURE 8.8. $E_5(tmf/\nu) = E_\infty(tmf/\nu)$ for $168 \le t - s \le 192$

$$
\begin{aligned}
h_0 \cdot (\alpha^3 g + h_0 w_1 w_2)\overline{g} &= w_1 \cdot h_0^2 w_2 \overline{g} \\
h_1 \cdot h_1 w_2^2 \overline{h_1} &= h_0 w_2^2 \overline{h_0 h_2} \\
h_0 \cdot (g^4 \overline{g} + i(w_1 w_2^2)) &= w_1 \cdot i(h_0 w_2^2) \\
h_1 \cdot (g^4 \overline{g} + i(w_1 w_2^2)) &= w_1 \cdot i(h_1 w_2^2) \\
h_1 \cdot (w_2^2 \overline{c_0} + i(\alpha w_2^2)) &= h_1 w_2^2 \overline{c_0} \\
h_1 \cdot w_1 w_2^2 \overline{h_1} &= w_1 \cdot h_1 w_2^2 \overline{h_1} \\
h_0 \cdot w_1 w_2^2 \overline{g} &= w_1 \cdot h_0 w_2^2 \overline{g} \\
h_1 \cdot w_1 w_2^2 \overline{g} &= w_1 \cdot h_1 w_2^2 \overline{g} \\
h_0 \cdot i(\alpha^3 g w_2^2 + h_0 w_1 w_2^3) &= w_1 \cdot i(h_0^2 w_2^3) \\
h_1 \cdot \gamma w_1 w_2^2 \overline{g} &= w_1 \cdot i(h_1^2 w_2^3) \\
h_0 \cdot (\alpha^3 g w_2^2 + h_0 w_1 w_2^3)\overline{g} &= w_1 \cdot h_0^2 w_2^3 \overline{g} \,.
\end{aligned}
$$

In addition, we have the relations

$$
\begin{aligned}
h_1 \cdot \delta' \overline{g} = h_1 \cdot i(c_0 w_2) &= g \cdot \delta' \overline{h_1} + d_2(w_2 \overline{h_2^2}) \\
h_1 \cdot \delta' w_2^2 \overline{g} = h_1 \cdot i(c_0 w_2^3) &= g \cdot \delta' w_2^2 \overline{h_1} + d_2(w_2^3 \overline{h_2^2}) \,.
\end{aligned}
$$

The $h_2$-multiplications are zero, because $\mathrm{Ext}_{A(2)}(M_4, \mathbb{F}_2)$ is a bigraded algebra by Lemma 1.39. □

THEOREM 8.12. *$E_5(tmf/\nu) = E_\infty(tmf/\nu)$.*

PROOF. To prove that the Adams spectral sequence for $tmf/\nu$ collapses at the $E_5$-term, we use Theorem 5.27 and show that each $E_5(tmf)$-module generator $x$ listed in Table 8.10 is an infinite cycle, i.e., that $d_r(x) = 0$ for each $r \geq 5$. In most cases this is clear because all target groups are trivial. Furthermore, all differentials on $\overline{h_1}$ must vanish by $h_0$-linearity.

The remaining cases are $x = \overline{\alpha\beta}$, $\gamma^2\overline{\gamma}$, $w_2^2\overline{h_0 h_2}$ and $w_2^2\overline{\alpha\beta}$, each of which is ($w_1$- or) $w_1^2$-torsion. If $d_r(x) = y$, then $w_1^2 y = 0$ at the $E_r$-term. Since $E_5(tmf/\nu)$ is trivial in the topological degree of $w_1^2 x$, this can only happen if $w_1^2 y = 0$ at the $E_5$-term. Since all possible targets $y$ are $w_1$-torsion free at $E_5$, this implies $y = 0$. □

# Part 3

# The abutment

CHAPTER 9

# The homotopy groups of *tmf*

In this chapter we use our mod 2 Adams spectral sequence calculations, together with the known image of $\pi_*(tmf)$ in the ring of modular forms, to determine the structure of $\pi_*(tmf)^\wedge_2 \cong \pi_*(tmf) \otimes \mathbb{Z}_2$ as a graded ring, or more precisely, as a graded $\mathbb{Z}_2$-algebra. All spectra are hereafter implicitly completed at the prime 2 (until Chapter 13), but we shall omit this from the notation.

The algebra generators fall into eight families, parameterized, roughly speaking, by the powers $\Delta^k$, $0 \le k \le 7$, of the discriminant $\Delta$. These are then made periodic with period 192 by an element $M \in \pi_{192}(tmf)$ detected by $\Delta^8$. We name "$M$" for Mark Mahowald, who first saw much of this structure. These generators are listed in Figure 9.1. The $k$-th term in a family is written with the subscript $k$, except that we usually omit the subscript 0 since the classes $\eta$, $\nu$, $\epsilon$, $\kappa$ and $\bar{\kappa}$ are the images under the unit map $S \to tmf$ of classes known by those names in $\pi_*(S)$. The remaining classes $\eta_k$, $\nu_k$, $\epsilon_k$ and $\kappa_k$ are higher analogs of these. The classes in the left part of Figure 9.1 have finite additive order, while those in the right part are of infinite additive order.

| $\eta$ | $\nu$ | $\epsilon$ | $\kappa$ | $\bar{\kappa}$ | $B$ | $C$ | $M$ |
|---|---|---|---|---|---|---|---|
| $\eta_1$ | $\nu_1$ | $\epsilon_1$ | | | $B_1$ | $C_1$ | $D_1$ |
| | $\nu_2$ | | | | $B_2$ | $C_2$ | $D_2$ |
| | | | | | $B_3$ | $C_3$ | $D_3$ |
| $\eta_4$ | $\nu_4$ | $\epsilon_4$ | $\kappa_4$ | | $B_4$ | $C_4$ | $D_4$ |
| | $\nu_5$ | $\epsilon_5$ | | | $B_5$ | $C_5$ | $D_5$ |
| | $\nu_6$ | | | | $B_6$ | $C_6$ | $D_6$ |
| | | | | | $B_7$ | $C_7$ | $D_7$ |

FIGURE 9.1. $\mathbb{Z}_2$-algebra generators of $\pi_*(tmf)$

These two sets of generators are intimately related. For example, in Chapter 10 we will see that the duality exhibited by *tmf* implies that the order of the subgroup $\langle \nu_k \rangle$ generated by $\nu_k$ is equal to the index of $\langle BD_{7-k} \rangle$ in $\langle B_{7-k} \rangle$. For this formula to apply for all $0 \le k \le 7$, we will introduce the notations $\nu_3 = \eta_1^3$ and $\nu_7 = 0$, even though these classes are not algebra generators. We also extend the notation above by the rule $x_{k+8} = x_k M$ for any generator $x$, for convenience in making general statements.

The classes $B = B_0$ and $C = C_0$ map to generators of $\pi_8(ko)$ and $\pi_{12}(ko)$ under a natural map $tmf \to ko$, see Proposition 9.21. Accordingly, we refer to the class $B$ as the "Bott element". The classes $\nu_k$, $\epsilon_k$, $\kappa_k$ and $\bar{\kappa}$ are $B$-power torsion, while the classes $\eta_k$, $B_k$, $C_k$, $D_k$ and $M$ are $B$-periodic. The classes $B_k$, $C_k$ and $D_k$ map to the modular forms $c_4\Delta^k$, $2c_6\Delta^k$ and an appropriate 2-power multiple of $\Delta^k$, respectively.

The classes $B_k$ have a kind of dual aspect. Many of them have a 2-torsion class in the same degree whose action on the $B$-power torsion submodule of $\pi_*(tmf)$ is the same. Mahowald referred to this with his statement that "$\epsilon$ tries to act like the Bott element." (See Proposition 9.40 for a precise statement of this.) For example, there is a hidden relation $\epsilon\kappa = B\kappa \in \pi_{22}(tmf)$. Since $B$ is not in the image of $\pi_*(S) \to \pi_*(tmf)$, this takes the subtler form $\epsilon\kappa = \{Pd_0\}$ in $\pi_*(S)$, but is already present there. Replacing $B$ by the sum $\widetilde{B} = B + \epsilon$ simplifies this to the relation $\widetilde{B}\kappa = 0$. Since $B\epsilon = \epsilon^2 = 0$, we have $\widetilde{B}^2 = B^2$, so that a class is $B$-power torsion if and only if it is $\widetilde{B}$-power torsion. Both $B$ and $\widetilde{B}$ map to $c_4$ in the ring of modular forms, and are hard to distinguish in the elliptic [**75**] and Adams–Novikov [**23**] spectral sequences. One of the virtues of the Adams spectral sequence for $\pi_*(tmf)$ is that the $B$-power torsion classes lie in low Adams filtration, making it relatively easy to establish relations that are difficult to detect with other tools.

The classes $B_k$ are the easiest to work with in the Adams spectral sequence, but in the end we will find that using the $\widetilde{B}_k$ gives the cleanest expression for the algebra structure. In particular, the $\mathbb{Z}[M]$-subalgebra generated by the $\widetilde{B}_k$, $C_k$ and $D_k$ is isomorphic to its image in the ring of modular forms. It also has an extremely simple action on the $B$-power torsion generators. For example, the $\widetilde{B}_k$ and $C_k$ annihilate them all, and the $C_k$ annihilate the $\eta_k$ as well.

The final section in this chapter gives a complete description of $\pi_*(tmf)$ as a $\mathbb{Z}_2$-algebra. This can be found in Theorems 9.51, 9.53 and 9.54, Figures 9.6 through 9.13, and Tables 9.8 and 9.9. In the end, there is one sign which we have not determined: $\nu_4\nu_6 = s\nu\nu_2 M$, where $s \in \{\pm 1\}$. This same sign appears in $\nu_4 D_4 = 2s\nu M$ and $\nu_6 D_4 = 2s\nu_2 M$.

The plan of the chapter is as follows.

In Section 9.1 we start by recalling the structure of $E_\infty(tmf)$ as a graded $\mathbb{F}_2$-algebra. It has the 43 generators shown in Table 9.1. As a preliminary definition, we specify the $\mathbb{Z}_2$-algebra generators of $\pi_*(tmf)$ displayed in Figure 9.1 by the classes in $E_\infty(tmf)$ that detect them, as shown in the $E_\infty(tmf)$ and $\pi_*(tmf)$ columns of Table 9.1. This determines the generators of $\pi_*(tmf)$ modulo higher Adams filtration. We also compute certain Massey products in the Adams spectral sequence $E_2$-term, which stem from multiplication by the discriminant $\Delta$. They show that our grouping into families is consistent, whether done in terms of the detecting classes at $E_\infty$ or in terms of the image in modular forms.

In the next section, we determine all the hidden 2-, $\eta$- and $\nu$-extensions in $E_\infty(tmf)$. Our calculations of the Adams spectral sequences for $tmf/2$ and $tmf/\nu$ are key to our determination of the hidden 2- and $\nu$-extensions. The $\eta$-extensions follow from these. In the process, we refine our specification of the algebra generators so that the $\eta_k$ and $\epsilon_k$ all have additive order 2, see Lemma 9.7. The last result in this section is the interesting relation

$$\nu^2\nu_4 = \eta\epsilon_4 + \eta_1\bar{\kappa}^4 .$$

It exhibits a hidden $\nu$-extension from the $E_\infty$-class detecting $\nu\nu_4$ to the $E_\infty$-class detecting $\eta\epsilon_4$. However, this is not the whole relation in homotopy: there is also the higher filtration term $\eta_1\bar{\kappa}^4$. A hidden extension is simply the lowest filtration part of a nonzero product that is zero at $E_\infty$.

In Section 9.3, we recall the homomorphism from $\pi_*(tmf)$ to the ring of integral modular forms. Next, in Section 9.4 we refine our definition of the 40 algebra generators of $\pi_*(tmf)$: For each generator we specify its image in modular forms, together with its detecting class in $E_\infty(tmf)$. This leaves some ambiguity in a number of cases. Where possible, we eliminate this immediately. For example, when we define $\eta_1$ it will be apparent that we can add a term of higher Adams filtration to ensure that $\eta_1 B = \eta B_1$, and we do this. Where indeterminacy remains we make it explicit, and note where in the succeeding sections it will be reduced or eliminated.

We determine the remaining multiplicative structure in Section 9.5, and in Section 9.6 we put our description of the algebra $\pi_*(tmf)$ in its final form.

## 9.1. Algebra generators for the $E_\infty$-term

The $E_\infty$-term of the Adams spectral sequence for *tmf* is generated as an $\mathbb{F}_2$-algebra by the 43 classes listed in Table 5.10. These are reproduced in Table 9.1. Each entry also lists an element of $\pi_*(tmf)$ which represents it, the image of that element in $mf_{*/2}$ (to be determined in Section 9.3), and the values of the Massey products $\Delta(x) = \langle h_2, g, x\rangle$ and $\Delta'(x) = \langle x, h_2, g\rangle$ (which we determine next). An entry "–" in the $\Delta$ or $\Delta'$ column means that the Massey product is not defined, while an entry "?" indicates that we have not calculated this particular Massey product.

Table 9.1: Algebra generators of $E_\infty(tmf)$ and $\pi_*(tmf)$

| $t-s$ | $s$ | $g$ | $E_\infty(tmf)$ | $\pi_*(tmf)$ | $mf_{*/2}$ | $\Delta$ | $\Delta'$ |
|---|---|---|---|---|---|---|---|
| 0 | 1 | 0 | $h_0$ | $2\iota = 2D$ | 2 | – | – |
| 1 | 1 | 1 | $h_1$ | $\eta$ | 0 | – | $\gamma$ |
| 3 | 1 | 2 | $h_2$ | $\nu$ | 0 | 0 | – |
| 8 | 3 | 2 | $c_0$ | $\epsilon$ | 0 | $\delta$ | $\{\delta, \delta'\}$ |
| 8 | 4 | 1 | $w_1$ | $B$ | $c_4$ | – | – |
| 12 | 6 | 4 | $h_0^3\alpha$ | $C$ | $2c_6$ | $h_0\alpha^3$ | ? |
| 14 | 4 | 4 | $d_0$ | $\kappa$ | 0 | – | – |
| 20 | 4 | 8 | $g$ | $\bar{\kappa}$ | 0 | – | 0 |
| 24 | 7 | 7 | $h_0\alpha^2$ | $D_1$ | $8\Delta$ | $h_0 \cdot h_0^2 w_2$ | ? |
| 25 | 5 | 11 | $\gamma$ | $\eta_1$ | 0 | – | $h_1 w_2$ |
| 27 | 6 | 10 | $\alpha\beta$ | $\nu_1$ | 0 | – | $h_0 \cdot h_2 w_2$ |
| 32 | 7 | 11 | $\delta$ | $B_1 + \epsilon_1$ | $c_4\Delta$ | $c_0 w_2$ | ? |
| 32 | 7 | 12 | $\delta'$ | $\epsilon_1$ | 0 | – | $c_0 w_2$ |
| 36 | 10 | 14 | $h_0\alpha^3$ | $C_1$ | $2c_6\Delta$ | $h_0^3\alpha w_2$ | ? |

Table 9.1: Algebra generators of $E_\infty(tmf)$ and $\pi_*(tmf)$ (cont.)

| $t-s$ | $s$ | $g$ | $E_\infty(tmf)$ | $\pi_*(tmf)$ | $mf_{*/2}$ | $\Delta$ | $\Delta'$ |
|---|---|---|---|---|---|---|---|
| 48 | 10 | 19 | $h_0^2 w_2$ | $D_2$ | $4\Delta^2$ | – | – |
| 51 | 9 | 23 | $h_2 w_2$ | $\nu_2$ | 0 | 0 | – |
| 56 | 11 | 24 | $c_0 w_2$ | $B_2$ | $c_4\Delta^2$ | $\delta w_2$ | $\{\delta w_2, \delta' w_2\}$ |
| 56 | 13 | $26+27$ | $\alpha^3 g + h_0 w_1 w_2$ | $2B_2$ | $2c_4\Delta^2$ | – | – |
| 60 | 14 | 28 | $h_0^3 \alpha w_2$ | $C_2$ | $2c_6\Delta^2$ | $h_0\alpha^3 w_2$ | ? |
| 72 | 15 | 36 | $h_0\alpha^2 w_2$ | $D_3$ | $8\Delta^3$ | $h_0^2 \cdot h_0 w_2^2$ | ? |
| 80 | 15 | 41 | $\delta w_2$ | $B_3$ | $c_4\Delta^3$ | $c_0 w_2^2$ | ? |
| 84 | 18 | 48 | $h_0\alpha^3 w_2$ | $C_3$ | $2c_6\Delta^3$ | $h_0^3\alpha w_2^2$ | ? |
| 96 | 17 | 58 | $h_0 w_2^2$ | $D_4$ | $2\Delta^4$ | – | – |
| 97 | 17 | 59 | $h_1 w_2^2$ | $\eta_4$ | 0 | – | $\gamma w_2^2$ |
| 99 | 17 | 60 | $h_2 w_2^2$ | $\nu_4$ | 0 | 0 | – |
| 104 | 19 | 62 | $c_0 w_2^2$ | $\epsilon_4$ | 0 | $\delta w_2^2$ | $\{\delta w_2^2, \delta' w_2^2\}$ |
| 104 | 20 | 69 | $w_1 w_2^2$ | $B_4$ | $c_4\Delta^4$ | – | – |
| 108 | 22 | 71 | $h_0^3\alpha w_2^2$ | $C_4$ | $2c_6\Delta^4$ | $h_0\alpha^3 w_2^2$ | ? |
| 110 | 20 | 74 | $d_0 w_2^2$ | $\kappa_4$ | 0 | – | – |
| 120 | 23 | 82 | $h_0\alpha^2 w_2^2$ | $D_5$ | $8\Delta^5$ | $h_0 \cdot h_0^2 w_2^3$ | ? |
| 123 | 22 | 82 | $\alpha\beta w_2^2$ | $\nu_5$ | 0 | – | $h_0 \cdot h_2 w_2^3$ |
| 128 | 23 | 87 | $\delta w_2^2$ | $B_5+\epsilon_5$ | $c_4\Delta^5$ | $c_0 w_2^3$ | ? |
| 128 | 23 | 88 | $\delta' w_2^2$ | $\epsilon_5$ | 0 | – | $c_0 w_2^3$ |
| 132 | 26 | 100 | $h_0\alpha^3 w_2^2$ | $C_5$ | $2c_6\Delta^5$ | $h_0^3\alpha w_2^3$ | ? |
| 144 | 26 | 107 | $h_0^2 w_2^3$ | $D_6$ | $4\Delta^6$ | – | – |
| 147 | 25 | 113 | $h_2 w_2^3$ | $\nu_6$ | 0 | 0 | – |
| 152 | 27 | 116 | $c_0 w_2^3$ | $B_6$ | $c_4\Delta^6$ | $\delta w_2^3$ | $\{\delta w_2^3, \delta' w_2^3\}$ |
| 152 | 29 | 131 | $\alpha^3 g w_2^2$ | $2B_6$ | $2c_4\Delta^6$ | – | – |
| | | $+\,132$ | $+\,h_0 w_1 w_2^3$ | | | | |
| 156 | 30 | 131 | $h_0^3\alpha w_2^3$ | $C_6$ | $2c_6\Delta^6$ | $h_0\alpha^3 w_2^3$ | ? |
| 168 | 31 | 144 | $h_0\alpha^2 w_2^3$ | $D_7$ | $8\Delta^7$ | $h_0^3 \cdot w_2^4$ | ? |
| 176 | 31 | 149 | $\delta w_2^3$ | $B_7$ | $c_4\Delta^7$ | $c_0 w_2^4$ | ? |
| 180 | 34 | 168 | $h_0\alpha^3 w_2^3$ | $C_7$ | $2c_6\Delta^7$ | $h_0^3\alpha w_2^4$ | ? |
| 192 | 32 | 172 | $w_2^4$ | $M$ | $\Delta^8$ | – | – |

Table 9.2: $\Delta$ and $\Delta'$ on certain decomposable elements of $E_\infty(tmf)$

| $t-s$ | $s$ | $g$ | $E_\infty(tmf)$ | $\pi_*(tmf)$ | $mf_{*/2}$ | $\Delta$ | $\Delta'$ |
|---|---|---|---|---|---|---|---|
| 0 | 3 | 0 | $h_0^3$ | $8\iota = 8D$ | 8 | $h_0\alpha^2$ | $h_0\alpha^2$ |
| 24 | 7 | 7 | $h_0\alpha^2$ | $D_1$ | $8\Delta$ | $h_0^3 w_2$ | ? |
| 48 | 11 | 19 | $h_0^3 w_2$ | $2D_2$ | $8\Delta^2$ | $h_0\alpha^2 w_2$ | $h_0\alpha^2 w_2$ |
| 72 | 15 | 36 | $h_0\alpha^2 w_2$ | $D_3$ | $8\Delta^3$ | $h_0^3 w_2^2$ | ? |
| 96 | 19 | 57 | $h_0^3 w_2^2$ | $4D_4$ | $8\Delta^4$ | $h_0\alpha^2 w_2^2$ | $h_0\alpha^2 w_2^2$ |
| 120 | 23 | 82 | $h_0\alpha^2 w_2^2$ | $D_5$ | $8\Delta^5$ | $h_0^3 w_2^3$ | ? |
| 144 | 27 | 111 | $h_0^3 w_2^3$ | $2D_6$ | $8\Delta^6$ | $h_0\alpha^2 w_2^3$ | ? |
| 168 | 31 | 144 | $h_0\alpha^2 w_2^3$ | $D_7$ | $8\Delta^7$ | $h_0^3 w_2^4$ | ? |
| 8 | 7 | 1 | $h_0^3 w_1$ | $8B$ | $8c_4$ | $h_0\alpha^2 w_1$ | ? |
| 32 | 11 | 10 | $h_0\alpha^2 w_1$ | $8B_1$ | $8c_4\Delta$ | $h_0^3 w_1 w_2$ | ? |
| 56 | 15 | 24 | $h_0^3 w_1 w_2$ | $8B_2$ | $8c_4\Delta^2$ | $h_0\alpha^2 w_1 w_2$ | ? |
| 80 | 19 | 43 | $h_0\alpha^2 w_1 w_2$ | $8B_3$ | $8c_4\Delta^3$ | $h_0^3 w_1 w_2^2$ | ? |
| 104 | 23 | 66 | $h_0^3 w_1 w_2^2$ | $8B_4$ | $8c_4\Delta^4$ | $h_0\alpha^2 w_1 w_2^2$ | ? |
| 128 | 27 | 94 | $h_0\alpha^2 w_1 w_2^2$ | $8B_5$ | $8c_4\Delta^5$ | $h_0^3 w_1 w_2^3$ | ? |
| 152 | 31 | 127 | $h_0^3 w_1 w_2^3$ | $8B_6$ | $8c_4\Delta^6$ | $h_0\alpha^2 w_1 w_2^3$ | ? |
| 176 | 35 | 164 | $h_0\alpha^2 w_1 w_2^3$ | $8B_7$ | $8c_4\Delta^7$ | $h_0^3 w_1 w_2^4$ | ? |

**9.1.1. The Massey products $\Delta$ and $\Delta'$.** The discriminant $\Delta = (c_6^2 - c_4^3)/1728$ is not in the image of the map $tmf_* \to mf_{*/2}$. There is a class in the $E_2$-term of the Adams–Novikov spectral sequence that would have mapped to the discriminant had it survived, but which supports a differential $d_5(\Delta) = h_2 g$ imposing the relation $\nu\bar{\kappa} = 0$. In the Adams spectral sequence, there is no such class at $E_2$, but precursor spectral sequences like the May spectral sequence (where $d_4(b_{30}^2) = h_{12}b_{21}^2$ by [**144**, §3.2] or [**81**, §3.2]) and the Davis–Mahowald spectral sequence (where $d_1(x_7^4) = h_2 g$ by Lemma 3.28) have differentials that impose the relation $h_2 g = 0$. Any such class gives rise to a Massey product at $E_2$ (Adams spectral sequence) or $E_6$ (Adams–Novikov spectral sequence) which, in favorable cases, detects the Toda bracket $\langle \nu, \bar{\kappa}, -\rangle$. In the Adams spectral sequence this Massey product is $\Delta(x) = \langle h_2, g, x\rangle$. Here, we compute this Massey product and a related one at $E_2$ of the Adams spectral sequence, and show that they group the generators of $\pi_*(tmf)$ into a small number of families, explaining and justifying our notations for these generators.

DEFINITION 9.1. Let $E_2 = \mathrm{Ext}_{A(2)}(\mathbb{F}_2, \mathbb{F}_2)$. For $x \in E_2$ satisfying $gx = 0$, let

$$\Delta(x) = \langle h_2, g, x\rangle \in E_2/(h_2 E_2 + E_2 x)\,.$$

For $x \in E_2$ satisfying $h_2 x = 0$, let

$$\Delta'(x) = \langle x, h_2, g\rangle \in E_2/(x E_2 + E_2 g)\,.$$

As is customary, we regard these as subsets of $E_2$, but when they are singletons $\{y\}$ we will write them simply as $y$. Since $E_2(tmf) = 0$ in bidegree $(t-s, s) = (24, 4)$, the contributions $E_2x$ and $xE_2$ to the indeterminacy always vanish. Hence the indeterminacy of $\Delta(x)$ is $h_2E_2$, and that of $\Delta'(x)$ is $E_2g$.

THEOREM 9.2. *The Massey products $\Delta(x)$ are shown for classes detecting the algebra generators of $E_\infty(tmf)$ in Table 9.1, and for the other elements here in Table 9.2. Repeated application of $\Delta$ gives classes detecting the following sequences of elements of $\pi_*(tmf)$:*

- $8D \mapsto D_1 \mapsto 2D_2 \mapsto D_3 \mapsto 4D_4 \mapsto D_5 \mapsto 2D_6 \mapsto D_7 \mapsto 8M$.
- $C \mapsto C_1 \mapsto C_2 \mapsto C_3 \mapsto C_4 \mapsto C_5 \mapsto C_6 \mapsto C_7 \mapsto CM$.
- $B{+}\epsilon \mapsto B_1{+}\epsilon_1 \mapsto B_2 \mapsto B_3 \mapsto B_4{+}\epsilon_4 \mapsto B_5{+}\epsilon_5 \mapsto B_6 \mapsto B_7 \mapsto (B{+}\epsilon)M$.
- $8B \mapsto 8B_1 \mapsto 8B_2 \mapsto 8B_3 \mapsto 8B_4 \mapsto 8B_5 \mapsto 8B_6 \mapsto 8B_7 \mapsto 8BM$.

PROOF. These calculations are a simple matter for `ext`: We compute the chain map for the cocycle $s_g$ = `s_g` and inspect the file `s_g/brackets.sym`. The lines which record the Massey products $\langle h_2, g, -\rangle$ are all of the form

```
s1_g1 in < h2, 8, s_g >.
```

where $s_1 - s = 4$. This difference in degrees tells us that the middle term is $g = 4_8$ = `4_8`, and the presence of this line says that the coefficient of `s1_g1` in the Massey product is nonzero, *if the Massey product is defined.* Thus, for example

```
7_7 in < h2, 8, 3_0 >
```

tells us that the coefficient of the generator $7_7$ = `7_7`, detecting $h_0\alpha^2$, in the Massey product $\langle h_2, g, h_0^3\rangle$ is nonzero. The Massey product is defined and there are no other lines indicating that other terms occur in the Massey product, so we conclude that $\Delta(h_0^3) = \langle h_2, g, h_0^3\rangle = h_0\alpha^2$ with 0 indeterminacy.

The $B$-, $C$- and $D$-families merit special attention. In each case, we have $\Delta(xw_2) = w_2\Delta(x)$, since there are no nonzero $h_2$-multiples in the bidegrees in which these 24 values of $\Delta$ lie. Therefore, it is necessary to calculate only the first two Massey products in each family of eight elements. In addition they are all defined with 0 indeterminacy.

For the "discriminant" family we have $\Delta(h_0^3) = h_0\alpha^2$ and $\Delta(h_0\alpha^2) = h_0^3w_2$. The coefficients 8, 4, 2 and 1 of the $D_k$ reflect the $h_0$-divisibility of these classes in $E_\infty(tmf)$.

Second, the "$2c_6$" family is much more uniform: the Massey products $\Delta(h_0^3\alpha) = h_0\alpha^3$ and $\Delta(h_0\alpha^3) = h_0^3\alpha w_2$ imply that $C_k \mapsto C_{k+1}$ for each $k$, with $C_0 = C$ and $C_{k+8} = C_kM$.

Third, we consider two sequences of elements of Ext which contain information about the "Bott", or "$c_4$" family. In degree 8, either $B$, detected by $w_1$, or $B+\epsilon$, detected by $c_0$, generates a $\mathbb{Z}$-summand and maps to $c_4 \in mf_{*/2}$. The Massey product $\Delta$ is not defined on $w_1$, but is defined on $c_0$, where we get $\Delta(c_0) = \delta$. In degree 32, similarly, either $B_1$, detected by $\alpha g = \delta + \delta'$, or $B_1 + \epsilon_1$, detected by $\delta$, generates a $\mathbb{Z}$-summand and maps to $c_4\Delta \in mf_{*/2}$. However, the Massey product $\Delta$ is not defined on $\alpha g$ or $\delta'$, but is defined on $\delta$, where it gives $\Delta(\delta) = c_0w_2$. The sequence of Massey products $c_0 \mapsto \delta \mapsto c_0w_2 \mapsto \cdots$ detects the $B+\epsilon \mapsto B_1+\epsilon_1 \mapsto B_2 \mapsto B_3 \mapsto B_4+\epsilon_4 \mapsto \cdots$ sequence.

Due to hidden extensions, the 2-power multiples of these classes are mostly not detected by their $h_0$-power multiples. Accordingly, we also calculate $\Delta$ on $E_2$-classes detecting 8 times each of these classes. The Massey products $\Delta(h_0^3w_1) = h_0\alpha^2w_1$

and $\Delta(h_0\alpha^2w_1) = h_0^3w_1w_2$, together with $w_2$-linearity, then give the sequence of Massey products detecting the very uniform sequence $8B \mapsto 8B_1 \mapsto 8B_2 \mapsto \cdots$. □

On $B$-power torsion classes the operator $\Delta$ is often undefined or zero modulo indeterminacy. For these classes, the closely related Massey product $\Delta'(x) = \langle g, h_2, x\rangle$ plays the role of tying them together into systematic families. It is not as simple to calculate, because the `ext` program as currently constituted only calculates the Massey products $\langle x, y, h_i\rangle = \langle h_i, y, x\rangle$. (It calculates this from the chain map lifting $x$.) Identities for the relevant Massey products are used to get around this.

Because of the identities $h_1^3 = h_0h_2^2$ and $h_1^2\gamma = h_0\alpha\beta$, we consider the elements in the $\eta$- and $\nu$-families together. At the $E_2$-term, the behavior is quite uniform. These results and their meaning for the $\eta_k$ and $\nu_k$ are shown in Figures 9.2 and 9.3.

THEOREM 9.3. *The following Massey products are defined in $E_2(tmf)$ with 0 indeterminacy:*

(1) $\Delta'(h_1w_2^i) = \gamma w_2^i$.
(2) $\Delta'(\gamma w_2^i) = h_1w_2^{i+1}$.
(3) $\Delta'(h_0h_2w_2^i) = \alpha\beta w_2^i$.
(4) $\Delta'(\alpha\beta w_2^i) = h_0h_2w_2^{i+1}$.

PROOF. The `ext` program can calculate that $\Delta'(h_1) = \langle h_1, h_2, g\rangle = \gamma$.

We can then write $\Delta'(\gamma) = a_0h_1w_2$ and $\Delta'(h_1\gamma) = \{a_1h_1^2w_2, a_1h_1^2w_2 + \beta^2g\}$ for some coefficients $a_i \in \mathbb{F}_2$. Since $\Delta'(h_1\gamma) \supseteq \Delta'(h_1)\gamma = \{\gamma^2\} = \{h_1^2w_2 + \beta^2g\}$, we conclude that $a_1 = 1$. Since $h_1\Delta'(\gamma) \subseteq \Delta'(h_1\gamma) = \{h_1^2w_2, \gamma^2\}$, we conclude that $a_0 = 1$ also. Now we will show that multiplication by $w_2$ is an isomorphism in the relevant degrees, establishing the first two formulas claimed. The elements $h_1w_2^i$ and $\gamma w_2^i$ lie in bidegrees $(s,t) = (1+4k, 2+28k)$, where $k = 2i$ or $k = 2i+1$, respectively. A monomial

$$h_0^{n_1}h_1^{n_2}h_2^{n_3}c_0^{n_4}\alpha^{n_5}\beta^{n_6}d_0^{n_7}e_0^{n_8}\gamma^{n_9}\delta^{n_{10}}g^{n_{11}}w_1^{n_{12}}w_2^{n_{13}}$$

has

$$s = (n_1+n_2+n_3) + 3(n_4+n_5+n_6) + 4(n_7+n_8+n_{11}+n_{12}) + 5n_9 + 7n_{10} + 8n_{13}$$

and

$$t = n_1 + 2n_2 + 4n_3 + 11n_4 + 15n_5 + 18n_6 + 18n_7 + \\ 21n_8 + 24n_{11} + 12n_{12} + 30n_9 + 39n_{10} + 56n_{13}$$

so that, if it lies in a bidegree $(s,t) = (1+4k, 2+28k)$, then

$$7s - t = 5 = 6n_1 + 5n_2 + 3n_3 + 10n_4 + 6n_5 + 3n_6 + 10n_7 + 7n_8 + 4n_{11} + 16n_{12} + 5n_9 + 10n_{10}.$$

Evidently $n_1 = n_4 = n_5 = n_7 = n_8 = n_{12} = n_{10} = 0$ and

$$5 = 5n_2 + 3n_3 + 3n_6 + 4n_{11} + 5n_9.$$

The only non-negative integer solutions have $n_2$ or $n_9$ equal to 1 and all other terms 0, corresponding to the $h_1w_2^i$ and $\gamma w_2^i$. Therefore $w_2$-multiplication acts isomorphically on these bidegrees, as claimed.

Similarly, we can write $\Delta'(h_0h_2) = b_0\alpha\beta$ and $\Delta'(h_0^2h_2) = \Delta'(h_1^3) = b_1h_0\alpha\beta$ for some coefficients $b_i \in \mathbb{F}_2$. From $\Delta'(h_1^3) \supseteq h_1^2\Delta'(h_1) = h_1^2\gamma = h_0\alpha\beta$ we conclude that $b_1 = 1$. Then $h_0\Delta'(h_0h_2) \subseteq \Delta'(h_0^2h_2) = h_0\alpha\beta$ implies that $b_0 = 1$ also.

We can write $\Delta'(\alpha\beta) = b_2 h_0 h_2 w_2$ for some coefficient $b_2 \in \mathbb{F}_2$. We then have $h_0\Delta'(\alpha\beta) \subseteq \Delta'(h_0\alpha\beta) = \Delta'(h_1^2\gamma)$, which must be either $h_1\gamma^2$ or 0. Since it contains $h_1\gamma\Delta'(h_1)$, it is $h_1\gamma^2$, and $\Delta'(\alpha\beta) = h_0 h_2 w_2$.

Next we can write $\Delta'(h_0 h_2 w_2) = b_3 \alpha\beta w_2$ and

$$\Delta'(h_0^2 h_2 w_2) = \{b_4 h_0 \alpha\beta w_2, b_4 h_0 \alpha\beta w_2 + \beta g^3\}$$

for some coefficients $b_i \in \mathbb{F}_2$. Then $\Delta'(h_0^2 h_2 w_2) = \Delta'(h_1\gamma^2) \supseteq \Delta'(h_1)\gamma^2 = \gamma^3 = h_0\alpha\beta w_2 + \beta g^3$ implies that $b_4 = 1$. Then $h_0\Delta'(h_0 h_2 w_2) \subseteq \Delta'(h_0^2 h_2 w_2)$ implies that $b_3 = 1$ also.

As above, multiplication by $w_2$ is an isomorphism in the relevant degrees, establishing the last two formulas claimed, as follows. The classes $h_0 h_2 w_2^i$ and $\alpha\beta w_2^i$ lie in bidegrees $(s,t) = (2+4k, 5+28k)$. A monomial lying in such a degree must satisfy

$$7s-t = 9 = 6n_1+5n_2+3n_3+10n_4+6n_5+3n_6+10n_7+7n_8+4n_{11}+16n_{12}+5n_9+10n_{10}.$$

Evidently $n_4 = n_7 = n_{12} = n_{10} = 0$ and

$$9 = 6n_1 + 5n_2 + 3n_3 + 6n_5 + 3n_6 + 7n_8 + 4n_{11} + 5n_9.$$

It is easily verified that there are six non-negative integer solutions to this, corresponding to $w_2^i$-multiples of the five elements $h_0h_2$, $h_0\beta = h_2\alpha$, $\alpha\beta$, $h_1 g$ and $\gamma g$. Of these, only $h_0 h_2$ and $\alpha\beta$ have $(s,t)$ of the form $(2+4k, 5+28k)$. □

The elements produced by iterating the Massey product $\Delta'$ on the set of $h_0$- and $h_1$-multiples of $h_1$ and $h_2$ then fit into the very simple pattern shown in Figures 9.2 and 9.3. We note that each of these classes in $\pi_*(tmf)$ for which $\Delta'$ of the detecting class at $E_2$ does not survive to $E_\infty$ supports a hidden $\nu$-multiplication. This could be used as a means to detect such hidden extensions, but we have been able to determine them all using primary information and induced maps.

THEOREM 9.4. *The following Massey products are defined in $E_2(tmf)$:*

(1) $\Delta'(c_0 w_2^i) = \{\delta w_2^i, \delta' w_2^i\}$.

(2) $\Delta'(\delta' w_2^i) = c_0 w_2^{i+1}$.

PROOF. We can write $\Delta'(c_0)$ as $\{a_0\delta, a_0\delta + \alpha g\}$ for some $a_0 \in \mathbb{F}_2$. Multiplying by $h_1$ gives that $a_0 h_1 \delta \in \Delta'(h_1 c_0)$, which is either $h_1\delta = c_0\gamma$ or 0. It must contain $\Delta'(h_1)c_0$, so $a_0 = 1$. Thus, $\Delta'(c_0) = \{\delta, \delta'\}$.

Next, we can write $\Delta'(\delta') = a_1 c_0 w_2$ for some $a_1 \in \mathbb{F}_2$. Multiplying by $h_1$ we get that $\Delta'(h_1\delta) = \{a_1 h_1 c_0 w_2, a_1 h_1 c_0 w_2 + e_0 g^2 \gamma g\}$. This must contain $\delta'\Delta'(h_1) = \delta'\gamma = h_1 c_0 w_2 + e_0 g^2$, so that $a_1 = 1$.

To finish the proof, we show that $c_0 w_2^i$, $\delta w_2^i$, $\delta' w_2^i$ and $\alpha g w_2^i$ are the only elements in bidegrees of the form $(s,t) = (3+4k, 11+28k)$. As in Theorem 9.3, we must solve

$$\begin{aligned} 7s - t = 10 = 6n_1 + 5n_2 + 3n_3 + 10n_4 + 6n_5 \\ + 3n_6 + 10n_7 + 7n_8 + 4n_{11} + 16n_{12} + 5n_9 + 10n_{10}. \end{aligned}$$

Evidently $n_{12} = 0$ and

$$10 = 6(n_1+n_5) + 5(n_2+n_9) + 3(n_3+n_6) + 7n_8 + 4n_{11} + 10(n_4+n_7+n_{10}).$$

It is easily verified that there are 13 non-negative integer solutions to this, corresponding to $w_2^i$-multiples of the elements $c_0$, $d_0$, $\delta$, $h_2 e_0 = h_0 g$, $\alpha g$, $h_2^2 g = 0$, $h_2\beta g = 0$, $\beta^2 g$, $h_1^2$, $\gamma^2$, $h_1\gamma$. Of these, only $c_0 w_2^i$, $\delta w_2^i$, $\delta' w_2^i$ and $\alpha g w_2^i$ have $(s,t)$ of

$$
\begin{array}{ccccc}
 & \xrightarrow{\;h_1\;} & \xrightarrow{\;h_1\;} & \xleftarrow{\;h_0\;} & \xleftarrow{\;h_0\;} \\
h_1 & h_1^2 & h_1^3 = h_0^2 h_2 & h_0 h_2 & h_2 \\
\downarrow & \downarrow & \downarrow & \downarrow & \\
\gamma & h_1\gamma & h_1^2\gamma = h_0\alpha\beta & \alpha\beta & \\
\downarrow & \downarrow & \downarrow & \downarrow & \\
h_1 w_2 & \{h_1^2 w_2, \gamma^2\} & h_1\gamma^2 = h_0^2 h_2 w_2 & h_0 h_2 w_2 & h_2 w_2 \\
\downarrow & \downarrow & \downarrow & \downarrow & \\
\gamma w_2 & h_1\gamma w_2 & \{h_1^2\gamma w_2 = h_0\alpha\beta w_2, \gamma^3\} & \alpha\beta w_2 & \\
\downarrow & \downarrow & \downarrow & \downarrow & \\
h_1 w_2^2 & \{h_1^2 w_2^2, \gamma^2 w_2\} & h_1\gamma^2 w_2 = h_0^2 h_2 w_2^2 & h_0 h_2 w_2^2 & h_2 w_2^2 \\
\downarrow & \downarrow & \downarrow & \downarrow & \\
\gamma w_2^2 & h_1\gamma w_2^2 & \{h_1^2\gamma w_2^2 = h_0\alpha\beta w_2^2, \gamma^3 w_2\} & \alpha\beta w_2^2 & \\
\downarrow & \downarrow & \downarrow & \downarrow & \\
h_1 w_2^3 & \{h_1^2 w_2^3, \gamma^2 w_2^2\} & h_1\gamma^2 w_2^2 = h_0^2 h_2 w_2^3 & h_0 h_2 w_2^3 & h_2 w_2^3 \\
\downarrow & \downarrow & \downarrow & \downarrow & \\
\gamma w_2^3 & h_1\gamma w_2^3 & \{h_1^2\gamma w_2^3 = h_0\alpha\beta w_2^3, \gamma^3 w_2^2\} & \alpha\beta w_2^3 & \\
\downarrow & \downarrow & \downarrow & \downarrow & \\
h_1 w_2^4 & \{h_1^2 w_2^4, \gamma^2 w_2^3\} & h_1\gamma^2 w_2^3 = h_0^2 h_2 w_2^4 & h_0 h_2 w_2^4 & h_2 w_2^4
\end{array}
$$

FIGURE 9.2. $\Delta'$ on $E_2$-classes, connecting the $\eta$- and $\nu$-families

the form $(3+4k, 11+28k)$, so that $w_2$-multiplication proves the cases $i > 0$ from the $i = 0$ case. $\square$

As with the $\eta$- and $\nu$-families, although iteration of $\Delta'$ produces the sequence

$$c_0 \mapsto \delta' \mapsto c_0 w_2 \mapsto \delta' w_2 \mapsto \cdots,$$

the process is interrupted in homotopy by the hidden $\nu$-extension $\nu\epsilon_1 = \nu_1 B \neq 0$, which we will show in the next section. Thus, in homotopy, we have only the elements $\epsilon \mapsto \epsilon_1$ and $\epsilon_4 \mapsto \epsilon_5$.

Finally, the classes detecting $\kappa$ and $\kappa_4$ are connected by multiplication by $w_2^2$, which justifies grouping these two elements into one family.

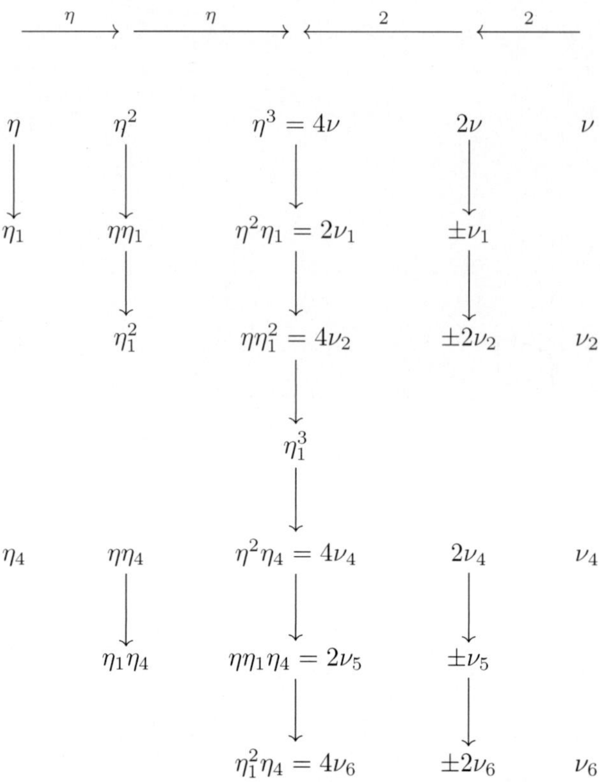


FIGURE 9.3. $\Delta'$ on homotopy classes, connecting the $\eta$- and $\nu$-families

## 9.2. Hidden extensions

In this section we determine all of the hidden 2-, $\nu$- and $\eta$-extensions, in turn. The results are displayed as dashed or dotted lines in Figures 9.6 through 9.13.

Isaksen [**82**, Def. 4.2] has given a precise clarification of the notion of a hidden extension in the Adams spectral sequence for a ring spectrum, such as *tmf*. The definition easily extends to the case of a pairing of spectra, such as the module action of *tmf* on *tmf*/2.

DEFINITION 9.5. Let $X \wedge Y \to Z$ be a pairing of spectra, with induced pairings $\pi_*(X) \otimes \pi_*(Y) \to \pi_*(Z)$ of homotopy groups and $E_\infty(X) \otimes E_\infty(Y) \to E_\infty(Z)$ of Adams $E_\infty$-terms. Let $\alpha \in \pi_*(X)$ be detected by $a \in E_\infty(X)$, and consider classes $b \in E_\infty(Y)$ and $c \in E_\infty(Z)$. We say that there is a hidden $\alpha$-extension from $b$ to $c$ if

(1) $ab = 0$,
(2) there is an element $\beta \in \pi_*(Y)$ detected by $b$ such that $\alpha\beta \in \pi_*(Z)$ is detected by $c$, and
(3) there is no element $\beta' \in \pi_*(Y)$ of higher Adams filtration than $\beta$ such that $\alpha\beta'$ is detected by $c$.

REMARK 9.6. If conditions (1) and (2) hold, but an element $\beta'$ exists that makes condition (3) fail, then we will say (in Chapter 12) that the $\alpha$-multiplication from $b$ (detecting $\beta$) to $c$ is eclipsed by the $\alpha$-multiplication from $b'$ (detecting $\beta'$) to $c$.

**9.2.1. Hidden 2-extensions.** We now determine the graded group structure of $\pi_*(tmf)$ using the structure of $i\colon E_\infty(tmf) \to E_\infty(tmf/2)$ obtained in Sections 5.5 and 6.5.

LEMMA 9.7. *We can (and do) choose $\eta_1$, $\epsilon_1$, $\eta_4$, $\epsilon_4$ and $\epsilon_5$ to have additive order* 2.

PROOF. We use the homotopy cofiber sequence

$$tmf \xrightarrow{2} tmf \xrightarrow{i} tmf/2 \xrightarrow{j} \Sigma tmf$$

and the associated short exact sequence

$$0 \to \pi_n(tmf)/2 \xrightarrow{i} \pi_n(tmf/2) \xrightarrow{j} {}_2\pi_{n-1}(tmf) \to 0\,, \tag{9.1}$$

where ${}_2A = \ker(2\colon A \to A)$ and $A/2 = \operatorname{cok}(2\colon A \to A)$ for any abelian group $A$. Consider classes $x$ in $\pi_*(tmf/2)$ that are detected by $\widetilde{\gamma}$, $\widetilde{\delta'}$, $\widetilde{w_2^2h_1}$, $w_2^2\widetilde{c_0}$ and $w_2^2\widetilde{\delta'}$ in $E_\infty(tmf/2)$. Their images $j(x) \in {}_2\pi_*(tmf)$ are of additive order 2, and are detected by $j(\widetilde{\gamma}) = \gamma$, $j(\widetilde{\delta'}) = \delta'$, $j(\widetilde{w_2^2h_1}) = h_1w_2^2$, $j(w_2^2\widetilde{c_0}) = c_0w_2^2$ and $j(w_2^2\widetilde{\delta'}) = \delta'w_2^2$, respectively. These images show that $\eta_1$, $\epsilon_1$, $\eta_4$, $\epsilon_4$ and $\epsilon_5$ can be chosen to have order 2. □

THEOREM 9.8. *The Adams spectral sequence for tmf contains precisely the following hidden 2-extensions. First we have six extensions that occur together with all their $w_1$- and $w_2^4$-power multiples:*

- (32) *From $h_0^2 \cdot \alpha g$ to $w_1 \cdot h_0\alpha^2$, with $4B_1 \in \{h_0^2\alpha g\}$ and $8B_1 \in \{h_0\alpha^2 w_1\}$.*
- (56) *From $c_0w_2$ to $\alpha^3 g + h_0w_1w_2$, with $B_2 \in \{c_0w_2\}$ and $2B_2 \in \{\alpha^3 g + h_0w_1w_2\}$.*
- (80) *From $h_0^2 \cdot \delta w_2$ to $w_1 \cdot h_0\alpha^2 w_2$, with $4B_3 \in \{h_0^2\alpha g w_2\}$ and $8B_3 \in \{h_0\alpha^2 w_1w_2\}$.*
- (128) *From $h_0^2 \cdot \alpha g w_2^2$ to $w_1 \cdot h_0\alpha^2 w_2^2$, detecting $4B_5 \in \{h_0^2\alpha g w_2^2\}$ and $8B_5 \in \{h_0\alpha^2 w_1 w_2^2\}$.*
- (152) *From $c_0w_2^3$ to $\alpha^3 g w_2^2 + h_0w_1w_2^3$, with $B_6 \in \{c_0w_2^3\}$ and $2B_6 \in \{\alpha^3 g w_2^2 + h_0w_1w_2^3\}$.*
- (176) *From $h_0^2 \cdot \delta w_2^3$ to $w_1 \cdot h_0\alpha^2 w_2^3$, detecting $4B_7 \in \{h_0^2\alpha g w_2^3\}$ and $8B_7 \in \{h_0\alpha^2 w_1 w_2^3\}$.*

*In addition we have seven extensions that only occur together with their $w_2^4$-power multiples:*

- (40) *From $g^2$ to $w_1 \cdot \delta'$, with $\bar\kappa^2 \in \{g^2\}$ and $2\bar\kappa^2 \in \{\delta' w_1\}$.*
- (54) *From $h_2 \cdot h_2 w_2$ to $d_0 g^2 = g^2 \cdot d_0$, with $\nu\nu_2 \in \{h_2^2 w_2\}$ and $2\nu\nu_2 \in \{d_0 g^2\}$.*
- (60) *From $g^3$ to $\delta' g w_1 = g w_1 \cdot \delta'$, with $\bar\kappa^3 \in \{g^3\}$ and $2\bar\kappa^3 \in \{\delta' g w_1\}$.*
- (110) *From $d_0 w_2^2$ to $\gamma^2 g^3 = g^3 \cdot \gamma^2$, with $\kappa_4 \in \{d_0 w_2^2\}$ and $2\kappa_4 \in \{\gamma^2 g^3\}$.*
- (130) *From $d_0 g w_2^2 = g \cdot d_0 w_2^2$ to $\gamma^2 g^4 = g^4 \cdot \gamma^2$, with $\kappa_4\bar\kappa \in \{d_0 g w_2^2\}$ and $2\kappa_4\bar\kappa \in \{\gamma^2 g^4\}$.*
- (150a) *From $h_2 \cdot h_2 w_2^3$ to $d_0 g^2 w_2^2 = g^2 \cdot d_0 w_2^2$, with $\nu\nu_6 \in \{h_2^2 w_2^3\}$ and $2\nu\nu_6 \in \{d_0 g^2 w_2^2\}$.*
- (150b) *From $g^2 \cdot d_0 w_2^2$ to $d_0\delta' w_1 w_2^2 = w_1 \cdot \alpha d_0 g w_2^2$, with $2\nu\nu_6$ and $\kappa_4\bar\kappa^2 \in \{d_0 g^2 w_2^2\}$ and $4\nu\nu_6 = 2\kappa_4\bar\kappa^2 \in \{d_0\delta' w_1 w_2^2\}$.*

PROOF. We start with the hidden 2-extensions between $w_1$-periodic classes.

(32) First we determine $\pi_{33}(tmf)$ from $E_\infty(tmf)$. From $2\eta_1 = 0$ (Lemma 9.7) we know that (the Adams filtration $\geq 5$ part of) $\pi_{25}(tmf)$ is isomorphic to $(\mathbb{Z}/2)^2$. Multiplying by $B$, and using the known action of $w_1$ on the $E_\infty$-term, it follows that the Adams filtration $\geq 9$ part of $\pi_{33}(tmf)$ is isomorphic to $(\mathbb{Z}/2)^2$. We can represent $h_1\delta$ by the $\eta$-multiple $\eta\epsilon_1$, so there is no hidden 2-extension from $h_1\delta$, since $2\eta = 0$. Hence $\pi_{33}(tmf) \cong (\mathbb{Z}/2)^3$.

From $E_\infty(tmf/2)$ we see that $\pi_{33}(tmf/2)$ has order $2^4$. Using the exact sequence (9.1) we deduce that ${}_2\pi_{32}(tmf) \cong \mathbb{Z}/2$. This shows that $\epsilon_1$ is the unique element of order 2 in $\pi_{32}(tmf)$.

Clearly $\pi_{31}(tmf) = 0$, so $\pi_{32}(tmf)/2 \cong \pi_{32}(tmf/2)$, which has order $2^3$. It follows that $\pi_{32}(tmf) \cong \mathbb{Z}^2 \oplus \mathbb{Z}/2$ (implicitly 2-completed), with $i\colon \mathbb{Z}^2 \oplus \mathbb{Z}/2 \to (\mathbb{Z}/2)^3$ surjective. The homotopy classes $B^4$, $B_1$ and $\epsilon_1$ in $\pi_{32}(tmf)$ map to homotopy classes in $\pi_{32}(tmf/2)$ detected by $i(w_1^4)$, $i(\alpha g) = i(\delta)+i(\delta')$ and $i(\delta')$. Since the latter three classes generate the $E_\infty$-term for $tmf/2$ in topological degree $t-s = 32$, the corresponding homotopy classes generate $\pi_{32}(tmf/2)$. It follows as a matter of algebra that $B^4$, $B_1$ and $\epsilon_1$ generate $\pi_{32}(tmf)$.

It also follows that there is a hidden 2-extension from $h_0^2\alpha g$ represented by $4B_1$ to $h_0\alpha^2 w_1$ represented by $8B_1$. To verify this, note that if $8B_1$ were not detected by $h_0\alpha^2 w_1$, then no linear combination of the three classes $B^4$, $B_1$ and $\epsilon_1$ would be detected by $h_0\alpha^2 w_1$, which contradicts the fact that these classes generate $\pi_{32}(tmf)$.

Regarding $w_1$- and $w_2^4$-power multiples, multiplication by $B^j M^\ell$ for $j \geq 0$ and $\ell \geq 0$ shows that there is a hidden 2-extension from $h_0^2\alpha g w_1^j w_2^{4\ell}$, represented by $4B_1 B^j M^\ell$, to $h_0\alpha^2 w_1^{1+j} w_2^{4\ell}$, represented by $8B_1 B^j M^\ell$. (This does not mean that every homotopy class detected by $h_0^2\alpha g w_1^j w_2^{4\ell}$ multiplies by 2 to be detected by $h_0\alpha^2 w_1^{1+j} w_2^{4\ell}$. For example, there are classes $\beta$ detected by $h_0^2\alpha g w_1^2$ for which $2\beta$ is detected by $h_0\alpha^2 w_1^3 + h_0^{11} w_2$.)

(56) We first calculate $\pi_{57}(tmf)$. Multiplying the Adams filtration $\geq 9$ part of $\pi_{33}(tmf)$ by $B^3$ shows that the Adams filtration $\geq 21$ part of $\pi_{57}$ is $(\mathbb{Z}/2)^2$. In Adams filtration 12 we can represent $\gamma\delta'$ and $h_1 c_0 w_2$ by the homotopy classes $\eta_1\epsilon_1$ and $\eta B_2$, both of which are of additive order 2. Hence $\pi_{57}(tmf) \cong (\mathbb{Z}/2)^4$.

We also know that $\pi_{57}(tmf/2)$ has order $2^4$. Using the exact sequence (9.1) we deduce that ${}_2\pi_{56}(tmf) = 0$, so that $\pi_{56}(tmf)$ is 2-torsion free.

Since $\pi_{55}(tmf) = 0$ and $\pi_{56}(tmf/2)$ has order $2^3$, it follows that $\pi_{56}(tmf) \cong \mathbb{Z}^3$, with $i\colon \mathbb{Z}^3 \to (\mathbb{Z}/2)^3$ surjective. The classes $B_2$, $B^3B_1$ and $B^7$ map to homotopy classes in $\pi_{56}(tmf/2)$ that are detected by the generators $i(c_0 w_2)$, $i(\alpha g w_1^3)$ and $i(w_1^7)$ of the $E_\infty$-term in this topological degree. It follows that $i(B_2)$, $i(B^3B_1)$ and $i(B^7)$ generate $\pi_{56}(tmf/2)$, and that $B_2$, $B^3B_1$ and $B^7$ generate $\pi_{56}(tmf)$.

This implies that there is a hidden 2-extension from $c_0 w_2$ detecting $B_2$ to $\alpha^3 g + h_0 w_1 w_2$, in addition to the previously known hidden 2-extension from $h_0^2\alpha g w_1^3$ to $h_0\alpha^2 w_1^4$. To see this, note that if $2B_2$ were not detected by $\alpha^3 g + h_0 w_1 w_2$, then no linear combination of $B_2$, $B^3B_1$ and $B^7$ would be detected by that class in the $E_\infty$-term for *tmf*.

As in the previous case, this hidden 2-extension propagates freely to all $w_1$- and $w_2^4$-power multiples.

(80) Multiplying (the Adams filtration $\geq 12$ part of) $\pi_{57}(tmf)$ by $B^3$, we see that the Adams filtration $\geq 24$ part of $\pi_{81}(tmf)$ is $(\mathbb{Z}/2)^3$. In Adams filtration 16 we can represent $h_1\delta w_2$ by $\eta B_3$, of additive order 2, so $\pi_{81}(tmf) \cong (\mathbb{Z}/2)^4$. Since $\pi_{81}(tmf/2)$ has order $2^5$, we deduce that ${}_2\pi_{80}(tmf) \cong \mathbb{Z}/2$. The element $\bar{\kappa}^4$ detected by $g^4$ has finite order, since $8\bar{\kappa} = 0$, hence is in fact the unique element of order 2 in $\pi_{80}(tmf)$.

Since $\pi_{79}(tmf) = 0$ and $\pi_{80}(tmf/2)$ has order $2^5$, it follows that $\pi_{80}(tmf) \cong \mathbb{Z}^4 \oplus \mathbb{Z}/2$, with $i\colon \mathbb{Z}^4 \oplus \mathbb{Z}/2 \to (\mathbb{Z}/2)^5$ surjective. The five homotopy classes $B_3$, $\bar{\kappa}^4$, $B^3B_2$, $B^6B_1$ and $B^{10}$ map to homotopy classes in $\pi_{80}(tmf/2)$ that are detected by $i(\delta w_2)$, $i(g^4)$, $i(c_0w_1^3w_2)$, $i(\alpha g w_1^6)$ and $i(w_1^{10})$, and which therefore generate this group. It follows that the given five classes in $\pi_{80}(tmf)$ generate that group. Hence there must be a hidden 2-extension from $h_0^2\alpha g w_2$ detecting $4B_3$ to $h_0\alpha^2 w_1 w_2$ detecting $8B_3$. It propagates freely to all $w_1$- and $w_2^4$-power multiples.

(128) This case is similar to the case $n = 32$. Multiplying $\pi_{81}(tmf)$ by $B^3$ we see that the Adams filtration $\geq 28$ part of $\pi_{105}(tmf)$ is $(\mathbb{Z}/2)^4$. We can represent $h_1w_1w_2^2$ and $\gamma g^4$ in Adams filtration 21 by the homotopy classes $\eta B_4$ and $\eta_1\bar{\kappa}^4$, both of which have order 2. Hence Adams filtration $\geq 21$ of $\pi_{105}(tmf)$ is $(\mathbb{Z}/2)^6$.

Multiplying by $B^3$ once more, we see that Adams filtration $\geq 33$ of $\pi_{129}(tmf)$ is $(\mathbb{Z}/2)^5$. In Adams filtration 24 and 25 we can represent $h_1 \cdot \delta' w_2^2$ and $\gamma w_1 w_2^2$ by the homotopy classes $\eta\epsilon_5$ and $\eta_1 B_4$, both of which have order 2. Hence $\pi_{129}(tmf) \cong (\mathbb{Z}/2)^7$. Since $\pi_{129}(tmf/2)$ has order $2^8$, we obtain ${}_2\pi_{128}(tmf) \cong \mathbb{Z}/2$. This shows that $\epsilon_5$ is the unique element of order 2 in $\pi_{128}(tmf)$.

Since $\pi_{127}(tmf) = 0$ and $\pi_{128}(tmf/2)$ has order $2^7$, we must have $\pi_{128}(tmf) \cong \mathbb{Z}^6 \oplus \mathbb{Z}/2$, with $i\colon \mathbb{Z}^6 \oplus \mathbb{Z}/2 \to (\mathbb{Z}/2)^7$ surjective. The seven homotopy classes $B_5$, $\epsilon_5$, $B^3B_4$, $B^6B_3$, $B^9B_2$, $B^{12}B_1$ and $B^{16}$ map to homotopy classes in $\pi_{128}(tmf/2)$ that are detected by $i(\alpha g w_2^2)$, $i(\delta' w_2^2)$, $i(w_1^4 w_2^2)$, $i(\delta w_1^6 w_2)$, $i(c_0 w_1^9 w_2)$, $i(\alpha g w_1^{12})$ and $i(w_1^{16})$, and which therefore generate this group. It follows that the given seven homotopy classes in $\pi_{128}(tmf)$ generate that group. As before, this implies that there must be a hidden 2-extension from $h_0^2\alpha g w_2^2$ detecting $4B_5$ to $h_0\alpha^2 w_1 w_2^2$ detecting $8B_5$. It propagates freely to all $w_1$- and $w_2^4$-power multiples.

(152) This case is very similar to the case $n = 56$. Multiplying by $B^3$ shows that Adams filtration $\geq 37$ of $\pi_{153}(tmf)$ is $(\mathbb{Z}/2)^6$. Since $h_1c_0w_2^3$ and $\gamma\delta' w_2^2$ in Adams filtration 28 are represented by $\eta B_6$ and $\eta_1\epsilon_5$, both of order 2, it follows that $\pi_{153}(tmf) \cong (\mathbb{Z}/2)^8$. Since $\pi_{153}(tmf/2)$ has order $2^8$, we deduce that $\pi_{152}(tmf)$ is 2-torsion free.

Since $\pi_{151}(tmf) = 0$ and $\pi_{152}(tmf/2)$ has order $2^7$, we must have $\pi_{152}(tmf) \cong \mathbb{Z}^7$. The seven homotopy classes $B_6$, $B^3B_5$, $B^6B_4$, $B^9B_3$, $B^{12}B_2$, $B^{15}B_1$ and $B^{19}$ map to classes that generate $\pi_{152}(tmf/2)$, because they are detected by $i(c_0w_2^3)$, $i(\alpha g w_1^3 w_2^2)$, $i(w_1^7 w_2^2)$, $i(\delta w_1^9 w_2)$, $i(c_0 w_1^{12} w_2)$, $i(\alpha g w_1^{15})$ and $i(w_1^{19})$, which generate the $E_\infty$-term in topological degree $t-s = 152$. Hence the seven homotopy classes generate $\pi_{152}(tmf)$, and there must be a hidden 2-extension from $c_0 w_2^3$ detecting $B_6$ to $\alpha^3 g w_2^2 + h_0 w_1 w_2^3$ detecting $2B_6$. It propagates freely to all $w_1$- and $w_2^4$-power multiples.

(176) This case is similar to that for $n = 80$. Multiplication by $B^3$ tells us that the Adams filtration $\geq 40$ part of $\pi_{177}(tmf)$ is $(\mathbb{Z}/2)^7$. The class $h_1 \delta w_2^3$ in Adams filtration 32 is represented by $\eta B_7$, of order 2, so $\pi_{177}(tmf) \cong (\mathbb{Z}/2)^8$. Since $\pi_{177}(tmf/2)$ has order $2^8$, it follows that $\pi_{176}(tmf)$ is 2-torsion free.

From $\pi_{175}(tmf) = 0$ and $\pi_{176}(tmf/2)$ having order $2^8$ we obtain $\pi_{176}(tmf) \cong \mathbb{Z}^8$. The eight homotopy classes $B_7$, $B^3B_6$, $B^6B_5$, $B^9B_4$, $B^{12}B_3$, $B^{15}B_2$, $B^{18}B_1$ and $B^{22}$ map to classes generating $\pi_{176}(tmf/2)$, since the detecting classes generate the $E_\infty$-term for $tmf/2$ for $t-s = 176$. Hence these eight classes generate $\pi_{176}(tmf)$, and there must be a hidden 2-extension from $h_0^2 \alpha g w_2^3$ detecting $4B_7$ to $h_0 \alpha^2 w_1 w_2^3$ detecting $8B_7$. It propagates freely to all $w_1$- and $w_2^4$-power multiples.

We now turn to the hidden 2-extensions between $w_1$-power torsion classes.

(54) From $E_\infty(tmf)$ we see that $\pi_{54}(tmf)$ has order $2^2 = 4$ and $\pi_{55}(tmf) = 0$, and from $E_\infty(tmf/2)$ we see that $\pi_{55}(tmf/2) \cong \mathbb{Z}/2$. Using (9.1) we deduce that $\pi_{54}(tmf) \cong \mathbb{Z}/4$. This group is generated by $\nu\nu_2$, detected by $h_2^2 w_2$, hence must encompass a hidden 2-extension to $d_0 g^2$, detecting $2\nu\nu_2 = \kappa\bar{\kappa}^2$. Regarding $w_2^4$-power multiples, multiplication by $M^\ell$ shows that there are hidden 2-extensions from $h_2^2 w_2^{1+4\ell}$ to $d_0 g^2 w_2^{4\ell}$, for all $\ell \geq 0$.

(150) From $E_\infty(tmf)$ we see that $\pi_{150}(tmf)$ has order $2^3 = 8$ and $\pi_{151}(tmf) = 0$, and from $E_\infty(tmf/2)$ we see that $\pi_{151}(tmf/2) \cong \mathbb{Z}/2$. Using (9.1) we deduce that $\pi_{150}(tmf) \cong \mathbb{Z}/8$. This group is generated by $\nu\nu_6$, detected by $h_2^2 w_2^3$, with a hidden 2-extension to $d_0 g^2 w_2^2$, and a second hidden 2-extension to $d_0 \delta' w_1 w_2^2$. Multiplication by $M^\ell$ for $\ell \geq 0$ shows that there are hidden 2-extensions from $h_2^2 w_2^{3+4\ell}$ to $d_0 g^2 w_2^{2+4\ell}$, and from $d_0 g^2 w_2^{2+4\ell}$ to $d_0 \delta' w_1 w_2^{2+4\ell}$.

(110) To show that there is a hidden 2-extension from $d_0 w_2^2$ to $\gamma^2 g^3$, we show that $2\kappa_4 \neq 0$. Multiplication by $\bar{\kappa}^2$ takes $\kappa_4$ to $\bar{\kappa}^2\kappa_4$, which is detected by $d_0 g^2 w_2^2$. From the case $n = 150$, we know that $2\bar{\kappa}^2\kappa_4 \neq 0$ is detected by $d_0 \delta' w_1 w_2^2$. Hence $2\kappa_4 \neq 0$, and the only $E_\infty$-class that can detect it is $\gamma^2 g^3$. This hidden 2-extension propagates freely to all $w_2^4$-power multiples.

(130) Multiplying the hidden 2-extension for $n = 110$ by $\bar{\kappa}$ we obtain a hidden 2-extension from $d_0 g w_2^2$ to $\gamma^2 g^4$, as asserted. It propagates freely to all $w_2^4$-power multiples.

(40) To show that there is a hidden 2-extension from $g^2$ to $\delta' w_1$, we show that $2\bar{\kappa}^2 \neq 0$. Multiplication by $\kappa_4$ takes $\bar{\kappa}^2$ to $\bar{\kappa}^2\kappa_4$, which is detected by $d_0 g^2 w_2^2$. From the case $n = 150$, we know that $2\bar{\kappa}^2\kappa_4 \neq 0$ is detected by $d_0 \delta' w_1 w_2^2$. Hence $2\bar{\kappa}^2 \neq 0$. Being a 2-power torsion class, $2\bar{\kappa}^2$ can only be detected by $\delta' w_1$, which establishes the claimed hidden 2-extension. It propagates freely to all $w_2^4$-power multiples.

(60) Multiplying the hidden 2-extension for $n = 40$ by $\bar{\kappa}$ we obtain a hidden 2-extension from $g^3$ to $\delta' g w_1$, as asserted. It propagates freely to all $w_2^4$-power multiples.

To finish the proof, we must check that there are no further hidden 2-extensions for *tmf* than those already mentioned. In all cases, this is easily seen by representing the possible source $b$ of a hidden 2-extension by a homotopy class $\beta$ that is known to be 2-torsion. For instance, in bidegree $(t-s,s)=(28,8)$ the class $gw_1=d_0^2$ is represented by $\kappa^2$, and $2\kappa=0$. As another example, in bidegree $(t-s,s)=(68,14)$ the class $h_0^2gw_2$ detects $\bar{\kappa}D_2$, and $8\bar{\kappa}=0$, so $2\bar{\kappa}D_2$ must be 2-power torsion. However, $\pi_{68}(tmf)$ is 2-torsion free above Adams filtration 14, so $2\bar{\kappa}D_2=0$. □

**9.2.2. The Bott torsion.** Recall the Bott element $B\in\pi_8(tmf)$, with $B\in\{w_1\}$. We now make precise how $w_1$-power torsion classes in $E_\infty(tmf)$ detect $B$-power torsion classes in $\pi_*(tmf)$.

LEMMA 9.9. *When viewed as an $\mathbb{F}_2[w_1,w_2^4]$-module, $E_\infty(tmf)$ splits as a direct sum of cyclic modules with annihilator ideals $(0)$, $(w_1)$ or $(w_1^2)$. All $w_1$-power torsion is $w_1$- or $w_1^2$-torsion.*

PROOF. The $\mathbb{F}_2[w_1,w_2^4]$-module structure on $E_\infty(tmf)$ is obtained by restriction from the $R_2$-module structure given in Tables 5.8 and 5.9. Examination of the annihilator ideals and the non-cyclic summands shows that its $w_1$-torsion free quotient is free as an $\mathbb{F}_2[w_1,w_2^4]$-module. Furthermore, the $w_1$-power torsion submodule of $E_\infty(tmf)$ splits as a sum of cyclic modules with annihilator ideals $(w_1)$ or $(w_1^2)$, as listed in Table 9.3. □

Table 9.3: $w_1$-power torsion in $E_\infty(tmf)$

| $t-s$ | $s$ | $g$ | $x$ | Ann$(x)$ | rep. |
|---|---|---|---|---|---|
| 3 | 1 | 2 | $h_2$ | $(w_1)$ | $\nu$ |
| 3 | 2 | 2 | $h_0h_2$ | $(w_1)$ | $2\nu$ |
| 3 | 3 | 1 | $h_0^2h_2$ | $(w_1)$ | $4\nu$ |
| 6 | 2 | 3 | $h_2^2$ | $(w_1)$ | $\nu^2$ |
| 8 | 3 | 2 | $c_0$ | $(w_1)$ | $\epsilon$ |
| 9 | 4 | 2 | $h_1c_0$ | $(w_1)$ | $\eta\epsilon$ |
| 14 | 4 | 4 | $d_0$ | $(w_1^2)$ | $\kappa$ |
| 15 | 5 | 6 | $h_1d_0$ | $(w_1)$ | $\eta\kappa$ |
| 17 | 5 | 7 | $h_2d_0$ | $(w_1)$ | $\nu\kappa$ |
| 20 | 4 | 8 | $g$ | $(w_1^2)$ | $\bar{\kappa}$ |
| 20 | 5 | 9 | $h_0g$ | $(w_1)$ | $2\bar{\kappa}$ |
| 20 | 6 | 7 | $h_0^2g$ | $(w_1)$ | $4\bar{\kappa}$ |
| 21 | 5 | 10 | $h_1g$ | $(w_1)$ | $\eta\bar{\kappa}$ |
| 27 | 6 | 10 | $\alpha\beta$ | $(w_1^2)$ | $\nu_1$ |
| 27 | 7 | 9 | $h_0\alpha\beta$ | $(w_1)$ | $2\nu_1$ |
| 32 | 7 | 12 | $\delta'$ | $(w_1^2)$ | $\epsilon_1$ |
| 33 | 8 | 15 | $h_1\delta$ | $(w_1)$ | $\eta\epsilon_1$ |
| 34 | 8 | 16 | $d_0g$ | $(w_1^2)$ | $\kappa\bar{\kappa}$ |
| 39 | 9 | 18 | $d_0\gamma$ | $(w_1)$ | $\eta_1\kappa$ |

Table 9.3: $w_1$-power torsion in $E_\infty(tmf)$ (cont.)

| $t-s$ | $s$ | $g$ | $x$ | $\mathrm{Ann}(x)$ | rep. |
|---|---|---|---|---|---|
| 40 | 8 | 18 | $g^2$ | $(w_1)$ | $\bar{\kappa}^2$ |
| 41 | 10 | 16 | $\alpha\beta d_0$ | $(w_1)$ | $\nu_1\kappa$ |
| 45 | 9 | 20 | $\gamma g$ | $(w_1^2)$ | $\eta_1\bar{\kappa}$ |
| 46 | 11 | 18 | $d_0\delta'$ | $(w_1)$ | $\epsilon_1\kappa$ |
| 51 | 9 | 23 | $h_2w_2$ | $(w_1^2)$ | $\nu_2$ |
| 51 | 10 | 22 | $h_0h_2w_2$ | $(w_1)$ | $2\nu_2$ |
| 51 | 11 | 21 | $h_0^2h_2w_2$ | $(w_1)$ | $4\nu_2$ |
| 52 | 11 | 22 | $\delta' g$ | $(w_1^2)$ | $\epsilon_1\bar{\kappa}$ |
| 54 | 10 | 23 | $h_2^2w_2$ | $(w_1)$ | $\nu\nu_2$ |
| 54 | 12 | 25 | $d_0g^2$ | $(w_1)$ | $\kappa\bar{\kappa}^2$ |
| 57 | 12 | $27+28$ | $\gamma\delta'$ | $(w_1)$ | $\eta_1\epsilon_1$ |
| 60 | 12 | 29 | $g^3$ | $(w_1)$ | $\bar{\kappa}^3$ |
| 65 | 13 | 35 | $\gamma g^2$ | $(w_1)$ | $\eta_1\bar{\kappa}^2$ |
| 65 | 13 | 36 | $h_2d_0w_2$ | $(w_1)$ | $\nu_2\kappa$ |
| 66 | 15 | 31 | $d_0\delta' g$ | $(w_1)$ | $\epsilon_1\kappa\bar{\kappa}$ |
| 68 | 14 | 34 | $h_2^2d_0w_2$ | $(w_1)$ | $\nu\nu_2\kappa$ |
| 70 | 14 | 35 | $\gamma^2 g$ | $(w_1)$ | $\eta_1^2\bar{\kappa}$ |
| 75 | 15 | $38+39$ | $\gamma^3$ | $(w_1)$ | $\eta_1^3=\nu_3$ |
| 80 | 16 | 48 | $g^4$ | $(w_1)$ | $\bar{\kappa}^4$ |
| 85 | 17 | 54 | $\gamma g^3$ | $(w_1)$ | $\eta_1\bar{\kappa}^3$ |
| 90 | 18 | 52 | $\gamma^2g^2$ | $(w_1)$ | $\eta_1^2\bar{\kappa}^2$ |
| 99 | 17 | 60 | $h_2w_2^2$ | $(w_1)$ | $\nu_4$ |
| 99 | 18 | 58 | $h_0h_2w_2^2$ | $(w_1)$ | $2\nu_4$ |
| 99 | 19 | 59 | $h_0^2h_2w_2^2$ | $(w_1)$ | $4\nu_4$ |
| 100 | 20 | 67 | $g^5$ | $(w_1)$ | $\bar{\kappa}^5$ |
| 102 | 18 | 59 | $h_2^2w_2^2$ | $(w_1)$ | $\nu\nu_4$ |
| 104 | 19 | 62 | $c_0w_2^2$ | $(w_1)$ | $\epsilon_4$ |
| 105 | 20 | 71 | $h_1c_0w_2^2$ | $(w_1)$ | $\eta\epsilon_4$ |
| 105 | 21 | 72 | $\gamma g^4$ | $(w_1)$ | $\eta_1\bar{\kappa}^4$ |
| 110 | 20 | 74 | $d_0w_2^2$ | $(w_1^2)$ | $\kappa_4$ |
| 110 | 22 | 73 | $\gamma^2g^3$ | $(w_1)$ | $\eta_1^2\bar{\kappa}^3$ |
| 111 | 21 | 79 | $h_1d_0w_2^2$ | $(w_1)$ | $\eta\kappa_4$ |

Table 9.3: $w_1$-power torsion in $E_\infty(tmf)$ (cont.)

| $t-s$ | $s$ | $g$ | $x$ | $\mathrm{Ann}(x)$ | rep. |
|---|---|---|---|---|---|
| 113 | 21 | 81 | $h_2d_0w_2^2$ | $(w_1)$ | $\nu\kappa_4$ |
| 116 | 21 | 83 | $h_0gw_2^2$ | $(w_1)$ | $\bar\kappa D_4$ |
| 116 | 22 | 78 | $h_0^2gw_2^2$ | $(w_1)$ | $2\bar\kappa D_4$ |
| 117 | 21 | 84 | $h_1gw_2^2$ | $(w_1^2)$ | $\eta_4\bar\kappa$ |
| 123 | 22 | 82 | $\alpha\beta w_2^2$ | $(w_1^2)$ | $\nu_5$ |
| 123 | 23 | 85 | $h_0\alpha\beta w_2^2$ | $(w_1)$ | $2\nu_5$ |
| 124 | 24 | 95 | $gw_1w_2^2$ | $(w_1)$ | $\kappa\kappa_4$ |
| 128 | 23 | 88 | $\delta' w_2^2$ | $(w_1^2)$ | $\epsilon_5$ |
| 129 | 24 | 101 | $h_1\delta w_2^2$ | $(w_1)$ | $\eta\epsilon_5$ |
| 130 | 24 | 102 | $d_0gw_2^2$ | $(w_1^2)$ | $\kappa_4\bar\kappa$ |
| 130 | 26 | 96 | $\gamma^2g^4$ | $(w_1)$ | $\eta_1^2\bar\kappa^4$ |
| 135 | 25 | 108 | $d_0\gamma w_2^2$ | $(w_1)$ | $\eta_1\kappa_4$ |
| 137 | 26 | 103 | $\alpha\beta d_0w_2^2$ | $(w_1)$ | $\nu_5\kappa$ |
| 142 | 27 | 109 | $d_0\delta' w_2^2$ | $(w_1^2)$ | $\epsilon_5\kappa$ |
| 147 | 25 | 113 | $h_2w_2^3$ | $(w_1^2)$ | $\nu_6$ |
| 147 | 26 | 110 | $h_0h_2w_2^3$ | $(w_1)$ | $2\nu_6$ |
| 147 | 27 | 113 | $h_0^2h_2w_2^3$ | $(w_1)$ | $4\nu_6$ |
| 148 | 27 | 114 | $\delta' gw_2^2$ | $(w_1^2)$ | $\epsilon_5\bar\kappa$ |
| 149 | 29 | 129 | $\gamma gw_1w_2^2$ | $(w_1)$ | $\eta_1\kappa\kappa_4$ |
| 150 | 26 | 111 | $h_2^2w_2^3$ | $(w_1)$ | $\nu\nu_6$ |
| 150 | 28 | 127 | $d_0g^2w_2^2$ | $(w_1)$ | $\kappa_4\bar\kappa^2$ |
| 153 | 28 | $129+130$ | $\gamma\delta' w_2^2$ | $(w_1)$ | $\eta_1\epsilon_5$ |
| 161 | 29 | 142 | $h_2d_0w_2^3$ | $(w_1)$ | $\nu_6\kappa$ |
| 162 | 31 | 138 | $d_0\delta' gw_2^2$ | $(w_1)$ | $\epsilon_5\kappa\bar\kappa$ |
| 164 | 30 | 138 | $h_2^2d_0w_2^3$ | $(w_1)$ | $\nu\nu_6\kappa$ |

Consider the $B$-power torsion submodule $\Gamma_B\pi_*(tmf) \subset \pi_*(tmf)$. The Adams filtration $F^s\pi_*(tmf)$ of $\pi_*(tmf)$ restricts to a filtration

$$F^s\Gamma_B\pi_*(tmf) = F^s\pi_*(tmf) \cap \Gamma_B\pi_*(tmf)$$

of $\Gamma_B\pi_*(tmf)$, and the filtration subquotient

$$\frac{F^s\Gamma_B\pi_*(tmf)}{F^{s+1}\Gamma_B\pi_*(tmf)} \subset \frac{F^s\pi_*(tmf)}{F^{s+1}\pi_*(tmf)}$$

corresponds under the isomorphism $F^s\pi_*(tmf)/F^{s+1}\pi_*(tmf) \cong E_\infty^{s,*}(tmf)$ to the classes that can be represented by $B$-power torsion elements. These classes are all $w_1$-power torsion. In the case of *tmf* the converse holds, so that the $w_1$-power torsion in $E_\infty(tmf)$ is precisely the associated graded of the restriction of the Adams filtration to $\Gamma_B\pi_*(tmf)$.

PROPOSITION 9.10. *Each $w_1$-torsion class in $E_\infty(tmf)$ is represented by a $B$-torsion class in $\pi_*(tmf)$, each $w_1^2$-torsion class is represented by a $B^2$-torsion class, and $E_\infty(tmf)$ is $w_2^4$-torsion free. Hence there are no hidden $B$- or $M$-extensions in the Adams spectral sequence for tmf.*

PROOF. Each class $b$ in the $x$-column of Table 9.3 is represented by a class $\beta \in \pi_*(tmf)$, as listed in the "rep."-column. We claim that if $w_1^k b = 0$ then $B^k\beta = 0$, for $k \in \{1, 2\}$.

In view of the multiplicative structure, it suffices to verify that $B$ annihilates $\beta = \nu$, $\epsilon$, $\eta\kappa$, $2\bar{\kappa}$, $\eta\bar{\kappa}$, $2\nu_1$, $\eta\epsilon_1$, $\eta_1\kappa$, $\bar{\kappa}^2$, $\nu_1\kappa$, $\epsilon_1\kappa$, $2\nu_2$, $\eta_1\epsilon_1$, $\nu_2\kappa$, $\eta_1^2\bar{\kappa}$, $\eta_1^3 = \nu_3$, $\nu_4$, $\epsilon_4$, $\eta\kappa_4$, $\bar{\kappa}D_4$, $2\nu_5$, $\kappa\kappa_4$, $\eta\epsilon_5$, $\eta_1\kappa_4$, $\nu_5\kappa$, $2\nu_6$, $\eta_1\epsilon_5$, $\nu_6\kappa$ and $\epsilon_5\kappa\bar{\kappa}$, and that $B^2$ annihilates $\beta = \kappa$, $\bar{\kappa}$, $\nu_1$, $\epsilon_1$, $\nu_2$, $\kappa_4$, $\nu_5$, $\epsilon_5$ and $\nu_6$. In most cases, this holds because $B^k\beta$ lies in a trivial Adams filtration. This is most easily seen from Figures 5.1 to 5.8 or Figures 9.6 through 9.13.

For $\beta = \epsilon$, $2\bar{\kappa}$, $\bar{\kappa}^2$, $\epsilon_4$, $\bar{\kappa}D_4$, $\kappa\kappa_4$, $\bar{\kappa}$, $\epsilon_1$ and $\epsilon_5$ it holds because $\beta$ is 2-power torsion (by our choices in Lemma 9.7), and $B^k\beta$ lies in a 2-torsion free Adams filtration.

In the remaining cases, $\beta = \eta\epsilon_1$, $\nu_1\kappa$, $\eta_1\epsilon_1$, $\nu_2\kappa$, $\eta\epsilon_5$, $\nu_5\kappa$, $\eta_1\epsilon_5$, $\nu_6\kappa$ and $\epsilon_5\kappa\bar{\kappa}$, it holds because $\beta$ is $B$-power torsion (by what we have already established for $\kappa$, $\epsilon_1$ and $\epsilon_5$), and $B\beta$ lies in a $B$-torsion free Adams filtration (because $w_1$ acts injectively on that part of the $E_\infty$-term). □

In the course of the previous proof, we also established the following lemma.

LEMMA 9.11. *The classes $\nu_k$, $\epsilon_k$, $\kappa_k$ and $\bar{\kappa}$ are $B$-power torsion. The minimal power of $B$ annihilating each is as follows:*

(1) $B \cdot \nu_k = 0$ *for* $k \in \{0, 3, 4\}$ *and* $B^2 \cdot \nu_k = 0$ *for* $k \in \{1, 2, 5, 6\}$.
(2) $B \cdot \epsilon_k = 0$ *for* $k \in \{0, 4\}$ *and* $B^2 \cdot \epsilon_k = 0$ *for* $k \in \{1, 5\}$.
(3) $B^2 \cdot \kappa = 0$ *and* $B^2 \cdot \kappa_4 = 0$.
(4) $B^2 \cdot \bar{\kappa} = 0$.

□

PROPOSITION 9.12. *The $B$-power torsion in $\pi_*(tmf)$ is the ideal*

$$\Gamma_B\pi_*(tmf) = (\nu_k, \epsilon_k, \kappa_k, \bar{\kappa})$$

*generated by the $\nu$-, $\epsilon$-, $\kappa$- and $\bar{\kappa}$-families, including $\nu_3 = \eta_1^3$. It is contained in the 2-power torsion ideal*

$$\Gamma_2\pi_*(tmf) = (\eta_k, \nu_k, \epsilon_k, \kappa_k, \bar{\kappa})$$

*generated by the $\eta$-, $\nu$-, $\epsilon$-, $\kappa$- and $\bar{\kappa}$-families.*

PROOF. The first claim follows from the "rep."-column of Table 9.3 and the previous lemma. The second claim follows because the $\nu$-, $\epsilon$-, $\kappa$- and $\bar{\kappa}$-families consist of 2-power torsion, and the remaining ($B$-periodic) 2-power torsion consists of multiples of $\eta$, $\eta_1$ and $\eta_4$. For example, the infinite cycles $\gamma w_1 w_2^2$ and $\gamma^2 w_1 w_2^2$ in bidegrees $(t-s, s) = (129, 25)$ and $(154, 30)$ detect $\eta_1 B_4$ and $\eta_1^2 B_4$, respectively. □

**9.2.3. Hidden $\nu$-extensions.** We proceed to determine the action of $\nu$ on $\pi_*(tmf)$ using the structure of $i\colon E_\infty(tmf) \to E_\infty(tmf/\nu)$ obtained in Sections 5.5 and 8.5.

LEMMA 9.13. *The relations $\nu \cdot (B_1 + \epsilon_1) = 0$, $\nu \cdot B_2 = 0$, $\nu \cdot (B_5 + \epsilon_5) = 0$ and $\nu \cdot B_6 = 0$ hold in $\pi_*(tmf)$.*

PROOF. We use the homotopy cofiber sequence

$$\Sigma^3 tmf \xrightarrow{\nu} tmf \xrightarrow{i} tmf/\nu \xrightarrow{j} \Sigma^4 tmf$$

and the associated short exact sequence

$$0 \to \pi_n(tmf)/\nu \xrightarrow{i} \pi_n(tmf/\nu) \xrightarrow{j} {}_\nu\pi_{n-4}(tmf) \to 0\,, \tag{9.2}$$

where ${}_\nu A = \ker(\nu\colon A \to \Sigma^{-3}A)$ and $A/\nu = \operatorname{cok}(\nu\colon \Sigma^3 A \to A)$. Consider classes $x \in \pi_*(tmf/\nu)$ detected by $\overline{\delta}$, $w_2\overline{c_0}$, $w_2^2\overline{\delta}$ and $w_2^3\overline{c_0}$ in $E_\infty(tmf/\nu)$. Their images $j(x) \in {}_\nu\pi_*(tmf)$ are detected by $\delta$, $c_0w_2$, $\delta w_2^2$ and $c_0w_2^3$, respectively.

In degree 32 we note that $\{\delta\}$ is the set of 2-adic units times $B_1 + \epsilon_1$, plus Adams filtration $\geq 16$. For filtration reasons, $\nu$ annihilates Adams filtration $\geq 16$ in $\pi_{32}(tmf)$, so from $\nu \cdot j(x) = 0$ for some $j(x) \in \{\delta\}$ we can deduce that $\nu$ annihilates $B_1 + \epsilon_1$.

In degree 56 we observe that $\{c_0w_2\}$ is the set of 2-adic units times $B_2$, plus Adams filtration $\geq 19$. For filtration reasons, $\nu$ annihilates Adams filtration $\geq 19$ in $\pi_{56}(tmf)$, so from $\nu \cdot j(x) = 0$ for some $j(x) \in \{c_0w_2\}$ we can deduce that $\nu$ annihilates $B_2$.

The proofs for $B_5 + \epsilon_5$ and $B_6$ are very similar. □

THEOREM 9.14. *The Adams spectral sequence for tmf contains precisely the following* 19 *hidden $\nu$-extensions, together with their $w_2^4$-power multiples.*

(6) *From $h_2^2$ to $h_1c_0$, with $\nu^2 \in \{h_2^2\}$ and $\nu^3 \in \{h_1c_0\}$.*
(25) *From $\gamma$ to $gw_1$, with $\eta_1 \in \{\gamma\}$ and $\eta_1\nu \in \{gw_1\}$.*
(32) *From $\delta'$ to $\alpha\beta w_1$, with $\epsilon_1 \in \{\delta'\}$ and $\nu\epsilon_1 \in \{\alpha\beta w_1\}$.*
(32) *From $\alpha g$ to $\alpha\beta w_1$, with $B_1 \in \{\alpha g\}$ and $\nu B_1 \in \{\alpha\beta w_1\}$.*
(39) *From $d_0\gamma$ to $d_0gw_1$, with $\eta_1\kappa \in \{d_0\gamma\}$ and $\eta_1\nu\kappa \in \{d_0gw_1\}$.*
(50) *From $\gamma^2$ to $\gamma gw_1$, with $\eta_1^2 \in \{\gamma^2\}$ and $\eta_1^2\nu \in \{\gamma gw_1\}$.*
(51) *From $h_0h_2w_2$ to $d_0g^2$, with $2\nu_2 \in \{h_0h_2w_2\}$ and $2\nu\nu_2 \in \{d_0g^2\}$.*
(54) *From $h_2^2w_2$ to $\gamma\delta'$, with $\nu\nu_2 \in \{h_2^2w_2\}$ and $\nu^2\nu_2 \in \{\gamma\delta'\}$.*
(57) *From $\gamma\delta'$ to $\delta' gw_1$, with $\nu^2\nu_2 = \eta_1\epsilon_1 \in \{\gamma\delta'\}$ and $\nu^3\nu_2 = \eta_1\nu\epsilon_1 \in \{\delta' gw_1\}$.*
(97) *From $h_1w_2^2$ to $g^5$, with $\eta_4 \in \{h_1w_2^2\}$ and $\eta_4\nu \in \{g^5\}$.*
(102) *From $h_2^2w_2^2$ to $h_1c_0w_2^2$, with $\nu\nu_4 \in \{h_2^2w_2^2\}$ and $\nu^2\nu_4 \in \{h_1c_0w_2^2\}$.*
(122) *From $h_1\gamma w_2^2$ to $h_1gw_1w_2^2$, with $\eta_1\eta_4 \in \{h_1\gamma w_2^2\}$ and $\eta_1\eta_4\nu \in \{h_1gw_1w_2^2\}$.*
(128) *From $\delta' w_2^2$ to $\alpha\beta w_1w_2^2$, with $\epsilon_5 \in \{\delta' w_2^2\}$ and $\nu\epsilon_5 \in \{\alpha\beta w_1w_2^2\}$.*
(128) *From $\alpha gw_2^2$ to $\alpha\beta w_1w_2^2$, with $B_5 \in \{\alpha gw_2^2\}$ and $\nu B_5 \in \{\alpha\beta w_1w_2^2\}$.*
(135) *From $d_0\gamma w_2^2$ to $d_0gw_1w_2^2$, with $\eta_1\kappa_4 \in \{d_0\gamma w_2^2\}$ and $\eta_1\nu\kappa_4 \in \{d_0gw_1w_2^2\}$.*
(147a) *From $h_0h_2w_2^3$ to $d_0g^2w_2^2$, with $2\nu_6 \in \{h_0h_2w_2^3\}$ and $2\nu\nu_6 \in \{d_0g^2w_2^2\}$.*
(147b) *From $h_0^2h_2w_2^3$ to $d_0\delta' w_1w_2^2$, with $4\nu_6 \in \{h_0^2h_2w_2^3\}$ and $4\nu\nu_6 \in \{d_0\delta' w_1w_2^2\}$.*
(150) *From $h_2^2w_2^3$ to $\gamma\delta' w_2^2$, with $\nu\nu_6 \in \{h_2^2w_2^3\}$ and $\nu^2\nu_6 \in \{\gamma\delta' w_2^2\}$.*
(153) *From $\gamma\delta' w_2^2$ to $\delta' gw_1w_2^2$, with $\nu^2\nu_6 = \eta_1\epsilon_5 \in \{\gamma\delta' w_2^2\}$ and $\nu^3\nu_6 = \eta_1\nu\epsilon_5 \in \{\delta' gw_1w_2^2\}$.*

REMARK 9.15. In Proposition 9.17 we will refine the hidden extension $\nu^2\nu_4 \in \{h_1c_0w_2^2\}$ to the relation $\nu^2\nu_4 = \eta\epsilon_4 + \eta_1\bar{\kappa}^4$.

PROOF. We first determine the hidden $\nu$-extensions from $n = 51$ and $n = 147$, and their multiplicative consequences.

(51) By Theorem 9.8 there is a hidden 2-extension from $h_2^2 w_2$ to $d_0 g^2$. The ordinary $\nu$-extension from $h_2 w_2$ to $h_2^2 w_2$ thus implies a hidden $\nu$-extension from $h_0 h_2 w_2$ detecting $2\nu_2$ to $d_0 g^2$ detecting $2\nu\nu_2$.

(147) By Theorem 9.8 there are hidden 2-extensions from $h_2^2 w_2^3$ to $d_0 g^2 w_2^2$, and from $d_0 g^2 w_2^2$ to $d_0 \delta' w_1 w_2^2$. The ordinary $\nu$-extension from $h_2 w_2^3$ to $h_2^2 w_2^3$ thus implies hidden $\nu$-extensions from $h_0 h_2 w_2^3$ to $d_0 g^2 w_2^2$, and from $h_0^2 h_2 w_2^3$ to $d_0 \delta' w_1 w_2^2$.

(50) Multiplying $\eta_1^2 \in \{\gamma^2\}$ by $\eta_4 \in \{h_1 w_2^2\}$ we obtain $\eta_1^2 \eta_4 = 4\nu_6 \in \{h_0^2 h_2 w_2^3\}$, since $h_1 \gamma^2 w_2^2 = h_0^2 h_2 w_2^3$ already at the $E_2$-term and there are no classes of higher Adams filtration in $\pi_{147}(tmf)$. Since $\nu \cdot 4\nu_6 \neq 0$ by case (147), it follows that $\nu \cdot \eta_1^2 \neq 0$, and this product can only be detected by $\gamma g w_1$.

(25) Multiplying $\eta_1 \in \{\gamma\}$ by $\eta_1$ we obtain $\eta_1^2 \in \{\gamma^2\}$. Since $\nu \cdot \eta_1^2 \neq 0$ by case (50), we deduce that $\nu \cdot \eta_1 \neq 0$, and being a homotopy class of order 2, this product can only be detected by $g w_1$.

(39) Multiplying $\eta_1 \in \{\gamma\}$ by $\kappa \in \{d_0\}$ we obtain $\eta_1 \kappa \in \{d_0 \gamma\}$. Since $\eta_1 \nu \in \{g w_1\}$ by case (25) we find that $\nu \cdot \eta_1 \kappa = \kappa \cdot \eta_1 \nu$ is detected by $d_0 g w_1 \neq 0$.

(57) Multiplying $\eta_1 \in \{\gamma\}$ by $\epsilon_1 \in \{\delta'\}$ we obtain $\eta_1 \epsilon_1 \in \{\gamma \delta'\}$. Since $\eta_1 \nu \in \{g w_1\}$, it follows that $\nu \cdot \eta_1 \epsilon_1$ is detected by $\delta' g w_1 \neq 0$.

(122) Multiplying $\eta_1 \in \{\gamma\}$ by $\eta_4 \in \{h_1 w_2^2\}$ we obtain $\eta_1 \eta_4 \in \{h_1 \gamma w_2^2\}$. Since $\eta_1 \nu \in \{g w_1\}$, we deduce that $\nu \cdot \eta_1 \eta_4$ is detected by $h_1 g w_1 w_2^2 \neq 0$.

(135) Multiplying $\eta_1 \in \{\gamma\}$ by $\kappa_4 \in \{d_0 w_2^2\}$ we obtain $\eta_1 \kappa_4 \in \{d_0 \gamma w_2^2\}$. Since $\eta_1 \nu \in \{g w_1\}$, we find that $\nu \cdot \eta_1 \kappa_4$ is detected by $d_0 g w_1 w_2^2 \neq 0$.

(153) Multiplying $\eta_1 \in \{\gamma\}$ by $\epsilon_5 \in \{\delta' w_2^2\}$ we obtain $\eta_1 \epsilon_5 \in \{\gamma \delta' w_2^2\}$. Since $\eta_1 \nu \in \{g w_1\}$, it follows that $\nu \cdot \eta_1 \epsilon_5$ is detected by $\delta' g w_1 w_2^2 \neq 0$.

(32) Multiplying $\epsilon_1 \in \{\delta'\}$ by $\eta_1 \in \{\gamma\}$ we obtain $\eta_1 \epsilon_1 \in \{\gamma \delta'\}$. Since $\nu \cdot \eta_1 \epsilon_1 \neq 0$ by case (57), we deduce that $\nu \cdot \epsilon_1 \neq 0$ must be detected by $\alpha \beta w_1$. Lemma 9.13 now shows that $B_1 \in \{\alpha g\}$ supports a hidden $\nu$-extension with the same target as $\epsilon_1 \in \{\delta'\}$.

(97) Multiplying $\eta_4 \in \{h_1 w_2^2\}$ by $\eta_1 \in \{\gamma\}$ we obtain $\eta_1 \eta_4 \in \{h_1 \gamma w_2^2\}$. Since $\nu \cdot \eta_1 \eta_4 \neq 0$ by case (122), we find that $\nu \cdot \eta_4 \neq 0$ must be detected by $g^5$, e.g. because the Adams filtration $\geq 26$ part of $\pi_{100}(tmf)$ is 2-torsion free.

(128) Multiplying $\epsilon_5 \in \{\delta' w_2^2\}$ by $\eta_1 \in \{\gamma\}$ we obtain $\eta_1 \epsilon_5 \in \{\gamma \delta' w_2^2\}$. Since $\nu \cdot \eta_1 \epsilon_5 \neq 0$ by case (153), it follows that $\nu \cdot \epsilon_5 \neq 0$ must be detected by $\alpha \beta w_1 w_2^2$. Lemma 9.13 now shows that $B_5 \in \{\alpha g w_2^2\}$ supports a hidden $\nu$-extension with the same target as $\epsilon_5 \in \{\delta' w_2^2\}$.

Next, we use the short exact sequence (9.2) to determine the hidden $\nu$-extensions on $h_2^2$ and its $w_2$-power multiples.

(6) We claim that $\nu$ times the generator $\nu^2 \in \{h_2^2\}$ of $\pi_6(tmf) \cong \mathbb{Z}/2$ is detected by $h_1 c_0$. The $E_\infty$-terms show that $\pi_9(tmf) \cong (\mathbb{Z}/2)^2$, $\pi_9(tmf/\nu) \cong \mathbb{Z}/2$, and $\pi_5(tmf) = 0$, which implies that $\nu\colon \pi_6(tmf) \to \pi_9(tmf)$ has image of order 2. Since $B \cdot \nu^2 = 0$, the image $\nu \cdot \nu^2 = \nu^3$ cannot be detected by the $w_1$-periodic class $h_1 w_1$. Hence $\nu^3$ must be detected by $h_1 c_0$.

(54) The $E_\infty$-terms show that $\pi_{57}(tmf) \cong (\mathbb{Z}/2)^4$, $\pi_{57}(tmf/\nu)$ has order $2^4$, and $\pi_{53}(tmf) \cong \mathbb{Z}/2$. Since $\pi_{56}(tmf)$ is 2-torsion free, $\pi_{53}(tmf)$ is $\nu$-torsion, so $\nu\colon \pi_{54}(tmf) \to \pi_{57}(tmf)$ has image of order 2. Since $\pi_{54}(tmf) \cong$

$\mathbb{Z}/4$ is generated by $\nu\nu_2$, and $B \cdot \nu\nu_2 = 0$, it follows that $\nu \cdot \nu\nu_2 = \nu^2\nu_2 \neq 0$ cannot be detected by a $w_1$-periodic class. Hence $\nu^2\nu_2 \in \{\gamma\delta'\}$.

(102) We claim that $\nu$ times the generator $\nu\nu_4 \in \{h_2^2 w_2^2\}$ of $\pi_{102}(tmf) \cong \mathbb{Z}/2$ is detected by $h_1 c_0 w_2^2$. The $E_\infty$-terms show that $\pi_{105}(tmf) \cong (\mathbb{Z}/2)^7$, $\pi_{105}(tmf/\nu)$ has order $2^6$, and $\pi_{101}(tmf) = 0$, which implies that the homomorphism $\nu\colon \pi_{102}(tmf) \to \pi_{105}(tmf)$ has image of order 2. Since $B^2 \cdot \nu\nu_4 = 0$, the image $\nu \cdot \nu\nu_4 = \nu^2\nu_4$ cannot be detected by a $w_1$-periodic class. It can also not be detected by $\gamma g^4$, since $i(\gamma g^4) = g^5\overline{h_1} \neq 0$ in $E_\infty(tmf/\nu)$. (This relation holds already in $E_2(tmf/\nu)$.) Hence $\nu^2\nu_4$ must be detected by $h_1 c_0 w_2^2$.

(150) The $E_\infty$-terms show that $\pi_{153}(tmf) \cong (\mathbb{Z}/2)^8$, $\pi_{153}(tmf/\nu)$ has order $2^8$, and $\pi_{149}(tmf) \cong \mathbb{Z}/2$. Since $\pi_{152}(tmf)$ is 2-torsion free, $\pi_{149}(tmf)$ is $\nu$-torsion, so $\nu\colon \pi_{150}(tmf) \to \pi_{153}(tmf)$ has image of order 2. Since $\pi_{150}(tmf) \cong \mathbb{Z}/8$ is generated by $\nu\nu_6$, and $B \cdot \nu\nu_6 = 0$, it follows that $\nu \cdot \nu\nu_6 = \nu^2\nu_6 \neq 0$ cannot be detected by a $w_1$-periodic class. Hence $\nu^2\nu_6 \in \{\gamma\delta' w_2^2\}$.

To finish the proof, we must show there are no further hidden $\nu$-extensions. In most degrees, the result is evident from the fact that $E_\infty(tmf) = 0$ in the relevant bidegrees (e.g., $\nu B_i = 0$ for $i = 0$, 3, 4 or 7), or the fact that $\nu$-multiples are $B$-torsion, hence cannot be detected by $w_1$-periodic classes. The remaining cases, $\nu \cdot B_2 = 0$ and $\nu \cdot B_6 = 0$, were handled in Lemma 9.13. □

**9.2.4. Hidden $\eta$-extensions.** We used the map of Adams spectral sequences induced by $i\colon tmf \to tmf/2$ to determine the hidden 2-extensions in the spectral sequence for $tmf$. This also determined most of the hidden $\nu$-extensions, except those on $\nu\nu_k$ for $k = 0, 2, 4, 6$, for which we used the spectral sequence map induced by $i\colon tmf \to tmf/\nu$. It turns out that this information also suffices to determine the hidden $\eta$-extensions.

THEOREM 9.16. *The Adams spectral sequence for tmf contains precisely the following hidden $\eta$-extensions. First we have four extensions that occur together with all their $w_1$- and $w_2^4$-power multiples:*

(32) *From $\alpha g$ to $\gamma w_1$, with $B_1 \in \{\alpha g\}$ and $\eta B_1 \in \{\gamma w_1\}$.*
(57) *From $h_1 c_0 w_2$ to $\gamma^2 w_1$, with $\eta B_2 \in \{h_1 c_0 w_2\}$ and $\eta^2 B_2 \in \{\gamma^2 w_1\}$.*
(128) *From $\alpha g w_2^2$ to $\gamma w_1 w_2^2$, with $B_5 \in \{\alpha g w_2^2\}$ and $\eta B_5 \in \{\gamma w_1 w_2^2\}$.*
(153) *From $h_1 c_0 w_2^3$ to $\gamma^2 w_1 w_2^2$, with $\eta B_6 \in \{h_1 c_0 w_2^3\}$ and $\eta^2 B_6 \in \{\gamma^2 w_1 w_2^2\}$.*

*In addition we have 24 extensions that only occur together with their $w_2^4$-power multiples:*

(21) *From $h_1 g$ to $d_0 w_1$, with $\eta\bar\kappa \in \{h_1 g\}$ and $\eta^2\bar\kappa \in \{d_0 w_1\}$.*
(27) *From $\alpha\beta$ to $g w_1$, with $\nu_1 \in \{\alpha\beta\}$ and $\eta\nu_1 \in \{g w_1\}$.*
(34) *From $d_0 g$ to $\alpha\beta w_1$, with $\kappa\bar\kappa \in \{d_0 g\}$ and $\eta\kappa\bar\kappa \in \{\alpha\beta w_1\}$.*
(39) *From $d_0\gamma$ to $\delta' w_1$, with $\eta_1\kappa \in \{d_0\gamma\}$ and $\eta\eta_1\kappa \in \{\delta' w_1\}$.*
(40) *From $g^2$ to $\alpha\beta d_0$, with $\bar\kappa^2 \in \{g^2\}$ and $\eta\bar\kappa^2 \in \{\alpha\beta d_0\}$.*
(41) *From $\alpha\beta d_0$ to $d_0 g w_1$, with $\eta\bar\kappa^2 = \nu_1\kappa \in \{\alpha\beta d_0\}$ and $\eta^2\bar\kappa^2 = \eta\nu_1\kappa \in \{d_0 g w_1\}$.*
(45) *From $\gamma g$ to $d_0\delta'$, with $\eta_1\bar\kappa \in \{\gamma g\}$ and $\eta\eta_1\bar\kappa \in \{d_0\delta'\}$.*
(51) *From $h_2 w_2$ to $\delta' g$, with $\nu_2 \in \{h_2 w_2\}$ and $\eta\nu_2 \in \{\delta' g\}$.*
(52) *From $\delta' g$ to $\gamma g w_1$, with $\eta\nu_2 = \epsilon_1\bar\kappa \in \{\delta' g\}$ and $\eta^2\nu_2 = \eta\epsilon_1\bar\kappa \in \{\gamma g w_1\}$.*

(59) *From* $h_2w_1w_2 = d_0\gamma g$ *to* $\delta' g w_1$*, with* $\nu_2 B = \eta_1\kappa\bar\kappa \in \{h_2w_1w_2\} = \{d_0\gamma g\}$ *and* $\eta\nu_2 B = \eta\eta_1\kappa\bar\kappa \in \{\delta' g w_1\}$.
(65a) *From* $\gamma g^2$ *to* $d_0\delta' g$*, with* $\eta_1\bar\kappa^2 \in \{\gamma g^2\}$ *and* $\eta\eta_1\bar\kappa^2 \in \{d_0\delta' g\}$.
(65b) *From* $h_2d_0w_2$ *to* $d_0\delta' g$*, with* $\nu_2\kappa \in \{h_2d_0w_2\}$ *and* $\eta\nu_2\kappa \in \{d_0\delta' g\}$.
(99) *From* $h_2w_2^2$ *to* $g^5$*, with* $\nu_4 \in \{h_2w_2^2\}$ *and* $\eta\nu_4 \in \{g^5\}$.
(117) *From* $h_1gw_2^2$ *to* $d_0w_1w_2^2$*, with* $\eta_4\bar\kappa \in \{h_1gw_2^2\}$ *and* $\eta\eta_4\bar\kappa \in \{d_0w_1w_2^2\}$.
(123) *From* $\alpha\beta w_2^2$ *to* $gw_1w_2^2$*, with* $\nu_5 \in \{\alpha\beta w_2^2\}$ *and* $\eta\nu_5 \in \{gw_1w_2^2\}$.
(129) *From* $h_1\delta w_2^2$ *to* $\gamma^2 g^4$ *with* $\eta\epsilon_5 \in \{h_1\delta w_2^2\}$ *and* $\eta^2\epsilon_5 \in \{\gamma^2 g^4\}$.
(130) *From* $d_0gw_2^2$ *to* $\alpha\beta w_1w_2^2$*, with* $\kappa_4\bar\kappa \in \{d_0gw_2^2\}$ *and* $\eta\kappa_4\bar\kappa \in \{\alpha\beta w_1w_2^2\}$.
(135) *From* $d_0\gamma w_2^2$ *to* $\delta' w_1w_2^2$*, with* $\eta_1\kappa_4 \in \{d_0\gamma w_2^2\}$ *and* $\eta\eta_1\kappa_4 \in \{\delta' w_1w_2^2\}$.
(137) *From* $\alpha\beta d_0w_2^2$ *to* $d_0gw_1w_2^2$*, with* $\nu_5\kappa \in \{\alpha\beta d_0w_2^2\}$ *and* $\eta\nu_5\kappa \in \{d_0gw_1w_2^2\}$.
(147) *From* $h_2w_2^3$ *to* $\delta' gw_2^2$*, with* $\nu_6 \in \{h_2w_2^3\}$ *and* $\eta\nu_6 \in \{\delta' gw_2^2\}$.
(148) *From* $\delta' gw_2^2$ *to* $\gamma gw_1w_2^2$*, with* $\eta\nu_6 = \epsilon_5\bar\kappa \in \{\delta' gw_2^2\}$ *and* $\eta^2\nu_6 = \eta\epsilon_5\bar\kappa \in \{\gamma gw_1w_2^2\}$.
(149) *From* $\gamma gw_1w_2^2$ *to* $d_0\delta' w_1w_2^2$*, with* $\eta^2\nu_6 = \eta_1\kappa\kappa_4 \in \{\gamma gw_1w_2^2\}$ *and* $\eta^3\nu_6 = \eta\eta_1\kappa\kappa_4 \in \{d_0\delta' w_1w_2^2\}$.
(155) *From* $h_2w_1w_2^3$ *to* $\delta' gw_1w_2^2$*, with* $\nu_6 B \in \{h_2w_1w_2^3\}$ *and* $\eta\nu_6 B \in \{\delta' gw_1w_2^2\}$.
(161) *From* $h_2d_0w_2^3$ *to* $d_0\delta' gw_2^2$*, with* $\nu_6\kappa \in \{h_2d_0w_2^3\}$ *and* $\eta\nu_6\kappa \in \{d_0\delta' gw_2^2\}$.

PROOF. The proof starts from the nontrivial $\eta^3$ on $\nu_6$ in degree 147, proved in Theorem 9.14, and deduces the majority of the $\eta$-extensions from this and its consequences.

(147-149) From Theorem 9.14 we have the relation $\eta^3\cdot\nu_6 = 4\nu\cdot\nu_6 \in \{d_0\delta' w_1w_2^2\} \neq 0$ in degree 150. This implies $\eta\cdot\nu_6 \in \{\delta' gw_2^2\}$ and $\eta^2\cdot\nu_6 \in \{\gamma gw_1w_2^2\}$, as these are the only classes of Adams filtration between 26 and 30 in these degrees.
(129) Next, $\eta\nu_6 = \epsilon_5\bar\kappa$ because both products are detected by $\delta' gw_2^2$, and from $E_\infty(tmf)$ we see that there is only one nonzero 2-torsion element in $\pi_{148}(tmf)$. Since $\eta^2\cdot\eta\nu_6 \neq 0$, by the previous case, we deduce that $\eta^2\epsilon_5 \neq 0$. It follows that $\eta^2\epsilon_5$ must be detected by $\gamma^2g^4$, since this class detects the unique $B$-power torsion element of order 2 in $\pi_{130}(tmf)$.
(21) From $\eta\nu_6 = \epsilon_5\bar\kappa$ and $\eta^2\cdot\eta\nu_6 \neq 0$ we also deduce that $\eta^2\bar\kappa \neq 0$, implying that there is a hidden $\eta$-extension from $h_1g$ to $d_0w_1$, detecting $\kappa B$.
(40,41) From $\eta^2\cdot\bar\kappa = \kappa B$ we get $\eta^2\cdot\bar\kappa^2 = \kappa\bar\kappa B$, detected by $d_0gw_1$. The intermediate class $\eta\bar\kappa^2$ must be detected by $\alpha\beta d_0$, since this is the only class of Adams filtration between 9 and 11 in degree 41.
(65a) Multiplying case (40) by $\eta_1 \in \{\gamma\}$ shows that $\eta$-multiplication takes $\eta_1\bar\kappa^2 \in \{\gamma g^2\}$ to $\eta\eta_1\bar\kappa^2 \in \{\gamma\cdot\alpha\beta d_0\} = \{\alpha d_0 g^2\} = \{d_0\delta' g\}$.
(45) Dividing the preceding case by $\bar\kappa$ gives that $\eta$-multiplication sends $\eta_1\bar\kappa$, detected by $\gamma g$, to $\eta\eta_1\bar\kappa$ detected by $d_0\delta' = \alpha d_0 g$.
(27) Dividing case (41) by $d_0$ shows that $\nu_1$, detected by $\alpha\beta$, is sent to $\eta\nu_1$, detected by $gw_1$.
(52) Multiplying the previous case by $\gamma$ shows that $\eta_1\nu_1$, detected by $\gamma\cdot\alpha\beta = \alpha g^2 = \delta' g$, is sent to $\eta\eta_1\nu_1$, detected by $\gamma gw_1 \neq 0$.
(32) In degree 32, filtration 7, there is a Klein 4-group with nonzero elements $\alpha g$, $\delta$ and $\delta'$, detecting $B_1$, $B_1+\epsilon_1$ and $\epsilon_1$, respectively. We see that $\eta B_1$ must be detected in Adams filtration at least 9, since $h_1\alpha g = 0$ in $E_2(tmf)$. Multiplying by $g$ gives $\alpha g^2 = \delta' g$, and we have just shown that $\eta$ times any class detected by this must be detected by $\gamma gw_1$ in Adams filtration 13.

The product $\eta B_1$ must therefore be detected in Adams filtration exactly 9, i.e., by $\gamma w_1$.

(59) In Adams filtration 13 of degree 59 we have $h_2 w_1 w_2$, which equals $\alpha^2\beta g$ in $E_3(tmf)$ from the differential $d_2(\alpha w_2) = d_0\gamma g + h_2 w_1 w_2$ and the relation $\alpha^2\beta = d_0\gamma$. Multiplying case (32) by $\alpha\beta$ we see that $\eta$-multiplication takes $\nu_1 \cdot B_1 = \eta_1\kappa\bar{\kappa} = \nu_2 B$, detected by $\alpha\beta\cdot\alpha g = d_0\gamma g = h_2 w_1 w_2$, to $\eta\nu_2 B$, detected by $\alpha\beta\cdot\gamma w_1 = \alpha g^2 w_1 = \delta' g w_1$.

(34) Dividing the preceding case by $\gamma$ gives that $\kappa\bar{\kappa}$, detected by $d_0 g$, is sent to $\eta\kappa\bar{\kappa}$, detected by $\alpha\beta w_1$.

(39) Dividing case (59) by $g$ gives that $\eta_1\kappa$, detected by $d_0\gamma$, is sent to $\eta\eta_1\kappa$, detected by $\delta' w_1$. Here we use the fact that $\kappa$ is $B$-power torsion to conclude that $\eta\eta_1\kappa$ is detected by $\delta' w_1$, rather than by $\alpha g w_1$ or $\delta w_1$, since the latter two classes are $w_1$-periodic.

(51) Dividing case (59) by $w_1$, we get that $\nu_2$, detected by $h_2 w_2$, is sent to $\eta\nu_2$, detected by $\delta' g$.

(65b) Multiplying by $d_0$ now shows that $\eta$-multiplication takes $\nu_2\kappa$, detected by $h_2 d_0 w_2$, to $\eta\nu_2\kappa$, detected by $d_0\delta' g$.

(57) In Adams filtration 12 of degree 57, we have a Klein 4-group with nonzero elements $\gamma\delta = h_1 c_0 w_2$, $\alpha\gamma g$ and $\gamma\delta'$, detecting $\eta B_2$, $\eta_1 B_1$ and $\eta_1\epsilon_1$, respectively. Now, $\eta_1\epsilon_1 = \nu^2\nu_2$ since $\gamma\delta'$ detects them both (by Theorem 9.14) and there is only one nonzero $B$-power torsion class in degree 57. Hence $\eta\eta_1\epsilon_1 = 0$. We have $\eta\cdot\eta_1 B_1 = \eta_1\cdot\eta B_1 \neq 0$ detected by $\gamma^2 w_1$, by case (32). Hence $\eta\cdot\eta B_2$ is also detected by $\gamma^2 w_1$.

(128) In degree 128, Adams filtration 23 is a Klein 4-group with nonzero elements $\alpha g w_2^2$, $\delta w_2^2$ and $\delta' w_2^2$. These classes are represented by $B_5$, $B_5+\epsilon_5$ and $\epsilon_5$, respectively. The class $\bar{\kappa}B_5$ is detected by $\alpha g^2 w_2^2 = \delta' g w_2^2$ in degree 148. We have shown that $\eta$ times any such class is detected in Adams filtration 29. It follows that $\eta B_5$ is detected in Adams filtration no more than 25. Since $h_1\alpha g w_2^2 = 0$, $\eta B_5$ must be detected in Adams filtration no less than 25. Hence $\eta B_5$ is detected by $\gamma w_1 w_2^2$, the unique class in Adams filtration 25. (It follows that $\eta^2 \colon \pi_{128}(tmf) \to \pi_{130}(tmf)$ maps $E_\infty^{23,23+128}(tmf)$ isomorphically to $E_\infty^{26,26+130}(tmf)$.)

(153) First, $\gamma\delta' w_2^2$ is represented by $\eta_1\epsilon_5$, which is $B$-power torsion. Since $\pi_{154}(tmf)$ has no $B$-power torsion, there is no hidden $\eta$-extension from $\gamma\delta' w_2^2$. Now, the identity $h_1 c_0 w_2^3 = \gamma\delta w_2^2 = \gamma\delta' w_2^2 + \alpha\gamma g w_2^2$ and the hidden $\eta$-extension from $\alpha g w_2^2$ to $\gamma w_1 w_2^2$ (case (128)) show that we have a hidden $\eta$-extension from $h_1 c_0 w_2^3$ to $\gamma^2 w_1 w_2^2$.

(123) Since $\nu_5 B^2 = 0$, $\eta\nu_5$ is either 0 or is detected by $g w_1 w_2^2$. It thus suffices to observe that $\eta$ times a class detected by $\gamma\cdot\alpha\beta w_2^2 = \delta' g w_2^2$ is nonzero, by case (148).

(137) Multiplying by $d_0$, we get that $\eta$ times a class detected by $\alpha\beta d_0 w_2^2$ must be detected by $d_0 g w_1 w_2^2$.

(155) Multiplying case (147) by $w_1$ proves this.

(161) Multiplying case (147) by $d_0$ proves this.

(130) Multiplying $\kappa_4\bar{\kappa}$, detected by $d_0 g w_2^2$, by $\eta$ must give either 0 or a class detected by $\alpha\beta w_1 w_2^2$. It must be nonzero because multiplying by $\gamma$ gives the product in case (155). This uses the equality $h_2 w_1 w_2^3 = d_0\gamma g w_2^2$ in $E_3(tmf)$ coming from $d_2(\alpha w_2^3)$.

(135) Similarly, multiplying by $g$ shows that this also follows from case (155).

(99) We have $\nu_4\kappa = \nu\kappa_4$, because there is only one $B$-power torsion class detected by $h_2 d_0 w_2^2$. By Theorem 9.14 the product $\eta_1\nu_4 \cdot \kappa = \eta_1\nu \cdot \kappa_4 = \bar{\kappa}B \cdot \kappa_4$ is detected by $d_0 g w_1 w_2^2 \neq 0$ in Adams filtration 28. It follows that $\eta_1\nu_4 \neq 0$ is detected in Adams filtration $\leq 24$, and $g w_1 w_2^2$ is the only nonzero class in sufficiently low Adams filtration. Hence $\eta \cdot \eta_1\nu_4$ is detected by $h_1 g w_1 w_2^2 \neq 0$, which implies that $\eta\nu_4 \neq 0$. Being a 2-torsion class, it can only be detected by $g^5$, so $\eta\nu_4 \in \{g^5\}$.

(117) We prove this by lifting to the top cell of $C2$. Consider the following commutative diagram. The elements we are interested in are named in the left hand and right hand columns; see Tables 6.10 and 6.11.

$$\begin{array}{ccccc}
h_1 g w_2^2 & \pi_{117}(tmf) & \xrightarrow{\ \eta\ } & \pi_{118}(tmf) & d_0 w_1 w_2^2 \\
 & \big\uparrow{\scriptstyle j} & & \big\uparrow{\scriptstyle j}\ \cong & \\
g \cdot w_2^2\widetilde{h_1} & \pi_{118}(tmf/2) & \xrightarrow{\ \eta\ } & \pi_{119}(tmf/2) & i(\beta w_1 w_2^2) \\
 & \big\downarrow{\scriptstyle \bar{\kappa}} & & \big\downarrow{\scriptstyle \bar{\kappa}} & \\
g^2 \cdot w_2^2\widetilde{h_1} & \pi_{138}(tmf/2) & \xrightarrow{\ \eta\ } & \pi_{139}(tmf/2) & g \cdot i(\beta w_1 w_2^2) \\
 & \big\downarrow{\scriptstyle j} & & \big\downarrow{\scriptstyle j} & \\
\alpha\beta d_0 w_2^2 & \pi_{137}(tmf) & \xrightarrow{\ \eta\ } & \pi_{138}(tmf) & d_0 g w_1 w_2^2
\end{array}$$

First note that $\pi_{117}(tmf)$ and $\pi_{118}(tmf)$ are both of order 2, and that $\pi_{119}(tmf) = 0$. Hence the map $j$ in the upper right of the diagram is an isomorphism. To show that $\eta$ times $\{h_1 g w_2^2\}$ is nonzero, it suffices to show that $\eta$ times any lift in $\{g w_2^2\widetilde{h_1}\}$ is nonzero. To show that, it suffices to show that $\eta \cdot j(\bar{\kappa} \cdot \{g w_2^2\widetilde{h_1}\}) \neq 0$. Clearly $\bar{\kappa}\{g w_2^2\widetilde{h_1}\} \subset \{g^2 w_2^2\widetilde{h_1}\}$. Since $\pi_{137}(tmf)$ has exponent 2, the map $j$ in the lower left of the diagram is an epimorphism. The class $\alpha\beta d_0 w_2^2$ has Adams filtration 26, and $g^2 w_2^2\widetilde{h_1}$ is the only class in $\pi_{138}(tmf/2)$ in filtration less than or equal to 26. Thus $j(\{g^2 w_2^2\widetilde{h_1}\}) = \{\alpha\beta d_0 w_2^2\}$. By case (137) above, $\eta$ acts nontrivially on any class in $\{\alpha\beta d_0 w_2^2\}$, and we are done. (Alternatively, this can be deduced from the vanishing of $\pi_{119}(tmf/\eta)$, which is clear from Figure 7.5.)

This exhausts the nonzero hidden $\eta$-extensions. In all other cases, a hidden $\eta$-extension would have to map from a $w_1$-power torsion class, which detects a $B$-power torsion class, to a $w_1$-periodic class, which can only detect $B$-periodic classes. □

PROPOSITION 9.17. $\nu^2\nu_4 = \eta\epsilon_4 + \eta_1\bar{\kappa}^4$.

PROOF. The $B$-power torsion subgroup of $\pi_{105}(tmf)$ is $(\mathbb{Z}/2)^2$, generated by $\eta\epsilon_4$ and $\eta_1\bar{\kappa}^4$, which are detected in adjacent Adams filtrations by $h_1 c_0 w_2^2$ and $\gamma g^4$. By Theorem 9.14, $\nu^2\nu_4$ is also detected by $h_1 c_0 w_2^2$, so $\nu^2\nu_4 - \eta\epsilon_4$ is either 0 or $\eta_1\bar{\kappa}^4$. We have $\eta_1\nu^2 = 0$ because $\pi_{31}(tmf) = 0$, and $\eta_1\epsilon_4 = \eta\epsilon_5$ since both are $B$-power torsion classes detected by $c_0\gamma w_2^2 = h_1\delta w_2^2$. Hence $\eta_1(\nu^2\nu_4 - \eta\epsilon_4) = \eta^2\epsilon_5 \neq 0$ by Theorem 9.16. It follows that $\nu^2\nu_4 - \eta\epsilon_4 = \eta_1\bar{\kappa}^4$, since it is not 0. □

## 9.3. The image of $\pi_*(tmf)$ in modular forms

To determine the ring structure on the torsion free quotient of $\pi_*(tmf)$, we make a comparison with the elliptic spectral sequence of [**75**, §4.3], with edge homomorphism

$$e\colon \pi_*(tmf) \longrightarrow mf_{*/2} = \mathbb{Z}[c_4, c_6, \Delta]/(c_4^3 - c_6^2 = 1728\Delta)\,.$$

Here $mf_{*/2}$ is the ring of integral modular forms, with $c_4$, $c_6$ and $\Delta$ in weights $*/2 =$ 4, 6 and 12, corresponding to topological degrees $* =$ 8, 12 and 24.

By [**75**, Prop. 4.6] and [**23**, §8], the image of the edge homomorphism is the subring of $mf_{*/2}$ given additively as

$$\mathbb{Z}\{a_{i,j,k}c_4^i c_6^j \Delta^k \mid i \geq 0, j \in \{0,1\}, k \geq 0\} \tag{9.3}$$

where

$$a_{i,j,k} = \begin{cases} 24/\gcd(k,24) & \text{for } i = j = 0, \\ 1 & \text{for } i \geq 1 \text{ and } j = 0, \\ 2 & \text{for } j = 1. \end{cases}$$

See also [**54**, §13.4] and [**89**, Thm. 1.2]. As stated this is an integral result, but following our standing conventions we are only concerned with its conclusion after implicit 2-completion.

DEFINITION 9.18. For $k \geq 0$ let $e_k = \max\{3 - \mathrm{ord}_2(k), 0\}$ and $d_k = 2^{e_k}$, so that

$$d_k = \begin{cases} 8 & \text{for } k \equiv 1,3,5,7 \mod 8, \\ 4 & \text{for } k \equiv 2,6 \mod 8, \\ 2 & \text{for } k \equiv 4 \mod 8, \\ 1 & \text{for } k \equiv 0 \mod 8 \end{cases}$$

is the 2-primary component of $a_{0,0,k}$. It follows that $8/d_k = \gcd(k,8)$.

PROPOSITION 9.19. *The kernel of the edge homomorphism $\pi_*(tmf) \to mf_{*/2}$ is equal to the 2-power torsion ideal in $\pi_*(tmf)$. Hence the torsion free quotient of $\pi_*(tmf)$ is isomorphic to the image of the edge homomorphism.*

1. *The generators $B_k$ in $\pi_*(tmf)$ can be chosen to map to $c_4\Delta^k$ in $mf_{*/2}$, for $0 \leq k \leq 7$.*
2. *The generators $C_k$ can be chosen to map to $2c_6\Delta^k$, for $0 \leq k \leq 7$.*
3. *The generators $D_k$ can be chosen to map to $d_k\Delta^k$, for $1 \leq k \leq 7$, where $d_k \in \{2, 4, 8\}$ is defined as above.*
4. *The generator $M$ can be chosen to map to $\Delta^8$.*
5. *The remaining algebra generators are 2-power torsion, and map to zero.*

REMARK 9.20. The modular form image in $mf_{*/2}$ and Adams detecting class in $E_\infty(tmf)$ uniquely determine each algebra generator $B_k$, $C_k$ and $D_k$ in $\pi_*(tmf)$, for $0 \leq k \leq 7$, with the following exceptions: $C_2$ is determined modulo $2\bar{\kappa}^3 = \nu^3\nu_2$, $B_3$ is determined modulo $\bar{\kappa}^4$, and $C_6$ is determined modulo $\nu^3\nu_6$. In each of the three exceptional cases the ambiguity is a class of order 2. A specific choice of $B_3$ will be made in Definition 9.22, but see also Definition 9.50.

Our proof of the proposition above makes use of the case $j = 1$ of (9.3). It also uses the construction in [**91**, Thm. 1.2] of an $E_\infty$ ring spectrum map $\iota'\colon tmf \to$

$tmf_1(3) \simeq BP\langle 2\rangle$, with an associated map of elliptic spectral sequences, yielding a commutative diagram

$$\begin{array}{ccc} \pi_*(tmf) & \xrightarrow{e} & mf_{*/2} \\ \downarrow{\iota'} & & \downarrow{\iota'} \\ \pi_*(tmf_1(3)) & \xrightarrow[\cong]{e} & mf_1(3)_{*/2} \end{array} \tag{9.4}$$

with horizontal edge homomorphisms. To justify the formulas connecting $mf_{*/2}$ to $mf_1(3)_{*/2}$ and $\pi_*(BP\langle 2\rangle)$, we first review some of the theory of $\Gamma_1(3)$-modular forms.

Following Mahowald and Rezk [**105**] we consider the moduli stack $\mathcal{M}_1(3)$ of elliptic curves with level structure of type $\Gamma_1(3)$, i.e., with a chosen point of order 3. There is an étale map $\mathcal{M}_1(3) \to \mathcal{M}_{ell}$ that represents forgetting the level structure, and the Goerss–Hopkins–Miller sheaf of $E_\infty$ ring spectra over $\mathcal{M}_{ell}$ pulls back to a similar sheaf over $\mathcal{M}_1(3)$. We let $TMF_1(3)$ be the global sections (= homotopy limit) of this sheaf, so that there is a canonical map $TMF \to TMF_1(3)$ of $E_\infty$ ring spectra. Since we are implicitly working locally at $p = 2$, each elliptic curve with $\Gamma_1(3)$ structure is uniquely strictly isomorphic, cf. [**105**, Prop. 3.2], to a non-singular Weierstrass curve of the form

$$y^2 + a_1xy + a_3y = x^3 ,$$

with $a_2 = a_4 = a_6 = 0$. This defines an elliptic curve with a flex point at $(x, y) = (0, 0)$, which gives the point of order 3. The classical expressions for $c_4$, $c_6$ and $\Delta$ of an elliptic curve in Weierstrass form, as given in Joseph Silverman's book [**157**, §III.1], then simplify to $c_4 = a_1(a_1^3 - 24a_3)$, $c_6 = -a_1^6 + 36a_1^3a_3 - 216a_3^2$ and $\Delta = a_3^3(a_1^3 - 27a_3)$. It follows that $\pi_*(TMF_1(3)) \cong MF_1(3)_{*/2} = \mathbb{Z}[a_1, a_3][1/\Delta]$. The 2-series of the associated formal group law can be calculated with the recipe of [**157**, §IV.1], and begins

$$[2](z) = 2z - a_1z^2 - 7a_3z^4 + \dots .$$

Hence the complex orientation $MU \to TMF_1(3)$ sends $v_1$ to $-a_1 \equiv a_1 \mod 2$ and $v_2$ to $-7a_3 \equiv a_3 \mod (2, a_1)$. Here we use that $v_n$ maps to the coefficient of $z^{2^n}$ in the 2-series, modulo $(2, \dots, v_{n-1})$, both for the Araki and the Hazewinkel generators [**144**, A2.2.4 and p. 371].

Using chromatic fracture squares, Lawson and Naumann [**91**, §3] proceed to construct a map $Tmf \to Tmf_1(3)$ of $E_\infty$ ring spectra, whose $K(2)$-localization agrees with the canonical map mentioned above. Passing to connective covers, they obtain the $E_\infty$ ring spectrum map $\iota'\colon tmf \to tmf_1(3)$, where $\pi_*(tmf_1(3)) \cong mf_1(3)_{*/2} = \mathbb{Z}[a_1, a_3]$. Furthermore, $tmf_1(3)$ is a (generalized) $BP\langle 2\rangle$, in the sense that the composite homomorphism

$$\mathbb{Z}[v_1, v_2] \to \pi_*(MU) \to \pi_*(tmf_1(3))$$

is an isomorphism. Moreover, they show in [**91**, Thm. 4.4] that $H^*(tmf_1(3)) \cong A//E(2)$, and $\iota'$ induces the evident surjection $A//E(2) \to A//A(2)$ in mod 2 cohomology.

Alternatively, one can follow the later work of Hill and Lawson [**70**, Thm. 5.17], who show that the Goerss–Hopkins–Miller étale sheaf over $\mathcal{M}_{ell}$ extends to a log-étale sheaf over the compactification $\overline{\mathcal{M}}_{ell}$. The direct image log structure from $\mathcal{M}_{ell}$ gives $\overline{\mathcal{M}}_{ell}$ the structure of a (Deligne–Mumford) log stack [**70**, Def. 3.1], and the extended sheaf can be pulled back along any log-étale cover of $\overline{\mathcal{M}}_{ell}$. In particular, there is a compactification $\overline{\mathcal{M}}_1(3)$ of $\mathcal{M}_1(3)$ classifying generalized elliptic curves with $\Gamma_1(3)$ level structure. When the compactification is equipped with the direct image log structure, the forgetful map $\overline{\mathcal{M}}_1(3) \to \overline{\mathcal{M}}_{ell}$ is log-étale. Passing to global sections, Hill and Lawson recover the map $Tmf \to Tmf_1(3)$ of $E_\infty$ ring spectra, and an associated map of descent spectral sequences [**70**, Thm. 6.1]. A presentation of $\overline{\mathcal{M}}_1(3)$ as a weighted projective space shows that the descent spectral sequence for $Tmf_1(3)$ collapses at the $E_2$-term, which is concentrated along the 0- and 1-lines. In particular, $\pi_*(Tmf_1(3))$ agrees with $mf_1(3)_{*/2} = \mathbb{Z}[a_1, a_3]$ in non-negative degrees.

Passing to connective covers, this leads to diagram (9.4), with $tmf_1(3)$ a generalized $BP\langle 2\rangle$. The edge homomorphism $e\colon \pi_*(BP\langle 2\rangle) \cong \pi_*(tmf_1(3)) \to mf_1(3)_{*/2}$ satisfies $e(v_1) \equiv a_1 \mod 2$ and $e(v_2) \equiv a_3 \mod (2, a_1)$, while the homomorphism $\iota'\colon mf_{*/2} \to mf_1(3)_{*/2}$ is given by

$$
\begin{aligned}
c_4 &\longmapsto a_1(a_1^3 - 3\cdot 2^3 a_3)\\
c_6 &\longmapsto -a_1^6 + 9\cdot 2^2 a_1^3 a_3 - 27\cdot 2^3 a_3^2\\
\Delta &\longmapsto a_3^3(a_1^3 - 27\cdot a_3)\,.
\end{aligned}
\tag{9.5}
$$

Here we have emphasized the powers of 2 that are present, in order to make it easier to recognize how (products of) the classes on the right hand side are detected in the Adams spectral sequence for $tmf_1(3)$.

PROOF OF PROPOSITION 9.19. We compare diagram (9.4) with the map of Adams spectral sequences

$$
\begin{array}{ccc}
E_2^{*,*}(tmf) & \Longrightarrow & \pi_*(tmf)\\
\big\downarrow{\scriptstyle \iota'} & & \big\downarrow{\scriptstyle \iota'}\\
E_2^{*,*}(tmf_1(3)) & \Longrightarrow & \pi_*(tmf_1(3))\,.
\end{array}
$$

The left hand homomorphism $\iota'\colon E_2^{*,*}(tmf) \to E_2^{*,*}(tmf_1(3))$ was calculated in Lemma 1.17 and given in Table 1.3.

In degree $* = 8$, $B \in \{w_1\}$ maps to a multiple $e(B) = xc_4$ of the generator of $mf_4 = \mathbb{Z}\{c_4\}$. Its image $\iota'(xc_4) = xa_1(a_1^3 - 3\cdot 2^3 a_3)$ must be detected by $\iota'(w_1) = v_1^4$ in $E_\infty(tmf_1(3))$. Here $a_1(a_1^3 - 3\cdot 2^3 a_3) \in \{v_1^4\}$, so $x$ is a 2-adic unit. Replacing $B$ by $B/x$ we may thus arrange that $e(B) = c_4$.

In degree $* = 12$, $C \in \{h_0^3\alpha\}$ maps to a multiple $e(C) = xc_6$ of the generator of $mf_6 = \mathbb{Z}\{c_6\}$. Its image $\iota'(xc_6) = x(-a_1^6 + 9\cdot 2^2 a_1^3 a_3 - 27\cdot 2^3 a_3^2)$ must be detected by $\iota'(h_0^3\alpha) = v_0^4 v_2^2$ in $E_\infty(tmf_1(3))$. Here $-a_1^6 + 9\cdot 2^2 a_1^3 a_3 - 27\cdot 2^3 a_3^2 \in \{v_0^3 v_2^2\}$, so $x$ is 2 times a unit. Dividing $C$ by this unit we obtain $e(C) = 2c_6$.

In degree $* = 24$, $D_1 \in \{h_0\alpha^2\}$ maps to a linear combination $e(D_1) = xc_4^3 + y\Delta$ of the generators of $mf_{12} = \mathbb{Z}\{c_4^3, \Delta\}$. Subtracting $xB^3$ from $D_1$ does not alter its detecting class in the Adams $E_\infty$-term, so we may assume that $x = 0$ and $e(D_1) = y\Delta$. The image $\iota'(y\Delta) = ya_3^3(a_1^3 - 27\cdot a_3)$ must be detected by $\iota'(h_0\alpha^2) = v_0^3 v_2^4$ in $E_\infty(tmf_1(3))$. Here $a_3^3(a_1^3 - 27\cdot a_3) \in \{v_2^4\}$, so $y$ is $d_1 = 2^3$ times a unit. Dividing $D_1$ by this unit we get $e(D_1) = 2^3\Delta$.

In degree $* = 32$, $B_1 \in \{\alpha g\}$ maps to a sum $e(B_1) = xc_4^4 + yc_4\Delta$ in $mf_{16}$. Subtracting $xB^4$ from $B_1$ we may assume that $x = 0$ and $e(B_1) = yc_4\Delta$. Due to the hidden 2-extension from $h_0^2\alpha g$ to $h_0\alpha^2 w_1$, we instead consider $8B_1 \in \{h_0\alpha^2 w_1\}$, with $e(8B_1) = 8yc_4\Delta$. The image

$$\iota'(8yc_4\Delta) = 8ya_1(a_1^3 - 3\cdot 2^3 a_3)a_3^3(a_1^3 - 27\cdot a_3)$$

must be detected by $\iota'(h_0\alpha^2 w_1) = v_0^3 v_1^4 v_2^4$ in $E_\infty(tmf_1(3))$. Here $\iota'(c_4\Delta) \in \{v_1^4 v_2^4\}$, so $8y$ is $2^3$ times a unit. Dividing $B_1$ by $y$ we get $e(B_1) = c_4\Delta$.

In degree $* = 36$, $C_1 \in \{h_0\alpha^3\}$ maps to a sum $e(C_1) = xc_4^3 c_6 + yc_6\Delta$ in $mf_{18}$. By formula (9.3), $x$ and $y$ are both even. Hence we can subtract $(x/2)B^3C$ from $C_1$ to arrange that $x = 0$ and $e(C_1) = yc_6\Delta$. The image $\iota'(yc_6\Delta)$ must be detected by $\iota'(h_0\alpha^3) = v_0^4 v_2^6$. Here $\iota'(c_6\Delta) \in \{v_0^3 v_2^6\}$, so $y$ is 2 times a unit. Dividing $C_1$ by that unit we get $e(C_1) = 2c_6\Delta$.

In degree $* = 48$, $D_2 \in \{h_0^2 w_2\}$ maps to a sum $e(D_2) = xc_4^6 + yc_4^3\Delta + z\Delta^2$ in $mf_{24} = \mathbb{Z}\{c_4^6, c_4^3\Delta, \Delta^2\}$. Subtracting $xB^6 + yB^2B_1$ from $D_2$ does not alter its detecting class in the Adams $E_\infty$-term, so we may assume that $x = 0$, $y = 0$ and $e(D_2) = z\Delta^2$. The image $\iota'(z\Delta^2)$ must be detected by $\iota'(h_0^2 w_2) = v_0^2 v_2^8$. Here $\iota'(\Delta^2) \in \{v_2^8\}$, so $z$ is $d_2 = 2^2$ times a unit. Dividing $D_2$ by this unit we get $e(D_2) = 2^2\Delta^2$.

In degree $* = 56$, $B_2 \in \{c_0 w_2\}$ maps to a sum $e(B_2) = xc_4^7 + yc_4^4\Delta + zc_4\Delta^2$ in $mf_{28}$. Subtracting $xB^7 + yB^3B_1$ from $B_2$ we may assume that $e(B_2) = zc_4\Delta^2$. Due to the hidden 2-extension from $c_0 w_2$ to $\alpha^3 g + h_0 w_1 w_2$, we instead consider $2B_2 \in \{\alpha^3 g + h_0 w_1 w_2\}$, with $e(2B_2) = 2zc_4\Delta^2$. The image $\iota'(2zc_4\Delta^2)$ must be detected by $\iota'(\alpha^3 g + h_0 w_1 w_2) = v_0 v_1^4 v_2^8$. Here $\iota'(c_4\Delta^2) \in \{v_1^4 v_2^8\}$, so $2z$ is 2 times a unit. Dividing $B_2$ by that unit we get $e(B_2) = c_4\Delta^2$.

The proofs for $C_2 \in \{h_0^3\alpha w_2\}$, $C_3 \in \{h_0\alpha^3 w_2\}$, $C_4 \in \{h_0^3\alpha w_2^2\}$, $C_5 \in \{h_0\alpha^3 w_2^2\}$, $C_6 \in \{h_0^3\alpha w_2^3\}$ and $C_7 \in \{h_0\alpha^3 w_2^3\}$ are very similar to the one for $C_1$. In each case we use that $e(C_k)$ is divisible by 2 in $mf_{*/2}$ by (9.3).

The proofs for $D_3 \in \{h_0\alpha^2 w_2\}$, $D_5 \in \{h_0\alpha^2 w_2^2\}$ and $D_7 \in \{h_0\alpha^2 w_2^3\}$ are very similar to the one for $D_1$.

The proofs for $B_3 \in \{\delta w_2\}$, $B_5 \in \{\alpha g w_2^2\}$ and $B_7 \in \{\delta w_2^3\}$ are very similar to the one for $B_1$.

In degree $* = 96$, $D_4 \in \{h_0 w_2^2\}$ maps to a sum $e(D_4) = xc_4^{12} + yc_4^9\Delta + zc_4^6\Delta^2 + sc_4^3\Delta^3 + t\Delta^4$ in $mf_{48}$. Subtracting $xB^{12} + yB^8B_1 + zB^5B_2 + sB^2B_3$ from $D_4$ we may assume that $e(D_4) = t\Delta^4$. The image $\iota'(t\Delta^4)$ must be detected by $\iota'(h_0 w_2^2) = v_0 v_2^{16}$. Here $\iota'(\Delta^4) \in \{v_2^{16}\}$, so $t$ is $d_4 = 2$ times a unit. Dividing by this unit we get $e(D_4) = 2\Delta^4$.

The proof for $B_4 \in \{w_1 w_2^2\}$ is very similar to that for $B$.

The proof for $D_6 \in \{h_0^2 w_2^3\}$ is very similar to that for $D_2$.

The proof for $B_6 \in \{c_0 w_2^3\}$ is very similar to that for $B_2$.

Finally, in degree $* = 192$, $M \in \{w_2^4\} \subset \pi_*(tmf)$ maps to a linear combination $e(M) \in mf_{*/2}$ of terms $c_4^i\Delta^j$ with $i + 3j = 24$ and $0 \le j \le 8$. Subtracting from $M$ the corresponding linear combination of terms $B^{i-1}B_j$ for $0 \le j \le 7$, we may assume that $e(M) = x\Delta^8$. The image $\iota'(x\Delta^8) \in \pi_*(tmf_1(3))$ must then be detected by $\iota'(w_2^4) = v_2^{32}$. Here $\iota'(\Delta^8) \in \{v_2^{32}\}$, so $x$ is a 2-adic unit. Dividing $M$ by that unit we get $e(M) = \Delta^8$, while still keeping $M \in \{w_2^4\}$.

In view of Theorem 9.8, the 2-torsion free quotient of $\pi_*(tmf)$ is generated as a $\mathbb{Z}[B, M]$-module (implicitly 2-completed) by $D_k$, $B_k$ and $C_k$ in degrees $24k$, $8 + 24k$ and $12 + 24k$, for $0 \le k \le 7$, subject to the relations $B \cdot D_k = d_k B_k$. These relations lift from $E_\infty(tmf)$ to $\pi_*(tmf)$ because all classes of higher Adams filtration in degree $8 + 24k$ are detected by the edge homomorphism, and the relations

$$c_4 \cdot d_k\Delta^k = d_k \cdot c_4\Delta^k$$

evidently hold in $mf_{*/2}$. Since the edge images $e(D_k) = d_k\Delta^k$, $e(B_k) = c_4\Delta^k$ and $e(C_k) = 2c_6\Delta^k$ satisfy no other $\mathbb{Z}[c_4, \Delta^8]$-module relations than these, it follows that $e$ maps the 2-torsion free quotient of $\pi_*(tmf)$ injectively to $mf_{*/2}$. This proves that the kernel of $e$ is precisely $\Gamma_2\pi_*(tmf)$. □

By [**91**, Thm. 1.2], the map $\iota' \colon tmf \to tmf_1(3) \simeq BP\langle 2\rangle$ sits in a commutative square

$$\begin{array}{ccc} tmf & \xrightarrow{q_0} & ko \\ {\scriptstyle \iota'}\downarrow & & \downarrow{\scriptstyle c} \\ tmf_1(3) & \xrightarrow{\tilde{c}} & ku \end{array}$$

of $E_\infty$ ring spectra, realizing the square of cyclic $A$-modules

$$\begin{array}{ccc} A/\!/A(2) & \longleftarrow & A/\!/A(1) \\ \uparrow & & \uparrow \\ A/\!/E(2) & \longleftarrow & A/\!/E(1) \end{array}$$

in cohomology. The Adams spectral sequences for $tmf_1(3) \simeq BP\langle 2\rangle$, $ko$ and $ku$ collapse at the $E_2$-term, and we have induced graded ring homomorphisms

$$\begin{array}{ccc} \pi_*(tmf) & \xrightarrow{q_0} & \dfrac{\mathbb{Z}[\eta, A, B]}{(2\eta, \eta^3, \eta A, A^2 - 4B)} \\ {\scriptstyle \iota'}\downarrow & & \downarrow{\scriptstyle c} \\ \mathbb{Z}[a_1, a_3] & \xrightarrow{\tilde{c}} & \mathbb{Z}[v_1] \end{array}$$

(implicitly 2-localized or 2-completed). The complexification map $c$ induces $\eta \mapsto 0$, $A \mapsto 2v_1^2$ and $B \mapsto v_1^4$, while the map $\tilde{c}$ is constructed [**91**, p. 2784] so as to induce $a_1 \mapsto -v_1$ and $a_3 \mapsto 0$.

Proposition 9.21. *The ring homomorphism $q_0 \colon \pi_*(tmf) \to \pi_*(ko)$ is given on the $B$-, $C$-, $D$- and $M$-families of generators by*

$$B \longmapsto B$$
$$C \longmapsto -AB$$

*while $B_k \mapsto 0$, $C_k \mapsto 0$, $D_k \mapsto 0$ and $M \mapsto 0$ for $1 \le k \le 7$.*

Proof. Since $c\colon \pi_*(ko) \to \pi_*(ku)$ is injective in degrees $* \equiv 0 \bmod 4$, it suffices to verify that $cq_0 = \tilde{c}\iota'$ is given by $B \mapsto v_1^4$, $C \mapsto -2v_1^6$, $B_k \mapsto 0$, $C_k \mapsto 0$, $D_k \mapsto 0$ and $M \mapsto 0$, where $1 \le k \le 7$. This follows from the choices of modular form images made in Proposition 9.19, together with the formulas $c_4 \mapsto v_1^4$, $c_6 \mapsto -v_1^6$ and $\Delta \mapsto 0$ for the composite $\tilde{c}\iota' \colon mf_{*/2} \to \pi_*(ku)$, which follow directly from (9.5). □

For degree reasons, it is clear that the 2-power torsion $\nu$-, $\epsilon$-, $\kappa$- and $\bar{\kappa}$-families map to 0 in $\pi_*(ko)$, but to determine the images of $\eta_1$ and $\eta_4$, more specific choices must be made. We do this in the following section.

## 9.4. Algebra generators for $\pi_*(tmf)$

We now aim to characterize the 40 homotopy classes from Figure 9.1, which generate $\pi_*(tmf)$ as a graded commutative ring, or more precisely (due to our implicit 2-completion), as a $\mathbb{Z}_2$-algebra. Each of these algebra generators will be detected in $E_\infty(tmf)$ by one of the 43 generators from Table 9.1, with the minor modification that $B_1 \in \{\alpha g\}$ and $B_5 \in \{\alpha g w_2^2\}$, where $\alpha g = \delta + \delta'$ and $\alpha g w_2^2 = \delta w_2^2 + \delta' w_2^2$ occur as sums of generators in that table. Due to the additive extensions found in Section 9.2, the remaining three generators (namely $h_0$, $\alpha^3 g + h_0 w_1 w_2$ and $\alpha^3 g w_2^2 + h_0 w_1 w_2^3$) from Table 9.1 are not needed to generate $\pi_*(tmf)$ as a $\mathbb{Z}_2$-algebra.

The detecting classes in $E_\infty(tmf)$ only determine these 40 algebra generators for $\pi_*(tmf)$ modulo classes of higher Adams filtration. By also specifying their images in $mf_{*/2}$ under the edge homomorphism to modular forms, as in Section 9.3, we eliminate most of the ambiguity in the definition of the generators of infinite additive order. Nonetheless, some ambiguity remains, which we account for on a case-by-case basis in the following definition.

DEFINITION 9.22.

(1) Let $B \in \pi_8(tmf)$, $C \in \pi_{12}(tmf)$ and $M \in \pi_{192}(tmf)$ be the classes detected by $w_1$, $h_0^3\alpha$ and $w_2^4$ in $E_\infty(tmf)$, and mapping to $c_4$, $2c_6$ and $\Delta^8$ in $mf_{*/2}$, respectively.

(2) Let $\eta \in \pi_1(tmf)$, $\nu \in \pi_3(tmf)$, $\epsilon \in \pi_8(tmf)$, $\kappa \in \pi_{14}(tmf)$ and $\bar\kappa \in \pi_{20}(tmf)$ be the images of the classes with the same names in $\pi_*(S)$. These satisfy $2\eta = 0$, $8\nu = 0$, $2\epsilon = 0$, $2\kappa = 0$ and $8\bar\kappa = 0$ in $\pi_*(S)$ (implicitly 2-completed), as well as in $\pi_*(tmf)$, and are detected in the $E_\infty$-term by $h_1$, $h_2$, $c_0$, $d_0$ and $g$, respectively.

(3) The classes $D = 1$, $B$, $C$, $\eta$, $\nu$, $\epsilon$ and $\kappa$ generate the remaining algebra generators for $\pi_*(tmf)$, up to scalars, by "formally multiplying by powers of $\Delta = v_2^4$." As discussed in Section 9.1, classes detecting these elements at the $E_2$-term are related by the Massey products $\Delta$ and $\Delta'$. For each class $x \in \pi_n(tmf)$ in the above list we write $x_k$ for the corresponding algebra generator in $\pi_{n+24k}(tmf)$, for some or all $1 \le k \le 7$. In some general formulas it is convenient to use the conventions that $x_0 = x$ and $x_{k+8} = x_k M$, but the latter products are not needed to generate $\pi_*(tmf)$.

(4) Hence, let $D_k \in \pi_{24k}(tmf)$ for $1 \le k \le 7$ be the classes detected by $h_0\alpha^2$, $h_0^2 w_2$, $h_0\alpha^2 w_2$, $h_0 w_2^2$, $h_0\alpha^2 w_2^2$, $h_0^2 w_2^3$ and $h_0\alpha^2 w_2^3$ in $E_\infty(tmf)$, respectively, and mapping to $d_k\Delta^k$ in $mf_{*/2}$, where $d_k = 2^{e_k}$ is as in Definition 9.18.

(5) Let $B_k \in \pi_{8+24k}(tmf)$ for $1 \le k \le 7$ be classes detected by $\alpha g$, $c_0 w_2$, $\delta w_2$, $w_1 w_2^2$, $\alpha g w_2^2$, $c_0 w_2^3$ and $\delta w_2^3$, respectively, and mapping to $c_4\Delta^k$ in each case. These conditions uniquely specify the $B_k$, except for $k = 3$: If $\bar B_3$ denotes a class detected by $\delta w_2$ and mapping to $c_4\Delta^3$, then $\bar B_3$ and $\bar B_3 + \bar\kappa^4$ are the two elements of $\pi_{80}(tmf)$ that meet these two conditions. Exactly one of $\bar B_3$ and $\bar B_3 + \bar\kappa^4$ satisfies

$$\bar\kappa B_3 = \bar\kappa^5 \,,$$

and we let $B_3$ be this one. The choices of classes $D_k$ and $B_k$ are compatible, in the sense that

$$B \cdot D_k = d_k B_k \,.$$

(6) Let $C_k \in \pi_{12+24k}(tmf)$ for $1 \le k \le 7$ be classes detected by $h_0\alpha^3$, $h_0^3\alpha w_2$, $h_0\alpha^3 w_2$, $h_0^3\alpha w_2^2$, $h_0\alpha^3 w_2^2$, $h_0^3\alpha w_2^3$ and $h_0\alpha^3 w_2^3$, respectively, and mapping to $2c_6\Delta^k$ in $mf_{*/2}$. These conditions uniquely specify the $C_k$, except for $k \in \{2,6\}$: each choice of $C_2$ or $C_6$ can be altered by adding $\nu^3\nu_2 = 2\bar{\kappa}^3$ or $\nu^3\nu_6$, respectively, without changing the detecting classes in $E_\infty(tmf)$ or the images in $mf_{*/2}$. We leave this additive indeterminacy in $C_2$ and $C_6$ unspecified.

(7) Let $\eta_k \in \pi_{1+24k}(tmf)$ for $k \in \{1,4\}$ be the classes detected by $\gamma$ and $h_1 w_2^2$, respectively, and subject to the condition

$$B \cdot \eta_k = \eta B_k\,.$$

This determines $\eta_1$ uniquely, since multiplication by $B$ maps $\pi_{25}(tmf) \cong (\mathbb{Z}/2)^2$ isomorphically to Adams filtration $\ge 9$ of $\pi_{33}(tmf)$, where $\eta B_1$ is detected by $\gamma w_1$. It also determines $\eta_4$ uniquely, since multiplication by $B$ maps $\pi_{97}(tmf) \cong (\mathbb{Z}/2)^5$ isomorphically to the part of Adams filtration $\ge 21$ of $\pi_{105}(tmf)$ that maps to $\{0, h_1 w_1 w_2^2\} \subset E_\infty^{21,21+105}(tmf)$, where $\eta B_4$ is detected by $h_1 w_1 w_2^2$. This definition is compatible with the earlier specification made in Lemma 9.7, namely that $2\eta_1 = 0$ and $2\eta_4 = 0$, since $\pi_{25}(tmf)$ and $\pi_{97}(tmf)$ both have exponent 2.

(8) Let $\nu_k \in \pi_{3+24k}(tmf)$ for $k \in \{1,2,4,5,6\}$ be classes detected by $\alpha\beta$, $h_2 w_2$, $h_2 w_2^2$, $\alpha\beta w_2^2$ and $h_2 w_2^3$, respectively. These are uniquely determined up to odd multiples, and satisfy $4\nu_1 = 0$, $8\nu_2 = 0$, $8\nu_4 = 0$, $4\nu_5 = 0$ and $8\nu_6 = 0$, as is easily seen from the $E_\infty$-term for *tmf*. This leaves $\mathbb{Z}/4^\times$ ambiguity in the choices of $\nu_1$ and $\nu_5$, and $\mathbb{Z}/8^\times$ ambiguity in the choices of $\nu_2$, $\nu_4$ and $\nu_6$. We shall see in Proposition 9.35 that $\nu_5$ and $\nu_6$ can be uniquely chosen to make

$$\nu_1\nu_5 = 2\nu\nu_6 \qquad \text{and} \qquad \nu_2\nu_4 = 3\nu\nu_6\,,$$

leaving only the multiplicative ambiguity in the choices of $\nu_1$, $\nu_2$ and $\nu_4$. In Theorem 9.54, we will see that we can choose $\nu_4$ so that

$$\nu D_4 = 2\nu_4\,,$$

and this further reduces the ambiguity in the choice of $\nu_4$ to a factor in $\{1,5\} \subset \mathbb{Z}/8^\times$. In some general formulas, it will be convenient to let $\nu_3 = \eta_1^3$, detected by $\gamma^3$, and $\nu_7 = 0 \in \pi_{171}(tmf)$, so that $\nu_k$ has order $d_{7-k}$ for each $0 \le k \le 7$.

(9) Let $\epsilon_k \in \pi_{8+24k}(tmf)$ for $k \in \{1,4,5\}$ be classes detected by $\delta'$, $c_0 w_2^2$ and $\delta' w_2^2$, respectively. We showed in Lemma 9.7 that we can choose these homotopy classes so that $2\epsilon_1 = 0$, $2\epsilon_4 = 0$ and $2\epsilon_5 = 0$. This uniquely determines these elements in $\pi_*(tmf)$, since in each case the 2-torsion subgroup is $\mathbb{Z}/2$.

(10) Finally, let $\kappa_4 \in \pi_{110}(tmf)$ be a class detected by $d_0 w_2^2$. It is easily seen from the $E_\infty$-term for *tmf* that $4\kappa_4 = 0$, and we saw in Theorem 9.8 that $2\kappa_4 \ne 0$. This determines $\kappa_4$ up to sign. In case (150a) of Theorem 9.8 we showed that $\kappa_4\bar{\kappa}^2 = \pm 2\nu\nu_6$, and we choose the sign of $\kappa_4$ to make

$$\kappa_4\bar{\kappa}^2 = 2\nu\nu_6\,.$$

While the preceding definition contains forward references to results which allow us to reduce or eliminate ambiguity, those results and the resulting specificity in our choices of generators are not used until the results have been proved. Their inclusion above is done simply to collect everything in one definition for the convenience of the reader.

REMARK 9.23. We note the following comparisons with other notations:

(1) The generators $B_k$ specified above are the most convenient for the calculations in this section and the next, but for our final description of the multiplicative structure in $\pi_*(tmf)$ we will find it best to replace them by generators $\widetilde{B}_k$, which sometimes have lower Adams filtration. See Definition 9.50.
(2) There is no relation between our classes $\eta_1$ and $\eta_4$ and Mahowald's classes $\eta_j \in \pi_{2^j}(S)$ detected by $h_1 h_j$, cf. [**101**]. The latter homotopy classes are decomposable for $j \leq 3$, and Mahowald's $\eta_4$ equals Toda's $\eta^*$ in $\pi_{16}(S)$, so the notation $\eta_j$ is mostly needed for $j \geq 5$, in which case there is no conflict of notation. The image of Mahowald's $\eta_4$ in $\pi_*(tmf)$ is zero, because $\pi_{16}(tmf)$ is 2-torsion free.
(3) Henriques [**54**, Ch. 13] writes $\{2\nu\Delta\}$ for our class $\nu_1$, and $\{\nu\Delta^5\}$ for our class $\nu_5$ (but $\{2\nu\Delta^5\}$ was intended). There are relations $\eta^2 \cdot \eta_1 = 2 \cdot \nu_1$ and $\nu \cdot \eta_1 = \eta \cdot \nu_1$. The first of these would look more familiar in Henriques' notation, but the second relation is more familiar in our notation, which is typographically simpler.
(4) As we will prove in Proposition 11.77, the element $\epsilon_1 \in \pi_{32}(tmf)$ is the image of a homotopy class $[q]$ in $\pi_{32}(S)$ detected by $q \in E_\infty^{6,6+32}(S)$, see Table 1.1. However, $[q]$ has Adams filtration 6 and $\epsilon_1$ has Adams filtration 7, so we prefer to keep separate notations. Further, as we have observed, all the $\epsilon_k$ play a similar role, making the more consistent notation preferable.

REMARK 9.24. The following indeterminacies remain in our choices of algebra generators for $\pi_*(tmf)$:

(1) The complex and quaternionic Hopf fibrations specify the classes $\eta$ and $\nu$ in $\pi_*(S)$, respectively, as well as their images in $\pi_*(tmf)$. The elements $\epsilon$ and $\kappa$ in $\pi_*(tmf)$ are characterized by being of order 2. The class $\bar{\kappa} \in \pi_{20}(S)$ was only defined up to a factor in $\mathbb{Z}/8^\times = \{1, 3, 5, 7\}$ in [**130**, Lemma 15.4]. A more precise choice can be made using fourfold Toda brackets $\langle \nu, \eta, 2, \kappa \rangle$ or $\langle \kappa, 2, \eta, \nu \rangle$, as in [**87**, Lemma 5.3.8] and [**23**, (8.1)], but in each case the indeterminacy $4\bar{\kappa} = \nu^2\kappa$ remains. The image of $\bar{\kappa}$ in $\pi_*(tmf)$ is then as uniquely specified as it is in $\pi_*(S)$. The products $\eta\bar{\kappa}$, $\nu\bar{\kappa} = 0$, $\epsilon\bar{\kappa}$, $\kappa\bar{\kappa}$ and $\bar{\kappa}^2$ are unambiguously defined.
(2) The classes $D_k$ for $1 \leq k \leq 7$ and $M$ are uniquely determined by their modular form images.
(3) The classes $B_k$ for $0 \leq k \leq 7$ are uniquely determined by their detecting $E_\infty$-classes and modular form images, except for $B_3$, which is unambiguously specified by the relation $\bar{\kappa}B_3 = \bar{\kappa}^5$. This choice is made so that the formula $\eta_i \nu_j = \bar{\kappa}B_{i+j-1}$ in Proposition 9.38 will hold for all $i$ and $j$.
(4) The classes $C_k$ for $0 \leq k \leq 7$ are uniquely determined by their modular form images, except for $C_2$ and $C_6$. We leave these two classes unspecified, with additive indeterminacy $2\bar{\kappa}^3 = \eta\nu_2\epsilon = \nu^3\nu_2 = \epsilon\epsilon_1\bar{\kappa}$ and $\eta\nu_6\epsilon = \nu^3\nu_6 = \epsilon\epsilon_5\bar{\kappa}$, respectively. See Proposition 9.41 for the factorizations involving $\epsilon$.

(5) The classes $\eta_k$ for $k \in \{1,4\}$ are uniquely determined by their detecting classes in $E_\infty(tmf)$ and the relations $\eta_k B = \eta B_k$.
(6) The $\nu_k$ for $k \in \{1,2,4,5,6\}$ are specified by their detecting classes in $E_\infty(tmf)$, together with the equations $\nu_1\nu_5 = 2\nu\nu_6$, $\nu_2\nu_4 = 3\nu\nu_6$ and $\nu D_4 = 2\nu_4$. This leaves multiplicative indeterminacy $\mathbb{Z}/4^\times$ for $\nu_1$, $\mathbb{Z}/8^\times$ for $\nu_2$, and $\{1,5\} \subset \mathbb{Z}/8^\times$ for $\nu_4$.
(7) The $\epsilon_k$ for $k \in \{1,4,5\}$ are uniquely determined by their detecting classes in $E_\infty(tmf)$, together with the fact that they have order 2. The latter clause could be replaced by the condition that they be $B$-power torsion.
(8) The class $\kappa_4$ is uniquely determined by its detecting class in $E_\infty(tmf)$ and the relation $\kappa_4\bar\kappa^2 = 2\nu\nu_6$.

To summarize: The classes that have not been uniquely specified are $C_2$, $C_6$, $\nu_1$, $\nu_2$ and $\nu_4$. The classes $\nu_5$, $\nu_6$ and $\kappa_4$ depend, in well-defined manner, on the choices of $\nu_1$, $\nu_2$ and $\nu_4$.

DEFINITION 9.25. Let $N_* \subset \pi_*(tmf)$ be the $\mathbb{Z}[B]$-submodule generated by all classes in degrees $0 \le * < 192$, and let $N = tmf/M$ be the homotopy cofiber of the map

$$M\colon \Sigma^{192} tmf \longrightarrow tmf\,.$$

THEOREM 9.26. *As a $\mathbb{Z}[B]$-module, $N_*$ is a split extension*

$$0 \to \Gamma_B N_* \longrightarrow N_* \longrightarrow N_*/\Gamma_B N_* \to 0\,.$$

*The $B$-power torsion submodule $\Gamma_B N_*$ is given in Table 9.4. It is concentrated in degrees $3 \le * \le 164$, and is finite in each degree. The action of $B$ is as indicated in the table, together with $2\bar\kappa^2 = \epsilon_1 B$, $2\bar\kappa^3 = \eta\nu_2 B$ and $4\nu\nu_6 = \epsilon_5\kappa B$.*

*The $B$-torsion free quotient of $N_*$ is the direct sum*

$$N_*/\Gamma_B N_* = \bigoplus_{k=0}^{7} ko[k]$$

*of the following eight (implicitly 2-completed) $\mathbb{Z}[B]$-modules, with $ko[k]$ concentrated in degrees $* \ge 24k$:*

$$\begin{aligned}
ko[0] &= \mathbb{Z}[B]\{1, C\} \oplus \mathbb{Z}/2[B]\{\eta, \eta^2\}\\
ko[1] &= \mathbb{Z}\{D_1\} \oplus \mathbb{Z}[B]\{B_1, C_1\} \oplus \mathbb{Z}/2[B]\{\eta_1, \eta\eta_1\}\\
ko[2] &= \mathbb{Z}\{D_2\} \oplus \mathbb{Z}[B]\{B_2, C_2\} \oplus \mathbb{Z}/2[B]\{\eta B_2, \eta_1^2\}\\
ko[3] &= \mathbb{Z}\{D_3\} \oplus \mathbb{Z}[B]\{B_3, C_3\} \oplus \mathbb{Z}/2[B]\{\eta B_3, \eta^2 B_3\}\\
ko[4] &= \mathbb{Z}\{D_4\} \oplus \mathbb{Z}[B]\{B_4, C_4\} \oplus \mathbb{Z}/2[B]\{\eta_4, \eta\eta_4\}\\
ko[5] &= \mathbb{Z}\{D_5\} \oplus \mathbb{Z}[B]\{B_5, C_5\} \oplus \mathbb{Z}/2[B]\{\eta B_5, \eta_1\eta_4\}\\
ko[6] &= \mathbb{Z}\{D_6\} \oplus \mathbb{Z}[B]\{B_6, C_6\} \oplus \mathbb{Z}/2[B]\{\eta B_6, \eta^2 B_6\}\\
ko[7] &= \mathbb{Z}\{D_7\} \oplus \mathbb{Z}[B]\{B_7, C_7\} \oplus \mathbb{Z}/2[B]\{\eta B_7, \eta^2 B_7\}\,.
\end{aligned}$$

*The $\mathbb{Z}[B]$-module structures are such that $B \cdot D_1 = 8B_1$, $B \cdot D_2 = 4B_2$, $B \cdot D_3 = 8B_3$, $B \cdot D_4 = 2B_4$, $B \cdot D_5 = 8B_5$, $B \cdot D_6 = 4B_6$ and $B \cdot D_7 = 8B_7$. In other words, $B \cdot D_k = d_k B_k$ for each $1 \le k \le 7$.*

PROOF. In view of Definition 9.22 this summarizes information from Tables 5.8 and 5.9, Theorems 9.8, 9.14 and 9.16, and Proposition 9.10. A splitting of the extension is provided by the chosen lifts $1, C, \eta, \dots, C_7, \eta B_7, \eta^2 B_7$ in $N_* \subset \pi_*(tmf)$ of the $\mathbb{Z}[B]$-module generators of the $ko[k]$. □

The $\alpha \in \pi_n(S)$ column of Table 9.4 will be explained in Section 11.11. We note that $N_*$ is not a direct sum of cyclic $\mathbb{Z}[B]$-modules. For instance, $B \cdot \epsilon_1 = 2 \cdot \bar{\kappa}^2$ and $B \cdot D_1 = 8 \cdot B_1$.

THEOREM 9.27. *As a $\mathbb{Z}[B, M]$-module, $\pi_*(tmf)$ is a split extension*

$$0 \to \Gamma_B \pi_*(tmf) \longrightarrow \pi_*(tmf) \longrightarrow \pi_*(tmf)/\Gamma_B \pi_*(tmf) \to 0\,.$$

*Here*

$$\Gamma_B \pi_*(tmf) \cong \Gamma_B N_* \otimes \mathbb{Z}[M]$$

*with $\Gamma_B N_*$ given in Table 9.4, and*

$$\pi_*(tmf)/\Gamma_B \pi_*(tmf) \cong \bigoplus_{k=0}^{7} ko[k] \otimes \mathbb{Z}[M]$$

*with $ko[k]$ given as above.*

PROOF. Since $w_2^4$ (detecting $M$) acts freely on the Adams $E_\infty$-term for *tmf*, the composite homomorphism

$$N_* \otimes \mathbb{Z}[M] \longrightarrow \pi_*(tmf) \otimes \pi_*(tmf) \stackrel{\cdot}{\longrightarrow} \pi_*(tmf)$$

is an isomorphism of $\mathbb{Z}[B, M]$-modules. This theorem therefore follows from the previous one. □

COROLLARY 9.28. *The composite*

$$N_* \subset \pi_*(tmf) \longrightarrow \pi_*(N)$$

*is an isomorphism of $\mathbb{Z}[B]$-modules.* □

REMARK 9.29. The submodule $N_* \subset \pi_*(tmf)$ is preserved by the action of $\eta$, $\nu$, $\epsilon$, $\kappa$ and $\bar{\kappa}$. To check this, note that the $B^2$-torsion classes $\kappa C_7$, $\bar{\kappa} B_7$ and $\bar{\kappa} C_7$ are zero. It follows that the isomorphisms $N_* \otimes \mathbb{Z}[M] \cong \pi_*(tmf)$ and $N_* \cong \pi_*(N)$ also respect the action by these elements.

Table 9.4: $B$-power torsion in $\pi_n(tmf)$ for $0 \leq n < 192$, with generators $\beta \in \{b\}$ and some lifts $\alpha \in \iota^{-1}(\beta) \subset \pi_n(S)$

| $n$ | $\Gamma_B \pi_n(tmf)$ | $\beta \in \pi_n(tmf)$ | $b \in E_\infty(tmf)$ | $\alpha \in \pi_n(S)$ |
|---|---|---|---|---|
| 3 | $\mathbb{Z}/8$ | $\nu$ | $h_2$ | $\nu$ |
| 6 | $\mathbb{Z}/2$ | $\nu^2$ | $h_2^2$ | $\nu^2$ |
| 8 | $\mathbb{Z}/2$ | $\epsilon$ | $c_0$ | $\epsilon + \eta\sigma$ |
| 9 | $\mathbb{Z}/2$ | $\eta\epsilon$ | $h_1 c_0$ | $\eta\epsilon + \eta^2\sigma$ |
| 14 | $\mathbb{Z}/2$ | $\kappa$ | $d_0$ | $\kappa$ |
| 15 | $\mathbb{Z}/2$ | $\eta\kappa$ | $h_1 d_0$ | $\eta\kappa$ |
| 17 | $\mathbb{Z}/2$ | $\nu\kappa$ | $h_2 d_0$ | $\nu\kappa$ |
| 20 | $\mathbb{Z}/8$ | $\bar{\kappa}$ | $g$ | $\bar{\kappa}$ |

Table 9.4: $B$-power torsion in $\pi_n(tmf)$ for $0 \le n < 192$, with generators $\beta \in \{b\}$ and some lifts $\alpha \in \iota^{-1}(\beta) \subset \pi_n(S)$ (cont.)

| $n$ | $\Gamma_B\pi_n(tmf)$ | $\beta \in \pi_n(tmf)$ | $b \in E_\infty(tmf)$ | $\alpha \in \pi_n(S)$ |
|---|---|---|---|---|
| 21 | $\mathbb{Z}/2$ | $\eta\bar{\kappa}$ | $h_1 g$ | $\eta\bar{\kappa}$ |
| 22 | $\mathbb{Z}/2$ | $\eta^2\bar{\kappa} = \kappa B$ | $d_0 w_1$ | $\eta^2\bar{\kappa}$ |
| 27 | $\mathbb{Z}/4$ | $\nu_1$ | $\alpha\beta$ | – |
| 28 | $\mathbb{Z}/2$ | $\eta\nu_1 = \bar{\kappa}B$ | $g w_1$ | $\epsilon\bar{\kappa}$ |
| 32 | $\mathbb{Z}/2$ | $\epsilon_1$ | $\delta'$ | $[q]$ |
| 33 | $\mathbb{Z}/2$ | $\eta\epsilon_1$ | $h_1\delta$ | $\eta[q]$ |
| 34 | $\mathbb{Z}/2$ | $\kappa\bar{\kappa}$ | $d_0 g$ | $\kappa\bar{\kappa}$ |
| 35 | $\mathbb{Z}/2$ | $\eta\kappa\bar{\kappa} = \nu_1 B$ | $\alpha\beta w_1$ | $\eta\kappa\bar{\kappa}$ |
| 39 | $\mathbb{Z}/2$ | $\eta_1\kappa$ | $d_0\gamma$ | $[u]$ |
| 40 | $\mathbb{Z}/4$ | $\bar{\kappa}^2$ | $g^2$ | $\bar{\kappa}^2$ |
| 41 | $\mathbb{Z}/2$ | $\eta\bar{\kappa}^2$ | $\alpha\beta d_0$ | $\eta\bar{\kappa}^2$ |
| 42 | $\mathbb{Z}/2$ | $\eta^2\bar{\kappa}^2 = \kappa\bar{\kappa}B$ | $d_0 g w_1$ | $\eta^2\bar{\kappa}^2$ |
| 45 | $\mathbb{Z}/2$ | $\eta_1\bar{\kappa}$ | $\gamma g$ | $\{w\}$ |
| 46 | $\mathbb{Z}/2$ | $\eta\eta_1\bar{\kappa}$ | $d_0\delta'$ | $\eta\{w\}$ |
| 51 | $\mathbb{Z}/8$ | $\nu_2$ | $h_2 w_2$ | – |
| 52 | $\mathbb{Z}/2$ | $\eta\nu_2$ | $\delta' g$ | $\bar{\kappa}[q]$ |
| 53 | $\mathbb{Z}/2$ | $\eta^2\nu_2 = \eta_1\bar{\kappa}B$ | $\gamma g w_1$ | $\eta\bar{\kappa}[q]$ |
| 54 | $\mathbb{Z}/4$ | $\nu\nu_2$ | $h_2^2 w_2$ | $\alpha_{54}$ |
| 57 | $\mathbb{Z}/2$ | $\nu^2\nu_2$ | $\gamma\delta'$ | $\nu\alpha_{54}$ |
| 59 | $\mathbb{Z}/2$ | $\nu_2 B$ | $h_2 w_1 w_2$ | $\kappa\{w\}$ |
| 60 | $\mathbb{Z}/4$ | $\bar{\kappa}^3$ | $g^3$ | $\bar{\kappa}^3$ |
| 65 | $(\mathbb{Z}/2)^2$ | $\eta_1\bar{\kappa}^2$ | $\gamma g^2$ | $\bar{\kappa}\{w\}$ |
| – | – | $\nu_2\kappa$ | $h_2 d_0 w_2$ | $\alpha_{65}$ |
| 66 | $\mathbb{Z}/2$ | $\eta\nu_2\kappa$ | $d_0\delta' g$ | $\eta\bar{\kappa}\{w\}$ |
| 68 | $\mathbb{Z}/2$ | $\nu\nu_2\kappa$ | $h_2^2 d_0 w_2$ | $\nu\alpha_{65}$ |
| 70 | $\mathbb{Z}/2$ | $\eta_1^2\bar{\kappa}$ | $\gamma^2 g$ | $\alpha_{70}$ |
| 75 | $\mathbb{Z}/2$ | $\eta_1^3$ | $\gamma^3$ | – |
| 80 | $\mathbb{Z}/2$ | $\bar{\kappa}^4$ | $g^4$ | $\bar{\kappa}^4$ |
| 85 | $\mathbb{Z}/2$ | $\eta_1\bar{\kappa}^3$ | $\gamma g^3$ | $\bar{\kappa}^2\{w\}$ |
| 90 | $\mathbb{Z}/2$ | $\eta_1^2\bar{\kappa}^2$ | $\gamma^2 g^2$ | $\{w\}^2$ |
| 99 | $\mathbb{Z}/8$ | $\nu_4$ | $h_2 w_2^2$ | – |

Table 9.4: $B$-power torsion in $\pi_n(tmf)$ for $0 \le n < 192$, with generators $\beta \in \{b\}$ and some lifts $\alpha \in \iota^{-1}(\beta) \subset \pi_n(S)$ (cont.)

| $n$ | $\Gamma_B\pi_n(tmf)$ | $\beta \in \pi_n(tmf)$ | $b \in E_\infty(tmf)$ | $\alpha \in \pi_n(S)$ |
|---|---|---|---|---|
| 100 | $\mathbb{Z}/2$ | $\eta\nu_4$ | $g^5$ | $\bar\kappa^5$ |
| 102 | $\mathbb{Z}/2$ | $\nu\nu_4$ | $h_2^2 w_2^2$ | (?) |
| 104 | $\mathbb{Z}/2$ | $\epsilon_4$ | $c_0 w_2^2$ | (?) |
| 105 | $(\mathbb{Z}/2)^2$ | $\eta\epsilon_4$ | $h_1 c_0 w_2^2$ | |
| – | – | $\eta_1\bar\kappa^4$ | $\gamma g^4$ | $\bar\kappa^3\{w\}$ |
| 110 | $\mathbb{Z}/4$ | $\kappa_4$ | $d_0 w_2^2$ | (?) |
| 111 | $\mathbb{Z}/2$ | $\eta\kappa_4$ | $h_1 d_0 w_2^2$ | |
| 113 | $\mathbb{Z}/2$ | $\nu\kappa_4$ | $h_2 d_0 w_2^2$ | |
| 116 | $\mathbb{Z}/4$ | $\bar\kappa D_4$ | $h_0 g w_2^2$ | (?) |
| 117 | $\mathbb{Z}/2$ | $\eta_4\bar\kappa$ | $h_1 g w_2^2$ | (?) |
| 118 | $\mathbb{Z}/2$ | $\eta\eta_4\bar\kappa = \kappa_4 B$ | $d_0 w_1 w_2^2$ | |
| 123 | $\mathbb{Z}/4$ | $\nu_5$ | $\alpha\beta w_2^2$ | – |
| 124 | $\mathbb{Z}/2$ | $\eta\nu_5$ | $g w_1 w_2^2$ | |
| 125 | $\mathbb{Z}/2$ | $\eta^2\nu_5 = \eta_4\bar\kappa B$ | $h_1 g w_1 w_2^2$ | $\bar\kappa^4\{w\}$ |
| 128 | $\mathbb{Z}/2$ | $\epsilon_5$ | $\delta' w_2^2$ | (?) |
| 129 | $\mathbb{Z}/2$ | $\eta\epsilon_5$ | $h_1 \delta w_2^2$ | |
| 130 | $\mathbb{Z}/4$ | $\kappa_4\bar\kappa$ | $d_0 g w_2^2$ | |
| 131 | $\mathbb{Z}/2$ | $\eta\kappa_4\bar\kappa = \nu_5 B$ | $\alpha\beta w_1 w_2^2$ | |
| 135 | $\mathbb{Z}/2$ | $\eta_1\kappa_4$ | $d_0\gamma w_2^2$ | (?) |
| 136 | $\mathbb{Z}/2$ | $\eta\eta_1\kappa_4 = \epsilon_5 B$ | $\delta' w_1 w_2^2$ | |
| 137 | $\mathbb{Z}/2$ | $\nu_5\kappa$ | $\alpha\beta d_0 w_2^2$ | |
| 138 | $\mathbb{Z}/2$ | $\eta\nu_5\kappa = \kappa_4\bar\kappa B$ | $d_0 g w_1 w_2^2$ | |
| 142 | $\mathbb{Z}/2$ | $\epsilon_5\kappa$ | $d_0\delta' w_2^2$ | |
| 147 | $\mathbb{Z}/8$ | $\nu_6$ | $h_2 w_2^3$ | – |
| 148 | $\mathbb{Z}/2$ | $\eta\nu_6$ | $\delta' g w_2^2$ | |
| 149 | $\mathbb{Z}/2$ | $\eta^2\nu_6$ | $\gamma g w_1 w_2^2$ | |
| 150 | $\mathbb{Z}/8$ | $\nu\nu_6$ | $h_2^2 w_2^3$ | (?) |
| 153 | $\mathbb{Z}/2$ | $\nu^2\nu_6$ | $\gamma\delta' w_2^2$ | |
| 155 | $\mathbb{Z}/2$ | $\nu_6 B$ | $h_2 w_1 w_2^3$ | |
| 156 | $\mathbb{Z}/2$ | $\eta\nu_6 B$ | $\delta' g w_1 w_2^2$ | |
| 161 | $\mathbb{Z}/2$ | $\nu_6\kappa$ | $h_2 d_0 w_2^3$ | (?) |

Table 9.4: $B$-power torsion in $\pi_n(tmf)$ for $0 \le n < 192$, with generators $\beta \in \{b\}$ and some lifts $\alpha \in \iota^{-1}(\beta) \subset \pi_n(S)$ (cont.)

| $n$ | $\Gamma_B\pi_n(tmf)$ | $\beta \in \pi_n(tmf)$ | $b \in E_\infty(tmf)$ | $\alpha \in \pi_n(S)$ |
|---|---|---|---|---|
| 162 | $\mathbb{Z}/2$ | $\eta\nu_6\kappa$ | $d_0\delta' g w_2^2$ | |
| 164 | $\mathbb{Z}/2$ | $\nu\nu_6\kappa$ | $h_2^2 d_0 w_2^3$ | |

DEFINITION 9.30. Let

$$T = \mathbb{Z}[\eta, \nu, B, M]/(2\eta, \eta^3 + 4\nu, \eta\nu, 2\nu^2, \nu B, \nu^4)$$

be the (implicitly 2-completed) subalgebra of $\pi_*(tmf)$ generated by $\eta$, $\nu$, $B$ and $M$.

PROPOSITION 9.31. *As a $T$-module, $\pi_*(tmf)$ is generated by the classes listed in the $x$-column of Table 9.5. Here $x \in \pi_n(tmf)$ is detected by the given class in $E_\infty^{s,s+n}(tmf)$ and maps to the given modular form in $mf_{n/2}$. Its annihilator ideal in $T$ is* $\mathrm{Ann}(x)$, *with radical* $\sqrt{\mathrm{Ann}(x)}$ *viewed as an ideal in $\mathbb{Z}[B, M] \cong T_{\mathrm{red}} = T/(\eta, \nu)$.*

PROOF. This summarizes information from Tables 5.8 and 5.9, Theorems 9.8, 9.14 and 9.16, and Propositions 9.10 and 9.17. The products $\eta \cdot D_k$ for $1 \le k \le 7$ are zero because $\eta B D_k = d_k \eta B_k$ and $d_k$ is even. □

REMARK 9.32. Note that $\pi_*(tmf)$ is not a direct sum of cyclic $T$-modules. For instance, $\eta \cdot \epsilon = \nu^3 \cdot 1$, $4 \cdot \bar\kappa = \nu^2 \cdot \kappa$ and $\eta^2 \cdot \bar\kappa = B \cdot \kappa$. These, and the other $T$-module relations, are visible in Figures 9.6 through 9.13

Table 9.5: $T$-module generators of $\pi_*(tmf)$

| $n$ | $s$ | $x$ | $E_\infty(tmf)$ | $mf$ | $\mathrm{Ann}(x)$ | $\sqrt{\mathrm{Ann}(x)}$ |
|---|---|---|---|---|---|---|
| 0 | 0 | 1 | 1 | 1 | $(0)$ | $(0)$ |
| 8 | 3 | $\epsilon$ | $c_0$ | 0 | $(2, \eta^2, \nu, B)$ | $(2, B)$ |
| 12 | 6 | $C$ | $h_0^3\alpha$ | $2c_6$ | $(\eta, \nu)$ | $(0)$ |
| 14 | 4 | $\kappa$ | $d_0$ | 0 | $(2, \eta^2, \nu^3, 2B, \eta B, B^2)$ | $(2, B)$ |
| 20 | 4 | $\bar\kappa$ | $g$ | 0 | $(8, \nu, 2B, \eta B, B^2)$ | $(2, B)$ |
| 24 | 7 | $D_1$ | $h_0\alpha^2$ | $8\Delta$ | $(\eta, \nu)$ | $(0)$ |
| 25 | 5 | $\eta_1$ | $\gamma$ | 0 | $(2, \nu^2, \eta^2 B)$ | $(2)$ |
| 27 | 6 | $\nu_1$ | $\alpha\beta$ | 0 | $(4, \eta^2, \nu, 2B, \eta B, B^2)$ | $(2, B)$ |
| 32 | 7 | $B_1$ | $\alpha g$ | $c_4\Delta$ | $(2\nu, \nu^2, \nu B)$ | $(0)$ |
| 32 | 7 | $\epsilon_1$ | $\delta'$ | 0 | $(2, \eta^2, \nu^2, \eta B, B^2)$ | $(2, B)$ |
| 34 | 8 | $\kappa\bar\kappa$ | $d_0 g$ | 0 | $(2, \eta^2, \nu, \eta B, B^2)$ | $(2, B)$ |
| 36 | 10 | $C_1$ | $h_0\alpha^3$ | $2c_6\Delta$ | $(\eta, \nu)$ | $(0)$ |
| 39 | 9 | $\eta_1\kappa$ | $d_0\gamma$ | 0 | $(2, \eta^2, \nu^2, B)$ | $(2, B)$ |
| 40 | 8 | $\bar\kappa^2$ | $g^2$ | 0 | $(4, \nu, B)$ | $(2, B)$ |

Table 9.5: $T$-module generators of $\pi_*(tmf)$ (cont.)

| $n$ | $s$ | $x$ | $E_\infty(tmf)$ | $mf$ | $\mathrm{Ann}(x)$ | $\sqrt{\mathrm{Ann}(x)}$ |
|---|---|---|---|---|---|---|
| 45 | 9 | $\eta_1\bar{\kappa}$ | $\gamma g$ | 0 | $(2, \eta^2, \nu, \eta B, B^2)$ | $(2, B)$ |
| 48 | 10 | $D_2$ | $h_0^2 w_2$ | $4\Delta^2$ | $(\eta, 2\nu, \nu^2)$ | $(0)$ |
| 50 | 10 | $\eta_1^2$ | $\gamma^2$ | 0 | $(2, \eta^2, \nu^2, \eta B)$ | $(2)$ |
| 51 | 9 | $\nu_2$ | $h_2 w_2$ | 0 | $(8, 4\nu, 2B, \nu^3 + \eta B, B^2)$ | $(2, B)$ |
| 56 | 11 | $B_2$ | $c_0 w_2$ | $c_4\Delta^2$ | $(\nu)$ | $(0)$ |
| 60 | 12 | $\bar{\kappa}^3$ | $g^3$ | 0 | $(4, \eta, \nu, B)$ | $(2, B)$ |
| 60 | 14 | $C_2$ | $h_0^3\alpha w_2$ | $2c_6\Delta^2$ | $(\eta, \nu)$ | $(0)$ |
| 65 | 13 | $\eta_1\bar{\kappa}^2$ | $\gamma g^2$ | 0 | $(2, \eta^2, \nu, B)$ | $(2, B)$ |
| 65 | 13 | $\nu_2\kappa$ | $h_2 d_0 w_2$ | 0 | $(2, \eta^2, \nu^2, B)$ | $(2, B)$ |
| 70 | 14 | $\eta_1^2\bar{\kappa}$ | $\gamma^2 g$ | 0 | $(2, \eta, \nu, B)$ | $(2, B)$ |
| 72 | 15 | $D_3$ | $h_0\alpha^2 w_2$ | $8\Delta^3$ | $(\eta, \nu)$ | $(0)$ |
| 75 | 15 | $\eta_1^3$ | $\gamma^3$ | 0 | $(2, \eta, \nu, B)$ | $(2, B)$ |
| 80 | 15 | $B_3$ | $\delta w_2$ | $c_4\Delta^3$ | $(\nu)$ | $(0)$ |
| 80 | 16 | $\bar{\kappa}^4$ | $g^4$ | 0 | $(2, \eta, \nu, B)$ | $(2, B)$ |
| 84 | 18 | $C_3$ | $h_0\alpha^3 w_2$ | $2c_6\Delta^3$ | $(\eta, \nu)$ | $(0)$ |
| 85 | 17 | $\eta_1\bar{\kappa}^3$ | $\gamma g^3$ | 0 | $(2, \eta, \nu, B)$ | $(2, B)$ |
| 90 | 18 | $\eta_1^2\bar{\kappa}^2$ | $\gamma^2 g^2$ | 0 | $(2, \eta, \nu, B)$ | $(2, B)$ |
| 96 | 17 | $D_4$ | $h_0 w_2^2$ | $2\Delta^4$ | $(\eta, \nu^2)$ | $(0)$ |
| 97 | 17 | $\eta_4$ | $h_1 w_2^2$ | 0 | $(2, \nu^2, \eta^2 B)$ | $(2)$ |
| 99 | 17 | $\nu_4$ | $h_2 w_2^2$ | 0 | $(8, \eta^2, 2\nu, B, \nu^3)$ | $(2, B)$ |
| 104 | 19 | $\epsilon_4$ | $c_0 w_2^2$ | 0 | $(2, \eta^2, \nu, B)$ | $(2, B)$ |
| 104 | 20 | $B_4$ | $w_1 w_2^2$ | $c_4\Delta^4$ | $(\nu)$ | $(0)$ |
| 108 | 22 | $C_4$ | $h_0^3\alpha w_2^2$ | $2c_6\Delta^4$ | $(\eta, \nu)$ | $(0)$ |
| 110 | 20 | $\kappa_4$ | $d_0 w_2^2$ | 0 | $(4, \eta^2, 2\nu, 2B, \nu^3, \eta B, B^2)$ | $(2, B)$ |
| 116 | 21 | $\bar{\kappa}D_4$ | $h_0 g w_2^2$ | 0 | $(4, \eta, \nu, B)$ | $(2, B)$ |
| 117 | 21 | $\eta_4\bar{\kappa}$ | $h_1 g w_2^2$ | 0 | $(2, \eta^2, \nu, \eta B, B^2)$ | $(2, B)$ |
| 120 | 23 | $D_5$ | $h_0\alpha^2 w_2^2$ | $8\Delta^5$ | $(\eta, \nu)$ | $(0)$ |
| 122 | 22 | $\eta_1\eta_4$ | $h_1\gamma w_2^2$ | 0 | $(2, \eta^2, \nu^2, \eta B)$ | $(2)$ |
| 123 | 22 | $\nu_5$ | $\alpha\beta w_2^2$ | 0 | $(4, \nu, 2B, \eta B, B^2)$ | $(2, B)$ |
| 128 | 23 | $B_5$ | $\alpha g w_2^2$ | $c_4\Delta^5$ | $(2\nu, \nu^2)$ | $(0)$ |
| 128 | 23 | $\epsilon_5$ | $\delta' w_2^2$ | 0 | $(2, \nu^2, \eta B, B^2)$ | $(2, B)$ |
| 130 | 24 | $\kappa_4\bar{\kappa}$ | $d_0 g w_2^2$ | 0 | $(4, \eta^2, \nu, 2B, \eta B, B^2)$ | $(2, B)$ |

Table 9.5: $T$-module generators of $\pi_*(tmf)$ (cont.)

| $n$ | $s$ | $x$ | $E_\infty(tmf)$ | $mf$ | $\mathrm{Ann}(x)$ | $\sqrt{\mathrm{Ann}(x)}$ |
|---|---|---|---|---|---|---|
| 132 | 26 | $C_5$ | $h_0\alpha^3 w_2^2$ | $2c_6\Delta^5$ | $(\eta, \nu)$ | $(0)$ |
| 135 | 25 | $\eta_1\kappa_4$ | $d_0\gamma w_2^2$ | 0 | $(2, \eta^2, \nu^2, B)$ | $(2, B)$ |
| 137 | 26 | $\nu_5\kappa$ | $\alpha\beta d_0 w_2^2$ | 0 | $(2, \eta^2, \nu, B)$ | $(2, B)$ |
| 142 | 27 | $\epsilon_5\kappa$ | $d_0\delta' w_2^2$ | 0 | $(2, \eta, \nu, B^2)$ | $(2, B)$ |
| 144 | 26 | $D_6$ | $h_0^2 w_2^3$ | $4\Delta^6$ | $(\eta, 2\nu, \nu^3)$ | $(0)$ |
| 147 | 25 | $\nu_6$ | $h_2 w_2^3$ | 0 | $(8, 2B, \nu^3 + \eta B, B^2)$ | $(2, B)$ |
| 152 | 27 | $B_6$ | $c_0 w_2^3$ | $c_4\Delta^6$ | $(\nu)$ | $(0)$ |
| 156 | 30 | $C_6$ | $h_0^3\alpha w_2^3$ | $2c_6\Delta^6$ | $(\eta, \nu)$ | $(0)$ |
| 161 | 29 | $\nu_6\kappa$ | $h_2 d_0 w_2^3$ | 0 | $(2, \eta^2, \nu^2, B)$ | $(2, B)$ |
| 168 | 31 | $D_7$ | $h_0\alpha^2 w_2^3$ | $8\Delta^7$ | $(\eta, \nu)$ | $(0)$ |
| 176 | 31 | $B_7$ | $\delta w_2^3$ | $c_4\Delta^7$ | $(\nu)$ | $(0)$ |
| 180 | 34 | $C_7$ | $h_0\alpha^3 w_2^3$ | $2c_6\Delta^7$ | $(\eta, \nu)$ | $(0)$ |

We have the following complement to Proposition 9.21.

PROPOSITION 9.33. *The ring homomorphism $q_0\colon \pi_*(tmf) \to \pi_*(ko)$ is given on the $\eta$-, $\nu$-, $\epsilon$-, $\kappa$- and $\bar\kappa$-families of generators by*

$$\eta \longmapsto \eta\,,$$

*while the remaining generators map to* 0.

PROOF. The map $E_2(tmf) = \mathrm{Ext}_{A(2)}(\mathbb{F}_2, \mathbb{F}_2) \to \mathrm{Ext}_{A(1)}(\mathbb{F}_2, \mathbb{F}_2) = E_2(ko)$ of Adams $E_2$-terms takes $h_1$ to $h_1$, so $q_0$ maps $\eta$ to $\eta$. For $k \in \{1, 4\}$, the relation $\eta_k B = \eta B_k$ implies that $q_0(\eta_k)B = \eta q_0(B_k) = 0$, so $q_0(\eta_k) = 0$ since $\pi_*(ko)$ is $B$-torsion free. The $B$-power torsion generators $\nu_k$, $\epsilon_k$, $\kappa_k$ and $\bar\kappa$ map to 0 for the same reason (or because the target is 2-torsion free in these degrees). □

We note that $q_0$ factors through $\pi_*(tmf)/\Gamma_B\pi_*(tmf) \cong \bigoplus_{k=0}^7 ko[k] \otimes \mathbb{Z}[M]$. On $ko[0]$ it is the injective $\mathbb{Z}[B]$-linear homomorphism given by

$$\begin{aligned} 1 &\longmapsto 1 \\ \eta &\longmapsto \eta \\ \eta^2 &\longmapsto \eta^2 \\ C &\longmapsto -AB\,, \end{aligned}$$

while it is zero on the $M$-multiples of $ko[0]$, and on the summands $ko[k] \otimes \mathbb{Z}[M]$ for $1 \le k \le 7$.

## 9.5. Relations in $\pi_*(tmf)$

Using the detecting classes in $E_\infty(tmf)$, the images in $mf_{*/2}$, and the hidden 2-, $\eta$- and $\nu$-extensions we have found, we are now able to compute nearly every product in the algebra $\pi_*(tmf)$. There is one sign we have not determined: $\nu_4\nu_6 =$

$s\nu\nu_2 M$, where $s \in \{\pm 1\}$. This same sign appears in the products $\nu_4 D_4 = 2s\nu M$ and $\nu_6 D_4 = 2s\nu_2 M$. All other products are completely known. We first make a systematic study of products involving the 2-power torsion classes in $\pi_*(tmf)$. Thereafter we turn to products of 2-power torsion classes and 2-torsion free classes. Finally we discuss the products of 2-torsion free classes.

Recall that we adopt the conventions that $x_0 = x$ and $x_{k+8} = x_k M$ for $x \in \{\eta, \nu, \epsilon, \kappa, \bar{\kappa}, B, C, D\}$. As a heuristic guide, note that one may expect a close relationship between all the elements $x_i y_j$ for a fixed value of $n = i + j$, on the grounds that in some spectral sequence these were all represented by $xy\Delta^n$, up to scalars for the $\nu$- and $D$-families. As stated, the heuristic fails for the $B$-family, as shown by the relation $\eta_1 B_1 = \eta B_2 + \nu^2 \nu_2$. However, it applies well with the modified $\widetilde{B}$-family, introduced in Definition 9.50, as we make precise in Corollary 9.56. Recall also the numerical function $d_k$ from Definition 9.18. We have

$$d_{7-n}/2 = \begin{cases} 4 & \text{for } n \equiv 0, 2, 4, 6 \mod 8, \\ 2 & \text{for } n \equiv 1, 5 \mod 8, \\ 1 & \text{for } n \equiv 3 \mod 8 \end{cases}$$

(omitting the case $n \equiv 7 \mod 8$) and

$$8/d_{n+2} = \begin{cases} 1 & \text{for } n \equiv 1, 3, 5, 7 \mod 8, \\ 2 & \text{for } n \equiv 0, 4 \mod 8, \\ 4 & \text{for } n \equiv 2 \mod 8, \\ 8 & \text{for } n \equiv 6 \mod 8. \end{cases}$$

We start with the products of the classes in the $\eta$-family, with two or more factors.

PROPOSITION 9.34.

(1) $\eta_i \eta_j = \eta_k \eta_\ell$ *only if* $\{i, j\} = \{k, \ell\}$, *for* $i, j, k, \ell \in \{0, 1, 4\}$.
(2) $\eta_4^2 = \eta^2 M$.
(3) $\eta_i \eta_j \eta_k = (d_{7-n}/2)\nu_n$ *where* $n = i + j + k$. *These are the unique classes of order* 2 *in their degree.*
(4) $\eta_1^4 = \bar{\kappa}^5 = \eta \nu_4 \neq 0$, *while all other* 4*-fold products of the* $\eta_i$ *are* 0.
(5) $\eta_1^5 = \eta^2 \nu_5 = \eta_1 \eta_4 \nu$ *is the unique nonzero element in* $\pi_{125}(tmf)$.
(6) $\eta_1^6 = 4\nu\nu_6$ *is the unique element of order* 2 *in* $\pi_{150}(tmf)$.
(7) $\eta_1^7 = 0$.

PROOF.

(1) The degree of $\eta_i \eta_j$ determines the set $\{i, j\}$ for $i, j \in \{0, 1, 4\}$.
(2) The normalization $\eta_4 B = \eta B_4$ in our choice of $\eta_4$ implies that $\eta_4^2 B^2 = \eta^2 B_4^2$. We have $B_4^2 = B^2 M$ since both are detected by $c_4^2 \Delta^8$ in the ring of modular forms and $\pi_{208}(tmf)$ is 2-torsion free. Hence $\eta_4^2 = \eta^2 M$, since $B^2$ acts monomorphically on $\pi_{194}(tmf)$.
(3) For $0 \le n \le 6$ these relations are visible at the $E_\infty$-term, since there are no classes of higher Adams filtration than $\eta_i \eta_j \eta_k$ in its degree. For $n = 2$ or 6 this depends on the relation $h_1 \gamma^2 = h_0^2 h_2 w_2$. The remaining cases, $n \in \{8, 9, 12\}$, follow from $\eta_4^2 = \eta^2 M$.
(4) The product $\eta_i \eta_j \eta_k$ is divisible by 2, so will annihilate $\eta_\ell$, unless $i = j = k = 1$. This applies equally to all 3-element subsets of the factors

| | $\nu$ | $\nu_1$ | $\nu_2$ | $\nu_3$ | $\nu_4$ | $\nu_5$ | $\nu_6$ |
|---|---|---|---|---|---|---|---|
| $\nu$ | $\nu^2$ | 0 | $\nu\nu_2$ | 0 | $\nu\nu_4$ | 0 | $\nu\nu_6$ |
| $\nu_1$ | 0 | $2\nu\nu_2$ | 0 | 0 | 0 | $2a\nu\nu_6$ | 0 |
| $\nu_2$ | $-\nu\nu_2$ | 0 | $\nu\nu_4$ | 0 | $b\nu\nu_6$ | 0 | $\nu^2 M$ |
| $\nu_3$ | 0 | 0 | 0 | $4\nu\nu_6$ | 0 | 0 | 0 |
| $\nu_4$ | $-\nu\nu_4$ | 0 | $-b\nu\nu_6$ | 0 | $\nu^2 M$ | 0 | $s\nu\nu_2 M$ |
| $\nu_5$ | 0 | $-2a\nu\nu_6$ | 0 | 0 | 0 | $2\nu\nu_2 M$ | 0 |
| $\nu_6$ | $-\nu\nu_6$ | 0 | $-\nu^2 M$ | 0 | $-s\nu\nu_2 M$ | 0 | $\nu\nu_4 M$ |

FIGURE 9.4. Products of the $\nu_i$, with $\nu_i$ chosen independently

of a 4-fold product. The relation $\gamma^4 = g^5$ holds in $E_2(tmf)$, and this survives to detect the only element of order 2 in degree 100, which is $\eta\nu_4$ by Theorem 9.16.

(5) The group $\pi_{125}(tmf) \cong \mathbb{Z}/2$ is generated by $\eta^2\nu_5 = \eta_1\eta_4\nu$, detected by $h_1 g w_1 w_2^2$. The element $\eta_1^5$ is detected by $\gamma g^5$, which equals $h_1 g w_1 w_2^2$ in $E_\infty(tmf)$ because $d_3(\beta^2 w_2^2) = h_1 g w_1 w_2^2 + \gamma g^5$.

(6) The group $\pi_{150}(tmf) \cong \mathbb{Z}/8$ is generated by $\nu\nu_6$. We have $\eta_1^6 = \eta_1\eta_1^5 = \eta_1^2\eta_4\nu = 4\nu\nu_6$ by cases (3) and (5).

(7) Use $\pi_{175}(tmf) = 0$, or note that $4\eta_1 = 0$. □

We continue with the products of the classes in the $\nu$-family, first with two factors.

PROPOSITION 9.35. *The product $\nu_i\nu_j$ with $i+j=n$ lies in $\pi_{6+24n}(tmf)$, which is a cyclic group of order $8/d_{n+2} = \gcd(n+2, 8)$. In particular, the group is trivial if $n$ is odd, so $\nu_i\nu_j = 0$ unless $i \equiv j \mod 2$. We can (and do) choose $\nu_5$ and $\nu_6$ so that $\nu_1\nu_5 = 2\nu\nu_6$ and $\nu_2\nu_4 = 3\nu\nu_6$. This completely determines $\nu_5$ and $\nu_6$, given $\nu_1$, $\nu_2$ and $\nu_4$, and makes the relations in Figure 9.5 hold.*

| | $\nu$ | $\nu_1$ | $\nu_2$ | $\nu_3$ | $\nu_4$ | $\nu_5$ | $\nu_6$ |
|---|---|---|---|---|---|---|---|
| $\nu$ | $\nu^2$ | 0 | $\nu\nu_2$ | 0 | $\nu\nu_4$ | 0 | $\nu\nu_6$ |
| $\nu_1$ | 0 | $2\nu\nu_2$ | 0 | 0 | 0 | $2\nu\nu_6$ | 0 |
| $\nu_2$ | $-\nu\nu_2$ | 0 | $\nu\nu_4$ | 0 | $3\nu\nu_6$ | 0 | $\nu^2 M$ |
| $\nu_3$ | 0 | 0 | 0 | $4\nu\nu_6$ | 0 | 0 | 0 |
| $\nu_4$ | $-\nu\nu_4$ | 0 | $-3\nu\nu_6$ | 0 | $\nu^2 M$ | 0 | $s\,\nu\nu_2 M$ |
| $\nu_5$ | 0 | $-2\nu\nu_6$ | 0 | 0 | 0 | $2\nu\nu_2 M$ | 0 |
| $\nu_6$ | $-\nu\nu_6$ | 0 | $-\nu^2 M$ | 0 | $-s\,\nu\nu_2 M$ | 0 | $\nu\nu_4 M$ |

FIGURE 9.5. Products of the $\nu_i$, with specified $\nu_5$ and $\nu_6$ and $s \in \{\pm 1\}$

*The ambiguity in $\nu_4\nu_6 = s\nu\nu_2 M$, $s \in \{\pm 1\}$, is not affected by the choices of the $\nu_1$, $\nu_2$ and $\nu_4$, and has not yet been determined.*

PROOF. We start by working out the multiplication table with arbitrary choices of generators $\nu_i$, recording the results in Figure 9.4. In it, $a, s \in \{\pm 1\}$ and $b \in \{1, 3, 5, 7\}$ are odd integers depending upon the choices of the $\nu_i$. Most entries have well-defined coefficients, independent of the choices of specific generators $\nu_i$, because of the orders of the groups. The matrix is antisymmetric because the $\nu_i$ are all in odd degrees. In particular, the elements along the main diagonal all have order 2. We have written the result in the simplest form possible for the most part, but do show the antisymmetry even for classes of order 2.

The products $\nu_i\nu_j$ for $0 \le i, j \le 6$ land in the groups $\pi_{6+24n}(tmf)$, for $0 \le n \le 12$. These are trivial when $n$ is odd, so we only need to discuss the cases when $n$ is even. Note that in degrees 165 and higher, the $B$-power torsion classes are all $M$-multiples. As a result, in positive degrees $k \equiv 3, 5, 6, 7 \mod 8$, multiplication by $M$ is an isomorphism $\pi_k(tmf) \cong \pi_{k+192}(tmf)$.

($n = 0$) $\pi_6(tmf) \cong \mathbb{Z}/2$ is generated by $\nu^2$.

(2) $\pi_{54}(tmf) \cong \mathbb{Z}/4$ is generated by $\nu\nu_2$. Theorem 9.8 shows that $\nu_1^2$, detected by $\alpha^2\beta^2 = d_0 g^2$, is twice the generator.

(4) $\pi_{102}(tmf) \cong \mathbb{Z}/2$ is generated by $\nu\nu_4$. Both $\nu\nu_4$ and $\nu_2^2$ are detected by $h_2^2 w_2^2$. The product $\nu_1\nu_3 = \eta_1^3\nu_1$ has Adams filtration $\ge 21$ and is therefore zero.

(6) $\pi_{150}(tmf) \cong \mathbb{Z}/8$ is generated by $\nu\nu_6$. Theorem 9.8 shows that $\nu_1\nu_5$, detected by $\alpha^2\beta^2 w_2^2 = d_0 g^2 w_2^2$, is a class of order 4, which we can write as $2a\nu\nu_6$ for some $a \in \{\pm 1\}$. The product $\nu_2\nu_4$ is detected by $h_2^2 w_2^3$ and can therefore be written as $b\nu\nu_6$ for some $b \in \{1, 3, 5, 7\}$. The product $\nu_3^2 = \eta_1^6$ is $4\nu\nu_6$, the unique class of order 2, by Proposition 9.34.

(8) $\pi_{198}(tmf) \cong \pi_6(tmf) \cong \mathbb{Z}/2$ is generated by $\nu^2 M$. The products $\nu_2\nu_6$ and $\nu_4^2$ are each detected by $h_2^2 w_2^4$ and are therefore equal to $\nu^2 M$. The product $\nu_3\nu_5 = \eta_1^3\nu_5$ has Adams filtration $\ge 37$, hence is zero.

(10) $\pi_{246}(tmf) \cong \pi_{54}(tmf) \cong \mathbb{Z}/4$ is generated by $\nu\nu_2 M$. The product $\nu_4\nu_6$ is detected by $h_2^2 w_2^5$, hence can be written as $s\nu\nu_2 M$ for some $s \in \{\pm 1\}$, while $\nu_5^2$ is detected by $\alpha^2\beta^2 w_2^4 = d_0 g^2 w_2^4$ and is the unique class of order 2.

(12) $\pi_{294}(tmf) \cong \pi_{102}(tmf) \cong \mathbb{Z}/2$ is generated by $\nu_6^2 = \nu\nu_4 M$, detected by $h_2^2 w_2^6$.

Multiplication by $\nu$ from $\pi_{147}(tmf) \cong \mathbb{Z}/8$ to $\pi_{150}(tmf)$ is an isomorphism, and multiplication by $\nu_1$ from $\pi_{123}(tmf) \cong \mathbb{Z}/4$ to $\pi_{150}(tmf)$ is injective. Given choices of $\nu_1$, $\nu_2$ and $\nu_4$ we can therefore specify unique choices of $\nu_6$ and $\nu_5$ by the requirements $\nu_2\nu_4 = 3\nu\nu_6$ and $\nu_1\nu_5 = 2\nu\nu_6$. With these choices, $a = 1$ and $b = 3$.

The relation $\nu_4\nu_6 = s\nu\nu_2 M$ cannot be altered by our remaining choices: multiplying $\nu_2$ or $\nu_4$ by an odd integer will modify $\nu_6$ by the same factor, and has no effect on the sign $s$. □

REMARK 9.36. If $s = 1$, then we can summarize Figure 9.5 by the curious formula

$$\nu_i \cdot \nu_j = (i+1)\nu\nu_{i+j} .$$

This expression is the reason for our choices $a = 1$ and $b = 3$ in the normalization of $\nu_5$ and $\nu_6$, and suggests that $s = 1$ is the correct value of the unresolved sign. If instead $s = -1$, then this formula fails for $\{i, j\} = \{4, 6\}$. We note that the formula is compatible with antisymmetry, because $\nu\nu_n$ has order dividing $n+2$ for each $n \geq 0$.

Because Proposition 9.35 tells us that products $\nu_i\nu_j$ are integer multiples of $\nu\nu_{i+j}$, we need only compute the products $\nu^2\nu_i$, $\nu^3\nu_i$, etc., in order to determine all products of the $\nu_i$. We do this next. Recall the convention $\nu_{n+8} = \nu_n M$ for $n \geq 0$.

PROPOSITION 9.37. *Products of three or more $\nu_i$ are as follows:*

(1) $\nu_i\nu_j\nu_k = 0$ *unless $i$, $j$ and $k$ are even.*
(2) *The three- and four-fold products with all even subscripts,*

$$\nu_i\nu_j\nu_k = \nu^2\nu_{i+j+k} \qquad \text{and} \qquad \nu_i\nu_j\nu_k\nu_\ell = \nu^3\nu_{i+j+k+\ell}\,,$$

*are classes of order 2, given by the following table.*

| $n$ | 0 | 2 | 4 | 6 |
|---|---|---|---|---|
| $\nu^2\nu_n$ | $\eta\epsilon$ | $\eta_1\epsilon_1$ | $\eta\epsilon_4 + \eta_1\bar{\kappa}^4$ | $\eta_1\epsilon_5$ |
| $\nu^3\nu_n$ | 0 | $\eta\nu_2 B = 2\bar{\kappa}^3$ | 0 | $\eta\nu_6 B$ |

(3) *Any five-fold product of the $\nu_i$ is 0.*

PROOF. This is simply a combination of the results from Proposition 9.35, Theorem 9.14 and Proposition 9.17. □

Next consider the interaction between the $\eta_i$ and the $\nu_j$. Recall the shorthand $\nu_3 = \eta_1^3$ and $\nu_7 = 0$.

PROPOSITION 9.38. *The products of the $\eta_i$ and $\nu_j$ depend only on the sum of the subscripts as follows:*

(1) $\eta_i\nu_j = \eta\nu_{i+j} = \bar{\kappa}B_{i+j-1}$, *which is nonzero if and only if* $i+j \equiv 1, 2, 4, 5, 6 \mod 8$. *(Here we treat $\bar{\kappa}B_{-1}$ as zero.)*
(2) $\eta_i\eta_j\nu_k = \eta^2\nu_{i+j+k}$, *which is nonzero if and only if* $i+j+k \equiv 2, 5, 6 \mod 8$.
(3) $\eta_i\eta_j\eta_k\nu_\ell = \eta^3\nu_{i+j+k+\ell}$, *which is nonzero if and only if* $i+j+k+\ell \equiv 6 \mod 8$.
(4) $\eta_i\eta_j\eta_k\eta_\ell\nu_m = 0$ *for all* $i, j, k, \ell, m$.
(5) $\eta_i\nu_j\nu_k = 0$ *for all* $i, j, k$.

REMARK 9.39. In particular, the nonzero $\eta_i\eta_j\eta_k\nu_\ell$ are all multiples of $\eta^3\nu_6 = \eta_1^6$ by powers of $M$.

PROOF. We start by considering the $\eta_i\nu_j$ for $i+j = n$. First, since $i \equiv 0, 1, 4 \mod 8$, if $n \geq 12$ then one of $i$ or $j$ is $\geq 8$. Hence, we may assume $0 \leq n \leq 11$, as the remaining cases can be reduced to these by considering $M$-multiples. We consider each value separately.

($n = 0$) We have $\eta\nu = 0$ in $\pi_*(S)$, and in any case $\pi_4(tmf) = 0$.
(1) Theorems 9.14 and 9.16 show that $\eta\nu_1 = \bar{\kappa}B = \eta_1\nu$.
(2) We must show $\eta\nu_2 = \eta_1\nu_1 = \bar{\kappa}B_1$. In $E_2(tmf)$, $\eta_1\nu_1$ and $\bar{\kappa}B_1$ are detected by $\alpha\beta\gamma = \alpha g^2 = \delta' g$, while Theorem 9.16 shows that $\eta\nu_2$ is detected by this as well. Hence these classes are all equal to the unique nonzero $B$-power torsion element in $\pi_{52}(tmf)$.
(3) Since there is no 2-power torsion in $\pi_{76}(tmf)$, the products $\eta\nu_3$, $\eta_1\nu_2$ and $\bar{\kappa}B_2$ are all zero.

(4) Theorems 9.14 and 9.16 show that $\eta\nu_4$, $\eta_4\nu$, $\bar{\kappa}^5 = \bar{\kappa}B_3$ and $\eta_1^4 = \eta_1\nu_3$ are all detected by $g^5 = \gamma^4 \neq 0$. Since these are 2-torsion classes in $\pi_{100}(tmf)$, they must all be equal.

(5) We must show $\eta\nu_5 = \eta_1\nu_4 = \eta_4\nu_1 = \bar{\kappa}B_4 \neq 0$. By Theorems 9.14 and 9.16 we have $\eta^2\nu_5 = \eta_1\eta_4\nu = \eta\bar{\kappa}B_4$, detected by $h_1 g w_1 w_2^2$. But $\eta_1\eta_4\nu = \eta\eta_4\nu_1$ by case (1) and $\eta_1\eta_4\nu = \eta\eta_1\nu_4$ by case (4). Since $x = \bar{\kappa}B_4$ is the unique 2-power torsion solution to $\eta x = \eta\bar{\kappa}B_4$, the claims follow.

(6) We must show $\eta\nu_6 = \eta_1\nu_5 = \eta_4\nu_2 = \bar{\kappa}B_5 \neq 0$. By Theorem 9.16, $\eta\nu_6$ is detected by $\delta' g w_2^2 = \alpha g^2 w_2^2 = \alpha\beta\gamma w_2^2$, which also detects $\eta_1\nu_5 = \bar{\kappa}B_5$. By case (2), we have $\eta\eta_4\nu_2 = \eta_1\eta_4\nu_1$, and by case (5) this is equal to $\eta\eta_1\nu_5$. Since $x = \bar{\kappa}B_5$ is the unique 2-power torsion solution to $\eta x = \eta\bar{\kappa}B_5$, the claims follow.

(7) We have $\eta\nu_7 = 0$ by definition, while $\eta_1\nu_6 = \eta_4\nu_3 = \bar{\kappa}B_6 = 0$ because there is no 2-power torsion in $\pi_{172}(tmf)$.

(8) Since there is no 2-power torsion in $\pi_{196}(tmf)$, the products $\eta\nu_8$, $\eta_1\nu_7$, $\eta_4\nu_4$, $\eta_8\nu$ and $\bar{\kappa}B_7$ are all zero.

(9) We have $\eta\nu_9 = \eta_8\nu_1 = \eta\nu_1 M$, which equals $\eta_1\nu_8 = \eta_9\nu = \eta_1\nu M = \bar{\kappa}B_8 = \bar{\kappa}BM$ by case (1). This is the only nonzero 2-power torsion class in $\pi_{220}(tmf)$, so it suffices to show that $\eta_4\nu_5$ is nonzero. Multiplying by $\eta_4$, and using that $\eta_4^2 = \eta^2 M$ from Proposition 9.34, we have $\eta_4^2\nu_5 = \eta^2\nu_5 M \neq 0$ by Theorem 9.16.

(10) Similarly, $\eta\nu_{10} = \eta_1\nu_9 = \eta_8\nu_2 = \eta_9\nu_1 = \bar{\kappa}B_9 \neq 0$ from case (2) by $M$-multiplication. It remains to show that $\eta_4\nu_6$ is $\eta\nu_2 M$, the unique nonzero 2-power torsion class in $\pi_{244}(tmf)$. Multiplying by $\eta_4$ works just as in the preceding case, since $\eta^2\nu_6 M \neq 0$.

(11) By $M$-multiplication from the case (3), $\eta\nu_{11} = \eta_1\nu_{10} = \eta_8\nu_3 = \eta_9\nu_2 = \bar{\kappa}B_2 M = 0$, while $\eta_4\nu_7 = 0$ because $\nu_7 = 0$.

We observe that the $\eta_i\nu_j$ with $i + j = n \equiv 1, 2, 4, 5, 6 \mod 8$ are nonzero by Theorem 9.16, while the relation $\eta_i\nu_j = \bar{\kappa}B_{i+j-1}$ also holds for $i + j = n \equiv 0, 3, 7 \mod 8$ since there is no 2-power torsion in degrees congruent to 4, 76 or 172 mod 192.

Next, $\eta_i\eta_j\nu_k = \eta_i\eta\nu_{j+k} = \eta^2\nu_{i+j+k}$, which is nonzero if and only if $i + j + k \equiv 2, 5, 6 \mod 8$, while $\eta_i\eta_j\eta_k\nu_\ell = \eta_i\eta^2\nu_{j+k+\ell} = \eta^3\nu_{i+j+k+\ell}$, which is nonzero precisely when $i + j + k + \ell \equiv 6 \mod 8$. Similarly, $\eta_i\eta_j\eta_k\eta_\ell\nu_m = \eta_i\eta^3\nu_{j+k+\ell+m} = \eta^4\nu_{i+j+k+\ell+m} = 0$.

Finally, by Proposition 9.35, $\nu_j\nu_k = c\nu\nu_{j+k}$ for some integer $c$, so that $\eta_i\nu_j\nu_k = c\eta_i\nu\nu_{j+k} = c\eta\nu\nu_{i+j+k} = 0$. □

Next we show that $\epsilon$ acts like $B$ on the $B$-power torsion classes. A key input is that fact that the product $\epsilon\kappa$ is nonzero in $\pi_{22}(S)$. This was proved using unstable methods by Mimura [**129**, Thm. B]. We give an independent proof, using only stable methods, in Theorems 11.71 and 11.61. Recall from Proposition 9.12 that the $B$-power torsion in $\pi_*(tmf)$ is the ideal generated by the $\nu$-, $\epsilon$- and $\kappa$-families, including $\nu_3$, and $\bar{\kappa}$.

PROPOSITION 9.40. *For $x \in \{\nu, \epsilon, \kappa, \bar{\kappa}\}$ we have $\epsilon \cdot x_k = B \cdot x_k$ for each $k$ for which $x_k$ is defined. In contrast, $\epsilon \cdot \eta_k \neq B \cdot \eta_k$ for $k = 0$, 1 and 4. Instead, we have $\epsilon \cdot \eta = \nu^3$, $\epsilon \cdot \eta_1 = \eta\epsilon_1$, and $\epsilon \cdot \eta_4 = \eta\epsilon_4$.*

PROOF. We calculate $\epsilon \cdot x_k$ for each $x_k$, in the following order:

($x_k = \eta$) We showed that $\nu \cdot \nu^2 = \eta\epsilon$ in Theorem 9.14, so $\epsilon \cdot \eta = \nu^3$.

($\nu$) Since $\pi_{11}(tmf)$ is trivial, $\epsilon \cdot \nu = 0 = B \cdot \nu$.

($\epsilon$) Since $\pi_{16}(tmf)$ contains no 2-torsion, $\epsilon \cdot \epsilon = 0 = B \cdot \epsilon$.

($\kappa$) By Mimura [**129**, Thm. B], or our Theorems 11.71 and 11.61, the product $\epsilon\kappa$ is nonzero of Adams filtration $\geq 7$ in $\pi_{22}(S)$, and must therefore be detected by $Pd_0$ in $E_\infty(S)$. The unit map $\iota\colon S \to tmf$ takes $Pd_0$ to $d_0 w_1$, by Proposition 1.14. Hence both $\epsilon \cdot \kappa$ and $B \cdot \kappa$ are detected by $d_0 w_1$ in $E_\infty(tmf)$, and must therefore be equal in $\pi_{22}(tmf)$.

($\bar\kappa$) From $\epsilon\kappa = \kappa B$ we get $\epsilon\kappa\bar\kappa = \kappa\bar\kappa B$. Since $d_0 \cdot g w_1 \neq 0$ in $E_\infty(tmf)$, multiplication by $\kappa$ acts injectively on the 2-power torsion in $\pi_{28}(tmf)$, so we can conclude that $\epsilon \cdot \bar\kappa = B \cdot \bar\kappa$.

($\eta_1$) The products $\epsilon \cdot \eta_1$ and $\eta\epsilon_1$ are detected by $c_0\gamma = h_1\delta$ in the Adams spectral sequence, hence are both equal to the unique nonzero $B$-power torsion class in $\pi_{33}(tmf)$.

($\nu_2$) From $\eta\epsilon = \nu^3$ and Proposition 9.37 we have $\epsilon \cdot \eta\nu_2 = \nu^3\nu_2 = B \cdot \eta\nu_2 \neq 0$, so $\epsilon \cdot \nu_2$ and $B \cdot \nu_2$ are both equal to the unique nonzero class in $\pi_{59}(tmf)$.

($\nu_1$) By Proposition 9.38 and the previous case, $\epsilon \cdot \eta_1\nu_1 = \epsilon \cdot \eta\nu_2 = B \cdot \eta\nu_2 = B \cdot \eta_1\nu_1 \neq 0$, so $\epsilon \cdot \nu_1$ and $B \cdot \nu_1$ are both equal to the unique nonzero element in $\pi_{35}(tmf)$.

($\epsilon_1$) Both $\epsilon_1\bar\kappa$ and $\bar\kappa B_1$ are detected by $\delta' g = \alpha g^2$, hence equal the unique nonzero $B$-power torsion class in $\pi_{52}(tmf)$. By Proposition 9.38 and case ($\nu_2$), $\epsilon \cdot \epsilon_1\bar\kappa = \epsilon \cdot \bar\kappa B_1 = \epsilon \cdot \eta\nu_2 \neq 0$. It follows that both $\epsilon \cdot \epsilon_1 \neq 0$ and $B \cdot \epsilon_1$ are equal to the unique $B$-torsion class of order 2 in $\pi_{40}(tmf)$.

($\nu_3$) Since $\pi_{83}(tmf)$ is trivial, $\epsilon \cdot \nu_3 = 0 = B \cdot \nu_3$.

($\eta_4$) The products $\eta_4\epsilon_1$ and $\eta\epsilon_5$ are both detected by $h_1\delta' w_2^2$ in the Adams spectral sequence, hence equal the unique nonzero $B$-power torsion class in $\pi_{129}(tmf)$. Furthermore, $\epsilon \cdot \eta_4$ and $\eta\epsilon_4$ are both detected by $h_1 c_0 w_2^2$, hence agree modulo a $B$-torsion class of Adams filtration $\geq 21$, i.e., modulo $\eta_1\bar\kappa^4$. Since $\eta_1\nu^2 = 0$ and $\eta\epsilon_4 + \eta_1\bar\kappa^4 = \nu^2\nu_4$ by Proposition 9.17, we cannot have $\epsilon \cdot \eta_4 = \nu^2\nu_4$, since $\eta_1\eta_4\epsilon = \eta\eta_4\epsilon_1 = \eta^2\epsilon_5 \neq 0$. Having eliminated the only alternative, we deduce that $\epsilon \cdot \eta_4 = \eta\epsilon_4$.

($\nu_4$) Since $\pi_{107}(tmf)$ is trivial, $\epsilon \cdot \nu_4 = 0 = B \cdot \nu_4$.

($\epsilon_4$) Since the $B$-torsion in $\pi_{112}(tmf)$ is zero, $\epsilon \cdot \epsilon_4 = 0 = B \cdot \epsilon_4$.

($\kappa_4$) From $\epsilon\bar\kappa = \bar\kappa B$ we have $\epsilon\kappa_4\bar\kappa = \kappa_4\bar\kappa B$, which is nonzero because it is detected by $d_0 g w_1 w_2^2 \neq 0$. Hence $\epsilon \cdot \kappa_4 = B \cdot \kappa_4$ is the unique nonzero element in $\pi_{118}(tmf)$.

($\nu_6$) As in case ($\nu_2$), we have $\epsilon \cdot \eta\nu_6 = \nu^3\nu_6 = B \cdot \eta\nu_6 \neq 0$, and there is a unique nonzero class in $\pi_{155}(tmf)$.

($\nu_5$) As in case ($\nu_1$), $\epsilon \cdot \eta_1\nu_5 = \epsilon \cdot \eta\nu_6 = B \cdot \eta\nu_6 = B \cdot \eta_1\nu_5 \neq 0$, and there is a unique nonzero class in $\pi_{131}(tmf)$.

($\epsilon_5$) As in case ($\epsilon_1$), both $\epsilon_5\bar\kappa$ and $\bar\kappa B_5$ are detected by $\delta' g w_2^2 = \alpha g^2 w_2^2$, hence equal the unique nonzero $B$-power torsion class in $\pi_{148}(tmf)$. Thus $\epsilon \cdot \epsilon_5\bar\kappa = \epsilon \cdot \bar\kappa B_5 = \epsilon \cdot \eta\nu_6 = \nu^3\nu_6 \neq 0$. It follows that $\epsilon \cdot \epsilon_5 \neq 0$ and $B \cdot \epsilon_5$ are both equal to the unique nonzero $B$-torsion class in $\pi_{136}(tmf)$. □

Most of the following relations involving the $\epsilon_k$ have already been established.

PROPOSITION 9.41. *$\epsilon_k\bar\kappa = \bar\kappa B_k = \eta\nu_{k+1}$ for $k = 0$, 1, 4 and 5.*

PROOF. We showed in Proposition 9.38 that $\bar\kappa B_k = \eta\nu_{k+1}$ for any $k$.

In the course of the proof of Proposition 9.40, we also showed that $\epsilon_k\bar{\kappa} = \bar{\kappa}B_k$ for $k = 0$, 1 and 5. It remains to prove that $\epsilon_4\bar{\kappa} = \bar{\kappa}B_4$. By Proposition 9.17,

$$\eta\epsilon_4\bar{\kappa} = (\nu^2\nu_4 + \eta_1\bar{\kappa}^4)\bar{\kappa} = \eta_1\bar{\kappa}^5 = \eta\bar{\kappa}B_4\,.$$

Dividing by $\eta$ is possible and proves the relation. □

We next consider products of the $\epsilon_k$.

PROPOSITION 9.42.

(1) *For two-fold products we have* $\epsilon_i\epsilon_j = 0$, *unless* $i+j \equiv 1 \mod 4$, *for which* $\epsilon_i\epsilon_j = \epsilon\epsilon_{i+j} \neq 0$.
(2) *All products* $\epsilon_i\epsilon_j\epsilon_k$ *are* 0.

PROOF. The products $\epsilon_i\epsilon_j$ lie in degrees with no $B$-power torsion unless $i+j \equiv 1 \mod 4$. We have just shown that $\epsilon\epsilon_j = \epsilon_j B \neq 0$ when $j \equiv 1 \mod 4$, so it remains only to consider $\epsilon_1\epsilon_4$ and $\epsilon_4\epsilon_5$.

In Proposition 9.41 we saw that $\epsilon_1\epsilon_4\bar{\kappa} = \epsilon_4\bar{\kappa}B_1 = \bar{\kappa}B_1B_4$, which is detected by $g \cdot \alpha g \cdot w_1w_2^2 = \alpha g^2w_1w_2^2 = \delta' g w_1 w_2^2$ and is therefore nonzero and equal to $\nu^3\nu_6 = \epsilon_5\bar{\kappa}B$. Since there is only one nonzero $B$-torsion class in $\pi_{136}(tmf)$, $\epsilon_1\epsilon_4 = \epsilon_5 B = \epsilon\epsilon_5$. For $\epsilon_4\epsilon_5$, $\epsilon_4\epsilon_5\bar{\kappa} = \bar{\kappa}B_4B_5$, detected by $\alpha g^2 w_1 w_2^4 = \delta' g w_1 w_2^4$. This is nonzero, and $\epsilon_1 BM = \epsilon_9 B$ is the unique $B$-torsion class of order 2 in $\pi_{232}(tmf)$, so $\epsilon_4\epsilon_5 = \epsilon_9 B = \epsilon\epsilon_9$.

The three-fold products of the $\epsilon_i$ are 0 because the two-fold products are multiples of either $B\epsilon_1$ or $B\epsilon_5$, and we have $B^2\epsilon_1 = 0$ and $B^2\epsilon_5 = 0$ since there are no 2-torsion classes in these degrees. □

PROPOSITION 9.43. *The products of the* $\eta_i$ *and* $\epsilon_j$ *depend only on the sum of the subscripts as follows:*

(1) $\eta_i\epsilon_j$ *is always nonzero and can be described as follows:*

$$\eta_i\epsilon_j = \begin{cases} \eta\epsilon_{i+j} & \text{for } i+j \equiv 0,1 \mod 4, \\ \eta_1\epsilon_{i+j-1} & \text{for } i+j \equiv 2 \mod 4. \end{cases}$$

*These elements satisfy the following relations:*

| $i+j$ | $\eta_i\epsilon_j$ |
|---|---|
| 0 | $\eta\epsilon = \nu^3$ |
| 1 | $\eta\epsilon_1$ |
| 2 | $\eta_1\epsilon_1 = \nu^2\nu_2$ |
| 4 | $\eta\epsilon_4 = \nu^2\nu_4 + \eta_1\bar{\kappa}^4$ |
| 5 | $\eta\epsilon_5$ |
| 6 | $\eta_1\epsilon_5 = \nu^2\nu_6$ |

(2) $\eta_i\eta_j\epsilon_k = 0$ *unless* $i+j+k \equiv 5 \mod 8$, *in which case it is the appropriate power of* $M$ *times* $\eta^2\epsilon_5 = 2\kappa_4\bar{\kappa} = \eta_1^2\bar{\kappa}^4$.
(3) $\eta_i\eta_j\eta_k\epsilon_\ell = 0$ *for for all* $i, j, k, \ell$.
(4) $\eta_i\epsilon_j\epsilon_k = 0$ *for all* $i, j, k$.

PROOF. (1) Products with $\epsilon$ have already been dealt with in Proposition 9.40. In particular, we showed that $\eta_1\epsilon = \eta\epsilon_1$ and that $\eta_4\epsilon = \eta\epsilon_4$, which we earlier saw equals $\nu^2\nu_4 + \eta_1\bar{\kappa}^4$.

In Theorem 9.14 we saw that $\nu^2\nu_2 \in \{\gamma\delta'\}$ and $\nu^2\nu_6 \in \{\gamma\delta' w_2^2\}$. Clearly $\eta_1\epsilon_1 \in \{\gamma\delta'\}$ and $\eta_1\epsilon_5 \in \{\gamma\delta' w_2^2\}$. Hence $\nu^2\nu_2 = \eta_1\epsilon_1$ are both equal to the unique nonzero $B$-power torsion class in $\pi_{57}(tmf)$, and $\nu^2\nu_6 = \eta_1\epsilon_5$ are both equal to the unique nonzero $B$-power torsion class in $\pi_{153}(tmf)$.

The relation $h_1\delta' = c_0\gamma$ in $E_2(tmf)$ implies that $\eta_1\epsilon_4$, $\eta_4\epsilon_1$ and $\eta\epsilon_5$ are all detected by the same class at $E_\infty$, and this detects the unique nonzero $B$-power torsion class in $\pi_{129}(tmf)$.

The products $\eta_4\epsilon_4$ and $\eta_4\epsilon_5$ are detected by $h_1c_0w_2^4$ and $h_1\delta' w_2^4$, detecting $\eta\epsilon M$ and $\eta\epsilon_1 M$, respectively. In both cases these are the unique nonzero $B$-power torsion classes in their degrees.

(2) The products $\eta_i\eta_j\epsilon_k$ are $B$-power torsion, hence are zero unless $n = i + j + k \equiv 1, 5 \mod 8$. They must also be zero when $n \equiv 1 \mod 8$ because they have Adams filtration at least $4n + 5$, and the only $B$-power torsion class in degree $10 + 24n$ when $n \equiv 1 \mod 8$ is in Adams filtration $4n + 4$.

By Theorems 9.16 and 9.8, $\eta^2\epsilon_5$, $2\kappa_4\bar{\kappa}$ and $\eta_1^2\bar{\kappa}^4$ are all detected by $\gamma^2 g^4$, hence are all equal to the unique $B$-power torsion class of order 2 in $\pi_{130}(tmf)$.

(3) Since $\eta^2\epsilon_5 = 2\kappa_4\bar{\kappa}$, while each $\eta_i$ has order 2, multiplying by another member of the $\eta$-family must produce 0.

(4) Similarly, products $\epsilon_j\epsilon_k$ are always multiples of either $B\epsilon_1$ or $B\epsilon_5$, and it is easily checked that their products with $\eta$, $\eta_1$ and $\eta_4$ all lie in Adams filtrations that have no $B$-torsion. □

PROPOSITION 9.44. *The product $\nu_i\epsilon_j$ only depends on the sum $i + j$, with the usual conventions that $\nu_3 = \eta_1^3$, $\nu_7 = 0$ and $\nu_{k+8} = \nu_k M$. These products can be expressed as follows:*

| $i+j$ | $\nu_i\epsilon_j$ |
|---|---|
| 0 | $\nu\epsilon = \nu B = 0$ |
| 1 | $\nu_1\epsilon = \nu_1 B = \eta\kappa\bar{\kappa}$ |
| 2 | $\nu_2\epsilon = \nu_2 B = \eta_1\kappa\bar{\kappa}$ |
| 3 | $\nu_3\epsilon = \nu_3 B = 0$ |
| 4 | $\nu_4\epsilon = \nu_4 B = 0$ |
| 5 | $\nu_5\epsilon = \nu_5 B = \eta\kappa_4\bar{\kappa} = \eta_4\kappa\bar{\kappa}$ |
| 6 | $\nu_6\epsilon = \nu_6 B = \eta_1\kappa_4\bar{\kappa}$ |
| 7 | $\nu_7\epsilon = \nu_7 B = 0$ |

PROOF. When $n = i + j \equiv 0, 3 \mod 4$, $\pi_{11+24n}(tmf) = 0$, proving the result in those cases.

When $i + j = 1$, the result is established in Theorems 9.14, 9.16 and Proposition 9.40.

When $i+j = 2$ we have $\nu_1\epsilon_1 = \eta_1\kappa\bar{\kappa}$ since both are detected by $\alpha\beta\delta' = \alpha^2\beta g = d_0\gamma g$. By Proposition 9.40 we have $\nu_2\epsilon = \nu_2 B$, detected by $h_2w_1w_2$. These are equal at $E_\infty$ because of the differential $d_2(\alpha w_2) = d_0\gamma g + h_2w_1w_2$.

For $i+j=5$, Propositions 9.35, 9.38 and 9.43 show that multiplying each of $\nu\epsilon_5$, $\nu_1\epsilon_4$, $\nu_4\epsilon_1$ and $\nu_5\epsilon$ in $\pi_{131}(tmf)=\mathbb{Z}/2$ by $\eta_1$ yields $\nu^3\nu_6\neq 0$, hence they must be equal. Theorem 9.16 shows this class is $\eta\kappa_4\bar{\kappa}$, and the equality $\eta\kappa_4=\eta_4\kappa$ is already true in $E_2(tmf)$.

Similarly, multiplying any of $\nu_1\epsilon_5$, $\nu_2\epsilon_4$, $\nu_5\epsilon_1$ or $\nu_6\epsilon$ in $\pi_{155}(tmf)=\mathbb{Z}/2$ by $\eta$ yields $\nu^3\nu_6\neq 0$, showing these are all equal. We have $\nu_5\epsilon_1=\eta_1\kappa_4\bar{\kappa}$ because the classes detecting them are $\alpha\beta w_2^2\cdot\delta'=\gamma\cdot d_0w_2^2\cdot g$.

Finally, $i+j=9$ is handled in the same way, multiplying by $\eta_1$, and $i+j=10$ is handled by multiplying by $\eta$. □

PROPOSITION 9.45. *The product $\epsilon_i\kappa_j$ equals $(\eta\eta)_{i+j}\bar{\kappa}$, where $(\eta\eta)_n$ denotes $\eta_k\eta_\ell$ for any $k$ and $\ell$ with $k+\ell=n$. These can be expressed as follows:*

| $i+j$ | $\epsilon_i\kappa_j$ |
|---|---|
| 0 | $\epsilon\kappa=\eta^2\bar{\kappa}$ |
| 1 | $\epsilon_1\kappa=\eta\eta_1\bar{\kappa}$ |
| 4 | $\epsilon_4\kappa=\epsilon\kappa_4=\eta\eta_4\bar{\kappa}$ |
| 5 | $\epsilon_5\kappa=\epsilon_1\kappa_4=\eta_1\eta_4\bar{\kappa}$ |
| 8 | $\epsilon_4\kappa_4=\eta^2\bar{\kappa}M$ |
| 9 | $\epsilon_5\kappa_4=\eta\eta_1\bar{\kappa}M$ |

PROOF. We consider these products for $n=i+j$ on a case by case basis.

($n=0$) Theorem 9.16 and Proposition 9.40 imply that $\eta^2\bar{\kappa}$ is the unique nonzero class $\epsilon\kappa=\kappa B$ in $\pi_{22}(tmf)$, but this also follows directly from case (22) of Theorem 11.61, which was ultimately used in the proofs of those two results.

(1) Theorem 9.16 shows that $\eta\cdot\eta_1\bar{\kappa}$ is detected by $d_0\delta'$, and hence equals $\epsilon_1\kappa$, since this is the unique nonzero class in $\pi_{46}(tmf)$.

(4) Theorem 9.16 and Proposition 9.40 show that $\eta\eta_4\bar{\kappa}=\epsilon\kappa_4=\kappa_4B$. By Propositions 9.41 and 9.38 and Theorem 9.16, $\bar{\kappa}\cdot\epsilon_4\kappa=\eta\nu_5\kappa=\eta_4\nu_1\kappa=\bar{\kappa}\cdot\eta\eta_4\bar{\kappa}$. Multiplication by $\bar{\kappa}$ is a monomorphism here and the result follows.

(5) By Theorem 9.16 the product $\eta\cdot\eta_4\bar{\kappa}$ is detected by $d_0w_1w_2^2$ while $\eta\cdot\nu_5\kappa$ is detected by $d_0gw_1w_2^2$. Hence $\eta\eta_4\bar{\kappa}^2=\eta\nu_5\kappa$, since there is a unique nonzero $B$-power torsion class in $\pi_{138}(tmf)$. Since multiplication by $\eta$ is a monomorphism on $\pi_{137}(tmf)$, $\eta_4\bar{\kappa}^2=\nu_5\kappa$. Thus $\bar{\kappa}\cdot\eta_1\eta_4\bar{\kappa}=\eta_1\nu_5\kappa=\eta\nu_6\kappa$ by Proposition 9.38. Now Theorem 9.16 in degree 161 shows this equals $\bar{\kappa}\cdot\epsilon_5\kappa\neq 0$. Hence $\epsilon_5\kappa=\eta_1\eta_4\bar{\kappa}$, since there is only one nonzero class in $\pi_{142}(tmf)$. The equation $\epsilon_5\kappa=\epsilon_1\kappa_4$ is already true in $E_2(tmf)$.

(8) We must show that $\epsilon_4\kappa_4=\eta^2\bar{\kappa}M$. Multiplication by $\bar{\kappa}$ is an isomorphism between the $B$-power torsion in $\pi_{22}(tmf)$, spanned by $\eta^2\bar{\kappa}$, and the $B$-power torsion in $\pi_{42}(tmf)$, spanned by $\eta^2\bar{\kappa}^2$, so it suffices to show the relation holds after multiplication by $\bar{\kappa}$. By Theorems 9.14 (in degree 39) and 9.16 (in degrees 40 and 41), we have $\bar{\kappa}\cdot\eta^2\bar{\kappa}=\eta_1\nu\kappa$. Note that $\eta\kappa_4=\eta_4\kappa$ because they are both detected by $h_1d_0w_2^2$ and $\pi_{111}(tmf)=\mathbb{Z}/2$. Then Propositions 9.41 and 9.38 show that $\bar{\kappa}\cdot\epsilon_4\kappa_4=\kappa_4\cdot\epsilon_4\bar{\kappa}=\kappa_4\cdot\eta\nu_5=\eta\kappa_4\cdot\nu_5=\eta_4\kappa\cdot\nu_5=\eta_4\nu_5\cdot\kappa=\eta_1\nu\kappa M$, as required.

(9) We have $\epsilon_5\kappa_4 = \epsilon_1\kappa M$ in $\pi_{46+192}(tmf) = \mathbb{Z}/2$, since both are detected by $\delta' w_2^2 \cdot d_0 w_2^2 = d_0\delta' w_2^4$. Since $\epsilon_1\kappa = \eta\eta_1\bar{\kappa}$ by the case $i + j = n = 1$ of this proposition, we are done. □

PROPOSITION 9.46. *$\kappa D_4 = 2\kappa_4$, $\bar{\kappa}D_4$, $\kappa\bar{\kappa}D_4 = 2\kappa_4\bar{\kappa}$, $\bar{\kappa}^2 D_4 = \eta\eta_1\kappa_4$, $\kappa\bar{\kappa}^2 D_4 = 4\nu\nu_6$ and $\bar{\kappa}^3 D_4 = \nu^3\nu_6$ are nonzero.*

PROOF. By Theorem 9.8, $\pi_{96}(tmf) \cong \mathbb{Z}^5$ is generated by classes in Adams filtrations 17, 23, 31, 39 and 48. Since $\pi_{95}(tmf) = 0$, these generators must map nontrivially onto $\pi_{96}(tmf/2)$, which is $(\mathbb{Z}/2)^5$, generated in Adams filtrations 19, 23, 31, 39 and 48. It follows that $D_4$, in Adams filtration 17, must map to a class detected by $19_{51} = \gamma^2 g\tilde{\gamma}$ in $E_\infty(tmf/2)$. This lies in the $R_2$-module summand

$$\langle \gamma^2\tilde{\gamma}, i(\delta' w_2^2)\rangle \cong \frac{\Sigma^{15,91}R_2 \oplus \Sigma^{23,151}R_2}{\langle (gw_1, 0), (g^3, w_1), (0, g^2)\rangle}$$

of $E_\infty(tmf/2)$. Since $g^3 \cdot \gamma^2 g\tilde{\gamma} \neq 0$ we see that $\bar{\kappa}^i D_4 \neq 0$ for $1 \le i \le 3$. These are $B$-power torsion elements, so $\bar{\kappa}^2 D_4 = \eta\eta_1\kappa_4$ and $\bar{\kappa}^3 D_4 = \nu^3\nu_6$. Multiplying the former by $\kappa$ gives $\kappa\bar{\kappa}^2 D_4 = \eta^3\nu_6 = 4\nu\nu_6 \neq 0$ by Theorem 9.16. This then implies that $\kappa D_4 = 2\kappa_4$ and $\kappa\bar{\kappa}D_4 = 2\kappa_4\bar{\kappa}$, as these are the only $B$-power torsion classes of order 2 in their respective degrees. □

With the preceding results in hand we can now give a nearly complete multiplicative description of $\pi_*(tmf)$. Based on this, we will give our final description of this graded algebra in the next section.

THEOREM 9.47. *The products of 2-power torsion classes and the $\mathbb{Z}[\eta, \nu, B, M]$-module generators of $\pi_*(tmf)$ are as given in Tables 9.6 and 9.7. The undetermined constants $s_i$ are $\pm 1$.*

PROOF. We saw in Table 9.3 and Proposition 9.12 that the 2-power torsion in $\pi_*(tmf)$ is generated over $\mathbb{Z}[B, M]$ by products of one or more elements $y$ in the $\eta$-, $\nu$-, $\epsilon$-, $\kappa$- and $\bar{\kappa}$-families, together with $\bar{\kappa}D_4$. The $\mathbb{Z}[\eta, \nu, B, M]$-module generators $x$ are the same as the $T$-module generators from Proposition 9.31 and Table 9.5. We calculate the products $xy$ in terms of this $T$-module structure, and list the results in the multiplication tables at the end of this section. This also determines the remaining products $x \cdot \bar{\kappa}D_4 = D_4 \cdot \bar{\kappa}x$, except for the case $D_4 \cdot \bar{\kappa}D_4 = \bar{\kappa}D_4^2 = 4\bar{\kappa}M$, which we shall account for in Theorem 9.48.

The superscripts in square brackets in the multiplication tables refer to the following set of arguments establishing these products. They are proved in the order listed here, so that we may, for example, use relations of type [$0d$] and [6] to prove a relation of type [$8a$].

- [$t$] The product is tautologous, as in $\kappa \cdot \bar{\kappa} = \kappa\bar{\kappa}$ or $\bar{\kappa} \cdot \kappa_4 = \bar{\kappa}\kappa_4 = \kappa_4\bar{\kappa}$.
- [$Z$] The product lies in a zero group, hence is 0.
- [$B$] The product is a $B$-power torsion class whose Adams filtration is greater than that of all nonzero $B$-power torsion classes in its degree.
- [$T$] The product is a 2-torsion class whose Adams filtration is greater than that of all nonzero 2-torsion classes in its degree.
- [$M$] Mimura [**129**, Thm. B] proved that $\epsilon\kappa \neq 0$ in $\pi_{22}(S)$, using unstable methods. This product has Adams filtration $\geq 7$, and can only be detected by $Pd_0 \in E_\infty(S)$. As explained in the proof of Theorem 11.61, it follows that $\eta\epsilon\kappa = \nu^3\kappa = 4\nu\bar{\kappa} = \eta^3\bar{\kappa} \neq 0$ is detected by $h_1 Pd_0 \in E_\infty(S)$,

so $\eta^2\bar{\kappa}$ in Adams filtration $\geq 6$ must also be detected by $Pd_0$, which implies that $\epsilon\kappa = \eta^2\bar{\kappa}$. Alternatively, we use only stable methods to prove in Theorem 11.71 that $\eta^2\bar{\kappa} \neq 0$ is detected by $Pd_0$. It follows that $\eta^3\bar{\kappa} = \eta\epsilon\kappa \neq 0$ is detected by $h_1Pd_0$, which implies Mimura's result $\epsilon\kappa \neq 0$. Either way, we can compose with $\iota\colon S \to tmf$ to deduce that $\epsilon\kappa = \eta^2\bar{\kappa} \neq 0$ in $\pi_{22}(tmf)$, detected by $d_0w_1$.

[*S*] We have $\nu_{2j}D_4 = \pm 2\nu_{2j+4}$ because both are detected by $h_0h_2w_2^{j+2}$ in $E_2(tmf)$. Similarly, we saw that $\nu_4\nu_6 = \pm\nu\nu_2M$ in Proposition 9.35. We call these signs $s, s_i \in \{\pm 1\}$ for $i = 0, 2, 4, 6$, with

- $\nu_4\nu_6 = s\nu\nu_2M$
- $\nu D_4 = 2s_0\nu_4$
- $\nu_2D_4 = 2s_2\nu_6$
- $\nu_4D_4 = 2s_4\nu M$
- $\nu_6D_4 = 2s_6\nu_2M$.

From $\nu_2(\nu D_4) = (\nu_2\nu)D_4$ and $\nu_2\nu_4 = 3\nu\nu_6$ we deduce that $s_2 = s_0$. From $\nu_4(\nu_6D_4) = (\nu_4\nu_6)D_4$ we deduce that $s_6 = ss_2$. From $\nu_6(\nu_4D_4) = (\nu_6\nu_4)D_4$ we deduce that $s_4 = ss_2$. It follows that $s$ and $s_0$ determine the other three signs, so that

$$s_2 = s_0 \qquad \text{and} \qquad s_4 = s_6 = ss_0\,.$$

[0] These are either results of normalization decisions made in our definitions of the homotopy generators, or the results of propositions proved earlier in this section. In detail:

[0*a*] Choosing $B_3$ to be detected by $\delta w_2$ and to project to $c_4\Delta^3$ in the ring of modular forms leaves two possible values for $\bar{\kappa}B_3$. In Definition 9.22 we chose $B_3$ to make $\bar{\kappa}B_3 = \bar{\kappa}^5 = \eta\nu_4$. In case (150a) of Theorem 9.8 we showed that $\kappa_4\bar{\kappa}^2 = \pm 2\nu\nu_6$. In Definition 9.22 we chose $\kappa_4$ to make $\kappa_4\bar{\kappa}^2 = 2\nu\nu_6$. In Proposition 9.35 we saw that we could choose $\nu_5$ and $\nu_6$ so that $\nu_1\nu_5 = 2\nu\nu_6$ and $\nu_2\nu_4 = 3\nu\nu_6$, and this is the choice we made in Definition 9.22.

[0*b*] See Proposition 9.34 for the products $\eta_i\eta_j$, $\eta_i\eta_j\eta_k$, and so on.

[0*c*] See Proposition 9.35 for the products $\nu_i\nu_j$.

[0*d*] See Proposition 9.38 for the products $\eta_i\nu_j$.

[0*e*] Proposition 9.40 shows that $\eta_1\epsilon = \eta\epsilon_1$, $\eta_4\epsilon = \eta\epsilon_4$ and $\epsilon\cdot x_k = B\cdot x_k$ for $x \in \{\nu, \epsilon, \kappa, \bar{\kappa}\}$. Theorems 9.8 and 9.16 then show that these $B$-multiples are the 2- and $\eta$-multiples given.

[0*f*] See Proposition 9.41.

[0*g*] Proposition 9.42 shows that $\epsilon_i\epsilon_j = \epsilon\epsilon_{i+j}$ when $i + j \equiv 1 \mod 4$, and 0 otherwise. Theorems 9.8 and 9.16 then show that these elements are the 2- and $\eta$-multiples given, taking into account the relation $\epsilon\epsilon_k = \epsilon_kB$.

[0*h*] See Proposition 9.43 for the products $\eta_i\epsilon_j$.

[0*i*] Proposition 9.44 shows that $\nu_i\epsilon_j = \nu_{i+j}\epsilon$ when $i + j \equiv 1, 2 \mod 4$, and 0 otherwise. Theorem 9.16 then shows that these elements are the $\eta$-multiples given, taking into account the relation $\nu_k\epsilon = \nu_kB$.

[0*j*] See Proposition 9.45 for the products $\epsilon_i\kappa_j$.

[0*k*] See Proposition 9.46.

[1] The product is correct in $E_2(tmf)$, hence in $E_\infty(tmf)$, with no need to use the relations in $\mathrm{Ext}_{A(2)}(\mathbb{F}_2, \mathbb{F}_2)$. It is $B$-power torsion, and there are no $B$-power torsion classes of higher Adams filtration in this degree.

[2] The product in $E_2(tmf)$ is the target of a hidden 2- or $\eta$-extension, with no need to use the relations in $\mathrm{Ext}_{A(2)}(\mathbb{F}_2, \mathbb{F}_2)$. It is $B$-power torsion, and there are no $B$-power torsion classes of higher Adams filtration in this degree. In detail:

- [2*a*] The product in $E_2(tmf)$ is the target of a hidden 2-extension (see Theorem 9.8).
- [2*b*] The product in $E_2(tmf)$ is the target of a hidden $\eta$-extension (see Theorem 9.16).

[3] This is proved in Proposition 9.17.

[4] The product in $E_2(tmf)$ can be rewritten using relations in $\mathrm{Ext}_{A(2)}(\mathbb{F}_2, \mathbb{F}_2)$, it is $B$-power torsion, and there are no $B$-power torsion classes of higher Adams filtration in this degree. The necessary relations are $h_2^2 d_0 = h_0^2 g$ and $\alpha d_0 g = d_0 \delta'$.

[5] The product in $E_2(tmf)$ can be rewritten using relations in $\mathrm{Ext}_{A(2)}(\mathbb{F}_2, \mathbb{F}_2)$ to be the target of a hidden 2- or $\eta$-extension. It is $B$-power torsion, and there are no $B$-power torsion classes of higher Adams filtration in this degree. In detail:

- [5*a*] The product in $E_2(tmf)$ is the target of a hidden 2-extension (see Theorem 9.8). The necessary relation is $d_0^2 = g w_1$.
- [5*b*] The product $E_2(tmf)$ is the target of a hidden $\eta$-extension (see Theorem 9.16). The necessary relations are $d_0^2 = g w_1$, $\alpha d_0 g = d_0 \delta'$, $\alpha\beta\gamma = \alpha g^2 = \delta' g$ and $\gamma^4 = g^5$.

[6] The product takes place in a degree congruent to 59 mod 96, which contains a unique nonzero element detected by $h_2 w_1 w_2$ times the appropriate power of $w_2^2$. The $w_2^2$-linear differential $d_2(\alpha w_2) = d_0 \gamma g + h_2 w_1 w_2$ from Table 5.1 shows that the asserted relation holds in $E_3(tmf)$, hence also in $E_\infty(tmf)$ and $\pi_*(tmf)$, using $\alpha^2 \beta = d_0 \gamma$ in $\mathrm{Ext}_{A(2)}(\mathbb{F}_2, \mathbb{F}_2)$.

[7] These formulas for $\eta_i B_j$ hold in $E_2(tmf)$, hence also in $E_\infty(tmf)$. Multiplication by $B$ acts monomorphically on higher Adams filtrations in these degrees, so it suffices to verify the relations after multiplication by $B$ or $B^2$. This follows from the relation $\eta_k B = \eta B_k$ (Definition 9.22) and the fact (Proposition 9.19) that $B_i B_j = B B_{i+j}$ modulo 2-power torsion. In detail:

- [7*a*] These products hold in $E_2(tmf)$, as consequences of the relations $\gamma\delta = h_1 c_0 w_2$ or $c_0 \gamma = h_1 \delta$. The relations in $\pi_*(tmf)$ hold after multiplication by $B$. In some cases we use $\eta_1 \epsilon_1 = \nu^2 \nu_2$ or $\eta_1 \epsilon_5 = \nu^2 \nu_6$ to rewrite the expressions in our preferred form.
- [7*b*] The products $\eta_1 B_4$ and $\eta B_5$ are both detected by $\gamma w_1 w_2^2$ (using Theorem 9.16). They must be equal in $\pi_*(tmf)$, since $B^2 \cdot \eta_1 B_4 = \eta B B_1 B_4 = B^2 \cdot \eta B_5$.
- [7*c*] The products $\eta_4 B_1$ and $\eta B_5$ have Adams filtration $\geq 25$, since $h_1 w_2^2 \cdot \alpha g = 0$ in $E_2(tmf)$. Hence $B^2$ acts injectively on their difference, which must be zero since $B^2 \cdot \eta_4 B_1 = \eta B B_1 B_4 = B^2 \cdot \eta B_5$. Similarly, $\eta_4 B_5$ and $\eta B_1 M$ have Adams filtration $\geq 41$, hence are equal, because $B^2 \cdot \eta_4 B_5 = \eta B B_4 B_5 = B^2 \cdot \eta B_1 M$.

[7*d*] The products $\eta_1 B_3$ and $\nu^2\nu_4$ are both detected by $\gamma\delta w_2 = h_1 c_0 w_2^2$ (using Theorem 9.14), hence agree modulo Adams filtration $\geq 21$. This remains true if we add the filtration 21 class $\eta B_4$ to $\nu^2\nu_4$. Moreover, $\eta_1 B_3$ and $\eta B_4 + \nu^2\nu_4$ agree after multiplication by $B$, since $B \cdot \eta_1 B_3 = \eta B_1 B_3 = B \cdot \eta B_4$ and $B \cdot \nu^2\nu_4 = 0$. Hence the two classes $\eta_1 B_3$ and $\eta B_4 + \nu^2\nu_4$ can at most differ by the $B$-torsion class $\eta_1\bar{\kappa}^4$. We can detect this after multiplication by $\bar{\kappa}$, since $\bar{\kappa} \cdot \eta_1\bar{\kappa}^4 = \eta_1^5 = \eta^2\nu_5 \neq 0$. Our choice of $B_3$ in Definition 9.22 gives $\bar{\kappa} \cdot \eta_1 B_3 = \eta_1\bar{\kappa}^5 = \eta^2\nu_5$. Furthermore, $\bar{\kappa} \cdot (\eta B_4 + \nu^2\nu_4) = \eta\bar{\kappa}B_4$, since $\nu\bar{\kappa} = 0$, and Proposition 9.38 shows that this is also equal to $\eta^2\nu_5$. Hence the two classes are equal.

[8] The remaining products $\nu_4 B_j$, for $j \equiv 1, 2 \mod 4$, lie in groups of order 2.

[8*a*] To show $\nu_4 B_1 = \eta\kappa_4\bar{\kappa}$ and $\nu_4 B_5 = \eta\kappa\bar{\kappa}M$, observe that in both cases the right hand side is the unique nonzero element in its degree. It suffices then to show the left hand sides are nonzero, which we do by computing their products with $\eta_1$. We have $\eta_1 \cdot \nu_4 B_1 = \eta\nu_5 B_1 = \eta\nu_6 B \neq 0$, by what we have already shown in cases [0*d*] and [6]. Similarly, $\eta_1 \cdot \nu_4 B_5 = \eta\nu_5 B_5$ is the nonzero element $\eta\nu_2 BM$.

[8*b*] The product $\nu_4 B_2$ is either 0 or $\nu_6 B$. By Theorem 9.16, $\eta \cdot \nu_6 B$ is detected by $\delta' g w_1 w_2^2$ in Adams filtration 31, while the product $\eta\nu_4$ is detected by $g^5$ in Adams filtration 20. Since $c_0 g = 0$ in $E_2(tmf)$, $\eta\nu_4 \cdot B_2$ has Adams filtration at least 32, and hence must be zero. Thus $\nu_4 B_2 \neq \nu_6 B$ must be zero. Similarly, $\eta\nu_4 \cdot B_6$ has Adams filtration at least 48, hence is zero, so $\nu_4 B_6 = 0$.

[*Dx*] Here $x$ is one of the elements $\eta$, $\eta_1$, $\bar{\kappa}$ or $B$. The product is correct after multiplying by $x$ and multiplication by $x$ acts monomorphically on ($B$-power torsion) elements whose Adams filtration is equal to or higher than that of the product in question. In order:

[*D*$\eta$] Multiplication by $\eta$ detects $\bar{\kappa}^2 \cdot \eta_4 = \eta_4\bar{\kappa} \cdot \bar{\kappa}$, and $\eta\eta_4\bar{\kappa}^2 = \epsilon\kappa_4\bar{\kappa} \neq 0$, so $\eta_4\bar{\kappa}^2 = \nu_5\kappa$.

[*D*$\eta_1$] Multiplication by $\eta_1$ detects $B_i \cdot \nu_j$ for $i \equiv 3 \mod 4$ and $j \equiv 2 \mod 4$, since $\eta_1\nu_1\epsilon = \eta\nu_2 B$ and $\eta_1\nu_5\epsilon = \eta\nu_6 B$ are nonzero, so $\eta_1\nu_j = 0$ implies $\nu_j B_i = 0$ in these cases.

[*D*$\bar{\kappa}$] Multiplication by $\bar{\kappa}$ detects $\eta_1\bar{\kappa} \cdot \eta_4 = \eta_4\bar{\kappa} \cdot \eta_1 = \eta_1\eta_4 \cdot \bar{\kappa}$, and $\eta_1\eta_4\bar{\kappa}^2 = \eta_1\nu_5\kappa = \eta\nu_6\kappa = \epsilon_5\kappa\bar{\kappa}$, where the first equality uses case [*D*$\eta$]. Hence $\eta_1\eta_4\bar{\kappa} = \epsilon_5\kappa$. The products $B_i \cdot \epsilon_j$ for $i \equiv 1 \mod 4$ and $j \equiv 0 \mod 4$ are $B$-power torsion, and multiplication by $\bar{\kappa}$ acts injectively on the $B$-power torsion elements in these degrees. From $\epsilon_j \cdot \bar{\kappa}B_i = \epsilon_j\epsilon_i\bar{\kappa} = \epsilon\epsilon_{i+j} \cdot \bar{\kappa}$, using Propositions 9.41 and 9.42, we deduce $\epsilon_j B_i = \epsilon\epsilon_{i+j}$, which is then rewritten using $\epsilon\epsilon_1 = 2\bar{\kappa}^2$ and $\epsilon\epsilon_5 = \eta\eta_1\kappa_4$.

[*DB*] Multiplication by $B$ detects the products $D_i \cdot \eta_j$, and $\eta_j B D_i = \eta_j d_i B_i = 0$ for $1 \leq i \leq 7$ since $d_i$ is even and $2\eta_j = 0$.

[*Px*] If $x$ is one of the generators $\eta_1$, $\nu_6$, $\kappa$, $\kappa_4$ or $\bar{\kappa}$ of $\pi_*(tmf)$, then this relation follows by multiplying an earlier relation by that generator. Otherwise, we have one of the following arguments.

[*Pa*] Using $\epsilon \cdot \epsilon_5 = \eta\eta_1\kappa_4$, $\kappa \cdot \kappa_4 = \eta\nu_5$ and $\eta_1 \cdot \nu_5 = \eta\nu_6$ we calculate $\epsilon_5\kappa \cdot \epsilon = \eta\eta_1\kappa\kappa_4 = \eta^2\eta_1\nu_5 = \eta^3\nu_6 = 4\nu\nu_6$.

[*Pb*] Using $\eta_4 \cdot \kappa_4 = \eta\kappa M$ it follows that $\eta_1\kappa_4 \cdot \eta_4 = \eta_1\eta_4 \cdot \kappa_4 = \eta\eta_1\kappa M$, and $\eta\eta_1\kappa = 2\bar{\kappa}^2$ by Theorems 9.8 and 9.16.

[*Pc*] Using $\nu_4 \cdot \kappa_4 = \nu\kappa M$ and $\nu_5 \cdot \eta_4 = \eta\nu_1 M$ we calculate $\eta_1\kappa_4 \cdot \nu_4 = \eta_1\nu\kappa M$ and $\nu_5\kappa \cdot \eta_4 = \eta\nu_1\kappa M$. Here $\eta_1\nu\kappa = \eta\nu_1\kappa = \eta^2\bar{\kappa}^2$ by Theorems 9.14 and 9.16.

[*Pd*] We have $\eta_1\eta_4 \cdot \epsilon = \eta\eta_1\epsilon_4 = \eta^2\epsilon_5$, which is $2\kappa_4\bar{\kappa}$ by Theorems 9.8 and 9.16. □

We next turn to the products of classes that are not 2-power torsion. Recall the numbers $e_k = \max\{3 - \mathrm{ord}_2(k), 0\} \in \{0, 1, 2, 3\}$ and $d_k = 2^{e_k} \in \{1, 2, 4, 8\}$ from Definition 9.18. The 2-torsion free generators have the following Adams filtrations ("AF"):

$$AF(B_k) = 4k + \begin{cases} 4 & \text{for } k \equiv 0 \mod 4 \\ 3 & \text{for } k \not\equiv 0 \mod 4 \end{cases}$$
$$AF(C_k) = 4k + 6$$
$$AF(D_k) = 4k + e_k\,.$$

THEOREM 9.48. *The products of elements in the $B$-, $C$- and $D$-families are as follows:*

- $B_iB_j = BB_{i+j}$, *except for*
  - $B_2B_3 = BB_5 + \eta\eta_1\kappa_4$
  - $B_2B_7 = B_3B_6 = BB_1M + 2\bar{\kappa}^2M$
  - $B_6B_7 = BB_5M + \eta\eta_1\kappa_4M$
- $B_iC_j = BC_{i+j}$
- $B_iD_j = d_jB_{i+j}$
- $C_iC_j = 4(B^2B_{i+j} - (1728/d_{i+j+1})D_{i+j+1})$
- $C_iD_j = d_jC_{i+j}$
- $D_iD_j = (d_id_j/d_{i+j})D_{i+j}$.

REMARK 9.49. With the exception of the four listed products of the form $B_iB_j$, these are exactly the relations which hold between the images of these classes in $mf_{*/2}$. In those four cases, the 2-torsion "error term" can be written uniformly as either $(\eta\eta\kappa)_{i+j}$, i.e.,

$$\eta\eta_1\kappa_4,\ \eta\eta_1\kappa M,\ \eta\eta_1\kappa_4M\,,$$

or as $(\epsilon\epsilon)_{i+j}$, i.e.,

$$\epsilon\epsilon_5,\ \epsilon\epsilon_1M,\ \epsilon\epsilon_5M\,.$$

PROOF. It is straightforward to check that the images of these relations hold in $mf_{*/2}$, using the relation $c_6^2 = c_4^3 - 1728\Delta$ to obtain the expression for $C_iC_j$. It remains to determine any additional terms in them which lie in the kernel of this homomorphism, i.e., in the 2-power torsion. We consider these according to the form of the product.

[*BB*] The products $B_iB_j$ and $BB_{i+j}$ lie in degree $16 + 24(i + j)$ in Adams filtrations $4(i + j)$ plus 6, 7 or 8. There are 2-power torsion classes in these degrees of this high or higher Adams filtration only when $i + j \equiv 1 \mod 4$. These are $\eta\eta_1\kappa = 2\bar{\kappa}^2$ and $\eta\eta_1\kappa_4$, times the appropriate power of $M$. Multiplication by $\bar{\kappa}$ acts monomorphically on the 2-power torsion in these bidegrees. The claims then follow from the calculations

$$\begin{aligned} \bar{\kappa} \cdot BB_1 &= \eta\nu_2 B \neq 0 \\ \bar{\kappa} \cdot BB_5 &= \bar{\kappa} \cdot B_1B_4 = \eta\nu_6 B \neq 0 \\ \bar{\kappa} \cdot B_2B_3 &= 0 \\ \bar{\kappa} \cdot B_2B_7 &= \bar{\kappa} \cdot B_3B_6 = 0 \\ \bar{\kappa} \cdot B_4B_5 &= \eta\nu_2 BM \neq 0 \\ \bar{\kappa} \cdot B_6B_7 &= 0\,, \end{aligned}$$

which can be read off from Tables 9.6 and 9.7.

[*BC*] The products $B_iC_j$ lie in degree $20+24(i+j)$ in Adams filtrations $4(i+j)$ plus 9 or 10. There is no 2-power torsion of such high Adams filtration in these degrees.

[*BD*] The products $B_iD_j$ lie in degree $8+24(i+j)$ in Adams filtrations $4(i+j)$ plus 4, 5, 6 or 7. The 2-power torsion in these degrees lies in Adams filtrations less than $4(i+j)+4$ unless $i+j \equiv 3 \mod 8$. In these cases, the Adams filtration of $B_iD_j$ is greater than $4(i+j)+4$, except for $B_7D_4$, where a possible additional term $\bar{\kappa}^4M$ could occur. However, $B_7D_4$ and $2B_3M$ are each detected in Adams filtration $16+32$ by $h_0\delta w_2 \cdot w_2^4$ rather than by the sum of this with $g^4w_2^4$. (Alternatively, the product with $\bar{\kappa}$ shows that the additional term does not occur.)

[*CC*] The products $C_iC_j$ lie in degree $24+24(i+j)$, which has no 2-power torsion.

[*CD*] The products $C_iD_j$ lie in degree $12+24(i+j)$ in Adams filtrations $4(i+j)$ plus 7, 8 or 9. There is 2-power torsion in Adams filtration $4(i+j)+7$ when $i+j \equiv 2 \mod 4$, and $C_iD_j$ has this same Adams filtration only when $j=4$. Checking $E_\infty(tmf)$ shows that $C_2D_4$ and $2C_6$ are detected by $h_0^4\alpha w_2^3$, while $C_6D_4$ and $2C_2M$ are detected by $h_0^4\alpha w_2^5$, with no contribution from the 2-power torsion classes in these degrees. Furthermore, the indeterminacy in the choice of $C_2$ and $C_6$ has no effect because it is divisible by $\nu^3$, which is annihilated by 2 and by $D_k$ when $1 \leq k \leq 7$.

[*DD*] The products $D_iD_j$ lie in degree $24(i+j)$, which has no 2-power torsion.

□

Table 9.6: Preliminary products in $\pi_*(tmf)$: the entry in row $x$ (found in the $x$-column) and column $y$ (found in the top row) gives $xy$. Part 1 of 2: $\eta_i$- and $\nu_i$-multiples. Signs $s$, $s_i \in \{\pm 1\}$.

| $n$ | $s$ | $x$ | $\eta_1$ | $\eta_4$ | $\nu_1$ | $\nu_2$ | $\nu_4$ | $\nu_5$ | $\nu_6$ |
|---|---|---|---|---|---|---|---|---|---|
| 8 | 3 | $\epsilon$ | $\eta\epsilon_1{}^{[0e]}$ | $\eta\epsilon_4{}^{[0e]}$ | $\eta\kappa\bar{\kappa}^{[0e]}$ | $\nu_2 B^{[0e]}$ | $0^{[Z]}$ | $\eta\kappa_4\bar{\kappa}^{[0e]}$ | $\nu_6 B^{[0e]}$ |
| 12 | 6 | $C$ | $0^{[Z]}$ | $0^{[Z]}$ | $0^{[B]}$ | $0^{[Z]}$ | $0^{[B]}$ | $0^{[B]}$ | $0^{[Z]}$ |
| 14 | 4 | $\kappa$ | $\eta_1\kappa^{[t]}$ | $\eta\kappa_4{}^{[1]}$ | $\eta\bar{\kappa}^{2\,[2b]}$ | $\nu_2\kappa^{[t]}$ | $\nu\kappa_4{}^{[1]}$ | $\nu_5\kappa^{[t]}$ | $\nu_6\kappa^{[t]}$ |
| 20 | 4 | $\bar{\kappa}$ | $\eta_1\bar{\kappa}^{[t]}$ | $\eta_4\bar{\kappa}^{[t]}$ | $0^{[Z]}$ | $0^{[Z]}$ | $0^{[Z]}$ | $0^{[Z]}$ | $0^{[Z]}$ |
| 24 | 7 | $D_1$ | $0^{[DB]}$ | $0^{[DB]}$ | $0^{[B]}$ | $0^{[B]}$ | $0^{[B]}$ | $0^{[B]}$ | $0^{[Z]}$ |
| 25 | 5 | $\eta_1$ | $\eta_1^{2\,[t]}$ | $\eta_1\eta_4{}^{[t]}$ | $\eta\nu_2{}^{[0d]}$ | $0^{[B]}$ | $\eta\nu_5{}^{[0d]}$ | $\eta\nu_6{}^{[0d]}$ | $0^{[B]}$ |
| 27 | 6 | $\nu_1$ | $\eta\nu_2{}^{[0d]}$ | $\eta\nu_5{}^{[0d]}$ | $2\nu\nu_2{}^{[0c]}$ | $0^{[Z]}$ | $0^{[Z]}$ | $2\nu\nu_6{}^{[0a]}$ | $0^{[Z]}$ |
| 32 | 7 | $B_1$ | $\eta B_2 + \nu^2\nu_2{}^{[7a]}$ | $\eta B_5{}^{[7c]}$ | $\nu_2 B^{[6]}$ | $0^{[Z]}$ | $\eta\kappa_4\bar{\kappa}^{[8a]}$ | $\nu_6 B^{[6]}$ | $0^{[Z]}$ |
| 32 | 7 | $\epsilon_1$ | $\nu^2\nu_2{}^{[0h]}$ | $\eta\epsilon_5{}^{[0h]}$ | $\nu_2 B^{[0i]}$ | $0^{[Z]}$ | $\eta\kappa_4\bar{\kappa}^{[0i]}$ | $\nu_6 B^{[0i]}$ | $0^{[Z]}$ |
| 34 | 8 | $\kappa\bar{\kappa}$ | $\nu_2 B^{[6]}$ | $\eta\kappa_4\bar{\kappa}^{[P\bar{\kappa}]}$ | $0^{[Z]}$ | $0^{[P\kappa]}$ | $0^{[Z]}$ | $0^{[Z]}$ | $0^{[Z]}$ |
| 36 | 10 | $C_1$ | $0^{[Z]}$ | $0^{[Z]}$ | $0^{[Z]}$ | $0^{[Z]}$ | $0^{[B]}$ | $0^{[Z]}$ | $0^{[Z]}$ |
| 39 | 9 | $\eta_1\kappa$ | $0^{[B]}$ | $\eta\eta_1\kappa_4{}^{[P\eta_1]}$ | $\eta\nu_2\kappa^{[P\kappa]}$ | $0^{[P\kappa]}$ | $\eta\nu_5\kappa^{[P\kappa]}$ | $\eta\nu_6\kappa^{[P\kappa]}$ | $0^{[B]}$ |
| 40 | 8 | $\bar{\kappa}^2$ | $\eta_1\bar{\kappa}^{2\,[t]}$ | $\nu_5\kappa^{[D\eta]}$ | $0^{[Z]}$ | $0^{[Z]}$ | $0^{[Z]}$ | $0^{[Z]}$ | $0^{[Z]}$ |
| 45 | 9 | $\eta_1\bar{\kappa}$ | $\eta_1^2\bar{\kappa}^{[t]}$ | $\epsilon_5\kappa^{[D\bar{\kappa}]}$ | $0^{[B]}$ | $0^{[B]}$ | $0^{[B]}$ | $0^{[B]}$ | $0^{[B]}$ |
| 48 | 10 | $D_2$ | $0^{[DB]}$ | $0^{[DB]}$ | $0^{[B]}$ | $4\nu_4{}^{[1]}$ | $4\nu_6{}^{[1]}$ | $0^{[Z]}$ | $4\nu M^{[1]}$ |
| 50 | 10 | $\eta_1^2$ | $\eta_1^{3\,[t]}$ | $4\nu_6{}^{[0b]}$ | $0^{[Z]}$ | $0^{[Z]}$ | $\eta^2\nu_6{}^{[0d]}$ | $0^{[Z]}$ | $0^{[Z]}$ |
| 51 | 9 | $\nu_2$ | $0^{[B]}$ | $\eta\nu_6{}^{[0d]}$ | $0^{[Z]}$ | $\nu\nu_4{}^{[0c]}$ | $3\nu\nu_6{}^{[0a]}$ | $0^{[Z]}$ | $\nu^2 M^{[0c]}$ |

Table 9.6: Preliminary products in $\pi_*(tmf)$ (Part 1, cont.)

| $n$ | $s$ | $x$ | $\eta_1$ | $\eta_4$ | $\nu_1$ | $\nu_2$ | $\nu_4$ | $\nu_5$ | $\nu_6$ |
|---|---|---|---|---|---|---|---|---|---|
| 56 | 11 | $B_2$ | $\eta B_3{}^{[7a]}$ | $\eta B_6{}^{[7a]}$ | $0^{[Z]}$ | $0^{[Z]}$ | $0^{[8b]}$ | $0^{[Z]}$ | $0^{[Z]}$ |
| 60 | 12 | $\bar\kappa^3$ | $\eta_1\bar\kappa^{3[t]}$ | $0^{[Z]}$ | $0^{[Z]}$ | $0^{[P\bar\kappa]}$ | $0^{[Z]}$ | $0^{[Z]}$ | $0^{[P\bar\kappa]}$ |
| 60 | 14 | $C_2$ | $0^{[T]}$ | $0^{[Z]}$ | $0^{[Z]}$ | $0^{[B]}$ | $0^{[Z]}$ | $0^{[Z]}$ | $0^{[B]}$ |
| 65 | 13 | $\eta_1\bar\kappa^2$ | $\eta_1^2\bar\kappa^{2[t]}$ | $\eta\nu_6\kappa^{[P\eta_1]}$ | $0^{[B]}$ | $0^{[P\eta_1]}$ | $0^{[P\eta_1]}$ | $0^{[B]}$ | $0^{[P\eta_1]}$ |
| 65 | 13 | $\nu_2\kappa$ | $0^{[P\kappa]}$ | $\eta\nu_6\kappa^{[P\kappa]}$ | $0^{[B]}$ | $2\bar\kappa D_4{}^{[4]}$ | $\nu\nu_6\kappa^{[1]}$ | $0^{[B]}$ | $4\bar\kappa M^{[4]}$ |
| 70 | 14 | $\eta_1^2\bar\kappa$ | $0^{[Z]}$ | $0^{[Z]}$ | $0^{[B]}$ | $0^{[B]}$ | $0^{[B]}$ | $0^{[B]}$ | $0^{[B]}$ |
| 72 | 15 | $D_3$ | $0^{[DB]}$ | $0^{[DB]}$ | $0^{[B]}$ | $0^{[B]}$ | $0^{[Z]}$ | $0^{[B]}$ | $0^{[B]}$ |
| 75 | 15 | $\eta_1^3$ | $\eta\nu_4{}^{[0b]}$ | $0^{[T]}$ | $0^{[B]}$ | $0^{[Z]}$ | $0^{[Z]}$ | $0^{[B]}$ | $0^{[Z]}$ |
| 80 | 15 | $B_3$ | $\eta B_4 + \nu^2\nu_4{}^{[7d]}$ | $\eta B_7{}^{[7a]}$ | $0^{[Z]}$ | $0^{[D\eta_1]}$ | $0^{[Z]}$ | $0^{[Z]}$ | $0^{[D\eta_1]}$ |
| 80 | 16 | $\bar\kappa^4$ | $\eta\epsilon_4 + \nu^2\nu_4{}^{[3]}$ | $0^{[B]}$ | $0^{[Z]}$ | $0^{[P\bar\kappa]}$ | $0^{[Z]}$ | $0^{[Z]}$ | $0^{[P\bar\kappa]}$ |
| 84 | 18 | $C_3$ | $0^{[Z]}$ | $0^{[Z]}$ | $0^{[B]}$ | $0^{[B]}$ | $0^{[Z]}$ | $0^{[B]}$ | $0^{[B]}$ |
| 85 | 17 | $\eta_1\bar\kappa^3$ | $2\kappa_4{}^{[2a]}$ | $0^{[Z]}$ | $0^{[B]}$ | $0^{[P\bar\kappa]}$ | $0^{[B]}$ | $0^{[B]}$ | $0^{[P\bar\kappa]}$ |
| 90 | 18 | $\eta_1^2\bar\kappa^2$ | $0^{[Z]}$ | $0^{[Z]}$ | $0^{[B]}$ | $0^{[Z]}$ | $0^{[Z]}$ | $0^{[B]}$ | $0^{[B]}$ |
| 96 | 17 | $D_4$ | $0^{[DB]}$ | $0^{[DB]}$ | $2\nu_5{}^{[1]}$ | $2s_2\nu_6{}^{[S]}$ | $2s_4\nu M^{[S]}$ | $2\nu_1 M^{[1]}$ | $2s_6\nu_2 M^{[S]}$ |
| 97 | 17 | $\eta_4$ | $\eta_1\eta_4{}^{[t]}$ | $\eta^2 M^{[0b]}$ | $\eta\nu_5{}^{[0d]}$ | $\eta\nu_6{}^{[0d]}$ | $0^{[B]}$ | $\eta\nu_1 M^{[0d]}$ | $\eta\nu_2 M^{[0d]}$ |
| 99 | 17 | $\nu_4$ | $\eta\nu_5{}^{[0d]}$ | $0^{[B]}$ | $0^{[Z]}$ | $-3\nu\nu_6{}^{[0a]}$ | $\nu^2 M^{[0c]}$ | $0^{[Z]}$ | $s\nu\nu_2 M^{[S]}$ |
| 104 | 19 | $\epsilon_4$ | $\eta\epsilon_5{}^{[0h]}$ | $\eta\epsilon M^{[0h]}$ | $\eta\kappa_4\bar\kappa^{[0i]}$ | $\nu_6 B^{[0i]}$ | $0^{[Z]}$ | $\eta\kappa\bar\kappa M^{[0i]}$ | $\nu_2 BM^{[0i]}$ |
| 104 | 20 | $B_4$ | $\eta B_5{}^{[7b]}$ | $\eta BM^{[7a]}$ | $\eta\kappa_4\bar\kappa^{[2b]}$ | $\nu_6 B^{[1]}$ | $0^{[Z]}$ | $\eta\kappa\bar\kappa M^{[2b]}$ | $\nu_2 BM^{[1]}$ |
| 108 | 22 | $C_4$ | $0^{[Z]}$ | $0^{[Z]}$ | $0^{[B]}$ | $0^{[Z]}$ | $0^{[B]}$ | $0^{[B]}$ | $0^{[Z]}$ |

Table 9.6: Preliminary products in $\pi_*(tmf)$ (Part 1, cont.)

| $n$ | $s$ | $x$ | $\eta_1$ | $\eta_4$ | $\nu_1$ | $\nu_2$ | $\nu_4$ | $\nu_5$ | $\nu_6$ |
|---|---|---|---|---|---|---|---|---|---|
| 110 | 20 | $\kappa_4$ | $\eta_1\kappa_4^{[t]}$ | $\eta\kappa M^{[1]}$ | $\nu_5\kappa^{[1]}$ | $\nu_6\kappa^{[1]}$ | $\nu\kappa M^{[1]}$ | $\eta\bar{\kappa}^2 M^{[2b]}$ | $\nu_2\kappa M^{[1]}$ |
| 116 | 21 | $\bar{\kappa}D_4$ | $0^{[Z]}$ | $0^{[B]}$ | $0^{[Z]}$ | $0^{[Z]}$ | $0^{[Z]}$ | $0^{[Z]}$ | $0^{[Z]}$ |
| 117 | 21 | $\eta_4\bar{\kappa}$ | $\epsilon_5\kappa^{[D\bar{\kappa}]}$ | $\eta^2\bar{\kappa}M^{[P\bar{\kappa}]}$ | $0^{[B]}$ | $0^{[B]}$ | $0^{[B]}$ | $0^{[B]}$ | $0^{[B]}$ |
| 120 | 23 | $D_5$ | $0^{[DB]}$ | $0^{[DB]}$ | $0^{[B]}$ | $0^{[Z]}$ | $0^{[B]}$ | $0^{[B]}$ | $0^{[B]}$ |
| 122 | 22 | $\eta_1\eta_4$ | $4\nu_6^{[0b]}$ | $2\nu_1 M^{[0b]}$ | $\eta^2\nu_6^{[0d]}$ | $0^{[Z]}$ | $0^{[Z]}$ | $\eta^2\nu_2 M^{[0d]}$ | $0^{[Z]}$ |
| 123 | 22 | $\nu_5$ | $\eta\nu_6^{[0d]}$ | $\eta\nu_1 M^{[0d]}$ | $-2\nu\nu_6^{[0a]}$ | $0^{[Z]}$ | $0^{[Z]}$ | $2\nu\nu_2 M^{[0c]}$ | $0^{[Z]}$ |
| 128 | 23 | $B_5$ | $\eta B_6 + \nu^2\nu_6^{[7a]}$ | $\eta B_1 M^{[7c]}$ | $\nu_6 B^{[6]}$ | $0^{[Z]}$ | $\eta\kappa\bar{\kappa}M^{[8a]}$ | $\nu_2 BM^{[6]}$ | $0^{[Z]}$ |
| 128 | 23 | $\epsilon_5$ | $\nu^2\nu_6^{[0h]}$ | $\eta\epsilon_1 M^{[0h]}$ | $\nu_6 B^{[0i]}$ | $0^{[Z]}$ | $\eta\kappa\bar{\kappa}M^{[0i]}$ | $\nu_2 BM^{[0i]}$ | $0^{[Z]}$ |
| 130 | 24 | $\kappa_4\bar{\kappa}$ | $\nu_6 B^{[6]}$ | $\eta\kappa\bar{\kappa}M^{[P\bar{\kappa}]}$ | $0^{[Z]}$ | $0^{[Z]}$ | $0^{[Z]}$ | $0^{[Z]}$ | $0^{[P\kappa_4]}$ |
| 132 | 26 | $C_5$ | $0^{[Z]}$ | $0^{[Z]}$ | $0^{[Z]}$ | $0^{[Z]}$ | $0^{[B]}$ | $0^{[Z]}$ | $0^{[Z]}$ |
| 135 | 25 | $\eta_1\kappa_4$ | $0^{[B]}$ | $2\bar{\kappa}^2 M^{[Pb]}$ | $\eta\nu_6\kappa^{[5b]}$ | $0^{[B]}$ | $\eta^2\bar{\kappa}^2 M^{[Pc]}$ | $\eta\nu_2\kappa M^{[5b]}$ | $0^{[P\kappa_4]}$ |
| 137 | 26 | $\nu_5\kappa$ | $\eta\nu_6\kappa^{[P\kappa]}$ | $\eta^2\bar{\kappa}^2 M^{[Pc]}$ | $0^{[B]}$ | $0^{[B]}$ | $0^{[B]}$ | $0^{[B]}$ | $0^{[B]}$ |
| 142 | 27 | $\epsilon_5\kappa$ | $0^{[Z]}$ | $0^{[Z]}$ | $0^{[B]}$ | $0^{[B]}$ | $0^{[B]}$ | $0^{[B]}$ | $0^{[B]}$ |
| 144 | 26 | $D_6$ | $0^{[DB]}$ | $0^{[DB]}$ | $0^{[Z]}$ | $4\nu M^{[1]}$ | $4\nu_2 M^{[1]}$ | $0^{[B]}$ | $4\nu_4 M^{[1]}$ |
| 147 | 25 | $\nu_6$ | $0^{[B]}$ | $\eta\nu_2 M^{[0d]}$ | $0^{[Z]}$ | $\nu^2 M^{[0c]}$ | $-s\nu\nu_2 M^{[S]}$ | $0^{[Z]}$ | $\nu\nu_4 M^{[0c]}$ |
| 152 | 27 | $B_6$ | $\eta B_7^{[7a]}$ | $\eta B_2 M^{[7a]}$ | $0^{[Z]}$ | $0^{[Z]}$ | $0^{[8b]}$ | $0^{[Z]}$ | $0^{[Z]}$ |
| 156 | 30 | $C_6$ | $0^{[Z]}$ | $0^{[Z]}$ | $0^{[Z]}$ | $0^{[B]}$ | $0^{[Z]}$ | $0^{[Z]}$ | $0^{[B]}$ |
| 161 | 29 | $\nu_6\kappa$ | $0^{[B]}$ | $\eta\nu_2\kappa M^{[P\nu_6]}$ | $0^{[B]}$ | $4\bar{\kappa}M^{[4]}$ | $\nu\nu_2\kappa M^{[1]}$ | $0^{[B]}$ | $2\bar{\kappa}D_4 M^{[4]}$ |
| 168 | 31 | $D_7$ | $0^{[DB]}$ | $0^{[DB]}$ | $0^{[B]}$ | $0^{[B]}$ | $0^{[B]}$ | $0^{[B]}$ | $0^{[B]}$ |

Table 9.6: Preliminary products in $\pi_*(tmf)$ (Part 1, cont.)

| $n$ | $s$ | $x$ | $\eta_1$ | $\eta_4$ | $\nu_1$ | $\nu_2$ | $\nu_4$ | $\nu_5$ | $\nu_6$ |
|---|---|---|---|---|---|---|---|---|---|
| 176 | 31 | $B_7$ | $\eta(B+\epsilon)M^{[7a]}$ | $\eta B_3 M^{[7a]}$ | $0^{[Z]}$ | $0^{[D\eta_1]}$ | $0^{[Z]}$ | $0^{[Z]}$ | $0^{[D\eta_1]}$ |
| 180 | 34 | $C_7$ | $0^{[Z]}$ | $0^{[T]}$ | $0^{[B]}$ | $0^{[B]}$ | $0^{[Z]}$ | $0^{[B]}$ | $0^{[B]}$ |

Table 9.7: Preliminary products in $\pi_*(tmf)$: the entry in row $x$ (found in the $x$-column) and column $y$ (found in the top row) gives $xy$. Part 2 of 2: $\epsilon_i$-, $\kappa_i$- and $\bar{\kappa}$-multiples.

| $n$ | $s$ | $x$ | $\epsilon$ | $\epsilon_1$ | $\epsilon_4$ | $\epsilon_5$ | $\kappa$ | $\kappa_4$ | $\bar{\kappa}$ |
|---|---|---|---|---|---|---|---|---|---|
| 8 | 3 | $\epsilon$ | $0^{[B]}$ | $2\bar{\kappa}^{2[0e]}$ | $0^{[B]}$ | $\eta\eta_1\kappa_4{}^{[0e]}$ | $\eta^2\bar{\kappa}^{[M]}$ | $\eta\eta_4\bar{\kappa}^{[0e]}$ | $\eta\nu_1{}^{[0e]}$ |
| 12 | 6 | $C$ | $0^{[B]}$ | $0^{[B]}$ | $0^{[B]}$ | $0^{[B]}$ | $0^{[B]}$ | $0^{[B]}$ | $0^{[B]}$ |
| 14 | 4 | $\kappa$ | $\eta^2\bar{\kappa}^{[M]}$ | $\eta\eta_1\bar{\kappa}^{[0j]}$ | $\eta\eta_4\bar{\kappa}^{[0j]}$ | $\epsilon_5\kappa^{[t]}$ | $\eta\nu_1{}^{[5b]}$ | $\eta\nu_5{}^{[5b]}$ | $\kappa\bar{\kappa}^{[t]}$ |
| 20 | 4 | $\bar{\kappa}$ | $\eta\nu_1{}^{[0e]}$ | $\eta\nu_2{}^{[0f]}$ | $\eta\nu_5{}^{[0f]}$ | $\eta\nu_6{}^{[0f]}$ | $\kappa\bar{\kappa}^{[t]}$ | $\kappa_4\bar{\kappa}^{[t]}$ | $\bar{\kappa}^{2[t]}$ |
| 24 | 7 | $D_1$ | $0^{[B]}$ | $0^{[B]}$ | $0^{[B]}$ | $0^{[B]}$ | $0^{[Z]}$ | $0^{[Z]}$ | $0^{[B]}$ |
| 25 | 5 | $\eta_1$ | $\eta\epsilon_1{}^{[0e]}$ | $\nu^2\nu_2{}^{[0h]}$ | $\eta\epsilon_5{}^{[0h]}$ | $\nu^2\nu_6{}^{[0h]}$ | $\eta_1\kappa^{[t]}$ | $\eta_1\kappa_4{}^{[t]}$ | $\eta_1\bar{\kappa}^{[t]}$ |
| 27 | 6 | $\nu_1$ | $\eta\kappa\bar{\kappa}^{[0e]}$ | $\nu_2 B^{[0i]}$ | $\eta\kappa_4\bar{\kappa}^{[0i]}$ | $\nu_6 B^{[0i]}$ | $\eta\bar{\kappa}^{2[2b]}$ | $\nu_5\kappa^{[1]}$ | $0^{[Z]}$ |
| 32 | 7 | $B_1$ | $2\bar{\kappa}^{2[D\bar{\kappa}]}$ | $0^{[B]}$ | $\eta\eta_1\kappa_4{}^{[D\bar{\kappa}]}$ | $0^{[B]}$ | $\eta\eta_1\bar{\kappa}^{[5b]}$ | $\epsilon_5\kappa^{[4]}$ | $\eta\nu_2{}^{[5b]}$ |
| 32 | 7 | $\epsilon_1$ | $2\bar{\kappa}^{2[0e]}$ | $0^{[B]}$ | $\eta\eta_1\kappa_4{}^{[0g]}$ | $0^{[B]}$ | $\eta\eta_1\bar{\kappa}^{[0j]}$ | $\epsilon_5\kappa^{[0j]}$ | $\eta\nu_2{}^{[0f]}$ |
| 34 | 8 | $\kappa\bar{\kappa}$ | $\eta^2\bar{\kappa}^{2[P\bar{\kappa}]}$ | $\eta\nu_2\kappa^{[P\kappa]}$ | $\eta\nu_5\kappa^{[P\kappa]}$ | $\eta\nu_6\kappa^{[P\kappa]}$ | $0^{[B]}$ | $0^{[B]}$ | $2\nu\nu_2{}^{[2a]}$ |

Table 9.7: Preliminary products in $\pi_*(tmf)$ (Part 2, cont.)

| $n$ | $s$ | $x$ | $\epsilon$ | $\epsilon_1$ | $\epsilon_4$ | $\epsilon_5$ | $\kappa$ | $\kappa_4$ | $\bar{\kappa}$ |
|---|---|---|---|---|---|---|---|---|---|
| 36 | 10 | $C_1$ | $0^{[B]}$ | $0^{[B]}$ | $0^{[B]}$ | $0^{[B]}$ | $0^{[B]}$ | $0^{[B]}$ | $0^{[B]}$ |
| 39 | 9 | $\eta_1\kappa$ | $0^{[Z]}$ | $0^{[Z]}$ | $0^{[Z]}$ | $0^{[Z]}$ | $\eta^2\nu_2{}^{[5b]}$ | $\eta^2\nu_6{}^{[5b]}$ | $\nu_2 B^{[6]}$ |
| 40 | 8 | $\bar{\kappa}^2$ | $0^{[B]}$ | $0^{[B]}$ | $0^{[B]}$ | $0^{[B]}$ | $2\nu\nu_2{}^{[2a]}$ | $2\nu\nu_6{}^{[0a]}$ | $\bar{\kappa}^{3[t]}$ |
| 45 | 9 | $\eta_1\bar{\kappa}$ | $\eta^2\nu_2{}^{[P\bar{\kappa}]}$ | $0^{[Z]}$ | $\eta^2\nu_6{}^{[P\bar{\kappa}]}$ | $0^{[Z]}$ | $\nu_2 B^{[6]}$ | $\nu_6 B^{[6]}$ | $\eta_1\bar{\kappa}^{2[t]}$ |
| 48 | 10 | $D_2$ | $0^{[B]}$ | $0^{[B]}$ | $0^{[B]}$ | $0^{[B]}$ | $0^{[Z]}$ | $0^{[Z]}$ | $\nu\nu_2\kappa^{[4]}$ |
| 50 | 10 | $\eta_1^2$ | $0^{[B]}$ | $0^{[B]}$ | $0^{[B]}$ | $0^{[B]}$ | $0^{[B]}$ | $0^{[B]}$ | $\eta_1^2\bar{\kappa}^{[t]}$ |
| 51 | 9 | $\nu_2$ | $\nu_2 B^{[0e]}$ | $0^{[Z]}$ | $\nu_6 B^{[0i]}$ | $0^{[Z]}$ | $\nu_2\kappa^{[t]}$ | $\nu_6\kappa^{[1]}$ | $0^{[Z]}$ |
| 56 | 11 | $B_2$ | $0^{[B]}$ | $0^{[B]}$ | $0^{[B]}$ | $0^{[B]}$ | $0^{[B]}$ | $0^{[Z]}$ | $0^{[B]}$ |
| 60 | 12 | $\bar{\kappa}^3$ | $0^{[B]}$ | $0^{[B]}$ | $0^{[B]}$ | $0^{[B]}$ | $0^{[B]}$ | $0^{[B]}$ | $\bar{\kappa}^{4[t]}$ |
| 60 | 14 | $C_2$ | $0^{[B]}$ | $0^{[B]}$ | $0^{[B]}$ | $0^{[B]}$ | $0^{[B]}$ | $0^{[B]}$ | $0^{[B]}$ |
| 65 | 13 | $\eta_1\bar{\kappa}^2$ | $0^{[B]}$ | $0^{[B]}$ | $0^{[B]}$ | $0^{[B]}$ | $0^{[Z]}$ | $0^{[Z]}$ | $\eta_1\bar{\kappa}^{3[t]}$ |
| 65 | 13 | $\nu_2\kappa$ | $0^{[B]}$ | $0^{[B]}$ | $0^{[B]}$ | $0^{[B]}$ | $0^{[Z]}$ | $0^{[Z]}$ | $0^{[P\kappa]}$ |
| 70 | 14 | $\eta_1^2\bar{\kappa}$ | $0^{[Z]}$ | $0^{[B]}$ | $0^{[Z]}$ | $0^{[B]}$ | $0^{[B]}$ | $0^{[B]}$ | $\eta_1^2\bar{\kappa}^{2[t]}$ |
| 72 | 15 | $D_3$ | $0^{[B]}$ | $0^{[B]}$ | $0^{[B]}$ | $0^{[B]}$ | $0^{[Z]}$ | $0^{[Z]}$ | $0^{[B]}$ |
| 75 | 15 | $\eta_1^3$ | $0^{[Z]}$ | $0^{[Z]}$ | $0^{[Z]}$ | $0^{[Z]}$ | $0^{[B]}$ | $0^{[B]}$ | $0^{[Z]}$ |
| 80 | 15 | $B_3$ | $0^{[B]}$ | $0^{[B]}$ | $0^{[B]}$ | $0^{[B]}$ | $0^{[Z]}$ | $0^{[Z]}$ | $\eta\nu_4{}^{[0a]}$ |
| 80 | 16 | $\bar{\kappa}^4$ | $0^{[B]}$ | $0^{[B]}$ | $0^{[B]}$ | $0^{[B]}$ | $0^{[Z]}$ | $0^{[Z]}$ | $\eta\nu_4{}^{[2b]}$ |
| 84 | 18 | $C_3$ | $0^{[B]}$ | $0^{[B]}$ | $0^{[B]}$ | $0^{[B]}$ | $0^{[B]}$ | $0^{[B]}$ | $0^{[B]}$ |
| 85 | 17 | $\eta_1\bar{\kappa}^3$ | $0^{[Z]}$ | $0^{[B]}$ | $0^{[Z]}$ | $0^{[B]}$ | $0^{[B]}$ | $0^{[B]}$ | $\eta\epsilon_4 + \nu^2\nu_4{}^{[3]}$ |

Table 9.7: Preliminary products in $\pi_*(tmf)$ (Part 2, cont.)

| $n$ | $s$ | $x$ | $\epsilon$ | $\epsilon_1$ | $\epsilon_4$ | $\epsilon_5$ | $\kappa$ | $\kappa_4$ | $\bar\kappa$ |
|---|---|---|---|---|---|---|---|---|---|
| 90 | 18 | $\eta_1^2\bar\kappa^2$ | $0^{[B]}$ | $0^{[B]}$ | $0^{[B]}$ | $0^{[B]}$ | $0^{[B]}$ | $0^{[B]}$ | $2\kappa_4{}^{[2a]}$ |
| 96 | 17 | $D_4$ | $0^{[B]}$ | $0^{[B]}$ | $0^{[B]}$ | $0^{[B]}$ | $2\kappa_4{}^{[0k]}$ | $0^{[B]}$ | $\bar\kappa D_4{}^{[t]}$ |
| 97 | 17 | $\eta_4$ | $\eta\epsilon_4{}^{[0e]}$ | $\eta\epsilon_5{}^{[0h]}$ | $\eta\epsilon M^{[0h]}$ | $\eta\epsilon_1 M^{[0h]}$ | $\eta\kappa_4{}^{[1]}$ | $\eta\kappa M^{[1]}$ | $\eta_4\bar\kappa^{[t]}$ |
| 99 | 17 | $\nu_4$ | $0^{[Z]}$ | $\eta\kappa_4\bar\kappa^{[0i]}$ | $0^{[Z]}$ | $\eta\kappa\bar\kappa M^{[0i]}$ | $\nu\kappa_4{}^{[1]}$ | $\nu\kappa M^{[1]}$ | $0^{[Z]}$ |
| 104 | 19 | $\epsilon_4$ | $0^{[B]}$ | $\eta\eta_1\kappa_4{}^{[0g]}$ | $0^{[B]}$ | $2\bar\kappa^2 M^{[0g]}$ | $\eta\eta_4\bar\kappa^{[0j]}$ | $\eta^2\bar\kappa M^{[0j]}$ | $\eta\nu_5{}^{[0f]}$ |
| 104 | 20 | $B_4$ | $0^{[B]}$ | $\eta\eta_1\kappa_4{}^{[2b]}$ | $0^{[B]}$ | $2\bar\kappa^2 M^{[2a]}$ | $\eta\eta_4\bar\kappa^{[2b]}$ | $\eta^2\bar\kappa M^{[2b]}$ | $\eta\nu_5{}^{[2b]}$ |
| 108 | 22 | $C_4$ | $0^{[B]}$ | $0^{[B]}$ | $0^{[B]}$ | $0^{[B]}$ | $0^{[B]}$ | $0^{[B]}$ | $0^{[B]}$ |
| 110 | 20 | $\kappa_4$ | $\eta\eta_4\bar\kappa^{[0e]}$ | $\epsilon_5\kappa^{[0j]}$ | $\eta^2\bar\kappa M^{[0j]}$ | $\eta\eta_1\bar\kappa M^{[0j]}$ | $\eta\nu_5{}^{[5b]}$ | $\eta\nu_1 M^{[5b]}$ | $\kappa_4\bar\kappa^{[t]}$ |
| 116 | 21 | $\bar\kappa D_4$ | $0^{[P\bar\kappa]}$ | $0^{[B]}$ | $0^{[P\bar\kappa]}$ | $0^{[B]}$ | $2\kappa_4\bar\kappa^{[0k]}$ | $0^{[B]}$ | $\eta\eta_1\kappa_4{}^{[0k]}$ |
| 117 | 21 | $\eta_4\bar\kappa$ | $\eta^2\nu_5{}^{[P\bar\kappa]}$ | $\eta^2\nu_6{}^{[P\bar\kappa]}$ | $0^{[Z]}$ | $\eta^2\nu_2 M^{[P\bar\kappa]}$ | $\eta\kappa_4\bar\kappa^{[P\bar\kappa]}$ | $\eta\kappa\bar\kappa M^{[P\bar\kappa]}$ | $\nu_5\kappa^{[D\eta]}$ |
| 120 | 23 | $D_5$ | $0^{[B]}$ | $0^{[B]}$ | $0^{[B]}$ | $0^{[B]}$ | $0^{[Z]}$ | $0^{[Z]}$ | $0^{[B]}$ |
| 122 | 22 | $\eta_1\eta_4$ | $2\kappa_4\bar\kappa^{[Pd]}$ | $0^{[B]}$ | $0^{[B]}$ | $0^{[B]}$ | $\eta\eta_1\kappa_4{}^{[P\eta_1]}$ | $2\bar\kappa^2 M^{[Pb]}$ | $\epsilon_5\kappa^{[D\bar\kappa]}$ |
| 123 | 22 | $\nu_5$ | $\eta\kappa_4\bar\kappa^{[0e]}$ | $\nu_6 B^{[0i]}$ | $\eta\kappa\bar\kappa M^{[0i]}$ | $\nu_2 B M^{[0i]}$ | $\nu_5\kappa^{[t]}$ | $\eta\bar\kappa^2 M^{[2b]}$ | $0^{[Z]}$ |
| 128 | 23 | $B_5$ | $\eta\eta_1\kappa_4{}^{[D\bar\kappa]}$ | $0^{[B]}$ | $2\bar\kappa^2 M^{[D\bar\kappa]}$ | $0^{[B]}$ | $\epsilon_5\kappa^{[4]}$ | $\eta\eta_1\bar\kappa M^{[5b]}$ | $\eta\nu_6{}^{[5b]}$ |
| 128 | 23 | $\epsilon_5$ | $\eta\eta_1\kappa_4{}^{[0e]}$ | $0^{[B]}$ | $2\bar\kappa^2 M^{[0g]}$ | $0^{[B]}$ | $\epsilon_5\kappa^{[t]}$ | $\eta\eta_1\bar\kappa M^{[0j]}$ | $\eta\nu_6{}^{[0f]}$ |
| 130 | 24 | $\kappa_4\bar\kappa$ | $\eta\nu_5\kappa^{[P\kappa_4]}$ | $\eta\nu_6\kappa^{[2b]}$ | $\eta^2\bar\kappa^2 M^{[P\kappa_4]}$ | $\eta\nu_2\kappa M^{[2b]}$ | $0^{[B]}$ | $0^{[B]}$ | $2\nu\nu_6{}^{[0a]}$ |
| 132 | 26 | $C_5$ | $0^{[B]}$ | $0^{[B]}$ | $0^{[B]}$ | $0^{[B]}$ | $0^{[B]}$ | $0^{[B]}$ | $0^{[B]}$ |
| 135 | 25 | $\eta_1\kappa_4$ | $0^{[Z]}$ | $0^{[Z]}$ | $0^{[Z]}$ | $0^{[Z]}$ | $\eta^2\nu_6{}^{[5b]}$ | $\eta^2\nu_2 M^{[5b]}$ | $\nu_6 B^{[6]}$ |
| 137 | 26 | $\nu_5\kappa$ | $0^{[B]}$ | $0^{[B]}$ | $0^{[B]}$ | $0^{[B]}$ | $0^{[Z]}$ | $0^{[Z]}$ | $0^{[Z]}$ |

Table 9.7: Preliminary products in $\pi_*(tmf)$ (Part 2, cont.)

| $n$ | $s$ | $x$ | $\epsilon$ | $\epsilon_1$ | $\epsilon_4$ | $\epsilon_5$ | $\kappa$ | $\kappa_4$ | $\bar{\kappa}$ |
|---|---|---|---|---|---|---|---|---|---|
| 142 | 27 | $\epsilon_5\kappa$ | $4\nu\nu_6^{[Pa]}$ | $0^{[Z]}$ | $0^{[B]}$ | $0^{[Z]}$ | $\eta\nu_6 B^{[5b]}$ | $2\bar{\kappa}^3 M^{[5a]}$ | $\eta\nu_6\kappa^{[P\kappa]}$ |
| 144 | 26 | $D_6$ | $0^{[B]}$ | $0^{[B]}$ | $0^{[B]}$ | $0^{[B]}$ | $0^{[Z]}$ | $0^{[Z]}$ | $\nu\nu_6\kappa^{[4]}$ |
| 147 | 25 | $\nu_6$ | $\nu_6 B^{[0e]}$ | $0^{[Z]}$ | $\nu_2 B M^{[0i]}$ | $0^{[Z]}$ | $\nu_6\kappa^{[t]}$ | $\nu_2\kappa M^{[1]}$ | $0^{[Z]}$ |
| 152 | 27 | $B_6$ | $0^{[B]}$ | $0^{[B]}$ | $0^{[B]}$ | $0^{[B]}$ | $0^{[Z]}$ | $0^{[B]}$ | $0^{[B]}$ |
| 156 | 30 | $C_6$ | $0^{[B]}$ | $0^{[B]}$ | $0^{[B]}$ | $0^{[B]}$ | $0^{[B]}$ | $0^{[B]}$ | $0^{[B]}$ |
| 161 | 29 | $\nu_6\kappa$ | $0^{[B]}$ | $0^{[B]}$ | $0^{[B]}$ | $0^{[B]}$ | $0^{[Z]}$ | $0^{[Z]}$ | $0^{[Z]}$ |
| 168 | 31 | $D_7$ | $0^{[B]}$ | $0^{[B]}$ | $0^{[B]}$ | $0^{[B]}$ | $0^{[Z]}$ | $0^{[Z]}$ | $0^{[B]}$ |
| 176 | 31 | $B_7$ | $0^{[B]}$ | $0^{[B]}$ | $0^{[B]}$ | $0^{[B]}$ | $0^{[Z]}$ | $0^{[Z]}$ | $0^{[B]}$ |
| 180 | 34 | $C_7$ | $0^{[B]}$ | $0^{[B]}$ | $0^{[B]}$ | $0^{[B]}$ | $0^{[B]}$ | $0^{[B]}$ | $0^{[B]}$ |

## 9.6. The algebra structure of $\pi_*(tmf)$

Now that we have the complete product structure of $\pi_*(tmf)$ in hand, we can make a choice of generators that is optimized for simplicity and the relation to $mf_{*/2}$. To this end, we give an alternative set of generators $\widetilde{B}_k \in \pi_{8+24k}(tmf)$ and make a more precise choice for $\nu_4 \in \pi_{3+24\cdot 4}(tmf)$.

The exceptional products of the form $B_iB_j$ found in Theorem 9.48 suggest the following change: let $\widetilde{B}_k = B_k + \epsilon_k$ if $k \equiv 0, 1 \mod 4$ and $\widetilde{B}_k = B_k$ otherwise. We then have $\bar{\kappa} \cdot \widetilde{B}_k = 0$ except when $k = 3$, so it also makes good sense to reverse our choice of $B_3$: let $\widetilde{B}_3 = B_3 + \bar{\kappa}^4$.

DEFINITION 9.50. Let

$$\widetilde{B}_k = \begin{cases} B_k + \epsilon_k & \text{for } k \equiv 0, 1 \mod 4, \\ B_3 + \bar{\kappa}^4 & \text{for } k = 3, \\ B_k & \text{otherwise.} \end{cases}$$

Then

$$AF(\widetilde{B}_k) = 4k + 3$$

in all cases, with $\widetilde{B}_k$ detected by $c_0 w_2^i$ in $E_\infty(tmf)$ for $k = 2i$, and by $\delta w_2^i$ for $k = 2i + 1$. As usual, we often abbreviate $\widetilde{B}_0$ to $\widetilde{B} = B + \epsilon$. Note that $\widetilde{B}^2 = B^2$, since $\epsilon^2 = 0$, so a class is $\widetilde{B}$-power torsion if and only if it is $B$-power torsion.

In Theorem 9.54 we finish our refinement of the generators, by choosing $\nu_4$ so that $\nu D_4 = 2\nu_4$. In the notation of case $[S]$ of the proof of Theorem 9.47, we set $s_0 = 1$, so that $s_2 = 1$, $s_4 = s$ and $s_6 = s$, where $s \in \{\pm 1\}$ is the remaining undetermined sign.

Having done this, we now describe the products in $\pi_*(tmf)$ in terms of our final, optimized, choice of generators. We break the result into three parts:

(1) The $\mathbb{Z}[\eta, \nu, B, M]$-module structure is given in Theorem 9.51 and Figures 9.6 through 9.13.
(2) The products among the 2-torsion free classes $\widetilde{B}_k$, $C_k$ and $D_k$ are given in Theorem 9.53.
(3) The products with the 2-power torsion classes $\eta_k$, $\nu_k$, $\epsilon_k$, $\kappa_k$ and $\bar{\kappa}$ are given in Theorem 9.54 and Tables 9.8 and 9.9.

THEOREM 9.51. *The $\mathbb{Z}[\eta, \nu, B, M]$-module structure of $\pi_*(tmf)$ is given in Figures 9.6 through 9.13. The $B$-periodic classes are shown in black, while the $B$-power torsion classes are red. The action of $B$ is as shown in those charts on the (black) classes $\eta_k$, $B_k$, $C_k$ and $D_k$ and agrees with that of $\epsilon$ on the (red) $B$-power torsion classes. The element $M$ acts monomorphically in $\pi_*(tmf)$.*

PROOF. These figures simply summarize what we have shown in Theorems 9.8, 9.14 and 9.16, Lemma 9.11 and Proposition 9.17. Note that the $B$-multiples of $B$-periodic classes $x$ are usually not labeled in Figures 9.6 to 9.13, but are recognizable by their location 8 degrees and 4 Adams filtrations higher than the element $x$. For $B$-power torsion classes $x$ the $B$- and $\epsilon$-multiples agree, by Proposition 9.40, and are usually labeled $\epsilon x$ when nonzero. □

REMARK 9.52. The charts in Figures 9.6 to 9.13 are not Adams spectral sequence $E_\infty$ charts, though we have placed elements at the location of their detecting class in $E_\infty$ to make the charts as easy to read as possible. Vertical lines denote multiplication by 2, lines to one degree higher denote multiplication by $\eta$, and lines (or curves) to three degrees higher denote multiplication by $\nu$. In particular, they are intended to indicate that $\nu D_4 = 2\nu_4$ (Theorem 9.54), not simply $\pm 2\nu_4$, as would be the case in an Adams $E_\infty$ chart.

To avoid congestion in these diagrams, we display the elements $B_k$ rather than the $\widetilde{B}_k$. This avoids the issue that $\widetilde{B}_k$ and $\epsilon_k$ for $k \equiv 0 \mod 4$ have the same detecting class in $E_\infty(tmf)$. The translation between the two is easily made by use of Definition 9.50.

THEOREM 9.53. *The products of elements in the $\widetilde{B}$-, $C$- and $D$-families are as follows:*

$$\begin{aligned} \widetilde{B}_i\widetilde{B}_j &= \widetilde{B}\widetilde{B}_{i+j} & C_iC_j &= 4(\widetilde{B}^2\widetilde{B}_{i+j} - (1728/d_{i+j+1})D_{i+j+1}) \\ \widetilde{B}_iC_j &= \widetilde{B}C_{i+j} & C_iD_j &= d_jC_{i+j} \\ \widetilde{B}_iD_j &= d_j\widetilde{B}_{i+j} & D_iD_j &= (d_id_j/d_{i+j})D_{i+j}\,. \end{aligned}$$

*The ring homomorphism from $\pi_*(tmf)$ onto its image in $mf_{*/2}$ has a section, which is also a ring homomorphism, sending*

$$\begin{aligned} \Delta^8 &\longmapsto M & c_4\Delta^k &\longmapsto \widetilde{B}_k \\ 2c_6\Delta^k &\longmapsto C_k & d_k\Delta^k &\longmapsto D_k\,. \end{aligned}$$

PROOF. To verify the relation $\widetilde{B}_i\widetilde{B}_j = \widetilde{B}\widetilde{B}_{i+j}$ we calculate for $0 \le i \le j \le 7$ that $\widetilde{B}_i\widetilde{B}_j = BB_{i+j} + \epsilon\epsilon_{i+j}$ for $i+j \equiv 1 \mod 4$, and $\widetilde{B}_i\widetilde{B}_j = BB_{i+j}$ otherwise, all of which follows from Theorem 9.48 and Table 9.7.

The remaining products rely upon the facts that the $\epsilon_i$ and $\bar\kappa^4$ are annihilated by 2, by $\widetilde{B}^2 = B^2$, by each $C_j$, and by the $D_k$ for $k \not\equiv 0 \mod 8$. Again, these properties can be read off from Table 9.7.

Let $\mathrm{im}(e)$ be the image of the edge homomorphism $e\colon \pi_*(tmf) \to mf_{*/2}$. It is the subring of $\mathbb{Z}[c_4, c_6, \Delta]/(c_4^3 - c_6^2 = 1728\Delta)$ generated by $\Delta^8$, $c_4\Delta^k$, $2c_6\Delta^k$ and $d_k\Delta^k$ for all $0 \le k \le 7$. These generators are subject only to the ideal of relations generated by the identities

$$\begin{aligned} c_4\Delta^i \cdot c_4\Delta^j &= c_4 \cdot c_4\Delta^{i+j} \\ c_4\Delta^i \cdot 2c_6\Delta^j &= c_4 \cdot 2c_6\Delta^{i+j} \\ c_4\Delta^i \cdot d_j\Delta^j &= d_j \cdot c_4\Delta^{i+j} \\ 2c_6\Delta^i \cdot 2c_6\Delta^j &= 4(c_4^2 \cdot c_4\Delta^{i+j} - (1728/d_{i+j+1}) \cdot d_{i+j+1}\Delta^{i+j+1}) \\ 2c_6\Delta^i \cdot d_j\Delta^j &= d_j \cdot 2c_6\Delta^{i+j} \\ d_i\Delta^i \cdot d_j\Delta^j &= (d_id_j/d_{i+j}) \cdot d_{i+j}\Delta^{i+j}\,. \end{aligned}$$

To see that no further relations are required, note that the associated quotient ring is generated as a $\mathbb{Z}[c_4, \Delta^8]$-module by $d_k\Delta^k$, $c_4\Delta^k$ and $2c_6\Delta^k$ for all $0 \le k \le 7$, subject only to the relations $c_4 \cdot d_k\Delta^k = d_k \cdot c_4\Delta^k$. It therefore maps isomorphically to $\mathrm{im}(e)$. Hence the first part of the theorem shows that the rules $\Delta^8 \mapsto M$, $c_4\Delta^k \mapsto \widetilde{B}_k$, $2c_6\Delta^k \mapsto C_k$ and $d_k\Delta^k \mapsto D_k$ specify a well-defined ring homomorphism $\sigma\colon \mathrm{im}(e) \to \pi_*(tmf)$, such that $e \circ \sigma$ is the inclusion $\mathrm{im}(e) \subset mf_{*/2}$. □

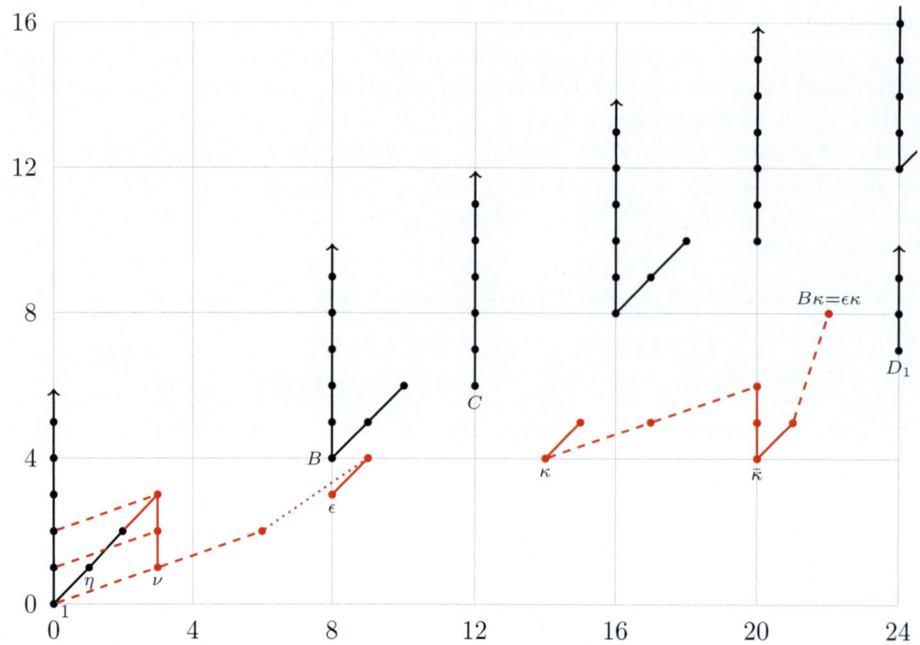


FIGURE 9.6. $\pi_n(tmf)$ for $0 \leq n \leq 24$

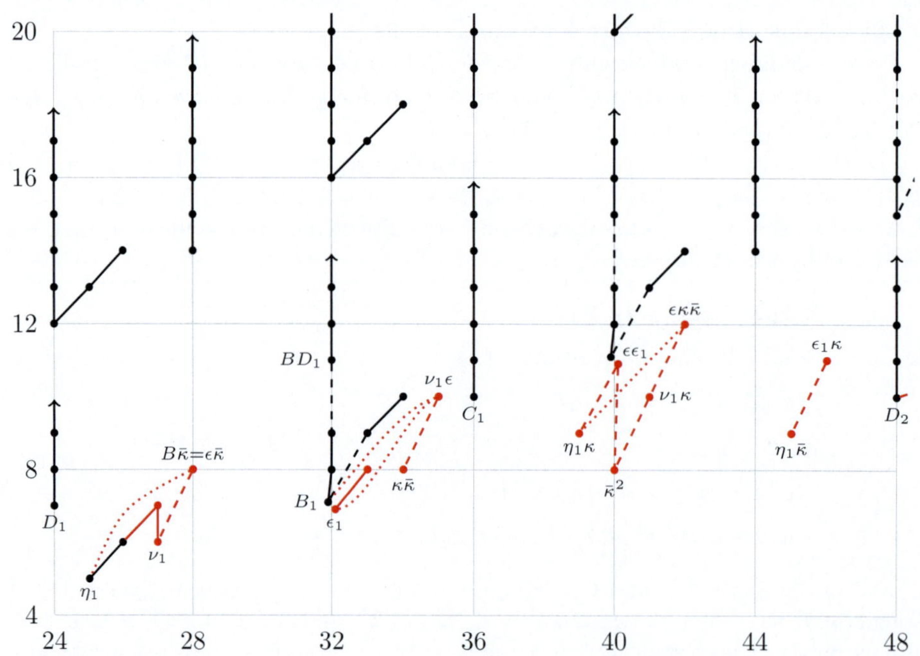


FIGURE 9.7. $\pi_n(tmf)$ for $24 \leq n \leq 48$

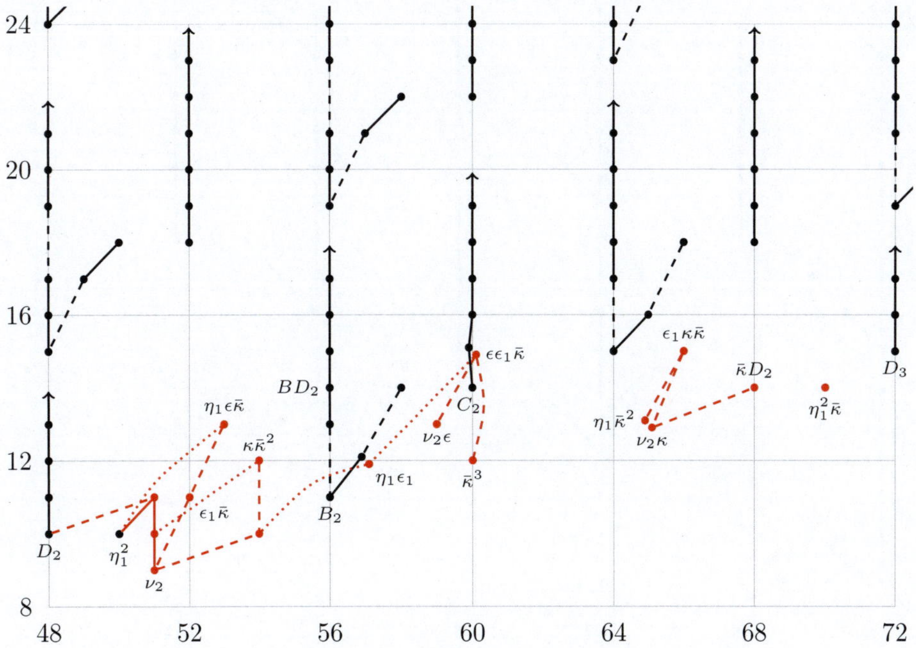


FIGURE 9.8. $\pi_n(tmf)$ for $48 \leq n \leq 72$

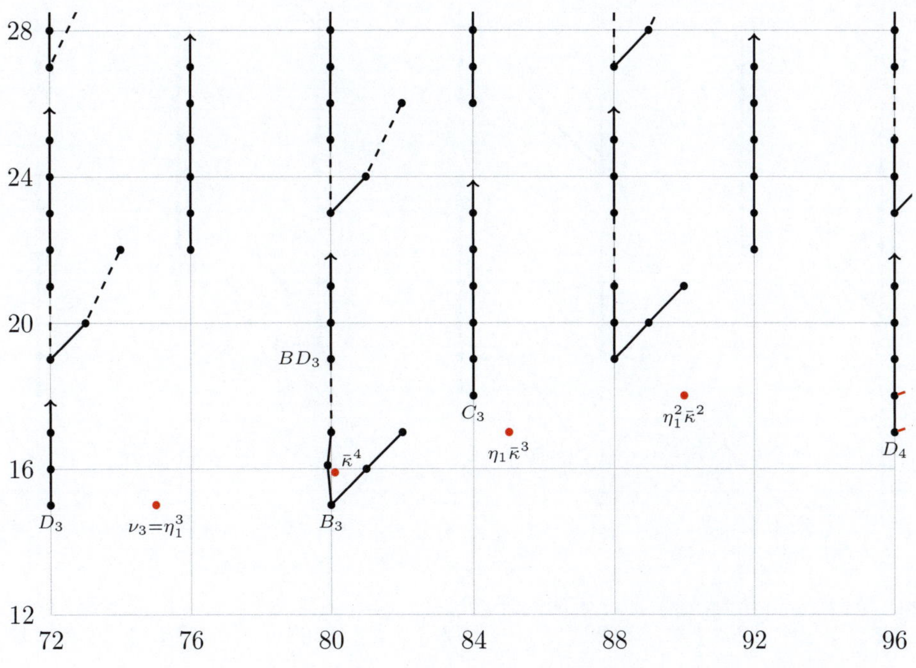


FIGURE 9.9. $\pi_n(tmf)$ for $72 \leq n \leq 96$

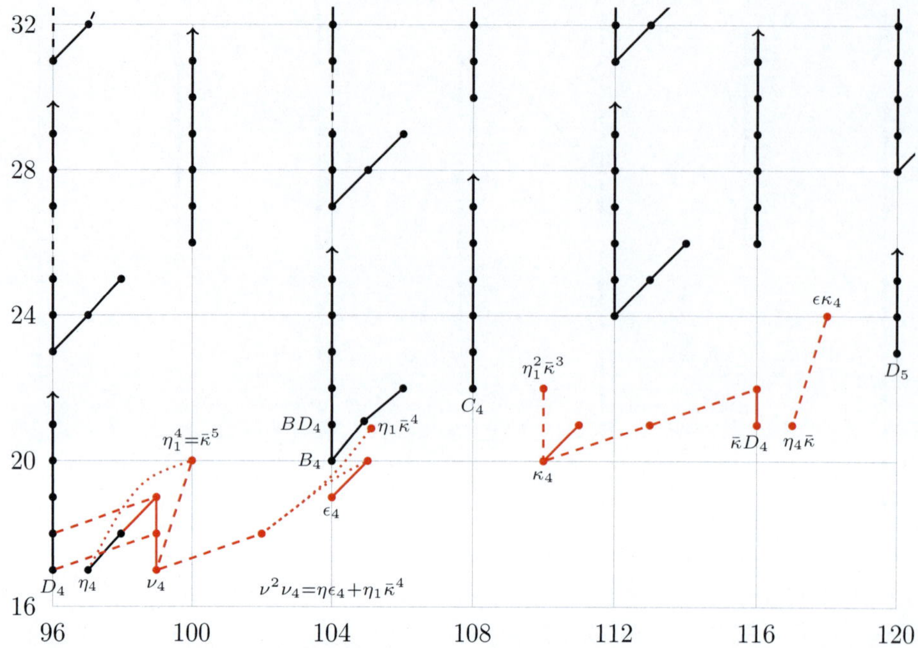

FIGURE 9.10. $\pi_n(tmf)$ for $96 \le n \le 120$

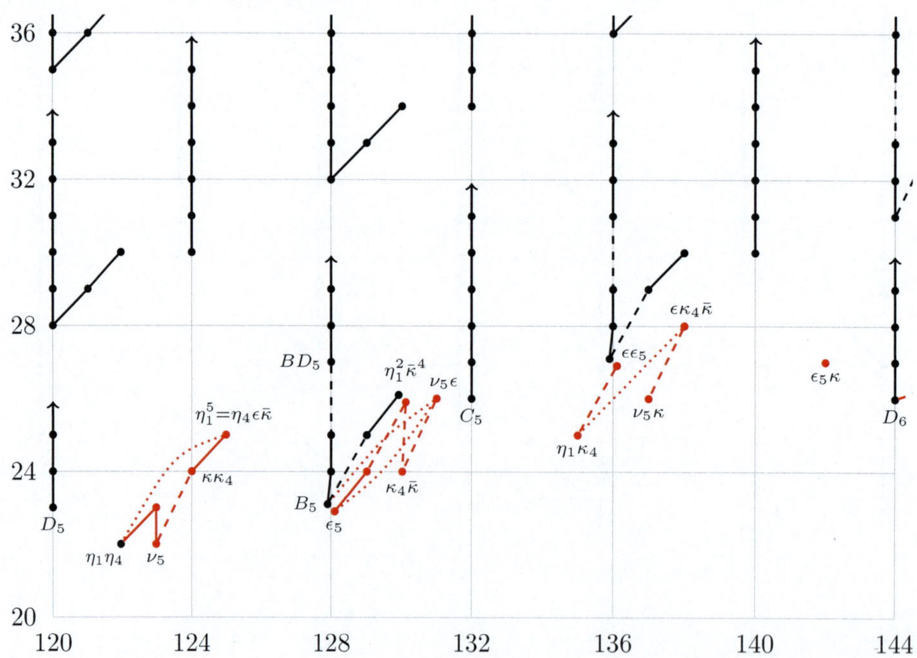

FIGURE 9.11. $\pi_n(tmf)$ for $120 \le n \le 144$

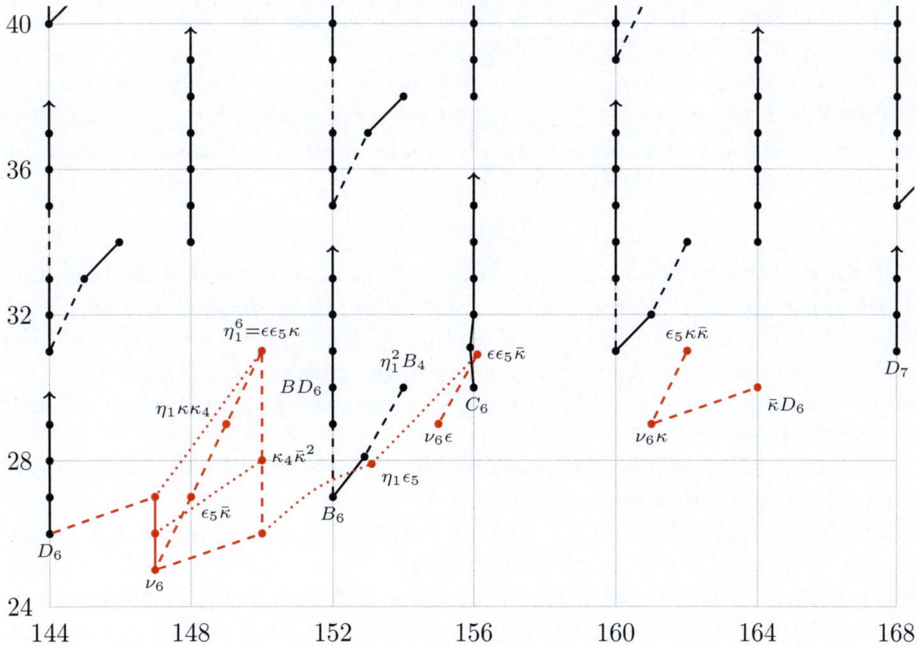


FIGURE 9.12. $\pi_n(tmf)$ for $144 \leq n \leq 168$

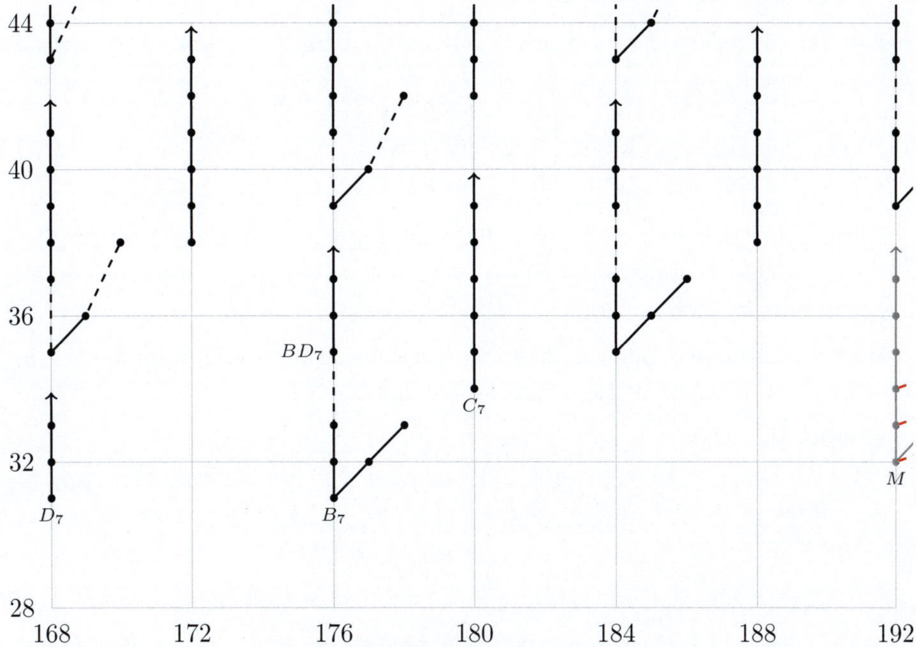


FIGURE 9.13. $\pi_n(tmf)$ for $168 \leq n \leq 192$

THEOREM 9.54. *We can (and do) choose $\nu_4$ so that $\nu D_4 = 2\nu_4$. This determines $\nu_4$ up to a factor in $\{1, 5\} \subset \mathbb{Z}/8^\times$.*

*The products with 2-power torsion elements in $\pi_*(tmf)$ are then as shown in Tables 9.8 and 9.9, where $s \in \{\pm 1\}$ is the sign in the product $\nu_4\nu_6 = s\nu\nu_2 M$. The rows for the $\widetilde{B}_k$, $C_k$ and odd-indexed $D_{2j+1}$ are omitted, because all products in these rows are zero with the exception of*

$$\eta_i \widetilde{B}_j = \eta \widetilde{B}_{i+j} \,.$$

PROOF. These tables are largely the same as those at the end of the preceding section. There are three sets of changes. First, each row of $B_k$-products is replaced by the corresponding row of $\widetilde{B}_k$-products. Having done this, we no longer need the rows we have omitted, as they contain only 0 entries with the exception of the $\eta_i\widetilde{B}_j$. We retain the rows for the products by the even-indexed $D_{2j}$ because a number of these products are nonzero.

Second, the $B$-multiples of $B$-power torsion classes are rewritten as $\epsilon$-multiples, which is justified by Proposition 9.40.

Third, the signs $s_i$ for $i \in \{0, 2, 4, 6\}$ are all replaced by the appropriate multiple of $s$: see case [$S$] of the proof of Theorem 9.47. There we saw that $\nu D_4 = 2s_0\nu_4$ with $s_0 \in \{\pm 1\}$. Thus far the generator $\nu_4$ of $\pi_{99}(tmf) \cong \mathbb{Z}/8$ was only determined by its detecting class $h_2 w_2^2$ in the Adams spectral sequence. Hence, by possibly changing the sign of $\nu_4$, we may arrange that $s_0 = 1$ and $\nu D_4 = 2\nu_4$. This then gives the remaining $\nu_{2j} D_4$ as in Theorem 9.47. □

This theorem contains the following generalization of Mahowald's dictum that $Bx = \epsilon x$ for $B$-power torsion classes $x$.

COROLLARY 9.55. *If $x$ is $B$-power torsion then $\widetilde{B}_k \cdot x = 0$ for all $k$. Equivalently,*

$$B_k x = \begin{cases} \epsilon_k x & \text{for } k = 0, 1, 4, 5, \\ \bar{\kappa}^4 x & \text{for } k = 3, \\ 0 & \text{for } k = 2, 6, 7. \end{cases}$$

PROOF. In other words, $\widetilde{B}_k \cdot x = 0$ for all $0 \le k \le 7$ and $x \in \Gamma_B \pi_*(tmf) = (\nu_k, \epsilon_k, \kappa_k, \bar{\kappa})$. The honorary case $\widetilde{B}_k \cdot \nu_3 = 0$ is not made explicit in our tables, but $\eta_1^3 \widetilde{B}_k = \eta^3 \widetilde{B}_{k+3} = 4\nu \widetilde{B}_{k+3} = 0$. □

It also confirms the following heuristic relationship between the products $x_i y_j$, including the $\widetilde{B}$-family but excluding the $D$-family.

COROLLARY 9.56.

(1) *When $x, y \in \{\eta, \nu, \epsilon, \kappa, \bar{\kappa}, \widetilde{B}, C\}$ and $x_i$ and $y_j$ are defined, then $x_i y_j$ depends only on $x$, $y$ and $i + j$, except when $x = y = \nu$.*

(2) *For $x = y = \nu$,*

$$\nu_i \nu_j = (i + 1)\nu\nu_{i+j} \,,$$

*except when $\{i, j\} = \{4, 6\}$ if $s = -1$.*

(3) *When $x, y, z \in \{\eta, \nu, \epsilon, \kappa, \bar{\kappa}, \widetilde{B}, C\}$ and $x_i$, $y_j$ and $z_k$ are defined, then $x_i y_j z_k$ depends only on $x$, $y$, $z$ and $i + j + k$, except when two or more of $x$, $y$ and $z$ equal $\nu$.*

(4) *$\eta_1^4 = \eta\nu_4 \neq 0$ while $\eta^3\eta_4 = 0$, so $x_i y_j z_k w_\ell$ does sometimes depend on more than $x$, $y$, $z$, $w$ and $i + j + k + \ell$.* □

Table 9.8: Products in $\pi_*(tmf)$: the entry in row $x$ (found in the $x$-column) and column $y$ (found in the top row) gives $xy$. Part 1 of 2: $\eta_i$- and $\nu_i$-multiples. Rows $\widetilde{B}_k$, $C_k$ and $D_{2j+1}$ omitted: see Theorem 9.54.

| $n$ | $s$ | $x$ | $\eta_1$ | $\eta_4$ | $\nu_1$ | $\nu_2$ | $\nu_4$ | $\nu_5$ | $\nu_6$ |
|---|---|---|---|---|---|---|---|---|---|
| 8 | 3 | $\epsilon$ | $\eta\epsilon_1$ | $\eta\epsilon_4$ | $\eta\kappa\bar{\kappa}$ | $\nu_2\epsilon$ | 0 | $\eta\kappa_4\bar{\kappa}$ | $\nu_6\epsilon$ |
| 14 | 4 | $\kappa$ | $\eta_1\kappa$ | $\eta\kappa_4$ | $\eta\bar{\kappa}^2$ | $\nu_2\kappa$ | $\nu\kappa_4$ | $\nu_5\kappa$ | $\nu_6\kappa$ |
| 20 | 4 | $\bar{\kappa}$ | $\eta_1\bar{\kappa}$ | $\eta_4\bar{\kappa}$ | 0 | 0 | 0 | 0 | 0 |
| 25 | 5 | $\eta_1$ | $\eta_1^2$ | $\eta_1\eta_4$ | $\eta\nu_2$ | 0 | $\eta\nu_5$ | $\eta\nu_6$ | 0 |
| 27 | 6 | $\nu_1$ | $\eta\nu_2$ | $\eta\nu_5$ | $2\nu\nu_2$ | 0 | 0 | $2\nu\nu_6$ | 0 |
| 32 | 7 | $\epsilon_1$ | $\nu^2\nu_2$ | $\eta\epsilon_5$ | $\nu_2\epsilon$ | 0 | $\eta\kappa_4\bar{\kappa}$ | $\nu_6\epsilon$ | 0 |
| 34 | 8 | $\kappa\bar{\kappa}$ | $\nu_2\epsilon$ | $\eta\kappa_4\bar{\kappa}$ | 0 | 0 | 0 | 0 | 0 |
| 39 | 9 | $\eta_1\kappa$ | 0 | $\eta\eta_1\kappa_4$ | $\eta\nu_2\kappa$ | 0 | $\eta\nu_5\kappa$ | $\eta\nu_6\kappa$ | 0 |
| 40 | 8 | $\bar{\kappa}^2$ | $\eta_1\bar{\kappa}^2$ | $\nu_5\kappa$ | 0 | 0 | 0 | 0 | 0 |
| 45 | 9 | $\eta_1\bar{\kappa}$ | $\eta_1^2\bar{\kappa}$ | $\epsilon_5\kappa$ | 0 | 0 | 0 | 0 | 0 |
| 48 | 10 | $D_2$ | 0 | 0 | 0 | $4\nu_4$ | $4\nu_6$ | 0 | $4\nu M$ |
| 50 | 10 | $\eta_1^2$ | $\eta_1^3$ | $4\nu_6$ | 0 | 0 | $\eta^2\nu_6$ | 0 | 0 |
| 51 | 9 | $\nu_2$ | 0 | $\eta\nu_6$ | 0 | $\nu\nu_4$ | $3\nu\nu_6$ | 0 | $\nu^2 M$ |
| 60 | 12 | $\bar{\kappa}^3$ | $\eta_1\bar{\kappa}^3$ | 0 | 0 | 0 | 0 | 0 | 0 |
| 65 | 13 | $\eta_1\bar{\kappa}^2$ | $\eta_1^2\bar{\kappa}^2$ | $\eta\nu_6\kappa$ | 0 | 0 | 0 | 0 | 0 |
| 65 | 13 | $\nu_2\kappa$ | 0 | $\eta\nu_6\kappa$ | 0 | $2\bar{\kappa}D_4$ | $\nu\nu_6\kappa$ | 0 | $4\bar{\kappa}M$ |
| 70 | 14 | $\eta_1^2\bar{\kappa}$ | 0 | 0 | 0 | 0 | 0 | 0 | 0 |
| 75 | 15 | $\eta_1^3$ | $\eta\nu_4$ | 0 | 0 | 0 | 0 | 0 | 0 |
| 80 | 16 | $\bar{\kappa}^4$ | $\eta\epsilon_4$ $+\nu^2\nu_4$ | 0 | 0 | 0 | 0 | 0 | 0 |
| 85 | 17 | $\eta_1\bar{\kappa}^3$ | $2\kappa_4$ | 0 | 0 | 0 | 0 | 0 | 0 |
| 90 | 18 | $\eta_1^2\bar{\kappa}^2$ | 0 | 0 | 0 | 0 | 0 | 0 | 0 |
| 96 | 17 | $D_4$ | 0 | 0 | $2\nu_5$ | $2\nu_6$ | $2s\nu M$ | $2\nu_1 M$ | $2s\nu_2 M$ |
| 97 | 17 | $\eta_4$ | $\eta_1\eta_4$ | $\eta^2 M$ | $\eta\nu_5$ | $\eta\nu_6$ | 0 | $\eta\nu_1 M$ | $\eta\nu_2 M$ |
| 99 | 17 | $\nu_4$ | $\eta\nu_5$ | 0 | 0 | $-3\nu\nu_6$ | $\nu^2 M$ | 0 | $s\nu\nu_2 M$ |
| 104 | 19 | $\epsilon_4$ | $\eta\epsilon_5$ | $\eta\epsilon M$ | $\eta\kappa_4\bar{\kappa}$ | $\nu_6\epsilon$ | 0 | $\eta\kappa\bar{\kappa}M$ | $\nu_2\epsilon M$ |
| 110 | 20 | $\kappa_4$ | $\eta_1\kappa_4$ | $\eta\kappa M$ | $\nu_5\kappa$ | $\nu_6\kappa$ | $\nu\kappa M$ | $\eta\bar{\kappa}^2 M$ | $\nu_2\kappa M$ |
| 116 | 21 | $\bar{\kappa}D_4$ | 0 | 0 | 0 | 0 | 0 | 0 | 0 |
| 117 | 21 | $\eta_4\bar{\kappa}$ | $\epsilon_5\kappa$ | $\eta^2\bar{\kappa}M$ | 0 | 0 | 0 | 0 | 0 |

Table 9.8: Products in $\pi_*(tmf)$ (Part 1, cont.)

| $n$ | $s$ | $x$ | $\eta_1$ | $\eta_4$ | $\nu_1$ | $\nu_2$ | $\nu_4$ | $\nu_5$ | $\nu_6$ |
|---|---|---|---|---|---|---|---|---|---|
| 122 | 22 | $\eta_1\eta_4$ | $4\nu_6$ | $2\nu_1 M$ | $\eta^2\nu_6$ | 0 | 0 | $\eta^2\nu_2 M$ | 0 |
| 123 | 22 | $\nu_5$ | $\eta\nu_6$ | $\eta\nu_1 M$ | $-2\nu\nu_6$ | 0 | 0 | $2\nu\nu_2 M$ | 0 |
| 128 | 23 | $\epsilon_5$ | $\nu^2\nu_6$ | $\eta\epsilon_1 M$ | $\nu_6\epsilon$ | 0 | $\eta\kappa\bar{\kappa}M$ | $\nu_2\epsilon M$ | 0 |
| 130 | 24 | $\kappa_4\bar{\kappa}$ | $\nu_6\epsilon$ | $\eta\kappa\bar{\kappa}M$ | 0 | 0 | 0 | 0 | 0 |
| 135 | 25 | $\eta_1\kappa_4$ | 0 | $2\bar{\kappa}^2 M$ | $\eta\nu_6\kappa$ | 0 | $\eta^2\bar{\kappa}^2 M$ | $\eta\nu_2\kappa M$ | 0 |
| 137 | 26 | $\nu_5\kappa$ | $\eta\nu_6\kappa$ | $\eta^2\bar{\kappa}^2 M$ | 0 | 0 | 0 | 0 | 0 |
| 142 | 27 | $\epsilon_5\kappa$ | 0 | 0 | 0 | 0 | 0 | 0 | 0 |
| 144 | 26 | $D_6$ | 0 | 0 | 0 | $4\nu M$ | $4\nu_2 M$ | 0 | $4\nu_4 M$ |
| 147 | 25 | $\nu_6$ | 0 | $\eta\nu_2 M$ | 0 | $\nu^2 M$ | $-s\nu\nu_2 M$ | 0 | $\nu\nu_4 M$ |
| 161 | 29 | $\nu_6\kappa$ | 0 | $\eta\nu_2\kappa M$ | 0 | $4\bar{\kappa}M$ | $\nu\nu_2\kappa M$ | 0 | $2\bar{\kappa}D_4 M$ |

Table 9.9: Products in $\pi_*(tmf)$: the entry in row $x$ (found in the $x$-column) and column $y$ (found in the top row) gives $xy$. Part 2 of 2: $\epsilon_i$-, $\kappa_i$- and $\bar{\kappa}$-multiples. Rows $\widetilde{B}_k$, $C_k$ and $D_{2j+1}$ omitted: see Theorem 9.54.

| $n$ | $s$ | $x$ | $\epsilon$ | $\epsilon_1$ | $\epsilon_4$ | $\epsilon_5$ | $\kappa$ | $\kappa_4$ | $\bar{\kappa}$ |
|---|---|---|---|---|---|---|---|---|---|
| 8 | 3 | $\epsilon$ | 0 | $2\bar{\kappa}^2$ | 0 | $\eta\eta_1\kappa_4$ | $\eta^2\bar{\kappa}$ | $\eta\eta_4\bar{\kappa}$ | $\eta\nu_1$ |
| 14 | 4 | $\kappa$ | $\eta^2\bar{\kappa}$ | $\eta\eta_1\bar{\kappa}$ | $\eta\eta_4\bar{\kappa}$ | $\epsilon_5\kappa$ | $\eta\nu_1$ | $\eta\nu_5$ | $\kappa\bar{\kappa}$ |
| 20 | 4 | $\bar{\kappa}$ | $\eta\nu_1$ | $\eta\nu_2$ | $\eta\nu_5$ | $\eta\nu_6$ | $\kappa\bar{\kappa}$ | $\kappa_4\bar{\kappa}$ | $\bar{\kappa}^2$ |
| 25 | 5 | $\eta_1$ | $\eta\epsilon_1$ | $\nu^2\nu_2$ | $\eta\epsilon_5$ | $\nu^2\nu_6$ | $\eta_1\kappa$ | $\eta_1\kappa_4$ | $\eta_1\bar{\kappa}$ |
| 27 | 6 | $\nu_1$ | $\eta\kappa\bar{\kappa}$ | $\nu_2\epsilon$ | $\eta\kappa_4\bar{\kappa}$ | $\nu_6\epsilon$ | $\eta\bar{\kappa}^2$ | $\nu_5\kappa$ | 0 |
| 32 | 7 | $\epsilon_1$ | $2\bar{\kappa}^2$ | 0 | $\eta\eta_1\kappa_4$ | 0 | $\eta\eta_1\bar{\kappa}$ | $\epsilon_5\kappa$ | $\eta\nu_2$ |
| 34 | 8 | $\kappa\bar{\kappa}$ | $\eta^2\bar{\kappa}^2$ | $\eta\nu_2\kappa$ | $\eta\nu_5\kappa$ | $\eta\nu_6\kappa$ | 0 | 0 | $2\nu\nu_2$ |
| 39 | 9 | $\eta_1\kappa$ | 0 | 0 | 0 | 0 | $\eta^2\nu_2$ | $\eta^2\nu_6$ | $\nu_2\epsilon$ |
| 40 | 8 | $\bar{\kappa}^2$ | 0 | 0 | 0 | 0 | $2\nu\nu_2$ | $2\nu\nu_6$ | $\bar{\kappa}^3$ |
| 45 | 9 | $\eta_1\bar{\kappa}$ | $\eta^2\nu_2$ | 0 | $\eta^2\nu_6$ | 0 | $\nu_2\epsilon$ | $\nu_6\epsilon$ | $\eta_1\bar{\kappa}^2$ |
| 48 | 10 | $D_2$ | 0 | 0 | 0 | 0 | 0 | 0 | $\nu\nu_2\kappa$ |
| 50 | 10 | $\eta_1^2$ | 0 | 0 | 0 | 0 | 0 | 0 | $\eta_1^2\bar{\kappa}$ |
| 51 | 9 | $\nu_2$ | $\nu_2\epsilon$ | 0 | $\nu_6\epsilon$ | 0 | $\nu_2\kappa$ | $\nu_6\kappa$ | 0 |
| 60 | 12 | $\bar{\kappa}^3$ | 0 | 0 | 0 | 0 | 0 | 0 | $\bar{\kappa}^4$ |
| 65 | 13 | $\eta_1\bar{\kappa}^2$ | 0 | 0 | 0 | 0 | 0 | 0 | $\eta_1\bar{\kappa}^3$ |

Table 9.9: Products in $\pi_*(tmf)$ (Part 2, cont.)

| $n$ | $s$ | $x$ | $\epsilon$ | $\epsilon_1$ | $\epsilon_4$ | $\epsilon_5$ | $\kappa$ | $\kappa_4$ | $\bar{\kappa}$ |
|---|---|---|---|---|---|---|---|---|---|
| 65 | 13 | $\nu_2\kappa$ | 0 | 0 | 0 | 0 | 0 | 0 | 0 |
| 70 | 14 | $\eta_1^2\bar{\kappa}$ | 0 | 0 | 0 | 0 | 0 | 0 | $\eta_1^2\bar{\kappa}^2$ |
| 75 | 15 | $\eta_1^3$ | 0 | 0 | 0 | 0 | 0 | 0 | 0 |
| 80 | 16 | $\bar{\kappa}^4$ | 0 | 0 | 0 | 0 | 0 | 0 | $\eta\nu_4$ |
| 85 | 17 | $\eta_1\bar{\kappa}^3$ | 0 | 0 | 0 | 0 | 0 | 0 | $\eta\epsilon_4 + \nu^2\nu_4$ |
| 90 | 18 | $\eta_1^2\bar{\kappa}^2$ | 0 | 0 | 0 | 0 | 0 | 0 | $2\kappa_4$ |
| 96 | 17 | $D_4$ | 0 | 0 | 0 | 0 | $2\kappa_4$ | 0 | $\bar{\kappa}D_4$ |
| 97 | 17 | $\eta_4$ | $\eta\epsilon_4$ | $\eta\epsilon_5$ | $\eta\epsilon M$ | $\eta\epsilon_1 M$ | $\eta\kappa_4$ | $\eta\kappa M$ | $\eta_4\bar{\kappa}$ |
| 99 | 17 | $\nu_4$ | 0 | $\eta\kappa_4\bar{\kappa}$ | 0 | $\eta\kappa\bar{\kappa}M$ | $\nu\kappa_4$ | $\nu\kappa M$ | 0 |
| 104 | 19 | $\epsilon_4$ | 0 | $\eta\eta_1\kappa_4$ | 0 | $2\bar{\kappa}^2 M$ | $\eta\eta_4\bar{\kappa}$ | $\eta^2\bar{\kappa}M$ | $\eta\nu_5$ |
| 110 | 20 | $\kappa_4$ | $\eta\eta_4\bar{\kappa}$ | $\epsilon_5\kappa$ | $\eta^2\bar{\kappa}M$ | $\eta\eta_1\bar{\kappa}M$ | $\eta\nu_5$ | $\eta\nu_1 M$ | $\kappa_4\bar{\kappa}$ |
| 116 | 21 | $\bar{\kappa}D_4$ | 0 | 0 | 0 | 0 | $2\kappa_4\bar{\kappa}$ | 0 | $\eta\eta_1\kappa_4$ |
| 117 | 21 | $\eta_4\bar{\kappa}$ | $\eta^2\nu_5$ | $\eta^2\nu_6$ | 0 | $\eta^2\nu_2 M$ | $\eta\kappa_4\bar{\kappa}$ | $\eta\kappa\bar{\kappa}M$ | $\nu_5\kappa$ |
| 122 | 22 | $\eta_1\eta_4$ | $2\kappa_4\bar{\kappa}$ | 0 | 0 | 0 | $\eta\eta_1\kappa_4$ | $2\bar{\kappa}^2 M$ | $\epsilon_5\kappa$ |
| 123 | 22 | $\nu_5$ | $\eta\kappa_4\bar{\kappa}$ | $\nu_6\epsilon$ | $\eta\kappa\bar{\kappa}M$ | $\nu_2\epsilon M$ | $\nu_5\kappa$ | $\eta\bar{\kappa}^2 M$ | 0 |
| 128 | 23 | $\epsilon_5$ | $\eta\eta_1\kappa_4$ | 0 | $2\bar{\kappa}^2 M$ | 0 | $\epsilon_5\kappa$ | $\eta\eta_1\bar{\kappa}M$ | $\eta\nu_6$ |
| 130 | 24 | $\kappa_4\bar{\kappa}$ | $\eta\nu_5\kappa$ | $\eta\nu_6\kappa$ | $\eta^2\bar{\kappa}^2 M$ | $\eta\nu_2\kappa M$ | 0 | 0 | $2\nu\nu_6$ |
| 135 | 25 | $\eta_1\kappa_4$ | 0 | 0 | 0 | 0 | $\eta^2\nu_6$ | $\eta^2\nu_2 M$ | $\nu_6\epsilon$ |
| 137 | 26 | $\nu_5\kappa$ | 0 | 0 | 0 | 0 | 0 | 0 | 0 |
| 142 | 27 | $\epsilon_5\kappa$ | $4\nu\nu_6$ | 0 | 0 | 0 | $\eta\nu_6\epsilon$ | $2\bar{\kappa}^3 M$ | $\eta\nu_6\kappa$ |
| 144 | 26 | $D_6$ | 0 | 0 | 0 | 0 | 0 | 0 | $\nu\nu_6\kappa$ |
| 147 | 25 | $\nu_6$ | $\nu_6\epsilon$ | 0 | $\nu_2\epsilon M$ | 0 | $\nu_6\kappa$ | $\nu_2\kappa M$ | 0 |
| 161 | 29 | $\nu_6\kappa$ | 0 | 0 | 0 | 0 | 0 | 0 | 0 |

REMARK 9.57. The only nonzero products between the 2-power torsion classes $\eta_i$, $\nu_i$, $\epsilon_i$, $\kappa_i$ and $\bar{\kappa}$ and the 2-torsion free classes $\widetilde{B}_j$, $C_j$ and $D_j$ (other than $D_0 = 1$) are the following:

- $\eta_i \cdot \widetilde{B}_j = \eta\widetilde{B}_{i+j}$.
- $\nu_i \cdot D_j = 4\nu_{i+j}$ for $i$ even and $j \in \{2, 6\}$.
- $\nu_i \cdot D_4 = \pm 2\nu_{i+4}$, with a sign depending on $i$.
- $\kappa \cdot D_4 = 2\kappa_4$.
- $\bar{\kappa} \cdot D_j = \nu\nu_j\kappa$ for $j \in \{2, 6\}$.
- $\bar{\kappa} \cdot D_4 = \bar{\kappa}D_4$, one of our generators.

REMARK 9.58. Most of this multiplicative structure was very concisely described by Henriques in [**54**, Ch. 13], on pages 190–192. We offer the following concordance between his presentation and our results.

(1) The (nonzero) 2- and $\eta$-multiplications, and almost all $\nu$-multiplications, are shown in the picture on page 190, which repeats $M$-periodically. The missing $\nu$-multiplications from degrees 0, 51, 96 and 147 are easily deduced from the ones shown, by means of the additive group structure.
(2) The action by $\widetilde{B}$ is trivial in the upper part of Henriques' picture (including all classes in degrees $* \equiv 3 \mod 24$), and is periodic in the lower part. The $\mathbb{Z}[\widetilde{B}]$-module generators of infinite order correspond to our $\widetilde{B}$-, $C$- and $D$-families.
(3) The products among 2-torsion free classes are determined up to 2-power torsion classes by the ring homomorphism to modular forms, as stated on page 191. The fact that a multiplicative section can be chosen so that there are no 2-power torsion correction terms is not made explicit, and may be new.
(4) Most $\epsilon$-, $\kappa$- and $\bar{\kappa}$-multiplications are also shown in the picture on page 190. The remaining degrees supporting nonzero products with these classes, as well as with $\eta_1 = \{\eta\Delta\}$, $\nu_1 = \{2\nu\Delta\}$, $\epsilon_1 = q$ and $\nu_2 = \{\nu\Delta^2\}$, are listed in the table on pages 191 and 192. We note the following deviations from our conclusions:
   ($\kappa$) A nonzero product from degree 3 is missing.
   ($\bar{\kappa}$) Some products from degrees 0, 20, 40 and 96 are not shown. The sign of the product from degree 130 is left undetermined.
   ($\eta_1$) A nonzero product from degree 17 is missing.
   ($\nu_1$) The sign of the product from degree 123 is left undetermined.
   ($\epsilon_1$) A nonzero product from degree 98 is missing.
   ($\nu_2$) Nonzero products from degrees 48, 144, 147, 150 and 161 are missing. The indicated product from degree 116 should be omitted. The sign of the product from degree 96, and the coefficient in $\mathbb{Z}/8^\times$ of the product from degree 99, are not determined.
(5) The products with $\eta_4$, $\nu_4$, $\nu_5$, $\nu_6$, $\epsilon_4$, $\epsilon_5$ and $\kappa_4$ are not listed.

Henriques also shows the Adams $E_\infty$-term for *tmf* on pages 196–197, with hidden 2-, $\eta$- and $\nu$-multiplications indicated. Our results appear to agree, except near degrees 32 and 128. Henriques indicates a class "$c_4\Delta + q$" in degree 32 of Adams filtration 7, equal to our class $B_1$, such that $\eta(c_4\Delta + q)$ has Adams filtration 9 and $\nu(c_4\Delta + q) = 0$. As our calculations show, the latter $\nu$-product should be nonzero. The same issue occurs in degree 128 and Adams filtration 23.

REMARK 9.59. The multiplicative structure in the Adams–Novikov spectral sequence for *tmf* does not seem to suffice to determine the common sign $s$ in the relations $\nu_4\nu_6 = s\nu\nu_2 M$, $\nu_4 D_4 = 2s\nu M$ and $\nu_6 D_4 = 2s\nu_2 M$. In the notation of [**23**, §8], the classes $\nu_4\nu_6$ and $\nu\nu_2 M$ are both detected by $h_2^2\Delta^{10}$, but this only tells us that they agree modulo the higher filtration class $2\nu\nu_2 M$. Likewise, the classes $\nu_4 D_4$ and $2\nu M$ are both detected by $2h_2\Delta^8$, and must agree modulo $4\nu M$, while $\nu_6 D_4$ and $2\nu_2 M$ are both detected by $2h_2\Delta^{10}$, and must agree modulo $4\nu_2 M$. Similarly, in the elliptic spectral sequence for $TMF$, and in the homotopy fixed point spectral sequence for $L_{K(2)}TMF = EO_2$, the sign in these products is invisible at the $E_\infty$-term.

CHAPTER 10

# Duality

## 10.1. Pontryagin duality in the $B$-power torsion of $\pi_*(tmf)$

The $B$-power torsion in $\pi_*(tmf)$ repeats 192-periodically, and is shown in red in Figures 9.6 through 9.13, and again in Figures 10.1 and 10.2. In the latter illustrations the groups in degrees $3 \le * \le 90$ are shown in the upper halves with degrees increasing toward the right, while the groups in degrees $75 \le * \le 164$ are shown in the lower halves with degrees increasing toward the left. As usual, 2-, $\eta$- and $\nu$-extensions are shown by solid or dashed lines increasing degree by 0, 1 and 3, respectively, but the vertical coordinate has no specific meaning. The mirror symmetry across the "fold line" in these pictures makes it clear that for $0 \le n < 192$ the $B$-power torsion in degree $n$ is abstractly isomorphic to the $B$-power torsion in degree $170 - n$, except in degrees $n \equiv 3 \mod 24$.

More precisely, we will see in Theorem 10.25 that these finite groups are naturally Pontryagin dual, so that there is a perfect pairing

$$(-,-)\colon \Theta\pi_n(tmf) \times \Theta\pi_{170-n}(tmf) \longrightarrow \mathbb{Q}/\mathbb{Z}$$

for $0 \le n < 192$. Here $\Theta\pi_n(tmf) \subset \Gamma_B\pi_n(tmf)$ denotes the self-dual part of the $B$-power torsion, i.e., the part in degrees $n \not\equiv 3 \mod 24$. A less ad hoc characterization of $\Theta\pi_*(tmf)$ is given in Definition 10.18, which makes it clear that this is a $\pi_*(tmf)$-submodule of $\Gamma_B\pi_*(tmf)$. The omitted groups in degrees $n \equiv 3 \mod 24$ are generated by the classes $\nu_k$ for $0 \le k \le 6$, and we will see that there is a more comprehensive spectral expression of the duality, for which the order $d_{7-k} \in \{2, 4, 8\}$ of the cyclic group $\langle\nu_k\rangle$ corresponds to the index of $\mathbb{Z}\{D_{7-k}\}$ in $\mathbb{Z}\{B_{7-k}/B\}$. The spectrum level statement

$$\Sigma^{20} tmf \simeq I(tmf/(2^\infty, B^\infty, M^\infty))$$

is given in Theorem 10.6, and the duality between $\langle\nu_k\rangle$ and $\mathbb{Z}\{B_{7-k}/B\}/\mathbb{Z}\{D_{7-k}\}$ appears in Theorem 10.25. We explain the notation $tmf/(2^\infty, B^\infty, M^\infty)$ in Section 10.2 and recall the Brown–Comenetz duality functor $I$ in Section 10.3, where we also establish the spectrum level duality by a descent argument along $\iota'\colon tmf \to tmf_1(3) \simeq BP\langle 2\rangle$.

The duality theorem can be re-expressed in terms of local cohomology spectra and Anderson duality, as we spell out in Proposition 10.12 of Section 10.4:

$$\Sigma^{22} tmf \simeq I_{\mathbb{Z}}(\Gamma_{(B,M)} tmf)\,.$$

By construction, $tmf$ is the connective cover of an $E_\infty$ ring spectrum $Tmf$, and the equivalence above extends to an Anderson self-duality of $Tmf$:

$$\Sigma^{21} Tmf \simeq I_{\mathbb{Z}}(Tmf)\,.$$

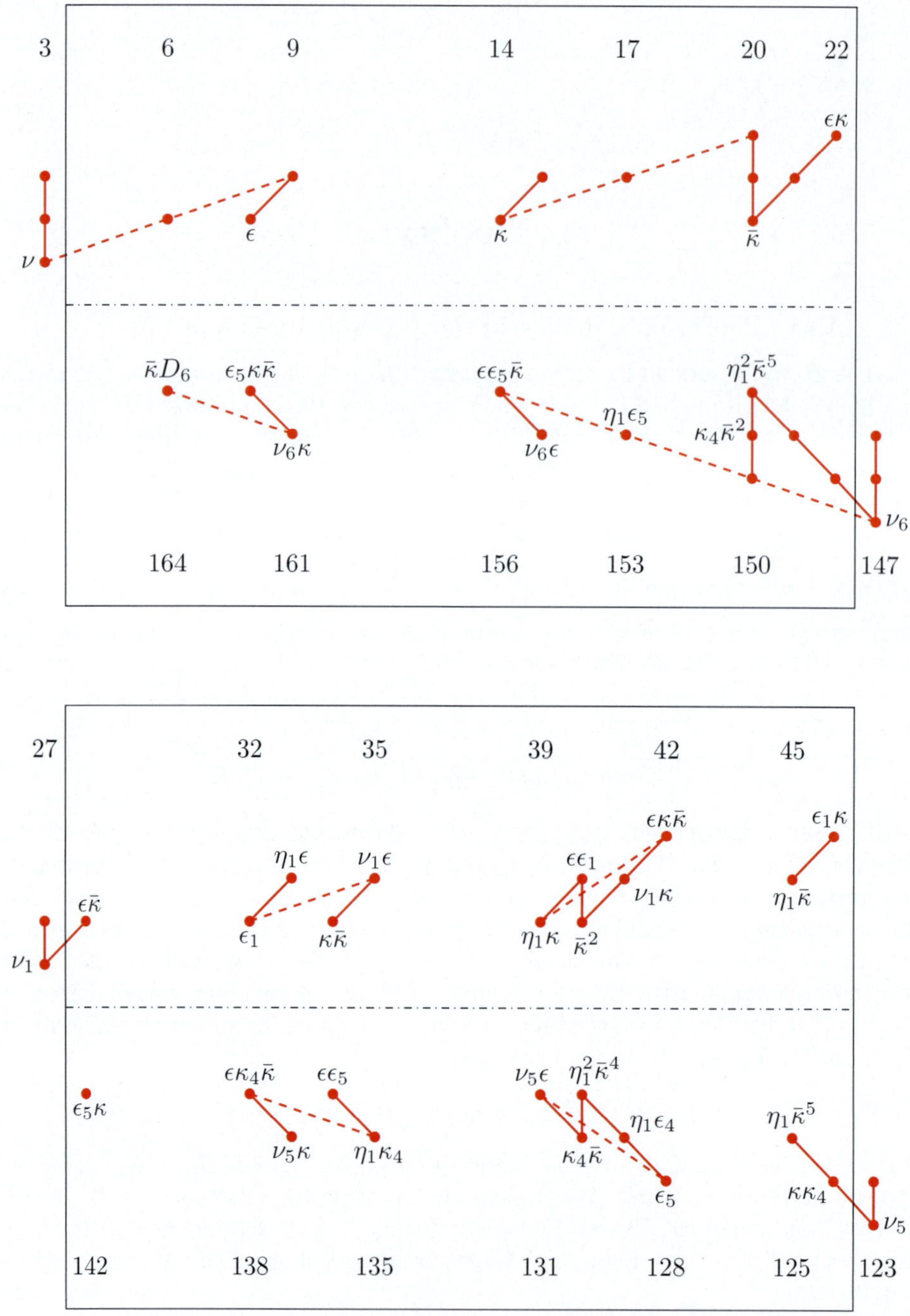


FIGURE 10.1. The self-dual submodule $\Theta\pi_n(tmf) \subset \Gamma_B\pi_n(tmf)$ for $4 \leq n, 170 - n \leq 46$, with $\eta_1\bar{\kappa}^5 = \eta_1^5$ and $\eta_1^2\bar{\kappa}^5 = \eta_1^6$

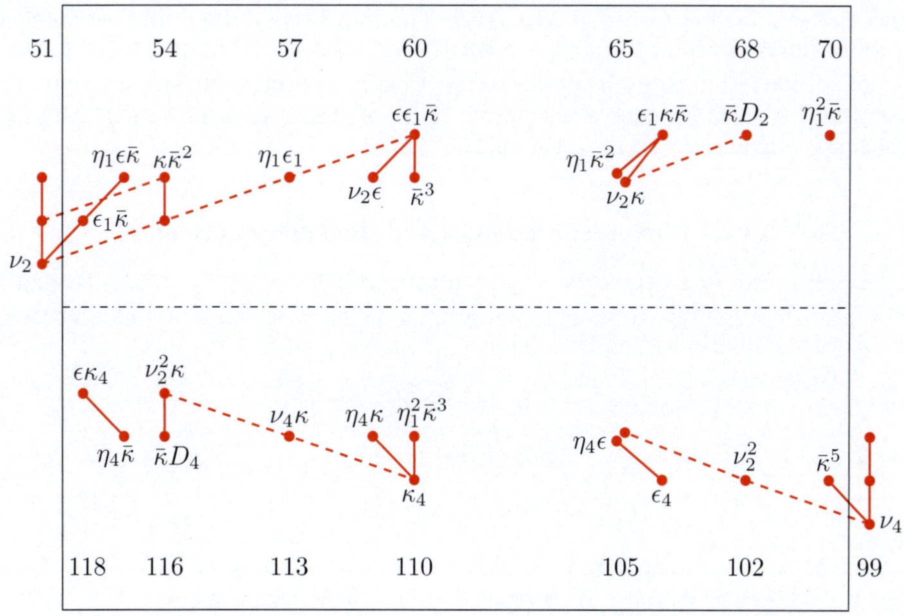


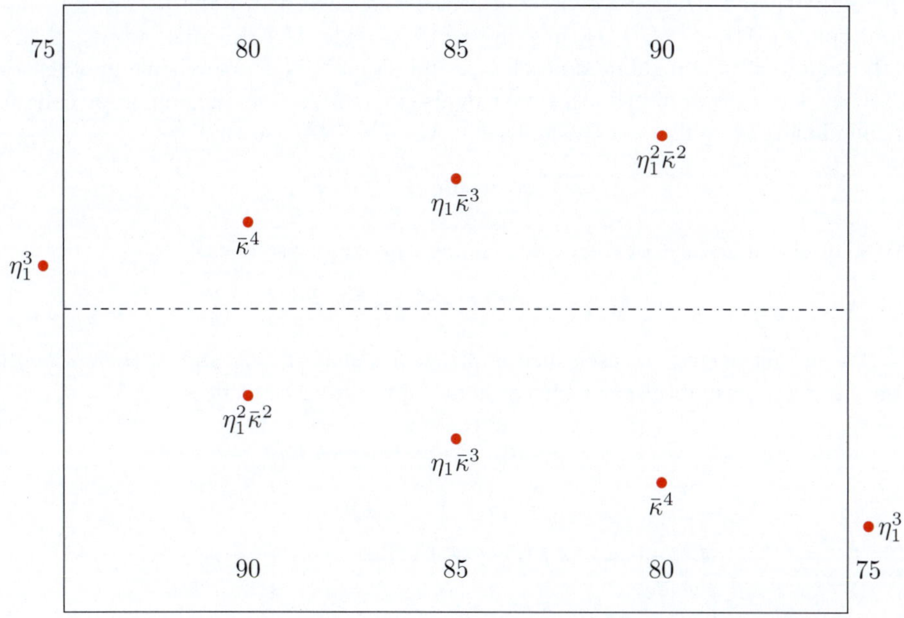


FIGURE 10.2. The self-dual submodule $\Theta\pi_n(tmf) \subset \Gamma_B\pi_n(tmf)$ for $52 \le n, 170 - n \le 94$, with $\eta_1\bar{\kappa}^4 = \eta\epsilon_4 + \nu^2\nu_4$ and $\bar{\kappa}^5 = \eta_1^4$

This theorem is due to Stojanoska [**161**, Thm. 13.1], but the published version assumes that the prime 2 has been inverted. See Theorem 10.13 for a proof following [**67**] of the 2-complete part of this result. Finally, in Section 10.5 we translate the spectrum level duality equivalence into a series of algebraic duality isomorphisms, which are summarized in Theorem 10.26.

## 10.2. Torsion submodules and divisible quotients

DEFINITION 10.1. Let $R$ be a commutative $S$-algebra ($=E_\infty$ ring spectrum), let $M$ be an $R$-module spectrum, and let $x \in \pi_d(R)$. Let $M/x$ be the homotopy cofiber of the multiplication-by-$x$ map

$$\Sigma^d M \xrightarrow{x} M ,$$

let $M[1/x]$ be the homotopy colimit of the sequence

$$M \xrightarrow{x} \Sigma^{-d} M \xrightarrow{x} \Sigma^{-2d} M \xrightarrow{x} \dots ,$$

and let $M/x^\infty$ be the homotopy cofiber of the structure map $M \to M[1/x]$. Note that $\pi_*(M[1/x]) = \pi_*(M)[1/x]$, so that there is a short exact sequence

$$0 \to \pi_*(M)/x^\infty \longrightarrow \pi_*(M/x^\infty) \longrightarrow \Gamma_x \pi_{*-1}(M) \to 0 , \tag{10.1}$$

where $\Gamma_x M_*$ and $M_*/x^\infty$ denote the kernel and cokernel of the localization homomorphism $M_* \to M_*[1/x]$, for any $\pi_*(R)$-module $M_*$. In other words, $\Gamma_x M_*$ is the $x$-power torsion submodule of $M_*$, and $M_*/x^\infty$ is an $x$-divisible quotient of $M_*[1/x]$. By reversal of priorities, we can also view $M/x^\infty$ as the homotopy colimit of the homotopy cofibers of the maps $x^n \colon M \to \Sigma^{-nd} M$, so that

$$M/x^\infty \simeq \operatorname*{hocolim}_n \Sigma^{-nd} M/x^n .$$

We shall also make use of the evident homotopy cofiber sequence

$$M/x \longrightarrow \Sigma^d M/x^\infty \xrightarrow{x} M/x^\infty . \tag{10.2}$$

We are interested in cases such as $M = R$, $M = R[1/x]$ and $M = R/x^\infty$. If also $y \in \pi_e(R)$, we obtain a square of homotopy cofiber sequences

$$\begin{array}{ccccc}
R & \longrightarrow & R[1/x] & \longrightarrow & R/x^\infty \\
\downarrow & & \downarrow & & \downarrow \\
R[1/y] & \longrightarrow & R[1/x, 1/y] & \longrightarrow & R/(x^\infty)[1/y] \\
\downarrow & & \downarrow & & \downarrow \\
R/y^\infty & \longrightarrow & R[1/x]/(y^\infty) & \longrightarrow & R/(x^\infty, y^\infty)
\end{array}$$

in the category of $R$-modules, where $R/(x^\infty, y^\infty) = R/x^\infty \wedge_R R/y^\infty$ is the iterated homotopy cofiber of the upper left hand square. Likewise, if $z \in \pi_f(R)$ then there is a cube of homotopy cofiber sequences, with

$$R/(x^\infty, y^\infty, z^\infty) = R/x^\infty \wedge_R R/y^\infty \wedge_R R/z^\infty$$

being the iterated homotopy cofiber of the initial cube

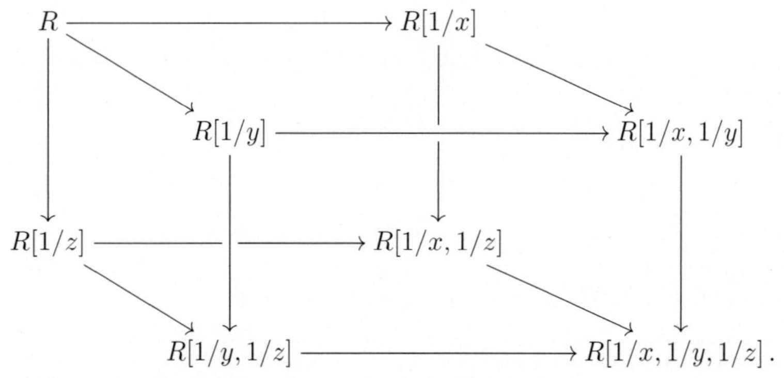


REMARK 10.2. Following Greenlees–May [**66**, §3], one can work with the homotopy $x$-power torsion (= local cohomology) spectrum $\Gamma_x M$, defined as the homotopy fiber of $M \to M[1/x]$, in place of the homotopy cofiber $M/x^\infty$. Since $\Sigma\Gamma_x M \simeq M/x^\infty$ this only amounts to a shift in grading, which, however, may be convenient for the discussion of multiplicative structure. There is a natural short exact sequence

$$0 \to \pi_{*+1}(M)/x^\infty \longrightarrow \pi_*(\Gamma_x M) \longrightarrow \Gamma_x \pi_*(M) \to 0\,.$$

Iterating, $\Gamma_{(x,y)}R = \Gamma_x(\Gamma_y R)$ is the double homotopy fiber of the initial square above, with $\Sigma^2\Gamma_{(x,y)}R \simeq R/(x^\infty, y^\infty)$. Similarly, $\Gamma_{(x,y,z)}R = \Gamma_x(\Gamma_y(\Gamma_z R))$ is the triple homotopy fiber of the displayed cube, with $\Sigma^3\Gamma_{(x,y,z)}R \simeq R/(x^\infty, y^\infty, z^\infty)$.

LEMMA 10.3 ([**66**]). *Let $x_1, \dots, x_n \in \pi_*(R)$. The homotopy type of the $R$-module $R/(x_1^\infty, \dots, x_n^\infty)$ only depends on $n$ and the radical $\sqrt{J}$ of the ideal $J = (x_1, \dots, x_n)$ in $\pi_*(R)$.*

PROOF. In view of the equivalence $R/(x_1^\infty, \dots, x_n^\infty) \simeq \Sigma^n\Gamma_{(x_1,\dots,x_n)}R$, this is a restatement of the fact that $\Gamma_{(x_1,\dots,x_n)}R$ only depends on the radical of $(x_1, \dots, x_n)$, which is explained in [**66**, p. 266]. □

## 10.3. Brown–Comenetz duality

Recall the Bott element $B \in \pi_8(tmf)$ with $B \in \{w_1\}$, and the Mahowald element $M \in \pi_{192}(tmf)$ with $M \in \{w_2^4\}$. We shall study $\mathbb{Z}[B, M]$-modules obtained by restriction along $\mathbb{Z}[B, M] \to \pi_*(tmf)$, or by induction along $\mathbb{Z}[B] \to \mathbb{Z}[B, M]$. Recall also the following notation from Section 9.4.

DEFINITION 10.4. Let $N_* \subset \pi_*(tmf)$ be the $\mathbb{Z}[B]$-submodule generated by the classes in degrees $0 \le * < 192$, or equivalently, by the classes in degrees $0 \le * \le 180$.

By the results of the previous chapter, cf. Theorem 9.26, the $B$-power torsion $\Gamma_B N_*$ is finite in degrees $3 \le * \le 164$ and is trivial outside this range. Furthermore, the $B$-divisible quotient $N_*/B^\infty$ is concentrated in degrees $\le 172$. The group in degree 172 is a copy of $\mathbb{Z}$ generated by $C_7/B$, where $C_7 \in \{h_0\alpha^3 w_2^3\}$, and the group in degree 171 is trivial.

Since $w_2^4$ acts freely on the Adams $E_\infty$-term for $tmf$, the composite homomorphism

$$N_* \otimes \mathbb{Z}[M] \longrightarrow \pi_*(tmf) \otimes \pi_*(tmf) \overset{\cdot}{\longrightarrow} \pi_*(tmf)$$

is an isomorphism of $\mathbb{Z}[B, M]$-modules. In particular, $\Gamma_M \pi_*(tmf) = 0$ and

$$\pi_*(tmf/M^\infty) \cong \pi_*(tmf)/M^\infty = N_* \otimes \mathbb{Z}[M]/M^\infty ,$$

where $\mathbb{Z}[M]/M^\infty = \mathbb{Z}[M, M^{-1}]/\mathbb{Z}[M] \cong \mathbb{Z}[M^{-1}]\{1/M\}$. Hence there is a short exact sequence

$$0 \to N_*/B^\infty \otimes \mathbb{Z}[M]/M^\infty \\ \longrightarrow \pi_*(tmf/(B^\infty, M^\infty)) \longrightarrow \Gamma_B N_{*-1} \otimes \mathbb{Z}[M]/M^\infty \to 0 .$$

It follows that $\pi_*(tmf/(B^\infty, M^\infty))$ is concentrated in degrees $* \le -20$, with the group in degree $-20$ being a copy of $\mathbb{Z}$ generated by $C_7/BM$, and the group in degree $-21$ being zero. Using the short exact sequence

$$0 \to \pi_*(tmf/(B^\infty, M^\infty))/2^\infty \\ \longrightarrow \pi_*(tmf/(2^\infty, B^\infty, M^\infty)) \longrightarrow \Gamma_2 \pi_{*-1}(tmf/(B^\infty, M^\infty)) \to 0$$

we conclude that $\pi_*(tmf/(2^\infty, B^\infty, M^\infty))$ is concentrated in degrees $* \le -20$, with the group in degree $-20$ being a copy of $\mathbb{Z}/2^\infty$.

DEFINITION 10.5. Let $I = I_{\mathbb{Q}/\mathbb{Z}}$ be the Brown–Comenetz dual of the sphere spectrum [**36**]. This is the spectrum representing the generalized cohomology theory

$$X \longmapsto I^n(X) = \mathrm{Hom}(\pi_n(X), \mathbb{Q}/\mathbb{Z}) .$$

Let $I(X) = F(X, I)$, so that $\pi_{-n} I(X) = I^n(X)$. If $M$ is an $R$-module spectrum, then $I(M) = F(M, I)$ is naturally an $R$-module spectrum.

Here is our formulation of the duality theorem.

THEOREM 10.6. *There is a duality equivalence of (implicitly 2-completed) tmf-modules*

$$\Sigma^{20} tmf \simeq I(tmf/(2^\infty, B^\infty, M^\infty)) .$$

PROOF. By the discussion at the beginning of this section, the homotopy groups of $tmf/(2^\infty, B^\infty, M^\infty)$ are concentrated in degrees $* \le -20$, with the group in degree $-20$ being a copy of $\mathbb{Z}/2^\infty$. Hence the homotopy groups of the Brown–Comenetz dual $I(tmf/(2^\infty, B^\infty, M^\infty))$ are concentrated in degrees $* \ge 20$, with the group in degree 20 being isomorphic to $\mathrm{Hom}(\mathbb{Z}/2^\infty, \mathbb{Q}/\mathbb{Z}) = \mathbb{Z}_2$. Representing a (2-adic) generator for this group by a map from $S^{20}$, we obtain a $tmf$-module map

$$a \colon \Sigma^{20} tmf \longrightarrow I(tmf/(2^\infty, B^\infty, M^\infty))$$

between 19-connected spectra, which induces an isomorphism on $\pi_{20}$. We will show that $a$ is in fact an equivalence, as a consequence of an easier duality result for the truncated Brown–Peterson spectrum $BP\langle 2\rangle$.

Recall from Remark 1.16 and Equation (9.5) in the proof of Proposition 9.19 that Lawson and Naumann [**91**] constructed a map of (implicitly 2-complete) commutative $S$-algebras $\iota' \colon tmf \to tmf_1(3) \simeq BP\langle 2\rangle$, where $\pi_*(BP\langle 2\rangle) = \mathbb{Z}[v_1, v_2]$ maps isomorphically to $\pi_*(tmf_1(3)) = \mathbb{Z}[a_1, a_3]$ by $v_1 \mapsto -a_1 \equiv a_1 \mod 2$ and $v_2 \mapsto -7a_3 \equiv a_3 \mod (2, a_1)$. The map $\iota'$ induces $B \mapsto c_4 = a_1(a_1^3 - 24a_3)$ and $M \mapsto \Delta^8$ with $\Delta = a_3^3(a_1^3 - 27a_3)$. It is straightforward to check that the radical of the ideal $J = (2, c_4, \Delta^8)$ in $\pi_*(tmf_1(3))$ equals $\sqrt{J} = (2, a_1, a_3)$, which corresponds to $(2, v_1, v_2)$ in $\pi_*(BP\langle 2\rangle)$.

The main step is to show that the coinduced $BP\langle 2\rangle$-module map

$$b = F_{tmf}(BP\langle 2\rangle, a)\colon F_{tmf}(BP\langle 2\rangle, \Sigma^{20} tmf) \longrightarrow F_{tmf}(BP\langle 2\rangle, I(tmf/(2^\infty, B^\infty, M^\infty)))$$

is an equivalence. We start with the target of $b$. Induction along $\iota'\colon tmf \to BP\langle 2\rangle$ takes $tmf/(2^\infty, B^\infty, M^\infty)$ to

$$\begin{aligned} BP\langle 2\rangle \wedge_{tmf} tmf/(2^\infty, B^\infty, M^\infty) &\cong BP\langle 2\rangle/(2^\infty, B^\infty, M^\infty) \\ &= BP\langle 2\rangle/(2^\infty, c_4^\infty, (\Delta^8)^\infty) \\ &\simeq BP\langle 2\rangle/(2^\infty, v_1^\infty, v_2^\infty)\,. \end{aligned}$$

The middle identity uses that $\iota'_*\colon \pi_*(tmf) \to \pi_*(BP\langle 2\rangle)$ maps $B$ and $M$ to $c_4$ and $\Delta^8$, respectively. The final equivalence uses that $BP\langle 2\rangle/(2^\infty, c_4^\infty, (\Delta^8)^\infty)$ and $BP\langle 2\rangle/(2^\infty, v_1^\infty, v_2^\infty)$ are equivalent as $BP\langle 2\rangle$-modules because $(2, c_4, \Delta^8)$ and $(2, v_1, v_2)$ have the same radical in $\pi_*(BP\langle 2\rangle) = \mathbb{Z}[v_1, v_2]$, cf. Lemma 10.3.

Applying the Brown–Comenetz duality functor $I$, we see that coinduction along $\iota'$ takes $I(tmf/(2^\infty, B^\infty, M^\infty))$ to

$$\begin{aligned} F_{tmf}(BP\langle 2\rangle, I(tmf/(2^\infty, B^\infty, M^\infty))) &\cong I(BP\langle 2\rangle \wedge_{tmf} tmf/(2^\infty, B^\infty, M^\infty)) \\ &\simeq I(BP\langle 2\rangle/(2^\infty, v_1^\infty, v_2^\infty))\,. \end{aligned}$$

The homotopy groups of $BP\langle 2\rangle/(2^\infty, v_1^\infty, v_2^\infty)$ are

$$\mathbb{Z}[v_1, v_2]/(2^\infty, v_1^\infty, v_2^\infty) = \mathbb{Z}/2^\infty[v_1^{-1}, v_2^{-1}]\{1/v_1v_2\}\,,$$

with $1/v_1v_2$ in degree $-8$. Hence the target

$$\pi_*(I(BP\langle 2\rangle/(2^\infty, v_1^\infty, v_2^\infty))) \cong \Sigma^8 \pi_*(BP\langle 2\rangle)$$

of $\pi_*(b)$ is a free module over $\pi_*(BP\langle 2\rangle)$, on a single generator in degree 8.

Next, we consider the source of $b$. Let $\Phi = \Phi A(1)$ be a finite (8-cell) 12-dimensional CW spectrum with cohomology realizing $A(2)//E(2)$, i.e., the double of $A(1) = \langle Sq^1, Sq^2\rangle$. We saw in Lemma 1.42 that such spectra exist. Then $tmf \wedge \Phi \simeq BP\langle 2\rangle$ as $tmf$-modules, because $A//A(2) \otimes A(2)//E(2) \cong A//E(2)$. The Spanier–Whitehead dual $D\Phi = F(\Phi, S)$ has cohomology realizing $\Sigma^{-12}A(2)//E(2)$ as an $A(2)$-module, so there is also an equivalence of $tmf$-modules $F(\Phi, tmf) \simeq \Sigma^{-12}BP\langle 2\rangle$. Coinduction along $\iota'$ therefore takes $\Sigma^{20}tmf$ to

$$F_{tmf}(BP\langle 2\rangle, \Sigma^{20}tmf) \simeq F(\Phi, \Sigma^{20}tmf) \simeq \Sigma^8 BP\langle 2\rangle\,,$$

in the category of $tmf$-modules. Hence the source of the homomorphism $\pi_*(b)$ is isomorphic to $\Sigma^8\pi_*(BP\langle 2\rangle)$ as a $\pi_*(tmf)$-module, and, in particular, as a graded abelian group.

The coinduced $BP\langle 2\rangle$-module map

$$b = F_{tmf}(BP\langle 2\rangle, a)\colon \Sigma^8 BP\langle 2\rangle \longrightarrow I(BP\langle 2\rangle/(2^\infty, v_1^\infty, v_2^\infty))$$

can be written as $F(\Phi, a)$, hence is a map between 7-connected spectra that induces an isomorphism on $\pi_8$. It follows that

$$\pi_*(b)\colon \Sigma^8\pi_*(BP\langle 2\rangle) \longrightarrow \Sigma^8\pi_*(BP\langle 2\rangle)$$

is surjective, since it is $\pi_*(BP\langle 2\rangle)$-linear and maps onto the $\pi_*(BP\langle 2\rangle)$-module generator of the target. Furthermore, its source and target are abstractly isomorphic and of finite type as (implicitly 2-completed) graded abelian groups, so the surjectivity implies that $\pi_*(b)$ is in fact an isomorphism.

It follows that $b$ is an equivalence and the homotopy cofiber $Cb$ is contractible. The Hurewicz theorem then implies that $Ca$ is contractible, and that $a$ is an equivalence, since $D\Phi \wedge Ca \simeq F(\Phi, Ca) \simeq Cb$ and $H_{-12}(D\Phi; \mathbb{Z}) \cong H^{12}(\Phi; \mathbb{Z}) \cong \mathbb{Z}$. □

REMARK 10.7. The theorem can be reformulated as saying that there is a perfect (Brown–Comenetz duality) pairing

$$\Sigma^{20} tmf \wedge tmf/(2^\infty, B^\infty, M^\infty) \longrightarrow I\,.$$

When smashed with the perfect (Spanier–Whitehead duality) pairing

$$D\Phi \wedge \Phi \longrightarrow S$$

it gives the perfect pairing

$$\Sigma^8 BP\langle 2\rangle \wedge BP\langle 2\rangle/(2^\infty, v_1^\infty, v_2^\infty) \longrightarrow I\,.$$

LEMMA 10.8. *The $\pi_*(tmf)$-module isomorphism*

$$\begin{aligned} a_* \colon \pi_*(\Sigma^{20} tmf) \xrightarrow{\cong} \pi_* I(tmf/(2^\infty, B^\infty, M^\infty)) \\ = \operatorname{Hom}(\pi_{-*}(tmf/(2^\infty, B^\infty, M^\infty)), \mathbb{Q}/\mathbb{Z}) \end{aligned}$$

*is adjoint to a perfect pairing*

$$\langle -, - \rangle \colon \pi_*(\Sigma^{20} tmf) \times \pi_{-*}(tmf/(2^\infty, B^\infty, M^\infty)) \longrightarrow \mathbb{Q}/\mathbb{Z}$$

*with $\langle x, y\rangle = a_*(x)(y)$, such that*

$$\langle r \cdot x, y\rangle = (-1)^{|r||x|} \langle x, r \cdot y\rangle$$

*for $r \in \pi_*(tmf)$, $x \in \pi_*(\Sigma^{20} tmf)$ and $y \in \pi_*(tmf/(2^\infty, B^\infty, M^\infty))$ with $|r| + |x| + |y| = 0$.*

PROOF. The formula $\langle r \cdot x, y\rangle = (-1)^{|r||x|} \langle x, r \cdot y\rangle$ follows by adjunction from $a_*(r \cdot x) = r \cdot a_*(x)$, where $(r \cdot a_*(x))(y) = (-1)^{|r||x|} a_*(x)(r \cdot y)$. □

REMARK 10.9. A similar argument proves that $\Sigma^4 ko \simeq I(ko/(2^\infty, B^\infty))$, using $ko \wedge C\eta \simeq ku$ and $\Sigma^2 ku \simeq I(ku/(2^\infty, v_1^\infty))$. The smash product of the perfect pairings

$$\Sigma^4 ko \wedge ko/(2^\infty, B^\infty) \longrightarrow I$$

and

$$DC\eta \wedge C\eta \longrightarrow S$$

gives the perfect pairing

$$\Sigma^2 ku \wedge ku/(2^\infty, v_1^\infty) \longrightarrow I\,.$$

The $\pi_*(ko)$-module isomorphism

$$\pi_*(\Sigma^4 ko) \xrightarrow{\cong} \pi_* I(ko/(2^\infty, B^\infty)) = \operatorname{Hom}(\pi_{-*}(ko/(2^\infty, B^\infty)), \mathbb{Q}/\mathbb{Z})$$

is adjoint to a perfect pairing

$$\langle -.- \rangle \colon \pi_*(\Sigma^4 ko) \times \pi_{-*}(ko/(2^\infty, B^\infty)) \longrightarrow \mathbb{Q}/\mathbb{Z}\,.$$

## 10.4. Anderson duality

The natural double duality map $\rho\colon X \to I(I(X))$ is an equivalence whenever each group $\pi_n(X)$ is finite. There is a modification $I_{\mathbb{Z}}(X)$ of the Brown–Comenetz dual, known as the Anderson dual [**180**], such that the natural map $\rho\colon X \to I_{\mathbb{Z}}(I_{\mathbb{Z}}(X))$ is an equivalence whenever each group $\pi_n(X)$ is finitely generated. The modification is defined by the homotopy fiber sequence

$$I_{\mathbb{Z}}(X) \longrightarrow I_{\mathbb{Q}}(X) \longrightarrow I_{\mathbb{Q}/\mathbb{Z}}(X)\,,$$

where $I_{\mathbb{Q}/\mathbb{Z}}(X) = I(X)$ is the Brown–Comenetz dual of $X$ and $I_{\mathbb{Q}}(X) = F(X, H\mathbb{Q})$. Here $H\mathbb{Q}$ represents ordinary rational cohomology

$$X \longmapsto H^n(X;\mathbb{Q}) \cong \mathrm{Hom}(\pi_n(X), \mathbb{Q})\,.$$

The associated long exact sequence of homotopy groups breaks up into short exact sequences

$$0 \to \mathrm{Ext}(\pi_{n-1}(X), \mathbb{Z}) \longrightarrow \pi_{-n} I_{\mathbb{Z}}(X) \longrightarrow \mathrm{Hom}(\pi_n(X), \mathbb{Z}) \to 0\,. \tag{10.3}$$

LEMMA 10.10. *For each prime p, there is a natural chain of equivalences*

$$I(X/p^\infty) \simeq I_{\mathbb{Z}}(X)^\wedge_p\,.$$

PROOF. Applying the contravariant duality functors to the homotopy cofiber sequence $X \xrightarrow{p^n} X \longrightarrow X/p^n$ we obtain a commutative diagram of horizontal and vertical homotopy (co-)fiber sequences

$$\begin{array}{ccccc}
I_{\mathbb{Z}}(X/p^n) & \longrightarrow & I_{\mathbb{Z}}(X) & \xrightarrow{p^n} & I_{\mathbb{Z}}(X) \\
\downarrow & & \downarrow & & \downarrow \\
I_{\mathbb{Q}}(X/p^n) & \longrightarrow & I_{\mathbb{Q}}(X) & \xrightarrow{p^n} & I_{\mathbb{Q}}(X) \\
\downarrow & & \downarrow & & \downarrow \\
I(X/p^n) & \longrightarrow & I(X) & \xrightarrow{p^n} & I(X)\,.
\end{array}$$

Here $I_{\mathbb{Q}}(X/p^n) \simeq *$, so

$$\begin{aligned}
I(X/p^\infty) &\simeq I(\operatorname*{hocolim}_n X/p^n) \simeq \operatorname*{holim}_n I(X/p^n) \\
&\simeq \operatorname*{holim}_n \Sigma I_{\mathbb{Z}}(X/p^n) \simeq \operatorname*{holim}_n I_{\mathbb{Z}}(X)/p^n \simeq I_{\mathbb{Z}}(X)^\wedge_p\,.
\end{aligned}$$

□

Using local cohomology spectra and/or Anderson duality, we can reformulate the duality equivalence of Theorem 10.6 in various ways.

DEFINITION 10.11. For brevity, let $tmf' = tmf/(B^\infty, M^\infty)$, so that $tmf' \simeq \Sigma^2 \Gamma_{(B,M)} tmf$.

PROPOSITION 10.12. *There are equivalences of (implicitly 2-completed) tmf-modules*

$$\begin{aligned}
\Sigma^{20} tmf &\simeq I_{\mathbb{Z}}(tmf/(B^\infty, M^\infty)) = I_{\mathbb{Z}}(tmf') \\
\Sigma^{22} tmf &\simeq I_{\mathbb{Z}}(\Gamma_{(B,M)} tmf) \\
\Sigma^{23} tmf &\simeq I(\Gamma_{(2,B,M)} tmf)\,.
\end{aligned}$$

PROOF. This follows directly from the $tmf$-module equivalences

$$\begin{aligned} I(tmf'/2^\infty) &\simeq I_{\mathbb{Z}}(tmf')^\wedge_2 \\ tmf' = tmf/(B^\infty, M^\infty) &\simeq \Sigma^2\Gamma_{(B,M)}tmf \\ tmf/(2^\infty, B^\infty, M^\infty) &\simeq \Sigma^3\Gamma_{(2,B,M)}tmf\,. \end{aligned}$$

□

The topological modular forms spectrum $tmf$ is defined as the connective cover of an $E_\infty$ ring spectrum $Tmf$, which can be constructed using Goerss–Hopkins–Miller obstruction theory [**62**], [**54**, Ch. 12], or Lurie's spectral orientation and deformation theories for $p$-divisible groups and formal groups [**96**, §4], [**97**] and [**98**]. In either case $Tmf$ is defined as the global sections in a sheaf of $E_\infty$ ring spectra over a compactified moduli stack $\overline{\mathcal{M}}_{ell}$ of generalized elliptic curves. This moduli stack is covered by the two open substacks of ordinary generalized elliptic curves (where $c_4$ and $B$ are invertible), and of non-generalized elliptic curves (where $\Delta$ and $M$ are invertible). It follows that there is a homotopy pullback square

$$\begin{array}{ccc} Tmf & \longrightarrow & Tmf[1/B] \\ \downarrow & & \downarrow \\ Tmf[1/M] & \longrightarrow & Tmf[1/B, 1/M]\,. \end{array}$$

Since the covering map $i\colon tmf \to Tmf$ induces equivalences after inverting $B$, $M$ or both, it also follows that we have a homotopy (co-)fiber sequence of $tmf$-modules

$$\Sigma^{-2}tmf/(B^\infty, M^\infty) \longrightarrow tmf \xrightarrow{i} Tmf \longrightarrow \Sigma^{-1}tmf/(B^\infty, M^\infty)\,,$$

which we can write in terms of local cohomology spectra as

$$\Gamma_{(B,M)}tmf \longrightarrow tmf \xrightarrow{i} Tmf \longrightarrow \Sigma\Gamma_{(B,M)}tmf\,.$$

Using the duality equivalence, we can rewrite this as

$$\Sigma^{-22}I_{\mathbb{Z}}(tmf) \xrightarrow{\partial} tmf \xrightarrow{i} Tmf \xrightarrow{j} \Sigma^{-21}I_{\mathbb{Z}}(tmf)\,. \tag{10.4}$$

Vesna Stojanoska [**161**, Thm. 13.1] showed that $Tmf$ is Anderson self-dual in the sense that there is an equivalence $\Sigma^{21}Tmf \simeq I_{\mathbb{Z}}(Tmf)$. More precisely, the cited reference shows this as an equivalence after inverting $p = 2$, while the corresponding 2-local calculations have not been fully published. In [**67**, Prop. 4.1], Greenlees and Stojanoska show how to deduce integral Anderson self-duality for $Tmf$ from Gorenstein duality for $tmf \to H\mathbb{Z}$, and later work [**43**] by Greenlees and the current two authors establishes this Gorenstein duality property, also at $p = 2$. Using our present notation, the 2-complete part of Stojanoska's theorem can be demonstrated as follows. The argument is essentially that of [**67**].

THEOREM 10.13. *There is a duality equivalence of (implicitly 2-completed) tmf-modules*

$$\Sigma^{21}Tmf \simeq I_{\mathbb{Z}}(Tmf)\,.$$

PROOF. Applying $\Sigma^{-21}I_{\mathbb{Z}}$ to (10.4) we obtain a homotopy (co-)fiber sequence of $tmf$-modules

$$\Sigma^{-22}I_{\mathbb{Z}}(tmf) \xrightarrow{\delta} tmf \longrightarrow \Sigma^{-21}I_{\mathbb{Z}}(Tmf) \longrightarrow \Sigma^{-21}I_{\mathbb{Z}}(tmf)$$

where $\delta = \Sigma^{-22} I_{\mathbb{Z}}(\partial)$. It suffices to prove that $\delta$ is homotopic to $\partial$, up to sign, since then $Tmf \simeq C\partial \simeq C\delta \simeq \Sigma^{-21} I_{\mathbb{Z}}(Tmf)$. The homotopy classes of $\partial$ and $\delta$ lie in the abelian group

$$G = [\Sigma^{-22} I_{\mathbb{Z}}(tmf), tmf]_0^{tmf} = \pi_0 F_{tmf}(\Sigma^{-22} I_{\mathbb{Z}}(tmf), tmf)$$

and the functor $\Sigma^{-22} I_{\mathbb{Z}}$ induces an involution on $G$, interchanging $\partial$ and $\delta$. For any $tmf$-module $X$ there are equivalences

$$\begin{aligned} F_{tmf}(\Sigma^{-22} I_{\mathbb{Z}}(tmf), X) &\simeq F_{tmf}(\Gamma_{(B,M)} tmf, X) \\ &\simeq F_{tmf}(\operatorname*{hocolim}_{k,\ell} \Sigma^{-8k-192\ell-2} tmf/(B^k, M^\ell), X) \\ &\simeq \operatorname*{holim}_{k,\ell} F_{tmf}(\Sigma^{-8k-192\ell-2} tmf/(B^k, M^\ell), X) \\ &\simeq \operatorname*{holim}_{k,\ell} X/(B^k, M^\ell) = X^\wedge_{(B,M)} . \end{aligned}$$

Here the final homotopy limit defines the $(B, M)$-completion of $X$. For $X = tmf$, the completion map $X \to \operatorname{holim}_{k,\ell} X/(B^k, M^\ell) = X^\wedge_{(B,M)}$ is an equivalence, since $tmf$ is bounded below and both $B$ and $M$ have positive degree. Hence $G = \pi_0 F_{tmf}(\Sigma^{-22} I_{\mathbb{Z}}(tmf), tmf) \cong \pi_0(tmf^\wedge_{(B,M)}) \cong \pi_0(tmf) \cong \mathbb{Z}$ (up to implicit 2-completion), and the only possible involutions on $G$ are given by multiplication by 1 or $-1$. This proves that $\delta \simeq \pm\partial$. □

## 10.5. Explicit formulas

We now turn the spectrum level duality equivalence from Theorem 10.6 into a series of algebraic duality statement about $\pi_*(tmf)$, or more precisely, about the subquotients of a filtration

$$0 \subset \Theta\pi_*(tmf) \subset \Gamma_B \pi_*(tmf) \subset \Gamma_2 \pi_*(tmf) \subset \pi_*(tmf) .$$

We will use the following variant of (10.3).

LEMMA 10.14. *For any $R$-module spectrum $M$ there is a natural short exact sequence of $\pi_*(R)$-modules*

$$0 \to \operatorname{Hom}(\Gamma_2 \pi_{*-1}(M), \mathbb{Q}/\mathbb{Z}) \longrightarrow \pi_{-*} I(M/2^\infty) \longrightarrow \operatorname{Hom}(\pi_*(M), \mathbb{Z}_2) \to 0 .$$

PROOF. The short exact sequence

$$0 \to \pi_*(M)/2^\infty \longrightarrow \pi_*(M/2^\infty) \longrightarrow \Gamma_2 \pi_{*-1}(M) \to 0$$

is Pontryagin dual to a short exact sequence of $\pi_*(R)$-modules

$$\begin{aligned} 0 \to \operatorname{Hom}(\Gamma_2 \pi_{*-1}(M), \mathbb{Q}/\mathbb{Z}) \\ \longrightarrow \operatorname{Hom}(\pi_*(M/2^\infty), \mathbb{Q}/\mathbb{Z}) \longrightarrow \operatorname{Hom}(\pi_*(M)/2^\infty, \mathbb{Q}/\mathbb{Z}) \to 0 . \end{aligned}$$

Here $\operatorname{Hom}(\pi_*(M/2^\infty), \mathbb{Q}/\mathbb{Z}) = \pi_{-*} I(M/2^\infty)$, by definition, and there is a natural chain of isomorphisms

$$\begin{aligned} \operatorname{Hom}(\pi_*(M)/2^\infty, \mathbb{Q}/\mathbb{Z}) &= \operatorname{Hom}(\operatorname*{colim}_n \pi_*(M)/2^n, \mathbb{Q}/\mathbb{Z}) \\ &\cong \lim_n \operatorname{Hom}(\pi_*(M)/2^n, \mathbb{Q}/\mathbb{Z}) \cong \lim_n \operatorname{Hom}(\pi_*(M), \mathbb{Z}/2^n) = \operatorname{Hom}(\pi_*(M), \mathbb{Z}_2) . \end{aligned}$$

□

Recall our notation $tmf' = tmf/(B^\infty, M^\infty)$.

THEOREM 10.15. *There are short exact sequences of $\pi_*(tmf)$-modules*

$$0 \to \pi_*(tmf)/(B^\infty, M^\infty) \longrightarrow \pi_*(tmf') \longrightarrow \Gamma_B\pi_{*-1}(tmf)/M^\infty \to 0$$

*and*

$$0 \to \mathrm{Hom}(\Gamma_2\pi_{*-1}(tmf'), \mathbb{Q}/\mathbb{Z}) \longrightarrow \pi_{-*}(\Sigma^{20}tmf) \longrightarrow \mathrm{Hom}(\pi_*(tmf'), \mathbb{Z}_2) \to 0\,.$$

PROOF. The first exact sequence follows from (10.1), since $\pi_*(tmf) \cong N_* \otimes \mathbb{Z}[M]$ implies $\Gamma_B\pi_*(tmf/M^\infty) \cong \Gamma_B N_* \otimes \mathbb{Z}[M]/M^\infty$.

The second exact sequence follows from the duality theorem, in the formulation $\Sigma^{20}tmf \simeq I(tmf'/2^\infty)$, and Lemma 10.14. □

Recall the $\mathbb{Z}[B,M]$-module identification $\pi_*(tmf) \cong N_* \otimes \mathbb{Z}[M]$. As we saw in Theorem 9.26, the $B$-torsion free image of $N_*$ in $N_*[1/B]$ is the direct sum

$$N_*/\Gamma_B N_* = \bigoplus_{k=0}^{7} ko[k]$$

of the following eight $\mathbb{Z}[B]$-modules, with $ko[k]$ concentrated in degrees $* \ge 24k$:

$$(10.5)\qquad \begin{aligned}
ko[0] &= \mathbb{Z}[B]\{1, C\} \oplus \mathbb{Z}/2[B]\{\eta, \eta^2\}\\
ko[1] &= \mathbb{Z}\{D_1\} \oplus \mathbb{Z}[B]\{B_1, C_1\} \oplus \mathbb{Z}/2[B]\{\eta_1, \eta\eta_1\}\\
ko[2] &= \mathbb{Z}\{D_2\} \oplus \mathbb{Z}[B]\{B_2, C_2\} \oplus \mathbb{Z}/2[B]\{\eta B_2, \eta_1^2\}\\
ko[3] &= \mathbb{Z}\{D_3\} \oplus \mathbb{Z}[B]\{B_3, C_3\} \oplus \mathbb{Z}/2[B]\{\eta B_3, \eta^2 B_3\}\\
ko[4] &= \mathbb{Z}\{D_4\} \oplus \mathbb{Z}[B]\{B_4, C_4\} \oplus \mathbb{Z}/2[B]\{\eta_4, \eta\eta_4\}\\
ko[5] &= \mathbb{Z}\{D_5\} \oplus \mathbb{Z}[B]\{B_5, C_5\} \oplus \mathbb{Z}/2[B]\{\eta B_5, \eta_1\eta_4\}\\
ko[6] &= \mathbb{Z}\{D_6\} \oplus \mathbb{Z}[B]\{B_6, C_6\} \oplus \mathbb{Z}/2[B]\{\eta B_6, \eta^2 B_6\}\\
ko[7] &= \mathbb{Z}\{D_7\} \oplus \mathbb{Z}[B]\{B_7, C_7\} \oplus \mathbb{Z}/2[B]\{\eta B_7, \eta^2 B_7\}\,.
\end{aligned}$$

The $\mathbb{Z}[B]$-module structures are specified by $B \cdot D_k = d_k B_k$ for each $1 \le k \le 7$, where the numbers $d_k \in \{2, 4, 8\}$ are as in Definition 9.18. In each case, $ko[k][1/B] \cong \pi_*(KO)$. It follows that

$$N_*/B^\infty = \bigoplus_{k=0}^{7} ko[k]/B^\infty$$

is the direct sum of the following eight $\mathbb{Z}[B]$-modules, with $ko[k]/B^\infty$ concentrated in degrees $* \le 4 + 24k$:

$$\begin{aligned}
ko[0]/B^\infty &= \mathbb{Z}[B^{-1}]\{1/B, C/B\} \oplus \mathbb{Z}/2[B^{-1}]\{\eta/B, \eta^2/B\}\\
ko[1]/B^\infty &= \mathbb{Z}[B^{-1}]\{B_1/B, C_1/B\}/(8B_1/B) \oplus \mathbb{Z}/2[B^{-1}]\{\eta_1/B, \eta\eta_1/B\}\\
ko[2]/B^\infty &= \mathbb{Z}[B^{-1}]\{B_2/B, C_2/B\}/(4B_2/B) \oplus \mathbb{Z}/2[B^{-1}]\{\eta B_2/B, \eta_1^2/B\}\\
ko[3]/B^\infty &= \mathbb{Z}[B^{-1}]\{B_3/B, C_3/B\}/(8B_3/B) \oplus \mathbb{Z}/2[B^{-1}]\{\eta B_3/B, \eta^2 B_3/B\}\\
ko[4]/B^\infty &= \mathbb{Z}[B^{-1}]\{B_4/B, C_4/B\}/(2B_4/B) \oplus \mathbb{Z}/2[B^{-1}]\{\eta_4/B, \eta\eta_4/B\}\\
ko[5]/B^\infty &= \mathbb{Z}[B^{-1}]\{B_5/B, C_5/B\}/(8B_5/B) \oplus \mathbb{Z}/2[B^{-1}]\{\eta B_5/B, \eta_1\eta_4/B\}\\
ko[6]/B^\infty &= \mathbb{Z}[B^{-1}]\{B_6/B, C_6/B\}/(4B_6/B) \oplus \mathbb{Z}/2[B^{-1}]\{\eta B_6/B, \eta^2 B_6/B\}\\
ko[7]/B^\infty &= \mathbb{Z}[B^{-1}]\{B_7/B, C_7/B\}/(8B_7/B) \oplus \mathbb{Z}/2[B^{-1}]\{\eta B_7/B, \eta^2 B_7/B\}\,.
\end{aligned}$$

The following lemma specifies a $\mathbb{Z}[B]$-module extension $N'_*$, uniquely up to isomorphism. It will play an important role in the following calculations. The notation $N'_*$ is chosen to parallel that of *tmf*$'$.

LEMMA 10.16. *The restriction of the $\pi_*(tmf)$-module extension*

$$0 \to \pi_*(tmf)/B^\infty \longrightarrow \pi_*(tmf/B^\infty) \longrightarrow \Gamma_B\pi_{*-1}(tmf) \to 0$$

*to a $\mathbb{Z}[B,M]$-module extension is induced up from a unique $\mathbb{Z}[B]$-module extension*

$$0 \to N_*/B^\infty \longrightarrow N'_* \longrightarrow \Gamma_B N_{*-1} \to 0\,.$$

*Hence $\pi_*(tmf/B^\infty) = N'_* \otimes \mathbb{Z}[M]$ and $\pi_*(tmf') = N'_* \otimes \mathbb{Z}[M]/M^\infty$ as $\mathbb{Z}[B,M]$-modules.*

PROOF. We claim that the induction homomorphism

$$\begin{aligned}\mathrm{Ext}^1_{\mathbb{Z}[B]}(\Gamma_B N_{*-1}, N_*/B^\infty) \longrightarrow \mathrm{Ext}^1_{\mathbb{Z}[B,M]}(\Gamma_B N_{*-1} \otimes \mathbb{Z}[M], N_*/B^\infty \otimes \mathbb{Z}[M])\\ \cong \mathrm{Ext}^1_{\mathbb{Z}[B]}(\Gamma_B N_{*-1}, N_*/B^\infty \otimes \mathbb{Z}[M])\end{aligned}$$

is bijective. This follows from the observation that

$$\mathrm{Ext}^s_{\mathbb{Z}[B]}(\Gamma_B N_{*-1}, N_*/B^\infty \otimes (\mathbb{Z}[M]/\mathbb{Z})) = 0$$

for $s \in \{0,1\}$, since $\Gamma_B N_{*-1}$ is concentrated in degrees $* \le 165$, and $N_*/B^\infty \otimes (\mathbb{Z}[M]/\mathbb{Z})$ agrees with $N_*[1/B] \otimes (\mathbb{Z}[M]/\mathbb{Z})$ in degrees $* < 192$. In more detail, the groups

$$\mathrm{Ext}^s_{\mathbb{Z}[B]}(\Gamma_B N_{*-1}, N_*[1/B] \otimes (\mathbb{Z}[M]/\mathbb{Z}))$$

vanish because $B$ acts nilpotently on $\Gamma_B N_{*-1}$ and invertibly on $N_*[1/B]$. The groups

$$\mathrm{Ext}^{s+1}_{\mathbb{Z}[B]}(\Gamma_B N_{*-1}, N_*/\Gamma_B N_* \otimes (\mathbb{Z}[M]/\mathbb{Z}))$$

vanish because $\Gamma_B N_{*-1}$ admits a projective $\mathbb{Z}[B]$-module resolution with generators in degrees $* \le 173$, and $N_*/\Gamma_B N_* \otimes (\mathbb{Z}[M]/\mathbb{Z})$ is concentrated in degrees $* \ge 192$. □

LEMMA 10.17. *$\pi_*(tmf')$ is bounded above and of finite type.*

PROOF. It is clear from the formulas for the $ko[k]/B^\infty$ that $N'_*$ is of finite type and bounded above, hence so is its tensor product with $\mathbb{Z}[M]/M^\infty$. □

We can now define the Pontryagin self-dual part of $\Gamma_B\pi_*(tmf) \subset \pi_*(tmf)$.

DEFINITION 10.18. Let the $\pi_*(tmf)$-module $\Theta\pi_{*-1}(tmf)$ be the image of the composite homomorphism

$$\Gamma_2\pi_*(tmf/B^\infty) \longrightarrow \pi_*(tmf/B^\infty) \longrightarrow \Gamma_B\pi_{*-1}(tmf)$$

and let the $\mathbb{Z}[B]$-module $\Theta N_{*-1}$ be the image of the composite homomorphism

$$\Gamma_2 N'_* \longrightarrow N'_* \longrightarrow \Gamma_B N_{*-1}\,.$$

There is an isomorphism

$$\Theta\pi_*(tmf) \cong \Theta N_* \otimes \mathbb{Z}[M]$$

of $\mathbb{Z}[B,M]$-modules. When we later use the notation $\Theta\pi_{-*}(\Sigma^{20}tmf)$, we mean the same as $\Theta\pi_{-*-20}(tmf)$.

LEMMA 10.19. *There is a filtration of $\pi_*(tmf)$-modules (= ideals)*

$$0 \subset \Theta\pi_*(tmf) \subset \Gamma_B\pi_*(tmf) \subset \Gamma_2\pi_*(tmf) \subset \pi_*(tmf)\,.$$

*When restricted to a filtration of $\mathbb{Z}[B,M]$-modules, it is induced up from the filtration*

$$0 \subset \Theta N_* \subset \Gamma_B N_* \subset \Gamma_2 N_* \subset N_*$$

*of $\mathbb{Z}[B]$-modules. Here*

$$\frac{\Gamma_2 N_*}{\Gamma_B N_*} = \bigoplus_{k=0}^{7} \Gamma_2 ko[k]$$

*and*

$$\frac{N_*}{\Gamma_2 N_*} = \bigoplus_{k=0}^{7} \frac{ko[k]}{\Gamma_2 ko[k]}\,.$$

PROOF. This is clear with the definitions above. Precise formulas for the $\Gamma_2 ko[k]$ and $ko[k]/\Gamma_2 ko[k]$ can be read off from the formulas (10.5) for $ko[k]$. □

LEMMA 10.20. *There is a filtration of $\pi_*(tmf)$-modules*

$$0 \subset (\Gamma_2\pi_*(tmf))/B^\infty \subset \Gamma_2(\pi_*(tmf)/B^\infty) \subset \Gamma_2\pi_*(tmf/B^\infty) \subset \pi_*(tmf/B^\infty)\,.$$

*When viewed as a filtration of $\mathbb{Z}[B,M]$-modules, it is induced up from the filtration*

$$0 \subset (\Gamma_2 N_*)/B^\infty \subset \Gamma_2(N_*/B^\infty) \subset \Gamma_2 N'_* \subset N'_*$$

*of $\mathbb{Z}[B]$-modules. Here*

$$(\Gamma_2 N_*)/B^\infty \cong \bigoplus_{k=0}^{7} (\Gamma_2 ko[k])/B^\infty\,,$$

$$\frac{\Gamma_2(N_*/B^\infty)}{(\Gamma_2 N_*)/B^\infty} \cong \bigoplus_{k=1}^{7} \langle B_k/B\rangle$$

*with $\langle B_k/B\rangle$ cyclic of order $d_k$,*

$$\frac{\Gamma_2 N'_*}{\Gamma_2(N_*/B^\infty)} \cong \Theta N_{*-1}$$

*and there is a short exact sequence*

$$0 \to \bigoplus_{k=0}^{7} \frac{ko[k]/B^\infty}{\Gamma_2(ko[k]/B^\infty)} \longrightarrow \frac{N'_*}{\Gamma_2 N'_*} \longrightarrow \frac{\Gamma_B N_{*-1}}{\Theta N_{*-1}} \to 0$$

*of $\mathbb{Z}[B]$-modules.*

PROOF. Formulas for

$$(\Gamma_2 N_*)/B^\infty \cong (\Gamma_2 N_*/\Gamma_B N_*)/B^\infty = \bigoplus_{k=0}^{7} (\Gamma_2 ko[k])/B^\infty$$

and

$$\Gamma_2(N_*/B^\infty) = \bigoplus_{k=0}^{7} \Gamma_2(ko[k]/B^\infty)$$

can be read off from the formulas for $ko[k]$ and $ko[k]/B^\infty$, respectively. This gives the stated seven-term sum for $\Gamma_2(N_*/B^\infty)$ modulo $(\Gamma_2 N_*)/B^\infty$.

By the definition of $\Theta N_{*-1}$, we have a $3 \times 3$ diagram of short exact sequences

$$\begin{array}{ccccccccc}
 & & 0 & & 0 & & 0 & & \\
 & & \downarrow & & \downarrow & & \downarrow & & \\
0 & \longrightarrow & \Gamma_2(N_*/B^\infty) & \longrightarrow & \Gamma_2 N'_* & \longrightarrow & \Theta N_{*-1} & \longrightarrow & 0 \\
 & & \downarrow & & \downarrow & & \downarrow & & \\
0 & \longrightarrow & N_*/B^\infty & \longrightarrow & N'_* & \longrightarrow & \Gamma_B N_{*-1} & \longrightarrow & 0 \\
 & & \downarrow & & \downarrow & & \downarrow & & \\
0 & \longrightarrow & \dfrac{N_*/B^\infty}{\Gamma_2(N_*/B^\infty)} & \longrightarrow & \dfrac{N'_*}{\Gamma_2 N'_*} & \longrightarrow & \dfrac{\Gamma_B N_{*-1}}{\Theta N_{*-1}} & \longrightarrow & 0 \\
 & & \downarrow & & \downarrow & & \downarrow & & \\
 & & 0 & & 0 & & 0 & &
\end{array}$$

and the remaining claims follow by inspection. □

COROLLARY 10.21. *The $\pi_*(tmf)$-module filtration*

$$0 \subset (\Gamma_2\pi_*(tmf))/(B^\infty, M^\infty) \subset \Gamma_2(\pi_*(tmf)/(B^\infty, M^\infty)) \subset \Gamma_2\pi_*(tmf') \subset \pi_*(tmf')$$

*is isomorphic to*

$$0 \subset (\Gamma_2 N_*)/B^\infty \otimes \mathbb{Z}[M]/M^\infty \subset \Gamma_2(N_*/B^\infty) \otimes \mathbb{Z}[M]/M^\infty \subset \Gamma_2 N'_* \otimes \mathbb{Z}[M]/M^\infty \subset N'_* \otimes \mathbb{Z}[M]/M^\infty$$

*when viewed as a filtration of $\mathbb{Z}[B, M]$-modules.* □

THEOREM 10.22. *The duality isomorphism*

$$a_* \colon \pi_{-*}(\Sigma^{20} tmf) \cong \mathrm{Hom}(\pi_*(tmf'/2^\infty), \mathbb{Q}/\mathbb{Z})$$

*of Theorem* 10.6 *specializes to isomorphisms of $\pi_*(tmf)$-modules*

$$\Gamma_2 a_* \colon \Gamma_2\pi_{-*}(\Sigma^{20} tmf) \cong \mathrm{Hom}(\Gamma_2\pi_{*-1}(tmf'), \mathbb{Q}/\mathbb{Z})$$

*and*

$$\frac{\pi_{-*}(\Sigma^{20} tmf)}{\Gamma_2\pi_{-*}(\Sigma^{20} tmf)} \cong \mathrm{Hom}(\pi_*(tmf'), \mathbb{Z}_2)\,.$$

*Hence there are isomorphisms*

$$\Gamma_2 N_{171-*} \cong \mathrm{Hom}(\Gamma_2 N'_*, \mathbb{Q}/\mathbb{Z})$$
$$\frac{N_{172-*}}{\Gamma_2 N_{172-*}} \cong \mathrm{Hom}(N'_*, \mathbb{Z}_2)$$

*and a short exact sequence*

$$0 \to \frac{N_{172-*}}{\Gamma_2 N_{172-*}} \longrightarrow \mathrm{Hom}(N_*/B^\infty, \mathbb{Z}_2) \longrightarrow \mathrm{Hom}\Big(\frac{\Gamma_B N_{*-1}}{\Theta N_{*-1}}, \mathbb{Q}/\mathbb{Z}\Big) \to 0\,,$$

*all in the category of $\mathbb{Z}[B]$-modules.*

PROOF. By Lemma 10.17, $\Gamma_2\pi_{*-1}(tmf')$ is finite in each degree, so the second short exact sequence in Theorem 10.15 specializes to a $\pi_*(tmf)$-isomorphism $\Gamma_2 a_*$ between $\mathrm{Hom}(\Gamma_2\pi_{*-1}(tmf'), \mathbb{Q}/\mathbb{Z})$ and the 2-power torsion in $\pi_{-*}(\Sigma^{20} tmf)$, as well as a $\pi_*(tmf)$-isomorphism between the 2-torsion free quotient of $\pi_{-*}(\Sigma^{20} tmf)$ and $\mathrm{Hom}(\pi_*(tmf'), \mathbb{Z}_2)$.

The first $\pi_*(tmf)$-isomorphism restricts to a $\mathbb{Z}[B, M]$-module isomorphism

$$\Gamma_2 N_{-*-20} \otimes \mathbb{Z}[M] \cong \mathrm{Hom}(\Gamma_2 N'_{*-1} \otimes \mathbb{Z}[M]/M^\infty, \mathbb{Q}/\mathbb{Z})$$

which is induced up from a $\mathbb{Z}[B]$-module isomorphism

$$\Gamma_2 N_{-*-20} \cong \mathrm{Hom}(\Gamma_2 N'_{*-1} \otimes \mathbb{Z}\{1/M\}, \mathbb{Q}/\mathbb{Z}) .$$

Here $\Gamma_2 N'_{*-1} \otimes \mathbb{Z}\{1/M\}$ in total degree $* - 1$ is isomorphic to $\Gamma_2 N'_{*+191}$, via multiplication by $M$, so the first asserted isomorphism follows after reindexing.

The second $\pi_*(tmf)$-isomorphism restricts to a $\mathbb{Z}[B, M]$-module isomorphism

$$\frac{N_{-*-20}}{\Gamma_2 N_{-*-20}} \otimes \mathbb{Z}[M] \cong \mathrm{Hom}(N'_* \otimes \mathbb{Z}[M]/M^\infty, \mathbb{Z}_2)$$

which is induced up from a $\mathbb{Z}[B]$-isomorphism

$$\frac{N_{-*-20}}{\Gamma_2 N_{-*-20}} \cong \mathrm{Hom}(N'_* \otimes \mathbb{Z}\{1/M\}, \mathbb{Z}_2) .$$

Here $N'_* \otimes \mathbb{Z}\{1/M\}$ in total degree $*$ is isomorphic to $N'_{*+192}$, so the second asserted isomorphism also follows after reindexing.

The short exact sequence

$$0 \to \frac{N_*/B^\infty}{\Gamma_2(N_*/B^\infty)} \longrightarrow \frac{N'_*}{\Gamma_2 N'_*} \longrightarrow \frac{\Gamma_B N_{*-1}}{\Theta N_{*-1}} \to 0$$

of Lemma 10.20, combined with the facts that $N'_*/\Gamma_2 N'_*$ is free in each degree and $\Gamma_B N_{*-1}/\Theta N_{*-1}$ is finite in each degree, leads to a short exact sequence

$$0 \to \mathrm{Hom}\Big(\frac{N'_*}{\Gamma_2 N'_*}, \mathbb{Z}_2\Big) \longrightarrow \mathrm{Hom}\Big(\frac{N_*/B^\infty}{\Gamma_2(N_*/B^\infty)}, \mathbb{Z}_2\Big) \longrightarrow \mathrm{Hom}\Big(\frac{\Gamma_B N_{*-1}}{\Theta N_{*-1}}, \mathbb{Q}/\mathbb{Z}\Big) \to 0 .$$

Substituting

$$\mathrm{Hom}\Big(\frac{N'_*}{\Gamma_2 N'_*}, \mathbb{Z}_2\Big) = \mathrm{Hom}(N'_*, \mathbb{Z}_2)$$

and

$$\mathrm{Hom}\Big(\frac{N_*/B^\infty}{\Gamma_2(N_*/B^\infty)}, \mathbb{Z}_2\Big) = \mathrm{Hom}(N_*/B^\infty, \mathbb{Z}_2)$$

yields the required short exact sequence. □

DEFINITION 10.23. For $0 \le k \le 6$ let $\langle \nu_k \rangle \subset \Gamma_B N_* \subset \Gamma_B \pi_*(tmf)$ denote the finite abelian group generated by the class $\nu_k$ in degree $3 + 24k$, subject to the interpretations $\nu_0 = \nu$ and $\nu_3 = \eta_1^3$. Note that $\langle \nu_k \rangle$ is cyclic of order $d_{7-k} \in \{2, 4, 8\}$.

PROPOSITION 10.24. *The $\pi_*(tmf)$-submodule $\Theta\pi_*(tmf)$ of $\Gamma_B \pi_*(tmf)$ consists precisely of the classes in degrees $* \not\equiv 3 \mod 24$. Likewise, the $\mathbb{Z}[B]$-submodule $\Theta N_*$ of $\Gamma_B N_*$ consists precisely of the classes in degrees $* \not\equiv 3 \mod 24$. Hence*

$$\frac{\Gamma_B \pi_*(tmf)}{\Theta\pi_*(tmf)} \cong \bigoplus_{k=0}^{6} \langle \nu_k \rangle \otimes \mathbb{Z}[M]$$

*as $\mathbb{Z}[B, M]$-modules and*

$$\frac{\Gamma_B N_*}{\Theta N_*} \cong \bigoplus_{k=0}^{6} \langle \nu_k \rangle$$

*as $\mathbb{Z}[B]$-modules (with trivial $B$-action).*

PROOF. For the moment we only prove that $\Theta\pi_*(tmf) \cong \Theta N_* \otimes \mathbb{Z}[M]$ is trivial in degrees $* \equiv 3 \mod 24$. By inspection of $E_\infty(tmf)$ (or $\pi_*(tmf)$), it is clear that $\pi_*(tmf) = 0$ for $* \equiv -1 \mod 24$. Hence $\Gamma_2 N_* = N_* = 0$ for $* \equiv -1 \mod 24$. By Theorem 10.22 it follows that $\Gamma_2 N'_* = 0$ for $* \equiv 4 \mod 24$. Thus the image $\Theta N_{*-1}$ of this group in $\Gamma_B N_{*-1}$ is also trivial, for each $* - 1 \equiv 3 \mod 24$.

The proof that all classes in degrees $* \not\equiv 3 \mod 24$ lie in $\Theta\pi_*(tmf)$ will be completed by a counting argument, in the course of the proof of Theorem 10.25. □

THEOREM 10.25. *The 2-power torsion isomorphism*

$$\Gamma_2 a_* \colon \Gamma_2\pi_{-*}(\Sigma^{20}tmf) \cong \operatorname{Hom}(\Gamma_2\pi_{*-1}(tmf'), \mathbb{Q}/\mathbb{Z})$$

*of Theorem* 10.22 *specializes to isomorphisms*

$$\begin{aligned} \Theta a_* \colon \Theta\pi_{-*}(\Sigma^{20}tmf) &\cong \operatorname{Hom}(\Theta\pi_{*-2}(tmf)/M^\infty, \mathbb{Q}/\mathbb{Z}) \\ \frac{\Gamma_B\pi_{-*}(\Sigma^{20}tmf)}{\Theta\pi_{-*}(\Sigma^{20}tmf)} &\cong \operatorname{Hom}\Big(\frac{\Gamma_2(\pi_{*-1}(tmf)/(B^\infty, M^\infty))}{(\Gamma_2\pi_{*-1}(tmf))/(B^\infty, M^\infty)}, \mathbb{Q}/\mathbb{Z}\Big) \\ \frac{\Gamma_2\pi_{-*}(\Sigma^{20}tmf)}{\Gamma_B\pi_{-*}(\Sigma^{20}tmf)} &\cong \operatorname{Hom}((\Gamma_2\pi_{*-1}(tmf))/(B^\infty, M^\infty), \mathbb{Q}/\mathbb{Z}) \end{aligned}$$

*of $\pi_*(tmf)$-modules. Hence there are isomorphisms*

$$\begin{aligned} \Theta N_{170-*} &\cong \operatorname{Hom}(\Theta N_*, \mathbb{Q}/\mathbb{Z}) \\ \frac{\Gamma_B N_{171-*}}{\Theta N_{171-*}} &\cong \operatorname{Hom}\Big(\frac{\Gamma_2(N_*/B^\infty)}{(\Gamma_2 N_*)/B^\infty}, \mathbb{Q}/\mathbb{Z}\Big) \\ \frac{\Gamma_2 N_{171-*}}{\Gamma_B N_{171-*}} &\cong \operatorname{Hom}((\Gamma_2 N_*)/B^\infty, \mathbb{Q}/\mathbb{Z}) \end{aligned}$$

*of $\mathbb{Z}[B]$-modules.*

PROOF. The 2-power torsion isomorphism specializes to an isomorphism

$$\Gamma_B a_* \colon \Gamma_B\pi_{-*}(\Sigma^{20}tmf) \cong \Gamma_B \operatorname{Hom}(\Gamma_2\pi_{*-1}(tmf'), \mathbb{Q}/\mathbb{Z})$$

between the $B$-power torsion submodules, and an isomorphism

$$\frac{\Gamma_2\pi_{-*}(\Sigma^{20}tmf)}{\Gamma_B\pi_{-*}(\Sigma^{20}tmf)} \cong \frac{\operatorname{Hom}(\Gamma_2\pi_{*-1}(tmf'), \mathbb{Q}/\mathbb{Z})}{\Gamma_B \operatorname{Hom}(\Gamma_2\pi_{*-1}(tmf'), \mathbb{Q}/\mathbb{Z})}$$

between the $B$-torsion free quotients. We now make the right hand sides more explicit.

The Pontryagin dual of the 2-power torsion part of the filtration in Corollary 10.21 is a sequence of surjective $\pi_*(tmf)$-module homomorphisms

$$\begin{aligned} &\operatorname{Hom}(\Gamma_2\pi_*(tmf'), \mathbb{Q}/\mathbb{Z}) \\ &\qquad \longrightarrow \operatorname{Hom}(\Gamma_2(\pi_*(tmf)/(B^\infty, M^\infty)), \mathbb{Q}/\mathbb{Z}) \\ &\qquad\qquad \longrightarrow \operatorname{Hom}((\Gamma_2\pi_*(tmf))/(B^\infty, M^\infty), \mathbb{Q}/\mathbb{Z}) \to 0\,. \end{aligned}$$

As a sequence of $\mathbb{Z}[B, M]$-modules, it is isomorphic to

$$\begin{aligned} &\operatorname{Hom}(\Gamma_2 N'_* \otimes \mathbb{Z}[M]/M^\infty, \mathbb{Q}/\mathbb{Z}) \\ &\qquad \longrightarrow \operatorname{Hom}(\Gamma_2(N_*/B^\infty) \otimes \mathbb{Z}[M]/M^\infty, \mathbb{Q}/\mathbb{Z}) \\ &\qquad\qquad \longrightarrow \operatorname{Hom}((\Gamma_2 N_*)/B^\infty \otimes \mathbb{Z}[M]/M^\infty, \mathbb{Q}/\mathbb{Z}) \to 0\,. \end{aligned}$$

In view of Lemma 10.20 and Corollary 10.21 the kernel of the first surjection is

$$\begin{aligned} K^1_* &= \mathrm{Hom}\Big(\frac{\Gamma_2\pi_*(tmf')}{\Gamma_2(\pi_*(tmf)/(B^\infty, M^\infty))}, \mathbb{Q}/\mathbb{Z}\Big) \\ &\cong \mathrm{Hom}(\Theta\pi_{*-1}(tmf)/M^\infty, \mathbb{Q}/\mathbb{Z}) \\ &\cong \mathrm{Hom}(\Theta N_{*-1} \otimes \mathbb{Z}[M]/M^\infty, \mathbb{Q}/\mathbb{Z}) \,, \end{aligned}$$

the kernel of the second surjection is

$$\begin{aligned} K^2_* &= \mathrm{Hom}\Big(\frac{\Gamma_2(\pi_*(tmf)/(B^\infty, M^\infty))}{(\Gamma_2\pi_*(tmf))/(B^\infty, M^\infty)}, \mathbb{Q}/\mathbb{Z}\Big) \\ &\cong \mathrm{Hom}(\bigoplus_{k=1}^{7} \langle B_k/B\rangle \otimes \mathbb{Z}[M]/M^\infty, \mathbb{Q}/\mathbb{Z}) \,, \end{aligned}$$

and both of these are $B$-power torsion. The kernel $K_*$ of the composite surjection thus sits in a short exact sequence

$$0 \to K^1_* \longrightarrow K_* \longrightarrow K^2_* \to 0$$

and is $B$-power torsion. On the other hand,

$$\begin{aligned} &\mathrm{Hom}((\Gamma_2 N_*)/B^\infty \otimes \mathbb{Z}[M]/M^\infty, \mathbb{Q}/\mathbb{Z}) \\ &\qquad\qquad \cong \mathrm{Hom}(\bigoplus_{k=0}^{7} (\Gamma_2 ko[k])/B^\infty \otimes \mathbb{Z}[M]/M^\infty, \mathbb{Q}/\mathbb{Z}) \end{aligned}$$

is $B$-torsion free. Hence $\mathrm{Hom}((\Gamma_2\pi_*(tmf))/(B^\infty, M^\infty), \mathbb{Q}/\mathbb{Z})$ is the $B$-torsion free quotient of $\mathrm{Hom}(\Gamma_2\pi_*(tmf'), \mathbb{Q}/\mathbb{Z})$. The isomorphism

$$\frac{\Gamma_2\pi_{-*}(\Sigma^{20}tmf)}{\Gamma_B\pi_{-*}(\Sigma^{20}tmf)} \cong \mathrm{Hom}((\Gamma_2\pi_{*-1}(tmf))/(B^\infty, M^\infty), \mathbb{Q}/\mathbb{Z})$$

is thus the specialization of the 2-power torsion isomorphism to the $B$-torsion free quotients.

The specialization of $\Gamma_2 a_*$ to the $B$-power torsion submodules takes the form

$$\Gamma_B a_* \colon \Gamma_B\pi_{-*}(\Sigma^{20}tmf) \cong K_{*-1} \,,$$

so that there is a short exact sequence

$$0 \to K^1_{*-1} \longrightarrow \Gamma_B\pi_{-*}(\Sigma^{20}tmf) \longrightarrow K^2_{*-1} \to 0$$

of $\pi_*(tmf)$-modules. Viewed as $\mathbb{Z}[B, M]$-modules, it can be rewritten as

$$\begin{aligned} 0 \to \mathrm{Hom}(\Theta N_{*-2} \otimes \mathbb{Z}[M]/M^\infty, \mathbb{Q}/\mathbb{Z}) &\longrightarrow \Gamma_B N_{-*-20} \otimes \mathbb{Z}[M] \\ &\longrightarrow \mathrm{Hom}(\bigoplus_{k=1}^{7} \langle B_k/B\rangle_{*-1} \otimes \mathbb{Z}[M]/M^\infty, \mathbb{Q}/\mathbb{Z}) \to 0 \,, \end{aligned}$$

hence is induced up, after regrading by $192 - 2 = 190$, from a short exact sequence

$$0 \to \mathrm{Hom}(\Theta N_*, \mathbb{Q}/\mathbb{Z}) \longrightarrow \Gamma_B N_{170-*} \longrightarrow \mathrm{Hom}(\bigoplus_{k=1}^{7} \langle B_k/B\rangle_{*+1}, \mathbb{Q}/\mathbb{Z}) \to 0$$

of $\mathbb{Z}[B]$-modules.

We now complete the proof of Proposition 10.24. The total order of the graded finite abelian group $\Theta N_*$ is equal to the total order of $\mathrm{Hom}(\Theta N_*, \mathbb{Q}/\mathbb{Z})$, and the total order of $\Gamma_B N_*$ is equal to the total order of $\Gamma_B N_{170-*}$. Hence the total order of $\Gamma_B N_*/\Theta N_*$ is equal to the total order of $\mathrm{Hom}(\bigoplus_{k=1}^{7}\langle B_k/B\rangle_{*+1}, \mathbb{Q}/\mathbb{Z})$, which is $8\cdot 4\cdot 8\cdot 2\cdot 8\cdot 4\cdot 8 = 2^{17}$. Since this is equal to the total order of $\bigoplus_{k=0}^{6}\langle \nu_k\rangle$, it follows that $\Theta N_*$ cannot be strictly smaller than the kernel of the surjection $\Gamma_B N_* \to \bigoplus_{k=0}^{6}\langle\nu_k\rangle$. Hence $\Theta N_*$ consists of all the classes in $\Gamma_B N_*$ in degrees $* \not\equiv 3 \mod 24$. Inducing up along $\mathbb{Z}[B] \subset \mathbb{Z}[B, M]$ it follows that $\Theta\pi_*(tmf)$ consists of all the classes in $\Gamma_B\pi_*(tmf)$ in degrees $* \not\equiv 3 \mod 24$. This concludes the delayed part of the proof of Proposition 10.24.

The two remaining $\pi_*(tmf)$-module isomorphisms are obtained by specializing the $B$-power torsion isomorphism $\Gamma_B a_*$ to degrees $-*-20 \not\equiv 3 \mod 24$ and degrees $-*-20 \equiv 3 \mod 24$, respectively. Since $K^1_{*-1}$ is concentrated in the degrees with $* \not\equiv 1 \mod 24$, and $K^2_{*-1}$ is concentrated in the degrees with $* \equiv 1 \mod 24$, it follows that

$$\Theta\pi_{-*}(\Sigma^{20}tmf) \cong K^1_{*-1}$$

and

$$\frac{\Gamma_B\pi_{-*}(\Sigma^{20}tmf)}{\Theta\pi_{-*}(\Sigma^{20}tmf)} \cong K^2_{*-1}\,.$$

When combined with the previous expressions for $K^1_*$ and $K^2_*$, this completes the proof of the three $\pi_*(tmf)$-module isomorphisms. The three $\mathbb{Z}[B]$-module isomorphisms follow easily from this. □

To emphasize how the previous results exhibit the spectrum level duality in algebraic terms, we formulate the following summary of the discussion of this section.

THEOREM 10.26. *(1) The graded ring $\pi_*(tmf)$ of topological modular forms is filtered by a sequence of ideals*

$$0 \subset \Theta\pi_*(tmf) \subset \Gamma_B\pi_*(tmf) \subset \Gamma_2\pi_*(tmf) \subset \pi_*(tmf)\,,$$

*where $\Theta\pi_*(tmf)$ is the image of the composite homomorphism*

$$\Gamma_2\pi_{*+1}(tmf/B^\infty) \longrightarrow \pi_{*+1}(tmf/B^\infty) \longrightarrow \Gamma_B\pi_*(tmf)\,.$$

*It consists precisely of the $B$-power torsion in $\pi_*(tmf)$ in degrees $* \not\equiv 3 \mod 24$.*

*(2) As a sequence of $\mathbb{Z}[B, M]$-modules, the filtration is induced up from the sequence of $\mathbb{Z}[B]$-modules*

$$0 \subset \Theta N_* \subset \Gamma_B N_* \subset \Gamma_2 N_* \subset N_*$$

*where $N_* \subset \pi_*(tmf)$ is the $\mathbb{Z}[B]$-submodule generated by the classes in degrees $0 \le * < 192$. The $B$-power torsion in $N_*$ is concentrated in degrees $3 \le * \le 164$, and is finite in each degree. The submodule $\Theta N_*$ is the part of $\Gamma_B N_*$ in degrees $* \not\equiv 3 \mod 24$, and is concentrated in degrees $6 \le * \le 164$.*

*(3) The duality equivalence $a\colon \Sigma^{20}tmf \simeq I(tmf/(2^\infty, B^\infty, M^\infty))$ specializes to a Pontryagin self-duality*

$$\Theta a_*\colon \Theta N_{170-*} \cong \mathrm{Hom}(\Theta N_*, \mathbb{Q}/\mathbb{Z})\,,$$

*illustrated in Figures 10.1 and 10.2.*

*(4) The remaining $B$-power torsion*

$$\frac{\Gamma_B N_*}{\Theta N_*} \cong \bigoplus_{k=0}^{6}\langle\nu_k\rangle$$

*is a direct sum of cyclic groups, with $\nu_k$ in degree $3+24k$ of order $d_{7-k} \in \{2,4,8\}$. The duality isomorphism specializes to an isomorphism*

$$\frac{\Gamma_B N_{171-*}}{\Theta N_{171-*}} \cong \operatorname{Hom}\Big(\frac{\Gamma_2(N_*/B^\infty)}{(\Gamma_2 N_*)/B^\infty}, \mathbb{Q}/\mathbb{Z}\Big)$$

*which is the direct sum of isomorphisms*

$$\Sigma^{-171}\langle \nu_{7-k}\rangle \cong \operatorname{Hom}(\langle B_k/B\rangle, \mathbb{Q}/\mathbb{Z})$$

*for $1 \le k \le 7$.*

*(5) The $B$-periodic 2-torsion is*

$$\frac{\Gamma_2 N_*}{\Gamma_B N_*} \cong \bigoplus_{k=0}^{7} \Gamma_2 ko[k] = \mathbb{Z}/2[B]\{\eta, \eta^2, \eta_1, \eta\eta_1, \eta B_2, \eta_1^2, \eta B_3, \eta^2 B_3, \\ \eta_4, \eta\eta_4, \eta B_5, \eta_1\eta_4, \eta B_6, \eta^2 B_6, \eta B_7, \eta^2 B_7\}.$$

*The duality equivalence specializes to an isomorphism*

$$\frac{\Gamma_2 N_{171-*}}{\Gamma_B N_{171-*}} \cong \operatorname{Hom}\Big(\Big(\frac{\Gamma_2 N_*}{\Gamma_B N_*}\Big)/B^\infty, \mathbb{Q}/\mathbb{Z}\Big),$$

*which is a direct sum of isomorphisms*

$$\Sigma^{-171}\Gamma_2 ko[7-k] \cong \operatorname{Hom}((\Gamma_2 ko[k])/B^\infty, \mathbb{Q}/\mathbb{Z})$$

*for $0 \le k \le 7$. Alternatively, writing $\Gamma_2 N_*/\Gamma_B N_* = \mathbb{Z}[B] \otimes H_*$ with $H_* = \mathbb{Z}/2\{\eta, \eta^2, \eta_1, \dots, \eta^2 B_6, \eta B_7, \eta^2 B_7\}$, the duality equivalence specializes to a Pontryagin self-duality*

$$H_{179-*} \cong \operatorname{Hom}(H_*, \mathbb{Q}/\mathbb{Z}).$$

*(6) The 2-torsion free quotient is*

$$\frac{N_*}{\Gamma_2 N_*} \cong \bigoplus_{k=0}^{7} \frac{ko[k]}{\Gamma_2 ko[k]} \cong \mathbb{Z}[B]\{1, C\} \oplus \bigoplus_{k=1}^{7} \Big(\mathbb{Z}\{D_k\} \oplus \mathbb{Z}[B]\{B_k, C_k\}\Big)$$

*where $B \cdot D_k = d_k B_k$ for each $1 \le k \le 7$. The duality equivalence induces a short exact sequence*

$$0 \to \frac{N_{172-*}}{\Gamma_2 N_{172-*}} \longrightarrow \operatorname{Hom}\Big(\Big(\frac{N_*}{\Gamma_2 N_*}\Big)/B^\infty, \mathbb{Z}_2\Big) \longrightarrow \operatorname{Hom}\Big(\frac{\Gamma_B N_{*-1}}{\Theta N_{*-1}}, \mathbb{Q}/\mathbb{Z}\Big) \to 0$$

*relating $N_*/\Gamma_2 N_*$ to its $\mathbb{Z}_2$-linear dual, with the Pontryagin dual of the remaining $B$-power torsion from (4) entering as a correction term. It is the direct sum of an isomorphism*

$$\Sigma^{-172}\frac{ko[0]}{\Gamma_2 ko[0]} \cong \operatorname{Hom}\Big(\Big(\frac{ko[7]}{\Gamma_2 ko[7]}\Big)/B^\infty, \mathbb{Z}_2\Big)$$

*and short exact sequences*

$$0 \to \Sigma^{-172}\frac{ko[7-k]}{\Gamma_2 ko[7-k]} \longrightarrow \operatorname{Hom}\Big(\Big(\frac{ko[k]}{\Gamma_2 ko[k]}\Big)/B^\infty, \mathbb{Z}_2\Big) \longrightarrow \operatorname{Hom}(\Sigma\langle \nu_k\rangle, \mathbb{Q}/\mathbb{Z}) \to 0$$

*for $0 \le k \le 6$.*

PROOF. (1) See Proposition 9.12, Definition 10.18 and Proposition 10.24.

(2) See Definition 10.4, Lemma 10.19, Table 9.4 and Proposition 10.24.

(3) See Theorem 10.25.

(4) See Definitions 9.18 and 9.22, and string together parts of Proposition 10.24, Theorem 10.25 and Lemma 10.20.

(5) See Equation (10.5), Lemma 10.19 and Theorem 10.25, keeping in mind that $(\Gamma_2 N_*)/B^\infty = (\Gamma_2 N_*/\Gamma_B N_*)/B^\infty$.

(6) See Equation (10.5), Lemma 10.19 and Theorem 10.22, keeping in mind that $N_*/B^\infty \to (N_*/\Gamma_2 N_*)/B^\infty$ induces an isomorphism under $\mathrm{Hom}(-, \mathbb{Z}_2)$. □

We note that $\Theta N_*$ in the $B$-power torsion $\Gamma_B N_*$ is Pontryagin 170-self dual, the $B$-periodic 2-torsion $\Gamma_2 N_*/\Gamma_B N_*$ is Pontryagin 171-dual to $(\Gamma_2 N_*/\Gamma_B N_*)/B^\infty$, and the 2-torsion free quotient $N_*/\Gamma_2 N_*$ is linearly 172-dual to $(N_*/\Gamma_2 N_*)/B^\infty$, modulo a correction term arising from $\Gamma_B N_*/\Theta N_*$. John Greenlees has pointed out how these three different degree shifts can be explained in terms of local cohomology , which we work out in a joint paper [**43**].

PROPOSITION 10.27. *The specialized $\pi_*(tmf)$-module isomorphism*

$$\Theta\pi_{-*}(\Sigma^{20}tmf) \xrightarrow{\cong} \mathrm{Hom}(\Theta\pi_{*-2}(tmf)/M^\infty, \mathbb{Q}/\mathbb{Z})$$

*is adjoint to a perfect pairing*

$$\langle -, - \rangle \colon \Theta\pi_{-*}(\Sigma^{20}tmf) \times \Theta\pi_{*-2}(tmf)/M^\infty \longrightarrow \mathbb{Q}/\mathbb{Z}$$

*such that*

$$\langle r \cdot x, y \rangle = (-1)^{|r||x|} \langle x, r \cdot y \rangle$$

*for $r \in \pi_*(tmf)$, $x \in \Theta\pi_{-*}(\Sigma^{20}tmf)$ and $y \in \Theta\pi_{*-2}(tmf)/M^\infty$ with $|r|+|x|+|y| = 0$. Similarly, the $\mathbb{Z}[B]$-module isomorphism*

$$\Theta N_{170-*} \xrightarrow{\cong} \mathrm{Hom}(\Theta N_*, \mathbb{Q}/\mathbb{Z})$$

*is adjoint to a perfect pairing*

$$(-, -) \colon \Theta N_{170-*} \times \Theta N_* \longrightarrow \mathbb{Q}/\mathbb{Z}\,.$$

*Under the isomorphisms $\Theta\pi_*(tmf) \cong \Theta N_* \otimes \mathbb{Z}[M]$ and $\Theta\pi_*(tmf)/M^\infty \cong \Theta N_* \otimes \mathbb{Z}[M]/M^\infty$, these pairings are related by*

$$\langle xM^\ell, y/M^{1+\ell} \rangle = (x, y)$$

*for $\ell \geq 0$ and $|x| + |y| = 170$.*

PROOF. Recall Lemma 10.8. The composite

$$\Theta\pi_{-*}(\Sigma^{20}tmf) \longrightarrow \pi_{-*}(\Sigma^{20}tmf) \xrightarrow{\cong} \mathrm{Hom}(\pi_*(tmf'/2^\infty, ), \mathbb{Q}/\mathbb{Z})$$

maps the source isomorphically to the homomorphisms that factor through the composite $\pi_*(tmf)$-module homomorphism

$$\pi_*(tmf'/2^\infty) \longrightarrow \Gamma_2\pi_{*-1}(tmf') \longrightarrow \Theta\pi_{*-2}(tmf)/M^\infty\,.$$

This leads to the specialized pairing $\langle -, - \rangle$. When restricted to $\mathbb{Z}[B, M]$, it takes the form

$$\langle -, - \rangle \colon \Theta N_{-*-20} \otimes \mathbb{Z}[M] \times \Theta N_{*-2} \otimes \mathbb{Z}[M]/M^\infty \longrightarrow \mathbb{Q}/\mathbb{Z}$$

and satisfies

$$\langle xM^\ell, y/M^{1+\ell} \rangle = \langle x, y/M \rangle$$

for all $\ell \geq 0$. Here $|x| + |y/M| = -22$, so $|x| + |y| = 170$. It follows that $(x, y) = \langle x, y/M \rangle$ is the pairing adjoint to the given $\mathbb{Z}[B]$-module isomorphism. □

REMARK 10.28. In the proof of Theorem 10.6 we made an arbitrary choice of a 2-adic generator of $\pi_{20}I(tmf/(2^\infty, B^\infty, M^\infty)) \cong \mathbb{Z}_2$. Multiplying by any 2-adic unit $u$ gives another generator, and would multiply the duality equivalence $a$ and the associated pairings $\langle -,-\rangle$ and $(-,-)$ by the same unit $u$. Hence we can only expect to give well-defined expressions for the pairings $(x,y) \in \mathbb{Q}/\mathbb{Z}$ when the values are 0 or $1/2 \mod 1$.

Once a choice of $a$ is fixed, it is possible to specify a choice of generator $\nu_6 \in \pi_{147}(tmf) \cong \mathbb{Z}/8$ in terms of a choice of $\bar\kappa \in \pi_{20}(tmf)$, e.g., by demanding that $(\bar\kappa, \nu\nu_6) \equiv 1/8 \mod 1$ (as opposed to 3/8, 5/8 or 7/8). In view of our conventions $\nu_2\nu_4 = 3\nu\nu_6$ and $\nu D_4 = 2\nu_4$ from Definition 9.22, this would reduce the combined multiplicative indeterminacy in $\nu_2$ and $\nu_4$ by a factor of $\mathbb{Z}/8^\times$.

THEOREM 10.29. *The values of the perfect pairing* $(-,-)\colon \Theta N_{170-*} \times \Theta N_* \to \mathbb{Q}/\mathbb{Z}$ *on classes* $x, y \in \Theta N_*$ *with* $|x| + |y| = 170$ *are given in Table* 10.1.

PROOF. Let $n = |x|$, so that $|y| = 170 - n$. When $\Theta N_n = \mathbb{Z}/2\{x\}$ and $\Theta N_{170-n} = \mathbb{Z}/2\{y\}$, the duality isomorphism implies that $(x,y) = 1/2 \mod 1$.

For $n = 20$, $\Theta N_{20} = \mathbb{Z}/8\{\bar\kappa\}$ is perfectly paired to $\Theta N_{150} = \mathbb{Z}/8\{\nu\nu_6\}$, so $(\bar\kappa, \nu\nu_6) \doteq 1/8 \mod 1$, up to an odd factor. Hence $(\bar\kappa, 4\nu\nu_6) = (\bar\kappa, \eta_1^2\bar\kappa^5) = 1/2 \mod 1$ and $(\nu^2\kappa, \nu\nu_6) = (4\bar\kappa, \nu\nu_6) = 1/2 \mod 1$.

For $n \in \{40, 54, 60\}$, $\Theta N_n = \mathbb{Z}/4\{x\}$ is perfectly paired to $\Theta N_{170-n} = \mathbb{Z}/4\{y\}$, for the appropriate $x \in \{\bar\kappa^2, \nu\nu_2, \bar\kappa^3\}$ and $y \in \{\kappa_4\bar\kappa, \bar\kappa D_4, \kappa_4\}$, so $(x,y) = \pm 1/4 \mod 1$ and $(x, 2y) = (2x, y) = 1/2 \mod 1$.

The case $n = 65$ remains, with Klein four-groups $\Theta N_{65} = \mathbb{Z}/2\{\nu_2\kappa, \eta_1\bar\kappa^2\}$ and $\Theta N_{105} = \mathbb{Z}/2\{\eta\epsilon_4, \nu^2\nu_4\}$. Using $\eta$- and $\nu$-linearity, we deduce from the cases $n \in \{66, 68\}$ that

$$
\begin{aligned}
(\nu_2\kappa, \eta\epsilon_4) &= (\eta\nu_2\kappa, \epsilon_4) = 1/2 \mod 1\\
(\nu_2\kappa, \nu^2\nu_4) &= (\nu\nu_2\kappa, \nu\nu_4) = 1/2 \mod 1\\
(\eta_1\bar\kappa^2, \eta\epsilon_4) &= (\eta\eta_1\bar\kappa^2, \epsilon_4) = 1/2 \mod 1\\
(\eta_1\bar\kappa^2, \nu^2\nu_4) &= (\eta_1\nu\bar\kappa^2, \nu\nu_4) = 0 \mod 1\,.
\end{aligned}
$$

It follows by bilinearity that $(\nu_2\kappa, \eta_1\bar\kappa^4) = 0 \mod 1$ and $(\eta_1\bar\kappa^2, \eta_1\bar\kappa^4) = 1/2 \mod 1$, since $\eta_1\bar\kappa^4 = \eta\epsilon_4 + \nu^2\nu_4$. □

REMARK 10.30. Heuristically, we have $(x,y) = 1/2 \mod 1$ when $x$ and $y$ formally multiply to

$$
(\eta\nu\epsilon\kappa)_6 = (\nu^4\kappa)_6 = (\eta^3\nu\bar\kappa)_6 = (\epsilon^2\kappa\bar\kappa)_5 = \eta_1^2\bar\kappa^6 = 2\kappa_4\bar\kappa^3 = \kappa\bar\kappa^3 D_4
$$

These identities follow formally from $\eta\epsilon = \nu^3$, $\epsilon\kappa = \eta^2\bar\kappa$, $\eta\nu_1 = \epsilon\bar\kappa$, $\epsilon\epsilon_5\kappa = \eta_1^2\bar\kappa^5$, $\eta_1^2\bar\kappa^3 = 2\kappa_4$ and $\kappa D_4 = 2\kappa_4$, but, of course, all of the displayed products actually evaluate to zero in $\pi_{170}(tmf)$.

By analogy, $\pi_*(ko) \cong N_*^1 \otimes \mathbb{Z}[B]$ as a $\mathbb{Z}[B]$-module, where $N_*^1 = \mathbb{Z}\{1, A\} \oplus \mathbb{Z}/2\{\eta, \eta^2\}$. The 2-power torsion $\Gamma_2 N_*^1 = \mathbb{Z}/2\{\eta, \eta^2\}$ is Pontryagin self-dual, with $(\eta, \eta^2) = (\eta^2, \eta) = 1/2 \mod 1$, but the product $\eta \cdot \eta^2 = \eta^3$ is zero in $\pi_3(ko)$.

REMARK 10.31. We spell out how the Pontryagin self-duality of $\Theta N_*$ arises from Theorem 10.6. Let $N = tmf/M$ be the homotopy cofiber of $M\colon \Sigma^{192}tmf \to tmf$, so that the composite homomorphism $N_* \subset \pi_*(tmf) \to \pi_*(N)$ is an isomorphism of $\mathbb{Z}[B]$-modules. Substituting $a\colon \Sigma^{20}tmf \simeq I(tmf/(2^\infty, B^\infty, M^\infty))$ in the

TABLE 10.1. Duality pairing in $\Theta N_*$

| $n$ | $x$ | $y$ | $(x, y)$ |
|---|---|---|---|
| 6 | $\nu^2$ | $\nu\nu_6\kappa$ | $1/2$ |
| 8 | $\epsilon$ | $\eta\nu_6\kappa$ | $1/2$ |
| 9 | $\eta\epsilon$ | $\nu_6\kappa$ | $1/2$ |
| 14 | $\kappa$ | $\eta\nu_6\epsilon$ | $1/2$ |
| 15 | $\eta\kappa$ | $\nu_6\epsilon$ | $1/2$ |
| 17 | $\nu\kappa$ | $\eta_1\epsilon_5$ | $1/2$ |
| 20 | $\bar\kappa$ | $\eta_1^2\bar\kappa^5$ | $1/2$ |
| 20 | $2\bar\kappa$ | $\kappa_4\bar\kappa^2$ | $1/2$ |
| 20 | $\nu^2\kappa$ | $\nu\nu_6$ | $1/2$ |
| 20 | $2\bar\kappa$ | $\nu\nu_6$ | $\pm 1/4$ |
| 20 | $\bar\kappa$ | $\kappa_4\bar\kappa^2$ | $\pm 1/4$ |
| 20 | $\bar\kappa$ | $\nu\nu_6$ | $?/8$ |
| 21 | $\eta\bar\kappa$ | $\eta^2\nu_6$ | $1/2$ |
| 22 | $\epsilon\kappa$ | $\eta\nu_6$ | $1/2$ |
| 28 | $\eta\nu_1$ | $\epsilon_5\kappa$ | $1/2$ |
| 32 | $\epsilon_1$ | $\eta\nu_5\kappa$ | $1/2$ |
| 33 | $\eta\epsilon_1$ | $\nu_5\kappa$ | $1/2$ |
| 34 | $\kappa\bar\kappa$ | $\epsilon\epsilon_5$ | $1/2$ |
| 35 | $\nu\epsilon_1$ | $\eta_1\kappa_4$ | $1/2$ |
| 39 | $\eta_1\kappa$ | $\nu_5\epsilon$ | $1/2$ |
| 40 | $\bar\kappa^2$ | $\eta_1^2\bar\kappa^4$ | $1/2$ |
| 40 | $\epsilon\epsilon_1$ | $\kappa_4\bar\kappa$ | $1/2$ |
| 40 | $\bar\kappa^2$ | $\kappa_4\bar\kappa$ | $\pm 1/4$ |

| $n$ | $x$ | $y$ | $(x, y)$ |
|---|---|---|---|
| 41 | $\nu_1\kappa$ | $\eta\epsilon_5$ | $1/2$ |
| 42 | $\eta\nu_1\kappa$ | $\epsilon_5$ | $1/2$ |
| 45 | $\eta_1\bar\kappa$ | $\eta_1\bar\kappa^5$ | $1/2$ |
| 46 | $\epsilon_1\kappa$ | $\eta\nu_5$ | $1/2$ |
| 52 | $\eta\nu_2$ | $\epsilon\kappa_4$ | $1/2$ |
| 53 | $\eta^2\nu_2$ | $\eta_4\bar\kappa$ | $1/2$ |
| 54 | $\nu\nu_2$ | $\nu^2\kappa_4$ | $1/2$ |
| 54 | $\kappa\bar\kappa^2$ | $\bar\kappa D_4$ | $1/2$ |
| 54 | $\nu\nu_2$ | $\bar\kappa D_4$ | $\pm 1/4$ |
| 57 | $\eta_1\epsilon_1$ | $\nu\kappa_4$ | $1/2$ |
| 59 | $\nu_2\epsilon$ | $\eta\kappa_4$ | $1/2$ |
| 60 | $\bar\kappa^3$ | $\eta_1^2\bar\kappa^3$ | $1/2$ |
| 60 | $\eta\nu_2\epsilon$ | $\kappa_4$ | $1/2$ |
| 60 | $\bar\kappa^3$ | $\kappa_4$ | $\pm 1/4$ |
| 65 | $\nu_2\kappa$ | $\eta\epsilon_4$ | $1/2$ |
| 65 | $\nu_2\kappa$ | $\nu^2\nu_4$ | $1/2$ |
| 65 | $\eta_1\bar\kappa^2$ | $\eta\epsilon_4$ | $1/2$ |
| 65 | $\eta_1\bar\kappa^2$ | $\nu^2\nu_4$ | $0$ |
| 66 | $\eta\nu_2\kappa$ | $\epsilon_4$ | $1/2$ |
| 68 | $\nu\nu_2\kappa$ | $\nu\nu_4$ | $1/2$ |
| 70 | $\eta_1^2\bar\kappa$ | $\bar\kappa^5$ | $1/2$ |
| 80 | $\bar\kappa^4$ | $\eta_1^2\bar\kappa^2$ | $1/2$ |
| 85 | $\eta_1\bar\kappa^3$ | $\eta_1\bar\kappa^3$ | $1/2$ |

homotopy cofiber sequence

$$\Sigma^{212} tmf \xrightarrow{M} \Sigma^{20} tmf \longrightarrow \Sigma^{20} N$$

and applying Brown–Comenetz duality, we obtain a homotopy cofiber sequence

$$I(\Sigma^{20} N) \longrightarrow tmf/(2^\infty, B^\infty, M^\infty) \xrightarrow{M} \Sigma^{-192} tmf/(2^\infty, B^\infty, M^\infty)\,.$$

The homotopy fiber of the right hand map is $\Sigma^{-192} N/(2^\infty, B^\infty)$, so we get an equivalence

$$\Sigma^{172} I(N) \simeq N/(2^\infty, B^\infty)$$

of *tmf*-modules. We can view each homomorphism $\phi\colon \pi_k(N) \to \mathbb{Q}/\mathbb{Z}$ as a homotopy class $\phi \in \pi_{-k} I(N)$, and $\Sigma^{172}\phi$ then corresponds under the equivalence above to a class $\psi \in \pi_{172-k}(N/(2^\infty, B^\infty))$. Its image $\partial^2(\psi)$ under the two connecting homomorphisms

$$\pi_{172-k}(N/(2^\infty, B^\infty)) \xrightarrow{\partial} \pi_{171-k}(N/B^\infty) \xrightarrow{\partial} \pi_{170-k}(N)$$

lies in $\Theta N_{170-k}$. Our analysis shows that $\partial^2(\psi)$ only depends on the restriction of $\phi$ to a homomorphism $\phi| \colon \Theta N_k \to \mathbb{Q}/\mathbb{Z}$, and Theorem 10.26(3) asserts that the correspondence $\phi| \leftrightarrow \partial^2(\psi)$ defines an isomorphism

$$\mathrm{Hom}(\Theta N_k, \mathbb{Q}/\mathbb{Z}) \cong \Theta N_{170-k} .$$

CHAPTER 11

# The Adams spectral sequence for the sphere

Our study of $\pi_*(tmf)$ relies on some initial information about $\pi_*(S)$, which we establish in this chapter. Since the literature on this subject is scattered through many sources, and since those sources have in many cases been subject to later corrections, we here attempt to give an account that is as succinct and comprehensive as possible, in the range we need. This has the virtue of shortening and clarifying many of the arguments, and makes our work reasonably self-contained. The three results that we will need about the Adams spectral sequence for $S$ are the following: the product $\eta^2\kappa$ is zero, the product $\eta\rho$ is detected by $Pc_0$, and the product $\eta^2\bar{\kappa}$ is detected by $Pd_0$. The first two facts are established in case (16) of Theorem 11.61, and the third is established in Theorem 11.71.

After completing the calculation of $\pi_*(tmf)$, we are then able to use the unit map $S \to tmf$ and its associated cofiber sequence to deduce further information about $\pi_*(S)$. In order to avoid splitting the statements about $\pi_*(S)$ into two disconnected sections, some parts of Theorems 11.52, 11.54, 11.56 and 11.59 are marked (*). Logically, we first only prove the statements without this mark. These suffice to give the necessary input for our calculations of the Adams spectral sequence and homotopy groups of *tmf*, given in Chapters 5 and 9. Thereafter we return to $S$ and use the results about *tmf* to prove the marked statements. We have chosen to break with the logical order for this presentation in order to have the results collected in one place, and to avoid repetition.

Our overall strategy in this chapter is thus as follows: The Adams spectral sequence for $S$ is a graded commutative algebra spectral sequence, with Steenrod operations $Sq^i$ acting on its initial term $E_2(S) = \mathrm{Ext}_A(\mathbb{F}_2, \mathbb{F}_2)$. The differential structure is therefore determined by the values of the $d_r$-differentials on the algebra indecomposables of the $E_r$-term, for each $r \geq 2$. The $H_\infty$ ring structure on $S$ implies a number of differentials on classes of the form $Sq^i(x)$. We summarize this theory in Section 11.1, and collect results specific to $S$ in Section 11.2. The proven Adams conjecture, about the ($d$- and) $e$-invariant map $e\colon S \to j$ to the image-of-$J$ spectrum, implies multiple differentials in $h_0$-towers leading up to the vanishing line of slope $1/2$ in the $(t-s, s)$-plane. We review this theory in Section 11.3, and are thereafter ready to determine the sequence of $(E_r, d_r)$-terms for $S$, up to topological degree $t - s = 48$, in Sections 11.4 through 11.7. The unit map $\iota\colon S \to tmf$ and our results on the Adams spectral sequence for *tmf* give simplified proofs of several differentials above or near a line of slope $1/6$. Some differentials below this line remain, principally $d_3(h_2h_5) = h_0p$ and $d_4(h_3h_5) = h_0x$, for which we follow [**107**] and [**22**], comparing $S$ with the finite CW spectra $C\nu$ and $C\sigma \cup_{2\sigma} e^{16}$, respectively. It is then mostly elementary to deduce the structure of $\pi_*(S)$ as a graded commutative ring for $* \leq 44$, but the details grow more complex as the degree increases. Our results are collected in the lengthy, but hopefully useful, Theorem 11.61. We give a purely stable proof in Section 11.9 of the key fact that $\eta^2\bar{\kappa} \in \pi_{22}(S)$ is detected by the nonzero class $Pd_0 \in E_\infty(S)$, and therefore equals the product $\epsilon\kappa$. In Theorem 11.89 we calculate the image of the $tmf$-Hurewicz homomorphism $\iota\colon \pi_*(S) \to \pi_*(tmf)$ in degrees $* \le 101$ and for $* = 125$, relying on results from [**83**] for degrees $* = 54$, 65 and 70. The multiplicative structure also allows us to deduce, in Proposition 11.83, that $\iota$ detects nonzero classes $\bar{\kappa}^3\{w\}$, $\bar{\kappa}\{w\}^2$, $\bar{\kappa}^2\{w\}^2$ and $\bar{\kappa}^3\{w\}^2 \in \pi_*(S)$, in degrees $* = 105$, 110, 130 and 150, respectively.

## 11.1. $H_\infty$ ring spectra

The principal spectra whose homotopy groups we are studying in this work, such as the sphere spectrum $S$ and the topological modular forms spectrum $tmf$, as well as the real $K$-theory spectrum $ko$ and the image-of-$J$ spectrum $j$, are all $E_\infty$ ring spectra [**121**, Ch. IV]. The presence of an $E_\infty$ ring structure on a spectrum $Y$ gives rise to power operations acting on its homotopy groups $\pi_*(Y)$, subject to suitable natural identities. There are also algebraic Steenrod operations acting on the Adams $E_2$-term for $Y$, and these will detect the power operations modulo Adams filtration. However, not every Steenrod operation comes from a homotopy operation, and not every element in $E_2(Y)$ comes from a homotopy class. In order to account for these two discrepancies, the Adams spectral sequence for any $E_\infty$ ring spectrum must contain certain universal differentials, for which we can give explicit formulas. We have already seen these formulas in action in our analysis of the Adams spectral sequence for $tmf$, in Chapter 5, and in the present chapter we will apply the same method to the Adams spectral sequence for the sphere spectrum. For the convenience of the reader, we here give a review of the main results regarding these power operations and the associated Adams differentials.

**11.1.1. Structured homotopy commutativity.** In terms of the Lewis–May category of spectra, an $E_\infty$ ring spectrum $Y$ comes equipped with an action by an $E_\infty$ operad $\mathcal{O} \to \mathcal{L}$ over the linear isometries operad [**92**, §VII.2]. Such an action is given by suitably compatible spectrum maps

$$\xi_j\colon \mathcal{O}(j) \ltimes_{\Sigma_j} (Y \wedge \cdots \wedge Y) \longrightarrow Y$$

for all integers $j \ge 0$, where $\Sigma_j$ denotes the symmetric group on $j$ letters, and there are $j$ copies of $Y$ in the source of the map. In terms of the categories of $S$-modules [**58**, §II.3 and §II.4] and orthogonal spectra [**111**, Ex. 4.4], [**110**, §1.1], each $E_\infty$ ring spectrum can be realized up to equivalence as a commutative monoid with respect to the symmetric monoidal smash product, i.e., as a commutative $S$-algebra and a commutative orthogonal ring spectrum, respectively. For the purpose of defining power operations in homotopy, as well as for the study of differentials in the Adams spectral sequence for $Y$, only a weakened form of the $E_\infty$ ring structure turns out to be needed. More precisely, we shall only make use of the structure maps $\xi_j$ in their relaxed incarnation as morphisms in the stable homotopy category. This "up-to-homotopy" image of an $E_\infty$ ring structure is known as an $H_\infty$ ring structure [**45**, §I.3], [**92**, §VII.2]. Taking the homotopy category of orthogonal spectra, equipped with the stable model structure of [**111**, §9], as our model for the stable homotopy category, we can write the $j$-th $H_\infty$ structure map as

$$\xi_j\colon E\Sigma_{j+} \wedge_{\Sigma_j} (Y \wedge \cdots \wedge Y) \longrightarrow Y\,,$$

where now $E\Sigma_j \simeq \mathcal{O}(j)$ is any free, contractible $\Sigma_j$-CW complex. If a product $\phi\colon Y \wedge Y \to Y$ makes $Y$ a commutative orthogonal ring spectrum, then $\xi_j$ can be taken to be the composite $E\Sigma_{j+} \wedge_{\Sigma_j} (Y \wedge \cdots \wedge Y) \to (Y \wedge \cdots \wedge Y)/\Sigma_j \to Y$, where the first map collapses $E\Sigma_j$ to a point and the second map is induced by the $j$-fold multiplication map $\phi_j : Y \wedge \cdots \wedge Y \to Y$, keeping in mind the hypothesis that $\phi$ is strictly commutative.

DEFINITION 11.1. For any orthogonal spectrum $X$ we call $D_j(X) = E\Sigma_{j+} \wedge_{\Sigma_j} (X \wedge \cdots \wedge X)$ the $j$-th extended power of $Y$. When $j = 2$ we also refer to $D_2(X) = E\Sigma_{2+} \wedge_{\Sigma_2} (X \wedge X)$ as the quadratic construction on $X$.

Suppose hereafter that $Y$ is an $H_\infty$ ring spectrum. The underlying ring spectrum pairing is given by the composite

$$\phi\colon Y \wedge Y \longrightarrow E\Sigma_{2+} \wedge_{\Sigma_2} (Y \wedge Y) \xrightarrow{\xi_2} Y\,,$$

where the first map is induced by any choice of point in $E\Sigma_2$. Since $E\Sigma_2$ is path connected, this pairing is homotopy commutative. Let $H_*(Y) = H(Y;\mathbb{F}_p)$ denote mod $p$ homology, for any prime $p$. It follows that the induced pairing

$$\phi_* : H_*(Y) \otimes H_*(Y) \cong H_*(Y \wedge Y) \longrightarrow H_*(Y)$$

makes $H_*(Y)$ a commutative algebra in the category of $A_*$-comodules, where $A_*$ denotes the dual of the mod $p$ Steenrod algebra $A$. Suppose also that $\pi_*(Y)$ is bounded below with $H_*(Y)$ of finite type, so that

$$\phi^* : H^*(Y) \longrightarrow H^*(Y \wedge Y) \cong H^*(Y) \otimes H^*(Y)$$

makes $H^*(Y)$ a cocommutative coalgebra in the category of $A$-modules. It was shown by Liulevicius [**95**, Ch. 2], see also May [**118**, §11], that there are Steenrod operations acting in the cohomology of any cocommutative Hopf algebra, such as $A_*$, including the $E_2$-term

$$\begin{aligned} E_2(Y) = \mathrm{Ext}_{A_*}(\mathbb{F}_p, H_*(Y)) &\cong \mathrm{Ext}_A(H^*(Y), \mathbb{F}_p) \\ &\Longrightarrow \pi_*(Y_p^\wedge) \end{aligned}$$

of the mod $p$ Adams spectral sequence for $Y$. These constructions were generalized by the first author to the cohomology of Hopf algebroids, in [**45**, Lem. IV.2.3], and play a corresponding role in the $E$-based Adams–Novikov spectral sequence for suitable ring spectra $E$.

When $p = 2$ we write the Steenrod operations in $\mathrm{Ext}_A(H^*(Y), \mathbb{F}_2)$ as cohomologically indexed Steenrod squares, viz.

$$Sq^i : \mathrm{Ext}_A^{s,t}(H^*(Y), \mathbb{F}_2) \longrightarrow \mathrm{Ext}_A^{s+i,2t}(H^*(Y), \mathbb{F}_2)\,.$$

Note that this operation increases the cohomological degree $s$ by $i$, and doubles the internal degree $t$. Only the operations with $0 \le i \le s$ can be nonzero, and $Sq^s(x) = x^2$ is given by the square with respect to the usual product on Ext.

For $p$ odd there are analogously defined Steenrod $p$-th powers, $P^i$ and $\beta P^i$, acting on $\mathrm{Ext}_A(H^*(Y), \mathbb{F}_p)$. This case can be found in [**45**, Ch. IV], which is also the definitive source for the material here. We shall concentrate on the 2-primary case.

REMARK 11.2. Alternatively, these operations can be homologically indexed by the change in the topologically significant degree $t - s$:

$$Q^j : \mathrm{Ext}_A^{s,t}(H^*(Y), \mathbb{F}_2) \longrightarrow \mathrm{Ext}_A^{s+t-j,2t}(H^*(Y), \mathbb{F}_2)\,.$$

Thus $Sq^i = Q^j$ on $\mathrm{Ext}_A^{s,t}$ for $i + j = t$. This homological indexing is compatible with that of the Dyer–Lashof operations present in the homology of an $H_\infty$ ring spectrum, under the edge and Hurewicz homomorphisms, hence the notation $Q^j$. In [**45**] the homological indexing was used, but was written $Sq^j$. We now find the cohomological indexing more convenient, and have therefore translated the discussion and the main theorems into the cohomological indexing just introduced.

To each element $\alpha \in \pi_N D_j(S^n)$ we can associate a $j$-th order power operation $\alpha^* \colon \pi_n(Y) \to \pi_N(Y)$, which is natural for $H_\infty$ ring spectra $Y$. We concentrate on the case $j = 2$, with $D_2(S^n) = E\Sigma_{2+} \wedge_{\Sigma_2} (S^n \wedge S^n)$, referring to [**45**, §IV.7] for the case when $j = p$ is an odd prime, and to [**42**] for the case of a general exponent $j$.

DEFINITION 11.3. For any $\alpha \in \pi_N D_2(S^n)$ and any $H_\infty$ ring spectrum $Y$ let

$$\alpha^* \colon \pi_n(Y) \longrightarrow \pi_N(Y)$$

be the natural power operation sending the homotopy class of a map $y \colon S^n \to Y$ to the homotopy class of the composite map

$$S^N \xrightarrow{\alpha} D_2(S^n) \xrightarrow{D_2(y)} D_2(Y) \xrightarrow{\xi_2} Y .$$

With these notations, the main results about power operations and differentials in the Adams spectral sequence for an $H_\infty$ ring spectrum $Y$ are the following: First, in Theorem 11.13, we show how the power operation $\alpha^*(y)$ on a homotopy class $y$ detected by an infinite cycle $x \in E_2(Y) = \mathrm{Ext}_A(H^*(Y), \mathbb{F}_2)$ is detected, modulo classes of higher Adams filtration, by a linear combination of Steenrod operations $Sq^i(x)$, where the coefficients lie in $E_2(S) = \mathrm{Ext}_A(\mathbb{F}_2, \mathbb{F}_2)$ and depend on $\alpha$. Second, in Theorem 11.22, we consider a class $x \in E_2(Y)$ that survives to the $E_r$-term, and identify the generically first Adams differential $d_*(Sq^i(x))$ on the class $Sq^i(x)$, in terms of $x$, $d_r(x)$, Steenrod operations on these classes, and coefficients in $E_2(S)$.

Let us outline the history of these results. By assumption, the spectrum $Y$ is bounded below with $H_*(Y)$ of finite type. It admits an Adams resolution

$$Y = Y_0 \longleftarrow Y_1 \longleftarrow Y_2 \longleftarrow \dots ,$$

where each homotopy cofiber $Y_{s,1} = \mathrm{cof}(Y_{s+1} \to Y_s)$ is equivalent to a wedge sum of suspensions of copies of $H = H\mathbb{F}_p$ and each homomorphism $H^*(Y_s) \to H^*(Y_{s+1})$ is zero. We may, and will, arrange that each $H^*(Y_{s,1})$ is of finite type. There is a smash power resolution

$$Y \wedge \cdots \wedge Y \longleftarrow (Y \wedge \cdots \wedge Y)_1 \longleftarrow (Y \wedge \cdots \wedge Y)_2 \longleftarrow \dots ,$$

with $j$ copies of $Y$ at each stage. The $j$-fold multiplication map $\phi_j \colon Y \wedge \cdots \wedge Y \to Y$ lifts to a weak map of Adams resolutions, where "weak" means that the evident squares are only required to commute up to homotopy. For $j = 2$ it induces the product pairing $\phi_r \colon E_r(Y) \otimes E_r(Y) \to E_r(Y)$ in the Adams spectral sequence for $Y$.

Daniel Kahn [**85**] and James Milgram [**122**] showed, in the case $Y = S$, that the lifts $(Y \wedge \cdots \wedge Y)_s \to Y_s$ can be gradually prolonged over the extended $j$-th powers $D_j(Y)$ to yield a collection of suitably compatible maps

$$\xi_{k,s} \colon E\Sigma_{j+}^{(k)} \wedge_{\Sigma_j} (Y \wedge \cdots \wedge Y)_s \longrightarrow Y_{s-k} ,$$

where $E\Sigma_j^{(k)}$ denotes the $\Sigma_j$-equivariant $k$-skeleton of $E\Sigma_j$. In particular, $\xi_{\infty,s}$ is given by the $H_\infty$ structure map $\xi_j$. The compatible maps $\xi_{k,s}$ give rise to a geometric construction of the Steenrod operations $Sq^i(x)$ and $\beta^\epsilon P^i(x)$ in the Adams

$E_2$-term for $Y$, as well as of the homotopy operations traditionally denoted $\cup_k(y)$. This allowed Kahn and Milgram to give formulas for Adams differentials on these Steenrod operations when applied to infinite cycles, or, by suitable truncations, to general elements within a range in which one can act as if one is operating on an infinite cycle. Jukka Mäkinen, in his thesis [**109**], showed how to remove this range restriction when $p = 2$, accounting for the contribution of the boundary $d_r(x)$, and obtaining much more extensive formulas for differentials on the values of the Steenrod operations. The first author (of the present book) showed in [**37**] and [**45**] how this could be done for all primes and for all $H_\infty$ ring spectra, as well as for many $E$-based Adams–Novikov spectral sequences.

**11.1.2. Extended powers of Adams resolutions.** We now review these constructions in the context of orthogonal spectra, to show how the compatibility conditions alluded to above can be clarified in terms of this symmetric monoidal and topologically enriched model for the stable homotopy category. Let $Y_\star$ denote an Adams resolution

$$Y = Y_0 \longleftarrow Y_1 \longleftarrow Y_2 \longleftarrow \dots$$

of $Y$, as above, with associated free resolution

$$0 \leftarrow H^*(Y) \xleftarrow{\epsilon} F_0 \xleftarrow{\partial} F_1 \xleftarrow{\partial} F_2 \xleftarrow{\partial} \dots$$

in the category of $A$-modules. Here $F_s = H^*(\Sigma^s Y_{s,1})$ where

$$Y_{s,r} = \mathrm{cof}(Y_{s+r} \to Y_s)\,, \tag{11.1}$$

and the homomorphisms $\epsilon$ and $\partial$ are induced by the evident maps $Y_0 \to Y_{0,1}$ and $Y_{s,1} \to \Sigma Y_{s+1} \to \Sigma Y_{s+1,1}$, respectively. This complex is exact, since each homomorphism $H^*(Y_s) \to H^*(Y_{s+1})$ is zero. Let us write $F_s^\vee = \mathrm{Hom}_A(F_s, \mathbb{F}_p)$, so that $\mathrm{Ext}_A(H^*(Y), \mathbb{F}_p)$ is the cohomology of the cocomplex

$$0 \to F_0^\vee \xrightarrow{\delta} F_1^\vee \xrightarrow{\delta} F_2^\vee \xrightarrow{\delta} \dots .$$

A typical element $[x] \in \mathrm{Ext}_A^{s,t}(H^*(Y), \mathbb{F}_p)$ is represented by a cocycle $x$ in

$$\begin{aligned} \mathrm{Hom}_A(F_s, \Sigma^t \mathbb{F}_p) &= \mathrm{Hom}_A(H^*(\Sigma^s Y_{s,1}), \Sigma^t \mathbb{F}_p) \\ &\cong \mathrm{Hom}_A(H^*(Y_{s,1}), H^*(S^{t-s})) \\ &\cong [S^{t-s}, Y_{s,1}] = \pi_{t-s}(Y_{s,1})\,. \end{aligned}$$

The last isomorphism uses our assumption that $H^*(Y_{s,1})$ is of finite type. We call $n = t - s$ the topological degree, $s$ the cohomological degree or filtration, and $t$ the internal degree of $x$. It will be convenient to extend the Adams resolution in the negative direction by letting $Y_s = Y$ for $s \le 0$, with $Y_{s+1} \to Y_s$ the identity map for each $s < 0$.

REMARK 11.4. The tensor square $F_* \otimes F_*$, with the diagonal $A$-module structure and the boundary operator $\partial \otimes 1 + 1 \otimes \partial$, is a free $A$-module resolution of $H^*(Y) \otimes H^*(Y) \cong H^*(Y \wedge Y)$, and can be realized as the algebraic resolution associated to an Adams resolution

$$Y \wedge Y = (Y \wedge Y)_0 \longleftarrow (Y \wedge Y)_1 \longleftarrow (Y \wedge Y)_2 \longleftarrow \dots .$$

Recalling that $Y$ is, in particular, a homotopy commutative ring spectrum, the product map $\phi\colon Y \wedge Y \to Y$ induces the cocommutative $A$-module coproduct $H^*(Y) \to H^*(Y) \otimes H^*(Y)$. It can be lifted to an $A$-module chain map $F_* \to F_* \otimes F_*$, which in turn can be realized by a (weak) map $(Y \wedge Y)_\star \to Y_\star$ of Adams resolutions. The twist isomorphisms $\tau\colon Y \wedge Y \to Y \wedge Y$ and $\tau\colon F_* \otimes F_* \to F_* \otimes F_*$ can be realized by a (weak) map $\tau\colon (Y \wedge Y)_\star \to (Y \wedge Y)_\star$ of Adams resolutions, but in this context there is no reason why $\tau^2$ should be equal to the identity, i.e., why the Adams resolution $(Y \wedge Y)_\star$ should be $\Sigma_2$-equivariant in any strict sense.

To obtain a $\Sigma_2$-equivariant Adams resolution of $Y \wedge Y$, we now assume that we are working in the context of orthogonal spectra, in the stable model structure [**111**, §9]. We assume that the spectra in the Adams resolution $Y_\star$ are all $q$-cofibrant and stably $q$-fibrant, and that each map $Y_{s+1} \to Y_s$ is a $q$-cofibration. In essence, we may assume that each $Y_s$ can be built from $Y_{s+1}$ or $*$ by attaching cells, and that each $Y_s$ is an $\Omega$-spectrum. We can then form the convolution product $(Y \wedge Y)_\star$ of two copies of $Y_\star$, by setting

$$(Y \wedge Y)_s = \bigcup_{s_1+s_2=s} Y_{s_1} \wedge Y_{s_2}\,.$$

By the pushout-product axiom for orthogonal spectra each $(Y \wedge Y)_s$ is $q$-cofibrant, and each inclusion $(Y \wedge Y)_{s+1} \to (Y \wedge Y)_s$ is a $q$-cofibration. (In general, we have no reason to expect that $(Y \wedge Y)_s$ is stably $q$-fibrant.) Hence there are natural equivalences

$$\begin{aligned}(Y \wedge Y)_{s,1} &\xrightarrow{\simeq} (Y \wedge Y)_s/(Y \wedge Y)_{s+1} \\ &\xleftarrow{\cong} \bigvee_{s_1+s_2=s} Y_{s_1}/Y_{s_1+1} \wedge Y_{s_2}/Y_{s_2+1} \\ &\xleftarrow{\simeq} \bigvee_{s_1+s_2=s} Y_{s_1,1} \wedge Y_{s_2,1}\end{aligned}$$

and the algebraic free resolution associated to $(Y \wedge Y)_\star$ is given by

$$H^*(\Sigma^s (Y \wedge Y)_{s,1}) \cong \bigoplus_{s_1+s_2=s} H^*(\Sigma^{s_1} Y_{s_1,1}) \otimes H^*(\Sigma^{s_2} Y_{s_2,1})$$

in cohomological degree $s$, i.e., by the tensor square $F_* \otimes F_*$. Moreover, the symmetric monoidal twist isomorphism $\tau\colon Y \wedge Y \to Y \wedge Y$ of orthogonal spectra now restricts to a well-defined $\Sigma_2$-action on $(Y \wedge Y)_\star$, inducing the algebraic twist isomorphism $\tau\colon F_* \otimes F_* \to F_* \otimes F_*$. Completely similar considerations define a $\Sigma_j$-equivariant Adams resolution of the $j$-fold smash power $Y \wedge \cdots \wedge Y$.

We can merge the $\Sigma_j$-equivariant skeleton filtration of $E\Sigma_j$ with this $\Sigma_j$-equivariant Adams resolution, to obtain a doubly-indexed filtration of $D_j(Y) = E\Sigma_{j+} \wedge_{\Sigma_j} (Y \wedge \cdots \wedge Y)$ by subspectra $E\Sigma_{j+}^{(k)} \wedge_{\Sigma_j} (Y \wedge \cdots \wedge Y)_s$ for $k \geq 0$ and $s \geq 0$. Furthermore, we can convolve this into a singly-indexed filtration $D_j(Y)_\star$. As before we concentrate on the case $j = 2$, and for definiteness, we shall work with the following concrete model for the free, contractible $\Sigma_2$-CW complex $E\Sigma_2 \simeq \mathcal{O}(2)$.

Definition 11.5. Let $S^\infty = S(\mathbb{R}^\infty)$ be the unit sphere in $\mathbb{R}^\infty$. The group $\Sigma_2 = \{1, T\}$ acts freely on $S^\infty \simeq E\Sigma_2$ by the antipodal action, with $T$ sending a unit vector $x$ to $-x$. Its $\Sigma_2$-equivariant $k$-skeleton $S^k = S(\mathbb{R}^{k+1})$ is the unit sphere in $\mathbb{R}^{k+1}$, and $S^\infty$ has precisely one $\Sigma_2$-free cell in each dimension $k \geq 0$. The associated cellular chain complex

$$\dots \xrightarrow{\partial} W_2 \xrightarrow{\partial} W_1 \xrightarrow{\partial} W_0 \xrightarrow{\epsilon} \mathbb{Z} \to 0$$

is the usual $\mathbb{Z}[\Sigma_2]$-free resolution of $\mathbb{Z}$: each $W_k = H_k(S^k/S^{k-1};\mathbb{Z})$ is free over $\mathbb{Z}[\Sigma_2]$ on one generator $e_k$, with boundary $\partial(e_k) = (T+1)e_{k-1}$ for $k \ge 2$ even and $\partial(e_k) = (T-1)e_{k-1}$ for $k \ge 1$ odd. The orbit space $P^\infty = S^\infty/\Sigma_2$ is the infinite-dimensional real projective space, with $k$-skeleton $P^k = S^k/\Sigma_2$ the $k$-dimensional projective space. Let $P_n^\infty = P^\infty/P^{n-1}$ denote the stunted projective space, with one cell in each dimension $k \ge n$ (together with the base point), and let $P_n^{n+k} = P^{n+k}/P^{n-1}$ denote its $n+k$-skeleton.

LEMMA 11.6 ([**16**, Prop. 4.3]). *Let $n \ge 0$. There is a natural homeomorphism*

$$D_2(S^n) = S_+^\infty \wedge_{\Sigma_2} (S^n \wedge S^n) \cong \Sigma^n P_n^\infty ,$$

*which is filtration-preserving in the sense that it sends $S_+^k \wedge_{\Sigma_2} (S^n \wedge S^n) \subset D_2(S^n)$ homeomorphically to $\Sigma^n P_n^{n+k}$. In particular, $\pi_N D_2(S^n) \cong \pi_N(\Sigma^n P_n^\infty)$.*

REMARK 11.7. This well-known identification shows that the operations $\alpha^*$ introduced in Definition 11.3 are precisely parameterized by the stable homotopy groups of stunted projective spaces. If $N = 2n+k$ and the mod 2 Hurewicz image $h(\alpha) \in H_N D_2(S^n) \cong H_N(\Sigma^n P_n^\infty) = \mathbb{F}_2\{\Sigma^n e_{n+k}\}$ is nonzero, then $\alpha$ splits off the top cell of the $N$-skeleton $\Sigma^n P_n^{n+k}$ of $\Sigma^n P_n^\infty$. Such operations $\alpha^*$ have traditionally been denoted $\cup_k \colon \pi_n(Y) \to \pi_{2n+k}(Y)$. Of course, $\alpha$ is not usually uniquely determined by its mod 2 Hurewicz image, and this can lead to some ambiguity in the meaning of $\cup_k$.

As shown by Liulevicius [**95**, Ch. 2] and May [**118**, §11], and already recalled in Section 1.3, the Steenrod squares in $\mathrm{Ext}_A(H^*(Y),\mathbb{F}_2)$ are induced by any choice of $\Sigma_2$-equivariant $A$-module chain map $\Delta\colon W_*\otimes F_* \to F_*\otimes F_*$ covering the coproduct $H^*(Y) \to H^*(Y)\otimes H^*(Y)$. Here $\Sigma_2$ acts freely upon $W_*$ and through the twist isomorphism on $F_*\otimes F_*$. Applying $\mathrm{Hom}_A(-,\mathbb{F}_2)$, we obtain a $\Sigma_2$-equivariant chain map $\Phi\colon W_*\otimes F_*^\vee \otimes F_*^\vee \to F_*^\vee$, graded so that we have homomorphisms

$$\Phi_{k,s_1,s_2}\colon W_k \otimes F_{s_1}^\vee \otimes F_{s_2}^\vee \longrightarrow F_{s_1+s_2-k}^\vee$$

for all $k, s_1, s_2 \ge 0$, compatible with the boundaries in $W_*$ and $F_*^\vee$. The Steenrod square

$$Sq^i\colon \mathrm{Ext}_A^{s,t}(H^*(Y),\mathbb{F}_2) \longrightarrow \mathrm{Ext}_A^{s+i,2t}(H^*(Y),\mathbb{F}_2)$$

is then defined by the formula $Sq^i([x]) = [\Phi_{s-i,s,s}(e_{s-i}\otimes x\otimes x)]$ for any cocycle $x \in F_s^\vee$. In the remainder of this subsection we will show how $\Phi$ and the $Sq^i$ admit a geometric realization, in terms of a filtration-preserving map $\Xi_\star\colon D_2(Y)_\star \to Y_\star$.

DEFINITION 11.8. Let

$$Z_{k,s} = S_+^k \wedge_{\Sigma_2} (Y\wedge Y)_s$$

for $k \ge 0$ and $s \ge 0$, and let $a\colon Y_{s+1} \to Y_s$, $b\colon Z_{k-1,s} \to Z_{k,s}$ and $c\colon Z_{k,s+1} \to Z_{k,s}$ denote the various inclusion maps associated to the Adams resolution $Y_\star$ of $Y$ and the bifiltration $Z_{\star,\star} = S_+^\star \wedge_{\Sigma_2} (Y\wedge Y)_\star$ of $D_2(Y) = S_+^\infty \wedge_{\Sigma_2} (Y\wedge Y)$. Note that $bc = cb$. Let

$$D_2(Y)_\ell = \bigcup_{s-k=\ell} Z_{k,s}$$

define the balanced convolution product $D_2(Y)_\star$ of the filtrations $S^\star$ and $(Y\wedge Y)_\star$.

PROPOSITION 11.9 ([**45**, Thm. IV.5.2]). *There are maps of orthogonal spectra*

$$\xi_{k,s}\colon Z_{k,s} = S^k_+ \wedge_{\Sigma_2} (Y \wedge Y)_s \longrightarrow Y_{s-k}$$

*for all* $k \geq 0$ *and* $s \geq 0$, *as well as "horizontal" homotopies*

$$H_{k,s}\colon a \circ \xi_{k,s} \simeq \xi_{k+1,s} \circ b$$

*of maps* $Z_{k,s} \to Y_{s-k-1}$ *and "vertical" homotopies*

$$V_{k,s+1}\colon a \circ \xi_{k,s+1} \simeq \xi_{k,s} \circ c$$

*of maps* $Z_{k,s+1} \to Y_{s-k}$. *Furthermore, these homotopies can be taken to be 2-categorically compatible, in the sense that there exists a 2-homotopy*

$$aV_{k,s+1} * H_{k,s}c \Longleftrightarrow aH_{k,s+1} * V_{k+1,s+1}b$$

*of maps* $Z_{k,s+1} \to Y_{s-k-1}$, *between the composite homotopies*

$$aV_{k,s+1} * H_{k,s}c\colon a^2 \circ \xi_{k,s+1} \simeq a \circ \xi_{k,s} \circ c \simeq \xi_{k+1,s} \circ bc$$

*and*

$$aH_{k,s+1} * V_{k+1,s+1}b\colon a^2 \circ \xi_{k,s+1} \simeq a \circ \xi_{k+1,s+1} \circ b \simeq \xi_{k+1,s} \circ cb\,.$$

*The maps* $\xi_{k,0}$ *are given by restriction of the* $H_\infty$ *structure map* $\xi_2\colon D_2(Y) \to Y$ *along* $S^k \subset S^\infty$. *The maps* $\xi_{0,s}$ *give a (weak) map* $(Y \wedge Y)_\star \to Y_\star$ *of Adams resolutions, lifting the ring spectrum product* $\phi\colon Y \wedge Y \to Y$.

$aV_{k,s+1} * H_{k,s}c$:

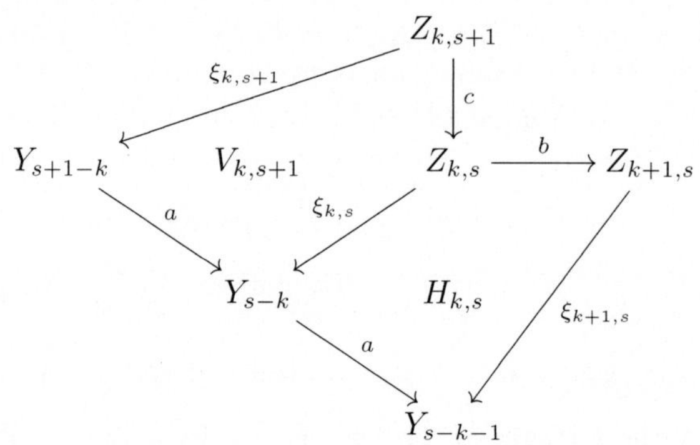


$aH_{k,s+1} * V_{k+1,s+1}b$:

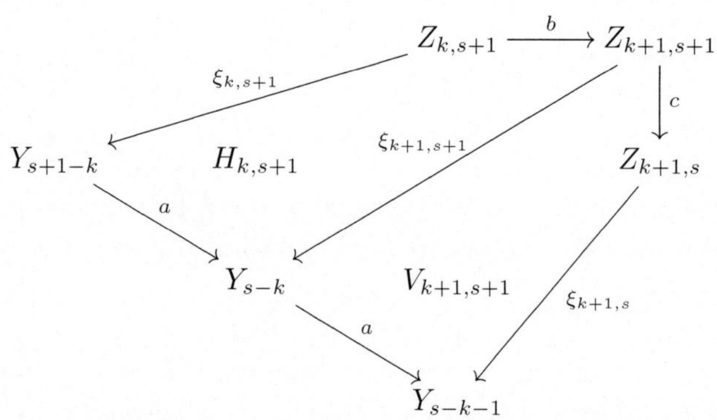

$$\begin{array}{ccc}
a^2 \circ \xi_{k,s+1} & \xrightarrow{aH_{k,s+1}} & a \circ \xi_{k+1,s+1} \circ b \\
{\scriptstyle aV_{k,s+1}} \downarrow & \Nearrow & \downarrow {\scriptstyle V_{k+1,s+1}b} \\
a \circ \xi_{k,s} \circ c & \xrightarrow{H_{k,s}c} & \xi_{k+1,s} \circ bc
\end{array}$$

PROOF. This is essentially the statement of Theorem IV.5.2 in [**45**], except for the assertion about 2-categorical compatibility of the commuting homotopies, which we will need when convolving the maps $\xi_{k,s}$. Fortunately, the proof given in that reference also justifies this slightly stronger statement, as we now outline.

The maps $\xi_{k,0}\colon S^k_+ \wedge_{\Sigma_2} (Y \wedge Y) \to Y_{-k} = Y$ and $\xi_{0,s}\colon S^0_+ \wedge_{\Sigma_2} (Y \wedge Y)_s \cong (Y \wedge Y)_s \to Y_s$ are given by restriction of $\xi_2\colon D_2(Y) \to Y$ and the (weak) map $(Y \wedge Y)_\star \to Y_\star$, as indicated. The horizontal homotopies $H_{k,0}$ are constant, but the vertical homotopies $V_{0,s+1}$ are generally not constant.

For $k \geq 0$ and $s \geq 0$ we inductively assume that $\xi_{k,s}$, $\xi_{k+1,s}$, $\xi_{k,s+1}$, $H_{k,s}$ and $V_{k,s+1}$ have been defined, and must construct $\xi_{k+1,s+1}$, $H_{k,s+1}$ and $V_{k+1,s+1}$, together with a commuting 2-homotopy. Contemplating the upper left hand square of mapping spaces in the diagram

$$\begin{array}{ccccc}
\mathrm{Map}(Z_{k+1,s+1}, Y_{s-k}) & \xrightarrow{b^*} & \mathrm{Map}(Z_{k,s+1}, Y_{s-k}) & & \\
{\scriptstyle a_*}\downarrow & & {\scriptstyle a_*}\downarrow & \overset{c^*}{\nwarrow} & \\
\mathrm{Map}(Z_{k+1,s+1}, Y_{s-k-1}) & \xrightarrow{b^*} & \mathrm{Map}(Z_{k,s+1}, Y_{s-k-1}) & & \mathrm{Map}(Z_{k,s}, Y_{s-k}) \\
& \overset{c^*}{\nwarrow} & & \overset{c^*}{\nwarrow} & \downarrow{\scriptstyle a_*} \\
& & \mathrm{Map}(Z_{k+1,s}, Y_{s-k-1}) & \xrightarrow{b^*} & \mathrm{Map}(Z_{k,s}, Y_{s-k-1})
\end{array}$$

we find that the obstruction to finding such data lies in

$$[\mathrm{cof}(Z_{k,s+1} \xrightarrow{b} Z_{k+1,s+1}), \mathrm{cof}(Y_{s-k} \xrightarrow{a} Y_{s-k-1})] \cong [Z_{k+1,s+1}/Z_{k,s+1}, Y_{s-k-1,1}] .$$

Furthermore, using that $c\colon Z_{k,s+1} \to Z_{k,s}$ is a $q$-cofibration and the fact that the stable model structure is topological, we find that the obstruction can be lifted over $c^*$ to come from $[Z_{k+1,s}/Z_{k,s}, Y_{s-k-1,1}]$. However, $c^*$ induces the zero homomorphism of obstruction groups, since the map $(Y \wedge Y)_{s+1} \to (Y \wedge Y)_s$ induces zero in cohomology. Hence the obstruction class vanishes, and we can construct $\xi_{k+1,s+1}$ and the required homotopies and 2-homotopy, as asserted. $\square$

The maps, homotopies and 2-homotopies of the previous proposition glue together to define a map

$$T = \mathrm{Tel}(S^\star)_+ \wedge_{\Sigma_2} \mathrm{Tel}((Y \wedge Y)_\star) \longrightarrow Y$$

from a double mapping telescope, where

$$\mathrm{Tel}(S^\star) = \bigcup_{k \geq 0} [k, k+1] \times S^k$$

is the mapping telescope of $S^0 \to S^1 \to S^2 \to \dots$ and

$$\mathrm{Tel}((Y \wedge Y)_\star) = \{0\}_+ \wedge (Y \wedge Y) \cup \bigcup_{s \geq 0} [s, s+1]_+ \wedge (Y \wedge Y)_{s+1}$$

is the mapping telescope of $\dots \to (Y \wedge Y)_2 \to (Y \wedge Y)_1 \to (Y \wedge Y)$. Here $\mathrm{Tel}(S^\star) \subset [0, \infty) \times S^\infty$ is filtered by letting $\mathrm{Tel}_k(S)$ be the part that meets $[0, k] \times S^\infty$, and $\mathrm{Tel}((Y \wedge Y)_\star) \subset [0, \infty)_+ \wedge (Y \wedge Y)$ is filtered by letting $\mathrm{Tel}_s(Y \wedge Y)$ be the part that meets $[s, \infty)_+ \wedge (Y \wedge Y)$. The double mapping telescope $T$ is filtered by setting

$$T_\ell = \bigcup_{s-k=\ell} \mathrm{Tel}_k(S)_+ \wedge_{\Sigma_2} \mathrm{Tel}_s(Y \wedge Y)\,,$$

and $T \to Y$ is then filtration-preserving in the sense that it maps $T_\ell$ to $Y_\ell$ for all integers $\ell$. The evident projections $\mathrm{Tel}_k(S) \to S^k$ and $\mathrm{Tel}_s(Y \wedge Y) \to (Y \wedge Y)_s$ are deformation retractions, and define a filtration-preserving equivalence $T \to D_2(Y)$. We obtain a zig-zag of filtration-preserving maps

$$\Xi_\star \colon D_2(Y)_\star \xleftarrow{\simeq} T_\star \longrightarrow Y_\star\,.$$

On each filtration quotient, this induces a zig-zag of maps

$$\bar{\Xi}_\ell \colon \frac{D_2(Y)_\ell}{D_2(Y)_{\ell+1}} \xleftarrow{\simeq} \frac{T_\ell}{T_{\ell+1}} \longrightarrow \frac{Y_\ell}{Y_{\ell+1}}$$

where

$$\frac{D_2(Y)_\ell}{D_2(Y)_{\ell+1}} \cong \bigvee_{s-k=\ell} \frac{S^k}{S^{k-1}} \wedge_{\Sigma_2} \bigvee_{s_1+s_2=s} \frac{Y_{s_1}}{Y_{s_1+1}} \wedge \frac{Y_{s_2}}{Y_{s_2+1}}\,.$$

Let $\bar{\Xi}_{k,s_1,s_2} \colon S^k/S^{k-1} \wedge Y_{s_1,1} \wedge Y_{s_2,2} \to Y_{s_1+s_2-k,1}$ denote the (weakly defined) components of $\bar{\Xi}_\ell$. Passing to cohomology, we get the components

$$\Delta_{k,\ell} \colon W_k \otimes F_\ell \longrightarrow \bigoplus_{k+\ell=s_1+s_2} F_{s_1} \otimes F_{s_2}$$

of a $\Sigma_2$-equivariant $A$-module chain map $\Delta$, as required for the definition of the Steenrod squares. The components of the dual $\Sigma_2$-equivariant chain map $\Phi \colon W_* \otimes F_*^\vee \otimes F_*^\vee \to F_*^\vee$ can therefore be calculated as the composites

(11.2) $\Phi_{k,s_1,s_2} \colon W_k \otimes F_{s_1,t_1}^\vee \otimes F_{s_2,t_2}^\vee$

$$\begin{aligned}
&\cong \mathrm{Hom}_A(H^*(S^k/S^{k-1}), \Sigma^k \mathbb{F}_2) \otimes \bigotimes_{i=1}^{2} \mathrm{Hom}_A(H^*(\Sigma^{s_i} Y_{s_i,1}), \Sigma^{t_i} \mathbb{F}_2) \\
&\xrightarrow{\otimes} \mathrm{Hom}_A(H^*(S^k/S^{k-1}) \otimes H^*(\Sigma^{s_1} Y_{s_1,1}) \otimes H^*(\Sigma^{s_2} Y_{s_2,1}), \Sigma^{k+t_1+t_2} \mathbb{F}_2) \\
&\xleftarrow{\cong} \mathrm{Hom}_A(H^*(S^k/S^{k-1} \wedge \Sigma^{s_1} Y_{s_1,1} \wedge \Sigma^{s_2} Y_{s_2,1}), \Sigma^{k+t} \mathbb{F}_2) \\
&\xrightarrow{(\bar{\Xi}^*_{k,s_1,s_2})^\vee} \mathrm{Hom}_A(H^*(\Sigma^s Y_{s-k,1}), \Sigma^{k+t} \mathbb{F}_2) = F_{s-k,t}^\vee\,.
\end{aligned}$$

Here $s = s_1 + s_2$ and $t = t_1 + t_2$, the first isomorphism uses the identification $W_k \otimes \mathbb{F}_2 \cong \mathrm{Hom}_A(H^*(S^k/S^{k-1}), \Sigma^k \mathbb{F}_2)$, the second homomorphism tensors together $A$-module homomorphisms, the next isomorphism is given by the Künneth theorem, and the final homomorphism is geometrically induced by $\bar{\Xi}_{k,s_1,s_2}$. This is a cohomological reformulation of Corollary IV.5.3 in [**45**].

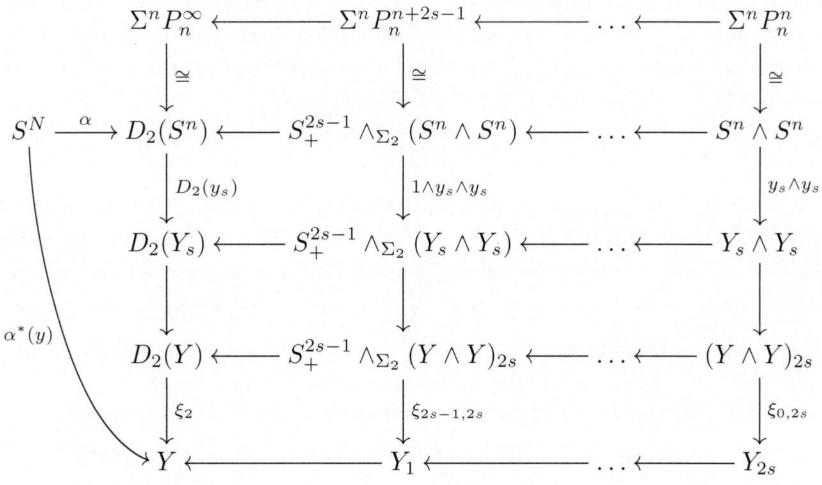

FIGURE 11.1. Factorization of power operation $\alpha^*(y)$

**11.1.3. A delayed Adams spectral sequence.** Our next goal is the detection result, Theorem 11.13, for power operations in the homotopy of $H_\infty$ ring spectra. Let $y \in \pi_n(Y)$ have Adams filtration $s$, so that it factors as $S^n \xrightarrow{y_s} Y_s \to Y$ and is detected by the class $x \in E_2(Y)$ of the cocycle in $F_s^\vee$ corresponding to the composite $S^n \xrightarrow{y_s} Y_s \to Y_{s,1}$. Let $\alpha \in \pi_N D_2(S^n)$. Using the maps $\xi_{k,s}$ from Proposition 11.9 we can piece together the homotopy commutative diagram shown in Figure 11.1. In particular, we have the following (weak) map of towers.

(11.3)

$$\begin{array}{ccccccccccccc} S^N & \xrightarrow{\alpha} & \Sigma^n P_n^\infty & \longleftarrow & \Sigma^n P_n^{n+2s-1} & \longleftarrow & \dots & \longleftarrow & \Sigma^n P_n^{n+k} & \longleftarrow \dots \longleftarrow & \Sigma^n P_n^n & \longleftarrow & * \\ & \searrow^{\alpha^*(y)} & \downarrow{\scriptstyle \xi_2 D_2(y)} & & \downarrow & & & & \downarrow & & \downarrow & & \downarrow \\ & & Y & \longleftarrow & Y_1 & \longleftarrow & \dots & \longleftarrow & Y_{2s-k} & \longleftarrow \dots \longleftarrow & Y_{2s} & \longleftarrow & Y_{2s+1} \end{array}$$

The Adams filtration of the composite $\alpha^*(y)$ will depend on a mixture of the cellular filtration and the Adams filtrations of the compressions of $\alpha$ through the skeleta of $D_2(S^n) \cong \Sigma^n P_n^\infty$. This can be neatly handled by the following construction.

DEFINITION 11.10. Let $Z_\star$ be a tower

$$Z = Z_0 \longleftarrow Z_1 \longleftarrow Z_2 \longleftarrow \dots$$

of orthogonal spectra. The delayed mod $p$ Adams spectral sequence for $Z_\star$ is the spectral sequence

$$E_1^{s,t}(Z_\star) = \pi_{t-s}((S \wedge Z)_{s,1}) \Longrightarrow \pi_{t-s}(Z_p^\wedge)$$

obtained by applying $\pi_*(-)$ to the convolution product

$$(S \wedge Z)_0 \longleftarrow (S \wedge Z)_1 \longleftarrow (S \wedge Z)_2 \longleftarrow \dots$$

of a mod $p$ Adams resolution $S = S_0 \leftarrow S_1 \leftarrow S_2 \leftarrow$ of the sphere spectrum with the tower $Z_\star$.

We may and will assume that each $S_i$ and $Z_j$ is $q$-cofibrant, and that each map $S_{i+1} \to S_i$ and $Z_{j+1} \to Z_j$ is a $q$-cofibration. We shall also assume that each $S_i$ is connective with $H^*(S_i)$ of finite type. The convolution product is then given in filtration $k$ by

$$(S \wedge Z)_k = \bigcup_{i+j=k} S_i \wedge Z_j \,,$$

and $(S \wedge Z)_{k,1} \simeq \bigvee_{i+j=k} S_{i,1} \wedge Z_{j,1}$. The delayed Adams spectral sequence is evidently natural in the tower $Z_\star$. Its name is meant to suggest that the resolution of $Z_k$ relative to $Z_{k+1}$ is delayed until Adams filtration $k$, cf. case (1) of the following theorem.

THEOREM 11.11. *Suppose that each $Z_k$ is bounded below, with $H^*(Z_k)$ of finite type.*

*(1) If $H^*(Z_k) \to H^*(Z_{k+1})$ is an epimorphism for each $k$, then*

$$E_2^{s,t}(Z_\star) = \bigoplus_{k \geq 0} \mathrm{Ext}_A^{s-k,t-k}(H^*(Z_k/Z_{k+1}), \mathbb{F}_p)\,.$$

*Furthermore, if $H^*(\mathrm{holim}_k Z_k) = 0$, then the delayed Adams spectral sequence converges conditionally and strongly to $\pi_*(Z_p^\wedge)$.*

*(2) If $H^*(Z_k) \to H^*(Z_{k+1})$ is zero for each $k$, then*

$$E_2^{s,t}(Z_\star) = \mathrm{Ext}_A^{s,t}(H^*(Z), \mathbb{F}_p)\,,$$

*and the delayed Adams spectral sequence for $Z_\star$ is equal to the ordinary Adams spectral sequence for $Z$ from the $E_2$-term and onward. In particular, it converges conditionally and strongly to $\pi_*(Z_p^\wedge)$.*

PROOF. This is a cohomological reformulation of [**39**, Thm. 5], where related results and their proofs can also be found.

In case (1) we offer the following variant of the convergence proof given in [**45**, Thm. IV.6.1]. Fix an integer $n_1$ so that $\pi_*(Z/p) = 0$ for $* < n_1$. Then $H_*(Z_k/p)$ is a quotient of $H_*(Z/p) = 0$ for $* < n_1$, so the $Z_k/p$ are uniformly $n_1$-connective. The vanishing of $H^*(\mathrm{holim}_k Z_k)$ can be rewritten as $H_*(\mathrm{holim}_k Z_k) = 0$. By Adams' Theorem 15.2 of [**9**, Part III], it follows that $\lim_k$ and $\mathrm{Rlim}_k$ of $H_*(Z_k)$ are both zero.

We have a $k$-indexed tower of short exact sequences

$$0 \to H_*(Z_k) \longrightarrow H_*((S \wedge Z)_k) \longrightarrow H_*((S \wedge Z)_k/Z_k) \to 0$$

and the bonding maps in the right hand tower are zero. Hence $\lim_k$ and $\mathrm{Rlim}_k$ for that tower are both zero. By the lim-Rlim exact sequence, $\lim_k$ and $\mathrm{Rlim}_k$ of $H_*((S \wedge Z)_k)$ are therefore both zero.

The spectra $(S \wedge Z)_k/p = (S \wedge Z/p)_k$ are uniformly $n_1$-connective. Hence $\mathrm{holim}_k(S \wedge Z)_k/p$ is bounded below, and $H_*(\mathrm{holim}_k(S \wedge Z)_k) = 0$ by another application of Adams' theorem. By the Hurewicz theorem, $\mathrm{holim}_k(S \wedge Z)_k/p$ is trivial. By induction on $n$, it follows that $\mathrm{holim}_k(S \wedge Z)_k/p^n$ is trivial for each $n$. Passing to the homotopy limit over $n$, we deduce that $\mathrm{holim}_k((S \wedge Z)_k)_p^\wedge$ is trivial. Therefore the homotopy spectral sequence associated to the $p$-completed tower

$$((S \wedge Z)_0)_p^\wedge \longleftarrow ((S \wedge Z)_1)_p^\wedge \longleftarrow ((S \wedge Z)_2)_p^\wedge \longleftarrow \dots$$

is conditionally convergent in the sense of Michael Boardman [**29**, Def. 5.10], with abutment $\pi_*(Z_p^\wedge)$. The completion map induces an isomorphism from the homotopy spectral sequence associated to the tower

$$(S \wedge Z)_0 \longleftarrow (S \wedge Z)_1 \longleftarrow (S \wedge Z)_2 \longleftarrow \dots$$

to the conditionally convergent one. This is the delayed mod $p$ Adams spectral sequence associated to $Z_\star$. Since the delayed Adams $E_2$-term is finite in each degree, we know that $RE_\infty(Z_\star) = 0$. Hence the spectral sequences are strongly convergent to $\pi_*(Z_p^\wedge)$, by [**29**, Thm. 7.3].

In case (2), the convolved tower $(S \wedge Z)_\star$ is itself an Adams resolution, so the delayed spectral sequence is an instance of the ordinary Adams spectral sequence, and has the usual convergence properties. □

REMARK 11.12. (1) The construction specializes to the ordinary Adams spectral sequence when the tower is "trivial": $Z_k = *$ for $k > 0$.

(2) The construction also specializes to the ordinary Adams spectral sequence when $Z_\star$ is itself an Adams resolution of $Z$.

(3) The vanishing of the homotopy limit (or microscope) $\operatorname{holim}_k Z_k$ is trivially satisfied if the tower is of finite length, with $Z_k = *$ for all sufficiently large $k$. This situation is adequate for our needs.

(4) A case of the delayed Adams spectral sequence was constructed in an ad hoc manner by Milgram in [**122**, Lem. 5.3.1], and used in the same way that we will use it.

(5) These results are phrased in terms of homology and proved for $E$-based Adams–Novikov spectral sequences in [**39**]. This reference also considers the spectral sequence obtained by applying $[X, -]_*$ in place of $\pi_*(-)$, for a fixed spectrum $X$.

(6) The paper [**39**] also treats a dual version of the theorem, in which one uses function spectra of maps from a direct sequence rather than smash products with an inverse sequence. This dual version was used by Adams (unpublished) to construct a spectral sequence $\operatorname{Ext}_{E_*E}(E_*X, E_*Y) \Longrightarrow [X, Y_E^\wedge]_*$, without the usual assumption that $E_*X$ be $\pi_*E$-projective. It was also used by Ravenel in his proof of the Segal conjecture for $C_{p^n}$, cf. [**143**, Def. 2.12], where he referred to it as the "modified Adams spectral sequence".

**11.1.4. Detection of power operations by Steenrod squares.** Let $Y$ be an $H_\infty$ ring spectrum, bounded below and with $H^*(Y)$ of finite type. For notational simplicity assume that $Y$ is $p$-complete, and that $p = 2$.

THEOREM 11.13. *Suppose that $y \in \pi_n(Y)$ is detected by $x \in \operatorname{Ext}_A^{s,t}(H^*(Y), \mathbb{F}_2)$, where $t = s + n$. Let $P_\star$ be the tower*

$$\Sigma^n P_n^\infty \longleftarrow \Sigma^n P_n^{n+2s-1} \longleftarrow \cdots \longleftarrow \Sigma^n P_n^{n+k} \longleftarrow \cdots \longleftarrow \Sigma^n P_n^n \longleftarrow *$$

*of stunted projective spaces, mapping as in (11.3) to an Adams resolution of $Y$, thereby inducing a morphism $E_r(P_\star) \to E_r(Y)$ of spectral sequences.*

*(1) The $E_2$-term $E_2(P_\star)$ of the delayed Adams spectral sequence for $P_\star$ is the direct sum of*

- *a free $\operatorname{Ext}_A(\mathbb{F}_2, \mathbb{F}_2)$-module on generators $\{\cup_0, \dots, \cup_{2s-1}\}$, and*
- *a copy of $\operatorname{Ext}_A(H^*(\Sigma^n P_{n+2s}^\infty), \mathbb{F}_2)$ with lowest degree class $\cup_{2s}$.*

*(2) For $0 \le k \le 2s$, the class $\cup_k$ lies in $E_2^{2s-k,2t}(P_\star)$ and maps to the class $Sq^{s-k}(x) \in E_2^{2s-k,2t}(Y)$.*

*(3) If $\alpha \in \pi_N(\Sigma^n P_n^\infty)$ is detected by $\sum_{k=0}^{2s} a_k \cup_k$ in $E_2(P_\star)$, with each $a_k \in \operatorname{Ext}_A(\mathbb{F}_2, \mathbb{F}_2)$, then $\alpha^*(y) \in \pi_N(Y)$ is weakly detected by $\sum_{k=0}^{2s} a_k Sq^{s-k}(x)$.*

PROOF. This is the content of Proposition 7.5, Theorem 7.6 and Corollary 7.7 of [**45**, §IV.7], translated to the case of ordinary mod 2 cohomology. The proof of Theorem 7.6 relies on the geometric description in (11.2) of the Steenrod squares in $E_2(Y)$. □

REMARK 11.14. (1) Saying that $y' \in \pi_*(Y)$ is "weakly detected" by $x' \in E_\infty^{s,*}(Y)$ means that $y'$ lifts to Adams filtration $s$ or higher, and if $x' \neq 0$ then it corresponds to $y'$ in $F^s\pi_*(Y)/F^{s+1}\pi_*(Y) \cong E_\infty^{s,*}(Y)$, while if $x' = 0$ then $y'$ lifts to Adams filtration $s+1$ or higher.

(2) For $\alpha$ detected in homology by $\Sigma^n e_{n+k}$, as in Remark 11.7, this theorem shows that $\alpha^* = \cup_k$ is detected in the Adams spectral sequence by the Steenrod operation $Sq^{s-k}$ acting on $\mathrm{Ext}_A^{s,t}(H^*(Y), \mathbb{F}_2)$.

(3) There are occasional homotopy classes of Adams filtration 0 and topological degree greater than $2t = 2s+2n$ detected in the summand $\mathrm{Ext}_A(H^*(\Sigma^n P_{n+2s}^\infty), \mathbb{F}_2)$. These can be considered to be instances of $\cup_k$, for $k > 2s$.

(4) There are also elements in the summand $\mathrm{Ext}_A(H^*(\Sigma^n P_{n+2s}^\infty), \mathbb{F}_2)$ that are not sums of classes of the form $a_k \cup_k$. In order to analyze their effect on the class $x$ we would need to express them in terms of the Atiyah–Hirzebruch spectral sequence for computing $\mathrm{Ext}_A(H^*(\Sigma^n P_{n+2s}^\infty), \mathbb{F}_2)$ that arises from filtering $H^*(\Sigma^n P_{n+2s}^\infty)$ by degree.

(5) The classes $\cup_k$ for $k > s$ always map to 0 in their bidegree of the Adams spectral sequence for $Y$, since $Sq^{s-k}(x) = 0$ in these cases. They do not necessarily map to 0 in homotopy; they simply map to classes of higher Adams filtration.

The preceding theorem does not encompass all the information that is available in the spectral sequence for the tower $P_\star$. In particular, we have the following consequence of the naturality of the delayed Adams spectral sequence.

COROLLARY 11.15. *Differentials and hidden extensions in the spectral sequence* $(E_r(P_\star), d_r)$ *for* $\Sigma^n P_n^\infty$ *map to differentials and hidden extensions in the Adams spectral sequence* $(E_r(Y), d_r)$ *for* $Y$. □

The first example of this is in the analysis of operations on a class of odd degree. If $y \in \pi_n(Y)$ with $n$ odd, then it is well known and elementary that $2y^2 = 0$. The preceding map of spectral sequences shows more.

PROPOSITION 11.16. *If* $y \in \pi_n(Y)$ *is an odd degree class detected by* $x$ *in Adams filtration* $s \geq 1$, *then* $d_2(Sq^{s-1}(x)) = h_0 x^2$ *and* $2y^2 = 0$. *There is a class* $y_1 \in \pi_{2n+2}(Y)$ *that is weakly detected by* $h_1 Sq^{s-1}(x)$ *and which satisfies* $2y_1 = \eta^2 y^2$. *This extension is hidden:* $h_0(h_1 Sq^{s-1}(x)) = 0$.

PROOF. The truncated tower $Z_\star$ with

$$\Sigma^n P_n^{n+1} \longleftarrow \Sigma^n P_n^n \longleftarrow *$$

in filtrations $2s-1$ through $2s+1$, extended by identity maps on either side, maps to the tower $P_\star$ of Theorem 11.13, which in turn maps as in (11.3) to the Adams resolution of $Y$. The delayed Adams spectral sequence for this truncated tower converges to the homotopy of the mod 2 Moore spectrum $\Sigma^n P_n^{n+1} \cong S^{2n} \cup_2 e^{2n+1}$. Its $E_2$-term is free over $\mathrm{Ext}_A(\mathbb{F}_2, \mathbb{F}_2)$ on classes $\cup_0$ and $\cup_1$ in Adams bidegrees $(2n, 2s)$ and $(2n+1, 2s-1)$, respectively, which map to $Sq^s(x)$ and $Sq^{s-1}(x)$ in $E_2(Y)$ by Theorem 11.13.

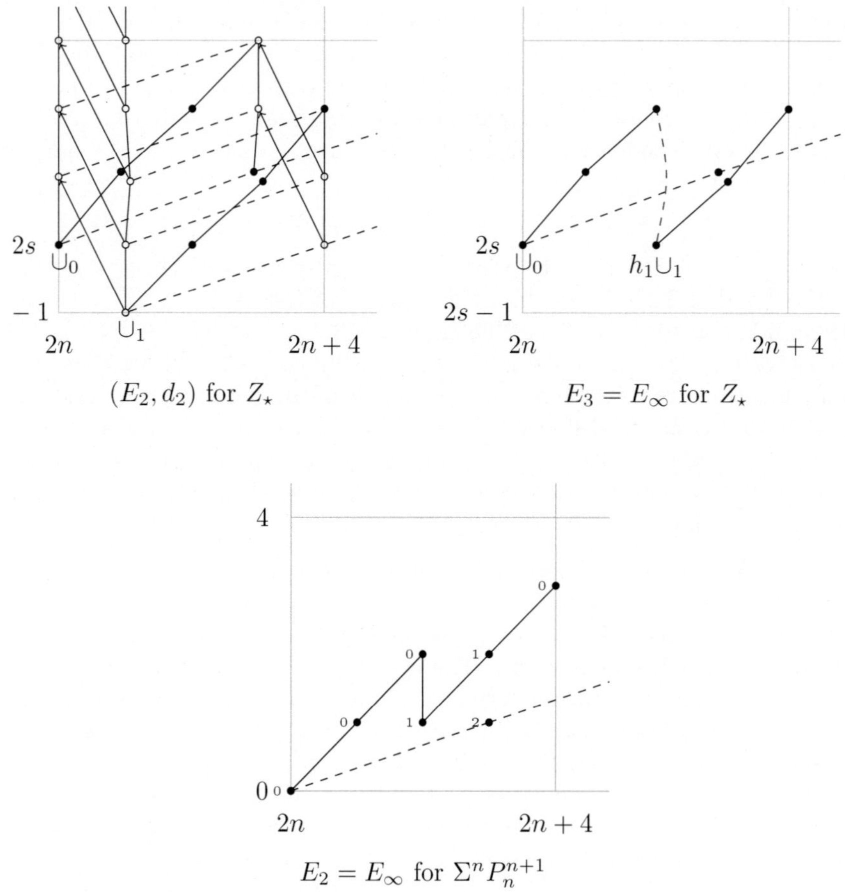


FIGURE 11.2. Delayed and ordinary Adams spectral sequences for $\pi_*(\Sigma^n P_n^{n+1})$, for $n$ odd

The ordinary Adams spectral sequence for the mod 2 Moore spectrum shows that $\pi_{2n}(\Sigma^n P_n^{n+1}) \cong \mathbb{Z}/2$ and $\pi_{2n+2}(\Sigma^n P_n^{n+1}) \cong \mathbb{Z}/4$. Since the spectral sequence for $Z_\star$ converges to the same abutment, we must have $d_2(\cup_1) = h_0\cup_0$ in $E_2(Z_\star)$. This differential propagates $h_0$- and $h_2$-linearly, and by sparsity the resulting $E_3$-term must be equal to the $E_\infty$-term in degrees less than $2n+6$. Furthermore, there must be a hidden 2-extension from $h_1\cup_1$ to $h_1^2\cup_0$. The relevant terms of these spectral sequences are shown in Figure 11.2. It follows by naturality that $d_2(Sq^{s-1}(x)) = h_0 Sq^s(x) = h_0 x^2$ in the Adams spectral sequence for $Y$.

The generator $\alpha$ of $\pi_{2n}(\Sigma^n P_n^{n+1})$ has order 2, and represents the operation $\alpha^*$ sending $y$ to $y^2$, so $2y^2 = 0$. A generator $\alpha_1$ of $\pi_{2n+2}(\Sigma^n P_n^{n+1})$ is detected by $h_1\cup_1$, hence represents an operation $\alpha_1^*$ sending $y$ to a class $y_1 \in \pi_{2n+2}(Y)$ that is weakly detected by $h_1 Sq^{s-1}(x)$. Since $2\alpha_1 = \eta^2\alpha$, it follows by naturality that $2y_1 = \eta^2 y^2$. Hence, if $h_1 Sq^{s-1}(x) \neq 0$ and $h_1^2 x^2 \neq 0$, then there is a hidden 2-extension from $h_1 Sq^{s-1}(x)$ to $h_1^2 x^2$ in the Adams spectral sequence for $Y$. □

REMARK 11.17. If we refine our hypotheses, we can be more precise. If $n \equiv 3 \mod 4$ then $\eta y^2 = 0$ and $2y_1 = 0$, while if $n \equiv 1 \mod 4$ then $y_1$ is itself divisible by 2: there is a class $y_2 = \cup_2(y) \in \pi_{2n+2}(Y)$, weakly detected by $Sq^{s-2}(x)$, such that $2y_2 = y_1$ and $\eta^2 y^2 = 4y_2$. These relations are computed by comparing the ordinary Adams spectral sequence for $\Sigma^n P_n^\infty$ to the delayed Adams spectral sequence for the tower $Z'_\star$ with

$$\Sigma^n P_n^{n+3} \longleftarrow \Sigma^n P_n^{n+2} \longleftarrow \Sigma^n P_n^{n+1} \longleftarrow \Sigma^n P_n^n \longleftarrow *$$

in filtrations $2s-3$ through $2s+1$, extended by identity maps on either side. See [**45**, Fig. V.3.2 and V.3.4]. There are evident maps of towers $Z_\star \to Z'_\star \to P_\star$, in the notation of Proposition 11.16 and Theorem 11.13.

The dashed line extending outside the boxes in Figure 11.2 indicates that the class $h_2^2 \cup_0$ remains nonzero in the spectral sequences of that figure. This class, and also classes shown within the displayed figure, may well map to zero or acquire new divisors as we add higher cells to the stunted projective spaces. The additional relations when $n \equiv 3 \mod 4$ and the additional class $y_2$ when $n \equiv 1 \mod 4$ are typical examples of this.

**11.1.5. Differentials on Steenrod squares.** To examine the implications of the $H_\infty$ ring structure on $Y$ for classes $x \in E_2(Y)$ that are not permanent cycles, it is simplest to focus on the consequences for Adams differentials on the $Sq^i(x)$. There are two kinds of contributions to them. The first comes from Steenrod operations on the boundary, giving terms of the form $Sq^{i+r-1}(d_r(x))$. The second comes from the geometry of the extended powers, and gives terms similar to those we saw in the discussion of homotopy operations above. These are of the form $\bar{a}\, x\, d_r(x)$ and $\bar{a}\, Sq^{i+v}(x)$, where $\bar{a}$ is a permanent cycle in the Adams spectral sequence for $S$. A partial statement of results, adapted to the case $Y = tmf$, was given in Section 5.2. The general statements are as follows.

DEFINITION 11.18 ([**45**, Def. V.2.15]). For $n \geq 0$ let $v = v(n)$ denote the "vector field number", i.e., the maximal number $v$ such that the attaching map of the $n$-cell in $P^n$ factors up to homotopy as

$$S^{n-1} \xrightarrow{\alpha} P^{n-v} \subset P^{n-1} .$$

Let $a = a(n) \in \pi_{v-1}(S)$ denote the top component

$$S^{n-1} \xrightarrow{\alpha} P^{n-v} \longrightarrow S^{n-v}$$

of a maximal compression. Let $\bar{a} \in E_\infty^{f,f+v-1}(S)$ be the infinite cycle that detects $a$ in the mod 2 Adams spectral sequence for $S$. Here $f$ is the Adams filtration of $a$.

REMARK 11.19. Strictly speaking, to be appropriate for small (or negative) values of $n$ this compression problem should be interpreted as taking place in the stunted projective spectrum $P^n_{-\infty}$. For $n = 2^i - 1$ with $i \in \{0, 1, 2, 3\}$ the attaching map $S^{n-1} \to P^{n-1}_{-\infty}$ factors through $P^{-1}_{-\infty}$, so that $v(n) = n + 1 = 2^i$ and $\bar{a} = h_i$. For all other positive $n$ the attaching map does not compress below $P^1_{-\infty}$, hence can equally well be studied at the space level.

Adams' solution of the vector-field problem for spheres [**5**] leads to the following formulas.

PROPOSITION 11.20 ([**45**, Prop. V.2.16 and V.2.17]). *Let the 2-adic valuation of $n+1$ be $4q+r$, with $0 \leq r \leq 3$. Then $v = v(n) = 8q + 2^r$.*

*If $n$ is even, then $v = 1$, $a = 2$ and $\bar{a} = h_0$. If $n$ is odd, then $v \ge 2$ and $a$ generates the image of the $J$-homomorphism in $\pi_{v-1}(S)^\wedge_2$. In particular,*

(1) *if* $n \equiv 1 \mod 4$ *then* $v = 2$, $a = \eta$ *and* $\bar{a} = h_1$,
(2) *if* $n \equiv 3 \mod 8$ *then* $v = 4$, $a \equiv \nu \mod 2\nu$ *and* $\bar{a} = h_2$,
(3) *if* $n \equiv 7 \mod 16$ *then* $v = 8$, $a \equiv \sigma \mod 2\sigma$ *and* $\bar{a} = h_3$,
(4) *if* $n \equiv 15 \mod 32$ *then* $v = 9$, $a = \eta\sigma$ *and* $\bar{a} = h_1 h_3$, *and*
(5) *if* $n \equiv 31 \mod 64$ *then* $v = 10$, $a = \eta^2\sigma$ *and* $\bar{a} = h_1^2 h_3$.

DEFINITION 11.21. Let $A \in E_2^{s,t}$, $B_1 \in E_2^{s+r_1,t+r_1-1}$ and $B_2 \in E_2^{s+r_2,t+r_2-1}$ be classes in a spectral sequence with differentials $d_r \colon E_r^{s,t} \to E_r^{s+r,t+r-1}$. The notation

$$d_*(A) = B_1 \dotplus B_2$$

means that $d_r(A) = 0$ for $2 \le r < \min\{r_1, r_2\}$, while

$$\begin{cases} d_{r_1}(A) = B_1 & \text{if } r_1 < r_2, \\ d_r(A) = B_1 + B_2 & \text{if } r_1 = r = r_2, \text{ and} \\ d_{r_2}(A) = B_2 & \text{if } r_1 > r_2. \end{cases}$$

THEOREM 11.22 ([**45**, Thm. VI.1.1 and VI.1.2]). *Let $E_r^{*,*}(Y)$ be the mod 2 Adams spectral sequence for an $H_\infty$ ring spectrum $Y$, and let $x \in E_2^{s,t}(Y)$ be an element that survives to the $E_r$-term, where $r \ge 2$. Let $0 \le i \le s$, and let $v = v(t-i)$, $a = a(t-i)$ and $\bar{a}$ be as just defined. Then*

$$d_*(Sq^i(x)) = Sq^{i+r-1}(d_r(x)) \dotplus \begin{cases} 0 & \textit{if } v > s-i+1, \\ \bar{a}\, x\, d_r(x) & \textit{if } v = s-i+1, \\ \bar{a}\, Sq^{i+v}(x) & \textit{if } v = s-i \textit{ or } v \le \min\{s-i, 10\}. \end{cases}$$

REMARK 11.23. If $r_1 < r_2$ and $B_1 = 0$, then $B_1 \dotplus B_2$ denotes the zero element in filtration $s + r_1$. In this case the theorem does not give information about $d_r(Sq^i(x))$ for $r > r_1$. Similar remarks apply if $r_1 > r_2$ and $B_2 = 0$. However, in the (first) case $v > s - i + 1$ of the theorem the summand $B_2 = 0$ should be interpreted as lying in arbitrarily high Adams filtration $s + r_2$, so that

$$d_{2r-1}(Sq^i(x)) = Sq^{i+r-1}(d_r(x)) .$$

We note the following special case.

COROLLARY 11.24. *With the notation of Theorem 11.22, let $n = t - s$ be the topological degree of $x$. If $n$ is odd then*

$$d_{2r-1}(x^2) = Sq^{s+r-1}(d_r(x)) ,$$

*while if $n$ is even and $r = 2$ then*

$$d_3(x^2) = Sq^{s+1}(d_2(x)) + h_0 x d_2(x) ,$$

*and if $n$ is even and $r > 2$ then*

$$d_{r+1}(x^2) = h_0 x d_r(x) .$$

PROOF. This is the $i = s$ case of Theorem 11.22. We then have $v = v(t-i) = v(n)$, so that $v > s - i + 1 = 1$ if $n$ is odd, giving the first case. If $n$ is even, then $v = s - i + 1 = 1$, so that

$$d_*(x^2) = Sq^{s+r-1}(d_r(x)) \dotplus h_0 x d_r(x) .$$

If $r = 2$, both terms are in filtration $2s + 3$, proving the second case, and if $r > 2$, the second term has the lower filtration, proving the final case. □

The analysis required to prove Theorem 11.22 is too involved to recount completely here, but we can give a quick overview as follows. An element $x$ of $E_r^{s,s+n}(Y)$ can be represented by a map $(y, \partial y)\colon (D^n, S^{n-1}) \to (Y_s, Y_{s+r})$. We define a $\Sigma_2$-equivariant filtration

$$S^{n-1} \wedge S^{n-1} = \Gamma_2 \subset \Gamma_1 \subset \Gamma_0 = D^n \wedge D^n$$

by letting $\Gamma_1 = S^{n-1} \wedge D^n \cup D^n \wedge S^{n-1}$. The homotopy orbits $S^k_+ \wedge_{\Sigma_2} \Gamma_i$ were analyzed in [**45**, §VI.2 and §VI.3]. In particular, we have the following identifications.

PROPOSITION 11.25.

(1) $\Gamma_0$ *is the cone* $C\Gamma_1$, *and* $S^k_+ \wedge_{\Sigma_2} \Gamma_0 \cong C(S^k_+ \wedge_{\Sigma_2} \Gamma_1)$.
(2) $S^k_+ \wedge_{\Sigma_2} (\Gamma_0/\Gamma_1) \cong \Sigma^n P_n^{n+k}$.
(3) $S^k_+ \wedge_{\Sigma_2} \Gamma_1 \simeq \Sigma^{n-1} P_n^{n+k}$.
(4) $S^k_+ \wedge_{\Sigma_2} \Gamma_2 \cong \Sigma^{n-1} P_{n-1}^{n-1+k}$.
(5) *The inclusion* $S^k_+ \wedge_{\Sigma_2} \Gamma_2 \longrightarrow S^k_+ \wedge_{\Sigma_2} \Gamma_1$ *is homotopic to the map*

$$\pi\colon \Sigma^{n-1} P_{n-1}^{n-1+k} \longrightarrow \Sigma^{n-1} P_n^{n-1+k}$$

*collapsing the bottom cell, followed by the inclusion*

$$\iota\colon \Sigma^{n-1} P_n^{n-1+k} \longrightarrow \Sigma^{n-1} P_n^{n+k} .$$

The maps $y$ and $\partial y$, and $\xi_{k,2s+ir}$ from Proposition 11.9, induce maps

$$S^k_+ \wedge_{\Sigma_2} \Gamma_i \longrightarrow S^k_+ \wedge_{\Sigma_2} (Y \wedge Y)_{2s+ir} \xrightarrow{\xi_{k,2s+ir}} Y_{2s+ir-k}$$

that are compatible as $i \in \{0, 1, 2\}$ and $k \geq 0$ vary, up to 2-coherent homotopy, with the maps in the Adams resolution of $Y$. Using these, we can show ([**45**, Lem. VI.4.2]) that the Steenrod operation $Sq^{s-k}(x)$ is represented by the induced map of pairs

$$(D^{2n+k}, S^{2n+k-1}) \simeq (S^k_+ \wedge_{\Sigma_2} \Gamma_0, S^{k-1}_+ \wedge_{\Sigma_2} \Gamma_0 \cup S^k_+ \wedge_{\Sigma_2} \Gamma_1) \longrightarrow (Y_{2s-k}, Y_{2s-k+1}) .$$

Let $i = s - k$. The Adams differential on $Sq^i(x)$ is then obtained by lifting the boundary map $S^{2n+k-1} \to S^{k-1}_+ \wedge_{\Sigma_2} \Gamma_0 \cup S^k_+ \wedge_{\Sigma_2} \Gamma_1 \to Y_{2s-k+1}$ into as high an Adams filtration as is possible. We do this by decomposing the boundary sphere into two hemispheres, which we analyze separately.

In Figure 11.3, $v = v(n + k) = v(t - i)$ and $\alpha$ is the maximally compressed attaching map of the top $(n - 1) + (n + k)$ cell of $S^{k+1}_+ \wedge_{\Sigma_2} \Gamma_2 \cong \Sigma^{n-1} P_{n-1}^{n+k}$. Since the top quadrangle gives the characteristic map of this cell, it maps to $Sq^{s+r-k-1}(d_r(x)) = Sq^{i+r-1}(d_r(x))$, no matter what values $v$ and $\alpha$ take on.

In the left hand quadrangle, since $\Gamma_0$ is the cone on $\Gamma_1$, and $S^{k-v}_+ \wedge_{\Sigma_2} \Gamma_1$ is $S^{k-v+1}_+ \wedge_{\Sigma_2} \Gamma_2$ modulo its bottom cell, we have $\bar{a}$ times the map carried by the top $(n - 1) + (n + k - v)$ cell of $S^{k-v+1}_+ \wedge_{\Sigma_2} \Gamma_2$. If $v \leq k$ this top cell maps by an equivalence to the top cell of $S^{k-v}_+ \wedge_{\Sigma_2} \Gamma_1$, giving $\bar{a}Sq^{i+v}(x)$, modulo "components" of $\alpha$ supported on cells below the $(n-1)+(n+k-v)$ cell. (This is made precise by the spectral sequence of Theorem 11.13, and accounts for the restriction to $v \leq 10$ in this case.) When $v = k+1$ this is the bottom cell of $S^{k-v+1}_+ \wedge_{\Sigma_2} \Gamma_2$, which we can show maps to $xd_r(x)$, contributing $\bar{a}xd_r(x)$ to the differential on $Sq^i(x)$. Finally, if

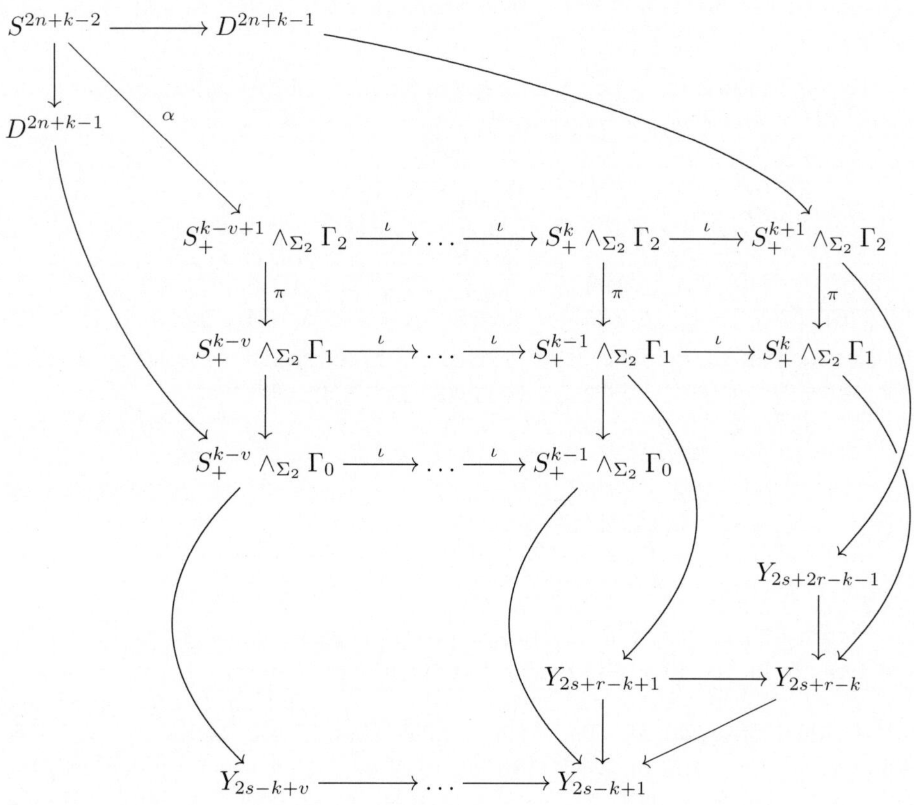

FIGURE 11.3. Maps from two hemispheres

$v > k+1$, then $\alpha = 0$ and this cell contributes nothing. This ends our overview of the arguments needed to prove Theorem 11.22.

## 11.2. Steenrod operations in $E_2(S)$

To use the results of the preceding section we need some information about the action of the Steenrod squares upon the $E_2$-term of the Adams spectral sequence for the sphere. We collect the results we shall use here.

Recall that in Proposition 1.4 we specified a basis for the algebra indecomposables of $E_2(S) = \mathrm{Ext}_A(\mathbb{F}_2, \mathbb{F}_2)$ in topological degrees $t - s \le 48$, together with their representing `ext`-cocycles. Let us write

$$Sq^*(x) = (Sq^s(x), Sq^{s-1}(x), \dots, Sq^1(x), Sq^0(x))$$

for the total Steenrod operation on a class $x \in \mathrm{Ext}_A^{s,t}(\mathbb{F}_2, \mathbb{F}_2)$. At one extreme, $Sq^s(x) = x^2$, which we can calculate by computing chain maps using `ext`. At the other extreme, $Sq^0(x)$ can also be easily calculated from the dual of the degree-doubling Frobenius homomorphism in the dual Steenrod algebra, i.e., from the degree-halving Verschiebung homomorphism in $A$. In the following proposition, we report the values in the range $t - s \le 48$ of $Sq^0$ on indecomposables, as well as a few values beyond this range, when it is possible to do so without having to introduce new notation other than $C = 6_{27}$ in bidegree $(t-s, s) = (50, 6)$ and $A'' = 6_{38}$ in bidegree $(t-s, s) = (64, 6)$.

PROPOSITION 11.26. *$Sq^0$ is an algebra homomorphism whose values on indecomposables, including all values in the range $t-s \leq 48$, satisfy*

(1) $Sq^0(h_i) = h_{i+1}$;
(2) $Sq^0(c_i) = c_{i+1}$, *with* $c_0 = 3_3$, $c_1 = 3_9$, $c_2 = 3_{19}$, $c_3 = 3_{34}$ *and* $c_4 = 3_{55}$;
(3) $Sq^0(d_i) = d_{i+1}$, *with* $d_0 = 4_3$, $d_1 = 4_{13}$, $d_2 = 4_{32}$ *and* $d_3 = 4_{65}$;
(4) $Sq^0(e_i) = e_{i+1}$, *with* $e_0 = 4_5$, $e_1 = 4_{16}$, $e_2 = 4_{40}$ *and* $e_3 = 4_{79}$;
(5) $Sq^0(f_i) = f_{i+1}$, *with* $f_0 = 4_6$, $f_1 = 4_{19}$, $f_2 = 4_{44}$ *and* $f_3 = 4_{84}$;
(6) $Sq^0(g_i) = g_{i+1}$, *with* $g = g_1 = 4_8$, $g_2 = 4_{22}$, $g_3 = 4_{48}$ *and* $g_4 = 4_{89}$.

*In each item above, the first element is defined by the specified cocycle $s_g$, while the remaining elements are calculated by applying $Sq^0$. In addition,*

(7) $Sq^0(Ph_1) = h_2g$, $Sq^0(Ph_2) = 0$, $Sq^0(Pc_0) = c_1g$, $Sq^0(Pd_0) = d_1g$ *and* $Sq^0(Pe_0) = 0$;
(8) $Sq^0(i) = h_2C$, $Sq^0(j) = 0$, $Sq^0(k) = h_2h_5n = h_4C$ *and* $Sq^0(\ell) = h_3A''$, *where* $C = 6_{27}$ *and* $A'' = 6_{38}$;
(9) $Sq^0(P^2h_1) = 0$, $Sq^0(P^2h_2) = 0$, $Sq^0(P^2c_0) = 0$, $Sq^0(P^2d_0) = d_1g^2$ *and* $Sq^0(P^2e_0) = 0$.

PROOF. That $Sq^0$ is an algebra homomorphism is immediate from the Cartan formula (1.1). In [**118**, Proposition 11.10], it is shown that the operation $Sq^0$ can be calculated by $Sq^0([a_1|\dots|a_s]) = [a_1^2|\dots|a_s^2]$ in the cobar complex for the dual Steenrod algebra. This implies that if $\Phi A_*$ is the double of the dual Steenrod algebra, in which the degrees of all the elements are doubled, then $Sq^0\colon \mathrm{Ext}_{A_*}^{s,t}(\mathbb{F}_2, \mathbb{F}_2) \to \mathrm{Ext}_{A_*}^{s,2t}(\mathbb{F}_2, \mathbb{F}_2)$ is induced by the degree-preserving Hopf algebra homomorphism $F\colon \Phi A_* \to A_*$ that sends $\xi_i$ to $\xi_i^2$ for each $i \geq 1$. Dually, it is induced by the degree-preserving Hopf algebra homomorphism $V\colon A \longrightarrow \Phi A$ that sends an "even" Milnor basis element $Sq^{(2r_1,\dots,2r_k)}$ to $Sq^{(r_1,\dots,r_k)}$, and other Milnor basis elements to 0. Restricting along this homomorphism gives

$$Sq^0\colon \mathrm{Ext}_A^{s,t}(\mathbb{F}_2, \mathbb{F}_2) \cong \mathrm{Ext}_{\Phi A}^{s,2t}(\mathbb{F}_2, \mathbb{F}_2) \longrightarrow \mathrm{Ext}_A^{s,2t}(\mathbb{F}_2, \mathbb{F}_2).$$

A slight modification of the computer code that calculates chain maps can compute this: a program `startsq0` computes the restriction $V_{s-1}(\partial(x))$ for each generator $x = s_g^*$ in the minimal $A$-module resolution $(C_*, \partial)$ of $\mathbb{F}_2$, and the same program that computes lifts for chain maps then solves for an element $V_s(x)$ satisfying $\partial(V_s(x)) = V_{s-1}(\partial(x))$. We recover $Sq^0$ as $\mathrm{Hom}_A(V_*, \mathbb{F}_2)$. This inductive calculation is begun by setting $V_0(0_0^*) = 0_0^*$, so that $Sq^0(1) = 1$. □

PROPOSITION 11.27. *The elements $h_i \in \mathrm{Ext}_A^{1,2^i}(\mathbb{F}_2, \mathbb{F}_2)$ dual to the $Sq^{2^i}$ satisfy the following relations:*

(1) $Sq^*(h_i) = (h_i^2, h_{i+1})$;
(2) $h_ih_{i+1} = 0$, $h_ih_{i+2}^2 = 0$, $h_i^2h_{i+2} = h_{i+1}^3$ *and, for* $i > 0$, $h_i^4 = 0$;
(3) $h_i^2h_{i+3}^2 = 0$, $h_0^{2^i}h_{i+2}^2 = 0$ *and, for* $i \neq 1$, $h_0^{2^i}h_{i+1} = 0$.

PROOF. The relations $h_0h_1 = 0$ and $h_0h_2^2 = 0$ are easily checked by hand. Applying $Sq^0$ repeatedly then gives $h_ih_{i+1} = 0$ and $h_ih_{i+2}^2 = 0$ for all $i$. By the Cartan formula, $0 = Sq^1(h_ih_{i+1}) = h_i^2h_{i+2} + h_{i+1}^3$. These relations then give $h_{i+1}^4 = h_{i+1}h_i^2h_{i+2} = 0$.

Next, $Sq^{2^i}(h_0^{2^i} h_{i+2}^2) = h_0^{2^{i+1}} h_{i+3}^2$ shows that $h_0^{2^i} h_{i+2}^2 = 0$ for all $i$. In particular, $h_0^2 h_3^2 = 0$. Repeatedly applying $Sq^0$ then shows that $h_i^2 h_{i+3}^2 = 0$ for all $i$. Applying $Sq^2$ to $h_0^2 h_2 = h_1^3$ gives $h_0^4 h_3 = 0$. Then $Sq^{2^i}(h_0^{2^i} h_{i+1}) = h_0^{2^{i+1}} h_{i+2}$ shows $h_0^{2^i} h_{i+1} = 0$ for all $i \neq 1$. □

REMARK 11.28. The first two items in Proposition 11.27 were shown by Adams in [**3**]. The last item was shown by Sergei Novikov [**138**], and can also be found in [**133**]. Novikov's third identity was incorrectly reported in [**45**, Ch. VI] to be $h_0^{2^n} h_n = 0$ (when $n > 0$). Novikov also established the identities $h_i h_{i+k}^2 h_{i+k+3} = 0$ and $h_i^2 h_{i+k+1}^2 h_{i+k+4} = 0$ for $i \geq 0$ and $k \geq 3$.

PROPOSITION 11.29.

(1) $Sq^*(c_0) = (c_0^2 = h_1^2 d_0, h_0 e_0, f_0, c_1)$ *with* $f_0 = 4_6$.
(2) $Sq^*(d_0) = (d_0^2, 0, r, 0, d_1)$.
(3) $Sq^*(e_0) = (e_0^2 = d_0 g, m, t, x, e_1)$.
(4) $Sq^*(f_0) = (0, h_3 r, y, 0, f_1)$ *with* $y = 6_{16}$.
(5) $Sq^*(g) = (g^2, h_1 h_5 Ph_1, h_5 Ph_2, 0, g_2)$.

PROOF. We have already discussed $Sq^s(x)$ and $Sq^0(x)$ for $x$ in cohomological degree $s$. The relations $c_0^2 = h_1^2 d_0$ and $e_0^2 = d_0 g$ can be verified with `ext`.

The Cartan formula and known relations in $\mathrm{Ext}_A(\mathbb{F}_2, \mathbb{F}_2)$ allow us to determine the remaining squaring operations on $d_0$ and $e_0$, as well as $Sq^2(c_0)$. First, we write

$$Sq^*(d_0) = (d_0^2\,,\, \alpha_3 \cdot k\,,\, \alpha_2 \cdot r\,,\, \alpha_1 \cdot n + \alpha_1' \cdot h_0^4 h_5\,,\, d_1)$$

for some coefficients $\alpha_i, \alpha_i' \in \mathbb{F}_2$. Since $c_0^2 = h_1^2 d_0$, we must have $0 = Sq^3(c_0^2) = Sq^3(h_1^2 d_0) = \alpha_3 \cdot h_2^2 k$. From $h_2^2 k \neq 0$ we determine that $\alpha_3 = 0$. Similarly, $0 = Sq^1(c_0^2) = Sq^1(h_1^2 d_0) = \alpha_1 \cdot h_2^2 n$ and $h_2^2 n \neq 0$ imply that $\alpha_1 = 0$. Using $h_0^4 d_0 = 0$ we get $Sq^5(h_0^4 d_0) = \alpha_1' \cdot h_0^{12} h_5 = 0$ with $h_0^{12} h_5 \neq 0$, and hence $\alpha_1' = 0$.

Second, we write

$$Sq^*(e_0) = (e_0^2\,,\, \beta_3 \cdot m\,,\, \beta_2 \cdot t\,,\, \beta_1 \cdot x\,,\, e_1)$$

for some coefficients $\beta_i \in \mathbb{F}_2$. The relation $h_2 d_0 = h_0 e_0$ then gives $h_3 d_0^2 = Sq^4(h_2 d_0) = Sq^4(h_0 e_0) = h_1 e_0^2 + \beta_3 \cdot h_0^2 m$ with $h_3 d_0^2 = 0$ and $h_1 e_0^2 = h_0^2 m \neq 0$, which implies $\beta_3 = 1$. The same relation gives $\alpha_2 \cdot h_3 r = Sq^2(h_2 d_0) = Sq^2(h_0 e_0) = \beta_1 \cdot h_0^2 x + \beta_2 \cdot h_1 t$. Then $h_3 r = 7_{13} + 7_{14}$, $h_0^2 x = 7_{14}$ and $h_1 t = 7_{13}$ imply that $\alpha_2 = \beta_1 = \beta_2$. Likewise, $h_2^2 d_1 = Sq^1(h_2 d_0) = Sq^1(h_0 e_0) = \beta_1 \cdot h_1 x + h_0^2 e_1$ with $h_2^2 d_1 \neq 0$ and $h_0^2 e_1 = 0$, which implies $\beta_1 = 1$, hence also $\alpha_2 = \beta_2 = 1$.

Third, applying $Sq^4$ to $c_0^2 = h_1^2 d_0$ we obtain $Sq^2(c_0)^2 = Sq^4(c_0^2) = Sq^4(h_1^2 d_0) = h_2^2 d_0^2 + h_1^4 Sq^2(d_0) = h_2^2 d_0^2 \neq 0$, so that $Sq^2(c_0)$ must be nonzero. The only possible value is $h_0 e_0$.

To continue, we need two key computational facts, namely that $Sq^1(c_0) = f_0$ and $Sq^2(f_0) = y$, where $c_0 = 3_3$, $f_0 = 4_6$ and $y = 6_{16}$ in the minimal resolution chosen by `ext`. These calculations are worked out in [**46**, Prop. 4 and 11] by the method of Nassau [**135**]. Recall from the proof of Proposition 1.4 that we fixed our choices of $f_0$ and $y$ to conform with these computations.

We can then use the Adem relations to determine the remaining squaring operations on $f_0$. First, $Sq^1(f_0) = Sq^1 Sq^1(c_0) = 0$, since $Sq^1 Sq^1 = 0$. Next, $Sq^3(f_0) = Sq^3 Sq^1(c_0) = Sq^2 Sq^2(c_0) = Sq^2(h_0 e_0) = h_1 Sq^2(e_0) + h_0^2 Sq^1(e_0) = h_1 t + h_0^2 x = h_3 r$, since $Sq^2 Sq^2 = Sq^3 Sq^1$. (The same results were obtained in [**46**] by direct calculation.)

To conclude the proof, we write

$$Sq^*(g) = (g^2\,,\ \gamma_3 \cdot h_1h_5Ph_1\,,\ \gamma_2 \cdot h_5Ph_2\,,\ 0\,,\ g_2)$$

for some coefficients $\gamma_i \in \mathbb{F}_2$. From $h_2e_0 = h_0g$ we get that $Sq^3(h_2e_0) = h_2^2t + h_3m = 0$ must equal $Sq^3(h_0g) = \gamma_2 \cdot h_0^2h_5Ph_2 + \gamma_3 \cdot h_1^2h_5Ph_1 = (\gamma_2 + \gamma_3) \cdot 8_{19}$, so that $\gamma_2 = \gamma_3$. Finally, $h_2f_0 = h_1g$ implies $h_3y = Sq^2(h_2f_0) = Sq^2(h_1g) = \gamma_2 \cdot h_2h_5Ph_2$, with $h_3y \neq 0$, so that $\gamma_2 = 1$. □

REMARK 11.30. The values of the squaring operations on $c_0$, $d_0$, $e_0$ and $f_0$ were calculated by Shunji Mukohda in [**133**, Prop. 4, 5 and 6] and by James Milgram in [**122**, §6]. More precisely, they both showed that $Sq^1(c_0)$ is an element of the Massey product $\langle h_0^2, h_3^2, h_2\rangle = \{f_0, f_0 + h_1^3h_4\} = \{4_6, 4_6 + 4_7\}$, and that $Sq^2(f_0)$ is an element of $\langle h_0^4, h_4^2, h_3\rangle = \{y, y + h_1x\} = \{6_{16}, 6_{16} + 6_{17}\}$. The result in [**46**] removes the indeterminacy in these two calculations.

COROLLARY 11.31.

(1) $Sq^*(c_i) = (c_i^2, h_ie_i, f_i, c_{i+1})$.
(2) $Sq^*(d_i) = (d_i^2, 0, r_i, 0, d_{i+1})$.
(3) $Sq^*(e_i) = (e_i^2, m_i, t_i, x_i, e_{i+1})$.
(4) $Sq^*(f_i) = (0, h_{i+3}r_i, y_i, 0, f_{i+1})$.

*The classes $a_i$ for $a \in \{r, m, t, x, y\}$ are inductively defined by $a_0 = a$ and $a_{i+1} = Sq^0(a_i)$.*

PROOF. This follows from Proposition 11.29 by repeatedly applying $Sq^0$, since $Sq^0Sq^i = Sq^iSq^0$ is one of the Adem relations (1.2) satisfied by the algebraic squaring operations. □

REMARK 11.32. In the range we are considering here we have full knowledge of the multiplicative relations from the machine calculation by `ext`. Outside that range, the squaring operations are a useful tool for extending them. For example, from $h_2f_0 = h_1g$ we know immediately that $h_{i+2}f_i = h_{i+1}g_{i+1}$ for all $i$. From the vanishing of $h_0^8c_4$, $h_0d_1$, $h_0^3e_2$, $h_0^8f_3$ and $h_0^3g_2$ we can inductively prove that $h_0^{2^i}c_{i+1} = 0$ for $i \geq 3$, $h_0^{2^i}d_{i+1} = 0$ for $i \geq 0$, $h_0^{2^i}e_i = 0$ for $i \geq 3$, $h_0^{2^i}f_i = 0$ for $i \geq 3$ and $h_0^{3\cdot 2^i}g_{i+2} = 0$ for $i \geq 0$.

PROPOSITION 11.33.

(1) $Sq^*(Ph_1) = (h_1P^2h_1, P^2h_2, 0, 0, 0, h_2g)$.
(2) $Sq^*(Ph_2) = (h_2P^2h_2, h_1Pd_0 + h_0^2i, 0, 0, h_2^2g, 0)$.
(3) $Sq^*(Pc_0) = (c_0P^2c_0, h_0P^2e_0, h_0Pj, 0, 0, \zeta_2 \cdot h_0e_0g, \zeta_1 \cdot f_0g, c_1g)$.
(4) $Sq^*(Pd_0) = (d_0P^2d_0, 0, i^2, 0, d_0^2g, 0, gr, 0, d_1g)$.

*The coefficients $\zeta_i \in \mathbb{F}_2$ of $h_0e_0g = h_0^4x$ and $f_0g = h_0^2y$ remain undetermined.*

PROOF. The values of the $Sq^0$ were computed by `ext` and recorded in Proposition 11.26. The other values can be computed as follows, using the Cartan formula and, in one case, the Adem relations.

We have $(Ph_1)^2 = h_1P^2h_1$, so we may write

$$Sq^*(Ph_1) = (h_1P^2h_1\,,\ \delta_4 \cdot P^2h_2\,,\ 0\,,\ 0\,,\ 0\,,\ h_2g)$$

for some coefficient $\delta_4 \in \mathbb{F}_2$. From $h_3Ph_1 = 6_5 = c_0^2$ we find that $Sq^4(h_3Ph_1) = \delta_4 \cdot h_4P^2h_2$ is equal to $Sq^4(c_0^2) = h_0^2e_0^2 = 10_{11} \neq 0$, so $\delta_4 = 1$.

We have $(Ph_2)^2 = h_2P^2h_2$, so we may write

$$Sq^*(Ph_2) = (h_2P^2h_2\,,\ \epsilon_4 \cdot h_1Pd_0 + \epsilon_4' \cdot h_0^2 i\,,\ 0\,,\ 0\,,\ \epsilon_1 \cdot h_2^2 g\,,\ 0)$$

for some coefficients $\epsilon_1, \epsilon_4, \epsilon_4' \in \mathbb{F}_2$. From $h_2Ph_2 = h_0^2d_0$ we get that $Sq^4(h_2Ph_2) = h_3(\epsilon_4 \cdot h_1Pd_0 + \epsilon_4' \cdot h_0^2 i) = \epsilon_4' \cdot h_0^2h_3 i$ is equal to $Sq^4(h_0^2d_0) = h_1^2d_0^2 + h_0^4 r = 10_8 \neq 0$. Hence $\epsilon_4' = 1$. Then, $h_1Ph_2 = 0$ gives $0 = Sq^5(h_1Ph_2) = h_1^2(\epsilon_4 \cdot h_1Pd_0 + h_0^2 i) + h_2^2P^2h_2 = (\epsilon_4 + 1) \cdot h_1^3Pd_0$, with $h_1^3Pd_0 \neq 0$. Hence $\epsilon_4 = 1$. To determine $\epsilon_1$ we use the Adem relations $Sq^1Sq^2 = Sq^3Sq^0 = Sq^0Sq^3$. On one hand, $Sq^1Sq^2(g) = Sq^1(h_5Ph_2) = h_6Sq^1(Ph_2) = \epsilon_1 \cdot h_2^2h_6g$. On the other hand, $Sq^0Sq^3(g) = Sq^0(h_1h_5Ph_1) = h_2h_6h_2g = h_2^2h_6g = 7_{75} \neq 0$. Hence $\epsilon_1 = 1$.

We have $(Pc_0)^2 = c_0P^2c_0$, so we may write

$$Sq^*(Pc_0) = (c_0P^2c_0\,,\ \zeta_6 \cdot h_0P^2e_0\,,\ \zeta_5 \cdot h_0Pj\,,\ 0\,,\ 0\,,\ \zeta_2 \cdot h_0e_0g\,,\ \zeta_1 \cdot f_0g\,,\ c_1g)$$

for some coefficients $\zeta_i \in \mathbb{F}_2$. From $h_1Pc_0 = c_0Ph_1$ we get that $Sq^5(h_1Pc_0) = \zeta_5 \cdot h_0h_2Pj$ is equal to $Sq^5(c_0Ph_1) = c_1h_1P^2h_1 + f_0P^2h_2 = h_0h_2Pj \neq 0$, so $\zeta_5 = 1$. Furthermore, $Sq^6(h_1Pc_0) = \zeta_6 \cdot h_2h_0P^2e_0 + h_1^2h_0Pj = \zeta_6 \cdot h_0h_2P^2e_0$ is equal to $Sq^6(c_0Ph_1) = f_0h_1P^2h_1 + h_0e_0P^2h_2 = h_0h_2P^2e_0 \neq 0$, so $\zeta_6 = 1$. We have $(Pd_0)^2 = d_0P^2d_0$, so we may write

$$\begin{aligned} Sq^*(Pd_0) = (d_0P^2d_0\,,\ &\eta_7 \cdot iPd_0\,,\ \eta_6 \cdot i^2\,,\ \eta_5 \cdot Q + \eta_5' \cdot Pu\,, \\ &\eta_4 \cdot d_0^2 g\,,\ \eta_3 \cdot gk\,,\ \eta_2 \cdot gr\,,\ \eta_1 \cdot gn\,,\ d_1g) \end{aligned}$$

for some coefficients $\eta_i, \eta_i' \in \mathbb{F}_2$. From $h_2Pd_0 = d_0Ph_2$, we get that $Sq^8(h_2Pd_0) = h_3d_0P^2d_0 + \eta_7 \cdot h_2^2 iPd_0 = \eta_7 \cdot 17_{15}$ is equal to $Sq^8(d_0Ph_2) = d_0^2(h_1Pd_0 + h_0^2 i) = 0$, so that $\eta_7 = 0$. Also, $Sq^6(h_2Pd_0) = \eta_6 \cdot h_3 i^2 + h_2^2(\eta_5 \cdot Q + \eta_5' \cdot Pu) = \eta_6 \cdot 15_{13}$ is equal to $Sq^6(d_0Ph_2) = r(h_1Pd_0 + h_0^2 i) = 15_{13}$, showing that $\eta_6 = 1$. Next, $Sq^5(h_2Pd_0) = h_3(\eta_5 \cdot Q + \eta_5' \cdot Pu) + h_2^2(\eta_4 \cdot d_0^2 g) = \eta_5 \cdot 14_{16}$ while $Sq^5(d_0Ph_2) = d_1h_2P^2h_2 + d_0^2h_2^2 g = 0$, showing that $\eta_5 = 0$.

The relation $h_0^2Pd_0 = (Ph_2)^2$ implies that $Sq^5(h_0^2Pd_0) = \eta_5' \cdot h_1^2Pu + \eta_3 \cdot h_0^4 gk = \eta_5' \cdot 15_{11}$ is equal to $Sq^5((Ph_2)^2) = 0$, so that $\eta_5' = 0$.

The relation $gPd_0 = d_0^3$ gives us the final four coefficients. First, $Sq^8(gPd_0) = g_2d_0P^2d_0 + h_5Ph_2 i^2 + \eta_4 \cdot g^2d_0^2 g = \eta_4 \cdot 20_{37}$ is equal to $Sq^8(d_0^3) = d_1d_0^4 + d_0^2r^2 = 20_{37}$, so that $\eta_4 = 1$. Next, $Sq^7(gPd_0) = h_1h_5Ph_1d_0^2 g + \eta_3 \cdot g^2 gk = \eta_3 \cdot 19_{43}$, while $Sq^7(d_0^3) = 0$, so that $\eta_3 = 0$. Similarly, $Sq^6(gPd_0) = g_2 i^2 + h_5Ph_2d_0^2 g + \eta_2 \cdot g^2 gr = 18_{53} + \eta_2 \cdot 18_{51}$, while $Sq^6(d_0^3) = rr^2 = 18_{51} + 18_{53}$, so that $\eta_2 = 1$. Finally, $Sq^5(gPd_0) = h_1h_5Ph_1gr + \eta_1 \cdot g^2 gn = \eta_1 \cdot 17_{54}$ is equal to $Sq^5(d_0^3) = 0$, letting us conclude that $\eta_1 = 0$. □

We now apply the $H_\infty$ ring structure on $S$ to construct classes in $\pi_*(S)$ using power operations, and to find permanent cycles in $E_\infty(S)$ detecting these classes. This also allows us to determine some relations in $\pi_*(S)$. It may be helpful to refer to Figures 11.10, 11.13 and 11.14.

PROPOSITION 11.34. *Let $\sigma^2$, $\eta^\circ$ and $\nu^\circ$ be given by the power operations $\alpha^*(\sigma)$, for classes $\alpha \in \pi_*D_2(S^7)$ detected by $\cup_0$, $h_1\cup_1$ and $h_0\cup_4$, respectively.*

*(1) The square $\sigma^2 \in \pi_{14}(S)$ is detected by $h_3^2 \in E_\infty(S)$, and satisfies $2\sigma^2 = 0$ and $\eta\sigma^2 = 0$.*

*(2) The class $\eta^\circ \in \pi_{16}(S)$ is detected by $h_1h_4 \in E_\infty(S)$, and satisfies $2\eta^\circ = 0$ and $\nu\eta^\circ = 0$.*

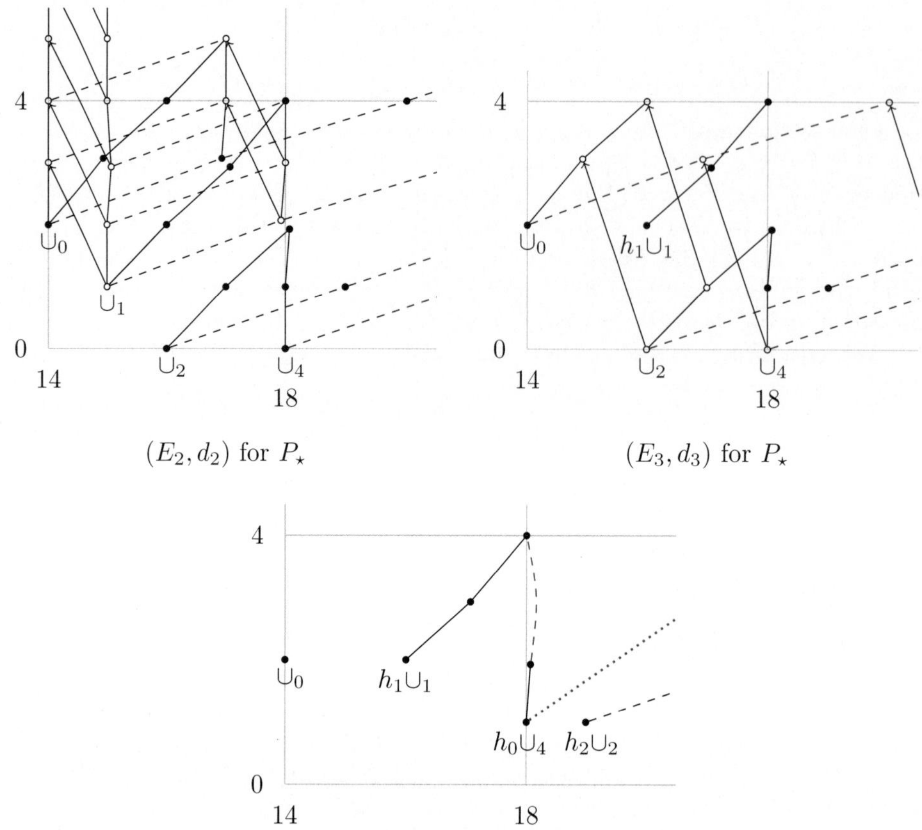


$(E_2, d_2)$ for $P_\star$ $(E_3, d_3)$ for $P_\star$

$E_4 = E_\infty$ for $P_\star$

FIGURE 11.4. Delayed Adams spectral sequence for $\pi_*(\Sigma^7 P_7^\infty)$

*(3) The class $\nu^\circ \in \pi_{18}(S)$ is defined up to multiplication by an odd integer, is detected by $h_2h_4 \in E_\infty(S)$, and satisfies $8\nu^\circ = 0$, $\eta\nu^\circ = 0$ and $4\nu^\circ = \eta^2\eta^\circ$. Furthermore, $\epsilon\nu^\circ$ is an $\eta^2$-multiple, possibly zero.*

PROOF. We apply Theorem 11.13 and Corollary 11.15 to $\sigma\colon S^7 \longrightarrow S$ detected by $h_3 \in \mathrm{Ext}_A^{1,8}(\mathbb{F}_2, \mathbb{F}_2)$, with $Sq^*(h_3) = (h_3^2, h_4)$. The tower $P_\star$ we must consider is

$$\Sigma^7 P_7^\infty \longleftarrow \Sigma^7 P_7^8 \longleftarrow \Sigma^7 P_7^7 \longleftarrow * ,$$

where $\Sigma^7 P_7^\infty \cong D_2(S^7)$. We have

$$E_2(P_\star) = \mathrm{Ext}_A(\mathbb{F}_2, \mathbb{F}_2)\{\cup_0, \cup_1\} \oplus \mathrm{Ext}_A(H^*(\Sigma^7 P_9^\infty), \mathbb{F}_2) ,$$

with classes $\cup_k$ in bidegrees $(t-s, s) = (14+k, 2-k)$ for $k = 0, 1, 2$. This $E_2$-term is shown in the upper left hand part of Figure 11.4. We have given the filtration 0 class in degree 18 the name $\cup_4$. This can be justified in terms of the spherical classes in $\pi_*(\Sigma^7 P_9^\infty)$, but is purely a notational convenience for us.

For comparison, the $E_2$-term of the ordinary Adams spectral sequence for $\Sigma^7 P_7^\infty$ is shown in Figure 11.5. In particular, $\pi_{14}(\Sigma^7 P_7^\infty) \cong \mathbb{Z}/2$ and $\pi_{15}(\Sigma^7 P_7^\infty) = 0$. Since $\cup_0$ maps to $\sigma^2$, it is immediate from naturality that $2\sigma^2 = 0$ and $\eta\sigma^2 = 0$.

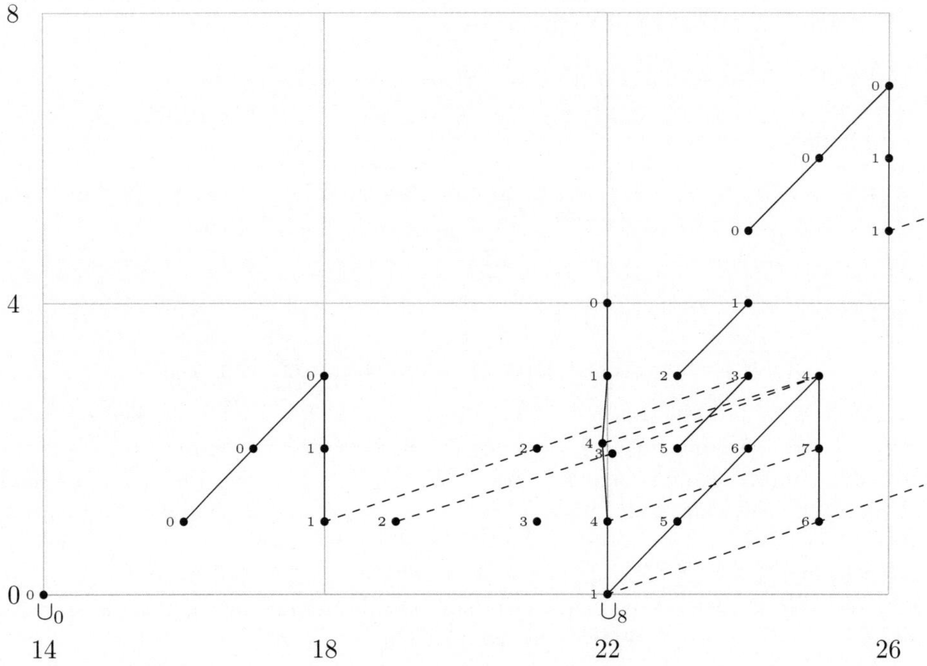

FIGURE 11.5. Adams spectral sequence for $\pi_*(\Sigma^7 P_7^\infty)$

Since $\pi_{14}(\Sigma^7 P_7^\infty) \cong \mathbb{Z}/2$ we must have $d_2(\cup_1) = h_0\cup_0$ in $E_2(P_\star)$. This differential extends $h_0$- and $h_2$-linearly, as illustrated. It follows easily that $d_2(\cup_2) = 0$ and $d_2(\cup_4) = 0$. This leads to the delayed Adams $E_3$-term shown in the upper right hand part of Figure 11.4. Since $\pi_{15}(\Sigma^7 P_7^\infty) = 0$ we must have $d_3(\cup_2) = h_1\cup_0$. Similarly, we must have $d_3(\cup_4) = h_2\cup_0$ because $\pi_{17}(\Sigma^7 P_7^\infty) \cong \mathbb{Z}/2$ and $h_1$-linearity precludes the term $h_1^2\cup_1$ from appearing in $d_3(\cup_4)$. These differentials extend $h_1$- and $h_2$-linearly, as shown. The resulting delayed Adams $E_4$-term is shown in the lower part of Figure 11.4. There is no room for further differentials, so $E_4(P_\star) = E_\infty(P_\star)$ in this range of degrees. It follows that $\pi_{18}(\Sigma^7 P_7^\infty) \cong \mathbb{Z}/8$.

When combined with the fact that the 15-cell $\cup_8$ is spherical in $P_7^{15} \subset P_7^\infty$, cf. Proposition 11.20, this also shows that $E_2 = E_\infty$ in the ordinary Adams spectral sequence, in the range of degrees shown.

The map of spectral sequences $E_2(P_\star) \to E_2(S)$ sends $\cup_0$ to $Sq^1(h_3) = h_3^2$ and $\cup_1$ to $Sq^0(h_3) = h_4$, while $\cup_2$ and $\cup_4$ map to 0 since these bidegrees are trivial in the Adams spectral sequence for the sphere. Hence $\eta^\circ \in \pi_{16}(S)$, the image of the class $\{h_1\cup_1\} \in \pi_{16}(\Sigma^7 P_7^\infty)$ detected by $h_1\cup_1$, is detected by $h_1h_4$ in $E_\infty(S)$. It satisfies $2\eta^\circ = 0$ and $\nu\eta^\circ = 0$, since $\{h_1\cup_1\}$ satisfies these relations in $\pi_*(\Sigma^7 P_7^\infty)$.

Let $\nu^\circ \in \pi_{18}(S)$ be the image of a class $\alpha \in \pi_{18}(\Sigma^7 P_7^\infty)$ detected by $h_0\cup_4$. Since $4\{h_0\cup_4\} = \eta^2\{h_1\cup_1\}$ and $\eta\{h_0\cup_4\} = 0$ we must have $4\nu^\circ = \eta^2\eta^\circ$ and $\eta\nu^\circ = 0$. This means that $\nu^\circ$ must be detected by $h_2h_4 \in E_\infty(S)$, since this is the only nonzero class in topological degree 18 and Adams filtration $s \leq 2$ in $E_2(S)$. It follows immediately that $8\nu^\circ = 0$. Finally, $\epsilon\alpha \in \pi_{26}(\Sigma^7 P_7^\infty)$ has order dividing 2, hence is either zero or detected by $7_0 = h_1^2 \cdot 5_0$ in the ordinary Adams spectral sequence for $\Sigma^7 P_7^\infty$. In either case, $\epsilon\nu^\circ$ is an $\eta^2$-multiple. (We will see in case (26) of Theorem 11.61 that this product is zero.) □

PROPOSITION 11.35. *Let $\sigma^\circ \in \pi_{19}(S)$ be given by the power operation $\alpha^*(\epsilon) = \cup_3(\epsilon)$, where $\alpha \in \pi_{19}D_2(S^8)$ is detected by $\cup_3$. Then $\sigma^\circ$ is detected by $c_1 \in E_\infty(S)$, and satisfies $\eta\sigma^\circ = 0$.*

PROOF. We apply Theorem 11.13 and Corollary 11.15 to $\epsilon\colon S^8 \to S$ detected by $c_0 \in \mathrm{Ext}_A^{3,11}(\mathbb{F}_2, \mathbb{F}_2)$, with $Sq^*(c_0) = (c_0^2, h_0 e_0, f_0, c_1)$. The tower $P_\star$ is

$$\Sigma^8 P_8^\infty \longleftarrow \Sigma^8 P_8^{13} \longleftarrow \Sigma^8 P_8^{12} \longleftarrow \Sigma^8 P_8^{11} \longleftarrow \Sigma^8 P_8^{10} \longleftarrow \Sigma^8 P_8^{9} \longleftarrow \Sigma^8 P_8^{8} \longleftarrow * ,$$

where $\Sigma^8 P_8^\infty \cong D_2(S^8)$. We have

$$E_2(P_\star) = \mathrm{Ext}_A(\mathbb{F}_2, \mathbb{F}_2)\{\cup_0, \dots, \cup_5\} \oplus \mathrm{Ext}_A(H^*(\Sigma^8 P_{14}^\infty), \mathbb{F}_2) ,$$

with classes $\cup_k$ in bidegrees $(t-s, s) = (16+k, 6-k)$ for $0 \le k \le 6$. The $E_\infty$-term of the delayed Adams spectral sequence for $P_\star$ is shown in Figure 11.6, while the ordinary Adams spectral sequence for $\Sigma^8 P_8^\infty$ is shown in Figure 11.7. We have $E_2 = E_\infty$ in this range, since $d_r(\cup_1) = 0$ and $d_2(\cup_3) = 0$ by $h_0$- and $h_1$-linearity, respectively. Any choice of class $\alpha \in \pi_{19}(\Sigma^8 P_8^\infty)$ detected by $\cup_3 \in E_\infty^{0,19}$ will be detected by $\cup_3 \in E_\infty^{3,22}(P_\star)$. Hence $\sigma^\circ = \alpha^*(\epsilon)$ will be detected by $Sq^0(c_0) = c_1$ in $E_\infty^{3,22}(S)$. (It is easy to see that $c_1$ cannot be a boundary in the Adams spectral sequence for $S$, hence $c_1$ remains nonzero at $E_\infty$.)

Since $h_1 \cup_3 = 0$ in Figure 11.7, we must have $\eta\alpha = 0$ in $\pi_{20}(\Sigma^8 P_8^\infty)$. It follows, by naturality, that $\eta\sigma^\circ = 0$ in $\pi_{20}(S)$. □

REMARK 11.36. Extensive Adams spectral sequence calculations for the stable homotopy of the stunted projective spaces $P_n^\infty$ were made by Mahowald in his memoir [**99**].

REMARK 11.37. The notations $\eta^\circ$, $\nu^\circ$ and $\sigma^\circ$ are meant to suggest connections to the homotopy classes $\eta^* \in \pi_{16}(S)$, $\nu^* \in \pi_{18}(S)$ and $\bar{\sigma} \in \pi_{19}(S)$ defined by Toda [**171**, Ch. XIV] in terms of the following secondary compositions (= Toda brackets):

$$\eta^* \in \langle \sigma, 2\sigma, \eta \rangle$$
$$\nu^* \in \langle \sigma, 2\sigma, \nu \rangle$$
$$\bar{\sigma} \in \langle \nu, \eta\sigma, \sigma \rangle .$$

These are known to be detected by $h_1 h_4$, $h_2 h_4$ and $c_1$, respectively, hence agree with $\eta^\circ$, $\nu^\circ$ and $\sigma^\circ$ modulo classes of higher Adams filtration. We outline these connections here, referring to Section 11.3 for a review of the Adams $d$- and $e$-invariants and Theorem 11.61 for the structure of $\pi_*(S)$ in this range of degrees.

In the first case, $\eta^\circ \equiv \eta^* \mod \eta\rho$, and this is as precise a comparison we can make, since the Toda bracket defining $\eta^*$ has indeterminacy $\{0, \eta\rho\}$. However, in Theorem 11.61 we will fix our choice of $\eta^*$ to have zero Adams $e$-invariant. In [**22**, p. 313] the authors suggest that this is the "natural" choice for $\eta^*$, which they call $\eta_3$, but which is now usually denoted $\eta_4$. We do not know if $e(\eta^\circ)$ is 0 or $\eta j_{15}$, corresponding to $\eta^\circ = \eta^*$ or $\eta^\circ = \eta^* + \eta\rho$, respectively.

In the second case, $\nu^\circ \doteq \nu^* \mod \eta\bar{\mu}$, meaning that $\nu^\circ$ is a 2-adic unit times $\nu^*$ plus a multiple of $\eta\bar{\mu}$. Here $\eta\bar{\mu}$ is detected by the Adams $d$-invariant, induced by the $E_\infty$ ring spectrum map $d\colon S \to ko$, and $d(\nu^\circ) = 0 = d(\nu^*)$ since $d(\sigma) = 0$.

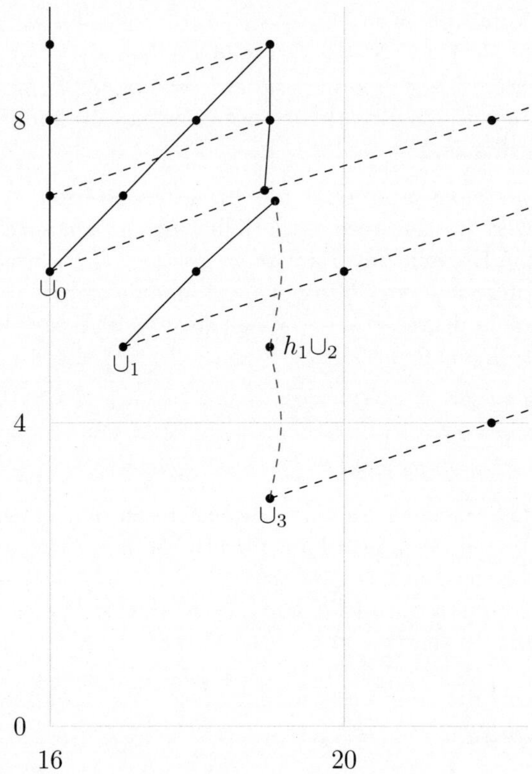


FIGURE 11.6. Delayed $E_\infty$-term for $\pi_*(\Sigma^8 P_8^\infty)$

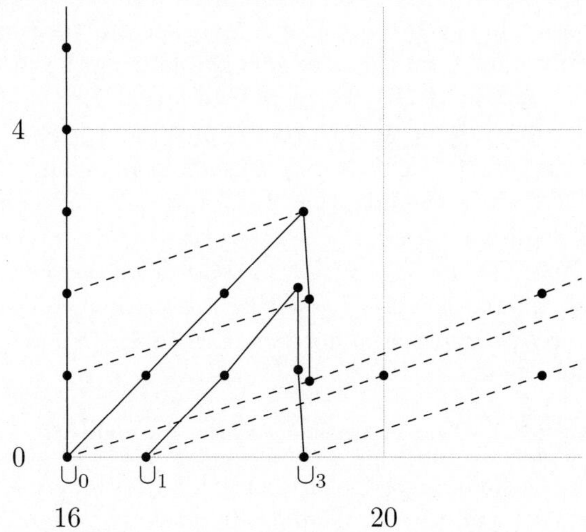


FIGURE 11.7. Adams spectral sequence for $\pi_*(\Sigma^8 P_8^\infty)$

Thus $\nu^\circ$ is an odd multiple of $\nu^*$. To specify this coefficient we would have to fix a generator of $\pi_{18}(\Sigma^7 P_7^\infty) \cong \mathbb{Z}/8$, so as to uniquely define $\nu^\circ$.

In the third case, $\sigma^\circ \equiv \bar{\sigma} \mod \bar{\zeta}$, and $e(\bar{\sigma}) = 0$ by [**8**, Thm. 5.3(v)]. We do not know the value of $e(\sigma^\circ) \in \mathbb{Z}/8\{j_{19}\}$, but note that $\eta\bar{\zeta} = 0$ implies $\eta\bar{\sigma} = \eta\sigma^\circ$, which we have just proved is zero.

We will give a simple proof of the following proposition, using *tmf*, in Theorem 11.54. However, we also record the following more classical argument, since the existence of the Kervaire invariant one class $\theta_4 \in \{h_4^2\}$ was a major result in [**107**], long predating the theory of topological modular forms. The proof presumes knowledge of $E_3(S)$ in degrees $t - s \le 30$, as given in Theorem 11.52, so the reader may wish to return here after reaching that result. See also Figure 11.11.

PROPOSITION 11.38.
*(1) $d_3(r) = h_1 d_0^2$ in $E_3(S)$.*
*(2) There is a homotopy class $\theta_4 \in \pi_{30}(S)$ detected by $h_4^2 \in E_\infty(S)$.*

PROOF. (1) Let $S_\star$ be a minimal Adams resolution of the sphere, and let $\kappa\colon S^{14} \to S$ be detected by $d_0$ in Adams filtration 4, admitting a lift $\kappa_4\colon S^{14} \to S_4$. Form the quadratic construction $\xi_2 D_2(\kappa)\colon \Sigma^{14} P_{14}^\infty \cong D_2(S^{14}) \to D_2(S) \to S$. By Proposition 11.9 its restriction to $\Sigma^{14} P_{14}^{16} \cong S^2_+ \wedge_{\Sigma_2} (S^{14} \wedge S^{14})$ factors through Adams filtration 6:

$$\begin{array}{ccccc} D_2(S^{14}) & \longleftarrow & \Sigma^{14} P_{14}^{16} & \longleftarrow & S^{14} \wedge S^{14} \\ {\scriptstyle \xi_2 D_2(\kappa)}\downarrow & & \downarrow{\scriptstyle \xi_{2,8}(1\wedge\kappa_4\wedge\kappa_4)} & & \downarrow{\scriptstyle \kappa_4^2} \\ S & \longleftarrow & S_6 & \longleftarrow & S_8 \,. \end{array}$$

Here $\Sigma^{14} P_{14}^{16} \simeq (S^{28} \vee S^{29}) \cup_{\eta+2} e^{30}$, as is clear from the action of $Sq^1$ and $Sq^2$ on $H^*(P_{14}^{16})$, so the extension over the 30-cell implies that $\eta\kappa_4^2 = 2y$ in $\pi_{29}(S_6)$, for some homotopy class $y$ in this group. The Adams spectral sequence for $S_6$, shifted up 6 filtrations, is obtained from the one for $S$ by omitting the rows $0 \le s < 6$. The differential $d_2(k) = h_0 d_0^2$ in $E_2(S)$ therefore shows that $\Sigma^{6,6} E_3(S_6)$ in topological degree 29 has only a single generator $h_1 d_0^2$, so $2y = 0$ in $\pi_{29}(S_6)$. Hence $\eta\kappa_4^2 = 0$ in $\pi_{29}(S_6)$, meaning that $h_1 d_0^2 \in \Sigma^{6,6} E_3(S_6)$ must be a boundary, and $d_3(r) = h_1 d_0^2$ is the only possibility, since the filtration 2 class $h_4^2 \in E_3(S)$ is not present in the truncated spectral sequence.

(2) It follows from (1) that $E_4(S) = 0$ in topological degree 29, so there are no possible targets for differentials on $h_4^2$. Neither are there any classes in low enough filtration to hit it, so $h_4^2$ survives as a nonzero class in $E_\infty(S)$, and therefore detects a nonzero homotopy class $\theta_4$. □

## 11.3. The Adams $d$- and $e$-invariants

To supplement the Adams spectral sequence information obtained from the $H_\infty$ ring structure on $S$, we will use two related refinements of the unit map $d\colon S \to ko$ to the connective real $K$-theory spectrum. This map $d$ induces the Adams $d$-invariant $d\colon \pi_*(S) \to \pi_*(ko)$. The first refinement is a lift through the homotopy fiber of a map $\tilde{p}\colon ko \to \prod_{i\ge1} \Sigma^{4i} H\mathbb{Z}$, related to Chern and Pontryagin characters. The second refinement is a lift $e\colon S \to j$, where $j$ is the homotopy fiber of a lift $\tilde{\psi}\colon ko \to bspin$ of the Adams operation $\psi^3 - 1\colon ko \to ko$. In the first case it will be more convenient to map the homotopy fiber of the Hurewicz map $h\colon S \to H\mathbb{Z}$ to the homotopy fiber of a map $p\colon ko \to \prod_{i\geq 0} \Sigma^{4i} H\mathbb{Z}$. In the second case, $e$ induces a homomorphism $e\colon \pi_*(S) \to \pi_*(j)$ that is equivalent on $\ker(d)$, up to a sign, to the Adams $e$-invariant. We call $j$ the connective image-of-$J$ spectrum. (This homomorphism $e$ is mostly unrelated to the edge homomorphism in the elliptic spectral sequence, cf. Section 9.3.)

We first recall work of Maunder [**115**], [**116**], building on the construction by Adams [**4**] of characteristic classes $ch_r \in H^{2r}(ku;\mathbb{Z})$, showing that classes of the form $P^k(h_0^j h_m)$ are never boundaries in the Adams spectral sequence for $S$. We outline a spectrum-level proof, similar to that of Ravenel [**144**, §3.4], in Theorem 11.39.

Maunder also applied the Adams $e$-invariant to obtain information about which classes $P^k(h_0^j h_m)$ must support Adams differentials. That information is contained in the statement that $e\colon \pi_*(S) \to \pi_*(j)$ is split surjective, see Theorem 11.47. This is a direct consequence of the Adams conjecture [**6**, Conj. 1.2], which was proved independently by Daniel Quillen [**141**] and Dennis Sullivan [**164**]. A more elementary proof was found later by James Becker and Daniel Gottlieb [**24**].

**11.3.1. Maunder's theorem.** For $k = 2^r \ell$ with $r \geq 0$ and $\ell$ odd, let $P^{2^r}(x) = \langle h_{r+3}, h_0^{2^{r+2}}, x\rangle$ and set $P^k = (P^{2^r})^\ell$, so that it has bidegree $(s,t) = (4k, 12k)$. In the following version of Maunder's results we consider the Adams periodicity operator $P^k$ to be defined in the region near the line $t-s = 2s$ where it is an isomorphism by [**7**, Thm. 1.2]. See also Theorem 4.9.

THEOREM 11.39 ([**116**, Thm. 2.4, Thm. 2.5]).
*(1) The elements $P^k(h_0^j h_m) \in E_2(S)$ are not $d_r$-boundaries, for any $r \geq 2$, $m \geq 2$, $0 \leq j < 2^{m-1}$ and $k$ such that $P^k$ is defined on $h_0^j h_m$.*
*(2) The elements $P^k(a)$, with*

$$a \in \{h_1, h_1^2, h_2, h_0h_2, h_0^2h_2 = h_1^3, h_0^2h_3, h_0^3h_3\}$$

*and $k \geq 0$, are $d_r$-cycles and not $d_r$-boundaries for all $r \geq 2$, hence survive as nonzero classes in $E_\infty(S)$.*

PROOF. We argue using the commutative diagram

$$\begin{array}{ccccc}
S & \xrightarrow{h} & H\mathbb{Z} & \longrightarrow & Ch \\
\Big\downarrow{\scriptstyle d} & & \Big\downarrow{\scriptstyle in_0} & & \Big\downarrow{\scriptstyle g} \\
ko & \xrightarrow{p} & \bigvee_{i\geq 0}\Sigma^{4i}H\mathbb{Z} & \longrightarrow & Cp\,,
\end{array}$$

with horizontal homotopy cofiber sequences. The unit maps $h\colon S \to H\mathbb{Z}$ and $d\colon S \to ko$ induce the Hurewicz homomorphism and Adams $d$-invariant, respectively. Recall from Section 2.6 that $H^*(ko) = A/\!/A(1) = A/A(Sq^1, Sq^2)$ and $H_*(ko) = \mathbb{F}_2[\xi_1^4, \bar\xi_2^2, \bar\xi_3, \bar\xi_4, \dots]$. The left action of $Sq^1$ on $H^*(ko)$ is dual to a right action on $H_*(ko)$, given by $\xi_1^4 \mapsto 0$, $\bar\xi_2^2 \mapsto 0$ and $\bar\xi_k \mapsto \bar\xi_{k-1}^2$ for $k \geq 3$, with Margolis homology

$$H(H_*(ko), Sq^1) = \mathbb{F}_2[\xi_1^4]\,.$$

The projection $H_*(ko) \to \mathbb{F}_2[\xi_1^4]$ sending $\xi_1^4$ to itself, and sending $\bar\xi_2^2$ and $\bar\xi_k$ for $k \geq 3$ to zero, is a homomorphism of $A(0)_*$-comodule algebras, adjoint to a homomorphism

$$H_*(ko) \longrightarrow A_* \,\square_{A(0)_*}\, \mathbb{F}_2[\xi_1^4] \cong H_*(H\mathbb{Z}) \otimes \mathbb{F}_2[\xi_1^4]$$

of $A_*$-comodule algebras. As shown by Adams [**4**, Thm. 2(5)] it is realized by a map of spectra

$$p\colon ko \longrightarrow \bigvee_{i\geq 0} \Sigma^{4i} H\mathbb{Z} \simeq \prod_{i\geq 0} \Sigma^{4i} H\mathbb{Z}$$

with $i$-th component $p_i\colon ko \to \Sigma^{4i} H\mathbb{Z}$ an integral lift of the map $ko \to \Sigma^{4i} H$ representing $\chi Sq^{4i} \in H^{4i}(ko)$, dual to $\xi_1^{4i}$ in the monomial basis generated by the conjugate classes $\xi_1^4$, $\bar{\xi}_2^2$, $\bar{\xi}_k$ for $k \geq 3$. (The maps $p_i$ are the composites of the natural map $ko \to ku$ and the maps constructed by Adams in degrees $4i$. They are related to, but not equal to, the Pontryagin classes in $H^{4i}(BO;\mathbb{Z})$.)

Let $in_0\colon H\mathbb{Z} \to \bigvee_{i\geq 0} \Sigma^{4i} H\mathbb{Z}$ be the inclusion of the 0-th summand. Then $p \circ d \simeq in_0 \circ h$, and we get an induced map $g\colon Ch \to Cp$ of homotopy cofibers, well-defined up to homotopy. The maps $h$ and $p$ induce surjections in cohomology, and the resulting map

$$\begin{array}{ccccccccc}
0 & \longleftarrow & \mathbb{F}_2 & \xleftarrow{h^*} & A/A(Sq^1) & \longleftarrow & \ker(h^*) & \longleftarrow & 0 \\
 & & \uparrow & & \uparrow{\scriptstyle pr_0} & & \uparrow{\scriptstyle g^*} & & \\
0 & \longleftarrow & A/A(Sq^1, Sq^2) & \xleftarrow{p^*} & \bigoplus_{i\geq 0} \Sigma^{4i} A/A(Sq^1) & \longleftarrow & \ker(p^*) & \longleftarrow & 0
\end{array}$$

of $A$-module extensions leads to a map

$$\begin{array}{ccccccccccc}
\dots & \longrightarrow & E_2(S) & \xrightarrow{h_*} & \mathbb{F}_2[h_0] & \longrightarrow & E_2(Ch) & \xrightarrow{\delta} & \Sigma^{-1,0} E_2(S) & \longrightarrow & \dots \\
 & & \downarrow & & \downarrow{\scriptstyle in_0} & & \downarrow{\scriptstyle g_*} & & \downarrow & & \\
\dots & \longrightarrow & E_2(ko) & \xrightarrow{p_*} & \bigoplus_{i\geq 0} \Sigma^{0,4i} \mathbb{F}_2[h_0] & \longrightarrow & E_2(Cp) & \longrightarrow & \Sigma^{-1,0} E_2(ko) & \longrightarrow & \dots
\end{array}$$

of long exact sequences of Adams $E_2$-terms. The suspensions refer to the $(s,t)$-bigrading, not the $(t-s,s)$-bigrading.

By the geometric boundary theorem [**38**, Prop.], the connecting homomorphism $\delta\colon E_2^{s,t}(Ch) \to E_2^{s+1,t}(S)$ is $h_0$- and $h_1$-linear and commutes with the differentials in these Adams spectral sequences. It is an isomorphism for $t-s>0$, mapping the class $\tilde{h}_m \in E_2^{0,2^m}(Ch)$ dual to the $A$-module indecomposable $\chi Sq^{2^m} \in \ker(h^*)$ to the standard generator $h_m \in E_2^{1,2^m}(S)$, for all $m \geq 1$. (Note that $\chi Sq^{2^m} \equiv Sq^{2^m}$ modulo decomposables.) It also maps the classes

$$\tilde{a} \in \{\tilde{h}_1, h_1\tilde{h}_1, \tilde{h}_2, h_0\tilde{h}_2, h_0^2\tilde{h}_2 = h_1^2\tilde{h}_1, h_0^2\tilde{h}_3, h_0^3\tilde{h}_3\}$$

in $E_2(Ch)$ to the corresponding classes $a$ in $E_2(S)$, as in the statement of the theorem. Hence it will suffice to prove that the elements $P^k(h_0^j\tilde{h}_m)$ and $P^k(\tilde{a})$ are not boundaries in the Adams spectral sequence for $Ch$, and that the classes $P^k(\tilde{a})$ are infinite cycles.

The Margolis homology of $Sq^1$ acting on $H_*(H\mathbb{Z}) = \mathbb{F}_2[\xi_1^2, \bar{\xi}_2, \bar{\xi}_3, \dots]$ is $\mathbb{F}_2$, and $h$ and $p$ induce isomorphisms

$$H(H_*(S), Sq^1) \xrightarrow{\cong} H(H_*(H\mathbb{Z}), Sq^1)$$
$$H(H_*(ko), Sq^1) \xrightarrow{\cong} H(H_*(\bigvee_{i\geq 0} \Sigma^{4i} H\mathbb{Z}), Sq^1)\,.$$

It follows [**11**, Thm. 2.1] that the $A_*$-comodules $\mathrm{cok}(h_*)$ and $\mathrm{cok}(p_*)$, and the dual $A$-modules $\ker(h^*)$ and $\ker(p^*)$, are all free as $A(0)$-modules. They are concentrated in degrees $* \geq 2$, so we are in a position to use the vanishing and periodicity theorems of Adams [**7**, Thm. 2.1 and Thm. 5.4]. First,

$$E_2^{s,t}(Ch) = \mathrm{Ext}_A^{s,t}(\ker(h^*), \mathbb{F}_2)$$
$$E_2^{s,t}(Cp) = \mathrm{Ext}_A^{s,t}(\ker(p^*), \mathbb{F}_2)$$

are both 0 in the region $t < 2 + T(s)$, where

$$\begin{aligned} T(4k) &= 12k \\ T(4k+1) &= 12k+2 \\ T(4k+2) &= 12k+4 \\ T(4k+3) &= 12k+7\,. \end{aligned}$$

It follows that each $P^k(\tilde{a})$ is an infinite cycle, for $k \geq 0$ and $\tilde{a}$ as above, since all Adams differentials on these classes land in trivial bidegrees. Second, the Adams periodicity operators

$$P\colon E_2^{s,t}(Ch) \longrightarrow E_2^{s+4,t+12}(Ch)$$
$$P\colon E_2^{s,t}(Cp) \longrightarrow E_2^{s+4,t+12}(Cp)$$

are both isomorphisms for $s > 0$ and $t < 2 + \min\{4s, 8 + T(s-1)\}$. Now $E_2(ko) = \mathrm{Ext}_{A(1)}(\mathbb{F}_2, \mathbb{F}_2) = \mathbb{F}_2[h_0, h_1, v, w_1]/(h_0h_1, h_1^3, h_1v, v^2 + h_0^2w_1)$ and $\bigoplus_{i\geq 0} \Sigma^{0,4i}\mathbb{F}_2[h_0]$ both have rank 1 in bidegrees $(t-s, s) = (4i, s)$ for $i \geq 0$ and $s$ large. Since $E_2(Cp)$ vanishes in these bidegrees, and the target of $p_*$ is $h_0$-torsion free, we must have $p_*(w_1^k) = \Sigma^{0,8k}h_0^{4k}$ and $p_*(vw_1^k) = \Sigma^{0,8k+4}h_0^{4k+3}$ for $k \geq 0$. It follows that $E_2(Cp)$ is an extension

$$0 \to \bigoplus_{k\geq 0}\Big(\Sigma^{0,8k}\mathbb{F}_2[h_0]/(h_0^{4k}) \oplus \Sigma^{0,8k+4}\mathbb{F}_2[h_0]/(h_0^{4k+3})\Big)$$
$$\longrightarrow E_2(Cp) \longrightarrow \Sigma^{-1,0}\mathbb{F}_2[w_1]\{h_1, h_1^2\} \to 0\,,$$

see Figure 11.8.

The $A$-module indecomposable $\chi Sq^{4i} + \Sigma^{4i}1$ in $\ker(p^*)$ maps under $g^*$ to $\chi Sq^{4i}$ in $\ker(h^*)$, which is indecomposable precisely if $4i = 2^m$ for some $m \geq 2$. It follows that, in these cases, the homomorphism $g_*$ maps $\tilde{h}_m \in E_2^{0,4i}(Ch)$ to $\Sigma^{0,4i}1 \in E_2^{0,4i}(Cp)$. By $h_0$-linearity, $g_*$ maps $h_0^j\tilde{h}_m$ to $\Sigma^{0,4i}h_0^j$ for each $j \geq 0$. In particular, $h_0^j\tilde{h}_m$ is nonzero for $0 \leq j < 2^{m-1}$ when $m \geq 3$, since $\Sigma^{0,4i}h_0^j \neq 0$ in these cases. Furthermore, $h_1^2\tilde{h}_1 = h_0^2\tilde{h}_2$ maps to $\Sigma^{0,4}h_0^2 \neq 0$. By $h_1$-linearity, this implies that $g_*$ maps $\tilde{h}_1$ to $\Sigma^{-1,0}h_1$ and $h_1\tilde{h}_1$ to $\Sigma^{-1,0}h_1^2$, and that $h_1 \cdot \Sigma^{-1,0}h_1^2 = \Sigma^{0,4}h_0^2$. By Adams periodicity it then follows that $h_1 \cdot \Sigma^{-1,0}h_1w_1^k = \Sigma^{0,8k+4}h_0^{4k+2} \neq 0$ for each $k \geq 0$, as indicated in Figure 11.8.

These $h_1$-extensions imply there is no room for any nonzero differentials in the Adams spectral sequence for $Cp$, so that $E_2(Cp) = E_\infty(Cp)$. The remaining conclusions now follow from this key vanishing result, by naturality for the map of spectral sequences $g_*\colon E_r(Ch) \to E_r(Cp)$, since any such map must take $d_r$-boundaries to $d_r$-boundaries.

In more detail, we have shown that $g_*$ maps the classes $\tilde{a}$ and $h_0^j\tilde{h}_m$ in $E_2(Ch)$, for $m \geq 2$ and $0 \leq j < 2^{m-1}$, to nonzero classes in $E_2(Cp)$. Applying the Adams periodicity operator $P^k$, in the range where it is known to act isomorphically, we deduce that $g_*$ maps the classes $P^k(\tilde{a})$ and $P^k(h_0^j\tilde{h}_m)$ to nonzero classes in $E_2(Cp)$,

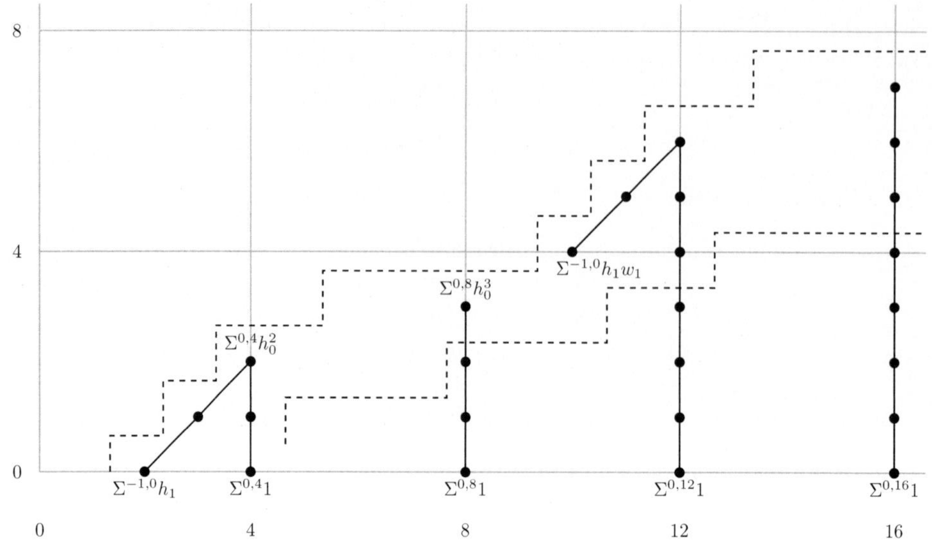


FIGURE 11.8. $E_2(Cp) = E_\infty(Cp)$ for $t - s \le 16$, with vanishing and periodicity range indicated by upper and lower dashed lines

for the same $m$ and $j$ as above. Since the target classes are not $d_r$-boundaries for any $r \ge 2$, it follows by naturality that the source classes $P^k(\tilde{a})$ and $P^k(h_0^j\tilde{h}_m)$ can also not be $d_r$-boundaries for any $r \ge 2$. □

**11.3.2. The image-of-$J$ spectrum.** We now review the construction of the connective image-of-$J$ spectrum $j$ and the map $e\colon S \to j$ merging the Adams $d$- and $e$-invariants, in the context of $E_\infty$ ring spectra.

Recall from Section 2.6 that the connective real $K$-theory spectrum $ko$ is an $E_\infty$ ring spectrum with homotopy ring $\pi_*(ko) = \mathbb{Z}[\eta, A, B]/(2\eta, \eta^3, \eta A, A^2 - 4B)$, where $A \in \pi_4(ko)$ and $B \in \pi_8(ko)$. The unit map $d\colon S \to ko$ induces homomorphisms $d\colon \pi_n(S) \to \pi_n(ko)$, which can only be nontrivial for $n = 0$ and for $n = 8k + r$ with $k \ge 0$ and $r \in \{1, 2\}$, since the groups $\pi_n(S)$ are finite for $n > 0$. Adams proved that $d$ is nontrivial in all of these degrees.

THEOREM 11.40 ([**8**, Thm. 1.2]). *There are unique classes $\mu_{8k+1} \in \pi_{8k+1}(S)$ and $\eta\mu_{8k+1} \in \pi_{8k+2}(S)$ detected by $P^kh_1$ and $h_1P^kh_1$ in the Adams spectral sequence. They are of order 2, with d-invariants $\eta B^k \in \pi_{8k+1}(ko)$ and $\eta^2B^k \in \pi_{8k+2}(ko)$, respectively.*

PROOF. We outline the proof. Adams constructs a map $\alpha\colon S^8/2 \to S/2$ inducing an isomorphism in $K$-theory. It has Adams filtration 4, so $\alpha^k\colon S^{8k}/2 \to S/2$ has Adams filtration $\ge 4k$. Let $\bar{\eta}\colon S^1/2 \to S$ be an extension of $\eta\colon S^1 \to S$ over $i\colon S^1 \to S^1/2$, and let $\mu_{8k+1}$ be the composite

$$S^{8k+1} \xrightarrow{i} S^{8k+1}/2 \xrightarrow{\alpha^k} S^1/2 \xrightarrow{\bar{\eta}} S\,.$$

By construction, $2 \cdot \mu_{8k+1} = 0$. Furthermore, $d(\mu_{8k+1}) = \eta B^k \neq 0$ and $\mu_{8k+1}$ has Adams filtration $\ge 4k + 1$. By Adams periodicity in $\mathrm{Ext}_A(\mathbb{F}_2, \mathbb{F}_2)$ (or by explicit calculation in the range of degrees we are considering), this means that $\mu_{8k+1}$ can only be detected by $P^k h_1$. In particular, this class in $E_2(S)$ survives to $E_\infty(S)$ and cannot be a boundary.

It follows that $d(\eta\mu_{8k+1}) = \eta^2 B^k$ and that $\eta\mu_{8k+1}$ has Adams filtration $\geq 4k+2$. Again, there is only one possible detecting class in the Adams $E_\infty$-term for $S$, namely $h_1 P^k h_1$. $\square$

REMARK 11.41. In Toda's notation [**171**], $\mu_1 = \eta$, $\mu_9 = \mu$ and $\mu_{17} = \bar{\mu}$.

The definition of the connective image-of-$J$ spectrum and the $e$-invariant map depends on the Adams operations. Let $q$ be a natural number and $X$ a finite cell complex. The complex Adams operation $\psi^q \colon KU(X) \to KU(X)$ is a natural ring homomorphism, satisfying $\psi^q(L) = L^q$ whenever $L$ is the class of a complex line bundle over $X$. The real Adams operation $\psi^q \colon KO(X) \to KO(X)$ is also a natural ring homomorphism, satisfies $\psi^q(L) = L^q$ when $L$ is the class of a real line bundle over $X$, and $\psi^q \circ c = c \circ \psi^q$ where $c \colon KO(X) \to KU(X)$ denotes complexification. See [**5**, Thm. 4.1]. After inverting $q$, these operations become stable, and can be represented by spectrum maps $\psi^q \colon KU[1/q] \to KU[1/q]$ and $\psi^q \colon KO[1/q] \to KO[1/q]$. In particular, we have maps $\psi^q \colon KU_p^\wedge \to KU_p^\wedge$ and $\psi^q \colon KO_p^\wedge \to KO_p^\wedge$ for each prime $p$ that does not divide $q$. Passing to connective covers, we get spectrum maps $\psi^q \colon ku_p^\wedge \to ku_p^\wedge$ and $\psi^q \colon ko_p^\wedge \to ko_p^\wedge$.

In fact, these can be realized as $E_\infty$ ring maps. One way to see this is to use the discrete models for topological $K$-theory discussed by May and Jørgen Tornehave in [**121**, Ch. VIII]. Suppose that $q$ is a prime power and let $\mathbb{F}_q$ be a field with $q$ elements and algebraic closure $\bar{\mathbb{F}}_q$. The bipermutative category $\mathcal{GL}(\bar{\mathbb{F}}_q)$ of finite dimensional $\bar{\mathbb{F}}_q$-vector spaces has a bipermutative subcategory $\mathcal{O}(\bar{\mathbb{F}}_q)$ in which the morphisms respect a standard inner product, i.e., are represented by orthogonal matrices. See [**121**, Ex. VI.5.3]. Let $K(\bar{\mathbb{F}}_q)$ and $KO(\bar{\mathbb{F}}_q)$ denote the associated $E_\infty$ ring spectra. As a consequence of Quillen's work on the algebraic $K$-theory of finite fields [**142**], there are equivalences $K(\bar{\mathbb{F}}_q)_p^\wedge \simeq ku_p^\wedge$ and $KO(\bar{\mathbb{F}}_q)_p^\wedge \simeq ko_p^\wedge$. The Frobenius automorphism $\phi^q \colon \bar{\mathbb{F}}_q \to \bar{\mathbb{F}}_q$, given by $\phi^q(x) = x^q$, induces functors $\phi^q \colon \mathcal{GL}(\bar{\mathbb{F}}_q) \to \mathcal{GL}(\bar{\mathbb{F}}_q)$ and $\phi^q \colon \mathcal{O}(\bar{\mathbb{F}}_q) \to \mathcal{O}(\bar{\mathbb{F}}_q)$ respecting the bipermutative structure. The induced maps $\phi^q \colon K(\bar{\mathbb{F}}_q)_p^\wedge \to K(\bar{\mathbb{F}}_q)_p^\wedge$ and $\phi^q \colon KO(\bar{\mathbb{F}}_q)_p^\wedge \to KO(\bar{\mathbb{F}}_q)_p^\wedge$ are then $E_\infty$ ring spectrum models for the $p$-adically completed Adams operations $\psi^q$. See [**121**, Thm. VIII.2.9].

Alternatively, we can appeal to the height 1 case of the Goerss–Hopkins–Miller theorem [**65**, Cor. 7.7], showing that $KU_p^\wedge = E(\hat{\mathbb{G}}_m, \mathbb{F}_p)$ is an $E_\infty$ ring spectrum, and the space of $E_\infty$ ring maps $KU_p^\wedge \to KU_p^\wedge$ has set of path components isomorphic to the automorphism group $\mathrm{Aut}(\hat{\mathbb{G}}_m, \mathbb{F}_p) \cong \mathbb{Z}_p^\times$, with each path component being contractible. Each $p$-adic unit $k \in \mathbb{Z}_p^\times$ then corresponds to an $E_\infty$ ring map $\psi^k \colon KU_p^\wedge \to KU_p^\wedge$, up to contractible choice. In particular, $\psi^{-1}$ acts as complex conjugation, and we can recover $KO_p^\wedge$ as the homotopy fixed points of the $C_2$-action on $KU_p^\wedge$ generated by $\psi^{-1}$. Furthermore, $\psi^k$ then induces an $E_\infty$ ring map $\psi^k \colon KO_p^\wedge \to KO_p^\wedge$, since $\psi^k$ and $\psi^{-1}$ commute up to contractible choice. Passing to connective covers, we have $E_\infty$ ring maps $\psi^k \colon ku_p^\wedge \to ku_p^\wedge$ and $\psi^k \colon ko_p^\wedge \to ko_p^\wedge$ for all $p$-adic units $k$.

The operations $\psi^q \colon \widetilde{KU}(S^2) \to \widetilde{KU}(S^2)$ and $\psi^q \colon \widetilde{KO}(S^1) \to \widetilde{KO}(S^1)$ are given by multiplication by $q$. Hence $\psi^k \colon ku_p^\wedge \to ku_p^\wedge$ acts on $\pi_*(ku_p^\wedge) \cong \mathbb{Z}_p[u]$ by $\psi^k(u) =$

$k \cdot u$, while $\psi^k \colon ko_p^\wedge \to ko_p^\wedge$ acts on $\pi_*(ko_p^\wedge)$ by $\psi^k(\eta) = k\eta$, $\psi^k(A) = k^2 A$ and $\psi^k(B) = k^4 B$.

We now concentrate on the operation $\psi^3 \colon ko_2^\wedge \to ko_2^\wedge$, with $\psi^3(\eta) = \eta$, $\psi^3(A) = 3^2 A$ and $\psi^3(B) = 3^4 B$. In the remainder of this section, all spectra are implicitly 2-completed.

DEFINITION 11.42. Let $ko^{h\psi^3} = \mathrm{hoeq}(\psi^3, 1)$ be the homotopy equalizer in the following diagram of $E_\infty$ ring spectra, and let $\bar{e} \colon S \to ko^{h\psi^3}$ be the unit map, lifting $d$. It exists because $\psi^3$ is unital, so that $\psi^3 \circ d = d$.

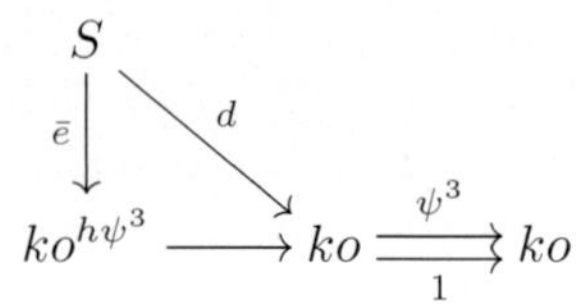

The forgetful functor from $E_\infty$ ring spectra to spectra respects the formation of homotopy equalizers. Hence there is a homotopy cofiber sequence

$$ko^{h\psi^3} \longrightarrow ko \xrightarrow{\psi^3 - 1} ko$$

of spectra. It follows that $\pi_n(ko^{h\psi^3}) \cong (\mathbb{Z}_2, \mathbb{Z}/2 \oplus \mathbb{Z}, \mathbb{Z}/2 \oplus \mathbb{Z}/2, \mathbb{Z}/2)$ for $-1 \le n \le 2$, whereas $\pi_n(S) \cong \pi_n(ko) \cong (0, \mathbb{Z}_2, \mathbb{Z}/2, \mathbb{Z}/2)$ in this range. We use Postnikov sections $X \to \tau_{\le 1} X$ for $X = S$ and $X = ko^{h\psi^3}$ to find a modification $j$ of $ko^{h\psi^3}$ that agrees with $S$, in this range of degrees. The original approach in [**121**, Def. VIII.3.1] used discrete models and the spinor norm [**121**, Ex. VI.5.7] to obtain this modification.

DEFINITION 11.43. Let $j$ be the homotopy pullback in the following diagram of $E_\infty$ ring spectra, and let $e \colon S \to j$ be the induced $E_\infty$ ring map to the homotopy pullback.

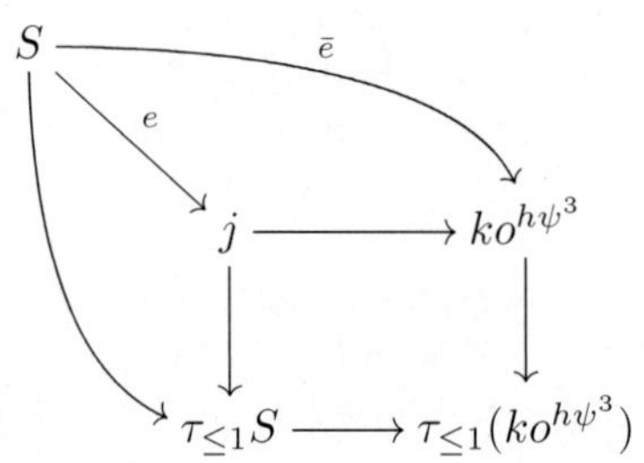

LEMMA 11.44. *There is a homotopy cofiber sequence of (implicitly* 2*-completed) spectra*

$$j \longrightarrow ko \xrightarrow{\tilde{\psi}} bspin \xrightarrow{\partial} \Sigma j\,,$$

*where bspin* $\to$ *ko is the* 3*-connected cover and* $\tilde{\psi}$ *lifts* $\psi^3 - 1 \colon ko \to ko$.

PROOF. The composite map $\tau_{\le 1} d \colon \tau_{\le 1} S \to \tau_{\le 1}(ko^{h\psi^3}) \to \tau_{\le 1} ko$ is an equivalence, so $\tau_{\le 1} S \to \tau_{\le 1}(ko^{h\psi^3})$ is 1-coconnected, i.e., induces an isomorphism on $\pi_n$ for $n > 1$ and an injection for $n = 1$. Hence its pullback $j \to ko^{h\psi^3}$ is also

1-coconnected. Consider the map of horizontal homotopy cofiber sequences

$$\begin{array}{ccccc} j & \longrightarrow & ko & \xrightarrow{\tilde{\psi}} & X \\ \downarrow & & \| & & \downarrow{\scriptstyle \pi} \\ ko^{h\psi^3} & \longrightarrow & ko & \xrightarrow{\psi^3-1} & ko \end{array}$$

associated to the factorization $j \to ko^{h\psi^3} \to ko$. Here $j \to ko$ is 3-connected, i.e., induces an isomorphism on $\pi_n$ for $n < 3$ and a surjection for $n = 3$, so its homotopy cofiber $X$ is 3-connected. Furthermore $\pi\colon X \to ko$ is 2-coconnected, so this map exhibits $X$ as the 3-connected cover $bspin$ of $ko$. □

DEFINITION 11.45. Let $j_{8k-1} = \partial(B^k) \in \pi_{8k-1}(j)$ for $k \geq 1$, and let $j_{8k+1} = e(\mu_{8k+1}) \in \pi_{8k+1}(j)$ and $j_{8k+3} = \partial(AB^k) \in \pi_{8k+3}(j)$ for $k \geq 0$.

LEMMA 11.46. *The map $e\colon S \to j$ is (at least) 2-connected, and for $n \geq 2$*

$$\pi_n(j) = \begin{cases} \mathbb{Z}_2/(16k)\{j_{8k-1}\} & \textit{for } n = 8k-1, \\ \mathbb{Z}/2\{\eta j_{8k-1}\} & \textit{for } n = 8k, \\ \mathbb{Z}/2\{\eta^2 j_{8k-1}\} \oplus \mathbb{Z}/2\{j_{8k+1}\} & \textit{for } n = 8k+1, \\ \mathbb{Z}/2\{\eta j_{8k+1}\} & \textit{for } n = 8k+2, \\ \mathbb{Z}/8\{j_{8k+3}\} & \textit{for } n = 8k+3, \\ 0 & \textit{otherwise}, \end{cases}$$

*with $\nu j_{8k-1} = 0$ and $\eta^2 j_{8k+1} = 4 j_{8k+3}$.*

PROOF. This is mostly clear from the long exact sequence in homotopy associated to the homotopy cofiber sequence in Lemma 11.44. The lift $\tilde{\psi}$ sends $B^k$ to $(3^{4k}-1)B^k$, and $\mathrm{ord}_2(3^{4k}-1) = 4 + \mathrm{ord}_2(k) = \mathrm{ord}_2(16k)$. It also sends $AB^k$ to $(3^{4k+2}-1)AB^k$, and $\mathrm{ord}_2(3^{4k+2}-1) = 3 = \mathrm{ord}_2(8)$. The short exact sequence

$$0 \to \pi_{8k+2}(bspin) \xrightarrow{\partial} \pi_{8k+1}(j) \longrightarrow \pi_{8k+1}(ko) \to 0$$

splits, because $j_{8k+1} = e(\mu_{8k+1})$ has order 2 and maps to $\eta B^k$ in $\pi_{8k+1}(ko)$.

Since $\nu B^k$ lies in $\pi_{8k+3}(ko) = 0$ we must have $\nu j_{8k-1} = \partial(\nu B^k) = 0$. It remains to show that $\eta^2 j_{8k+1} \neq 0$, since $4 j_{8k+3}$ in the only element of order 2 in $\pi_{8k+3}(j)$. To see this, we use the commutative diagram

$$\begin{array}{ccccc} Y & \longrightarrow & \Sigma ko & \longrightarrow & Z \\ \downarrow & & \downarrow{\scriptstyle \eta} & & \downarrow \\ j & \longrightarrow & ko & \xrightarrow{\tilde{\psi}} & bspin \\ \downarrow & & \downarrow{\scriptstyle c} & & \downarrow \\ ju & \longrightarrow & ku & \xrightarrow{\tilde{\psi}} & bu \end{array}$$

of horizontal and vertical homotopy cofiber sequences. Here $ju$ denotes the complex image-of-$J$ spectrum, defined as the connective cover of $ku^{h\psi^3}$. The middle vertical cofiber sequence expresses part of the real Bott periodicity theorem. The lift $\tilde{\psi}\colon ku \to bu$ of $\psi^3 - 1$ multiplies by $3^k - 1$ in degree $2k \geq 0$, so $\pi_n(ju) = 0$ for $n \geq 2$ even. Hence $\pi_{8k+2}(Y) \to \pi_{8k+2}(j)$ is surjective and $\pi_{8k+3}(Y) \to \pi_{8k+3}(j)$ is injective. Let $x \in \pi_{8k+2}(Y)$ be a lift of $\eta j_{8k+1} \in \pi_{8k+2}(j)$. Their common image in

$\pi_{8k+2}(ko)$ is $\eta d(\mu_{8k+1}) = \eta^2 B^k \neq 0$, so the image of $x$ in $\pi_{8k+2}(\Sigma ko)$ must be the nonzero class $\Sigma\eta B^k$. Multiplying by $\eta$, the image of $\eta x$ in $\pi_{8k+3}(\Sigma ko)$ is $\Sigma\eta^2 B^k \neq 0$, so $\eta x \neq 0$. Due to the injectivity noted above, the image $\eta \cdot \eta j_{8k+1} \in \pi_{8k+3}(j)$ of $\eta x$ is also nonzero. □

We can now formulate a key consequence of the confirmed Adams conjecture. We shall appeal to this theorem to determine some of the differentials originating in topological degree $8k-1$ of the Adams spectral sequence for the sphere.

THEOREM 11.47 ([**6**, Conj. 1.2], [**141**], [**164**]). *The ring map $e\colon S \to j$ induces a surjective ring homomorphism*

$$e\colon \pi_*(S) \longrightarrow \pi_*(j)$$

*which admits an additive section.*

REMARK 11.48. This is a substantial theorem, and we just mention the role of the main references. Let $SO$ denote the infinite special orthogonal group. Adams [**8**, Thm. 7.16] showed that Whitehead's [**177**] $J$-homomorphism $J\colon \pi_n(SO) \to \pi_n(S)$ creates enough elements in $\pi_*(S)$ to ensure that $e\colon \pi_n(S) \to \pi_n(j)$ is surjective. More precisely, he showed that $e$ is split surjective for all $n$, up to a possible factor of 2 in the cases when $n = 8k-1$. The subsequent proofs of the Adams conjecture, first by Quillen [**141**] and by Sullivan [**164**], thereafter by Becker and Gottlieb [**24**], eliminated the remaining factor of 2. Davis and Mahowald [**100**], [**53**, Thm. 1.1] determined the precise Adams filtration of the classes in the image of the $J$-homomorphism, but we shall not rely on this stronger result. However, see Proposition 11.88 for partial information in this direction.

This now allows us to compute the ring structure in $\pi_*(j)$. Since it is generated over $\pi_*(S)$ by the $j_n$, it suffices to determine the products of these generators.

PROPOSITION 11.49 (cf. [**8**, Prop. 12.14 and Ex. 12.15]). *The products of the $j_n$ are given as follows: $j_{8k-1} \cdot j_{8\ell+1} = j_{8k+1} \cdot j_{8\ell-1} = \eta j_{8(k+\ell)-1}$, $j_{8k+1} \cdot j_{8\ell+1} = \eta j_{8(k+\ell)+1}$, and the remaining products are zero.*

PROOF. The products in degrees $* \equiv 4, 6 \mod 8$ are trivially zero. The products in degrees $* \equiv 2 \mod 8$ are detected by their images in $\pi_*(ko)$, since $\pi_*(j) \to \pi_*(ko)$ is an isomorphism in these degrees. Hence $j_{8k-1} \cdot j_{8\ell+3} = 0$, since $j_{8k-1}$ and $j_{8\ell+3}$ both map to zero in $\pi_*(ko)$. On the other hand, $j_{8k+1} \cdot j_{8\ell+1}$ maps to $\eta B^k \cdot \eta B^\ell = \eta^2 B^{k+\ell} \neq 0$, so $j_{8k+1} \cdot j_{8\ell+1} = \eta j_{8(k+\ell)+1}$. In degrees $* \equiv 0 \mod 8$ we calculate $j_{8k+1} \cdot j_{8\ell-1} = e(\mu_{8k+1}) \cdot \partial(B^\ell) = \mu_{8k+1} \cdot \partial(B^\ell) = \partial(\mu_{8k+1} \cdot B^\ell) = \partial(\eta B^k \cdot B^\ell) = \eta j_{8(k+\ell)-1}$, since the $S$-action on $j$ factors through $e$ and the $S$-action on *bspin* factors through the action by $ko$. □

REMARK 11.50. The homotopy cofiber sequence in Lemma 11.44 induces a long exact sequence of $A$-modules

$$\cdots \longrightarrow H^*(bspin) \xrightarrow{\tilde{\psi}^*} H^*(ko) \longrightarrow H^*(j) \xrightarrow{\partial^*} \Sigma^{-1} H^*(bspin) \longrightarrow \dots$$

with $H^*(ko) = A/A(Sq^1, Sq^2)$, $H^*(bspin) = \Sigma^4 A/A(Sq^1, Sq^2 Sq^3)$ and $\tilde{\psi}^*$ mapping $\Sigma^4 1$ to $Sq^4$, cf. [**104**]. Hence there is a (nontrivial) $A$-module extension

$$0 \to C \longrightarrow H^*(j) \longrightarrow K \to 0\,,$$

where

$$C = \mathrm{cok}(\tilde{\psi}^*) \cong A/A(Sq^1, Sq^2, Sq^4)$$

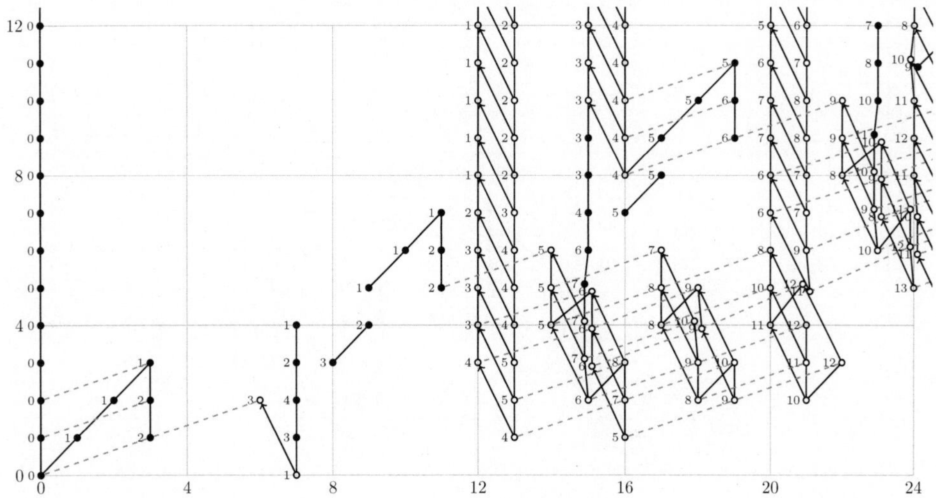

FIGURE 11.9. $(E_2(j), d_2)$ for $t - s \leq 24$

and

$$K = \Sigma^{-1} \ker(\tilde{\psi}^*) \cong \Sigma^7 A/A(Sq^1, Sq^7, Sq^4Sq^6 + Sq^6Sq^4)\,.$$

See [**50**, Thm. 1] or [**15**, Lem. 7.10(c)]. In spite of the isomorphism $C \cong H^*(tmf)$, the monomorphism $C \to H^*(j)$ not induced by a map $j \to tmf$ under $S$, since $\nu^2 \in \pi_6(S)$ is detected by $tmf$ but not by $j$.

The $(E_2, d_2)$-term of the Adams spectral sequence for $j$ is shown, for $t-s \leq 24$, in Figure 11.9. We have recently confirmed [**44**] the first author's conjecture that this spectral sequence collapses at the $E_3$-term, except for a regular pattern of later differentials connecting the $h_0$-towers in topological degrees $32i$ and $32i-1$ for $i \geq 1$. The classes detecting $\eta\rho_{8k-1} \mapsto \eta j_{8k-1}$, $\eta^2\rho_{8k-1} \mapsto \eta^2 j_{8k-1}$, $\mu_{8k+1} \mapsto j_{8k+1}$, $\eta\mu_{8k+1} \mapsto \eta j_{8k+1}$, and $\zeta_{8k+3} \mapsto j_{8k+3}$ map isomorphically at $E_\infty$, while the $h_0$-towers detecting $\langle \rho_{8k-1} \rangle \mapsto \langle j_{8k-1} \rangle$ undergo an Adams filtration shift equal to the 2-adic valuation of $k$: in $\pi_*(S)$, the $h_0$-tower on $\rho_{8k-1}$ ends in Adams filtration $4k$, while in $\pi_*(j)$, the corresponding $h_0$-tower on $j_{8k-1}$ starts in Adams filtration $4k-3$.

## 11.4. Some $d_2$-differentials for $S$

We now reach the main object of study in this chapter: the mod 2 Adams spectral sequence for the sphere spectrum. Its $E_2$-term

$$E_2^{s,t}(S) = \mathrm{Ext}_A^{s,t}(\mathbb{F}_2, \mathbb{F}_2) \Longrightarrow \pi_{t-s}(S)^\wedge_2$$

is given for $t \leq 200$ in Figures 1.1 to 1.8. A list of algebra generators for $t-s \leq 48$ is given in Table 1.1. In this section we will justify the values of the $d_2$-differentials listed in that table.

REMARK 11.51. In Theorems 11.52, 11.54, 11.56 and 11.59 some statements or proofs are marked with an asterisk (*). Logically, we first only prove the statements without this mark. These suffice to give the necessary input for our calculations of the Adams spectral sequence and homotopy groups of $tmf$, given in Chapters 5 and 9. After this we can return to $S$ and use the results about $tmf$ to prove the

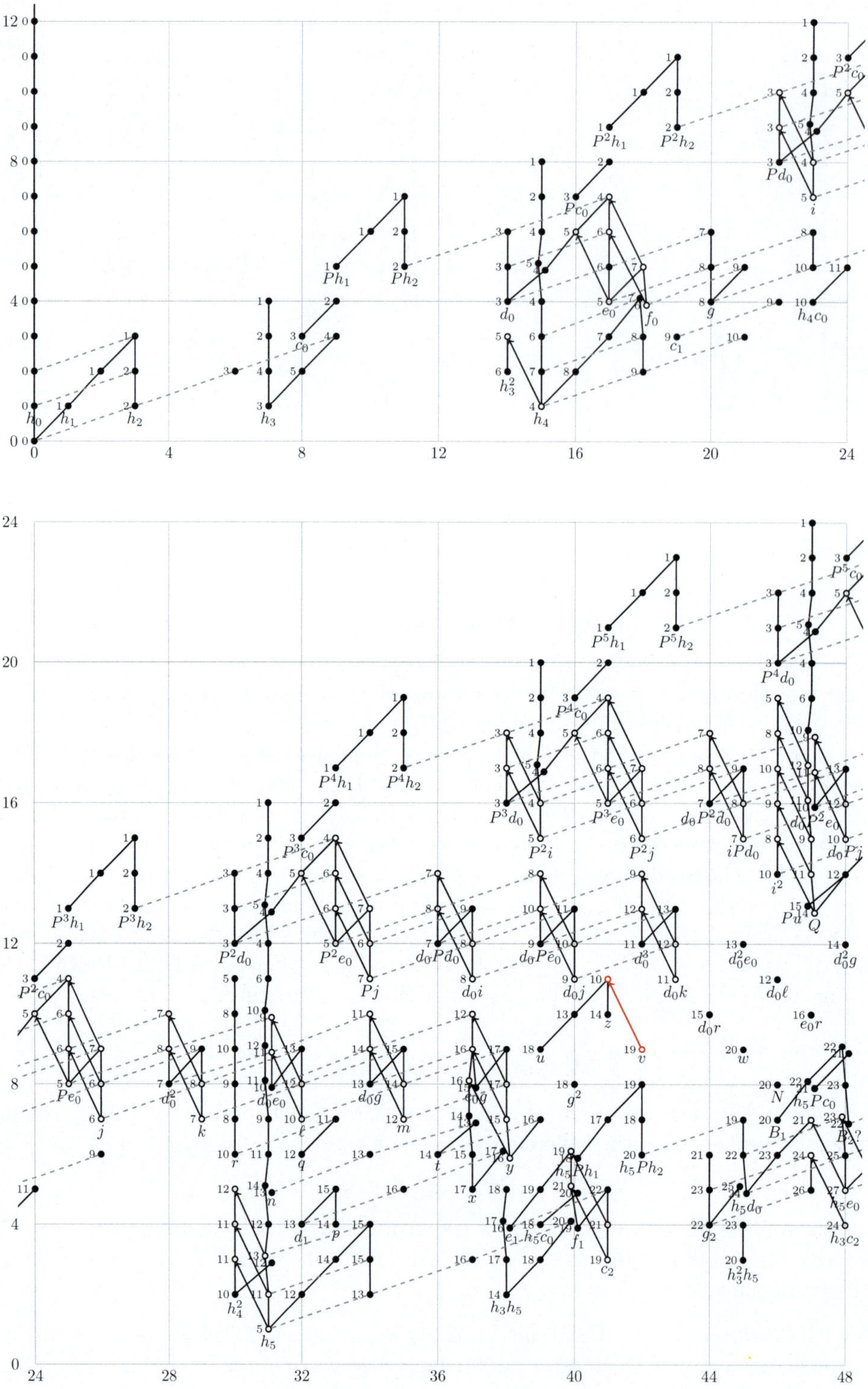

FIGURE 11.10. $(E_2(S), d_2)$ for $t - s \leq 48$

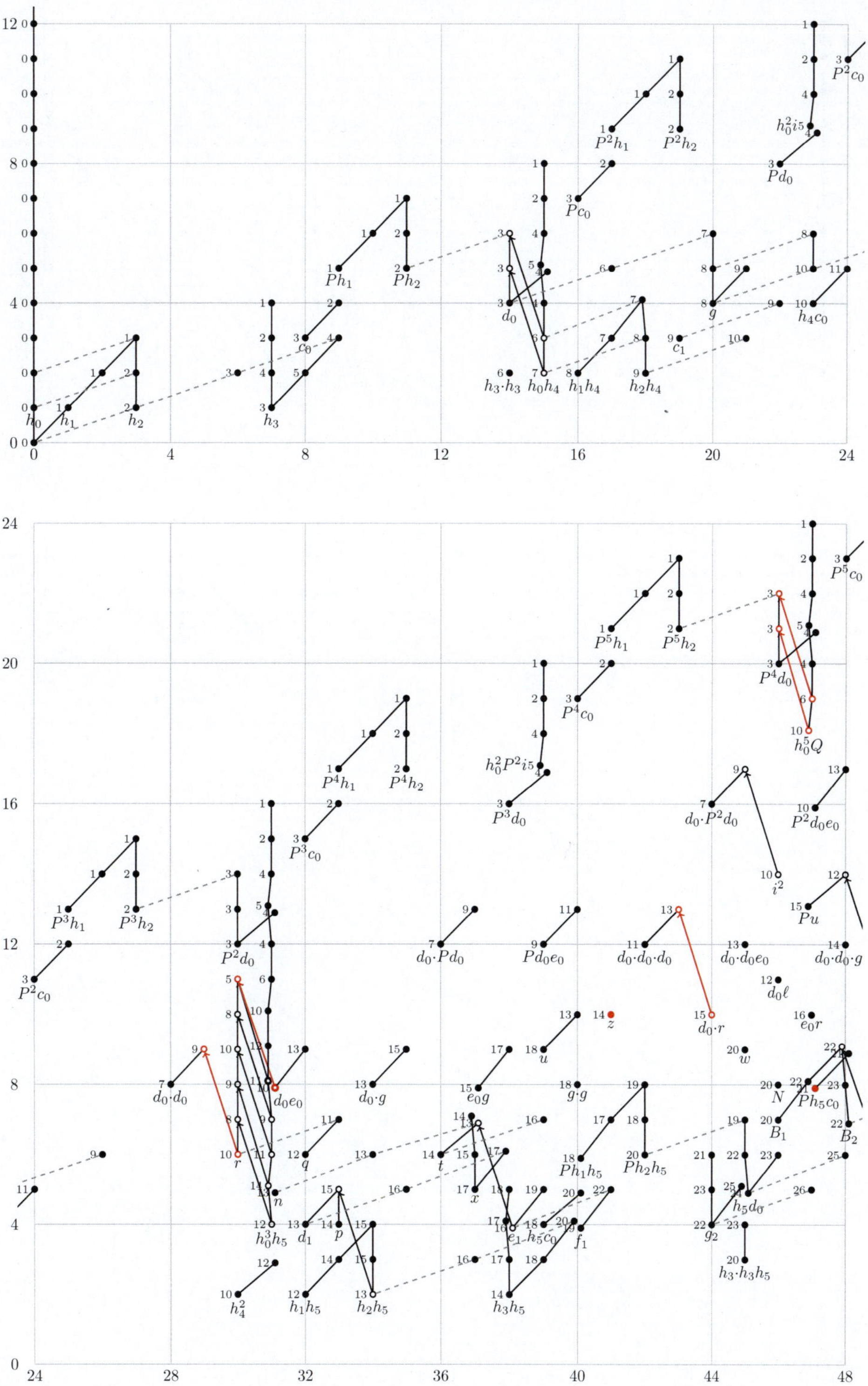

FIGURE 11.11. $(E_3(S), d_3)$ for $t - s \leq 48$

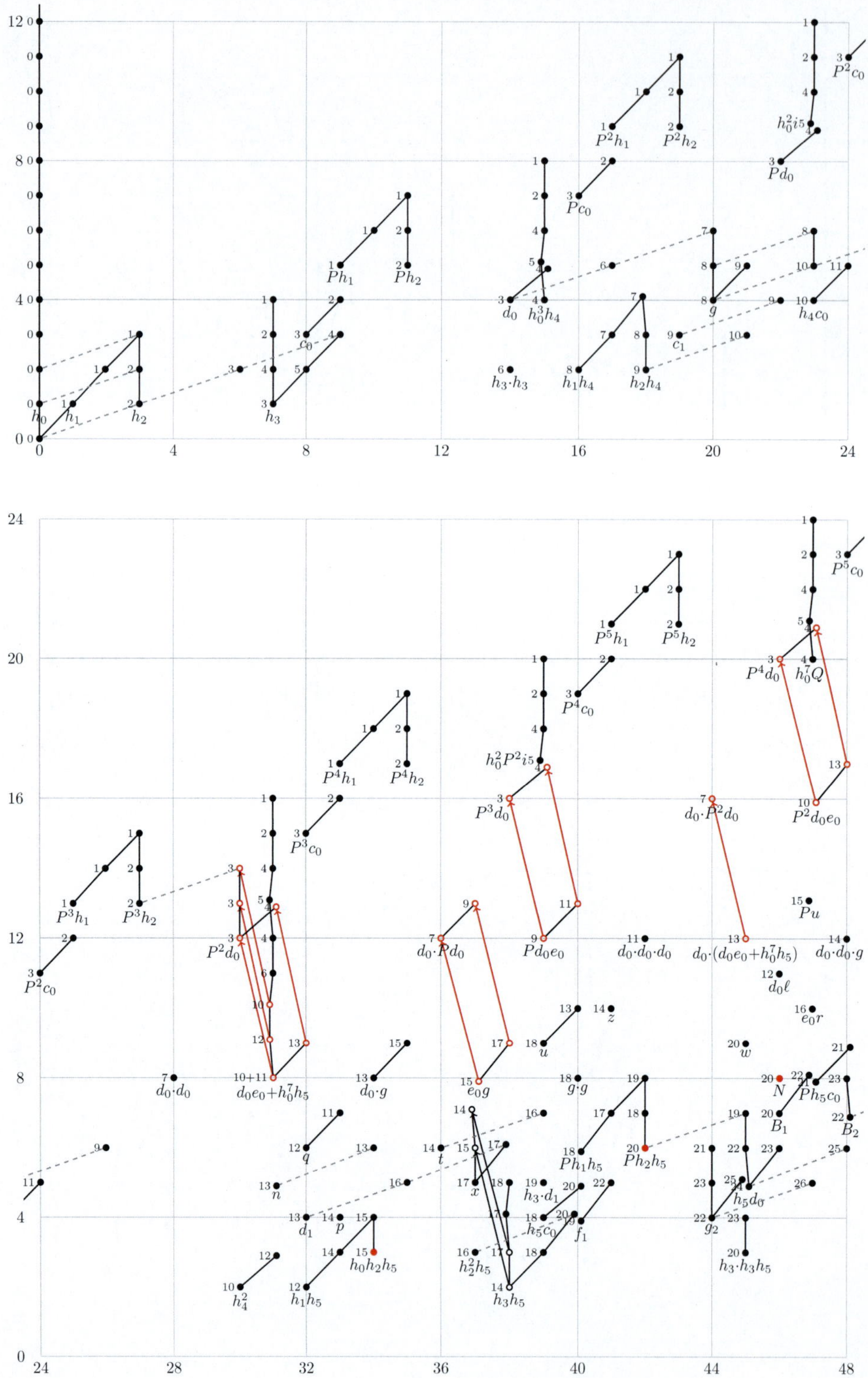

FIGURE 11.12. $(E_4(S), d_4)$ for $t - s \leq 48$

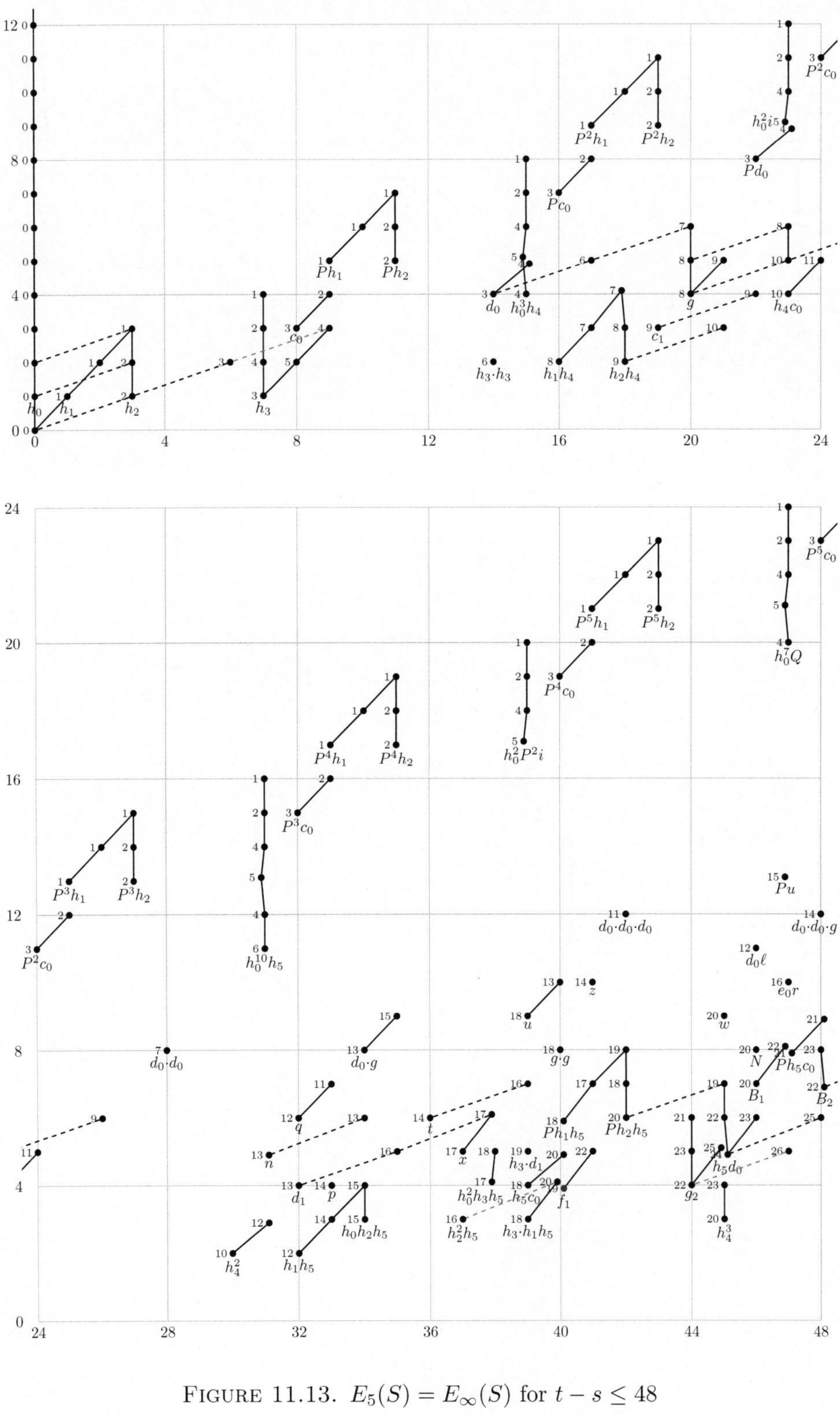

FIGURE 11.13. $E_5(S) = E_\infty(S)$ for $t - s \leq 48$

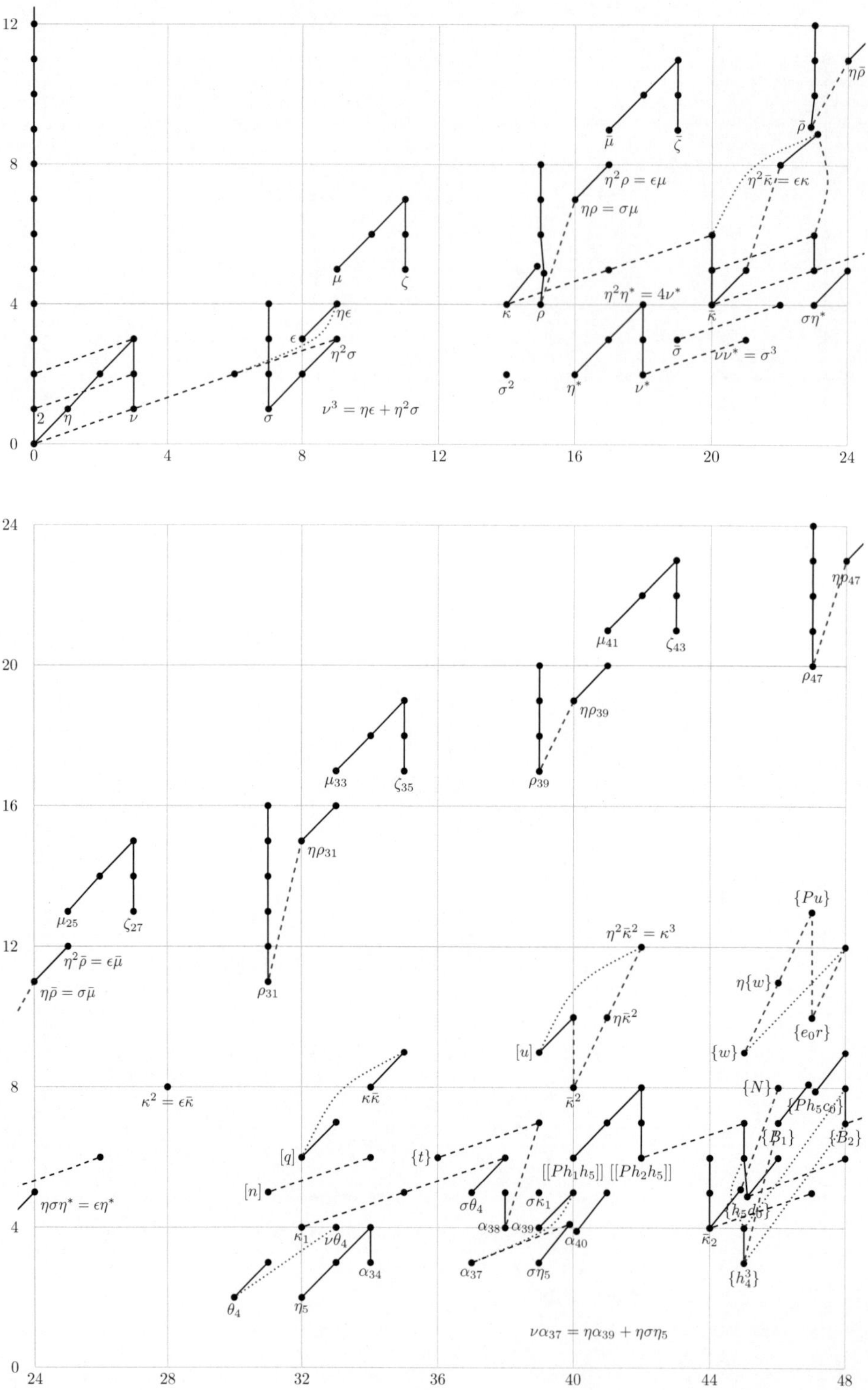

FIGURE 11.14. $\pi_n(S)$ for $n \leq 48$

marked statements. We have chosen to break with the logical order in this presentation in order to have the results collected in one place, and to avoid repetition. The red arrows and red dots in Figures 11.10 through 11.13 indicate nonzero and zero differentials, respectively, for which our most straightforward argument relies on the use of *tmf*.

THEOREM 11.52. *In the mod 2 Adams spectral sequence for $S$:*

(1) $d_2(a) = 0$ *for* $a = h_0$, $h_1$, $h_2$, $h_3$, $c_0$, $Ph_1$, $Ph_2$, $d_0$, $Pc_0$, $P^2h_1$, $P^2h_2$, $g$, $Pd_0$, $P^2c_0$, $P^3h_1$, $P^3h_2$, $P^2d_0$, $n$, $d_1$, $q$, $P^3c_0$, $p$, $P^4h_1$, $P^4h_2$, $t$, $x$, $e_1$, $P^3d_0$, $u$, $f_1$, $P^4c_0$, $z$, $P^5h_1$, $P^5h_2$, $g_2$, $w$, $N$, $P^4d_0$, $Pu$, $B_2$ *and* $P^5c_0$.
(2) $d_2(h_4) = h_0h_3^2$ *and* $d_2(h_5) = h_0h_4^2$.
(3) $d_2(e_0) = h_1^2d_0$, $d_2(f_0) = h_0^2e_0$, $d_2(c_1) = 0$, $d_2(i) = h_0Pd_0$, $d_2(Pe_0) = h_1^2Pd_0$, $d_2(j) = h_0Pe_0$, $d_2(k) = h_0d_0^2$, $d_2(r) = 0$, $d_2(\ell) = h_0d_0e_0$, $d_2(P^2e_0) = h_1^2P^2d_0$, $d_2(Pj) = h_0P^2e_0$, $d_2(m) = h_0d_0g$, $d_2(y) = h_0^3x$, $d_2(P^2i) = h_0P^3d_0$, $d_2(P^3e_0) = h_1^2P^3d_0$ *and* $d_2(P^2j) = h_0P^3e_0$.
(4) $d_2(c_2) = h_0f_1$.
(5) *(*)* $d_2(v) = h_0z$.
(6) $d_2(B_1) = 0$.
(7) $d_2(Q) = h_0i^2$.

The Adams $(E_2, d_2)$-term is shown in Figure 11.10, and the algebra generators for $t - s \le 48$ of the resulting $E_3$-term are listed in Table 11.1.

PROOF. (1) By inspection of $E_2(S)$, it is clear that $d_2(a) = 0$ for the algebra generators $a = h_0$, $h_2$, $h_3$, $c_0$, $Ph_1$, $Ph_2$, $d_0$, $Pc_0$, $P^2h_1$, $P^2h_2$, $g$, $Pd_0$, $P^2c_0$, $P^3h_1$, $P^3h_2$, $P^2d_0$, $P^3c_0$, $P^4h_1$, $P^4h_2$, $x$, $P^3d_0$, $u$, $f_1$, $P^4c_0$, $P^5h_1$, $P^5h_2$, $g_2$, $w$, $N$, $P^4d_0$, $B_2$ and $P^5c_0$ because the target groups are trivial. Furthermore, $d_2(a) = 0$ for $a = h_1$, $n$, $d_1$, $q$, $t$, $e_1$ and $Pu$ by $h_0$-linearity, for $a = p$ by $h_1$-linearity, and for $a = z$ by $h_2$-linearity.

(2) The Adams differentials $d_2(h_{i+1}) = h_0h_i^2$ follow from the $H_\infty$ ring structure on $S$. We apply Theorem 11.22 for $Y = S$ and $x = h_i$, with $Sq^0(h_i) = h_{i+1}$ and $Sq^1(h_i) = h_i^2$ as in Proposition 11.27, to obtain the formula

$$d_*(h_{i+1}) = d_*(Sq^0(h_i)) = Sq^1(d_2(h_i)) \dotplus h_0Sq^1(h_i)\,.$$

Here $Sq^1(d_2(h_i))$ has Adams filtration 4 and $h_0Sq^1(h_i)$ has Adams filtration 3, so the expression simplifies to $d_2(h_{i+1}) = h_0h_i^2$.

(3) The Adams $d_2$-differential on $f_0$ follows from the $H_\infty$ ring structure on $S$. We apply Theorem 11.22 for $x = c_0$, with $Sq^1(c_0) = f_0$ and $Sq^2(c_0) = h_0e_0$ as in Proposition 11.29, to obtain the formula

$$d_*(f_0) = d_*(Sq^1(c_0)) = Sq^2(d_2(c_0)) \dotplus h_0Sq^2(c_0)\,,$$

which simplifies to $d_2(f_0) = h_0^2e_0$. Then $h_i$-linearity gives $d_2(e_0) = h_1^2d_0$, while the relation $d_0f_0 = h_2k$ and $h_i$-linearity give $d_2(k) = h_0d_0^2$, $d_2(j) = h_0Pe_0$, $d_2(Pe_0) = h_1^2Pd_0$, $d_2(i) = h_0Pd_0$, $d_2(\ell) = h_0d_0e_0$, $d_2(m) = h_0d_0g$ and $d_2(h_0y) = h_0^4x$. The $H_\infty$ differential for $Sq^2(f_0) = y$ obtained from Theorems 11.22 and 11.29 improves the last of these to

$$d_*(y) = d_*(Sq^2(f_0)) = Sq^3(h_0^2e_0) \dotplus h_0Sq^3(f_0)\,,$$

which simplifies to $d_2(y) = h_0h_3r = h_0(h_1t + h_0^2x) = h_0^3x$. The relation $h_0d_0i = h_2Pj$ and $h_i$-linearity then give $d_2(Pj) = h_0P^2e_0$ and $d_2(P^2e_0) = h_1^2P^2d_0$. One more application of $d_0$- and $h_i$-linearity then gives $d_2(P^2j) = h_0P^3e_0$, $d_2(P^3e_0) =$

$h_1^2P^3d_0$ and $d_2(P^2i) = h_0P^3d_0$. Finally, $d_2(c_1) \neq h_0f_0$ and $d_2(r) \neq h_0k$ follow from the fact that $d_2 \circ d_2 = 0$, since $d_2(h_0f_0) = h_0^3e_0 \neq 0$ and $d_2(h_0k) = h_0^2d_0^2 \neq 0$. Hence $d_2(c_1)$ and $d_2(r)$ must both be zero.

(Alternatively, we can deduce $d_2(y) = h_0^3x$ from $d_2(h_0y) = h_0^4x$ and $d_2(t) = 0$ using $h_1$- and $h_2$-linearity.)

(4) The Adams $d_2$-differential on $c_2$ follows from the $H_\infty$ ring structure on $S$. We apply Theorem 11.22 for $x = c_1$, with $Sq^0(c_1) = c_2$ and $Sq^1(c_1) = f_1$ as in Corollary 11.31, to obtain

$$d_*(c_2) = d_*(Sq^0(c_1)) = Sq^1(d_2(c_1)) \dotplus h_0Sq^1(c_1)\,,$$

which simplifies to $d_2(c_2) = h_0f_1$.

(This differential was overlooked in Mahowald and Tangora's pioneering 1967 paper [**107**]. It was found by Milgram [**122**] in his systematic application of the differentials implied by the geometric construction of the Steenrod operations in $\mathrm{Ext}_A(\mathbb{F}_2, \mathbb{F}_2)$.)

(5) We prove this using *tmf*. In Lemma 1.15 we showed that the morphism $\iota\colon E_2(S) \to E_2(tmf)$ maps $v$ to $e_0\gamma$, and in Theorem 5.18 we showed that $d_3(e_0\gamma) = w_1 \cdot h_1\delta \neq 0$. If $d_2(v) = 0$, then $d_3(v) = 0$ because the target group is trivial, implying $d_3(e_0\gamma) = 0$ by naturality. This contradiction shows that $d_2(v) \neq 0$, which can only mean that $d_2(v) = h_0z$.

(The original proof [**107**, Prop. 6.1.5] for this differential uses $C\sigma$.)

(6) We use $d_0$- and $h_1$-linearity of $d_2$ to show that $d_2(B_1) \neq w$. This involves classes in $\mathrm{Ext}_A(\mathbb{F}_2, \mathbb{F}_2)$ beyond the range $t-s \leq 48$, for which we refer to Figure 1.3. The relation $d_0B_1 = 11_{22} = h_1B_{21}$ is readily verified by `ext`, where $B_{21} = 10_{24}$. Here $d_2(B_{21}) = 0$ lives in a trivial group, so $d_0 \cdot d_2(B_1) = d_2(d_0B_1) = d_2(h_1B_{21}) = h_1 \cdot 0 = 0$. On the other hand, $d_0 \cdot w = 13_{22} \neq 0$. Hence $d_2(B_1) \neq w$, and $d_2(B_1) = 0$ is the only alternative.

(Alternatively, we can deduce this using *tmf*, since $w$ maps to $\gamma g \in E_2(tmf)$, which is not a $d_2$-boundary.)

(7) We use $h_0^2h_3$- and $h_0d_0^2$-linearity of $d_2$ to show that $d_2(Q) \neq 0$, and $d_2(Q) = h_0i^2$ is the only alternative. Using `ext` we calculate $h_0^2h_3Q = 16_{17} = h_0d_0^2j$, see Figure 1.3. Hence $h_0^2h_3 \cdot d_2(Q) = d_2(h_0^2h_3Q) = d_2(h_0d_0^2j) = h_0d_0^2 \cdot d_2(j) = h_0^2d_0^2Pe_0 = 18_{12} \neq 0$. This implies $d_2(Q) \neq 0$. □

REMARK 11.53. The nonzero $d_2$-differentials landing in topological degree 48 are $d_2(5_{28}) = d_2(h_5f_0) = h_0^2h_5e_0 = 7_{23}$, and $d_2(20_5) = d_2(P^4e_0) = h_1^2P^4d_0 = 22_5$. The $d_2$-differential on $14_{13} = ij$ is zero by $h_2$-linearity.

## 11.5. Some $d_3$-differentials for $S$

THEOREM 11.54. *In the mod 2 Adams spectral sequence for $S$:*

(1) $d_3(a) = 0$ *for* $a = h_0,\ h_1,\ h_2,\ h_3,\ c_0,\ Ph_1,\ Ph_2,\ d_0,\ Pc_0,\ P^2h_1,\ c_1,\ P^2h_2,\ g,\ Pd_0,\ h_4c_0,\ h_0^2i,\ P^2c_0,\ P^3h_1,\ P^3h_2,\ h_4^2,\ P^2d_0,\ n,\ h_1h_5,\ d_1,\ q,\ P^3c_0,\ p,\ P^4h_1,\ P^4h_2,\ x,\ e_0g,\ P^3d_0,\ h_5c_0,\ u,\ Pd_0e_0,\ h_0^2P^2i,\ P^4c_0,\ P^5h_1,\ Ph_2h_5,\ P^5h_2,\ g_2,\ h_5d_0,\ w,\ B_1,\ N,\ d_0\ell,\ P^4d_0,\ e_0r,\ Pu,\ P^2d_0e_0$ *and* $P^5c_0$.

(2) $d_3(h_0h_4) = h_0d_0$.

(3) $d_3(h_1h_4) = 0$.

(4) $d_3(h_2h_4) = 0$.

(5) *(*)* $d_3(r) = h_1d_0^2$.

TABLE 11.1. Algebra indecomposables in $E_3(S)$ for $t-s \leq 48$

| $t-s$ | $s$ | $g$ | $x$ | $d_3(x)$ |
|---|---|---|---|---|
| 0 | 1 | 0 | $h_0$ | 0 |
| 1 | 1 | 1 | $h_1$ | 0 |
| 3 | 1 | 2 | $h_2$ | 0 |
| 7 | 1 | 3 | $h_3$ | 0 |
| 8 | 3 | 3 | $c_0$ | 0 |
| 9 | 5 | 1 | $Ph_1$ | 0 |
| 11 | 5 | 2 | $Ph_2$ | 0 |
| 14 | 4 | 3 | $d_0$ | 0 |
| 15 | 2 | 7 | $h_0h_4$ | $h_0d_0$ |
| 16 | 2 | 8 | $h_1h_4$ | 0 |
| 16 | 7 | 3 | $Pc_0$ | 0 |
| 17 | 9 | 1 | $P^2h_1$ | 0 |
| 18 | 2 | 9 | $h_2h_4$ | 0 |
| 19 | 3 | 9 | $c_1$ | 0 |
| 19 | 9 | 2 | $P^2h_2$ | 0 |
| 20 | 4 | 8 | $g$ | 0 |
| 22 | 8 | 3 | $Pd_0$ | 0 |
| 23 | 4 | 10 | $h_4c_0$ | 0 |
| 23 | 9 | 5 | $h_0^2i$ | 0 |
| 24 | 11 | 3 | $P^2c_0$ | 0 |
| 25 | 13 | 1 | $P^3h_1$ | 0 |
| 27 | 13 | 2 | $P^3h_2$ | 0 |
| 30 | 2 | 10 | $h_4^2$ | 0 |
| 30 | 6 | 10 | $r$ | $h_1d_0^2$ |
| 30 | 12 | 3 | $P^2d_0$ | 0 |
| 31 | 4 | 12 | $h_0^3h_5$ | $h_0r$ |
| 31 | 5 | 13 | $n$ | 0 |
| 31 | 8 | 10 | $d_0e_0$ | $h_0^5r$ |
| 32 | 2 | 12 | $h_1h_5$ | 0 |
| 32 | 4 | 13 | $d_1$ | 0 |
| 32 | 6 | 12 | $q$ | 0 |
| 32 | 15 | 3 | $P^3c_0$ | 0 |
| 33 | 4 | 14 | $p$ | 0 |
| 33 | 17 | 1 | $P^4h_1$ | 0 |

| $t-s$ | $s$ | $g$ | $x$ | $d_3(x)$ |
|---|---|---|---|---|
| 34 | 2 | 13 | $h_2h_5$ | $h_0p$ |
| 35 | 17 | 2 | $P^4h_2$ | 0 |
| 36 | 6 | 14 | $t$ | 0 |
| 37 | 5 | 17 | $x$ | 0 |
| 37 | 8 | 15 | $e_0g$ | 0 |
| 38 | 2 | 14 | $h_3h_5$ | 0 |
| 38 | 4 | 16 | $e_1$ | $h_1t$ |
| 38 | 16 | 3 | $P^3d_0$ | 0 |
| 39 | 4 | 18 | $h_5c_0$ | 0 |
| 39 | 9 | 18 | $u$ | 0 |
| 39 | 12 | 9 | $Pd_0e_0$ | 0 |
| 39 | 17 | 5 | $h_0^2P^2i$ | 0 |
| 40 | 4 | 19 | $f_1$ | 0 |
| 40 | 6 | 18 | $Ph_1h_5$ | 0 |
| 40 | 19 | 3 | $P^4c_0$ | 0 |
| 41 | 10 | 14 | $z$ | 0 |
| 41 | 21 | 1 | $P^5h_1$ | 0 |
| 42 | 6 | 20 | $Ph_2h_5$ | 0 |
| 43 | 21 | 2 | $P^5h_2$ | 0 |
| 44 | 4 | 22 | $g_2$ | 0 |
| 45 | 5 | 24 | $h_5d_0$ | 0 |
| 45 | 9 | 20 | $w$ | 0 |
| 46 | 7 | 20 | $B_1$ | 0 |
| 46 | 8 | 20 | $N$ | 0 |
| 46 | 11 | 12 | $d_0\ell$ | 0 |
| 46 | 14 | 10 | $i^2$ | $h_1(Pd_0)^2$ |
| 46 | 20 | 3 | $P^4d_0$ | 0 |
| 47 | 8 | 21 | $Ph_5c_0$ | 0 |
| 47 | 10 | 16 | $e_0r$ | 0 |
| 47 | 13 | 15 | $Pu$ | 0 |
| 47 | 16 | 10 | $P^2d_0e_0$ | 0 |
| 47 | 18 | 10 | $h_0^5Q$ | $h_0P^4d_0$ |
| 48 | 7 | 22 | $B_2$ | 0 |
| 48 | 23 | 3 | $P^5c_0$ | 0 |

(6) $d_3(h_0^3h_5) = h_0r$.
(7) (*) $d_3(d_0e_0) = h_0^5r$.
(8) $d_3(h_2h_5) = h_0p$.
(9) $d_3(t) = 0$.
(10) $d_3(h_3h_5) = 0$.
(11) $d_3(e_1) = h_1t$.
(12) $d_3(f_1) = 0$.
(13) $d_3(Ph_1h_5) = 0$.
(14) (*) $d_3(z) = 0$.
(15) $d_3(i^2) = h_1(Pd_0)^2$.
(16) (*) $d_3(Ph_5c_0) = 0$.
(17) (*) $d_3(h_0^5Q) = h_0P^4d_0$.
(18) $d_3(B_2) = 0$.

The Adams $(E_3, d_3)$-term is shown in Figure 11.11, and the algebra generators for $t - s \leq 48$ of the resulting $E_4$-term are listed in Table 11.2.

PROOF. (1) By inspection of $E_3(S)$ as a subquotient of $E_2(S)$, it is clear that $d_3(a) = 0$ for the algebra generators $a = h_0$, $h_2$, $h_3$, $c_0$, $Ph_1$, $Ph_2$, $d_0$, $Pc_0$, $P^2h_1$, $c_1$, $P^2h_2$, $g$, $Pd_0$, $h_4c_0$, $h_0^2i$, $P^2c_0$, $P^3h_1$, $P^3h_2$, $h_4^2$, $P^2d_0$, $P^3c_0$, $p$, $P^4h_1$, $P^4h_2$, $x$, $e_0g$, $P^3d_0$, $h_5c_0$, $u$, $Pd_0e_0$, $h_0^2P^2i$, $P^4c_0$, $P^5h_1$, $Ph_2h_5$, $P^5h_2$, $g_2$, $h_5d_0$, $w$, $B_1$, $N$, $d_0\ell$, $P^4d_0$, $e_0r$, $Pu$, $P^2d_0e_0$ and $P^5c_0$ because the target groups are trivial. Furthermore, $d_3(a) = 0$ for $a = h_1$, $n$, $d_1$ and $q$, by $h_0$-linearity, and for $a = h_1h_5$, by combined $h_0$- and $h_2$-linearity.

(2) By Maunder's Theorem 11.39, the classes $h_0^ih_4$ for $1 \leq i \leq 7$ are not boundaries. By Theorem 11.47 the homomorphism

$$e\colon \pi_{15}(S) \longrightarrow \pi_{15}(j) = \mathbb{Z}/32$$

is split surjective. Suppose, for a contradiction, that $d_3(h_0h_4) = 0$. Then $h_0^ih_4$ survives to $E_\infty(S)$ for each $2 \leq i \leq 7$, and there must be a class $x \in \{h_0^2h_4\}$ of order 64. This is impossible, since $\ker(e) \subset \pi_{15}(S)$ has order dividing 8, in view of the total dimension of the $E_3^{s,t}(S)$ with $t - s = 15$. Hence $d_3(h_0h_4)$ must be nonzero, i.e., equal to $h_0d_0$.

(In this degree, we can instead give an elementary argument using $C\sigma$. We suppose known that $\pi_6(S) = \mathbb{Z}/2\{\nu^2\}$, $\pi_7(S) = \mathbb{Z}/16\{\sigma\}$ and $\pi_{13}(S) = 0$, all of which are evident from $E_2(S)$ in this range. The classes $h_3^2$ and $d_0$ in $E_3(S)$ cannot support or be hit by differentials for bidegree reasons, hence detect independent homotopy classes $\sigma^2$ and $\kappa$ in $\pi_{14}(S)$. By graded commutativity $2\sigma^2 = 0$, so that the image of $\sigma\colon \pi_7(S) \to \pi_{14}(S)$ is $\mathbb{Z}/2\{\sigma^2\}$. Hence we have an exact sequence

$$0 \to \mathbb{Z}/2\{\sigma^2\} \longrightarrow \pi_{14}(S) \xrightarrow{i} \pi_{14}(C\sigma) \xrightarrow{j} \mathbb{Z}/2\{\nu^2\} \to 0\,.$$

Let $M_8 = H^*(C\sigma)$. The map $i\colon E_2(S) \to E_2(C\sigma)$ of Adams spectral sequences sends $d_2(f_0) = h_0^2e_0$ to $d_2(4_8) = 6_6$, with the generators of $\mathrm{Ext}_A(M_8, \mathbb{F}_2)$ chosen by `ext`. See Figure 11.15, where $\overline{\overline{a}}$ denotes a class with $j(\overline{\overline{a}}) = a$. By $h_2$- and $h_0$-linearity, $d_2(3_4) = 5_4$ and $d_2(4_4) = 6_4$. Hence $E_3(C\sigma)$ has at most two generators for $t - s = 14$, proving that $\pi_{14}(C\sigma)$ has order dividing 4. It follows from the exact sequence above that $\pi_{14}(S)$ also has order dividing 4. Hence $2\kappa = 0$, so $h_0d_0 \in E_3(S)$ must be a boundary, and $d_3(h_0h_4) = h_0d_0$ is the only possibility.)

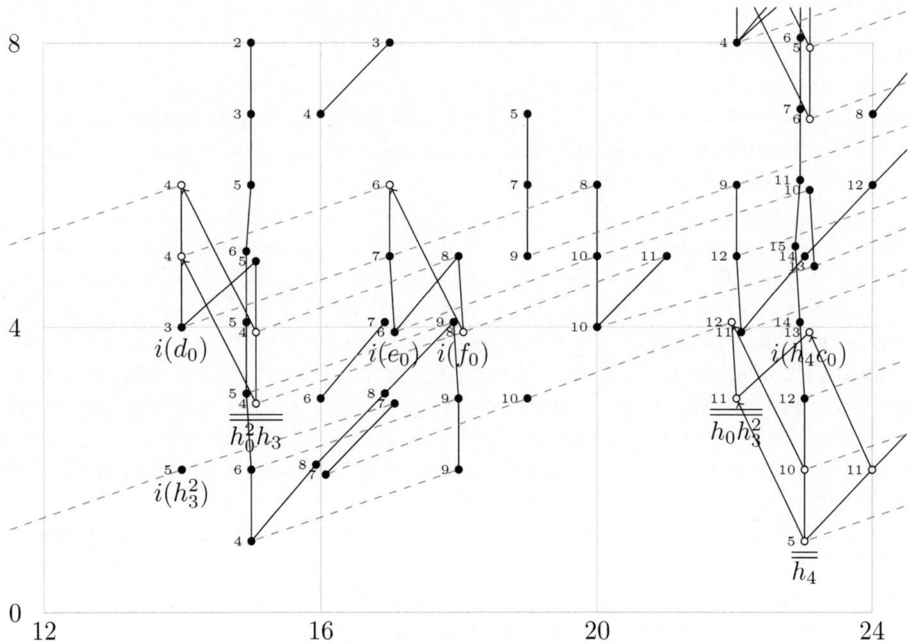

FIGURE 11.15. $(E_2(C\sigma), d_2)$ for $12 \le t - s \le 24$

(3) The $H_\infty$ ring structure gives

$$d_*(h_1h_4) = d_*(Sq^0(h_0h_3)) = Sq^2(d_3(h_0h_3)) \dotplus h_1Sq^2(h_0h_3)\,,$$

which simplifies to $d_3(h_1h_4) = h_1(h_0h_3)^2 = 0$.

(Alternatively, we proved in Proposition 11.34 that there is a homotopy class $\eta^\circ$ detected by $h_1h_4$, which ensures that $d_r(h_1h_4) = 0$ for all $r$.)

(4) In view of the relation $h_3^3 = h_2^2h_4$ we have $h_2 \cdot d_3(h_2h_4) = d_3(h_3^3)$, which is zero because $d_3(h_3) = 0$. It follows that $d_3(h_2h_4) = 0$, since $h_2$ acts injectively on the target group of this differential.

(Alternatively, we proved in Proposition 11.34 that there is a homotopy class $\nu^\circ$ detected by $h_2h_4$, which ensures that $d_r(h_2h_4) = 0$ for all $r$.)

(5) We can easily prove this using *tmf*. The map $E_3(S) \to E_3(tmf)$ takes $r$ to $\beta^2$, with $d_3(\beta^2) \neq 0$ for *tmf* by Proposition 5.8. Hence $d_3(r)$ cannot be zero, and $h_1d_0^2$ is the only possible value.

(We gave a different proof of this in Proposition 11.38, using the quadratic construction on $\kappa$.)

(6) We prove that $d_3(h_0^3h_5) \neq 0$ by a counting argument based on the proven Adams conjecture, Theorem 11.47, asserting that the homomorphism

$$e\colon \pi_{31}(S) \longrightarrow \pi_{31}(j) = \mathbb{Z}/64$$

is split surjective. The image of such a splitting is then a subgroup $\langle y\rangle$ of $\pi_{31}(S)$, mapping isomorphically to $\mathbb{Z}/64$ under $e$. (One such subgroup is the image of the $J$-homomorphism $J\colon \pi_{31}(SO) \to \pi_{31}(S)$, but we will not use this fact.) The elements $2^ky$ of this subgroup, for $0 \le k \le 5$, must be detected by nonzero classes $x_k \in E_\infty(S)$, with $x_{k+1}$ in higher Adams filtration than its predecessor $x_k$, and with $x_{k+1} = h_0 x_k$ if the Adams filtrations only differ by 1. Furthermore, $h_0^i x_k = 0$ for $i + k = 6$, since $2^6 y = 0$.

We have $d_3(h_0^8 h_5) = h_0^5 d_3(h_0^3 h_5) = 0$ since $h_0^5 \cdot h_0 r = 0$, and $d_4(h_0^{10} h_5) = h_0^2 d_4(h_0^8 h_5) = 0$ since $h_0^2 \cdot h_0 P^2 d_0 = 0$. Hence $h_0^i h_5$ is an infinite cycle for each $10 \le i \le 15$. By Maunder's Theorem 11.39, none of these classes are boundaries, hence they remain nonzero in $E_\infty(S)$.

Suppose, for a contradiction, that $d_4(h_0^9 h_5) = 0$, so that $h_0^9 h_5$ survives to $E_\infty(S)$ and detects a homotopy class $z \in \pi_{31}(S)$. Then $2^6 z$ will be detected by $h_0^{15} h_5 \neq 0$. We cannot have $2^6 z = 2^5 y$, since $2^6 z$ maps to zero in $\pi_{31}(j) = \mathbb{Z}/64$ whereas $2^5 y$ maps to the element of order 2. This implies that $x_5 \neq h_0^{15} h_5$, so $x_5 \in E_\infty(S)$ must be the class of one of the other $h_0$-torsion classes, $h_1 P^2 d_0$, $d_0 e_0$, $n$ or $h_1 h_4^2$, in topological degree 31 of $E_3(S)$. Since $h_0^i x_k = 0$ for $i + k = 6$ it follows that $x_4$ must be one of the $h_0^2$-torsion classes $d_0 e_0$, $n$ or $h_1 h_4^2$, that $x_3$ must be one of the $h_0^3$-torsion classes $n$ or $h_1 h_4^2$, and that $x_2$ must be $h_1 h_4^2$. At this point we obtain a contradiction, since there is no nonzero class in lower Adams filtration than $h_1 h_4^2$ that could be equal to $x_1$.

This proves that $x_5 = h_0^{15} h_5$ and $d_4(h_0^9 h_5) = h_0^2 P^2 d_0 \neq 0$. If $d_3(h_0^3 h_5)$ were 0, then $d_4(h_0^9 h_5) = h_0^6 d_4(h_0^3 h_5)$ would be a multiple of $h_0^6 \cdot h_0^2 r = 0$, and this gives another contradiction. The only remaining possibility is $d_3(h_0^3 h_5) = h_0 r \neq 0$.

(7) We prove this using *tmf*. If $d_3(d_0 e_0) = 0$ in $E_3(S)$ then $\iota\colon E_4(S) \to E_4(tmf)$ takes $d_0 e_0$ to $d_0 e_0$ with $d_4(d_0 e_0) = d_0 w_1^2 \neq 0$ in $E_4(tmf)$, as we showed in Corollary 5.13. Hence $d_4(d_0 e_0) = P^2 d_0 \neq 0$ in $E_4(S)$. However, this is impossible by $h_0$-linearity, since $h_0 \cdot d_0 e_0 = 0$ at $E_3(S)$ and $h_0 \cdot P^2 d_0 \neq 0$ at $E_4(S)$. The only alternative is $d_3(d_0 e_0) = h_0^5 r$.

(The original proof [**107**, Prop. 4.3.1] used a similar deduction from $d_4(d_0^2 e_0) \neq 0$ in $E_4(S)$, which they obtained using $C\sigma$.)

(8) We have not found an $H_\infty$-based proof of the differential $d_3(h_2 h_5) = h_0 p$, in spite of the operation $Sq^0(h_1 h_4) = h_2 h_5$, due to the intervening class $p$. We therefore reproduce the argument of Barratt, Mahowald and Tangora [**22**, Prop. 3.3.7], who deduced this differential from a comparison with $C\nu$. They showed that there is a hidden $\nu$-extension from $h_4^2$ to $p$, detecting $\theta_4 \in \pi_{30}(S)$ and $\nu\theta_4 \in \pi_{33}(S)$, respectively. Since $2\theta_4$ has high Adams filtration (in fact, is zero), it follows that $h_0 p$ must be a boundary, and $h_2 h_5$ is the only possible source of this differential. This proof relies on the nonzero $d_2$-differential on $c_2$, which was missed in [**107**], which explains why the nonzero $d_3$-differential on $h_2 h_5$ was also missing in that reference.

See Figure 11.16 for a part of $E_2(C\nu)$, as calculated by `ext`, showing only some of the $d_2$-differentials. There is a unique lift $\overline{h_4^2} = 2_{16}$ in $E_2(C\nu)$ of $h_4^2 \in E_2(S)$, and $h_3 \cdot \overline{h_4^2} = 3_{26} = i(c_2)$. From $d_2(c_2) = h_0 f_1$ we deduce $h_3 \cdot d_2(\overline{h_4^2}) = d_2(i(c_2)) = i(h_0 f_1) = 5_{26} \neq 0$. Hence $d_2(\overline{h_4^2}) \neq 0$ in $E_2(C\nu)$, and $4_{19} = i(p)$ is the only possible value, where $p \in E_2^{4,37}(S)$. Thus $d_2(\overline{h_4^2}) = i(p)$ in $E_2(C\nu)$.

We have not yet shown that $p$ is an infinite cycle for the sphere spectrum, but we can limit our attention to the classes in Adams filtration $\le 4$ by mapping $S = S_0$ to $S_{0,5} = \mathrm{cof}(S_5 \to S_0)$, where $S_\star$ is a minimal Adams resolution of $S$ and we are using the notation of (11.1). A truncated (non-Adams) spectral sequence converging to $\pi_*(S_{0,5})$ is obtained from the Adams spectral sequence for $S$ by omitting the rows $s \ge 5$ from the $E_1$-term, which equals the $E_2$-term by minimality. The tower $S_\star \wedge C\nu$ is a (non-minimal) Adams resolution of $C\nu$, and in the same way we

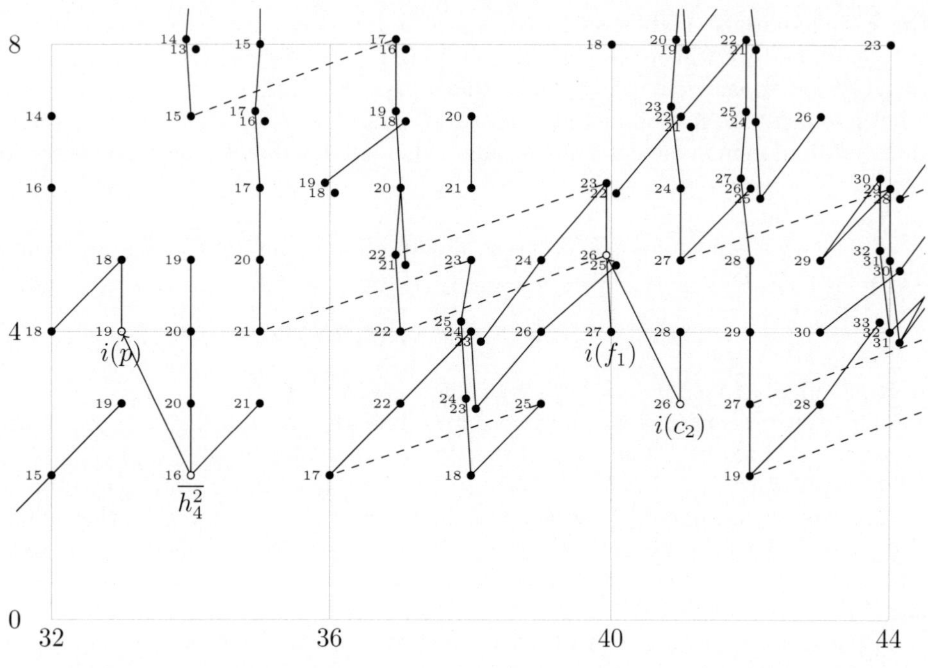

FIGURE 11.16. $E_2(C\nu)$ for $32 \leq t-s \leq 44$, with some $d_2$-differentials

can get a truncated spectral sequence converging to $\pi_*(S_{0,5} \wedge C\nu)$, with $E_r$-terms concentrated in the rows $0 \leq s \leq 4$.

$$\begin{array}{ccc} E_2(S) & \xrightarrow{\quad i \quad} & E_2(C\nu) \\ \downarrow & & \downarrow \\ E_2(S_{0,5}) & \xrightarrow{i \wedge 1} & E_2(S_{0,5} \wedge C\nu) \end{array}$$

Since $d_2(h_2h_5) = 0$, the class $p \in E_2(S_{0,5})$ survives to the $E_\infty$-term and detects a nonzero homotopy class $\gamma \in \pi_{33}(S_{0,5})$. Since $i(p)$ is a $d_2$-boundary in $E_2(C\nu)$, the image $i(\gamma) \in \pi_{33}(S_{0,5} \wedge C\nu)$ must be zero, meaning that $\gamma = \nu \cdot \beta$ for some nonzero $\beta \in \pi_{30}(S_{0,5})$. (This follows from the long exact sequence

$$\cdots \longrightarrow \pi_{n-3}(X) \xrightarrow{\nu} \pi_n(X) \xrightarrow{i} \pi_n(C\nu \wedge X) \longrightarrow \ldots$$

for $X = S_{0,5}$.) Due to the Adams differential $d_2(h_5) = h_0h_4^2$ and its $h_0$-multiple, the homotopy class $\beta$ can only be detected by $h_4^2 \in E_\infty(S_{0,5})$. In other words, $\beta$ is the image under $S \to S_{0,5}$ of a class $\theta_4 \in \pi_{30}(S)$ detected by $h_4^2 \in E_\infty(S)$, and $\gamma = \nu\beta$ is the image of $\nu\theta_4$.

Since $h_2h_4^2 = 0$, the product $\nu\theta_4$ must have Adams filtration $\geq 4$. Since it maps to $\gamma \in \pi_{33}(S_{0,5})$, detected by $p$ in filtration 4 for $S_{0,5}$, it follows that $\nu\theta_4$ is detected by $p \in E_\infty(S)$ in the Adams spectral sequence for $S$. This proves that $p$ is an infinite cycle, and that there is a hidden $\nu$-extension from $h_4^2$ to $p$, in the sense of Definition 9.5.

If the product $h_0p$ remains nonzero at $E_\infty(S)$, then it will detect $2 \cdot \nu\theta_4$. However, $2\theta_4$ must have Adams filtration $\geq 6$, as we see by inspection of $E_3(S)$. (Using the $d_3$-differentials established in cases (5) and (6), it must have Adams filtration $\geq 12$, and we will see in Theorem 11.56 that no other classes than $h_4^2$ survive to $E_\infty(S)$ in topological degree 30, so that $\theta_4$ is a well-defined class with $2\theta_4 = 0$, but these facts are not needed at this stage.) Hence $2\nu\theta$ must have Adams filtration $\geq 7$, and cannot be detected by $h_0 p$. This implies that $h_0 p$ is a boundary in the Adams spectral sequence for $S$. By inspection of $E_3(S)$, the only possibility is $d_3(h_2 h_5) = h_0 p$.

(9) The vanishing of the $d_3$-differential on $t$ follows from the $H_\infty$ ring structure. Theorem 11.22 for $x = e_0$, with $Sq^2(e_0) = t$, gives

$$d_*(t) = d_*(Sq^2(e_0)) = Sq^3(d_2(e_0)) \dot{+} 0\,,$$

which simplifies to $d_3(t) = Sq^3(c_0^2) = 0$, since $d_2(e_0) = c_0^2$.

(10) With the aid of `ext`, we can use $h_1 h_6$-linearity to show that $d_3(h_3 h_5)$ must be zero. See Figures 1.3 and 1.5. From the Adams differential $d_2(h_6) = h_0 h_5^2$ it follows that $d_2(h_0^4 h_6) = h_0^5 h_5^2 \neq 0$, so $d_3(h_1 h_6) = 0$ because the target group is trivial. Now $h_1 h_6 \cdot h_3 h_5 = 0$, so $h_1 h_6 \cdot d_3(h_3 h_5) = d_3(0) = 0$. On the other hand, $h_1 h_6 \cdot x = 7_{89} \neq 0$ in bidegree $(t-s, s) = (101, 7)$ of $E_2(S)$, as calculated by `ext`, and this class cannot be a $d_2$-boundary for degree reasons. Hence $d_3(h_3 h_5) \neq x$, and 0 is the only alternative.

(The original argument [**107**, Cor. 7.3.6] for $d_3(h_3 h_5) = 0$ uses a comparison with $C\sigma$ and $C\sigma \cup_{2\sigma} e^{16}$. It involves steps, such as their Proposition 7.3.5, that build on the mistaken assertion that $d_3(e_1) = 0$. These are therefore incorrect as stated, but can probably be rectified.)

(11) The differential $d_3(e_1) = h_1 t \neq 0$ follows from the $H_\infty$ ring structure. Theorem 11.22 for $x = e_0$, with $Sq^0(e_0) = e_1$ and $Sq^2(e_0) = t$, gives

$$d_*(e_1) = d_*(Sq^0(e_0)) = Sq^1(d_2(e_0)) \dot{+} h_1 Sq^2(e_0)$$

which simplifies to $d_3(e_1) = Sq^1(c_0^2) + h_1 t = h_1 t$.

(This differential was argued to be zero in [**107**, §8.6], and the error persisted in [**22**]. It was corrected by the first author in [**40**, Thm. 4.1], using the proof just given.)

(12) The differential $d_3(f_1) = 0$ follows from the $H_\infty$ ring structure. Theorem 11.22 for $x = c_1$, with $Sq^1(c_1) = f_1$, gives

$$d_*(f_1) = d_*(Sq^1(c_1)) = Sq^2(d_2(c_1)) \dot{+} h_1 Sq^3(c_1)\,,$$

which simplifies to $d_3(f_1) = 0 + h_1 c_1^2 = h_1^2 x = 0$.

(13) We can show that $d_3(Ph_1 h_5) = 0$ using the $H_\infty$ ring structure and $h_1$-linearity. Theorem 11.22 for $x = g$, with $Sq^3(g) = h_1 h_5 P h_1$ according to Proposition 11.29, gives

$$d_*(h_1 h_5 P h_1) = d_*(Sq^3(g)) = Sq^4(d_2(g)) \dot{+} h_1\, g\, d_2(g)\,,$$

which simplifies to $h_1 d_3(Ph_1 h_5) = d_3(h_1 h_5 P h_1) = Sq^4(d_2(g)) = 0$. It follows that we cannot have $d_3(Ph_1 h_5) = u$, since $h_1 u \neq 0$ in $E_3(S)$. The only alternative is $d_3(Ph_1 h_5) = 0$.

(We can also prove this using *tmf*. The map $\iota\colon E_3(S) \to E_3(tmf)$ takes $u$ to $d_0\gamma$, which remains nonzero at the $E_4$- and $E_\infty$-terms for *tmf*. Hence $u$ cannot be $d_3(Ph_1 h_5)$.)

(14) We prove this using *tmf*. The map $E_3(S) \to E_3(tmf)$ takes $h_1Pd_0e_0$ to $h_1d_0e_0w_1 = h_0^2\alpha g w_1$, which remains nonzero at the $E_4$-term for *tmf*, as we can read off from Table 5.5. Hence $h_1Pd_0e_0$ cannot be $d_3(z)$.

(15) Corollary 11.24 and $d_2(i) = h_0Pd_0$ give

$$d_3(i^2) = Sq^8(h_0Pd_0),$$

which is $h_1(Pd_0)^2$ by Proposition 11.33.

(We can also prove this using *tmf*. The map $E_3(S) \to E_3(tmf)$ takes $i^2$ to $\beta^2w_1^2$, with $d_3(\beta^2w_1^2) \neq 0$, so $d_3(i^2) \neq 0$.)

(16) We prove this using *tmf*. The map $E_3(S) \to E_3(tmf)$ takes $d_0\ell$ to $\alpha d_0 g$, which remains nonzero at the $E_4$-term for *tmf*, as we can read off from Table 5.5. Hence $d_0\ell$ cannot be $d_3(Ph_5c_0)$.

(This was proved in [**21**, p. 541] using the Toda bracket $\langle\theta_4, 2, \eta\rho\rangle$ and Moss' theorem [**132**].)

(17) We prove this using *tmf*. The map $E_3(S) \to E_3(tmf)$ takes $P^2d_0e_0$ to $d_0e_0w_1^2$, with $d_4(d_0e_0w_1^2) = d_0w_1^4 \neq 0$ in $E_4(tmf)$. Hence $d_4(P^2d_0e_0) = P^4d_0 \neq 0$ in $E_4(S)$. Now $h_0 \cdot P^2d_0e_0 = 0$ in $E_3(S)$, which implies $h_0 \cdot P^4d_0 = 0$ in $E_4(S)$ by $h_0$-linearity of $d_4$. Thus $h_0P^4d_0 \in E_3(S)$ must be a $d_3$-boundary, and $h_0^5Q$ is the only possible source of that differential.

(It is possible to give a counting argument for $d_3(h_0^5Q) = h_0P^4d_0$, using the Adams conjecture as in case (6), but a bit of work is needed to see why a lifting of $\pi_{47}(j) = \mathbb{Z}/32$ to $\pi_{47}(S)$ cannot be detected by some of the $h_0$-torsion classes $h_1P^4d_0$, $P^2d_0e_0$, $Pu$, $e_0r$, $Ph_5c_0$, $h_1B_1$ and $h_2g_2$.)

(18) This is Lemma 3.67 in Isaksen's memoir [**82**]. We have $d_0 \cdot B_2 = 11_{27} = h_2 \cdot B_{21}$, with $B_{21} = 10_{24}$. See Figure 1.3. Here $d_2(B_{21}) = 0$ and $d_3(B_{21}) = 0$ because the target groups are trivial, so $d_0 \cdot d_3(B_2) = h_2 \cdot 0 = 0$. On the other hand, $d_0 \cdot e_0r = 14_{20}$ remains nonzero at $E_3(S)$ by $h_0$-linearity. Hence $d_3(B_2) \neq e_0r$. □

REMARK 11.55. The nonzero $d_3$-differentials landing in topological degree 48 are $d_3(6_{26}) = d_3(h_0h_5f_0) = h_0^2B_2 = 9_{22}$ and $d_3(11_{13}) = d_3(gk) = h_1Pu = 14_{12}$, according to [**21**, Diag. A]. The latter differential is only possible because we also had $d_2(13_{16}) = d_2(Pv) = h_1^2Pu = 15_{11}$ landing in topological degree 49. It, and the differential $d_3(14_{13}) = d_3(ij) = 0$, are easily shown by comparison with *tmf*.

## 11.6. Some $d_4$-differentials for $S$

THEOREM 11.56. *In the mod 2 Adams spectral sequence for $S$:*

(1) $d_4(a) = 0$ *for* $a = h_0$, $h_1$, $h_2$, $h_3$, $c_0$, $Ph_1$, $Ph_2$, $d_0$, $h_0^3h_4$, $h_1h_4$, $Pc_0$, $P^2h_1$, $h_2h_4$, $c_1$, $P^2h_2$, $g$, $Pd_0$, $h_4c_0$, $h_0^2i$, $P^2c_0$, $P^3h_1$, $P^3h_2$, $h_4^2$, $P^2d_0$, $n$, $h_1h_5$, $d_1$, $q$, $P^3c_0$, $p$, $P^4h_1$, $P^4h_2$, $t$, $h_2^2h_5$, $x$, $P^3d_0$, $h_5c_0$, $u$, $h_0^2P^2i$, $f_1$, $Ph_1h_5$, $P^4c_0$, $z$, $P^5h_1$, $P^5h_2$, $g_2$, $h_5d_0$, $w$, $B_1$, $d_0\ell$, $P^4d_0$, $Ph_5c_0$, $e_0r$, $Pu$, $h_0^7Q$, $B_2$ *and* $P^5c_0$.

(2) *(*)* $d_4(d_0e_0 + h_0^7h_5) = P^2d_0$, $d_4(Pd_0e_0) = P^3d_0$ *and* $d_4(P^2d_0e_0) = P^4d_0$.

(3) *(*)* $d_4(e_0g) = d_0Pd_0$.

(4) *(*)* $d_4(h_0h_2h_5) = 0$.

(5) $d_4(h_3h_5) = h_0x$.

(6) *(*)* $d_4(Ph_2h_5) = 0$.

(7) *(*)* $d_4(N) = 0$.

TABLE 11.2. Algebra indecomposables in $E_4(S)$ for $t-s \leq 48$

| $t-s$ | $s$ | $g$ | $x$ | $d_4(x)$ |
|---|---|---|---|---|
| 0 | 1 | 0 | $h_0$ | 0 |
| 1 | 1 | 1 | $h_1$ | 0 |
| 3 | 1 | 2 | $h_2$ | 0 |
| 7 | 1 | 3 | $h_3$ | 0 |
| 8 | 3 | 3 | $c_0$ | 0 |
| 9 | 5 | 1 | $Ph_1$ | 0 |
| 11 | 5 | 2 | $Ph_2$ | 0 |
| 14 | 4 | 3 | $d_0$ | 0 |
| 15 | 4 | 4 | $h_0^3h_4$ | 0 |
| 16 | 2 | 8 | $h_1h_4$ | 0 |
| 16 | 7 | 3 | $Pc_0$ | 0 |
| 17 | 9 | 1 | $P^2h_1$ | 0 |
| 18 | 2 | 9 | $h_2h_4$ | 0 |
| 19 | 3 | 9 | $c_1$ | 0 |
| 19 | 9 | 2 | $P^2h_2$ | 0 |
| 20 | 4 | 8 | $g$ | 0 |
| 22 | 8 | 3 | $Pd_0$ | 0 |
| 23 | 4 | 10 | $h_4c_0$ | 0 |
| 23 | 9 | 5 | $h_0^2i$ | 0 |
| 24 | 11 | 3 | $P^2c_0$ | 0 |
| 25 | 13 | 1 | $P^3h_1$ | 0 |
| 27 | 13 | 2 | $P^3h_2$ | 0 |
| 30 | 2 | 10 | $h_4^2$ | 0 |
| 30 | 12 | 3 | $P^2d_0$ | 0 |
| 31 | 5 | 13 | $n$ | 0 |
| 31 | 8 | $10+11$ | $d_0e_0+h_0^7h_5$ | $P^2d_0$ |
| 32 | 2 | 12 | $h_1h_5$ | 0 |
| 32 | 4 | 13 | $d_1$ | 0 |
| 32 | 6 | 12 | $q$ | 0 |
| 32 | 15 | 3 | $P^3c_0$ | 0 |
| 33 | 4 | 14 | $p$ | 0 |
| 33 | 17 | 1 | $P^4h_1$ | 0 |
| 34 | 3 | 15 | $h_0h_2h_5$ | 0 |
| 35 | 17 | 2 | $P^4h_2$ | 0 |
| 36 | 6 | 14 | $t$ | 0 |
| 37 | 3 | 16 | $h_2^2h_5$ | 0 |
| 37 | 5 | 17 | $x$ | 0 |
| 37 | 8 | 15 | $e_0g$ | $d_0Pd_0$ |
| 38 | 2 | 14 | $h_3h_5$ | $h_0x$ |
| 38 | 16 | 3 | $P^3d_0$ | 0 |
| 39 | 4 | 18 | $h_5c_0$ | 0 |
| 39 | 9 | 18 | $u$ | 0 |
| 39 | 12 | 9 | $Pd_0e_0$ | $P^3d_0$ |
| 39 | 17 | 5 | $h_0^2P^2i$ | 0 |
| 40 | 4 | 19 | $f_1$ | 0 |
| 40 | 6 | 18 | $Ph_1h_5$ | 0 |
| 40 | 19 | 3 | $P^4c_0$ | 0 |
| 41 | 10 | 14 | $z$ | 0 |
| 41 | 21 | 1 | $P^5h_1$ | 0 |
| 42 | 6 | 20 | $Ph_2h_5$ | 0 |
| 43 | 21 | 2 | $P^5h_2$ | 0 |
| 44 | 4 | 22 | $g_2$ | 0 |
| 45 | 5 | 24 | $h_5d_0$ | 0 |
| 45 | 9 | 20 | $w$ | 0 |
| 46 | 7 | 20 | $B_1$ | 0 |
| 46 | 8 | 20 | $N$ | 0 |
| 46 | 11 | 12 | $d_0\ell$ | 0 |
| 46 | 20 | 3 | $P^4d_0$ | 0 |
| 47 | 8 | 21 | $Ph_5c_0$ | 0 |
| 47 | 10 | 16 | $e_0r$ | 0 |
| 47 | 13 | 15 | $Pu$ | 0 |
| 47 | 16 | 10 | $P^2d_0e_0$ | $P^4d_0$ |
| 47 | 20 | 4 | $h_0^7Q$ | 0 |
| 48 | 7 | 22 | $B_2$ | 0 |
| 48 | 23 | 3 | $P^5c_0$ | 0 |

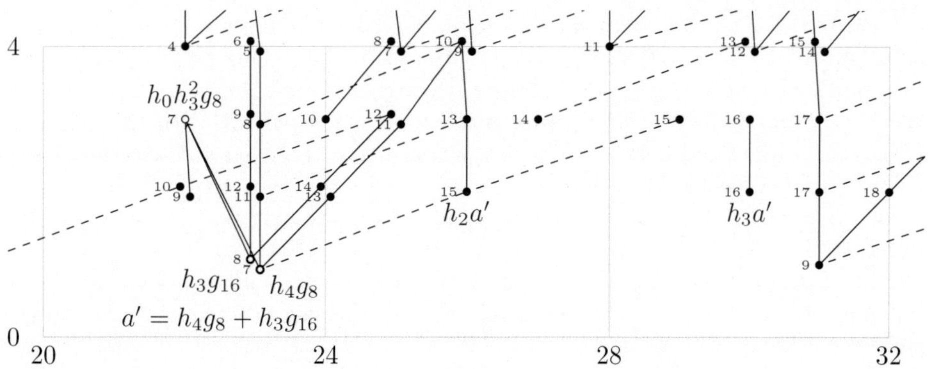

FIGURE 11.17. $E_2(\Sigma^8 C(2\sigma))$ for $20 \le t-s \le 32$, with some $d_2$-differentials

The Adams $(E_4, d_4)$-term is shown in Figure 11.12, and the algebra generators for $t-s \le 48$ of the resulting $E_5$-term are listed in Table 11.3.

PROOF. (1) The target group of $d_4(a)$ is trivial for $a = h_0$, $h_2$, $h_3$, $c_0$, $Ph_1$, $Ph_2$, $d_0$, $h_0^3 h_4$, $Pc_0$, $P^2h_1$, $h_2h_4$, $c_1$, $P^2h_2$, $g$, $Pd_0$, $h_0^2 i$, $P^2c_0$, $P^3h_1$, $P^3h_2$, $h_4^2$, $P^2d_0$, $n$, $h_1h_5$, $P^3c_0$, $p$, $P^4h_1$, $P^4h_2$, $t$, $h_2^2h_5$, $x$, $P^3d_0$, $h_5c_0$, $u$, $h_0^2P^2i$, $f_1$, $Ph_1h_5$, $P^4c_0$, $z$, $P^5h_1$, $P^5h_2$, $g_2$, $h_5d_0$, $w$, $B_1$, $d_0\ell$, $P^4d_0$, $Ph_5c_0$, $e_0r$, $Pu$, $h_0^7Q$, $B_2$ and $P^5c_0$, so these differentials are zero. Furthermore, $d_4(a) = 0$ by $h_0$-linearity for $a = h_1$, $h_1h_4$, $d_1$ and $q$. Finally, $d_4(h_4c_0) = 0$ by $h_1$-linearity, since $h_1 \cdot h_4c_0 = h_1h_4 \cdot c_0$ is a product of $d_4$-cycles, and $h_1 \cdot Pd_0 \neq 0$. Hence $d_4(h_4c_0) \neq Pd_0$.

(2) We prove this using *tmf*. The map $E_4(S) \to E_4(tmf)$ takes $d_0e_0 + h_0^7h_5$ to $d_0e_0$, with $d_4(d_0e_0) = d_0w_1^2 \neq 0$ in $E_4(tmf)$. Hence $d_4(d_0e_0 + h_0^7h_5) \neq 0$ in $E_4(S)$, and $P^2d_0$ is the only possible value. Similarly, $E_4(S) \to E_4(tmf)$ takes $P^id_0e_0$ to $d_0e_0w_1^i$ with $d_4(d_0e_0w_1^i) = d_0w_1^{2+i} \neq 0$, so $d_4(P^id_0e_0) \neq 0$ must be equal to $P^{2+i}d_0$ for $i = 1$ and $i = 2$.

(Alternatively, we can deduce the first of these from the differential $d_4(h_0^9h_5) = h_0^2P^2d_0$ that we established in the proof of case (6) of Theorem 11.54, since $h_0^2 \cdot (d_0e_0 + h_0^7h_5) = h_0^9h_5$ in $E_4(S)$, so that $h_0^2 \cdot d_4(d_0e_0 + h_0^7h_5) = h_0^2 \cdot P^2d_0$, and $d_4(d_0e_0 + h_0^7h_5) = P^2d_0$ is the only possibility.)

(3) We prove this using *tmf*. The map $E_4(S) \to E_4(tmf)$ takes $e_0g$ to $e_0g$, with $d_4(e_0g) = gw_1^2 \neq 0$ for *tmf*. Hence $d_4(e_0g) \neq 0$ for $S$, and $d_0Pd_0$ is the only possible value.

(The original proof in [**107**, Thm. 4.2.1] uses $C\eta$.)

(4) We prove that $d_4(h_0h_2h_5) = 0$ using the homotopy cofiber sequence $S \to tmf \to tmf/S$ and a harmless forward reference. By Proposition 11.77, which only depends on the known differential $d_2(Q) = h_0 i^2$ for $S$, machine calculations by `ext` of $E_2(tmf/S)$, and our results on the Adams spectral sequence for *tmf*, the map $\iota\colon S \to tmf$ takes each element $\alpha \in \{q\}$ in $\pi_{32}(S)$ to $\epsilon_1 \in \{\delta'\}$ in $\pi_{32}(tmf)$, shifting Adams filtration from 6 to 7. It follows that $\eta\alpha$ maps to $\eta\epsilon_1 \in \{h_1\delta\}$, which has Adams filtration 8. Hence $\eta\alpha$ must have Adams filtration 7 or 8. The only possible detecting class is $h_1q$, which therefore is not a boundary, and $d_4(h_0h_2h_5) = 0$.

(See Remark 11.57 concerning the proof of this fact in [**22**].)

(5) We have not found an $H_\infty$-based proof of the differential $d_4(h_3h_5) = h_0x$, in spite of the operation $Sq^0(h_2h_4) = h_3h_5$, due to the intervening class $x$. We therefore reproduce, with some modifications, the argument of Mahowald and Tangora [**107**, §7], who deduced this $d_4$-differential from a comparison with the cell complexes $C\sigma$, $C(2\sigma)$ and $C\sigma \cup_{2\sigma} e^{16}$. They showed that there is a hidden $\sigma$-extension from $h_4^2$ to $x$, detecting $\theta_4 \in \pi_{30}(S)$ and $\sigma\theta_4 \in \pi_{37}(S)$, respectively. Since $2\theta_4 = 0$ it follows that $h_0x$ must be a boundary, and $h_3h_5$ is the only possible source of this differential.

Let $C\sigma \cup_{2\sigma} e^{16} = S \cup_\sigma e^8 \cup_{2\sigma} e^{16}$ be a 3-cell spectrum with nontrivial action by $Sq^8$ and $Sq^{16}$ on the 0-th cohomology class. It can be constructed as the homotopy cofiber of a map $S^8 \to C\sigma \wedge C\sigma$ of degree $+1$ and $-1$, respectively, on the two 8-cells. Consider the following diagram of horizontal and vertical homotopy cofiber sequences.

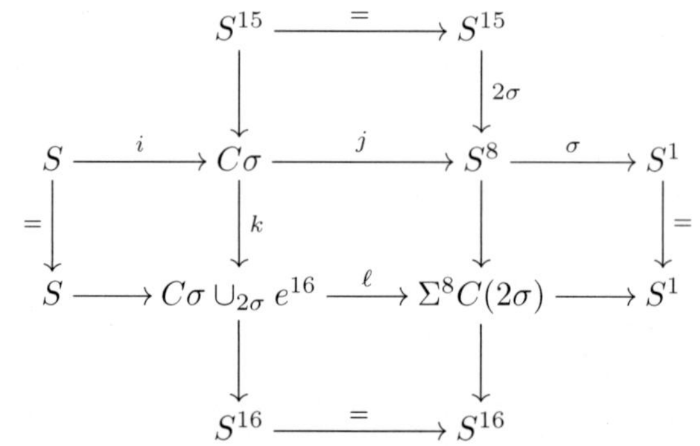


We start with the Adams spectral sequence for $\Sigma^8 C(2\sigma) = S^8 \cup_{2\sigma} e^{16}$. The Steenrod action on $H^*(\Sigma^8 C(2\sigma))$ is trivial, so $E_2(\Sigma^8 C(2\sigma)) = E_2(S)\{g_8, g_{16}\}$ is a free module on two generators $g_8 = 0_0$ and $g_{16} = 0_1$, in bidegrees $(t-s, s) = (8, 0)$ and $(16, 0)$, respectively. Since $2\sigma$ is detected by $h_0h_3$ we have $d_2(g_{16}) = h_0h_3 \cdot g_8$, which implies $d_2(h_3 \cdot g_{16}) = h_0h_3^2 \cdot g_8$ by $h_3$-linearity. Furthermore, $d_2(h_4 \cdot g_8) = h_0h_3^2 \cdot g_8$ by naturality with respect to $S^8 \to \Sigma^8 C(2\sigma)$. See Figure 11.17. It follows that $a' = h_4 \cdot g_8 + h_3 \cdot g_{16}$ is a $d_2$-cycle in bidegree $(t-s, s) = (23, 1)$ of $E_2(\Sigma^8 C(2\sigma))$. Using `ext` we can verify that $h_2 a' = 2_{15} \neq 0$ and $h_3 a' = 2_{16} \neq 0$.

Next we show that $a'$ lifts to a $d_2$-cycle $a$ in the Adams spectral sequence for $C\sigma \cup_{2\sigma} e^{16}$. See Figure 11.18. In view of the long exact sequence

$$\cdots \longrightarrow E_2(S) \longrightarrow E_2(C\sigma \cup_{2\sigma} e^{16}) \xrightarrow{\ell} E_2(\Sigma^8 C(2\sigma)) \longrightarrow \dots$$

the map $\ell$ of $E_2$-terms is an isomorphism in bidegrees $(23, 1)$, $(22, 3)$, $(26, 2)$ and $(30, 2)$, so there is a unique $d_2$-cycle $a$ in bidegree $(23, 1)$ of $E_2(C\sigma \cup_{2\sigma} e^{16})$ such that $\ell(a) = a'$, with $h_2 \cdot a = 2_{12} \neq 0$ and $h_3 \cdot a = 2_{13} \neq 0$. We can calculate $h_2 \cdot 1_4 = 2_{12} = h_2 \cdot 1_5$, $h_3 \cdot 1_4 = 2_{13}$ and $h_3 \cdot 1_5 = 0$ using `ext`, and deduce that $a = 1_4$.

Moreover, we claim that $d_3(a) = 0$. Since $d_2(1_5) = 3_8$, which implies $d_2(2_9) = 4_{10}$, the only alternative target at the $E_3$-term is $b = 4_9$. However, calculation with `ext` shows that $h_1 d_0 \cdot b = 5_4 \cdot 4_9 = 9_{24}$, which remains nonzero at the $E_3$-term by $h_0$-linearity, while $h_1 d_0 \cdot a = 5_4 \cdot 1_4 = 0$. It follows that $d_3(a) \neq b$, so $a$ is a $d_3$-cycle. Multiplying $a$ by $d_3(h_0h_4) = h_0 d_0$ in $E_3(S)$ we obtain a differential $d_3(h_0h_4 \cdot a) = h_0 d_0 \cdot a$ in $E_3(C\sigma \cup_{2\sigma} e^{16})$. Here $h_0h_4 \cdot a = 2_7 \cdot 1_4 = 3_{25}$ and

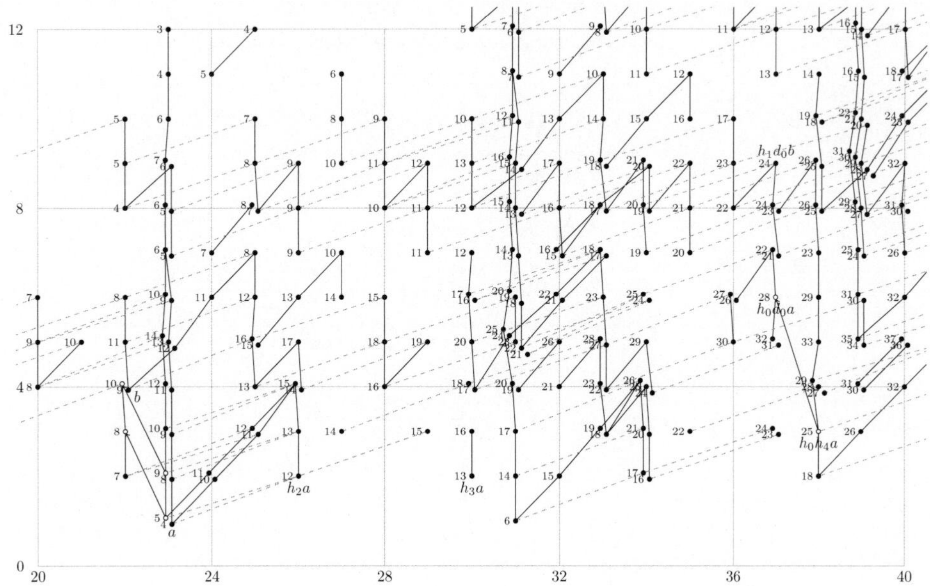

FIGURE 11.18. $E_2(C\sigma \cup_{2\sigma} e^{16})$ for $20 \le t-s \le 40$, with some $d_2$- and $d_3$-differentials

$h_0 d_0 \cdot a = 5_3 \cdot 1_4 = 6_{28}$. The latter class cannot be a $d_2$-boundary, by $h_0$-linearity, hence is nonzero at the $E_3$-term.

We proceed to lift this nonzero $d_3$-differential over $k$ to the Adams spectral sequence for $C\sigma$. Let $\overline{\overline{h_4^2}} = 2_{17}$ denote the unique class in $E_2(C\sigma)$ with $j(\overline{\overline{h_4^2}}) = h_4^2$ in $E_2(S)$. See Figure 11.19. Trivially $d_2(\overline{\overline{h_4^2}}) = 0$, so $h_0\overline{\overline{h_4^2}} = 3_{23}$ survives to the $E_3$-term. In view of the long exact sequence

$$\cdots \longrightarrow E_2(C\sigma) \xrightarrow{k} E_2(C\sigma \cup_{2\sigma} e^{16}) \longrightarrow E_2(S^{16}) \longrightarrow \dots,$$

the latter class maps under $k$ to $h_0 h_4 \cdot a = 3_{25}$. By naturality, $d_3(h_0\overline{\overline{h_4^2}})$ maps under $k$ to $d_3(h_0 h_4 \cdot a) = h_0 d_0 \cdot a \ne 0$, hence is itself nonzero. It follows by $h_0$-linearity that $d_3(\overline{\overline{h_4^2}})$ is nonzero, and the only possible value is $i(x) = 5_{27}$. Hence $d_3(\overline{\overline{h_4^2}}) = i(x)$ in $E_3(C\sigma)$.

We now conclude the proof as in Theorem 11.54, case (8). Let $S_\star$ be a minimal Adams resolution of $S$, and let $S_{0,6} = \mathrm{cof}(S_6 \to S_0)$ be its truncation to filtrations $0 \le s \le 5$. Since $d_3(h_3h_5) = 0$, the class $x \in E_2(S_{0,6})$ survives to the $E_\infty$-term and detects a nonzero homotopy class $\gamma \in \pi_{37}(S_{0,6})$. Since $i(x)$ is a $d_3$-boundary in $E_2(C\sigma)$, the image $i(\gamma) \in \pi_{37}(S_{0,6} \wedge C\sigma)$ must be zero, meaning that $\gamma = \sigma \cdot \beta$ for some nonzero $\beta \in \pi_{30}(S_{0,6})$. The only possibility is that $\beta$ is detected by $h_4^2 \in E_\infty(S_{0,6})$, so that $\beta$ is the image of $\theta_4 \in \pi_{30}(S)$. Hence $\gamma$ is the image of $\sigma\theta_4$. Since $h_3h_4^2 = 0$, it follows that $\sigma\theta_4$ is detected by $x \in E_\infty(S)$. This proves that $x$ is an infinite cycle, and that there is a hidden $\sigma$-extension from $h_4^2$ to $x$. Finally, $2\theta_4 = 0$ implies that $2\sigma\theta_4 = 0$, so $h_0x$ must be a boundary, and $d_4(h_3h_5) = h_0x$ is the only remaining possibility.

(6) We prove this using *tmf*. By Lemma 1.15 and Table 5.5 the map $E_4(S) \to E_4(tmf)$ takes $z$ to $\alpha^2 e_0$, which remains nonzero at the $E_4$-term. On the other

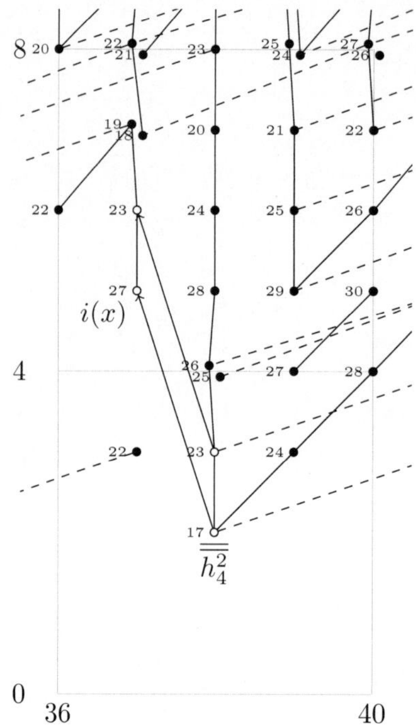

FIGURE 11.19. $E_2(C\sigma)$ for $36 \le t-s \le 40$, with some $d_3$-differentials

hand, $Ph_2h_5$ maps to zero, since $E_4(tmf)$ is trivial in bidegree $(t-s,s) = (42,6)$. Hence we cannot have $d_4(Ph_2h_5) = z$, and $d_4(Ph_2h_5) = 0$ is the only alternative.

(The original proof in [**107**, §8.5] uses $C\sigma$.)

(7) We prove this using $tmf$. Multiplying $d_4(d_0e_0 + h_0^7h_5) = P^2d_0$ by $d_0$ we obtain $d_4(d_0^2e_0) = d_0P^2d_0 = (Pd_0)^2 \neq 0$. Since $d_4 \circ d_4 = 0$ we cannot have $d_4(N) = d_0^2e_0$, and this implies $d_4(N) = 0$. □

REMARK 11.57. The claim that $h_2h_5$ is a permanent cycle, in [**107**, §8.3], would have implied $d_4(h_0h_2h_5) = 0$. With the corrected claim about $d_3(h_2h_5)$ in [**22**, Prop. 3.3.7], a new argument is needed for why $d_4(h_0h_2h_5)$ vanishes. This is implicit in [**22**, Cor. 3.2.3], stating that the elements of the Toda bracket $\langle\eta, 2, \eta_5\rangle$ are detected by $h_0h_2h_5$. Here $\eta_5 \in \pi_{32}(S)$ is specified to be an element of $\langle\eta, 2, \theta_4\rangle$, which is detected by $h_1h_5$ and satisfies $\nu\eta_5 = 0$. To make sense of the first Toda bracket one must know that $2\eta_5 = 0$, i.e., that there is no hidden 2-extension from $h_1h_5$ to $q$, with an accompanying hidden $\eta$-extension from $h_1h_4^2$ to $q$. This can be deduced from the subsequent result [**22**, Prop. 3.3.1], which uses the inclusions

$$\eta\kappa\bar{\kappa} \in \eta\kappa\langle\kappa, 2, \eta, \nu\rangle \subset \eta\langle\kappa^2, 2, \eta, \nu\rangle = \langle\eta, \kappa^2, 2, \eta\rangle\nu$$

to show that $\eta\kappa\bar{\kappa}$, which is detected by $h_1d_0g$, has the form $\nu\{q\}$, with $\{q\} \subset \langle\eta, \kappa^2, 2, \eta\rangle$. See Sections 2.2 and 2.3 of Kochman's book [**87**] for a definition of these four-fold Toda brackets and proofs of the requisite properties. Note that the class that is now commonly referred to as $\eta_5$ is denoted by $\eta_4$ in [**22**].

REMARK 11.58. According to [**21**, Diag. A], there are no nonzero $d_r$-differentials for $r \ge 4$ landing in topological degree 48.

## 11.7. Collapse at $E_5$

THEOREM 11.59. *$d_r(a) = 0$ for all $a \in E_r^{s,t}(S)$ with $r \geq 5$ and $t - s \leq 48$, so $E_5(S) = E_\infty(S)$ in this range.*

The Adams $E_5 = E_\infty$-term is shown in Figure 11.13. The algebra generators in Table 11.3 are thus also algebra generators for the $E_\infty$-term, in this range of degrees.

PROOF. (1) For $a = h_0, h_2, h_3, c_0, Ph_1, Ph_2, d_0, h_0^3h_4, Pc_0, P^2h_1, P^2h_2, Pd_0, h_4c_0, h_0^2i, P^2c_0, P^3h_1, P^3h_2, h_4^2, n, h_0^{10}h_5, P^3c_0, P^4h_1, P^4h_2, h_2^2h_5, x, h_0^2h_3h_5, h_5c_0, u, h_0^2P^2i, P^4c_0, P^5h_1, P^5h_2, h_4^3, h_5d_0, w, B_1, N, d_0\ell, Ph_5c_0, e_0r, Pu, h_0^7Q$ and $P^5c_0$ all $d_r$-differentials for $r \geq 5$ land in trivial groups.

(2) For $a = h_1$ all later differentials vanish by $h_0$-linearity.

(3) For $a = h_1h_4, c_1, g, h_1h_5, d_1, q, t, Ph_1h_5$ and $g_2$ all later differentials vanish because the possible targets are not boundaries by Maunder's Theorem 11.39.

(We already showed in Propositions 11.34 and 11.35 that $h_1h_4$ and $c_1$, respectively, are infinite cycles, since they detect $\eta^\circ$ and $\sigma^\circ$.)

(4) We showed in Proposition 11.34 that $h_2h_4$ is an infinite cycle, since it detects $\nu^\circ$.

(Alternatively, we can argue that the remaining possible targets for a differential on $h_2h_4$, namely $h_1Pc_0$ and $P^2h_1$, must detect independent classes $\eta^2\rho$ and $\bar{\mu}$, respectively, hence cannot be boundaries. This follows from Adams' Theorem 11.40 and Theorem 11.47.)

(5) We showed in the proof of case (8) of Theorem 11.54 that $p$ is an infinite cycle detecting $\nu\theta_4$.

Alternatively, we can argue that the remaining possible target for a differential on $p$, namely $P^3c_0$, must detect a nonzero class $\eta\rho_{31}$, hence cannot be a boundary. We give the details, since they are useful for the next case. Let $\rho_{31} \in \pi_{31}(S)$ be detected by $h_0^{10}h_5$. Then $2^5\rho_{31}$ is the unique class detected by $h_0^{15}h_5$, so $e(2^5\rho_{31}) = 2^5j_{31}$ is the order 2 class in $\pi_{31}(j) = \mathbb{Z}/64$, by our proof of case (6) of Theorem 11.54. Hence $e(\rho_{31}) \doteq j_{31}$, meaning that these agree up to a 2-adic unit, and $e(\eta\rho_{31}) = \eta j_{31} \neq 0$. Since $\eta\rho_{31}$ has Adams filtration $\geq 12$, it can only be detected by $P^3c_0$.

(6) Choosing $\rho_{31} \in \{h_0^{10}h_5\}$ as in the previous case, all later differentials on $a = h_0h_2h_5$ vanish because the possible targets $h_1P^3c_0$ and $P^4h_1$ must detect independent classes $\eta^2\rho_{31}$ and $\mu_{33}$, respectively, hence cannot be boundaries.

(7) (*) Using *tmf* we can show that $d_5(f_1) \neq u$, since $u \mapsto d_0\gamma \neq 0$ at $E_\infty(tmf)$. Then all later differentials on $f_1$ vanish, by Theorem 11.39.

(The original proof in [**107**, §8.7] uses $C\eta$.)

(8) The only possible later differential on $a = z$, with target $P^4c_0$, must vanish by $h_1$-linearity.

Alternatively, we can argue that $P^4c_0$ must detect a nonzero class $\eta\rho_{39}$, hence cannot be a boundary. We give the details, since they are useful for the next case. By Theorem 11.47 there is a class $y \in \pi_{39}(S)$ of order 16 with $e(y) \doteq j_{39}$ generating $\pi_{39}(j) = \mathbb{Z}/16$. Let $x_k \in E_\infty(S)$ detect $2^ky$ for $0 \leq k \leq 3$. Then $x_0$ must have Adams filtration $\geq 3$, $x_1$ must have Adams filtration $\geq 5$, $x_2$ must have Adams filtration $\geq 7$, and $x_3$ must be a nonzero $h_0$-torsion class of Adams filtration $\geq 9$, by inspection of $E_5(S)$. If $x_3$ were $u$ then $\eta \cdot 2^3y = 0$ would be detected by $h_1u$, but we now know that $h_1u \neq 0$ in $E_\infty(S)$, so this is impossible. Therefore $x_3 = h_0^5P^2i = P^4(h_0^3h_3)$ has Adams filtration 20, and detects the class $2^3y$, mapping to

TABLE 11.3. Algebra indecomposables in $E_5(S) = E_\infty(S)$ for $t-s \leq 48$

| $t-s$ | $s$ | $g$ | $x$ | $y \in \{x\}$ |
|---|---|---|---|---|
| 0 | 1 | 0 | $h_0$ | 2 |
| 1 | 1 | 1 | $h_1$ | $\eta$ |
| 3 | 1 | 2 | $h_2$ | $\nu$ |
| 7 | 1 | 3 | $h_3$ | $\sigma$ |
| 8 | 3 | 3 | $c_0$ | $\epsilon$ |
| 9 | 5 | 1 | $Ph_1$ | $\mu$ |
| 11 | 5 | 2 | $Ph_2$ | $\zeta$ |
| 14 | 4 | 3 | $d_0$ | $\kappa$ |
| 15 | 4 | 4 | $h_0^3h_4$ | $\rho$ |
| 16 | 2 | 8 | $h_1h_4$ | $\eta^*$ |
| 16 | 7 | 3 | $Pc_0$ | $\eta\rho$ |
| 17 | 9 | 1 | $P^2h_1$ | $\bar{\mu}$ |
| 18 | 2 | 9 | $h_2h_4$ | $\nu^*$ |
| 19 | 3 | 9 | $c_1$ | $\bar{\sigma}$ |
| 19 | 9 | 2 | $P^2h_2$ | $\bar{\zeta}$ |
| 20 | 4 | 8 | $g$ | $\bar{\kappa}$ |
| 22 | 8 | 3 | $Pd_0$ | $\eta^2\bar{\kappa}$ |
| 23 | 4 | 10 | $h_4c_0$ | $\sigma\eta^*$ |
| 23 | 9 | 5 | $h_0^2i$ | $\bar{\rho}$ |
| 24 | 11 | 3 | $P^2c_0$ | $\eta\bar{\rho}$ |
| 25 | 13 | 1 | $P^3h_1$ | $\mu_{25}$ |
| 27 | 13 | 2 | $P^3h_2$ | $\zeta_{27}$ |
| 30 | 2 | 10 | $h_4^2$ | $\theta_4$ |
| 31 | 5 | 13 | $n$ | $[n]$ |
| 31 | 11 | 6 | $h_0^{10}h_5$ | $\rho_{31}$ |
| 32 | 2 | 12 | $h_1h_5$ | $\eta_5$ |
| 32 | 4 | 13 | $d_1$ | $\kappa_1$ |
| 32 | 6 | 12 | $q$ | $[q]$ |
| 32 | 15 | 3 | $P^3c_0$ | $\eta\rho_{31}$ |
| 33 | 4 | 14 | $p$ | $\nu\theta_4$ |
| 33 | 17 | 1 | $P^4h_1$ | $\mu_{33}$ |
| 34 | 3 | 15 | $h_0h_2h_5$ | $\alpha_{34}$ |
| 35 | 17 | 2 | $P^4h_2$ | $\zeta_{35}$ |
| 36 | 6 | 14 | $t$ | $\{t\}$ |
| 37 | 3 | 16 | $h_2^2h_5$ | $\alpha_{37}$ |
| 37 | 5 | 17 | $x$ | $\sigma\theta_4$ |
| 38 | 4 | 17 | $h_0^2h_3h_5$ | $\alpha_{38}$ |
| 39 | 4 | 18 | $h_5c_0$ | $\alpha_{39}$ |
| 39 | 9 | 18 | $u$ | $[u]$ |
| 39 | 17 | 5 | $h_0^2P^2i$ | $\rho_{39}$ |
| 40 | 4 | 19 | $f_1$ | $\alpha_{40}$ |
| 40 | 6 | 18 | $Ph_1h_5$ | $[[Ph_1h_5]]$ |
| 40 | 19 | 3 | $P^4c_0$ | $\eta\rho_{39}$ |
| 41 | 10 | 14 | $z$ | $\eta\bar{\kappa}^2$ |
| 41 | 21 | 1 | $P^5h_1$ | $\mu_{41}$ |
| 42 | 6 | 20 | $Ph_2h_5$ | $[[Ph_2h_5]]$ |
| 43 | 21 | 2 | $P^5h_2$ | $\zeta_{43}$ |
| 44 | 4 | 22 | $g_2$ | $\bar{\kappa}_2$ |
| 45 | 3 | 20 | $h_4^3$ | |
| 45 | 5 | 24 | $h_5d_0$ | |
| 45 | 9 | 20 | $w$ | $\{w\}$ |
| 46 | 7 | 20 | $B_1$ | |
| 46 | 8 | 20 | $N$ | $\eta^2\bar{\kappa}_2$ |
| 46 | 11 | 12 | $d_0\ell$ | $\eta\{w\}$ |
| 47 | 8 | 21 | $Ph_5c_0$ | |
| 47 | 10 | 16 | $e_0r$ | $[e_0r]$ |
| 47 | 13 | 15 | $Pu$ | $2[e_0r]$ |
| 47 | 20 | 4 | $h_0^7Q$ | $\rho_{47}$ |
| 48 | 7 | 22 | $B_2$ | |
| 48 | 23 | 3 | $P^5c_0$ | $\eta\rho_{47}$ |

the element $2^3 j_{39}$ of order 2 in $\pi_{39}(j)$. Let $\rho_{39} \in \pi_{39}(S)$ be detected by $h_0^2 P^2 i$. Then $2^3\rho_{39}$ is the unique class $2^3 y$ detected by $h_0^5 P^2 i$, so $e(2^3\rho_{39}) = 2^3 j_{39}$, $e(\rho_{39}) \doteq j_{39}$ and $e(\eta\rho_{39}) = \eta j_{39} \neq 0$. Since $\eta\rho_{39}$ has Adams filtration $\geq 18$, it must be detected by $P^4 c_0$.

(9) Choosing $\rho_{39} \in \{h_0^2 P^2 i\}$ as in the previous case, all later differentials on $a = Ph_2h_5$ vanish because the possible targets $h_1 P^4 c_0$ and $P^5 h_1$ must detect independent classes $\eta^2\rho_{39}$ and $\mu_{41}$, respectively, hence cannot be boundaries.

(10) To show that $B_2$ is an infinite cycle we follow Isaksen [**82**, Lem. 4.93] and use Moss' Theorem 1.2 of [**132**], relating the Toda bracket $\langle \nu, \sigma, 2\sigma\theta_4 \rangle$ in $\pi_*(S)$ to the Massey product $\langle h_2, h_3, h_0 x\rangle$ in $E_2(S)$. The infinite cycles $h_2 \in E_2^{1,4}(S)$, $h_3 \in E_2^{1,8}(S)$ and $h_0 x \in E_2^{6,43}(S)$ satisfy $h_2 \cdot h_3 = 0$ and $h_3 \cdot h_0 x = 0$ in $E_2(S)$, with $h_0 x = d_4(h_3h_5)$. They detect homotopy classes $\nu \in \pi_3(S)$, $\sigma \in \pi_7(S)$ and $2\sigma\theta_4 = 0 \in \pi_{37}(S)$, respectively, which satisfy $\nu \cdot \sigma = 0$ and $\sigma \cdot 0 = 0$ in $\pi_*(S)$. The hypotheses that must hold for Moss' theorem to apply are that the groups $E_3^{0,11}(S)$ and $E_{8-n}^{n,45+n}(S)$, for $0 \leq n \leq 5$, consist of infinite cycles. This is readily verified from Figures 11.11, 11.12 and 11.13. The Massey product $\langle h_2, h_3, h_0 x\rangle$ contains $B_2 = 7_{22}$, as can be calculated with `ext`, and has indeterminacy spanned by $h_2 \cdot h_0 h_5 d_0 = h_0^2 h_5 e_0 = 7_{23} = d_2(h_5 f_0)$. Moss' theorem thus asserts that the Toda bracket $\langle \nu, \sigma, 0\rangle$ contains an element $\beta_{48} \in \pi_{48}(S)$ that is detected by $B_2$ or $B_2 + h_0^2 h_5 e_0$. Hence both of these elements are infinite cycles, with the same image in $E_\infty(S)$, and $\beta_{48} \in \{B_2\}$. Since $B_2$ cannot be a boundary, $\beta_{48} \neq 0$. Isaksen also notes that $\langle \nu, \sigma, 0\rangle \subset \pi_{48}(S)$ contains 0, hence is equal to $\nu \cdot \pi_{45}(S)$, so that $\beta_{48} = \nu \cdot \beta_{45}$ for some nonzero class $\beta_{45} \in \pi_{45}(S)$. A look at the $h_0$-, $h_1$- and $h_2$-multiplications in Figure 11.13 shows that $\beta_{45}$ must be detected by $h_4^3$, so there is a hidden $\nu$-extension from $h_4^3$ to $B_2$.

(See Remark 11.60 concerning the treatment in [**166**] and [**21**] of this fact.) □

REMARK 11.60. We have chosen to write the element at bidegree $(46, 11)$ as $d_0\ell$ rather than $gj$ to focus on the important role of $\kappa$, detected by $d_0$. Early work in the subject called it $gj$ and the following historical remark conforms to this usage.

The proof in [**166**, p. 583] that $\eta\{gj\} \neq 0$, which implies that there is a hidden $\eta$-extension from $gj$ to $Pu$, and that $d_6(B_2) \neq Pu$, is, unfortunately, circular. Tangora assumes that $\eta\{gj\} = 0$ and appeals to Moss' Theorem 1.2 of [**132**], relating the Toda bracket $\langle \{gj\}, \eta, \nu\rangle$ in $\pi_*(S)$ to the Massey product $\langle gj, h_1, h_2\rangle = d_0 e_0 g$ in $E_2(S)$, to obtain a contradiction. However, one of the hypotheses needed for this case of Moss' theorem is that $B_2$ is an infinite cycle, in particular that $d_6(B_2) = 0$, and one cannot apply the theorem before this has been established. Barratt, Jones and Mahowald [**21**, Thm. 3.1] cite Tangora's paper [**166**] for this result, and rely on it for their account of the Adams differentials from topological degrees 48 and 49. Fortunately, Isaksen's argument reproduced above circumvents this hole in the logic.

## 11.8. Some homotopy groups of $S$

We can now calculate the graded commutative ring structure of the stable homotopy groups of spheres, in an interesting range of degrees, using principally the Adams spectral sequence methods that are available for the $H_\infty$ ring spectra $S$ and *tmf*. In Theorem 11.61 we calculate the additive and multiplicative structure of $\pi_*(S)$ for $* \leq 44$, implicitly completed at 2. To pass from $E_\infty(S)$ to $\pi_*(S)$ we must solve the extension problems, i.e., identify the additive and multiplicative relations that are not directly visible in the associated graded of the Adams filtration. A chart showing the hidden 2-, $\eta$- and $\nu$-extensions for $\pi_*(S)$ is given in Figure 11.14.

In Theorem 11.61, for each degree $n \le 44$ we first give the structure of $\pi_n(S)$ as an abelian group. We then give the conditions defining the additive generators that we have chosen. Thereafter we specify the products of the form $a \cdot x$ for $a$ indecomposable and $x$ in our generating set. We omit explicit mention of products in trivial groups, such as $\eta\nu$ in $\pi_4(S)$. We also omit $a \cdot xy$ if the result is evident from the values of $a \cdot x$ or $a \cdot y$. Thus, for example, in $\pi_{23}(S)$ we do not mention $\mu \cdot \sigma^2 = 0$, because we have already stated that $\sigma \cdot \mu = \eta\rho$ and $\sigma \cdot \rho = 0$, so that $\sigma^2\mu = \eta\sigma\rho = 0$. However, we do not require the reader to reverse relations from earlier degrees to extract this information. Thus, for example, in $\pi_{31}(S)$ we do note that $\nu \cdot \kappa^2 = 0$, even though this is implied by the products $\epsilon \cdot \bar{\kappa} = \kappa^2$ and $\nu \cdot \epsilon = 0$.

In Remark 11.62 we give a quick overview of the history of stable stem calculations. We stop our detailed work at $\pi_{44}(S)$ because the group structure of $\pi_{45}(S)$ involves a hidden 4-extension from $h_4^3$ to $h_0h_5d_0$, with a delicate proof [**166**, Part 2], and relatively soon thereafter there is a $d_2$-differential landing in degree 51 for which the known proof [**84**] relies on motivic methods.

Our overall strategy is to specify classes in $\pi_*(S)$ by their detecting classes in $E_\infty(S)$, together with their images under the Adams $e$-invariant $e\colon \pi_*(S) \to \pi_*(j)$ and the *tmf*-Hurewicz homomorphism $\iota\colon \pi_*(S) \to \pi_*(tmf)$, both of which are ring homomorphisms. Table 11.3 lists representing homotopy classes in $\pi_*(S)$ for the algebra indecomposables in $E_\infty(S)$, in our range. We use the customary notation $\{x\}$ to denote the set of all $y \in \pi_*(S)$ that are detected by a given infinite cycle $x \in E_\infty(S)$, and refine this in two steps by setting $[x] = \{x\} \cap \ker(e)$ and $[[x]] = [x] \cap \ker(\iota)$. We write $\mathbb{Z}/n\{y\}$ for the cyclic group of order $n$ generated by a class $y$, and write $y \doteq z$ to indicate that $y$ is a multiple of $z$, and vice versa, so that $\langle y\rangle = \langle z\rangle$ as $\mathbb{Z}_2$-modules.

THEOREM 11.61. *The (implicitly 2-completed) stable homotopy groups $\pi_n(S)$ for $0 \le n \le 44$ have the following presentations. They satisfy the listed multiplicative relations.*

(0) $\pi_0(S) = \mathbb{Z}$;
$2 \in \{h_0\}$.

(1) $\pi_1(S) = \mathbb{Z}/2\{\eta\}$;
$\eta = \{h_1\}$;
$e(\eta) = j_1$.

(2) $\pi_2(S) = \mathbb{Z}/2\{\eta^2\}$.

(3) $\pi_3(S) = \mathbb{Z}/8\{\nu\}$;
$\nu \in \{h_2\}$;
$e(\nu) \doteq j_3$, $\eta \cdot \eta^2 = 4\nu$.

(4) $\pi_4(S) = 0$.

(5) $\pi_5(S) = 0$.

(6) $\pi_6(S) = \mathbb{Z}/2\{\nu^2\}$.

(7) $\pi_7(S) = \mathbb{Z}/16\{\sigma\}$;
$\sigma \in \{h_3\}$;
$e(\sigma) \doteq j_7$.

(8) $\pi_8(S) = \mathbb{Z}/2\{\epsilon\} \oplus \mathbb{Z}/2\{\eta\sigma\}$;
$\epsilon = \{c_0\}$;
$e(\epsilon) = \eta j_7$.

(9) $\pi_9(S) = \mathbb{Z}/2\{\mu\} \oplus \mathbb{Z}/2\{\eta\epsilon\} \oplus \mathbb{Z}/2\{\eta^2\sigma\}$;
$\mu = \{Ph_1\}$;
$e(\mu) = j_9$, $\nu \cdot \nu^2 = \eta\epsilon + \eta^2\sigma$.

(10) $\pi_{10}(S) = \mathbb{Z}/2\{\eta\mu\}$;
$\eta \cdot \eta\epsilon = 0$, $\nu \cdot \sigma = 0$.

(11) $\pi_{11}(S) = \mathbb{Z}/8\{\zeta\}$;
$\zeta \in \{Ph_2\}$;
$e(\zeta) \doteq j_{11}$, $\eta \cdot \eta\mu = 4\zeta$, $\nu \cdot \epsilon = 0$.

(12) $\pi_{12}(S) = 0$.

(13) $\pi_{13}(S) = 0$.

(14) $\pi_{14}(S) = \mathbb{Z}/2\{\kappa\} \oplus \mathbb{Z}/2\{\sigma^2\}$;
$\kappa = \{d_0\}$;
$\nu \cdot \zeta = 0$.

(15) $\pi_{15}(S) = \mathbb{Z}/2\{\eta\kappa\} \oplus \mathbb{Z}/32\{\rho\}$;
$\rho \in \{h_0^3 h_4\}$ *with* $\epsilon\rho = 0$;
$e(\rho) \doteq j_{15}$, $\eta \cdot \sigma^2 = 0$, $\sigma \cdot \epsilon = 0$.

(16) $\pi_{16}(S) = \mathbb{Z}/2\{\eta\rho\} \oplus \mathbb{Z}/2\{\eta^*\}$;
$\eta\rho = \{Pc_0\}$, $\eta^* \in \{h_1 h_4\}$, $e(\eta^*) = 0$;
$\eta \cdot \eta\kappa = 0$, $\sigma \cdot \mu = \eta\rho$, $\epsilon \cdot \epsilon = 0$.

(17) $\pi_{17}(S) = \mathbb{Z}/2\{\bar{\mu}\} \oplus \mathbb{Z}/2\{\eta^2\rho\} \oplus \mathbb{Z}/2\{\nu\kappa\} \oplus \mathbb{Z}/2\{\eta\eta^*\}$;
$\bar{\mu} = \{P^2 h_1\}$;
$e(\bar{\mu}) = j_{17}$, $\epsilon \cdot \mu = \eta^2\rho$.

(18) $\pi_{18}(S) = \mathbb{Z}/2\{\eta\bar{\mu}\} \oplus \mathbb{Z}/8\{\nu^*\}$;
$\nu^* \in \{h_2 h_4\}$, $e(\nu^*) = 0$;
$\eta \cdot \eta\eta^* = 4\nu^*$, $\nu \cdot \rho = 0$, $\sigma \cdot \zeta = 0$, $\mu \cdot \mu = \eta\bar{\mu}$.

(19) $\pi_{19}(S) = \mathbb{Z}/8\{\bar{\zeta}\} \oplus \mathbb{Z}/2\{\bar{\sigma}\}$;
$\bar{\zeta} \in \{P^2 h_2\}$, $\bar{\sigma} \in \{c_1\}$, $e(\bar{\sigma}) = 0$;
$e(\bar{\zeta}) \doteq j_{19}$, $\eta \cdot \eta\bar{\mu} = 4\bar{\zeta}$, $\eta \cdot \nu^* = 0$, $\nu \cdot \eta^* = 0$, $\epsilon \cdot \zeta = 0$.

(20) $\pi_{20}(S) = \mathbb{Z}/8\{\bar{\kappa}\}$;
$\bar{\kappa} \in \{g\}$;
$\eta \cdot \bar{\zeta} = 0$, $\eta \cdot \bar{\sigma} = 0$, $\nu \cdot \bar{\mu} = 0$, $\nu \cdot \nu\kappa = 4\bar{\kappa}$, $\mu \cdot \zeta = 0$.

(21) $\pi_{21}(S) = \mathbb{Z}/2\{\eta\bar{\kappa}\} \oplus \mathbb{Z}/2\{\nu\nu^*\}$;
$\sigma \cdot \kappa = 0$, $\sigma \cdot \sigma^2 = \nu\nu^*$.

(22) $\pi_{22}(S) = \mathbb{Z}/2\{\eta^2\bar{\kappa}\} \oplus \mathbb{Z}/2\{\nu\bar{\sigma}\}$;
$\eta^2\bar{\kappa} = \{Pd_0\}$;
$\nu \cdot \bar{\zeta} = 0$, $\sigma \cdot \rho = 0$, $\epsilon \cdot \kappa = \eta^2\bar{\kappa}$, $\zeta \cdot \zeta = 0$.

(23) $\pi_{23}(S) = \mathbb{Z}/16\{\bar{\rho}\} \oplus \mathbb{Z}/8\{\nu\bar{\kappa}\} \oplus \mathbb{Z}/2\{\sigma\eta^*\}$;
$\bar{\rho} \in \{h_0^2 i\}$, $4\nu\bar{\kappa} \in \{h_1 P d_0\}$, $\sigma\eta^* \in \{h_4 c_0\}$;
$e(\bar{\rho}) \doteq j_{23}$, $\epsilon \cdot \rho = 0$, $\mu \cdot \kappa = 4\nu\bar{\kappa}$.

(24) $\pi_{24}(S) = \mathbb{Z}/2\{\eta\bar{\rho}\} \oplus \mathbb{Z}/2\{\epsilon\eta^*\}$;
$\eta\bar{\rho} = \{P^2c_0\}$;
$\eta \cdot \sigma\eta^* = \epsilon\eta^*$, $\nu \cdot \nu\nu^* = 0$, $\sigma \cdot \bar{\mu} = \eta\bar{\rho}$, $\mu \cdot \rho = \eta\bar{\rho}$.

(25) $\pi_{25}(S) = \mathbb{Z}/2\{\mu_{25}\} \oplus \mathbb{Z}/2\{\eta^2\bar{\rho}\}$;
$\mu_{25} = \{P^3h_1\}$;
$e(\mu_{25}) = j_{25}$, $\eta \cdot \epsilon\eta^* = 0$, $\nu \cdot \nu\bar{\sigma} = 0$, $\sigma \cdot \nu^* = 0$, $\epsilon \cdot \bar{\mu} = \eta^2\bar{\rho}$, $\mu \cdot \eta^* = 0$, $\zeta \cdot \kappa = 0$.

(26) $\pi_{26}(S) = \mathbb{Z}/2\{\eta\mu_{25}\} \oplus \mathbb{Z}/2\{\nu^2\bar{\kappa}\}$;
$\nu \cdot \bar{\rho} = 0$, $\sigma \cdot \bar{\zeta} = 0$, $\sigma \cdot \bar{\sigma} = 0$, $\epsilon \cdot \nu^* = 0$, $\mu \cdot \bar{\mu} = \eta\mu_{25}$, $\zeta \cdot \rho = 0$.

(27) $\pi_{27}(S) = \mathbb{Z}/8\{\zeta_{27}\}$;
$\zeta_{27} \in \{P^3h_2\}$;
$e(\zeta_{27}) \doteq j_{27}$, $\eta \cdot \eta\mu_{25} = 4\zeta_{27}$, $\sigma \cdot \bar{\kappa} = 0$, $\epsilon \cdot \bar{\zeta} = 0$, $\epsilon \cdot \bar{\sigma} = 0$, $\mu \cdot \nu^* = 0$, $\zeta \cdot \eta^* = 0$.

(28) $\pi_{28}(S) = \mathbb{Z}/2\{\kappa^2\}$;
$\eta \cdot \zeta_{27} = 0$, $\nu \cdot \mu_{25} = 0$, $\epsilon \cdot \bar{\kappa} = \kappa^2$, $\mu \cdot \bar{\zeta} = 0$, $\mu \cdot \bar{\sigma} = 0$, $\zeta \cdot \bar{\mu} = 0$.

(29) $\pi_{29}(S) = 0$.

(30) $\pi_{30}(S) = \mathbb{Z}/2\{\theta_4\}$;
$\theta_4 = \{h_4^2\}$;
$\nu \cdot \zeta_{27} = 0$, $\sigma \cdot \bar{\rho} = 0$, $\sigma \cdot \sigma\eta^* = 0$, $\zeta \cdot \bar{\zeta} = 0$, $\zeta \cdot \bar{\sigma} = 0$, $\kappa \cdot \eta^* = 0$, $\rho \cdot \rho = 0$.

(31) $\pi_{31}(S) = \mathbb{Z}/64\{\rho_{31}\} \oplus \mathbb{Z}/2\{[n]\} \oplus \mathbb{Z}/2\{\eta\theta_4\}$;
$\rho_{31} \in \{h_0^{10}h_5\}$, $[n] \in \{n\}$, $e([n]) = 0$;
$e(\rho_{31}) \doteq j_{31}$, $\nu \cdot \kappa^2 = 0$, $\epsilon \cdot \bar{\rho} = 0$, $\zeta \cdot \bar{\kappa} = 0$, $\kappa \cdot \bar{\mu} = 0$, $\rho \cdot \eta^* = 0$.

(32) $\pi_{32}(S) = \mathbb{Z}/2\{\eta\rho_{31}\} \oplus \mathbb{Z}/2\{[q]\} \oplus \mathbb{Z}/2\{\kappa_1\} \oplus \mathbb{Z}/2\{\eta_5\}$;
$\eta\rho_{31} = \{P^3c_0\}$, $[q] \in \{q\}$, $\kappa_1 \in \{d_1\}$, $\eta_5 \in \{h_1h_5\}$, $e([q]) = e(\kappa_1) = e(\eta_5) = 0$, $\iota(\kappa_1) = \iota(\eta_5) = 0$, $\nu \cdot \eta_5 = 0$;
$\eta \cdot [n] = 0$, $\eta \cdot \eta\theta_4 = 0$, $\sigma \cdot \mu_{25} = \eta\rho_{31}$, $\mu \cdot \bar{\rho} = \eta\rho_{31}$, $\kappa \cdot \nu^* = 0$, $\rho \cdot \bar{\mu} = \eta\rho_{31}$, $\eta^* \cdot \eta^* = 0$.

(33) $\pi_{33}(S) = \mathbb{Z}/2\{\mu_{33}\} \oplus \mathbb{Z}/2\{\eta^2\rho_{31}\} \oplus \mathbb{Z}/2\{\eta[q]\} \oplus \mathbb{Z}/2\{\nu\theta_4\} \oplus \mathbb{Z}/2\{\eta\eta_5\}$;
$\mu_{33} = \{P^4h_1\}$, $\nu\theta_4 \in \{p\}$;
$e(\mu_{33}) = j_{33}$, $\eta \cdot \kappa_1 = 0$, $\epsilon \cdot \mu_{25} = \eta^2\rho_{31}$, $\kappa \cdot \bar{\zeta} = 0$, $\kappa \cdot \bar{\sigma} = 0$, $\rho \cdot \nu^* = 0$, $\eta^* \cdot \bar{\mu} = 0$.

(34) $\pi_{34}(S) = \mathbb{Z}/2\{\eta\mu_{33}\} \oplus \mathbb{Z}/2\{\kappa\bar{\kappa}\} \oplus \mathbb{Z}/2\{\nu[n]\} \oplus \mathbb{Z}/4\{\alpha_{34}\}$;
$\alpha_{34} \in \{h_0h_2h_5\}$, $e(\alpha_{34}) = 0$, $\eta \cdot \alpha_{34} = 0$;
$\eta \cdot \eta[q] = 0$, $\eta \cdot \eta\eta_5 = 2\alpha_{34}$, $\nu \cdot \rho_{31} = 0$, $\sigma \cdot \zeta_{27} = 0$, $\mu \cdot \mu_{25} = \eta\mu_{33}$, $\zeta \cdot \bar{\rho} = 0$, $\rho \cdot \bar{\zeta} = 0$, $\rho \cdot \bar{\sigma} = 0$, $\eta^* \cdot \nu^* = 0$, $\bar{\mu} \cdot \bar{\mu} = \eta\mu_{33}$.

(35) $\pi_{35}(S) = \mathbb{Z}/8\{\zeta_{35}\} \oplus \mathbb{Z}/2\{\eta\kappa\bar{\kappa}\} \oplus \mathbb{Z}/2\{\nu\kappa_1\}$;
$\zeta_{35} \in \{P^4h_2\}$;
$e(\zeta_{35}) \doteq j_{35}$, $\eta \cdot \eta\mu_{33} = 4\zeta_{35}$, $\eta \cdot \alpha_{34} = 0$, $\nu \cdot [q] = \eta\kappa\bar{\kappa}$, $\nu \cdot \eta_5 = 0$, $\epsilon \cdot \zeta_{27} = 0$, $\rho \cdot \bar{\kappa} = 0$, $\eta^* \cdot \bar{\zeta} = 0$, $\eta^* \cdot \bar{\sigma} = 0$, $\bar{\mu} \cdot \nu^* = 0$.

(36) $\pi_{36}(S) = \mathbb{Z}/2\{\{t\}\}$;
$\eta \cdot \zeta_{35} = 0$, $\nu \cdot \mu_{33} = 0$, $\nu \cdot \nu\theta_4 = 0$, $\mu \cdot \zeta_{27} = 0$, $\zeta \cdot \mu_{25} = 0$, $\eta^* \cdot \bar{\kappa} = 0$, $\bar{\mu} \cdot \bar{\zeta} = 0$, $\bar{\mu} \cdot \bar{\sigma} = 0$, $\nu^* \cdot \nu^* = 0$.

(37) $\pi_{37}(S) = \mathbb{Z}/2\{\sigma\theta_4\} \oplus \mathbb{Z}/2\{\alpha_{37}\}$;
$\sigma\theta_4 = \{x\}$, $\alpha_{37} \in \{h_2^2 h_5\}$, $\eta \cdot \alpha_{37} = 0$;
$\eta \cdot \{t\} = 0$, $\nu \cdot \kappa\bar{\kappa} = 0$, $\nu \cdot \nu[n] = 0$, $\nu \cdot \alpha_{34} = 0$, $\kappa \cdot \bar{\rho} = 0$, $\bar{\mu} \cdot \bar{\kappa} = 0$, $\nu^* \cdot \bar{\zeta} = 0$, $\nu^* \cdot \bar{\sigma} = 0$.

(38) $\pi_{38}(S) = \mathbb{Z}/2\{\eta\sigma\theta_4\} \oplus \mathbb{Z}/4\{\alpha_{38}\}$;
$\eta\sigma\theta_4 = \{h_1 x\}$, $\alpha_{38} \in \{h_0^2 h_3 h_5\}$;
$\eta \cdot \alpha_{37} = 0$, $\nu \cdot \zeta_{35} = 0$, $\nu \cdot \nu\kappa_1 = \eta\sigma\theta_4$, $\sigma \cdot \rho_{31} = 0$, $\sigma \cdot [n] = \eta\sigma\theta_4$, $\epsilon \cdot \theta_4 = \eta\sigma\theta_4$, $\zeta \cdot \zeta_{27} = 0$, $\rho \cdot \bar{\rho} = 0$, $\nu^* \cdot \bar{\kappa} = \eta\sigma\theta_4$, $\bar{\zeta} \cdot \bar{\zeta} = 0$, $\bar{\zeta} \cdot \bar{\sigma} = 0$, $\bar{\sigma} \cdot \bar{\sigma} = \eta\sigma\theta_4$.

(39) $\pi_{39}(S) = \mathbb{Z}/16\{\rho_{39}\} \oplus \mathbb{Z}/2\{[u]\} \oplus \mathbb{Z}/2\{\nu\{t\}\} \oplus \mathbb{Z}/2\{\sigma\kappa_1\} \oplus \mathbb{Z}/2\{\alpha_{39}\} \oplus \mathbb{Z}/2\{\sigma\eta_5\}$;
$\rho_{39} \in \{h_0^2 P^2 i\}$, $[u] \in \{u\}$, $\alpha_{39} \in \{h_5 c_0\}$, $e([u]) = e(\alpha_{39}) = 0$, $\iota(\alpha_{39}) = 0$;
$e(\rho_{39}) \doteq j_{39}$, $\eta \cdot \alpha_{38} = \nu\{t\}$, $\sigma \cdot [q] = \nu\{t\}$, $\epsilon \cdot \rho_{31} = 0$, $\epsilon \cdot [n] = 0$, $\mu \cdot \theta_4 = 0$, $\kappa \cdot \mu_{25} = 0$, $\eta^* \cdot \bar{\rho} = 0$, $\bar{\zeta} \cdot \bar{\kappa} = 0$, $\bar{\sigma} \cdot \bar{\kappa} = \nu\{t\}$.

(40) $\pi_{40}(S) = \mathbb{Z}/2\{\eta\rho_{39}\} \oplus \mathbb{Z}/4\{\bar{\kappa}^2\} \oplus \mathbb{Z}/2\{[[Ph_1h_5]]\} \oplus \mathbb{Z}/2\{\eta\alpha_{39}\} \oplus \mathbb{Z}/2\{\alpha_{40}\} \oplus \mathbb{Z}/2\{\eta\sigma\eta_5\}$;
$\eta\rho_{39} = \{P^4 c_0\}$, $2\bar{\kappa}^2 \in \{h_1 u\}$, $[[Ph_1h_5]] \in \{Ph_1h_5\}$, $\alpha_{40} \in \{f_1\}$, $e([[Ph_1h_5]]) = e(\alpha_{40}) = 0$, $\iota([[Ph_1h_5]]) = \iota(\alpha_{40}) = 0$, $\eta^2 \cdot \alpha_{40} = 0$;
$\eta \cdot [u] = 2\bar{\kappa}^2$, $\nu \cdot \alpha_{37} = \eta\alpha_{39} + \eta\sigma\eta_5$, $\sigma \cdot \mu_{33} = \eta\rho_{39}$, $\epsilon \cdot [q] = 2\bar{\kappa}^2$, $\epsilon \cdot \kappa_1 = 0$, $\epsilon \cdot \eta_5 = \eta\alpha_{39}$, $\mu \cdot \rho_{31} = \eta\rho_{39}$, $\mu \cdot [n] = 0$, $\rho \cdot \mu_{25} = \eta\rho_{39}$, $\bar{\mu} \cdot \bar{\rho} = \eta\rho_{39}$.

(41) $\pi_{41}(S) = \mathbb{Z}/2\{\mu_{41}\} \oplus \mathbb{Z}/2\{\eta^2\rho_{39}\} \oplus \mathbb{Z}/2\{\eta\bar{\kappa}^2\} \oplus \mathbb{Z}/2\{\eta[[Ph_1h_5]]\} \oplus \mathbb{Z}/2\{\eta\alpha_{40}\}$;
$\mu_{41} = \{P^5 h_1\}$, $\eta\bar{\kappa}^2 \in \{z\}$;
$e(\mu_{41}) = j_{41}$, $\eta \cdot \eta\alpha_{39} = 0$, $\eta \cdot \eta\sigma\eta_5 = 0$, $\nu \cdot \alpha_{38} = 0$, $\sigma \cdot \alpha_{34} = \eta\alpha_{40}$, $\epsilon \cdot \mu_{33} = \eta^2\rho_{39}$, $\mu \cdot [q] = 0$, $\mu \cdot \kappa_1 = 0$, $\mu \cdot \eta_5 = \eta[[Ph_1h_5]]$, $\zeta \cdot \theta_4 = 0$, $\kappa \cdot \zeta_{27} = 0$, $\eta^* \cdot \mu_{25} = 0$, $\nu^* \cdot \bar{\rho} = 0$.

(42) $\pi_{42}(S) = \mathbb{Z}/2\{\eta\mu_{41}\} \oplus \mathbb{Z}/2\{\kappa^3\} \oplus \mathbb{Z}/8\{[[Ph_2h_5]]\}$;
$[[Ph_2h_5]] \in \{Ph_2h_5\}$, $e([[Ph_2h_5]]) = 0$, $\iota([[Ph_2h_5]]) = 0$;
$\eta \cdot \eta\bar{\kappa}^2 = \kappa^3$, $\eta \cdot \eta[[Ph_1h_5]] = 4[[Ph_2h_5]]$, $\eta \cdot \eta\alpha_{40} = 0$, $\nu \cdot \rho_{39} = 0$, $\nu \cdot [u] = \kappa^3$, $\nu \cdot \nu\{t\} = 0$, $\nu \cdot \alpha_{39} = 0$, $\sigma \cdot \zeta_{35} = 0$, $\epsilon \cdot \alpha_{34} = 0$, $\mu \cdot \mu_{33} = \eta\mu_{41}$, $\zeta \cdot \rho_{31} = 0$, $\zeta \cdot [n] = 0$, $\rho \cdot \zeta_{27} = 0$, $\bar{\mu} \cdot \mu_{25} = \eta\mu_{41}$, $\bar{\zeta} \cdot \bar{\rho} = 0$, $\bar{\sigma} \cdot \bar{\rho} = 0$.

(43) $\pi_{43}(S) = \mathbb{Z}/8\{\zeta_{43}\}$;
$\zeta_{43} \in \{P^5 h_2\}$;
$e(\zeta_{43}) \doteq j_{43}$, $\eta \cdot \eta\mu_{41} = 4\zeta_{43}$, $\eta \cdot \kappa^3 = 0$, $\eta \cdot [[Ph_2h_5]] = 0$, $\nu \cdot \bar{\kappa}^2 = 0$, $\nu \cdot [[Ph_1h_5]] = 0$, $\nu \cdot \alpha_{40} = 0$, $\sigma \cdot \{t\} = 0$, $\epsilon \cdot \zeta_{35} = 0$, $\mu \cdot \alpha_{34} = 0$, $\zeta \cdot [q] = 0$, $\zeta \cdot \kappa_1 = 0$, $\zeta \cdot \eta_5 = 0$, $\eta^* \cdot \zeta_{27} = 0$, $\nu^* \cdot \mu_{25} = 0$, $\bar{\kappa} \cdot \bar{\rho} = 0$.

(44) $\pi_{44}(S) = \mathbb{Z}/8\{\bar{\kappa}_2\}$;
$\bar{\kappa}_2 \in \{g_2\}$;
$\eta \cdot \zeta_{43} = 0$, $\nu \cdot \mu_{41} = 0$, $\sigma \cdot \sigma\theta_4 = 4\bar{\kappa}_2$, $\sigma \cdot \alpha_{37} = 4\bar{\kappa}_2$, $\epsilon \cdot \{t\} = 0$, $\mu \cdot \zeta_{35} = 0$, $\zeta \cdot \mu_{33} = 0$, $\kappa \cdot \theta_4 = 0$, $\bar{\mu} \cdot \zeta_{27} = 0$, $\bar{\zeta} \cdot \mu_{25} = 0$, $\bar{\sigma} \cdot \mu_{25} = 0$.

REMARK 11.62. Building on L.E.J. Brouwer's notion of degree [**34**], Heinz Hopf [**71**] showed that the homotopy classes of maps $S^m \to S^m$ are in one-to-one correspondence with the integers, for each $m \geq 1$. It follows that the homotopy groups $\pi_{n+m}(S^m)$, as defined by Eduard Čech [**48**] and Witold Hurewicz [**79**], are trivial for $n < 0$ and isomorphic to $\mathbb{Z}$ for $n = 0$. Hopf [**72**] also introduced his invariant $\pi_{2m-1}(S^m) \to \mathbb{Z}$, for even $m$, showing that the fibrations $\eta\colon S^3 \to S^2$,

$\nu\colon S^7 \to S^4$ and $\sigma\colon S^{15} \to S^8$ are essential. Hans Freudenthal [**60**] recognized the role of the stable group $\pi_n(S) = \operatorname{colim}_m \pi_{n+m}(S^m)$, known as the $n$-stem, and calculated $\pi_1(S)$. Lev Pontryagin [**140**] and George Whitehead [**178**] determined $\pi_2(S)$. Hirosi Toda [**168**] and Jean-Pierre Serre [**152**], [**153**] calculated $\pi_n(S)$ for $3 \le n \le 5$, using composition methods and cohomology of Postnikov systems, respectively, while Vladimir Rokhlin [**147**] obtained $\pi_3(S)$ by manifold-geometric methods. Thereafter Serre [**154**] calculated the groups $\pi_n(S)$ for $6 \le n \le 8$, and Toda [**169**], [**170**] obtained the groups for $9 \le n \le 13$. Toda's results were extended to the range $n \le 19$ in his book [**171**]. Mamoru Mimura and Toda [**130**] calculated $\pi_{20}(S)$, and Mimura [**129**] then obtained $\pi_{21}(S)$ and $\pi_{22}(S)$, including the salient fact that $\epsilon\kappa \ne 0$ in the latter group.

By this time the Adams spectral sequence [**2**] was available as a new tool, and Peter May [**117**] calculated enough of its $E_2$-term to obtain the correct groups for $n \le 28$, except for $n = 23$. Mark Mahowald and Martin Tangora [**107**] showed how Mimura's fact implied hidden 2-, $\eta$- and $\nu$-extensions in the range $20 \le n \le 23$. They proceeded to calculate the groups for $n \le 37$ and $n \in \{39, 42, 43, 44\}$, except that they missed the three differentials $d_2(c_2) = h_0 f_1$, $d_3(h_2h_5) = h_0 p$ and $d_3(e_1) = h_1 t$, which affected the results for $n \in \{33, 34, 37, 38, 40, 41\}$. The first two of these differentials were corrected by Michael Barratt, Mahowald and Tangora in [**22**, Cor. 3.3.6], and by Milgram in [**122**, Cor. 6.5.2], giving a calculation for $n \le 44$, except for $n \in \{37, 38\}$. Tangora [**166**, Thm. on p. 583] determined the group structure of $\pi_{45}(S)$, which entails a hidden 4-extension from $h_4^3$ to $h_0h_5d_0$, cf. [**21**, Thm. 3.3]. The third differential was corrected by the first author in [**40**], giving $\pi_{37}(S)$ and $\pi_{38}(S)$.

For odd primes $p$, with $q = 2p - 2$, Toda [**172**], [**173**] introduced extended powers to calculate the $p$-primary torsion in $\pi_n(S)$ for $n \le (p^2 + 2p)q - 4$. At $p = 3$ these results were extended to $n \le 103$ by Osamu Nakamura [**134**] and Tangora [**167**], and to $n \le 108$ by Douglas Ravenel [**144**] using the Adams–Novikov spectral sequence, but see [**28**, Add. on p. 12] for a possible inconsistency. For $p \ge 5$, Marc Aubry [**17**] obtained a full calculation for $n < (3p^2 + 4p)q$, and Ravenel [**144**] extended this range to $n < 1000$ for $p = 5$. Our odd-primary understanding of $\pi_n(S)$ thus goes well beyond our 2-primary knowledge.

Stanley Kochman [**87**] used computer calculations with an Atiyah–Hirzebruch spectral sequence to calculate the 2-primary part of $\pi_n(S)$ in the range $46 \le n \le 53$, except for $n = 51$, and for $58 \le n \le 60$. A mistake for $n = 55$ was resolved with Mahowald in [**88**], giving the correct groups $\pi_{54}(S)$ and $\pi_{55}(S)$. Adams differentials $d_2(D_1) = h_0^2 h_3 g_2$ and $d_3(Q_2) = gt$, landing in the 51- and 56-stems, respectively, were established by Daniel Isaksen and Zhouli Xu in [**84**], and Isaksen in [**82**], using the motivic weight grading as a new ingredient. Guozhen Wang and Xu then obtained an Adams differential $d_3(D_3) = B_3$ landing in the 60-stem [**174**], and resolved the group structure of $\pi_{51}(S)$ in [**175**]. At this point, the group structure of $\pi_n(S)$ was known for all $n \le 61$. Combining the motivic method with machine calculation of the Adams–Novikov $E_2$-term, Isaksen, Wang and Xu [**83**] have recently made extensive new calculations in the range $62 \le n \le 90$. This brings them close to degree 93, where one can optimistically hope to prove that the Toda bracket $\langle \theta_5, 2, \theta_4 \rangle$ contains zero, which, according to [**174**, Rem. 1.11], would suffice to prove the existence of $\theta_6 \in \{h_6^2\} \subset \pi_{126}(S)$, the last potential Kervaire invariant one element [**69**].

Proof of Theorem 11.61. (0) The $E_\infty$-term for $t-s=0$ is additively generated by $h_0^i$ for $i \ge 0$. Hence $\pi_0(S) \cong \mathbb{Z}$ is generated by the identity map $1\colon S \to S$, and the stable class $2\colon S \to S$ of the real Hopf fibration $S^1 \to S^1$ is detected by $h_0$.

(1) The $E_\infty$-term for $t-s=1$ is generated by $h_1$. Hence $\pi_1(S) = \mathbb{Z}/2\{\eta\}$ is generated by the stable class of the complex Hopf fibration $S^3 \to S^2$, which is detected by $h_1$. Its $e$-invariant $j_1$ generates $\pi_1(j) = \mathbb{Z}/2$. The relation $h_0h_1 = 0$ implies that $2\eta$ and $0$ agree modulo Adams filtration $\ge 3$, hence these classes are equal.

(2) The $E_\infty$-term for $t-s=2$ is generated by $h_1^2$. Hence $\pi_2(S) = \mathbb{Z}/2\{\eta^2\}$ is generated by $\eta^2$, which is detected by $h_1^2$. Its $e$-invariant $\eta j_1$ generates $\pi_2(j) = \mathbb{Z}/2$.

(3) The $E_\infty$-term for $t-s=3$ is generated by $h_2$, $h_0h_2$ and $h_0^2h_2$. Hence $\pi_3(S) = \mathbb{Z}/8\{\nu\}$ is generated by the stable class of the quaternionic Hopf fibration $S^7 \to S^4$, which is detected by $h_2$. Its $e$-invariant is an odd multiple of $j_3$, and generates $\pi_3(j) = \mathbb{Z}/8$. The relation $h_1^3 = h_0^2h_2$ implies that $\eta^3$ and $4\nu$ agree modulo Adams filtration $\ge 4$, hence these classes are equal.

(4) The $E_\infty$-term for $t-s=4$ is trivial. Hence $\pi_4(S) = 0$ and $\eta\nu = 0$.

(5) The $E_\infty$-term for $t-s=5$ is trivial. Hence $\pi_5(S) = 0$.

(6) The $E_\infty$-term for $t-s=6$ is generated by $h_2^2$. Hence $\pi_6(S) = \mathbb{Z}/2\{\nu^2\}$ is generated by $\nu^2$, which is detected by $h_2^2$. Clearly $e(\nu^2) = 0$, since $\pi_6(j) = 0$.

(7) The $E_\infty$-term for $t-s=7$ is generated by $h_0^kh_3$ for $k \in \{0,1,2,3\}$. Hence $\pi_7(S) = \mathbb{Z}/16\{\sigma\}$ is generated by the stable class of the octonionic Hopf fibration $S^{15} \to S^8$, which is detected by $h_3$. Its $e$-invariant is an odd multiple of $j_7$, and generates $\pi_7(j) = \mathbb{Z}/16$. The product $\eta \cdot \nu^2$ is zero, since $\eta\nu = 0$.

(8) The $E_\infty$-term for $t-s=8$ is generated by $h_1h_3$ and $c_0$, detecting $\eta\sigma$ and $\epsilon$, respectively. The Adams filtration splits, since $2 \cdot \eta\sigma = 0$, so that $\pi_8(S) = \mathbb{Z}/2\{\epsilon\} \oplus \mathbb{Z}/2\{\eta\sigma\}$. Here $e(\eta\sigma) = \eta j_7$ generates $\pi_8(j) = \mathbb{Z}/2$. We postpone the proof that $e(\epsilon) = \eta j_7$ to the next case. Once that is established, we know that $\ker(e) = \mathbb{Z}/2\{\epsilon + \eta\sigma\} \subset \pi_8(S)$. (Toda [**171**] uses the notation $\bar{\nu}$ for $\epsilon + \eta\sigma$.)

(9) The $E_\infty$-term for $t-s=9$ is generated by $h_1^2h_3$, $h_1c_0$ and $Ph_1$, detecting $\eta^2\sigma$, $\eta\epsilon$ and $\mu = \mu_9$, respectively. Since $2 \cdot \eta\epsilon = 0$ and $2 \cdot \eta^2\sigma = 0$ we have $\pi_9(S) = \mathbb{Z}/2\{\mu\} \oplus \mathbb{Z}/2\{\eta\epsilon\} \oplus \mathbb{Z}/2\{\eta^2\sigma\}$. By construction $e(\mu) = j_9$, while $e(\eta^2\sigma) = \eta^2 j_7$, and these classes generate $\pi_9(j) = \mathbb{Z}/2\{j_9\} \oplus \mathbb{Z}/2\{\eta^2 j_7\}$. The relation $h_2^3 = h_1^2h_3$ implies that $\nu^3$ and $\eta^2\sigma$ agree modulo Adams filtration $\ge 4$, so that $\nu^3 = x\mu + y\eta\epsilon + \eta^2\sigma$, for some $x, y \in \{0,1\}$. Since $e(\nu^3) = \nu e(\nu^2) = 0$, we must have $e(x\mu + y\eta\epsilon + \eta^2\sigma) = xj_9 + y\eta e(\epsilon) + \eta^2 j_7 = 0$. It follows that $x = 0$, $y = 1$ and $\eta e(\epsilon) = \eta^2 j_7$. Hence $e(\epsilon) = \eta j_7$, and $\nu \cdot \nu^2 = \eta\epsilon + \eta^2\sigma$.

(10) The $E_\infty$-term for $t-s=10$ is generated by $h_1Ph_1$. Hence $\pi_{10}(S) = \mathbb{Z}/2\{\eta\mu\}$ is generated by $\eta\mu$, which is detected by $h_1Ph_1$. Its $e$-invariant $\eta j_9$ generates $\pi_{10}(j) = \mathbb{Z}/2$. Since $e(\eta^2\epsilon) = \eta^3 j_7 = 0$ and $e(\nu\sigma) \doteq \nu j_7 = 0$, cf. Lemma 11.46, we must have $\eta \cdot \eta\epsilon = 0$ and $\nu \cdot \sigma = 0$.

(11) The $E_\infty$-term for $t-s=11$ is generated by $Ph_2$, $h_0Ph_2$ and $h_0^2Ph_2$, detecting $\zeta$, $2\zeta$ and $4\zeta$, respectively. Hence $\pi_{11}(S) = \mathbb{Z}/8\{\zeta\}$. This determines $\zeta$ up to an odd multiple. Its $e$-invariant is $j_{11}$, up to an odd multiple, and generates $\pi_{11}(j) = \mathbb{Z}/8$. The relation $h_1^2Ph_1 = h_0^2Ph_2$ implies that $\eta^2\mu$ and $4\zeta$ agree modulo Adams filtrations $\ge 8$, hence are equal. It follows that $\nu \cdot \epsilon = 0$, since $e(\nu\epsilon) = \eta\nu j_7 = 0$ and $e\colon \pi_{11}(S) \to \pi_{11}(j)$ is an isomorphism.

(12) The $E_\infty$-term for $t-s=12$ is trivial. Hence $\pi_{12}(S) = 0$, so that $\eta\zeta = 0$ and $\nu\mu = 0$.

(13) The $E_\infty$-term for $t-s=13$ is trivial. Hence $\pi_{13}(S)=0$.

(14) The $E_\infty$-term for $t-s=14$ is generated by $h_3^2$ and $d_0$, detecting $\sigma^2$ and $\kappa$, respectively. Since $2\sigma^2=0$ by the graded commutativity of $\pi_*(S)$ (a simple consequence of its $H_\infty$ ring structure), we have $\pi_{14}(S)=\mathbb{Z}/2\{\kappa\}\oplus\mathbb{Z}/2\{\sigma^2\}$. Clearly $e(\kappa)=0$, since $\pi_{14}(j)=0$. The product $\nu\cdot\zeta$ is detected by $h_2Ph_2=0$ modulo Adams filtration $\geq 7$, hence is zero.

(15) The $E_\infty$-term for $t-s=15$ is generated by $h_0^kh_4$ for $k\in\{3,4,5,6,7\}$ and $h_1d_0$, with $h_0^3h_4$ detecting $\rho$ and $h_1d_0$ detecting $\eta\kappa$. Since $2\cdot\eta\kappa=0$ we have $\pi_{15}(S)=\mathbb{Z}/2\{\eta\kappa\}\oplus\mathbb{Z}/32\{\rho\}$. This determines $\rho$ modulo $\eta\kappa$ and even multiples of $\rho$. We shall fix a more specific choice of $\rho$ by asking that $\epsilon\cdot\rho=0$ in $\pi_{23}(S)$, or by asking that $\iota(\rho)=0$ in $\pi_{15}(tmf)\cong\mathbb{Z}/2\{\eta\kappa\}$. These conditions determine $\rho$ up to an odd multiple, and are equivalent.

To see that a choice of $\rho$ in $\ker(\iota)$ satisfies $\epsilon\rho=0$ (and $\rho\bar\kappa=0$), we can use the homotopy cofiber sequence

$$S\xrightarrow{\iota} tmf\xrightarrow{i} tmf/S\xrightarrow{j}\Sigma S$$

and the calculation of $E_2(tmf/S)$ and $E_3(tmf/S)$ given in Figures 11.27 and 11.28. The Adams differential $d_2(h_4)=h_0h_3^2$ lifts to differentials $d_2(h_0^k\widetilde{h_4})=h_0^{1+k}h_3\widetilde{h_3}$ for $0\leq k\leq 2$, so that a generator $\alpha$ of $\pi_{16}(tmf/S)\cong\mathbb{Z}$ is detected by $h_0^3\widetilde{h_4}$, with $j(\alpha)\in\pi_{15}(S)$ detected by $h_0^3h_4$. Setting $\rho=j(\alpha)$ we obtain $\iota(\rho)=0$. Furthermore, $\epsilon\cdot\alpha=0$ and $\bar\kappa\cdot\alpha=0$, since these products have finite order and Adams filtration $\geq 6$, and $E_\infty(tmf/S)$ contains no $h_0$-power torsion in topological degrees 24 and 36 and Adams filtrations $\geq 5$.

Conversely, once we know that $\eta\epsilon\kappa\neq 0$ in $\pi_{23}(S)$, cf. case (23) below, it will be clear that any choice of $\rho$ with $\epsilon\rho=0$ will also satisfy $\iota(\rho)=0$. (Alternatively, a specific choice can be made using the classical $J$-homomorphism $J\colon\pi_*(SO)\to\pi_*(S)$, by declaring $\rho$ to be the image of a generator of $\pi_{15}(SO)$. This is consistent with the condition $\epsilon\cdot\rho=0$, because $\epsilon$ can be realized unstably to act naturally on $\pi_{15}$ of spaces, and will then map a generator of $\pi_{15}(SO)$ to zero.)

We have $e(\eta\kappa)=\eta e(\kappa)=0$, so by the surjectivity of the $e$-invariant, cf. Remark 11.48, the class $e(\rho)$ generates $\pi_{15}(j)=\mathbb{Z}/32$, hence is an odd multiple of $j_{15}$. We showed that $\eta\cdot\sigma^2=0$ in Proposition 11.34, using the quadratic construction on $\sigma\colon S^7\to S$. From $e(\sigma\epsilon)=\eta\sigma j_7=0$ in $\pi_{15}(j)$ we deduce that $\sigma\cdot\epsilon\in\ker(e)=\{0,\eta\kappa\}$. One way to show that $\sigma\epsilon\neq\eta\kappa$ is to use that multiplication by $\bar\kappa\in\{g\}$ satisfies $\bar\kappa\cdot\sigma=0$, since $e(\sigma\bar\kappa)=0$, and $\bar\kappa\cdot\eta\kappa\in\{h_1d_0g\}\neq 0$. Another way is to use that $\iota\colon\pi_*(S)\to\pi_*(tmf)$ satisfies $\iota(\sigma)=0$ and $\iota(\eta\kappa)=\{h_1d_0\}\neq 0$. Either way the conclusion is that $\sigma\epsilon=0$.

(16) The $E_\infty$-term for $t-s=16$ is generated by $h_1h_4$ and $Pc_0$. We know that $\eta\rho$ is nonzero, since $e(\eta\rho)=\eta j_{15}$ generates $\pi_{16}(j)=\mathbb{Z}/2$, and that it is detected modulo Adams filtration $\geq 6$ by $h_1\cdot h_0^3h_4=0$. Thus $Pc_0$ must detect $\eta\rho$, and there is a hidden $\eta$-extension from $h_0^3h_4$ to $Pc_0$. The class $\eta^*=\eta_4\in\{h_1h_4\}$ is determined modulo $\eta\rho$, and we choose $\eta^*$ so that $e(\eta^*)=0$, cf. Remark 11.37. Since $e$ is split surjective, it follows that $\pi_{16}(S)=\mathbb{Z}/2\{\eta\rho\}\oplus\mathbb{Z}/2\{\eta^*\}$, with $2\eta^*=0$.

The products $\eta\cdot\eta\kappa$, $\sigma\cdot\mu$ and $\epsilon\cdot\epsilon$ lie in Adams filtration $\geq 6$, hence are detected by $e$. We have $e(\eta^2\kappa)=\eta^2e(\kappa)=0$, $e(\sigma\mu)=j_7j_9=\eta j_{15}$, and $e(\epsilon^2)=\eta^2j_7^2=0$, the latter two by Proposition 11.49. Hence $\eta^2\kappa=0$, $\sigma\mu=\eta\rho$ and $\epsilon^2=0$.

(17) The $E_\infty$-term for $t-s=17$ is generated by $h_1^2h_4$, $h_2d_0$, $h_1Pc_0$ and $P^2h_1$, detecting $\eta\eta^*=\eta\eta_4$, $\nu\kappa$, $\eta^2\rho$ and $\bar\mu=\mu_{17}$, respectively. Here $e(\eta\eta^*)=0$,

$e(\nu\kappa) = 0$ and $e(\eta^2\rho) = \eta^2 j_{15}$, by $\eta$- and $\nu$-linearity of $e$, while $e(\bar{\mu}) = j_{17}$ by the construction of $\bar{\mu}$. Hence $\ker(e) = \mathbb{Z}/2\{\nu\kappa\} \oplus \mathbb{Z}/2\{\eta\eta^*\}$, so that $\pi_{17}(S) = \mathbb{Z}/2\{\bar{\mu}\} \oplus \mathbb{Z}/2\{\eta^2\rho\} \oplus \mathbb{Z}/2\{\nu\kappa\} \oplus \mathbb{Z}/2\{\eta\eta^*\}$. Clearly $\nu \cdot \sigma^2 = 0$, $\sigma \cdot \eta\mu = \eta \cdot \sigma\mu = \eta^2\rho$, $\epsilon \cdot \eta\epsilon = \eta \cdot \epsilon^2 = 0$ and $\epsilon \cdot \eta^2\sigma = 0$. The product $\epsilon \cdot \mu$ in Adams filtration $\geq 8$ is detected by $e$, with $e(\epsilon \cdot \mu) = \eta j_7 j_9 = \eta^2 j_{15}$, hence equals $\eta^2\rho$.

(18) The $E_\infty$-term for $t - s = 18$ is generated by $h_2h_4$, $h_0h_2h_4$, $h_0^2h_2h_4$ and $h_1P^2h_1$, detecting $\nu^*$, $2\nu^*$, $4\nu^*$ and $\eta\bar{\mu}$, respectively. Since $e(\eta\bar{\mu}) = \eta j_{17}$ generates $\pi_{18}(j)$ we can and will choose $\nu^*$ so that $e(\nu^*) = 0$. This determines $\nu^*$ up to an odd multiple, $\ker(e) = \mathbb{Z}/8\{\nu^*\}$, and $\pi_{18}(S) = \mathbb{Z}/2\{\eta\bar{\mu}\} \oplus \mathbb{Z}/8\{\nu^*\}$. The Adams filtration $\geq 5$ part of $\pi_{18}(S)$ is thus detected by $e$. From $h_1^3h_4 = h_0^2h_2h_4$ we deduce that the difference between $\eta^2\eta^*$ and $4\nu^*$ is detected by $e$. Since $e(\eta^*) = 0$ and $e(\nu^*) = 0$ it follows that this difference is zero. From $e(\nu \cdot \rho) = \nu j_{15} = 0$, $e(\sigma \cdot \zeta) = j_7 j_{11} = 0$, $e(\mu^2) = j_9^2 = \eta j_{17} = e(\eta\bar{\mu})$ and $e(\eta \cdot \eta^2\rho) = \eta^3 j_{15} = 0$, we see that $\nu \cdot \rho = \sigma \cdot \zeta = \eta \cdot \eta^2\rho = 0$ while $\mu^2 = \eta\bar{\mu}$.

(19) The $E_\infty$-term for $t-s = 19$ is generated by $c_1$, $P^2h_2$, $h_0P^2h_2$ and $h_0^2P^2h_2$. Let $\bar{\zeta}$ in $\pi_{19}(S)$ be detected by $P^2h_2$. Then $4\bar{\zeta}$ is detected by $h_0^2P^2h_2 = h_1^2P^2h_1$, hence is equal to $\eta^2\bar{\mu}$, with $e(4\bar{\zeta}) = \eta^2 j_{17} = 4j_{19}$ in $\pi_{19}(j) = \mathbb{Z}/8$. It follows that $e(\bar{\zeta}) \doteq j_{19}$, so we can and will choose $\bar{\sigma} \in \pi_{19}$ to be detected by $c_1$ and to satisfy $e(\bar{\sigma}) = 0$. This uniquely determines $\bar{\sigma}$, with $\ker(e) = \mathbb{Z}/2\{\bar{\sigma}\}$ and $\pi_{19}(S) = \mathbb{Z}/8\{\bar{\zeta}\} \oplus \mathbb{Z}/2\{\bar{\sigma}\}$. The Adams filtration $\geq 4$ part of $\pi_{19}(S)$ is detected by $e\colon S \to j$. Hence $\eta \cdot \nu^* = 0$, since $h_1 \cdot h_2h_4 = 0$ and $e(\nu^*) = 0$. Similarly, $\nu \cdot \eta^* = 0$ because $h_2 \cdot h_1h_4 = 0$ and $e(\eta^*) = 0$. Clearly $\nu \cdot \eta\rho = 0$. Finally, $\epsilon \cdot \zeta = 0$ because $e(\epsilon \cdot \zeta) = \eta j_7 j_{11} = 0$.

(20) The $E_\infty$-term for $t-s = 20$ is generated by $g$, $h_0g$ and $h_0^2g$. Let $\bar{\kappa} \in \pi_{20}(S)$ be detected by $g$. This determines $\bar{\kappa}$ up to an odd multiple. Trivially $e(\bar{\kappa}) = 0$, since $\pi_{20}(j) = 0$. The product $\eta \cdot \bar{\zeta}$ lies in Adams filtration $\geq 10$, hence is zero. We showed that $\eta \cdot \sigma^\circ = 0$ in Proposition 11.35, using the quadratic construction on $\epsilon\colon S^8 \to S$. Both $\bar{\sigma}$ and $\sigma^\circ$ are detected by $c_1$, so $\bar{\sigma} \equiv \sigma^\circ \mod \bar{\zeta}$, which implies $\eta \cdot \bar{\sigma} = \eta \cdot \sigma^\circ = 0$. Multiplication by $\nu$ is clearly trivial on $\eta\eta^*$ and $\eta^2\rho$. The relation $h_2^2d_0 = h_0^2g$ in $E_2(S)$, and the fact that $\pi_{20}(S)$ is trivial in Adams filtrations $\geq 7$, imply that $\nu \cdot \nu\kappa = 4\bar{\kappa}$. Similarly $\nu \cdot \bar{\mu} = 0$ and $\mu \cdot \zeta = 0$, because these products land in Adams filtration $\geq 10$.

(21) The $E_\infty$-term for $t - s = 21$ is generated by $h_2^2h_4$ and $h_1g$, detecting $\nu\nu^*$ and $\eta\bar{\kappa}$, respectively. In view of the relation $h_3^3 = h_2^2h_4$, this class also detects $\sigma^3$, with $2\sigma^3 = 2\sigma^2 \cdot \sigma = 0$. Hence $\pi_{21}(S) = \mathbb{Z}/2\{\eta\bar{\kappa}\} \oplus \mathbb{Z}/2\{\nu\nu^*\}$. The product $\sigma \cdot \kappa$ is detected modulo Adams filtrations $\geq 6$ by $h_3d_0 = 0$. These filtrations are trivial, so $\sigma\kappa = 0$. We postpone the proof that $\sigma \cdot \sigma^2 = \nu\nu^*$ until we have established $\eta^2\bar{\kappa} \neq 0$, in the next case.

(22) The $E_\infty$-term for $t - s = 22$ is generated by $h_2c_1$ and $Pd_0$, with $\nu\bar{\sigma}$ detected by $h_2c_1$. By Theorem 11.71 below, due to Mimura [**129**] and Mahowald–Tangora [**107**], the product $\eta^2\bar{\kappa}$ is detected by $Pd_0$. Since $2\bar{\sigma} = 0$, we must have $2 \cdot \nu\bar{\sigma} = 0$ and $\pi_{22}(S) = \mathbb{Z}/2\{\eta^2\bar{\kappa}\} \oplus \mathbb{Z}/2\{\nu\bar{\sigma}\}$. There is thus a hidden $\eta$-extension from $h_1g$ to $Pd_0$. Clearly $\eta \cdot \eta\bar{\kappa} = \eta^2\bar{\kappa}$, $\eta \cdot \nu\nu^* = 0$ and $\nu \cdot \bar{\sigma} = \nu\bar{\sigma}$. The products $\nu \cdot \bar{\zeta}$ and $\zeta \cdot \zeta$ lie in Adams filtration $\geq 10$, hence vanish. We can prove that $\sigma \cdot \rho = 0$ using $\iota\colon S \to tmf$. The product has Adams filtration $\geq 5$, hence is either 0 or $\eta^2\bar{\kappa}$. The latter class remains nonzero in $\pi_{22}(tmf)$, cf. Theorem 9.16, while $\iota(\sigma) = 0$. Hence $\sigma\rho$ cannot be $\eta^2\bar{\kappa}$. (Alternatively, one can prove that $\sigma \cdot \rho = 0$ using the $J$-homomorphism $J\colon \pi_*(SO) \to \pi_*(S)$, since $\sigma$ acts naturally on $\pi_n$ of spaces for

$n \geq 8$ by composition with the Hopf fibration $S^{15} \to S^8$, $J$ maps a generator of $\pi_{15}(SO)$ to $\rho$, and $\pi_{22}(SO) = 0$.)

We postpone the proof that $\epsilon \cdot \kappa = \eta^2\bar{\kappa}$, giving a hidden $\epsilon$-extension from $d_0$ to $Pd_0$, until the next case.

Returning to $\pi_{21}(S)$, the relation $h_3^3 = h_2^2 h_4$ implies that the difference between $\sigma^3$ and $\nu\nu^*$ has Adams filtration $\geq 4$, i.e., is either 0 or $\eta\bar{\kappa}$. From $\eta \cdot \sigma^2 = 0$, $\eta \cdot \nu = 0$ and $\eta \cdot \eta\bar{\kappa} \neq 0$ it follows that the difference is zero, so that $\sigma \cdot \sigma^2 = \nu\nu^*$.

(23) The $E_\infty$-term for $t - s = 23$ is generated by $h_4c_0$, $h_2g$, $h_0h_2g$, $h_1Pd_0$ and $h_0^k i$ for $k \in \{2, 3, 4, 5\}$, with $h_2g$ detecting $\nu\bar{\kappa}$ and $h_0^2 i$ detecting $\bar{\rho}$. It follows that $2\nu\bar{\kappa}$ is detected by $h_0h_2g$. Furthermore, $4\nu\bar{\kappa} = \eta \cdot \eta^2\bar{\kappa}$ is detected by $h_1 \cdot Pd_0 \neq 0$. Hence there is a hidden 2-extension from $h_0h_2g$ to $h_1Pd_0$, and a hidden $\nu$-extension from $h_0^2 g$ to $h_1Pd_0$.

We claim that $\sigma\eta^*$ is detected by $h_4c_0$, so that there is a hidden $\sigma$-extension from $h_1h_4$ to $h_4c_0$. The proof is similar to that of Theorem 11.54, case (8), using the Adams spectral sequence for $C\sigma$. See Figure 11.15. The Adams differential $d_2(h_4) = h_0h_3^2$ for $S$ lifts to $d_2(\overline{h_4}) = \overline{h_0h_3^2}$ for $C\sigma$. Multiplying by $h_1$ gives $d_2(h_1\overline{h_4}) = h_1\overline{h_0h_3^2}$. A calculation with `ext` shows that $h_1\overline{h_0h_3^2} = i(h_4c_0) = 4_{13} \neq 0$. We now compare with Adams filtrations $\leq 4$.

$$\begin{array}{ccc} E_2(S) & \xrightarrow{\quad i \quad} & E_2(C\sigma) \\ \downarrow & & \downarrow \\ E_2(S_{0,5}) & \xrightarrow{i \wedge 1} & E_2(S_{0,5} \wedge C\sigma) \end{array}$$

The infinite cycle $h_4c_0 \in E_2(S_{0,5})$ detects a nonzero class $\gamma \in \pi_{23}(S_{0,5})$. Since $i(h_4c_0)$ is a boundary in $E_2(S_{0,5} \wedge C\sigma)$, the image $i(\gamma) \in \pi_{23}(S_{0,5} \wedge C\sigma)$ must be zero, so that $\gamma = \sigma \cdot \beta$ for some nonzero class $\beta \in \pi_{16}(S_{0,5})$. The only possibility is that $\beta$ is the image of $\eta^* \in \pi_{16}(S)$, detected by $h_1h_4$. Hence $\sigma \cdot \eta^* \in \pi_{23}(S)$ maps to $\gamma$, and must be detected by $h_4c_0$. The hidden $\sigma$-extension from $h_1h_4$ to $h_4c_0$ follows. From $2\eta^* = 0$ we deduce $2 \cdot \sigma\eta^* = 0$. Thus $\pi_{23}(S) = \mathbb{Z}/16\{\bar{\rho}\} \oplus \mathbb{Z}/8\{\nu\bar{\kappa}\} \oplus \mathbb{Z}/2\{\sigma\eta^*\}$.

The class $\bar{\rho} \in \{h_0^2 i\}$ is well-defined up to an odd multiple. (A more specific choice can be made using the $J$-homomorphism, by taking $\bar{\rho}$ to be the image of a generator of $\pi_{23}(SO)$.) We have $e(\nu\bar{\kappa}) = \nu e(\bar{\kappa}) = 0$ and $e(\sigma\eta^*) = \sigma e(\eta^*) = 0$, so by the surjectivity of $e$, cf. Remark 11.48, $e(\bar{\rho})$ generates $\pi_{23}(j) = \mathbb{Z}/16$. We chose $\rho \in \pi_{15}(S)$ so as to satisfy $\epsilon\rho = 0$. The product $\mu \cdot \kappa$ is detected by $Ph_1 \cdot d_0 = h_1Pd_0 \neq 0$ (verified by `ext`), and $e(\mu \cdot \kappa) = \mu e(\kappa) = 0$, which together imply $\mu \cdot \kappa = 4\nu\bar{\kappa}$. Returning to $\pi_{22}(S)$, it follows from $\nu^2\kappa = 4\bar{\kappa}$ that $\nu^3 \cdot \kappa = 4\nu\bar{\kappa}$ is detected by $h_1Pd_0$. Since $\eta^2\sigma \cdot \kappa = 0$, it follows from $\nu^3 = \eta\epsilon + \eta^2\sigma$ that $\eta\epsilon \cdot \kappa = 4\nu\bar{\kappa} \neq 0$. In particular, $\epsilon\kappa \neq 0$. Since this product lives in Adams filtration $\geq 7$, it can only be detected by $Pd_0$, hence is equal to $\eta^2\bar{\kappa}$.

Finally, let us note that since $\epsilon \cdot \eta\kappa \neq 0$, the condition $\epsilon\rho = 0$ from case (15) characterizes $\rho$, up to an odd multiple, in the same way as the condition $\iota(\rho) = 0$.

(24) The $E_\infty$-term for $t - s = 24$ is generated by $h_1h_4c_0$ and $P^2c_0$, detecting $\epsilon\eta^*$ and $\eta\bar{\rho}$, respectively. The latter claim holds since $e(\eta\bar{\rho}) = \eta j_{23} \neq 0$, and $\eta\bar{\rho}$ has Adams filtration $\geq 10$. Thus $\pi_{24}(S) = \mathbb{Z}/2\{\eta\bar{\rho}\} \oplus \mathbb{Z}/2\{\epsilon\eta^*\}$.

To see that $\eta \cdot \sigma\eta^* = \epsilon\eta^*$ note that both homotopy classes are detected by $h_1h_4c_0$ and lie in $\ker(e)$. This implies the claim, since $e$ detects their possible difference, $\eta\bar{\rho} = \{P^2c_0\}$. We have $\nu \cdot \nu\nu^* = 0$, since $\nu\nu^* = \sigma^3$ and $\nu\sigma = 0$. The products $\sigma \cdot \bar{\mu}$

and $\mu \cdot \rho$ have Adams filtration $\ge 9$, hence are detected by their $e$-invariants. Since $e(\sigma\bar{\mu}) = j_7 j_{17} = \eta j_{23}$ and $e(\mu\rho) = j_9 j_{15} = \eta j_{23}$, both of these products equal $\eta\bar{\rho}$.

(25) The $E_\infty$-term for $t-s=25$ is generated by $h_1 P^2 c_0$ and $P^3 h_1$, detecting $\eta^2\bar{\rho}$ and $\mu_{25}$, respectively. Hence $\pi_{25}(S) = \mathbb{Z}/2\{\mu_{25}\} \oplus \mathbb{Z}/2\{\eta^2\bar{\rho}\}$. By construction $e(\mu_{25}) = j_{25}$, while $e(\eta^2\bar{\rho}) = \eta^2 j_{23}$. Thus $\ker(e) = 0$ in degree 25.

We have $\eta \cdot \epsilon\eta^* = 0$, $\nu \cdot \nu\bar{\sigma} = 0$, $\sigma \cdot \nu^* = 0$, $\epsilon \cdot \bar{\mu} = \eta\sigma \cdot \bar{\mu} = \eta^2\bar{\rho}$, $\mu \cdot \eta^* = 0$ and $\zeta \cdot \kappa = 0$ since $e(\eta^*) = 0$, $e(\bar{\sigma}) = 0$, $e(\nu^*) = 0$, $e(\epsilon) = e(\eta\sigma)$, $e(\eta^*) = 0$ and $e(\kappa) = 0$, respectively.

(26) The $E_\infty$-term for $t-s=26$ is generated by $h_2^2 g$ and $h_1 P^3 h_1$, detecting $\nu^2\bar{\kappa}$ and $\eta\mu_{25}$, respectively. Thus $\pi_{26}(S) = \mathbb{Z}/2\{\eta\mu_{25}\} \oplus \mathbb{Z}/2\{\nu^2\bar{\kappa}\}$. Here $e(\eta\mu_{25}) = \eta j_{25}$ and $e(\nu^2\bar{\kappa}) = 0$. Products in Adams filtration $\ge 7$ are detected by the $e$-invariant. Thus $\nu \cdot \bar{\rho} = 0$, $\sigma \cdot \bar{\zeta} = 0$, $\mu \cdot \bar{\mu} = \eta\mu_{25}$ and $\zeta \cdot \rho = 0$, since $e(\nu \cdot \bar{\rho}) = \nu j_{23} = 0$, $e(\sigma \cdot \bar{\zeta}) = j_7 j_{19} = 0$, $e(\mu \cdot \bar{\mu}) = j_9 j_{17} = \eta j_{25}$ and $e(\zeta \cdot \rho) = j_{11} j_{15} = 0$.

To show that the product $\sigma \cdot \bar{\sigma}$ is zero we use Toda brackets and Moss' theorem. As can be verified with `ext`, the Massey product $\langle h_2, h_1, h_3^2\rangle$ is $c_1$ with no indeterminacy. The groups $E_2^{s,t}(S)$ vanish for $(s,t) = (0,5)$, $(0,16)$ and $(1,17)$, so [**132**, Thm. 1.2] applies to show that the Toda bracket $\langle \nu, \eta, \sigma^2\rangle$, which has no indeterminacy, is an element in $\{c_1\}$. Since $\sigma\bar{\zeta} = 0$, it follows that $\sigma\bar{\sigma} = \sigma\langle \nu, \eta, \sigma^2\rangle$. By the shuffle relation [**171**, (3.6)] for Toda brackets, $\sigma\langle \nu, \eta, \sigma^2\rangle = \langle \sigma, \nu, \eta\rangle\sigma^2 = 0$, since $\langle \sigma, \nu, \eta\rangle \in \pi_{12}(S) = 0$.

We showed in Proposition 11.34 that $\epsilon\nu^\circ$ is an $\eta^2$-multiple. Since $\nu^* \doteq \nu^\circ$, this shows that $\epsilon \cdot \nu^* \in \eta^2 \cdot \pi_{24}(S) = 0$, which we now know only contains zero. Hence $\epsilon\nu^* = 0$. (Alternatively, this can be deduced from Moss' theorem: The Massey product $\langle h_3, h_2, h_3\rangle$ is $h_2 h_4$ with no indeterminacy. The group $E_3^{0,11}(S)$ vanishes, so $h_2 h_4$ detects an element of $\langle \sigma, \nu, \sigma\rangle$. Since $\epsilon \cdot \eta\bar{\mu} = 0$ and $2\epsilon = 0$ it follows that $\epsilon\nu^* = \epsilon\langle \sigma, \nu, \sigma\rangle$. The shuffle relation $\epsilon\langle \sigma, \nu, \sigma\rangle = -\langle \epsilon, \sigma, \nu\rangle\sigma$ then implies that $\epsilon\nu^*$ is a $\sigma$-multiple, hence must be 0.)

(27) The $E_\infty$-term for $t-s=27$ is generated by $P^3 h_2$, $h_0 P^3 h_2$ and $h_0^2 P^3 h_2$, with $\zeta_{27}$ detected by $P^3 h_2$. Thus $\pi_{27}(S) = \mathbb{Z}/8\{\zeta_{27}\}$ and $e(\zeta_{27}) \doteq j_{27}$, so that $\ker(e) = 0$ in degree 27. It follows that $\eta \cdot \eta\mu_{25} = 4\zeta_{27}$, $\sigma \cdot \bar{\kappa} = 0$, $\epsilon \cdot \bar{\zeta} = \eta\sigma \cdot \bar{\zeta} = 0$, $\epsilon \cdot \bar{\sigma} = 0$, $\mu \cdot \nu^* = 0$ and $\zeta \cdot \eta^* = 0$, because $e(\eta^2\mu_{25}) = \eta^2 j_{25} = 4j_{27}$, $e(\bar{\kappa}) = 0$, $e(\epsilon) = e(\eta\sigma)$, $e(\bar{\sigma}) = 0$, $e(\nu^*) = 0$ and $e(\eta^*) = 0$, respectively.

(28) The $E_\infty$-term for $t-s=28$ is generated by $d_0^2$, detecting $\kappa^2$. Hence $\pi_{28}(S) = \mathbb{Z}/2\{\kappa^2\}$. We have $\eta \cdot \zeta_{27} = 0$, $\nu \cdot \mu_{25} = 0$, $\mu \cdot \bar{\zeta} = 0$, $\mu \cdot \bar{\sigma} = 0$ and $\zeta \cdot \bar{\mu} = 0$, since these products have Adams filtration $\ge 9$. The case of $\mu \cdot \bar{\sigma}$ uses that $Ph_1 \cdot c_1 = 0$, as can be checked with `ext`.

On the other hand, $\epsilon \cdot \bar{\kappa} = \kappa^2$, since $\epsilon\kappa = \eta^2\bar{\kappa} = \{Pd_0\}$ implies $\epsilon\kappa\bar{\kappa} = \eta^2\bar{\kappa}^2 \in \{Pd_0 g\}$ where $Pd_0 \cdot g = d_0^3 \neq 0$ in $E_\infty(S)$. Hence $\epsilon\bar{\kappa} \neq 0$, and $\kappa^2$ is the only possible value. This calculation also shows that $\eta\bar{\kappa}^2$ must have Adams filtration between 9 and 11, since $\bar{\kappa}^2 \in \{g^2\}$. The only possible detecting class is $z$, which proves that there are hidden $\eta$-extensions from $g^2$ to $z$ and from $z$ to $d_0^3$.

(29) The $E_\infty$-term for $t-s=29$ is trivial. Hence $\pi_{29}(S) = 0$.

(30) The $E_\infty$-term for $t-s=30$ is generated by $h_4^2$, detecting the Kervaire invariant one class $\theta_4$. Each product of lower-degree classes landing in $\pi_{30}(S)$ has Adams filtration $\ge 3$, hence is zero.

(31) The $E_\infty$-term for $t-s=31$ is generated by $h_1 h_4^2$, $n$ and $h_0^k h_5$ for $k \in \{10, 11, \dots, 15\}$, with $h_1 h_4^2$ detecting $\eta\theta_4$ and $h_0^{10} h_5$ detecting $\rho_{31}$. The class $\rho_{31}$ is determined up to an odd multiple. (A more specific choice can be made using

the $J$-homomorphism, by taking $\rho_{31}$ to be the image of a generator of $\pi_{31}(SO)$.) In the proof of Theorem 11.54, case (6), we showed that $\{x_5\} = \{h_0^{15}h_5\}$ maps to $2^5 j_{31} \in \pi_{31}(j) = \mathbb{Z}/64$, which implies that $e(\rho_{31}) \doteq j_{31}$. It follows that we can choose an element $[n] \in \{n\}$ with $e([n]) = 0$, and this uniquely determines $[n]$ in $\ker(e) \subset \pi_{31}(S)$. Clearly $\eta\theta_4$ and $[n]$ have order 2, while $\rho_{31}$ has order 64.

The products $\nu \cdot \kappa^2$, $\epsilon \cdot \bar{\rho}$, $\zeta \cdot \bar{\kappa}$, $\kappa \cdot \bar{\mu}$ and $\rho \cdot \eta^*$ vanish because they have Adams filtration $\geq 6$, hence are detected by $e$. In the second case, $e(\epsilon \cdot \bar{\rho}) = \eta j_7 j_{23} = 0$, and in the other cases the $e$-invariant of one of the factors is zero.

(32) The $E_\infty$-term for $t - s = 32$ is generated by $h_1h_5$, $d_1$, $q$ and $P^3c_0$. Since $e(\eta\rho_{31}) = \eta j_{31} \neq 0$ in $\pi_{32}(j) = \mathbb{Z}/2$ we see that $\eta\rho_{31} \neq 0$ must be detected by a class in Adams filtration $\geq 12$, and $P^3c_0$ is the only possibility. The remaining three generators are therefore represented by the elements of $[q] = \{q\} \cap \ker(e)$, $[d_1] = \{d_1\} \cap \ker(e)$ and $[h_1h_5] = \{h_1h_5\} \cap \ker(e)$. Here $[q]$ consists of a single element, with $2[q] = 0$. The indeterminacy of $[d_1]$ is generated by $[q]$, and the indeterminacy of $[h_1h_5]$ is generated by $[q]$ and $[d_1]$. Since $h_1q \neq 0$ in $E_\infty(S)$ we have $\eta[q] \neq 0$, so $2[d_1] \neq [q]$, which implies $2[d_1] = 0$. Similarly, $2[h_1h_5] \neq [q]$. Furthermore, $h_2d_1 \neq 0$ in $E_\infty(S)$, so $\nu[d_1] \neq 0$ is detected in Adams filtration 5. If $2[h_1h_5] = [d_1]$ then $\nu[d_1] = 2\nu[h_1h_5]$, but there is no class in Adams filtration $\leq 4$ that could detect $\nu[h_1h_5]$. Hence $2[h_1h_5] = 0$, so that $\pi_{32}(S) = \mathbb{Z}/2\{\eta\rho_{31}\} \oplus \mathbb{Z}/2\{[q]\} \oplus \mathbb{Z}/2\{[d_1]\} \oplus \mathbb{Z}/2\{[h_1h_5]\}$.

We will show in Proposition 11.77 that the Hurewicz homomorphism $\iota\colon \pi_*(S) \to \pi_*(tmf)$ takes $\{q\} \subset \pi_{32}(S)$ to $\epsilon_1 \in \{\delta'\} \subset \pi_{32}(tmf)$, increasing Adams filtration from 6 to 7. Here $\epsilon_1$ generates the 2-power torsion in $\pi_{32}(tmf)$, see Table 9.3 and Theorem 9.26. It follows that we can make refined choices $\kappa_1 \in [[d_1]] = [d_1] \cap \ker(\iota)$ and $\eta_5 \in [[h_1h_5]] = [h_1h_5] \cap \ker(\iota)$ of representatives for $d_1$ and $h_1h_5$. This uniquely determines $\kappa_1$, and specifies $\eta_5$ modulo $\kappa_1$. Finally, $\nu\kappa_1$ is detected by $h_2d_1 \neq 0$, so we can fix a single element $\eta_5 \in [[h_1h_5]]$ by insisting that $\nu\eta_5 = 0$. (Alternatively, one can define $\eta_5 \in \{h_1h_5\}$ to be an element of the Toda bracket $\langle \eta, 2, \theta_4 \rangle \subset \pi_{32}(S)$, as in [**22**, §3.2]. The bracket has indeterminacy $\eta \cdot \pi_{31}(S)$, so we can choose $\eta_5$ in $[h_1h_5] \subset \ker(e)$. The image of the Toda bracket in $\pi_{32}(tmf)$ is zero, since $\pi_{30}(tmf) = \pi_{31}(tmf) = 0$, so $\eta_5 \in [[h_1h_5]]$. Furthermore, $\nu\langle \eta, 2, \theta_4 \rangle = \langle \nu, \eta, 2 \rangle \theta_4 = 0$, so this $\eta_5$ is equal to the one we have specified above. We will use this Toda bracket description of $\eta_5$ in our discussion of $\pi_{40}(S)$.)

The product $\eta \cdot [n]$ has Adams filtration $\geq 7$, since $h_1n = 0$ at $E_2(S)$, hence is detected by $e$, and $e$ is zero on $[n]$, so $\eta[n] = 0$. Similarly, the product $\eta \cdot \eta\theta_4$ has Adams filtration $\geq 5$, since $h_1 \cdot h_1h_4^2 = 0$. It cannot be detected by $q$, because $\eta^3\theta_4 = 4\nu\theta_4$ would then be detected by $h_1q \neq 0$, but $\nu\theta_4$ is detected in Adams filtration at least 4, and there is no class in $E_\infty^{s,t}(S)$ for $t - s = 33$ and $5 \leq s \leq 6$ that could detect $2\nu\theta_4$. Hence $\eta \cdot \eta\theta_4$ is detected by $e$. Since $e(\theta_4) = 0$, we have $\eta^2\theta_4 = 0$. The products $\sigma \cdot \mu_{25}$, $\mu \cdot \bar{\rho}$ and $\rho \cdot \bar{\mu}$ all have Adams filtration $\geq 13$, hence are detected by $e$. Here $e(\sigma \cdot \mu_{25}) = j_7 j_{25}$, $e(\mu \cdot \bar{\rho}) = j_9 j_{23}$ and $e(\rho \cdot \bar{\mu}) = j_{15} j_{17}$, each of which equals $e(\eta\rho_{31}) = \eta j_{31}$. The product $\kappa \cdot \nu^*$ represents $d_0 \cdot h_2h_4 = 0$ modulo Adams filtration $\geq 7$, hence is detected by $e$, which is zero on both factors, so that $\kappa\nu^* = 0$.

We use $tmf$ to show that $\eta^* \cdot \eta^* = 0$. This product has Adams filtration $\geq 5$, since $h_1h_4 \cdot h_1h_4 = 0$. It cannot be detected by $q$, since $\eta^* \mapsto 0$ in $\pi_{16}(tmf) \cong \mathbb{Z}$, while $[q] \mapsto \epsilon_1 \neq 0$ in $\pi_{32}(tmf)$, by Proposition 11.77. Hence $\eta^* \cdot \eta^*$ is detected by $e$, and $e(\eta^*) = 0$, so $(\eta^*)^2 = 0$.

(33) The $E_\infty$-term for $t-s=33$ is generated by $h_1^2h_5$, $p$, $h_1q$, $h_1P^3c_0$ and $P^4h_1$, detecting $\eta\eta_5$, $\nu\theta_4$, $\eta[q]$, $\eta^2\rho_{31}$ and $\mu_{33}$, respectively. See the proof of Theorem 11.54, case (8) for the hidden $\nu$-extension from $h_4^2$ to $p$. Since $2\eta=0$ and $2\theta_4=0$ it follows that $\pi_{33}(S)\cong(\mathbb{Z}/2)^5$. By construction, $e(\mu_{33})=j_{33}$.

We use *tmf* to show that $\eta\cdot\kappa_1=0$. This product has Adams filtration $\geq 6$ since $h_1\cdot d_1=0$ in $E_\infty(S)$. It cannot be detected by $h_1q$, because $\iota(\kappa_1)=0$ and $\iota(\eta[q])=\eta\epsilon_1\neq 0$ in $\pi_{33}(tmf)$. Hence the product is detected by $e$, and $e(\kappa_1)=0$, so $\eta\kappa_1=0$. Similarly, $\rho\cdot\nu^*=0$. The product has Adams filtration $\geq 6$, and cannot be detected by $h_1q$, because $d=q_0\iota\colon S\to tmf\to ko$ maps $\nu^*$ to zero, so that $\iota(\nu^*)=0$, whereas $\iota(\eta[q])=\eta\epsilon_1\neq 0$, as recalled above. Hence $e$ detects $\rho\cdot\nu^*$, and $e(\nu^*)=0$, so that $\rho\nu^*=0$.

The products $\epsilon\cdot\mu_{25}$, $\kappa\cdot\bar{\zeta}$, $\kappa\cdot\bar{\sigma}$ and $\eta^*\cdot\bar{\mu}$ have Adams filtration $\geq 8$, since $d_0\cdot c_1=0$ in $E_2(S)$, hence are detected by $e$. Here $e(\epsilon\cdot\mu_{25})=\eta j_7j_{25}=\eta^2j_{31}$, so that $\epsilon\mu_{25}=\eta^2\rho_{31}$. Also $e(\kappa\cdot\bar{\zeta})=0$, $e(\kappa\cdot\bar{\sigma})=0$ and $e(\eta^*\cdot\bar{\mu})=0$, since $e(\kappa)=0$ and $e(\eta^*)=0$, so that $\kappa\bar{\zeta}=0$, $\kappa\bar{\sigma}=0$ and $\eta^*\bar{\mu}=0$.

(34) The $E_\infty$-term for $t-s=34$ is generated by $h_0h_2h_5$, $h_0^2h_2h_5$, $h_2n$, $d_0g$ and $h_1P^4h_1$, with $\eta\mu_{33}$ detected by $h_1P^4h_1$, $\kappa\bar{\kappa}$ detected by $d_0g$, $\nu[n]$ detected by $h_2n$ and $\eta^2\eta_5$ detected by $h_1^3h_5=h_0^2h_2h_5$. Since $e(\eta\mu_{33})=\eta j_{33}$ generates $\pi_{34}(j)$ we can represent $h_0h_2h_5$ by an element $\alpha_{34}$ in $[h_0h_2h_5]=\{h_0h_2h_5\}\cap\ker(e)$. Then $\eta\mu_{33}$, $\kappa\bar{\kappa}$ and $\nu[n]$ have order 2, and $2\alpha_{34}=\eta^2\eta_5$ modulo Adams filtration $\geq 6$, so that $4\alpha_{34}=0$.

The indeterminacy of $[h_0h_2h_5]$ is generated by $\kappa\bar{\kappa}$, $\nu[n]$ and $2\alpha_{34}$. We can remove the indeterminacy generated by $\kappa\bar{\kappa}$ in two equivalent ways. First, we can insist that $\eta\alpha_{34}=0$. Here $\eta\alpha_{34}\in\ker(e)$ cannot be detected by $h_2d_1$ since $h_2^2d_1\neq 0$ would then detect $\nu\cdot\eta\alpha_{34}=0$, and if $\eta\alpha_{34}$ is detected by $h_1d_0g$ then we can subtract $\kappa\bar{\kappa}$ from $\alpha_{34}$ to make $\eta\alpha_{34}=0$. The remaining indeterminacy of $\alpha_{34}$ is generated by $\nu[n]$ and $2\alpha_{34}$. Second, in view of Proposition 11.82, $\iota$ maps $\ker(e)\subset\pi_*(S)$ into the $B$-power torsion in $\pi_*(tmf)$, so $\iota(\alpha_{34})$ is 0 or $\kappa\bar{\kappa}$, with $\eta\kappa\bar{\kappa}\neq 0$ in $\pi_{35}(tmf)$. See Figure 9.7. Hence, for $\alpha_{34}\in[h_0h_2h_5]$ the conditions $\eta\alpha_{34}=0$ and $\iota(\alpha_{34})=0$ are equivalent. We set $[[h_0h_2h_5]]=[h_0h_2h_5]\cap\ker(\iota)=[h_0h_2h_5]\cap\ker(\eta)$. Moreover, we can use a Toda bracket to remove the indeterminacy generated by $\nu[n]$. Following [**22**, §4] we may form the Toda bracket $\langle\eta,2,\eta_5\rangle\subset\pi_{34}(S)$ with indeterminacy $\eta\cdot\pi_{33}(S)$. By Moss' theorem and an `ext`-calculation the Massey product $\langle h_1,h_0,h_1h_5\rangle=h_0h_2h_5$ detects one, hence each, element of this Toda bracket, and we may choose $\alpha_{34}\in\langle\eta,2,\eta_5\rangle\cap\ker(e)$. We will see in the next paragraph that this reduces the indeterminacy of $\alpha_{34}$ to $\mathbb{Z}/2\{\eta^2\eta_5\}=\mathbb{Z}/2\{2\alpha_{34}\}$. Furthermore, $\alpha_{34}\in[h_0h_2h_5]$ and $\iota(\alpha_{34})\in\langle\eta,2,0\rangle\subset\pi_{34}(tmf)$. The latter Toda bracket contains zero, hence equals $\eta\cdot\pi_{33}(tmf)$, which only contains $B$-periodic classes. See Figure 9.7, again. Since $\iota(\alpha_{34})$ is $B$-power torsion, it follows that $\iota(\alpha_{34})=0$, as required by the previous specification.

The product $\eta\cdot\eta[q]$ is detected by $e$, since $h_1^2q=0$, and $e([q])=0$, so $\eta^2[q]=0$. The third (Toda bracket) specification of $\alpha_{34}$ lets us calculate that $2\alpha_{34}=2\langle\eta,2,\eta_5\rangle=-\langle 2,\eta,2\rangle\eta_5=\eta^2\eta_5$, since $\langle 2,\eta,2\rangle=\eta^2$.

The products $\nu\cdot\rho_{31}$, $\sigma\cdot\zeta_{27}$, $\mu\cdot\mu_{25}$, $\zeta\cdot\bar{\rho}$, $\rho\cdot\bar{\zeta}$ and $\bar{\mu}^2$ lie in Adams filtrations detected by $e$. Here $e(\nu\cdot\rho_{31})=\nu j_{31}=0$, $e(\sigma\cdot\zeta_{27})=j_7j_{27}=0$, $e(\mu\cdot\mu_{25})=j_9j_{25}=\eta j_{33}$, $e(\zeta\cdot\bar{\rho})=j_{11}j_{23}=0$, $e(\rho\cdot\bar{\zeta})=j_{15}j_{19}=0$ and $e(\bar{\mu}^2)=j_{17}^2=\eta j_{33}$, so $\nu\rho_{31}=0$, $\sigma\zeta_{27}=0$, $\zeta\bar{\rho}=0$, and $\rho\bar{\zeta}=0$, while $\mu\mu_{25}=\eta\mu_{33}=\bar{\mu}^2$. (The products $\mu\cdot\mu_{25}$ and $\bar{\mu}\cdot\bar{\mu}$ are also detected by $Ph_1\cdot P^3h_1=h_1P^4h_1=h_1P^2h_1\cdot h_1P^2h_1$.)

We use *tmf* to show that $\rho \cdot \bar{\sigma} = 0$. This product has Adams filtration $\geq 7$. It cannot be detected by $\kappa\bar{\kappa}$, since $\iota(\bar{\sigma})$ must vanish in $\pi_{19}(tmf) = 0$, while $\iota(\kappa\bar{\kappa}) \neq 0$ in $\pi_{34}(tmf)$. Likewise, the product cannot be detected by $\eta\mu_{33}$, since $e(\bar{\sigma}) = 0$.

The product $\eta^* \cdot \nu^*$ has Adams filtration $\geq 5$, since $h_1h_4 \cdot h_2h_4 = 0$ in $E_2(S)$. To eliminate the possibility that it is detected by $\nu[n]$, we use the Toda bracket presentation $\eta^* \in \langle \sigma, 2\sigma, \eta \rangle$ and the relation $\eta\nu^* = 0$ to see that $\eta^*\nu^* \in \langle \sigma, 2\sigma, \eta\rangle\nu^* \subset \langle \sigma, 2\sigma, \eta\nu^*\rangle = \sigma \cdot \pi_{27}(S)$. Since $\sigma\zeta_{27} = 0$ we must have $\eta^*\nu^* = 0$.

(35) The $E_\infty$-term for $t - s = 35$ is generated by $h_2d_1$, $h_1d_0g$ and $h_0^kP^4h_2$ for $k \in \{0, 1, 2\}$, detecting $\nu\kappa_1$, $\eta\kappa\bar{\kappa}$ and $2^k\zeta_{35}$, respectively. We have $2 \cdot \nu\kappa_1 = 0$ since $2\kappa_1 = 0$.

As usual, $\eta \cdot \eta\mu_{33} = 4\zeta_{35}$ because $h_1^2P^4h_1 = h_0^2P^4h_2$. Hence $4e(\zeta_{35}) = \eta^2 j_{33} = 4j_{35}$ and $e(\zeta_{35}) \doteq j_{35}$. We chose $\alpha_{34}$ so that $\eta\alpha_{34} = 0$. Similarly, we required $\eta_5$ to satisfy $\nu\eta_5 = 0$. The products $\epsilon \cdot \zeta_{27}$, $\eta^* \cdot \bar{\zeta}$ and $\bar{\mu} \cdot \nu^*$ are detected by $e$, and $e(\epsilon \cdot \zeta_{27}) = \eta j_7 j_{27} = 0$, $e(\eta^*) = 0$ and $e(\nu^*) = 0$, so $\epsilon\zeta_{27} = 0$, $\eta^*\bar{\zeta} = 0$ and $\bar{\mu}\nu^* = 0$. The product $\eta^* \cdot \bar{\sigma}$ lies in Adams filtration $\geq 6$, because $h_1h_4 \cdot c_1 = 0$ in $E_2(S)$. Since $e(\eta^*) = 0$ and $\iota(\eta^*) = 0$ the product cannot be detected by $h_0^2P^4h_2$ or by $h_1d_0g$, which map to the nonzero elements $4j_{35} \in \pi_{35}(j)$ and $\eta\kappa\bar{\kappa} \in \pi_{35}(tmf)$, respectively, hence $\eta^*\bar{\sigma} = 0$.

Using *tmf*, we see that there is a hidden $\nu$-extension from $q$ to $h_1d_0g$, since $\iota$ maps $[q]$ to $\epsilon_1$ with $\nu \cdot \epsilon_1 = \eta\kappa\bar{\kappa} \neq 0$ in $\pi_{35}(tmf)$, so that $\nu \cdot [q] \neq 0$ in $\pi_{35}(S)$. Since $e([q]) = 0$, only $h_1d_0g$ can detect $\nu[q] = \eta\kappa\bar{\kappa}$. Also using *tmf*, we showed in case (15) that $\rho \cdot \bar{\kappa} = 0$ for our choice of $\rho$ with $\iota(\rho) = 0$, or equivalently, with $\epsilon\rho = 0$. (Alternatively, one can prove that $\rho\bar{\kappa} = 0$ using the $J$-homomorphism $J\colon \pi_*(SO) \to \pi_*(S)$. The class $\bar{\kappa}$ can be realized unstably by a map $\bar{\kappa}_7\colon S^{27} \to S^7$ with $8\bar{\kappa}_7 = 0$, according to [**130**, Lem. 15.4]. Composition with $\bar{\kappa}_7$ acts naturally on $\pi_n$ of spaces for $\geq 7$, and takes the generator of $\pi_{15}(SO)$ to zero in $\pi_{35}(SO)$. Since $J$ maps this generator to $\rho$, it follows that $\rho\bar{\kappa}$ is zero.)

(36) The $E_\infty$-term for $t - s = 36$ is generated by $t$, so $\{t\}$ consists of a single element.

The products $\eta \cdot \zeta_{35}$, $\nu \cdot \mu_{33}$, $\mu \cdot \zeta_{27}$, $\zeta \cdot \mu_{25}$, $\bar{\mu} \cdot \bar{\zeta}$ and $\bar{\mu} \cdot \bar{\sigma}$ lie in Adams filtration $\geq 12$, hence are zero. Furthermore, $h_1h_4 \cdot g = 0$ in $E_2(S)$, so $\eta^* \cdot \bar{\kappa}$ has Adams filtration $\geq 7$, and is also zero. We claim that $\nu \cdot \nu\theta_4 = 0$. Recall the relation $\nu^3 = \eta\epsilon + \eta^2\sigma$ from case (9). If $\nu^2\theta_4$ were detected by $t$, then $\nu^3\theta_4 = (\eta\epsilon + \eta^2\sigma)\theta_4$ would be detected by $h_2t \neq 0$ in $E_\infty(S)$. Here $\epsilon\theta_4$ and $\eta\sigma\theta_4$ have Adams filtration $\geq 6$, since $c_0 \cdot h_4^2 = 0$ in $E_2(S)$. It follows that $\eta\epsilon\theta_4$ and $\eta^2\sigma\theta_4$ have Adams filtration $\geq 8$, since $h_1 \cdot h_1x = 0$. Hence their sum cannot be detected in Adams filtration 7, showing that $\nu^2\theta_4 = 0$. Finally, if $\nu^* \cdot \nu^*$ were detected by $t$, then $\nu \cdot (\nu^*)^2$ would be detected by $h_2t$, but $\nu \cdot (\nu^*)^2 = \nu^* \cdot \sigma^3 = 0$ because $\nu\nu^* = \sigma^3$ and $\sigma\nu^* = 0$. Hence $(\nu^*)^2 = 0$.

(37) The $E_\infty$-term for $t-s = 37$ is generated by $h_2^2h_5$ and $x$. We proved that $\sigma\theta_4$ is detected by $x$ in Theorem 11.56, case (5). There cannot be a hidden 2-extension from $h_2^2h_5$ to $x$, since $h_1x \neq 0$ in $E_\infty(S)$ detects $\eta\sigma\theta_4 \neq 0$ in $\pi_{38}(S)$. Following [**22**, §4], we can form the Toda bracket $\langle \nu^2, 2, \theta_4\rangle \subset \pi_{37}(S)$, with indeterminacy $\langle \sigma\theta_4\rangle$. Recall that $h_2^2 \cdot h_0 = 0$ and $h_0 \cdot h_4^2 = d_2(h_5)$ in $E_2(S)$. Moss' theorem for the $E_3$-term applies, and shows that the $E_3$-Massey product $\langle h_2^2, h_0, h_4^2\rangle = h_2^2h_5$ detects one, hence both, elements in this Toda bracket, so that $\langle \nu^2, 2, \theta_4\rangle = \{h_2^2h_5\}$. The Toda shuffle relation $\eta\langle \nu^2, 2, \theta_4\rangle = \langle \eta, \nu^2, 2\rangle\theta_4 \subset \pi_8(S)\cdot\theta_4$, and our observation from case (36) that $\epsilon\theta_4$ and $\eta\sigma\theta_4$ both have Adams filtration $\geq 6$, prove that $\eta\{h_2^2h_5\}$

has Adams filtration at least 6. Hence there is no hidden $\eta$-extension from $h_2^2h_5$ to $h_0^3h_3h_5$. We can therefore uniquely specify an element $\alpha_{37} \in \{h_2^2h_5\} = \langle\nu^2, 2, \theta_4\rangle$ by the condition $\eta\cdot\alpha_{37} = 0$. (We will use this Toda bracket description of $\alpha_{37}$ when discussing $\pi_{40}(S)$.)

The products $\eta \cdot \{t\}$, $\nu \cdot \kappa\bar{\kappa}$, $\nu \cdot \nu[n]$, $\kappa \cdot \bar{\rho}$, $\bar{\mu} \cdot \bar{\kappa}$, $\nu^* \cdot \bar{\zeta}$ and $\nu^* \cdot \bar{\sigma}$ have Adams filtration $\geq 6$, since $h_2h_4 \cdot c_1 = 0$ in $E_2(S)$, hence are zero. Furthermore $\nu \cdot \alpha_{34}$ cannot be $\sigma\theta_4$, since $\eta\sigma\theta_4 \neq 0$, so $\nu\alpha_{34} = 0$.

(38) The $E_\infty$-term for $t-s = 38$ is generated by $h_0^2h_3h_5$, $h_0^3h_3h_5$ and $h_1x$. The latter class detects $\eta\sigma\theta_4$. Let $\alpha_{38}$ be any class detected by $h_0^2h_3h_5$. Then $2\alpha_{38}$ is detected by $h_0^3h_3h_5$, and $4\alpha_{38} = 0$.

We defined $\alpha_{37}$ so that $\eta\alpha_{37} = 0$. The products $\nu \cdot \zeta_{35}$, $\sigma \cdot \rho_{31}$, $\zeta \cdot \zeta_{27}$, $\rho \cdot \bar{\rho}$, $\bar{\zeta} \cdot \bar{\zeta}$ and $\bar{\zeta} \cdot \bar{\sigma}$ lie in Adams filtration $\geq 12$, hence are zero. We have factorizations $\eta\sigma\theta_4 = \nu \cdot \nu\kappa_1 = \sigma \cdot [n] = \nu^* \cdot \bar{\kappa} = \bar{\sigma} \cdot \bar{\sigma}$, since $h_1x = h_2^2d_1 = h_3n = h_2h_4g = c_1^2$ in $E_2(S)$. In particular, there is not a hidden $\sigma$-extension from $h_1h_4^2$ to $h_1x$, in the strict sense of Definition 9.5, though there is a hidden $\eta\sigma$-extension from $h_4^2$ to $h_1x$.

The product $\epsilon \cdot \theta_4 = \eta\sigma\theta_4$ was calculated by Tangora [**166**, Prop. 1.3], using four-fold Toda brackets. We are grateful to Daniel Isaksen for pointing out this reference. Since Tangora's paper is not easily available, we review the argument in Remark 11.63.

(39) The $E_\infty$-term for $t-s = 39$ is generated by $h_1h_3h_5$, $h_5c_0$, $h_3d_1$, $h_2t$, $u$ and $h_0^kP^2i$ for $k \in \{2,3,4,5\}$. Let $\rho_{39}$ be detected by $h_0^2P^2i$. We know that $h_2t$ detects $\nu\{t\}$, $h_3d_1$ detects $\sigma\kappa_1$ and $h_1h_3h_5$ detects $\sigma\eta_5$. The homomorphism $e\colon \pi_{39}(S) \to \pi_{39}(j) = \mathbb{Z}/16$ maps $\nu\{t\}$, $\sigma\kappa_1$ and $\sigma\eta_5$ to zero, hence can only be surjective if $e(\rho_{39}) \doteq j_{39}$. The intersection $[u] = \{u\} \cap \ker(e)$ therefore consists of a single element. The homomorphism $\iota\colon \pi_{39}(S) \to \pi_{39}(tmf) \cong \mathbb{Z}/2$ maps $\rho_{39}$, $\nu\{t\}$, $\sigma\kappa_1$ and $\sigma\eta_5$ to zero, and sends $[u]$ to the nonzero class $\eta_1\kappa$ detected by $d_0\gamma$, cf. Lemma 1.15 and Table 1.1. Hence we can choose $\alpha_{39} \in \{h_5c_0\} \cap \ker(e) \cap \ker(\iota)$, with indeterminacy generated by $\nu\{t\}$ and $\sigma\kappa_1$. We can remove the indeterminacy in $\alpha_{39}$ by means of a Toda bracket. Following [**22**, Prop. 3.2.4(b)], we can form the $E_3$-Massey product $\langle c_0, h_0, h_4^2\rangle = h_5c_0$ with zero indeterminacy. Moss' theorem for the $E_3$-term applies to show that $\langle\epsilon, 2, \theta_4\rangle$ meets $\{h_5c_0\}$. To see that this Toda bracket has no indeterminacy, we use the fact that $Ph_1 \cdot h_4^2 = 0$ in $E_2(S)$, and the discussion in case (36), to see that the products $\mu\theta_4$, $\eta\epsilon\theta_4$, $\eta^2\sigma\theta_4$, $\epsilon\rho_{31}$ and $\epsilon[n]$ have Adams filtration $\geq 8$, hence are detected by $e$ and $\iota$, and are therefore zero. The homomorphism $e$ maps $\langle\epsilon, 2, \theta_4\rangle$ into $\langle\eta j_7, 2, 0\rangle = \eta j_7 \cdot \pi_{31}(j) = 0$. The homomorphism $\iota$ maps $\langle\epsilon, 2, \theta_4\rangle$ into $\langle\epsilon, 2, 0\rangle = \epsilon \cdot \pi_{31}(tmf) = 0$. Hence we can consistently refine the definition above by setting $\alpha_{39} = \langle\epsilon, 2, \theta_4\rangle$, with zero indeterminacy. By Toda shuffling, $2\langle\epsilon, 2, \theta_4\rangle = -\langle 2, \epsilon, 2\rangle\theta_4$, which lies in $\pi_9(S)\cdot\theta_4 = 0$, as we just saw. This proves that there is no hidden 2-extension from $h_5c_0$ to $h_2t$. Since $[u]$, $\{t\}$, $\kappa_1$ and $\eta_5$ have order 2, and $\eta[u] \neq 0$ is detected by $h_1u$, it follows that $\ker(e) \cong (\mathbb{Z}/2)^5$ is elementary abelian.

The products $\epsilon \cdot \rho_{31}$, $\kappa \cdot \mu_{25}$, $\eta^* \cdot \bar{\rho}$ and $\bar{\zeta} \cdot \bar{\kappa}$ have Adams filtration $\geq 11$, hence are detected by $e$, and are all zero. Furthermore, $\epsilon \cdot [n]$ and $\mu \cdot \theta_4$ have Adams filtration $\geq 8$, because $Ph_1 \cdot h_2^2 = 0$ in $E_2(S)$, hence are detected by $e$ and $\iota$, and must vanish because $\pi_*(tmf)$ is trivial in degrees 30 and 31. On the other hand, $\sigma \cdot [q]$ and $\bar{\sigma} \cdot \bar{\kappa}$ are detected by $h_3 \cdot q = h_2t = c_1 \cdot g$, hence agree with $\nu\{t\}$ modulo Adams filtration $\geq 8$. The differences are detected by $e$ and $\iota$, and vanish because $e([q])$, $e(\bar{\sigma})$, $e(\{t\})$, $\iota(\sigma)$, $\iota(\bar{\sigma})$ and $\iota(\{t\})$ are all zero. Hence $\sigma[q] = \nu\{t\} = \bar{\sigma}\bar{\kappa}$.

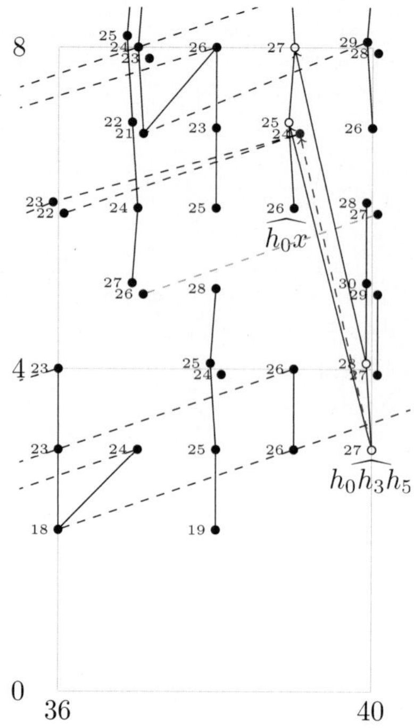


FIGURE 11.20. $E_2(C\eta)$ for $36 \le t-s \le 40$, with some $d_4$-differentials

We use $C\eta$ to prove that there is a hidden $\eta$-extension from $h_0^2h_3h_5$ to $h_2t$. This will prove that $\eta\cdot\alpha_{38} = \nu\{t\}$, since the difference between these classes must have Adams filtration $\ge 8$, hence be detected by $e$ and $\iota$, and these homomorphisms vanish on both classes.

There is a long exact sequence of Adams $E_2$-terms

$$\cdots \longrightarrow E_2^{s-1,t-2}(S) \xrightarrow{h_1} E_2^{s,t}(S) \xrightarrow{i} E_2^{s,t}(C\eta) \xrightarrow{j} E_2^{s,t-2}(S) \longrightarrow \dots,$$

where a part of $E_2(C\eta)$ is shown in Figure 11.20. The infinite cycle $h_2t = h_1y = 7_{16}$ detects $\nu\{t\} \in \pi_{39}(S)$. Since $i(h_1y) = 0$ in $E_2(C\eta)$, it follows that $i(\nu\{t\}) \in \pi_{39}(C\eta)$ has Adams filtration $\ge 8$. Consider the lift $\widehat{h_0h_3h_5} = 3_{27}$ in $E_2(C\eta)$ of $h_0h_3h_5$ in $E_2(S)$. Trivially $d_2(\widehat{h_0h_3h_5}) = 0$. By naturality with respect to $j$ we cannot have $d_3(\widehat{h_0h_3h_5}) = \widehat{h_0x} = 6_{26}$, since $d_3(h_0h_3h_5) = 0$, while $j(\widehat{h_0x}) = h_0x \ne 0$ in $E_3(S)$. Hence $d_3(\widehat{h_0h_3h_5}) = 0$. On the other hand, we must have $d_4(\widehat{h_0h_3h_5}) \equiv h_0\widehat{h_0x}$ mod $h_2^2\widehat{n} = 7_{25}$ mod $7_{24}$, since $d_4(h_0h_3h_5) = h_0^2x \ne 0$ in $E_4(S)$. (This ambiguity is indicated in blue.) Multiplying by $h_0$ we obtain $d_4(h_0\widehat{h_0h_3h_5}) = h_0^2\widehat{h_0x} = 8_{27}$. Hence $E_\infty(C\eta) = 0$ in bidegree $(t-s,s) = (39,8)$, proving that $i(\nu\{t\})$ has Adams filtration $\ge 9$.

Let $S_\star$ be a minimal Adams resolution of $S$, and let $S_{0,9} = \mathrm{cof}(S_9 \to S_0)$ be its truncation to filtrations $0 \le s \le 8$. The image of $i(\nu\{t\})$ in $\pi_{39}(S_{0,9} \wedge C\eta)$ must then be zero, so the nonzero image $\gamma$ of $\nu\{t\}$ in $\pi_{39}(S_{0,9})$ must be of the form $\gamma = \eta\cdot\beta$, with $\beta \in \pi_{38}(S_{0,9})$ of filtration $\le 6$. From $d_2(y) \ne 0$, $h_1^2x = 0$ and $2\eta = 0$ we see that the only class in $E_\infty(S_{0,9})$ that can detect $\beta$ is $h_0^2h_3h_5$. It follows that $\eta\{h_0^2h_3h_5\} \subset \{h_2t\}$ in $\pi_{39}(S)$.

(40) The $E_\infty$-term for $t-s=40$ is generated by $f_1$, $h_1^2h_3h_5$, $h_1h_5c_0$, $Ph_1h_5$, $g^2$, $h_1u$ and $P^4c_0$. Since $e(\eta\rho_{39}) = \eta j_{39} \neq 0$ the product $\eta\rho_{39}$ must be detected by $P^4c_0$. The products $\eta[u]$, $\bar{\kappa}^2$, $\eta\alpha_{39}$ and $\eta\sigma\eta_5$ are detected by $h_1u$, $g^2$, $h_1h_5c_0$ and $h_1^2h_3h_5$, respectively. The $tmf$-Hurewicz homomorphism takes $\bar{\kappa}^2 \in \pi_{40}(S)$ to $\bar{\kappa}^2 \in \pi_{40}(tmf)$, with $2\bar{\kappa}^2 = \epsilon\epsilon_1 \neq 0$, cf. Theorem 9.8. Here $2\bar{\kappa}^2$ is detected by $\delta' w_1$ in Adams filtration 11, so $2\bar{\kappa}^2 \neq 0$ in $\pi_{40}(S)$ can only be detected by $h_1u$. Hence there is a hidden 2-extension from $g^2$ to $h_1u$ in $E_\infty(S)$. Since $\bar{\kappa}^2$ generates the 2-power torsion in $\pi_{40}(tmf)$, the intersections $[[Ph_1h_5]] = \{Ph_1h_5\} \cap \ker(e) \cap \ker(\iota)$ and $[[f_1]] = \{f_1\} \cap \ker(e) \cap \ker(\iota)$ are nonempty. The first contains a single element, while the second has indeterminacy of order four spanned by $[[Ph_1h_5]]$ and $\eta\alpha_{39}$. Multiplication by $\eta^2$ takes any element in $[[f_1]]$ to a 2-torsion element in $\ker(e) \cap \ker(\iota) \subset \pi_{42}(S)$, which must be 0 or detected by $h_1^2Ph_1h_5$. Since $\eta^2 \cdot [[Ph_1h_5]]$ is detected by $h_1^2Ph_1h_5$, we can choose $\alpha_{40} \in [[f_1]] \cap \ker(\eta^2)$, with indeterminacy of order two. In a moment we shall see that $\eta^2 \cdot \eta\alpha_{39} = 0$, so that the indeterminacy left in $\alpha_{40}$ is $\mathbb{Z}/2\{\eta\alpha_{39}\}$. The $e$-invariant splits off $\mathbb{Z}/2\{\eta\rho_{39}\}$ from $\pi_{40}(S)$, and $\iota$ splits off $\mathbb{Z}/4\{\bar{\kappa}^2\}$ from $\ker(e)$. There can be no hidden 2-extensions within $\ker(e) \cap \ker(\iota)$, since $h_1 \cdot Ph_1h_5 \neq 0$, meaning that $\ker(e) \cap \ker(\iota) \cong (\mathbb{Z}/2)^4$ is elementary abelian.

The products $\eta \cdot [u]$ and $2 \cdot \bar{\kappa}^2$ agree modulo $\eta\rho_{39}$, and both map to 0 under $e$, so $\eta[u] = 2\bar{\kappa}^2$. The products $\sigma \cdot \mu_{33}$, $\mu \cdot \rho_{31}$, $\rho \cdot \mu_{25}$ and $\bar{\mu} \cdot \bar{\rho}$ have Adams filtration $\geq 16$, hence are detected by $e$, and $e(\sigma \cdot \mu_{33}) = j_7 j_{33} = \eta j_{39}$, $e(\mu \cdot \rho_{31}) = j_9 j_{31} = \eta j_{39}$, $e(\rho \cdot \mu_{25}) = j_{15} j_{25} = \eta j_{39}$ and $e(\bar{\mu} \cdot \bar{\rho}) = j_{17} j_{23} = \eta j_{39}$, so $\sigma\mu_{33} = \mu\rho_{31} = \rho\mu_{25} = \bar{\mu}\bar{\rho} = \eta\rho_{39}$. The product $\epsilon \cdot [q]$ in Adams filtration $\geq 9$ has trivial $e$-invariant and maps under $\iota$ to $\epsilon \cdot \epsilon_1 = 2\bar{\kappa}^2$ in $\pi_{40}(tmf)$, hence must be equal to $2\bar{\kappa}^2$ in $\pi_{40}(S)$. Similarly, $\epsilon \cdot \kappa_1$ and $\mu \cdot [n]$ have Adams filtration $\geq 7$, hence are detected by $e$ and $\iota$. Since $\kappa_1$ and $[n]$ lie in $\ker(e) \cap \ker(\iota)$, these products are zero.

The product $\epsilon \cdot \eta_5$ is detected by $c_0 \cdot h_1h_5 = h_1 \cdot h_5c_0 \neq 0$, hence agrees with $\eta \cdot \alpha_{39}$ modulo Adams filtration $\geq 6$. Both products map to zero under $e$ and $\iota$, so they are equal modulo $[[Ph_1h_5]]$. Since $h_1^2 \cdot Ph_1h_5 \neq 0$, they are exactly equal because $\eta^2 \cdot \epsilon\eta_5 = 0$ and, as promised above, $\eta^2 \cdot \eta\alpha_{39} = \eta^3\langle\epsilon, 2, \theta_4\rangle = \langle\eta^3, \epsilon, 2\rangle\theta_4$ is also 0, because $\pi_{12}(S) = 0$.

Finally, the relation $\nu\alpha_{37} = \eta\alpha_{39} + \eta\sigma\eta_5$ is the image under the Toda bracket $\langle -, 2, \theta_4\rangle$ of the relation $\nu^3 = \eta\epsilon + \eta^2\sigma$, by virtue of the Toda brackets $\nu\alpha_{37} \in \langle\nu^3, 2, \theta_4\rangle$, $\eta\alpha_{39} \in \langle\eta\epsilon, 2, \theta_4\rangle$ and $\eta\sigma\eta_5 \in \langle\eta^2\sigma, 2, \theta_4\rangle$, each with zero indeterminacy.

(41) The $E_\infty$-term for $t-s=41$ is generated by $h_1f_1$, $h_1Ph_1h_5$ $z$, $h_1P^4c_0$ and $P^5h_1$, with $\mu_{41}$ detected by $P^5h_1$, $\eta^2\rho_{39}$ detected by $h_1P^4c_0$, $\eta[[Ph_1h_5]]$ detected by $h_1Ph_1h_5$ and $\eta\alpha_{40}$ detected by $h_1f_1$. As we noted in case (28) there is a hidden $\eta$-extension from $g^2$ to $z$, so that $\eta\bar{\kappa}^2$ is detected by $z$. By construction, $e(\mu_{41}) = j_{41}$. The $e$-invariant splits off $\mathbb{Z}/2\{\mu_{41}\} \oplus \mathbb{Z}/2\{\eta^2\rho_{39}\}$ from $\pi_{41}(S)$, and there cannot be hidden 2-extensions from the $h_1$-multiples $h_1Ph_1h_5$ and $h_1f_1$, since $2\eta = 0$, so $\ker(e) \cong (\mathbb{Z}/2)^3$.

The products $\eta \cdot \eta\alpha_{39}$ and $\eta \cdot \eta\sigma\eta_5$ lie in $\ker(e) \cap \ker(\iota)$. Since $h_1 \cdot h_1^2h_3h_5 = 0$, they are either zero or detected by $h_1Ph_1h_5$. In the latter case, $\eta^2 \cdot \eta\alpha_{39} = \eta^2 \cdot \epsilon\eta_5$ and $\eta^2 \cdot \eta\sigma\eta_5$ would be detected by $h_1^2Ph_1h_5 \neq 0$, but $\eta^2\epsilon = \eta^3\sigma = 0$, so this is impossible. Hence $\eta \cdot \eta\alpha_{39} = \eta \cdot \eta\sigma\eta_5 = 0$.

The product $\nu \cdot \alpha_{38}$ lies in $\ker(e) \cap \ker(\iota)$. Since $h_2 \cdot h_0^2h_3h_5 = 0$, it is either zero or detected by $h_1Ph_1h_5$. Since $h_1^2Ph_1h_5 \neq 0$ and $\eta\nu = 0$, the latter is impossible. Hence $\nu\alpha_{38} = 0$.

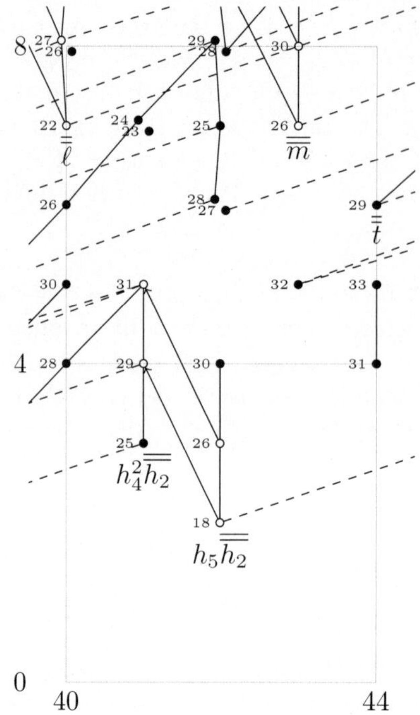

FIGURE 11.21. $(E_2(C\sigma), d_2)$ for $40 \le t-s \le 44$

We show that $\sigma \cdot \alpha_{34} = \eta\alpha_{40}$ using $j\colon C\sigma \to S^8$. See Figure 11.21, where $\bar{\bar{a}} \in j^{-1}(a) \subset E_2^{s,t}(C\sigma)$ denotes a lift of $a \in E_2^{s,t}(S^8) = E_2^{s,t-8}(S)$. In $E_2(C\sigma)$ we have $d_2(h_5\overline{\overline{h_2}}) = h_0h_4^2\overline{\overline{h_2}} \neq 0$ and $d_2(h_0h_5\overline{\overline{h_2}}) = h_0^2h_4^2\overline{\overline{h_2}} \neq 0$, so $\pi_{42}(C\sigma)$ is concentrated in Adams filtrations $\geq 4$. Hence $\mathrm{im}(j) = \ker(\sigma) \subset \pi_{34}(S)$ is also concentrated in filtrations $\geq 4$. Since $\alpha_{34} \in [[h_0h_2h_5]] = \{h_0h_2h_5\} \cap \ker(e) \cap \ker(\iota)$ lies in filtration $s = 3$, we must have $\sigma\alpha_{34} \neq 0$, in the span of $\eta[[Ph_1h_5]]$ and $\eta\alpha_{40}$. Since $\eta\alpha_{34} = 0$ we have $\eta\sigma\alpha_{34} = 0$, leaving $\sigma\alpha_{34} = \eta\alpha_{40}$ as the only possibility.

The products $\epsilon \cdot \mu_{33}$, $\mu \cdot [q]$, $\kappa \cdot \zeta_{27}$, $\eta^* \cdot \mu_{25}$ and $\nu^* \cdot \bar{\rho}$ have Adams filtration $\geq 11$ and are detected by $e$, with $e(\epsilon \cdot \mu_{33}) = \eta j_7 j_{33} = \eta^2 j_{39}$, $e([q]) = 0$, $e(\kappa) = 0$, $e(\eta^*) = 0$ and $e(\nu^*) = 0$, so that $\epsilon\mu_{33} = \eta^2\rho_{39}$, $\mu[q] = 0$, $\kappa\zeta_{27} = 0$, $\eta^*\mu_{25} = 0$ and $\nu^*\bar{\rho} = 0$.

The product $\mu \cdot \kappa_1$ has Adams filtration $\geq 9$ and is detected by $e$ and $\iota$, with $e(\mu \cdot \kappa_1) = 0$ and $\iota(\mu \cdot \kappa_1) = 0$, so that $\mu\kappa_1 = 0$. Similarly, $\mu \cdot \eta_5$ and $\eta[[Ph_1h_5]]$ are both detected by $Ph_1 \cdot h_1h_5 = h_1 \cdot Ph_1h_5$, so their difference lies in Adams filtration $\geq 8$ and is detected by $e$ and $\iota$. Since $\eta_5$ and $[[Ph_1h_5]]$ lie in $\ker(e) \cap \ker(\iota)$, that difference is zero, so $\mu\eta_5 = \eta[[Ph_1h_5]]$. Likewise, $\zeta \cdot \theta_4$ lies in Adams filtration $\geq 8$, because $Ph_2 \cdot h_4^2 = 0$, and is therefore detected by $e$ and $\iota$. Since $e(\theta_4)$ and $\iota(\theta_4)$ lie in trivial groups, we conclude that $\zeta\theta_4 = 0$.

(42) The $E_\infty$-term for $t-s = 42$ is generated by $Ph_2h_5$, $h_0Ph_2h_5$, $h_0^2Ph_2h_5$, $d_0^3$ and $h_1P^5h_1$. Here $\eta\mu_{41}$ is detected by $h_1P^5h_1$, $\kappa^3$ is detected by $d_0^3$, and $2^k[[Ph_2h_5]]$ is detected by $h_0^kPh_2h_5$ for $k \in \{0,1,2\}$, where $[[Ph_2h_5]] \in \{Ph_2h_5\}$ can and will be chosen to lie in $\ker(e) \cap \ker(\iota)$. (Here we use that $\iota$ maps $\kappa^3 \in \pi_{42}(S)$ to $\kappa^3 = \epsilon\kappa\bar{\kappa}$, which generates the $B$-power torsion in $\pi_{42}(tmf)$, and that $\iota([[Ph_2h_5]])$ must be

$B$-power torsion since $[[Ph_2h_5]] \in \ker(e)$, cf. Proposition 11.82.) There cannot be a hidden 2-extension from $h_0^2Ph_2h_5$ to $d_0^3$, since $h_0^2Ph_2h_5$ is an $h_1$-multiple.

As discussed in case (28), $\eta^2\bar{\kappa}^2 = \epsilon\kappa\bar{\kappa} = \kappa^3$ in $\ker(e) \subset \pi_{42}(S)$. Hence there is a hidden $\eta$-extension from $z$ to $d_0^3$, and a hidden $\epsilon$-extension from $d_0g$ to $d_0^3$. We chose $\alpha_{40} \in [[f_1]] = \{f_1\} \cap \ker(e) \cap \ker(\iota)$ so that $\eta^2\alpha_{40} = 0$. The relation $h_1^2 \cdot Ph_1h_5 = h_0^2Ph_2h_5$ in $E_2(S)$ implies that $\eta^2[[Ph_1h_5]] = 4[[Ph_2h_5]]$ modulo Adams filtration $\geq 9$, but these filtrations are detected by $e$ and $\iota$, and both $[[Ph_1h_5]]$ and $[[Ph_2h_5]]$ lie in $\ker(e) \cap \ker(\iota)$, so this identity holds strictly. Similarly, $\zeta \cdot [n]$ has Adams filtration $\geq 10$, hence is detected by $e$ and $\iota$. Since $e([n]) = 0$ and $\iota([n]) = 0$ we must have $\zeta[n] = 0$.

The products $\nu{\cdot}\rho_{39}$, $\sigma{\cdot}\zeta_{35}$, $\zeta{\cdot}\rho_{31}$, $\rho{\cdot}\zeta_{27}$, $\bar{\zeta}{\cdot}\bar{\rho}$ and $\bar{\sigma}{\cdot}\bar{\rho}$ lie in Adams filtration $\geq 13$, because $c_1 \cdot h_0^2 i = 0$ in $E_2(S)$, hence are detected by $e$, and $e(\nu \cdot \rho_{39}) = \nu j_{39} = 0$, $e(\sigma{\cdot}\zeta_{35}) = j_7j_{35} = 0$, $e(\zeta{\cdot}\rho_{31}) = j_{11}j_{31} = 0$, $e(\rho{\cdot}\zeta_{27}) = j_{15}j_{27} = 0$, $e(\bar{\zeta}{\cdot}\bar{\rho}) = j_{19}j_{23} = 0$ and $e(\bar{\sigma}) = 0$, so $\nu\rho_{39} = 0$, $\sigma\zeta_{35} = 0$, $\zeta\rho_{31} = 0$, $\rho\zeta_{27} = 0$, $\bar{\zeta}\bar{\rho} = 0$ and $\bar{\sigma}\bar{\rho} = 0$. The products $\mu{\cdot}\mu_{33}$ and $\bar{\mu}{\cdot}\mu_{25}$ are detected by $Ph_1 \cdot P^4h_1 = P^2h_1 \cdot P^3h_1 = h_1P^5h_1 \neq 0$, hence are both equal to $\eta\mu_{41}$.

We use $\iota\colon S \to tmf$ to detect a hidden $\nu$-extension from $u$ to $d_0^3$. From Lemma 1.15 and Table 1.1 we know that $\iota([u])$ lies in $\{d_0\gamma\}$, hence is equal to $\eta_1\kappa$ in $\pi_{39}(tmf)$. By Theorem 9.14, $\nu \cdot \eta_1\kappa$ in $\pi_{42}(tmf)$ is detected by $d_0gw_1 \neq 0$ in filtration 12 of $E_\infty(tmf)$. See Figure 9.7. It follows that $\nu \cdot [u]$, in Adams filtration $\geq 10$ of $\pi_{42}(S)$, cannot have Adams filtration $\geq 13$, and must therefore be detected by $d_0^3$. Thus $\nu[u] \equiv \kappa^3 \mod \eta\mu_{41}$, and a comparison of $e$-invariants shows that $\nu[u] = \kappa^3$.

The known relations $\eta\alpha_{38} = \nu\{t\}$ and $\eta\nu = 0$ imply $\nu \cdot \nu\{t\} = 0$.

We showed in case (39) that $\alpha_{39} = \langle\epsilon, 2, \theta_4\rangle$. By shuffling, $\nu \cdot \alpha_{39} = \nu\langle\epsilon, 2, \theta_4\rangle = \langle\nu, \epsilon, 2\rangle\theta_4 = 0$, since $\pi_{12}(S) = 0$.

To show that $\epsilon \cdot \alpha_{34} = 0$ we use that $\epsilon \in \langle\nu^2, 2, \eta\rangle$, by [**171**, Ch. XIV] or `ext` and Moss' theorem. By shuffling, $\epsilon\alpha_{34} \in \langle\nu^2, 2, \eta\rangle\alpha_{34} = -\nu^2\langle 2, \eta, \alpha_{34}\rangle$, which is zero because $\nu^2\{t\} = 0$.

(43) The $E_\infty$-term for $t - s = 43$ is generated by $P^5h_2$, $h_0P^5h_2$ and $h_0^2P^5h_2$, detecting $\zeta_{43}$, $2\zeta_{43}$ and $4\zeta_{43}$, respectively. Hence $\pi_{43}(S) \cong \mathbb{Z}/8$ maps isomorphically by $e$ to $\pi_{43}(j)$, and $e(\zeta_{43}) \doteq j_{43}$.

The relation $h_1^2P^5h_1 = h_0^2P^5h_2$ shows that $\eta \cdot \eta\mu_{41} = 4\zeta_{43}$. The products $\eta \cdot \kappa^3$, $\eta \cdot [[Ph_2h_5]]$, $\nu \cdot \bar{\kappa}^2$, $\nu \cdot [[Ph_1h_5]]$, $\nu \cdot \alpha_{40}$, $\sigma \cdot \{t\}$, $\mu \cdot \alpha_{34}$, $\zeta \cdot [q]$, $\zeta \cdot \kappa_1$, $\zeta \cdot \eta_5$, $\eta^* \cdot \zeta_{27}$, $\nu^* \cdot \mu_{25}$ and $\bar{\kappa} \cdot \bar{\rho}$ are zero because $\kappa$, $\eta^*$, $\nu^*$, $\bar{\kappa}$, $[q]$, $\kappa_1$, $\eta_5$, $\alpha_{34}$, $\{t\}$, $[[Ph_1h_5]]$, $\alpha_{40}$ and $[[Ph_2h_5]]$ lie in $\ker(e)$. Likewise, $e(\epsilon \cdot \zeta_{35}) = \eta j_7 j_{35} = 0$, so $\epsilon\zeta_{35} = 0$.

(44) The $E_\infty$-term for $t - s = 44$ is generated by $g_2$, $h_0g_2$ and $h_0^2g_2$, detecting $\bar{\kappa}_2$, $2\bar{\kappa}_2$ and $4\bar{\kappa}_2$, respectively.

The products $\eta \cdot \zeta_{43}$, $\nu \cdot \mu_{41}$, $\epsilon \cdot \{t\}$, $\mu \cdot \zeta_{35}$, $\zeta \cdot \mu_{33}$, $\kappa \cdot \theta_4$, $\bar{\mu} \cdot \zeta_{27}$, $\bar{\zeta} \cdot \mu_{25}$ and $\bar{\sigma} \cdot \mu_{25}$ lie in Adams filtration $\geq 7$, since $d_0 \cdot h_4^2 = 0$, hence are zero. We have $\sigma \cdot \sigma\theta_4 = 4\bar{\kappa}_2$, since $h_3x = h_0^2g_2$ has maximal Adams filtration in $E_\infty(S)$.

Finally, $\sigma{\cdot}\alpha_{37} = 4\bar{\kappa}_2$, as we learned from Isaksen and Xu. See Lemma 11.64. □

REMARK 11.63. We recall Tangora's proof from [**166**, Part 1] that $\bar{\nu}\theta_4 = 0$, where $\bar{\nu} = \epsilon + \eta\sigma$. This follows from $\theta_4 \in \langle\sigma, 2\sigma, \sigma, 2\sigma\rangle$, proved in [**107**, Thm. 8.1.1], and the shuffling relation

$$\bar{\nu}\langle\sigma, 2\sigma, \sigma, 2\sigma\rangle \subset \langle\langle\bar{\nu}, \sigma, 2\sigma\rangle, \sigma, 2\sigma\rangle\,,$$

proved in [**87**, Thm. 2.3.6(a)], once one has shown that the latter iterated Toda bracket only contains 0. First, $\langle \bar\nu, \sigma, 2\sigma\rangle \subset \pi_{23}(S)$ is defined with indeterminacy $4\nu\bar\kappa$. Here $2\langle \bar\nu, \sigma, 2\sigma\rangle \subset \pi_{16}(S)\cdot 2\sigma = \{0\}$ and $\langle \bar\nu, \sigma, 2\sigma\rangle\eta = -\bar\nu\langle \sigma, 2\sigma, \eta\rangle$ equals $\bar\nu\eta^* = 0$ with zero indeterminacy, because $\eta^* \in \langle \sigma, 2\sigma, \eta\rangle$ with indeterminacy $\{0, \eta\rho\}$ and $\bar\nu \cdot \eta\rho = 0$. Hence $\langle \bar\nu, \sigma, 2\sigma\rangle$ contains either 0 or $8\bar\rho$, modulo $4\nu\bar\kappa$. Finally, $\langle 8\bar\rho, \sigma, 2\sigma\rangle$ and $\langle 4\nu\bar\kappa, \sigma, 2\sigma\rangle$ are both 0 with no indeterminacy, since they contain $8\langle \bar\rho, \sigma, 2\sigma\rangle \subset 8\cdot\pi_{38}(S) = \{0\}$ and $4\langle \nu\bar\kappa, \sigma, 2\sigma\rangle \subset 4\cdot\pi_{38}(S) = \{0\}$, respectively, and since $8\bar\rho\cdot\pi_{15}(S) = 4\nu\bar\kappa\cdot\pi_{15}(S) = \pi_{31}(S)\cdot 2\sigma = 0$.

LEMMA 11.64 (Isaksen–Xu). *$\sigma\cdot\alpha_{37} = 4\bar\kappa_2$.*

PROOF. Consider the Toda bracket $\langle \nu, \eta, \kappa_1\rangle \subset \pi_{37}(S)$, which has zero indeterminacy. On one hand, $\eta\langle \nu, \eta, \kappa_1\rangle = \langle \eta, \nu, \eta\rangle\kappa_1 = \nu^2\kappa_1 = \eta\sigma\theta_4$, since $\langle \eta, \nu, \eta\rangle = \nu^2$. On the other hand, the differential $d_3(h_2h_5) = h_0p = h_1d_1$ and Moss' theorem for $E_4$-Massey products shows that $\langle \nu, \eta, \kappa_1\rangle$ is detected by $h_2^2h_5$. It follows that $\langle \nu, \eta, \kappa_1\rangle = \sigma\theta_4 + \alpha_{37}$. Finally, $\sigma\langle \nu, \eta, \kappa_1\rangle = \langle \sigma, \nu, \eta\rangle\kappa_1 = 0$ since $\langle \sigma, \nu, \eta\rangle = 0$, so $\sigma\cdot\alpha_{37} = \sigma^2\theta_4 = 4\bar\kappa_2$. □

REMARK 11.65. For ease of reference, we summarize our definitions of the multiplicative generators for $\pi_*(S)$ in degrees $* \le 44$.

- $\eta = \{h_1\}$ is well-defined.
- $\nu \in \{h_2\}$ is defined up to (multiplication by) a unit in $\mathbb{Z}/8$. The Hopf fibration gives a specific choice.
- $\sigma \in \{h_3\}$ is defined up to a unit in $\mathbb{Z}/16$. The Hopf fibration gives a specific choice.
- $\epsilon = \{c_0\}$ is well-defined.
- $\mu = \{Ph_1\}$ is well-defined.
- $\zeta \in \{Ph_2\}$ is defined up to a unit in $\mathbb{Z}/8$. The $J$-homomorphism gives a specific choice.
- $\kappa = \{d_0\}$ is well-defined.
- $\rho \in \{h_0^3h_4\}$ is defined up to a unit in $\mathbb{Z}/32$ by the condition $\epsilon\rho = 0$, or equivalently, by $\iota(\rho) = 0$. The $J$-homomorphism gives a specific choice.
- $\eta^* \in \{h_1h_4\}$ is well-defined by the condition $e(\eta^*) = 0$.
- $\bar\mu = \{P^2h_1\}$ is well-defined.
- $\nu^* \in \{h_2h_4\}$ is defined up to a unit in $\mathbb{Z}/8$ by the condition $e(\nu^*) = 0$.
- $\bar\zeta \in \{P^2h_2\}$ is defined up to a unit in $\mathbb{Z}/8$. The $J$-homomorphism gives a specific choice.
- $\bar\sigma \in \{c_1\}$ is well-defined by the condition $e(\bar\sigma) = 0$.
- $\bar\kappa \in \{g\}$ is defined up to a unit in $\mathbb{Z}/8$.
- $\bar\rho \in \{h_0^2i\}$ is defined up to a unit in $\mathbb{Z}/16$. The $J$-homomorphism gives a specific choice.
- $\mu_{25} = \{P^3h_1\}$ is well-defined.
- $\zeta_{27} \in \{P^3h_2\}$ is defined up to a unit in $\mathbb{Z}/8$. The $J$-homomorphism gives a specific choice.
- $\theta_4 = \{h_4^2\}$ is well-defined.
- $\rho_{31} \in \{h_0^{10}h_5\}$ is defined up to a unit in $\mathbb{Z}/64$. The $J$-homomorphism gives a specific choice.
- $[n] \in \{n\}$ is well-defined by the condition $e([n]) = 0$.
- $[q] \in \{q\}$ is well-defined by the condition $e([q]) = 0$.
- $\kappa_1 \in \{d_1\}$ is well-defined by the conditions $e(\kappa_1) = 0$ and $\iota(\kappa_1) = 0$.

- $\eta_5 \in \{h_1h_5\}$ is well-defined by the conditions $e(\eta_5) = 0$, $\iota(\eta_5) = 0$ and $\nu\eta_5 = 0$. It is the unique element in $\langle \eta, 2, \theta_4 \rangle = \{\eta_5, \eta_5 + \eta\rho_{31}\}$ satisfying $e(\eta_5) = 0$.
- $\mu_{33} = \{P^4h_1\}$ is well-defined.
- $\alpha_{34} \in \{h_0h_2h_5\}$ is defined up to a unit in $\mathbb{Z}/4$, as an element of $\langle \eta, 2, \eta_5 \rangle$ with $e(\alpha_{34}) = 0$. Less precisely, it is defined up to the same unit, modulo $\mathbb{Z}/2\{\nu[n]\}$, by the conditions $e(\alpha_{34}) = 0$ and $\eta\alpha_{34} = 0$, or equivalently, by the conditions $e(\alpha_{34}) = 0$ and $\iota(\alpha_{34}) = 0$.
- $\zeta_{35} \in \{P^4h_2\}$ is defined up to a unit in $\mathbb{Z}/8$. The $J$-homomorphism gives a specific choice.
- $\{t\}$ is well-defined.
- $\alpha_{37} \in \{h_2^2h_5\} = \langle \nu^2, 2, \theta_4 \rangle$ is well-defined by the condition $\eta\alpha_{37} = 0$.
- $\alpha_{38} \in \{h_0^2h_3h_5\}$ is defined up to a unit in $\mathbb{Z}/4$, modulo $\mathbb{Z}/2\{\eta\sigma\theta_4\}$.
- $\rho_{39} \in \{h_0^2P^2i\}$ is defined up to a unit in $\mathbb{Z}/16$. The $J$-homomorphism gives a specific choice.
- $[u] \in \{u\}$ is well-defined by the condition $e([u]) = 0$.
- $\alpha_{39} \in \{h_5c_0\}$ is well-defined as the single element of $\langle \epsilon, 2, \theta_4 \rangle$. Less precisely, it is defined modulo $\mathbb{Z}/2\{\nu\{t\}\} \oplus \mathbb{Z}/2\{\sigma\kappa_1\}$ by the conditions $e(\alpha_{39}) = 0$ and $\iota(\alpha_{39}) = 0$.
- $[[Ph_1h_5]] \in \{Ph_1h_5\}$ is well-defined by the conditions $e([[Ph_1h_5]]) = 0$ and $\iota([[Ph_1h_5]]) = 0$.
- $\alpha_{40} \in \{f_1\}$ is defined modulo $\mathbb{Z}/2\{\eta\alpha_{39}\}$ by the conditions $e(\alpha_{40}) = 0$, $\iota(\alpha_{40}) = 0$ and $\eta^2\alpha_{40} = 0$.
- $\mu_{41} = \{P^5h_1\}$ is well-defined.
- $[[Ph_2h_5]] \in \{Ph_2h_5\}$ is defined up to a unit in $\mathbb{Z}/8$ by the conditions $e([[Ph_2h_5]]) = 0$ and $\iota([[Ph_2h_5]]) = 0$.
- $\zeta_{43} \in \{P^5h_2\}$ is defined up to a unit in $\mathbb{Z}/8$. The $J$-homomorphism gives a specific choice.
- $\bar{\kappa}_2 \in \{g_2\}$ is defined up to a unit in $\mathbb{Z}/8$.

The notations $\eta$ and $\nu$ were used by Toda in [**168**], with $\eta$ being associated to Hopf. The remaining notations in degrees $\leq 19$ are those used in [**171**], while $\bar{\kappa}$ is from [**130**]. Several notational schemes are in use for the generators of the image of the $J$-homomorphism; we continue Toda's pattern $\zeta, \rho, \bar{\zeta}$ with $\bar{\rho}, \zeta_{8k+3}, \rho_{8k-1}$. Adams [**8**] introduced the classes $\mu_{8k+1} = \{P^kh_1\}$, extending Toda's $\mu$ and $\bar{\mu}$. The notation $\theta_j \in \{h_j^2\}$ appeared for $j = 4$ in Barratt–Mahowald–Tangora [**22**], presumably due to the connection to the Kervaire–Milnor [**86**] group $\Theta_n$. The notation $\eta_j \in \{h_1h_j\}$ is that of [**101**], extending Toda's $\eta^*$. The notations $[n]$, $[q]$, $\{t\}$, $[u]$, $[[Ph_1h_5]]$ and $[[Ph_2h_5]]$ are inherited from the May spectral sequence calculation of $E_2(S)$ [**117**], [**165**]. We allow ourselves to write $\kappa_i \in \{d_i\}$ and $\bar{\kappa}_i \in \{g_i\}$, extending $\kappa \in \{d_0\}$ and $\bar{\kappa} \in \{g\}$, even if the Steenrod operations $Sq^0(d_0) = d_1$ and $Sq^0(g) = g_2$ cannot immediately be lifted to homotopy operations. Keep in mind that $g = g_1$ and $\bar{\kappa} = \bar{\kappa}_1$; there is no class $g_0$ in $E_2(S)$. The remaining ad hoc notations $\alpha_n \in \pi_n(S)$ for $n \in \{34, 37, 38, 39, 40\}$ are only introduced here for typesetting convenience, and illustrate the limitations of the existing nomenclature for the stable homotopy groups of spheres.

The following five lemmas account for the hidden 2-, $\eta$- and $\nu$-extensions shown in Figure 11.14, in the region where $45 \leq t - s \leq 48$ and $s \geq 9$. For the remaining hidden extensions, in the range $45 \le t-s \le 48$ and $s \le 8$, we refer to the literature, in particular to [**166**], [**21**] and [**87**].

LEMMA 11.66. *There is a hidden $\eta$-extension from $w$ to $d_0\ell$.*

PROOF. This is detected by $\iota\colon S \to tmf$, which maps $w$ to $\gamma g$ detecting $\eta_1\bar{\kappa}$. Since $\eta\cdot\eta_1\bar{\kappa} = \epsilon_1\kappa$ is nonzero in $\pi_{46}(tmf)$, there must be a hidden $\eta$-extension on $w$, and $d_0\ell$ is the only possible target. □

LEMMA 11.67. *There is a hidden $\eta$-extension from $d_0\ell$ to $Pu$.*

PROOF. We prove this using the homotopy cofiber sequence

$$S^1 \xrightarrow{\eta} S \xrightarrow{i} C\eta \xrightarrow{j} S^2 .$$

The differential $d_2(\ell) = h_0d_0e_0$ for $S$ lifts to a differential $d_2(\widehat{\ell}) = d_0e_0\widehat{h_0}$ in the Adams spectral sequence for $C\eta$. Multiplying by $d_0$ we obtain a nonzero differential $d_2(d_0\widehat{\ell}) = d_0^2e_0\widehat{h_0} = 13_{21} = i(Pu)$ for $C\eta$, as verified by `ext`. It follows that $\eta$ times a class detected by $j(d_0\widehat{\ell}) = d_0\ell$ is detected by $Pu$. □

LEMMA 11.68. *There is a hidden 2-extension from $e_0r$ to $Pu$, and a hidden $\eta$-extension from $e_0r$ to $d_0^2g$.*

PROOF. This follows from the homotopy cofiber sequence

$$\Sigma^{-1}tmf/S \xrightarrow{j} S \xrightarrow{\iota} tmf \xrightarrow{i} tmf/S .$$

See Figure 11.30 for the $E_\infty$-term of $tmf/S$. The differentials $d_2(w_2) = \alpha\beta g$, $d_3(h_1w_2) = g^2w_1$ and $d_4(h_0w_2) = d_0\gamma w_1$ for $tmf$ imply that $j\colon \pi_n(tmf/S) \to \pi_{n-1}(S)$ maps homotopy classes detected by $i(w_2) = \overline{w_2}$, $i(h_1w_2) = h_1\overline{w_2}$ and $i(h_0w_2) = h_0\overline{w_2}$ to homotopy classes detected by $e_0r$, $d_0^2g$ and $Pu$, respectively, since $\iota(e_0r) = \alpha\beta g$, $\iota(d_0^2g) = g^2w_1$ and $\iota(Pu) = d_0\gamma w_1$. Since 2 times each class detected by $\overline{w_2}$ is detected by $h_0\overline{w_2}$, it follows that 2 times a class detected by $e_0r$ is detected by $Pu$. Similarly, since $\eta$ times each class detected by $\overline{w_2}$ is detected by $h_1\overline{w_2}$, it follows that $\eta$ times a class detected by $e_0r$ is detected by $d_0^2g$. □

LEMMA 11.69. *There is a hidden $\nu$-extension from $w$ to $d_0^2g$.*

PROOF. We prove this using the homotopy cofiber sequence

$$S^3 \xrightarrow{\nu} S \xrightarrow{i} C\nu \xrightarrow{j} S^4 .$$

The hidden $\nu$-extension from $u$ to $d_0^3$ corresponds to a differential $d_3(\overline{u}) = i(d_0^3)$ in the Adams spectral sequence for $C\nu$. Multiplying by $g$ we obtain a differential $d_3(g\overline{u}) = g\cdot i(d_0^3) = 16_{24}$, where $g\cdot\overline{u} = 13_{36} = d_0\cdot\overline{w}$, as can be verified with `ext`. The class $16_{24}$ remains nonzero at the $E_3$-term, by $h_0$-linearity. Hence $d_0\cdot d_3(\overline{w}) = 16_{24} \ne 0$, which implies $d_3(\overline{w}) = 12_{17} = i(d_0^2g)$. In turn, this differential corresponds to a hidden $\nu$-extension from $j(\overline{w}) = w$ to $d_0^2g$, as claimed. □

LEMMA 11.70. *There is a hidden $\eta$-extension from $h_0^7Q$ to $P^5c_0$.*

PROOF. Let $\rho_{47}$ be detected by $h_0^7Q$. The only way that $e\colon \pi_{47}(S) \to \pi_{47}(j) = \mathbb{Z}/32$ can be surjective is that $e(\rho_{47}) \doteq j_{47}$. Hence $e(\eta\rho_{47}) = \eta j_{47} \ne 0$, so $\eta\rho_{47}$ is nonzero and must be detected by $P^5c_0$. □

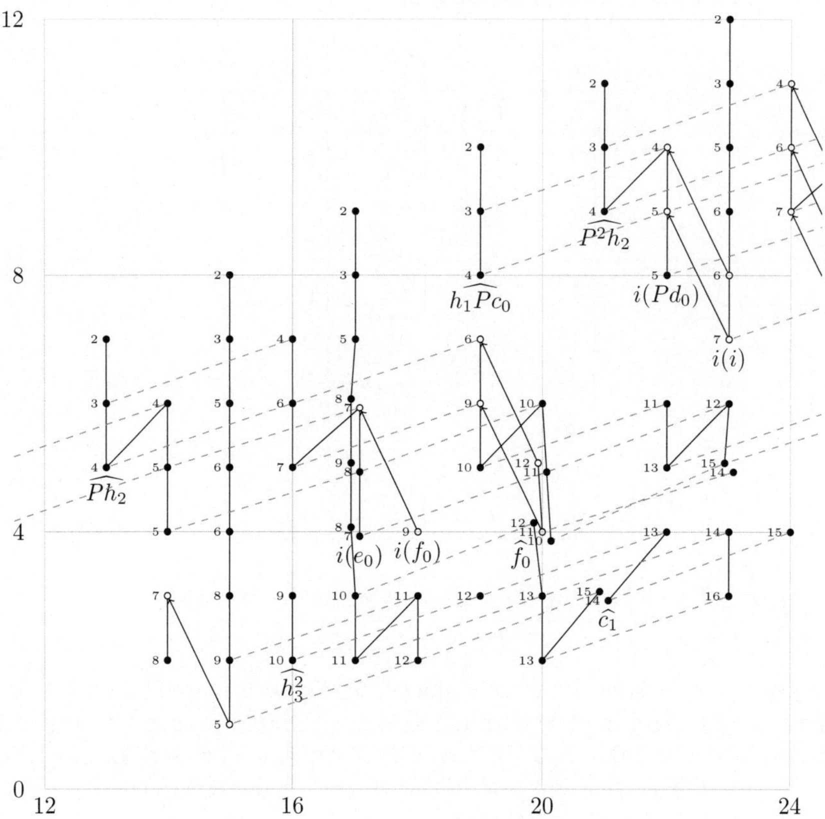

FIGURE 11.22. $(E_2(C\eta), d_2)$ for $12 \le t-s \le 24$

## 11.9. A hidden $\eta$-extension

Using space-level (unstable) methods, Mimura [**129**, Thm. B] showed that $\epsilon\kappa$ is nonzero in $\pi_{22}(S)$. This product has Adams filtration $\geq 7$, hence can only be detected by $Pd_0$ in $E_\infty(S)$. Mahowald and Tangora [**107**, Thm. 2.1.1] used a Toda bracket calculation due to Barratt to deduce that $\eta^2\bar{\kappa}$ is also detected by $Pd_0$, so that there is a hidden $\eta$-extension from $h_1 g$ to $Pd_0$. Other proofs of these results have been given by Bauer [**23**, p. 30], using the elliptic spectral sequence to show that the image of $\eta^2\bar{\kappa}$ in $\pi_{22}(tmf)$ is nonzero, and by Daniel Dugger and Isaksen [**55**, Prop. 8.9], using a hidden $\tau$-extension in a motivic Adams spectral sequence. We provide a classical spectrum-level (stable) proof of this hidden $\eta$-extension, from which Mimura's theorem follows as in cases (22) and (23) of the proof of Theorem 11.61.

THEOREM 11.71 (Mimura [**129**], Mahowald–Tangora [**107**]). *The product $\eta^2\bar{\kappa}$ is detected by $Pd_0$.*

PROOF. We argue using the maps of Adams spectral sequences induced by the maps of spectra

$$S \xrightarrow{i} C\eta \xrightarrow{1\wedge i} C\eta \wedge C\nu\,.$$

TABLE 11.4. $E_2(S)$-module generators of $E_2(C\eta)$ for $t-s \le 24$

| $t-s$ | $s$ | $g$ | $x$ | $d_2(x)$ |
|---|---|---|---|---|
| 0 | 0 | 0 | $i(1)$ | 0 |
| 2 | 1 | 1 | $\widehat{h_0}$ | 0 |
| 5 | 1 | 3 | $\widehat{h_2}$ | 0 |
| 11 | 4 | 4 | $\widehat{h_1 c_0}$ | 0 |
| 13 | 5 | 4 | $\widehat{Ph_2}$ | 0 |
| 16 | 2 | 10 | $\widehat{h_3^2}$ | 0 |
| 19 | 8 | 4 | $\widehat{h_1 P c_0}$ | 0 |
| 20 | 4 | 11 | $\widehat{f_0}$ | $h_0 e_0 \widehat{h_0}$ |
| 21 | 3 | 14 | $\widehat{c_1}$ | 0 |
| 21 | 9 | 4 | $\widehat{P^2 h_2}$ | 0 |

We start with $C\eta$, defined by the homotopy cofiber sequence

$$S^1 \xrightarrow{\eta} S \xrightarrow{i} C\eta \xrightarrow{j} S^2 .$$

The $E_2$-term of the Adams spectral sequence for $C\eta$ is displayed for $12 \le t-s \le 24$ in Figure 11.22. As a module over $E_2(S)$ it is generated in degrees $t-s \le 24$ by the classes listed in Table 11.9. In each case $\widehat{x}$ denotes a lift of $x$, i.e., a class with $j(\widehat{x}) = x$. The differential structure in this range follows by $h_0$-linearity, naturality with respect to $j$, and the fact that $d_2 \circ d_2 = 0$.

The resulting $E_3$-term of the Adams spectral sequence for $C\eta$ is displayed for $12 \le t-s \le 24$ in Figure 11.23. As a module over $E_3(S)$ it is generated in degrees $t-s \le 24$ by the classes listed in Table 11.9. Most of the $d_3$-differentials follow by $h_0$-linearity and naturality with respect to $i$ or $j$. For completeness, we show in Lemmas 11.72 and 11.73 that $d_3$ vanishes on $i(e_0)$ and $\widehat{c_1}$, but these results are not necessary for the proof of the theorem.

The key differential, $d_3(h_2\widehat{f_0}) = i(Pd_0)$, is established in Lemma 11.74. It implies that Adams filtration $\ge 7$ in $\pi_{22}(C\eta)$ is trivial. Letting $\gamma = \{Pd_0\}$ denote the unique class in $\pi_{22}(S)$ that is detected by $Pd_0$, it follows that $i(\gamma) = 0$. Hence $\gamma = \eta \cdot \beta$ for some class $\beta \in \pi_{21}(S) = \mathbb{Z}/2\{\eta\bar{\kappa}\} \oplus \mathbb{Z}/2\{\nu\nu^*\}$. Since $\eta \cdot \nu\nu^* = 0$ we must have $\gamma = \eta \cdot \eta\bar{\kappa}$. □

LEMMA 11.72. *$d_3(i(e_0)) = 0$ in $E_3(C\eta)$.*

PROOF. (This is implicit in [**107**, Lem. 3.1.4].) If $d_3(i(e_0)) = 0$ then $i(e_0)$ survives to $E_\infty(C\eta)$ in bidegree $(t-s, s) = (17, 4)$, otherwise $i(e_0) + h_0^2 h_4 \widehat{h_0}$ is the surviving class. Let $\alpha \in \pi_{17}(C\eta)$ be detected in Adams filtration 4. Consider the exact sequence

$$\dots \xrightarrow{\eta} \pi_{17}(S) \xrightarrow{i} \pi_{17}(C\eta) \xrightarrow{j} \pi_{15}(S) \xrightarrow{\eta} \dots .$$

Since $i(\eta\eta^*) = 0$ the image of $i$ lies in Adams filtration $\ge 5$, hence $\alpha$ maps non-trivially by $j$ to $\ker(\eta) \subset \pi_{15}(S)$. If $\alpha$ were detected by $i(e_0) + h_0^2 h_4 \widehat{h_0}$ then $j(\alpha)$ would be detected by $h_0^3 h_4$ modulo Adams filtration $\ge 5$. But then $j(\alpha) = \rho$

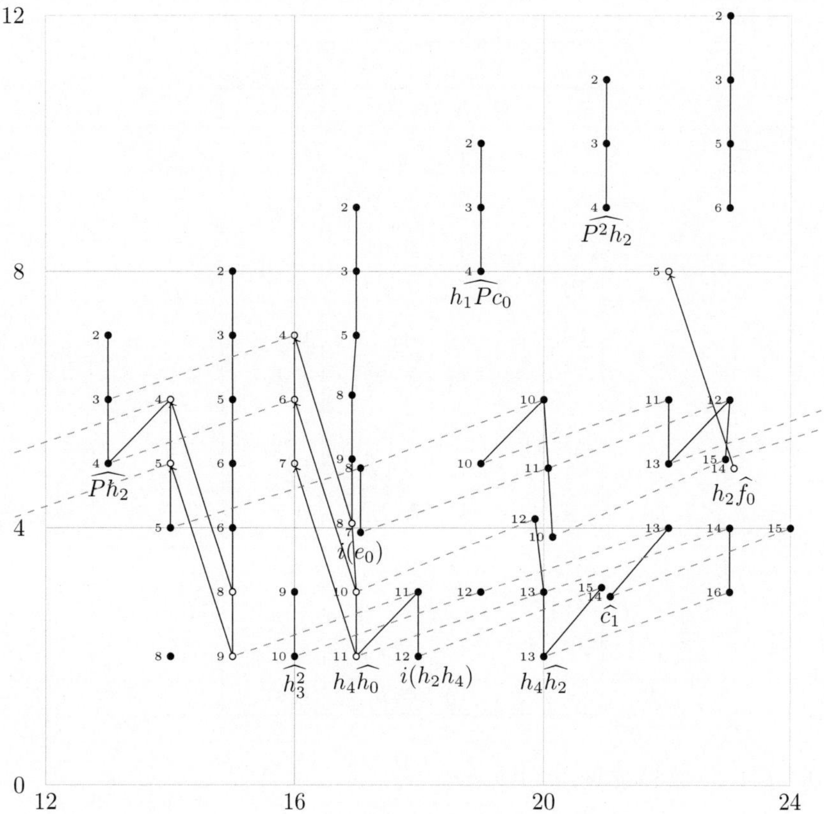

FIGURE 11.23. $(E_3(C\eta), d_3)$ for $12 \le t - s \le 24$

modulo Adams filtration $\geq 5$. Since $\eta\rho \neq 0$ and $\eta^2\kappa = 0$, this means that $j(\alpha)$ is not in $\ker(\eta)$, which contradicts exactness. Hence $\alpha$ is detected by $i(e_0)$, so $d_3(i(e_0)) = 0$. □

LEMMA 11.73. *$d_3(\widehat{c_1}) = 0$ in $E_3(C\eta)$.*

PROOF. From $\eta\bar{\sigma} = 0$ in $\pi_{20}(S)$ we deduce that there must be a class $\beta \in \pi_{21}(C\eta)$ with $j(\beta) = \bar{\sigma}$ in Adams filtration 3. Any such lift $\beta$ must have Adams filtration $\leq 3$, hence be detected by a nonzero element in $\mathbb{F}_2\{i(h_2^2 h_4), \widehat{c_1}\}$. If $d_3(\widehat{c_1})$ were nonzero then $\beta$ would have to be detected by $i(h_2^2 h_4)$. However, $i(\nu\nu^*)$ is detected by $i(h_2^2 h_4)$, so modifying the choice of $\beta$ by $i(\nu\nu^*)$ would give a lift of $\bar{\sigma}$ of Adams filtration $\geq 4$. This contradiction implies that $d_3(\widehat{c_1}) = 0$. □

LEMMA 11.74. *$d_3(h_2\widehat{f_0}) = i(Pd_0)$ in $E_3(C\eta)$.*

PROOF. Suppose for a contradiction that $d_3(h_2\widehat{f_0}) = 0$. Then $i(Pd_0)$ would survive to a nonzero class in $E_\infty(C\eta)$, since $d_r(\widehat{h_4 h_2}) = 0$ for $r \in \{4, 5\}$. Let $\gamma' \in \{i(Pd_0)\} \subset \pi_{22}(C\eta)$. Consider the homotopy cofiber sequence

$$\Sigma^3 C\eta \xrightarrow{1\wedge\nu} C\eta \xrightarrow{1\wedge i} C\eta \wedge C\nu \xrightarrow{1\wedge j} \Sigma^4 C\eta$$

TABLE 11.5. $E_3(S)$-module generators of $E_3(C\eta)$ for $t-s \leq 24$

| $t-s$ | $s$ | $g$ | $x$ | $d_3(x)$ |
|---|---|---|---|---|
| 0 | 0 | 0 | $i(1)$ | 0 |
| 2 | 1 | 1 | $\widehat{h_0}$ | 0 |
| 5 | 1 | 3 | $\widehat{h_2}$ | 0 |
| 11 | 4 | 4 | $\widehat{h_1 c_0}$ | 0 |
| 13 | 5 | 4 | $\widehat{Ph_2}$ | 0 |
| 16 | 2 | 10 | $\widehat{h_3^2}$ | 0 |
| 17 | 2 | 11 | $h_4\widehat{h_0}$ | $d_0\widehat{h_0}$ |
| 17 | 4 | 7 | $i(e_0)$ | 0 |
| 18 | 2 | 12 | $i(h_2h_4)$ | 0 |
| 19 | 8 | 4 | $\widehat{h_1Pc_0}$ | 0 |
| 20 | 2 | 13 | $h_4\widehat{h_2}$ | 0 |
| 21 | 3 | 14 | $\widehat{c_1}$ | 0 |
| 21 | 9 | 4 | $\widehat{P^2h_2}$ | 0 |
| 23 | 5 | 14 | $h_2\widehat{f_0}$ | $i(Pd_0)$ |

and the associated long exact sequence

$$\cdots \longrightarrow \pi_{19}(C\eta) \xrightarrow{\nu} \pi_{22}(C\eta) \xrightarrow{1\wedge i} \pi_{22}(C\eta \wedge C\nu) \longrightarrow \dots .$$

We prove in Lemma 11.75 below that $(1\wedge i)(\gamma') = 0$ in $\pi_{22}(C\eta \wedge C\nu)$, so that $\gamma' = \nu \cdot \beta'$ is a $\nu$-multiple. However, $E_\infty(C\eta)$ in topological degree $t-s=19$ is generated by $h_2\widehat{h_3^2} = 3_{12}$, $d_0\widehat{h_2} = 5_{10}$ and classes in Adams filtration $\geq 8$. Since $h_2 \cdot h_2\widehat{h_3^2} = 4_{13}$ and $h_2 \cdot d_0\widehat{h_2} = 6_{11}$ are linearly independent, it follows that there is no class $\beta' \in \pi_{19}(C\eta)$ such that $\nu\beta'$ is detected by $i(Pd_0)$. □

The $(E_2, d_2)$-term of the Adams spectral sequence for $C\eta \wedge C\nu$ is displayed for $12 \leq t-s \leq 24$ in Figure 11.24. Most $d_2$-differentials follow by $h_0$- and $h_3$-linearity, and naturality with respect to $i$ or $j$. The first nontrivial case is that handled by the following lemma.

LEMMA 11.75. *(1) There is a differential $d_2(a) = b$ in $E_2(C\eta \wedge C\nu)$, where*

$$a = \overline{h_0 e_0 \widehat{h_0}} = 6_{13}$$

*has bidegree $(t-s, s) = (23, 6)$, and*

$$b = (1\wedge i)i(Pd_0) = 8_9$$

*has bidegree $(t-s, s) = (22, 8)$.*

*(2) If $\gamma' \in \{i(Pd_0)\} \subset \pi_{22}(C\eta)$, then $(1\wedge i)(\gamma') = 0$ in $\pi_{22}(C\eta \wedge C\nu)$.*

PROOF. (1) The differential $d_2(f_0) = h_0^2 e_0$ in $E_2(S)$ pushes forward along $i\colon S \to C\eta$ to give $d_2(i(f_0)) = i(h_0^2 e_0)$ in $E_2(C\eta)$. See Figure 11.22. This lifts

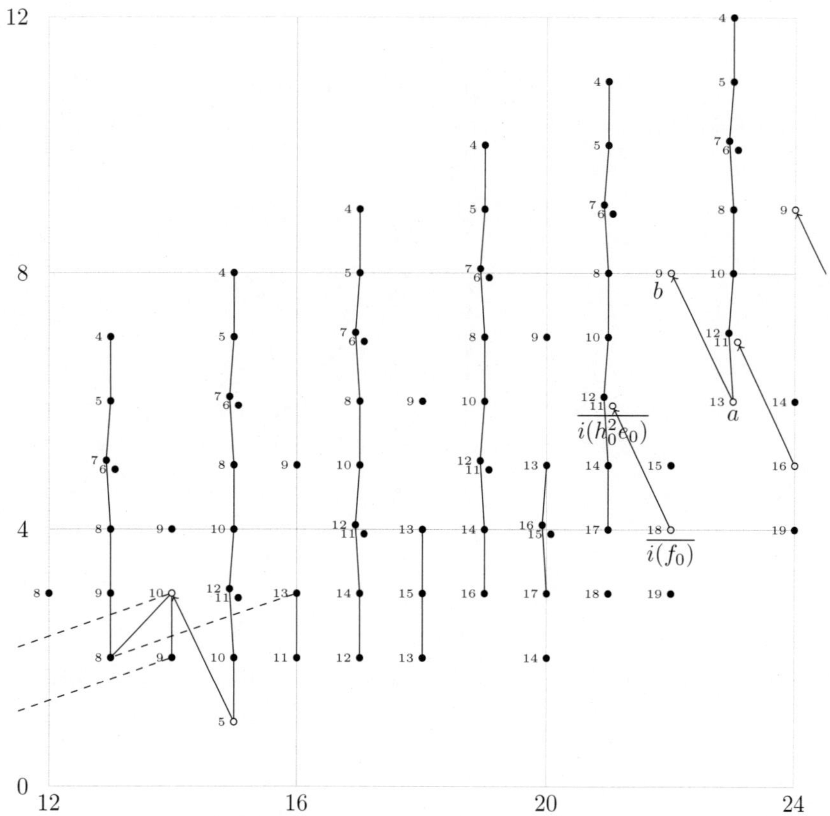

FIGURE 11.24. $(E_2(C\eta \wedge C\nu), d_2)$ for $12 \leq t - s \leq 24$

over $1 \wedge j \colon C\eta \wedge C\nu \to \Sigma^4 C\eta$ to give

$$d_2(\overline{i(f_0)}) = \overline{i(h_0^2 e_0)}$$

in $E_2(C\eta \wedge C\nu)$. Here $\overline{i(f_0)} = 4_{18}$ maps by $1 \wedge j$ to $i(f_0) = 4_9$, and $\overline{i(h_0^2 e_0)} = 6_{11}$ maps by $1 \wedge j$ to $i(h_0^2 e_0) = 6_7$. See Figure 11.24, and note that, by $h_0$-linearity, $d_2(\overline{i(f_0)})$ cannot involve $6_{12}$. Next, multiply by $d_0 e_0 \in E_2(S)$, with $d_2(d_0 e_0) = 0$, to get

$$d_2(d_0 e_0 \cdot \overline{i(f_0)}) = d_0 e_0 \cdot \overline{i(h_0^2 e_0)}$$

in $E_2(C\eta \wedge C\nu)$. Here $d_0 e_0 \cdot \overline{i(f_0)} = 12_{34}$ and $d_0 e_0 \cdot \overline{i(h_0^2 e_0)} = 14_{28}$ in the minimal resolution calculated by `ext`. See Figure 11.25. There is a second generator $c = 12_{35}$ in the same bidegree of $E_2(C\eta \wedge C\nu)$ as $d_0 e_0 \cdot \overline{i(f_0)} = 12_{34}$, but $d_2(c) = 0$ by $h_0$-linearity. Recall the class $r = 6_{10}$ in $E_2(S)$, with $d_2(r) = 0$. The identity

$$r \cdot a = 6_{10} \cdot 6_{13} = 12_{34} + 12_{35} = d_0 e_0 \cdot \overline{i(f_0)} + c$$

can be verified with `ext`. It follows that

$$r \cdot d_2(a) = d_0 e_0 \cdot \overline{i(h_0^2 e_0)} + 0 \neq 0\,,$$

so $d_2(a) \neq 0$ in $E_2(C\eta \wedge C\nu)$. The only possible target is $b = (1 \wedge i)i(Pd_0) = 8_9$.

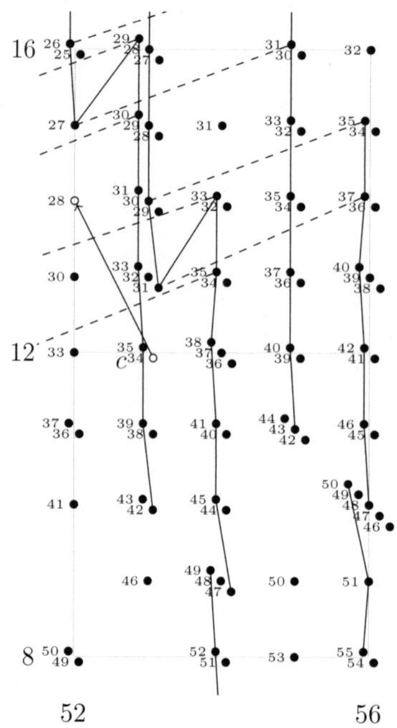

FIGURE 11.25. $E_2(C\eta\wedge C\nu)$ for $52 \le t-s \le 56$, $8 \le s \le 16$, with one $d^2$-differential

(2) If $\gamma' \in \{i(Pd_0)\}$ then $(1\wedge i)(\gamma')$ is detected by $(1\wedge i)i(Pd_0) = 0$ in $E_\infty(C\eta\wedge C\nu)$, modulo classes of higher Adams filtration. But Adams filtration $\ge 9$ of $\pi_{22}(C\eta\wedge C\nu)$ is trivial, as is evident from Figure 11.24. Hence $(1\wedge i)(\gamma') = 0$. □

## 11.10. The *tmf*-Hurewicz homomorphism

Consider the homotopy cofiber sequence

$$S \xrightarrow{\iota} tmf \xrightarrow{i} tmf/S \xrightarrow{j} \Sigma S$$

and the induced long exact sequence of Adams spectral sequence $E_2$-terms

$$\cdots \longrightarrow E_2^{s,t}(S) \xrightarrow{\iota} E_2^{s,t}(tmf) \xrightarrow{i} E_2^{s,t}(tmf/S) \xrightarrow{j} E_2^{s+1,t}(S) \longrightarrow \dots.$$

Here $H^*(tmf/S)$ is the positive-degree part of $H^*(tmf)$, and we can calculate

$$E_2^{s,t}(tmf/S) = \mathrm{Ext}_A^{s,t}(H^*(tmf/S), \mathbb{F}_2)$$

in a finite range using `ext`. This requires producing module definition files for $A//A(2)$ and $IA//A(2) = \ker(A//A(2) \to \mathbb{F}_2)$, together with map definition files for the homomorphisms $\iota \in \mathrm{Ext}_A^{0,0}(A//A(2), \mathbb{F}_2)$ and $i \in \mathrm{Ext}_A^{0,0}(IA//A(2), A//A(2))$ and for the 1-cocycle $j \in \mathrm{Ext}_A^{1,0}(\mathbb{F}_2, IA//A(2))$, in the requisite formats. This can be achieved with computer algebra software such as `MAGMA` together with a bit of hand work.

The $E_2(S)$-module generators of $E_2(tmf/S)$ for $t-s \le 48$ are listed in Table 11.6. Here $\overline{a} = i(a)$ denotes the image of a class $a \in E_2(tmf)$, and $\widetilde{a} \in j^{-1}(a)$ denotes a lift of a class $a \in E_2(S)$. Most of the $d_2$-differentials in that table are

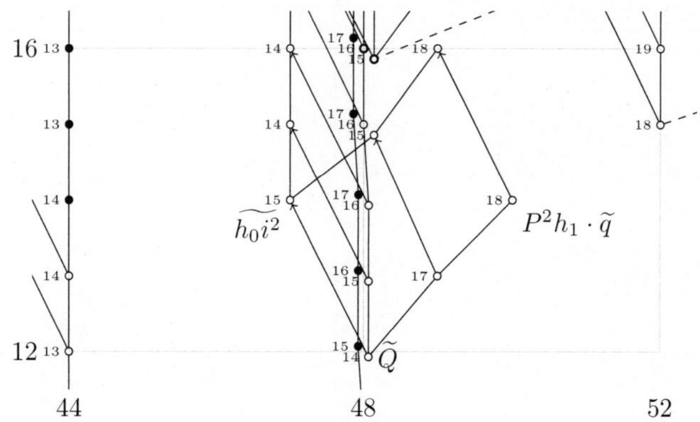


FIGURE 11.26. $(E_2(tmf/S), d_2)$ for $44 \le t-s \le 52$, $12 \le s \le 16$

determined by $h_0$-linearity or $j$-naturality, or vanish because the target group is zero. The following lemma accounts for the two remaining cases.

LEMMA 11.76. *$d_2(\widetilde{q}) = \overline{\delta'} = h_1 \cdot \widetilde{h_0 r}$ and $d_2(\widetilde{h_1 u}) = \overline{\delta' w_1} = Ph_1 \cdot \widetilde{h_0 r}$.*

PROOF. We lift $d_2(Q) = h_0 i^2$ in $E_2(S)$ to $d_2(\widetilde{Q}) = \widetilde{h_0 i^2}$ in $E_2(tmf/S)$. See Figure 11.26. Multiplying by $h_1^2$ we obtain $d_2(h_1^2 \cdot \widetilde{Q}) = h_1^2 \cdot \widetilde{h_0 i^2} = 16_{18}$ in $E_2(tmf/S)$. Here $h_1^2 \cdot \widetilde{Q} = 14_{18} = P^2h_1 \cdot \widetilde{q} = Ph_1 \cdot \widetilde{h_1 u}$, as verified by `ext`. Thus $P^2h_1 \cdot d_2(\widetilde{q}) = Ph_1 \cdot d_2(\widetilde{h_1 u}) = 16_{18} \ne 0$, which implies $d_2(\widetilde{q}) \ne 0$ and $d_2(\widetilde{h_1 u}) \ne 0$. The only possible values are $\overline{\delta'} = h_1 \cdot \widetilde{h_0 r}$ and $\overline{\delta' w_1} = Ph_1 \cdot \widetilde{h_0 r}$, respectively. □

These differentials for $tmf/S$ correspond to filtration shifts for $\iota\colon S \to tmf$.

PROPOSITION 11.77. *The homomorphism $\iota\colon \pi_*(S) \to \pi_*(tmf)$ takes the classes in $\{q\} \subset \pi_{32}(S)$ to $\epsilon_1 \in \{\delta'\} \subset \pi_{32}(tmf)$, increasing Adams filtration from* 6 *to* 7. *Similarly, it takes the classes in $\{h_1 u\} \subset \pi_{40}(S)$ to $B\epsilon_1 \in \{\delta' w_1\} \subset \pi_{40}(tmf)$, increasing Adams filtration from* 10 *to* 11. *Here $B\epsilon_1 = \epsilon\epsilon_1 = 2\bar\kappa^2$.*

PROOF. Let $S_{5,3} = \mathrm{cof}(S_8 \to S_5)$, where $S_\star$ is a minimal Adams resolution of $S$. Note that $S_\star \wedge tmf$ and $S_\star \wedge tmf/S$ are then (non-minimal) Adams resolutions of $tmf$ and $tmf/S$, respectively. Consider the following vertical maps of horizontal homotopy cofiber sequences.

$$\begin{array}{ccccc}
S & \xrightarrow{\ \iota\ } & tmf & \xrightarrow{\ i\ } & tmf/S \\
\uparrow & & \uparrow & & \uparrow \\
S_5 & \xrightarrow{1\wedge\iota} & S_5 \wedge tmf & \xrightarrow{1\wedge i} & S_5 \wedge tmf/S \\
\downarrow & & \downarrow & & \downarrow \\
S_{5,3} & \xrightarrow{1\wedge\iota} & S_{5,3} \wedge tmf & \xrightarrow{1\wedge i} & S_{5,3} \wedge tmf/S
\end{array}$$

The class $\epsilon_1 \in \pi_{32}(tmf)$ is detected by $\delta'$ in Adams filtration 7, hence comes from a class $\epsilon_1' \in \pi_{32}(S_5 \wedge tmf)$, also detected by $\delta'$, which maps to a well-defined class $\epsilon_1'' = \{\delta'\}$ in $\pi_{32}(S_{5,3} \wedge tmf)$. The latter class $\epsilon_1''$ maps to zero in $\pi_{32}(S_{5,3} \wedge tmf/S)$ under $1 \wedge i$, because $i(\delta') = d_2(\widetilde{q})$ by the previous lemma, hence is the image under

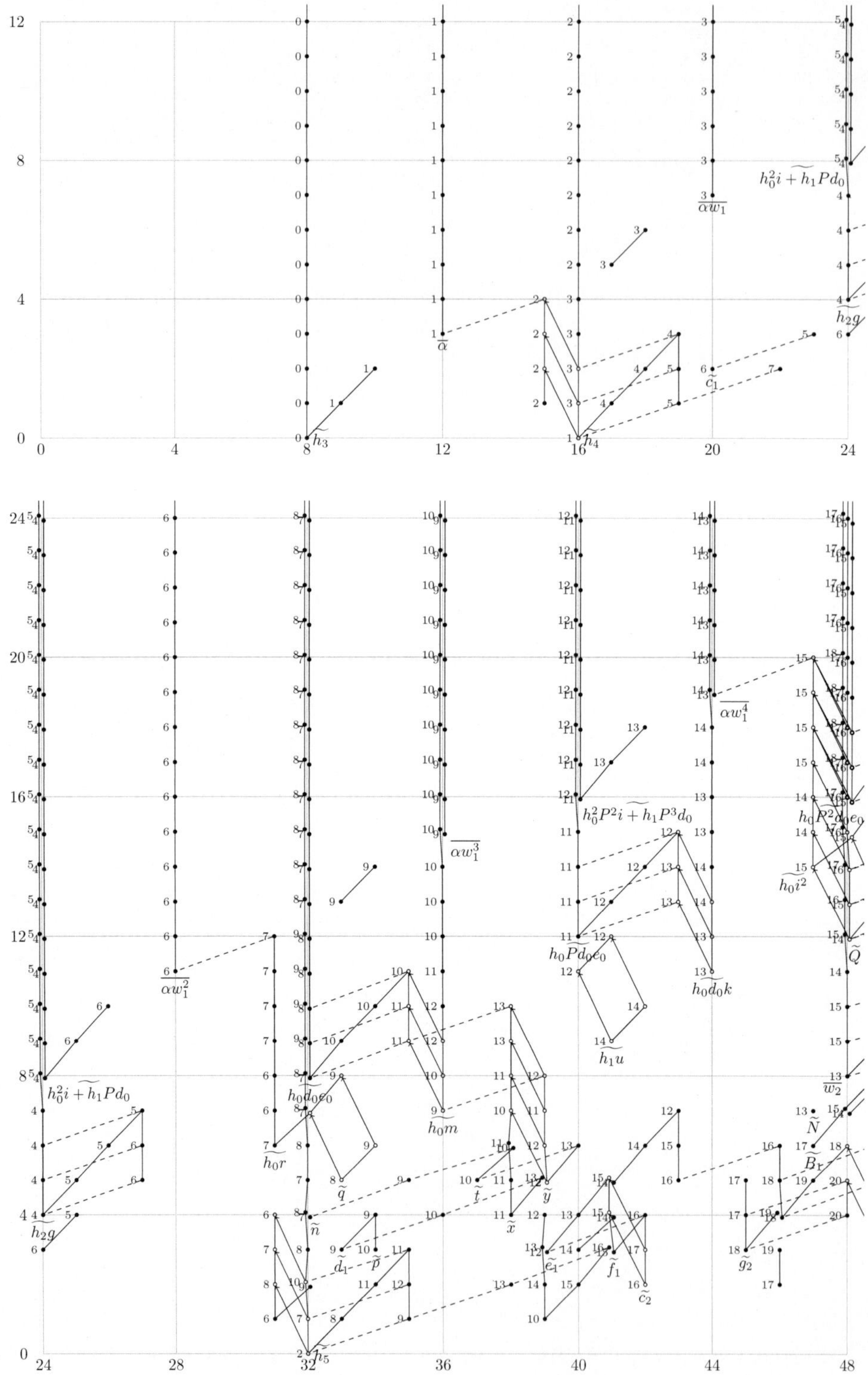

FIGURE 11.27. $(E_2(tmf/S), d_2)$ for $t - s \leq 48$

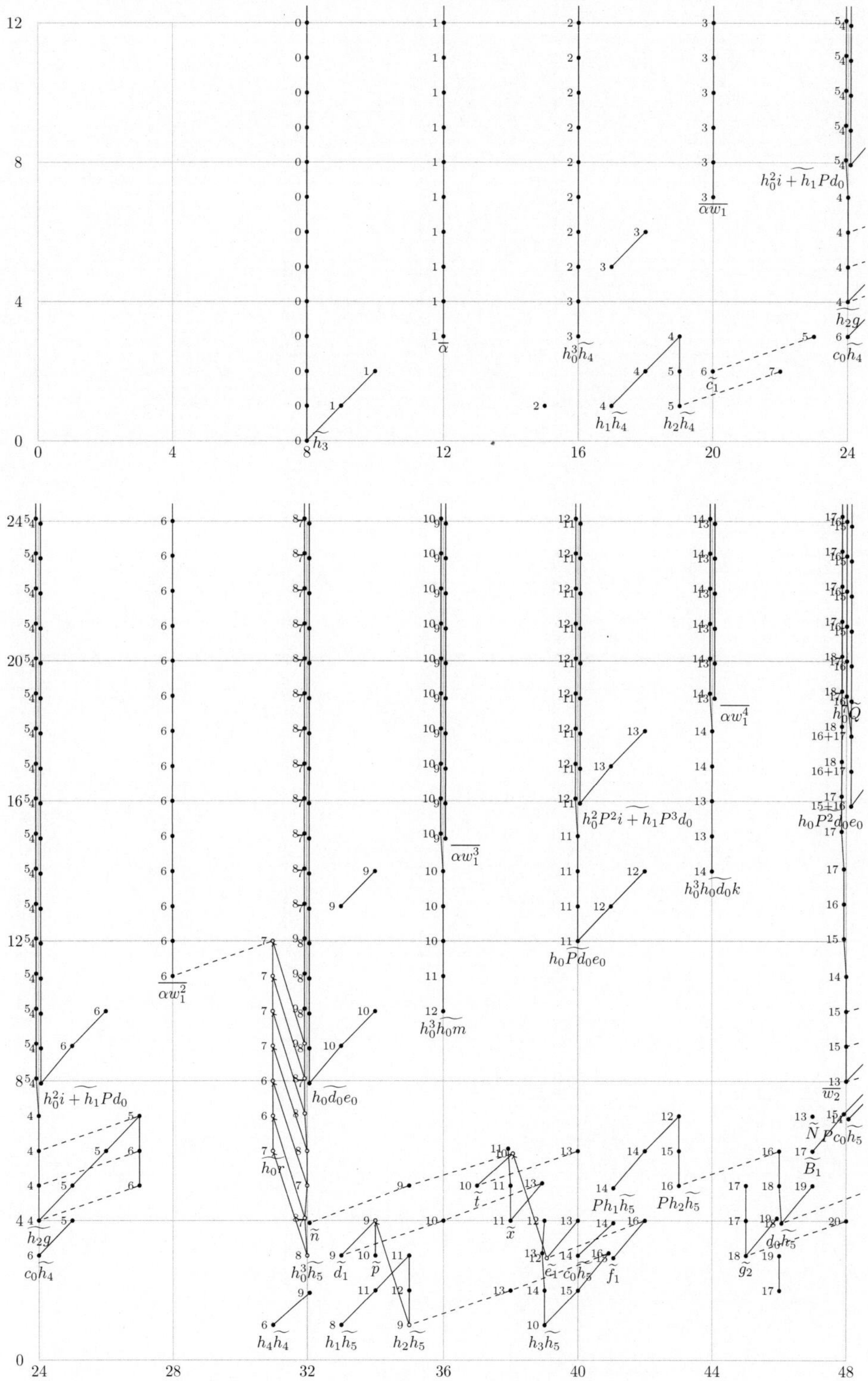

FIGURE 11.28. $(E_3(tmf/S), d_3)$ for $t - s \leq 48$

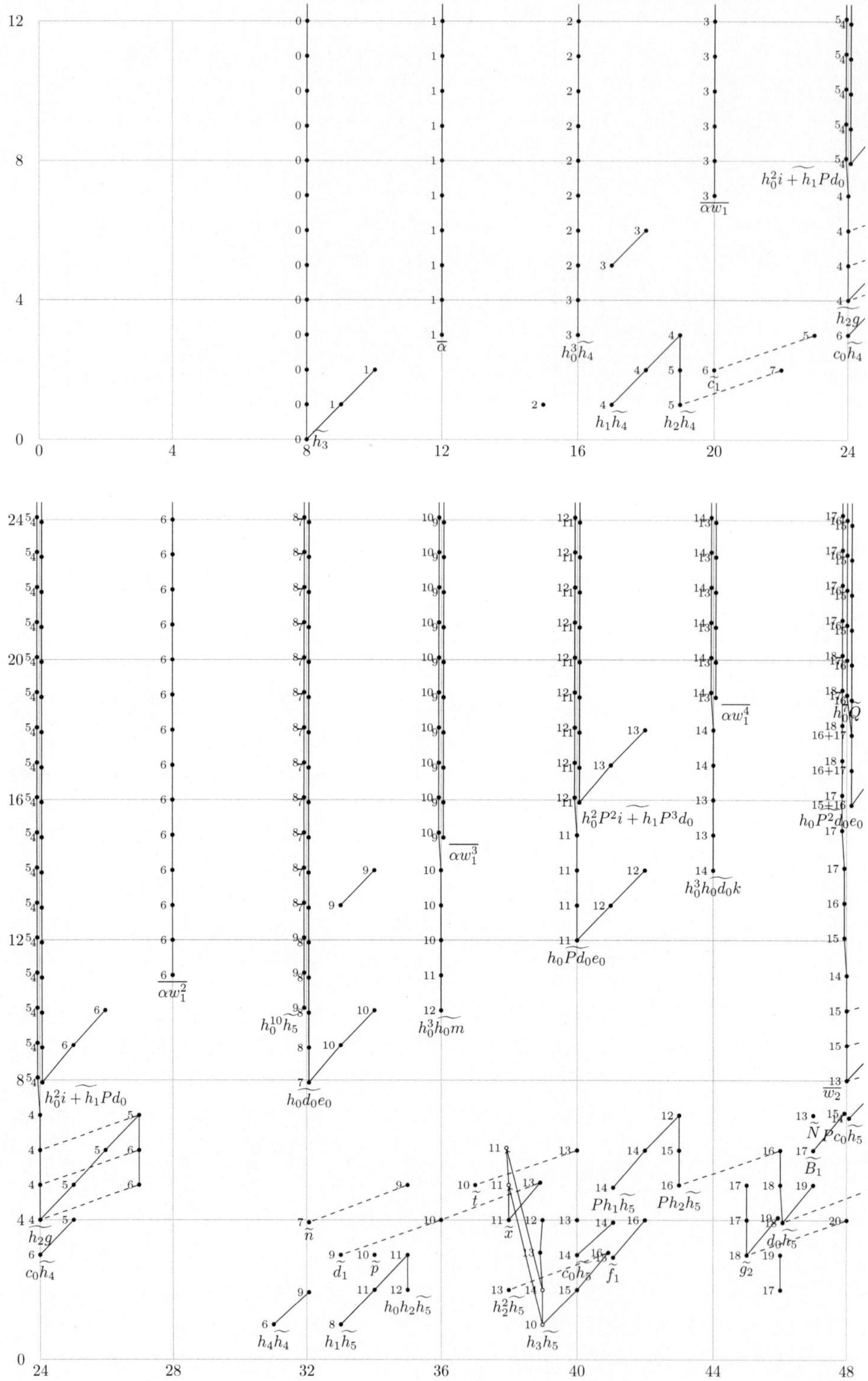

FIGURE 11.29. $(E_4(tmf/S), d_4)$ for $t - s \leq 48$

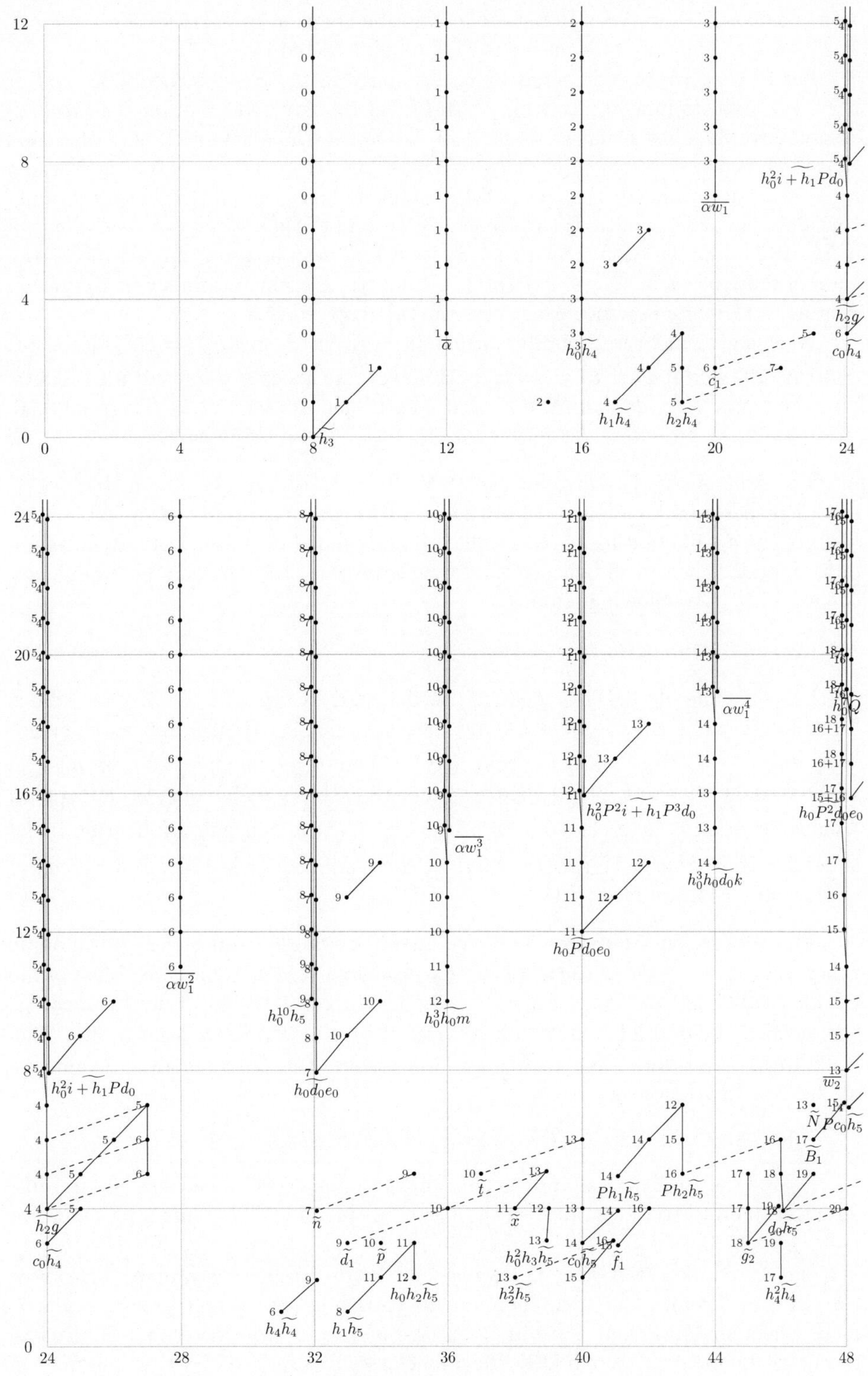


FIGURE 11.30. $E_5(tmf/S) = E_\infty(tmf/S)$ for $t - s \leq 48$

$1 \wedge \iota$ of a class $\beta'' \in \pi_{32}(S_{5,3})$. Here $\beta''$ cannot be detected by $\ell$, since $\iota(\ell) = \alpha g \neq \delta'$ in $E_\infty(tmf)$. The only alternative is that $\beta''$ is detected by $q$.

Let $\beta' \in \pi_{32}(S_5)$ be detected by $q$. Its image in $\pi_{32}(S_{5,3})$ is then either $\beta''$ or $\beta'' + \{\ell\}$, and its image in $\pi_{32}(S_{5,3} \wedge tmf)$ will be detected by $\delta'$ or $\delta' + \alpha g = \delta$, respectively. It follows that $(1 \wedge \iota)(\beta')$ in $\pi_{32}(S_5 \wedge tmf)$ is detected by $\delta'$ or $\delta$, and that the image $\beta \in \pi_{32}(S)$ of $\beta'$ maps by $\iota$ to a class $\iota(\beta) \in \pi_{32}(tmf)$ that is detected by $\delta'$ or $\delta$, according to the case. Now $\pi_{32}(S)$ is finite, and $\{\delta\} \subset \pi_{32}(tmf)$ only contains classes of infinite order. Hence the second of the two cases is excluded, $\beta'$ maps to $\beta''$, and we have shown that at least one class in $\{q\} \subset \pi_{32}(S)$ maps to a class of finite order in $\{\delta'\} \subset \pi_{32}(tmf)$, i.e., to $\epsilon_1$. The indeterminacy in $\{q\}$ lies in Adams filtration $\geq 8$, hence every class in $\{q\}$ maps to $\epsilon_1$.

We can give an entirely similar argument for $\{\widetilde{h_1 u}\}$, using $S_{9,3} = \mathrm{cof}(S_{12} \to S_9)$ and the differential $d_2(\widetilde{h_1 u}) = i(\delta' w_1)$. However, in this case the result also follows directly from $\iota(u) = d_0\gamma$ and the hidden $\eta$-extension in case (39) of Theorem 9.16. □

The Adams $(E_2, d_2)$-term for $tmf/S$ is shown in Figure 11.27, and the $E_3(S)$-module generators for $t - s \leq 48$ of the resulting $E_3$-term are listed in Table 11.7. Most of the $d_3$-differentials in that table are determined by $h_0$-linearity, $h_2$-linearity or $j$-naturality, or vanish because the target group is zero. The one remaining case is covered by the following lemma.

LEMMA 11.78. $d_3(\widetilde{h_0 d_0 e_0}) = 0$.

PROOF. The class $B_1 \in \pi_{32}(tmf)$ is detected by $\alpha g = 7_{11} + 7_{12}$ in Adams filtration 7, while $8B_1$ is detected by $h_0\alpha^2 w_1 = 11_{10}$. Using `ext` to calculate $i\colon E_2(tmf) \to E_2(tmf/S)$, we see that $i(\alpha g) = 0$ and $i(h_0\alpha^2 w_1) = 11_8 = h_0^3 \cdot \widetilde{h_0 d_0 e_0}$, which must survive to $E_\infty(tmf/S)$ by $h_0$-linearity. Hence $i(B_1)$ must be detected in Adams filtration 8, by a class $b$ with $h_0^3 \cdot b = h_0^3 \cdot \widetilde{h_0 d_0 e_0}$. Since multiplication by $h_0^3$ acts injectively from bidegree $(t-s, s) = (32, 8)$, the only possibility is $b = \widetilde{h_0 d_0 e_0}$, so this class is an infinite cycle. □

The Adams $(E_3, d_3)$-term is shown in Figure 11.28, and the $E_4(S)$-module generators for $t - s \leq 48$ of the resulting $E_4$-term are listed in Table 11.8. Most of the $d_4$-differentials in this range vanish because the target is trivial, or by $h_0$-linearity. The nonzero differential on $h_3\widetilde{h_5}$ follows from the one in $E_4(S)$ by $j$-naturality. This leads to the $E_5$-term shown in Figure 11.30, where the $E_5(S)$-module generators for $t - s \leq 48$ are labeled.

PROPOSITION 11.79. *$E_5(tmf/S) = E_\infty(tmf/S)$ for $t - s \leq 48$.*

PROOF. Most $E_5(S)$-module generators are infinite cycles because all later differentials land in trivial groups, or by $h_0$- or $h_1$-linearity. The remaining cases are $h_2\widetilde{h_4}$, $h_0 h_2\widetilde{h_5}$ and $Ph_2\widetilde{h_5}$. By Proposition 11.82 the classes $\nu^*$, $\alpha_{34}$ and $[[Ph_2h_5]]$ in $\ker(e) \subset \pi_*(S)$, detected by $h_2h_4$, $h_0h_2h_5$ and $Ph_2h_5$, respectively, map to $B$-power torsion in $\pi_*(tmf)$. They can therefore be chosen so as to map to zero under $\iota$, as in Theorem 11.61, hence are in the image of $j\colon \pi_{*+1}(tmf/S) \to \pi_*(S)$. By the geometric boundary theorem [**38**], the only classes in $\pi_{*+1}(tmf/S)$ that can map to $\nu^*$, $\alpha_{34}$ and $[[Ph_2h_5]]$ must be detected by $h_2\widetilde{h_4}$, $h_0h_2\widetilde{h_5}$ and $Ph_2\widetilde{h_5}$, respectively. Hence the latter three classes are infinite cycles. □

THEOREM 11.80. *The tmf-Hurewicz homomorphism* $\iota\colon \pi_*(S) \to \pi_*(tmf)$ *maps the algebra generators* $\alpha$ *in degrees* $* \le 44$ *to* $\iota(\alpha)$, *as in the following table.*

| $\alpha$ | $\eta$ | $\nu$ | $\sigma$ | $\epsilon$ | $\mu$ | $\zeta$ | $\kappa$ | $\rho$ | $\eta^*$ | $\bar{\mu}$ | $\nu^*$ | $\bar{\zeta}$ | $\bar{\sigma}$ | $\bar{\kappa}$ | $\bar{\rho}$ | $\mu_{25}$ |
|---|---|---|---|---|---|---|---|---|---|---|---|---|---|---|---|---|
| $\iota(\alpha)$ | $\eta$ | $\nu$ | 0 | $\epsilon$ | $\eta B$ | 0 | $\kappa$ | 0 | 0 | $\eta B^2$ | 0 | 0 | 0 | $\bar{\kappa}$ | 0 | $\eta B^3$ |

| $\alpha$ | $\zeta_{27}$ | $\theta_4$ | $\rho_{31}$ | $[n]$ | $[q]$ | $\kappa_1$ | $\eta_5$ | $\mu_{33}$ | $\alpha_{34}$ | $\zeta_{35}$ | $\{t\}$ | $\alpha_{37}$ | $\alpha_{38}$ |
|---|---|---|---|---|---|---|---|---|---|---|---|---|---|
| $\iota(\alpha)$ | 0 | 0 | 0 | 0 | $\epsilon_1$ | 0 | 0 | $\eta B^4$ | 0 | 0 | 0 | 0 | 0 |

| $\alpha$ | $\rho_{39}$ | $[u]$ | $\alpha_{39}$ | $[[Ph_1h_5]]$ | $\alpha_{40}$ | $\mu_{41}$ | $[[Ph_2h_5]]$ | $\zeta_{43}$ | $\bar{\kappa}_2$ |
|---|---|---|---|---|---|---|---|---|---|
| $\iota(\alpha)$ | 0 | $\eta_1\kappa$ | 0 | 0 | 0 | $\eta B^5$ | 0 | 0 | 0 |

*Furthermore,* $\iota(\{w\}) = \eta_1\bar{\kappa}$, *and the remaining algebra generators in degrees* $45 \le * \le 50$ *can be chosen to map to zero.*

PROOF. The claims for $\eta$, $\nu$, $\epsilon$, $\kappa$ and $\bar{\kappa}$ are clear from Definition 9.22.

The classes $\sigma$, $\zeta$, $\eta^*$, $\bar{\zeta}$, $\bar{\sigma}$, $\bar{\rho}$, $\zeta_{27}$, $\theta_4$, $\rho_{31}$, $[n]$, $\zeta_{35}$, $\{t\}$, $\alpha_{37}$, $\alpha_{38}$, $\rho_{39}$, $\zeta_{43}$, $\bar{\kappa}_2$, $\rho_{47}$ and $\{e_0r\}$ map to zero, because the corresponding Adams filtration of the target group is 2-torsion free. See Figures 9.6 and 9.7.

The classes $\mu$, $\bar{\mu}$, $\mu_{25}$, $\mu_{33}$ and $\mu_{41}$ are detected by $P^kh_1$ for $k \ge 0$, hence map to classes detected by $\iota(P^kh_1) = h_1w_1^k$, which uniquely characterizes the $\eta B^k$. The classes $[u]$ and $\{w\}$ map to classes detected by $\iota(u) = d_0\gamma$ and $\gamma g$, see Table 1.1, which uniquely characterizes $\eta_1\kappa$ and $\eta_1\bar{\kappa}$, respectively.

We chose $\rho$ to be in $j(\{h_0^3\widetilde{h_4}\})$, so that $\epsilon\rho = 0$ and $\iota(\rho) = 0$, cf. the proof of cases (15) and (23) of Theorem 11.61. We chose $\nu^*$ to be in $\ker(e)$, hence $\iota(\nu^*)$ is $B$-power torsion by Proposition 11.82, of which there is none in $\pi_{18}(tmf)$.

We proved that $\iota([q]) = \epsilon_1$, with a filtration shift from 6 to 7, in Proposition 11.77.

We could, and did, choose $\kappa_1$, $\eta_5$, $\alpha_{34}$, $\alpha_{39}$, $[[Ph_1h_5]]$, $\alpha_{40}$ and $[[Ph_2h_5]]$ in $\ker(e)$ to also lie in $\ker(\iota)$, since in each case there are classes in higher Adams filtration whose images under $\iota$ span the $B$-power torsion in the relevant degree of $\pi_*(tmf)$. The same argument applies to the remaining algebra generators in degree $* \in \{45, 46\}$ of filtration $\le 8$. Finally, the generators in degrees $47 \le * \le 50$ must map to zero, since $\pi_*(tmf)$ contains no $B$-power torsion in these degrees. $\square$

Table 11.6: $E_2(S)$-module generators of $E_2(tmf/S)$ for $t - s \le 48$

| $t-s$ | $s$ | $g$ | $x$ | $d_2(x)$ |
|---|---|---|---|---|
| 8 | 0 | 0 | $\widetilde{h_3}$ | 0 |
| 12 | 3 | 1 | $\overline{\alpha}$ | 0 |
| 16 | 0 | 1 | $\widetilde{h_4}$ | $h_0h_3 \cdot \widetilde{h_3}$ |
| 20 | 2 | 6 | $\widetilde{c_1}$ | 0 |
| 20 | 7 | 3 | $\overline{\alpha w_1}$ | 0 |
| 24 | 4 | 4 | $\widetilde{h_2g}$ | 0 |
| 24 | 8 | 4 | $\widetilde{h_0^2i + h_1Pd_0}$ | 0 |

Table 11.6: $E_2(S)$-module generators of $E_2(tmf/S)$ for $t-s \leq 48$ (cont.)

| $t-s$ | $s$ | $g$ | $x$ | $d_2(x)$ |
|---|---|---|---|---|
| 28 | 11 | 6 | $\overline{\alpha w_1^2}$ | 0 |
| 31 | 6 | 7 | $\widetilde{h_0 r}$ | 0 |
| 32 | 0 | 2 | $\widetilde{h_5}$ | $h_0 h_4 \cdot \widetilde{h_4}$ |
| 32 | 4 | 7 | $\widetilde{n}$ | 0 |
| 32 | 8 | 7 | $\widetilde{h_0 d_0 e_0}$ | 0 |
| 33 | 3 | 9 | $\widetilde{d_1}$ | 0 |
| 33 | 5 | 8 | $\widetilde{q}$ | $\overline{\delta'} = h_1 \cdot \widetilde{h_0 r}$ |
| 34 | 3 | 10 | $\widetilde{p}$ | 0 |
| 36 | 7 | 9 | $\widetilde{h_0 m}$ | $h_2 \cdot \widetilde{h_0 d_0 e_0}$ |
| 36 | 15 | 9 | $\overline{\alpha w_1^3}$ | 0 |
| 37 | 5 | 10 | $\widetilde{t}$ | 0 |
| 38 | 4 | 11 | $\widetilde{x}$ | 0 |
| 39 | 3 | 12 | $\widetilde{e_1}$ | 0 |
| 39 | 5 | 12 | $\widetilde{y}$ | $h_0^3 \cdot \widetilde{x}$ |
| 40 | 12 | 11 | $h_0 \widetilde{P d_0 e_0}$ | 0 |
| 40 | 16 | 11 | $h_0^2 \widetilde{P^2 i + h_1 P^3 d_0}$ | 0 |
| 41 | 3 | 15 | $\widetilde{f_1}$ | 0 |
| 41 | 9 | 14 | $\widetilde{h_1 u}$ | $\overline{\delta' w_1} = P h_1 \cdot \widetilde{h_0 r}$ |
| 42 | 2 | 16 | $\widetilde{c_2}$ | $h_0 \cdot \widetilde{f_1}$ |
| 44 | 11 | 13 | $\widetilde{h_0 d_0 k}$ | $h_2 \cdot h_0 \widetilde{P d_0 e_0}$ |
| 44 | 19 | 13 | $\overline{\alpha w_1^4}$ | 0 |
| 45 | 3 | 18 | $\widetilde{g_2}$ | 0 |
| 47 | 6 | 17 | $\widetilde{B_1}$ | 0 |
| 47 | 7 | 13 | $\widetilde{N}$ | 0 |
| 47 | 14 | 15 | $\widetilde{h_0 i^2}$ | 0 |
| 48 | 8 | 13 | $\overline{w_2}$ | 0 |
| 48 | 12 | 14 | $\widetilde{Q}$ | $\widetilde{h_0 i^2}$ |
| 48 | 16 | $15+16$ | $h_0 \widetilde{P^2 d_0 e_0}$ | 0 |

Table 11.7: $E_3(S)$-module generators of $E_3(tmf/S)$ for $t-s \leq 48$

| $t-s$ | $s$ | $g$ | $x$ | $d_3(x)$ |
|---|---|---|---|---|
| 8 | 0 | 0 | $\widetilde{h_3}$ | 0 |
| 12 | 3 | 1 | $\overline{\alpha}$ | 0 |
| 16 | 3 | 3 | $h_0^3\widetilde{h_4}$ | 0 |
| 17 | 1 | 4 | $h_1\widetilde{h_4}$ | 0 |
| 19 | 1 | 5 | $h_2\widetilde{h_4}$ | 0 |
| 20 | 2 | 6 | $\widetilde{c_1}$ | 0 |
| 20 | 7 | 3 | $\overline{\alpha w_1}$ | 0 |
| 24 | 3 | 6 | $c_0\widetilde{h_4}$ | 0 |
| 24 | 4 | 4 | $\widetilde{h_2 g}$ | 0 |
| 24 | 8 | 4 | $h_0^2\widetilde{i + h_1 P}d_0$ | 0 |
| 28 | 11 | 6 | $\overline{\alpha w_1^2}$ | 0 |
| 31 | 1 | 6 | $h_4\widetilde{h_4}$ | 0 |
| 31 | 6 | 7 | $\widetilde{h_0 r}$ | 0 |
| 32 | 3 | 8 | $h_0^3\widetilde{h_5}$ | $\widetilde{h_0 r}$ |
| 32 | 4 | 7 | $\widetilde{n}$ | 0 |
| 32 | 8 | 7 | $\widetilde{h_0 d_0 e_0}$ | 0 |
| 33 | 1 | 8 | $h_1\widetilde{h_5}$ | 0 |
| 33 | 3 | 9 | $\widetilde{d_1}$ | 0 |
| 34 | 3 | 10 | $\widetilde{p}$ | 0 |
| 35 | 1 | 9 | $h_2\widetilde{h_5}$ | $h_0\widetilde{p}$ |
| 36 | 10 | 12 | $h_0^3\widetilde{h_0 m}$ | 0 |
| 36 | 15 | 9 | $\overline{\alpha w_1^3}$ | 0 |
| 37 | 5 | 10 | $\widetilde{t}$ | 0 |
| 38 | 4 | 11 | $\widetilde{x}$ | 0 |
| 39 | 1 | 10 | $h_3\widetilde{h_5}$ | 0 |
| 39 | 3 | 12 | $\widetilde{e_1}$ | $h_1\widetilde{t}$ |
| 40 | 3 | 14 | $c_0\widetilde{h_5}$ | 0 |
| 40 | 12 | 11 | $h_0\widetilde{P d_0 e_0}$ | 0 |
| 40 | 16 | 11 | $h_0^2 P^2\widetilde{i + h_1 P^3} d_0$ | 0 |
| 41 | 3 | 15 | $\widetilde{f_1}$ | 0 |
| 41 | 5 | 14 | $P h_1\widetilde{h_5}$ | 0 |
| 43 | 5 | 16 | $P h_2\widetilde{h_5}$ | 0 |

Table 11.7: $E_3(S)$-module generators of $E_3(tmf/S)$ for $t-s \leq 48$ (cont.)

| $t-s$ | $s$ | $g$ | $x$ | $d_3(x)$ |
|---|---|---|---|---|
| 44 | 14 | 14 | $h_0^3\widetilde{h_0 d_0 k}$ | 0 |
| 44 | 19 | 13 | $\overline{\alpha w_1^4}$ | 0 |
| 45 | 3 | 18 | $\widetilde{g_2}$ | 0 |
| 46 | 4 | 18 | $d_0\widetilde{h_5}$ | 0 |
| 47 | 6 | 17 | $\widetilde{B_1}$ | 0 |
| 47 | 7 | 13 | $\widetilde{N}$ | 0 |
| 48 | 7 | 14 | $Pc_0\widetilde{h_5}$ | 0 |
| 48 | 8 | 13 | $\overline{w_2}$ | 0 |
| 48 | 16 | $15+16$ | $h_0\widetilde{P^2 d_0 e_0}$ | 0 |
| 48 | 19 | 17 | $h_0^7\widetilde{Q}$ | 0 |

Table 11.8: $E_4(S)$-module generators of $E_4(tmf/S)$ for $t-s \leq 48$

| $t-s$ | $s$ | $g$ | $x$ | $d_4(x)$ |
|---|---|---|---|---|
| 8 | 0 | 0 | $\widetilde{h_3}$ | 0 |
| 12 | 3 | 1 | $\overline{\alpha}$ | 0 |
| 16 | 3 | 3 | $h_0^3\widetilde{h_4}$ | 0 |
| 17 | 1 | 4 | $h_1\widetilde{h_4}$ | 0 |
| 19 | 1 | 5 | $h_2\widetilde{h_4}$ | 0 |
| 20 | 2 | 6 | $\widetilde{c_1}$ | 0 |
| 20 | 7 | 3 | $\overline{\alpha w_1}$ | 0 |
| 24 | 3 | 6 | $c_0\widetilde{h_4}$ | 0 |
| 24 | 4 | 4 | $\widetilde{h_2 g}$ | 0 |
| 24 | 8 | 4 | $h_0^2 \widetilde{i + h_1 P d_0}$ | 0 |
| 28 | 11 | 6 | $\overline{\alpha w_1^2}$ | 0 |
| 31 | 1 | 6 | $h_4\widetilde{h_4}$ | 0 |
| 32 | 4 | 7 | $\widetilde{n}$ | 0 |
| 32 | 8 | 7 | $\widetilde{h_0 d_0 e_0}$ | 0 |
| 32 | 10 | 9 | $h_0^{10}\widetilde{h_5}$ | 0 |
| 33 | 1 | 8 | $h_1\widetilde{h_5}$ | 0 |
| 33 | 3 | 9 | $\widetilde{d_1}$ | 0 |

Table 11.8: $E_4(S)$-module generators of $E_4(tmf/S)$ for $t-s \le 48$ (cont.)

| $t-s$ | $s$ | $g$ | $x$ | $d_4(x)$ |
|---|---|---|---|---|
| 34 | 3 | 10 | $\widetilde{p}$ | 0 |
| 35 | 2 | 12 | $h_0h_2\widetilde{h_5}$ | 0 |
| 36 | 10 | 12 | $h_0^3\widetilde{h_0m}$ | 0 |
| 36 | 15 | 9 | $\overline{\alpha w_1^3}$ | 0 |
| 37 | 5 | 10 | $\widetilde{t}$ | 0 |
| 38 | 2 | 13 | $h_2^2\widetilde{h_5}$ | 0 |
| 38 | 4 | 11 | $\widetilde{x}$ | 0 |
| 39 | 1 | 10 | $h_3\widetilde{h_5}$ | $h_0\widetilde{x}$ |
| 40 | 3 | 14 | $c_0\widetilde{h_5}$ | 0 |
| 40 | 12 | 11 | $\widetilde{h_0Pd_0e_0}$ | 0 |
| 40 | 16 | 11 | $h_0^2\widetilde{P^2i+h_1P^3d_0}$ | 0 |
| 41 | 3 | 15 | $\widetilde{f_1}$ | 0 |
| 41 | 5 | 14 | $Ph_1\widetilde{h_5}$ | 0 |
| 43 | 5 | 16 | $Ph_2\widetilde{h_5}$ | 0 |
| 44 | 14 | 14 | $h_0^3\widetilde{h_0d_0k}$ | 0 |
| 44 | 19 | 13 | $\overline{\alpha w_1^4}$ | 0 |
| 45 | 3 | 18 | $\widetilde{g_2}$ | 0 |
| 46 | 4 | 18 | $d_0\widetilde{h_5}$ | 0 |
| 47 | 6 | 17 | $\widetilde{B_1}$ | 0 |
| 47 | 7 | 13 | $\widetilde{N}$ | 0 |
| 48 | 7 | 14 | $Pc_0\widetilde{h_5}$ | 0 |
| 48 | 8 | 13 | $\overline{w_2}$ | 0 |
| 48 | 16 | $15+16$ | $\widetilde{h_0P^2d_0e_0}$ | 0 |
| 48 | 19 | 17 | $h_0^7\widetilde{Q}$ | 0 |

## 11.11. The *tmf*-Hurewicz image

The image of the *tmf*-Hurewicz homomorphism $\iota\colon \pi_*(S) \to \pi_*(tmf)$ lies mostly in the Pontryagin self-dual part. Working integrally, for a moment, the image contains $\pi_0(tmf) \cong \mathbb{Z}\{\iota\}$, $\pi_3(tmf) \cong \mathbb{Z}/24\{\nu\}$ and the groups $\mathbb{Z}/2\{\eta B^k\} \subset \pi_{8k+1}(tmf)$ and $\mathbb{Z}/2\{\eta^2 B^k\} \subset \pi_{8k+2}(tmf)$ for $k \ge 0$. The remainder of the 2-primary Hurewicz image was conjectured by Mahowald [**54**, §13.5] to be equal to the part of the $B$-power torsion that we refer to as $\Theta\pi_*(tmf)^\wedge_2$, cf. Chapter 10. A proof of this conjecture was announced by Behrens (ca. 2012), in joint work with Mahowald.

See Remark 11.84. Similarly, the remainder of the 3-primary Hurewicz image is asserted in [**54**, §13.1] to be equal to the part of the $B$-power torsion that we denote $\Theta\pi_*(tmf)^\wedge_3$. In this section we outline calculations leading toward these conclusions at $p = 2$. See Section 13.7 for calculations at $p = 3$.

Returning to the implicitly 2-complete setting, let the cokernel-of-$J$ spectrum $c$ be defined by the homotopy cofiber sequence

$$c \longrightarrow S \xrightarrow{e} j\,,$$

where $e\colon S \to j$ is the unit map representing the combined Adams $d$- and $e$-invariants. Adams [**8**] proved that $e\colon \pi_*(S) \to \pi_*(j)$ is surjective, which implies that $\pi_*(c) \cong \ker(e)$.

As a consequence of Mahowald's work on $bo$-resolutions and $v_1$-periodic homotopy [**102**, Thm. 1.1], Bousfield [**33**, Thm. 4.3] deduced that the map $e$ is a $KU$-equivalence, so that $c$ is $KU$-acyclic. A simpler proof of this fact can be given, following Stephen Mitchell [**131**, p. 201], by noting that $e$ induces an isomorphism $e^*\colon KU^*(j) \to KU^*(S)$. Here $KU^*(KO)$ is $\mathbb{Z}_2[[\mathbb{Z}_2^\times/\langle -1\rangle]]$ for $* = 0$ and $0$ for $* = 1$. The Adams operation $\psi^k\colon KO \to KO$ induces multiplication by $k$ in $\mathbb{Z}_2^\times/\langle -1\rangle$, so $KU^*(j)$ is $\mathbb{Z}_2[[\mathbb{Z}_2^\times/\langle -1, 3\rangle]] \cong \mathbb{Z}_2$ for $* = 0$ and $0$ for $* = 1$. Since $d^*\colon KU^0(KO) \to KU^0(S) \cong \mathbb{Z}_2$ is surjective and $KU^1(S) = 0$, it follows that $e^*$ is an isomorphism, which implies that $e$ is a $KU$-equivalence.

PROPOSITION 11.81. *$tmf[1/B]$ is Bousfield $KU$-local.*

PROOF. By Bousfield's criterion [**33**, Thm. 4.8] it suffices to check that

$$tmf[1/B] \wedge Z \simeq *$$

for $Z = M(1,4)$. Here $M(1,4) = S/(2, v_1^4)$ is the mapping cone of an Adams map $v_1^4\colon \Sigma^8 S/2 \to S/2$. By the Hopkins–Smith thick subcategory theorem [**78**], we may equally well verify the condition for $Z = \Phi \wedge M(1,4)$, since both $M(1,4)$ and $\Phi \wedge M(1,4)$ are type 2 finite CW spectra. Here $\Phi = \Phi A(1)$ is as in Lemma 1.42. In view of the equivalence $tmf \wedge \Phi \simeq BP\langle 2\rangle$ from Proposition 1.44, it suffices to prove that $BP\langle 2\rangle \wedge M(1,4)$ becomes trivial after inverting $B$. Here $\pi_*(BP\langle 2\rangle \wedge M(1,4)) = \mathbb{Z}_2[v_1, v_2]/(2, v_1^4) = \mathbb{Z}/2[v_1, v_2]/(v_1^4)$, where $B$ acts nilpotently, so this claim is clear. □

PROPOSITION 11.82. *If $x \in \ker(e) \subset \pi_n(S)$, then $\iota(x)$ lies in $\Theta\pi_n(tmf) \subset \Gamma_B\pi_n(tmf) \subset \pi_n(tmf)$.*

PROOF. The composite $c \to S \to tmf \to tmf[1/B]$ is null-homotopic, since $c$ is $KU$-acyclic and $tmf[1/B]$ is $KU$-local, so there is a commutative diagram

$$\begin{array}{ccccc}
c & \longrightarrow & S & \xrightarrow{e} & j \\
\downarrow & & \downarrow{\scriptstyle \iota} & & \downarrow \\
\Sigma^{-1} tmf/B^\infty & \longrightarrow & tmf & \longrightarrow & tmf[1/B]
\end{array}$$

with horizontal homotopy cofiber sequences. If $e(x) = 0$ then $x$ admits a lift $\tilde{x} \in \pi_n(c)$. Its image $\tilde{y} \in \pi_{n+1}(tmf/B^\infty)$ must be 2-power torsion, since $\pi_n(c)$ is finite. Hence its image $y = \iota(x) \in \pi_n(tmf)$ lies in the image of the composite homomorphism

$$\Gamma_2\pi_{n+1}(tmf/B^\infty) \subset \pi_{n+1}(tmf/B^\infty) \longrightarrow \Gamma_B\pi_n(tmf) \subset \pi_n(tmf)\,.$$

By Definition 10.18, this means that $\iota(x)$ lies in $\Theta\pi_n(tmf)$. In particular, $\iota(x) = 0$ if $n \equiv 3 \mod 24$. □

PROPOSITION 11.83. *The tmf-Hurewicz image of* $\ker(e) \subset \pi_n(S)$ *is equal to* $\Theta\pi_n(tmf)$ *for* $n \leq 101$ *and for* $n = 125$. *It also contains the nonzero elements* $\eta_1\bar{\kappa}^4$, $2\kappa_4 = \eta_1^2\bar{\kappa}^3$, $2\kappa_4\bar{\kappa} = \eta_1^2\bar{\kappa}^4$ *and* $4\nu\nu_6 = \eta_1^6$, *in degrees* 105, 110, 130 *and* 150, *respectively. In particular, the products* $\bar{\kappa}^5$, $\bar{\kappa}^4\{w\}$ *and* $\bar{\kappa}^3\{w\}^2$ *are nonzero in* $\pi_*(S)$.

PROOF. In Table 9.4 we have listed classes $\alpha \in \ker(e) \subset \pi_n(S)$ with $\iota(\alpha) = \beta$, for many $B$-power torsion classes $\beta \in \pi_n(tmf)$. We have also noted that $\iota(\nu) = \nu$ for $n = 3$, while for the other $n \equiv 3 \mod 24$ no spherical lift $\alpha$ exists, since $\Theta\pi_n(tmf) = 0$ for these $n$. The values of $\iota$ on the multiplicative generators of $\pi_*(S)$ in degrees $n \leq 44$ are given in Theorem 11.80, and these suffice to determine the *tmf*-Hurewicz image for $n \leq 53$, as well as in some higher degrees, including $n = 125$. We appeal to the work of Isaksen, Wang and Xu [**83**] to prove Propositions 11.85, 11.86 and 11.87, which suffice to determine the image for $n \leq 101$. Granting these results it is mostly trivial to see that a given $\alpha$ maps to the stated $\beta$. The following factorizations in $\pi_*(tmf)$, from Tables 9.8 and 9.9, handle the remaining cases:

- $\eta\nu_1 = \epsilon\bar{\kappa} = \iota(\epsilon\bar{\kappa})$
- $\eta\nu_2 = \epsilon_1\bar{\kappa} = \iota(\bar{\kappa}[q])$
- $\nu_2 B = \eta_1\kappa\bar{\kappa} = \iota(\kappa\{w\})$
- $\eta\nu_2\kappa = \eta\eta_1\bar{\kappa}^2 = \iota(\eta\bar{\kappa}\{w\})$
- $\eta\nu_4 = \eta_1^4 = \bar{\kappa}^5 = \iota(\bar{\kappa}^5)$
- $\eta_1\bar{\kappa}^4 = \iota(\bar{\kappa}^3\{w\})$
- $2\kappa_4 = \eta_1^2\bar{\kappa}^3 = \iota(\bar{\kappa}\{w\}^2)$
- $\eta^2\nu_5 = \eta_1^5 = \eta_1\bar{\kappa}^5 = \iota(\bar{\kappa}^4\{w\})$
- $2\kappa_4\bar{\kappa} = \eta_1^2\bar{\kappa}^4 = \iota(\bar{\kappa}^2\{w\}^2)$
- $4\nu\nu_6 = \eta_1^6 = \eta_1^2\bar{\kappa}^5 = \iota(\bar{\kappa}^3\{w\}^2)$. □

REMARK 11.84. As mentioned above, Mahowald effectively conjectured that $\iota(\ker(e)) = \Theta\pi_*(tmf)$ holds in all degrees, and a recent preprint [**27**] by Behrens, Mahowald and Quigley affirms this conjecture. In outline, their proof is obtained by first constructing enough classes in $\pi_*(S)$ to generate $\Theta\pi_*(tmf)$ as a $\pi_*(S)$-module in degrees $0 \leq * < 192$, and then to apply a variant for $M(3,8) = S/(8, v_1^8)$ of the $v_2^{32}$-self map of $M(1,4)$ from [**26**] to extend this 192-periodically. Proposition 11.83 accounts for a little over half the initial range of degrees. Using Tables 9.8 and 9.9 we see that the classes $\nu\nu_4$, $\epsilon_4$, $\kappa_4$, $\bar{\kappa}D_4$, $\eta_4\bar{\kappa}$, $\epsilon_5$, $\eta_1\kappa_4$, $\nu\nu_6$ and $\nu_6\kappa$ suffice to generate the remainder of $\Theta\pi_*(tmf)$, up to degree 192. These nine classes were emphasized with question-marks in the $\alpha$-column in Table 9.4, and part of the work in [**27**] is to verify that these classes are in the *tmf*-Hurewicz image from $\pi_*(S)$.

PROPOSITION 11.85 ([**82**], [**83**]). *The class* $h_0h_5i \in E_2(S)$ *survives to the* $E_\infty$-*term. Let* $\alpha_{54} \in \{h_0h_5i\}$, *in Adams filtration* 9. *Then* $\iota(\alpha_{54}) \doteq \nu\nu_2$, *in Adams filtration* 10.

SKETCH PROOF. According to [**82**, Lem. 4.56], the class $h_0h_5i = 9_{25}$ in bidegree $(t-s, s) = (54, 9)$ survives to the $E_\infty$-term in the Adams spectral sequence for $S$, and corresponds to $\beta_{10/2}$ in the Adams–Novikov spectral sequence for $S$. The latter class maps to $\Delta^2h_2^2$ in the Adams–Novikov spectral sequence for *tmf*, which detects a generator of $\pi_{54}(tmf) \cong \mathbb{Z}/4\{\nu\nu_2\}$, with $2\nu\nu_2 = \kappa\bar{\kappa}^2$. Hence $\iota$ maps $\{h_0h_5i\}$ to $\{h_2^2w_2\} = \{\pm\nu\nu_2\}$ in bidegree $(54, 10)$. □

Isaksen's subsequent argument for why there must be a hidden 2-extension from $h_0h_5i$ to $d_0g^2 = e_0^2g$ in bidegree $(54, 12)$ of $E_\infty(S)$ is incomplete, due to an intervening class $h_1x'$ in bidegree $(54, 11)$. See [**83**, Rem. 7.11] and the recent preprint [**47**] by Robert Burklund.

Proposition 11.86 ([**83**]). *The class $Ph_5j \in E_2(S)$ survives to the $E_\infty$-term. We can choose $\alpha_{65} \in \{Ph_5j\}$, in Adams filtration* 12, *with $\eta\alpha_{65} \neq 0$. Then $\iota(\alpha_{65}) = \nu_2\kappa$, in Adams filtration* 13.

Proof. According to [**83**, Cor. 1.2] and its accompanying chart, $Ph_5j = 12_{29} + 12_{30}$ in bidegree $(t-s, s) = (65, 12)$ of $E_2(S)$ survives to the $E_\infty$-term. Furthermore, $h_2 \cdot Ph_5j = 13_{31} = d_0 \cdot h_0h_5i$ is nonzero in $E_\infty(S)$, of maximal filtration in topological degree 68. Hence, letting $\alpha_{65}$ be detected by $Ph_5j$ in Adams filtration 12, we must have $\nu\alpha_{65} = \kappa\alpha_{54}$. Here $\iota(\kappa\alpha_{54}) = \nu\nu_2\kappa$, so $\iota(\alpha_{65}) \equiv \nu_2\kappa \mod \eta_1\bar{\kappa}^2$.

This suffices to prove that $\iota$ maps onto the $B$-power torsion in $\pi_{65}(tmf)$, but we can refine the choice of $\alpha_{65}$ to ensure that $\iota(\alpha_{65}) = \nu_2\kappa$, as follows. The class $\bar{\kappa}\{w\} \in \{gw\}$ in Adams filtration 13 maps to $\eta_1\bar{\kappa}^2 \in \{\gamma g^2\}$, with $\eta \cdot \eta_1\bar{\kappa}^2 = \eta \cdot \nu_2\kappa \in \{d_0\delta' g\}$. Hence $\eta\bar{\kappa}\{w\}$ must be nonzero of Adams filtration $\geq 14$, and by [**83**, Cor. 1.2] the only possible detecting class is $d_0e_0m = 15_{24} = g^2j$. If $\eta\alpha_{65} = 0$ we can therefore add $\bar{\kappa}\{w\}$ to the choice of $\alpha_{65} \in \{Ph_5j\}$, with no change in its detecting class, to arrange that $\eta\alpha_{65} = \eta\bar{\kappa}\{w\} \neq 0$, with $\eta\iota(\alpha_{65}) = \eta\nu_2\kappa \neq 0$, and this implies $\iota(\alpha_{65}) = \nu_2\kappa$. (It might be more natural to fix a choice of $\alpha_{65}$ with $\eta\alpha_{65} = 0$, but this is typographically less convenient in Table 9.4.) □

Proposition 11.87 ([**83**]). *The class $m^2 + h_1a \in E_2(S)$ survives to the $E_\infty$-term, where $a = 13_{32}$ denotes the nonzero class in bidegree $(t-s, s) = (69, 13)$. Let $\alpha_{70} \in \{m^2 + h_1a\}$. Then $\iota(\alpha_{70}) = \eta_1^2\bar{\kappa}$.*

Proof. Isaksen, Wang and Xu write $\Delta^2h_1g$ for the class $a = 13_{32} \in E_2(S)$. By [**83**, Cor. 1.2] and its accompanying chart, $m^2 + h_1a = m^2 + \Delta^2h_1^2g = 14_{29} + 14_{31}$ in bidegree $(t-s, s) = (70, 14)$ survives to $E_\infty(S)$. Furthermore, `ext` calculates that $\iota\colon E_2(S) \to E_2(tmf)$ maps $m^2 + h_1a$ to $14_{35} = \gamma^2g$, which has maximal filtration in its topological degree and detects $\eta_1^2\bar{\kappa}$. Hence $\alpha_{70} \in \{m^2 + h_1a\}$ satisfies $\iota(\alpha_{70}) = \eta_1^2\bar{\kappa}$. □

In order to complete our discussion of the image of the *tmf*-Hurewicz homomorphism $\iota$ we need some information about the classes complementary to $\ker(e)$ in $\pi_*(S)$, i.e., about the image of the $J$-homomorphism $J\colon \pi_*(SO) \to \pi_*(S)$ and the classes detected by the Adams $d$-invariant, represented by the unit map $d\colon S \to ko$. We recall that $J$ is induced by a space-level map

$$J\colon SO \longrightarrow Q_1S^0 \simeq Q_0S^0 \subset QS^0\,,$$

obtained by stabilizing maps $J_m\colon SO(m) \to \Omega_1^mS^m \simeq \Omega_0^mS^m \subset \Omega^mS^m$. Here $J_m$ takes an orientation-preserving isometry $\mathbb{R}^m \to \mathbb{R}^m$ to the induced degree $+1$ map of one-point compactifications $S^m \to S^m$, followed by loop sum with a fixed degree $-1$ map, cf. [**177**]. In degrees $* = 8k-1$ the precise Adams filtration of the elements in the image of the $J$-homomorphism is determined by Davis and Mahowald in [**53**, Thm. 1.1], with significant effort, but for our purposes the much more elementary estimate from [**53**, Prop. 2.5] suffices, and its proof can readily be extended to also account for degrees of the form $* = 8k+3$.

PROPOSITION 11.88 ([**53**, Prop. 2.5]).

*(1) If $n = 8k-1 < 2^\ell - 1$, then the image of $J\colon \pi_n(SO) \to \pi_n(S)$ lies in Adams filtration $\geq 4k+1-\ell$.*

*(2) If $n = 8k+3 < 2^\ell - 1$, then the image of $J\colon \pi_n(SO) \to \pi_n(S)$ lies in Adams filtration $\geq 4k+4-\ell$.*

PROOF. Let $X[n]$ denote the $(n-1)$-connected cover of a space $X$. Robert Stong [**163**, Thm. A] calculated the mod 2 cohomology $H^*(BO[8q+2^r])$ for $q \geq 0$ and $0 \leq r \leq 3$, showing, in particular, that this cohomology is polynomial and that the projection in the homotopy fiber sequence

$$BO[8q+2^r+1] \longrightarrow BO[8q+2^r] \longrightarrow K(\pi_{8q+2^r}(BO), 8q+2^r)$$

induces a surjection in cohomology in degrees $* < 2^\ell$, where $\ell = 4q+r+1$. It follows by the Eilenberg–Moore spectral sequence that the projection $p$ in the homotopy fiber sequence

$$SO[8q+2^r] \xrightarrow{i} SO[8q+2^r-1] \xrightarrow{p} K(\pi_{8q+2^r-1}(SO), 8q+2^r-1)$$

induces a surjection in cohomology in degrees $* < 2^\ell - 1$, so that the inclusion $i$ induces the zero homomorphism in reduced cohomology in the same degrees. Hence for $n < 2^\ell - 1$ each map $f\colon S^n \to SO$ factors as a composite

$$S^n \to SO[n] \xrightarrow{i} SO[n-1] \xrightarrow{i} \dots \xrightarrow{i} SO[8q+2^r] \xrightarrow{i} SO[8q+2^r-1] \to SO\,,$$

where about half of the maps $i$ induce the zero homomorphisms in degrees $\leq n$, and the remaining maps $i$ are equivalences. More precisely, for $n = 8k-1$ and $n = 8k+3$ there are $(4k-1)-(\ell-1)$ and $(4k+2)-(\ell-1)$ maps of the first kind, respectively. Passing to suspension spectra, each map $i$ of the first kind induces homomorphisms $\pi_*(\Sigma^\infty i)$ that increase Adams filtration by at least 1 for $* \leq n$, since $E_2^{s,t}(i) = 0$ for $t-s \leq n$. The composite $Jf\colon S^n \to SO \to QS^0$ is adjoint to a map

$$\Sigma^\infty S^n \xrightarrow{\Sigma^\infty f} \Sigma^\infty SO \xrightarrow{\tilde{J}} \Sigma^\infty S^0 = S$$

of suspension spectra, and its homotopy class in $[\Sigma^\infty S^n, \Sigma^\infty S^0] \cong \pi_n(S)$ corresponds to the homotopy class of $Jf$ in $\pi_n(QS^0)$. Since $\tilde{J}$ induces zero in cohomology, it follows that $Jf$ has Adams filtration $\geq (4k-1)-(\ell-1)+1$ (resp. $\geq (4k+2)-(\ell-1)+1$) for $n = 8k-1$ (resp. $n = 8k+3$), where $n < 2^\ell - 1$. □

THEOREM 11.89. *The image of the Hurewicz homomorphism*

$$\iota\colon \pi_*(S) \longrightarrow \pi_*(tmf)\,,$$

*implicitly completed at $p = 2$, is the direct sum of the following terms:*

1. *The group $\mathbb{Z}\{\iota\} \cong \pi_0(tmf)$ and the subgroups $\mathbb{Z}/2\{\eta B^k\} \subset \pi_{8k+1}(tmf)$ and $\mathbb{Z}/2\{\eta^2 B^k\} \subset \pi_{8k+2}(tmf)$ for $k \geq 0$.*
2. *The group $\mathbb{Z}/8\{\nu\} \cong \pi_3(tmf)$.*
3. *The groups $\Theta\pi_n(tmf) \subset \pi_n(tmf)$ for $n \leq 101$ and $n = 125$.*
4. *A subgroup of $\Theta\pi_n(tmf) \subset \pi_n(tmf)$ for the remaining $n \geq 102$.*

See Remark 11.84 for Mahowald's conjecture that the subgroups in case (4) are always the whole of $\Theta\pi_n(tmf)$.

PROOF. Let $d\colon S \to ko$ be the unit map representing the Adams $d$-invariant. We have inclusions

$$\ker(e) \subset \ker(d) \subset \pi_n(S)\,.$$

By Proposition 11.82 the image of $\iota$ on $\ker(e)$ is contained in $\Theta\pi_n(tmf)$, and by Proposition 11.83 this containment is an equality for $n \le 101$ and $n = 125$.

The image $\mathrm{im}(J)$ of $J\colon \pi_n(SO) \to \pi_n(S)$ gives a complementary summand in $\ker(d)$ to $\ker(e)$, for $n \ge 3$. We claim that $\iota(\mathrm{im}(J)) = 0$, except when $n = 3$. When $n = 8k - 1$ for $k \ge 1$, the image lies in Adams filtration $\ge 4k + 1 - \ell$ by Proposition 11.88, which for $k \ne 2$ is strictly larger than the maximal Adams filtration of the classes in $\pi_n(tmf)$. Since $\iota$ cannot decrease Adams filtration, it follows that $\iota(\mathrm{im}(J)) = 0$ in degrees $8k-1 \ne 15$. Furthermore, $\iota(\rho) = 0$, cf. the proof of case (15) of Theorem 11.61, which accounts for the case $k = 2$. Multiplying by $\eta$ (resp. $\eta^2$), it follows that $\iota(\mathrm{im}(J)) = 0$ in degrees $n = 8k \ge 8$ (resp. $n = 8k+1 \ge 9$). When $n = 8k + 3$ for $k \ge 0$, the image $\mathrm{im}(J)$ lies in Adams filtration $\ge 4k + 4 - \ell$. This Adams filtration of $\pi_n(tmf)$ is trivial, except for $k = 0$, so $\iota(\mathrm{im}(J)) = 0$ in degrees $8k + 3 \ge 11$. On the other hand, $\iota$ maps $\mathrm{im}(J) = \pi_3(S) = \mathbb{Z}/8\{\nu\}$ isomorphically to $\pi_3(tmf) = \mathbb{Z}/8\{\nu\}$ in degree 3.

Finally, the groups $\pi_0(S)$, $\mathbb{Z}/2\{\mu_{8k+1}\}$ and $\mathbb{Z}/2\{\eta\mu_{8k+1}\}$ for $k \ge 0$ give complementary summands to $\ker(d)$, which $\iota$ maps isomorphically to summands $\pi_0(tmf)$, $\mathbb{Z}/2\{\eta B^k\}$ and $\mathbb{Z}/2\{\eta^2 B^k\}$ in $\pi_*(tmf)$, as in the proof of Theorem 11.80. □

CHAPTER 12

# Homotopy of some finite cell *tmf*-modules

In this chapter we study the homotopy groups of $tmf/2 \simeq tmf \wedge C2$, $tmf/\eta \simeq tmf \wedge C\eta$ and $tmf/\nu \simeq tmf \wedge C\nu$, whose Adams $E_\infty$-terms were determined in Chapters 6, 7 and 8. We also study $tmf/B$ and $tmf/(B, M)$, where the latter is Anderson self-dual, as well as $tmf/(2, B) \simeq tmf \wedge M(1, 4)$ and $tmf/(2, B, M) \simeq tmf \wedge M(1, 4, 32)$, where the latter is Brown–Comenetz self-dual. Here $M(1, 4) = \mathrm{cof}(v_1^4 \colon \Sigma^8 S/2 \to S/2)$ and $M(1, 4, 32) = \mathrm{cof}(v_2^{32} \colon \Sigma^{192} M(1, 4) \to M(1, 4))$ are type 2 and 3 finite CW spectra shown to exist in [**8**, §12] and [**26**], respectively.

## 12.1. Homotopy of *tmf*/2

We study $\pi_*(tmf/2)$ using the short exact sequence

$$0 \to \pi_*(tmf)/2 \xrightarrow{i} \pi_*(tmf/2) \xrightarrow{j} {}_2\pi_{*-1}(tmf) \to 0$$

of $\pi_*(tmf)$-modules. Here ${}_2\pi_{n-1}(tmf) = \ker(2 \colon \pi_{n-1}(tmf) \to \pi_{n-1}(tmf))$. We do not fully describe the $\pi_*(tmf)$-module structure on $\pi_*(tmf/2)$, but aim to determine the 2-, $\eta$-, $\nu$-, $B$- and $M$-action on this extension. As tools we use the maps

$$E_\infty^{*,*}(tmf) \xrightarrow{i} E_\infty^{*,*}(tmf/2) \xrightarrow{j} E_\infty^{*,*-1}(tmf)$$

of $E_\infty(tmf)$-modules, calculated in Chapters 5 and 6, and our knowledge from Chapter 9 of the graded ring $\pi_*(tmf)$. We determine the hidden 2-, $\eta$- and $\nu$-extensions in $E_\infty(tmf/2)$, except for some unresolved $\eta$-extensions, and show that there are no hidden $B$- and $M$-extensions.

The $E_\infty$-term for $tmf/2$ is displayed in Figures 12.1 to 12.8. A label $i(x)$ denotes the class of an infinite cycle in the image under $i \colon E_2^{s,t}(tmf) \to E_2^{s,t}(tmf/2)$ of $x \in E_2(tmf)$. A label $\widetilde{x}$ denotes the class of an infinite cycle mapping to $x \in E_2(tmf)$ under $j \colon E_2^{s,t}(tmf/2) \to E_2^{s,t-1}(tmf)$. We omit to label the classes that are $h_0$-, $h_1$-, $h_2$- or $w_1$-multiples, and this specifies the $w_1$-action on $E_\infty(tmf/2)$.

LEMMA 12.1. *There are no hidden $B$- or $M$-power extensions in $E_\infty(tmf/2)$.*

PROOF. For each $b \in E_\infty(tmf/2)$ with $w_1 b = 0$, each class in the same topological degree but in higher filtration than $w_1 b$ is a $w_1$-multiple. Hence each such $b$ can be represented by a homotopy class $\beta$ with $B\beta = 0$. Similarly, for almost every $b \in E_\infty(tmf/2)$ with $w_1^2 b = 0$, each class in the same topological degree but in higher filtration than $w_1^2 b$ is a $w_1^2$-multiple. Hence each such $b$ can be represented by a homotopy class $\beta$ with $B^2\beta = 0$.

There is one exceptional case, namely $b = i(h_2 w_2^3)$ in bidegree $(t - s, s) = (147, 25)$ detecting $i(\nu_6)$. Here $w_1^2 b = 0$ and the class $c = w_1 \cdot \gamma w_1 w_2^2 \widetilde{\gamma}$ is not a $w_1^2$-multiple, so we must exclude the possibility of a hidden $B^2$-extension from $b$ to $c$. Each class $\beta \in \{b\}$ has the form $i(\nu_6) + \beta'$ with $\beta'$ in Adams filtration $\geq 34$. It

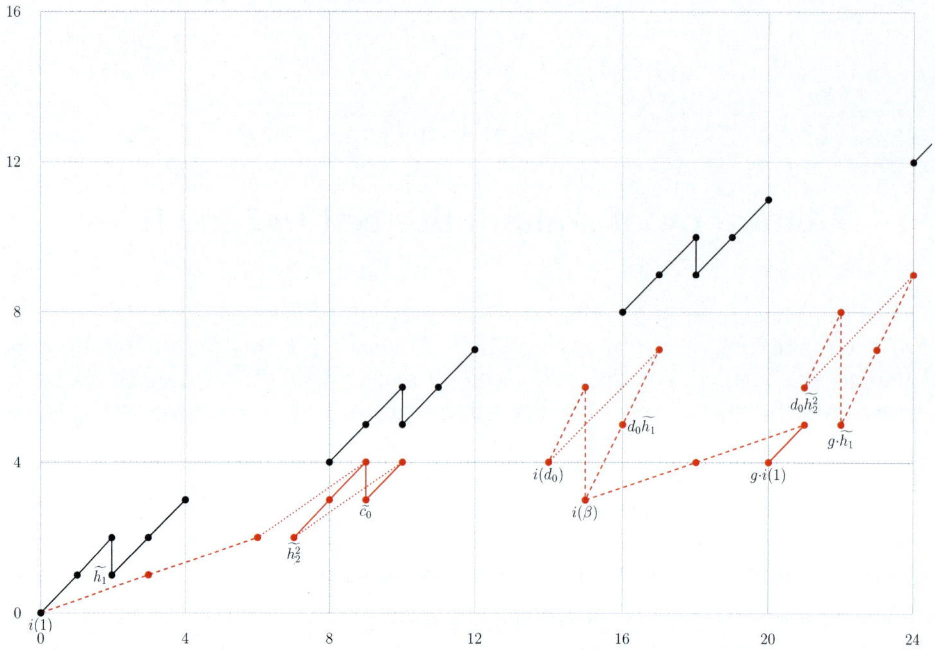


FIGURE 12.1. $E_\infty(tmf/2)$ for $0 \leq t-s \leq 24$, with all hidden 2-, $\eta$- and $\nu$-extensions

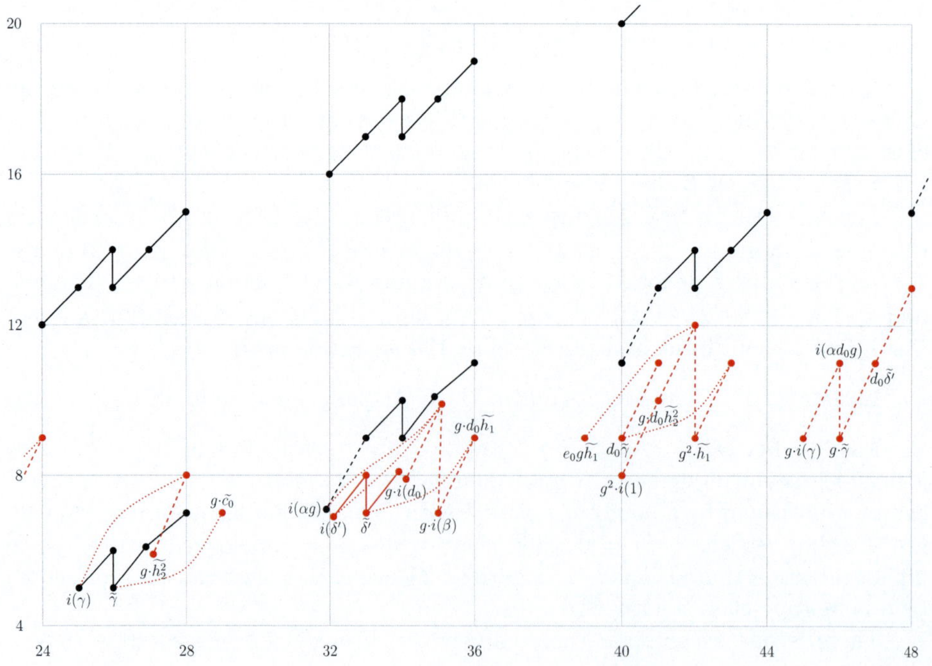


FIGURE 12.2. $E_\infty(tmf/2)$ for $24 \leq t-s \leq 48$, with all hidden 2-, $\eta$- and $\nu$-extensions

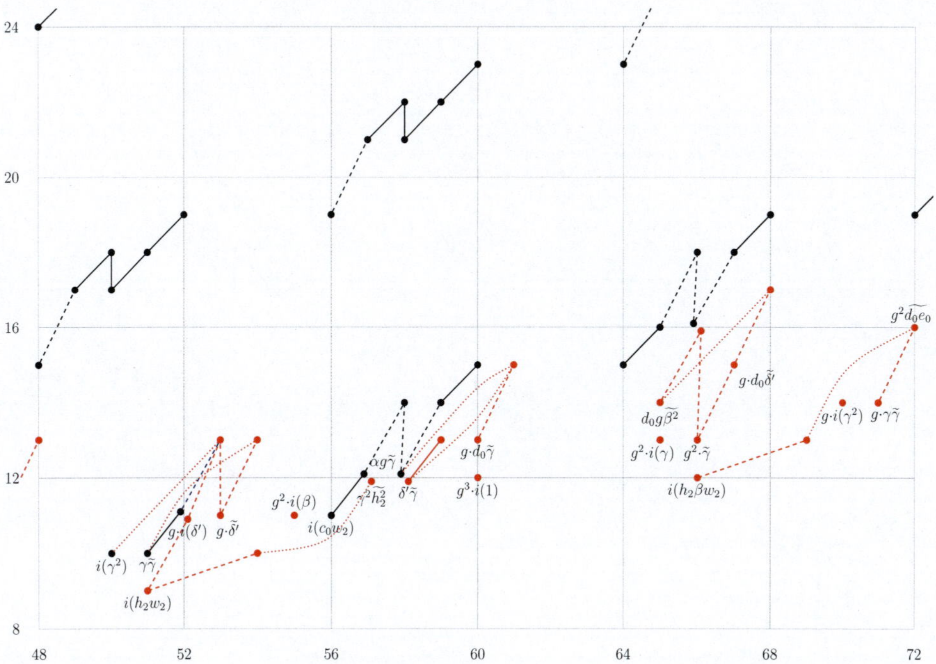


FIGURE 12.3. $E_\infty(tmf/2)$ for $48 \le t-s \le 72$, with all (potential) hidden 2-, $\eta$- and $\nu$-extensions

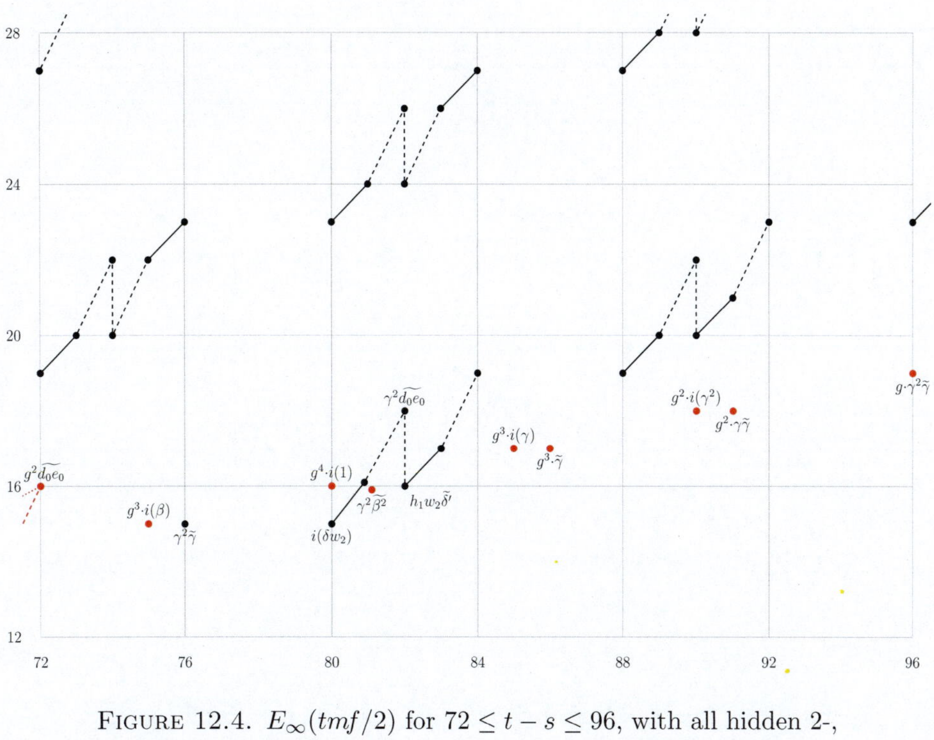


FIGURE 12.4. $E_\infty(tmf/2)$ for $72 \le t-s \le 96$, with all hidden 2-, $\eta$- and $\nu$-extensions

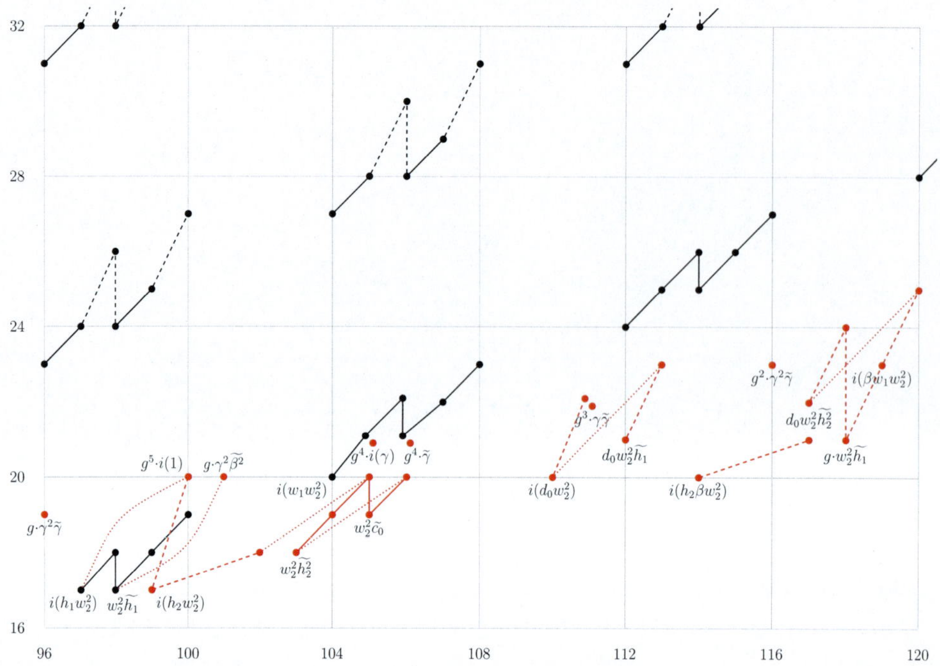


FIGURE 12.5. $E_\infty(tmf/2)$ for $96 \leq t-s \leq 120$, with all hidden 2-, $\eta$- and $\nu$-extensions

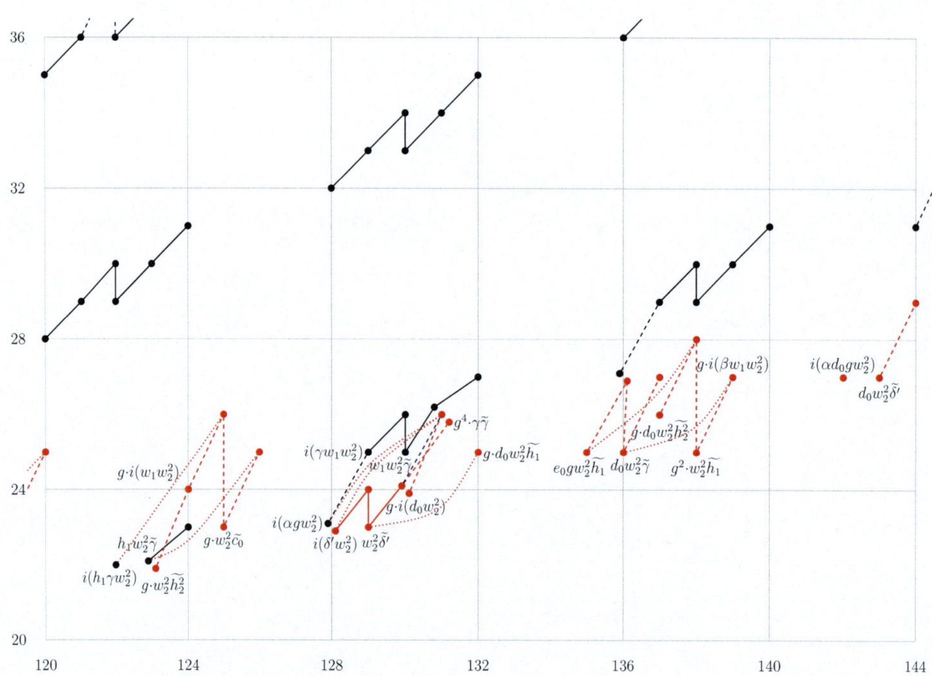


FIGURE 12.6. $E_\infty(tmf/2)$ for $120 \leq t-s \leq 144$, with all (potential) hidden 2-, $\eta$- and $\nu$-extensions

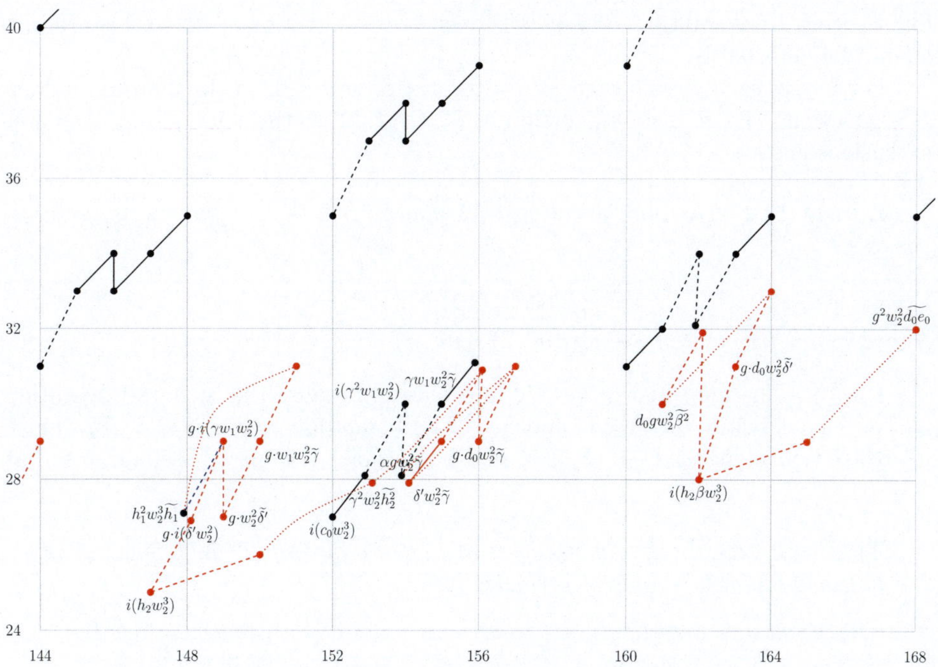


FIGURE 12.7. $E_\infty(tmf/2)$ for $144 \le t - s \le 168$, with all (potential) hidden 2-, $\eta$- and $\nu$-extensions

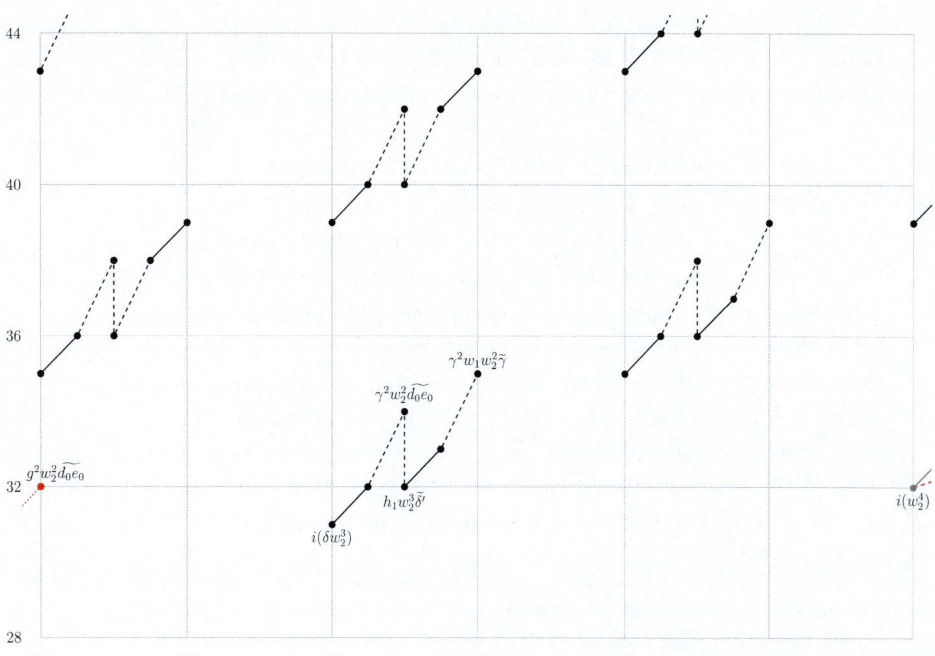


FIGURE 12.8. $E_\infty(tmf/2)$ for $168 \le t - s \le 192$, with all hidden 2-, $\eta$- and $\nu$-extensions

follows from $B^2\nu_6 = 0$ that $B^2\beta = B^2\beta'$ has Adams filtration $\geq 42$, so this product cannot be detected by $c$.

For $k \geq 3$, each class $b$ with $w_1^k b = 0$ satisfies $w_1^2 b = 0$, so there are no hidden $B^k$-extensions. There is no $w_2^4$-torsion in $E_\infty(tmf/2)$, so the claim about $M$-power extensions is clear. □

LEMMA 12.2. *The multiplication-by-2 map $2\colon S/2 \to S/2$ factors as the composite*

$$S/2 \xrightarrow{j} S^1 \xrightarrow{\eta} S \xrightarrow{i} S/2 .$$

*Hence $2 \cdot \widetilde{y} = i(\eta \cdot y)$ for $\widetilde{y} \in \pi_*(tmf/2)$ with $j(\widetilde{y}) = y$.*

PROOF. The map $2\colon S/2 \to S/2$ is essential, because the Steenrod operation $Sq^2$ acts nontrivially in the cohomology of its homotopy cofiber $S/2 \wedge S/2$. Since its restriction along $i\colon S \to S/2$ is null-homotopic, the only possibility is that $2 = i\eta j$. □

THEOREM 12.3. *In the Adams spectral sequence for $tmf/2$, the following hidden 2-extensions repeat $w_1$- and $w_2^4$-periodically:*

(58) *From $\alpha g\widetilde{\gamma}$ detecting $\widetilde{\eta_1 B_1}$ to $w_1 \cdot i(\gamma^2)$ detecting $i(\eta\eta_1 B_1) = i(\eta^2 B_2)$.*
(82) *From $h_1 w_2 \widetilde{\delta'}$ detecting $\widetilde{\eta B_3}$ to $\gamma^2 \widetilde{d_0 e_0}$ detecting $i(\eta^2 B_3)$.*
(154) *From $\alpha g w_2^2 \widetilde{\gamma}$ detecting $\widetilde{\eta_1 B_5}$ to $i(\gamma^2 w_1 w_2^2)$ detecting $i(\eta\eta_1 B_5) = i(\eta^2 B_6)$.*
(178) *From $h_1 w_2^3 \widetilde{\delta'}$ detecting $\widetilde{\eta B_7}$ to $\gamma^2 w_2^2 \widetilde{d_0 e_0}$ detecting $i(\eta^2 B_7)$.*

*The following hidden 2-extensions repeat $w_2^4$-periodically:*

(15) *From $i(\beta)$ detecting $\widetilde{\kappa}$ to $w_1 \cdot \widetilde{h_2^2}$ detecting $i(\eta\kappa)$.*
(22) *From $g \cdot \widetilde{h_1}$ detecting $\widetilde{\eta\bar{\kappa}}$ to $w_1 \cdot i(d_0)$ detecting $i(\eta^2\bar{\kappa})$.*
(35) *From $g \cdot i(\beta)$ detecting $\widetilde{\kappa\bar{\kappa}}$ to $g w_1 \cdot \widetilde{h_2^2}$ detecting $i(\eta\kappa\bar{\kappa})$.*
(42) *From $g^2 \cdot \widetilde{h_1}$ detecting $\widetilde{\eta\bar{\kappa}^2}$ to $g w_1 \cdot i(d_0)$ detecting $i(\eta^2\bar{\kappa}^2)$.*
(46) *From $g \cdot \widetilde{\gamma}$ detecting $\widetilde{\eta_1\bar{\kappa}}$ to $i(\alpha d_0 g)$ detecting $i(\eta\eta_1\bar{\kappa})$.*
(53) *From $g \cdot \widetilde{\delta'}$ detecting $\widetilde{\eta\nu_2}$ to $g w_1 \cdot i(\gamma)$ detecting $i(\eta^2\nu_2)$.*
(66) *From $g^2 \cdot \widetilde{\gamma}$ detecting $\widetilde{\eta_1\bar{\kappa}^2}$ to $w_1 \cdot \delta'\widetilde{\gamma}$ detecting $i(\eta\eta_1\bar{\kappa}^2)$.*
(118) *From $g \cdot w_2^2 \widetilde{h_1}$ detecting $\widetilde{\eta_4\bar{\kappa}}$ to $w_1 \cdot i(d_0 w_2^2)$ detecting $i(\eta\eta_4\bar{\kappa})$.*
(125) *From $g \cdot w_2^2 \widetilde{c_0}$ detecting $\widetilde{\eta\nu_5}$ to $w_1 \cdot d_0 w_2^2 \widetilde{h_2^2}$ detecting $i(\eta^2\nu_5)$.*
(136) *From $d_0 w_2^2 \widetilde{\gamma}$ detecting $\widetilde{\eta_1\kappa_4}$ to $w_1 \cdot i(\delta' w_2^2)$ detecting $i(\eta\eta_1\kappa_4)$.*
(138) *From $g^2 \cdot w_2^2 \widetilde{h_1}$ detecting $\widetilde{\nu_5\kappa}$ to $g w_1 \cdot i(d_0 w_2^2)$ detecting $i(\eta\nu_5\kappa)$.*
(149) *From $g \cdot w_2^2 \widetilde{\delta'}$ detecting $\widetilde{\eta\nu_6}$ to $g \cdot i(\gamma w_1 w_2^2)$ detecting $i(\eta^2\nu_6)$.*
(156) *From $g \cdot d_0 w_2^2 \widetilde{\gamma}$ detecting $\widetilde{\nu_6\epsilon}$ to $g w_1 \cdot i(\delta' w_2^2)$ detecting $i(\eta\nu_6\epsilon)$.*
(162) *From $i(h_2\beta w_2^3)$ detecting $\widetilde{\nu_6\kappa}$ to $w_1 \cdot \delta' w_2^2 \widetilde{\gamma}$ detecting $i(\eta\nu_6\kappa)$.*

*There are no other hidden 2-extensions in this spectral sequence.*

PROOF. By Lemma 12.1 there can be no hidden 2-extensions from $w_1$-power torsion classes to $w_1$-periodic classes. Furthermore, there are no hidden 2-extensions on classes detecting elements of the form $i(y)$ with $y \in \pi_*(tmf)$.

By Lemma 12.2 there is a hidden 2-extension from $b$ to $c$ if $b$ detects $\widetilde{y}$, $c$ detects $i(\eta \cdot y) \neq 0$, and there is no shorter 2-extension to $c$. All the nonzero hidden 2-extensions for $tmf/2$ arise in this way. □

To determine the action of $\eta$ on $\pi_*(tmf/2)$ we sometimes compare the two short exact sequences

$$0 \to \pi_n(tmf/2)/\eta \xrightarrow{i} \pi_n(tmf/(2,\eta)) \xrightarrow{j} {}_\eta\pi_{n-2}(tmf/2) \to 0$$

$$0 \to \pi_n(tmf/\eta)/2 \xrightarrow{i} \pi_n(tmf/(2,\eta)) \xrightarrow{j} {}_2\pi_{n-1}(tmf/\eta) \to 0\,,$$

using our knowledge of $E_\infty(tmf/\eta)$ to obtain information about $\pi_n(tmf/(2,\eta))$. Here $tmf/(2,\eta) = (tmf/2)/\eta \simeq (tmf/\eta)/2$, $\pi_n(tmf/2)/\eta = \mathrm{cok}(\eta\colon \pi_{n-1}(tmf/2) \to \pi_n(tmf/2))$ and ${}_\eta\pi_{n-2}(tmf/2) = \ker(\eta\colon \pi_{n-2}(tmf/2) \to \pi_{n-1}(tmf/2))$. Logically, Theorem 12.9 precedes this result.

THEOREM 12.4. *In the Adams spectral sequence for $tmf/2$, the following hidden $\eta$-extensions repeat $w_1$- and $w_2^4$-periodically:*

(32) *From $i(\alpha g)$ detecting $i(B_1)$ to $w_1 \cdot i(\gamma)$ detecting $i(\eta B_1)$.*
(57) *From $h_1 \cdot i(c_0 w_2)$ detecting $i(\eta B_2)$ to $w_1 \cdot i(\gamma^2)$ detecting $i(\eta^2 B_2)$.*
(58) *From $\alpha g\widetilde{\gamma}$ detecting $\widetilde{\eta_1 B_1}$ to $w_1 \cdot \gamma\widetilde{\gamma}$ detecting a lift $\widetilde{\eta^2 B_2}$.*
(81) *From $h_1 \cdot i(\delta w_2)$ detecting $i(\eta B_3)$ to $\gamma^2\widetilde{d_0 e_0}$ detecting $i(\eta^2 B_3)$.*
(83) *From $h_1 \cdot h_1 w_2\widetilde{\delta'}$ detecting $\widetilde{\eta^2 B_3}$ to $w_1 \cdot \gamma^2\widetilde{\gamma}$ detecting $i(C_3)$.*
(128) *From $i(\alpha g w_2^2)$ detecting $i(B_5)$ to $i(\gamma w_1 w_2^2)$ detecting $i(\eta B_5)$.*
(153) *From $h_1 \cdot i(c_0 w_2^3)$ detecting $i(\eta B_6)$ to $i(\gamma^2 w_1 w_2^2)$ detecting $i(\eta^2 B_6)$.*
(154) *From $\alpha g w_2^2\widetilde{\gamma}$ detecting $\widetilde{\eta_1 B_5}$ to $\gamma w_1 w_2^2\widetilde{\gamma}$ detecting a lift $\widetilde{\eta^2 B_6}$.*
(177) *From $h_1 \cdot i(\delta w_2^3)$ detecting $i(\eta B_7)$ to $\gamma^2 w_2^2\widetilde{d_0 e_0}$ detecting $i(\eta^2 B_7)$.*
(179) *From $h_1 \cdot h_1 w_2^3\widetilde{\delta'}$ detecting $\widetilde{\eta^2 B_7}$ to $\gamma^2 w_1 w_2^2\widetilde{\gamma}$ detecting $i(C_7)$.*

*The following hidden $\eta$-extensions repeat $w_2^4$-periodically:*

(14) *From $i(d_0)$ detecting $i(\kappa)$ to $w_1 \cdot \widetilde{h_2^2}$ detecting $i(\eta\kappa)$.*
(15) *From $i(\beta)$ detecting $\widetilde{\kappa}$ to $d_0\widetilde{h_1}$ detecting $\widetilde{\eta\kappa}$.*
(16) *From $d_0\widetilde{h_1}$ detecting $\widetilde{\eta\kappa}$ to $w_1 \cdot \widetilde{c_0}$ detecting $i(\nu\kappa)$.*
(21) *From $d_0\widetilde{h_2^2}$ detecting a lift $\widetilde{4\bar\kappa}$ to $w_1 \cdot i(d_0)$ detecting $i(\eta^2\bar\kappa)$.*
(22) *From $g \cdot \widetilde{h_1}$ detecting $\widetilde{\eta\bar\kappa}$ to $w_1 \cdot i(\beta)$ detecting $\widetilde{\eta^2\bar\kappa}$.*
(23) *From $w_1 \cdot i(\beta)$ detecting $\widetilde{\eta^2\bar\kappa}$ to $w_1 \cdot d_0\widetilde{h_1}$ detecting $i(D_1)$.*
(27) *From $g \cdot \widetilde{h_2^2}$ detecting $i(\nu_1)$ to $g w_1 \cdot i(1)$ detecting $i(\eta\nu_1)$.*
(34a) *From $g \cdot i(d_0)$ detecting $i(\kappa\bar\kappa)$ to $g w_1 \cdot \widetilde{h_2^2}$ detecting $i(\eta\kappa\bar\kappa)$.*
(35) *From $g \cdot i(\beta)$ detecting $\widetilde{\kappa\bar\kappa}$ to $g \cdot d_0\widetilde{h_1}$ detecting $\widetilde{\eta\kappa\bar\kappa}$.*
(40a) *From $g^2 \cdot i(1)$ detecting $i(\bar\kappa^2)$ to $g \cdot d_0\widetilde{h_2^2}$ detecting $i(\eta\bar\kappa^2)$.*
(40b) *From $d_0\widetilde{\gamma}$ detecting a lift $\widetilde{\eta_1\kappa}$ to $w_1 \cdot \widetilde{\delta'}$ detecting a lift $\widetilde{2\bar\kappa^2}$.*
(41) *From $g \cdot d_0\widetilde{h_2^2}$ detecting $i(\eta\bar\kappa^2)$ to $g w_1 \cdot i(d_0)$ detecting $i(\eta^2\bar\kappa^2)$.*
(42) *From $g^2 \cdot \widetilde{h_1}$ detecting $\widetilde{\eta\bar\kappa^2}$ to $g w_1 \cdot i(\beta)$ detecting $\widetilde{\eta^2\bar\kappa^2}$.*
(45) *From $g \cdot i(\gamma)$ detecting $i(\eta_1\bar\kappa)$ to $i(\alpha d_0 g)$ detecting $i(\eta\eta_1\bar\kappa)$.*
(46) *From $g \cdot \widetilde{\gamma}$ detecting $\widetilde{\eta_1\bar\kappa}$ to $d_0\widetilde{\delta'}$ detecting $\widetilde{\eta\eta_1\bar\kappa}$.*
(47) *From $d_0\widetilde{\delta'}$ detecting $\widetilde{\eta\eta_1\bar\kappa}$ to $w_1 \cdot d_0\widetilde{\gamma}$ detecting $i(D_2)$.*
(51) *From $i(h_2 w_2)$ detecting $i(\nu_2)$ to $g \cdot i(\delta')$ detecting $i(\eta\nu_2)$.*
(52a) *From $g \cdot i(\delta')$ detecting $i(\eta\nu_2)$ to $g w_1 \cdot i(\gamma)$ detecting $i(\eta^2\nu_2)$.*
(53) *From $g \cdot \widetilde{\delta'}$ detecting $\widetilde{\eta\nu_2}$ to $g w_1 \cdot \widetilde{\gamma}$ detecting a lift $\widetilde{\eta^2\nu_2}$.*
(60) *From $g \cdot d_0\widetilde{\gamma}$ detecting a lift $\widetilde{\nu_2}\epsilon$ to $g w_1 \cdot \widetilde{\delta'}$ detecting $\widetilde{2\bar\kappa^3}$.*

(65) *From* $d_0g\widetilde{\beta^2}$ *detecting* $i(\nu_2\kappa)$ *to* $w_1\cdot\delta'\widetilde{\gamma}$ *detecting* $i(\eta\nu_2\kappa)$.
(66) *From* $g^2\cdot\widetilde{\gamma}$ *detecting* $\widetilde{\eta_1\bar\kappa^2}$ *to* $g\cdot d_0\widetilde{\delta'}$ *detecting* $\widetilde{\eta\eta_1\bar\kappa^2}$.
(67) *From* $g\cdot d_0\widetilde{\delta'}$ *detecting* $\widetilde{\eta\eta_1\bar\kappa^2}$ *to* $gw_1\cdot d_0\widetilde{\gamma}$ *detecting* $i(\bar\kappa D_2)$.
(71) *From* $g\cdot\gamma\widetilde{\gamma}$ *detecting* $\widetilde{\eta_1^2\bar\kappa}$ *to* $g^2\widetilde{d_0e_0}$ *detecting* $i(D_3)$.
(99) *From* $i(h_2w_2^2)$ *detecting* $i(\nu_4)$ *to* $g^5\cdot i(1)$ *detecting* $i(\eta\nu_4)$.
(110) *From* $i(d_0w_2^2)$ *detecting* $i(\kappa_4)$ *to* $w_1\cdot w_2^2\widetilde{h_2^2}$ *detecting* $i(\eta\kappa_4)$.
(112) *From* $d_0w_2^2\widetilde{h_1}$ *detecting* $\widetilde{\eta\kappa_4}$ *to* $w_1\cdot w_2^2\widetilde{c_0}$ *detecting* $i(\nu\kappa_4)$.
(117) *From* $d_0w_2^2\widetilde{h_2^2}$ *detecting a lift* $\widetilde{2\bar\kappa D_4}$ *to* $w_1\cdot i(d_0w_2^2)$ *detecting* $i(\eta\eta_4\bar\kappa)$.
(118) *From* $g\cdot w_2^2\widetilde{h_1}$ *detecting* $\widetilde{\eta_4\bar\kappa}$ *to* $i(\beta w_1w_2^2)$ *detecting* $\widetilde{\eta\eta_4\bar\kappa}$.
(119) *From* $i(\beta w_1w_2^2)$ *detecting* $\widetilde{\eta\eta_4\bar\kappa}$ *to* $w_1\cdot d_0w_2^2\widetilde{h_1}$ *detecting* $i(D_5)$.
(123) *From* $g\cdot w_2^2\widetilde{h_2^2}$ *detecting* $i(\nu_5)$ *to* $g\cdot i(w_1w_2^2)$ *detecting* $i(\eta\nu_5)$.
(124) *From* $g\cdot i(w_1w_2^2)$ *detecting* $i(\eta\nu_5)$ *to* $w_1\cdot d_0w_2^2\widetilde{h_2^2}$ *detecting* $i(\eta^2\nu_5)$.
(125) *From* $g\cdot w_2^2\widetilde{c_0}$ *detecting* $\widetilde{\eta\nu_5}$ *to* $gw_1\cdot w_2^2\widetilde{h_1}$ *detecting* $\widetilde{\eta^2\nu_5}$.
(130a) *From* $g\cdot i(d_0w_2^2)$ *detecting* $i(\kappa_4\bar\kappa)$ *to* $gw_1\cdot w_2^2\widetilde{h_2^2}$ *detecting* $i(\eta\kappa_4\bar\kappa)$.
(135) *From* $e_0gw_2^2\widetilde{h_1}$ *detecting* $i(\eta_1\kappa_4)$ *to* $w_1\cdot i(\delta' w_2^2)$ *detecting* $i(\eta\eta_1\kappa_4)$.
(136) *From* $d_0w_2^2\widetilde{\gamma}$ *detecting* $\widetilde{\eta_1\kappa_4}$ *to* $w_1\cdot w_2^2\widetilde{\delta'}$ *detecting a lift* $\widetilde{\eta\eta_1\kappa_4}$.
(137) *From* $g\cdot d_0w_2^2\widetilde{h_2^2}$ *detecting* $i(\nu_5\kappa)$ *to* $gw_1\cdot i(d_0w_2^2)$ *detecting* $i(\eta\nu_5\kappa)$.
(138) *From* $g^2\cdot w_2^2\widetilde{h_1}$ *detecting* $\widetilde{\nu_5\kappa}$ *to* $g\cdot i(\beta w_1w_2^2)$ *detecting* $\widetilde{\eta\nu_5\kappa}$.
(143) *From* $d_0w_2^2\widetilde{\delta'}$ *detecting* $\widetilde{\epsilon_5\kappa}$ *to* $w_1\cdot d_0w_2^2\widetilde{\gamma}$ *detecting* $i(D_6)$.
(147) *From* $i(h_2w_2^3)$ *detecting* $i(\nu_6)$ *to* $g\cdot i(\delta' w_2^2)$ *detecting* $i(\eta\nu_6)$.
(148a) *From* $g\cdot i(\delta' w_2^2)$ *detecting* $i(\eta\nu_6)$ *to* $g\cdot i(\gamma w_1w_2^2)$ *detecting* $i(\eta^2\nu_6)$.
(149) *From* $g\cdot w_2^2\widetilde{\delta'}$ *detecting* $\widetilde{\eta\nu_6}$ *to* $g\cdot w_1w_2^2\widetilde{\gamma}$ *detecting a lift* $\widetilde{\eta^2\nu_6}$.
(150) *From* $g\cdot w_1w_2^2\widetilde{\gamma}$ *detecting a lift* $\widetilde{\eta^2\nu_6}$ *to* $w_1\cdot d_0w_2^2\widetilde{\delta'}$ *detecting* $\widetilde{4\nu\nu_6}$.
(155) *From* $h_1\cdot\delta' w_2^2\widetilde{\gamma}$ *detecting* $i(\nu_6\epsilon)$ *to* $gw_1\cdot i(\delta' w_2^2)$ *detecting* $i(\eta\nu_6\epsilon)$.
(156) *From* $g\cdot d_0w_2^2\widetilde{\gamma}$ *detecting* $\widetilde{\nu_6\epsilon}$ *to* $gw_1\cdot w_2^2\widetilde{\delta'}$ *detecting* $\widetilde{\eta\nu_6\epsilon}$.
(161) *From* $d_0gw_2^2\widetilde{\beta^2}$ *detecting* $i(\nu_6\kappa)$ *to* $w_1\cdot\delta' w_2^2\widetilde{\gamma}$ *detecting* $i(\eta\nu_6\kappa)$.
(162) *From* $i(h_2\beta w_2^3)$ *detecting* $\widetilde{\nu_6\kappa}$ *to* $g\cdot d_0w_2^2\widetilde{\delta'}$ *detecting* $\widetilde{\eta\nu_6\kappa}$.
(163) *From* $g\cdot d_0w_2^2\widetilde{\delta'}$ *detecting* $\widetilde{\eta\nu_6\kappa}$ *to* $gw_1\cdot d_0w_2^2\widetilde{\gamma}$ *detecting* $i(\bar\kappa D_6)$.

*The following potential hidden* $\eta$*-extensions repeat* $w_2^4$*-periodically, but remain to be precisely determined.*

(34b) *From* $h_1\widetilde{\delta'}$ *detecting a lift* $\widetilde{\eta\epsilon_1}$ *to zero, or to* $gw_1\cdot\widetilde{h_2^2}$ *detecting* $i(\eta\kappa\bar\kappa)$. *(We prove in Lemma* 12.26 *that this* $\eta$*-multiple is zero.)*
(52b) *From* $h_1\cdot\gamma\widetilde{\gamma}$ *detecting a lift* $\eta\widetilde{\eta_1^2}$ *to zero, or to* $gw_1\cdot i(\gamma)$ *detecting* $i(\eta^2\nu_2)$.
(130b) *From* $h_1\cdot w_2^2\widetilde{\delta'}$ *detecting a lift* $\widetilde{\eta\epsilon_5}$ *to* $g^4\cdot\gamma\widetilde{\gamma}$ *or to* $g^4\cdot\gamma\widetilde{\gamma}+gw_1\cdot w_2^2\widetilde{h_2^2}$, *detecting a lift* $\widetilde{2\kappa_4\bar\kappa}$.
(148b) *From* $h_1^2w_2^3\widetilde{h_1}$ *detecting a lift* $\widetilde{4\nu_6}$ *to zero, or to* $g\cdot i(\gamma w_1w_2^2)$ *detecting* $i(\eta^2\nu_6)$.

*There are no other hidden* $\eta$*-extensions in this spectral sequence.*

PROOF. By Lemma 12.1 there can be no hidden $\eta$-extensions from $w_1$-power torsion classes to $w_1$-periodic classes.

By naturality with respect to $i$ there is a hidden $\eta$-extension from $b$ to $c$ if $b$ detects $i(y)$, $h_1b=0$, $c$ detects $i(\eta y)\neq 0$, and there is no shorter $\eta$-extension to $c$.

By naturality with respect to $j$ there is also a hidden $\eta$-extension from $b$ to $c$ if $b$ detects a lift $\widetilde{y}$ of $y$, $h_1 b = 0$, $c$ detects $\eta\widetilde{y} \neq 0$, and there is no shorter $\eta$-extension to $c$.

(16) From $E_\infty(tmf/\eta)$ we see that $\pi_{17}(tmf/(2,\eta))$ has order 2. Since the $\eta$-torsion subgroup ${}_\eta\pi_{15}(tmf/2)$ contains $i(\eta\kappa)$ detected by $w_1 \cdot \widetilde{h_2^2}$, it follows that $\pi_{17}(tmf/2)/\eta = 0$. Hence $i(\nu\kappa)$ detected by $w_1 \cdot \widetilde{c_0}$ must be $\eta$ times some class in $\pi_{16}(tmf/2)$, and $\widetilde{\eta\kappa}$ detected by $d_0\widetilde{h_1}$ is the only possibility.

(21) We know that $\eta$ times $\widetilde{\nu^2}$ detected by $\widetilde{h_2^2}$ is $i(\epsilon)$ detected by $i(c_0)$. Multiplying by $\kappa$ we find that $\eta$ times $\kappa\widetilde{\nu^2} = \widetilde{4\bar\kappa}$ detected by $d_0\widetilde{h_2^2}$ is $i(\epsilon\kappa) = i(\eta^2\bar\kappa)$ detected by $w_1 \cdot i(d_0)$.

(23) We multiply the hidden $\eta$-extension from case (15) by $B$ to obtain this hidden $\eta$-extension.

(65) Since $i(\nu_2)$ is detected by $i(h_2 w_2)$, and $d_0 \cdot i(h_2 w_2) = 0$ in $E_2(tmf/2)$, we see that $i(\nu_2\kappa)$ must be detected in Adams filtration $\geq 14$, i.e., by $d_0 g\widetilde{\beta^2}$. Hence $i(\eta\nu_2\kappa)$ must be detected in Adams filtration $\geq 15$, i.e., by $w_1 \cdot \delta'\widetilde{\gamma}$.

(67) From $E_\infty(tmf/\eta)$ we see that $\pi_{69}(tmf/(2,\eta))$ has order 2. Since $\pi_{69}(tmf/2)$ contains $\bar\kappa D_2$ detected by $h_2 \cdot i(h_2\beta w_2)$, which cannot be an $\eta$-multiple, it follows that $\eta$ acts injectively on $\pi_{67}(tmf/2)$. Hence $\eta \cdot \widetilde{\eta\eta_1\bar\kappa^2}$ is nonzero, and must be detected by $gw_1 \cdot d_0\widetilde{\gamma}$.

(47) We divide the hidden $\eta$-extension from case (67) by $\bar\kappa$ to obtain this hidden $\eta$-extension.

(83) From $E_\infty(tmf/\eta)$ we see that $\pi_{85}(tmf/(2,\eta))$ has order 2. Since $\pi_{85}(tmf/2)$ contains $i(\eta_1\bar\kappa^3)$ detected by $g^3 \cdot i(\gamma)$, which cannot be an $\eta$-multiple, we must have that $\eta$ acts injectively on $\pi_{83}(tmf/2)$. Hence $\eta \cdot \widetilde{\eta^2 B_3}$ is nonzero, and must be detected by $w_1 \cdot \gamma^2\widetilde{\gamma}$.

(110) To see that $i(\eta\kappa_4)$ is detected by $w_1 \cdot w_2^2\widetilde{h_2^2}$, we note that $j$ maps $w_1 \cdot w_2^2\widetilde{h_2^2}$ to $h_2^2 w_1 w_2^2 = 0$ in $E_\infty(tmf)$, while it maps $g^3 \cdot \gamma\widetilde{\gamma}$ to $\gamma^2 g^3 \neq 0$. The conclusion follows, since $ji = 0$.

(117) We know that $\eta$ times $\widetilde{\nu\nu_4}$ detected by $w_2^2\widetilde{h_2^2}$ is $i(\epsilon_4)$ detected by $i(c_0 w_2^2)$. Multiplying by $\kappa$ we find that $\eta$ times $\kappa\widetilde{\nu\nu_4} = \widetilde{2\bar\kappa D_4}$ detected by $d_0 w_2^2\widetilde{h_2^2}$ is $i(\epsilon_4\kappa) = i(\eta\eta_4\bar\kappa)$ detected by $w_1 \cdot i(d_0 w_2^2)$.

(130a) The product $\eta \cdot \kappa_4\bar\kappa$ is detected by $w_1 \cdot \alpha\beta w_2^2$, which maps by $i$ to $gw_1 \cdot w_2^2\widetilde{h_2^2}$. Hence this is the class detecting $i(\eta\kappa_4\bar\kappa)$.

(163) From $E_\infty(tmf/\eta)$ we see that $\pi_{165}(tmf/(2,\eta))$ has order 2. Furthermore, $\pi_{165}(tmf/2)$ contains $\bar\kappa D_6$ detected by $h_2 \cdot i(h_2\beta w_2^3)$, which cannot be an $\eta$-multiple. It follows that $\eta$ acts injectively on $\pi_{163}(tmf/2)$. Hence $\eta \cdot \widetilde{\eta\nu_6\kappa}$ is nonzero, and must be detected by $gw_1 \cdot d_0 w_2^2\widetilde{\gamma}$.

(143) We divide the hidden $\eta$-extension from case (163) by $\bar\kappa$ to obtain this hidden $\eta$-extension.

(179) From $E_\infty(tmf/\eta)$ we see that $\pi_{181}(tmf/(2,\eta))$ is trivial, which implies that $\eta$ acts injectively on $\pi_{179}(tmf/2)$. Hence $\eta \cdot \widetilde{\eta^2 B_7}$ must be detected by $\gamma^2 w_1 w_2^2\widetilde{\gamma}$, which also detects $i(C_7)$.

For the next three cases, we will make use of the knowledge established in Theorem 12.9 of the hidden 2-extensions in degrees 72, 112 and 120 in the Adams spectral sequence for $tmf/\eta$. The proofs of these 2-extensions do not rely on the results of the present section, so there is no circularity.

(71) From $E_\infty(tmf/\eta)$ and case (32) of Theorem 12.9, we see that ${}_2\pi_{72}(tmf/\eta) = 0$ and $\pi_{73}(tmf/\eta)/2 = 0$, so that $\pi_{73}(tmf/(2,\eta)) = 0$. Hence ${}_\eta\pi_{71}(tmf/2) = 0$, so $\eta$ times the class $\widetilde{\eta_1^2\bar\kappa}$ detected by $g \cdot \gamma\widetilde{\gamma}$ must be nonzero. Since it is a $B$-torsion class in $\pi_{72}(tmf/2)$, it can only be detected by the $w_1$-torsion class $g^2\widetilde{d_0e_0}$, and be equal to $i(D_3)$, which establishes the asserted hidden $\eta$-extension.

(112) From $E_\infty(tmf/\eta)$ and cases (32) and (80) of Theorem 12.9, we see that ${}_2\pi_{112}(tmf/\eta) = \mathbb{Z}/2$ and $\pi_{113}(tmf/\eta)/2 = \mathbb{Z}/2$, so $\pi_{113}(tmf/(2,\eta))$ has order $2^2 = 4$. Since ${}_\eta\pi_{111}(tmf/2) = (\mathbb{Z}/2)^2$, it follows that $\pi_{113}(tmf/2)/\eta = 0$. In particular, $i(\nu\kappa_4)$ detected by $w_1 \cdot w_2^2\widetilde{c_0}$ must be an $\eta$-multiple, and the only possible source of this $\eta$-extension is $d_0\widetilde{w_2^2h_1}$, detecting $\widetilde{\eta\kappa_4}$.

(119) From $E_\infty(tmf/\eta)$ and cases (32) and (80) of Theorem 12.9, we see that ${}_2\pi_{120}(tmf/\eta)$ and $\pi_{121}(tmf/\eta)/2$ are trivial, so that $\pi_{121}(tmf/(2,\eta)) = 0$. Hence ${}_\eta\pi_{119}(tmf/2) = 0$. In particular, $\eta$ times $\widetilde{\eta\eta_4\bar\kappa}$ must be nonzero, and the only possible value is $i(D_5)$ detected by $w_1 \cdot d_0\widetilde{w_2^2h_1}$. □

To determine the action of $\nu$ on $\pi_*(tmf/2)$ we sometimes compare the two short exact sequences

$$0 \to \pi_n(tmf/2)/\nu \xrightarrow{i} \pi_n(tmf/(2,\nu)) \xrightarrow{j} {}_\nu\pi_{n-4}(tmf/2) \to 0$$

$$0 \to \pi_n(tmf/\nu)/2 \xrightarrow{i} \pi_n(tmf/(2,\nu)) \xrightarrow{j} {}_2\pi_{n-1}(tmf/\nu) \to 0,$$

using our knowledge of $E_\infty(tmf/\nu)$ to obtain information about $\pi_n(tmf/(2,\nu))$. Here $tmf/(2,\nu) = (tmf/2)/\nu \simeq (tmf/\nu)/2$, $\pi_n(tmf/2)/\nu = \mathrm{cok}(\nu\colon \pi_{n-3}(tmf/2) \to \pi_n(tmf/2))$ and ${}_\nu\pi_{n-4}(tmf/2) = \ker(\nu\colon \pi_{n-4}(tmf/2) \to \pi_{n-1}(tmf/2))$. Logically, Theorem 12.15 precedes this result.

THEOREM 12.5. *In the Adams spectral sequence for tmf/2, the following hidden $\nu$-extensions repeat $w_2^4$-periodically:*

(6) *From $h_2^2 \cdot i(1)$ detecting $i(\nu^2)$ to $h_0 \cdot \widetilde{c_0}$ detecting $i(\eta\epsilon)$.*
(7) *From $\widetilde{h_2^2}$ detecting $\widetilde{\nu^2}$ to $h_1 \cdot \widetilde{c_0}$ detecting $\widetilde{\eta\epsilon}$.*
(14) *From $i(d_0)$ detecting $i(\kappa)$ to $w_1 \cdot \widetilde{c_0}$ detecting $i(\nu\kappa)$.*
(21) *From $d_0\widetilde{h_2^2}$ detecting a lift $\widetilde{4\bar\kappa}$ to $w_1 \cdot d_0\widetilde{h_1}$ detecting $i(D_1)$.*
(25) *From $i(\gamma)$ detecting $i(\eta_1)$ to $gw_1 \cdot i(1)$ detecting $i(\eta\nu_1)$.*
(26) *From $\widetilde{\gamma}$ detecting $\widetilde{\eta_1}$ to $g \cdot \widetilde{c_0}$ detecting $\widetilde{\eta\nu_1}$.*
(32a) *From $i(\delta')$ detecting $i(\epsilon_1)$ to $gw_1 \cdot h_2^2$ detecting $i(\eta\kappa\bar\kappa)$.*
(32b) *From $i(\alpha g)$ detecting $i(B_1)$ to $gw_1 \cdot \widetilde{h_2^2}$ detecting $i(\eta\kappa\bar\kappa)$.*
(33) *From $\widetilde{\delta'}$ detecting $\widetilde{\epsilon_1}$ to $g \cdot d_0\widetilde{h_1}$ detecting $\widetilde{\eta\kappa\bar\kappa}$.*
(39) *From $e_0g\widetilde{h_1}$ detecting $i(\eta_1\kappa)$ to $gw_1 \cdot i(d_0)$ detecting $i(\eta^2\bar\kappa^2)$.*
(40) *From $d_0\widetilde{\gamma}$ detecting a lift $\widetilde{\eta_1\kappa}$ to $gw_1 \cdot i(\beta)$ detecting $\eta^2\bar\kappa^2$.*
(50) *From $i(\gamma^2)$ detecting $i(\eta_1^2)$ to $gw_1 \cdot i(\gamma)$ detecting $i(\eta^2\nu_2)$.*
(51) *From $\gamma\widetilde{\gamma}$ detecting a lift $\widetilde{\eta_1^2}$ to $gw_1 \cdot \widetilde{\gamma}$ detecting a lift $\widetilde{\eta^2\nu_2}$.*
(54) *From $h_2 \cdot i(h_2w_2)$ detecting $i(\nu\nu_2)$ to $\gamma^2\widetilde{h_2^2}$ detecting $i(\nu^2\nu_2)$.*
(58a) *From $\delta'\widetilde{\gamma}$ detecting $\widetilde{\nu^2\nu_2}$ to $gw_1 \cdot \widetilde{\delta'}$ detecting $\widetilde{2\bar\kappa^3}$.*
(58b) *From $\alpha g\widetilde{\gamma}$ detecting $\widetilde{\eta_1B_1}$ to $gw_1 \cdot \widetilde{\delta'}$ detecting $\widetilde{2\bar\kappa^3}$.*
(65) *From $d_0g\beta^2$ detecting $i(\nu_2\kappa)$ to $gw_1 \cdot d_0\widetilde{\gamma}$ detecting $i(\bar\kappa D_2)$.*
(69) *From $h_2 \cdot i(h_2\beta w_2)$ detecting $\widetilde{\bar\kappa D_2}$ to $g^2\widetilde{d_0e_0}$ detecting $i(D_3)$.*

(97) *From* $i(h_1 w_2^2)$ *detecting* $i(\eta_4)$ *to* $g^5 \cdot i(1)$ *detecting* $i(\eta\nu_4)$.
(98) *From* $\widetilde{w_2^2 h_1}$ *detecting* $\widetilde{\eta_4}$ *to* $g \cdot \gamma^2\beta^2$ *detecting* $\widetilde{\eta\nu_4}$.
(102) *From* $h_2 \cdot i(h_2 w_2^2)$ *detecting* $i(\nu\nu_4)$ *to* $h_0 \cdot w_2^2\widetilde{c_0}$ *detecting* $i(\eta\epsilon_4)$.
(103) *From* $\widetilde{w_2^2 h_2^2}$ *detecting* $\widetilde{\nu\nu_4}$ *to* $h_1 \cdot w_2^2\widetilde{c_0}$ *detecting* $\widetilde{\eta\epsilon_4}$.
(110) *From* $i(d_0 w_2^2)$ *detecting* $i(\kappa_4)$ *to* $w_1 \cdot w_2^2\widetilde{c_0}$ *detecting* $i(\nu\kappa_4)$.
(117) *From* $d_0\widetilde{w_2^2 h_2^2}$ *detecting a lift* $2\bar\kappa D_4$ *to* $w_1 \cdot d_0\widetilde{w_2^2 h_1}$ *detecting* $i(D_5)$.
(122) *From* $i(h_1\gamma w_2^2)$ *detecting* $i(\eta_1\eta_4)$ *to* $w_1 \cdot d_0\widetilde{w_2^2 h_2^2}$ *detecting* $i(\eta^2\nu_5)$.
(123) *From* $h_1\widetilde{w_2^2\gamma}$ *detecting* $\widetilde{\eta_1\eta_4}$ *to* $gw_1 \cdot \widetilde{w_2^2 h_1}$ *detecting* $\eta^2\nu_5$.
(128a) *From* $i(\delta' w_2^2)$ *detecting* $i(\epsilon_5)$ *to* $gw_1 \cdot \widetilde{w_2^2 h_2^2}$ *detecting* $i(\eta\kappa_4\bar\kappa)$.
(128b) *From* $i(\alpha g w_2^2)$ *detecting* $i(B_5)$ *to* $gw_1 \cdot \widetilde{w_2^2 h_2^2}$ *detecting* $i(\eta\kappa_4\bar\kappa)$.
(129) *From* $\widetilde{w_2^2\delta'}$ *detecting* $\widetilde{\epsilon_5}$ *to* $g \cdot d_0\widetilde{w_2^2 h_1}$ *detecting* $\widetilde{\eta\kappa_4\bar\kappa}$.
(135) *From* $e_0 g\widetilde{w_2^2 h_1}$ *detecting* $i(\eta_1\kappa_4)$ *to* $gw_1 \cdot i(d_0 w_2^2)$ *detecting* $i(\eta\nu_5\kappa)$.
(136) *From* $d_0\widetilde{w_2^2\gamma}$ *detecting* $\widetilde{\eta_1\kappa_4}$ *to* $g \cdot i(\beta w_1 w_2^2)$ *detecting* $\widetilde{\eta\nu_5}\kappa$.
(148) *From* $h_1^2\widetilde{w_2^3 h_1}$ *detecting* $\widetilde{4\nu_6}$ *to* $w_1 \cdot d_0\widetilde{w_2^2\delta'}$ *detecting* $\widetilde{4\nu\nu_6}$.
(150) *From* $h_2 \cdot i(h_2 w_2^3)$ *detecting* $i(\nu\nu_6)$ *to* $\gamma^2\widetilde{w_2^2 h_2^2}$ *detecting* $i(\nu^2\nu_6)$.
(153) *From* $\gamma^2\widetilde{w_2^2 h_2^2}$ *detecting* $i(\nu^2\nu_6)$ *to* $gw_1 \cdot i(\delta' w_2^2)$ *detecting* $i(\eta\nu_6\epsilon)$.
(154a) *From* $\delta'\widetilde{w_2^2\gamma}$ *detecting* $\widetilde{\nu^2\nu_6}$ *to* $gw_1 \cdot \widetilde{w_2^2\delta'}$ *detecting* $\widetilde{\eta\nu_6}\epsilon$.
(154b) *From* $\alpha g\widetilde{w_2^2\gamma}$ *detecting* $\widetilde{\eta_1 B_5}$ *to* $gw_1 \cdot \widetilde{w_2^2\delta'}$ *detecting* $\widetilde{\eta\nu_6}\epsilon$.
(161) *From* $d_0 g\widetilde{w_2^2\beta^2}$ *detecting* $i(\nu_6\kappa)$ *to* $gw_1 \cdot d_0\widetilde{w_2^2\gamma}$ *detecting* $i(\bar\kappa D_6)$.
(165) *From* $h_2 \cdot i(h_2\beta w_2^3)$ *detecting* $\widetilde{\bar\kappa D_6}$ *to* $g^2\widetilde{w_2^2 d_0 e_0}$ *detecting* $i(D_7)$.

*There are no other hidden $\nu$-extensions in this spectral sequence. In particular, there is no hidden $\nu$-extension on* $g \cdot \widetilde{w_2^2 h_2^2}$.

PROOF. Most cases are readily deduced from the known action of $\nu$ on $\pi_*(tmf)$, and naturality with respect to $i\colon tmf \to tmf/2$ and $j\colon tmf/2 \to \Sigma tmf$. The following notes account for the remaining cases.

(21) Multiplying the $\nu$-extension in case (7) by $\kappa$ shows that $\nu$ times $\widetilde{\kappa\nu^2}$ is $\eta\kappa$ times $\widetilde{\epsilon}$, which is $\eta$ times $\widetilde{\epsilon\kappa} = \widetilde{\eta^2\bar\kappa}$. We saw in Theorem 12.4 that this equals $i(D_1)$. (Alternatively, use $E_\infty(tmf/\nu)$ to see that $\pi_{24}(tmf/(2,\nu)) = (\mathbb{Z}/2)^3$, so $i(D_1)$ must be a $\nu$-multiple.)

(65) See case (65) of Theorem 12.4 for why $i(\nu_2\kappa)$ is detected by $d_0 g\widetilde{\beta^2}$.

(102) The product $\nu \cdot i(\nu\nu_4)$ equals $i(\eta\epsilon_4) + i(\eta_1\bar\kappa^4)$, but is detected by the same class as $i(\eta\epsilon_4)$.

(103) The product $\nu \cdot \widetilde{\nu\nu_4}$ is a lift of $\eta\epsilon_4 + \eta_1\bar\kappa^4$, but is detected by the same class as $\widetilde{\eta\epsilon_4}$.

(117) Multiplying the $\nu$-extension in case (103) by $\kappa$ (and noting that $d_0 \cdot g^4 \cdot \widetilde{\gamma} = 0$ implies $\kappa \cdot \widetilde{\eta_1\bar\kappa^4} = 0$) shows that $\nu$ times $\kappa\widetilde{\nu\nu_4}$ is $\eta\kappa$ times $\widetilde{\epsilon_4}$, which is $\eta$ times $\widetilde{\epsilon_4\kappa} = \widetilde{\eta\eta_4\bar\kappa}$. We saw in Theorem 12.4 that this equals $i(D_5)$.

For the last two cases, we will make use of the hidden 2-extensions in degrees 72 and 168 in the Adams spectral sequence for $tmf/\nu$, which we establish in Theorem 12.15. The proofs of these 2-extensions do not rely on the results of the present or next section, so there is no circularity.

(69) From $E_\infty(tmf/\nu)$ calculated in Section 8.5, and case (56) of Theorem 12.15, we see that ${}_2\pi_{72}(tmf/\nu) = 0$ and $\pi_{73}(tmf/\nu)/2 = (\mathbb{Z}/2)^3$, so that $\pi_{73}(tmf/(2,\nu)) = (\mathbb{Z}/2)^3$. Furthermore, $\pi_{73}(tmf/2)/\nu = (\mathbb{Z}/2)^3$, since $\nu$ acts trivially on the class

$i(\eta_1^2\bar\kappa)$ detected by $g \cdot i(\gamma^2)$. Hence ${}_\nu\pi_{69}(tmf/2) = 0$, so that $\nu$ times $\widetilde{\bar\kappa D_2}$ is nonzero. The only possible value is $i(D_3)$ detected by $g^2\widetilde{d_0 e_0}$.

(165) From $E_\infty(tmf/\nu)$ and cases (56), (80) and (152) of Theorem 12.15 we see that ${}_2\pi_{168}(tmf/\nu) = 0$ and $\pi_{169}(tmf/\nu)$ has order $2^7$. Furthermore, $\pi_{169}(tmf/2) \cong \pi_{169}(tmf/2)/\nu$ has order equal to $2^7$, which implies that $\pi_{169}(tmf/(2,\nu)) = (\mathbb{Z}/2)^7$ and ${}_\nu\pi_{165}(tmf/2) = 0$, so that $\nu$ times $\widetilde{\bar\kappa D_6}$ is nonzero. The only possible value is $i(D_7)$ detected by $g^2 w_2^2\widetilde{d_0 e_0}$. □

DEFINITION 12.6. We fix representatives $y$ in $\pi_*(tmf/2)$ of the twelve generators $x$, listed in Table 6.12, of $E_\infty(tmf/2)$ as a module over $E_\infty(tmf)$.

| $y$ | $i(1)$ | $\widetilde{\eta}$ | $\widetilde{\nu^2}$ | $\widetilde{\epsilon}$ | $\widetilde{\kappa}$ | $\widetilde{\eta_1}$ | $\widetilde{\epsilon_1}$ | $\widetilde{\bar\kappa^4}$ | $\widetilde{\eta_4}$ | $\widetilde{\nu\nu_4}$ | $\widetilde{\epsilon_4}$ | $\widetilde{\epsilon_5}$ |
|---|---|---|---|---|---|---|---|---|---|---|---|---|
| $n$ | 0 | 2 | 7 | 9 | 15 | 26 | 33 | 81 | 98 | 103 | 105 | 129 |
| $x$ | $i(1)$ | $\widetilde{h_1}$ | $\widetilde{h_2^2}$ | $\widetilde{c_0}$ | $i(\beta)$ | $\widetilde{\gamma}$ | $\widetilde{\delta'}$ | $\gamma^2\widetilde{\beta^2}$ | $w_2^2\widetilde{h_1}$ | $w_2^2\widetilde{h_2^2}$ | $w_2^2\widetilde{c_0}$ | $w_2^2\widetilde{\delta'}$ |

We may assume that the representatives of the $w_1$-power torsion classes are chosen as $B$-power torsion classes. Moreover, we may assume that each representative of the form $y = \widetilde{z}$ in $\pi_n(tmf/2)$ maps under $j$ to $z \in \pi_{n-1}(tmf)$. (This involves a shift in Adams filtration only for $z = \kappa$.) Then $i(1)$, $\widetilde{\nu^2}$, $\widetilde{\bar\kappa^4}$, $\widetilde{\nu\nu_4}$ are well-defined, and $\widetilde{\eta}$, $\widetilde{\epsilon}$, $\widetilde{\kappa}$, $\widetilde{\epsilon_1}$ and $\widetilde{\epsilon_5}$ are defined up to sign, whereas $\widetilde{\epsilon_4}$ is defined modulo a sign and $i(\eta_1\bar\kappa^4)$. Having chosen $\widetilde{\eta}$, we can fix $\widetilde{\eta_1}$ and $\widetilde{\eta_4}$ by demanding that $B \cdot \widetilde{\eta_1} = B_1\widetilde{\eta}$ and $B \cdot \widetilde{\eta_4} = B_4\widetilde{\eta}$.

It is immediate from Proposition 6.11 that the twelve classes in Definition 12.6 generate $\pi_*(tmf/2)$ as a module over $\pi_*(tmf)$. Let $(N/2)_* \subset \pi_*(tmf/2)$ denote the $\mathbb{Z}[B]$-submodule generated by all classes in degrees $0 \le * < 192$. There is an isomorphism

$$(N/2)_* \otimes \mathbb{Z}[M] \cong \pi_*(tmf/2)$$

of $\mathbb{Z}[B, M]$-modules. The submodule $(N/2)_*$ is preserved by the action of $\eta$, $\nu$, $\epsilon$, $\kappa$ and $\bar\kappa$ (since $\bar\kappa \cdot i(B_7) = 0$), and the isomorphism respects these actions.

In most degrees it is straightforward to read off the group structure of $(N/2)_*$, together with its $\eta$- and $\nu$-action, from $E_\infty(tmf/2)$ with the hidden 2-, $\eta$- and $\nu$-extensions, keeping in mind that the $w_1$-power torsion classes form the associated graded of the restriction to $\Gamma_B(N/2)_*$ of the Adams filtration, cf. the discussion before Proposition 9.10. The next result summarizes what we know about the less obvious cases.

PROPOSITION 12.7.

(21) $\pi_{21}(tmf/2) \cong (\mathbb{Z}/2)^2$ *is generated by* $\kappa\widetilde{\nu^2}$ *and* $i(\eta\bar\kappa)$, *which are detected by* $d_0\widetilde{h_2^2}$ *and* $h_1 g \cdot i(1)$, *respectively. The relations* $\nu^2 \cdot \widetilde{\kappa} = \kappa\widetilde{\nu^2} + i(\eta\bar\kappa)$, $\eta \cdot i(\bar\kappa) = i(\eta\bar\kappa)$, $\eta \cdot i(\eta\bar\kappa) = i(\eta^2\bar\kappa)$ *and* $\nu \cdot i(\eta\bar\kappa) = 0$ *hold. Hence* $\nu^3 \cdot \widetilde{\kappa} = i(D_1)$.

(35) *The product* $\eta \cdot \eta\widetilde{\epsilon_1}$ *is zero.*

(40) $\Gamma_B\pi_{40}(tmf/2) \cong (\mathbb{Z}/2)^2$ *is generated by* $\kappa\widetilde{\eta_1}$ *and* $i(\bar\kappa^2)$, *which are detected by* $d_0\widetilde{\gamma}$ *and* $g^2 \cdot i(1)$, *respectively. The product* $\nu \cdot i(\bar\kappa^2)$ *is zero.*

(53) *The product of* $\eta$ *with* $\eta\eta_1\widetilde{\eta_1}$, *detected by* $h_1 \cdot \gamma\widetilde{\gamma}$, *is zero or* $i(\eta^2\nu_2)$.

(60) $\Gamma_B\pi_{60}(tmf/2) \cong (\mathbb{Z}/2)^2$ *is generated by* $\nu_2\widetilde{\epsilon}$ *and* $i(\bar\kappa^3)$, *which are detected by* $g \cdot d_0\widetilde{\gamma}$ *and* $g^3 \cdot i(1)$, *respectively. The product* $\eta \cdot i(\bar\kappa^3)$ *is zero.*

(65) $\Gamma_B\pi_{65}(tmf/2) \cong (\mathbb{Z}/2)^2$ *is generated by* $i(\nu_2\kappa)$ *and* $i(\eta_1\bar\kappa^2)$*, which are detected by* $d_0 g\widetilde{\beta^2}$ *and* $g^2 \cdot i(\gamma)$*, respectively. The relations* $\eta \cdot i(\eta_1\bar\kappa^2) = i(\eta\nu_2\kappa)$ *and* $\nu \cdot i(\eta_1\bar\kappa^2) = 0$ *hold.*

(66) $\Gamma_B\pi_{66}(tmf/2) \cong \mathbb{Z}/4 \oplus \mathbb{Z}/2$ *is generated by* $\bar\kappa^2\widetilde{\eta_1}$ *of order* 4 *and* $\nu_2\widetilde{\kappa} + \bar\kappa^2\widetilde{\eta_1}$ *of order* 2*, which are detected by* $g^2 \cdot \widetilde{\gamma}$ *and* $i(h_2\beta w_2)$*, respectively. The product* $\eta \cdot (\nu_2\widetilde{\kappa} + \bar\kappa^2\widetilde{\eta_1})$ *is zero.*

(105) $\Gamma_B\pi_{105}(tmf/2) \cong \mathbb{Z}/2 \oplus \mathbb{Z}/4$ *is generated by* $i(\eta_1\bar\kappa^4)$ *of order* 2 *and* $\widetilde{\epsilon_4}$ *of order* 4*, detected by* $g^4 \cdot i(\gamma)$ *and* $w_2^2\widetilde{c_0}$*, respectively. The relations* $\eta \cdot i(\epsilon_4) = 2\widetilde{\epsilon_4}$ *and* $\nu^2 \cdot i(\nu_4) = 2\widetilde{\epsilon_4} + i(\eta_1\bar\kappa^4)$ *hold.*

(106) $\Gamma_B\pi_{106}(tmf/2) \cong (\mathbb{Z}/2)^2$ *is generated by* $\bar\kappa^4\widetilde{\eta_1}$ *and* $\eta\widetilde{\epsilon_4}$*, which are detected by* $g^4 \cdot \widetilde{\gamma}$ *and* $h_1 \cdot w_2^2\widetilde{c_0}$*, respectively. The relation* $\nu \cdot \widetilde{\nu\nu_4} = \eta\widetilde{\epsilon_4} + \bar\kappa^4\widetilde{\eta_1}$ *holds.*

(117) $\pi_{117}(tmf/2) \cong (\mathbb{Z}/2)^2$ *is generated by* $\kappa_4\widetilde{\nu^2}$ *and* $i(\eta_4\bar\kappa)$*, which are detected by* $d_0 w_2^2 h_2^2$ *and* $h_2 \cdot i(h_2\beta w_2^2)$*, respectively. The relations* $\nu \cdot \nu_4\widetilde{\kappa} = \kappa_4\widetilde{\nu^2} + i(\eta_4\bar\kappa)$*,* $\eta \cdot i(\eta_4\bar\kappa) = i(\eta\eta_4\bar\kappa)$ *and* $\nu \cdot i(\eta_4\bar\kappa) = 0$ *hold. Hence* $\nu^2 \cdot \nu_4\widetilde{\kappa} = i(D_5)$*.*

(125) *The product of* $\eta$ *with* $\eta\eta_4\widetilde{\eta_1}$*, detected by* $h_1 \cdot h_1 w_2^2\widetilde{\gamma}$*, is zero or* $i(\eta^2\nu_5)$*.*

(131) *The product of* $\eta$ *with* $\eta\widetilde{\epsilon_5}$*, detected by* $h_1 \cdot w_2^2\widetilde{\delta'}$*, is* $\eta_1\bar\kappa^4\widetilde{\eta_1}$ *or* $\eta_1\bar\kappa^4\widetilde{\eta_1} + i(\eta\kappa_4\bar\kappa)$*.*

(149) *The product of* $\eta$ *with* $\eta_1\eta_4\widetilde{\eta_1}$*, detected by* $h_1^2 w_2^3\widetilde{h_1}$*, is zero or* $i(\eta^2\nu_6)$*.*

PROOF. (21) We know that $\eta \cdot i(\eta\bar\kappa) = i(\eta^2\bar\kappa)$ is detected by $w_1 \cdot i(d_0)$, and $\eta\nu = 0$. Hence $\nu^2\widetilde{\kappa}$ is detected by $h_1 g \cdot i(1)$, but is not equal to $i(\eta\bar\kappa)$. Their difference must be the higher-filtration class $\kappa\widetilde{\nu^2}$.

(35) We prove this in Lemma 12.26.

(40) This is clear from $\nu\bar\kappa = 0$.

(60) The product $\nu_2\widetilde{\epsilon}$ must be detected by $g \cdot d_0\widetilde{\gamma}$, because $h_2 w_2 \cdot \widetilde{c_0} = 0$. The $\eta$-product vanishes because $\eta\bar\kappa^3 = 0$.

(65) As previously noted, $i(\nu_2\kappa)$ must be detected by $d_0 g\widetilde{\beta^2}$ because $i(h_2 w_2 \cdot d_0) = 0$. The relations already hold before applying $i$.

(66) The classes $\bar\kappa^2\widetilde{\eta_1}$ and $\nu_2\widetilde{\kappa}$ are detected by $g^2 \cdot \widetilde{\gamma}$ and $i(h_2\beta w_2)$, with $2 \cdot \bar\kappa^2\widetilde{\eta_1} = i(\eta\eta_1\bar\kappa^2) = i(\eta\nu_2\kappa) = 2 \cdot \nu_2\widetilde{\kappa}$. Furthermore $\eta \cdot \bar\kappa^2\widetilde{\eta_1}$ and $\eta \cdot \nu_2\widetilde{\kappa}$ are both lifts of $\eta\eta_1\bar\kappa^2 = \eta\nu_2\kappa$, hence they are equal.

(105) This follows from Lemma 12.2 and Proposition 9.17.

(106) The relation lifts that of Proposition 9.17.

(117) We know that $\eta \cdot i(\eta_4\bar\kappa) = i(\eta\eta_4\bar\kappa)$ is detected by $w_1 \cdot i(d_0 w_2^2)$, and $\eta\nu = 0$. Hence $\nu\nu_4\widetilde{\kappa}$ is detected by $h_2 \cdot i(h_2\beta w_2^2)$, but is not equal to $i(\eta_4\bar\kappa)$. Their difference must be $\kappa_4\widetilde{\nu^2}$. □

## 12.2. Homotopy of $tmf/\eta$

We describe $\pi_*(tmf/\eta)$ using the short exact sequence

$$0 \to \pi_*(tmf)/\eta \xrightarrow{i} \pi_*(tmf/\eta) \xrightarrow{j} {}_\eta\pi_{*-2}(tmf) \to 0$$

of $\pi_*(tmf)$-modules, where

$$\pi_n(tmf)/\eta = \operatorname{cok}(\eta\colon \pi_{n-1}(tmf) \to \pi_n(tmf))$$
$${}_\eta\pi_{n-2}(tmf) = \ker(\eta\colon \pi_{n-2}(tmf) \to \pi_{n-1}(tmf))\,.$$

To achieve this we use the maps

$$E_\infty^{*,*}(tmf) \xrightarrow{i} E_\infty^{*,*}(tmf/\eta) \xrightarrow{j} E_\infty^{*,*-2}(tmf)$$

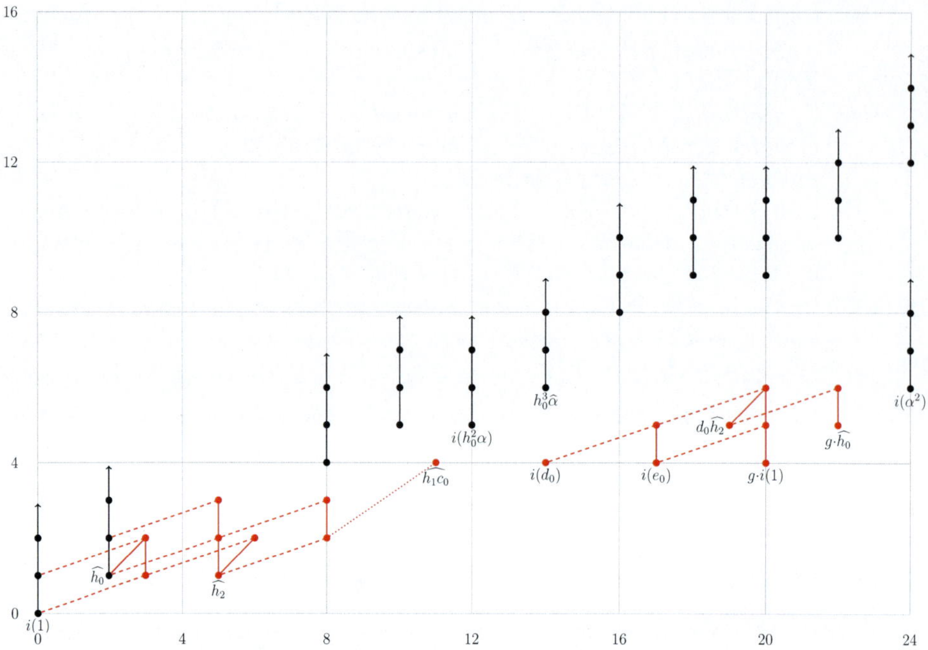

FIGURE 12.9. $E_\infty(tmf/\eta)$ for $0 \le t-s \le 24$, with all hidden 2-, $\eta$- and $\nu$-extensions

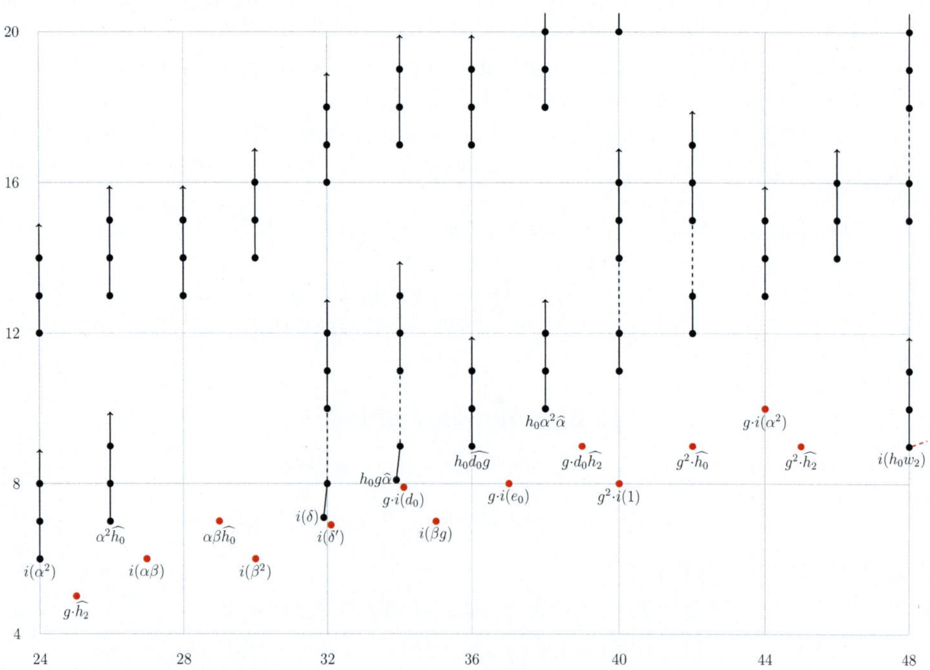

FIGURE 12.10. $E_\infty(tmf/\eta)$ for $24 \le t-s \le 48$, with all hidden 2-, $\eta$- and $\nu$-extensions

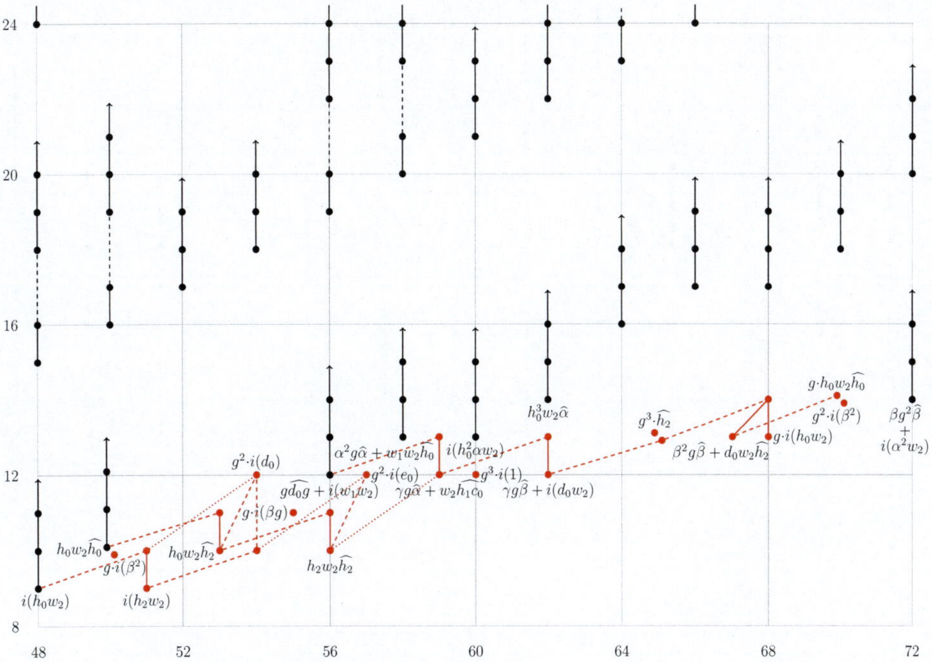


FIGURE 12.11. $E_\infty(tmf/\eta)$ for $48 \leq t-s \leq 72$, with all hidden 2-, $\eta$- and $\nu$-extensions

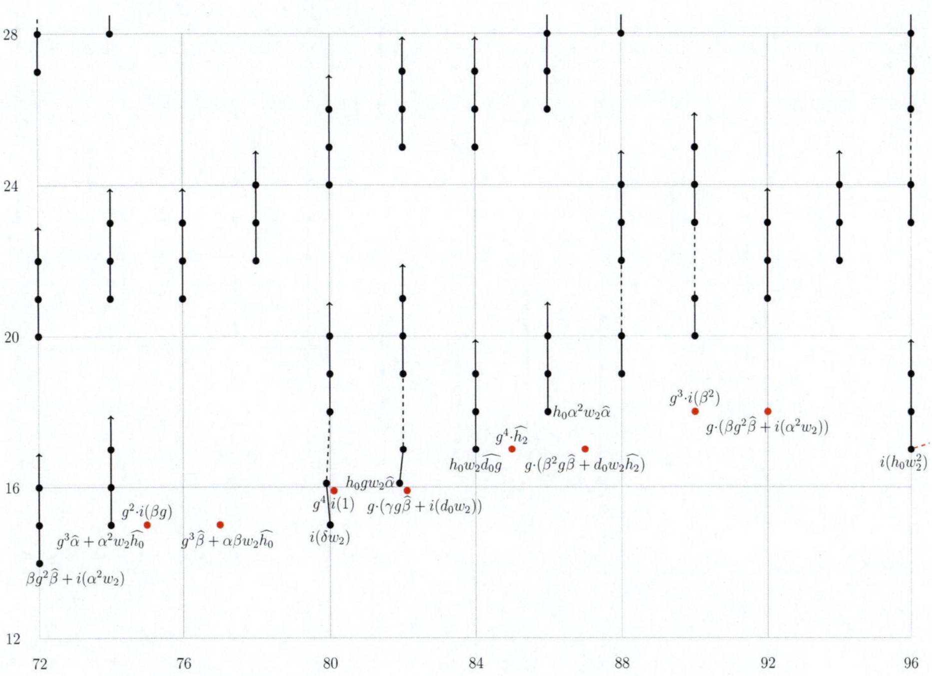


FIGURE 12.12. $E_\infty(tmf/\eta)$ for $72 \leq t-s \leq 96$, with all hidden 2-, $\eta$- and $\nu$-extensions

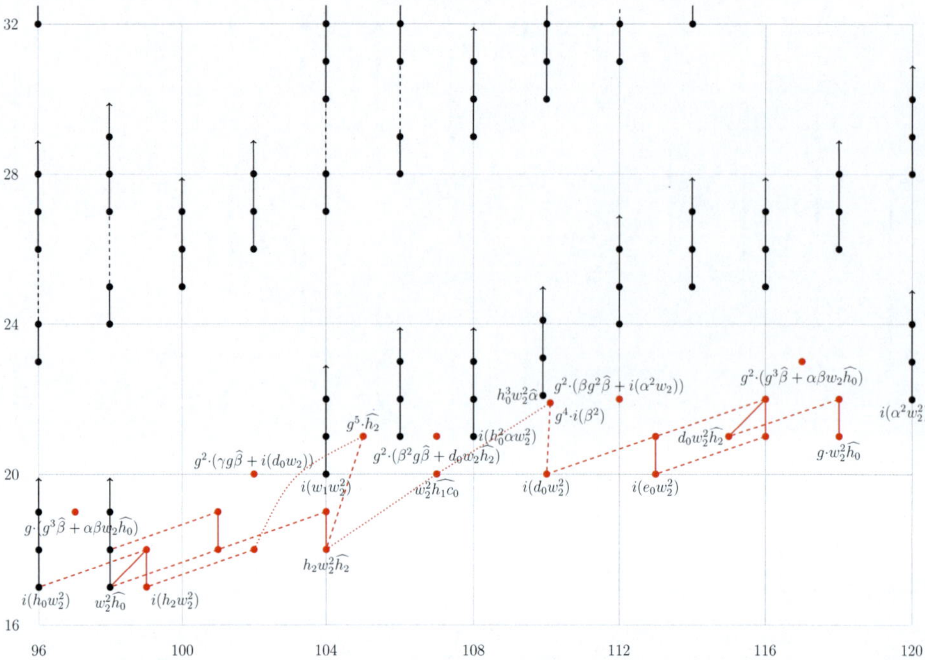

FIGURE 12.13. $E_\infty(tmf/\eta)$ for $96 \leq t-s \leq 120$, with all hidden 2-, $\eta$- and $\nu$-extensions

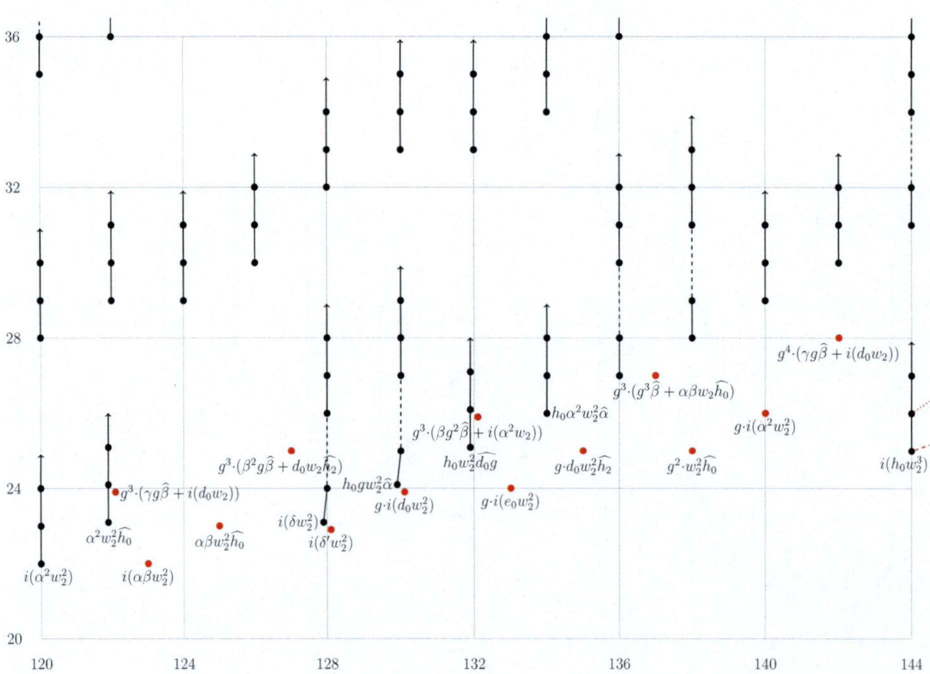

FIGURE 12.14. $E_\infty(tmf/\eta)$ for $120 \leq t-s \leq 144$, with all hidden 2-, $\eta$- and $\nu$-extensions

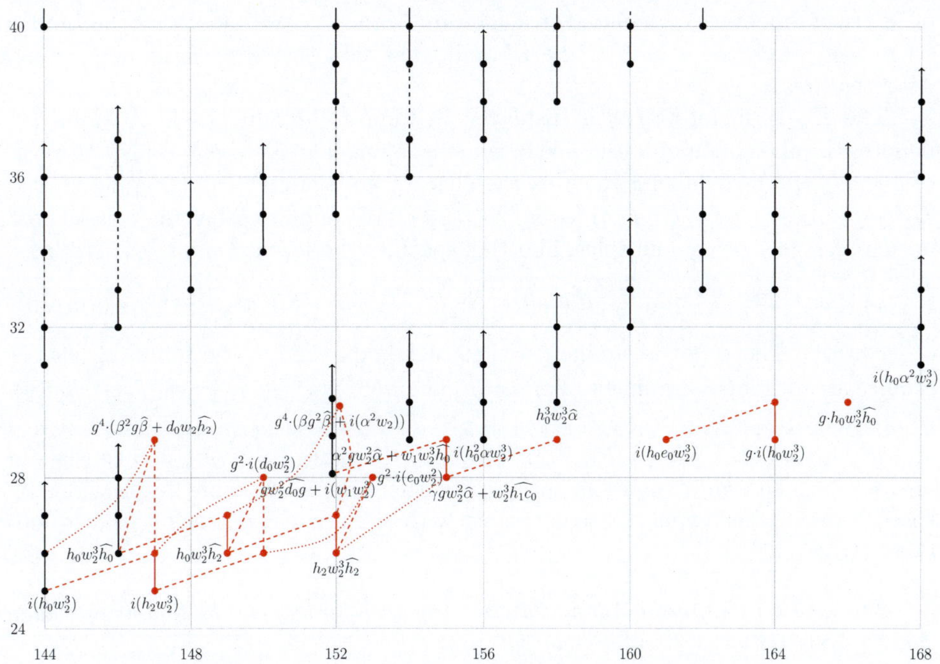


FIGURE 12.15. $E_\infty(tmf/\eta)$ for $144 \leq t-s \leq 168$, with all hidden 2-, $\eta$- and $\nu$-extensions

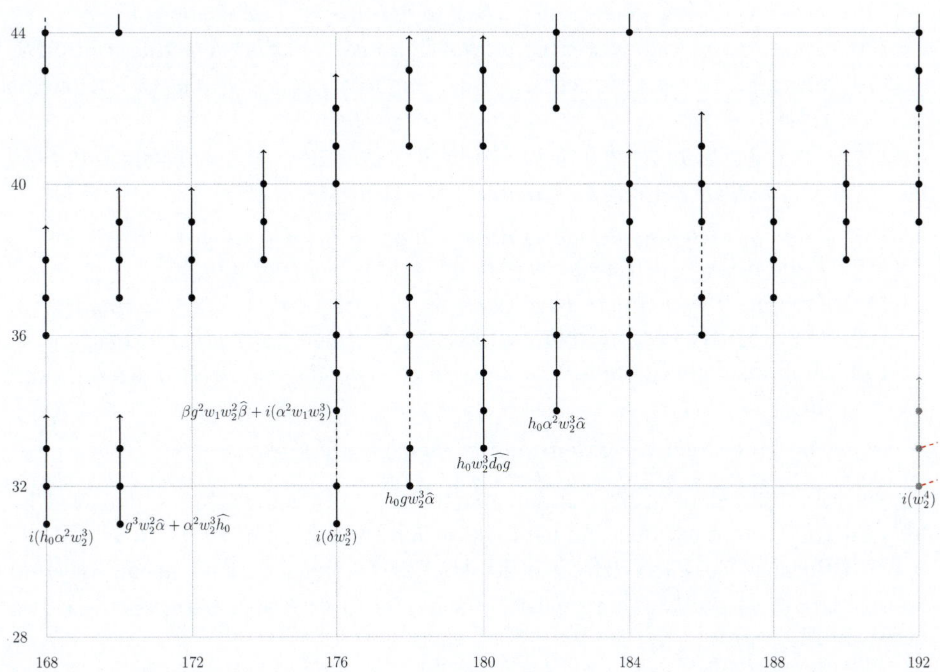


FIGURE 12.16. $E_\infty(tmf/\eta)$ for $168 \leq t-s \leq 192$, with all hidden 2-, $\eta$- and $\nu$-extensions

of $E_\infty(tmf)$-modules, calculated in Chapters 5 and 7. We determine the hidden 2-, $\eta$- and $\nu$-extensions in $E_\infty(tmf/\eta)$, and show that there are no hidden $B$- and $M$-extensions.

The $E_\infty$-term for $tmf/\eta$ is displayed in Figures 12.9 to 12.16. A label $i(x)$ denotes the class of an infinite cycle in the image under $i\colon E_2^{s,t}(tmf) \to E_2^{s,t}(tmf/\eta)$ of $x \in E_2(tmf)$. A label $\widehat{x}$ denotes the class of an infinite cycle mapping to $x \in E_2(tmf)$ under $j\colon E_2^{s,t}(tmf/\eta) \to E_2^{s,t-2}(tmf)$. We omit to label the classes that are $h_0$-, $h_1$-, $h_2$- or $w_1$-multiples, and this specifies the $w_1$-action on $E_\infty(tmf/\eta)$.

LEMMA 12.8. *There are no hidden $B$- or $M$-power extensions in $E_\infty(tmf/\eta)$.*

PROOF. The proof is similar to that of Lemma 12.1. The following classes require an additional argument. For $b = h_1 \cdot \widehat{h_2}$, $i(\beta^2)$, $g^2 \cdot i(1)$, $g^2 \cdot \widehat{h_0}$, $h_2 \cdot i(h_2w_2)$, $g^2 \cdot i(d_0)$, $h_2 \cdot i(h_2w_2^2)$, $g^2 \cdot (\gamma g\widehat{\beta} + i(d_0w_2))$, $g^2 \cdot (\beta g^2\widehat{\beta} + i(\alpha^2 w_2))$, $g^2 \cdot w_2^2\widehat{h_0}$, $h_2 \cdot i(h_2w_2^3)$ and $g^2 \cdot i(d_0w_2^2)$ the part of $E_\infty(tmf/\eta)$ above the bidegree of $w_1 b = 0$ consists of $w_1$-multiples and $h_0$-torsion free towers, but there are no possible 2-extensions on these classes $b$ that would be compatible with a hidden $B$-extension from $b$ into these $h_0$-towers. □

THEOREM 12.9. *In the Adams spectral sequence for tmf/$\eta$, the following hidden 2-extensions repeat $w_1$- and $w_2^4$-periodically:*

(32) *From $h_0 \cdot i(\delta)$ detecting $i(2B_1)$ to $w_1 \cdot i(\alpha^2)$ detecting $i(4B_1)$.*
(34) *From $h_0 \cdot h_0 g\widehat{\alpha}$ detecting a lift $\widehat{4B_1}$ to $w_1 \cdot \alpha^2\widehat{h_0}$ detecting a lift $\widehat{8B_1}$.*
(80) *From $h_0 \cdot i(\delta w_2)$ detecting $i(2B_3)$ to $w_1 \cdot (\beta g^2\widehat{\beta} + i(\alpha^2 w_2))$ detecting $i(4B_3)$.*
(82) *From $h_0 \cdot h_0 g w_2\widehat{\alpha}$ detecting $\widehat{4B_3}$ to $w_1 \cdot (g^3\widehat{\alpha} + \alpha^2 w_2\widehat{h_0})$ detecting $\widehat{8B_3}$.*
(128) *From $h_0 \cdot i(\delta w_2^2)$ detecting $i(2B_5)$ to $w_1 \cdot i(\alpha^2 w_2^2)$ detecting $i(4B_5)$.*
(130) *From $h_0 \cdot h_0 g w_2^2\widehat{\alpha}$ detecting a lift $\widehat{4B_5}$ to $w_1 \cdot \alpha^2 w_2^2\widehat{h_0}$ detecting a lift $\widehat{8B_5}$.*
(176) *From $h_0 \cdot i(\delta w_2^3)$ detecting $i(2B_7)$ to $\beta g^2 w_1 w_2^2\widehat{\beta} + i(\alpha^2 w_1 w_2^3)$ detecting $i(4B_7)$.*
(178) *From $h_0 \cdot h_0 g w_2^3\widehat{\alpha}$ detecting $\widehat{4B_7}$ to $w_1 \cdot (g^3 w_2^2\widehat{\alpha} + \alpha^2 w_2^3\widehat{h_0})$ detecting $\widehat{8B_7}$.*

*The following hidden 2-extensions repeat $w_2^4$-periodically:*

(54) *From $h_2 \cdot i(h_2w_2)$ detecting $i(\nu\nu_2)$ to $g^2 \cdot i(d_0)$ detecting $i(2\nu\nu_2)$.*
(110) *From $i(d_0w_2^2)$ detecting $i(\kappa_4)$ to $g^4 \cdot i(\beta^2)$ detecting $i(2\kappa_4)$.*
(147) *From $h_0 \cdot i(h_2w_2^3)$ detecting $i(2\nu_6)$ to $g^4 \cdot (\beta^2 g\widehat{\beta} + d_0 w_2\widehat{h_2})$ detecting $i(4\nu_6)$.*
(150) *From $h_2 \cdot i(h_2w_2^3)$ detecting $i(\nu\nu_6)$ to $g^2 \cdot i(d_0w_2^2)$ detecting $i(2\nu\nu_6)$.*
(152) *From $h_0 \cdot h_2 w_2^3\widehat{h_2}$ detecting $\widehat{2\nu\nu_6}$ to $g^4 \cdot (\beta g^2\widehat{\beta} + i(\alpha^2 w_2))$ detecting a lift $\widehat{4\nu\nu_6}$.*

*There are no other hidden 2-extensions in this spectral sequence.*

PROOF. (32) Since $i(2B_1)$ is detected by $h_0 \cdot i(\delta)$ and $i(8B_1)$ is detected by $h_0 w_1 \cdot i(\alpha^2)$, there must be a hidden 2-extension from $h_0 \cdot i(\delta)$ to $w_1 \cdot i(\alpha^2)$.

(34) Because $h_0 \cdot h_0 g\widehat{\alpha}$ detects a lift $\widehat{4B_1}$, and 2 times that lift lies in $\widehat{8B_1}$ and is detected by $w_1 \cdot \alpha^2\widehat{h_0}$, there must be a hidden 2-extension between these two classes in $E_\infty(tmf/\nu)$.

(80) Since $i(2B_3)$ is detected by $h_0 \cdot i(\delta w_2)$ and $i(8B_3)$ is detected by $h_0 w_1 \cdot (\beta g^2\widehat{\beta} + i(\alpha^2 w_2))$, there must be a hidden 2-extension from $h_0 \cdot i(\delta w_2)$ to $w_1 \cdot (\beta g^2\widehat{\beta} + i(\alpha^2 w_2))$.

(82) Because $h_0 \cdot h_0 g w_2 \widehat{\alpha}$ detects $\widehat{4B_3}$, and 2 times that lift lies in $\widehat{8B_3}$ and is detected by $w_1 \cdot (g^3\widehat{\alpha} + \alpha^2 w_2 \widehat{h_0})$, there must be a hidden 2-extension between these two classes.

(128) Since $i(2B_5)$ is detected by $h_0 \cdot i(\delta w_2^2)$ and $i(8B_5)$ is detected by $h_0 w_1 \cdot i(\alpha^2 w_2^2)$, there must be a hidden 2-extension from $h_0 \cdot i(\delta w_2^2)$ to $w_1 \cdot i(\alpha^2 w_2^2)$.

(130) Because $h_0 \cdot h_0 g w_2^2 \widehat{\alpha}$ detects a lift $\widehat{4B_5}$, and 2 times that lift lies in $\widehat{8B_5}$ and is detected by $w_1 \cdot \alpha^2 w_2^2 \widehat{h_0}$, there must be a hidden 2-extension between these two classes in $E_\infty(tmf/\nu)$.

(176) Since $i(2B_7)$ is detected by $h_0 \cdot i(\delta w_2^3)$ and $i(8B_7)$ is detected by $h_0 \cdot (\beta g^2 w_1 w_2^2 \widehat{\beta} + i(\alpha^2 w_1 w_2^3))$, there must be a hidden 2-extension from $h_0 \cdot i(\delta w_2^3)$ to $\beta g^2 w_1 w_2^2 \widehat{\beta} + i(\alpha^2 w_1 w_2^3)$.

(178) Because $h_0 \cdot h_0 g w_2^3 \widehat{\alpha}$ detects $\widehat{4B_7}$, and 2 times that lift lies in $\widehat{8B_7}$ and is detected by $w_1 \cdot (g^3 w_2^2 \widehat{\alpha} + \alpha^2 w_2^3 \widehat{h_0})$, there must be a hidden 2-extension between these two classes. □

LEMMA 12.10. *The multiplication-by-$\eta$ map $\eta\colon \Sigma S/\eta \to S/\eta$ factors as the composite*

$$\Sigma S/\eta \xrightarrow{j} S^3 \xrightarrow{\nu} S \xrightarrow{i} S/\eta\,.$$

*Hence $\eta \cdot \widehat{y} = i(\nu \cdot y)$ for $\widehat{y} \in \pi_*(tmf/\eta)$ with $j(\widehat{y}) = y$.*

PROOF. The map $\eta\colon \Sigma S/\eta \to S/\eta$ is essential, because $Sq^4$ acts nontrivially in the cohomology of its homotopy cofiber $S/\eta \wedge S/\eta$, and $i\nu j$ is the only nontrivial such map. □

THEOREM 12.11. *In the Adams spectral sequence for $tmf/\eta$, the following hidden $\eta$-extensions repeat $w_2^4$-periodically:*

(53) *From $h_0 w_2 \widehat{h_2}$ detecting $\widehat{2\nu_2}$ to $g^2 \cdot i(d_0)$ detecting $i(2\nu\nu_2)$.*
(56) *From $h_2 w_2 \widehat{h_2}$ detecting $\widehat{\nu\nu_2}$ to $g^2 \cdot i(e_0)$ detecting $i(\nu^2\nu_2)$.*
(104) *From $h_2 w_2^2 \widehat{h_2}$ detecting $\widehat{\nu\nu_4}$ to $g^5 \cdot \widehat{h_2}$ detecting $i(\nu^2\nu_4) = i(\eta_1\bar{\kappa}^4)$.*
(146) *From $h_0 w_2^3 \widehat{h_0}$ detecting $\widehat{D_6}$ to $g^4 \cdot (\beta^2 g \widehat{\beta} + d_0 w_2 \widehat{h_2})$ detecting $i(4\nu_6)$.*
(149) *From $h_0 w_2^3 \widehat{h_2}$ detecting $\widehat{2\nu_6}$ to $g^2 \cdot i(d_0 w_2^3)$ detecting $i(2\nu\nu_6)$.*
(152) *From $h_2 w_2^3 \widehat{h_2}$ detecting $\widehat{\nu\nu_6}$ to $g^2 \cdot i(e_0 w_2^3)$ detecting $i(\nu^2\nu_6)$.*

*There are no other hidden $\eta$-extensions in this spectral sequence.*

PROOF. This follows directly from Lemma 12.10, using the known action of $\nu$ on $\pi_*(tmf)$. □

Logically, Theorem 12.16 precedes the following result.

THEOREM 12.12. *In the Adams spectral sequence for $tmf/\eta$, the following hidden $\nu$-extensions repeat $w_2^4$-periodically:*

(8) *From $h_2 \cdot \widehat{h_2}$ detecting $\widehat{\nu^2}$ to $\widehat{h_1 c_0}$ detecting $\widehat{\eta\epsilon}$.*
(51) *From $h_0 \cdot i(h_2 w_2)$ detecting $i(2\nu_2)$ to $g^2 \cdot i(d_0)$ detecting $i(2\nu\nu_2)$.*
(54) *From $h_2 \cdot i(h_2 w_2)$ detecting $i(\nu\nu_2)$ to $g^2 \cdot i(e_0)$ detecting $i(\nu^2\nu_2)$.*
(56) *From $h_2 w_2 \widehat{h_2}$ detecting $\widehat{\nu\nu_2}$ to $\gamma g \widehat{\alpha} + w_2 \widehat{h_1 c_0}$ detecting $\widehat{\nu^2\nu_2}$.*
(102) *From $h_2 \cdot i(h_2 w_2^2)$ detecting $i(\nu\nu_4)$ to $g^5 \cdot \widehat{h_2}$ detecting $i(\nu^2\nu_4) = i(\eta_1\bar{\kappa}^4)$.*
(104) *From $h_2 w_2^2 \widehat{h_2}$ detecting $\widehat{\nu\nu_4}$ to $w_2^2 \widehat{h_1 c_0}$ detecting $\widehat{\nu^2\nu_4}$.*
(107) *From $w_2^2 \widehat{h_1 c_0}$ detecting $\widehat{\nu^2\nu_4}$ to $g^4 \cdot i(\beta^2)$ detecting $i(2\kappa_4)$.*

(144) *From* $h_0 \cdot i(h_0 w_2^3)$ *detecting* $i(D_6)$ *to* $g^4 \cdot (\beta^2 g\widehat{\beta} + d_0 w_2 \widehat{h_2})$ *detecting* $i(4\nu_6)$.
(147) *From* $h_0 \cdot i(h_2 w_2^3)$ *detecting* $i(2\nu_6)$ *to* $g^2 \cdot i(d_0 w_2^2)$ *detecting* $i(2\nu\nu_6)$.
(149) *From* $h_0 \cdot h_0 w_2^2 \widehat{h_2}$ *detecting* $\widehat{4\nu_6}$ *to* $g^4 \cdot (\beta g^2 \widehat{\beta} + i(\alpha^2 w_2))$ *detecting* $\widehat{4\nu\nu_6}$.
(150) *From* $h_2 \cdot i(h_2 w_2^3)$ *detecting* $i(\nu\nu_6)$ *to* $g^2 \cdot i(e_0 w_2^2)$ *detecting* $i(\nu^2\nu_6)$.
(152) *From* $h_2 w_2^3 \widehat{h_2}$ *detecting* $\widehat{\nu\nu_6}$ *to* $\gamma g w_2^2 \widehat{\alpha} + w_2^3 \widehat{h_1 c_0}$ *detecting* $\widehat{\nu^2\nu_6}$.

*There are no other hidden* $\nu$*-extensions in this spectral sequence.*

PROOF. The following case relies on a hidden $\eta$-extension for $tmf/\nu$, determined in Theorem 12.16. The remaining cases are routine.

(107) From $E_\infty(tmf/\nu)$ and case (109) of Theorem 12.16 we see that $i(2\kappa_4)$ in $\pi_{110}(tmf/\nu)$, which is detected by $g^4 \cdot \gamma\overline{h_1}$, is an $\eta$-multiple and therefore maps to zero in $\pi_{110}(tmf/(\eta,\nu))$. Hence $i(2\kappa_4)$ in $\pi_{110}(tmf/\eta)$, which is detected by $g^4 \cdot i(\beta^2)$, must be a $\nu$-multiple. Since $h_2$ acts trivially on $g^2 \cdot (\beta^2 g\widehat{\beta} + d_0 w_2 \widehat{h_2})$, it follows that $i(2\kappa_4)$ equals $\nu$ times the class $\widehat{\nu^2\nu_4}$ detected by $w_2^2 \widehat{h_1 c_0}$. □

It follows from Proposition 7.7 that $\pi_*(tmf/\eta)$ is generated as a $\pi_*(tmf)$-module by elements detected by the classes listed in Table 7.7, where we may assume that the $w_1$-power torsion classes are represented by $B$-power torsion elements. We omit to enumerate 45 such elements.

Let $(N/\eta)_* \subset \pi_*(tmf/\eta)$ denote the $\mathbb{Z}[B]$-submodule generated by all classes in degrees $0 \le * < 192$. There is an isomorphism

$$(N/\eta)_* \otimes \mathbb{Z}[M] \cong \pi_*(tmf/\eta)$$

of $\mathbb{Z}[B,M]$-modules. The submodule $(N/\eta)_*$ is preserved by the action of $\eta$, $\nu$, $\epsilon$, $\kappa$ and $\bar\kappa$ (since $\kappa$- and $\bar\kappa$-multiples are 2-power torsion), and the isomorphism respects these actions.

In most degrees it is straightforward to read off the group structure of $(N/\eta)_*$, together with its $\eta$- and $\nu$-action, from $E_\infty(tmf/\eta)$ with the hidden 2-, $\eta$- and $\nu$-extensions, keeping in mind that the $w_1$-power torsion classes form the associated graded of the restriction to $\Gamma_B(N/\eta)_*$ of the Adams filtration. The next result summarizes some not quite obvious cases.

PROPOSITION 12.13.

(5) $\pi_2(tmf/\eta) \cong \mathbb{Z}$ *is generated by* $\widehat{2}$ *detected by* $\widehat{h_0}$, *and* $\pi_5(tmf/\eta) \cong \mathbb{Z}/8$ *is generated by* $\widehat{\nu}$ *detected by* $\widehat{h_2}$. *The relation* $\nu \cdot \widehat{2} = 2 \cdot \widehat{\nu}$ *holds.*

(20) $\pi_{17}(tmf/\eta) \cong \mathbb{Z}/4$ *is generated by* $\widehat{\eta\kappa}$ *detected by* $i(e_0)$, *and the* $B$*-power torsion* $\Gamma_B\pi_{20}(tmf/\eta) \cong \mathbb{Z}/8$ *is generated by* $i(\bar\kappa)$ *detected by* $g \cdot i(1)$. *We can choose* $\widehat{\eta\kappa}$ *so that* $\nu \cdot \widehat{\eta\kappa} = 2 \cdot i(\bar\kappa)$.

(59) *The* $B$*-power torsion* $\Gamma_B\pi_{56}(tmf/\eta) \cong \mathbb{Z}/4$ *is generated by* $\widehat{\nu\nu_2}$ *detected by* $h_2 w_2 \widehat{h_2}$, *and* $\pi_{59}(tmf/\eta) \cong \mathbb{Z}/4$ *is generated by* $\widehat{\nu^2\nu_2}$ *detected by* $\gamma g\widehat{\alpha} + w_2 \widehat{h_1 c_0}$. *The relation* $\nu \cdot 2\widehat{\nu\nu_2} = 2 \cdot \widehat{\nu^2\nu_2}$ *holds.*

(107) $\pi_{107}(tmf/\eta) \cong (\mathbb{Z}/2)^2$ *is generated by* $\widehat{\eta_1\bar\kappa^4}$ *and* $\widehat{\eta\epsilon_4}$, *which are detected by* $g^2 \cdot (\beta^2 g\widehat{\beta} + d_0 w_2 \widehat{h_2})$ *and* $w_2^2 \widehat{h_1 c_0}$, *respectively. The relation* $\nu \cdot \widehat{\nu\nu_4} = \widehat{\eta\epsilon_4} + \widehat{\eta_1\bar\kappa^4}$ *holds.*

(147) $\pi_{144}(tmf/\eta) \cong \mathbb{Z}^7$ *has one generator* $\widehat{\epsilon_5\kappa}$ *detected by* $i(h_0 w_2^3)$ *and six others, and* $\pi_{147}(tmf/\eta) \cong \mathbb{Z}/8$ *is generated by* $i(\nu_6)$ *detected by* $i(h_2 w_2^3)$. *We can choose* $\widehat{\epsilon_5\kappa}$ *so that* $\nu \cdot \widehat{\epsilon_5\kappa} = 2 \cdot i(\nu_6)$.

(152) $\pi_{149}(tmf/\eta) \cong \mathbb{Z}/4$ *is generated by* $\widehat{2\nu_6}$ *detected by* $h_0 w_2^3 \widehat{h_2}$, *and the* $B$*-power torsion* $\Gamma_B \pi_{152}(tmf/\eta) \cong \mathbb{Z}/8$ *is generated by* $\widehat{\nu\nu_6}$ *detected by* $h_2 w_2^3 \widehat{h_2}$. *The relation* $\nu \cdot \widehat{2\nu_6} = 2 \cdot \widehat{\nu\nu_6}$ *holds.*

(155) $\pi_{155}(tmf/\eta) \cong \mathbb{Z}/4$ *is generated by* $\widehat{\nu^2\nu_6}$ *detected by* $\gamma g w_2^2 \widehat{\alpha} + w_2^3 \widehat{h_1 c_0}$, *and* $\nu \cdot 2\widehat{\nu\nu_6} = 2 \cdot \widehat{\nu^2\nu_6}$.

PROOF. (5) The two lifts of $\nu 2 = 2\nu$ must agree, because $\pi_5(tmf)/\eta = 0$.

(20) Adding $i(\nu\kappa)$ to $\widehat{\eta\kappa}$ changes the sign in $\nu \cdot \widehat{\eta\kappa} = \pm 2 \cdot i(\bar{\kappa})$.

(59) This expresses how the $\nu$-extension from $h_0 \cdot h_2 w_2 \widehat{h_2}$ to $h_0 \cdot (\gamma g \widehat{\alpha} + w_2 \widehat{h_1 c_0})$ is eclipsed by the $h_2$-multiplication from $\widehat{g d_0 g} + i(w_1 w_2)$.

(107) This lifts the relation $\nu^2 \nu_4 = \eta \epsilon_4 + \eta_1 \bar{\kappa}^4$ in $\pi_{105}(tmf)$.

(147) Adding $i(D_6)$ to $\widehat{\epsilon_5 \kappa}$ changes the sign in $\nu \cdot \widehat{\epsilon_5\kappa} = \pm 2 \cdot i(\nu_6)$.

(152) The two lifts of $2\nu\nu_6$ must agree, because $j$ maps the $B$-power torsion in $\pi_{152}(tmf/\eta)$ isomorphically to $\pi_{150}(tmf)$.

(155) The $\nu$-extension from $h_0 \cdot h_2 w_2^3 \widehat{h_2}$ to $h_0 \cdot (\gamma g w_2^2 \widehat{\alpha} + w_2^3 \widehat{h_1 c_0})$ is eclipsed by the $h_2$-multiplication from $g w_2^2 \widehat{d_0 g} + i(w_1 w_2^3)$. □

## 12.3. Homotopy of $tmf/\nu$

We describe $\pi_*(tmf/\nu)$ using the short exact sequence

$$0 \to \pi_*(tmf)/\nu \xrightarrow{i} \pi_*(tmf/\nu) \xrightarrow{j} {}_\nu\pi_{*-4}(tmf) \to 0$$

of $\pi_*(tmf)$-modules, where

$$\pi_n(tmf)/\nu = \mathrm{cok}(\nu\colon \pi_{n-3}(tmf) \to \pi_n(tmf))$$
$${}_\nu\pi_{n-4}(tmf) = \ker(\nu\colon \pi_{n-4}(tmf) \to \pi_{n-1}(tmf))\,.$$

To achieve this we use the maps

$$E_\infty^{*,*}(tmf) \xrightarrow{i} E_\infty^{*,*}(tmf/\nu) \xrightarrow{j} E_\infty^{*,*-4}(tmf)$$

of $E_\infty(tmf)$-modules, calculated in Chapters 5 and 8. We determine the hidden 2- and $\eta$-extensions in $E_\infty(tmf/\nu)$, and show that there are no hidden $B$- and $M$-extensions.

The $E_\infty$-term for $tmf/\nu$ is displayed in Figures 12.17 to 12.24. A label $i(x)$ denotes the class of an infinite cycle in the image under $i\colon E_2^{s,t}(tmf) \to E_2^{s,t}(tmf/\nu)$ of $x \in E_2(tmf)$. A label $\overline{x}$ denotes the class of an infinite cycle mapping to $x \in E_2(tmf)$ under $j\colon E_2^{s,t}(tmf/\nu) \to E_2^{s,t-4}(tmf)$. (This may look peculiar when $x$ involves $\bar{\kappa}$, but no real ambiguity should occur.) We omit to label the classes that are $h_0$-, $h_1$-, $h_2$- or $w_1$-multiples, and this specifies the $w_1$-action on $E_\infty(tmf/\nu)$.

LEMMA 12.14. *There are no hidden* $B$*- or* $M$*-power extensions in* $E_\infty(tmf/\nu)$.

PROOF. The proof is similar to that of Lemma 12.1. Again, several classes require an additional argument. For $b = h_1 \cdot \overline{h_0 h_2}$, $\overline{c_0} + i(\alpha)$, $g \cdot i(1)$, $h_1 \cdot \overline{\alpha\beta}$, $g \cdot (\overline{c_0} + i(\alpha))$, $g^2 \cdot i(1)$, $g \cdot \overline{g}$, $i(\alpha^2 g)$, $h_1 \cdot g \cdot \overline{\alpha\beta}$, $\delta' \overline{g}$, $g^2 \cdot \overline{g}$, $g^4 \cdot i(1)$, $h_1 \cdot w_2^2 \overline{h_0 h_2}$, $g^4 \cdot \overline{g}$, $w_2^2 \overline{c_0} + i(\alpha w_2^2)$, $h_1 \cdot d_0 w_2^2 \overline{h_1}$, $g^5 \cdot \overline{g}$, $h_1 \cdot w_2^2 \overline{\alpha\beta}$, $g \cdot (w_2^2 \overline{c_0} + i(\alpha w_2^2))$, $h_1 \cdot g \cdot d_0 w_2^2 \overline{h_1}$, $g \cdot w_1 w_2^2 \overline{g}$, $\delta' w_2^2 \overline{g}$ and $h_1 \cdot d_0 g w_2^2 \overline{\gamma}$ the part of $E_\infty(tmf/\nu)$ above the bidegree of $w_1 b = 0$ (or $w_1^2 b = 0$) consists of $w_1$-multiples (or $w_1^2$-multiples) and $h_0$-torsion free towers, and no possible 2-extension on $b$ would be compatible with a hidden $B$-extension (or $B^2$-extension) from $b$ into these $h_0$-towers. □

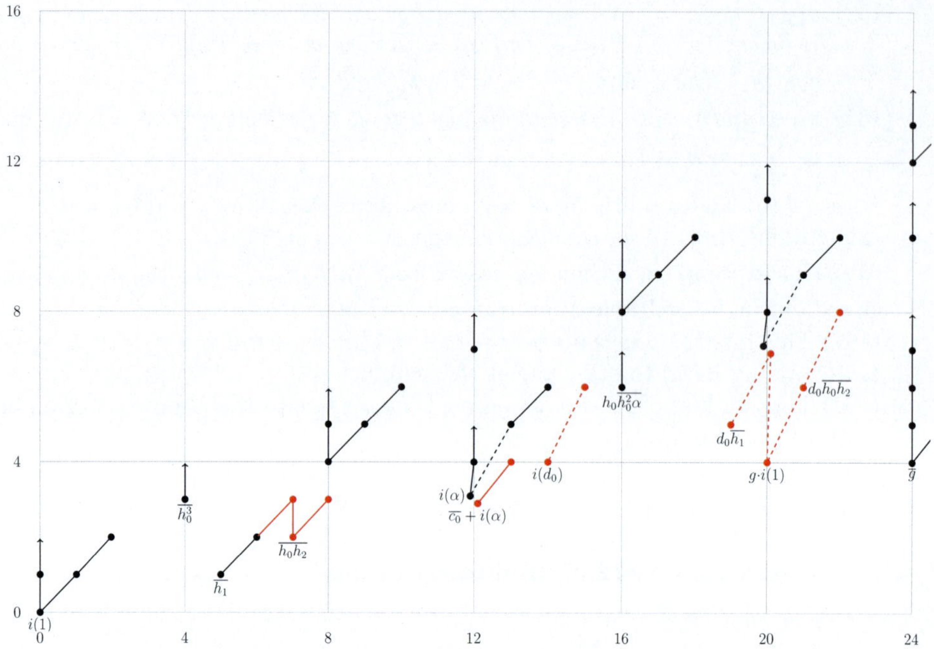

FIGURE 12.17. $E_\infty(tmf/\nu)$ for $0 \le t - s \le 24$, with all hidden 2- and $\eta$-extensions

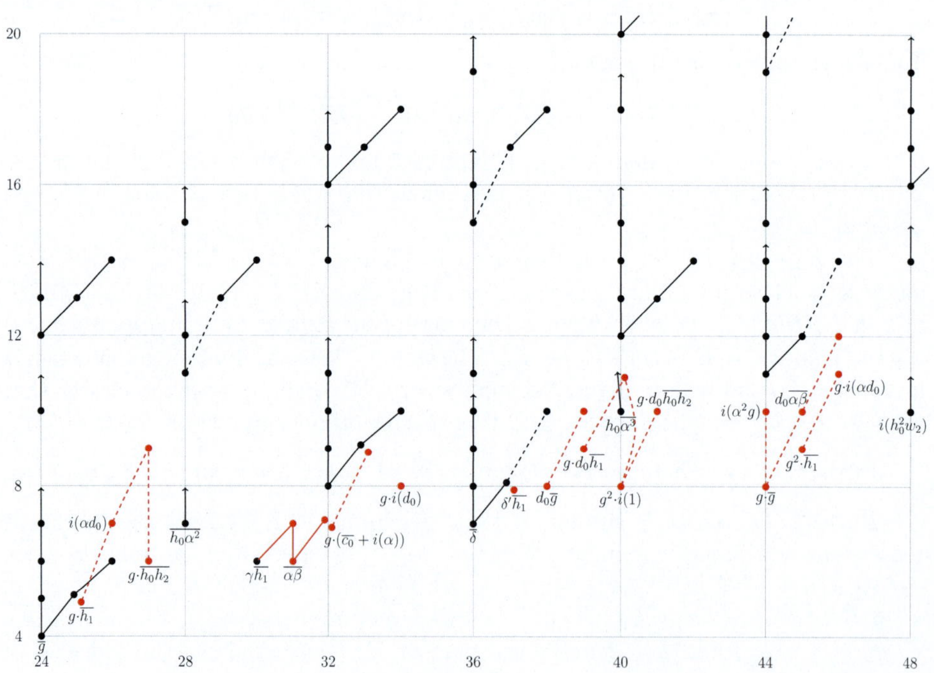

FIGURE 12.18. $E_\infty(tmf/\nu)$ for $24 \le t - s \le 48$, with all hidden 2- and $\eta$-extensions

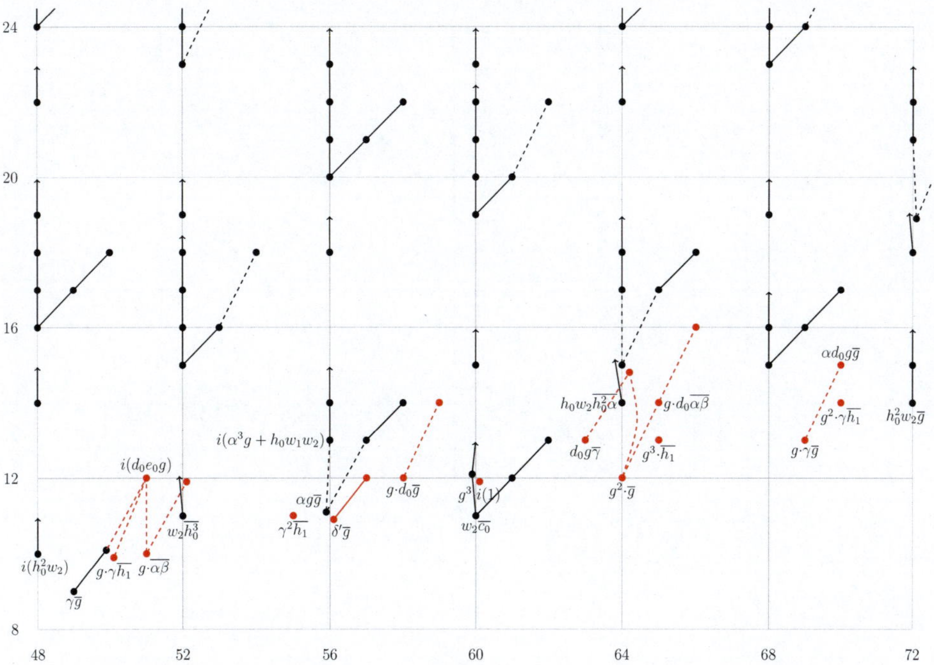

FIGURE 12.19. $E_\infty(tmf/\nu)$ for $48 \le t-s \le 72$, with all hidden 2- and $\eta$-extensions

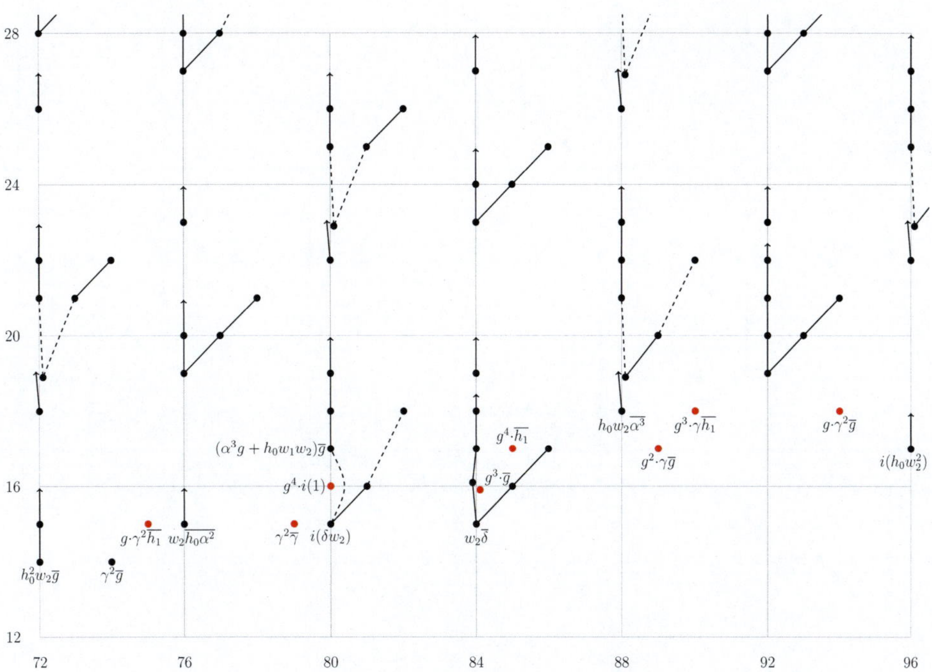

FIGURE 12.20. $E_\infty(tmf/\nu)$ for $72 \le t-s \le 96$, with all hidden 2- and $\eta$-extensions

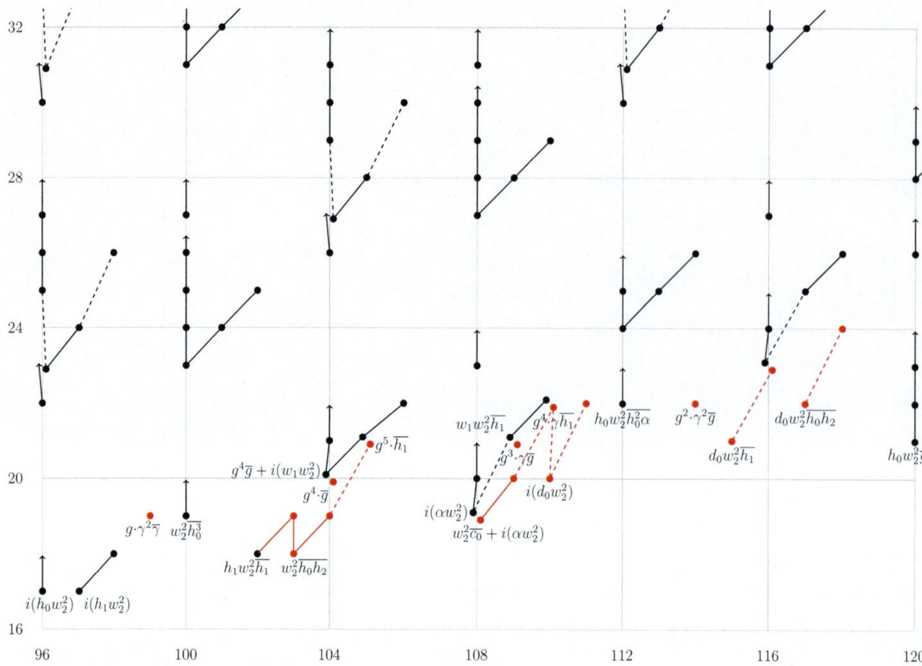

FIGURE 12.21. $E_\infty(tmf/\nu)$ for $96 \leq t-s \leq 120$, with all hidden 2- and $\eta$-extensions

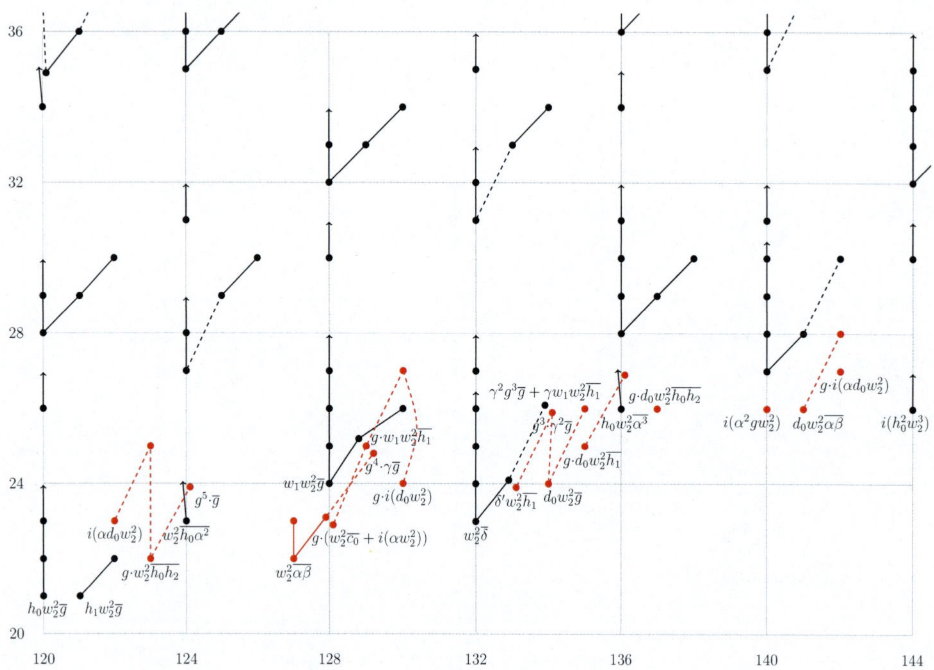

FIGURE 12.22. $E_\infty(tmf/\nu)$ for $120 \leq t-s \leq 144$, with all hidden 2- and $\eta$-extensions

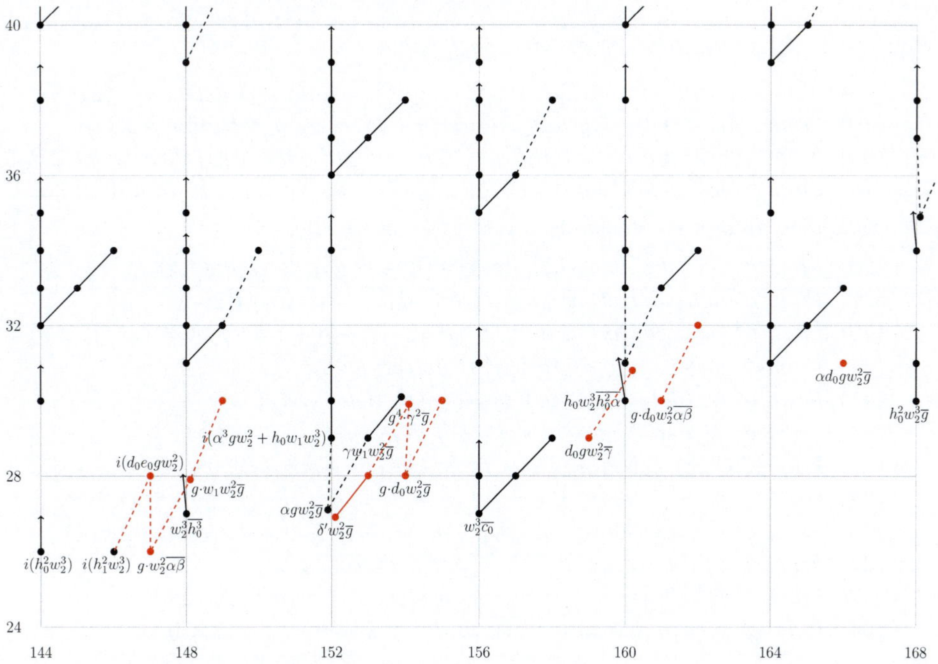

FIGURE 12.23. $E_\infty(tmf/\nu)$ for $144 \le t-s \le 168$, with all hidden 2- and $\eta$-extensions

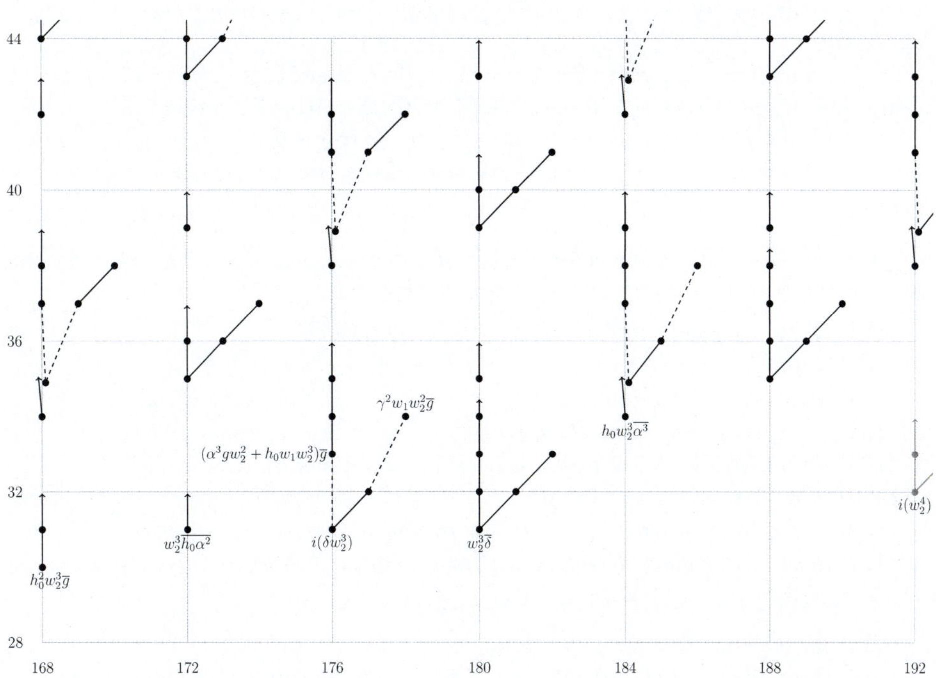

FIGURE 12.24. $E_\infty(tmf/\nu)$ for $168 \le t-s \le 192$, with all hidden 2- and $\eta$-extensions

THEOREM 12.15. *In the Adams spectral sequence for* $tmf/\nu$*, the following hidden* 2*-extensions repeat* $w_1$*- and* $w_2^4$*-periodically:*

(56) *From* $\alpha g\overline{g}$ *detecting* $i(B_2)+\epsilon_1\overline{\kappa}$ *to* $i(\alpha^3 g+h_0w_1w_2)$ *detecting* $i(2B_2)$.
(80) *From* $i(\delta w_2)$ *detecting* $i(B_3)$ *to* $(\alpha^3 g+h_0w_1w_2)\overline{g}$ *detecting* $i(2B_3)$.
(152) *From* $\alpha gw_2^2\overline{g}$ *detecting* $i(B_6)+\epsilon_5\overline{\kappa}$ *to* $i(\alpha^3 gw_2^2+h_0w_1w_2^3)$ *detecting* $i(2B_6)$.
(176) *From* $i(\delta w_2^3)$ *detecting* $i(B_7)$ *to* $(\alpha^3 gw_2^2+h_0w_1w_2^3)\overline{g}$ *detecting* $i(2B_7)$.

*The following hidden* 2*-extensions repeat* $w_2^4$*-periodically:*

(20) *From* $g\cdot i(1)$ *detecting* $i(\overline{\kappa})$ *to* $w_1\cdot(\overline{c_0}+i(\alpha))$ *detecting* $i(2\overline{\kappa})$.
(27) *From* $g\cdot\overline{h_0h_2}$ *detecting* $i(\nu_1)$ *to* $w_1\cdot d_0\overline{h_1}$ *detecting* $i(2\nu_1)$.
(40) *From* $g^2\cdot i(1)$ *detecting* $i(\overline{\kappa}^2)$ *to* $gw_1\cdot(\overline{c_0}+i(\alpha))$ *detecting* $i(2\overline{\kappa}^2)$.
(44) *From* $g\cdot\overline{g}$ *detecting* $\overline{\overline{\kappa}^2}$ *to* $i(\alpha^2 g)$ *detecting* $\overline{2\overline{\kappa}^2}$.
(51) *From* $g\cdot\overline{\alpha\beta}$ *detecting* $i(\nu_2)$ *to* $i(d_0e_0g)$ *detecting* $i(2\nu_2)$.
(64) *From* $g^2\cdot\overline{g}$ *detecting* $\overline{\overline{\kappa}^3}$ *to* $w_1\cdot\delta'\overline{g}$ *detecting a lift* $\overline{2\overline{\kappa}^3}$.
(110) *From* $i(d_0w_2^2)$ *detecting* $i(\kappa_4)$ *to* $g^4\cdot\gamma\overline{h_1}$ *detecting* $i(2\overline{\kappa}^4)$.
(123) *From* $g\cdot w_2^2\overline{h_0h_2}$ *detecting* $i(\nu_5)$ *to* $w_1\cdot d_0w_2^2\overline{h_1}$ *detecting* $i(2\nu_5)$.
(130) *From* $g\cdot i(d_0w_2^2)$ *detecting* $i(\kappa_4\overline{\kappa})$ *to* $w_1\cdot i(\alpha d_0w_2^2)$ *detecting* $i(2\kappa_4\overline{\kappa})$.
(134) *From* $d_0w_2^2\overline{g}$ *detecting* $\overline{\kappa_4\overline{\kappa}}$ *to* $g^3\cdot\gamma^2\overline{g}$ *detecting* $\overline{2\kappa_4\overline{\kappa}}$.
(147) *From* $g\cdot w_2^2\overline{\alpha\beta}$ *detecting* $i(\nu_6)$ *to* $i(d_0e_0gw_2^2)$ *detecting* $i(2\nu_6)$.
(154) *From* $g\cdot d_0w_2^2\overline{g}$ *detecting* $\overline{2\nu\nu_6}$ *to* $g^4\cdot\gamma^2\overline{g}$ *detecting a lift* $\overline{4\nu\nu_6}$.

*There are no other hidden* 2*-extensions in this spectral sequence.*

PROOF. (56) The image under $i$ of $c_0w_2=11_{24}$ detecting $B_2$ is $\delta\overline{g}=11_{40}$, which is the sum of the classes $\alpha g\overline{g}$ and $\delta'\overline{g}$, with the latter detecting $\epsilon_1\overline{\kappa}$.

(80) Since $i(B_3)$ is detected by $i(\delta w_2)$ and $i(8B_3)$ is detected by $h_0^2\cdot(\alpha^3 g+h_0w_1w_2)\overline{g}$ there must be a hidden 2-extension from the former class to $(\alpha^3 g+h_0w_1w_2)\overline{g}$.

(152) The image under $i$ of $c_0w_2^3=27_{116}$ detecting $B_6$ is $\delta w_2^2\overline{g}=27_{212}$, which is the sum of the classes $\alpha gw_2^2\overline{g}$ and $\delta'w_2^2\overline{g}$, with the latter detecting $\epsilon_5\overline{\kappa}$.

(176) Since $i(B_7)$ is detected by $i(\delta w_2^3)$ and $i(8B_7)$ is detected by $h_0^2\cdot(\alpha^3 gw_2^2+h_0w_1w_2^3)\overline{g}$ there must be a hidden 2-extension from the former class to $(\alpha^3 gw_2^2+h_0w_1w_2^3)\overline{g}$. □

THEOREM 12.16. *In the Adams spectral sequence for* $tmf/\nu$*, the following hidden* $\eta$*-extensions repeat* $w_1$*- and* $w_2^4$*-periodically:*

(12) *From* $i(\alpha)$ *detecting* $\overline{B}$ *to* $w_1\cdot\overline{h_1}$ *detecting* $\overline{\eta B}$.
(37) *From* $h_1\cdot\overline{\delta}$ *detecting* $\overline{\eta(B_1+\epsilon_1)}$ *to* $w_1\cdot\gamma\overline{h_1}$ *detecting* $\overline{\eta^2 B_1}$.
(56) *From* $\alpha g\overline{g}$ *detecting a lift* $\overline{\eta\nu_2}$ *to* $w_1\cdot\gamma\overline{g}$ *detecting a lift* $\overline{\eta^2\nu_2}$.
(81) *From* $h_1\cdot i(\delta w_2)$ *detecting* $i(\eta B_3)$ *to* $w_1\cdot\gamma^2\overline{g}$ *detecting* $i(\eta^2 B_3)$.
(108) *From* $i(\alpha w_2^2)$ *detecting* $\overline{B_4}$ *to* $w_1w_2^2\overline{h_1}$ *detecting* $\overline{\eta B_4}$.
(133b) *From* $h_1\cdot w_2^2\overline{\delta}$ *detecting* $\overline{\eta B_5+\epsilon_5}$ *to* $\gamma^2 g^3\overline{g}+\gamma w_1w_2^2\overline{h_1}$ *detecting* $\overline{\eta^2 B_5+\epsilon_5}$.
(152) *From* $\alpha gw_2^2\overline{g}$ *detecting a lift* $\overline{\eta\nu_6}$ *to* $\gamma w_1w_2^2\overline{g}$ *detecting a lift* $\overline{\eta^2\nu_6}$.
(177) *From* $h_1\cdot i(\delta w_2^3)$ *detecting* $i(\eta B_7)$ *to* $\gamma^2 w_1w_2^2\overline{g}$ *detecting* $i(\eta^2 B_7)$.

*The following hidden* $\eta$*-extensions repeat* $w_2^4$*-periodically:*

(14) *From* $i(d_0)$ *detecting* $i(\kappa)$ *to* $w_1\cdot\overline{h_0h_2}$ *detecting* $i(\eta\kappa)$.
(19) *From* $d_0\overline{h_1}$ *detecting* $\overline{\eta\kappa}$ *to* $w_1\cdot(\overline{c_0}+i(\alpha))$ *detecting* $i(2\overline{\kappa})$.
(20) *From* $g\cdot i(1)$ *detecting* $i(\overline{\kappa})$ *to* $d_0\overline{h_0h_2}$ *detecting* $i(\eta\overline{\kappa})$.
(21) *From* $d_0\overline{h_0h_2}$ *detecting* $i(\eta\overline{\kappa})$ *to* $w_1\cdot i(d_0)$ *detecting* $i(\eta^2\overline{\kappa})$.

(25) *From* $g \cdot \overline{h_1}$ *detecting a lift* $\overline{\eta\bar\kappa}$ *to* $i(\alpha d_0)$ *detecting a lift* $\overline{\eta^2\bar\kappa}$.
(26) *From* $i(\alpha d_0)$ *detecting a lift* $\overline{\eta^2\bar\kappa}$ *to* $w_1 \cdot d_0\overline{h_1}$ *detecting* $i(2\nu_1)$.
(32) *From* $g \cdot (\overline{c_0} + i(\alpha))$ *detecting* $y_{32} = i(B_1) + B\bar\kappa$ *to* $gw_1 \cdot \overline{h_1}$ *detecting* $\eta y_{32}$.
(38) *From* $d_0\overline{g}$ *detecting* $\overline{\kappa\bar\kappa}$ *to* $w_1 \cdot \overline{\alpha\beta}$ *detecting* $\overline{\eta\kappa\bar\kappa}$.
(39) *From* $g \cdot d_0\overline{h_1}$ *detecting* $i(\eta_1\kappa)$ *to* $gw_1 \cdot (\overline{c_0} + i(\alpha))$ *detecting* $i(2\bar\kappa^2)$.
(40) *From* $g^2 \cdot i(1)$ *detecting* $i(\bar\kappa^2)$ *to* $g \cdot d_0\overline{h_0h_2}$ *detecting* $i(\eta\bar\kappa^2)$.
(44) *From* $g \cdot \overline{g}$ *detecting* $\overline{\kappa}^2$ *to* $d_0\overline{\alpha\beta}$ *detecting a lift* $\overline{\eta\bar\kappa^2}$.
(45a) *From* $g^2 \cdot \overline{h_1}$ *detecting* $i(\eta_1\bar\kappa)$ *to* $g \cdot i(\alpha d_0)$ *detecting* $i(\eta\eta_1\bar\kappa)$.
(45b) *From* $d_0\overline{\alpha\beta}$ *detecting a lift* $\overline{\eta\bar\kappa^2}$ *to* $w_1 \cdot d_0\overline{g}$ *detecting a lift* $\overline{\eta^2\bar\kappa^2}$.
(50a) *From* $g \cdot \gamma\overline{h_1}$ *detecting* $\bar\kappa\overline{\eta\eta_1}$ *to* $i(d_0e_0g)$ *detecting* $i(2\nu_2)$.
(50b) *From* $h_1 \cdot \gamma\overline{g}$ *detecting* $\eta\overline{\eta_1\bar\kappa}$ *to* $i(d_0e_0g)$ *detecting* $i(2\nu_2)$.
(51) *From* $g \cdot \overline{\alpha\beta}$ *detecting* $i(\nu_2)$ *to* $gw_1 \cdot \overline{g}$ *detecting* $i(\eta\nu_2)$.
(58) *From* $g \cdot d_0\overline{g}$ *detecting* $\overline{2\nu\nu_2}$ *to* $gw_1 \cdot \overline{\alpha\beta}$ *detecting* $i(\nu_2\epsilon)$.
(63) *From* $d_0g\overline{\gamma}$ *detecting* $\overline{\nu_2\epsilon}$ *to* $w_1 \cdot \delta'\overline{g}$ *detecting* $\overline{2\bar\kappa^3}$.
(64) *From* $g^2 \cdot \overline{g}$ *detecting* $\overline{\kappa}^3$ *to* $g \cdot d_0\overline{\alpha\beta}$ *detecting* $i(\nu_2\kappa)$.
(65) *From* $g \cdot d_0\overline{\alpha\beta}$ *detecting* $i(\nu_2\kappa)$ *to* $gw_1 \cdot d_0\overline{g}$ *detecting* $i(\eta\nu_2\kappa)$.
(69) *From* $g \cdot \gamma\overline{g}$ *detecting* $\overline{\eta_1\bar\kappa^2}$ *to* $\alpha d_0g\overline{g}$ *detecting a lift* $\overline{\eta\eta_1\bar\kappa^2}$.
(104) *From* $h_1 \cdot w_2^2\overline{h_0h_2}$ *detecting* $i(\epsilon_4)$ *to* $g^5 \cdot \overline{h_1}$ *detecting* $i(\eta\epsilon_4) = i(\eta_1\bar\kappa^4)$.
(109) *From* $h_1 \cdot (w_2^2\overline{c_0} + i(\alpha w_2^2))$ *detecting* $\eta\overline{\epsilon_4}$ *to* $g^4 \cdot \gamma\overline{h_1}$ *detecting* $i(2\kappa_4)$.
(110) *From* $i(d_0w_2^2)$ *detecting* $i(\kappa_4)$ *to* $w_1 \cdot w_2^2\overline{h_0h_2}$ *detecting* $i(\eta\kappa_4)$.
(115) *From* $d_0w_2^2\overline{h_1}$ *detecting* $\overline{\eta\kappa_4}$ *to* $w_1 \cdot (w_2^2\overline{c_0} + i(\alpha w_2^2))$ *detecting* $i(\bar\kappa D_4)$.
(117) *From* $d_0w_2^2\overline{h_0h_2}$ *detecting* $i(\eta_4\bar\kappa)$ *to* $w_1 \cdot i(d_0w_2^2)$ *detecting* $i(\eta\eta_4\bar\kappa)$.
(122) *From* $i(\alpha d_0w_2^2)$ *detecting a lift* $\overline{\eta\eta_4\bar\kappa}$ *to* $w_1 \cdot d_0w_2^2\overline{h_1}$ *detecting* $i(2\nu_5)$.
(123) *From* $g \cdot w_2^2\overline{h_0h_2}$ *detecting* $i(\nu_5)$ *to* $g^5 \cdot \overline{g}$ *detecting* $i(\eta\nu_5)$.
(128a) *From* $g \cdot (w_2^2\overline{c_0} + i(\alpha w_2^2))$ *detecting* $y_{128} = i(B_5) + B_4\bar\kappa$ *to* $g \cdot w_1w_2^2\overline{h_1}$ *detecting* $\eta y_{128}$.
(128b) *From* $h_1 \cdot w_2^2\overline{\alpha\beta}$ *detecting* $\eta\overline{\nu_5}$ *to* $g^4 \cdot \gamma\overline{g}$ *detecting* $\eta^2\overline{\nu_5}$.
(129) *From* $g \cdot w_1w_2^2\overline{h_1}$ *detecting* $\eta y_{128}$ *to* $w_1 \cdot i(\alpha d_0w_2^2)$ *detecting* $i(2\kappa_4\bar\kappa)$.
(133a) *From* $\delta'w_2^2\overline{h_1}$ *detecting* $\overline{\eta\epsilon_5}$ *to* $g^3 \cdot \gamma^2\overline{g}$ *detecting* $\overline{2\kappa_4\bar\kappa}$.
(134) *From* $d_0w_2^2\overline{g}$ *detecting* $\overline{\kappa_4\bar\kappa}$ *to* $w_1 \cdot w_2^2\overline{\alpha\beta}$ *detecting a lift* $\overline{\eta\kappa_4\bar\kappa}$.
(135) *From* $g \cdot d_0w_2^2\overline{h_1}$ *detecting* $i(\eta_1\kappa_4)$ *to* $gw_1 \cdot (w_2^2\overline{c_0} + i(\alpha w_2^2))$ *detecting* $i(\eta\eta_1\kappa_4)$.
(141) *From* $d_0w_2^2\overline{\alpha\beta}$ *detecting* $\overline{\nu_5\kappa}$ *to* $w_1 \cdot d_0w_2^2\overline{g}$ *detecting a lift* $\overline{\eta\nu_5\kappa}$.
(146) *From* $i(h_1^2w_2^3)$ *detecting* $\overline{\epsilon_5\kappa}$ *to* $i(d_0e_0gw_2^2)$ *detecting* $i(2\nu_6)$.
(147) *From* $g \cdot w_2^2\overline{\alpha\beta}$ *detecting* $i(\nu_6)$ *to* $g \cdot w_1w_2^2\overline{g}$ *detecting* $i(\eta\nu_6)$.
(148) *From* $g \cdot w_1w_2^2\overline{g}$ *detecting* $i(\eta\nu_6)$ *to* $w_1 \cdot d_0w_2^2\overline{\alpha\beta}$ *detecting* $i(\eta^2\nu_6)$.
(153) *From* $h_1 \cdot \delta'w_2^2\overline{g}$ *detecting* $\overline{\eta^2\nu_6}$ *to* $g^4 \cdot \gamma^2\overline{g}$ *detecting a lift* $\overline{4\nu\nu_6}$.
(154) *From* $g \cdot d_0w_2^2\overline{g}$ *detecting* $\overline{2\nu\nu_6}$ *to* $gw_1 \cdot w_2^2\overline{\alpha\beta}$ *detecting* $i(\nu_6\epsilon)$.
(159) *From* $d_0gw_2^2\overline{\gamma}$ *detecting* $\overline{\nu_6\epsilon}$ *to* $w_1 \cdot \delta'w_2^2\overline{g}$ *detecting* $\overline{\eta\nu_6\epsilon}$.
(161) *From* $g \cdot d_0w_2^2\overline{\alpha\beta}$ *detecting* $i(\nu_6\kappa)$ *to* $gw_1 \cdot d_0w_2^2\overline{g}$ *detecting* $i(\eta\nu_6\kappa)$.

*There are no other hidden* $\eta$*-extensions in this spectral sequence. In particular, there are no hidden* $\eta$*-extensions on* $h_1 \cdot \overline{\alpha\beta}$ *or on* $g^4 \cdot \gamma\overline{g}$.

PROOF. (12) To see that $i(\alpha)$ detects $\overline{B}$ we can use that $d_2(\alpha) = h_2 \cdot w_1$ in $E_2(tmf)$, or note that $j$ maps $\overline{c_0}$ and $\overline{c_0} + i(\alpha)$ to $c_0$, so only $i(\alpha)$ can detect a class mapping to $B$.

(19) From $E_\infty(tmf/\eta)$ we see that $\pi_{21}(tmf/(\eta,\nu)) = 0$, so that $\eta$ acts injectively on $\pi_{19}(tmf/\nu)$. Hence $\eta \cdot \overline{\eta\bar\kappa}$ is nonzero, and must be equal to $i(2\bar\kappa)$.

(25) Since $h_1$ times $g \cdot \overline{h_1}$ is zero, $\eta$ times the lift of $\eta\bar{\kappa}$ detected by this class has Adams filtration 7 or 8, hence must be detected by $i(\alpha d_0)$.

(26) We multiply case (12) by $\kappa$ to deduce this extension.

(32) The class $B_1$ detected by $\alpha g = 7_{11} + 7_{12}$ maps to $i(B_1)$ in $\pi_{32}(tmf/\nu)$ detected by $i(\alpha g) = 7_{16} = g \cdot (\overline{c_0} + i(\alpha))$ in Adams filtration 7. Its $\eta$-multiple $i(\eta B_1) = Bi(\eta_1)$ is detected by $w_1 \cdot i(\gamma) = 9_{17} + 9_{18} = gw_1 \cdot \overline{h_1} + h_1 w_1 \cdot \overline{g}$. The filtration 8 class $B\overline{\kappa}$ is detected by $w_1 \cdot \overline{g}$, and $\eta B\overline{\kappa}$ is detected by $h_1 w_1 \cdot \overline{g} = 9_{18}$. The sum $y_{32} = i(B_1) + B\overline{\kappa}$ is also detected by $g \cdot (\overline{c_0} + i(\alpha))$, with $\eta y_{32}$ detected by $gw_1 \cdot \overline{h_1} = 9_{17}$. Hence there is a hidden $\eta$-extension from $g \cdot (\overline{c_0} + i(\alpha))$ to $gw_1 \cdot \overline{h_1}$.

Furthermore, the class $\epsilon_1$ detected by $\delta' = 7_{12}$ maps to $i(\epsilon_1)$ in $\pi_{32}(tmf/\nu)$ detected by $i(\delta') = 7_{16} + 7_{17} = g \cdot (\overline{c_0} + i(\alpha)) + h_1 \cdot \overline{\alpha\beta}$. Its $\eta$-multiple $i(\eta\epsilon_1)$ is nonzero and $B$-power torsion, hence must be detected by $gw_1 \cdot \overline{h_1} = 9_{17}$ in $E_\infty(tmf/\nu)$. It follows that there is no hidden $\eta$-extension from $h_1 \cdot \overline{\alpha\beta}$.

(50a) Multiplying the relation $\eta \cdot \overline{\eta\eta_1} = 2 \cdot \overline{\nu_1}$ in $\pi_{31}(tmf/\nu)$ by $\bar{\kappa}$, we see that $\eta$ times the class $\bar{\kappa}\overline{\eta\eta_1}$ detected by $g \cdot \gamma\overline{h_1}$ is 2 times the class $\bar{\kappa}\overline{\nu_1}$ detected by $g \cdot \overline{\alpha\beta}$. This common value must be equal to $i(2\nu_2)$ detected by $i(d_0 e_0 g)$.

(50b) The class $\eta_1^2$ detected by $\gamma^2 = 10_{20} + 10_{21}$ maps to $i(\eta_1^2)$ in $\pi_{50}(tmf/\nu)$ detected by $i(\gamma^2) = 10_{32} + 10_{33} = g \cdot \gamma\overline{h_1} + h_1 \cdot \gamma\overline{g}$. The $\eta$-multiple $\eta \cdot \eta_1^2 = \nu D_2$ maps to zero in $\pi_{51}(tmf/\nu)$. Hence $\eta$ times $\eta\eta_1\bar{\kappa}$ detected by $h_1 \cdot \gamma\overline{g}$ is equal to $\eta$ times $\bar{\kappa}\overline{\eta\eta_1}$ detected by $g \cdot \gamma\overline{h_1}$, i.e., $i(2\nu_2)$ detected by $i(d_0 e_0 g)$.

(56) The class $B_2$ detected by $c_0 w_2 = 11_{24}$ maps to $i(B_2)$ in $\pi_{56}(tmf/\nu)$ detected by $i(c_0 w_2) = 11_{40} = \delta\overline{g}$. Hence $\alpha g\overline{g}$ and $\delta'\overline{g}$ both detect lifts of $\eta\nu_2$ in ${}_\nu\pi_{52}(tmf)$. Since $\eta \cdot \eta\nu_2$ is nonzero of Adams filtration 13, $\eta$ times the lift detected by $\alpha g\overline{g}$ is nonzero of Adams filtration exactly 13, hence is detected by $w_1 \cdot \gamma\overline{g}$.

(58) We multiply case (38) by $\bar{\kappa}$ to deduce this extension.

(64) We multiply case (44) by $\bar{\kappa}$ to deduce that there is a hidden $\eta$-extension from $g^2 \cdot \overline{g}$ detecting $\overline{\kappa^3}$ to $g \cdot d_0\overline{\alpha\beta}$. The latter is the class that detects $i(\nu_2\kappa)$ because $i(h_2 w_2 \cdot d_0) = 0$.

(108) To see that $i(\alpha w_2^2)$ detects $\overline{B_4}$ we can use that $d_2(\alpha w_2^2) = h_2 \cdot w_1 w_2^2$ in $E_2(tmf)$, or note that $j$ maps $w_2^2\overline{c_0}$ and $w_2^2\overline{c_0} + i(\alpha w_2^2)$ to $c_0 w_2^2$, so only $i(\alpha w_2^2)$ can detect a class mapping to $B_4$.

(115) From $E_\infty(tmf/\eta)$ we see that $\pi_{117}(tmf/(\eta, \nu)) = \mathbb{Z}/2$. Since $d_0 w_2^2\overline{h_0 h_2}$ detects a class in $\pi_{117}(tmf/\nu)$ that cannot be an $\eta$-multiple, we see that $\eta$ acts injectively on $\pi_{115}(tmf/\nu)$. Hence $\eta \cdot \overline{\eta\kappa_4}$ is nonzero, and must be equal to $i(\bar{\kappa}D_4)$.

(122) We multiply case (108) by $\kappa$ to deduce this extension.

(128a) The class $B_5$ detected by $\alpha g w_2^2 = 23_{87} + 23_{88}$ maps by $i$ to $i(B_5)$ in $\pi_{128}(tmf/\nu)$ detected by $i(\alpha g w_2^2) = 23_{156} = g \cdot (w_2^2\overline{c_0} + i(\alpha w_2^2))$ in Adams filtration 23. Its $\eta$-multiple $i(\eta B_5)$ is detected by $i(\gamma w_1 w_2^2) = 25_{168} + 25_{169} = g \cdot w_1 w_2^2\overline{h_1} + h_1 \cdot w_1 w_2^2\overline{g}$. The filtration 24 class $B_4\overline{\kappa}$ is detected by $w_1 w_2^2\overline{g}$, and $\eta B_4\overline{\kappa}$ is detected by $h_1 \cdot w_1 w_2^2\overline{g} = 25_{169}$. The sum $y_{128} = i(B_5) + B_4\overline{\kappa}$ is also detected by $g \cdot (w_2^2\overline{c_0} + i(\alpha w_2^2))$, with $\eta y_{128}$ detected by $g \cdot w_1 w_2^2\overline{h_1}$. Hence there is a hidden $\eta$-extension from $g \cdot (w_2^2\overline{c_0} + i(\alpha w_2^2))$ to $g \cdot w_1 w_2^2\overline{h_1}$.

(128b) From $E_\infty(tmf/\eta)$ we see that $\pi_{129}(tmf/(\eta, \nu)) = \mathbb{Z}/2$, and $\eta$ acts trivially on $\overline{2\nu_5}$ in $\pi_{127}(tmf/\nu)$ detected by $h_0 \cdot w_2^2\overline{\alpha\beta}$, so each class in $\pi_{129}(tmf/\nu)$ is an $\eta$-multiple. By the previous case $\eta B_4\overline{\kappa}$ is detected by $h_1 \cdot w_1 w_2^2\overline{g}$ and $\eta y_{128}$ is detected by $g \cdot w_1 w_2^2\overline{h_1}$. Hence $\eta^2\overline{\nu_5}$ must be detected by $g^4 \cdot \gamma\overline{g}$, modulo these two classes. Since $\eta^2\overline{\nu_5}$ is $B$-power torsion and a lift of $\eta^2\nu_5$, and $j$ maps $g \cdot w_1 w_2^2\overline{h_1}$ and

$g^4 \cdot \gamma\overline{g}$ to $h_1 g w_1 w_2^2 = \gamma g^5$ in $E_\infty(tmf)$ detecting $\eta^2\nu_5$, it follows that $\eta^2\overline{\nu_5}$ must be detected by precisely $g^4 \cdot \gamma\overline{g}$.

(129) There is no hidden $\eta$-extension on $g^4 \cdot \gamma\overline{g}$, because $\nu \cdot \overline{\nu_5}$ must have Adams filtration $\geq 24$, so that $2\nu \cdot \overline{\nu_5}$ must have Adams filtration $\geq 27$, and there is no $w_1$-power torsion class in Adams filtration $\geq 28$ that can detect $\eta^3 = 4\nu$ times $\overline{\nu_5}$. Alternatively, we can use that there is no visible or hidden $\eta$-multiplication on $g^3 \cdot \gamma\overline{g}$, and multiply by $\bar\kappa$.

Furthermore, from $E_\infty(tmf/\eta)$ we see that $\pi_{131}(tmf/(\eta,\nu)) = \mathbb{Z}/2$, while $\pi_{131}(tmf/\nu) = 0$, so ${}_\eta\pi_{129}(tmf/\nu) = \mathbb{Z}/2$. It follows that $\eta$ times $\eta y_{128}$ is nonzero, so that there is a hidden $\eta$-extension from $g \cdot w_1 w_2^2 \overline{h_1}$ to $w_1 \cdot i(\alpha d_0 w_2^2)$.

(109) The class $w_2^2\overline{c_0} + i(\alpha w_2^2)$ detects a $B^2$-torsion lift $y_{108}$ in $\pi_{108}(tmf/\nu)$ of $\epsilon_4$, and $g \cdot (w_2^2\overline{c_0} + i(\alpha w_2^2))$ detects $\bar\kappa \cdot y_{108}$. By cases (128a) and (129) there is a hidden $\eta^2$-extension from $g \cdot (w^2\overline{c_0} + i(\alpha w_2^2))$ to $w_1 \cdot i(\alpha d_0 w_2^2)$. Since $\eta^2\bar\kappa \cdot y_{108}$ is nonzero, it follows that $\eta^2 \cdot y_{108}$ is nonzero, and only $g^4 \cdot \gamma\overline{h_1}$ can detect this product. Hence there is also a hidden $\eta$-extension from $h_1 \cdot (w_2^2\overline{c_0} + i(\alpha w_2^2))$ to $g^4 \cdot \gamma\overline{h_1}$.

(133b) The class $w_2^2\overline{\delta}$ detects $\overline{B_5 + \epsilon_5}$ in $\pi_{132}(tmf/\nu)$, so $h_1 \cdot w_2^2\overline{\delta}$ detects $\eta\overline{B_5 + \epsilon_5}$ and $\eta^2 \cdot \overline{B_5 + \epsilon_5}$ must be detected by a lift over $j$ of $h_1\gamma w_1 w_2^2 + \gamma^2 g^4$, i.e., by $\gamma^2 g^3\overline{g} + \gamma w_1 w_2^2\overline{h_1}$.

(146) From $E_\infty(tmf/\eta)$ and case (147) of Theorem 12.15 we can read off that $\pi_{147}(tmf/(\eta,\nu)) = \mathbb{Z}/2$. Since $\eta$ acts injectively on $\pi_{145}(tmf/\nu)$, by case (81) above, it follows that $i(d_0 e_0 g w_2^2)$ must detect an $\eta$-multiple. The only possible source of this multiplication is $\overline{\epsilon_5\kappa}$ detected by $i(h_1^2 w_2^3)$.

(152) The class $B_6$ detected by $c_0 w_2^3 = 27_{116}$ maps to $i(B_6)$ in $\pi_{152}(tmf/\nu)$ detected by $i(c_0 w_2^3) = 27_{212} = \delta w_2^2\overline{g}$. Hence $\alpha g w_2^2\overline{g}$ and $\delta' w_2^2\overline{g}$ both detect lifts of $\eta\nu_6$ in ${}_\nu\pi_{148}(tmf)$. Since $\eta \cdot \eta\nu_6$ is nonzero of Adams filtration 29, $\eta$ times the lift detected by $\alpha g w_2^2\overline{g}$ is nonzero of Adams filtration exactly 29, hence is detected by $\gamma w_1 w_2^2\overline{g}$.

(154) We multiply case (134) by $\bar\kappa$ to deduce this extension. □

It follows from Proposition 8.10 that $\pi_*(tmf/\nu)$ is generated as a $\pi_*(tmf)$-module by elements detected by the classes listed in Table 8.10, where we may assume that the $w_1$-power torsion classes are represented by $B$-power torsion elements. We omit to enumerate 34 such elements.

Let $(N/\nu)_* \subset \pi_*(tmf/\nu)$ denote the $\mathbb{Z}[B]$-submodule generated by all classes in degrees $0 \leq * < 192$. There is an isomorphism

$$(N/\nu)_* \otimes \mathbb{Z}[M] \cong \pi_*(tmf/\nu)$$

of $\mathbb{Z}[B, M]$-modules. The submodule $(N/\nu)_*$ is preserved by the action of $\eta$, $\nu$, $\epsilon$, $\kappa$ and $\bar\kappa$ (because $\bar\kappa \cdot \overline{B_7} = 0$, which follows from $g \cdot w_2^3\overline{\delta} = 0$ in $E_\infty(tmf/\nu)$), and the isomorphism respects these actions.

In most degrees it is straightforward to read off the group structure of $(N/\nu)_*$, together with its $\eta$-action, from $E_\infty(tmf/\nu)$ with the hidden 2- and $\eta$-extensions, keeping in mind that the $w_1$-power torsion classes form the associated graded of the restriction to $\Gamma_B(N/\nu)_*$ of the Adams filtration. The next result summarizes some less obvious cases.

PROPOSITION 12.17.

(27) *The product of $\eta$ with $\eta^2\bar\kappa$, detected by $h_1^2 \cdot \overline{g}$, is zero.*

(66) *The product of* $\eta$ *with* $i(\eta_1\bar\kappa^2)$, *detected by* $g^3 \cdot \overline{h_1}$, *is* $i(\eta\nu_2\kappa)$ *detected by* $gw_1 \cdot d_0\overline{g}$.

(123) *The product of* $\eta$ *with* $\eta\overline{\eta_4\bar\kappa}$, *detected by* $h_1 \cdot h_1w_2^2\overline{g}$, *is zero.*

(149) *A lift* $\overline{2D_6}$, *detected by* $w_2^3h_0^3$, *can be chosen so that* $\eta \cdot \overline{2D_6}$ *is zero.*

PROOF. (27) Four times $\nu\bar\kappa$ is zero in $\pi_{27}(tmf/\nu)$.

(66) The detection follows from $i(\gamma g^2) = g^3 \cdot \overline{h_1}$.

(123) Using [**171**, Prop. 1.8], $\eta^2 \cdot \overline{\eta_4\bar\kappa} = \pm i(z)$ where $z \in \langle \eta^2, \eta_4\bar\kappa, \nu\rangle$. By Moss' theorem [**132**, Thm. 1.2], this Toda bracket is weakly detected by the Massey product $\langle h_1^2, h_1gw_2^2, h_2\rangle$, which `ext` calculates is zero. It follows that $z = 0$.

(149) There are two lifts of $2D_6$ over $j$, differing by $i(\eta\nu_6)$, and multiplication by $\eta$ annihilates precisely one of them. □

## 12.4. Homotopy of $tmf/B$

We study $\pi_*(tmf/B)$ using the short exact sequence

$$(12.1) \qquad 0 \to \pi_*(tmf)/B \xrightarrow{i} \pi_*(tmf/B) \xrightarrow{j} {}_B\pi_{*-9}(tmf) \to 0$$

of $\pi_*(tmf)$-modules, where

$$\pi_n(tmf)/B = \mathrm{cok}(B\colon \pi_{n-8}(tmf) \to \pi_n(tmf))$$
$${}_B\pi_{n-9}(tmf) = \ker(B\colon \pi_{n-9}(tmf) \to \pi_{n-1}(tmf))\,.$$

Since $B\colon \Sigma^8 tmf \to tmf$ has Adams filtration 4, there is a split extension of Adams $E_2$-terms

$$0 \to E_2^{*,*}(tmf) \xrightarrow{i} E_2^{*,*}(tmf/B) \xrightarrow{j} E_2^{*,*-9}(tmf) \to 0$$

which persists to the $E_4$-term. However, at this stage the differentials in $E_4(tmf)$ and the action of $B$ will interact. To avoid this interference, we instead consider the delayed Adams spectral sequence $(E_r(Z_\star), d_r)$ of the $tmf$-module tower $Z_\star$ given by

$$tmf/B \xleftarrow{i} tmf \longleftarrow * \,.$$

See [**45**, §VI.6] and Definition 11.10. We set $Z_k = *$ for $k \geq 2$, $Z_1 = tmf$ and $Z_0 = tmf/B$. The nontrivial filtration quotients are then $Z_{1,1} \simeq tmf$ and $Z_{0,1} \simeq \Sigma^9 tmf$. Letting $(S_\star, \alpha)$ be an Adams resolution for $S$, the delayed Adams spectral sequence $(E_r(Z_\star), d_r)$ is associated to the convolved filtration $(S \wedge Z)_\star$. There is a homotopy cofiber sequence

$$S_s \wedge \Sigma^8 tmf \overset{\alpha\wedge B}{\longrightarrow} S_{s-1} \wedge tmf \longrightarrow (S \wedge Z)_s \longrightarrow S_s \wedge \Sigma^9 tmf$$

of filtered spectra, where the structure map $\alpha\colon S_s \to S_{s-1}$ has Adams filtration 1. The associated homotopy cofiber sequence of filtration quotients induces a long exact sequence in homotopy, which breaks up into split short exact sequences

$$(12.2) \qquad 0 \to E_r^{*-1,*-1}(tmf) \longrightarrow E_r^{*,*}(Z_\star) \longrightarrow E_r^{*,*-9}(tmf) \to 0$$

for $r = 1$ and $r = 2$, as in [**45**, Thm. VI.6.1(i)] and Theorem 11.11. Furthermore, the connecting map

$$\bar\alpha \wedge B\colon S_{s,r} \wedge \Sigma^8 tmf \longrightarrow S_{s-1,r} \wedge tmf$$

induces zero in homotopy for $r \leq 5$, which by the proof of Proposition 5.4 of [**148**] implies that (12.2) remains short exact for $r \leq 5$. The module structure over

$E_r(tmf)$ ensures that the sequence remains split, and that the $d_5$-differential is given by multiplication with $w_1$ detecting $B$. Hence we have a short exact sequence

$$0 \to E_\infty^{*-1,*-1}(tmf)/w_1 \longrightarrow E_6^{*,*}(Z_\star) \longrightarrow {}_{w_1}E_\infty^{*,*-9}(tmf) \to 0$$

of $E_\infty(tmf)$-modules, where $E_5(tmf) = E_\infty(tmf)$. The resulting $E_6$-term is displayed in Figures 12.25 to 12.32, and it follows by inspection that there is no room for any further differentials, so that $E_6(Z_\star) = E_\infty(Z_\star)$ in the delayed Adams spectral sequence converging to $\pi_*(tmf/B)$ (implicitly 2-completed).

In these charts, the filled (black) circles show the image of the cokernel of $w_1$, offset by $(t-s, s) = (0, 1)$ bidegrees, while the open (white) circles show lifts of the kernel of $w_1$, offset by $(9, 0)$ bidegrees. Hence the class labeled $c_0$ in bidegree $(8, 4)$ is the image of $c_0$ in $E_\infty(tmf)/w_1$, while the class labeled $c_0$ in bidegree $(17, 3)$ is the unique lift of $c_0$ in ${}_{w_1}E_\infty(tmf)$. The black lines (solid, dashed or dotted) show 2-, $\eta$- and $\nu$-extensions within the image of $E_\infty(tmf)/w_1$ and the lift of ${}_{w_1}E_\infty(tmf)$.

Hidden extensions from the lift to the image are shown in red (dashed or dotted). These will be determined in Theorems 12.19, 12.20 and 12.21. Using these, we can specify

(51) the lift of $d_0gw_1$ to detect $\eta^2$ times a class detected by the lift of $g^2$,
(99) the lift of $\gamma^2g^2$ to detect $\bar\kappa$ times the class detected by the lift of $\gamma^2g$, and
(147) the lift of $d_0gw_1w_2^2$ to detect $\eta$ times the class detected by the lift of $\alpha\beta d_0w_2^2$.

Let $(N/B)_* \subset \pi_*(tmf/B)$ denote the graded subgroup of classes in degrees $0 \le * < 192$. Since $w_2^4$ acts freely on the delayed $E_\infty(tmf/B)$, we have an isomorphism

$$(N/B)_* \otimes \mathbb{Z}[M] \cong \pi_*(tmf/B)$$

of $\mathbb{Z}[M]$-modules. The subgroup $(N/B)_*$ is preserved by the action of $\eta$, $\nu$, $\epsilon$, $\kappa$, $\bar\kappa$ and $B$, and the isomorphism respects these actions. We have isomorphisms

$$(N/B)_* \cong \pi_*(tmf/B)/M \cong \pi_*(tmf/(B, M))$$

and a short exact sequence

$$0 \to N_*/B \xrightarrow{i} (N/B)_* \xrightarrow{j} {}_BN_{*-9} \to 0\,,$$

where $N_*$ is as in Definition 9.25. Recall the Anderson duality functor $I_\mathbb{Z}$ from Section 10.4.

PROPOSITION 12.18. *The spectrum $tmf/(B, M)$ is Anderson self-dual, in the sense that there is an equivalence of $tmf$-modules*

$$tmf/(B, M) \simeq \Sigma^{180} I_\mathbb{Z}(tmf/(B, M))\,.$$

*Hence there is a short exact sequence*

$$0 \to \mathrm{Ext}((N/B)_{n-1}, \mathbb{Z}) \longrightarrow (N/B)_{180-n} \longrightarrow \mathrm{Hom}((N/B)_n, \mathbb{Z}) \to 0$$

*for each integer $n$.*

PROOF. By Proposition 10.12 we have an equivalence

$$\Sigma^{20} tmf \simeq I_\mathbb{Z}(tmf/(B^\infty, M^\infty))\,.$$

The homotopy cofiber sequences

$$tmf/(B, M^\infty) \longrightarrow \Sigma^8 tmf/(B^\infty, M^\infty) \xrightarrow{B} tmf/(B^\infty, M^\infty)$$
$$tmf/(B, M) \longrightarrow \Sigma^{192} tmf/(B, M^\infty) \xrightarrow{M} tmf/(B, M^\infty)$$

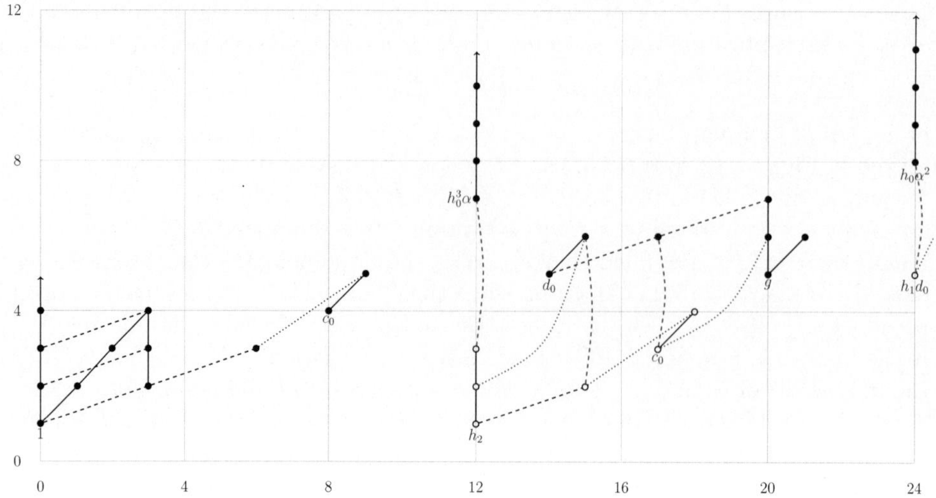

FIGURE 12.25. Delayed $E_\infty(tmf/(B,M))$ for $0 \le t-s \le 24$, with all hidden 2-, $\eta$- and $\nu$-extensions

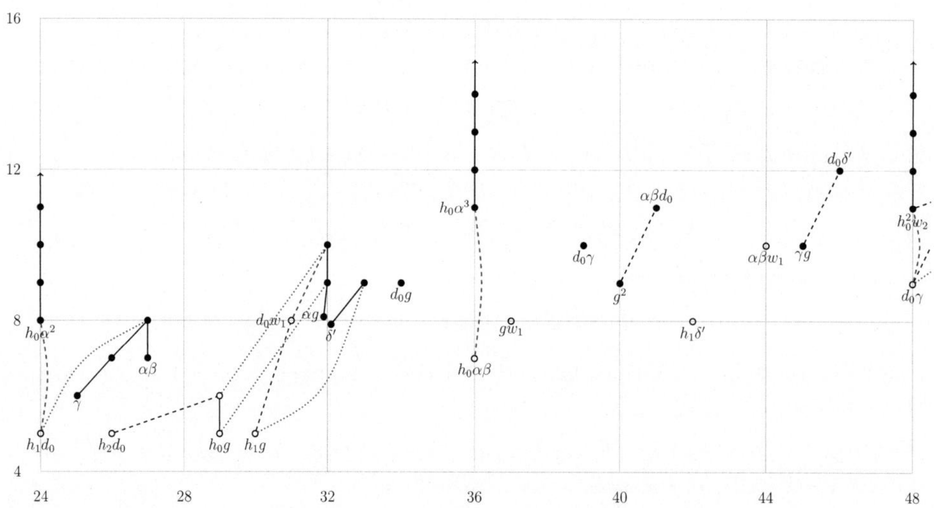

FIGURE 12.26. Delayed $E_\infty(tmf/(B,M))$ for $24 \le t-s \le 48$, with all hidden 2-, $\eta$- and $\nu$-extensions

from (10.2) dualize to homotopy cofiber sequences

$$I_{\mathbb{Z}}(tmf/(B^\infty, M^\infty)) \xrightarrow{B} \Sigma^{-8} I_{\mathbb{Z}}(tmf/(B^\infty, M^\infty)) \longrightarrow I_{\mathbb{Z}}(tmf/(B, M^\infty))$$
$$I_{\mathbb{Z}}(tmf/(B, M^\infty)) \xrightarrow{M} \Sigma^{-192} I_{\mathbb{Z}}(tmf/(B, M^\infty)) \longrightarrow I_{\mathbb{Z}}(tmf/(B, M)),$$

which translate to equivalences

$$\Sigma^{12} tmf/B \simeq I_{\mathbb{Z}}(tmf/(B, M^\infty)$$
$$\Sigma^{-180} tmf/(B,M) \simeq I_{\mathbb{Z}}(tmf/(B,M)).$$

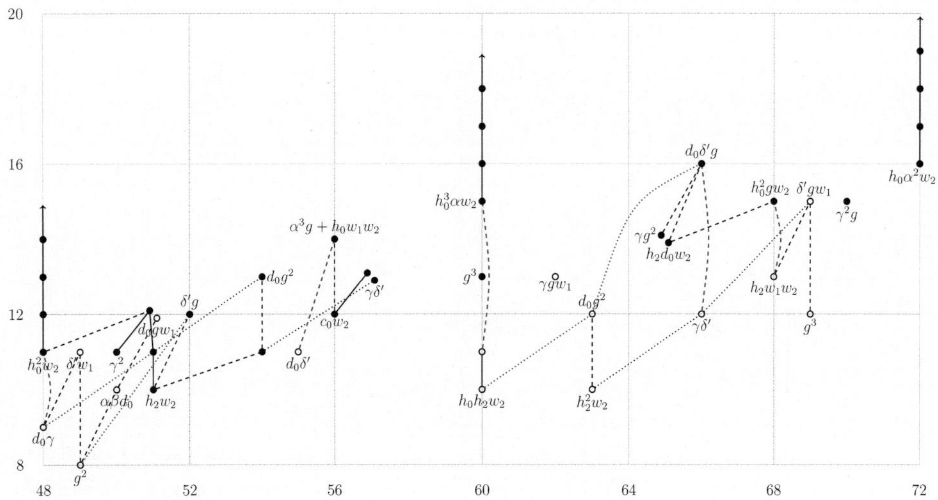

FIGURE 12.27. Delayed $E_\infty(tmf/(B,M))$ for $48 \leq t-s \leq 72$, with all hidden 2-, $\eta$- and $\nu$-extensions

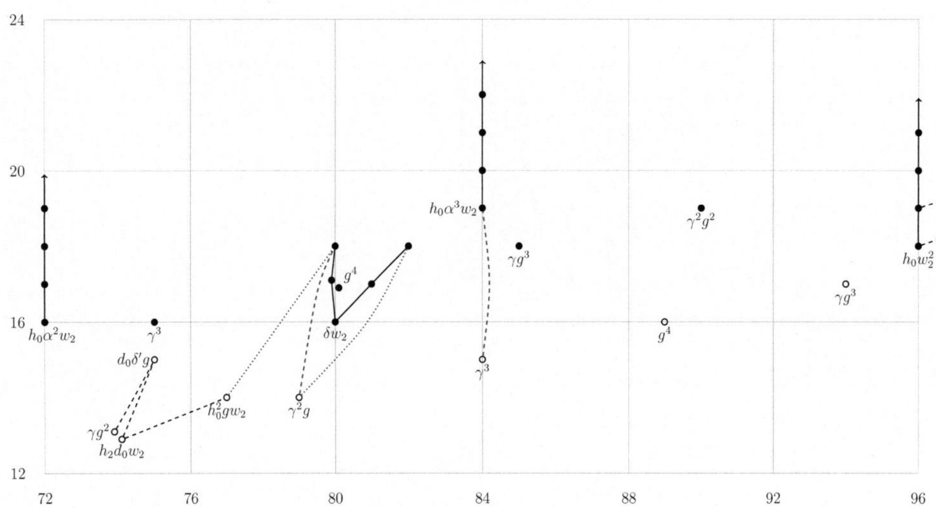

FIGURE 12.28. Delayed $E_\infty(tmf/(B,M))$ for $72 \leq t-s \leq 96$, with all hidden 2-, $\eta$- and $\nu$-extensions

The short exact sequence is a special case of (10.3). □

THEOREM 12.19. *In the delayed Adams spectral sequence for tmf/B, the following hidden 2-extensions repeat $w_2^4$-periodically:*

(12) *From the lift of $h_0^2h_2$ to the image of $h_0^3\alpha$.*
(15) *From the lift of $h_2^2$ to the image of $h_1d_0$.*
(17) *From the lift of $c_0$ to the image of $h_2d_0$.*
(24) *From the lift of $h_1d_0$ to the image of $h_0\alpha^2$.*
(36) *From the lift of $h_0\alpha\beta$ to the image of $h_0\alpha^3$.*

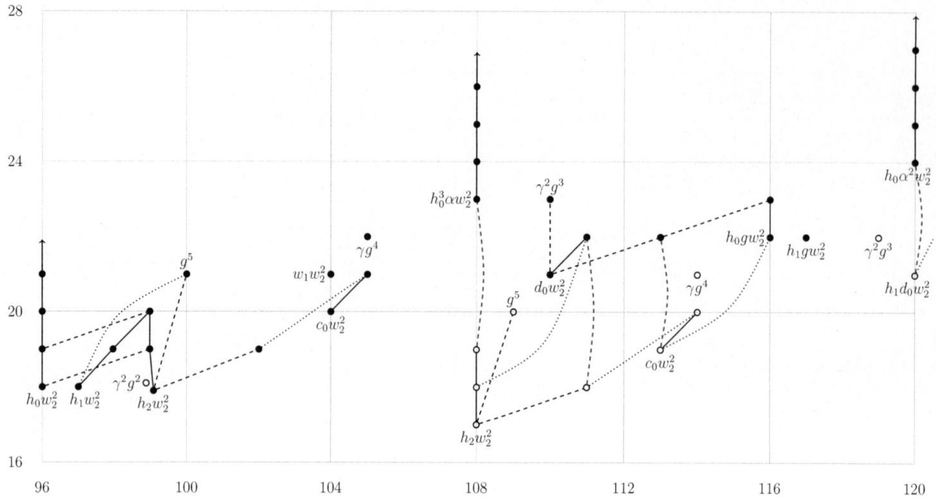

FIGURE 12.29. Delayed $E_\infty(tmf/(B,M))$ for $96 \le t-s \le 120$, with all hidden 2-, $\eta$- and $\nu$-extensions

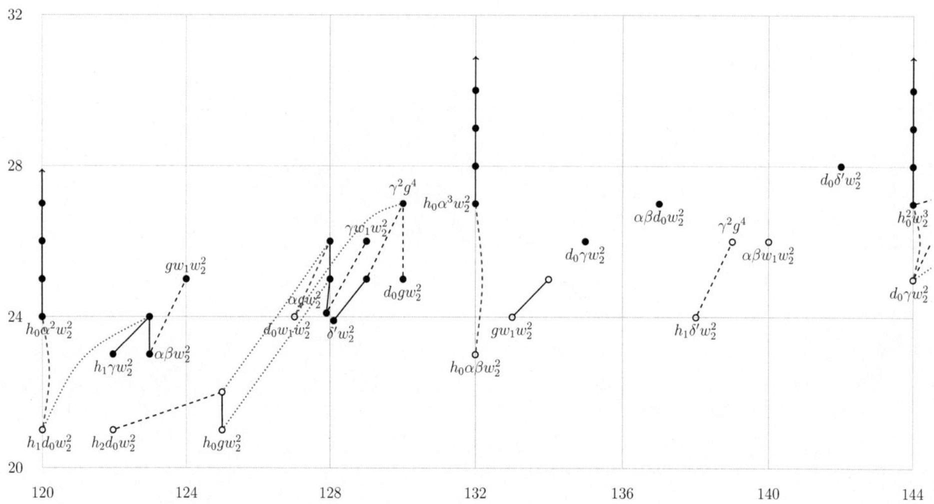

FIGURE 12.30. Delayed $E_\infty(tmf/(B,M))$ for $120 \le t-s \le 144$, with all hidden 2-, $\eta$- and $\nu$-extensions

(48) *From the lift of $d_0\gamma$ to the image of $h_0^2 w_2$.*
(49) *From the lift of $g^2$ to the lift of $\delta' w_1$.*
(54) *From the image of $h_2^2 w_2$ to the image of $d_0 g^2$.*
(56) *From the image of $c_0 w_2$ to the image of $\alpha^3 g + h_0 w_1 w_2$.*
(60) *From the lift of $h_0^2 h_2 w_2$ to the image of $h_0^3 \alpha w_2$.*
(63) *From the lift of $h_2^2 w_2$ to the lift of $d_0 g^2$.*
(66) *From the lift of $\gamma\delta'$ to the image of $d_0\delta' g$.*
(68) *From the lift of $h_2 w_1 w_2$ to the image of $h_0^2 g w_2$.*

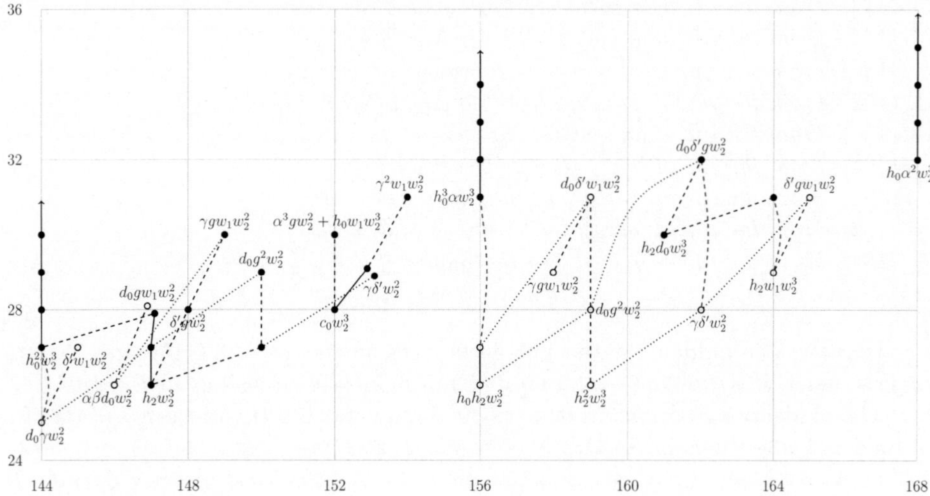


FIGURE 12.31. Delayed $E_\infty(tmf/(B, M))$ for $144 \leq t - s \leq 168$, with all hidden 2-, $\eta$- and $\nu$-extensions

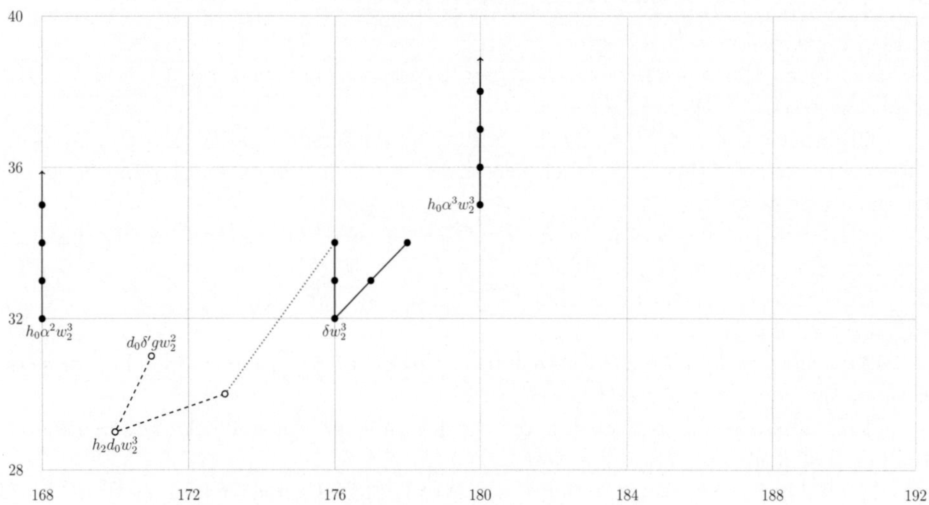


FIGURE 12.32. Delayed $E_\infty(tmf/(B, M))$ for $168 \leq t - s \leq 192$, with all hidden 2-, $\eta$- and $\nu$-extensions

(69) *From the lift of* $g^3$ *to the lift of* $\delta' g w_1$.
(84) *From the lift of* $\gamma^3$ *to the image of* $h_0\alpha^3 w_2$.
(108) *From the lift of* $h_0^2 h_2 w_2^2$ *to the image of* $h_0^3 \alpha w_2^2$.
(110) *From the image of* $d_0 w_2^2$ *to the image of* $\gamma^2 g^3$.
(111) *From the lift of* $h_2^2 w_2^2$ *to the image of* $h_1 d_0 w_2^2$.
(113) *From the lift of* $c_0 w_2^2$ *to the image of* $h_2 d_0 w_2^2$.
(120) *From the lift of* $h_1 d_0 w_2^2$ *to the image of* $h_0 \alpha^2 w_2^2$.
(130) *From the image of* $d_0 g w_2^2$ *to the image of* $\gamma^2 g^4$.

(132) *From the lift of* $h_0\alpha\beta w_2^2$ *to the image of* $h_0\alpha^3 w_2^2$.
(144) *From the lift of* $d_0\gamma w_2^2$ *to the image of* $h_0^2 w_2^3$.
(150) *From the image of* $h_2^2 w_2^3$ *to the image of* $d_0 g^2 w_2^2$.
(152) *From the image of* $c_0 w_2^3$ *to the image of* $\alpha^3 g w_2^2 + h_0 w_1 w_2^3$.
(156) *From the lift of* $h_0^2 h_2 w_2^3$ *to the image of* $h_0^3 \alpha w_2^3$.
(159a) *From the lift of* $h_2^2 w_2^3$ *to the lift of* $d_0 g^2 w_2^2$.
(159b) *From the lift of* $d_0 g^2 w_2^2$ *to the lift of* $d_0\delta' w_1 w_2^2$.
(162) *From the lift of* $\gamma\delta' w_2^2$ *to the image of* $d_0\delta' g w_2^2$.
(164) *From the lift of* $h_2 w_1 w_2^3$ *to the image of* $h_2^2 d_0 w_2^3$.

*There are no other hidden 2-extensions in this spectral sequence.*

PROOF. The hidden 2-extensions in degrees 54, 56, 110, 130, 150 and 152 are images under $E_\infty(tmf)/w_1 \to E_\infty(tmf/B)$ of hidden 2-extensions in $E_\infty(tmf)$.

The hidden 2-extensions in degrees 49, 63, 69 and 159 (two instances) are lifts of hidden 2-extensions in $E_\infty(tmf)$ along $E_\infty(tmf/B) \to {}_{w_1}E_\infty(tmf)$.

For $n = 12$, 24, 36, 48, 84, 108, 120, 132, 144 and 156 we see from $E_\infty(tmf/B)$ that $\pi_{179-n}(tmf/B) = 0$. By Anderson duality it follows that $\pi_n(tmf/B)$ is 2-torsion free. This implies that there must be a hidden 2-extension from the $h_0$-torsion class to the beginning of the $h_0$-tower in each of these degrees.

(15) From the $w_1$-action on $E_\infty(tmf/2)$ we see that $\pi_{15}(tmf/(2,B))$ has order $2^2 = 4$. Since ${}_2\pi_{14}(tmf/B) = \mathbb{Z}/2$ it follows that $\pi_{15}(tmf/B)/2 = \mathbb{Z}/2$, implying a hidden 2-extension in this degree.

(17) From the $w_1$-action on $E_\infty(tmf/2)$ we see that $\pi_{17}(tmf/(2,B)) = \mathbb{Z}/2$, implying $\pi_{17}(tmf/B)/2 = \mathbb{Z}/2$.

(60) Since $\pi_{119}(tmf/B) = \mathbb{Z}/2$, it follows by Anderson duality that the 2-power torsion in $\pi_{60}(tmf/B)$ is $\mathbb{Z}/2$, which necessarily must be generated by $i(\bar\kappa^3)$ detected by the image of $g^3$.

(66) From the $w_1$-action on $E_\infty(tmf/2)$ we see that $\pi_{66}(tmf/(2,B))$ has order $2^3 = 8$, so $\pi_{66}(tmf/B)/2 = \mathbb{Z}/2$.

(68) From the $w_1$-action on $E_\infty(tmf/2)$ we see that $\pi_{68}(tmf/(2,B)) = \mathbb{Z}/2$, so $\pi_{68}(tmf/B)/2 = \mathbb{Z}/2$.

(99) There cannot be a 2-extension on the lift of $\gamma^2 g^2$, since this class detects $\bar\kappa$ times the generator of $\pi_{79}(tmf/B) = \mathbb{Z}/2$.

(122) There is no 2-extension on the lift of $h_2 d_0 w_2^2$, since Anderson duality implies that $\pi_{122}(tmf/B) \cong \pi_{57}(tmf/B) \cong (\mathbb{Z}/2)^2$.

The hidden 2-extensions in degrees $n = 111$, 113, 162 and 164 follow from those in degree $179 - n$ by Anderson duality. □

THEOREM 12.20. *In the delayed Adams spectral sequence for* $tmf/B$*, the following hidden* $\eta$*-extensions repeat* $w_2^4$*-periodically:*

(30) *From the lift of* $h_1 g$ *to the lift of* $d_0 w_1$.
(31) *From the lift of* $d_0 w_1$ *to the image of* $h_0^2 \alpha g$.
(40) *From the image of* $g^2$ *to the image of* $\alpha\beta d_0$.
(45) *From the image of* $\gamma g$ *to the image of* $d_0\delta'$.
(48) *From the lift of* $d_0\gamma$ *to the lift of* $\delta' w_1$.
(49) *From the lift of* $g^2$ *to the lift of* $\alpha\beta d_0$.
(50) *From the lift of* $\alpha\beta d_0$ *to the specified lift of* $d_0 g w_1$.
(51) *From the image of* $h_2 w_2$ *to the image of* $\delta' g$.
(55) *From the lift of* $d_0\delta'$ *to the image of* $\alpha^3 g + h_0 w_1 w_2$.

(65a) *From the image of $h_2 d_0 w_2$ to the image of $d_0\delta' g$.*
(65b) *From the image of $\gamma g^2$ to the image of $d_0\delta' g$.*
(68) *From the lift of $h_2 w_1 w_2$ to the lift of $\delta' g w_1$.*
(74a) *From the lift of $h_2 d_0 w_2$ to the lift of $d_0\delta' g$.*
(74b) *From the lift of $\gamma g^2$ to the lift of $d_0\delta' g$.*
(79) *From the lift of $\gamma^2 g$ to the image of $h_0^2\delta w_2$.*
(99a) *From the image of $h_2 w_2^2$ to the image of $g^5$.*
(108) *From the lift of $h_2 w_2^2$ to the lift of $g^5$.*
(123) *From the image of $\alpha\beta w_2^2$ to the image of $g w_1 w_2^2$.*
(127) *From the lift of $d_0 w_1 w_2^2$ to the image of $h_0^2\alpha g w_2^2$.*
(128) *From the image of $\alpha g w_2^2$ to the image of $\gamma w_1 w_2^2$.*
(129) *From the image of $h_1\delta' w_2^2$ to the image of $\gamma^2 g^4$.*
(138) *From the lift of $h_1\delta' w_2^2$ to the lift of $\gamma^2 g^4$.*
(144) *From the lift of $d_0\gamma w_2^2$ to the lift of $\delta' w_1 w_2^2$.*
(146) *From the lift of $\alpha\beta d_0 w_2^2$ to the specified lift of $d_0 g w_1 w_2^2$.*
(147) *From the image of $h_2 w_2^3$ to the image of $\delta' g w_2^2$.*
(148) *From the image of $\delta' g w_2^2$ to the image of $\gamma g w_1 w_2^2$.*
(153) *From the image of $h_1 c_0 w_2^3$ to the image of $\gamma^2 w_1 w_2^2$.*
(158) *From the lift of $\gamma g w_1 w_2^2$ to the lift of $d_0\delta' w_1 w_2^2$.*
(161) *From the image of $h_2 d_0 w_2^3$ to the image of $d_0\delta' g w_2^2$.*
(164) *From the lift of $h_2 w_1 w_2^3$ to the lift of $\delta' g w_1 w_2^2$.*
(170) *From the lift of $h_2 d_0 w_2^3$ to the lift of $d_0\delta' g w_2^2$.*

*There are no other hidden $\eta$-extensions in this spectral sequence. In particular, there is no hidden $\eta$-extension on the lift of $h_2 d_0 w_2 + \gamma g^2$ or on the specified lift of $\gamma^2 g^2$.*

PROOF. The hidden $\eta$-extensions from degrees 40, 45, 51, 65 (two cases), 99, 123, 128, 129, 147, 148, 153 and 161 are images under $E_\infty(tmf)/w_1 \to E_\infty(tmf/B)$ of hidden $\eta$-extensions in $E_\infty(tmf)$.

The hidden $\eta$-extensions from degrees 30, 48, 49, 50, 68, 74 (two cases), 108, 138, 144, 146, 158, 164 and 170 are lifts of hidden $\eta$-extensions in $E_\infty(tmf)$ along $E_\infty(tmf/B) \to {}_{w_1}E_\infty(tmf)$. In cases (50) and (146) the target classes are the preferred lifts of $d_0 g w_1$ and $d_0 g w_1 w_2^2$, respectively, by how those lifts were specified.

(69) There is no hidden $\eta$-extension from the lift of $g^3$ to the image of $\gamma^2 g$. We use the $w_1$-action on $E_\infty(tmf/\eta)$ to see that $\pi_{70}(tmf/(\eta, B)) = (\mathbb{Z}/2)^2$. Since ${}_\eta\pi_{68}(tmf/B) = \mathbb{Z}/2$ it follows that $\pi_{70}(tmf/B)/\eta = \mathbb{Z}/2$, generated by $i(\eta_1^2\bar\kappa)$ detected by the image of $\gamma^2 g$.

(74) There is no hidden $\eta$-extension from the lift of $h_2 d_0 w_2 + \gamma g^2$ to the image of $\gamma^3$. This follows by Anderson duality, since multiplication by $\eta$ from $\pi_{104}(tmf/B)$ is not surjective.

(84) There is no hidden $\eta$-extension from the lift of $\gamma^3$ to the image of $\gamma g^3$. We use the $w_1$-action on $E_\infty(tmf/\eta)$ to see that $\pi_{85}(tmf/(\eta, B)) = \mathbb{Z}/2$. Since $\pi_{83}(tmf/B) = 0$ we must have $\pi_{85}(tmf/B)/\eta = \mathbb{Z}/2$, generated by $i(\eta_1\bar\kappa^3)$ detected by the image of $\gamma g^3$.

(89) We multiply case (69) by $\bar\kappa$ to see that there is no hidden $\eta$-extension from the lift of $g^4$ to the image of $\gamma^2 g^2$.

(79) By Anderson duality from case (99a) there is a hidden $\eta$-extension on the lift of $\gamma^2 g$. Let $y_{79} \in \pi_{79}(tmf/B)$ be the class detected by this lift. If $\eta y_{79}$ were detected by the image of $g^4$ then $\eta\eta_1 y_{79}$ would be detected by the image of $\gamma g^4$, i.e.,

be equal to $i(\eta_1\bar{\kappa}^4)$. However, $i(\eta_1\bar{\kappa}^4)$ is not an $\eta$-multiple, so this is impossible. Therefore $\eta y_{79}$ is detected by $h_0^2\delta w_2$.

(99b) We multiply case (79) by $\bar{\kappa}$ to see that there is no hidden $\eta$-extension on the lift of $\gamma^2 g^2$ given by $g$ times the lift of $\gamma^2 g$. (The other lift is given by adding the image of $h_2 w_2^2$, and does support a nontrivial $\eta$-extension by case (99a).)

(109) We multiply case (89) by $\bar{\kappa}$ to see that there is no hidden $\eta$-extension from the lift of $g^5$ to the image of $\gamma^2 g^3$.

(134) There is no $\eta$-extension from the lift of $h_1 g w_1 w_2^2$ to the image of $d_0\gamma w_2^2$, by Anderson duality.

The hidden $\eta$-extensions from degrees $n = 31$, 55 and 127 follow by Anderson duality from those from degree $178 - n$. □

THEOREM 12.21. *In the delayed Adams spectral sequence for $tmf/B$, the following hidden $\nu$-extensions repeat $w_2^4$-periodically:*

(6) *From the image of $h_2^2$ to the image of $h_1 c_0$.*
(12) *From the lift of $h_0 h_2$ to the image of $h_1 d_0$.*
(15) *From the lift of $h_2^2$ to the lift of $h_1 c_0$.*
(17) *From the lift of $c_0$ to the image of $h_0 g$.*
(24) *From the lift of $h_1 d_0$ to the image of $h_0\alpha\beta$.*
(29a) *From the lift of $h_0 g$ to the image of $h_0\alpha g$.*
(29b) *From the lift of $h_0^2 g$ to the image of $h_0^2\alpha g$.*
(30) *From the lift of $h_1 g$ to the image of $h_1\delta'$.*
(48) *From the lift of $d_0\gamma$ to the image of $h_0 h_2 w_2$.*
(49) *From the lift of $g^2$ to the image of $\delta' g$.*
(51) *From the image of $h_0 h_2 w_2$ to the image of $d_0 g^2$.*
(54) *From the image of $h_2^2 w_2$ to the image of $\gamma\delta'$.*
(60) *From the lift of $h_0 h_2 w_2$ to the lift of $d_0 g^2$.*
(63a) *From the lift of $h_2^2 w_2$ to the lift of $\gamma\delta'$.*
(63b) *From the lift of $d_0 g^2$ to the image of $d_0\delta' g$.*
(65) *From the image of $h_2 d_0 w_2$ to the image of $h_0^2 g w_2$.*
(66) *From the lift of $\gamma\delta'$ to the lift of $\delta' g w_1$.*
(74) *From the lift of $h_2 d_0 w_2$ to the lift of $h_0^2 g w_2$.*
(77) *From the lift of $h_0^2 g w_2$ to the image of $h_0^2\delta w_2$.*
(79) *From the lift of $\gamma^2 g$ to the image of $h_1^2\delta w_2$.*
(97) *From the image of $h_1 w_2^2$ to the image of $g^5$.*
(102) *From the image of $h_2^2 w_2^2$ to the image of $h_1 c_0 w_2^2$.*
(108) *From the lift of $h_0 h_2 w_2^2$ to the image of $h_1 d_0 w_2^2$.*
(111) *From the lift of $h_2^2 w_2^2$ to the lift of $h_1 c_0 w_2^2$.*
(113) *From the lift of $c_0 w_2^2$ to the image of $h_0 g w_2^2$.*
(120) *From the lift of $h_1 d_0 w_2^2$ to the image of $h_0\alpha\beta w_2^2$.*
(125a) *From the lift of $h_0 g w_2^2$ to the image of $h_0\alpha g w_2^2$.*
(125b) *From the lift of $h_0^2 g w_2^2$ to the image of $h_0^2\alpha g w_2^2$.*
(127) *From the lift of $d_0 w_1 w_2^2$ to the image of $\gamma^2 g^4$.*
(144) *From the lift of $d_0\gamma w_2^2$ to the image of $h_0 h_2 w_2^3$.*
(146) *From the lift of $\alpha\beta d_0 w_2^2$ to the image of $\gamma g w_1 w_2^2$.*
(147) *From the image of $h_0 h_2 w_2^3$ to the image of $d_0 g^2 w_2^2$.*
(150) *From the image of $h_2^2 w_2^3$ to the image of $\gamma\delta' w_2^2$.*
(156a) *From the lift of $h_0 h_2 w_2^3$ to the lift of $d_0 g^2 w_2^2$.*
(156b) *From the lift of $h_0^2 h_2 w_2^3$ to the lift of $d_0\delta' w_1 w_2^2$.*

(159a) *From the lift of $h_2^2 w_2^3$ to the lift of $\gamma\delta' w_2^2$.*
(159b) *From the lift of $d_0 g^2 w_2^2$ to the image of $d_0\delta' g w_2^2$.*
(162) *From the lift of $\gamma\delta' w_2^2$ to the lift of $\delta' g w_1 w_2^2$.*
(173) *From the lift of $h_2^2 d_0 w_2^3$ to the image of $h_0^2\delta w_2^3$.*

*There are no other hidden $\nu$-extensions in this spectral sequence.*

PROOF. The hidden $\nu$-extensions from degrees 6, 51, 54, 65, 97, 102, 147 and 150 are images under $E_\infty(tmf)/w_1 \to E_\infty(tmf/B)$ of hidden $\nu$-extensions in $E_\infty(tmf)$.

The hidden $\nu$-extensions from degrees 15, 60, 63 (case (a)), 66, 74, 111, 156 (two cases), 159 (case (a)) and 162 are lifts of hidden $\nu$-extensions in $E_\infty(tmf)$ along $E_\infty(tmf/B) \to {}_{w_1}E_\infty(tmf)$.

The hidden $\nu$-extensions from degrees $n = 12$, 17, 48, 63 (case (b)), 108, 113, 144 and 159 (case (b)) follow from the relation $2\nu = \nu 2$, previously known $\nu$-multiplications from degree $n$ to degree $n+3$, and hidden 2-extensions established in Theorem 12.19 in one or both of these degrees. In cases (48) and (144) these hidden $\nu$-extensions eclipse lifts of hidden $\nu$-extension from $d_0\gamma$ to $d_0 g w_1$, and from $d_0\gamma w_2^2$ to $d_0 g w_1 w_2^2$, respectively.

(24) From the $w_1$-action on $E_\infty(tmf/\nu)$ we see that $\pi_{27}(tmf/(\nu, B)) = \mathbb{Z}/2$, so $\pi_{27}(tmf/B)/\nu = \mathbb{Z}/2$ and the image of $h_0\alpha\beta$ must detect a $\nu$-multiple.

(29a, 29b, 30) From the $w_1$-action on $E_\infty(tmf/\nu)$ we see that $\pi_{33}(tmf/(\nu, B)) = 0$, so that $\nu$ acts injectively on $\pi_{29}(tmf/B)$ and maps onto $\pi_{33}(tmf/B)$. Since $\eta\nu = 0$, this implies the asserted hidden $\nu$-extensions.

(36) From the $w_1$-action on $E_\infty(tmf/\nu)$ we see that $\pi_{39}(tmf/(\nu, B)) = \mathbb{Z}/2$, so $\nu$ acts trivially on $\pi_{36}(tmf/B)$.

(49) From the $w_1$-action on $E_\infty(tmf/\nu)$ we see that $\pi_{53}(tmf/(\nu, B)) = \mathbb{Z}/2$, so ${}_\nu\pi_{49}(tmf/B) = \mathbb{Z}/2$. Hence the lift of $g^2$ must support a hidden $\nu$-extension.

(62) From the $w_1$-action on $E_\infty(tmf/\nu)$ we see that $\pi_{65}(tmf/(\nu, B)) = (\mathbb{Z}/2)^2$, so $\nu$ acts trivially on $\pi_{62}(tmf/B)$.

(77) From the $w_1$-action on $E_\infty(tmf/\nu)$ we see that $\pi_{80}(tmf/(\nu, B)) = \mathbb{Z}/4 \oplus \mathbb{Z}/2$, generated by the images of $B_3$ and $\bar\kappa^4$ in $\pi_{80}(tmf)$, with $4B_3$ mapping to zero. Hence $i(4B_3)$ in $\pi_{80}(tmf/B)$, detected by $h_0^2\delta w_2$, is a $\nu$-multiple. This implies that there is a hidden $\nu$-extension from the lift of $h_0^2 g w_2$ to the image of $h_0^2\delta w_2$.

(79) This follows by Anderson duality from case (97).

(99) Multiplying case (79) by $\bar\kappa$ confirms that there is no (hidden) $\nu$-extension on the specified lift of $\gamma^2 g^2$.

(114) There are no hidden $\nu$-extensions on the lifts of $h_1 c_0 w_2^2$ and $\gamma g^4$, e.g. by Anderson duality from case (62) of the proof.

(119) There is no hidden $\nu$-extension on the lift of $\gamma^2 g^3$, because $\eta\nu = 0$.

(120) From the $w_1$-action on $E_\infty(tmf/\nu)$ we see that $\pi_{123}(tmf/(\nu, B))$ has order $2^2 = 4$. From case (119) it follows that $\pi_{123}(tmf/B)/\nu = \mathbb{Z}/2$, so that the image of $h_0\alpha\beta w_2^2$ detects a $\nu$-multiple.

(125a, 125b) These follow by Anderson duality from case (51), since $\eta\nu = 0$.

(127) This follows by Anderson duality from case (49).

(132) From the $w_1$-action on $E_\infty(tmf/\nu)$ we see that $\pi_{135}(tmf/(\nu, B)) = \mathbb{Z}/2$, so $\nu$ acts trivially on $\pi_{132}(tmf/B)$.

(146) This follows by Anderson duality from case (30).

(173) This follows by Anderson duality from the nontrivial $\nu$-multiplication on $\pi_3(tmf/B)$. □

In most degrees it is straightforward to read off the group structure of $(N/B)_*$, together with its $\eta$- and $\nu$-actions, from the delayed $E_\infty(tmf/B)$ with its hidden 2-, $\eta$- and $\nu$-extensions. The next result summarizes some less obvious cases. We write $\overline{y} \in \pi_n(tmf/B)$ for the image of $y \in \pi_n(tmf)$, and $\widetilde{y} \in \pi_n(tmf/B)$ for lifts of $y \in \pi_{n-9}(tmf)$, with respect to the maps $i$ and $j$ in (12.1).

PROPOSITION 12.22.

(20) *$\pi_{20}(tmf/B) \cong \mathbb{Z}/8$ is generated by $\overline{\bar\kappa}$, which is detected by the image of $g$. There is a lift $\widetilde{\epsilon}$, detected by the lift of $c_0$, with $\nu \cdot \widetilde{\epsilon} = 2\overline{\bar\kappa}$.*

(32) *$\pi_{32}(tmf/B) \cong \mathbb{Z}/8 \oplus \mathbb{Z}/2$ is generated by $\overline{B_1}$ of order 8 and $\overline{\epsilon_1}$ of order 2, detected by the images of $\alpha g$ and $\delta'$, respectively. A relation $\nu \cdot \widetilde{2\bar\kappa} = \pm 2\overline{B_1}$ holds, but we have not determined the sign.*

(51) *$\pi_{51}(tmf/B) \cong \mathbb{Z}/8 \oplus \mathbb{Z}/2$ is generated by $\overline{\nu_2}$ of order 8 and $\eta^2 \cdot \widetilde{\bar\kappa^2}$ of order 2, detected by the image of $h_2 w_2$ and the specified lift of $d_0 g w_1$, respectively. A relation $\nu \cdot \widetilde{\eta_1 \kappa} = \pm 2\overline{\nu_2} + \eta^2 \widetilde{\bar\kappa^2}$ holds, but we have not determined the sign.*

(66) *$\pi_{66}(tmf/B) \cong \mathbb{Z}/4$ is generated by $\widetilde{\eta_1 \epsilon_1} = \nu \cdot \widetilde{\nu\nu_2}$, which is detected by the lift of $\gamma\delta'$. Here the lift $\widetilde{\nu\nu_2}$ is detected by the lift of $h_2^2 w_2$.*

(75) *$\pi_{75}(tmf/B) \cong (\mathbb{Z}/2)^2$ is generated by $\overline{\eta_1^3}$ and $\eta\widetilde{\nu_2\kappa}$, which are detected by the image of $\gamma^3$ and the lift of $d_0\delta' g$. The relation $\eta \cdot \overline{\eta_1 \bar\kappa^2} = \eta\widetilde{\nu_2\kappa}$ holds.*

(99) *$\pi_{99}(tmf/B) \cong \mathbb{Z}/8 \oplus \mathbb{Z}/2$ is generated by $\overline{\nu_4}$ of order 8 and $\widetilde{\bar\kappa\eta_1^2\bar\kappa}$ of order 2, detected by the image of $h_2 w_2^2$ and the specified lift of $\gamma^2 g^2$.*

(105) *$\pi_{105}(tmf/B) \cong (\mathbb{Z}/2)^2$ is generated by $\overline{\eta_1 \bar\kappa^4}$ and $\eta\overline{\epsilon_4}$, which are detected by the images of $\gamma g^4$ and $h_1 c_0 w_2^2$, respectively. The relation $\nu^2 \cdot \overline{\nu_4} = \eta\overline{\epsilon_4} + \overline{\eta_1 \bar\kappa^4}$ holds.*

(114) *$\pi_{114}(tmf/B) \cong (\mathbb{Z}/2)^2$ is generated by $\widetilde{\eta_1 \bar\kappa^4}$ and $\eta\widetilde{\epsilon_4}$, which are detected by the lifts of $\gamma g^4$ and $h_1 c_0 w_2^2$, respectively. The relation $\nu^2 \cdot \widetilde{\nu_4} = \eta\widetilde{\epsilon_4} + \widetilde{\eta_1 \bar\kappa^4}$ holds.*

(116) *$\pi_{116}(tmf/B) \cong \mathbb{Z}/4$ is generated by $\overline{\bar\kappa D_4}$, which is detected by the image of $h_0 g w_2^2$. There is a lift $\widetilde{\epsilon_4}$, detected by the lift of $c_0 w_2^2$, with $\nu \cdot \widetilde{\epsilon_4} = \overline{\bar\kappa D_4}$.*

(122) *$\pi_{122}(tmf/B) \cong (\mathbb{Z}/2)^2$ is generated by $\overline{\eta_1 \eta_4}$ and a lift $\widetilde{\nu\kappa_4}$, which are detected by the image of $h_1 \gamma w_2^2$ and the lift of $h_2 d_0 w_2^2$, respectively. The lift can be chosen so that $\eta \cdot \widetilde{\nu\kappa_4} = 0$.*

(128) *$\pi_{128}(tmf/B) \cong \mathbb{Z}/8 \oplus \mathbb{Z}/2$ is generated by $\overline{B_5}$ of order 8 and $\overline{\epsilon_5}$ of order 2, detected by the images of $\alpha g w_2^2$ and $\delta' w_2^2$, respectively. A relation $\nu \cdot \widetilde{\bar\kappa D_4} = \pm 2\overline{B_5}$ holds, but we have not determined the sign.*

(147) *$\pi_{147}(tmf/B) \cong \mathbb{Z}/8 \oplus \mathbb{Z}/2$ is generated by $\overline{\nu_6}$ of order 8 and $\eta \cdot \widetilde{\nu_5\kappa}$ of order 2, detected by the image of $h_2 w_2^3$ and the specified lift of $d_0 g w_1 w_2^2$, respectively. A relation $\nu \cdot \widetilde{\eta_1 \kappa_4} = \pm 2\overline{\nu_6} + \eta\widetilde{\nu_5\kappa}$ holds, but we have not determined the sign.*

(162) *$\pi_{162}(tmf/B) \cong \mathbb{Z}/4$ is generated by $\widetilde{\eta_1 \epsilon_5} = \nu \cdot \widetilde{\nu\nu_6}$, which is detected by the lift of $\gamma\delta' w_2^2$. Here the lift $\widetilde{\nu\nu_6}$ is detected by the lift of $h_2^2 w_2^3$.*

PROOF. (20) Adding $\overline{\nu\kappa}$ to a choice of $\widetilde{\epsilon}$ changes the sign of $\nu \cdot \widetilde{\epsilon}$.

(32) Recall that $B_1$ is detected by $\alpha g$ and satisfies $8B_1 = BD_1$, while $\epsilon_1$ is detected by $\delta'$ and satisfies $2\epsilon_1 = 0$.

(51) The hidden $\nu$-extension on the lift of $d_0\gamma$ implies that $\nu \cdot \widetilde{\eta_1\kappa} \equiv 2\overline{\nu_2}$ modulo $4\overline{\nu_2}$ and $\eta^2\widetilde{\bar\kappa^2}$. Since $\nu \cdot \eta_1\kappa = \eta^2\bar\kappa^2$ in $\pi_{42}(tmf)$, a summand $\eta^2\widetilde{\bar\kappa^2}$ must be present in $\nu \cdot \widetilde{\eta_1\kappa}$.

(66) We choose the lift $\widetilde{\eta_1\epsilon_1}$ to be the given $\nu$-multiple.

(75) The relation $\eta \cdot \eta_1\bar\kappa^2 = \eta \cdot \nu_2\kappa$ in $\pi_{66}(tmf)$ lifts to the stated relation in $\pi_{75}(tmf/B)$ because $\eta\colon \pi_{74}(tmf/B) \to \pi_{75}(tmf/B)$ is not surjective, e.g. by Anderson duality.

(99) The lift $\widetilde{\eta_1^2\bar\kappa}$ has order 2, hence so does its $\bar\kappa$-multiple.

(105) This is the image of the relation in Proposition 9.17.

(114) This is the lift of the relation in Proposition 9.17.

(116) Adding $\overline{\nu\kappa_4}$ to a choice of $\widetilde{\epsilon_4}$ changes the sign of $\nu \cdot \widetilde{\epsilon_4}$.

(122) Adding $\overline{\eta_1\eta_4}$ to a lift $\widetilde{\nu\kappa_4}$ changes $\eta \cdot \widetilde{\nu\kappa_4}$ by $\eta\overline{\eta_1\eta_4} = 2\overline{\nu_5}$.

(128) This is similar to case (32).

(147) The hidden $\nu$-extension on the lift of $d_0\gamma w_2^2$ implies that $\nu \cdot \widetilde{\eta_1\kappa_4} \equiv 2\overline{\nu_6}$ modulo $4\overline{\nu_6}$ and $\eta\widetilde{\nu_5\kappa}$. Since $\nu \cdot \eta_1\kappa_4 = \eta\nu_5\kappa$ in $\pi_{138}(tmf)$, a summand $\eta\widetilde{\nu_5\kappa}$ must be present in $\nu \cdot \widetilde{\eta_1\kappa_4}$.

(162) This is similar to case (66). □

There are additive extensions $\overline{C} \doteq 8\widetilde{\nu}$ in $\pi_{12}(tmf/B)$, $\overline{D_1} \doteq 2\widetilde{\eta\kappa}$ in $\pi_{24}(tmf/B)$, and so on. We have not determined the 2-adic units implicit in these identities.

## 12.5. Homotopy of $tmf/(2, B)$

We study $\pi_*(tmf/(2, B))$ using the short exact sequence

$$0 \to \pi_*(tmf/2)/B \xrightarrow{i} \pi_*(tmf/(2, B)) \xrightarrow{j} {}_B\pi_{*-9}(tmf/2) \to 0 \tag{12.3}$$

of $\pi_*(tmf)$-modules, where

$$\pi_n(tmf/2)/B = \mathrm{cok}(B\colon \pi_{n-8}(tmf/2) \to \pi_n(tmf/2))$$
$${}_B\pi_{n-9}(tmf/2) = \ker(B\colon \pi_{n-9}(tmf/2) \to \pi_{n-1}(tmf/2))\,.$$

Let $(E_r(Z_\star), d_r)$ denote the delayed Adams spectral sequence for the $tmf$-module tower $Z_\star$ given by

$$tmf/(2, B) \xleftarrow{i} tmf/2 \longleftarrow *\,.$$

Here $Z_k = *$ for $k \geq 2$, $Z_1 = tmf/2$ and $Z_0 = tmf/(2, B)$. The nontrivial filtration quotients are $Z_{1,1} \simeq tmf/2$ and $Z_{0,1} \simeq \Sigma^9 tmf/2$. The delayed Adams spectral sequence $(E_r(Z_\star), d_r)$ is associated to the convolved filtration $(S \wedge Z)_\star$, and there is a homotopy cofiber sequence

$$S_s \wedge \Sigma^8 tmf/2 \overset{\alpha\wedge B}{\longrightarrow} S_{s-1} \wedge tmf/2 \longrightarrow (S \wedge Z)_s \longrightarrow S_s \wedge \Sigma^9 tmf/2$$

of filtered spectra. Since $\alpha \wedge B$ has Adams filtration 5 we have short exact sequences

$$0 \to E_r^{*-1,*-1}(tmf/2) \longrightarrow E_r^{*,*}(Z_\star) \longrightarrow E_r^{*,*-9}(tmf/2) \to 0$$

for $r \leq 5$, as in [**45**, Thm. VI.6.1(i)], Theorem 11.11 and [**148**, Prop. 5.4]. The $d_5$-differential is given by multiplication by $w_1$ detecting $B$, as can be checked case-by-case for the twelve $E_5(tmf)$-module generators of the quotient copy of $E_5(tmf/2)$, cf. Table 6.12. Hence we have a short exact sequence

$$0 \to E_\infty^{*-1,*-1}(tmf/2)/w_1 \longrightarrow E_6^{*,*}(Z_\star) \longrightarrow {}_{w_1}E_\infty^{*,*-9}(tmf/2) \to 0$$

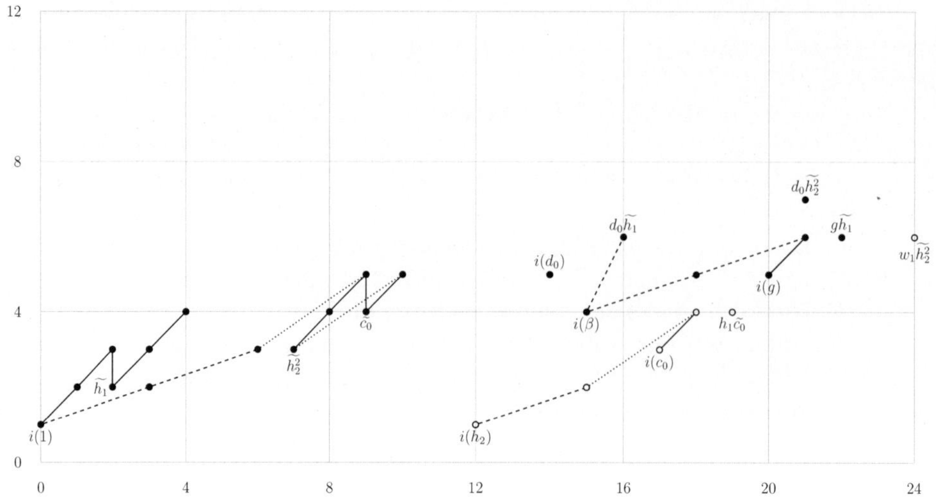

FIGURE 12.33. Delayed $E_\infty(tmf/(2,B,M))$ for $0 \le t-s \le 24$, with all hidden 2-, $\eta$- and $\nu$-extensions

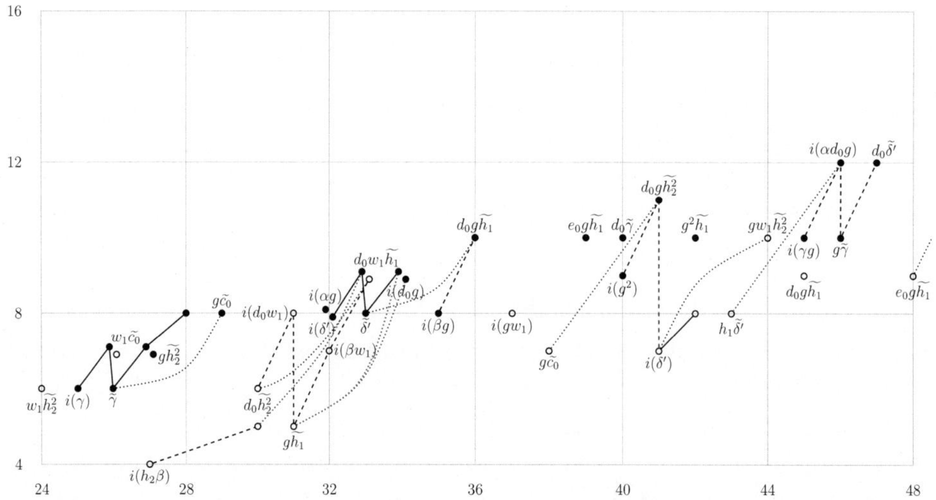

FIGURE 12.34. Delayed $E_\infty(tmf/(2,B,M))$ for $24 \le t-s \le 48$, with all (potential) hidden 2-, $\eta$- and $\nu$-extensions

of $E_\infty(tmf)$-modules. The resulting $E_6$-term is displayed in Figures 12.33 to 12.40. There is no room for any further differentials, and therefore $E_6(Z_\star) = E_\infty(Z_\star)$ in the delayed Adams spectral sequence converging to $\pi_*(tmf/(2,B))$.

In these charts, the filled (black) circles show the image of the cokernel of $w_1$, offset by $(t-s,s) = (0,1)$ bidegrees, while the open (white) circles show lifts of the kernel of $w_1$, offset by $(9,0)$ bidegrees. The black lines (solid, dashed or dotted) show 2-, $\eta$- and $\nu$-extensions within the image of $E_\infty(tmf/2)/w_1$ and the lift of $w_1E_\infty(tmf/2)$. Hidden extensions from the lift to the image are shown in red

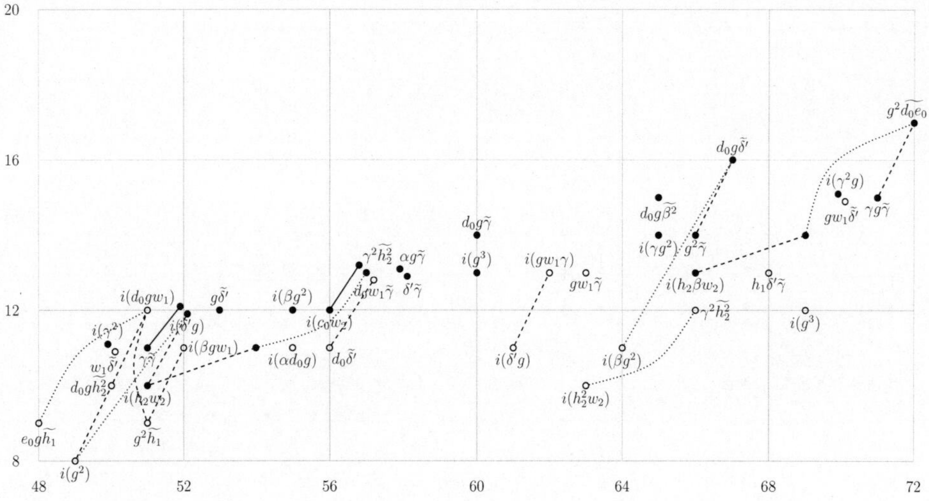


FIGURE 12.35. Delayed $E_\infty(tmf/(2, B, M))$ for $48 \le t - s \le 72$, with all hidden 2-, $\eta$- and $\nu$-extensions

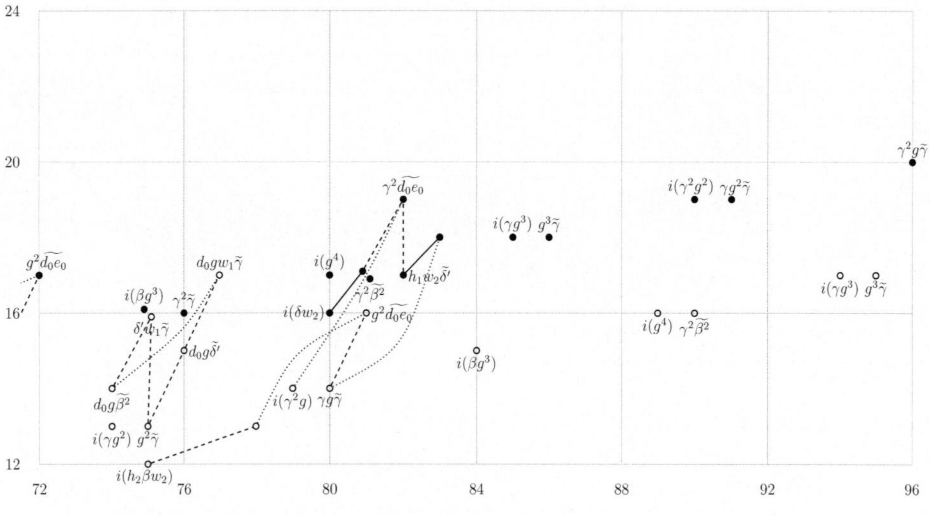


FIGURE 12.36. Delayed $E_\infty(tmf/(2, B, M))$ for $72 \le t - s \le 96$, with all hidden 2-, $\eta$- and $\nu$-extensions

(dashed or dotted). These will be determined in Theorems 12.24, 12.25 and 12.27. Using these, we can specify

(26) the lift of $w_1\widetilde{c_0}$ to detect $\kappa$ times the class detected by the lift of $i(\underline{h_2})$,
(33) the lift of $d_0w_1\widetilde{h_1}$ to detect $\eta^2$ times a class detected by the lift of $g\widetilde{h_1}$,
(50) the lift of $w_1\widetilde{\delta'}$ to detect a class annihilated by $\bar{\kappa}^2$,
(57) the lift of $d_0w_1\widetilde{\gamma}$ (modulo the image of $h_1 \cdot i(c_0w_2)$) to detect $\eta$ times a class detected by the lift of $d_0\widetilde{\delta'}$,
(70) the lift of $gw_1\widetilde{\delta'}$ to detect a class annihilated by $\bar{\kappa}$,

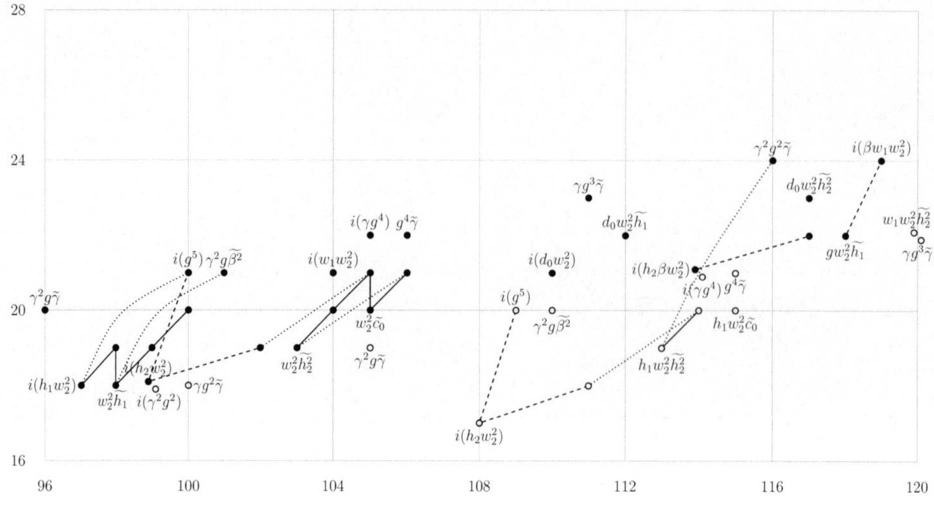

FIGURE 12.37. Delayed $E_\infty(tmf/(2,B,M))$ for $96 \leq t-s \leq 120$, with all hidden 2-, $\eta$- and $\nu$-extensions

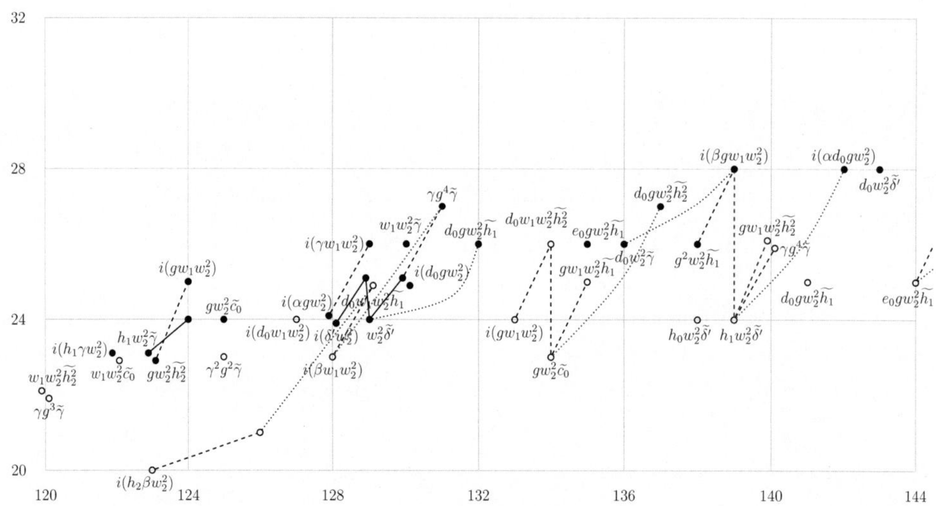

FIGURE 12.38. Delayed $E_\infty(tmf/(2,B,M))$ for $120 \leq t-s \leq 144$, with all (potential) hidden 2-, $\eta$- and $\nu$-extensions

(75) the lift of $\delta' w_1 \widetilde{\gamma}$ to detect $\eta$ times the class detected by the lift of $\widetilde{d_0 g\beta^2}$,
(99) the lift of $i(\gamma^2 g^2)$ to detect $\bar{\kappa}$ times the class detected by the lift of $i(\gamma^2 g)$,
(114) the lift of $i(\gamma g^4)$ to detect $\bar{\kappa}$ times the class detected by the lift of $i(\gamma g^3)$,
(122) the lift of $w_1 w_2^2 \widetilde{c_0}$ to detect $\kappa$ times the class detected by the lift of $i(h_2 w_2^2)$,
(129) the lift of $\widetilde{d_0 w_1 w_2^2 h_1}$ to detect $\nu$ times the class detected by the lift of $i(h_2^2 \beta w_2^2)$, and
(153) the lift of $d_0 w_1 w_2^2 \widetilde{\gamma}$ to detect a class that is annihilated by $\eta$.

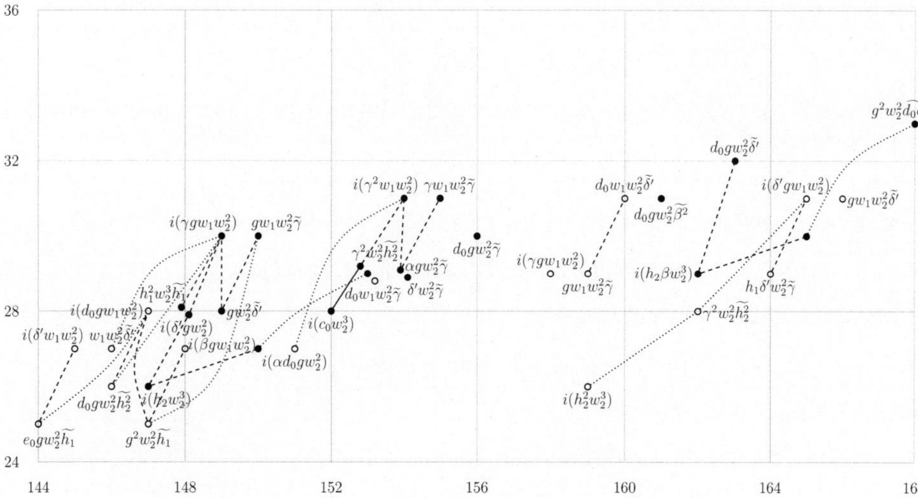


FIGURE 12.39. Delayed $E_\infty(tmf/(2, B, M))$ for $144 \leq t-s \leq 168$, with all (potential) hidden 2-, $\eta$- and $\nu$-extensions

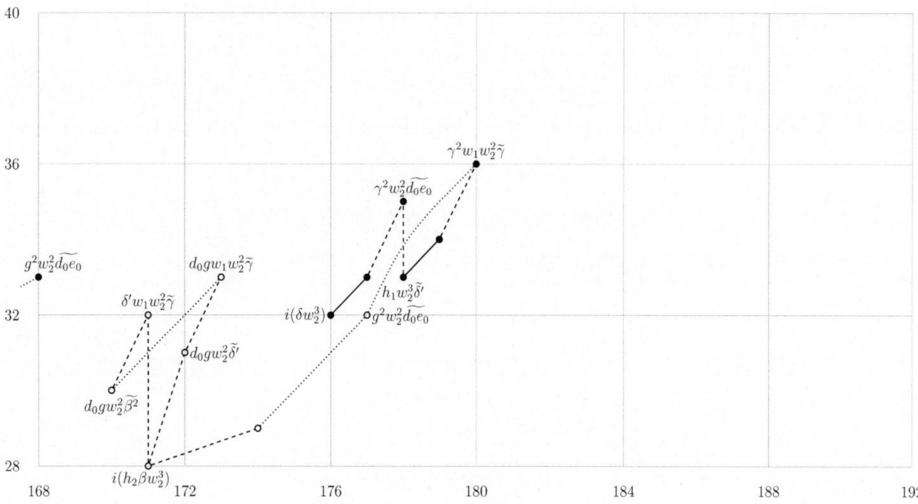


FIGURE 12.40. Delayed $E_\infty(tmf/(2, B, M))$ for $168 \leq t-s \leq 192$, with all hidden 2-, $\eta$- and $\nu$-extensions

Let $N/(2, B)_* \subset \pi_*(tmf/(2, B))$ denote the graded subgroup of classes in degrees $0 \leq * < 192$. Since $w_2^4$ acts freely on the delayed $E_\infty(tmf/(2, B))$, we have an isomorphism

$$N/(2, B)_* \otimes \mathbb{Z}[M] \cong \pi_*(tmf/(2, B))$$

of $\mathbb{Z}[M]$-modules. The subgroup $N/(2, B)_*$ is preserved by the action of $\eta$, $\nu$, $\epsilon$, $\kappa$, $\bar{\kappa}$ (since $\bar{\kappa} \cdot i(B_7) = 0$ in $\pi_*(tmf/2)$) and $B$, and the isomorphism respects these actions. We have isomorphisms

$$N/(2, B)_* \cong \pi_*(tmf/(2, B))/M \cong \pi_*(tmf/(2, B, M))$$

and a short exact sequence

$$0 \to (N/2)_*/B \xrightarrow{i} N/(2,B)_* \xrightarrow{j} {}_B(N/2)_{*-9} \to 0\,,$$

where $(N/2)_*$ is as in Section 12.1. Recall the Brown–Comenetz duality functor $I$ from Section 10.3.

PROPOSITION 12.23. *The spectrum $tmf/(2,B,M)$ is Brown–Comenetz self-dual, in the sense that there is an equivalence of $tmf$-modules*

$$tmf/(2,B,M) \simeq \Sigma^{180} I(tmf/(2,B,M))\,.$$

*Hence there is an isomorphism*

$$N/(2,B)_{180-n} \cong \mathrm{Hom}(N/(2,B)_n, \mathbb{Q}/\mathbb{Z})$$

*for each integer $n$.*

PROOF. By Theorem 10.6 we have an equivalence

$$\Sigma^{20} tmf \simeq I(tmf/(2^\infty, B^\infty, M^\infty))\,.$$

The homotopy cofiber sequences

$$tmf/(2,B^\infty,M^\infty) \longrightarrow tmf/(2^\infty,B^\infty,M^\infty) \xrightarrow{2} tmf/(2^\infty,B^\infty,M^\infty)$$
$$tmf/(2,B,M^\infty) \longrightarrow \Sigma^8 tmf/(2,B^\infty,M^\infty) \xrightarrow{B} tmf/(2,B^\infty,M^\infty)$$
$$tmf/(2,B,M) \longrightarrow \Sigma^{192} tmf/(2,B,M^\infty) \xrightarrow{M} tmf/(2,B,M^\infty)$$

from (10.2) dualize to homotopy cofiber sequences that translate to equivalences

$$\Sigma^{20} tmf/2 \simeq I(tmf/(2,B^\infty,M^\infty))$$
$$\Sigma^{12} tmf/(2,B) \simeq I(tmf/(2,B,M^\infty))$$
$$\Sigma^{-180} tmf/(2,B,M) \simeq I(tmf/(2,B,M))\,.$$

□

THEOREM 12.24. *In the delayed Adams spectral sequence for $tmf/(2,B)$, the following hidden 2-extensions repeat $w_2^4$-periodically:*

(31) *From the lift of $\widetilde{gh_1}$ to the lift of $i(d_0w_1)$.*
(41) *From the lift of $i(\delta')$ to the image of $\widetilde{d_0gh_2^2}$.*
(46) *From the image of $g\widetilde{\gamma}$ to the image of $i(\alpha d_0 g)$.*
(51) *From the lift of $\widetilde{g^2h_1}$ to the lift of $i(d_0gw_1)$.*
(75) *From the lift of $g^2\widetilde{\gamma}$ to the specified lift of $\delta' w_1\widetilde{\gamma}$.*
(82) *From the image of $h_1w_2\widetilde{\delta'}$ to the image of $\widetilde{\gamma^2 d_0 e_0}$.*
(134) *From the lift of $gw_2^2\widetilde{c_0}$ to the lift of $\widetilde{d_0w_1w_2^2h_2^2}$.*
(139) *From the lift of $h_1w_2^2\widetilde{\delta'}$ to the image of $i(\beta g w_1 w_2^2)$.*
(147) *From the lift of $\widetilde{g^2w_2^2h_1}$ to the lift of $i(d_0gw_1w_2^2)$.*
(149) *From the image of $gw_2^2\widetilde{\delta'}$ to the image of $i(\gamma g w_1 w_2^2)$.*
(154) *From the image of $\alpha g w_2^2\widetilde{\gamma}$ to the image of $i(\gamma^2 w_1 w_2^2)$.*
(171) *From the lift of $i(h_2\beta w_2^3)$ to the lift of $\delta' w_1 w_2^2\widetilde{\gamma}$.*
(178) *From the image of $h_1w_2^3\widetilde{\delta'}$ to the image of $\widetilde{\gamma^2 w_2^2 d_0 e_0}$.*

*There are no other hidden 2-extensions in this spectral sequence.*

PROOF. The hidden 2-extensions in degrees 46, 82, 149, 154 and 178 are images of known hidden 2-extensions in $E_\infty(tmf/2)/w_1$, and the hidden 2-extensions in degrees 31, 134 and 171 are lifts of known hidden 2-extensions in ${}_{w_1}E_\infty(tmf/2)$. The ambiguous lifts in degrees 51, 75 and 147 are treated separately below.

(41) The hidden 2-extension from the lift of $i(\delta')$ to the image of $\widetilde{d_0 g h_2^2}$ follows as in Lemma 12.2 from the hidden $\eta$-extension in $E_\infty(tmf/B)$ from the image of $g^2$ to the image of $\alpha\beta d_0$. In more detail, a class $\widetilde{y} \in \pi_{41}(tmf/(2, B))$ detected by the lift of $i(\delta')$ is mapped by $j$ to $y \in \pi_{40}(tmf/B)$ detected by the image of $g^2$, so that $2 \cdot \widetilde{y} = i(\eta y) = \eta \cdot i(y)$ is detected by the image of $i(\alpha\beta d_0) = \widetilde{d_0 g h_2^2}$.

(51) There is no 2-extension from the lift of $g^2\widetilde{h_1}$ to the image of $i(h_2 w_2)$ or the image of $\gamma\widetilde{\gamma}$, because $\eta$ acts nontrivially on these potential targets. Hence the hidden 2-extension from $g^2\widetilde{h_1}$ to $i(d_0 g w_1)$ in $E_\infty(tmf/2)$ lifts to the delayed $E_\infty(tmf/(2, B))$.

(75) There is a nontrivial hidden 2-extension in degree 75. The lift of $g^2\widetilde{\gamma}$ detects a class $\widetilde{y} \in \pi_{75}(tmf/(2, B))$ mapping by $j$ to a class $y \in \pi_{74}(tmf/B)$ detected by the lift of $\gamma g^2$. By Theorem 12.20, case (74b), the product $\eta y$ is detected by the lift of $d_0\delta' g$. We know that $d_0\delta' g$ detects $\eta\nu_2\kappa$ in $\pi_{66}(tmf)$, which maps by $i$ to $\eta i(\nu_2\kappa)$ detected by $\delta' w_1\widetilde{\gamma}$. Hence the lift of $d_0\delta' g$ maps by $i$ to $2 \cdot \widetilde{y} = i(\eta y) = \eta \cdot i(y)$ in $\pi_{75}(tmf/(2, B))$, which must be detected by the specified lift of $\delta' w_1\widetilde{\gamma}$.

(139) There is a hidden 2-extension from the lift of $h_1 w_2^2\widetilde{\delta'}$ to the image of $i(\beta g w_1 w_2^2)$, either because of the hidden $\eta$-extension from the lift of $h_1\delta' w_2^2$ to the lift of $\gamma^2 g^4$ in the delayed $E_\infty(tmf/B)$, or by Brown–Comenetz duality from case (41).

(147) There is no 2-extension from the lift of $g^2 w_2^2\widetilde{h_1}$ to the image of $i(h_2 w_2^3)$, because of the hidden $\eta$-extension on the latter class. Hence the hidden 2-extension from $g^2 w_2^2\widetilde{h_1}$ to $i(d_0 g w_1 w_2^2)$ in $E_\infty(tmf/2)$ lifts to the delayed $E_\infty(tmf/(2, B))$.

There are no hidden 2-extensions in degrees 15, 45, 56, 123, 128, 138, 162 or 177, because $\eta$ acts nontrivially on the possible targets and $\eta 2 = 0$.

Similarly, there are no hidden 2-extensions in degrees 18, 32, 42, 50, 52, 76, 81, 114, 129, 135 or 148, because the possible sources detect $\eta$-multiples and $2\eta = 0$.

Using Lemma 12.2, there are no 2-extensions in degrees $n = 27$, 55, 66, 69, 80, 90, 99, 110, 111 or 125 because the $\eta$-multiples in $\pi_n(tmf/B)$ are divisible by 2, hence map to zero under $i$.

(100) There is no hidden 2-extension in degree 100 by Brown–Comenetz duality from case (80).

(105) Finally, there is no hidden 2-extension on the lift of $\gamma^2 g\widetilde{\gamma}$, since $i$ maps the $\eta$-multiple in $\pi_{105}(tmf/B)$ to a class detected by the image of $\widetilde{h_1^2 w_2^2 h_2^2} = h_0 w_2^2 \widetilde{c_0}$, which is already an $h_0$-multiple. □

To determine the $\eta$- and $\nu$-action on $\pi_*(tmf/(2, B))$ we shall make use of the evident morphisms

$$\begin{aligned} i\colon E_r(tmf/B) &\longrightarrow E_r(tmf/(2, B)) \\ j\colon E_r(tmf/(2, B)) &\longrightarrow E_r(\Sigma tmf/B) \end{aligned}$$

of delayed Adams spectral sequences. We shall also make use of another variant of the Adams spectral sequence, which we call the hastened Adams spectral sequence, and which we discuss in Section 12.6. See in particular Proposition 12.38 and the accompanying figures, which do not depend on the work in the present section.

THEOREM 12.25. *In the delayed Adams spectral sequence for $tmf/(2,B)$, the following hidden $\eta$-extensions repeat $w_2^4$-periodically:*

(15) *From the image of $i(\beta)$ to the image of $d_0\widetilde{h_1}$.*
(30) *From the lift of $d_0\widetilde{h_2^2}$ to the lift of $i(d_0w_1)$.*
(31) *From the lift of $g\widetilde{h_1}$ to the lift of $i(\beta w_1)$.*
(32) *From the lift of $i(\beta w_1)$ to the specified lift of $d_0w_1\widetilde{h_1}$.*
(35) *From the image of $i(\beta g)$ to the image of $d_0g\widetilde{h_1}$.*
(40) *From the image of $i(g^2)$ to the image of $d_0g\widetilde{h_2^2}$.*
(45) *From the image of $i(\gamma g)$ to the image of $i(\alpha d_0 g)$.*
(46) *From the image of $g\widetilde{\gamma}$ to the image of $d_0\widetilde{\delta'}$.*
(49) *From the lift of $i(g^2)$ to the lift of $d_0g\widetilde{h_2^2}$.*
(50) *From the lift of $d_0g\widetilde{h_2^2}$ to the lift of $i(d_0gw_1)$.*
(51a) *From the lift of $g^2\widetilde{h_1}$ to the lift of $i(\beta gw_1)$.*
(51b) *From the image of $i(h_2w_2)$ to the image of $i(\delta' g)$.*
(56) *From the lift of $d_0\widetilde{\delta'}$ to the specified lift of $d_0w_1\widetilde{\gamma}$.*
(61) *From the lift of $i(\delta' g)$ to the lift of $i(gw_1\gamma)$.*
(66) *From the image of $g^2\widetilde{\gamma}$ to the image of $d_0g\widetilde{\delta'}$.*
(71) *From the image of $\gamma g\widetilde{\gamma}$ to the image of $g^2\widetilde{d_0e_0}$.*
(74) *From the lift of $d_0g\widetilde{\beta^2}$ to the specified lift of $\delta' w_1\widetilde{\gamma}$.*
(75) *From the lift of $g^2\widetilde{\gamma}$ to the lift of $d_0g\widetilde{\delta'}$.*
(76) *From the lift of $d_0g\widetilde{\delta'}$ to the lift of $d_0gw_1\widetilde{\gamma}$.*
(80) *From the lift of $\gamma g\widetilde{\gamma}$ to the lift of $g^2\widetilde{d_0e_0}$.*
(81) *From the image of $i(h_1\delta w_2)$ to the image of $\gamma^2\widetilde{d_0e_0}$.*
(99) *From the image of $i(h_2w_2^2)$ to the image of $i(g^5)$.*
(108) *From the lift of $i(h_2w_2^2)$ to the lift of $i(g^5)$.*
(118) *From the image of $gw_2^2\widetilde{h_1}$ to the image of $i(\beta w_1w_2^2)$.*
(123) *From the image of $gw_2^2\widetilde{h_2^2}$ to the image of $i(gw_1w_2^2)$.*
(128a) *From the lift of $i(\beta w_1w_2^2)$ to the specified lift of $d_0w_1w_2^2\widetilde{h_1}$.*
(128b) *From the image of $i(\alpha gw_2^2)$ to the image of $i(\gamma w_1w_2^2)$.*
(130) *From the image of $h_1w_2^2\widetilde{\delta'}$ to the image of $\gamma g^4\widetilde{\gamma}$.*
(133) *From the lift of $i(gw_1w_2^2)$ to the lift of $d_0w_1w_2^2\widetilde{h_2^2}$.*
(134) *From the lift of $gw_2^2\widetilde{c_0}$ to the lift of $gw_1w_2^2\widetilde{h_1}$.*
(138) *From the image of $g^2w_2^2\widetilde{h_1}$ to the image of $i(\beta gw_1w_2^2)$.*
(144) *From the lift of $e_0gw_2^2\widetilde{h_1}$ to the lift of $i(\delta' w_1w_2^2)$.*
(146) *From the lift of $d_0gw_2^2\widetilde{h_2^2}$ to the lift of $i(d_0gw_1w_2^2)$.*
(147a) *From the lift of $g^2w_2^2\widetilde{h_1}$ to the lift of $i(\beta gw_1w_2^2)$.*
(147b) *From the image of $i(h_2w_2^3)$ to the image of $i(\delta' gw_2^2)$.*
(148a) *From the image of $i(\delta' gw_2^2)$ to the image of $i(\gamma gw_1w_2^2)$.*
(149) *From the image of $gw_2^2\widetilde{\delta'}$ to the image of $gw_1w_2^2\widetilde{\gamma}$.*
(153) *From the image of $i(h_1c_0w_2^3)$ to the image of $i(\gamma^2w_1w_2^2)$.*
(154) *From the image of $\alpha gw_2^2\widetilde{\gamma}$ to the image of $\gamma w_1w_2^2\widetilde{\gamma}$.*
(159) *From the lift of $gw_1w_2^2\widetilde{\gamma}$ to the lift of $d_0w_1w_2^2\widetilde{\delta'}$.*
(162) *From the image of $i(h_2\beta w_2^3)$ to the image of $d_0gw_2^2\widetilde{\delta'}$.*

(164) *From the lift of* $h_1\delta' w_2^2\widetilde{\gamma}$ *to the lift of* $i(\delta' g w_1 w_2^2)$.
(170) *From the lift of* $d_0 g w_2^2 \widetilde{\beta^2}$ *to the lift of* $\delta' w_1 w_2^2 \widetilde{\gamma}$.
(171) *From the lift of* $i(h_2\beta w_2^3)$ *to the lift of* $d_0 g w_2^2 \widetilde{\delta'}$.
(172) *From the lift of* $d_0 g w_2^2 \widetilde{\delta'}$ *to the lift of* $d_0 g w_1 w_2^2 \widetilde{\gamma}$.
(177) *From the image of* $i(h_1\delta w_2^3)$ *to the image of* $\gamma^2 w_2^2 \widetilde{d_0 e_0}$.
(179) *From the image of* $h_1^2 w_2^3 \widetilde{\delta'}$ *to the image of* $\gamma^2 w_1 w_2^2 \widetilde{\gamma}$.

*The following potential hidden* $\eta$*-extensions repeat* $w_2^4$*-periodically, but remain to be precisely determined.*

(139) *From the lift of* $h_1 w_2^2 \widetilde{\delta'}$ *to the lift of* $\gamma g^4 \widetilde{\gamma}$, *or to the lift of* $\gamma g^4 \widetilde{\gamma} + g w_1 w_2^2 \widetilde{h_2^2}$.
(148b) *From the image of* $h_1^2 w_2^3 \widetilde{h_1}$ *to zero, or to the image of* $i(\gamma g w_1 w_2^2)$.

*There are no other hidden* $\eta$*-extensions in this spectral sequence.*

PROOF. The hidden $\eta$-extensions between pairs of image classes are images of known or potential hidden $\eta$-extensions in $E_\infty(tmf/2)/w_1$, and the hidden $\eta$-extensions between pairs of lifted classes are lifts of known or potential hidden $\eta$-extensions in ${}_{w_1}E_\infty(tmf/2)$. The lifted $\eta$-extension targets in degrees 33, 57, 75 and 129 are ambiguous, but we have specified the lifts of $d_0 w_1 \widetilde{h_1}$, $d_0 w_1 \widetilde{\gamma}$ and $\delta' w_1 \widetilde{\gamma}$ to detect the appropriate $\eta$-multiples. There is no early $\eta$-extension from the lift of $d_0 g \widetilde{h_2^2}$ to the image of $\gamma\widetilde{\gamma}$ in degree 51, because $\eta^3 = 4\nu$ must act trivially on $\pi_{49}(tmf/(2, B))$. In degree 129 the ambiguity in the lift of $d_0 w_1 w_2^2 \widetilde{h_1}$ is an $h_1$-multiple (the image of $i(h_1\delta' w_2^2)$), and does therefore not affect the presence of a hidden $\eta$-extension. We still have to argue that there are no other hidden $\eta$-extensions.

There is no $\eta$-extension from the class in degree 15 that detects a $\nu$-multiple, because $\eta\nu = 0$.

There are no $\eta$-extensions from the classes in degrees 32, 45, and 128 to classes with nonzero 2-multiples, since $2\eta = 0$.

There are no $\eta$-extensions from the classes in degrees 68 and 164 to classes supporting nonzero $\nu$-multiplications, because $\nu\eta = 0$.

(38) There is no hidden $\eta$-extension from the lift of $g\widetilde{c_0}$ to the image of $e_0 g \widetilde{h_1}$, by comparison with the hastened Adams spectral sequence for $tmf/(2, B)$, where $h_1$-multiplication is trivial from degree 38 and there is no room for hidden $\eta$-extensions. See Section 12.6, and Figure 12.50 in particular.

(64) There is no hidden $\eta$-extension from the lift of $i(\beta g^2)$ to the image of $i(\gamma g^2)$, nor to the image of $d_0 g \widetilde{\beta^2}$, by comparison with the hastened Adams spectral sequence, see Figure 12.51, where $h_1$-multiplication is trivial from degree 65 and there is no room for hidden $\eta$-extensions.

(69) A class in $\pi_{69}(tmf/B)$ detected by the lift of $g^3$ maps by $i$ to a class in $\pi_{69}(tmf/(2, B))$ detected by the lift of $i(g^3)$. Since $\eta$ acts trivially on $\pi_{69}(tmf/B)$, it also acts trivially on this class in $\pi_{69}(tmf/(2, B))$, so there is no hidden $\eta$-extension from the lift of $i(g^3)$.

(74) There is no hidden $\eta$-extension from the lift of $i(\gamma g^2)$ to the image of $i(\beta g^3)$, because multiplication by $\eta$ does not act injectively on $\pi_{74}(tmf/B) \cong (\mathbb{Z}/2)^2$, hence it also does not act injectively on $\pi_{74}(tmf/(2, B))$.

(75) There is no hidden $\eta$-extension from the lift of $i(h_2\beta w_2)$ to the image of $\gamma^2\widetilde{\gamma}$, because multiplication by $\eta$ does not map surjectively to $\pi_{75}(tmf/B) \cong (\mathbb{Z}/2)^2$, hence it also does not map surjectively to $\pi_{76}(tmf/(2, B))$.

(79) The class $y_{79}$ in $\pi_{79}(tmf/B)$ detected by the lift of $\gamma^2 g$ maps by $i$ to the class $i(y_{79})$ in $\pi_{79}(tmf/(2,B))$ detected by the lift of $i(\gamma^2 g)$. Since $\eta y_{79}$ is divisible by 2, it maps trivially under $i$. Hence multiplication by $\eta$ acts trivially on $i(y_{79})$.

(84) A class $y_{84}$ in $\pi_{84}(tmf/B)$ detected by the lift of $\gamma^3$ maps by $i$ to the class $i(y_{84})$ in $\pi_{84}(tmf/(2,B))$ detected by the lift of $i(\beta g^3)$. Since $\eta$ acts trivially on $\pi_{84}(tmf/B)$, it also acts trivially on $i(y_{84})$, so there is no hidden $\eta$-extension on the lift of $i(\beta g^3)$.

(89) A class $y_{89}$ in $\pi_{89}(tmf/B)$ detected by the lift of $g^4$ maps by $i$ to the class $i(y_{89})$ in $\pi_{89}(tmf/(2,B))$ detected by the lift of $i(g^4)$. Since $\eta$ acts trivially on $\pi_{89}(tmf/B)$, it also acts trivially on $i(y_{89})$, so there is no hidden $\eta$-extension on the lift of $i(g^4)$.

(90) The class in $\pi_{91}(tmf/(2,B))$ detected by the image of $\gamma g^2\widetilde{\gamma}$ maps by $j$ to the class in $\pi_{90}(tmf/B)$ detected by the image of $\gamma^2 g^2$. Since the latter is not an $\eta$-multiple, the $\eta$-multiplication on $\pi_{90}(tmf/(2,B))$ must be trivial.

(99) Since $\eta$ acts trivially on the class detected by the lift of $i(\gamma^2 g)$, it also acts trivially on $\bar{\kappa}$ times this class, which is detected by our specified lift of $i(\gamma^2 g^2)$.

(105) Multiplication by $\eta$ does not map surjectively to $\pi_{105}(tmf/B) \cong (\mathbb{Z}/2)^2$, hence it also does not map surjectively to $\pi_{106}(tmf/(2,B))$. It follows that there is no room for a hidden $\eta$-extension on the lift of $\gamma^2 g\widetilde{\gamma}$ to the image of $g^4\widetilde{\gamma}$.

(153) We have specified the lift of $d_0 w_1 w_2^2 \widetilde{\gamma}$ to detect a class that is annihilated by $\eta$.

There are no $\eta$-extensions from degree $n$ to $n+1$ for $n = 19$, 43, 52, 55, 95, 100, 109, 110, 111, 115, 141 or 151, by Brown–Comenetz duality and the vanishing of $\eta$-multiplication from degree $179-n$ to $180-n$. See also Lemma 12.26 for more detail on the case $n = 43$, which resolves a case that was left open in Theorem 12.4. □

LEMMA 12.26. *There is no hidden $\eta$-extension on $h_1\widetilde{\delta'}$ in $E_\infty(tmf/2)$, nor on the lift of $h_1\widetilde{\delta'}$ in the delayed $E_\infty(tmf/(2,B))$.*

PROOF. Multiplication by $\eta$ is trivial from degree 136 in $\pi_*(tmf/2)/B$ and $\pi_*(tmf/(2,B))$. Hence, by Brown–Comenetz duality, it is trivial from degree 43 in $\pi_*(tmf/(2,B))$ and from degree 34 in ${}_B\pi_*(tmf/2)$. □

THEOREM 12.27. *In the delayed Adams spectral sequence for $tmf/(2,B)$, the following hidden $\nu$-extensions repeat $w_2^4$-periodically:*

(6) *From the image of $i(h_2^2)$ to the image of $h_0\widetilde{c_0}$.*
(7) *From the image of $\widetilde{h_2^2}$ to the image of $h_1\widetilde{c_0}$.*
(15) *From the lift of $i(h_2^2)$ to the lift of $i(h_1c_0)$.*
(26) *From the image of $\widetilde{\gamma}$ to the image of $g\widetilde{c_0}$.*
(30a) *From the lift of $i(h_2^2\beta)$ to the image of $h_0\widetilde{\delta'}$.*
(30b) *From the lift of $d_0\widetilde{h_2^2}$ to the specified lift of $d_0w_1\widetilde{h_1}$.*
(33) *From the image of $\widetilde{\delta'}$ to the image of $d_0g\widetilde{h_1}$.*
(41) *From the lift of $i(\delta')$ to the lift of $gw_1\widetilde{h_2^2}$.*
(48) *From the lift of $e_0g\widetilde{h_1}$ to the lift of $i(d_0gw_1)$.*
(49) *From the lift of $i(g^2)$ to the image of $i(\delta' g)$.*
(54) *From the image of $i(h_2^2w_2)$ to the image of $\gamma^2\widetilde{h_2^2}$.*
(63) *From the lift of $i(h_2^2w_2)$ to the lift of $\gamma^2\widetilde{h_2^2}$.*
(64) *From the lift of $i(\beta g^2)$ to the image of $d_0g\widetilde{\delta'}$.*

(69) *From the image of* $i(h_2^2\beta w_2)$ *to the image of* $g^2\widetilde{d_0e_0}$.

(74) *From the lift of* $\widetilde{d_0g\beta^2}$ *to the lift of* $d_0gw_1\widetilde{\gamma}$.

(78) *From the lift of* $i(h_2^2\beta w_2)$ *to the lift of* $g^2\widetilde{d_0e_0}$.

(79) *From the lift of* $i(\gamma^2 g)$ *to the image of* $\gamma^2\widetilde{d_0e_0}$.

(80) *From the lift of* $\gamma g\widetilde{\gamma}$ *to the image of* $h_1^2w_2\widetilde{\delta}'$.

(97) *From the image of* $i(h_1w_2^2)$ *to the image of* $i(g^5)$.

(98) *From the image of* $w_2^2\widetilde{h_1}$ *to the image of* $\gamma^2 g\widetilde{\beta^2}$.

(102) *From the image of* $i(h_2^2w_2^2)$ *to the image of* $h_0w_2^2\widetilde{c_0}$.

(103) *From the image of* $w_2^2\widetilde{h_2^2}$ *to the image of* $h_1w_2^2\widetilde{c_0}$.

(111) *From the lift of* $i(h_2^2w_2^2)$ *to the lift of* $h_1^2w_2^2\widetilde{h_2^2}$.

(113) *From the lift of* $h_1w_2^2\widetilde{h_2^2}$ *to the image of* $\gamma^2g^2\widetilde{\gamma}$.

(126) *From the lift of* $i(h_2^2\beta w_2^2)$ *to the specified lift of* $d_0w_1w_2^2\widetilde{h_1}$.

(128) *From the lift of* $i(\beta w_1w_2^2)$ *to the image of* $\gamma g^4\widetilde{\gamma}$.

(129) *From the image of* $w_2^2\widetilde{\delta}'$ *to the image of* $d_0gw_2^2\widetilde{h_1}$.

(136) *From the image of* $d_0w_2^2\widetilde{\gamma}$ *to the image of* $i(\beta gw_1w_2^2)$.

(144) *From the lift of* $e_0gw_2^2\widetilde{h_1}$ *to the lift of* $i(d_0gw_1w_2^2)$.

(147) *From the lift of* $g^2w_2^2\widetilde{h_1}$ *to the image of* $gw_1w_2^2\widetilde{\gamma}$.

(150) *From the image of* $i(h_2^2w_2^3)$ *to the image of* $\gamma^2w_2^2\widetilde{h_2^2}$.

(151) *From the lift of* $i(\alpha d_0gw_2^2)$ *to the image of* $i(\gamma^2w_1w_2^2)$.

(159) *From the lift of* $i(h_2^2w_2^3)$ *to the lift of* $\gamma^2w_2^2\widetilde{h_2^2}$.

(162) *From the lift of* $\gamma^2w_2^2\widetilde{h_2^2}$ *to the lift of* $i(\delta' gw_1w_2^2)$.

(165) *From the image of* $i(h_2^2\beta w_2^3)$ *to the image of* $g^2w_2^2\widetilde{d_0e_0}$.

(170) *From the lift of* $d_0gw_2^2\widetilde{\beta^2}$ *to the lift of* $d_0gw_1w_2^2\widetilde{\gamma}$.

(174) *From the lift of* $i(h_2^2\beta w_2^3)$ *to the lift of* $g^2w_2^2\widetilde{d_0e_0}$.

(177) *From the lift of* $g^2w_2^2\widetilde{d_0e_0}$ *to the image of* $\gamma^2w_1w_2^2\widetilde{\gamma}$.

*The following potential hidden* $\nu$*-extensions repeat* $w_2^4$*-periodically, but remain to be precisely determined.*

(31) *From the lift of* $g\widetilde{h_1}$ *to the image of* $h_1\widetilde{\delta}'$, *or to the image of* $i(d_0g) + h_1\widetilde{\delta}'$.

(38) *From the lift of* $g\widetilde{c_0}$ *to zero, or to the image of* $d_0g\widetilde{h_2^2}$.

(43) *From the lift of* $h_1\widetilde{\delta}'$ *to zero, or to the image of* $i(\alpha d_0g)$.

(134) *From the lift of* $gw_2^2\widetilde{c_0}$ *to zero, or to the image of* $d_0gw_2^2\widetilde{h_2^2}$.

(139) *From the lift of* $h_1w_2^2\widetilde{\delta}'$ *to zero, or to the image of* $i(\alpha d_0gw_2^2)$.

(146a) *From the lift of* $d_0gw_2^2\widetilde{h_2^2}$ *to the image of* $i(\gamma gw_1w_2^2)$. *(This* $\nu$*-extension may be eclipsed by case (146b).)*

(146b) *From the lift of* $w_1w_2^2\widetilde{\delta}'$ *to zero, or to the image of* $i(\gamma gw_1w_2^2)$.

*There are no other hidden* $\nu$*-extensions in this spectral sequence.*

PROOF. The hidden $\nu$-extensions between pairs of image classes are images of known hidden $\nu$-extensions in $E_\infty(tmf/2)/w_1$, and the hidden $\nu$-extensions between pairs of lifted classes are mostly lifts of known hidden $\nu$-extensions in ${}_{w_1}E_\infty(tmf/2)$. The following two cases are exceptional.

(30b-1) The hidden $\nu$-extension in $E_\infty(tmf/2)$ from $d_0\widetilde{h_2^2}$ to $d_0w_1\widetilde{h_1}$ lifts to the delayed $E_\infty(tmf/(2, B))$, but it is ambiguous whether $\nu$ times the class detected by the lift of $d_0\widetilde{h_2^2}$ is detected by the specified lift of $d_0 w_1 \widetilde{h_1}$ or its sum with the image of $h_0\widetilde{\delta}'$. We continue this case in (30b-2) below.

(126) The hidden $\nu$-extension for $tmf/2$ from $d_0\widetilde{w_2^2h_2^2}$ to $d_0w_1\widetilde{w_2^2h_1}$ does not lift to $tmf/(2,B)$, because $d_0\widetilde{w_2^2h_2^2}$ is not $w_1$-torsion. This allows the nontrivial $\nu^2$-extension from $i(h_2\beta w_2^2)$ to $d_0w_1\widetilde{w_2^2h_1}$ to lift to a hidden $\nu^2$-extension from the lift of $i(h_2\beta w_2^2)$ to a lift of $d_0w_1\widetilde{w_2^2h_1}$, which in turn contributes a hidden $\nu$-extension from the lift of $i(h_2^2\beta w_2^2)$, with the same target. This hidden $\nu$-extension corresponds to a primary $h_2$-multiplication in the hastened Adams spectral sequence for $tmf/(2,B)$, cf. Figure 12.54. We have specified the lift of $d_0w_1\widetilde{w_2^2h_1}$ to be the class detecting this $\nu$-multiple.

Next we use the morphisms $i$ and $j$ of delayed Adams spectral sequences.

(30a) The hidden $\nu$-extension from the lift of $h_1g$ to the image of $h_1\delta'$ in the delayed $E_\infty(tmf/B)$ maps by $i$ to a hidden $\nu$-extension from the lift of $i(h_1g) = i(h_2^2\beta)$ to the image of $i(h_1\delta') = i(h_0\widetilde{\delta}')$ in the delayed $E_\infty(tmf/(2,B))$.

(31) Multiplication by $\nu$ on a class detected by the lift of $g\widetilde{h_1}$ maps by $j$ to the hidden $\nu$-extension from the lift of $h_1g$ to the image of $h_1\delta'$ in the delayed $E_\infty(\Sigma tmf/B)$, hence there must be a hidden $\nu$-extension in the delayed $E_\infty(tmf/(2,B))$ from the lift of $g\widetilde{h_1}$ to the image of $h_1\widetilde{\delta}'$ modulo $i(d_0g)$.

(49) The hidden $\nu$-extension from the lift of $g^2$ to the image of $\delta' g$ in the delayed $E_\infty(tmf/B)$ maps by $i$ to a hidden $\nu$-extension from the lift of $i(g^2)$ to the image of $i(\delta' g)$ in the delayed $E_\infty(tmf/(2,B))$.

(64) Multiplication by $\nu$ on the class detected by the lift of $i(\beta g^2)$ maps by $j$ to the hidden $\nu$-extension from the lift of $d_0g^2$ to the image of $d_0\delta' g$ in the delayed $E_\infty(tmf/B)$, hence there must be a hidden $\nu$-extension in the delayed $E_\infty(tmf/(2,B))$ from the lift of $i(\beta g^2)$ to the image of $d_0g\widetilde{\delta}'$.

In some cases we can use our known results about the $B$-action on $\pi_*(tmf/\nu)$ from Section 12.3, and the long exact sequences

$$\cdots \longrightarrow \pi_{n-3}(tmf/(2,B)) \xrightarrow{\nu} \pi_n(tmf/(2,B)) \xrightarrow{i} \pi_n(tmf/(2,\nu,B)) \xrightarrow{j} \pi_{n-4}(tmf/(2,B)) \xrightarrow{\nu} \pi_{n-1}(tmf/(2,B)) \longrightarrow \dots$$

and

$$\cdots \longrightarrow \pi_n(tmf/(\nu,B)) \xrightarrow{2} \pi_n(tmf/(\nu,B)) \xrightarrow{i} \pi_n(tmf/(2,\nu,B)) \xrightarrow{j} \pi_{n-1}(tmf/(\nu,B)) \xrightarrow{2} \pi_{n-1}(tmf/(\nu,B)) \longrightarrow \dots ,$$

with the group $\pi_n(tmf/(2,\nu,B))$ in common, to deduce information about the $\nu$-action on $\pi_*(tmf/(2,B))$.

(30b-2) We see from Figures 12.17 and 12.18 that $\pi_{32}(tmf/(\nu,B)) = (\mathbb{Z}/2)^2$ and $\pi_{33}(tmf/(\nu,B)) = 0$, so that $\pi_{33}(tmf/(2,\nu,B)) = (\mathbb{Z}/2)^2$. From Figure 12.34 we see that $\nu$ acts trivially on $\pi_{29}(tmf/(2,B)) = \mathbb{Z}/2$ for filtration reasons. Hence the cokernel of $\nu\colon \pi_{30}(tmf/(2,B)) = (\mathbb{Z}/2)^2 \to \pi_{33}(tmf/(2,B)) = \mathbb{Z}/4 \oplus \mathbb{Z}/2$ is $\mathbb{Z}/2$, which implies that $\nu$ acts monomorphically on $\pi_{30}(tmf/(2,B))$. We finish this case in (30b-3) below.

(147) We showed in case (30b-2) that $\nu$ acts injectively from $\pi_{30}(tmf/(2,B))$. Hence, by Brown–Comenetz duality it must map surjectively to $\pi_{150}(tmf/(2,B))$.

It follows that there must be a hidden $\nu$-extension from the lift of $g^2 w_2^2 \widetilde{h_1}$ to the image of $g w_1 w_2^2 \widetilde{\gamma}$.

We also deduce from Brown–Comenetz duality that $\nu$ must act nontrivially from degree $n$ to $n+3$, for $n = 79, 80, 113, 128, 151$ and $177$, since $\nu$ acts nontrivially from degree $177-n$ to $180-n$ by our other results. Since $\eta\nu = 0$, there is only one possible source and target for the corresponding hidden $\nu$-extensions.

(30b-3) We see from Figure 12.34 that the nonzero 2-multiple and the nonzero $\eta^2$-multiple in $\pi_{33}(tmf/(2,B))$ are not equal. By comparison with Figure 12.50, it follows that this difference is a $\nu^2$-multiple. In other words, the 2-, $\eta^2$- and $\nu^2$-multiples give the three nonzero 2-torsion elements in $\pi_{33}(tmf/(2,B))$, and their sum is zero. Let $y_{27} \in \pi_{27}(tmf/(2,B))$ be any class detected by the lift of $i(h_2\beta)$. Then $\nu^2 y_{27}$ is detected by the sum of the specified lift of $d_0 w_1 \widetilde{h_1}$ and the image of $h_0\widetilde{\delta}'$. It follows that $\nu y_{27} \in \pi_{30}(tmf/(2,B))$ is not the image $i(y_{30})$ of the generator $y_{30} \in \pi_{30}(tmf/B)$, but differs from it by the class detected by the lift of $d_0\widetilde{h_2^2}$. By case (30a) the product $\nu \cdot i(y_{30})$ is detected by the image of $h_0\widetilde{\delta}'$. Hence the hidden $\nu$-extension on the lift of $d_0\widetilde{h_2^2}$ must map to the specified lift of $d_0 w_1 \widetilde{h_1}$.

It remains to argue that there are no other hidden $\nu$-extensions.

There are no hidden $\nu$-extensions from degrees 17, 18, 32, 42, 52, 57, 62, 68, 100, 109, 115, 120, 125, 135, 145 or 160, because $\eta\nu = 0$. For the same reason there is no early $\nu$-extension from the lift of $d_0 w_1 \widetilde{h_1}$ in degree 33, no early $\nu$-extension from degree 48 to the images of $i(h_2 w_2)$ or $i(\gamma\widetilde{\gamma})$, no $\nu$-extension on the lift of $g w_1 \widetilde{\gamma}$ in degree 63, no early $\nu$-extension from the lift of $d_0 w_1 w_2^2 \widetilde{h_1}$ in degree 129, no early $\nu$-extension from degree 144 to the image of $i(h_2 w_2^3)$, and no early $\nu$-extension from the lift of $i(\delta' g w_1 w_2^2)$ in degree 165.

There are no hidden $\nu$-extensions from degrees 19, 24, 37, 44, 50, 55, 94, 95, 133, 140, 153 or 158, by filtration considerations in the hastened Adams spectral sequence for $tmf/(2,B)$, cf. Figures 12.49 to 12.56.

(26) By the hastened spectral sequence for $tmf/(2,B)$, see Figure 12.50, we have $\ker(\eta^2) \subset \ker(\nu)$ inside $\pi_{26}(tmf/(2,B))$. Since the lift of $w_1\widetilde{c_0}$ detects a class in $\ker(\eta^2)$, there cannot be a nonzero $\nu$-extension from it.

(99) There is no $\nu$-extension on the specified lift of $i(\gamma^2 g^2)$, which detects a $\bar{\kappa}$-multiple, since $\nu\bar{\kappa} = 0$.

(114) There is no $\nu$-extension on the lift of $h_1^2 w_2^2 \widetilde{h_2^2}$ since $\eta\nu = 0$, and no $\nu$-extension on the specified lift of $i(\gamma g^4)$, since $\nu\bar{\kappa} = 0$.

Finally, there is no $\nu$-extension from degree 122 by Brown–Comenetz duality from case (55), and no $\nu$-extension from degree 127 by Brown–Comenetz duality from case (50). □

REMARK 12.28. In view of the self-duality of $tmf/(2,B,M)$, the $\nu$-extensions from degree 38 and degree 139 are either both zero or both nonzero. Likewise, the $\nu$-extensions from degree 43 and degree 134 are either both zero or both nonzero. The hidden $\nu$-extension from degree 31 maps to the image of $h_1\widetilde{\delta}'$ if and only if the extension in case (146b) is zero, so that the hidden $\nu$-extension in case (146a) is present.

In most degrees it is straightforward to read off the group structure of $N/(2,B)_*$, together with its $\eta$- and $\nu$-actions, from the delayed $E_\infty(tmf/(2,B))$ with its hidden 2-, $\eta$- and $\nu$-extensions. The next result summarizes some less obvious cases. We write $\overline{y} \in \pi_n(tmf/(2,B))$ for the image of $y \in \pi_n(tmf/2)$, and $\widetilde{y} \in \pi_n(tmf/(2,B))$ for lifts of $y \in \pi_{n-9}(tmf/2)$, with respect to the maps $i$ and $j$ in (12.3).

PROPOSITION 12.29.

(18) $\pi_{18}(tmf/(2,B)) \cong (\mathbb{Z}/2)^2$ *is generated by* $\overline{\nu\widetilde{\kappa}}$ *and* $\eta\widetilde{i(\epsilon)}$, *which are detected by the image of* $i(h_2\beta)$ *and the lift of* $i(h_1c_0)$, *respectively. The relation* $\nu^2 \cdot \widetilde{i(\nu)} = \eta\widetilde{i(\epsilon)}$ *holds.*

(21) $\pi_{21}(tmf/(2,B)) \cong (\mathbb{Z}/2)^2$ *is generated by* $\overline{\kappa\widetilde{\nu^2}}$ *and* $\eta\overline{i(\bar\kappa)}$, *which are detected by the image of* $d_0\widetilde{h_2^2}$ *and the image of* $i(h_1g)$, *respectively. The relation* $\nu^2 \cdot \overline{\widetilde{\kappa}} = \overline{\kappa\widetilde{\nu^2}} + \eta\overline{i(\bar\kappa)}$ *holds.*

(81) $\pi_{81}(tmf/(2,B)) \cong (\mathbb{Z}/2)^3$ *is generated by* $\eta\overline{i(B_3)}$, $\overline{\widetilde{\bar\kappa^4}}$ *and a lift* $\widetilde{i(D_3)}$, *which are detected by the image of* $i(h_1\delta w_2)$, *the image of* $\gamma^2\widetilde{\beta^2}$, *and the lift of* $g^2\widetilde{d_0e_0}$, *respectively. We can choose* $\widetilde{i(D_3)}$ *to be equal to* $\eta \cdot \widetilde{\eta_1\bar\kappa\widetilde{\eta_1}} = \nu \cdot \widetilde{\nu\nu_2\widetilde{\kappa}}$, *for a choice of* $\widetilde{\eta_1\bar\kappa\widetilde{\eta_1}}$ *detected by the lift of* $\gamma g\widetilde{\gamma}$.

(105) $\pi_{105}(tmf/(2,B)) \cong \mathbb{Z}/4 \oplus (\mathbb{Z}/2)^2$ *is generated by* $\overline{i(\eta_1\bar\kappa^4)}$ *of order* 2, $\overline{\widetilde{\epsilon_4}}$ *of order* 4, *and a lift* $\widetilde{\eta_1^2\bar\kappa\widetilde{\eta_1}}$ *of order* 2, *detected by the image of* $i(\gamma g^4)$, *the image of* $w_2^2\widetilde{c_0}$ *and the lift of* $\gamma^2 g\widetilde{\gamma}$, *respectively. The relations* $\eta \cdot \overline{i(\epsilon_4)} = 2\overline{\widetilde{\epsilon_4}}$ *and* $\nu^2 \cdot \overline{i(\nu_4)} = 2\overline{\widetilde{\epsilon_4}} + \overline{i(\eta_1\bar\kappa^4)}$ *hold.*

(106) $\pi_{106}(tmf/(2,B)) \cong (\mathbb{Z}/2)^2$ *is generated by* $\overline{\bar\kappa^4\widetilde{\eta_1}}$ *and* $\eta\overline{\widetilde{\epsilon_4}}$, *which are detected by the image of* $g^4\widetilde{\gamma}$ *and the image of* $h_1w_2^2\widetilde{c_0}$, *respectively. The relation* $\nu \cdot \overline{\widetilde{\nu\nu_4}} = \eta\overline{\widetilde{\epsilon_4}} + \overline{\bar\kappa^4\widetilde{\eta_1}}$ *holds.*

(114) $\pi_{114}(tmf/(2,B)) \cong (\mathbb{Z}/2)^3$ *is generated by* $\overline{\nu_4\widetilde{\kappa}}$, $\bar\kappa\widetilde{i(\eta_1\bar\kappa^3)}$ *and* $\eta\widetilde{i(\epsilon_4)}$, *which are detected by the image of* $i(h_2\beta w_2^2)$, *the specified lift of* $i(\gamma g^4)$ *and the lift of* $h_1^2w_2^2\widetilde{h_2^2}$, *respectively. The relation* $\nu^2 \cdot \widetilde{i(\nu_4)} = \eta\widetilde{i(\epsilon_4)} + \bar\kappa\widetilde{i(\eta_1\bar\kappa^3)}$ *holds.*

(129) $\pi_{129}(tmf/(2,B)) \cong \mathbb{Z}/4 \oplus (\mathbb{Z}/2)^2$ *is generated by* $\eta\overline{i(B_5)}$ *of order* 2, $\nu^2\widetilde{\nu_4\widetilde{\kappa}}$ *of order* 2, *and* $\overline{\widetilde{\epsilon_5}}$ *of order* 4, *detected by the image of* $i(\gamma w_1w_2^2)$, *the specified lift of* $d_0w_1w_2^2\widetilde{h_1}$ *and the image of* $w_2^2\widetilde{\delta}'$, *respectively. The relations* $\eta \cdot \overline{i(\epsilon_5)} = 2\overline{\widetilde{\epsilon_5}}$ *and* $\eta \cdot \widetilde{\eta\bar\kappa\widetilde{\eta_4}} = \nu^2\widetilde{\nu_4\widetilde{\kappa}}$ *hold, for one choice of class* $\widetilde{\eta\bar\kappa\widetilde{\eta_4}}$ *detected by the lift of* $i(\beta w_1w_2^2)$.

PROOF. (18) The relation holds modulo $\overline{\nu\widetilde{\kappa}}$, by the delayed Adams spectral sequence for $tmf/(2,B,M)$. To see that the error term is zero, we can compare with the hastened Adams spectral sequence with the same abutment, see Figure 12.49. In this case both products must be detected by the same class in maximal filtration, hence they are equal.

Cases (21), (105) and (106) follow from the corresponding cases of Proposition 12.7, by applying the map $i \colon tmf/2 \to tmf/(2,B)$.

(81) For any choice of lift $\widetilde{\eta_1\bar\kappa\widetilde{\eta_1}}$ we can set $\widetilde{i(D_3)} = \eta\cdot\widetilde{\eta_1\bar\kappa\widetilde{\eta_1}}$. The classes $\eta\cdot\widetilde{\eta_1\bar\kappa\widetilde{\eta_1}}$ and $\nu \cdot \widetilde{\nu_4\widetilde{\kappa}}$ have the same image under $j$ in $\pi_{80}(tmf/B)$, cf. Figure 12.28. Hence they differ at most by the class detected by the image of $i(h_1\delta w_2)$. If necessary, we can add a class detected by the image of $i(\delta w_2)$ to the chosen lift $\widetilde{\eta_1\bar\kappa\widetilde{\eta_1}}$ to make this difference vanish.

(114) The relation holds modulo $\overline{\nu_4\widetilde{\kappa}}$, by case (105) of Proposition 12.7. Furthermore, $\nu$ annihilates the right hand side of the relation, because $\eta\nu = 0$ and

$\nu\bar{\kappa} = 0$. Since $\nu \cdot \overline{\nu_4\bar{\kappa}}$ is detected by the image of $i(h_2^2\beta w_2^2)$, hence is nonzero, it suffices to argue that $\nu^3 \cdot \widetilde{i(\nu_4)}$ vanishes. However, $\nu^3 = \eta\epsilon$ in $\pi_9(tmf)$, and $\eta$ acts trivially on $\pi_{116}(tmf/(2,B))$, which implies this fact.

(129) We can (uniquely) modify the class detected by the lift of $i(\beta w_1 w_2^2)$ to make its $\eta$-multiple equal to the nonzero $\nu$-multiple in $\pi_{129}(tmf/(2,B))$, due to the independent $\eta$-extensions on the images of $i(\delta' w_2^2)$ and $i(\alpha g w_2^2)$. □

## 12.6. Modified Adams spectral sequences

Let $f\colon X \to Y$ be a map of spectra, and consider the homotopy cofiber sequence

$$X \xrightarrow{f} Y \xrightarrow{i} Cf \xrightarrow{j} \Sigma X .$$

Also fix a homotopy cofiber sequence

$$\Sigma^{-1}\bar{H} \xrightarrow{\alpha} S \xrightarrow{\beta} H \xrightarrow{\gamma} \bar{H}$$

and form the canonical Adams resolution

$$\begin{array}{ccccccc}
S & \xleftarrow{\alpha} & S_1 & \xleftarrow{\alpha} & S_2 & \xleftarrow{\alpha} & \dots \\
\downarrow{\scriptstyle\beta} & & \downarrow{\scriptstyle\beta} & & \downarrow{\scriptstyle\beta} & & \\
H \wedge S & & H \wedge S_1 & & H \wedge S_2 & &
\end{array}$$

of $S = S_0$, with $S_s = (\Sigma^{-1}\bar{H})^{\wedge s}$ and $S_{s,1} = H \wedge (\Sigma^{-1}\bar{H})^{\wedge s}$. There are several ways to combine the canonical Adams resolutions $X_\star = S_\star \wedge X$ and $Y_\star = S_\star \wedge Y$ into a tower ending at $Cf$. The resulting spectral sequences are often referred to as "modified Adams spectral sequences", but as we shall clarify below there are a couple of different modifications involved. We therefore begin by reviewing the "ordinary" and "delayed" approaches that we have already made use of in this work, and then discuss a "hastened" modification of the Adams resolution.

First, the canonical Adams resolution $(Cf)_\star = S_\star \wedge Cf$ sits in a diagram

$$\begin{array}{ccccccc}
X & \xleftarrow{\alpha} & X_1 & \xleftarrow{\alpha} & X_2 & \xleftarrow{\alpha} & \dots \\
\downarrow{\scriptstyle f} & & \downarrow{\scriptstyle f_1} & & \downarrow{\scriptstyle f_2} & & \\
Y & \xleftarrow{\alpha} & Y_1 & \xleftarrow{\alpha} & Y_2 & \xleftarrow{\alpha} & \dots \\
\downarrow{\scriptstyle i} & & \downarrow{\scriptstyle i_1} & & \downarrow{\scriptstyle i_2} & & \\
Cf & \xleftarrow{\alpha} & (Cf)_1 & \xleftarrow{\alpha} & (Cf)_2 & \xleftarrow{\alpha} & \dots \\
\downarrow{\scriptstyle j} & & \downarrow{\scriptstyle j_1} & & \downarrow{\scriptstyle j_2} & & \\
\Sigma X & \xleftarrow{\alpha} & \Sigma X_1 & \xleftarrow{\alpha} & \Sigma X_2 & \xleftarrow{\alpha} & \dots
\end{array} \tag{12.4}$$

with vertical homotopy cofiber sequences, where $f_s = S_s \wedge f$ and $(Cf)_s = C(f_s)$. If $f_* = H_*(f) = 0$, so that $f$ has Adams filtration $\geq 1$, then the associated Adams $E_1$-terms for $Y$, $Cf$ and $\Sigma X$ form a short exact sequence

$$0 \to C^*_{A_*}(\mathbb{F}_2, H_*(Y)) \xrightarrow{i} C^*_{A_*}(\mathbb{F}_2, H_*(Cf)) \xrightarrow{j} C^*_{A_*}(\mathbb{F}_2, \Sigma H_*(X)) \to 0$$

of cobar complexes, as in Definition 2.12. (Strictly speaking, each $(E_1, d_1)$-term is most directly identified with the $A_*$-comodule primitives in the version for left $A_*$-comodules of the canonical injective resolution of [**45**, Def. IV.1.1], but the latter is isomorphic to the cobar resolution, with primitives given by the cobar complex.) The associated long exact sequence of $E_2$-terms takes the form

$$\cdots \longrightarrow \mathrm{Ext}^{s-1}_{A_*}(\mathbb{F}_2, \Sigma H_*(X)) \xrightarrow{\delta} \mathrm{Ext}^s_{A_*}(\mathbb{F}_2, H_*(Y)) \xrightarrow{i} \mathrm{Ext}^s_{A_*}(\mathbb{F}_2, H_*(Cf)) \xrightarrow{j} \mathrm{Ext}^s_{A_*}(\mathbb{F}_2, \Sigma H_*(X)) \longrightarrow \dots,$$

where the connecting homomorphism $\delta$ is given by Yoneda composition with the class in $\mathrm{Ext}^1_{A_*}(\Sigma H_*(X), H_*(Y))$ of the extension $H_*(Y) \to H_*(Cf) \to \Sigma H_*(X)$. Similar considerations apply (under the usual finite type hypotheses) (a) for the opposite variance, replacing homology and $A_*$-comodules with cohomology and $A$-modules, (b) for maps of *tmf*-modules, replacing $A_*$ with $A(2)_*$, and (c) at other primes, replacing $\mathbb{F}_2$ with $\mathbb{F}_p$. We leave to the reader to make the notational substitutions needed in these cases. We made use of this long exact sequence of $E_2$-terms in Chapters 4, 6, 7, 8 and 11. If $f$ has Adams filtration $\sigma \ge 2$, then

$$0 \to E_r(Y) \xrightarrow{i} E_r(Cf) \xrightarrow{j} E_r(\Sigma X) \to 0$$

remains short exact up to and including the case $r = \sigma$, so that the $d_r$-differentials for $Cf$ are more-or-less determined by those for $X$ and $Y$ when $r < \sigma$, and only the differentials with $r \ge \sigma$ are directly affected by $f$.

Second, we have seen in Chapter 11, cf. Definition 11.10, and in Sections 12.4 and 12.5, that we can delay the effect of $f$ on the differentials in the spectral sequence by $d \ge 1$ terms, by the device of replacing the canonical Adams resolution of $Cf$ with the convolution product $(S \wedge Z)_\star$ of $S_\star$ and $Z_\star$, where now $Z_\star$ is the tower

$$Cf \xleftarrow{i} Y \xleftarrow{=} \dots \xleftarrow{=} Y \longleftarrow *$$

with $Z_0 = Cf$, $Z_k = Y$ for $1 \le k \le d$, and $Z_k = *$ for $k > d$. We then have a diagram

(12.5)
$$\begin{array}{ccccccc}
X & \xleftarrow{\alpha} \dots \xleftarrow{\alpha} & X_d & \xleftarrow{\alpha} & X_{d+1} & \xleftarrow{\alpha} & \dots \\
\downarrow f & & \downarrow f\alpha^d & & \downarrow f_1\alpha^d & & \\
Y & \xleftarrow{=} \dots \xleftarrow{=} & Y & \xleftarrow{\alpha} & Y_1 & \xleftarrow{\alpha} & \dots \\
\downarrow i & & \downarrow i' & & \downarrow i'_1 & & \\
(S\wedge Z)_0 & \xleftarrow{\alpha} \dots \xleftarrow{\alpha} & (S\wedge Z)_d & \xleftarrow{\alpha} & (S\wedge Z)_{d+1} & \xleftarrow{\alpha} & \dots \\
\downarrow j & & \downarrow j' & & \downarrow j'_1 & & \\
\Sigma X & \xleftarrow{\alpha} \dots \xleftarrow{\alpha} & \Sigma X_d & \xleftarrow{\alpha} & \Sigma X_{d+1} & \xleftarrow{\alpha} & \dots
\end{array}$$

with vertical homotopy cofiber sequences, so that $(S \wedge Z)_s = C(f\alpha^s)$ for $0 \le s \le d$ and $(S \wedge Z)_s = C(f_{s-d}\alpha^d)$ for $s \ge d$. The associated $E_1$-terms form a short exact sequence

$$0 \to C^{*-d}_{A_*}(\mathbb{F}_2, \Sigma^d H_*(Y)) \xrightarrow{i'} E_1^*((S \wedge Z)_\star) \xrightarrow{j'} C^*_{A_*}(\mathbb{F}_2, \Sigma H_*(X)) \to 0\,.$$

If $f$ has Adams filtration $\sigma \ge 1$, then the associated long exact sequence of $E_2$-terms splits into short exact sequences

$$0 \to \mathrm{Ext}^{s-d}_{A_*}(\mathbb{F}_2, \Sigma^d H_*(Y)) \xrightarrow{i'} E_2^s((S \wedge Z)_\star) \xrightarrow{j'} \mathrm{Ext}^s_{A_*}(\mathbb{F}_2, \Sigma H_*(X)) \to 0\,,$$

and

$$0 \to E_r^{s-d}(\Sigma^d Y) \xrightarrow{i'} E_r^s((S \wedge Z)_\star) \xrightarrow{j'} E_r^s(\Sigma X) \to 0$$

remains short exact as long as $r \le \sigma + d$, so that only the $d_r$-differentials with $r \ge \sigma + d$ are directly affected by $f$. In this sense, the interaction between $f$ and the differentials internal to $E_r(X)$ and $E_r(Y)$ is delayed to only influence the differentials in $E_r((S \wedge Z)_\star)$ for $r \ge \sigma + d$. This can be advantageous if one already has a good understanding of the differential structure and hidden extensions in the ordinary Adams spectral sequences for $X$ and $Y$.

LEMMA 12.30. *The maps $\alpha^d \colon Y_s \to Y_{s-d}$ induce a morphism from the ordinary Adams spectral sequence for $Cf$ to the delayed one.*

PROOF. This is the morphism of spectral sequences induced by the following map of towers, where the top row is the canonical Adams resolution of $Cf$.

$$\begin{array}{ccccccccc}
Cf & \xleftarrow{\alpha} & C(f_1) & \xleftarrow{\alpha} \cdots \xleftarrow{\alpha} & C(f_d) & \xleftarrow{\alpha} & C(f_{d+1}) & \xleftarrow{\alpha} & \cdots \\
{\scriptstyle =}\downarrow & & \downarrow & & \downarrow & & \downarrow & & \\
Cf & \xleftarrow{\alpha} & C(f\alpha) & \xleftarrow{\alpha} \cdots \xleftarrow{\alpha} & C(f\alpha^d) & \xleftarrow{\alpha} & C(f_1\alpha^d) & \xleftarrow{\alpha} & \cdots
\end{array}$$

□

Third, we come to a modification of the Adams spectral sequence for $Cf$ where the effect of $f$ on the differentials is hastened by $e \ge 1$ terms. This is the "modified Adams spectral sequence" of Behrens, Hill, Hopkins and Mahowald [**26**]. Suppose that we have factored $f = \alpha^e g$ for some map $g \colon X \to Y_e = S_e \wedge Y$, and that $e = \sigma$ equals the Adams filtration of $f$. This implies that the composite $\beta g \colon X \to H \wedge Y_e$ is essential, so that $g_* \colon H_*(X) \to H_*(Y_e)$ is nonzero. We extend $g$ to a map of canonical Adams resolutions with $g_s = S_s \wedge g$, and obtain the diagram

$$\begin{array}{ccccccc}
X & \xleftarrow{=} \cdots \xleftarrow{=} & X & \xleftarrow{\alpha} & X_1 & \xleftarrow{\alpha} & \cdots \\
{\scriptstyle f}\downarrow & & {\scriptstyle g}\downarrow & & {\scriptstyle g_1}\downarrow & & \\
Y & \xleftarrow{\alpha} \cdots \xleftarrow{\alpha} & Y_e & \xleftarrow{\alpha} & Y_{e+1} & \xleftarrow{\alpha} & \cdots \\
{\scriptstyle i}\downarrow & & {\scriptstyle i''}\downarrow & & {\scriptstyle i''_1}\downarrow & & \\
Cf & \xleftarrow{\alpha} \cdots \xleftarrow{\alpha} & Cg & \xleftarrow{\alpha} & (Cg)_1 & \xleftarrow{\alpha} & \cdots \\
{\scriptstyle j}\downarrow & & {\scriptstyle j''}\downarrow & & {\scriptstyle j''_1}\downarrow & & \\
\Sigma X & \xleftarrow{=} \cdots \xleftarrow{=} & \Sigma X & \xleftarrow{\alpha} & \Sigma X_1 & \xleftarrow{\alpha} & \cdots
\end{array} \tag{12.6}$$

with vertical homotopy cofiber sequences.

DEFINITION 12.31. Let $W_\star$ denote the tower ending at $Cf$, displayed above, with $W_s = C(\alpha^{e-s} g)$ for $0 \le s \le e$ and $W_s = S_{s-e} \wedge Cg$ for $s \ge e$. We call the spectral sequence obtained by applying $\pi_*(-)$ the hastened Adams spectral sequence for $Cf$.

LEMMA 12.32. *The maps $\alpha^e \colon X_s \to X_{s-e}$ induce a morphism from the ordinary Adams spectral sequence for $Cf$ to the hastened one.*

PROOF. This is the morphism of spectral sequences induced by the following map of towers.

$$\begin{array}{ccccccccccc}
Cf & \xleftarrow{\alpha} & C(\alpha^{e-1}g\alpha) & \xleftarrow{\alpha} & \dots & \xleftarrow{\alpha} & C(g\alpha^e) & \xleftarrow{\alpha} & C(g_1\alpha^e) & \xleftarrow{\alpha} & \dots \\
{\scriptstyle =}\downarrow & & \downarrow & & & & \downarrow & & \downarrow & & \\
Cf & \xleftarrow{\alpha} & C(\alpha^{e-1}g) & \xleftarrow{\alpha} & \dots & \xleftarrow{\alpha} & Cg & \xleftarrow{\alpha} & C(g_1) & \xleftarrow{\alpha} & \dots
\end{array}$$

The top row is an Adams resolution of $Cf$, but not the canonical one. Its associated spectral sequence therefore agrees with the ordinary Adams spectral sequence from the $E_2$-term and onward. □

In general there is no direct connection between the delayed and hastened Adams spectral sequences. Returning to diagram (12.6), the associated $E_1$-terms fit into vertical long exact sequences, as in the following diagram.

$$\begin{array}{ccccccccc}
\downarrow & & & & \downarrow & & \downarrow & & \\
0 & \xrightarrow{=} & \dots & \longrightarrow & \Sigma^e H_*(X) & \xrightarrow{d_1^0} & \bar{A}_* \otimes \Sigma^e H_*(X) & \xrightarrow{d_1^1} & \dots \\
{\scriptstyle f_*}\downarrow & & & & {\scriptstyle g_*}\downarrow & & {\scriptstyle \bar{A}\otimes g_*}\downarrow & & \\
H_*(Y) & \xrightarrow{d_1^0} & \dots & \xrightarrow{d_1^{e-1}} & \bar{A}_*^{\otimes e} \otimes H_*(Y) & \xrightarrow{d_1^e} & \bar{A}_*^{\otimes e+1} \otimes H_*(Y) & \xrightarrow{d_1^{e+1}} & \dots \\
{\scriptstyle i_*}\downarrow & & & & {\scriptstyle i''_*}\downarrow & & {\scriptstyle \bar{A}_*\otimes i''_*}\downarrow & & \\
E_1^0(W_\star) & \xrightarrow{d_1^0} & \dots & \xrightarrow{d_1^{e-1}} & E_1^e(W_\star) & \xrightarrow{d_1^e} & E_1^{e+1}(W_\star) & \xrightarrow{d_1^{e+1}} & \dots \\
{\scriptstyle j_*}\downarrow & & & & {\scriptstyle j''_*}\downarrow & & {\scriptstyle \bar{A}_*\otimes j''_*}\downarrow & & \\
0 & \xrightarrow{=} & \dots & \longrightarrow & \Sigma^{e+1} H_*(X) & \xrightarrow{d_1^0} & \bar{A}_* \otimes \Sigma^{e+1} H_*(X) & \xrightarrow{d_1^1} & \dots \\
\downarrow & & & & \downarrow & & \downarrow & &
\end{array}$$

We now make the additional assumption that $g_* \colon H_*(X) \to H_*(Y_e)$ is a monomorphism. This ensures that the vertical long exact sequences break up into a short exact sequence

$$0 \to C^{*-e}_{A_*}(\mathbb{F}_2, \Sigma^e H_*(X)) \xrightarrow{g} C^*_{A_*}(\mathbb{F}_2, H_*(Y)) \xrightarrow{i''} E_1^*(W_\star) \to 0$$

of cochain complexes. In the induced long exact sequence of $E_2$-terms

$$\dots \longrightarrow \mathrm{Ext}^{s-e}_{A_*}(\mathbb{F}_2, \Sigma^e H_*(X)) \xrightarrow{g} \mathrm{Ext}^s_{A_*}(\mathbb{F}_2, H_*(Y))$$
$$\xrightarrow{i''} E_2^s(W_\star) \xrightarrow{\delta} \mathrm{Ext}^{s+1-e}_{A_*}(\mathbb{F}_2, \Sigma^e H_*(X)) \longrightarrow \dots$$

the homomorphism $g$ is given by Yoneda composition with the infinite cycle in $\mathrm{Ext}^{e,e}_{A_*}(H_*(X), H_*(Y))$ detecting the lift $g$ of $f$. In the spectral sequence $E_r(W_\star)$, the role of the map $f$ of Adams filtration $e$ has thus been hastened to have a direct effect on the $E_1$-term, rather than affecting the $d_e$-differential and $E_{e+1}$-term, as for the ordinary Adams spectral sequence $E_r(Cf)$. This has the advantage that information about $f$ enters at a stage where the determination of the $E_r$-term is still a purely algebraic problem.

This algebraic connection can be made more explicit. The homotopy cofiber $W_{s,1} = \mathrm{cof}(W_{s+1} \to W_s)$ has the form $H \wedge Y_s$ for $0 \le s < e$, and the form $H \wedge W_s$ for $s \ge e$, so that its homotopy

$$E_1^s(W_\star) = \pi_{*-s}(W_{s,1}) \cong \mathrm{Hom}_{A_*}(\mathbb{F}_2, H_{*-s}(W_{s,1}))$$

is given by the $A_*$-comodule primitives in its homology. Our assumption that $g_* \colon H_*(X) \to H_*(Y_e)$ is a monomorphism implies that the homologies of $X_{s-e,1}$, $Y_{s,1}$ and $W_{s,1}$ form a short exact sequence

$$0 \to C^{*-e}_{A_*}(A_*, \Sigma^e H_*(X)) \xrightarrow{g} C^*_{A_*}(A_*, H_*(Y)) \xrightarrow{i''} Q^* \to 0$$

of $A_*$-comodule cochain complexes, where $Q^s = H_{*-s}(W_{s,1})$ is extended, hence injective, and $\delta \colon Q^s \to Q^{s+1}$ is the usual composite homomorphism $H_{*-s}(W_{s,1}) \to H_{*-s-1}(W_{s+1}) \to H_{*-s-1}(W_{s+1,1})$. Furthermore, $\eta \colon H_*(Y) \to C^*_{A_*}(A_*, H_*(Y))$ is an injective resolution of the $A_*$-comodule $H_*(Y)$, whereas $\eta \colon \Sigma^e H_*(X)[-e] \to C^{*-e}_{A_*}(A_*, \Sigma^e H_*(X))$ is an injective resolution of $\Sigma^e H_*(X)$ shifted to cohomological degree $e$. In the derived category $\mathcal{D}(A_*)$ of $A_*$-comodules, we thus have a distinguished triangle

$$\Sigma^e H_*(X)[-e] \xrightarrow{g} H_*(Y) \xrightarrow{i''} Q^* \xrightarrow{j''} \Sigma^e H_*(X)[1-e]\,.$$

PROPOSITION 12.33 ([**26**, §3]). *Let $f = \alpha^e g \colon X \to Y_e \to Y$, and assume that $g_* \colon H_*(X) \to H_*(Y_e)$ is a monomorphism. The hastened Adams spectral sequence for $Cf$ has $E_2$-term*

$$E_2^s(W_\star) \cong \mathrm{Ext}^s_{\mathcal{D}(A_*)}(\mathbb{F}_2, Q^*)\,,$$

*where $Q^*$ is the homotopy cofiber in $\mathcal{D}(A_*)$ of the morphism $g \colon \Sigma^e H_*(X)[-e] \to H_*(Y)$ corresponding to the class in $\mathrm{Ext}^{e,e}_{A_*}(H_*(X), H_*(Y))$ that detects the lift $g$ of $f$.*

PROOF. We have arranged that $Q^*$ is injective in each cohomological degree. Therefore the hastened $E_2$-term

$$E_2^s(W_\star) \cong H^s(\mathrm{Hom}_{A_*}(\mathbb{F}_2, Q^*))$$

calculates, by definition, the hyper-Ext groups

$$\mathrm{Ext}^s_{\mathcal{D}(A_*)}(\mathbb{F}_2, Q^*) = \mathcal{D}(A_*)(\mathbb{F}_2, Q^*[s])$$

of the $A_*$-comodule complex $Q^*$. □

It follows that we can calculate the hastened $E_2$-term as the hyper-Ext of any other $A_*$-comodule complex that is isomorphic in $\mathcal{D}(A_*)$ to $Q^*$. Thus, let $\eta \colon H_*(Y) \to P^*$ be any injective $A_*$-comodule resolution of $H_*(Y)$, pick a chain equivalence $\theta \colon C^*_{A_*}(A_*, H_*(Y)) \to P^*$ under $H_*(Y)$, and suppose that the composite $e$-cocycle

$$h = \theta g \eta \colon \Sigma^e H_*(X) \longrightarrow C^0_{A_*}(A_*, \Sigma^e H_*(X)) \longrightarrow C^e_{A_*}(A_*, H_*(Y)) \longrightarrow P^e$$

in $\mathrm{Hom}_{A_*}(\Sigma^e H_*(X), P^*)$ is also a monomorphism. Clearly $h$ represents the same class in $\mathrm{Ext}^{e,e}_{A_*}(H_*(X), H_*(Y))$ as $g\eta$. Let

$$\tilde{h} \colon \Sigma^e H_*(X) \longrightarrow J^e = \mathrm{im}(\delta \colon P^{e-1} \to P^e) = \ker(\delta \colon P^e \to P^{e+1})$$

be the unique lift of $h$ through the $e$-coboundaries in $P^*$, and form the short exact sequence

$$0 \to \Sigma^e H_*(X) \xrightarrow{\tilde{h}} J^e \longrightarrow \bar{J}^e \longrightarrow 0\,.$$

We then have a zig-zag of quasi-isomorphisms connecting $Q^*$ to the $A_*$-comodule complex

$$0 \to P^0 \xrightarrow{\delta} P^1 \xrightarrow{\delta} \dots \xrightarrow{\delta} P^{e-1} \longrightarrow \bar{J}^e \to 0$$

with cohomology concentrated in degrees 0 and $e-1$. If $\eta\colon \bar{J}^e \to \bar{P}^{*+e}$ is another injective $A_*$-comodule resolution, then we can form an injective complex $\bar{P}^*$ by splicing $P^{*<e}$ with $\bar{P}^{*\geq e}$, using the composite

$$P^{e-1} \longrightarrow J^e \longrightarrow \bar{J}^e \longrightarrow \bar{P}^e$$

to connect the two parts of the complex. Then $Q^*$ and $\bar{P}^*$ are isomorphic in the derived category, and we can calculate the hastened $E_2$-term as

$$E_2^s(W_\star) \cong \mathrm{Ext}^s_{\mathcal{D}(A_*)}(\mathbb{F}_2, Q^*) \cong \mathrm{Ext}^s_{\mathcal{D}(A_*)}(\mathbb{F}_2, \bar{P}^*) = H^s(\mathrm{Hom}_{A_*}(\mathbb{F}_2, \bar{P}^*)) .$$

In particular, if $P^*$ and $\bar{P}^{*+e}$ were chosen as minimal resolutions, then $\bar{P}^*$ is also a minimal complex, which makes it trivial to pass to cohomology in the right hand term above.

We illustrate this method by two examples, working in cohomology and in the context of *tmf*-modules. In each case we use a method discovered by Mahowald, which tricks `ext` into calculating a minimal complex $\tilde{P}_*$ by carefully interrupting the machine computation of a minimal free $A(2)$-module resolution $\epsilon\colon P_* \to H^*(Y)$ and adjusting the boundary homomorphism $\partial\colon P_e \to P_{e-1}$ to have image $\tilde{J}_e = \ker(\bar{h}\colon J_e \to \Sigma^e H^*(X))$ in place of $J_e = \mathrm{im}(\partial\colon P_e \to P_{e-1}) \cong \mathrm{cok}(\partial\colon P_{e+1} \to P_e)$.

EXAMPLE 12.34. Let $f = B\colon \Sigma^8 tmf \to tmf$ be given by multiplication with the Bott element $B \in \pi_8(tmf)$. This map has Adams filtration $e=4$, and we can pick a lift $g\colon \Sigma^8 tmf \to tmf_4$ with $B \simeq \alpha^4 g$, which is detected by $w_1 \in \mathrm{Ext}^{4,12}_{A(2)}(\mathbb{F}_2, \mathbb{F}_2)$. The hastened Adams spectral sequence for $Cf = tmf/B$ has $E_2$-term

$$E_2^s(W_\star) = \mathrm{Ext}^s_{\mathcal{D}(A(2))}(Q_*, \mathbb{F}_2) ,$$

where $Q_*$ is the homotopy fiber in $\mathcal{D}(A(2))$ of the morphism $g\colon \mathbb{F}_2 \to \Sigma^{12}\mathbb{F}_2[4]$ corresponding to $w_1$. The inclusion $i''\colon Y_\star = tmf_\star \to W_\star$ induces a map of spectral sequences from the Adams spectral sequence for *tmf* to the hastened spectral sequence for *tmf*/$B$, which appears in the long exact sequence of $E_2$-terms

$$\begin{aligned} \dots \longrightarrow \mathrm{Ext}^{s-4,t-12}_{A(2)}(\mathbb{F}_2, \mathbb{F}_2) &\xrightarrow{w_1} \mathrm{Ext}^{s,t}_{A(2)}(\mathbb{F}_2, \mathbb{F}_2) \\ &\xrightarrow{i''} E_2^{s,t}(W_\star) \xrightarrow{j''} \mathrm{Ext}^{s-3,t-12}_{A(2)}(\mathbb{F}_2, \mathbb{F}_2) \longrightarrow \dots . \end{aligned}$$

In this case we know that $w_1$ acts injectively on $E_2(tmf) = \mathrm{Ext}_{A(2)}(\mathbb{F}_2, \mathbb{F}_2)$, so that $j'' = 0$ and $\mathrm{Ext}_{A(2)}(\mathbb{F}_2, \mathbb{F}_2)/w_1 \cong E_2(W_\star)$.

To use `ext` to calculate a minimal free complex $\tilde{P}_*$ that is derived isomorphic to $Q_*$, we work in the context of Remark 1.9, call on `newmodule tmfmodB tmf.def` in the directory `A2`, and execute `dims 0 12` in `A2/tmfmodB` to compute a minimal free $A(2)$-module resolution $P_* \to \mathbb{F}_2$ in internal degrees $t \leq 12$. At this point the file `Diff.4` specifies $\partial(4_g^*) \in P_3$ for the generators $4_g^* \in P_4$ in these degrees, in a machine readable format. To explain the method we display this data file in its "humanly readable" form `hDiff.4`, which is obtained from `Diff.4` by means of the command `convert Diff.4 hDiff.4 2 1 1 i`. It appears as follows:

```
              2           12
4


1
0 1 1 i(1).

12


2
0 9 4 i(6,1).
1 6 3 i(3,1).
```

This tells us that $P_4$ has two generators in this range, namely $4_0^*$ and $4_1^*$ in internal degrees 4 and 12, respectively. Furthermore, $\partial(4_0^*) = Sq^1 3_0^*$ and $\partial(4_1^*) = Sq^{(6,1)}3_0^* + Sq^{(3,1)}3_1^*$, where $3_0^*$ and $3_1^*$ are generators of $P_3$, and the coefficients in $A(2)$ written in terms of the Milnor basis. It follows that only the $A(2)$-linear 4-cocycle $h = 4_1 \colon P_4 \to \Sigma^{12}\mathbb{F}_2$ represents $w_1$. It is clear that $h$ is an epimorphism. We let

$$\bar{h} \colon J_4 = \operatorname{im}(\partial \colon P_4 \to P_3) \longrightarrow \Sigma^{12}\mathbb{F}_2$$

be its unique factorization through the 3-boundaries in $P_*$, and form the short exact sequence

$$0 \to \tilde{J}_4 \longrightarrow J_4 \xrightarrow{\bar{h}} \Sigma^{12}\mathbb{F}_2 \to 0\,.$$

We get an isomorphism in $\mathcal{D}(A(2))$ between the homotopy fiber $Q_*$ and the minimal complex

$$0 \leftarrow P_0 \xleftarrow{\partial} P_1 \xleftarrow{\partial} P_2 \xleftarrow{\partial} P_3 \longleftarrow \tilde{J}_4 \leftarrow 0\,.$$

Following Mahowald, we now edit the file `Diff.4`, changing $\partial \colon P_4 \to P_3$ to $\partial \colon \tilde{P}_4 \to P_3$ in degree 12 so that $\operatorname{im}(\partial \colon \tilde{P}_4 \to P_3) = \tilde{J}_4$. In this case, $\partial$ already maps the $A(2)$-module generated by $4_0^*$ to $\tilde{J}_4$, while $\bar{h}\partial$ is nonzero on $4_1^*$, so we obtain the desired modification by removing the generator $4_1^*$, together with the value of $\partial(4_1^*)$, and adjusting the total number of generators. The resulting file appears as follows:

```
              1           12
4


1
0 1 1 i(1).
```

The change in degree 12 from $P_4$ to $\tilde{P}_4$ is the only difference between $P_*$ and $\tilde{P}_*$ in this range of degrees. Running `dims 13 240` now has the effect of calculating the rest of $P_{*<4}$, and simultaneously to extend $\tilde{P}_{*+4}$ to a minimal free resolution of $\tilde{J}_4$, in degrees 13 and above. Since the resulting complex $\tilde{P}_*$ is minimal, we can call on `report` and

```
chart 0 40 0 200 Shape himults E2-tmfmodB.tex E2-tmfmodB
pdflatex E2-tmfmodB.tex
```

to obtain the hastened $E_2$-term for *tmf/B*, as shown in Figures 12.41 to 12.48.

PROPOSITION 12.35. *The differential structure in the hastened Adams spectral sequence* $E_2(W_\star) \Longrightarrow \pi_*(tmf/B)_2^\wedge$ *is as displayed in Figures* 12.41 *to* 12.48*, repeated* $w_2^4$*-periodically.*

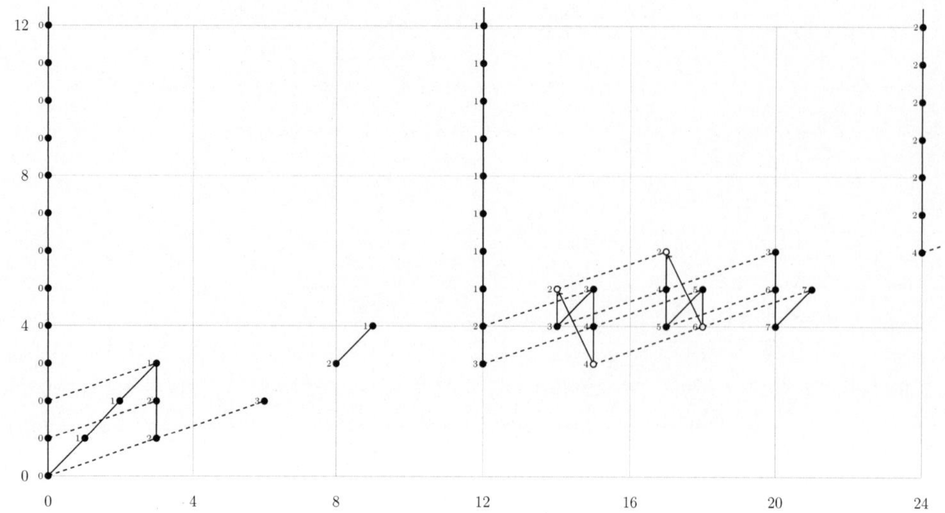

FIGURE 12.41. Hastened $(E_r(tmf/B), d_r)$ for $0 \leq t - s \leq 24$

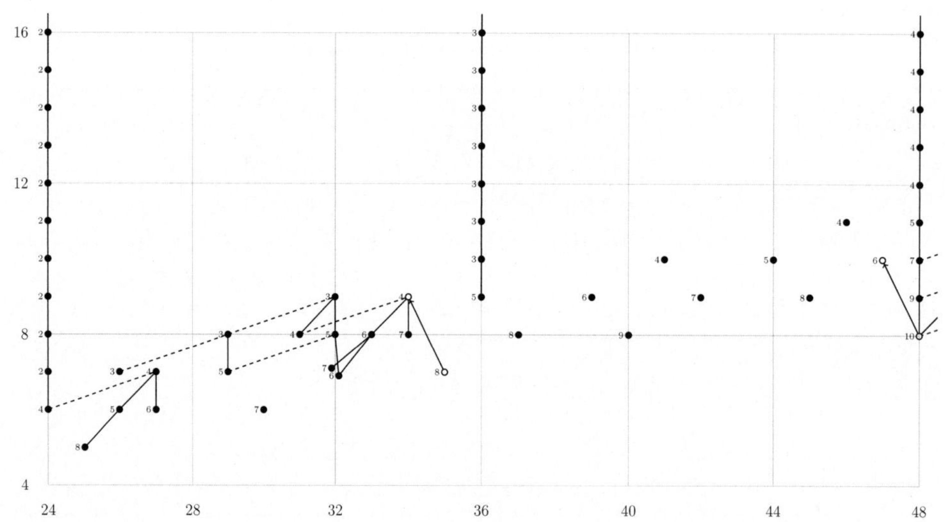

FIGURE 12.42. Hastened $(E_r(tmf/B), d_r)$ for $24 \leq t - s \leq 48$

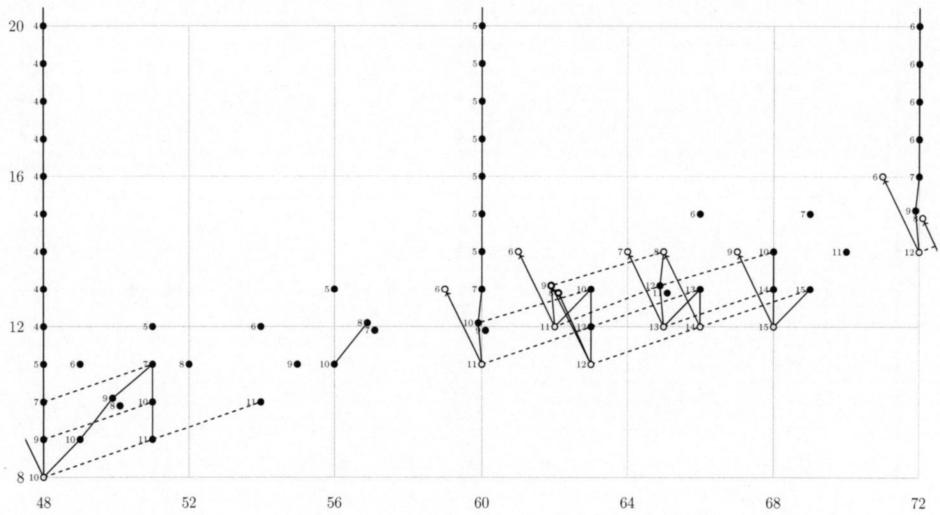

FIGURE 12.43. Hastened $(E_r(tmf/B), d_r)$ for $48 \le t - s \le 72$

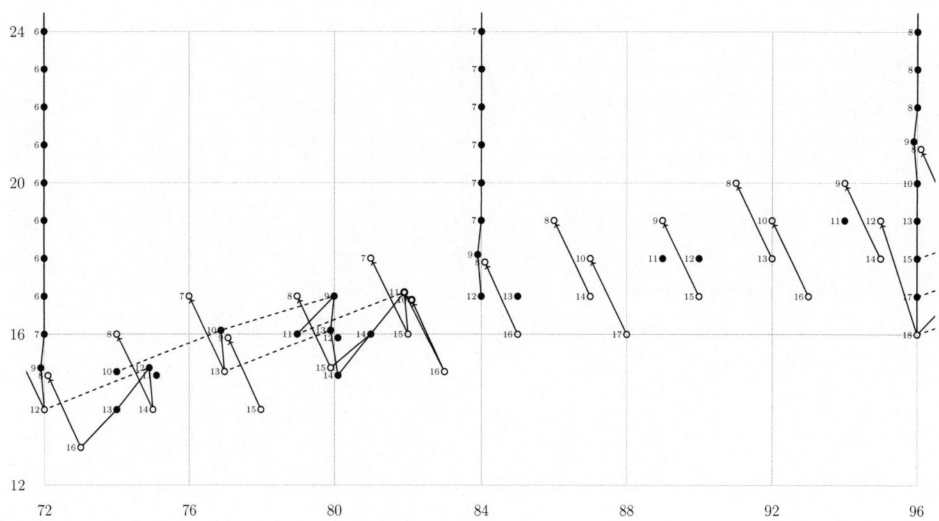

FIGURE 12.44. Hastened $(E_r(tmf/B), d_r)$ for $72 \le t - s \le 96$

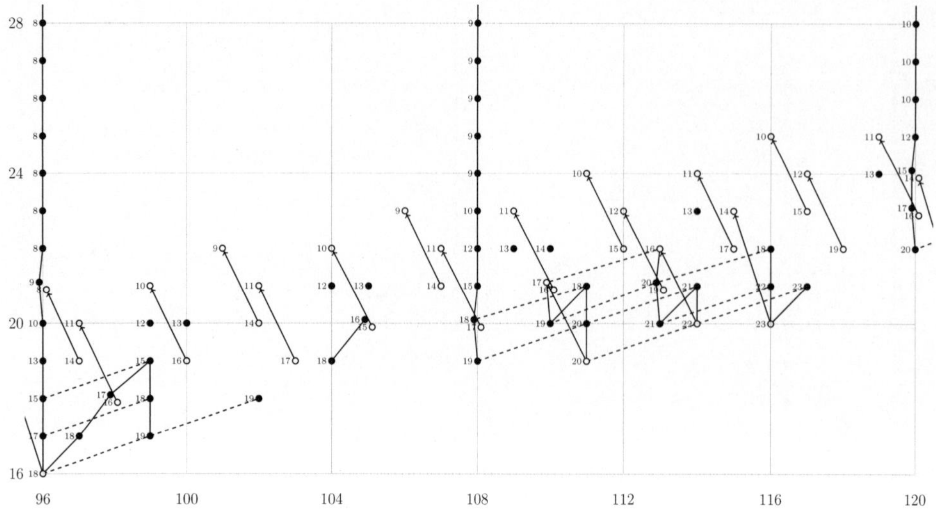

FIGURE 12.45. Hastened $(E_r(tmf/B), d_r)$ for $96 \le t - s \le 120$

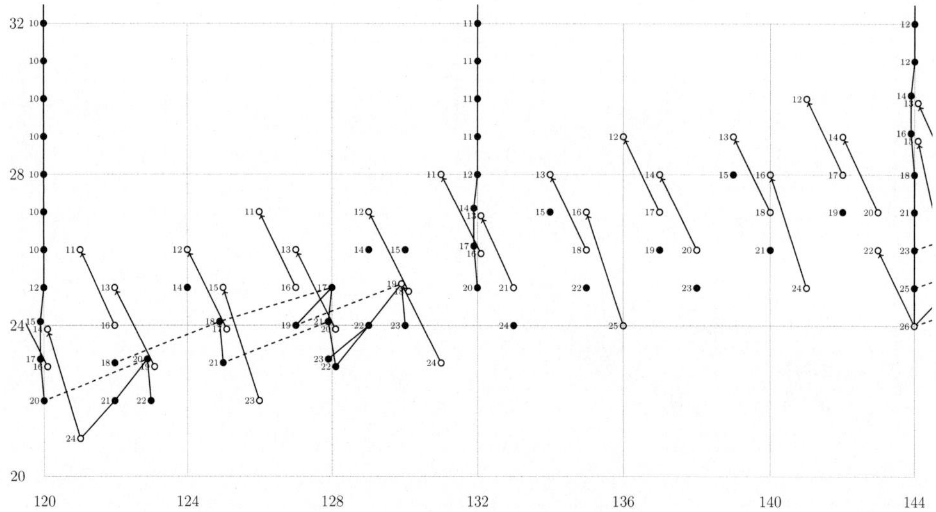

FIGURE 12.46. Hastened $(E_r(tmf/B), d_r)$ for $120 \le t - s \le 144$

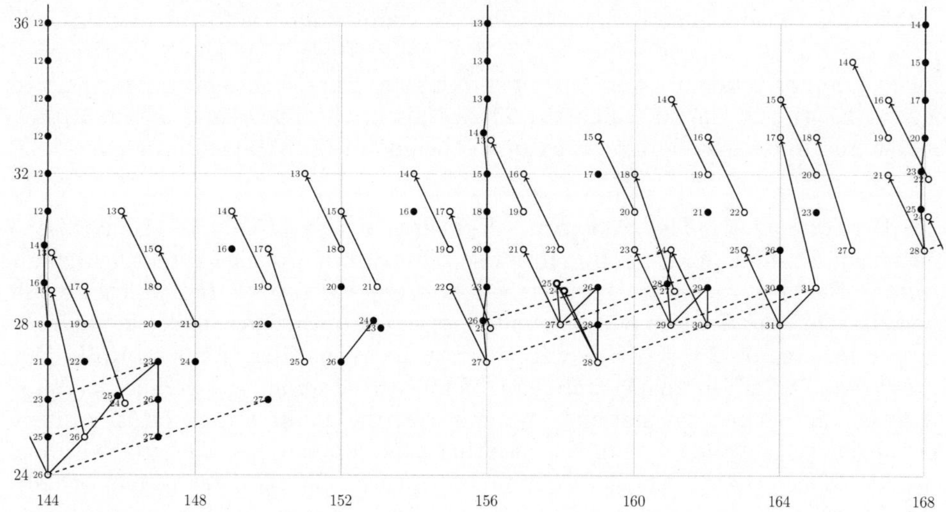

FIGURE 12.47. Hastened $(E_r(tmf/B), d_r)$ for $144 \leq t - s \leq 168$

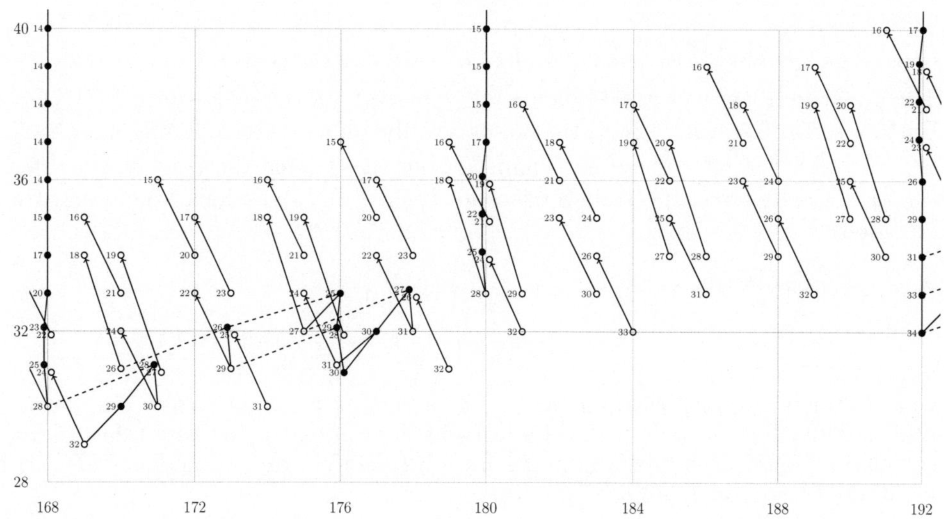

FIGURE 12.48. Hastened $(E_r(tmf/B), d_r)$ for $168 \leq t - s \leq 192$

PROOF. The $d_2$-differentials all follow by compatibility with the morphism $i''$ from the Adams spectral sequence for *tmf*, using Table 5.1. The $d_3$-differentials follow from the action of $w_2^2$ in $E_3(tmf)$ on this spectral sequence, using the Leibniz rule with $d_3(w_2^2) = \beta g^4$. The final $d_4$-differentials, on the images of $h_1 w_2^3$ and $h_1 g w_2^3$, follow from the known group structure of the abutment, obtained in Section 12.4. □

REMARK 12.36. The Anderson self-duality of $tmf/(B, M)$ is visible in the hastened $E_\infty$-term for $tmf/B$. Likewise, the Brown–Comenetz self-duality for $tmf/(2, B, M)$ is visible in the hastened $E_\infty$-term for $tmf/(2, B)$, which we will now discuss. The $E_2$-term and differential structure of the latter spectral sequence were calculated in [**26**, §8]. However, some of the $h_1$-multiplications shown in Figures 8.1 and 8.2 of that paper were based on incorrect arguments, and do not agree with our automated calculations. Nonetheless, the additive rank of each differential shown, and the order of the resulting homotopy groups, all agree with our results. Hence these mistakes have no consequences for the later results of [**26**]. On the other hand, for our proof of Theorem 12.25, concerning the action of $\eta$ on $\pi_*(tmf/(2, B))$, it is crucial to work with the correct $h_1$-multiplications.

EXAMPLE 12.37. Let $f = B\colon \Sigma^8 tmf/2 \to tmf/2$ be given by multiplication by $B$. We pick a lift $g\colon \Sigma^8 tmf \to (tmf/2)_4$, which is detected by the nonzero class $v_1^4$ in

$$\mathrm{Ext}_{A(2)}^{4,12}(M_1, M_1) \xrightarrow{\cong} \mathrm{Ext}_{A(2)}^{4,12}(M_1, \mathbb{F}_2) = \mathbb{F}_2\{i(w_1)\}\,.$$

The hastened Adams spectral sequence for $Cf = tmf/(2, B)$ has $E_2$-term $E_2^s(W_\star) = \mathrm{Ext}^s_{\mathcal{D}(A(2))}(Q_*, \mathbb{F}_2)$, where $Q_*$ is the homotopy fiber of the corresponding morphism $v_1^4\colon M_1 \to \Sigma^{12} M_1[4]$. There is a map $i''$ from the Adams spectral sequence for $tmf/2$ to the hastened spectral sequence for $tmf/(2, B)$, and a long exact sequence of $E_2$-terms

$$\cdots \longrightarrow \mathrm{Ext}_{A(2)}^{s-4,t-12}(M_1, \mathbb{F}_2) \xrightarrow{v_1^4} \mathrm{Ext}_{A(2)}^{s,t}(M_1, \mathbb{F}_2)$$
$$\xrightarrow{i''} E_2^{s,t}(W_\star) \xrightarrow{j''} \mathrm{Ext}_{A(2)}^{s-3,t-12}(M_1, \mathbb{F}_2) \longrightarrow \dots\,,$$

where $v_1^4$ acts as multiplication by $w_1$. To calculate a minimal free complex $\tilde{P}_*$ that is derived isomorphic to $Q_*$ we work in the context of Remark 1.26. Those calculations show that in the minimal free $A(2)$-module resolution $P_*$ of $M_1$, the file `Diff.4` begins as follows:

```
           7        240
12

1
0 5 2 i(2,1).

13

3
0 6 3 i(6)(0,2).
1 2 1 i(2).
2 1 1 i(1).

[...]
```

This means that $P_4$ in internal degrees $t \le 13$ has two free $A(2)$-module generators $4_0^*$ and $4_1^*$, in degrees 12 and 13, respectively. Furthermore, in these degrees $J_4 = \operatorname{im}(\partial\colon P_4 \to P_3)$ is generated by $\partial(4_0^*) = Sq^{(2,1)}3_0^*$ in degree 12 and $\partial(4_1^*) = (Sq^6 + Sq^{(0,2)})3_0^* + Sq^2 3_1^* + Sq^1 3_2^*$ in degree 13. To obtain an $\mathbb{F}_2$-basis for $J_4$ in this range of degrees we must adjoin $Sq^1\partial(4_0^*) = Sq^1 Sq^{(2,1)}3_0^* = Sq^{(3,1)}3_0^*$.

Let $h\colon P_4 \to \Sigma^{12}M_1$ be a 4-cocycle representing $v_1^4$. The composite $P_4 \to \Sigma^{12}M_1 \to \Sigma^{12}\mathbb{F}_2$ must then be the cocycle $4_0$ representing $i(w_1)$, meaning that $h(4_0^*)$ is nonzero in degree 12. It follows that $h(Sq^1 4_0^*)$ is nonzero in degree 13, so that $h$ is an epimorphism. We can choose whether $h(4_1^*)$ is to be zero or nonzero, but the two choices give cohomologous 4-cocycles, and therefore give equivalent resolutions. For convenience we choose $h$ so that $h(4_1^*) = 0$. We let $\bar{h}\colon J_4 \to \Sigma^{12}M_1$ be the unique factorization of $h$ through the 3-boundaries in $P_*$, and define $\tilde{J}_4 = \ker(\bar{h}) \subset \tilde{J}_4$. It follows that $\partial(4_1^*)$ gives an $\mathbb{F}_2$-basis for $\tilde{J}_4$ in degrees $\le 13$.

We now want to use Mahowald's trick to modify $P_*$ to a complex $\tilde{P}_*$, such that $P_s = \tilde{P}_s$ for $0 \le s \le 3$, with $\operatorname{im}(\partial\colon \tilde{P}_4 \to P_3) = \tilde{J}_4$, and such that $\tilde{P}_{*+4}$ is a minimal free resolution of $\tilde{J}_4$. This requires altering the image of $\partial\colon P_4 \to P_3$ both in degree 12 and in degree 13, and must therefore be performed in two steps. To start we call on `newmodule tmfC2modB tmfC2.def` and execute `dims 0 12`, to calculate a minimal free $A(2)$-module resolution $P_* \to M_1$ in degrees $t \le 12$. The file `Diff.4` then has the following content:

```
          1        12
12


1
0 5 2 i(2,1).
```

Since $\tilde{J}_4$ is trivial in degree 12, the first modification we must make is to delete the generator $4_0^*$ from `Diff.4`, leaving the following result:

```
          0        12
```

Let $\hat{P}_*$ denote the resulting subcomplex of $P_*$, in degrees $t \le 12$. We now run `dims 13 13` to extend $\hat{P}_*$ to internal degree 13. Thereafter, `Diff.4` appears as follows:

```
          2        13
13


1
0 6 3 i(3,1).


13


3
0 6 3 i(6)(0,2).
1 2 1 i(2).
2 1 1 i(1).
```

This means that $\hat{P}_4$ has two generators $\hat{4}_0^*$ and $\hat{4}_1^*$ in degree 13, with $\partial(\hat{4}_0^*) = Sq^{(3,1)}3_0^* = Sq^1\partial(4_0^*)$ and $\partial(\hat{4}_1^*) = (Sq^6 + Sq^{(0,2)})3_0^* + Sq^2 3_1^* + Sq^1 3_2^* = \partial(4_1^*)$. The second of these gives an $\mathbb{F}_2$-basis for $\tilde{J}_4$ in this degree, so as the second modification we delete the generator $\hat{4}_0^*$ from $\hat{P}_4$, leaving the following file `Diff.4`:

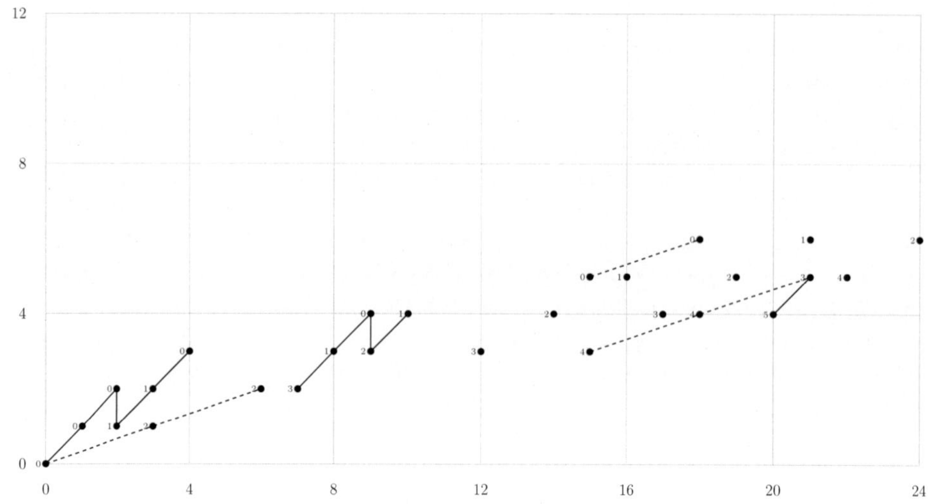

FIGURE 12.49. Hastened $(E_r(tmf/(2,B)), d_r)$ for $0 \leq t-s \leq 24$

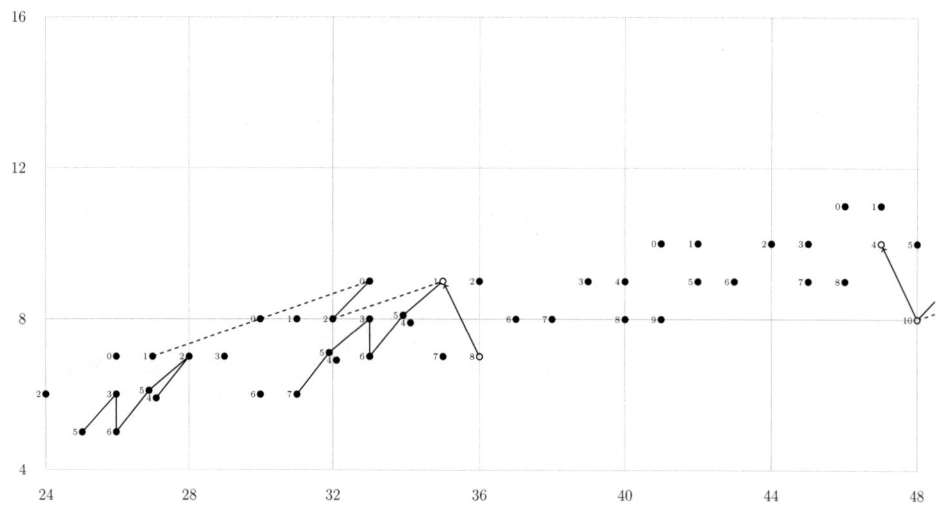

FIGURE 12.50. Hastened $(E_r(tmf/(2,B)), d_r)$ for $24 \leq t-s \leq 48$

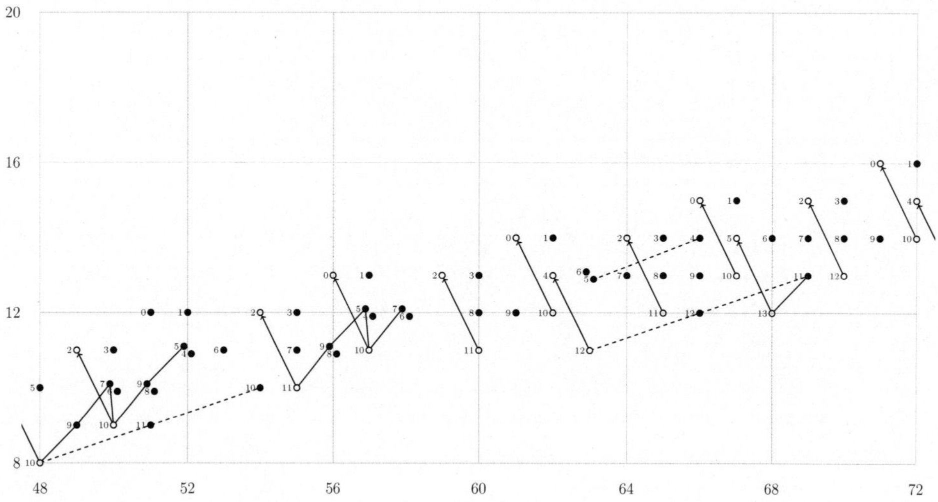

FIGURE 12.51. Hastened $(E_r(tmf/(2,B)), d_r)$ for $48 \le t - s \le 72$

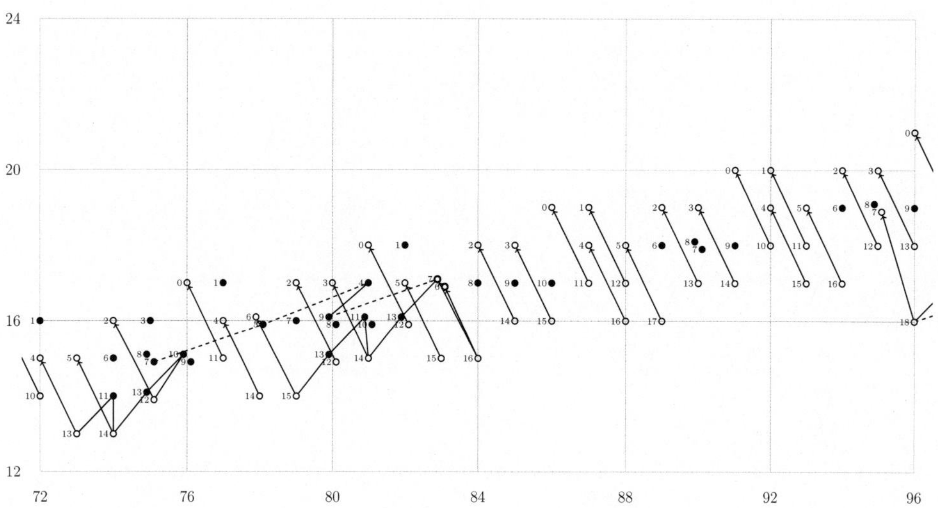

FIGURE 12.52. Hastened $(E_r(tmf/(2,B)), d_r)$ for $72 \le t - s \le 96$

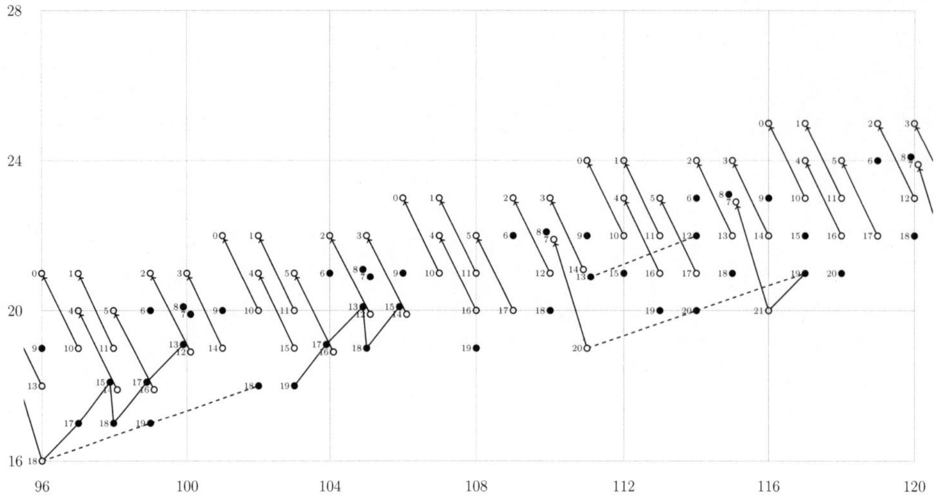

FIGURE 12.53. Hastened $(E_r(tmf/(2,B)), d_r)$ for $96 \leq t-s \leq 120$

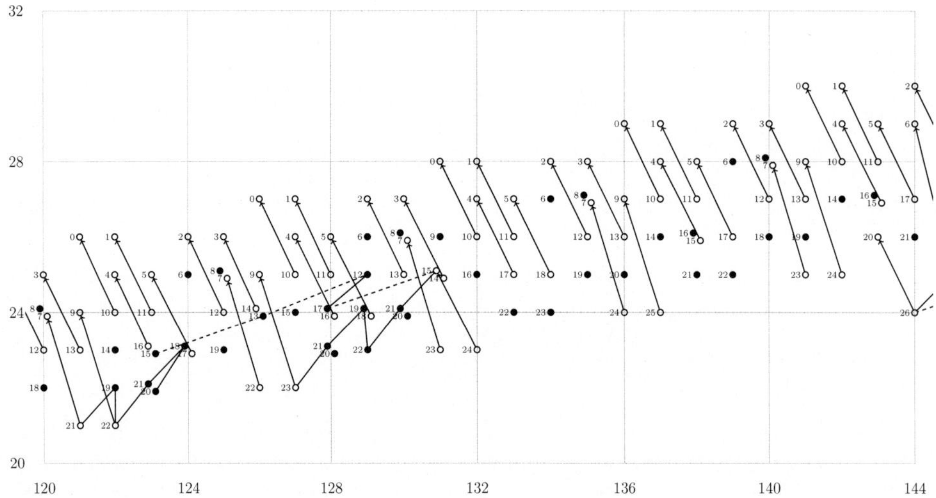

FIGURE 12.54. Hastened $(E_r(tmf/(2,B)), d_r)$ for $120 \leq t-s \leq 144$

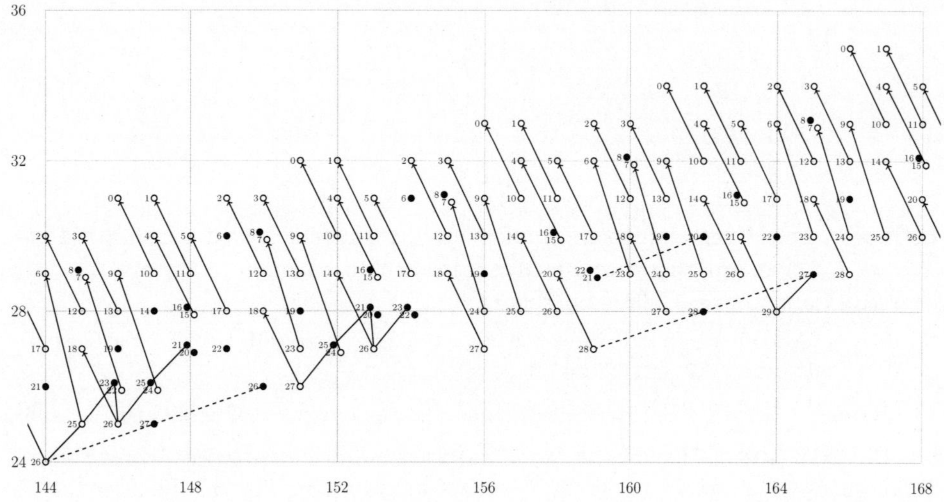

FIGURE 12.55. Hastened $(E_r(tmf/(2,B)),d_r)$ for $144 \le t-s \le 168$

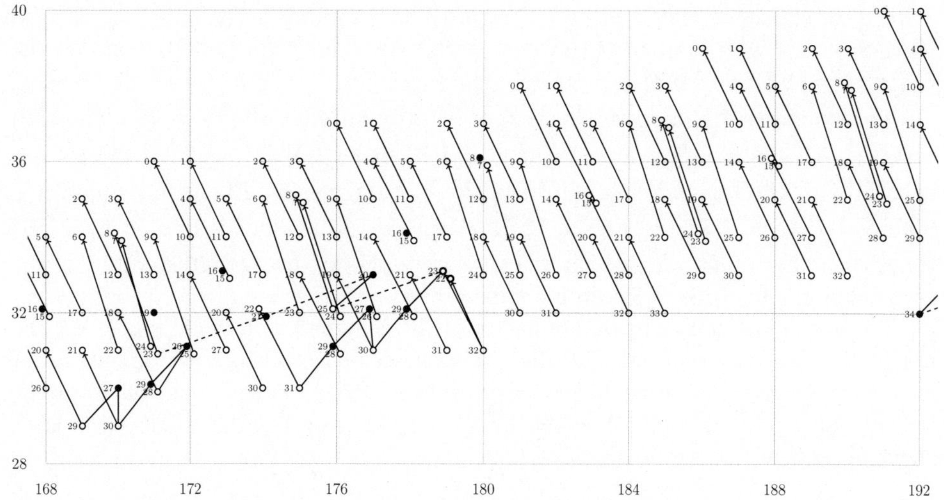

FIGURE 12.56. Hastened $(E_r(tmf/(2,B)),d_r)$ for $168 \le t-s \le 192$

```
          1         13
13

3
0 6 3 i(6)(0,2).
1 2 1 i(2).
2 1 1 i(1).
```

Letting $\tilde{P}_*$ be the remaining subcomplex of $\hat{P}_*$, we have achieved that $\mathrm{im}(\partial\colon \tilde{P}_4 \to P_3) = \tilde{J}_4$. We can therefore run `dims 14 240` to calculate $P_{*<4}$ and $\tilde{P}_{*+4}$ in degrees 14 and above. Calling on `report` and

```
chart 0 40 0 200 Shape himults E2-tmfC2modB.tex E2-tmfC2modB
pdflatex E2-tmfC2modB.tex
```

we obtain the hastened $E_2$-term for $tmf/(2,B)$, as shown in Figures 12.49 to 12.56.

PROPOSITION 12.38. *The differential structure in the hastened Adams spectral sequence $E_2(W_\star) \Longrightarrow \pi_*(tmf/(2,B))$ is as displayed in Figures* 12.49 *to* 12.56*, repeated $w_2^4$-periodically.*

PROOF. The $d_2$-differentials all follow by compatibility with the morphism $i''$ from the Adams spectral sequence for $tmf/2$, using Table 6.2. Most $d_3$-differentials also follow this way, using Table 6.4. The remaining possible $d_3$-differentials are defined on classes of the form $w_2^2 \cdot x$, where $x$ is an $E_3$-cycle for bidegree reasons. In these cases we calculate $d_3(w_2^2 \cdot x) = \beta g^4 \cdot x$, using the Leibniz rule for the $E_3(tmf)$-module structure.

For example, the image in $E_2(W_\star)$ of $d_0\widetilde{w_2^3h_1}$ in bidegree $(t-s,s) = (160,29)$ survives to $E_3$, since $d_2(d_0\widetilde{w_2^3h_1}) = g^2w_1\widetilde{w_2^2i(\beta)}$ in $E_2(tmf/2)$, which maps to zero in $E_2(W_\star)$. The image $x = 13_7$ of $d_0\widetilde{w_2h_1}$ in bidegree $(64,13)$ cannot support a differential, as is visible in Figure 12.51. Hence $d_3$ on the image of $d_0\widetilde{w_2^3h_1}$ is $\beta g^4$ times the image of $d_0\widetilde{w_2h_1}$, which we calculate with `ext` in `A2/tmfC2modB` to be the nonzero class $19_{56} \cdot 13_7 = 32_6$ in bidegree $(159,32)$.

At this point the only remaining possible hastened differential is $d_4$ on the image of $i(h_1w_2^3)$ in bidegree $(145,25)$. This must be nonzero by the known order of $\pi_*(tmf/(2,B))$ in degree 144 (or degree 145), which we can read off from the known $B$-action in $\pi_*(tmf/2)$ given in Section 12.1, or from the delayed Adams spectral sequence shown in Figure 12.39. □

CHAPTER 13

# Odd primes

After inverting $2 \cdot 3 = 6$, the edge homomorphism $e$ from connective topological modular forms to integral modular forms becomes an isomorphism

$$e \colon \pi_*(tmf)[1/6] \xrightarrow{\cong} mf_{*/2}[1/6] = \mathbb{Z}[1/6][c_4, c_6]\,,$$

with $\Delta = (c_4^3 - c_6^2)/1728$, and the spectral enrichment from algebra to topology carries little new information. At primes $p \geq 5$ the Hurewicz image of $\pi_*(S)_{(p)}$ in $\pi_*(tmf)_{(p)}$ is necessarily trivial in positive degrees. The main interest in the study of topological modular forms at odd primes is therefore concentrated at $p = 3$. At this prime, the homotopy of *tmf* (in its $K(2)$-localized form $EO_{p-1}$) was first calculated by Hopkins and Miller, cf. [**64**, Thm. 3.7] and [**137**, Thm. 2.1].

We shall follow Hill [**68**] and study $\pi_*(tmf)^\wedge_3$ by means of the Baker–Lazarev [**20**] mod 3 Adams spectral sequence built in the category of *tmf*-modules, as opposed to the classical Adams spectral sequence built in the category of spectra (= $S$-modules). Its $E_2$-term

$$E_2^{s,t} = \mathrm{Ext}^{s,t}_{A_*^{tmf}}(\mathbb{F}_3, \mathbb{F}_3) \Longrightarrow_s \pi_{t-s}(tmf)^\wedge_3$$

is given by the comodule Ext-groups for the *tmf*-module analogue

$$A_*^{tmf} = H_*^{tmf}(H) = \pi_*(H \wedge_{tmf} H)$$

of the dual Steenrod algebra, with coefficients in the *tmf*-module mod 3 homology groups

$$H_*^{tmf}(tmf) = \pi_*(H \wedge_{tmf} tmf) = \mathbb{F}_3$$

of *tmf*. Here $H = H\mathbb{F}_3$ denotes the mod 3 Eilenberg–Mac Lane spectrum, with its unique commutative *tmf*-algebra structure. Our first task is to determine the structure of $A_*^{tmf}$. Next, we shall use the Davis–Mahowald spectral sequence to calculate the *tmf*-module Adams $E_2$-term above. Thereafter, we use the $H_\infty$ ring structure on *tmf* to determine the differential structure in this Adams spectral sequence. Finally, we shall resolve the extension questions to pass from $E_\infty(tmf)$ to $\pi_*(tmf)$, implicitly completed at $p = 3$.

The calculation of $\Gamma_* = A_*^{tmf}$ is due to Henriques and Hill, using results of Hopkins–Mahowald (unpublished) and Behrens [**25**]. We supplement these arguments with later work of Hill and Lawson [**70**], and Mathew [**114**]. The calculation of $\mathrm{Ext}_{\Gamma_*}(\mathbb{F}_3, \mathbb{F}_3)$ was done by Hill using the May spectral sequence. We offer an alternative argument using the Davis–Mahowald spectral sequence of Chapter 2, based on a surjection $\Gamma_* \to \Lambda_*$ of commutative Hopf algebras, with $\Lambda_* = \mathbb{F}_3[\xi_1]/(\xi_1^3)$ dual to the subalgebra $\Lambda = \langle P^1 \rangle \subset A$ of the mod 3 Steenrod algebra. Hill determined the differentials in the *tmf*-module Adams spectral sequence for *tmf* at $p = 3$ by a comparison with the calculation of the corresponding Adams–Novikov spectral sequence made by Bauer [**23**]. We instead give direct arguments for these differentials in the style of our discussion at $p = 2$, starting from the $H_\infty$ ring structure on *tmf*. Bauer determined the hidden $\nu$-extensions in the Adams–Novikov spectral sequence by Toda bracket arguments. We shall give a different argument for these hidden extensions, based on our understanding of $tmf \wedge \Psi$, where $\Psi = S \cup_\nu e^4 \cup_\nu e^8$ is a 3-local CW spectrum with cohomology realizing the subalgebra $\Lambda = \langle P^1 \rangle$ mentioned above.

We conclude in Section 13.4 that $\pi_*(tmf)^\wedge_3$ is generated as a $\mathbb{Z}_3$-algebra by the three 3-torsion classes

| $x$ | $\nu$ | $\nu_1$ | $\beta$ |
|---|---|---|---|
| $n$ | 3 | 27 | 10 |
| $E_\infty(tmf)$ | $h_0$ | $h_0\Delta$ | $b_0$ |

and the following nine 3-torsion free classes.

| $x$ | $B$ | $B_1$ | $B_2$ | $C$ | $C_1$ | $C_2$ | $D_1$ | $D_2$ | $H$ |
|---|---|---|---|---|---|---|---|---|---|
| $n$ | 8 | 32 | 56 | 12 | 36 | 60 | 24 | 48 | 72 |
| $E_\infty(tmf)$ | $c_4$ | $c_4\Delta$ | $c_4\Delta^2$ | $c_6$ | $c_6\Delta$ | $c_6\Delta^2$ | $a_0\Delta$ | $a_0\Delta^2$ | $\Delta^3$ |
| $mf_{*/2}$ | $c_4$ | $c_4\Delta$ | $c_4\Delta^2$ | $c_6$ | $c_6\Delta$ | $c_6\Delta^2$ | $3\Delta$ | $3\Delta^2$ | $\Delta^3$ |

The 3-torsion is equal to the $B$-torsion, where $B$ is the Bott element, and the (Hopkins–Miller) element $H$ acts freely, so that $\pi_*(tmf) \cong N_* \otimes \mathbb{Z}[H]$, where $N_* \subset \pi_*(tmf)$ is the $\mathbb{Z}[B]$-submodule generated by the classes in degrees $0 \le * < 72$. We express $N_*$ as a $\mathbb{Z}[B]$-module, hence also $\pi_*(tmf)$ as a $\mathbb{Z}[B, H]$-module, in Theorem 13.18, and evaluate the product in $\pi_*(tmf)$ in Theorem 13.19. Only one product remains uncertain: we know that

$$B_2^2 = BB_1H + t\beta^4 H$$

for some number $t \in \mathbb{Z}/3$, but we do not know this coefficient. If $t = 0$ then the surjection $\pi_*(tmf) \to \mathrm{im}(e)$ admits a multiplicative section, otherwise it does not.

We also show that *tmf* satisfies duality at $p = 3$: there are equivalences of 3-completed *tmf*-modules

$$a\colon \Sigma^{20}tmf \xrightarrow{\simeq} I(tmf/(3^\infty, B^\infty, H^\infty))$$

and $\Sigma^{21}Tmf \simeq I_{\mathbb{Z}}(Tmf)$, where the latter was proved earlier by Stojanoska. In particular, the $B$-torsion in degrees $* \not\equiv 3 \mod 24$ in $N_*$ is

$$\Theta N_* = \mathbb{Z}/3\{\beta, \nu\beta, \beta^2, \beta^3, \nu_1\beta, \beta^4\}$$

and $a$ induces a perfect pairing

$$\Theta N_{50-n} \otimes \Theta N_n \longrightarrow \mathbb{Q}/\mathbb{Z}$$

for all $n$. Finally we show in Theorem 13.32 that for $* > 3$ the Hurewicz image of $\pi_*(S)$ in $\pi_*(tmf)$ is contained in $\Theta\pi_*(tmf) = \Theta N_* \otimes \mathbb{Z}[H]$, and that this upper bound is achieved at least up to degree 154.

REMARK 13.1. At the prime $p = 2$, the *tmf*-module mod 2 Adams spectral sequence for *tmf* agrees with the classical mod 2 Adams spectral sequence, since

$H_*(tmf) = A_* \,\square_{A(2)_*}\, \mathbb{F}_2$ and $A_*^{tmf} = \pi_*(H \wedge_{tmf} H) \cong A(2)_*$ at $p = 2$, so that the natural map

$$\mathrm{Ext}^{s,t}_{A_*}(\mathbb{F}_2, H_*(tmf)) \xrightarrow{\cong} \mathrm{Ext}^{s,t}_{A_*^{tmf}}(\mathbb{F}_2, \mathbb{F}_2)$$

is equal to the coalgebra version of the change-of-rings isomorphism along $A(2) \subset A$. Hence our discussion in the previous chapters can be interpreted to be all about the Baker–Lazarev *tmf*-module Adams spectral sequence for *tmf*, also for $p = 2$.

REMARK 13.2. The classical mod 3 Adams spectral sequence for *tmf* has $E_2$-term

$$E_2^{s,t} = \mathrm{Ext}^{s,t}_{A_*}(\mathbb{F}_3, H_*(tmf)) \Longrightarrow_s \pi_{t-s}(tmf)^\wedge_3 \,.$$

Dominic Culver [**49**] has worked out the rather elaborate structure of this spectral sequence, deducing several differentials and extensions from the known structure of the abutment. This approach does not seem to determine the unknown coefficient $t \in \mathbb{Z}/3$, since $\beta^4 H$ has higher classical Adams filtration than $B_2^2$ and $BB_1H$.

## 13.1. The *tmf*-module Steenrod algebra and its dual

We implicitly localize all spectra, abelian groups and stacks at 3 in this section. Hence $S$, $\mathbb{Z}$ and $\mathcal{M}_{ell}$ denote $S_{(3)}$, $\mathbb{Z}_{(3)}$ and $(\mathcal{M}_{ell})_{(3)}$, respectively. In particular, $3\nu = 0$ and $\nu^2 = 0$ in $\pi_*(S)$, where $\nu$ is the stable homotopy class of the quaternionic Hopf fibration.

DEFINITION 13.3. Let $P(0) = \langle P^1 \rangle$, $E(1) = \langle \beta, Q_1 \rangle$ and $A(1) = \langle \beta, P^1 \rangle$ be sub Hopf algebras of the mod 3 Steenrod algebra $A$, each generated by the listed elements. As usual, $Q_1 = [P^1, \beta]$. These are dual to the quotient Hopf algebras $P(0)_* = \mathbb{F}_3[\xi_1]/(\xi_1^3)$, $E(1)_* = E(\tau_0, \tau_1)$ and $A(1)_* = \mathbb{F}_3[\xi_1]/(\xi_1^3) \otimes E(\tau_0, \tau_1)$, respectively, of the mod 3 dual Steenrod algebra $A_*$. Let

$$\Psi = S \cup_\nu e^4 \cup_\nu e^8$$

be a 3-cell CW spectrum with $H^*(\Psi) = P(0)$ and $H_*(\Psi) = P(0)_*$, and let

$$V(1) = \mathrm{cof}(\Sigma^4 S/3 \xrightarrow{v_1} S/3) = S \cup_3 e^1 \cup_\nu e^5 \cup_3 e^6$$

be a type 2 Smith–Toda complex with $H^*(V(1)) = E(1)$ and $H_*(V(1)) = E(1)_*$. The smash product $V(1) \wedge \Psi$ has cohomology $E(1) \otimes P(0)$, which is free of rank 1 as an $A(1)$-module.

Hill [**68**, Prop. 2.3] credits the following result to Hopkins–Mahowald and Behrens [**25**]. Our outline of proof also relies on later work by Hill–Lawson [**70**] and Mathew [**114**].

THEOREM 13.4. *There is a map* $tmf \to tmf_0(2) = tmf_1(2)$ *of connective* $E_\infty$ *ring spectra, with* $\pi_*(tmf_0(2)) \cong \mathbb{Z}[a_2, a_4]$, *and an equivalence of tmf-modules*

$$tmf \wedge \Psi \simeq tmf_0(2) \,.$$

*The complex orientation associated to the Weierstrass curve* $y^2 = x^3 + a_2x^2 + a_4x$ *induces* $v_1 \mapsto a_2 \mod 3$ *and* $v_2 \mapsto 2a_4^2 \mod (3, a_2)$.

SKETCH PROOF. Following Behrens [**25**, §1.2.1] and Mahowald–Rezk [**105**], we first consider the moduli stack $\mathcal{M}_0(2)$ of elliptic curves with level structure of type $\Gamma_0(2) = \Gamma_1(2)$, i.e., with a chosen subgroup of order 2. There is an étale map $\mathcal{M}_0(2) \to \mathcal{M}_{ell}$ that represents forgetting the level structure, and the Goerss–Hopkins–Miller sheaf of $E_\infty$ ring spectra over $\mathcal{M}_{ell}$ pulls back to a similar sheaf over $\mathcal{M}_0(2)$. We let $TMF_0(2) = TMF_1(2)$ be the global sections (= homotopy limit) of this sheaf, so that there is a canonical map $TMF \to TMF_0(2)$ of $E_\infty$ ring spectra.

Since we are working locally at 3, each elliptic curve with $\Gamma_0(2)$ structure is uniquely strictly isomorphic to a non-singular Weierstrass curve of the form

$$y^2 = x^3 + a_2x^2 + a_4x$$

with $a_1 = a_3 = a_6 = 0$. This defines an elliptic curve with a vertical tangent at $(x, y) = (0, 0)$, which gives the point of order 2. The classical modular invariants are $c_4 = 16(a_2^2 - 3a_4)$, $c_6 = 32a_2(9a_4 - 2a_2^2)$ and $\Delta = 16a_4^2(a_2^2 - 4a_4)$. Hence $\pi_*(TMF_0(2)) \cong MF_0(2)_{*/2} = \mathbb{Z}[a_2, a_4][1/\Delta]$.

The 3-series of the associated formal group law begins

$$[3](z) = 3z - 8a_2z^3 + (24a_2^2 - 96a_4)z^5 - (72a_2^3 - 288a_2a_4)z^7 \\ + (216a_2^4 - 1472a_2^2a_4 + 2432a_4^2)z^9 + \dots$$

as can be verified with a computer algebra system such as sage, so that $v_1 \mapsto -8a_2 \equiv a_2 \mod 3$ and $v_2 \mapsto 216a_2^4 - 1472a_2^2a_4 + 2432a_4^2 \equiv 2a_4^2 \mod (3, a_2)$. Here we use that $v_n$ maps to the coefficient of $z^{3^n}$ in the 3-series, modulo $(3, \dots, v_{n-1})$, both for the Araki and the Hazewinkel generators [**144**, A2.2.4 and p. 371]. In particular, $\pi_*(TMF_0(2) \wedge V(1)) \cong \mathbb{Z}/3[a_4^{\pm 1}]$ is a quadratic extension of $\pi_*(K(2)) = \mathbb{Z}/3[v_2^{\pm 1}]$.

The unit map $S \to TMF_0(2)$ extends over $\Psi$, since $\pi_*(TMF_0(2))$ is concentrated in even degrees, so we obtain a $TMF$-module map

$$TMF \wedge \Psi \xrightarrow{\simeq} TMF_0(2) .$$

The descent spectral sequence for $\pi_*(TMF \wedge \Psi)$ based on the étale cover $TMF \to TMF_0(2)$ collapses at the $E_2$-term, which is concentrated on the 0-line, and implies that the map above is an equivalence. See [**23**, §5] and [**25**, p. 383].

Next, we follow Hill and Lawson [**70**, Thm. 5.17], who show that the Goerss–Hopkins–Miller étale sheaf over $\mathcal{M}_{ell}$ extends to a log-étale sheaf over the compactification $\overline{\mathcal{M}}_{ell}$. The direct image log structure from $\mathcal{M}_{ell}$ gives $\overline{\mathcal{M}}_{ell}$ the structure of a (Deligne–Mumford) log stack [**70**, Def. 3.1], and the extended sheaf can be pulled back along any log-étale cover of $\overline{\mathcal{M}}_{ell}$.

In particular, there is a compactification $\overline{\mathcal{M}}_0(2)$ of $\mathcal{M}_0(2)$ classifying generalized elliptic curves with $\Gamma_0(2)$ level structure. When the compactification is equipped with the direct image log structure, the forgetful map $\overline{\mathcal{M}}_0(2) \to \overline{\mathcal{M}}_{ell}$ is log-étale. Passing to global sections, Hill and Lawson obtain a map $Tmf \to Tmf_0(2)$ of $E_\infty$ ring spectra. Its localization away from (a power of) $\Delta$ is the map $TMF \to TMF_0(2)$ discussed above. The log stack $\overline{\mathcal{M}}_0(2)$ is equivalent to the subscheme of $\operatorname{Spec} \mathbb{Z}[a_2, a_4]$ given by the union of the two open subschemes $\operatorname{Spec} \mathbb{Z}[a_2, a_4][1/c_4]$ and $\operatorname{Spec} \mathbb{Z}[a_2, a_4][1/\Delta]$, equipped with the direct image log structure from the latter subscheme. Since the radical of $(c_4, \Delta)$ is $(a_2, a_4)$, the associated descent spectral sequence collapses at the $E_2$-term, which is concentrated along the 0- and 1-lines, and $\pi_*(Tmf_0(2))$ agrees with $\mathbb{Z}[a_2, a_4]$ in non-negative degrees.

By definition, $tmf_0(2)$ is the connective cover of $Tmf_0(2)$. Hence $\pi_*(tmf_0(2)) \cong mf_0(2)_{*/2} = \mathbb{Z}[a_2, a_4]$, and it follows that $\pi_*(tmf_0(2) \wedge V(1)) \cong \mathbb{Z}/3[a_4]$ is a quadratic extension of $\pi_*(k(2)) = \mathbb{Z}/3[v_2]$. The unit maps $S \to tmf_0(2) \to Tmf_0(2)$ also extend over $\Psi$, and Mathew [**114**, §4.6] shows that the equivalence $TMF \wedge \Psi \simeq$

$TMF_0(2)$ globalizes to an equivalence

$$Tmf \wedge \Psi \xrightarrow{\simeq} Tmf_0(2)$$

of $Tmf$-modules. By the Gap Theorem for $\pi_*(Tmf)$, it follows that there is also an equivalence

$$tmf \wedge \Psi \xrightarrow{\simeq} tmf_0(2)$$

of $tmf$-modules. □

The following two results are very similar to [**68**, Thm. 2.2]. Our proofs are perhaps a little more direct.

LEMMA 13.5. *The unit map $\iota\colon S \to tmf$ is 7-connected, and $\pi_7(tmf) = 0$.*

PROOF. The $v_1$-map $\Sigma^4 S/3 \to S/3$ acts on $\pi_*(tmf_0(2) \wedge S/3) = \mathbb{Z}/3[a_2, a_4]$ as multiplication by $a_2$, so $\pi_*(tmf_0(2) \wedge V(1)) = \mathbb{Z}/3[a_4]$. Hence multiplication by $a_4$ induces a homotopy cofiber sequence of $tmf_0(2)$-modules

$$\Sigma^8 tmf_0(2) \wedge V(1) \xrightarrow{a_4} tmf_0(2) \wedge V(1) \xrightarrow{i} H \xrightarrow{j} \Sigma^9 tmf_0(2) \wedge V(1)\,, \tag{13.1}$$

where $tmf_0(2)/a_4 \wedge V(1) \simeq H$ is characterized by its single nontrivial homotopy group. Note that $i_*\colon \mathbb{Z}/3[a_4] \to \mathbb{Z}/3 \cong \mathbb{F}_3$ is 8-connected.

Smashing the unit map $\iota\colon S \to tmf$ with $\Psi \wedge V(1)$ yields the left hand map in the following diagram:

$$\Psi \wedge V(1) \xrightarrow{\iota \wedge id} tmf \wedge \Psi \wedge V(1) \simeq tmf_0(2) \wedge V(1) \xrightarrow{i} H\,.$$

The composite $i \circ (\iota \wedge id)$ induces the monomorphism $A(1)_* \cong H_*(\Psi \wedge V(1)) \to H_*(H) = A_*$ in homology, which is 11-connected since the lowest-degree class in its cokernel is $\xi_1^3$. Hence $\iota \wedge id$ must be 7-connected. It follows that $\iota\colon S \to tmf$ is 7-connected, as a map of 3-local connective spectra of finite type.

We take as known that

$$\pi_*(S) = (\mathbb{Z}, 0, 0, \mathbb{Z}/3\{\alpha_1\}, 0, 0, 0, \mathbb{Z}/3\{\alpha_2\}, 0, 0, \mathbb{Z}/3\{\beta_1\}, \dots)$$

with $\alpha_1\alpha_2 = 0$, see e.g. [**144**, Fig. 1.2.15]. It remains to prove that the surjection $\pi_7(\iota)\colon \pi_7(S) \to \pi_7(tmf)$ maps $\alpha_2$ to zero. Consider the Atiyah–Hirzebruch spectral sequence

$$\begin{aligned} E^2_{s,t} = H_s(\Psi; \pi_t(tmf)) &\cong H_s(\Psi; \mathbb{Z}) \otimes \pi_t(tmf) \\ &\Longrightarrow_s \pi_{s+t}(tmf \wedge \Psi) \cong \pi_{s+t}(tmf_0(2))\,. \end{aligned}$$

Here $H_*(\Psi; \mathbb{Z}) = \mathbb{Z}\{g_0, g_4, g_8\}$ with $|g_s| = s$. We have $d^2 = d^3 = 0$ and

$$d^4(g_s \otimes x) = g_{s-4} \otimes \alpha_1 \cdot x$$

for $s \in \{4, 8\}$, since the 4- and 8-cells are attached by $\nu = \alpha_1$ to the 0- and 4-cells, respectively. Since the abutment $\pi_*(tmf_0(2)) \cong \mathbb{Z}[a_2, a_4]$ is concentrated in degrees $* \equiv 0 \mod 4$, the $E^\infty$-term must be trivial in the remaining total degrees. Hence, if $\iota(\alpha_2) \neq 0$ then $g_4 \otimes \iota(\alpha_2) \in E^2_{4,7}$ must either support a nonzero differential, or be hit by a nonzero differential. The latter is impossible, since $E^4_{8,4} = 0$. Thus $d_4(g_4 \otimes \iota(\alpha_2)) = g_0 \otimes \alpha_1 \cdot \iota(\alpha_2)$ must be nonzero in $E^4_{0,10}$. This contradicts the relation $\alpha_1\alpha_2 = 0$, so we can deduce that $\iota(\alpha_2) = 0$. □

THEOREM 13.6 (Henriques–Hill). *The tmf-module dual Steenrod algebra* $A^{tmf}_* = \pi_*(H \wedge_{tmf} H)$ *is isomorphic to the commutative Hopf algebra*

$$\mathbb{F}_3[\xi_1]/(\xi_1^3) \otimes E(\tau_0, \tau_1, \theta_2)$$

*with coproducts*

$$\begin{aligned}
\psi(\xi_1) &= 1 \otimes \xi_1 + \xi_1 \otimes 1 \\
\psi(\tau_0) &= 1 \otimes \tau_0 + \tau_0 \otimes 1 \\
\psi(\tau_1) &= 1 \otimes \tau_1 + \xi_1 \otimes \tau_0 + \tau_1 \otimes 1 \\
\psi(\theta_2) &= 1 \otimes \theta_2 + \xi_1 \otimes \tau_1 - \xi_1^2 \otimes \tau_0 + \theta_2 \otimes 1 \,.
\end{aligned}$$

*Here* $|\xi_1| = 4$, $|\tau_0| = 1$, $|\tau_1| = 5$ *and* $|\theta_2| = 9$. *The unit map* $\iota \colon S \to tmf$ *induces an 8-connected Hopf algebra homomorphism*

$$k_* \colon A_* = \pi_*(H \wedge H) \longrightarrow \pi_*(H \wedge_{tmf} H) = A^{tmf}_*$$

*taking* $\xi_1$, $\tau_0$ *and* $\tau_1 \in A_*$ *to the classes with the same names in* $A^{tmf}_*$, *and sending* $\xi_i$ *and* $\tau_i$ *to zero for each* $i \geq 2$. *Hence the image of* $k_*$ *is the sub Hopf algebra* $A(1)_* = \mathbb{F}_3[\xi_1]/(\xi_1^3) \otimes E(\tau_0, \tau_1)$ *of* $A^{tmf}_*$.

PROOF. Lemma 13.5 implies that

$$k = id \wedge_\iota id \colon H \wedge H = H \wedge_S H \longrightarrow H \wedge_{tmf} H$$

is at least 8-connected, since the homotopy type of $H \wedge_{tmf} H$ can be calculated using the 2-sided bar construction $B_\bullet(H, tmf, H)$. The induced Hopf algebra homomorphism $k_* \colon A_* \to A^{tmf}_*$ therefore maps the classes $\tau_0$, $\xi_1$ and $\tau_1$ in $A_*$ to indecomposable classes in $A^{tmf}_*$, with coproducts given by the usual formulas.

Applying $H \wedge_{tmf} (-)$ to (13.1) we obtain a homotopy cofiber sequence

$$\Sigma^8 H \wedge \Psi \wedge V(1) \xrightarrow{a_4} H \wedge \Psi \wedge V(1) \xrightarrow{i} H \wedge_{tmf} H \xrightarrow{j} \Sigma^9 H \wedge \Psi \wedge V(1)$$

in the category of module spectra over the $E_\infty$ ring spectrum $H \wedge_{tmf} tmf_0(2) \simeq H \wedge \Psi$. We claim that $a_{4*} = 0$ in the associated long exact sequence

$$\cdots \longrightarrow \Sigma^8 H_*(\Psi \wedge V(1)) \xrightarrow{a_{4*}} H_*(\Psi \wedge V(1)) \xrightarrow{i_*} A^{tmf}_* \xrightarrow{j_*} \Sigma^9 H_*(\Psi \wedge V(1)) \longrightarrow \dots .$$

To see this, note that $H_*(\Psi \wedge V(1)) \cong P(0)_* \otimes E(1)_*$ has dimension 12, and degree considerations show that $a_{4*}$ has rank $\leq 4$. Hence $A^{tmf}_*$ is a commutative Hopf algebra over $\mathbb{F}_3$, of dimension between 16 and 24, with indecomposable classes in degrees 1, 4 and 5. By Armand Borel's classification [**32**, §6] it follows that the dimension is 24, the rank of $a_{4*}$ is 0, and there is one more indecomposable class in degree 9. Thus

$$A^{tmf}_* \cong \mathbb{F}_3[\xi_1]/(\xi_1^3) \otimes E(\tau_0, \tau_1, \theta_2)$$

as an algebra, with $\theta_2$ in degree 9 defined modulo the image of $k_*$, i.e., modulo $\mathbb{F}_3\{\xi_1\tau_1, \xi_1^2\tau_0\}$. (Hill writes $a_2$ for the class we denote $\theta_2$.)

To show that we can choose $\theta_2$ so that

$$\psi(\theta_2) = 1 \otimes \theta_2 + \xi_1 \otimes \tau_1 - \xi_1^2 \otimes \tau_0 + \theta_2 \otimes 1$$

we use the fact that $\iota \colon S \to tmf$ maps $\alpha_2 \in \pi_7(S)$ to zero in $\pi_7(tmf)$. Here $\alpha_2 = \langle \alpha_1, \alpha_1, 3 \rangle$ is detected in the classical Adams spectral sequence for $S$ by $\Pi_0 a_0 = \langle h_0, h_0, a_0 \rangle \in \mathrm{Ext}^{2,9}_{A_*}(\mathbb{F}_3, \mathbb{F}_3)$ [**95**, Table 1], [**144**, Thm. 3.4.2]. Recall that

$d_1([\gamma]) = -\sum[\gamma'|\gamma'']$ in the cobar complex, where $\psi(\gamma) = 1\otimes\gamma+\sum\gamma'\otimes\gamma''+\gamma\otimes 1$. Hence $\Pi_0 a_0$ is represented by the 2-cocycle

$$y = [\xi_1^2|\tau_0] - [\xi_1|\tau_1]$$

in the cobar complex for $A_*$. Applying base change along $\iota\colon S \to tmf$, it follows that $k_*(y)$ detects zero in $\pi_7(tmf)$ in the $tmf$-module Adams spectral sequence for $tmf$, meaning that it is a coboundary in the cobar complex for $A_*^{tmf}$. Hence there is a class $x \in A_*^{tmf}$ with $d_1([x]) = k_*(y) = [\xi_1^2|\tau_0] - [\xi_1|\tau_1]$, and we let $\theta_2 = x$.

For degree reasons the Hopf algebra homomorphism $k_*\colon A_* \to A_*^{tmf}$ maps $\xi_i$ and $\tau_i$ to zero for $i \geq 2$, with the possible exception of $\tau_2$, which maps to a multiple of $\xi_1^2\theta_2$. In $A_*$ we have

$$\psi(\tau_2) = 1\otimes\tau_2 + \xi_1^3\otimes\tau_1 + \xi_2\otimes\tau_0 + \tau_2\otimes 1\,.$$

Since $\xi_1^3$ and $\xi_2$ map to zero in $A_*^{tmf}$ this implies that $k_*(\tau_2)$ must be primitive, which eliminates the nonzero multiples of $\xi_1^2\theta_2$. □

COROLLARY 13.7 ([**54**, §13.3]). *The tmf-module Steenrod algebra $A_{tmf}$ is generated by classes $\beta$ and $P^1$ in degrees 1 and 4, respectively, subject to the relations*

$$\beta^2 = 0$$
$$\beta(P^1)^2\beta = (\beta P^1)^2 + (P^1\beta)^2$$
$$(P^1)^3 = 0\,.$$

*The image of $k^*\colon A_{tmf} = H^*_{tmf}(H) \to H^*(H) = A$ is $A(1)$. The surjection $A_{tmf} \to A(1)$ introduces the Adem relation*

$$\beta(P^1)^2 + P^1\beta P^1 + (P^1)^2\beta = 0$$

*for $P^1\beta P^1$, which implies and replaces the relation for $\beta(P^1)^2\beta$.*

PROOF. By dualization, the Hopf algebra homomorphism $k^*\colon A_{tmf} \to A$ has image $A(1)$, and $A_{tmf} \to A(1)$ is an isomorphism in degrees $* \leq 8$. Let $\beta$ and $P^1$ in $A_{tmf}$ map to the classes with the same names in $A(1)$. Then $\beta^2 = 0$ and $(P^1)^3 = 0$, since $A_*^{tmf}$, hence also $A_{tmf}$, is trivial in degrees 2 and 12. A calculation with the coproduct in $A_*^{tmf}$ shows that $(\beta P^1)^2 + (P^1\beta)^2 - \beta(P^1)^2\beta$ evaluates to zero on $\xi_1\tau_0\tau_1$ and $\tau_0\theta_2$, hence is zero in $A_{tmf}$. It is then elementary to verify that these two generators and three relations give a presentation of $A_{tmf}$. The Adem relation is of course known to hold in $A(1) \subset A$. □

One can check that $\beta(P^1)^2 + P^1\beta P^1 + (P^1)^2\beta$ evaluates nontrivially on $\theta_2$. It squares to zero in $A_{tmf}$, so $A_{tmf} \to A(1)$ is a square-zero extension.

## 13.2. The Adams $E_2$-term

For brevity, let $\Gamma_* = A_*^{tmf}$. We use the Davis–Mahowald spectral sequence from Chapter 2 to calculate the $E_2$-term of the $tmf$-module Adams spectral sequence

$$E_2^{s,t}(tmf) = \mathrm{Ext}_{\Gamma_*}^{s,t}(\mathbb{F}_3, \mathbb{F}_3) \Longrightarrow_s \pi_{t-s}(tmf)\,,$$

where $tmf$ is now implicitly completed at $p = 3$.

Let

$$\Lambda_* = P(0)_* = \mathbb{F}_3[\xi_1]/(\xi_1^3)\,.$$

There is an evident surjection $\Gamma_* \to \Lambda_*$ of commutative Hopf algebras, and

$$\Omega_* = \Gamma_* \,\square_{\Lambda_*} \mathbb{F}_3 = E(\tau_0, \tau_1, \theta_2)$$

is a $\Gamma_*$-comodule subalgebra of $\Gamma_*$. There is a $\Gamma_*$-comodule algebra resolution

$$\mathbb{F}_3 \xrightarrow{\simeq} (\Omega_* \otimes R^*, d)\,,$$

where $R^* = \mathbb{F}_3[x_1, x_5, x_9]$ with $|x_1| = 1$, $|x_5| = 5$, $|x_9| = 9$ has $\Gamma_*$-coaction

$$\begin{aligned}\nu(x_1) &= 1 \otimes x_1 \\ \nu(x_5) &= 1 \otimes x_5 + \xi_1 \otimes x_1 \\ \nu(x_9) &= 1 \otimes x_9 + \xi_1 \otimes x_5 - \xi_1^2 \otimes x_1\end{aligned}$$

and differential $d(\tau_0) = x_1$, $d(\tau_1) = x_5$, $d(\theta_2) = x_9$. The associated Davis–Mahowald spectral sequence is

$$E_1^{\sigma,s,t} = \mathrm{Ext}_{\Lambda_*}^{s-\sigma,t}(\mathbb{F}_3, R^\sigma) \Longrightarrow_\sigma \mathrm{Ext}_{\Gamma_*}^{s,t}(\mathbb{F}_3, \mathbb{F}_3)\,,$$

cf. Definition 2.15. To analyze this $E_1$-term we note that $\Delta = x_9^3$ is $\Gamma_*$-comodule primitive, and let

$$\bar{R}^* = R^*/(x_9^3) = \mathbb{F}_3[x_1, x_5, x_9]/(x_9^3)\,.$$

There is then a $\Gamma_*$-comodule algebra extension

$$\mathbb{F}_3[\Delta] \longrightarrow E_1^* \longrightarrow \bar{E}_1^*\,,$$

where $\bar{E}_1^{\sigma,s,t} = \mathrm{Ext}_{\Lambda_*}^{s-\sigma,t}(\mathbb{F}_3, \bar{R}^\sigma)$.

Here $R^0 = \bar{R}^0 = \mathbb{F}_3$ with

$$\bar{E}_1^0 = \mathrm{Ext}_{\Lambda_*}(\mathbb{F}_3, \bar{R}^0) = E(h_0) \otimes \mathbb{F}_3[b_0]\,,$$

where $h_0 = [\xi_1]$ lies in bidegree $(t-s, s) = (3, 1)$ and $b_0 = [\xi_1|\xi_1^2] + [\xi_1^2|\xi_1]$ lies in bidegree $(t-s, s) = (10, 2)$; see [**144**, Lem. 3.2.4] for these cobar representatives. The Massey products and Steenrod operations in Ext relate these two generators, so that $\langle h_0, h_0, h_0\rangle = b_0$ and $\beta P^2(h_0) = -b_0$, with the sign conventions of [**118**, Rem. 11.11].

On the other hand,

$$\begin{aligned}R^1 = \bar{R}^1 &\cong \Sigma\Lambda_* \\ R^2 = \bar{R}^2 &\cong \Sigma^2\Lambda_* \oplus \Sigma^{10}\Lambda_* \\ \bar{R}^3 &\cong \Sigma^3\Lambda_* \oplus \Sigma^{11}\Lambda_* \oplus \Sigma^{15}\Lambda_*\end{aligned}$$

are dual to free modules, with

$$\begin{aligned}\mathrm{Ext}_{\Lambda_*}(\mathbb{F}_3, \bar{R}^1) &= \mathbb{F}_3\{x_1\} \\ \mathrm{Ext}_{\Lambda_*}(\mathbb{F}_3, \bar{R}^2) &= \mathbb{F}_3\{x_1^2, x_5^2 + x_1x_9\} \\ \mathrm{Ext}_{\Lambda_*}(\mathbb{F}_3, \bar{R}^3) &= \mathbb{F}_3\{x_1^3, x_1x_5^2 + x_1^2x_9, x_5^3\}\end{aligned}$$

all concentrated in $\mathrm{Ext}^0$. Letting $a_0 = x_1$, $c_4 = x_5^2 + x_1x_9$, $c_6 = x_5^3$ the pattern continues $a_0$- and $c_4$-periodically, as one can prove by filtering each $\bar{R}^\sigma$ by the powers of $x_1$ present. It follows that

$$\bar{E}_1^* = \mathrm{Ext}_{\Lambda_*}(\mathbb{F}_3, \bar{R}^*) \equiv \mathbb{F}_3[a_0, c_4, c_6]/(c_4^3 = c_6^2)$$

modulo the term with $\sigma = 0$. The relation $c_4^3 = c_6^2$ in $\bar{E}_1^*$ lifts to $c_4^3 = c_6^2 + a_0^3\Delta$ in $E_1^*$, since $(x_5^2 + x_1x_9)^3 = (x_5^3)^2 + x_1^3x_9^3$ in $R^*$. The Davis–Mahowald spectral sequence collapses at $E_1 = E_\infty$ for degree reasons, leading to the Adams $E_2$-term shown in Figure 13.1.

PROPOSITION 13.8. *$E_2(tmf) = \mathrm{Ext}_{\Gamma_*}(\mathbb{F}_3, \mathbb{F}_3)$ is generated by $h_0 \in E_2^{1,4}$, $b_0 \in E_2^{2,12}$, $a_0 \in E_2^{1,1}$, $c_4 \in E_2^{2,10}$, $c_6 \in E_2^{3,15}$ and $\Delta \in E_2^{3,27}$, subject to the relations $h_0^2 = 0$, $yz = 0$ for $y \in \{h_0, b_0\}$ and $z \in \{a_0, c_4, c_6\}$, and $c_4^3 - c_6^2 = a_0^3\Delta$.*

## 13.3. The Adams differentials

DEFINITION 13.9. Let $\nu \in \pi_3(tmf) \cong \mathbb{Z}/3$ and $\beta \in \pi_{10}(tmf) \cong \mathbb{Z}/3$ be the classes detected by $h_0$ and $b_0$, respectively.

These are the images of the classes $\nu = \alpha_1$ and $\beta_1$ in $\pi_*(S)$, as can be checked by a comparison of cobar representatives for $E_2(S)$ and $E_2(tmf)$.

PROPOSITION 13.10. *$d_2(\Delta) = \pm h_0 b_0^2$.*

PROOF. This follows from the $H_\infty$ ring structure on $tmf$. Since $3 \cdot \beta = 0$ it follows as in [**173**, Thm. 3] or [**45**, Cor. V.1.15] that $\nu\beta^3 = 0$. Hence $h_0 b_0^3$ is a boundary in $E_r(tmf)$, and for bidegree reasons (see Figure 13.1) the only possibility is $d_2(b_0\Delta) = \pm h_0 b_0^3$. Since $d_2(b_0) = 0$ it follows that $d_2(\Delta) = \pm h_0 b_0^2$. □

PROPOSITION 13.11. *$d_3(h_0\Delta^2) = \pm b_0^5$.*

PROOF. This also follows from the $H_\infty$ ring structure on $tmf$. Since $\nu \cdot \beta^2 = 0$ (by the previous proposition) it follows as in [**173**, Thm. 4] or [**45**, Cor. V.1.15] that $\beta(\beta^2)^3 = 0$. Hence $b_0^7$ is a boundary in $E_r(tmf)$, and for bidegree reasons the only possibility is $d_3(h_0 b_0^2 \Delta^2) = \pm b_0^7$. Since $d_3(b_0) = 0$ it follows that $d_3(h_0\Delta^2) = \pm b_0^5$. □

PROPOSITION 13.12. *$E_4(tmf) = E_\infty(tmf)$ is generated as an algebra by $a_0$, $h_0$, $c_4$, $b_0$, $c_6$, $a_0\Delta$, $h_0\Delta$, $c_4\Delta$, $c_6\Delta$, $a_0\Delta^2$, $c_4\Delta^2$, $c_6\Delta^2$ and $\Delta^3$.*

PROOF. There is no room for further differentials, as can be seen by inspection of Figure 13.1. □

## 13.4. The graded ring $\pi_*(tmf)$

We can now specify the remaining algebra generators for $\pi_*(tmf)$, implicitly completed at 3. As at the prime 2, they occur in families whose members are formally related by scalar multiples of the discriminant.

| $\nu$ | $\beta$ | $B$ | $C$ | $H$ |
|---|---|---|---|---|
| $\nu_1$ | | $B_1$ | $C_1$ | $D_1$ |
| | | $B_2$ | $C_2$ | $D_2$ |

DEFINITION 13.13.

(1) Let $\nu_1 \in \pi_{3+24}(tmf) \cong \mathbb{Z}/3$ be the class detected by $h_0\Delta$ in $E_\infty(tmf)$.
(2) Let $B_k \in \pi_{8+24k}(tmf)$ for $k \in \{0, 1, 2\}$ be the classes detected by $c_4\Delta^k$ in $E_\infty(tmf)$ and mapping to $c_4\Delta^k$ in $mf_{*/2}$. We call $B = B_0$ the Bott element.
(3) Let $C_k \in \pi_{12+24k}(tmf)$ for $k \in \{0, 1, 2\}$ be the classes detected by $c_6\Delta^k$ in $E_\infty(tmf)$ and mapping to $c_6\Delta^k$ in $mf_{*/2}$.

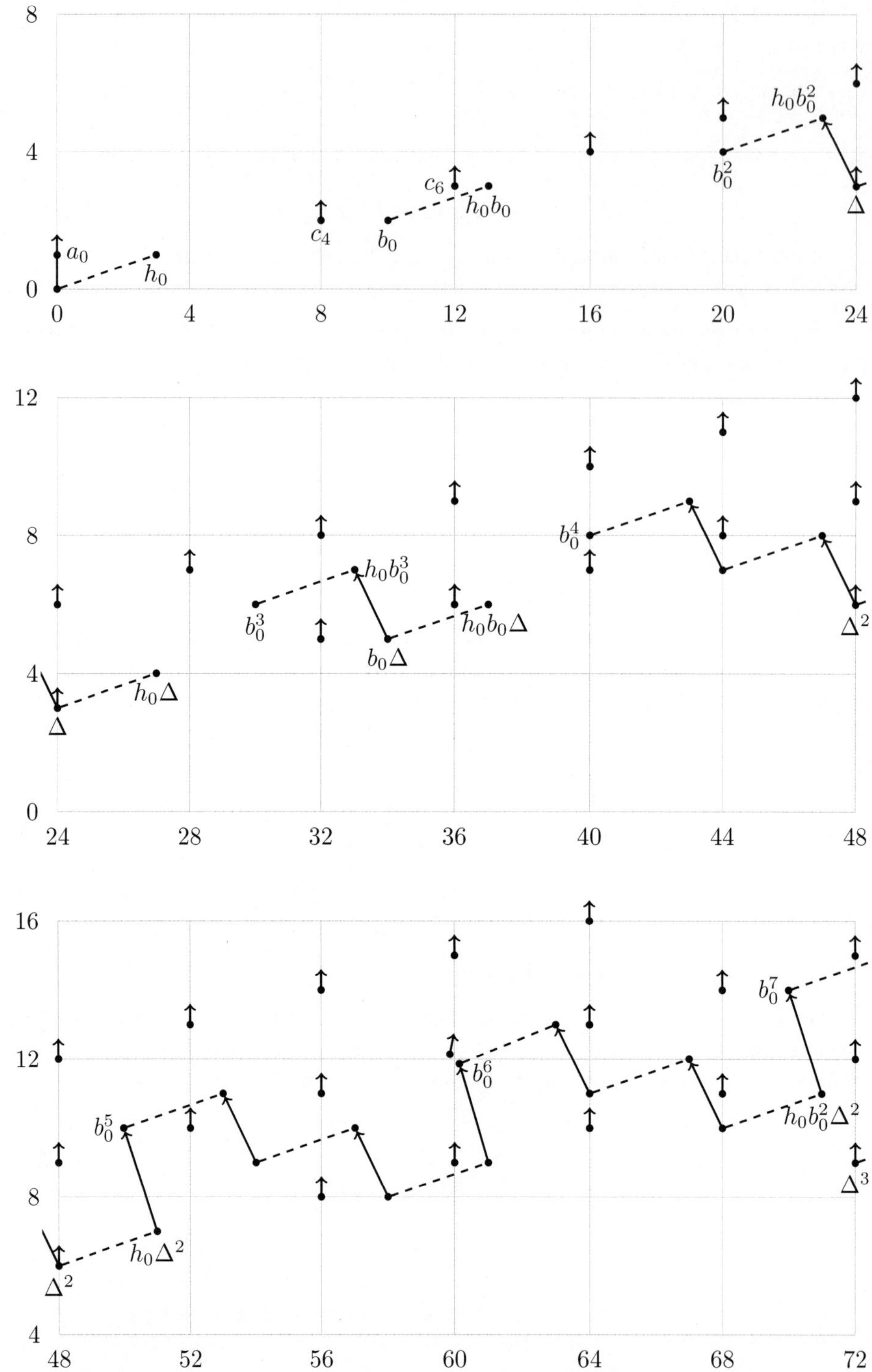


FIGURE 13.1. $E_2^{s,t}(tmf) \Longrightarrow_s \pi_{t-s}(tmf)$ at $p = 3$ for $0 \leq t - s \leq 72$

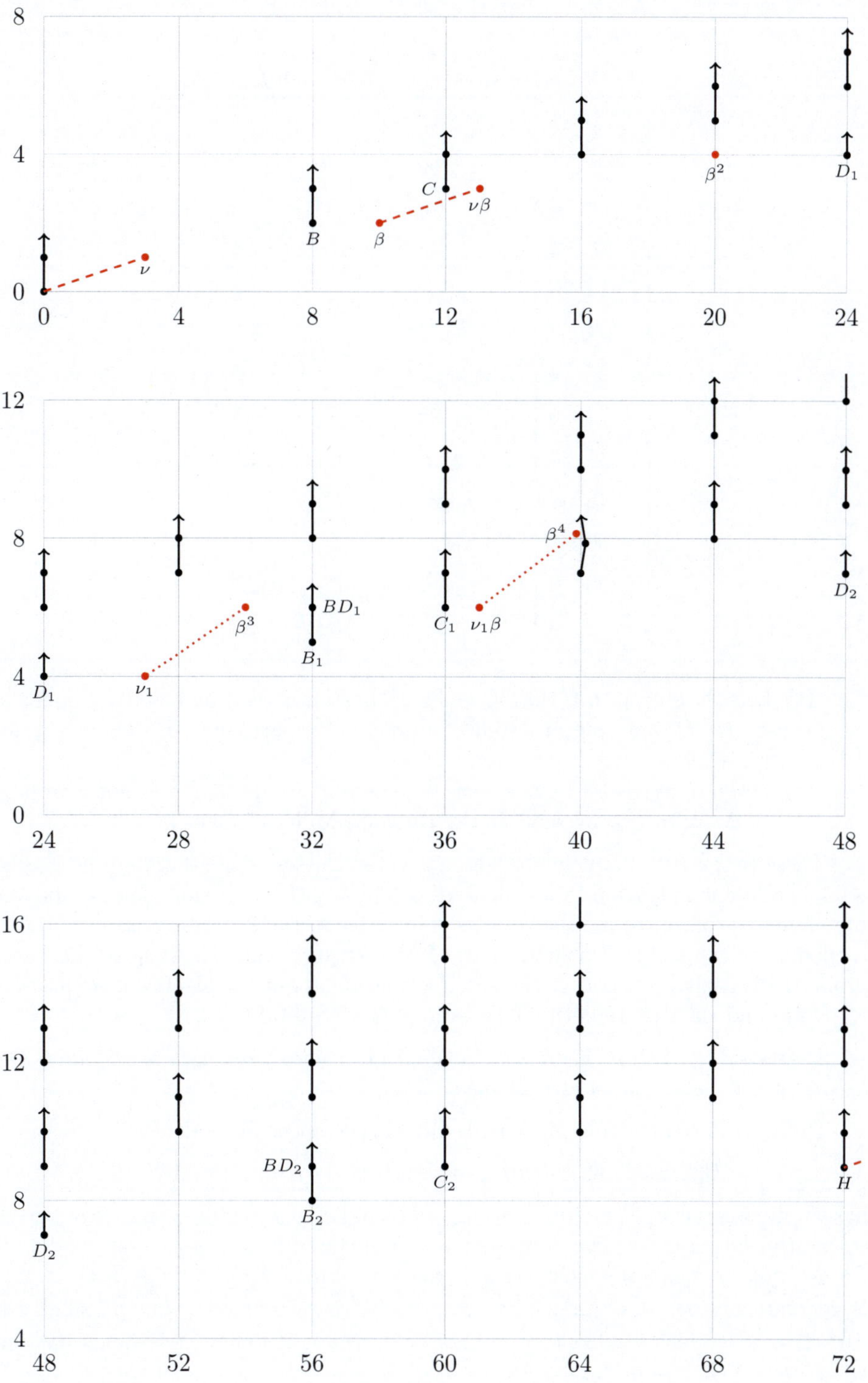


FIGURE 13.2. $\pi_n(tmf)$ at $p = 3$ for $0 \leq n \leq 72$

TABLE 13.1. Algebra generators of $E_\infty(tmf)$ and $\pi_*(tmf)$ at $p = 3$

| $t-s$ | $s$ | $E_\infty(tmf)$ | $\pi_*(tmf)$ | $mf_{*/2}$ |
|---|---|---|---|---|
| 0 | 1 | $a_0$ | 3 | 3 |
| 3 | 1 | $h_0$ | $\nu$ | 0 |
| 8 | 2 | $c_4$ | $B$ | $c_4$ |
| 10 | 2 | $b_0$ | $\beta$ | 0 |
| 12 | 3 | $c_6$ | $C$ | $c_6$ |
| 24 | 4 | $a_0\Delta$ | $D_1$ | $3\Delta$ |
| 27 | 4 | $h_0\Delta$ | $\nu_1$ | 0 |
| 32 | 5 | $c_4\Delta$ | $B_1$ | $c_4\Delta$ |
| 36 | 6 | $c_6\Delta$ | $C_1$ | $c_6\Delta$ |
| 48 | 7 | $a_0\Delta^2$ | $D_2$ | $3\Delta^2$ |
| 56 | 8 | $c_4\Delta^2$ | $B_2$ | $c_4\Delta^2$ |
| 60 | 9 | $c_6\Delta^2$ | $C_2$ | $c_6\Delta^2$ |
| 72 | 9 | $\Delta^3$ | $H$ | $\Delta^3$ |

(4) Let $D_k \in \pi_{24k}(tmf)$ for $k \in \{1, 2\}$ be the classes detected by $a_0\Delta^k$ in $E_\infty(tmf)$ and mapping to $3\Delta^k$ in $mf_{*/2}$. In particular, $B \cdot D_k = 3B_k$ for $k \in \{1, 2\}$.

(5) Let $H \in \pi_{72}(tmf)$ be the class detected by $\Delta^3$ in $E_\infty(tmf)$ and mapping to $\Delta^3$ in $mf_{*/2}$. We call $H$ the Hopkins–Miller element .

These properties uniquely characterize these classes, since there are no classes of higher Adams filtration in the degrees containing the 3-torsion classes, and the edge homomorphism to modular forms is injective in the degrees containing the 3-torsion free classes. The 3-primary part of (9.3) ensures that classes satisfying these properties do exist. We collect the key information about the algebra generators of $E_\infty(tmf)$ and their representing classes in $\pi_*(tmf)$ in Table 13.1.

PROPOSITION 13.14. *There are hidden $\nu$-extensions from $h_0\Delta$ to $\pm b_0^3$ and from $h_0b_0\Delta$ to $\pm b_0^4$, which propagate $\Delta^3$-periodically.*

PROOF. The three-column Atiyah–Hirzebruch spectral sequence

$$E^2_{*,*} = H_*(\Psi; \pi_*(tmf)) \Longrightarrow \pi_*(tmf \wedge \Psi) \cong \pi_*(tmf_0(2))$$

has $d^4$-differentials $d^4_{4,t} \colon \pi_t(tmf) \to \pi_{t+3}(tmf)$ and $d^4_{8,t} \colon \pi_t(tmf) \to \pi_{t+3}(tmf)$ given by multiplication by $\nu$. The abutment is concentrated in degrees $* \equiv 0 \mod 4$, so $\beta^3 \in E^2_{0,30} \cong \pi_{30}(tmf)$ must be a boundary. Since $E^2_{4,27} \cong \pi_{27}(tmf) \cong \mathbb{Z}/3$ is generated by $\nu_1 = \{h_0\Delta\}$, and $E^2_{8,23} \cong \pi_{23}(tmf) = 0$, the only possibility is $d_4(\nu_1) = \pm\beta^3$, which implies $\nu \cdot \nu_1 = \pm\beta^3$ in $\pi_*(tmf)$. It follows by $\beta$-linearity that $\nu \cdot \nu_1\beta = \pm\beta^4$ in $\pi_*(tmf)$.

By inspection of bidegrees, there is no room for any other hidden $\nu$-extensions. The extensions we have found propagate $\Delta^3$-periodically, since this class detects $H$ and acts freely on the $E_\infty$-term. □

DEFINITION 13.15. Let $N_* \subset \pi_*(tmf)$ be the $\mathbb{Z}[B]$-submodule generated by the classes in degrees $0 \le * < 72$.

LEMMA 13.16. *$\pi_*(tmf) \cong N_* \otimes \mathbb{Z}[H]$ as a $\mathbb{Z}[B,H]$-module.*

PROOF. This follows since $\Delta^3$ detects $H$ and acts freely on $E_\infty(tmf)$, with basis the classes detecting $N_*$. □

LEMMA 13.17. *The 3-power torsion and $B$-power torsion in $\pi_*(tmf)$ are both equal to the ideal*

$$\Gamma_3\pi_*(tmf) = \Gamma_B\pi_*(tmf) = (\nu, \beta, \nu_1) .$$

PROOF. It is clear from $E_\infty(tmf)$ in degrees $0 \le * < 72$ that

$$\Gamma_3 N_* = \Gamma_B N_* = \mathbb{Z}/3\{\nu, \beta, \nu\beta, \beta^2, \nu_1, \beta^3, \nu_1\beta, \beta^4\} .$$

This uses the fact that there are no hidden 3-extensions in $\pi_*(tmf)$, which follows easily from the multiplicative structure. It also uses the fact that $B \cdot D_k = 3B_k$ for $k \in \{1,2\}$, which follows from the observation that the edge homomorphism $\pi_*(tmf) \to mf_{*/2}$ is injective in these degrees. Hence $\Gamma_3\pi_*(tmf) = \Gamma_3 N_* \otimes \mathbb{Z}[H]$ and $\Gamma_B\pi_*(tmf) = \Gamma_B N_* \otimes \mathbb{Z}[H]$ are both equal to the ideal $(\nu, \beta, \nu_1)$ in $\pi_*(tmf)$. □

THEOREM 13.18. *There is a split extension*

$$0 \to \Gamma_B N_* \longrightarrow N_* \longrightarrow N_*/\Gamma_B N_* \to 0$$

*of $\mathbb{Z}[B]$-modules, where*

$$N_*/\Gamma_B N_* = ko[0] \oplus ko[1] \oplus ko[2]$$

*is the direct sum of the following three torsion-free $\mathbb{Z}[B]$-modules:*

$$\begin{aligned} ko[0] &= \mathbb{Z}[B]\{1, C\} \\ ko[1] &= \mathbb{Z}\{D_1\} \oplus \mathbb{Z}[B]\{B_1, C_1\} \\ ko[2] &= \mathbb{Z}\{D_2\} \oplus \mathbb{Z}[B]\{B_2, C_2\} . \end{aligned}$$

*The $\mathbb{Z}[B]$-module structures are such that $B \cdot D_1 = 3B_1$ and $B \cdot D_2 = 3B_2$.*

PROOF. This is now clear by inspection. □

THEOREM 13.19. *The products $xy$ in $\pi_*(tmf)$ (implicitly completed at 3) of the $\mathbb{Z}[B,H]$-module generators*

$$x \in \{\nu, \beta, C, \nu\beta, \beta^2, D_1, \nu_1, \beta^3, B_1, C_1, \nu_1\beta, \beta^4, D_2, B_2, C_2\}$$

*(omitting 1) and ring generators*

$$y \in \{\nu, \beta, C, D_1, \nu_1, B_1, C_1, D_2, B_2, C_2\}$$

*(omitting $B$ and $H$) are given in Table 13.2, except for the products $C_i \cdot C_j$, which are*

$$\begin{aligned} C \cdot C &= B^3 - 576D_1 \\ C \cdot C_1 = C_1 \cdot C &= B^2B_1 - 576D_2 \\ C \cdot C_2 = C_1 \cdot C_1 = C_2 \cdot C &= B^2B_2 - 1728H \\ C_1 \cdot C_2 = C_2 \cdot C_1 &= B^3H - 576D_1H \\ C_2 \cdot C_2 &= B^2B_1H - 576D_2H , \end{aligned} \tag{13.2}$$

*and the product*

$$B_2 \cdot B_2 = BB_1H + t\beta^4 H\,, \tag{13.3}$$

*where we have not determined the coefficient* $t \in \{0, 1, 2\}$.

PROOF. Many products are zero because they land in trivial groups. Many other products are determined by their image in $mf_{*/2}$, because the edge homomorphism $\pi_*(tmf) \to mf_{*/2}$ is injective in their degree. The product $\nu \cdot D_1$ is zero because it has higher Adams filtration than $\nu_1$, and $\nu_1 \cdot D_2$ is zero because it has higher Adams filtration than $\nu H$. The products $B_1 \cdot C_2 = B_2 \cdot C_1 = BCH$ have no contribution from $\beta^2 H$, because they have higher Adams filtration than that class. The relation $c_4^3 - c_6^2 = 1728\Delta$ in $mf_{*/2}$ implies $C \cdot C = B^3 - 576D_1$, and the other products $C_i \cdot C_j$ are treated similarly. The only difficult product is $B_2 \cdot B_2$ in $\pi_{112}(tmf)$, which is detected by $c_4^2\Delta^4$ in Adams filtration 16. The class $\beta^4 H$ in the same degree has Adams filtration 17, so these methods do not determine whether $B_2^2 = BB_1H$ or $B_2^2 = BB_1H \pm \beta^4 H$. In the former case, the surjection $\pi_*(tmf) \to \mathrm{im}(e)$ admits a multiplicative section, but not in the latter case. □

Table 13.2: Products in $\pi_*(tmf)$

| $n$ | $s$ | $x$ | $\nu$ | $\beta$ | $C$ | $D_1$ | $\nu_1$ | $B_1$ | $C_1$ | $D_2$ | $B_2$ | $C_2$ |
|---|---|---|---|---|---|---|---|---|---|---|---|---|
| 3 | 1 | $\nu$ | 0 | $\nu\beta$ | 0 | 0 | $\pm\beta^3$ | 0 | 0 | 0 | 0 | 0 |
| 10 | 2 | $\beta$ | $\nu\beta$ | $\beta^2$ | 0 | 0 | $\nu_1\beta$ | 0 | 0 | 0 | 0 | 0 |
| 12 | 3 | $C$ | 0 | 0 | (13.2) | $3C_1$ | 0 | $BC_1$ | (13.2) | $3C_2$ | $BC_2$ | (13.2) |
| 13 | 3 | $\nu\beta$ | 0 | 0 | 0 | 0 | $\pm\beta^4$ | 0 | 0 | 0 | 0 | 0 |
| 20 | 4 | $\beta^2$ | 0 | $\beta^3$ | 0 | 0 | 0 | 0 | 0 | 0 | 0 | 0 |
| 24 | 4 | $D_1$ | 0 | 0 | $3C_1$ | $3D_2$ | 0 | $3B_2$ | $3C_2$ | $9H$ | $3BH$ | $3CH$ |
| 27 | 4 | $\nu_1$ | $\mp\beta^3$ | $\nu_1\beta$ | 0 | 0 | 0 | 0 | 0 | 0 | 0 | 0 |
| 30 | 6 | $\beta^3$ | 0 | $\beta^4$ | 0 | 0 | 0 | 0 | 0 | 0 | 0 | 0 |
| 32 | 5 | $B_1$ | 0 | 0 | $BC_1$ | $3B_2$ | 0 | $BB_2$ | $BC_2$ | $3BH$ | $B^2H$ | $BCH$ |
| 36 | 6 | $C_1$ | 0 | 0 | (13.2) | $3C_2$ | 0 | $BC_2$ | (13.2) | $3CH$ | $BCH$ | (13.2) |
| 37 | 6 | $\nu_1\beta$ | $\mp\beta^4$ | 0 | 0 | 0 | 0 | 0 | 0 | 0 | 0 | 0 |
| 40 | 8 | $\beta^4$ | 0 | 0 | 0 | 0 | 0 | 0 | 0 | 0 | 0 | 0 |
| 48 | 7 | $D_2$ | 0 | 0 | $3C_2$ | $9H$ | 0 | $3BH$ | $3CH$ | $3D_1H$ | $3B_1H$ | $3C_1H$ |
| 56 | 8 | $B_2$ | 0 | 0 | $BC_2$ | $3BH$ | 0 | $B^2H$ | $BCH$ | $3B_1H$ | (13.3) | $BC_1H$ |
| 60 | 9 | $C_2$ | 0 | 0 | (13.2) | $3CH$ | 0 | $BCH$ | (13.2) | $3C_1H$ | $BC_1H$ | (13.2) |

## 13.5. Brown–Comenetz and Anderson duality

THEOREM 13.20. *There is an (implicitly 3-complete) equivalence of tmf-modules*

$$\Sigma^{20} tmf \simeq I(tmf/(3^\infty, B^\infty, H^\infty)) .$$

*Hence there is a perfect Brown–Comenetz duality pairing*

$$\Sigma^{20} tmf \wedge tmf/(3^\infty, B^\infty, H^\infty) \longrightarrow I .$$

PROOF. The proof is, of course, similar to that of Theorem 10.6. The $B$-power torsion $\Gamma_B N_*$ is finite and concentrated in degrees $3 \le * \le 40$. The $B$-divisible quotient $N_*/B^\infty$ is the direct sum of

$$\begin{aligned} ko[0]/B^\infty &= \mathbb{Z}[B^{-1}]\{1/B, C/B\} \\ ko[1]/B^\infty &= \mathbb{Z}[B^{-1}]\{B_1/B, C_1/B\}/(3B_1/B) \\ ko[2]/B^\infty &= \mathbb{Z}[B^{-1}]\{B_2/B, C_2/B\}/(3B_2/B) \end{aligned}$$

and is concentrated in degrees $* \le 52$. The group in degree 52 is a copy of $\mathbb{Z}$ generated by $C_2/B$, and the group in degree 51 is trivial. Hence the homotopy groups of $tmf/(3^\infty, B^\infty, H^\infty)$ are concentrated in degrees $* \le -20$, with the group in degree $-20$ being a copy of $\mathbb{Z}/3^\infty$. Passing to Brown–Comenetz duals, $I(tmf/(3^\infty, B^\infty, H^\infty))$ is a 19-connected $tmf$-module with homotopy group in degree 20 isomorphic to $\mathrm{Hom}(\mathbb{Z}/3^\infty, \mathbb{Q}/\mathbb{Z}) = \mathbb{Z}_3$. We represent a generator of this group by a $tmf$-module map

$$a \colon \Sigma^{20} tmf \longrightarrow I(tmf/(3^\infty, B^\infty, H^\infty)) .$$

We show that $a$ is an equivalence, by first showing that the coinduced $tmf_0(2)$-module map

$$\begin{aligned} b = F_{tmf}(tmf_0(2), a) \colon F_{tmf}(tmf_0(2), \Sigma^{20} tmf) \\ \longrightarrow F_{tmf}(tmf_0(2), I(tmf/(3^\infty, B^\infty, H^\infty))) \end{aligned}$$

is an equivalence. The target of $b$ is equivalent to

$$\begin{aligned} I(tmf_0(2) \wedge_{tmf} tmf/(3^\infty, B^\infty, H^\infty)) &\simeq I(tmf_0(2)/(3^\infty, B^\infty, H^\infty)) \\ &\simeq I(tmf_0(2)/(3^\infty, a_2^\infty, a_4^\infty)) , \end{aligned}$$

where we use that the ideal $J = (3, c_4, \Delta^8)$ generated by the images of 3, $B$ and $H$ in $\pi_*(tmf_0(2)) \cong \mathbb{Z}[a_2, a_4]$ has the same radical as $(3, a_2, a_4)$. The homotopy groups of $tmf_0(2)/(3^\infty, a_2^\infty, a_4^\infty)$ are

$$\mathbb{Z}[a_2, a_4]/(3^\infty, a_2^\infty, a_4^\infty) = \mathbb{Z}/3^\infty[a_2^{-1}, a_4^{-1}]\{1/a_2 a_4\}$$

with $1/a_2 a_4$ in degree $-12$, so

$$\pi_* I(tmf_0(2)/(3^\infty, a_2^\infty, a_4^\infty)) \cong \Sigma^{12} \pi_*(tmf_0(2))$$

is a free module over $\pi_*(tmf_0(2))$ on one generator in degree 12. The source of $b$ is equivalent as a $tmf$-module to

$$F_{tmf}(tmf \wedge \Psi, \Sigma^{20} tmf) \simeq F(\Psi, \Sigma^{20} tmf) \simeq \Sigma^{12} \Psi \wedge tmf \simeq \Sigma^{12} tmf_0(2) ,$$

where the finite 3-local CW spectrum $\Psi = S \cup_\nu e^4 \cup_\nu e^8$ is Spanier–Whitehead self-dual in the sense that $F(\Psi, S) = D\Psi \simeq \Sigma^{-8}\Psi$. Hence $\pi_*(b)$ is a surjective $\pi_*(tmf_0(2))$-module homomorphism, with abstractly isomorphic source and target (as graded abelian groups). It follows that $b$ is an equivalence. Since $b$ is obtained by smashing $a$ with $D\Psi$, with lowest homology group $H_{-8}(D\Psi) \cong H^8(\Psi) \cong \mathbb{Z}$, it follows that $a$ is an equivalence. $\square$

As in the case $p = 2$ we can rewrite the duality theorem in terms of local cohomology spectra and/or Anderson duality. Let $tmf' = tmf/(B^\infty, H^\infty)$.

PROPOSITION 13.21. *There are equivalences of tmf-modules*

$$\begin{aligned} \Sigma^{20} tmf &\simeq I_{\mathbb{Z}}(tmf/(B^\infty, H^\infty)) = I_{\mathbb{Z}}(tmf') \\ \Sigma^{22} tmf &\simeq I_{\mathbb{Z}}(\Gamma_{(B,H)} tmf) \\ \Sigma^{23} tmf &\simeq I(\Gamma_{(3,B,H)} tmf) . \end{aligned}$$

PROOF. This follows from $I(tmf'/3^\infty) \simeq I_{\mathbb{Z}}(tmf')^\wedge_3$, $tmf' = tmf/(B^\infty, H^\infty) \simeq \Sigma^2\Gamma_{(B,H)} tmf$ and $tmf/(3^\infty, B^\infty, H^\infty) \simeq \Sigma^3\Gamma_{(3,B,H)} tmf$. $\square$

The proof of Theorem 10.13 carries over with minor adjustments, replacing $M$, 192 and 2-completion by $H$, 72 and 3-completion, respectively, to recover the following 3-complete version of the theorem of Stojanoska [**161**, Thm. 13.1].

THEOREM 13.22. *There is a duality equivalence of (implicitly 3-completed) tmf-modules*

$$\Sigma^{21} Tmf \simeq I_{\mathbb{Z}}(Tmf) .$$

## 13.6. Explicit formulas

LEMMA 13.23. *The $\mathbb{Z}[B, H]$-module extension*

$$0 \to \pi_*(tmf)/B^\infty \longrightarrow \pi_*(tmf/B^\infty) \longrightarrow \Gamma_B \pi_{*-1}(tmf) \to 0$$

*is induced up from a unique $\mathbb{Z}[B]$-module extension*

$$0 \to N_*/B^\infty \longrightarrow N'_* \longrightarrow \Gamma_B N_{*-1} \to 0 .$$

PROOF. The induction homomorphism

$$\begin{aligned} \mathrm{Ext}^1_{\mathbb{Z}[B]}(\Gamma_B N_{*-1}, N_*/B^\infty) \longrightarrow \mathrm{Ext}^1_{\mathbb{Z}[B,H]}(\Gamma_B N_{*-1} \otimes \mathbb{Z}[H], N_*/B^\infty \otimes \mathbb{Z}[H]) \\ \cong \mathrm{Ext}^1_{\mathbb{Z}[B]}(\Gamma_B N_{*-1}, N_*/B^\infty \otimes \mathbb{Z}[H]) \end{aligned}$$

is bijective, because $\mathrm{Ext}^s_{\mathbb{Z}[B]}(\Gamma_B N_{*-1}, N_*/B^\infty \otimes (\mathbb{Z}[H]/\mathbb{Z})) = 0$ for $s \in \{0, 1\}$. This follows because $\Gamma_B N_{*-1}$ is concentrated in degrees $* \le 41$ and $N_*/B^\infty \otimes (\mathbb{Z}[H]/\mathbb{Z})$ agrees with $N_*[1/B] \otimes (\mathbb{Z}[H]/\mathbb{Z})$ in degrees $* < 72$. $\square$

DEFINITION 13.24. Let the $\pi_*(tmf)$-module $\Theta\pi_{*-1}(tmf)$ be the image of the composite homomorphism

$$\Gamma_3\pi_*(tmf/B^\infty) \longrightarrow \pi_*(tmf/B^\infty) \longrightarrow \Gamma_B\pi_{*-1}(tmf)$$

and let the $\mathbb{Z}[B]$-module $\Theta N_{*-1}$ be the image of the composite

$$\Gamma_3 N'_* \longrightarrow N'_* \longrightarrow \Gamma_B N_{*-1} ,$$

so that $\Theta\pi_*(tmf) \cong \Theta N_* \otimes \mathbb{Z}[H]$ as $\mathbb{Z}[B, H]$-modules.

The proof of Theorem 10.26 carries over to give the following algebraic consequences of the spectrum level duality equivalence $\Sigma^{20} tmf \simeq I(tmf/(3^\infty, B^\infty, H^\infty))$.

THEOREM 13.25. *(1) The graded ring $\pi_*(tmf)$ is filtered by a sequence of ideals*

$$0 \subset \Theta\pi_*(tmf) \subset \Gamma_B\pi_*(tmf) \subset \pi_*(tmf)\,,$$

*where $\Theta\pi_*(tmf)$ equals the part of $\Gamma_B\pi_*(tmf)$ in degrees $* \not\equiv 3 \mod 24$.*

*(2) The underlying sequence of $\mathbb{Z}[B, H]$-modules is induced up from the sequence of $\mathbb{Z}[B]$-modules*

$$0 \subset \Theta N_* \subset \Gamma_B N_* \subset N_*\,.$$

*The submodule*

$$\Theta N_* = \mathbb{Z}/3\{\beta, \nu\beta, \beta^2, \beta^3, \nu_1\beta, \beta^4\}$$

*is the part of $\Gamma_B N_*$ in degrees $* \not\equiv 3 \mod 24$, and is concentrated in degrees $10 \le * \le 40$.*

*(3) The duality equivalence specializes to a Pontryagin self-duality*

$$\Theta N_{50-*} \cong \mathrm{Hom}(\Theta N_*, \mathbb{Q}/\mathbb{Z})$$

*of part of the $B$-power torsion.*

*(4) The remaining $B$-power torsion*

$$\frac{\Gamma_B N_{51-*}}{\Theta N_{51-*}} \cong \mathbb{Z}/3\{\nu, \nu_1\}$$

*is Pontryagin dual to*

$$\frac{\Gamma_3(N_*/B^\infty)}{(\Gamma_3 N_*)/B^\infty} \cong \mathbb{Z}/3\{B_1/B, B_2/B\}\,.$$

*(5) The $B$-torsion free quotient*

$$\frac{N_*}{\Gamma_B N_*} = ko[0] \oplus ko[1] \oplus ko[2]$$

*participates in a short exact sequence*

$$0 \to \frac{N_{52-*}}{\Gamma_B N_{52-*}} \longrightarrow \mathrm{Hom}(\Big(\frac{N_*}{\Gamma_B N_*}\Big)/B^\infty, \mathbb{Z}_3) \longrightarrow \mathrm{Hom}(\frac{\Gamma_B N_{*-1}}{\Theta N_{*-1}}, \mathbb{Q}/\mathbb{Z}) \to 0\,.$$

PROPOSITION 13.26. *(1) The $\pi_*(tmf)$-module isomorphism*

$$\Theta\pi_{-*}(\Sigma^{20}tmf) \xrightarrow{\cong} \mathrm{Hom}(\Theta\pi_{*-2}(tmf)/H^\infty, \mathbb{Q}/\mathbb{Z})$$

*is adjoint to a perfect pairing*

$$\langle -, - \rangle \colon \Theta\pi_{-*}(\Sigma^{20}tmf) \times \Theta\pi_{*-2}(tmf)/H^\infty \longrightarrow \mathbb{Q}/\mathbb{Z}\,.$$

*(2) The $\mathbb{Z}[B]$-module isomorphism*

$$\Theta N_{50-*} \xrightarrow{\cong} \mathrm{Hom}(\Theta N_*, \mathbb{Q}/\mathbb{Z})$$

*is adjoint to a perfect pairing*

$$(-, -) \colon \Theta N_{50-*} \times \Theta N_* \longrightarrow \mathbb{Q}/\mathbb{Z}\,.$$

*Under the isomorphisms $\Theta\pi_*(tmf) \cong \Theta N_* \otimes \mathbb{Z}[H]$ and $\Theta\pi_*(tmf)/H^\infty \cong \Theta N_* \otimes \mathbb{Z}[H]/H^\infty$, these pairings are related by*

$$\langle xH^\ell, y/H^{1+\ell} \rangle = (x, y)$$

*for $\ell \ge 0$ and $|x| + |y| = 50$.*

*(3) The perfect pairing $(-, -)$ is given by*

$$(\beta, \beta^4) = \pm 1/3$$

$$(\nu\beta, \nu_1\beta) = \pm 1/3$$

$$(\beta^2, \beta^3) = \pm 1/3\,.$$

*In other words, $(x, y) = \pm 1/3$ if $x$ and $y$ formally multiply to $\beta^5$.*

REMARK 13.27. Here is how Pontryagin self-duality of $\Theta N_*$ at $p = 3$ arises from Theorem 13.20. Let $N = tmf/H$ be the homotopy cofiber of $H\colon \Sigma^{72}tmf \to tmf$, so that the composite homomorphism $N_* \subset \pi_*(tmf) \to \pi_*(N)$ is an isomorphism of $\mathbb{Z}[B]$-modules. Substituting $a\colon \Sigma^{20}tmf \simeq I(tmf/(3^\infty, B^\infty, H^\infty))$ in the homotopy cofiber sequence

$$\Sigma^{92}tmf \xrightarrow{H} \Sigma^{20}tmf \longrightarrow \Sigma^{20}N$$

and applying Brown–Comenetz duality, we obtain a homotopy cofiber sequence

$$I(\Sigma^{20}N) \longrightarrow tmf/(3^\infty, B^\infty, H^\infty) \xrightarrow{H} \Sigma^{-72}tmf/(3^\infty, B^\infty, H^\infty)\,.$$

The homotopy fiber of the right hand map is $\Sigma^{-72}N/(3^\infty, B^\infty)$, so we get an equivalence

$$\Sigma^{52}I(N) \simeq N/(3^\infty, B^\infty)$$

of $tmf$-modules. We can view each homomorphism $\phi\colon \pi_k(N) \to \mathbb{Q}/\mathbb{Z}$ as a homotopy class $\phi \in \pi_{-k}I(N)$, and $\Sigma^{52}\phi$ then corresponds under the equivalence above to a class $\psi \in \pi_{52-k}(N/(3^\infty, B^\infty))$. Its image $\partial^2(\psi)$ under the two connecting homomorphisms

$$\pi_{52-k}(N/(3^\infty, B^\infty)) \xrightarrow{\partial} \pi_{51-k}(N/B^\infty) \xrightarrow{\partial} \pi_{50-k}(N)$$

lies in $\Theta N_{50-k}$. As for $p = 2$, $\partial^2(\psi)$ only depends on the restriction $\phi|\colon \Theta N_k \to \mathbb{Q}/\mathbb{Z}$, and the correspondence $\phi| \leftrightarrow \partial^2(\psi)$ defines an isomorphism

$$\mathrm{Hom}(\Theta N_k, \mathbb{Q}/\mathbb{Z}) \cong \Theta N_{50-k}\,.$$

## 13.7. The *tmf*-Hurewicz image

The image of the Hurewicz homomorphism $\iota\colon \pi_*(S) \to \pi_*(tmf)$ lies mostly in the Pontryagin self-dual part. Integrally, it contains $\pi_0(tmf) \cong \mathbb{Z}\{\iota\}$ and $\pi_3(tmf) \cong \mathbb{Z}/24\{\nu\}$. The remainder of the 3-primary Hurewicz image is asserted in [**54**, §13.1] to be equal to the part of the $B$-power torsion that we denote $\Theta\pi_*(tmf)^\wedge_3$. In this section we show that the former is contained in the latter, and that the two agree in degrees $* < 154$. See Remark 13.33 for a discussion of what remains to be proved.

We proceed in the 3-complete setting. Let $j$ be the connective image-of-$J$ spectrum, which can be defined as the homotopy fiber of a lift $\tilde{\psi}\colon ku \to bu$ of $\psi^r - 1\colon ku \to ku$, where $r$ is any topological generator of the 3-adic units. Let $e\colon S \to j$ be the unit map representing the Adams $e$-invariant, and let the cokernel-of-$J$ spectrum $c$ be defined as its homotopy fiber. Adams [**8**, Thm. 7.16] proved that $e\colon \pi_*(S) \to \pi_*(j)$ is surjective, so that $\pi_*(c) \cong \ker(e)$. As a consequence of a theorem of Miller [**124**, Thm. 4.11], Bousfield [**33**, Thm. 4.3] showed that the map $e$ is a $KU$-equivalence, so $c$ is $KU$-acyclic. As for $p = 2$, a simpler proof can be given by calculating that $e^*\colon KU^*(j) \to KU^*(S)$ is an isomorphism [**131**, p. 201]. The 3-primary analogue of Proposition 11.81 is also true, with a similar proof.

PROPOSITION 13.28. *$tmf[1/B]$ is Bousfield $KU$-local.*

PROOF. Recall our notations from Definition 13.3. By Bousfield's criterion [**33**, Thm. 4.8] it suffices to check that

$$tmf[1/B] \wedge Z \simeq *$$

for $Z = V(1)$. By the Hopkins–Smith thick subcategory theorem [**78**], this is equivalent to verifying the condition for $Z = \Psi \wedge V(1)$, since both $V(1)$ and $\Psi \wedge V(1)$ are type 2 finite CW spectra. In view of the equivalence $tmf \wedge \Psi \simeq tmf_0(2)$ from Theorem 13.4, it suffices to prove that $tmf_0(2) \wedge V(1)$ becomes trivial after inverting $B$. Here $\pi_*(tmf_0(2) \wedge V(1)) = \mathbb{Z}[a_2, a_4]/(3, a_2) \cong \mathbb{Z}/3[a_4]$ with $B$ acting as multiplication by $c_4 \equiv 0 \mod (3, a_2)$, so this is clear. □

Proposition 11.82 is also valid as stated for $p = 3$, replacing 2-power torsion and $\Gamma_2$ by 3-power torsion and $\Gamma_3$, respectively, in its proof.

PROPOSITION 13.29. *For $0 \le n < 154$, the tmf-Hurewicz image of* $\ker(e) \subset \pi_n(S)$ *is equal to* $\Theta\pi_n(tmf)$.

PROOF. Since $\iota\colon \pi_*(S) \to \pi_*(tmf)$ is a graded ring homomorphism, mapping $\nu = \alpha_1$ to $\nu$ and $\beta_1$ to $\beta$, it is clear that $\nu\beta$, $\beta^2$, $\beta^3$ and $\beta^4$ are also in the Hurewicz image.

According to the 3-primary version of Proposition 11.82, $\nu_1 \in \pi_{27}(tmf)$ is not in the Hurewicz image of $\ker(e)$, since any lift of it in $\pi_{28}(tmf/B^\infty)$ has infinite order. Similarly, $\nu H \in \pi_{75}(tmf)$, $\nu_1 H \in \pi_{99}(tmf)$ and $\nu H^2 \in \pi_{147}(tmf)$ are not in the Hurewicz image.

By [**144**, §4.4, §7.4, A3.4], there is a 3-primary homotopy class traditionally denoted $\epsilon' = \langle \alpha_1, \alpha_1, \beta_1^3 \rangle \in \pi_{37}(S)$, with

$$\alpha_1 \epsilon' = \alpha_1 \langle \alpha_1, \alpha_1, \beta_1^3 \rangle = \langle \alpha_1, \alpha_1, \alpha_1 \rangle \beta_1^3 = \beta_1^4$$

by the shuffling formula [**171**, (3.6)]. The Toda bracket for $\epsilon'$ is well defined, since $\alpha_1 \beta_1^3 = 0$ in $\pi_*(S)$. Hence $\nu\iota(\epsilon') = \beta^4$, meaning that $\iota(\epsilon')$ must be $\pm\nu_1\beta$.

By the same calculations, there is a class in $\pi_{82}(S)$ detected by $\beta_{6/3} \in E_\infty^{2,84}$ in the Adams–Novikov spectral sequence

$$E_2(S) = \mathrm{Ext}_{BP_*BP}(BP_*, BP_*) \Longrightarrow \pi_*(S)_{(p)}$$

for $p = 3$. We claim that $\iota(\beta_{6/3}) = \pm\beta H$ is nonzero in $\pi_{82}(tmf)$. Granting this, it is clear that $\iota$ maps $\nu\beta_{6/3}$ and $\beta^i\beta_{6/3}$ to $\pm\nu\beta H$ and $\pm\beta^{i+1}H$ for $i \in \{1, 2, 3\}$.

Furthermore, by [**144**, A3.4] the product $\alpha_1 \cdot \beta_1^2\beta_{6/3} \in \pi_{105}(S)$ lies in a trivial group, so the Toda bracket $\epsilon'' = \langle \alpha_1, \alpha_1, \beta_1^2\beta_{6/3} \rangle$ is well-defined in $\pi_{109}(S)$, and

$$\alpha_1 \epsilon'' = \alpha_1 \langle \alpha_1, \alpha_1, \beta_1^2\beta_{6/3} \rangle = \langle \alpha_1, \alpha_1, \alpha_1 \rangle \beta_1^2\beta_{6/3} = \beta_1^3\beta_{6/3} \,.$$

Hence $\nu\iota(\epsilon'') = \pm\beta^4 H$, which proves that $\iota(\epsilon'') = \pm\nu_1\beta H$. It follows that all of $\Theta\pi_*(tmf)$ for $0 \le * < 154$ is in the image of $\iota$ on $\ker(e) \subset \pi_*(S)$.

It remains to verify the claim that $\beta_{6/3} \in \pi_{82}(S)$ has nontrivial Hurewicz image in $\pi_{82}(tmf)$. This can be deduced from calculations of Katsumi Shimomura [**156**, Lem. 2.4], showing that the image of $\beta_{6/3}$ is detected by a class $-v_2^3 b_{11}$ in the Adams–Novikov spectral sequence for $L_2V(1)$, combined with calculations of Goerss, Henn and Mahowald [**63**, Thm. 9, Pf. of Lem. 17, Cor. 19], showing that $v_2^3 b_{11}$ maps to $\pm v_2^{9/2}\beta$ in their spectral sequence for $\pi_*(E_2^{hN} \wedge V(1))$, which remains nonzero in their spectral sequence for $\pi_*(EO_2 \wedge V(1))$. Since the map $S \to EO_2 \wedge V(1)$ factors through $\iota\colon S \to tmf$, it follows that $\beta_{6/3}$ is also detected in $\pi_*(tmf)$. We refer to these papers for further explanation of the notations used. □

REMARK 13.30. We are grateful to Paul Goerss, Hans-Werner Henn, Mike Hill and Guozhen Wang for near-simultaneous help with finding a reference for the proof that $\iota(\beta_{6/3}) = \pm\beta H$. A more direct proof may be possible, tracing the Miller–Ravenel–Wilson [**126**, Thm. 1.1], [**125**, Thm. 2.6] construction of $\beta_{6/3}$ in $\mathrm{Ext}^2_{BP_*BP}(BP_*, BP_*)$ via $\mathrm{Ext}^2_{MU_*MU}(MU_*, MU_*)$ to $\mathrm{Ext}^2_\Gamma(A, A)$, where $(A, \Gamma)$ denotes the Weierstrass curve Hopf algebroid of [**23**, §3].

Let $U$ be the infinite unitary group. Localized at an odd prime $p$, the realification map $r\colon U \to SO$ induces an isomorphism $\pi_*(U) \to \pi_*(SO)$ in degrees $* = 4k - 1$, so the image of the $J$-homomorphism is equal to the image of the complex $J$-homomorphism $Jr\colon \pi_*(U) \to \pi_*(S)$ in these degrees.

PROPOSITION 13.31. *If $n \ge 2(p-1)k - 1 = |\alpha_k|$ and $n < 2p^\ell - 1$ then the image of $Jr\colon \pi_n(U) \to \pi_n(S)$ lies in Adams filtration $\ge k + 1 - \ell$.*

PROOF. Let $X[n]$ denote the $(n-1)$-connected cover of a space $X$. William Singer [**158**, Thm. 4.1] calculated the mod $p$ cohomology of each $U[2m+1]$, showing that the $(p-1)$-fold fiber inclusion

$$i^{p-1}\colon U[2(p-1)(\ell+1) - 1] \longrightarrow U[2(p-1)\ell - 1]$$

induces the zero homomorphism in reduced cohomology in all degrees $* < 2p^\ell - 1$. Hence, for integers $k$ and $\ell$ such that $n \ge 2(p-1)k - 1$ and $n < 2p^\ell - 1$, each map $f\colon S^n \to U$ factors as a composite

$$S^n \longrightarrow U[n] \longrightarrow U[2(p-1)k - 1] \xrightarrow{i^{p-1}} \dots$$
$$\dots \xrightarrow{i^{p-1}} U[2(p-1)(\ell+1) - 1] \xrightarrow{i^{p-1}} U[2(p-1)\ell - 1] \longrightarrow U\,,$$

where each map $i^{p-1}$ induces the zero homomorphism in degrees $\le n$. There are $k - \ell$ such maps, and each induced homomorphism $\pi_*(\Sigma^\infty i^{p-1})$ increases Adams filtration by at least 1 in degrees $\le n$. The composite $Jrf\colon S^n \to U \to SO \to QS^0$ is adjoint to

$$\Sigma^\infty S^n \xrightarrow{\Sigma^\infty f} \Sigma^\infty U \xrightarrow{\Sigma^\infty r} \Sigma^\infty SO \xrightarrow{\tilde J} S\,,$$

and $\tilde J$ has Adams filtration 1. It follows that $Jrf$ has Adams filtration at least $k + 1 - \ell$. □

THEOREM 13.32. *The image of the Hurewicz homomorphism*

$$\iota\colon \pi_*(S) \longrightarrow \pi_*(tmf)\,,$$

*implicitly completed at $p = 3$, is the direct sum of the following terms:*

1. *The group $\mathbb{Z}\{\iota\} \cong \pi_0(tmf)$.*
2. *The group $\mathbb{Z}/3\{\nu\} \cong \pi_3(tmf)$.*
3. *The groups $\Theta\pi_n(tmf) \subset \pi_n(tmf)$ for $n < 154$.*
4. *A subgroup of $\Theta\pi_n(tmf) \subset \pi_n(tmf)$ for the remaining $n \ge 154$.*

PROOF. Let $h\colon S \to H\mathbb{Z}$ be the unit map. We have inclusions

$$\ker(e) \subset \ker(h) \subset \pi_n(S)\,.$$

By the 3-primary version of Proposition 11.82 the image of $\iota$ on $\ker(e)$ is contained in $\Theta\pi_n(tmf)$, and by Proposition 13.29 this containment is an equality for $n < 154$. The image of $J\colon \pi_n(SO) \to \pi_n(S)$ gives a complementary summand $\mathrm{im}(J)$ in $\ker(h)$ to $\ker(e)$. We claim that $\iota(\mathrm{im}(J)) = 0$, except when $n = 3$. When $n = 4k - 1$ and $n < 2 \cdot 3^\ell - 1$ the image of $J$ lies in Adams filtration $\ge k + 1 - \ell$ in $\pi_n(S)$, by Proposition 13.31. Hence this is also a lower bound on the filtration of $\iota(\mathrm{im}(J))$ in the classical ($S$-module) Adams spectral sequence for *tmf*, as well as in the *tmf*-module Adams spectral sequence calculated in Figure 13.2. Since there are no infinite cycles in topological degree $n$ and filtration $\geq k+1-\ell$, except for $n=3$, the conclusion follows. □

REMARK 13.33. In [**54**, §13.1 (2)] it is stated that the 3-primary Hurewicz image of $\pi_*(S)$ in $\pi_*(tmf)$ is equal to $\mathbb{Z}\{\iota\} \oplus \mathbb{Z}/3\{\nu\} \oplus \Theta\pi_*(tmf)$ (using our notation for the 3- and $B$-power torsion in degrees $* \neq 3 \mod 24$). Our Theorem 13.32 confirms that this is an upper bound for the Hurewicz image, and shows that the bound is attained in the first 154 degrees. In order to extend this to all degrees, it is tempting to appeal to the self-map $v_2^9 \colon \Sigma^{144}V(1) \to V(1)$ constructed by Behrens and Pemmaraju [**28**], where $V(1) = S/(3, v_1)$ as above. To show that a class $x \colon S^n \to S$ with nontrivial *tmf*-Hurewicz image repeats periodically, one would like to extend $x$ over $\Sigma^n V(1)$, and then compose with iterates of $v_2^9$. Each of the key classes $\beta_1$, $\epsilon'$, $\beta_{6/3}$ and $\epsilon''$ has additive order 3, hence extends over $S \to S/3$, but in the case of $\beta_{6/3}$ there is no further extension over $S/3 \to V(1)$, essentially because $\beta_{6/2} \in \pi_{86}(S)$ is nonzero. In order to propagate $\beta_{6/3}$ periodically, it would therefore seem necessary to extend the work of Behrens and Pemmaraju to construct a $v_2^9$ self-map of $S/(3, v_1^3)$, i.e., the mapping cone of $v_1^3 \colon \Sigma^{12}S/3 \to S/3$. This appears to be an open problem.

APPENDIX A

# Calculation of $E_r(tmf)$ for $r = 3, 4, 5$

Recall from Definition 5.1 that $R_0 = \mathbb{F}_2[g, w_1, w_2]$, $R_1 = \mathbb{F}_2[g, w_1, w_2^2]$ and $R_2 = \mathbb{F}_2[g, w_1, w_2^4]$. Our calculations show that $E_2(tmf)$ is a complex of $R_1$-modules with $d_2(w_2) = \alpha\beta g$, $E_3(tmf)$ is a complex of $R_2$-modules with $d_3(w_2^2) = \beta g^4$, and $E_4(tmf)$ is a complex of $R_2$-modules with $d_4(w_2^4) = 0$.

## A.1. Calculation of $E_3(tmf) = H(E_2(tmf), d_2)$

The $(E_2, d_2)$-term of the Adams spectral sequence for *tmf* splits as a direct sum of 26 $R_1$-module complexes of length two or three, labeled (A) to (Z), plus a large summand with trivial differential, labeled 0. The Type-columns in Table A.1 give the labels of the complexes containing the $R_1$-module generators $x$ and $xw_2$. For each complex we discuss the passage to homology with respect to the $d_2$-differential, giving the transition from the $E_2$-term to the $E_3$-term.

Table A.1: Summands in $(E_2(tmf), d_2)$

| $t-s$ | $s$ | $g$ | $x$ | $\mathrm{Ann}(x)$ | Type$(x)$ | Type$(xw_2)$ |
|---|---|---|---|---|---|---|
| 0 | 0 | 0 | $1$ | $(0)$ | (A) | (O) |
| 0 | 1 | 0 | $h_0$ | $(g^2)$ | (B) | (P) |
| 0 | 2 | 0 | $h_0^2$ | $(g^2)$ | (C) | (V) |
| 0 | $3+i$ | 0 | $h_0^{3+i}$ | $(g)$ | 0 | 0 |
| 1 | 1 | 1 | $h_1$ | $(g^2)$ | 0 | 0 |
| 2 | 2 | 1 | $h_1^2$ | $(g)$ | 0 | 0 |
| 3 | 1 | 2 | $h_2$ | $(g)$ | (D) | (S) |
| 3 | 2 | 2 | $h_0h_2$ | $(g)$ | (E) | (W) |
| 3 | 3 | 1 | $h_0^2h_2$ | $(g)$ | (F) | (X) |
| 6 | 2 | 3 | $h_2^2$ | $(g)$ | (G) | (Y) |
| 8 | 3 | 2 | $c_0$ | $(g)$ | 0 | 0 |
| 9 | 4 | 2 | $h_1c_0$ | $(g)$ | 0 | 0 |
| 12 | 3 | 3 | $\alpha$ | $(0)$ | (D) | (S) |
| 12 | 4 | 3 | $h_0\alpha$ | $(g^2)$ | (E) | (W) |
| 12 | 5 | 4 | $h_0^2\alpha$ | $(g^2)$ | (F) | (X) |
| 12 | $6+i$ | 4 | $h_0^{3+i}\alpha$ | $(g)$ | 0 | 0 |

Table A.1: Summands in $(E_2(tmf), d_2)$ (cont.)

| $t-s$ | $s$ | $g$ | $x$ | Ann$(x)$ | Type$(x)$ | Type$(xw_2)$ |
|---|---|---|---|---|---|---|
| 14 | 4 | 4 | $d_0$ | (0) | (H) | (T) |
| 14 | 5 | 5 | $h_0d_0$ | $(g^2)$ | (I) | (U) |
| 15 | 3 | 4 | $\beta$ | (0) | (I) | (U) |
| 15 | 4 | 5 | $h_0\beta$ | $(g)$ | (G) | (Y) |
| 15 | 5 | 6 | $h_1d_0$ | $(g)$ | 0 | 0 |
| 17 | 4 | 6 | $e_0$ | (0) | (J) | (M) |
| 17 | 5 | 7 | $h_0e_0$ | $(g)$ | (K) | (N) |
| 17 | 6 | 6 | $h_0^2e_0$ | $(g)$ | (L) | (Z) |
| 18 | 4 | 7 | $h_2\beta$ | $(g)$ | (L) | (Z) |
| 18 | 5 | 8 | $h_1e_0$ | $(g)$ | 0 | 0 |
| 24 | 6 | 8 | $\alpha^2$ | (0) | (M) | (R) |
| 24 | $7+i$ | 7 | $h_0^{1+i}\alpha^2$ | $(g)$ | 0 | 0 |
| 25 | 5 | 11 | $\gamma$ | (0) | (N) | (D) |
| 26 | 6 | 9 | $h_1\gamma$ | $(g)$ | 0 | 0 |
| 26 | 7 | 8 | $\alpha d_0$ | (0) | (K) | (N) |
| 27 | 6 | 10 | $\alpha\beta$ | (0) | (O) | (H) |
| 27 | 7 | 9 | $h_1^2\gamma$ | $(g)$ | (P) | (I) |
| 29 | 7 | 10 | $\alpha e_0$ | (0) | (B) | (P) |
| 29 | 8 | 12 | $h_0\alpha e_0$ | $(g)$ | (C) | (V) |
| 30 | 6 | 11 | $\beta^2$ | (0) | (Q) | (J) |
| 31 | 8 | 13 | $d_0e_0$ | (0) | (R) | (Q) |
| 32 | 7 | 11 | $\delta$ | $(g)$ | 0 | 0 |
| 33 | 8 | 15 | $h_1\delta$ | $(g)$ | 0 | 0 |
| 36 | 9 | 17 | $\alpha^3$ | (0) | (P) | (I) |
| 36 | $10+i$ | 14 | $h_0^{1+i}\alpha^3$ | $(g)$ | 0 | 0 |
| 39 | 9 | 18 | $d_0\gamma$ | (0) | (S) | (K) |
| 41 | 10 | 16 | $\alpha^2e_0$ | (0) | (T) | (A) |
| 42 | 9 | 19 | $e_0\gamma$ | (0) | (U) | (B) |

Complex (A) is

$$\begin{array}{ccc} \langle \alpha^2 e_0 w_2 \rangle & \xrightarrow{g^4 w_1} & \langle 1 \rangle \\ \| & & \| \\ R_1 & & R_1 \end{array}$$

(implicitly extended with trivial groups at both sides). The class $\alpha^2 e_0 w_2$ does not survive (is not a $d_2$-cycle), leaving the cyclic module

$$\langle 1 \rangle \cong R_1/(g^4 w_1)$$

at $E_3$. Complex (B) is

$$\begin{array}{ccccc} \langle e_0\gamma w_2\rangle & \xrightarrow{g^3} & \langle \alpha e_0\rangle & \xrightarrow{gw_1} & \langle h_0\rangle \\ \| & & \| & & \| \\ R_1 & & R_1 & & R_1/(g^2) \end{array}$$

The class $e_0\gamma w_2$ does not survive, and $\alpha e_0$ is replaced by $\alpha e_0 g$, leaving the direct sum of the cyclic modules

$$\langle h_0\rangle \cong R_1/(g^2, gw_1)$$
$$\langle \alpha e_0 g\rangle \cong R_1/(g^2)$$

at $E_3$. (More precisely, $\langle h_0\rangle \cong \Sigma^{1,1} R_1/(g^2, gw_1)$, but we will omit the $(s,t)$-bidegree shifts in these formulas.) Complex (C) is

$$\begin{array}{ccc} \langle h_0\alpha e_0\rangle & \xrightarrow{gw_1} & \langle h_0^2\rangle \\ \| & & \| \\ R_1/(g) & & R_1/(g^2) \end{array}$$

The class $h_0\alpha e_0$ does not survive, leaving

$$\langle h_0^2\rangle \cong R_1/(g^2, gw_1)$$

at $E_3$. Complex (D) is

$$\begin{array}{ccccc} \langle \gamma w_2\rangle & \xrightarrow{g^3} & \langle \alpha\rangle & \xrightarrow{w_1} & \langle h_2\rangle \\ \| & & \| & & \| \\ R_1 & & R_1 & & R_1/(g) \end{array}$$

The class $\gamma w_2$ does not survive, and $\alpha$ is replaced by $\alpha g$, leaving

$$\langle h_2\rangle \cong R_1/(g, w_1)$$
$$\langle \alpha g\rangle \cong R_1/(g^2)\,.$$

Complex (E) is

$$\begin{array}{ccc} \langle h_0\alpha\rangle & \xrightarrow{w_1} & \langle h_0 h_2\rangle \\ \| & & \| \\ R_1/(g^2) & & R_1/(g) \end{array}$$

The class $h_0\alpha$ is replaced by $h_0\alpha g$, leaving

$$\langle h_0 h_2\rangle \cong R_1/(g, w_1)$$
$$\langle h_0\alpha g\rangle \cong R_1/(g)\,.$$

Complex (F) is

$$\begin{array}{ccc} \langle h_0^2\alpha\rangle & \xrightarrow{w_1} & \langle h_0^2 h_2\rangle \\ \| & & \| \\ R_1/(g^2) & & R_1/(g) \end{array}$$

The class $h_0^2\alpha$ is replaced by $h_0^2\alpha g$, leaving

$$\langle h_0^2 h_2\rangle \cong R_1/(g, w_1)$$
$$\langle h_0^2 \alpha g\rangle \cong R_1/(g)\,.$$

Complex (G) is

$$\begin{array}{ccc} \langle h_0\beta\rangle & \xrightarrow{w_1} & \langle h_2^2\rangle \\ \| & & \| \\ R_1/(g) & & R_1/(g) \end{array}$$

The class $h_0\beta$ does not survive, leaving

$$\langle h_2^2\rangle \cong R_1/(g, w_1)\,.$$

Complex (H) is

$$\begin{array}{ccc} \langle \alpha\beta w_2\rangle & \xrightarrow{g^3} & \langle d_0\rangle \\ \| & & \| \\ R_1 & & R_1 \end{array}$$

The class $\alpha\beta w_2$ does not survive, leaving

$$\langle d_0\rangle \cong R_1/(g^3)\,.$$

Complex (I) is

$$\begin{array}{ccccc} \langle \alpha^3 w_2\rangle & \xrightarrow{\binom{g^3 w_1}{w_1}} & \langle \beta\rangle \oplus \langle h_1^2\gamma w_2\rangle & \xrightarrow{(1\ 0)} & \langle h_0 d_0\rangle \\ \| & & \| & & \| \\ R_1 & & R_1 \oplus R_1/(g) & & R_1/(g^2) \end{array}$$

The class $\alpha^3 w_2$ does not survive, $\beta$ is replaced by $\beta g^2$, and $h_0 d_0$ becomes zero, leaving the non-cyclic module

$$\langle \beta g^2, h_1^2\gamma w_2\rangle \cong \frac{R_1 \oplus R_1}{\langle (g w_1, w_1), (0, g)\rangle}\,.$$

(For typographical reasons, we write the elements of $R_1 \oplus R_1$ as pairs $(x, y)$ rather than as column vectors.) Complex (J) is

$$\begin{array}{ccc} \langle \beta^2 w_2\rangle & \xrightarrow{g^3} & \langle e_0\rangle \\ \| & & \| \\ R_1 & & R_1 \end{array}$$

The class $\beta^2 w_2$ does not survive, leaving

$$\langle e_0\rangle \cong R_1/(g^3)\,.$$

Complex (K) is

$$\begin{array}{ccccc} \langle d_0\gamma w_2\rangle & \xrightarrow{g^3} & \langle \alpha d_0\rangle & \xrightarrow{w_1} & \langle h_0 e_0\rangle \\ \| & & \| & & \| \\ R_1 & & R_1 & & R_1/(g) \end{array}$$

The class $d_0\gamma w_2$ does not survive, and $\alpha d_0$ is replaced by $\alpha d_0 g$, leaving

$$\langle h_0 e_0\rangle \cong R_1/(g, w_1)$$
$$\langle \alpha d_0 g\rangle \cong R_1/(g^2)\,.$$

Complex (L) is

$$\begin{array}{ccc} \langle h_2\beta\rangle & \xrightarrow{1} & \langle h_0^2 e_0\rangle \\ \| & & \| \\ R_1/(g) & & R_1/(g) \end{array}$$

The class $h_2\beta$ does not survive, and $h_0^2 e_0$ becomes zero. Complex (M) is

$$\begin{array}{ccc} \langle e_0 w_2\rangle & \xrightarrow{g^2} & \langle \alpha^2\rangle \\ \| & & \| \\ R_1 & & R_1 \end{array}$$

The class $e_0 w_2$ does not survive, leaving

$$\langle \alpha^2\rangle \cong R_1/(g^2)\,.$$

Complex (N) is

$$\begin{array}{ccc} \langle \alpha d_0 w_2\rangle & \xrightarrow{\binom{g^2 w_1}{w_1}} & \langle \gamma\rangle \oplus \langle h_0 e_0 w_2\rangle \\ \| & & \| \\ R_1 & & R_1 \oplus R_1/(g) \end{array}$$

The class $\alpha d_0 w_2$ does not survive, leaving the non-cyclic module

$$\langle \gamma, h_0 e_0 w_2\rangle \cong \frac{R_1 \oplus R_1}{\langle (g^2 w_1, w_1), (0, g)\rangle}\,.$$

Complex (O) is

$$\begin{array}{ccc} \langle w_2\rangle & \xrightarrow{g} & \langle \alpha\beta\rangle \\ \| & & \| \\ R_1 & & R_1 \end{array}$$

The class $w_2$ does not survive, leaving

$$\langle \alpha\beta\rangle \cong R_1/(g)\,.$$

Complex (P) is

$$\begin{array}{ccccc} \langle \alpha e_0 w_2\rangle & \xrightarrow{\binom{g^2}{g w_1}} & \langle \alpha^3\rangle \oplus \langle h_0 w_2\rangle & \xrightarrow{(w_1\ 0)} & \langle h_1^2\gamma\rangle \\ \| & & \| & & \| \\ R_1 & & R_1 \oplus R_1/(g^2) & & R_1/(g) \end{array}$$

The class $\alpha e_0 w_2$ does not survive, and $\alpha^3$ is replaced by $\alpha^3 g + h_0 w_1 w_2$, leaving

$$\begin{aligned} \langle h_1^2\gamma\rangle &\cong R_1/(g, w_1) \\ \langle h_0 w_2\rangle &\cong R_1/(g^2) \\ \langle \alpha^3 g + h_0 w_1 w_2\rangle &\cong R_1/(g)\,. \end{aligned}$$

(We choose this replacement of $\alpha^3$ in order to present the $R_1$-module generated by $\alpha^3 g$ and $h_0 w_2$ as a direct sum of cyclic modules.) Complex (Q) is

$$\begin{array}{ccc} \langle d_0 e_0 w_2\rangle & \xrightarrow{g^2 w_1} & \langle \beta^2\rangle \\ \| & & \| \\ R_1 & & R_1 \end{array}$$

The class $d_0e_0w_2$ does not survive, leaving

$$\langle \beta^2 \rangle \cong R_1/(g^2w_1) .$$

Complex (R) is

$$\begin{array}{ccc} \langle \alpha^2 w_2 \rangle & \xrightarrow{g^2} & \langle d_0 e_0 \rangle \\ \| & & \| \\ R_1 & & R_1 \end{array}$$

The class $\alpha^2 w_2$ does not survive, leaving

$$\langle d_0 e_0 \rangle \cong R_1/(g^2) .$$

Complex (S) is

$$\begin{array}{ccc} \langle \alpha w_2 \rangle & \xrightarrow{\binom{g}{w_1}} & \langle d_0 \gamma \rangle \oplus \langle h_2 w_2 \rangle \\ \| & & \| \\ R_1 & & R_1 \oplus R_1/(g) \end{array}$$

The class $\alpha w_2$ does not survive, leaving the non-cyclic module

$$\langle d_0\gamma, h_2w_2 \rangle \cong \frac{R_1 \oplus R_1}{\langle (g, w_1), (0, g) \rangle} .$$

Complex (T) is

$$\begin{array}{ccc} \langle d_0 w_2 \rangle & \xrightarrow{g} & \langle \alpha^2 e_0 \rangle \\ \| & & \| \\ R_1 & & R_1 \end{array}$$

The class $d_0w_2$ does not survive, leaving

$$\langle \alpha^2 e_0 \rangle \cong R_1/(g) .$$

Complex (U) is

$$\begin{array}{ccc} \langle \beta w_2 \rangle & \xrightarrow{\binom{g}{1}} & \langle e_0 \gamma \rangle \oplus \langle h_0 d_0 w_2 \rangle \\ \| & & \| \\ R_1 & & R_1 \oplus R_1/(g^2) \end{array}$$

The class $\beta w_2$ does not survive, and $h_0d_0w_2$ becomes equal to $g \cdot e_0\gamma$, leaving

$$\langle e_0 \gamma \rangle \cong R_1/(g^3) .$$

Complex (V) is

$$\begin{array}{ccc} \langle h_0 \alpha e_0 w_2 \rangle & \xrightarrow{g w_1} & \langle h_0^2 w_2 \rangle \\ \| & & \| \\ R_1/(g) & & R_1/(g^2) \end{array}$$

The class $h_0\alpha e_0 w_2$ does not survive, leaving

$$\langle h_0^2 w_2 \rangle \cong R_1/(g^2, gw_1) .$$

Complex (W) is

$$\begin{array}{ccc} \langle h_0 \alpha w_2 \rangle & \xrightarrow{w_1} & \langle h_0 h_2 w_2 \rangle \\ \| & & \| \\ R_1/(g^2) & & R_1/(g) \end{array}$$

The class $h_0\alpha w_2$ is replaced by $h_0\alpha g w_2$, leaving

$$\langle h_0h_2w_2\rangle \cong R_1/(g, w_1)$$
$$\langle h_0\alpha g w_2\rangle \cong R_1/(g)\,.$$

Complex (X) is

$$\begin{array}{ccc} \langle h_0^2\alpha w_2\rangle & \xrightarrow{w_1} & \langle h_0^2h_2w_2\rangle \\ \| & & \| \\ R_1/(g^2) & & R_1/(g) \end{array}$$

The class $h_0^2\alpha w_2$ is replaced by $h_0^2\alpha g w_2$, leaving

$$\langle h_0^2h_2w_2\rangle \cong R_1/(g, w_1)$$
$$\langle h_0^2\alpha g w_2\rangle \cong R_1/(g)\,.$$

Complex (Y) is

$$\begin{array}{ccc} \langle h_0\beta w_2\rangle & \xrightarrow{w_1} & \langle h_2^2w_2\rangle \\ \| & & \| \\ R_1/(g) & & R_1/(g) \end{array}$$

The class $h_0\beta w_2$ does not survive, leaving

$$\langle h_2^2w_2\rangle \cong R_1/(g, w_1)\,.$$

Complex (Z) is

$$\begin{array}{ccc} \langle h_2\beta w_2\rangle & \xrightarrow{1} & \langle h_0^2e_0w_2\rangle \\ \| & & \| \\ R_1/(g) & & R_1/(g) \end{array}$$

The class $h_2\beta w_2$ does not survive, and $h_0^2e_0w_2$ becomes zero at the $E_3$-term.

## A.2. Calculation of $E_4(tmf) = H(E_3(tmf), d_3)$

The $(E_3, d_3)$-term of the Adams spectral sequence for *tmf* splits as a direct sum of 14 $R_2$-module complexes of length two or three, labeled (A) to (N), plus a large summand with trivial differential. The Type-columns in Table A.2 give the labels of the complexes containing the $R_2$-module generators $x$ and $xw_2^2$.

Table A.2: Summands in $(E_3(tmf), d_3)$

| $t-s$ | $s$ | $g$ | $x$ | Ann$(x)$ | Type$(x)$ | Type$(xw_2^2)$ |
|---|---|---|---|---|---|---|
| 0 | 0 | 0 | 1 | $(g^4w_1)$ | (A) | (I) |
| 0 | 1 | 0 | $h_0$ | $(g^2, gw_1)$ | 0 | 0 |
| 0 | 2 | 0 | $h_0^2$ | $(g^2, gw_1)$ | 0 | 0 |
| 0 | $3+i$ | 0 | $h_0^{3+i}$ | $(g)$ | 0 | 0 |
| 1 | 1 | 1 | $h_1$ | $(g^2)$ | (B) | (G) |
| 2 | 2 | 1 | $h_1^2$ | $(g)$ | 0 | 0 |
| 3 | 1 | 2 | $h_2$ | $(g, w_1)$ | 0 | 0 |
| 3 | 2 | 2 | $h_0h_2$ | $(g, w_1)$ | 0 | 0 |

Table A.2: Summands in $(E_3(tmf), d_3)$ (cont.)

| $t-s$ | $s$ | $g$ | $x$ | $\mathrm{Ann}(x)$ | $\mathrm{Type}(x)$ | $\mathrm{Type}(xw_2^2)$ |
|---|---|---|---|---|---|---|
| 3 | 3 | 1 | $h_0^2h_2$ | $(g, w_1)$ | 0 | 0 |
| 6 | 2 | 3 | $h_2^2$ | $(g, w_1)$ | 0 | 0 |
| 8 | 3 | 2 | $c_0$ | $(g)$ | (C) | (J) |
| 9 | 4 | 2 | $h_1c_0$ | $(g)$ | (D) | (K) |
| 12 | $6+i$ | 4 | $h_0^{3+i}\alpha$ | $(g)$ | 0 | 0 |
| 14 | 4 | 4 | $d_0$ | $(g^3)$ | (E) | (L) |
| 15 | 5 | 6 | $h_1d_0$ | $(g)$ | (F) | (M) |
| 17 | 4 | 6 | $e_0$ | $(g^3)$ | (C) | (J) |
| 17 | 5 | 7 | $h_0e_0$ | $(g, w_1)$ | 0 | 0 |
| 18 | 5 | 8 | $h_1e_0$ | $(g)$ | (D) | (K) |
| 24 | 6 | 8 | $\alpha^2$ | $(g^2)$ | (F) | (M) |
| 24 | $7+i$ | 7 | $h_0^{1+i}\alpha^2$ | $(g)$ | 0 | 0 |
| 25 | 5 | 11 | $\gamma$ | $-$ | (G) | (A) |
| 26 | 6 | 9 | $h_1\gamma$ | $(g)$ | 0 | 0 |
| 27 | 6 | 10 | $\alpha\beta$ | $(g)$ | 0 | 0 |
| 27 | 7 | 9 | $h_1^2\gamma$ | $(g, w_1)$ | 0 | 0 |
| 30 | 6 | 11 | $\beta^2$ | $(g^2w_1)$ | (B) | (G) |
| 31 | 8 | 13 | $d_0e_0$ | $(g^2)$ | 0 | 0 |
| 32 | 7 | 11 | $\delta$ | $(g)$ | 0 | 0 |
| 32 | 7 | $11+12$ | $\alpha g$ | $(g^2)$ | 0 | 0 |
| 32 | 8 | 14 | $h_0\alpha g$ | $(g)$ | 0 | 0 |
| 32 | 9 | 14 | $h_0^2\alpha g$ | $(g)$ | 0 | 0 |
| 33 | 8 | 15 | $h_1\delta$ | $(g)$ | (H) | (N) |
| 36 | $10+i$ | 14 | $h_0^{1+i}\alpha^3$ | $(g)$ | 0 | 0 |
| 39 | 9 | 18 | $d_0\gamma$ | $-$ | 0 | 0 |
| 41 | 10 | 16 | $\alpha^2e_0$ | $(g)$ | 0 | 0 |
| 42 | 9 | 19 | $e_0\gamma$ | $(g^3)$ | (H) | (N) |
| 46 | 11 | 18 | $\alpha d_0g$ | $(g^2)$ | 0 | 0 |
| 48 | 9 | 21 | $h_0w_2$ | $(g^2)$ | 0 | 0 |
| 48 | 10 | 19 | $h_0^2w_2$ | $(g^2, gw_1)$ | 0 | 0 |
| 48 | $11+i$ | 19 | $h_0^{3+i}w_2$ | $(g)$ | 0 | 0 |
| 49 | 9 | 22 | $h_1w_2$ | $(g^2)$ | (A) | (I) |

Table A.2: Summands in $(E_3(tmf), d_3)$ (cont.)

| $t-s$ | $s$ | $g$ | $x$ | $\mathrm{Ann}(x)$ | $\mathrm{Type}(x)$ | $\mathrm{Type}(xw_2^2)$ |
|---|---|---|---|---|---|---|
| 49 | 11 | 20 | $\alpha e_0 g$ | $(g^2)$ | 0 | 0 |
| 50 | 10 | 21 | $h_1^2 w_2$ | $(g)$ | 0 | 0 |
| 51 | 9 | 23 | $h_2 w_2$ | $-$ | 0 | 0 |
| 51 | 10 | 22 | $h_0 h_2 w_2$ | $(g, w_1)$ | 0 | 0 |
| 51 | 11 | 21 | $h_0^2 h_2 w_2$ | $(g, w_1)$ | 0 | 0 |
| 54 | 10 | 23 | $h_2^2 w_2$ | $(g, w_1)$ | 0 | 0 |
| 55 | 11 | 23 | $\beta g^2$ | $-$ | (I) | (B) |
| 56 | 11 | 24 | $c_0 w_2$ | $(g)$ | 0 | 0 |
| 56 | 13 | $26+27$ | $\alpha^3 g + h_0 w_1 w_2$ | $(g)$ | 0 | 0 |
| 57 | 12 | 28 | $h_1 c_0 w_2$ | $(g)$ | (C) | (J) |
| 60 | $14+i$ | 28 | $h_0^{3+i} \alpha w_2$ | $(g)$ | 0 | 0 |
| 63 | 13 | 34 | $h_1 d_0 w_2$ | $(g)$ | (E) | (L) |
| 65 | 13 | 36 | $h_0 e_0 w_2$ | $-$ | (G) | (A) |
| 66 | 13 | 37 | $h_1 e_0 w_2$ | $(g)$ | (C) | (J) |
| 72 | $15+i$ | 36 | $h_0^{1+i} \alpha^2 w_2$ | $(g)$ | 0 | 0 |
| 74 | 14 | 37 | $h_1 \gamma w_2$ | $(g)$ | (G) | (A) |
| 75 | 15 | 39 | $h_1^2 \gamma w_2$ | $-$ | (I) | (B) |
| 80 | 15 | 41 | $\delta w_2$ | $(g)$ | 0 | 0 |
| 80 | 16 | 49 | $h_0 \alpha g w_2$ | $(g)$ | 0 | 0 |
| 80 | 17 | 49 | $h_0^2 \alpha g w_2$ | $(g)$ | 0 | 0 |
| 81 | 16 | 50 | $h_1 \delta w_2$ | $(g)$ | 0 | 0 |
| 84 | $18+i$ | 48 | $h_0^{1+i} \alpha^3 w_2$ | $(g)$ | 0 | 0 |

Complex (A) is

$$\begin{array}{ccccc} \langle h_1\gamma w_2^3\rangle & \xrightarrow{\begin{pmatrix} 0 \\ g^2 w_1 \\ 0 \end{pmatrix}} & \langle h_1 w_2\rangle \oplus \langle \gamma w_2^2, h_0 e_0 w_2^3\rangle & \xrightarrow{(\, g^2 w_1 \;\; g^6 \;\; 0\,)} & \langle 1\rangle \\ \| & & \| & & \| \\ R_2/(g) & & R_2/(g^2) \oplus \dfrac{R_2 \oplus R_2}{\langle (g^2 w_1, w_1), (0, g)\rangle} & & R_2/(g^4 w_1) \end{array}$$

The classes $h_1\gamma w_2^3$ and $h_1 w_2$ do not survive, and $\gamma w_2^2$ is replaced by $\gamma w_1 w_2^2$, leaving

$$\begin{aligned} \langle 1\rangle &\cong R_2/(g^6, g^2 w_1) \\ \langle \gamma w_1 w_2^2\rangle &\cong R_2/(g^2) \\ \langle h_0 e_0 w_2^3\rangle &\cong R_2/(g, w_1) \end{aligned}$$

at $E_4$. Complex (B) is

$$\begin{array}{ccccc} \langle \beta g^2 w_2^2, h_1^2\gamma w_2^3\rangle & \xrightarrow{(g^6\ 0)} & \langle \beta^2\rangle & \xrightarrow{gw_1} & \langle h_1\rangle \\ \| & & \| & & \| \\ \dfrac{R_2 \oplus R_2}{\langle (gw_1, w_1), (0, g)\rangle} & & R_2/(g^2 w_1) & & R_2/(g^2) \end{array}$$

The classes $\beta^2$ and $\beta g^2 w_2^2$ are replaced by $\beta^2 g$ and $\beta g^2 w_1 w_2^2$, respectively, leaving

$$\langle h_1\rangle \cong R_2/(g^2, gw_1)$$
$$\langle \beta^2 g\rangle \cong R_2/(g^5, gw_1)$$

and

$$\langle \beta g^2 w_1 w_2^2, h_1^2\gamma w_2^3\rangle \cong \frac{R_2 \oplus R_2}{\langle (g, w_1), (0, g)\rangle}$$

at $E_4$. Complex (C) is

$$\begin{array}{ccccc} \langle h_1 e_0 w_2\rangle & \xrightarrow{\binom{g^2 w_1}{w_1}} & \langle e_0\rangle \oplus \langle h_1 c_0 w_2\rangle & \xrightarrow{(w_1\ 0)} & \langle c_0\rangle \\ \| & & \| & & \| \\ R_2/(g) & & R_2/(g^3) \oplus R_2/(g) & & R_2/(g) \end{array}$$

The class $h_1 e_0 w_2$ does not survive, and $e_0$ and $h_1 c_0 w_2$ are replaced by $e_0 g$ and

$$\gamma\delta' = 12_{27} + 12_{28} = e_0 g^2 + h_1 c_0 w_2\,,$$

respectively, leaving

$$\langle c_0\rangle \cong R_2/(g, w_1)$$
$$\langle e_0 g\rangle \cong R_2/(g^2)$$
$$\langle \gamma\delta'\rangle \cong R_2/(g, w_1)$$

at $E_4$. Complex (D) is

$$\begin{array}{ccc} \langle h_1 e_0\rangle & \xrightarrow{w_1} & \langle h_1 c_0\rangle \\ \| & & \| \\ R_2/(g) & & R_2/(g) \end{array}$$

The class $h_1 e_0$ does not survive, leaving

$$\langle h_1 c_0\rangle \cong R_2/(g, w_1)\,.$$

Complex (E) is

$$\begin{array}{ccc} \langle h_1 d_0 w_2\rangle & \xrightarrow{g^2 w_1} & \langle d_0\rangle \\ \| & & \| \\ R_2/(g) & & R_2/(g^3) \end{array}$$

The class $h_1 d_0 w_2$ does not survive, leaving

$$\langle d_0\rangle \cong R_2/(g^3, g^2 w_1)\,.$$

Complex (F) is

$$\begin{array}{ccc}
\langle \alpha^2 \rangle & \xrightarrow{w_1} & \langle h_1 d_0 \rangle \\
\| & & \| \\
R_2/(g^2) & & R_2/(g)
\end{array}$$

The class $\alpha^2$ is replaced by $\alpha^2 g$, leaving

$$\langle h_1 d_0 \rangle \cong R_2/(g, w_1)$$
$$\langle \alpha^2 g \rangle \cong R_2/(g)\,.$$

Complex (G) is

$$\begin{array}{ccc}
\langle h_1 \gamma w_2 \rangle \oplus \langle \beta^2 w_2^2 \rangle & \xrightarrow{\begin{pmatrix} g^2 w_1 & g^5 \\ 0 & 0 \\ 0 & g w_1 \end{pmatrix}} & \langle \gamma, h_0 e_0 w_2 \rangle \oplus \langle h_1 w_2^2 \rangle \\
\| & & \| \\
R_2/(g) \oplus R_2/(g^2 w_1) & & \dfrac{R_2 \oplus R_2}{\langle (g^2 w_1, w_1), (0, g) \rangle} \oplus R_2/(g^2)
\end{array}$$

The class $h_1 \gamma w_2$ does not survive, and $\beta^2 w_2^2$ is replaced by $\beta^2 g w_1 w_2^2$, leaving

$$\langle \gamma, h_1 w_2^2 \rangle \cong \frac{R_2 \oplus R_2}{\langle (g^2 w_1, 0), (g^5, g w_1), (0, g^2) \rangle}$$

and

$$\langle h_0 e_0 w_2 \rangle \cong R_2/(g, w_1)$$
$$\langle \beta^2 g w_1 w_2^2 \rangle \cong R_2/(g)\,.$$

Complex (H) is

$$\begin{array}{ccc}
\langle e_0 \gamma \rangle & \xrightarrow{w_1} & \langle h_1 \delta \rangle \\
\| & & \| \\
R_2/(g^3) & & R_2/(g)
\end{array}$$

The class $e_0 \gamma$ is replaced by $e_0 \gamma g$, leaving

$$\langle h_1 \delta \rangle \cong R_2/(g, w_1)$$
$$\langle e_0 \gamma g \rangle \cong R_2/(g^2)\,.$$

Complex (I) is

$$\begin{array}{ccccc}
\langle h_1 w_2^3 \rangle & \xrightarrow{g^2 w_1} & \langle w_2^2 \rangle & \xrightarrow{\binom{g^2}{0}} & \langle \beta g^2, h_1^2 \gamma w_2 \rangle \\
\| & & \| & & \| \\
R_2/(g^2) & & R_2/(g^4 w_1) & & \dfrac{R_2 \oplus R_2}{\langle (g w_1, w_1), (0, g) \rangle}
\end{array}$$

The class $h_1 w_2^3$ does not survive, and $w_2^2$ and $h_1^2 \gamma w_2$ are replaced by $w_1 w_2^2$ and

$$\gamma^3 = 15_{38} + 15_{39} = \beta g^3 + h_1^2 \gamma w_2\,,$$

respectively, leaving

$$\langle \beta g^2 \rangle \cong R_2/(g^2)$$
$$\langle \gamma^3 \rangle \cong R_2/(g, w_1)$$

$$\langle w_1 w_2^2 \rangle \cong R_2/(g^2)\,.$$

Complex (J) is $w_2^2$ times complex (C). (We omit to display it.) The class $h_1 e_0 w_2^3$ does not survive, and $e_0 w_2^2$ and $h_1 c_0 w_2^3$ are replaced by $e_0 g w_2^2$ and

$$\gamma\delta' w_2^2 = 28_{129} + 28_{130} = e_0 g^2 w_2^2 + h_1 c_0 w_2^3\,,$$

respectively, leaving

$$\langle c_0 w_2^2 \rangle \cong R_2/(g, w_1)$$
$$\langle e_0 g w_2^2 \rangle \cong R_2/(g^2)$$
$$\langle \gamma\delta' w_2^2 \rangle \cong R_2/(g, w_1)\,.$$

Complex (K) is $w_2^2$ times complex (D). The class $h_1 e_0 w_2^2$ does not survive, leaving

$$\langle h_1 c_0 w_2^2 \rangle \cong R_2/(g, w_1)\,.$$

Complex (L) is $w_2^2$ times complex (E). The class $h_1 d_0 w_2^3$ does not survive, leaving

$$\langle d_0 w_2^2 \rangle \cong R_2/(g^3, g^2 w_1)\,.$$

Complex (M) is $w_2^2$ times complex (F). The class $\alpha^2 w_2^2$ is replaced by $\alpha^2 g w_2^2$, leaving

$$\langle h_1 d_0 w_2^2 \rangle \cong R_2/(g, w_1)$$
$$\langle \alpha^2 g w_2^2 \rangle \cong R_2/(g)\,.$$

Complex (N) is $w_2^2$ times complex (H). The class $e_0 \gamma w_2^2$ is replaced by $e_0 \gamma g w_2^2$, leaving

$$\langle h_1 \delta w_2^2 \rangle \cong R_2/(g, w_1)$$
$$\langle e_0 \gamma g w_2^2 \rangle \cong R_2/(g^2)$$

at the $E_4$-term.

## A.3. Calculation of $E_5(tmf) = H(E_4(tmf), d_4)$

The $(E_4, d_4)$-term of the Adams spectral sequence for *tmf* splits as a direct sum of 16 $R_2$-module complexes of length two, labeled (A) to (P), plus a large summand with trivial differential. The Type-column in Table A.3 gives the label of the complex containing the $R_2$-module generator $x$.

Table A.3: Summands in $(E_4(tmf), d_4)$

| $t-s$ | $s$ | $g$ | $x$ | Ann($x$) | Type($x$) |
|---|---|---|---|---|---|
| 0 | 0 | 0 | 1 | $(g^6, g^2 w_1)$ | (A) |
| 0 | 1 | 0 | $h_0$ | $(g^2, g w_1)$ | 0 |
| 0 | 2 | 0 | $h_0^2$ | $(g^2, g w_1)$ | 0 |
| 0 | $3+i$ | 0 | $h_0^{3+i}$ | $(g)$ | 0 |
| 1 | 1 | 1 | $h_1$ | $(g^2, g w_1)$ | 0 |
| 2 | 2 | 1 | $h_1^2$ | $(g)$ | 0 |
| 3 | 1 | 2 | $h_2$ | $(g, w_1)$ | 0 |
| 3 | 2 | 2 | $h_0 h_2$ | $(g, w_1)$ | 0 |

Table A.3: Summands in $(E_4(tmf), d_4)$ (cont.)

| $t-s$ | $s$ | $g$ | $x$ | Ann$(x)$ | Type$(x)$ |
|---|---|---|---|---|---|
| 3 | 3 | 1 | $h_0^2 h_2$ | $(g, w_1)$ | 0 |
| 6 | 2 | 3 | $h_2^2$ | $(g, w_1)$ | 0 |
| 8 | 3 | 2 | $c_0$ | $(g, w_1)$ | 0 |
| 9 | 4 | 2 | $h_1 c_0$ | $(g, w_1)$ | 0 |
| 12 | $6+i$ | 4 | $h_0^{3+i}\alpha$ | $(g)$ | 0 |
| 14 | 4 | 4 | $d_0$ | $(g^3, g^2 w_1)$ | (B) |
| 15 | 5 | 6 | $h_1 d_0$ | $(g, w_1)$ | 0 |
| 17 | 5 | 7 | $h_0 e_0$ | $(g, w_1)$ | 0 |
| 24 | $7+i$ | 7 | $h_0^{1+i}\alpha^2$ | $(g)$ | 0 |
| 25 | 5 | 11 | $\gamma$ | – | (C) |
| 26 | 6 | 9 | $h_1\gamma$ | $(g)$ | 0 |
| 27 | 6 | 10 | $\alpha\beta$ | $(g)$ | (D) |
| 27 | 7 | 9 | $h_1^2\gamma$ | $(g, w_1)$ | 0 |
| 31 | 8 | 13 | $d_0 e_0$ | $(g^2)$ | (B) |
| 32 | 7 | 11 | $\delta$ | $(g)$ | (E) |
| 32 | 7 | $11+12$ | $\alpha g$ | $(g^2)$ | (E) |
| 32 | 8 | 14 | $h_0\alpha g$ | $(g)$ | 0 |
| 32 | 9 | 14 | $h_0^2\alpha g$ | $(g)$ | 0 |
| 33 | 8 | 15 | $h_1\delta$ | $(g, w_1)$ | 0 |
| 36 | $10+i$ | 14 | $h_0^{1+i}\alpha^3$ | $(g)$ | 0 |
| 37 | 8 | 17 | $e_0 g$ | $(g^2)$ | (A) |
| 39 | 9 | 18 | $d_0\gamma$ | – | (F) |
| 41 | 10 | 16 | $\alpha^2 e_0$ | $(g)$ | (G) |
| 44 | 10 | 17 | $\alpha^2 g$ | $(g)$ | (D) |
| 46 | 11 | 18 | $\alpha d_0 g$ | $(g^2)$ | (H) |
| 48 | 9 | 21 | $h_0 w_2$ | $(g^2)$ | (F) |
| 48 | 10 | 19 | $h_0^2 w_2$ | $(g^2, g w_1)$ | 0 |
| 48 | $11+i$ | 19 | $h_0^{3+i} w_2$ | $(g)$ | 0 |
| 49 | 11 | 20 | $\alpha e_0 g$ | $(g^2)$ | (E) |
| 50 | 10 | 20 | $\beta^2 g$ | $(g^5, g w_1)$ | (G) |
| 50 | 10 | 21 | $h_1^2 w_2$ | $(g)$ | (G) |
| 51 | 9 | 23 | $h_2 w_2$ | – | (F) |

Table A.3: Summands in $(E_4(tmf), d_4)$ (cont.)

| $t-s$ | $s$ | $g$ | $x$ | $\mathrm{Ann}(x)$ | $\mathrm{Type}(x)$ |
|---|---|---|---|---|---|
| 51 | 10 | 22 | $h_0h_2w_2$ | $(g, w_1)$ | 0 |
| 51 | 11 | 21 | $h_0^2h_2w_2$ | $(g, w_1)$ | 0 |
| 54 | 10 | 23 | $h_2^2w_2$ | $(g, w_1)$ | 0 |
| 55 | 11 | 23 | $\beta g^2$ | $(g^2)$ | (H) |
| 56 | 11 | 24 | $c_0w_2$ | $(g)$ | 0 |
| 56 | 13 | $26+27$ | $\alpha^3g + h_0w_1w_2$ | $(g)$ | 0 |
| 57 | 12 | $27+28$ | $\gamma\delta'$ | $(g, w_1)$ | 0 |
| 60 | $14+i$ | 28 | $h_0^{3+i}\alpha w_2$ | $(g)$ | 0 |
| 62 | 13 | 32 | $e_0\gamma g$ | $(g^2)$ | (C) |
| 65 | 13 | 36 | $h_0e_0w_2$ | $(g, w_1)$ | 0 |
| 72 | $15+i$ | 36 | $h_0^{1+i}\alpha^2w_2$ | $(g)$ | 0 |
| 75 | 15 | $38+39$ | $\gamma^3$ | $(g, w_1)$ | 0 |
| 80 | 15 | 41 | $\delta w_2$ | $(g)$ | 0 |
| 80 | 16 | 49 | $h_0\alpha g w_2$ | $(g)$ | 0 |
| 80 | 17 | 49 | $h_0^2\alpha g w_2$ | $(g)$ | 0 |
| 81 | 16 | 50 | $h_1\delta w_2$ | $(g)$ | 0 |
| 84 | $18+i$ | 48 | $h_0^{1+i}\alpha^3w_2$ | $(g)$ | 0 |
| 96 | 17 | 58 | $h_0w_2^2$ | $(g^2, gw_1)$ | 0 |
| 96 | 18 | 55 | $h_0^2w_2^2$ | $(g^2, gw_1)$ | 0 |
| 96 | $19+i$ | 57 | $h_0^{3+i}w_2^2$ | $(g)$ | 0 |
| 97 | 17 | 59 | $h_1w_2^2$ | $-$ | (C) |
| 98 | 18 | 57 | $h_1^2w_2^2$ | $(g)$ | 0 |
| 99 | 17 | 60 | $h_2w_2^2$ | $(g, w_1)$ | 0 |
| 99 | 18 | 58 | $h_0h_2w_2^2$ | $(g, w_1)$ | 0 |
| 99 | 19 | 59 | $h_0^2h_2w_2^2$ | $(g, w_1)$ | 0 |
| 102 | 18 | 59 | $h_2^2w_2^2$ | $(g, w_1)$ | 0 |
| 104 | 19 | 62 | $c_0w_2^2$ | $(g, w_1)$ | 0 |
| 104 | 20 | 69 | $w_1w_2^2$ | $(g^2)$ | (I) |
| 105 | 20 | 71 | $h_1c_0w_2^2$ | $(g, w_1)$ | 0 |
| 108 | $22+i$ | 71 | $h_0^{3+i}\alpha w_2^2$ | $(g)$ | 0 |
| 110 | 20 | 74 | $d_0w_2^2$ | $(g^3, g^2w_1)$ | (J) |
| 111 | 21 | 79 | $h_1d_0w_2^2$ | $(g, w_1)$ | 0 |

Table A.3: Summands in $(E_4(tmf), d_4)$ (cont.)

| $t-s$ | $s$ | $g$ | $x$ | Ann($x$) | Type($x$) |
|---|---|---|---|---|---|
| 113 | 21 | 81 | $h_0 e_0 w_2^2$ | $(g, w_1)$ | 0 |
| 120 | $23+i$ | 82 | $h_0^{1+i}\alpha^2 w_2^2$ | $(g)$ | 0 |
| 122 | 22 | 81 | $h_1 \gamma w_2^2$ | $(g)$ | 0 |
| 123 | 22 | 82 | $\alpha\beta w_2^2$ | $(g)$ | (K) |
| 123 | 23 | 85 | $h_1^2 \gamma w_2^2$ | $(g, w_1)$ | 0 |
| 127 | 24 | 98 | $d_0 e_0 w_2^2$ | $(g^2)$ | (J) |
| 128 | 23 | 87 | $\delta w_2^2$ | $(g)$ | (L) |
| 128 | 23 | $87+88$ | $\alpha g w_2^2$ | $(g^2)$ | (L) |
| 128 | 24 | 100 | $h_0 \alpha g w_2^2$ | $(g)$ | 0 |
| 128 | 25 | 102 | $h_0^2 \alpha g w_2^2$ | $(g)$ | 0 |
| 129 | 24 | 101 | $h_1 \delta w_2^2$ | $(g, w_1)$ | 0 |
| 129 | 25 | 103 | $\gamma w_1 w_2^2$ | $(g^2)$ | (M) |
| 132 | $26+i$ | 100 | $h_0^{1+i}\alpha^3 w_2^2$ | $(g)$ | 0 |
| 133 | 24 | 103 | $e_0 g w_2^2$ | $(g^2)$ | (I) |
| 135 | 25 | 108 | $d_0 \gamma w_2^2$ | – | (N) |
| 137 | 26 | 103 | $\alpha^2 e_0 w_2^2$ | $(g)$ | (O) |
| 140 | 26 | 105 | $\alpha^2 g w_2^2$ | $(g)$ | (K) |
| 142 | 27 | 109 | $\alpha d_0 g w_2^2$ | $(g^2)$ | (P) |
| 144 | 25 | 111 | $h_0 w_2^3$ | $(g^2)$ | (N) |
| 144 | 26 | 107 | $h_0^2 w_2^3$ | $(g^2, g w_1)$ | 0 |
| 144 | $27+i$ | 111 | $h_0^{3+i} w_2^3$ | $(g)$ | 0 |
| 145 | 27 | 112 | $\alpha e_0 g w_2^2$ | $(g^2)$ | (L) |
| 146 | 26 | 109 | $h_1^2 w_2^3$ | $(g)$ | (O) |
| 147 | 25 | 113 | $h_2 w_2^3$ | – | (N) |
| 147 | 26 | 110 | $h_0 h_2 w_2^3$ | $(g, w_1)$ | 0 |
| 147 | 27 | 113 | $h_0^2 h_2 w_2^3$ | $(g, w_1)$ | 0 |
| 150 | 26 | 111 | $h_2^2 w_2^3$ | $(g, w_1)$ | 0 |
| 152 | 27 | 116 | $c_0 w_2^3$ | $(g)$ | 0 |
| 152 | 29 | $131+132$ | $\alpha^3 g w_2^2 + h_0 w_1 w_2^3$ | $(g)$ | 0 |
| 153 | 28 | $129+130$ | $\gamma\delta' w_2^2$ | $(g, w_1)$ | 0 |
| 154 | 30 | 127 | $\beta^2 g w_1 w_2^2$ | $(g)$ | (O) |
| 156 | $30+i$ | 131 | $h_0^{3+i}\alpha w_2^3$ | $(g)$ | 0 |

Table A.3: Summands in $(E_4(tmf), d_4)$ (cont.)

| $t-s$ | $s$ | $g$ | $x$ | Ann$(x)$ | Type$(x)$ |
|---|---|---|---|---|---|
| 158 | 29 | 138 | $e_0\gamma g w_2^2$ | $(g^2)$ | (M) |
| 159 | 31 | 135 | $\beta g^2 w_1 w_2^2$ | – | (P) |
| 161 | 29 | 142 | $h_0 e_0 w_2^3$ | $(g, w_1)$ | 0 |
| 168 | $31+i$ | 144 | $h_0^{1+i}\alpha^2 w_2^3$ | $(g)$ | 0 |
| 171 | 31 | 147 | $h_1^2\gamma w_2^3$ | – | (P) |
| 176 | 31 | 149 | $\delta w_2^3$ | $(g)$ | 0 |
| 176 | 32 | 167 | $h_0\alpha g w_2^3$ | $(g)$ | 0 |
| 176 | 33 | 171 | $h_0^2\alpha g w_2^3$ | $(g)$ | 0 |
| 177 | 32 | 168 | $h_1\delta w_2^3$ | $(g)$ | 0 |
| 180 | $34+i$ | 168 | $h_0^{1+i}\alpha^3 w_2^3$ | $(g)$ | 0 |

Complex (A) is

$$\begin{array}{ccc} \langle e_0 g\rangle & \xrightarrow{\;gw_1^2\;} & \langle 1\rangle \\ \| & & \| \\ R_2/(g^2) & & R_2/(g^6, g^2w_1) \end{array}$$

The class $e_0 g$ is replaced by $e_0 g^2$, leaving

$$\langle 1\rangle \cong R_2/(g^6, g^2w_1, gw_1^2)$$
$$\langle e_0 g^2\rangle \cong R_2/(g)$$

at $E_5$. Complex (B) is

$$\begin{array}{ccc} \langle d_0 e_0\rangle & \xrightarrow{\;w_1^2\;} & \langle d_0\rangle \\ \| & & \| \\ R_2/(g^2) & & R_2/(g^3, g^2w_1) \end{array}$$

The class $d_0 e_0$ does not survive, leaving

$$\langle d_0\rangle \cong R_2/(g^3, g^2w_1, w_1^2)$$

at $E_5$. Complex (C) is

$$\begin{array}{ccc} \langle e_0\gamma g\rangle & \xrightarrow{\;\begin{pmatrix} gw_1^2 \\ 0 \end{pmatrix}\;} & \langle \gamma, h_1 w_2^2\rangle \\ \| & & \| \\ R_2/(g^2) & & \dfrac{R_2 \oplus R_2}{\langle (g^2w_1, 0), (g^5, gw_1), (0, g^2)\rangle} \end{array}$$

The class $e_0\gamma g$ is replaced by $e_0\gamma g^2$, leaving

$$\langle \gamma, h_1 w_2^2\rangle \cong \frac{R_2 \oplus R_2}{\langle (g^2w_1, 0), (gw_1^2, 0), (g^5, gw_1), (0, g^2)\rangle}$$

and

$$\langle e_0\gamma g^2\rangle \cong R_2/(g)$$

at $E_5$. Complex (D) is

$$\begin{array}{ccc} \langle \alpha^2 g\rangle & \xrightarrow{w_1^2} & \langle \alpha\beta\rangle \\ \| & & \| \\ R_2/(g) & & R_2/(g) \end{array}$$

The class $\alpha^2 g$ does not survive, leaving

$$\langle \alpha\beta\rangle \cong R_2/(g, w_1^2)\,.$$

Complex (E) is

$$\begin{array}{ccc} \langle \alpha e_0 g\rangle & \xrightarrow{\binom{w_1^2}{w_1^2}} & \langle \delta\rangle \oplus \langle \alpha g\rangle \\ \| & & \| \\ R_2/(g^2) & & R_2/(g) \oplus R_2/(g^2) \end{array}$$

The class $\alpha e_0 g$ does not survive, and $\alpha g$ is replaced by $\delta' = \delta + \alpha g$, leaving

$$\langle \delta\rangle \cong R_2/(g)$$
$$\langle \delta'\rangle \cong R_2/(g^2, w_1^2)\,.$$

Complex (F) is

$$\begin{array}{ccc} \langle h_0 w_2\rangle & \xrightarrow{\binom{w_1}{0}} & \langle d_0\gamma, h_2 w_2\rangle \\ \| & & \| \\ R_2/(g^2) & & \dfrac{R_2 \oplus R_2}{\langle (g, w_1), (0, g)\rangle} \end{array}$$

The class $h_0 w_2$ does not survive, leaving

$$\langle d_0\gamma, h_2 w_2\rangle \cong \frac{R_2 \oplus R_2}{\langle (w_1, 0), (g, w_1), (0, g)\rangle}\,.$$

Complex (G) is

$$\begin{array}{ccc} \langle \beta^2 g\rangle \oplus \langle h_1^2 w_2\rangle & \xrightarrow{(\,w_1\ w_1\,)} & \langle \alpha^2 e_0\rangle \\ \| & & \| \\ R_2/(g^5, g w_1) \oplus R_2/(g) & & R_2/(g) \end{array}$$

The classes $\beta^2 g$ and $h_1^2 w_2$ are replaced by

$$\gamma^2 = 10_{20} + 10_{21} = \beta^2 g + h_1^2 w_2\,,$$

leaving

$$\langle \alpha^2 e_0\rangle \cong R_2/(g, w_1)$$
$$\langle \gamma^2\rangle \cong R_2/(g^5, g w_1)\,.$$

Complex (H) is

$$\begin{array}{ccc} \langle \beta g^2\rangle & \xrightarrow{w_1} & \langle \alpha d_0 g\rangle \\ \| & & \| \\ R_2/(g^2) & & R_2/(g^2) \end{array}$$

The class $\beta g^2$ does not survive, leaving

$$\langle \alpha d_0 g\rangle \cong R_2/(g^2, w_1)\,.$$

Complex (I) is

$$\begin{array}{ccc} \langle e_0 g w_2^2\rangle & \xrightarrow{g w_1} & \langle w_1 w_2^2\rangle \\ \| & & \| \\ R_2/(g^2) & & R_2/(g^2) \end{array}$$

The class $e_0 g w_2^2$ is replaced by $e_0 g^2 w_2^2$, leaving

$$\langle w_1 w_2^2\rangle \cong R_2/(g^2, g w_1)$$
$$\langle e_0 g^2 w_2^2\rangle \cong R_2/(g)\,.$$

Complex (J) is $w_2^2$ times complex (B). The class $d_0 e_0 w_2^2$ does not survive, leaving

$$\langle d_0 w_2^2\rangle \cong R_2/(g^3, g^2 w_1, w_1^2)\,.$$

Complex (K) is $w_2^2$ times complex (D). The class $\alpha^2 g w_2^2$ does not survive, leaving

$$\langle \alpha\beta w_2^2\rangle \cong R_2/(g, w_1^2)\,.$$

Complex (L) is $w_2^2$ times complex (E). The class $\alpha e_0 g w_2^2$ does not survive, and $\alpha g w_2^2$ is replaced by $\delta' w_2^2 = \delta w_2^2 + \alpha g w_2^2$, leaving

$$\langle \delta w_2^2\rangle \cong R_2/(g)$$
$$\langle \delta' w_2^2\rangle \cong R_2/(g^2, w_1^2)\,.$$

Complex (M) is $\gamma$ times complex (I). The class $e_0\gamma g w_2^2$ is replaced by $e_0\gamma g^2 w_2^2$, leaving

$$\langle \gamma w_1 w_2^2\rangle \cong R_2/(g^2, g w_1)$$
$$\langle e_0\gamma g^2 w_2^2\rangle \cong R_2/(g)\,.$$

Complex (N) is $w_2^2$ times complex (F). The class $h_0 w_2^3$ does not survive, leaving

$$\langle d_0\gamma w_2^2, h_2 w_2^3\rangle \cong \frac{R_2 \oplus R_2}{\langle (w_1, 0), (g, w_1), (0, g)\rangle}\,.$$

Complex (O) is

$$\begin{array}{ccc} \langle \beta^2 g w_1 w_2^2\rangle \oplus \langle h_1^2 w_2^3\rangle & \xrightarrow{(\,w_1^2 \;\; w_1\,)} & \langle \alpha^2 e_0 w_2^2\rangle \\ \| & & \| \\ R_2/(g) \oplus R_2/(g) & & R_2/(g) \end{array}$$

The classes $\beta^2 g w_1 w_2^2$ and $h_1^2 w_2^3$ are replaced by

$$\gamma^2 w_1 w_2^2 = 30_{127} + 30_{128} = \beta^2 g w_1 w_2^2 + h_1^2 w_1 w_2^3\,,$$

leaving

$$\langle \alpha^2 e_0 w_2^2\rangle \cong R_2/(g, w_1)$$
$$\langle \gamma^2 w_1 w_2^2\rangle \cong R_2/(g)\,.$$

Complex (P) is

$$\begin{array}{ccc}
\langle \beta g^2 w_1 w_2^2, h_1^2 \gamma w_2^3 \rangle & \xrightarrow{\left( w_1^2 \; gw_1 \right)} & \langle \alpha d_0 g w_2^2 \rangle \\
\| & & \| \\
\dfrac{R_2 \oplus R_2}{\langle (g, w_1), (0, g) \rangle} & & R_2/(g^2)
\end{array}$$

The classes $\beta g^2 w_1 w_2^2$ and $h_1^2 \gamma w_2^3$ do not survive, leaving

$$\langle \alpha d_0 g w_2^2 \rangle \cong R_2/(g^2, gw_1, w_1^2)$$

at the $E_5$-term.

APPENDIX B

# Calculation of $E_r(tmf/2)$ for $r = 3, 4, 5$

Recall from Definition 5.1 that $R_0 = \mathbb{F}_2[g, w_1, w_2]$, $R_1 = \mathbb{F}_2[g, w_1, w_2^2]$ and $R_2 = \mathbb{F}_2[g, w_1, w_2^4]$. Our calculations show that $E_2(tmf/2)$ is a complex of $R_1$-modules, while $E_3(tmf/2)$ and $E_4(tmf/2)$ are complexes of $R_2$-modules.

## B.1. Calculation of $E_3(tmf/2) = H(E_2(tmf/2), d_2)$

When regarded as a complex of $R_1$-modules, the $(E_2, d_2)$-term of the Adams spectral sequence for $tmf/2$ splits as a direct sum of 24 two-term complexes of the form

$$R_1\{x\} \xrightarrow{a} R_1\{y\},$$

eight other complexes labeled (A) to (H), and a large summand with trivial differential. Table B.1 gives, for each $R_0$-module generator $x$ of $E_2(tmf/2)$, the summands to which $x$ and $xw_2$ belong. Table B.2 describes the two-term complexes, numbered $n = 1$ to 24. In each case, $R_1\{x\}$ does not survive to $E_3$, while $\langle y\rangle = R_1/(a)$ in $E_3$. The remaining complexes and their homology are described individually following these tables.

Table B.1: Summands in $(E_2(tmf/2), d_2)$

| $t-s$ | $s$ | $g$ | $x$ | $\mathrm{Ann}(x)$ | Type$(x)$ | Type$(xw_2)$ |
|---|---|---|---|---|---|---|
| 0 | 0 | 0 | $i(1)$ | $(0)$ | 1 | 3 |
| 1 | 1 | 0 | $i(h_1)$ | $(g^2, gw_1)$ | 0 | 0 |
| 2 | 1 | 1 | $\widetilde{h_1}$ | $(0)$ | 2 | 4 |
| 2 | 2 | 0 | $i(h_1^2)$ | $(g)$ | 0 | 0 |
| 3 | 1 | 2 | $i(h_2)$ | $(g)$ | (A) | (D) |
| 3 | 2 | 1 | $h_1\widetilde{h_1}$ | $(g)$ | 0 | 0 |
| 4 | 3 | 0 | $h_1^2\widetilde{h_1}$ | $(g)$ | 0 | 0 |
| 6 | 2 | 2 | $i(h_2^2)$ | $(g, w_1)$ | 0 | 0 |
| 7 | 2 | 3 | $\widetilde{h_2^2}$ | $(0)$ | 3 | 5 |
| 8 | 3 | 1 | $i(c_0)$ | $(g)$ | 0 | 0 |
| 9 | 3 | 2 | $\widetilde{c_0}$ | $(0)$ | 4 | 7 |
| 9 | 4 | 1 | $i(h_1c_0)$ | $(g)$ | (B) | (C) |
| 10 | 4 | 2 | $h_1\widetilde{c_0}$ | $(g)$ | 0 | 0 |
| 12 | 3 | 3 | $i(\alpha)$ | $(0)$ | (A) | (D) |

Table B.1: Summands in $(E_2(tmf/2), d_2)$ (cont.)

| $t-s$ | $s$ | $g$ | $x$ | $\mathrm{Ann}(x)$ | $\mathrm{Type}(x)$ | $\mathrm{Type}(xw_2)$ |
|---|---|---|---|---|---|---|
| 14 | 4 | 3 | $i(d_0)$ | $(0)$ | 5 | 8 |
| 15 | 3 | 4 | $i(\beta)$ | $(0)$ | 6 | 2 |
| 16 | 5 | 3 | $d_0\widetilde{h_1}$ | $(0)$ | 7 | 6 |
| 17 | 4 | 4 | $i(e_0)$ | $(0)$ | (C) | 9 |
| 18 | 4 | 5 | $i(h_2\beta)$ | $(g, w_1)$ | 0 | 0 |
| 18 | 6 | 3 | $\widetilde{h_0^2 e_0}$ | – | (B) | (C) |
| 19 | 5 | 4 | $e_0\widetilde{h_1}$ | $(0)$ | (D) | 12 |
| 21 | 6 | 4 | $d_0\widetilde{h_2^2}$ | $(0)$ | 8 | 1 |
| 24 | 6 | 5 | $i(\alpha^2)$ | $(0)$ | 9 | 14 |
| 25 | 5 | 7 | $i(\gamma)$ | $(0)$ | 10 | (A) |
| 26 | 5 | 8 | $\widetilde{\gamma}$ | $(0)$ | 11 | 16 |
| 26 | 6 | 6 | $i(h_1\gamma)$ | $(g)$ | 0 | 0 |
| 26 | 7 | 5 | $i(\alpha d_0)$ | $(0)$ | 12 | 10 |
| 27 | 6 | 8 | $h_1\widetilde{\gamma}$ | $(g)$ | 0 | 0 |
| 28 | 7 | 6 | $h_1^2\widetilde{\gamma}$ | $(g)$ | 0 | 0 |
| 30 | 6 | 9 | $i(\beta^2)$ | – | (B) | (C) |
| 31 | 6 | 10 | $\widetilde{\beta^2}$ | $(0)$ | 13 | 17 |
| 31 | 8 | 6 | $i(d_0 e_0)$ | $(0)$ | 14 | (B) |
| 32 | 7 | 9 | $i(\delta)$ | $(g)$ | 0 | 0 |
| 32 | 8 | 7 | $\widetilde{d_0 e_0}$ | $(0)$ | 15 | 13 |
| 33 | 7 | 10 | $\widetilde{\delta}'$ | $(0)$ | 16 | 18 |
| 33 | 8 | 8 | $i(h_1\delta)$ | $(g)$ | (E) | (G) |
| 33 | 9 | 7 | $h_1\widetilde{d_0 e_0}$ | $(g)$ | 0 | 0 |
| 34 | 8 | 10 | $h_1\widetilde{\delta}'$ | $(g)$ | 0 | 0 |
| 35 | 9 | 9 | $h_1^2\widetilde{\delta}'$ | $(g)$ | (F) | (H) |
| 36 | 7 | 12 | $\widetilde{\beta g}$ | $(0)$ | (F) | (H) |
| 38 | 8 | 12 | $\alpha\widetilde{\gamma}$ | $(0)$ | 17 | 19 |
| 40 | 9 | 12 | $d_0\widetilde{\gamma}$ | $(0)$ | 18 | 20 |
| 41 | 8 | 14 | $\beta\widetilde{\gamma}$ | $(0)$ | (G) | 21 |
| 42 | 10 | 12 | $\widetilde{\alpha^2 e_0}$ | $(0)$ | (E) | (G) |
| 43 | 9 | 14 | $e_0\widetilde{\gamma}$ | $(0)$ | (H) | 22 |
| 45 | 10 | 14 | $d_0\widetilde{\beta^2}$ | $(0)$ | 19 | 15 |

Table B.1: Summands in $(E_2(tmf/2), d_2)$ (cont.)

| $t-s$ | $s$ | $g$ | $x$ | Ann$(x)$ | Type$(x)$ | Type$(xw_2)$ |
|---|---|---|---|---|---|---|
| 47 | 11 | 14 | $d_0\widetilde{\delta'}$ | (0) | 20 | 11 |
| 48 | 10 | 16 | $e_0\widetilde{\beta^2}$ | (0) | 21 | 23 |
| 50 | 11 | 16 | $d_0\widetilde{\beta g}$ | (0) | 22 | 24 |
| 55 | 12 | 18 | $\alpha^2\widetilde{\beta^2}$ | (0) | 23 | (E) |
| 57 | 13 | 18 | $d_0e_0\widetilde{\gamma}$ | (0) | 24 | (F) |

Table B.2: Two-term complexes $a\colon R_1\{x\} \to R_1\{y\}$ in $E_2(tmf/2)$

| $n$ | $x$ | $a$ | $y$ |
|---|---|---|---|
| 1 | $d_0w_2\widetilde{h_2^2}$ | $g^3w_1$ | $i(1)$ |
| 2 | $i(\beta w_2)$ | $g^3$ | $\widetilde{h_1}$ |
| 3 | $i(w_2)$ | $g^2$ | $\widetilde{h_2^2}$ |
| 4 | $w_2\widetilde{h_1}$ | $g^2$ | $\widetilde{c_0}$ |
| 5 | $w_2\widetilde{h_2^2}$ | $g^2$ | $i(d_0)$ |
| 6 | $d_0w_2\widetilde{h_1}$ | $g^2w_1$ | $i(\beta)$ |
| 7 | $w_2\widetilde{c_0}$ | $g^2$ | $d_0\widetilde{h_1}$ |
| 8 | $i(d_0w_2)$ | $g^2$ | $d_0\widetilde{h_2^2}$ |
| 9 | $i(e_0w_2)$ | $g^2$ | $i(\alpha^2)$ |
| 10 | $i(\alpha d_0w_2)$ | $g^2w_1$ | $i(\gamma)$ |
| 11 | $d_0w_2\widetilde{\delta'}$ | $g^3w_1$ | $\widetilde{\gamma}$ |
| 12 | $e_0w_2\widetilde{h_1}$ | $g^2$ | $i(\alpha d_0)$ |
| 13 | $w_2\widetilde{d_0e_0}$ | $g^2w_1$ | $\widetilde{\beta^2}$ |
| 14 | $i(\alpha^2w_2)$ | $g^2$ | $i(d_0e_0)$ |
| 15 | $d_0w_2\widetilde{\beta^2}$ | $g^3$ | $\widetilde{d_0e_0}$ |
| 16 | $w_2\widetilde{\gamma}$ | $g^2$ | $\widetilde{\delta'}$ |
| 17 | $w_2\widetilde{\beta^2}$ | $g^2$ | $\alpha\widetilde{\gamma}$ |
| 18 | $w_2\widetilde{\delta'}$ | $g^2$ | $d_0\widetilde{\gamma}$ |
| 19 | $\alpha w_2\widetilde{\gamma}$ | $g^2$ | $d_0\widetilde{\beta^2}$ |
| 20 | $d_0w_2\widetilde{\gamma}$ | $g^2$ | $d_0\widetilde{\delta'}$ |
| 21 | $\beta w_2\widetilde{\gamma}$ | $g^2$ | $e_0\widetilde{\beta^2}$ |
| 22 | $e_0w_2\widetilde{\gamma}$ | $g^2$ | $d_0\widetilde{\beta g}$ |
| 23 | $e_0w_2\widetilde{\beta^2}$ | $g^2$ | $\alpha^2\widetilde{\beta^2}$ |
| 24 | $d_0w_2\widetilde{\beta g}$ | $g^2$ | $d_0e_0\widetilde{\gamma}$ |

We treat the remaining eight complexes individually. Complex (A) is

$$\begin{array}{ccccc} \langle i(\gamma w_2)\rangle & \xrightarrow{g^3} & \langle i(\alpha)\rangle & \xrightarrow{w_1} & \langle h_2\rangle \\ \| & & \| & & \| \\ R_1 & & R_1 & & R_1/(g) \end{array}$$

The class $i(\gamma w_2)$ does not survive to $E_3$, while $i(\alpha)$ is replaced by $i(\alpha g)$, leaving

$$\langle i(h_2)\rangle \cong R_1/(g, w_1)$$
$$\langle i(\alpha g)\rangle \cong R_1/(g^2)$$

at $E_3$. Complex (B) is

$$\begin{array}{ccccc} \langle i(d_0e_0w_2)\rangle & \xrightarrow{\begin{pmatrix} 0 \\ g^2w_1 \end{pmatrix}} & \langle \widetilde{h_0^2e_0}, i(\beta^2)\rangle & \xrightarrow{(w_1\ 0)} & \langle i(h_1c_0)\rangle \\ \| & & \| & & \| \\ R_1 & & \dfrac{R_1 \oplus R_1}{\langle (g, w_1)\rangle} & & R_1/(g) \end{array}$$

The classes $\widetilde{h_0^2e_0}$ and $i(d_0e_0w_2)$ do not survive to $E_3$, leaving

$$\langle i(h_1c_0)\rangle \cong R_1/(g, w_1)$$
$$\langle i(\beta^2)\rangle \cong R_1/(g^2w_1)$$

at $E_3$. Complex (C) is

$$\begin{array}{ccc} \langle w_2\widetilde{h_0^2e_0}, i(\beta^2w_2)\rangle & \xrightarrow{\begin{pmatrix} g^2w_1 & g^3 \\ w_1 & 0 \end{pmatrix}} & \langle i(e_0)\rangle \oplus \langle i(h_1c_0w_2)\rangle \\ \| & & \| \\ \dfrac{R_1 \oplus R_1}{\langle (g, w_1)\rangle} & & R_1 \oplus R_1/(g) \end{array}$$

The classes $i(\beta^2w_2)$ and $w_2\widetilde{h_0^2e_0}$ do not survive to $E_3$, while $i(e_0)$ and $i(h_1c_0w_2)$ do. Replacing $i(h_1c_0w_2)$ by the sum $i(h_1c_0w_2 + e_0g^2)$ gives a description of the result at $E_3$ as a sum of cyclic modules. We then use the relations $i(e_0g^2) = 12_{20} = \beta^2g\widetilde{h_2^2}$, $i(h_1c_0w_2) = 12_{21} = h_1^2w_2\widetilde{h_2^2}$ and $\gamma^2 = \beta^2g + h_1^2w_2$ to shorten the name of this second generator from $i(h_1c_0w_2 + e_0g^2)$ to $\gamma^2\widetilde{h_2^2}$. This gives the summands

$$\langle i(e_0)\rangle \cong R_1/(g^3)$$
$$\langle \gamma^2\widetilde{h_2^2}\rangle \cong R_1/(g, w_1)$$

at $E_3$. Complex (D) is

$$\begin{array}{ccc} \langle i(\alpha w_2)\rangle & \xrightarrow{\begin{pmatrix} g^2 \\ w_1 \end{pmatrix}} & \langle \widetilde{e_0h_1}\rangle \oplus \langle i(h_2w_2)\rangle \\ \| & & \| \\ R_1 & & R_1 \oplus R_1/(g) \end{array}$$

The class $i(\alpha w_2)$ does not survive to $E_3$, while $e_0\widetilde{h_1}$ and $i(h_2 w_2)$ generate the non-cyclic summand

$$\langle e_0\widetilde{h_1}, i(h_2 w_2)\rangle \cong \frac{R_1 \oplus R_1}{\langle (g^2, w_1), (0, g)\rangle}.$$

Complex (E) is

$$\begin{array}{ccccc} \langle \alpha^2 w_2 \widetilde{\beta^2}\rangle & \xrightarrow{g^3} & \langle \widetilde{\alpha^2 e_0}\rangle & \xrightarrow{w_1} & \langle i(h_1\delta)\rangle \\ \| & & \| & & \| \\ R_1 & & R_1 & & R_1/(g) \end{array}$$

The class $\alpha^2 w_2\widetilde{\beta^2}$ does not survive to $E_3$, while $\widetilde{\alpha^2 e_0}$ is replaced by $g\widetilde{\alpha^2 e_0}$, leaving

$$\langle i(h_1\delta)\rangle \cong R_1/(g, w_1)$$
$$\langle g\widetilde{\alpha^2 e_0}\rangle \cong R_1/(g^2).$$

Complex (F) is

$$\begin{array}{ccccc} \langle d_0 e_0 w_2 \widetilde{\gamma}\rangle & \xrightarrow{g^3 w_1} & \langle \widetilde{\beta g}\rangle & \xrightarrow{1} & \langle h_1^2\widetilde{\delta'}\rangle \\ \| & & \| & & \| \\ R_1 & & R_1 & & R_1/(g) \end{array}$$

The class $d_0 e_0 w_2\widetilde{\gamma}$ does not survive to $E_3$, while $\widetilde{\beta g}$ is replaced by $g\widetilde{\beta g}$ and $h_1^2\widetilde{\delta'}$ becomes 0. We use the identity $g\widetilde{\beta g} = 11_{21} = \beta^2\widetilde{\gamma}$ to rewrite the element $g\widetilde{\beta g}$ as $\beta^2\widetilde{\gamma}$ henceforth, to simplify the rest of the calculation. This leaves only the summand

$$\langle \beta^2\widetilde{\gamma}\rangle \cong R_1/(g^2 w_1).$$

Complex (G) is

$$\begin{array}{ccc} \langle w_2\widetilde{\alpha^2 e_0}\rangle & \xrightarrow{\binom{g^2 w_1}{w_1}} & \langle \beta\widetilde{\gamma}\rangle \oplus \langle i(h_1\delta w_2)\rangle \\ \| & & \| \\ R_1 & & R_1 \oplus R_1/(g) \end{array}$$

The class $w_2\widetilde{\alpha^2 e_0}$ does not survive to $E_3$, while $\beta\widetilde{\gamma}$ and $i(h_1\delta w_2)$ generate the non-cyclic summand

$$\langle \beta\widetilde{\gamma}, i(h_1\delta w_2)\rangle \cong \frac{R_1 \oplus R_1}{\langle (g^2 w_1, w_1), (0, g)\rangle}.$$

Complex (H) is

$$\begin{array}{ccc} \langle w_2\widetilde{\beta g}\rangle & \xrightarrow{\binom{g^2}{1}} & \langle e_0\widetilde{\gamma}\rangle \oplus \langle h_1^2 w_2\widetilde{\delta'}\rangle \\ \| & & \| \\ R_1 & & R_1 \oplus R_1/(g) \end{array}$$

The class $w_2\widetilde{\beta g}$ does not survive to $E_3$, while $h_1^2 w_2\widetilde{\delta'}$ becomes equal to $g^2 \cdot e_0\widetilde{\gamma}$, leaving only the summand

$$\langle e_0\widetilde{\gamma}\rangle \cong R_1/(g^3)$$

at $E_3$.

## B.2. Calculation of $E_4(tmf/2) = H(E_3(tmf/2), d_3)$

When regarded as a complex of $R_2$-modules, the $(E_3, d_3)$-term of the Adams spectral sequence for $tmf/2$ splits as a direct sum of six two-term complexes of the form

$$R_2/(g^2)\{x\} \xrightarrow{gw_1} R_2/(g^2)\{y\},$$

14 other complexes labeled (A) to (N), and a large summand with trivial differential. Table B.3 gives, for each $R_1$-module generator $x$ of $E_3(tmf/2)$, the summands to which $x$ and $xw_2^2$ belong. Table B.4 describes the two-term complexes, numbered $n = 1$ to 6. In these, $x$ does not survive to $E_4$, but $gx$ and $y$ do, and generate $R_2$-summands

$$\langle gx\rangle = R_2/(g)$$
$$\langle y\rangle = R_2/(g^2, gw_1).$$

The remaining complexes and their homology are described individually following these tables.

Note that we replace $i(\alpha g)$ in Table 6.4 by $i(\delta') = i(\delta + \alpha g)$ here, since this simplifies $d_3$.

Table B.3: Summands in $(E_3(tmf/2), d_3)$

| $t-s$ | $s$ | $g$ | $x$ | $\mathrm{Ann}(x)$ | $\mathrm{Type}(x)$ | $\mathrm{Type}(xw_2^2)$ |
|---|---|---|---|---|---|---|
| 0 | 0 | 0 | $i(1)$ | $(g^3w_1)$ | (A) | (G) |
| 1 | 1 | 0 | $i(h_1)$ | $(g^2, gw_1)$ | 0 | 0 |
| 2 | 1 | 1 | $\widetilde{h_1}$ | $(g^3)$ | (B) | (J) |
| 2 | 2 | 0 | $i(h_1^2)$ | $(g)$ | 0 | 0 |
| 3 | 1 | 2 | $i(h_2)$ | $(g, w_1)$ | 0 | 0 |
| 3 | 2 | 1 | $h_1\widetilde{h_1}$ | $(g)$ | 0 | 0 |
| 4 | 3 | 0 | $h_1^2\widetilde{h_1}$ | $(g)$ | 0 | 0 |
| 6 | 2 | 2 | $i(h_2^2)$ | $(g, w_1)$ | 0 | 0 |
| 7 | 2 | 3 | $\widetilde{h_2^2}$ | $(g^2)$ | 0 | 0 |
| 8 | 3 | 1 | $i(c_0)$ | $(g)$ | (C) | (D) |
| 9 | 3 | 2 | $\widetilde{c_0}$ | $(g^2)$ | 1 | 2 |
| 9 | 4 | 1 | $i(h_1c_0)$ | $(g, w_1)$ | 0 | 0 |
| 10 | 4 | 2 | $h_1\widetilde{c_0}$ | $(g)$ | (E) | (F) |
| 14 | 4 | 3 | $i(d_0)$ | $(g^2)$ | 0 | 0 |
| 15 | 3 | 4 | $i(\beta)$ | $(g^2w_1)$ | (G) | (H) |
| 16 | 5 | 3 | $d_0\widetilde{h_1}$ | $(g^2)$ | 3 | 4 |
| 17 | 4 | 4 | $i(e_0)$ | $(g^3)$ | (C) | (D) |
| 18 | 4 | 5 | $i(h_2\beta)$ | $(g, w_1)$ | 0 | 0 |
| 19 | 5 | 4 | $e_0\widetilde{h_1}$ | – | (E) | (F) |

Table B.3: Summands in $(E_3(tmf/2), d_3)$ (cont.)

| $t-s$ | $s$ | $g$ | $x$ | Ann$(x)$ | Type$(x)$ | Type$(xw_2^2)$ |
|---|---|---|---|---|---|---|
| 21 | 6 | 4 | $d_0\widetilde{h_2^2}$ | $(g^2)$ | 0 | 0 |
| 24 | 6 | 5 | $i(\alpha^2)$ | $(g^2)$ | 0 | 0 |
| 25 | 5 | 7 | $i(\gamma)$ | $(g^2w_1)$ | (I) | (A) |
| 26 | 5 | 8 | $\widetilde{\gamma}$ | $(g^3w_1)$ | (J) | (K) |
| 26 | 6 | 6 | $i(h_1\gamma)$ | $(g)$ | 0 | 0 |
| 26 | 7 | 5 | $i(\alpha d_0)$ | $(g^2)$ | 5 | 6 |
| 27 | 6 | 8 | $h_1\widetilde{\gamma}$ | $(g)$ | 0 | 0 |
| 28 | 7 | 6 | $h_1^2\widetilde{\gamma}$ | $(g)$ | 0 | 0 |
| 30 | 6 | 9 | $i(\beta^2)$ | $(g^2w_1)$ | (H) | (I) |
| 31 | 6 | 10 | $\widetilde{\beta^2}$ | $(g^2w_1)$ | (B) | (J) |
| 31 | 8 | 6 | $i(d_0e_0)$ | $(g^2)$ | 0 | 0 |
| 32 | 7 | $8+9$ | $i(\delta')$ | $(g^2)$ | (K) | (N) |
| 32 | 7 | 9 | $i(\delta)$ | $(g)$ | 0 | 0 |
| 32 | 8 | 7 | $\widetilde{d_0e_0}$ | $(g^3)$ | (G) | (H) |
| 33 | 7 | 10 | $\widetilde{\delta'}$ | $(g^2)$ | 0 | 0 |
| 33 | 8 | 8 | $i(h_1\delta)$ | $(g, w_1)$ | 0 | 0 |
| 33 | 9 | 7 | $h_1\widetilde{d_0e_0}$ | $(g)$ | 0 | 0 |
| 34 | 8 | 10 | $h_1\widetilde{\delta'}$ | $(g)$ | (L) | (M) |
| 38 | 8 | 12 | $\alpha\widetilde{\gamma}$ | $(g^2)$ | 1 | 2 |
| 40 | 9 | 12 | $d_0\widetilde{\gamma}$ | $(g^2)$ | 0 | 0 |
| 41 | 8 | 14 | $\beta\widetilde{\gamma}$ | – | (K) | (N) |
| 43 | 9 | 14 | $e_0\widetilde{\gamma}$ | $(g^3)$ | (L) | (M) |
| 45 | 10 | 14 | $d_0\widetilde{\beta^2}$ | $(g^2)$ | 3 | 4 |
| 47 | 11 | 14 | $d_0\widetilde{\delta'}$ | $(g^2)$ | 0 | 0 |
| 48 | 10 | 16 | $e_0\widetilde{\beta^2}$ | $(g^2)$ | (E) | (F) |
| 49 | 9 | 17 | $i(h_1w_2)$ | $(g^2, gw_1)$ | (A) | (G) |
| 50 | 10 | 18 | $i(h_1^2w_2)$ | $(g)$ | 0 | 0 |
| 50 | 11 | 16 | $d_0\widetilde{\beta g}$ | $(g^2)$ | 0 | 0 |
| 51 | 9 | 19 | $i(h_2w_2)$ | – | (E) | (F) |
| 51 | 10 | 20 | $h_1w_2\widetilde{h_1}$ | $(g)$ | (B) | (J) |
| 52 | 11 | 18 | $h_1^2w_2\widetilde{h_1}$ | $(g)$ | 0 | 0 |
| 54 | 10 | 21 | $i(h_2^2w_2)$ | $(g, w_1)$ | 0 | 0 |

Table B.3: Summands in $(E_3(tmf/2), d_3)$ (cont.)

| $t-s$ | $s$ | $g$ | $x$ | Ann$(x)$ | Type$(x)$ | Type$(xw_2^2)$ |
|---|---|---|---|---|---|---|
| 55 | 12 | 18 | $\alpha^2\widetilde{\beta^2}$ | $(g^2)$ | 5 | 6 |
| 56 | 11 | 21 | $\beta^2\widetilde{\gamma}$ | $(g^2w_1)$ | (N) | (B) |
| 56 | 11 | 22 | $i(c_0w_2)$ | $(g)$ | 0 | 0 |
| 57 | 12 | $20+21$ | $\gamma^2\widetilde{h_2^2}$ | $(g, w_1)$ | 0 | 0 |
| 57 | 13 | 18 | $d_0e_0\widetilde{\gamma}$ | $(g^2)$ | 0 | 0 |
| 58 | 12 | 23 | $h_1w_2\widetilde{c_0}$ | $(g)$ | 0 | 0 |
| 62 | 14 | 22 | $\widetilde{g\alpha^2e_0}$ | $(g^2)$ | (I) | (A) |
| 66 | 12 | 28 | $i(h_2\beta w_2)$ | $(g, w_1)$ | 0 | 0 |
| 74 | 14 | 33 | $i(h_1\gamma w_2)$ | $(g)$ | 0 | 0 |
| 75 | 14 | 35 | $h_1w_2\widetilde{\gamma}$ | $(g)$ | (J) | (K) |
| 76 | 15 | 35 | $h_1^2w_2\widetilde{\gamma}$ | $(g)$ | 0 | 0 |
| 80 | 15 | 38 | $i(\delta w_2)$ | $(g)$ | 0 | 0 |
| 81 | 16 | 39 | $i(h_1\delta w_2)$ | $-$ | (K) | (N) |
| 81 | 17 | 36 | $h_1w_2\widetilde{d_0e_0}$ | $(g)$ | (G) | (H) |
| 82 | 16 | 41 | $h_1w_2\widetilde{\delta'}$ | $(g)$ | 0 | 0 |

Table B.4: Two-term complexes $gw_1\colon R_2/(g^2)\{x\} \to R_2/(g^2)\{y\}$ in $E_3(tmf/2)$

| $n$ | $x$ | $y$ |
|---|---|---|
| 1 | $\alpha\widetilde{\gamma}$ | $\widetilde{c_0}$ |
| 2 | $\alpha w_2^2\widetilde{\gamma}$ | $w_2^2\widetilde{c_0}$ |
| 3 | $d_0\widetilde{\beta^2}$ | $d_0\widetilde{h_1}$ |
| 4 | $d_0w_2^2\widetilde{\beta^2}$ | $d_0w_2^2\widetilde{h_1}$ |
| 5 | $\alpha^2\widetilde{\beta^2}$ | $i(\alpha d_0)$ |
| 6 | $\alpha^2w_2^2\widetilde{\beta^2}$ | $i(\alpha d_0w_2^2)$ |

We treat the remaining 14 complexes individually. Complex (A) is

$$
\begin{array}{ccccc}
\langle gw_2^2\widetilde{\alpha^2e_0}\rangle & \xrightarrow{\begin{pmatrix} 0 \\ gw_1^2 \end{pmatrix}} & \langle i(h_1w_2)\rangle \oplus \langle i(\gamma w_2^2)\rangle & \xrightarrow{(g^2w_1\ g^6)} & \langle i(1)\rangle \\
\| & & \| & & \| \\
R_2/(g^2) & & R_2/(g^2, gw_1) \oplus R_2/(g^2w_1) & & R_2/(g^3w_1)
\end{array}
$$

The classes $gw_2^2\widetilde{\alpha^2 e_0}$, $i(h_1w_2)$ and $i(\gamma w_2^2)$ are replaced by $g^2w_2^2\widetilde{\alpha^2 e_0}$, $i(h_1gw_2)$ and $i(\gamma w_1 w_2^2)$, respectively, leaving

$$\begin{aligned} \langle i(1)\rangle &\cong R_2/(g^6, g^2w_1) \\ \langle i(h_1gw_2)\rangle &\cong R_2/(g, w_1) \\ \langle i(\gamma w_1 w_2^2)\rangle &\cong R_2/(g^2, gw_1) \\ \langle g^2w_2^2\widetilde{\alpha^2 e_0}\rangle &\cong R_2/(g) \end{aligned}$$

at $E_4$. Complex (B) is

$$\begin{array}{ccccc} \langle \beta^2 w_2^2\widetilde{\gamma}\rangle & \xrightarrow{\binom{g^6}{0}} & \langle\widetilde{\beta^2}\rangle \oplus \langle h_1w_2\widetilde{h_1}\rangle & \xrightarrow{(\,gw_1\ g^2w_1\,)} & \langle\widetilde{h_1}\rangle \\ \| & & \| & & \| \\ R_2/(g^2w_1) & & R_2/(g^2w_1)\oplus R_2/(g) & & R_2/(g^3) \end{array}$$

The classes $\beta^2w_2^2\widetilde{\gamma}$, $\widetilde{\beta^2}$ and $h_1w_2\widetilde{h_1}$ do not individually survive to $E_4$, being replaced by $\beta^2w_1w_2^2\widetilde{\gamma}$ and $\gamma\widetilde{\gamma} = 10_{19} + 10_{20} = g\widetilde{\beta^2} + h_1w_2\widetilde{h_1}$, leaving

$$\begin{aligned} \langle\widetilde{h_1}\rangle &\cong R_2/(g^3, gw_1) \\ \langle\gamma\widetilde{\gamma}\rangle &\cong R_2/(g^5, gw_1) \\ \langle\beta^2w_1w_2^2\widetilde{\gamma}\rangle &\cong R_2/(g^2) \end{aligned}$$

at $E_4$. Complex (C) is

$$\begin{array}{ccc} \langle i(e_0)\rangle & \xrightarrow{w_1} & \langle i(c_0)\rangle \\ \| & & \| \\ R_2/(g^3) & & R_2/(g) \end{array}$$

The class $i(e_0)$ does not survive to $E_4$, being replaced by $i(e_0g)$, leaving

$$\begin{aligned} \langle i(c_0)\rangle &\cong R_2/(g, w_1) \\ \langle i(e_0g)\rangle &\cong R_2/(g^2) \end{aligned}$$

at $E_4$. Complex (D) is isomorphic to complex (C) under multiplication by $w_2^2$. Therefore, the class $i(e_0w_2^2)$ does not survive to $E_4$, being replaced by $i(e_0gw_2^2)$, leaving

$$\begin{aligned} \langle i(c_0w_2^2)\rangle &\cong R_2/(g, w_1) \\ \langle i(e_0gw_2^2)\rangle &\cong R_2/(g^2)\,. \end{aligned}$$

Complex (E) is

$$\begin{array}{ccccc} \langle e_0\widetilde{\beta^2}\rangle & \xrightarrow{\binom{gw_1}{0}} & \langle e_0\widetilde{h_1}, i(h_2w_2)\rangle & \xrightarrow{(\,w_1\ 0\,)} & \langle h_1\widetilde{c_0}\rangle \\ \| & & \| & & \| \\ R_2/(g^2) & & \dfrac{R_2\oplus R_2}{\langle (g^2, w_1), (0, g)\rangle} & & R_2/(g) \end{array}$$

The class $e_0\widetilde{\beta^2}$ does not survive to $E_4$, while $e_0\widetilde{h_1}$ is replaced by $e_0g\widetilde{h_1}$, leaving

$$\langle h_1\widetilde{c_0}\rangle \cong R_2/(g, w_1)$$

and

$$\langle e_0 g\widetilde{h_1}, i(h_2w_2)\rangle \cong \frac{R_2 \oplus R_2}{\langle (w_1, 0), (g, w_1), (0, g)\rangle}.$$

Complex (F) is isomorphic to complex (E) under multiplication by $w_2^2$. Therefore, the class $e_0 w_2^2 \widetilde{\beta^2}$ does not survive to $E_4$, while $e_0 w_2^2 \widetilde{h_1}$ is replaced by $e_0 g w_2^2 \widetilde{h_1}$, leaving

$$\langle h_1 w_2^2 \widetilde{c_0}\rangle \cong R_2/(g, w_1)$$

and

$$\langle e_0 g w_2^2 \widetilde{h_1}, i(h_2 w_2^3)\rangle \cong \frac{R_2 \oplus R_2}{\langle (w_1, 0), (g, w_1), (0, g)\rangle}.$$

Complex (G) is

$$\begin{array}{ccccc} \langle h_1 w_2 \widetilde{d_0 e_0}\rangle \oplus \langle i(h_1 w_2^3)\rangle & \xrightarrow{\begin{pmatrix} g^2 w_1 & 0 \\ 0 & g^2 w_1 \end{pmatrix}} & \langle \widetilde{d_0 e_0}\rangle \oplus \langle i(w_2^2)\rangle & \xrightarrow{\begin{pmatrix} w_1^2 & g^4 \end{pmatrix}} & \langle i(\beta)\rangle \\ \| & & \| & & \| \\ R_2/(g) \oplus R_2/(g^2, g w_1) & & R_2/(g^3) \oplus R_2/(g^3 w_1) & & R_2/(g^2 w_1) \end{array}$$

The class $h_1 w_2 \widetilde{d_0 e_0}$ does not survive to $E_4$, while $i(h_1 w_2^3)$, $\widetilde{d_0 e_0}$ and $i(w_2^2)$ are replaced by $i(h_1 g w_2^3)$, $g^2 \widetilde{d_0 e_0}$, and $i(w_1 w_2^2)$, respectively, leaving

$$\begin{aligned} \langle i(\beta)\rangle &\cong R_2/(g^4, g^2 w_1, w_1^2) \\ \langle g^2 \widetilde{d_0 e_0}\rangle &\cong R_2/(g, w_1) \\ \langle i(w_1 w_2^2)\rangle &\cong R_2/(g^2) \\ \langle i(h_1 g w_2^3)\rangle &\cong R_2/(g, w_1). \end{aligned}$$

Complex (H) is

$$\begin{array}{ccccccc} \langle h_1 w_2^3 \widetilde{d_0 e_0}\rangle & \xrightarrow{g^2 w_1} & \langle w_2^2 \widetilde{d_0 e_0}\rangle & \xrightarrow{w_1^2} & \langle i(\beta w_2^2)\rangle & \xrightarrow{g^4} & \langle i(\beta^2)\rangle \\ \| & & \| & & \| & & \| \\ R_2/(g) & & R_2/(g^3) & & R_2/(g^2 w_1) & & R_2/(g^2 w_1) \end{array}$$

The class $h_1 w_2^3 \widetilde{d_0 e_0}$ does not survive to $E_4$, while $w_2^2 \widetilde{d_0 e_0}$ is replaced by $g^2 w_2^2 \widetilde{d_0 e_0}$ and $i(\beta w_2^2)$ is replaced by $i(\beta w_1 w_2^2)$, leaving

$$\begin{aligned} \langle i(\beta^2)\rangle &\cong R_2/(g^4, g^2 w_1) \\ \langle i(\beta w_1 w_2^2)\rangle &\cong R_2/(g^2, w_1) \\ \langle g^2 w_2^2 \widetilde{d_0 e_0}\rangle &\cong R_2/(g, w_1). \end{aligned}$$

Complex (I) is

$$\begin{array}{ccc} \langle g\widetilde{\alpha^2 e_0}\rangle \oplus \langle i(\beta^2 w_2^2)\rangle & \xrightarrow{\begin{pmatrix} g w_1^2 & g^5 \end{pmatrix}} & \langle i(\gamma)\rangle \\ \| & & \| \\ R_2/(g^2) \oplus R_2/(g^2 w_1) & & R_2/(g^2 w_1) \end{array}$$

The classes $g\widetilde{\alpha^2 e_0}$ and $i(\beta^2 w_2^2)$ are replaced by $g^2 \widetilde{\alpha^2 e_0}$ and $i(\beta^2 w_1 w_2^2)$, respectively, leaving

$$\langle i(\gamma)\rangle \cong R_2/(g^5, g^2 w_1, g w_1^2)$$

$$\langle g^2\widetilde{\alpha^2 e_0}\rangle \cong R_2/(g)$$
$$\langle i(\beta^2 w_1 w_2^2)\rangle \cong R_2/(g^2)\,.$$

Complex (J) is

$$\begin{array}{ccc} \langle h_1 w_2\widetilde{\gamma}\rangle \oplus \langle w_2^2\widetilde{\beta^2}\rangle \oplus \langle h_1 w_2^3\widetilde{h_1}\rangle & \xrightarrow{\begin{pmatrix} g^2w_1 & g^5 & 0 \\ 0 & gw_1 & g^2w_1 \end{pmatrix}} & \langle \widetilde{\gamma}\rangle \oplus \langle w_2^2\widetilde{h_1}\rangle \\ \| & & \| \\ R_2/(g) \oplus R_2/(g^2w_1) \oplus R_2/(g) & & R_2/(g^3w_1) \oplus R_2/(g^3) \end{array}$$

The class $h_1 w_2\widetilde{\gamma}$ does not survive to $E_4$, while $w_2^2\widetilde{\beta^2}$ and $h_1 w_2^3\widetilde{h_1}$ are replaced by $gw_1w_2^2\widetilde{\beta^2} + h_1w_1w_2^3\widetilde{h_1}$. As in complex (B), we use the relation $\gamma\widetilde{\gamma} = g\widetilde{\beta^2} + h_1 w_2\widetilde{h_1}$ to rewrite this sum as $\gamma w_1 w_2^2\widetilde{\gamma} = gw_1w_2^2\widetilde{\beta^2} + h_1w_1w_2^3\widetilde{h_1}$. The homology is then the sum of the non-cyclic summand

$$\langle \widetilde{\gamma}, w_2^2\widetilde{h_1}\rangle \cong \frac{R_2 \oplus R_2}{\langle (g^2w_1, 0), (g^5, gw_1), (0, g^3), (0, g^2w_1)\rangle}$$

and

$$\langle \gamma w_1 w_2^2\widetilde{\gamma}\rangle \cong R_2/(g)\,.$$

Complex (K) is

$$\begin{array}{ccccccc} \langle h_1 w_2^3\widetilde{\gamma}\rangle & \xrightarrow{g^2w_1} & \langle w_2^2\widetilde{\gamma}\rangle & \xrightarrow{\binom{g^4}{0}} & \langle \beta\widetilde{\gamma}, i(h_1\delta w_2)\rangle & \xrightarrow{(w_1\ 0)} & \langle i(\delta')\rangle \\ \| & & \| & & \| & & \| \\ R_2/(g) & & R_2/(g^3w_1) & & \dfrac{R_2 \oplus R_2}{\langle (g^2w_1, w_1), (0, g)\rangle} & & R_2/(g^2) \end{array}$$

The class $h_1 w_2^3\widetilde{\gamma}$ does not survive to $E_4$, while $\beta\widetilde{\gamma}$ and $w_2^2\widetilde{\gamma}$ are replaced by $\beta g^2\widetilde{\gamma}$ and $w_1 w_2^2\widetilde{\gamma}$, respectively, leaving

$$\langle i(\delta')\rangle \cong R_2/(g^2, w_1)$$
$$\langle w_1 w_2^2\widetilde{\gamma}\rangle \cong R_2/(g^2)$$

and

$$\langle \beta g^2\widetilde{\gamma}, i(h_1\delta w_2)\rangle \cong \frac{R_2 \oplus R_2}{\langle (g^2, 0), (w_1, w_1), (0, g)\rangle}\,.$$

Changing generators and using the relation $\gamma^2\widetilde{\beta^2} = 16_{38} + 16_{39} = \beta g^2\widetilde{\gamma} + i(h_1\delta w_2)$, the apparently non-cyclic module is the direct sum of

$$\langle \gamma^2\widetilde{\beta^2}\rangle \cong R_2/(g^2, w_1)$$

and

$$\langle i(h_1\delta w_2)\rangle \cong R_2/(g)\,.$$

Complex (L) is

$$\begin{array}{ccc} \langle e_0\widetilde{\gamma}\rangle & \xrightarrow{w_1} & \langle h_1\widetilde{\delta'}\rangle \\ \| & & \| \\ R_2/(g^3) & & R_2/(g) \end{array}$$

The class $e_0\widetilde{\gamma}$ is replaced by $e_0 g\widetilde{\gamma}$, leaving

$$\langle h_1\widetilde{\delta'}\rangle \cong R_2/(g, w_1)$$

$$\langle e_0 g\widetilde{\gamma}\rangle \cong R_2/(g^2)\,.$$

Complex (M) is isomorphic to complex (L) under multiplication by $w_2^2$. Therefore, the class $e_0 w_2^2\widetilde{\gamma}$ is replaced by $e_0 g w_2^2\widetilde{\gamma}$, leaving

$$\langle h_1 w_2^2\widetilde{\delta'}\rangle \cong R_2/(g, w_1)$$
$$\langle e_0 g w_2^2\widetilde{\gamma}\rangle \cong R_2/(g^2)\,.$$

Complex (N) is

$$\begin{array}{ccc} \langle \beta w_2^2\widetilde{\gamma}, i(h_1\delta w_2^3)\rangle & \xrightarrow{\begin{pmatrix} g^4 & 0 \\ w_1 & 0\end{pmatrix}} & \langle \beta^2\widetilde{\gamma}\rangle \oplus \langle i(\delta' w_2^2)\rangle \\ \| & & \| \\ \dfrac{R_2\oplus R_2}{\langle (g^2w_1, w_1), (0, g)\rangle} & & R_2/(g^2w_1)\oplus R_2/(g^2) \end{array}$$

The class $\beta w_2^2\widetilde{\gamma}$ does not survive to $E_4$, though $g^2w_1\cdot\beta w_2^2\widetilde{\gamma} = w_1\cdot i(h_1\delta w_2^3)$ does, leaving the non-cyclic summand

$$\langle \beta^2\widetilde{\gamma}, i(\delta' w_2^2)\rangle \cong \frac{R_2\oplus R_2}{\langle (g^2w_1, 0), (g^4, w_1), (0, g^2)\rangle}$$

and

$$\langle i(h_1\delta w_2^3)\rangle \cong R_2/(g)$$

at $E_4$.

### B.3. Calculation of $E_5(tmf/2) = H(E_4(tmf/2), d_4)$

The $(E_4, d_4)$-term of the Adams spectral sequence for $tmf/2$ splits as a direct sum of 24 $R_2$-module complexes of length two, plus a large summand with trivial differential. The length two complexes are of 11 types labeled (A) to (K), and the Type-column in Table B.5 gives the label of the complex containing the $R_2$-module generator $x$. If there is more than one complex of a given type, indices are added to distinguish them, as in (B1), …, (B8) for the 8 complexes of isomorphism type (B).

Table B.5: Summands in $(E_4(tmf/2), d_4)$

| $t-s$ | $s$ | $g$ | $x$ | $\mathrm{Ann}(x)$ | $\mathrm{Type}(x)$ |
|---|---|---|---|---|---|
| 0 | 0 | 0 | $i(1)$ | $(g^6, g^2w_1)$ | (A) |
| 1 | 1 | 0 | $i(h_1)$ | $(g^2, gw_1)$ | 0 |
| 2 | 1 | 1 | $\widetilde{h_1}$ | $(g^3, gw_1)$ | 0 |
| 2 | 2 | 0 | $i(h_1^2)$ | $(g)$ | 0 |
| 3 | 1 | 2 | $i(h_2)$ | $(g, w_1)$ | 0 |
| 3 | 2 | 1 | $h_1\widetilde{h_1}$ | $(g)$ | 0 |
| 4 | 3 | 0 | $h_1^2\widetilde{h_1}$ | $(g)$ | 0 |
| 6 | 2 | 2 | $i(h_2^2)$ | $(g, w_1)$ | 0 |
| 7 | 2 | 3 | $\widetilde{h_2^2}$ | $(g^2)$ | (B1) |
| 8 | 3 | 1 | $i(c_0)$ | $(g, w_1)$ | 0 |

Table B.5: Summands in $(E_4(tmf/2), d_4)$ (cont.)

| $t-s$ | $s$ | $g$ | $x$ | $\mathrm{Ann}(x)$ | $\mathrm{Type}(x)$ |
|---|---|---|---|---|---|
| 9 | 3 | 2 | $\widetilde{c_0}$ | $(g^2, gw_1)$ | (C1) |
| 9 | 4 | 1 | $i(h_1c_0)$ | $(g, w_1)$ | 0 |
| 10 | 4 | 2 | $h_1\widetilde{c_0}$ | $(g, w_1)$ | 0 |
| 14 | 4 | 3 | $i(d_0)$ | $(g^2)$ | (B2) |
| 15 | 3 | 4 | $i(\beta)$ | $(g^4, g^2w_1, w_1^2)$ | 0 |
| 16 | 5 | 3 | $d_0\widetilde{h_1}$ | $(g^2, gw_1)$ | (D1) |
| 18 | 4 | 5 | $i(h_2\beta)$ | $(g, w_1)$ | 0 |
| 21 | 6 | 4 | $d_0\widetilde{h_2^2}$ | $(g^2)$ | (E) |
| 24 | 6 | 5 | $i(\alpha^2)$ | $(g^2)$ | (B1) |
| 25 | 5 | 7 | $i(\gamma)$ | $(g^5, g^2w_1, gw_1^2)$ | 0 |
| 26 | 5 | 8 | $\widetilde{\gamma}$ | – | (F) |
| 26 | 6 | 6 | $i(h_1\gamma)$ | $(g)$ | 0 |
| 26 | 7 | 5 | $i(\alpha d_0)$ | $(g^2, gw_1)$ | (C1) |
| 27 | 6 | 8 | $h_1\widetilde{\gamma}$ | $(g)$ | 0 |
| 28 | 7 | 6 | $h_1^2\widetilde{\gamma}$ | $(g)$ | 0 |
| 30 | 6 | 9 | $i(\beta^2)$ | $(g^4, g^2w_1)$ | (E) |
| 31 | 8 | 6 | $i(d_0e_0)$ | $(g^2)$ | (B2) |
| 32 | 7 | $8+9$ | $i(\delta')$ | $(g^2, w_1)$ | 0 |
| 32 | 7 | 9 | $i(\delta)$ | $(g)$ | 0 |
| 33 | 7 | 10 | $\widetilde{\delta'}$ | $(g^2)$ | (B3) |
| 33 | 8 | 8 | $i(h_1\delta)$ | $(g, w_1)$ | 0 |
| 33 | 9 | 7 | $h_1\widetilde{d_0e_0}$ | $(g)$ | (D1) |
| 34 | 8 | 10 | $h_1\widetilde{\delta'}$ | $(g, w_1)$ | 0 |
| 37 | 8 | 11 | $i(e_0g)$ | $(g^2)$ | (A) |
| 39 | 9 | 11 | $e_0g\widetilde{h_1}$ | – | 0 |
| 40 | 9 | 12 | $d_0\widetilde{\gamma}$ | $(g^2)$ | (B4) |
| 47 | 11 | 14 | $d_0\widetilde{\delta'}$ | $(g^2)$ | (G) |
| 50 | 10 | 18 | $i(h_1^2w_2)$ | $(g)$ | (E) |
| 50 | 11 | 16 | $d_0\widetilde{\beta g}$ | $(g^2)$ | (B3) |
| 51 | 9 | 19 | $i(h_2w_2)$ | – | 0 |
| 51 | 10 | $19+20$ | $\gamma\widetilde{\gamma}$ | $(g^5, gw_1)$ | 0 |
| 52 | 11 | 18 | $h_1^2w_2\widetilde{h_1}$ | $(g)$ | 0 |

Table B.5: Summands in $(E_4(tmf/2), d_4)$ (cont.)

| $t-s$ | $s$ | $g$ | $x$ | $\mathrm{Ann}(x)$ | $\mathrm{Type}(x)$ |
|---|---|---|---|---|---|
| 54 | 10 | 21 | $i(h_2^2 w_2)$ | $(g, w_1)$ | 0 |
| 56 | 11 | 21 | $\beta^2\widetilde{\gamma}$ | – | (G) |
| 56 | 11 | 22 | $i(c_0 w_2)$ | $(g)$ | 0 |
| 57 | 12 | $20+21$ | $\gamma^2\widetilde{h_2^2}$ | $(g, w_1)$ | 0 |
| 57 | 13 | 18 | $d_0 e_0\widetilde{\gamma}$ | $(g^2)$ | (B4) |
| 58 | 12 | $22+23$ | $\delta'\widetilde{\gamma}$ | $(g)$ | (H1) |
| 58 | 12 | 23 | $\delta\widetilde{\gamma}$ | $(g)$ | 0 |
| 63 | 13 | 25 | $e_0 g\widetilde{\gamma}$ | $(g^2)$ | (F) |
| 65 | 14 | 25 | $d_0 g\widetilde{\beta^2}$ | $(g)$ | (I1) |
| 66 | 12 | 28 | $i(h_2\beta w_2)$ | $(g, w_1)$ | 0 |
| 69 | 13 | 30 | $i(h_1 g w_2)$ | $(g, w_1)$ | 0 |
| 72 | 16 | 28 | $g^2\widetilde{d_0 e_0}$ | $(g, w_1)$ | 0 |
| 74 | 14 | 33 | $i(h_1\gamma w_2)$ | $(g)$ | (I1) |
| 75 | 16 | 31 | $\alpha^2 g\widetilde{\beta^2}$ | $(g)$ | (H1) |
| 76 | 15 | 35 | $h_1^2 w_2\widetilde{\gamma}$ | $(g)$ | (G) |
| 80 | 15 | 38 | $i(\delta w_2)$ | $(g)$ | 0 |
| 81 | 16 | $38+39$ | $\gamma^2\widetilde{\beta^2}$ | $(g^2, w_1)$ | 0 |
| 81 | 16 | 39 | $i(h_1\delta w_2)$ | $(g)$ | 0 |
| 82 | 16 | 41 | $h_1 w_2\widetilde{\delta'}$ | $(g)$ | 0 |
| 82 | 18 | 34 | $g^2\widetilde{\alpha^2 e_0}$ | $(g)$ | (I1) |
| 97 | 17 | 50 | $i(h_1 w_2^2)$ | $(g^2, g w_1)$ | 0 |
| 98 | 17 | 51 | $w_2^2\widetilde{h_1}$ | – | (F) |
| 98 | 18 | 53 | $i(h_1^2 w_2^2)$ | $(g)$ | 0 |
| 99 | 17 | 52 | $i(h_2 w_2^2)$ | $(g, w_1)$ | 0 |
| 99 | 18 | 55 | $h_1 w_2^2\widetilde{h_1}$ | $(g)$ | 0 |
| 100 | 19 | 55 | $h_1^2 w_2^2\widetilde{h_1}$ | $(g)$ | 0 |
| 102 | 18 | 56 | $i(h_2^2 w_2^2)$ | $(g, w_1)$ | 0 |
| 103 | 18 | 57 | $w_2^2\widetilde{h_2^2}$ | $(g^2)$ | (B5) |
| 104 | 19 | 59 | $i(c_0 w_2^2)$ | $(g, w_1)$ | 0 |
| 104 | 20 | 58 | $i(w_1 w_2^2)$ | $(g^2)$ | (J1) |
| 105 | 19 | 60 | $w_2^2\widetilde{c_0}$ | $(g^2, g w_1)$ | (C2) |
| 105 | 20 | 60 | $i(h_1 c_0 w_2^2)$ | $(g, w_1)$ | 0 |

Table B.5: Summands in $(E_4(tmf/2), d_4)$ (cont.)

| $t-s$ | $s$ | $g$ | $x$ | $\mathrm{Ann}(x)$ | $\mathrm{Type}(x)$ |
|---|---|---|---|---|---|
| 106 | 20 | 62 | $h_1 w_2^2 \widetilde{c_0}$ | $(g, w_1)$ | 0 |
| 110 | 20 | 65 | $i(d_0 w_2^2)$ | $(g^2)$ | (B6) |
| 112 | 21 | 67 | $d_0 w_2^2 \widetilde{h_1}$ | $(g^2, gw_1)$ | (D2) |
| 114 | 20 | 67 | $i(h_2 \beta w_2^2)$ | $(g, w_1)$ | 0 |
| 117 | 22 | 72 | $d_0 w_2^2 \widetilde{h_2^2}$ | $(g^2)$ | (K1) |
| 119 | 23 | 74 | $i(\beta w_1 w_2^2)$ | $(g^2, w_1)$ | 0 |
| 120 | 22 | 75 | $i(\alpha^2 w_2^2)$ | $(g^2)$ | (B5) |
| 122 | 22 | 76 | $i(h_1 \gamma w_2^2)$ | $(g)$ | 0 |
| 122 | 23 | 77 | $i(\alpha d_0 w_2^2)$ | $(g^2, gw_1)$ | (C2) |
| 123 | 22 | 78 | $h_1 w_2^2 \widetilde{\gamma}$ | $(g)$ | 0 |
| 124 | 23 | 80 | $h_1^2 w_2^2 \widetilde{\gamma}$ | $(g)$ | 0 |
| 127 | 24 | 82 | $i(d_0 e_0 w_2^2)$ | $(g^2)$ | (B6) |
| 128 | 23 | $82+83$ | $i(\delta' w_2^2)$ | – | (G) |
| 128 | 23 | 83 | $i(\delta w_2^2)$ | $(g)$ | 0 |
| 129 | 23 | 84 | $w_2^2 \widetilde{\delta'}$ | $(g^2)$ | (B7) |
| 129 | 24 | 86 | $i(h_1 \delta w_2^2)$ | $(g, w_1)$ | 0 |
| 129 | 25 | $84+85$ | $i(\gamma w_1 w_2^2)$ | $(g^2, gw_1)$ | 0 |
| 129 | 25 | 85 | $h_1 w_2^2 \widetilde{d_0 e_0}$ | $(g)$ | (D2) |
| 130 | 24 | 88 | $h_1 w_2^2 \widetilde{\delta'}$ | $(g, w_1)$ | 0 |
| 130 | 25 | 87 | $w_1 w_2^2 \widetilde{\gamma}$ | $(g^2)$ | (J2) |
| 133 | 24 | 89 | $i(e_0 g w_2^2)$ | $(g^2)$ | (J1) |
| 134 | 26 | 91 | $i(\beta^2 w_1 w_2^2)$ | $(g^2)$ | (K1) |
| 135 | 25 | 93 | $e_0 g w_2^2 \widetilde{h_1}$ | – | 0 |
| 136 | 25 | 94 | $d_0 w_2^2 \widetilde{\gamma}$ | $(g^2)$ | (B8) |
| 143 | 27 | 103 | $d_0 w_2^2 \widetilde{\delta'}$ | $(g^2)$ | (K2) |
| 146 | 26 | 104 | $i(h_1^2 w_2^3)$ | $(g)$ | (K1) |
| 146 | 27 | 106 | $d_0 w_2^2 \widetilde{\beta g}$ | $(g^2)$ | (B7) |
| 147 | 25 | 101 | $i(h_2 w_2^3)$ | – | 0 |
| 148 | 27 | 108 | $h_1^2 w_2^3 \widetilde{h_1}$ | $(g)$ | 0 |
| 150 | 26 | 107 | $i(h_2^2 w_2^3)$ | $(g, w_1)$ | 0 |
| 152 | 27 | 112 | $i(c_0 w_2^3)$ | $(g)$ | 0 |
| 153 | 28 | $114+115$ | $\gamma^2 w_2^2 \widetilde{h_2^2}$ | $(g, w_1)$ | 0 |

Table B.5: Summands in $(E_4(tmf/2), d_4)$ (cont.)

| $t-s$ | $s$ | $g$ | $x$ | $\mathrm{Ann}(x)$ | Type$(x)$ |
|---|---|---|---|---|---|
| 153 | 29 | 115 | $d_0e_0w_2^2\widetilde{\gamma}$ | $(g^2)$ | (B8) |
| 154 | 28 | $116+117$ | $\delta' w_2^2\widetilde{\gamma}$ | $(g)$ | (H2) |
| 154 | 28 | 117 | $\delta w_2^2\widetilde{\gamma}$ | $(g)$ | 0 |
| 155 | 30 | $118+119$ | $\gamma w_1 w_2^2\widetilde{\gamma}$ | $(g)$ | 0 |
| 159 | 29 | 123 | $e_0gw_2^2\widetilde{\gamma}$ | $(g^2)$ | (J2) |
| 160 | 31 | 124 | $\beta^2w_1w_2^2\widetilde{\gamma}$ | $(g^2)$ | (K2) |
| 161 | 30 | 127 | $d_0gw_2^2\widetilde{\beta^2}$ | $(g)$ | (I2) |
| 162 | 28 | 122 | $i(h_2\beta w_2^3)$ | $(g, w_1)$ | 0 |
| 165 | 29 | 128 | $i(h_1gw_2^3)$ | $(g, w_1)$ | 0 |
| 168 | 32 | 137 | $g^2w_2^2\widetilde{d_0e_0}$ | $(g, w_1)$ | 0 |
| 170 | 30 | 135 | $i(h_1\gamma w_2^3)$ | $(g)$ | (I2) |
| 171 | 32 | 141 | $\alpha^2gw_2^2\widetilde{\beta^2}$ | $(g)$ | (H2) |
| 172 | 31 | 141 | $h_1^2w_2^3\widetilde{\gamma}$ | $(g)$ | (K2) |
| 176 | 31 | 144 | $i(\delta w_2^3)$ | $(g)$ | 0 |
| 177 | 32 | 149 | $i(h_1\delta w_2^3)$ | $(g)$ | 0 |
| 178 | 32 | 151 | $h_1w_2^3\widetilde{\delta'}$ | $(g)$ | 0 |
| 178 | 34 | 151 | $g^2w_2^2\widetilde{\alpha^2e_0}$ | $(g)$ | (I2) |

The complex of type (A) is

$$\begin{array}{ccc} \langle i(e_0g)\rangle & \xrightarrow{\ gw_1^2\ } & \langle i(1)\rangle \\ \| & & \| \\ R_2/(g^2) & & R_2/(g^6, g^2w_1) \end{array}$$

The class $i(e_0g)$ is replaced by $i(e_0g^2)$, leaving

$$\langle i(1)\rangle \cong R_2/(g^6, g^2w_1, gw_1^2)$$
$$\langle i(e_0g^2)\rangle \cong R_2/(g)$$

at $E_5$. Type (B) complexes have the form

$$\begin{array}{ccc} \langle x\rangle & \xrightarrow{\ w_1^2\ } & \langle y\rangle \\ \| & & \| \\ R_2/(g^2) & & R_2/(g^2) \end{array}$$

There are eight such summands in $(E_4(tmf/2), d_4)$, with $x$ and $y$ as in Table B.6, each leaving

$$\langle y\rangle \cong R_2/(g^2, w_1^2)$$

at $E_5$.

Table B.6: Summands of type (B)

| Type | $x$ | $y$ |
|---|---|---|
| (B1) | $i(\alpha^2)$ | $\widetilde{h_2^2}$ |
| (B2) | $i(d_0e_0)$ | $i(d_0)$ |
| (B3) | $d_0\widetilde{\beta g}$ | $\widetilde{\delta}'$ |
| (B4) | $d_0e_0\widetilde{\gamma}$ | $d_0\widetilde{\gamma}$ |
| (B5) | $i(\alpha^2 w_2^2)$ | $w_2^2\widetilde{h_2^2}$ |
| (B6) | $i(d_0e_0w_2^2)$ | $i(d_0w_2^2)$ |
| (B7) | $d_0w_2^2\widetilde{\beta g}$ | $w_2^2\widetilde{\delta}'$ |
| (B8) | $d_0e_0w_2^2\widetilde{\gamma}$ | $d_0w_2^2\widetilde{\gamma}$ |

The type (C) complexes are

$$\begin{array}{ccc} \langle i(\alpha d_0)\rangle & \xrightarrow{w_1^2} & \langle \widetilde{c_0}\rangle \\ \| & & \| \\ R_2/(g^2, gw_1) & & R_2/(g^2, gw_1) \end{array}$$

and its $w_2^2$-multiple, leaving

$$\begin{aligned} \langle \widetilde{c_0}\rangle &\cong R_2/(g^2, gw_1, w_1^2) \\ \langle i(\alpha d_0 g)\rangle &\cong R_2/(g, w_1) \\ \langle w_2^2\widetilde{c_0}\rangle &\cong R_2/(g^2, gw_1, w_1^2) \\ \langle i(\alpha d_0 g w_2^2)\rangle &\cong R_2/(g, w_1) \end{aligned}$$

at $E_5$. The type (D) complexes are

$$\begin{array}{ccc} \langle h_1\widetilde{d_0e_0}\rangle & \xrightarrow{w_1^2} & \langle d_0\widetilde{h_1}\rangle \\ \| & & \| \\ R_2/(g) & & R_2/(g^2, gw_1) \end{array}$$

and its $w_2^2$-multiple, leaving

$$\begin{aligned} \langle d_0\widetilde{h_1}\rangle &\cong R_2/(g^2, gw_1, w_1^2) \\ \langle d_0w_2^2\widetilde{h_1}\rangle &\cong R_2/(g^2, gw_1, w_1^2)\,. \end{aligned}$$

The complex of type (E) is

$$\begin{array}{ccc} \langle i(\beta^2)\rangle \oplus \langle i(h_1^2w_2)\rangle & \xrightarrow{(\,w_1\ gw_1\,)} & \langle d_0\widetilde{h_2^2}\rangle \\ \| & & \| \\ R_2/(g^4, g^2w_1) \oplus R_2/(g) & & R_2/(g^2) \end{array}$$

The classes $i(\beta^2)$ and $i(h_1^2 w_2)$ are replaced by $g \cdot i(\beta^2) + i(h_1^2 w_2) = 10_{17} + 10_{18} = i(\gamma^2)$. We choose the latter name for this class, leaving

$$\begin{aligned} \langle d_0 \widetilde{h_2^2} \rangle &\cong R_2/(g^2, w_1) \\ \langle i(\gamma^2) \rangle &\cong R_2/(g^3, gw_1) \,. \end{aligned}$$

The complex of type (F) is

$$\begin{array}{ccc} \langle e_0 g\widetilde{\gamma} \rangle & \xrightarrow{\binom{gw_1^2}{0}} & \langle \widetilde{\gamma}, \widetilde{w_2^2 h_1} \rangle \\ \| & & \| \\ R_2/(g^2) & & \dfrac{R_2 \oplus R_2}{\langle (g^2 w_1, 0), (g^5, gw_1), (0, g^3), (0, g^2 w_1) \rangle} \end{array}$$

The class $e_0 g\widetilde{\gamma}$ is replaced by $e_0 g^2 \widetilde{\gamma}$ leaving

$$\langle \widetilde{\gamma}, \widetilde{w_2^2 h_1} \rangle \cong \frac{R_2 \oplus R_2}{\langle (g^2 w_1, 0), (gw_1^2, 0), (g^5, gw_1), (0, g^3), (0, g^2 w_1) \rangle}$$

and

$$\langle e_0 g^2 \widetilde{\gamma} \rangle \cong R_2/(g) \,.$$

The complex of type (G) is

$$\begin{array}{ccc} \langle \beta^2 \widetilde{\gamma}, i(\delta' w_2^2) \rangle \oplus \langle h_1^2 w_2 \widetilde{\gamma} \rangle & \xrightarrow{(\, w_1 \;\; 0 \;\; gw_1 \,)} & \langle d_0 \widetilde{\delta'} \rangle \\ \| & & \| \\ \dfrac{R_2 \oplus R_2}{\langle (g^2 w_1, 0), (g^4, w_1), (0, g^2) \rangle} \oplus R_2/(g) & & R_2/(g^2) \end{array}$$

The classes $\beta^2 \widetilde{\gamma}$ and $h_1^2 w_2 \widetilde{\gamma}$ are replaced by $g \cdot \beta^2 \widetilde{\gamma} + h_1^2 w_2 \widetilde{\gamma} = 15_{34} + 15_{35} = \gamma^2 \widetilde{\gamma}$. We choose the shorter name for this class, leaving

$$\langle d_0 \widetilde{\delta'} \rangle \cong R_2/(g^2, w_1)$$

and

$$\langle \gamma^2 \widetilde{\gamma}, i(\delta' w_2^2) \rangle \cong \frac{R_2 \oplus R_2}{\langle (gw_1, 0), (g^3, w_1), (0, g^2) \rangle} \,.$$

The complexes of type (H) are

$$\begin{array}{ccc} \langle \alpha^2 g \widetilde{\beta^2} \rangle & \xrightarrow{w_1^2} & \langle \delta' \widetilde{\gamma} \rangle \\ \| & & \| \\ R_2/(g) & & R_2/(g) \end{array}$$

and its $w_2^2$-multiple, leaving

$$\begin{aligned} \langle \delta' \widetilde{\gamma} \rangle &\cong R_2/(g, w_1^2) \\ \langle \delta' w_2^2 \widetilde{\gamma} \rangle &\cong R_2/(g, w_1^2) \end{aligned}$$

at $E_5$. The complexes of type (I) are

$$\begin{array}{ccc} \langle i(h_1 \gamma w_2) \rangle \oplus \langle \widetilde{g^2 \alpha^2 e_0} \rangle & \xrightarrow{(\, w_1 \;\; w_1^2 \,)} & \langle d_0 g \widetilde{\beta^2} \rangle \\ \| & & \| \\ R_2/(g) \oplus R_2/(g) & & R_2/(g) \end{array}$$

and its $w_2^2$-multiple. The individual classes $i(h_1\gamma w_2)$ and $g^2\widetilde{\alpha^2 e_0}$ are replaced by the sum $w_1 \cdot i(h_1\gamma w_2) + g^2\widetilde{\alpha^2 e_0} = 18_{34} + 18_{35} = \gamma^2\widetilde{d_0 e_0}$. We use the latter, shorter, name for this class, leaving

$$\begin{aligned} \langle d_0 g\widetilde{\beta^2}\rangle &\cong R_2/(g, w_1) \\ \langle \gamma^2 \widetilde{d_0 e_0}\rangle &\cong R_2/(g) \\ \langle d_0 g w_2^2 \widetilde{\beta^2}\rangle &\cong R_2/(g, w_1) \\ \langle \gamma^2 w_2^2 \widetilde{d_0 e_0}\rangle &\cong R_2/(g)\,. \end{aligned}$$

The complexes of type (J) are $i$ applied to

$$\begin{array}{ccc} \langle e_0 g w_2^2\rangle & \xrightarrow{g w_1} & \langle w_1 w_2^2\rangle \\ \| & & \| \\ R_2/(g^2) & & R_2/(g^2) \end{array}$$

and its product with $\widetilde{\gamma}$, leaving

$$\begin{aligned} \langle i(w_1 w_2^2)\rangle &\cong R_2/(g^2, g w_1) \\ \langle i(e_0 g^2 w_2^2)\rangle &\cong R_2/(g) \\ \langle w_1 w_2^2 \widetilde{\gamma}\rangle &\cong R_2/(g^2, g w_1) \\ \langle e_0 g^2 w_2^2 \widetilde{\gamma}\rangle &\cong R_2/(g)\,. \end{aligned}$$

There are two complexes of type (K). They have the form

$$\begin{array}{ccc} \langle x\rangle \oplus \langle y\rangle & \xrightarrow{\left(\begin{smallmatrix} w_1^2 & g w_1 \end{smallmatrix}\right)} & \langle z\rangle \\ \| & & \| \\ R_2/(g^2) \oplus R_2/(g) & & R_2/(g^2) \end{array}$$

with homology

$$\begin{aligned} \langle z\rangle &\cong R_2/(g^2, g w_1, w_1^2) \\ \langle g x + w_1 y\rangle &\cong R_2/(g) \end{aligned}$$

at $E_5$. These are shown in Table B.7 together with shorter descriptions of the homology class $gx + w_1 y$ in each case, stemming from the relation $\gamma^2 = \beta^2 g + h_1^2 w_2$.

Table B.7: Summands of type (K)

| Type | $x$ | $y$ | $z$ | $gx + w_1 y$ |
|---|---|---|---|---|
| (K1) | $i(\beta^2 w_1 w_2^2)$ | $i(h_1^2 w_2^3)$ | $d_0 w_2^2 \widetilde{h_2^2}$ | $i(\gamma^2 w_1 w_2^2) = 30_{115} + 30_{116}$ |
| (K2) | $\beta^2 w_1 w_2^2 \widetilde{\gamma}$ | $h_1^2 w_2^3 \widetilde{\gamma}$ | $d_0 w_2^2 \widetilde{\delta'}$ | $\gamma^2 w_1 w_2^2 \widetilde{\gamma} = 35_{153} + 35_{154}$ |

APPENDIX C

# Calculation of $E_r(tmf/\eta)$ for $r = 3, 4$

Recall from Definition 5.1 that $R_0 = \mathbb{F}_2[g, w_1, w_2]$, $R_1 = \mathbb{F}_2[g, w_1, w_2^2]$ and $R_2 = \mathbb{F}_2[g, w_1, w_2^4]$. Our calculations show that $E_2(tmf/\eta)$ is a complex of $R_1$-modules, and $E_3(tmf/\eta)$ is a complex of $R_2$-modules.

## C.1. Calculation of $E_3(tmf/\eta) = H(E_2(tmf/\eta), d_2)$

When regarded as a complex of $R_1$-modules, the $(E_2, d_2)$-term of the Adams spectral sequence for $tmf/\eta$ splits as a direct sum of twelve complexes of the form

$$R_1 \longrightarrow R_1 \oplus R_1 \longrightarrow R_1\,,$$

labeled (A1)–(A12), twenty-eight other complexes labeled (B1) to (I4), and a summand with trivial differential. Table C.1 gives, for each $R_0$-module generator $x$ of $E_2(tmf/\eta)$, the summands to which $x$ and $xw_2$ belong. Tables C.2 and C.3 describe the complexes (A1) to (A12) and their homology, respectively. The remaining complexes and their homology are described following this.

Table C.1: Summands in $(E_2(tmf/\eta), d_2)$

| $t-s$ | $s$ | $g$ | $x$ | Ann$(x)$ | Type$(x)$ | Type$(xw_2)$ |
|---|---|---|---|---|---|---|
| 0 | 0 | 0 | $i(1)$ | $(0)$ | (A1) | (A2) |
| 0 | 1 | 0 | $i(h_0)$ | $(g^2)$ | (B1) | (B2) |
| 0 | 2 | 0 | $i(h_0^2)$ | $(g^2)$ | (B3) | (B4) |
| 0 | $3+i$ | 0 | $i(h_0^{3+i})$ | $(g)$ | 0 | 0 |
| 2 | 1 | 1 | $\widehat{h_0}$ | $(0)$ | (C) | (A3) |
| 2 | 2 | 1 | $h_0\widehat{h_0}$ | $(g^2)$ | (B5) | (B6) |
| 2 | $3+i$ | 1 | $h_0^{2+i}\widehat{h_0}$ | $(g)$ | 0 | 0 |
| 3 | 1 | 2 | $i(h_2)$ | $(g)$ | (D) | (E) |
| 3 | 2 | 2 | $i(h_0h_2)$ | $(g)$ | (F1) | (F2) |
| 5 | 1 | 3 | $\widehat{h_2}$ | $(0)$ | (A4) | (D) |
| 5 | 2 | 3 | $h_0\widehat{h_2}$ | $(g)$ | (F3) | (F4) |
| 5 | 3 | 2 | $h_0^2\widehat{h_2}$ | $(g)$ | (F5) | (F6) |
| 6 | 2 | 4 | $i(h_2^2)$ | $(g)$ | (G1) | (G2) |
| 8 | 2 | 5 | $h_2\widehat{h_2}$ | $(g)$ | (G3) | (G4) |
| 8 | 3 | 3 | $i(c_0)$ | $(g)$ | (G5) | (G6) |

Table C.1: Summands in $(E_2(tmf/\eta), d_2)$ (cont.)

| $t-s$ | $s$ | $g$ | $x$ | $\mathrm{Ann}(x)$ | $\mathrm{Type}(x)$ | $\mathrm{Type}(xw_2)$ |
|---|---|---|---|---|---|---|
| 11 | 4 | 3 | $\widehat{h_1 c_0}$ | $(0)$ | (A5) | (A6) |
| 12 | 3 | 4 | $i(\alpha)$ | $(0)$ | (D) | (E) |
| 12 | 4 | 4 | $i(h_0\alpha)$ | $(g^2)$ | (F1) | (F2) |
| 12 | $5+i$ | 5 | $i(h_0^{2+i}\alpha)$ | $(g)$ | 0 | 0 |
| 14 | 3 | 5 | $\widehat{\alpha}$ | $(0)$ | (A4) | (D) |
| 14 | 4 | 5 | $i(d_0)$ | $(0)$ | (A7) | (A8) |
| 14 | 4 | 6 | $h_0\widehat{\alpha}$ | $(g^2)$ | (F3) | (F4) |
| 14 | 5 | 7 | $i(h_0 d_0)$ | $(g)$ | (H) | (C) |
| 14 | 5 | 8 | $h_0^2\widehat{\alpha}$ | $(g^2)$ | (F5) | (F6) |
| 14 | $6+i$ | 8 | $h_0^{3+i}\widehat{\alpha}$ | $(g)$ | 0 | 0 |
| 15 | 3 | 6 | $i(\beta)$ | $(0)$ | (H) | (C) |
| 15 | 4 | 7 | $i(h_0\beta)$ | $(g)$ | (G1) | (G2) |
| 16 | 5 | 9 | $d_0\widehat{h_0}$ | $(0)$ | (A9) | (H) |
| 17 | 3 | 7 | $\widehat{\beta}$ | $(0)$ | (A9) | (H) |
| 17 | 4 | $8+9$ | $i(e_0)$ | $(0)$ | (A10) | (A11) |
| 17 | 4 | 9 | $h_0\widehat{\beta}$ | $(g)$ | (G3) | (G4) |
| 17 | 5 | $10+11$ | $i(h_0 e_0)$ | $(g)$ | (G7) | (G8) |
| 17 | 5 | 11 | $h_0^2\widehat{\beta}$ | $(g)$ | (G5) | (G6) |
| 17 | 6 | 10 | $i(h_0^2 e_0)$ | $(g)$ | (I1) | (I2) |
| 18 | 4 | 10 | $i(h_2\beta)$ | $(g)$ | (I1) | (I2) |
| 19 | 5 | 12 | $d_0\widehat{h_2}$ | $(0)$ | (E) | (A12) |
| 19 | 6 | 11 | $h_0 d_0\widehat{h_2}$ | $(g)$ | (I3) | (I4) |
| 20 | 4 | 12 | $h_2\widehat{\beta}$ | $(g)$ | (I3) | (I4) |
| 20 | 5 | 14 | $h_0 h_2\widehat{\beta}$ | $(g)$ | 0 | 0 |
| 23 | 5 | 16 | $h_2^2\widehat{\beta}$ | $(g)$ | 0 | 0 |
| 24 | 6 | 14 | $i(\alpha^2)$ | $(0)$ | (A11) | (A5) |
| 24 | $7+i$ | 11 | $i(h_0^{1+i}\alpha^2)$ | $(g)$ | 0 | 0 |
| 26 | 6 | 15 | $\alpha\widehat{\alpha}$ | $(0)$ | (A10) | (A11) |
| 26 | 7 | 13 | $\alpha^2\widehat{h_0}$ | $(0)$ | (A12) | (A4) |
| 26 | 7 | 14 | $h_0\alpha\widehat{\alpha}$ | $(g)$ | (G7) | (G8) |
| 26 | $8+i$ | 15 | $h_0^{1+i}\alpha^2\widehat{h_0}$ | $(g)$ | 0 | 0 |
| 27 | 6 | 16 | $i(\alpha\beta)$ | $(0)$ | (A2) | (A7) |

Table C.1: Summands in $(E_2(tmf/\eta), d_2)$ (cont.)

| $t-s$ | $s$ | $g$ | $x$ | $\mathrm{Ann}(x)$ | Type$(x)$ | Type$(xw_2)$ |
|---|---|---|---|---|---|---|
| 28 | 7 | 15 | $d_0\widehat{\alpha}$ | (0) | (E) | (A12) |
| 29 | 6 | 17 | $\alpha\widehat{\beta}$ | (0) | (A1) | (A2) |
| 29 | 7 | 16 | $\alpha\beta\widehat{h_0}$ | (0) | (A3) | (A9) |
| 29 | 7 | 17 | $h_0\alpha\widehat{\beta}$ | $(g)$ | (B1) | (B2) |
| 29 | 8 | 19 | $h_0^2\alpha\widehat{\beta}$ | $(g)$ | (B3) | (B4) |
| 30 | 6 | 18 | $i(\beta^2)$ | (0) | (A6) | (A10) |
| 31 | 7 | 18 | $d_0\widehat{\beta}$ | (0) | (C) | (A3) |
| 31 | 8 | 21 | $h_0d_0\widehat{\beta}$ | $(g)$ | (B5) | (B6) |
| 32 | 6 | 19 | $\dot{\beta}\widehat{\beta}$ | (0) | (A5) | (A6) |
| 32 | 7 | 20 | $i(\delta)$ | $(g)$ | (D) | (E) |
| 36 | 8 | 25 | $\widehat{d_0g}$ | (0) | (A2) | (A7) |
| 36 | $9+i$ | 26 | $h_0^{1+i}\widehat{d_0g}$ | $(g)$ | 0 | 0 |
| 38 | 9 | 27 | $\alpha^2\widehat{\alpha}$ | (0) | (A3) | (A9) |
| 38 | $10+i$ | 26 | $h_0^{1+i}\alpha^2\widehat{\alpha}$ | $(g)$ | 0 | 0 |
| 39 | 8 | 27 | $\gamma\widehat{\alpha}$ | (0) | (A6) | (A10) |
| 41 | 9 | 29 | $\alpha^2\widehat{\beta}$ | (0) | (D) | (E) |
| 41 | 10 | 28 | $i(\alpha^2e_0)$ | (0) | (A8) | (A1) |
| 42 | 8 | 29 | $\gamma\widehat{\beta}$ | (0) | (A8) | (A1) |
| 43 | 10 | 29 | $\alpha d_0\widehat{\beta}$ | (0) | (A7) | (A8) |
| 44 | 9 | 31 | $\alpha\beta\widehat{\beta}$ | (0) | (H) | (C) |
| 47 | 9 | 33 | $\beta^2\widehat{\beta}$ | (0) | (A12) | (A4) |
| 53 | 12 | 41 | $d_0\gamma\widehat{\alpha}$ | (0) | (A11) | (A5) |

The complexes (A1) to (A12) have the form

$$R_1\{z\} \xrightarrow{\binom{b_0}{b_1}} R_1\{y_0\} \oplus R_1\{y_1\} \xrightarrow{(a_0\ a_1)} R_1\{x\}$$

with the $a_i$ and $b_i$ being monomials in $g$ and $w_1$. If we write $a_i = d \cdot a_i'$, with $d$ the greatest common divisor of $a_0$ and $a_1$, then we must have

$$\binom{b_0}{b_1} = c\binom{a_1'}{a_0'}$$

for some $c \in R_1$. The homology of the complex is then the sum of

$$\frac{R_1}{(a_0, a_1)}\{x\}$$

and

$$\frac{R_1}{(c)}\{a_1' y_0 + a_0' y_1\}\,.$$

Of course, this second summand is 0 when $c = 1$. When this occurs, the generator $x$ remains at the $E_3$-term and the generators $y_0$, $y_1$ and $z$ disappear. When $c \neq 1$, the generator $x$ remains at $E_3$, the generators $y_0$ and $y_1$ are replaced by $a_1' y_0 + a_0' y_1$, and $z$ disappears. In Table C.3, the superfluous entry $a_1' y_0 + a_0' y_1$ and its annihilator ideal (1) are replaced by dashes when $c = 1$. Similarly, in cases (A8) and (A9), the superfluous generators $x = i(\alpha^2 e_0)$ and $d_0\widehat{h_0}$, with annihilator ideal (1), are replaced by dashes. Note also that $\mathrm{Ann}(x) = (a_0, a_1)$ often has a simpler description, which we give in Table C.3.

Table C.2: Complexes (A1)–(A12) in $E_2(tmf/\eta)$

| | $z$ | $\binom{b_0}{b_1}$ | $y_0$ | $y_1$ | $(a_0\ a_1)$ | $x$ |
|---|---|---|---|---|---|---|
| (A1) | $\gamma w_2\widehat{\beta}$ | $\binom{g^3}{1}$ | $\alpha\widehat{\beta}$ | $i(\alpha^2 e_0 w_2)$ | $(gw_1\ g^4 w_1)$ | $i(1)$ |
| (A2) | $\alpha w_2\widehat{\beta}$ | $\binom{g^2}{gw_1}$ | $\widehat{d_0 g}$ | $i(w_2)$ | $(w_1\ g)$ | $i(\alpha\beta)$ |
| (A3) | $d_0 w_2\widehat{\beta}$ | $\binom{g^2}{gw_1}$ | $\alpha^2\widehat{\alpha}$ | $w_2\widehat{h_0}$ | $(w_1\ g)$ | $\alpha\beta\widehat{h_0}$ |
| (A4) | $\beta^2 w_2\widehat{\beta}$ | $\binom{g^4}{g}$ | $\widehat{\alpha}$ | $\alpha^2 w_2\widehat{h_0}$ | $(w_1\ g^3 w_1)$ | $\widehat{h_2}$ |
| (A5) | $d_0\gamma w_2\widehat{\alpha}$ | $\binom{g^3 w_1}{gw_1}$ | $\beta\widehat{\beta}$ | $i(\alpha^2 w_2)$ | $(g\ g^3)$ | $\widehat{h_1 c_0}$ |
| (A6) | $\beta w_2\widehat{\beta}$ | $\binom{g^2}{g}$ | $\gamma\widehat{\alpha}$ | $w_2\widehat{h_1 c_0}$ | $(w_1\ gw_1)$ | $i(\beta^2)$ |
| (A7) | $w_2\widehat{d_0 g}$ | $\binom{g^2}{w_1}$ | $\alpha d_0\widehat{\beta}$ | $i(\alpha\beta w_2)$ | $(gw_1\ g^3)$ | $i(d_0)$ |
| (A8) | $\alpha d_0 w_2\widehat{\beta}$ | $\binom{g^2 w_1}{gw_1}$ | $\gamma\widehat{\beta}$ | $i(d_0 w_2)$ | $(1\ g)$ | $i(\alpha^2 e_0)$ |
| (A9) | $\alpha^2 w_2\widehat{\alpha}$ | $\binom{g^3 w_1}{w_1}$ | $\widehat{\beta}$ | $\alpha\beta w_2\widehat{h_0}$ | $(1\ g^3)$ | $d_0\widehat{h_0}$ |
| (A10) | $\gamma w_2\widehat{\alpha}$ | $\binom{g^3}{w_1}$ | $\alpha\widehat{\alpha}$ | $i(\beta^2 w_2)$ | $(w_1\ g^3)$ | $i(e_0)$ |
| (A11) | $\alpha w_2\widehat{\alpha}$ | $\binom{g}{w_1}$ | $d_0\gamma\widehat{\alpha}$ | $i(e_0 w_2)$ | $(gw_1\ g^2)$ | $i(\alpha^2)$ |
| (A12) | $d_0 w_2\widehat{\alpha}$ | $\binom{gw_1}{w_1}$ | $\beta^2\widehat{\beta}$ | $d_0 w_2\widehat{h_2}$ | $(g\ g^2)$ | $\alpha^2\widehat{h_0}$ |

Table C.3: Nonzero homology of the complexes (A1)–(A12) in $E_2(tmf/\eta)$

| | $x$ | $\mathrm{Ann}(x)$ | $y = a_1' y_0 + a_0' y_1$ | $\mathrm{Ann}(y)$ |
|---|---|---|---|---|
| (A1) | $i(1)$ | $(gw_1)$ | – | – |
| (A2) | $i(\alpha\beta)$ | $(g, w_1)$ | $g\widehat{d_0 g} + i(w_1 w_2)$ | $(g)$ |
| (A3) | $\alpha\beta\widehat{h_0}$ | $(g, w_1)$ | $\alpha^2 g\widehat{\alpha} + w_1 w_2\widehat{h_0}$ | $(g)$ |
| (A4) | $\widehat{h_2}$ | $(w_1)$ | $g^3\widehat{\alpha} + \alpha^2 w_2\widehat{h_0}$ | $(g)$ |
| (A5) | $\widehat{h_1 c_0}$ | $(g)$ | $\beta g^2\widehat{\beta} + i(\alpha^2 w_2)$ | $(gw_1)$ |
| (A6) | $i(\beta^2)$ | $(w_1)$ | $\gamma g\widehat{\alpha} + w_2\widehat{h_1 c_0}$ | $(g)$ |

Table C.3: Nonzero homology of the complexes (A1)–(A12) in $E_2(tmf/\eta)$ (cont.)

| | $x$ | $\mathrm{Ann}(x)$ | $y = a_1'y_0 + a_0'y_1$ | $\mathrm{Ann}(y)$ |
|---|---|---|---|---|
| (A7) | $i(d_0)$ | $(g^3, gw_1)$ | – | – |
| (A8) | – | – | $\gamma g\widehat{\beta} + i(d_0w_2)$ | $(gw_1)$ |
| (A9) | – | – | $g^3\widehat{\beta} + \alpha\beta w_2\widehat{h_0}$ | $(w_1)$ |
| (A10) | $i(e_0)$ | $(g^3, w_1)$ | – | – |
| (A11) | $i(\alpha^2)$ | $(g^2, gw_1)$ | – | – |
| (A12) | $\alpha^2\widehat{h_0}$ | $(g)$ | $\beta^2 g\widehat{\beta} + d_0w_2\widehat{h_2}$ | $(w_1)$ |

Complexes (B1)–(B6) have the form

$$\begin{array}{ccc} \langle x\rangle & \xrightarrow{gw_1} & \langle y\rangle \\ \| & & \| \\ R_1/(g) & & R_1/(g^2) \end{array}$$

for the $x$ and $y$ in Table C.4.

Table C.4: Complexes (B1)–(B6) in $E_2(tmf/\eta)$

| | $x$ | $y$ |
|---|---|---|
| (B1) | $h_0\alpha\widehat{\beta}$ | $i(h_0)$ |
| (B2) | $h_0\alpha w_2\widehat{\beta}$ | $i(h_0w_2)$ |
| (B3) | $h_0^2\alpha\widehat{\beta}$ | $i(h_0^2)$ |
| (B4) | $h_0^2\alpha w_2\widehat{\beta}$ | $i(h_0^2w_2)$ |
| (B5) | $h_0d_0\widehat{\beta}$ | $h_0\widehat{h_0}$ |
| (B6) | $h_0d_0w_2\widehat{\beta}$ | $h_0w_2\widehat{h_0}$ |

The classes $x$ do not survive to $E_3$, while the classes $y$ remain, leaving

$$\begin{aligned} \langle i(h_0)\rangle &\cong R_1/(g^2, gw_1) \\ \langle i(h_0w_2)\rangle &\cong R_1/(g^2, gw_1) \\ \langle i(h_0^2)\rangle &\cong R_1/(g^2, gw_1) \\ \langle i(h_0^2w_2)\rangle &\cong R_1/(g^2, gw_1) \\ \langle h_0\widehat{h_0}\rangle &\cong R_1/(g^2, gw_1) \\ \langle h_0w_2\widehat{h_0}\rangle &\cong R_1/(g^2, gw_1) \end{aligned}$$

at $E_3$. Complex (C) is

$$\begin{array}{ccccc}
\langle \alpha\beta w_2\widehat{\beta}\rangle & \xrightarrow{\left(\begin{smallmatrix} g^3 \\ gw_1 \end{smallmatrix}\right)} & \langle d_0\widehat{\beta}\rangle \oplus \langle i(\beta w_2)\rangle & \xrightarrow{\left(\begin{smallmatrix} gw_1 & g^3 \\ 0 & 1 \end{smallmatrix}\right)} & \langle \widehat{h_0}\rangle \oplus \langle i(h_0 d_0 w_2)\rangle \\
\| & & \| & & \| \\
R_1 & & R_1 \oplus R_1 & & R_1 \oplus R_1/(g)
\end{array}$$

This is exact except at the right hand end, so that the classes $\alpha\beta w_2\widehat{\beta}$, $d_0\widehat{\beta}$ and $i(\beta w_2)$ do not survive to $E_3$, leaving

$$\langle \widehat{h_0}\rangle \cong R_1/(g^4, gw_1)$$

at $E_3$, together with a new relation $i(h_0 d_0 w_2) = g^3\widehat{h_0}$. Complex (D) is

$$\begin{array}{ccccccc}
\langle w_2\widehat{\alpha}\rangle & \xrightarrow{\left(\begin{smallmatrix} g \\ w_1 \end{smallmatrix}\right)} & \langle \alpha^2\widehat{\beta}\rangle \oplus \langle w_2\widehat{h_2}\rangle & \xrightarrow{\left(\begin{smallmatrix} gw_1 & g^2 \\ w_1 & 0 \end{smallmatrix}\right)} & \langle i(\alpha)\rangle \oplus \langle i(\delta)\rangle & \xrightarrow{(\,w_1\ 0\,)} & \langle i(h_2)\rangle \\
\| & & \| & & \| & & \| \\
R_1 & & R_1 \oplus R_1 & & R_1 \oplus R_1/(g) & & R_1/(g)
\end{array}$$

The complex is exact at the left two modules, so that the classes $w_2\widehat{\alpha}$, $\alpha^2\widehat{\beta}$ and $w_2\widehat{h_2}$ do not survive to $E_3$, while $i(\alpha)$ is replaced by $i(\alpha g)$, leaving

$$\begin{aligned}
\langle i(h_2)\rangle &\cong R_1/(g, w_1) \\
\langle i(\alpha g)\rangle &\cong R_1/(g) \\
\langle i(\delta')\rangle &\cong R_1/(g, w_1)
\end{aligned}$$

at $E_3$. Here $\delta' = \alpha g + \delta$, as in $E_3(tmf)$. Complex (E) is

$$\begin{array}{ccccc}
\langle \alpha^2 w_2\widehat{\beta}\rangle & \xrightarrow{\left(\begin{smallmatrix} g^3 \\ gw_1 \\ w_1 \end{smallmatrix}\right)} & \langle d_0\widehat{\alpha}\rangle \oplus \langle i(\alpha w_2)\rangle \oplus \langle i(\delta w_2)\rangle & \xrightarrow{\left(\begin{smallmatrix} w_1 & g^2 & 0 \\ 0 & w_1 & 0 \end{smallmatrix}\right)} & \langle d_0\widehat{h_2}\rangle \oplus \langle i(h_2 w_2)\rangle \\
\| & & \| & & \| \\
R_1 & & R_1 \oplus R_1 \oplus R_1/(g) & & R_1 \oplus R_1/(g)
\end{array}$$

The classes $\alpha^2 w_2\widehat{\beta}$, $d_0\widehat{\alpha}$ and $i(\alpha w_2)$ do not survive to $E_3$, leaving a non-cyclic summand and a cyclic summand

$$\begin{aligned}
\langle d_0\widehat{h_2}, i(h_2 w_2)\rangle &\cong \frac{R_1 \oplus R_1}{\langle (w_1, 0), (g^2, w_1), (0, g)\rangle} \\
\langle i(\delta w_2)\rangle &\cong R_1/(g)
\end{aligned}$$

at $E_3$. Complexes (F1)–(F6) have the form

$$\begin{array}{ccc}
\langle x\rangle & \xrightarrow{\ \ w_1\ \ } & \langle y\rangle \\
\| & & \| \\
R_1/(g^2) & & R_1/(g)
\end{array}$$

for the $x$ and $y$ in Table C.5. The classes $x$ are replaced by $gx$ at $E_3$, while the classes $y$ remain, leaving summands

$$\begin{aligned}
\langle y\rangle &\cong R_1/(g, w_1) \\
\langle gx\rangle &\cong R_1/(g)
\end{aligned}$$

for the classes shown in Table C.6 at $E_3$.

Table C.5: Complexes (F1)–(F6) in $E_2(tmf/\eta)$

| | $x$ | $y$ |
|---|---|---|
| (F1) | $i(h_0\alpha)$ | $i(h_0h_2)$ |
| (F2) | $i(h_0\alpha w_2)$ | $i(h_0h_2w_2)$ |
| (F3) | $h_0\widehat{\alpha}$ | $h_0\widehat{h_2}$ |
| (F4) | $h_0w_2\widehat{\alpha}$ | $h_0w_2\widehat{h_2}$ |
| (F5) | $h_0^2\widehat{\alpha}$ | $h_0^2\widehat{h_2}$ |
| (F6) | $h_0^2w_2\widehat{\alpha}$ | $h_0^2w_2\widehat{h_2}$ |

Table C.6: Generators of the homology of the complexes (F1)–(F6) in $E_2(tmf/\eta)$

| | $gx$ | $y$ |
|---|---|---|
| (F1) | $i(h_0\alpha g)$ | $i(h_0h_2)$ |
| (F2) | $i(h_0\alpha gw_2)$ | $i(h_0h_2w_2)$ |
| (F3) | $h_0g\widehat{\alpha}$ | $h_0\widehat{h_2}$ |
| (F4) | $h_0gw_2\widehat{\alpha}$ | $h_0w_2\widehat{h_2}$ |
| (F5) | $h_0^2g\widehat{\alpha}$ | $h_0^2\widehat{h_2}$ |
| (F6) | $h_0^2gw_2\widehat{\alpha}$ | $h_0^2w_2\widehat{h_2}$ |

Complexes (G1)–(G8) have the form

$$\begin{array}{ccc} \langle x\rangle & \xrightarrow{w_1} & \langle y\rangle \\ \| & & \| \\ R_1/(g) & & R_1/(g) \end{array}$$

for the $x$ and $y$ in Table C.7. The classes $x$ do not survive to $E_3$, while the classes $y$ remain, leaving summands

$$\langle y\rangle \cong R_1/(g, w_1)$$

for each $y$ in Table C.7 at $E_3$.

Table C.7: Complexes (G1)–(G8) in $E_2(tmf/\eta)$

| | $x$ | $y$ |
|---|---|---|
| (G1) | $i(h_0\beta)$ | $i(h_2^2)$ |
| (G2) | $i(h_0\beta w_2)$ | $i(h_2^2w_2)$ |
| (G3) | $h_0\widehat{\beta}$ | $h_2\widehat{h_2}$ |
| (G4) | $h_0w_2\widehat{\beta}$ | $h_2w_2\widehat{h_2}$ |

Table C.7: Complexes (G1)–(G8) in $E_2(tmf/\eta)$ (cont.)

| | $x$ | $y$ |
|---|---|---|
| (G5) | $h_0^2\widehat{\beta}$ | $i(c_0)$ |
| (G6) | $h_0^2 w_2\widehat{\beta}$ | $i(c_0 w_2)$ |
| (G7) | $h_0\alpha\widehat{\alpha}$ | $i(h_0 e_0)$ |
| (G8) | $h_0\alpha w_2\widehat{\alpha}$ | $i(h_0 e_0 w_2)$ |

Complex (H) is

$$\begin{array}{ccccccc} \langle w_2\widehat{\beta}\rangle & \xrightarrow{\binom{g}{1}} & \langle \alpha\beta\widehat{\beta}\rangle \oplus \langle d_0 w_2\widehat{h_0}\rangle & \xrightarrow{(gw_1\ g^2w_1)} & \langle i(\beta)\rangle & \xrightarrow{(1)} & \langle i(h_0 d_0)\rangle \\ \| & & \| & & \| & & \| \\ R_1 & & R_1 \oplus R_1 & & R_1 & & R_1/(g) \end{array}$$

The classes $w_2\widehat{\beta}$, $\alpha\beta\widehat{\beta}$ and $d_0 w_2\widehat{h_0}$ do not survive to $E_3$, while $i(\beta)$ is replaced by $i(\beta g)$, leaving

$$\langle i(\beta g)\rangle \cong R_1/(w_1)$$

at $E_3$. The acyclic complexes (I1) to (I4) are

$$\begin{array}{ccc} \langle i(h_2\beta)\rangle & \xrightarrow{1} & \langle i(h_0^2 e_0)\rangle \\ \| & & \| \\ R_1/(g) & & R_1/(g) \end{array}$$

and

$$\begin{array}{ccc} \langle h_2\widehat{\beta}\rangle & \xrightarrow{1} & \langle h_0 d_0\widehat{h_2}\rangle \\ \| & & \| \\ R_1/(g) & & R_1/(g) \end{array}$$

and their isomorphic images under multiplication by $w_2$. The classes $i(h_2\beta)$, $i(h_2\beta w_2)$, $h_2\widehat{\beta}$ and $h_2 w_2\widehat{\beta}$ do not survive to $E_3$, while the classes $i(h_0^2 e_0)$, $i(h_0^2 e_0 w_2)$, $h_0 d_0\widehat{h_2}$ and $h_0 d_0 w_2\widehat{h_2}$ become 0, leaving no terms contributing to $E_3$.

## C.2. Calculation of $E_4(tmf/\eta) = H(E_3(tmf/\eta), d_3)$

The $E_3$-term for $tmf/\eta$ is the direct sum of a large complex with trivial differential together with fourteen complexes of seven types, which we label (A1) to (G).

Table C.8: Summands in $(E_3(tmf/\eta), d_3)$

| $t-s$ | $s$ | $g$ | $x$ | $\mathrm{Ann}(x)$ | $\mathrm{Type}(x)$ | $\mathrm{Type}(xw_2^2)$ |
|---|---|---|---|---|---|---|
| 0 | 0 | 0 | $i(1)$ | $(gw_1)$ | (A1) | (E1) |
| 0 | 1 | 0 | $i(h_0)$ | $(g^2, gw_1)$ | 0 | 0 |
| 0 | 2 | 0 | $i(h_0^2)$ | $(g^2, gw_1)$ | 0 | 0 |

Table C.8: Summands in $(E_3(tmf/\eta), d_3)$ (cont.)

| $t-s$ | $s$ | $g$ | $x$ | $\mathrm{Ann}(x)$ | Type($x$) | Type($xw_2^2$) |
|---|---|---|---|---|---|---|
| 0 | $3+i$ | 0 | $i(h_0^{3+i})$ | $(g)$ | 0 | 0 |
| 2 | 1 | 1 | $\widehat{h_0}$ | $(g^4, gw_1)$ | 0 | 0 |
| 2 | 2 | 1 | $h_0\widehat{h_0}$ | $(g^2, gw_1)$ | 0 | 0 |
| 2 | $3+i$ | 1 | $h_0^{2+i}\widehat{h_0}$ | $(g)$ | 0 | 0 |
| 3 | 1 | 2 | $i(h_2)$ | $(g, w_1)$ | 0 | 0 |
| 3 | 2 | 2 | $i(h_0h_2)$ | $(g, w_1)$ | 0 | 0 |
| 5 | 1 | 3 | $\widehat{h_2}$ | $(w_1)$ | (B1) | (A1) |
| 5 | 2 | 3 | $h_0\widehat{h_2}$ | $(g, w_1)$ | 0 | 0 |
| 5 | 3 | 2 | $h_0^2\widehat{h_2}$ | $(g, w_1)$ | 0 | 0 |
| 6 | 2 | 4 | $i(h_2^2)$ | $(g, w_1)$ | 0 | 0 |
| 8 | 2 | 5 | $h_2\widehat{h_2}$ | $(g, w_1)$ | 0 | 0 |
| 8 | 3 | 3 | $i(c_0)$ | $(g, w_1)$ | 0 | 0 |
| 11 | 4 | 3 | $\widehat{h_1c_0}$ | $(g)$ | (C1) | (C3) |
| 12 | $5+i$ | 5 | $i(h_0^{2+i}\alpha)$ | $(g)$ | 0 | 0 |
| 14 | 4 | 5 | $i(d_0)$ | $(g^3, gw_1)$ | (D1) | (D2) |
| 14 | $6+i$ | 8 | $h_0^{3+i}\widehat{\alpha}$ | $(g)$ | 0 | 0 |
| 17 | 4 | $8+9$ | $i(e_0)$ | $(g^3, w_1)$ | 0 | 0 |
| 17 | 5 | $10+11$ | $i(h_0e_0)$ | $(g, w_1)$ | 0 | 0 |
| 19 | 5 | 12 | $d_0\widehat{h_2}$ | $-$ | 0 | 0 |
| 20 | 5 | 14 | $h_0h_2\widehat{\beta}$ | $(g)$ | (C1) | (C3) |
| 23 | 5 | 16 | $h_2^2\widehat{\beta}$ | $(g)$ | (D1) | (D2) |
| 24 | 6 | 14 | $i(\alpha^2)$ | $(g^2, gw_1)$ | 0 | 0 |
| 24 | $7+i$ | 11 | $i(h_0^{1+i}\alpha^2)$ | $(g)$ | 0 | 0 |
| 26 | $7+i$ | 13 | $h_0^i\alpha^2\widehat{h_0}$ | $(g)$ | 0 | 0 |
| 27 | 6 | 16 | $i(\alpha\beta)$ | $(g, w_1)$ | 0 | 0 |
| 29 | 7 | 16 | $\alpha\beta\widehat{h_0}$ | $(g, w_1)$ | 0 | 0 |
| 30 | 6 | 18 | $i(\beta^2)$ | $(w_1)$ | (B2) | (B1) |
| 32 | 7 | $19+20$ | $i(\alpha g)$ | $(g)$ | 0 | 0 |
| 32 | 7 | 19 | $i(\delta')$ | $(g, w_1)$ | 0 | 0 |
| 32 | 8 | 22 | $i(h_0\alpha g)$ | $(g)$ | 0 | 0 |
| 34 | 8 | 24 | $h_0g\widehat{\alpha}$ | $(g)$ | 0 | 0 |
| 34 | 9 | 24 | $h_0^2g\widehat{\alpha}$ | $(g)$ | 0 | 0 |

Table C.8: Summands in $(E_3(tmf/\eta), d_3)$ (cont.)

| $t-s$ | $s$ | $g$ | $x$ | $\mathrm{Ann}(x)$ | $\mathrm{Type}(x)$ | $\mathrm{Type}(xw_2^2)$ |
|---|---|---|---|---|---|---|
| 35 | 7 | 22 | $i(\beta g)$ | $(w_1)$ | (E1) | (B2) |
| 36 | $9+i$ | 26 | $h_0^{1+i}\widehat{d_0 g}$ | $(g)$ | 0 | 0 |
| 38 | $10+i$ | 26 | $h_0^{1+i}\alpha^2\widehat{\alpha}$ | $(g)$ | 0 | 0 |
| 48 | 9 | 34 | $i(h_0 w_2)$ | $(g^2, gw_1)$ | 0 | 0 |
| 48 | 10 | 33 | $i(h_0^2 w_2)$ | $(g^2, gw_1)$ | 0 | 0 |
| 48 | $11+i$ | 34 | $i(h_0^{3+i} w_2)$ | $(g)$ | 0 | 0 |
| 50 | 10 | 36 | $h_0 w_2 \widehat{h_0}$ | $(g^2, gw_1)$ | 0 | 0 |
| 50 | $11+i$ | 36 | $h_0^{2+i} w_2 \widehat{h_0}$ | $(g)$ | 0 | 0 |
| 51 | 9 | 36 | $i(h_2 w_2)$ | $-$ | 0 | 0 |
| 51 | 10 | 37 | $i(h_0 h_2 w_2)$ | $(g, w_1)$ | 0 | 0 |
| 53 | 10 | 39 | $h_0 w_2 \widehat{h_2}$ | $(g, w_1)$ | 0 | 0 |
| 53 | 11 | 39 | $h_0^2 w_2 \widehat{h_2}$ | $(g, w_1)$ | 0 | 0 |
| 54 | 10 | 40 | $i(h_2^2 w_2)$ | $(g, w_1)$ | 0 | 0 |
| 56 | 10 | 41 | $h_2 w_2 \widehat{h_2}$ | $(g, w_1)$ | 0 | 0 |
| 56 | 11 | 42 | $i(c_0 w_2)$ | $(g, w_1)$ | 0 | 0 |
| 56 | 12 | $43+44$ | $g\widehat{d_0 g} + i(w_1 w_2)$ | $(g)$ | 0 | 0 |
| 58 | 13 | $46+47$ | $\alpha^2 g\widehat{\alpha} + w_1 w_2 \widehat{h_0}$ | $(g)$ | 0 | 0 |
| 59 | 12 | $46+47$ | $\gamma g\widehat{\alpha} + w_2\widehat{h_1 c_0}$ | $(g)$ | (C2) | (C4) |
| 60 | $13+i$ | 50 | $i(h_0^{2+i}\alpha w_2)$ | $(g)$ | 0 | 0 |
| 62 | 12 | $50+51$ | $\gamma g\widehat{\beta} + i(d_0 w_2)$ | $(gw_1)$ | (F) | (G) |
| 62 | $14+i$ | 53 | $h_0^{3+i} w_2\widehat{\alpha}$ | $(g)$ | 0 | 0 |
| 65 | 13 | $59+60$ | $i(h_0 e_0 w_2)$ | $(g, w_1)$ | 0 | 0 |
| 67 | 13 | $61+62$ | $\beta^2 g\widehat{\beta} + d_0 w_2 \widehat{h_2}$ | $(w_1)$ | (E2) | (F) |
| 68 | 13 | 64 | $h_0 h_2 w_2 \widehat{\beta}$ | $(g)$ | (C2) | (C4) |
| 71 | 13 | 66 | $h_2^2 w_2 \widehat{\beta}$ | $(g)$ | (F) | (G) |
| 72 | 14 | $65+66$ | $\beta g^2\widehat{\beta} + i(\alpha^2 w_2)$ | $(gw_1)$ | (A2) | (E2) |
| 72 | $15+i$ | 64 | $i(h_0^{1+i}\alpha^2 w_2)$ | $(g)$ | 0 | 0 |
| 74 | 15 | $66+67$ | $g^3\widehat{\alpha} + \alpha^2 w_2\widehat{h_0}$ | $(g)$ | 0 | 0 |
| 74 | $16+i$ | 72 | $h_0^{1+i}\alpha^2 w_2 \widehat{h_0}$ | $(g)$ | 0 | 0 |
| 77 | 15 | $71+72$ | $g^3\widehat{\beta} + \alpha\beta w_2 \widehat{h_0}$ | $(w_1)$ | (G) | (A2) |
| 80 | 15 | 76 | $i(\delta w_2)$ | $(g)$ | 0 | 0 |
| 80 | 16 | 83 | $i(h_0 \alpha g w_2)$ | $(g)$ | 0 | 0 |

Table C.8: Summands in $(E_3(tmf/\eta), d_3)$ (cont.)

| $t-s$ | $s$ | $g$ | $x$ | $\mathrm{Ann}(x)$ | Type$(x)$ | Type$(xw_2^2)$ |
|---|---|---|---|---|---|---|
| 82 | 16 | 86 | $h_0gw_2\widehat{\alpha}$ | $(g)$ | 0 | 0 |
| 82 | 17 | 87 | $h_0^2gw_2\widehat{\alpha}$ | $(g)$ | 0 | 0 |
| 84 | $17+i$ | 90 | $h_0^{1+i}w_2\widehat{d_0g}$ | $(g)$ | 0 | 0 |
| 86 | $18+i$ | 90 | $h_0^{1+i}\alpha^2w_2\widehat{\alpha}$ | $(g)$ | 0 | 0 |

Complexes (A1) and (A2) are

$$\begin{array}{ccc} \langle w_2^2\widehat{h_2}\rangle & \xrightarrow{g^5} & \langle i(1)\rangle \\ \| & & \| \\ R_2/(w_1) & & R_2/(gw_1) \end{array}$$

and

$$\begin{array}{ccc} \langle g^3w_2^2\widehat{\beta} + \alpha\beta w_2^3\widehat{h_0}\rangle & \xrightarrow{g^5} & \langle \beta g^2\widehat{\beta} + i(\alpha^2w_2)\rangle \\ \| & & \| \\ R_2/(w_1) & & R_2/(gw_1) \end{array}$$

The domain classes do not survive to $E_4$, while the targets persist, leaving

$$\langle i(1)\rangle \cong R_2/(g^5, gw_1)$$
$$\langle \beta g^2\widehat{\beta} + i(\alpha^2w_2)\rangle \cong R_2/(g^5, gw_1)$$

at $E_4$. Complexes (B1) and (B2) are

$$\begin{array}{ccc} \langle i(\beta^2w_2^2)\rangle & \xrightarrow{g^6} & \langle\widehat{h_2}\rangle \\ \| & & \| \\ R_2/(w_1) & & R_2/(w_1) \end{array}$$

and

$$\begin{array}{ccc} \langle i(\beta gw_2^2)\rangle & \xrightarrow{g^5} & \langle i(\beta^2)\rangle \\ \| & & \| \\ R_2/(w_1) & & R_2/(w_1) \end{array}$$

The domain classes do not survive to $E_4$, while the targets persist, leaving

$$\langle\widehat{h_2}\rangle \cong R_2/(g^6, w_1)$$
$$\langle i(\beta^2)\rangle \cong R_2/(g^5, w_1)$$

at $E_4$. Complexes (C1)–(C4) have the form

$$\begin{array}{ccc} \langle x\rangle & \xrightarrow{w_1} & \langle y\rangle \\ \| & & \| \\ R_2/(g) & & R_2/(g) \end{array}$$

for the $x$ and $y$ in Table C.9.

Table C.9: Complexes (C1)–(C4) in $E_3(tmf/\eta)$

| | $x$ | $y$ |
|---|---|---|
| (C1) | $h_0h_2\widehat{\beta}$ | $\widehat{h_1c_0}$ |
| (C2) | $h_0h_2w_2\widehat{\beta}$ | $\gamma g\widehat{\alpha} + w_2\widehat{h_1c_0}$ |
| (C3) | $h_0h_2w_2^2\widehat{\beta}$ | $w_2^2\widehat{h_1c_0}$ |
| (C4) | $h_0h_2w_2^3\widehat{\beta}$ | $\gamma gw_2^2\widehat{\alpha} + w_2^3\widehat{h_1c_0}$ |

The classes $x$ do not survive to $E_4$, while the classes $y$ remain, leaving

$$\begin{aligned}\langle\widehat{h_1c_0}\rangle &\cong R_2/(g, w_1)\\ \langle\gamma g\widehat{\alpha} + w_2\widehat{h_1c_0}\rangle &\cong R_2/(g, w_1)\\ \langle w_2^2\widehat{h_1c_0}\rangle &\cong R_2/(g, w_1)\\ \langle\gamma gw_2^2\widehat{\alpha} + w_2^3\widehat{h_1c_0}\rangle &\cong R_2/(g, w_1)\end{aligned}$$

at $E_4$. Complexes (D1) and (D2) are

$$\begin{array}{ccc}\langle h_2^2\widehat{\beta}\rangle & \xrightarrow{w_1} & \langle i(d_0)\rangle\\ \| & & \|\\ R_2/(g) & & R_2/(g^3, gw_1)\end{array}$$

and its $w_2^2$-multiple. The domain classes do not survive to $E_4$, while the targets persist, leaving

$$\begin{aligned}\langle i(d_0)\rangle &\cong R_2/(g^3, w_1)\\ \langle i(d_0w_2^2)\rangle &\cong R_2/(g^3, w_1)\end{aligned}$$

at $E_4$. Complexes (E1) and (E2) are

$$\begin{array}{ccc}\langle i(w_2^2)\rangle & \xrightarrow{g^3} & \langle i(\beta g)\rangle\\ \| & & \|\\ R_2/(gw_1) & & R_2/(w_1)\end{array}$$

and

$$\begin{array}{ccc}\langle \beta g^2w_2^2\widehat{\beta} + i(\alpha^2w_2^3)\rangle & \xrightarrow{g^5} & \langle\beta^2g\widehat{\beta} + d_0w_2\widehat{h_2}\rangle\\ \| & & \|\\ R_2/(gw_1) & & R_2/(w_1)\end{array}$$

The domain generators $x$ are replaced by $w_1x$ while the targets persist, leaving

$$\begin{aligned}\langle i(\beta g)\rangle &\cong R_2/(g^3, w_1)\\ \langle i(w_1w_2^2)\rangle &\cong R_2/(g)\end{aligned}$$

$$\langle \beta^2 g\widehat{\beta} + d_0 w_2 \widehat{h_2} \rangle \cong R_2/(g^5, w_1)$$
$$\langle \beta g^2 w_1 w_2^2 \widehat{\beta} + i(\alpha^2 w_1 w_2^3) \rangle \cong R_2/(g)$$

at $E_4$. Complex (F) is

$$\begin{array}{ccc} \langle \beta^2 g w_2^2 \widehat{\beta} + d_0 w_2^3 \widehat{h_2} \rangle \oplus \langle h_2^2 w_2 \widehat{\beta} \rangle & \xrightarrow{(g^5 \ w_1)} & \langle \gamma g \widehat{\beta} + i(d_0 w_2) \rangle \\ \| & & \| \\ R_2/(w_1) \oplus R_2/(g) & & R_2/(g w_1) \end{array}$$

The classes $\beta^2 g w_2^2 \widehat{\beta} + d_0 w_2^3 \widehat{h_2}$ and $h_2^2 w_2 \widehat{\beta}$ do not survive to $E_4$, leaving

$$\langle \gamma g \widehat{\beta} + i(d_0 w_2) \rangle \cong R_2/(g^5, w_1)$$

at $E_4$. Complex (G) is

$$\begin{array}{ccccc} \langle h_2^2 w_2^3 \widehat{\beta} \rangle & \xrightarrow{w_1} & \langle \gamma g w_2^2 \widehat{\beta} + i(d_0 w_2^3) \rangle & \xrightarrow{g^4} & \langle g^3 \widehat{\beta} + \alpha \beta w_2 \widehat{h_0} \rangle \\ \| & & \| & & \| \\ R_2/(g) & & R_2/(g w_1) & & R_2/(w_1) \end{array}$$

The classes $h_2^2 w_2^3 \widehat{\beta}$ and $\gamma g w_2^2 \widehat{\beta} + i(d_0 w_2^3)$ do not survive to $E_4$, leaving

$$\langle g^3 \widehat{\beta} + \alpha \beta w_2 \widehat{h_0} \rangle \cong R_2/(g^4, w_1)$$

at $E_4$.

APPENDIX D

# Calculation of $E_r(tmf/\nu)$ for $r = 3, 4, 5$

Recall from Definition 5.1 that $R_0 = \mathbb{F}_2[g, w_1, w_2]$, $R_1 = \mathbb{F}_2[g, w_1, w_2^2]$ and $R_2 = \mathbb{F}_2[g, w_1, w_2^4]$. Our calculations show that $E_2(tmf/\nu)$ is a complex of $R_1$-modules, while $E_3(tmf/\nu)$ and $E_4(tmf/\nu)$ are complexes of $R_2$-modules.

## D.1. Calculation of $E_3(tmf/\nu) = H(E_2(tmf/\nu), d_2)$

The $(E_2, d_2)$-term of the Adams spectral sequence for $tmf/\nu$ splits as a direct sum of 32 $R_1$-module complexes of length two or three, labeled (A1-4), (B1-18) and (C) to (L), plus a large summand with trivial differential. The Type-columns in Table D.1 give the labels of the complexes containing the $R_1$-module generators $x$ and $xw_2$. For each complex we discuss the passage to homology with respect to the $d_2$-differential, giving the transition from the $E_2$-term to the $E_3$-term.

Table D.1: Summands in $(E_2(tmf/\nu), d_2)$

| $t-s$ | $s$ | $g$ | $x$ | Ann$(x)$ | Type$(x)$ | Type$(xw_2)$ |
|---|---|---|---|---|---|---|
| 0 | 0 | 0 | $i(1)$ | $(0)$ | (A1) | (B1) |
| 0 | $1+i$ | 0 | $i(h_0^{1+i})$ | $(g)$ | 0 | 0 |
| 1 | 1 | 1 | $i(h_1)$ | $(g)$ | 0 | 0 |
| 2 | 2 | 1 | $i(h_1^2)$ | $(g)$ | 0 | 0 |
| 4 | $3+i$ | 1 | $h_0^i\overline{h_0^3}$ | $(g)$ | 0 | 0 |
| 5 | 1 | 2 | $\overline{h_1}$ | $(0)$ | (A2) | (B2) |
| 6 | 2 | 2 | $h_1\overline{h_1}$ | $(g)$ | 0 | 0 |
| 7 | 2 | 3 | $\overline{h_0h_2}$ | $(0)$ | (B1) | (B3) |
| 7 | 3 | 2 | $h_0\overline{h_0h_2}$ | $(g)$ | (C) | (E) |
| 8 | 3 | 3 | $i(c_0)$ | $(g)$ | 0 | 0 |
| 9 | 4 | 3 | $i(h_1c_0)$ | $(g)$ | (D) | (F) |
| 10 | 2 | 4 | $\overline{h_2^2}$ | $(0)$ | (D) | (F) |
| 12 | 3 | 4 | $\overline{c_0}$ | $(g)$ | 0 | 0 |
| 12 | 3 | $4+5$ | $i(\alpha)$ | $(0)$ | (B2) | (B4) |
| 12 | $4+i$ | 4 | $i(h_0^{1+i}\alpha)$ | $(g)$ | 0 | 0 |
| 13 | 4 | 5 | $h_1\overline{c_0}$ | $(g)$ | 0 | 0 |
| 14 | 4 | 6 | $i(d_0)$ | $(0)$ | (B3) | (B5) |

Table D.1: Summands in $(E_2(tmf/\nu), d_2)$ (cont.)

| $t-s$ | $s$ | $g$ | $x$ | Ann$(x)$ | Type$(x)$ | Type$(xw_2)$ |
|---|---|---|---|---|---|---|
| 14 | 5 | 6 | $i(h_0d_0)$ | $(g)$ | (E) | (G) |
| 15 | 3 | 6 | $i(\beta)$ | $(0)$ | (E) | (G) |
| 16 | 5 | 7 | $\overline{h_0^2\alpha}$ | $(0)$ | (C) | (E) |
| 16 | $6+i$ | 7 | $h_0^{1+i}\overline{h_0^2\alpha}$ | $(g)$ | 0 | 0 |
| 17 | 4 | 7 | $i(e_0)$ | $(0)$ | (F) | (B6) |
| 19 | 5 | 8 | $d_0\overline{h_1}$ | $(0)$ | (B4) | (B7) |
| 21 | 6 | 9 | $d_0\overline{h_0h_2}$ | $(0)$ | (B5) | (A1) |
| 22 | 5 | 9 | $e_0\overline{h_1}$ | $(0)$ | (G) | (B8) |
| 24 | 4 | 9 | $\overline{g}$ | $(0)$ | (A3) | (B9) |
| 24 | 5 | 10 | $h_0\overline{g}$ | $(g)$ | 0 | 0 |
| 24 | 6 | $10+11$ | $i(\alpha^2)$ | $(0)$ | (B6) | (B10) |
| 24 | 6 | 11 | $h_0^2\overline{g}$ | $(g)$ | 0 | 0 |
| 24 | $7+i$ | 11 | $i(h_0^{1+i}\alpha^2)$ | $(g)$ | 0 | 0 |
| 25 | 5 | 12 | $h_1\overline{g}$ | $(g)$ | 0 | 0 |
| 26 | 6 | 12 | $i(h_1\gamma)$ | $(g)$ | 0 | 0 |
| 26 | 7 | 12 | $i(\alpha d_0)$ | $(0)$ | (B7) | (A2) |
| 28 | $7+i$ | 13 | $h_0^i\overline{h_0\alpha^2}$ | $(g)$ | 0 | 0 |
| 29 | 5 | 13 | $\overline{\gamma}$ | $(0)$ | (A4) | (B11) |
| 29 | 7 | 14 | $i(\alpha e_0)$ | $(0)$ | (B8) | (C) |
| 30 | 6 | 15 | $h_1\overline{\gamma}$ | $(g)$ | 0 | 0 |
| 31 | 6 | 16 | $\overline{\alpha\beta}$ | $(0)$ | (B9) | (B12) |
| 31 | 7 | 15 | $h_0\overline{\alpha\beta}$ | $(g)$ | (H) | (J) |
| 31 | 8 | 15 | $i(d_0e_0)$ | $(0)$ | (B10) | (D) |
| 32 | 7 | 17 | $i(\delta)$ | $(g)$ | 0 | 0 |
| 33 | 8 | 17 | $i(h_1\delta)$ | $(g)$ | (I) | (K) |
| 34 | 6 | 17 | $\overline{\beta^2}$ | $(0)$ | (I) | (K) |
| 36 | 7 | 19 | $\overline{\delta}$ | $(g)$ | 0 | 0 |
| 36 | 7 | $19+20$ | $\alpha\overline{g}$ | $(0)$ | (B11) | (B13) |
| 36 | 8 | 19 | $h_0\overline{\delta}$ | $(g)$ | 0 | 0 |
| 36 | 9 | 20 | $h_0^2\overline{\delta}$ | $(g)$ | 0 | 0 |
| 36 | $10+i$ | 20 | $i(h_0^{1+i}\alpha^3)$ | $(g)$ | 0 | 0 |
| 37 | 8 | 21 | $h_1\overline{\delta}$ | $(g)$ | 0 | 0 |

Table D.1: Summands in $(E_2(tmf/\nu), d_2)$ (cont.)

| $t-s$ | $s$ | $g$ | $x$ | $\mathrm{Ann}(x)$ | Type($x$) | Type($xw_2$) |
|---|---|---|---|---|---|---|
| 38 | 8 | 22 | $d_0\overline{g}$ | (0) | (B12) | (B14) |
| 38 | 9 | 22 | $h_0d_0\overline{g}$ | $(g)$ | (J) | (L) |
| 39 | 7 | 21 | $\beta\overline{g}$ | (0) | (J) | (L) |
| 40 | 9 | 24 | $\overline{\alpha^3}$ | (0) | (H) | (J) |
| 40 | $10+i$ | 24 | $h_0^{1+i}\overline{\alpha^3}$ | $(g)$ | 0 | 0 |
| 41 | 8 | 24 | $e_0\overline{g}$ | (0) | (K) | (B15) |
| 43 | 9 | 26 | $d_0\overline{\gamma}$ | (0) | (B13) | (B16) |
| 45 | 10 | 28 | $d_0\overline{\alpha\beta}$ | (0) | (B14) | (A3) |
| 46 | 9 | 28 | $e_0\overline{\gamma}$ | (0) | (L) | (B17) |
| 48 | 10 | $30+31$ | $\alpha^2\overline{g}$ | (0) | (B15) | (B18) |
| 50 | 11 | 33 | $\alpha d_0\overline{g}$ | (0) | (B16) | (A4) |
| 53 | 11 | 36 | $\alpha^2\overline{\gamma}$ | (0) | (B17) | (H) |
| 55 | 12 | 38 | $d_0e_0\overline{g}$ | (0) | (B18) | (I) |

Type (A) complexes have the form

$$\begin{array}{ccc} \langle x\rangle & \xrightarrow{g^3w_1} & \langle y\rangle \\ \| & & \| \\ R_1 & & R_1 \end{array}$$

There are four such summands in $(E_2(tmf/\nu), d_2)$, with $x$ and $y$ as in Table D.2. The class $x$ does not survive, leaving the cyclic module

$$\langle y\rangle \cong R_1/(g^3w_1)$$

at $E_3$.

Table D.2: Summands of type (A)

| $n$ | $x$ | $y$ |
|---|---|---|
| 1 | $d_0w_2\overline{h_0h_2}$ | $i(1)$ |
| 2 | $i(\alpha d_0w_2)$ | $\overline{h_1}$ |
| 3 | $d_0w_2\overline{\alpha\beta}$ | $\overline{g}$ |
| 4 | $\alpha d_0w_2\overline{g}$ | $\overline{\gamma}$ |

Type (B) complexes have the form

$$\begin{array}{ccc} \langle x \rangle & \xrightarrow{g^2} & \langle y \rangle \\ \| & & \| \\ R_1 & & R_1 \end{array}$$

There are 18 such summands in $(E_2(tmf/\nu), d_2)$, with $x$ and $y$ as in Table D.3. The class $x$ does not survive, leaving the cyclic module

$$\langle y \rangle \cong R_1/(g^2)$$

at $E_3$.

Table D.3: Summands of type (B)

| $n$ | $x$ | $y$ |
|---|---|---|
| 1 | $i(w_2)$ | $\overline{h_0 h_2}$ |
| 2 | $w_2\overline{h_1}$ | $i(\alpha)$ |
| 3 | $w_2\overline{h_0 h_2}$ | $i(d_0)$ |
| 4 | $i(\alpha w_2)$ | $d_0\overline{h_1}$ |
| 5 | $i(d_0 w_2)$ | $d_0\overline{h_0 h_2}$ |
| 6 | $i(e_0 w_2)$ | $i(\alpha^2)$ |
| 7 | $d_0 w_2\overline{h_1}$ | $i(\alpha d_0)$ |
| 8 | $e_0 w_2\overline{h_1}$ | $i(\alpha e_0)$ |
| 9 | $w_2\overline{g}$ | $\overline{\alpha\beta}$ |
| 10 | $i(\alpha^2 w_2)$ | $i(d_0 e_0)$ |
| 11 | $w_2\overline{\gamma}$ | $\alpha\overline{g}$ |
| 12 | $w_2\overline{\alpha\beta}$ | $d_0\overline{g}$ |
| 13 | $\alpha w_2\overline{g}$ | $d_0\overline{\gamma}$ |
| 14 | $d_0 w_2\overline{g}$ | $d_0\overline{\alpha\beta}$ |
| 15 | $e_0 w_2\overline{g}$ | $\alpha^2\overline{g}$ |
| 16 | $d_0 w_2\overline{\gamma}$ | $\alpha d_0\overline{g}$ |
| 17 | $e_0 w_2\overline{\gamma}$ | $\alpha^2\overline{\gamma}$ |
| 18 | $\alpha^2 w_2\overline{g}$ | $d_0 e_0\overline{g}$ |

Complex (C) is

$$\begin{array}{ccccc} \langle i(\alpha e_0 w_2) \rangle & \xrightarrow{g^3} & \langle \overline{h_0^2\alpha} \rangle & \xrightarrow{w_1} & \langle h_0\overline{h_0 h_2} \rangle \\ \| & & \| & & \| \\ R_1 & & R_1 & & R_1/(g) \end{array}$$

The class $i(\alpha e_0 w_2)$ does not survive, and $\overline{h_0^2\alpha}$ is replaced by $g\overline{h_0^2\alpha}$, leaving

$$\langle h_0\overline{h_0h_2}\rangle \cong R_1/(g, w_1)$$
$$\langle g\overline{h_0^2\alpha}\rangle \cong R_1/(g^2)\,.$$

Complex (D) is

$$\begin{array}{ccccc} \langle i(d_0e_0w_2)\rangle & \xrightarrow{g^3w_1} & \langle\overline{h_2^2}\rangle & \xrightarrow{1} & \langle i(h_1c_0)\rangle \\ \| & & \| & & \| \\ R_1 & & R_1 & & R_1/(g) \end{array}$$

The classes $i(h_1c_0)$ and $i(d_0e_0w_2)$ do not survive, and $\overline{h_2^2}$ is replaced by $g\overline{h_2^2}$, leaving

$$\langle g\overline{h_2^2}\rangle \cong R_1/(g^2w_1)\,.$$

Complex (E) is

$$\begin{array}{ccccc} \langle w_2\overline{h_0^2\alpha}\rangle & \xrightarrow{\binom{g^2w_1}{w_1}} & \langle i(\beta)\rangle \oplus \langle h_0w_2\overline{h_0h_2}\rangle & \xrightarrow{(1\ 0)} & \langle i(h_0d_0)\rangle \\ \| & & \| & & \| \\ R_1 & & R_1 \oplus R_1/(g) & & R_1/(g) \end{array}$$

The classes $i(h_0d_0)$ and $w_2\overline{h_0^2\alpha}$ do not survive, and $i(\beta)$ is replaced by $i(\beta g)$, leaving the non-cyclic module

$$\langle i(\beta g), h_0w_2\overline{h_0h_2}\rangle \cong \frac{R_1 \oplus R_1}{\langle (gw_1, w_1), (0, g)\rangle}\,.$$

Complex (F) is

$$\begin{array}{ccc} \langle w_2\overline{h_2^2}\rangle & \xrightarrow{\binom{g^2}{1}} & \langle i(e_0)\rangle \oplus \langle i(h_1c_0w_2)\rangle \\ \| & & \| \\ R_1 & & R_1 \oplus R_1/(g) \end{array}$$

The class $w_2\overline{h_2^2}$ does not survive, and $i(h_1c_0w_2)$ becomes equal to $g^2 \cdot i(e_0)$, leaving

$$\langle i(e_0)\rangle \cong R_1/(g^3)\,.$$

Complex (G) is

$$\begin{array}{ccc} \langle i(\beta w_2)\rangle & \xrightarrow{\binom{g^2}{1}} & \langle e_0\overline{h_1}\rangle \oplus \langle i(h_0d_0w_2)\rangle \\ \| & & \| \\ R_1 & & R_1 \oplus R_1/(g) \end{array}$$

The class $i(\beta w_2)$ does not survive, and $i(h_0d_0w_2)$ becomes equal to $g^2 \cdot e_0\overline{h_1}$, leaving

$$\langle e_0\overline{h_1}\rangle \cong R_1/(g^3)\,.$$

Complex (H) is

$$\begin{array}{ccccc} \langle \alpha^2w_2\overline{\gamma}\rangle & \xrightarrow{g^3} & \langle\overline{\alpha^3}\rangle & \xrightarrow{w_1} & \langle h_0\overline{\alpha\beta}\rangle \\ \| & & \| & & \| \\ R_1 & & R_1 & & R_1/(g) \end{array}$$

The class $\alpha^2 w_2\overline{\gamma}$ does not survive, and $\overline{\alpha^3}$ is replaced by $g\overline{\alpha^3}$, leaving

$$\langle h_0\overline{\alpha\beta}\rangle \cong R_1/(g, w_1)$$
$$\langle g\overline{\alpha^3}\rangle \cong R_1/(g^2)\,.$$

Complex (I) is

$$\begin{array}{ccccc}\langle d_0e_0w_2\overline{g}\rangle & \xrightarrow{g^3w_1} & \langle\overline{\beta^2}\rangle & \xrightarrow{1} & \langle i(h_1\delta)\rangle \\ \| & & \| & & \| \\ R_1 & & R_1 & & R_1/(g)\end{array}$$

The classes $i(h_1\delta)$ and $d_0e_0w_2\overline{g}$ do not survive, and $\overline{\beta^2}$ is replaced by $g\overline{\beta^2}$, leaving

$$\langle g\overline{\beta^2}\rangle \cong R_1/(g^2w_1)\,.$$

Complex (J) is

$$\begin{array}{ccccc}\langle w_2\overline{\alpha^3}\rangle & \xrightarrow{\binom{g^2w_1}{w_1}} & \langle\beta\overline{g}\rangle \oplus \langle h_0w_2\overline{\alpha\beta}\rangle & \xrightarrow{(1\ 0)} & \langle h_0d_0\overline{g}\rangle \\ \| & & \| & & \| \\ R_1 & & R_1 \oplus R_1/(g) & & R_1/(g)\end{array}$$

The classes $h_0d_0\overline{g}$ and $w_2\overline{\alpha^3}$ do not survive, and $\beta\overline{g}$ is replaced by $\beta g\overline{g}$, leaving the non-cyclic module

$$\langle\beta g\overline{g}, h_0w_2\overline{\alpha\beta}\rangle \cong \frac{R_1 \oplus R_1}{\langle(gw_1, w_1), (0, g)\rangle}\,.$$

Complex (K) is

$$\begin{array}{ccc}\langle w_2\overline{\beta^2}\rangle & \xrightarrow{\binom{g^2}{1}} & \langle e_0\overline{g}\rangle \oplus \langle i(h_1\delta w_2)\rangle \\ \| & & \| \\ R_1 & & R_1 \oplus R_1/(g)\end{array}$$

The class $w_2\overline{\beta^2}$ does not survive, and $i(h_1\delta w_2)$ becomes equal to $g^2 \cdot e_0\overline{g}$, leaving

$$\langle e_0\overline{g}\rangle \cong R_1/(g^3)\,.$$

Complex (L) is

$$\begin{array}{ccc}\langle\beta w_2\overline{g}\rangle & \xrightarrow{\binom{g^2}{1}} & \langle e_0\overline{\gamma}\rangle \oplus \langle h_0d_0w_2\overline{g}\rangle \\ \| & & \| \\ R_1 & & R_1 \oplus R_1/(g)\end{array}$$

The class $\beta w_2\overline{g}$ does not survive, and $h_0d_0w_2\overline{g}$ becomes equal to $g^2 \cdot e_0\overline{\gamma}$, leaving

$$\langle e_0\overline{\gamma}\rangle \cong R_1/(g^3)\,.$$

## D.2. Calculation of $E_4(tmf/\nu) = H(E_3(tmf/\nu), d_3)$

The $(E_3, d_3)$-term of the Adams spectral sequence for $tmf/\nu$ splits as a direct sum of 20 $R_2$-module complexes of length two, three or four, labeled (A) to (H), plus a large summand with trivial differential. The Type-columns in Table D.4 give the labels of the complexes containing the $R_2$-module generators $x$ and $xw_2^2$.

Table D.4: Summands in $(E_3(tmf/\nu), d_3)$

| $t-s$ | $s$ | $g$ | $x$ | $\mathrm{Ann}(x)$ | $\mathrm{Type}(x)$ | $\mathrm{Type}(xw_2^2)$ |
|---|---|---|---|---|---|---|
| 0 | 0 | 0 | $i(1)$ | $(g^3 w_1)$ | (A) | (F) |
| 0 | $1+i$ | 0 | $i(h_0^{1+i})$ | $(g)$ | 0 | 0 |
| 1 | 1 | 1 | $i(h_1)$ | $(g)$ | 0 | 0 |
| 2 | 2 | 1 | $i(h_1^2)$ | $(g)$ | 0 | 0 |
| 4 | $3+i$ | 1 | $h_0^i \overline{h_0^3}$ | $(g)$ | 0 | 0 |
| 5 | 1 | 2 | $\overline{h_1}$ | $(g^3 w_1)$ | (B) | (A) |
| 6 | 2 | 2 | $h_1 \overline{h_1}$ | $(g)$ | 0 | 0 |
| 7 | 2 | 3 | $\overline{h_0 h_2}$ | $(g^2)$ | (C1) | (C6) |
| 7 | 3 | 2 | $h_0 \overline{h_0 h_2}$ | $(g, w_1)$ | 0 | 0 |
| 8 | 3 | 3 | $i(c_0)$ | $(g)$ | (D) | (H) |
| 12 | 3 | 4 | $\overline{c_0}$ | $(g)$ | 0 | 0 |
| 12 | 3 | $4+5$ | $i(\alpha)$ | $(g^2)$ | 0 | 0 |
| 12 | $4+i$ | 4 | $i(h_0^{1+i}\alpha)$ | $(g)$ | 0 | 0 |
| 13 | 4 | 5 | $h_1 \overline{c_0}$ | $(g)$ | (E1) | (E3) |
| 14 | 4 | 6 | $i(d_0)$ | $(g^2)$ | (C2) | (C7) |
| 16 | $6+i$ | 7 | $h_0^{1+i} \overline{h_0^2 \alpha}$ | $(g)$ | 0 | 0 |
| 17 | 4 | 7 | $i(e_0)$ | $(g^3)$ | (D) | (H) |
| 19 | 5 | 8 | $d_0 \overline{h_1}$ | $(g^2)$ | 0 | 0 |
| 21 | 6 | 9 | $d_0 \overline{h_0 h_2}$ | $(g^2)$ | (C3) | (C8) |
| 22 | 5 | 9 | $e_0 \overline{h_1}$ | $(g^3)$ | (E1) | (E3) |
| 24 | 4 | 9 | $\overline{g}$ | $(g^3 w_1)$ | (F) | (G) |
| 24 | 5 | 10 | $h_0 \overline{g}$ | $(g)$ | 0 | 0 |
| 24 | 6 | $10+11$ | $i(\alpha^2)$ | $(g^2)$ | (C4) | (C9) |
| 24 | 6 | 11 | $h_0^2 \overline{g}$ | $(g)$ | 0 | 0 |
| 24 | $7+i$ | 11 | $i(h_0^{1+i}\alpha^2)$ | $(g)$ | 0 | 0 |
| 25 | 5 | 12 | $h_1 \overline{g}$ | $(g)$ | 0 | 0 |
| 26 | 6 | 12 | $i(h_1 \gamma)$ | $(g)$ | 0 | 0 |
| 26 | 7 | 12 | $i(\alpha d_0)$ | $(g^2)$ | 0 | 0 |
| 28 | $7+i$ | 13 | $h_0^i \overline{h_0 \alpha^2}$ | $(g)$ | 0 | 0 |
| 29 | 5 | 13 | $\overline{\gamma}$ | $(g^3 w_1)$ | (A) | (F) |
| 29 | 7 | 14 | $i(\alpha e_0)$ | $(g^2)$ | 0 | 0 |
| 30 | 6 | 14 | $g \overline{h_2^2}$ | $(g^2 w_1)$ | (G) | (B) |

Table D.4: Summands in $(E_3(tmf/\nu), d_3)$ (cont.)

| $t-s$ | $s$ | $g$ | $x$ | $\mathrm{Ann}(x)$ | $\mathrm{Type}(x)$ | $\mathrm{Type}(xw_2^2)$ |
|---|---|---|---|---|---|---|
| 30 | 6 | 15 | $h_1\overline{\gamma}$ | $(g)$ | 0 | 0 |
| 31 | 6 | 16 | $\overline{\alpha\beta}$ | $(g^2)$ | 0 | 0 |
| 31 | 7 | 15 | $h_0\overline{\alpha\beta}$ | $(g, w_1)$ | 0 | 0 |
| 31 | 8 | 15 | $i(d_0e_0)$ | $(g^2)$ | (C5) | (C10) |
| 32 | 7 | 17 | $i(\delta)$ | $(g)$ | (E2) | (E4) |
| 35 | 7 | 18 | $i(\beta g)$ | – | (F) | (G) |
| 36 | 7 | 19 | $\overline{\delta}$ | $(g)$ | 0 | 0 |
| 36 | 7 | $19+20$ | $\alpha\overline{g}$ | $(g^2)$ | (C1) | (C6) |
| 36 | 8 | 19 | $h_0\overline{\delta}$ | $(g)$ | 0 | 0 |
| 36 | 9 | 19 | $g\overline{h_0^2\alpha}$ | $(g^2)$ | 0 | 0 |
| 36 | 9 | 20 | $h_0^2\overline{\delta}$ | $(g)$ | 0 | 0 |
| 36 | $10+i$ | 20 | $i(h_0^{1+i}\alpha^3)$ | $(g)$ | 0 | 0 |
| 37 | 8 | 21 | $h_1\overline{\delta}$ | $(g)$ | (D) | (H) |
| 38 | 8 | 22 | $d_0\overline{g}$ | $(g^2)$ | 0 | 0 |
| 40 | $10+i$ | 24 | $h_0^{1+i}\overline{\alpha^3}$ | $(g)$ | 0 | 0 |
| 41 | 8 | 24 | $e_0\overline{g}$ | $(g^3)$ | (E2) | (E4) |
| 43 | 9 | 26 | $d_0\overline{\gamma}$ | $(g^2)$ | (C2) | (C7) |
| 45 | 10 | 28 | $d_0\overline{\alpha\beta}$ | $(g^2)$ | 0 | 0 |
| 46 | 9 | 28 | $e_0\overline{\gamma}$ | $(g^3)$ | (D) | (H) |
| 48 | $9+i$ | 29 | $i(h_0^{1+i}w_2)$ | $(g)$ | 0 | 0 |
| 48 | 10 | $30+31$ | $\alpha^2\overline{g}$ | $(g^2)$ | 0 | 0 |
| 49 | 9 | 31 | $i(h_1w_2)$ | $(g)$ | (A) | (F) |
| 50 | 10 | 33 | $i(h_1^2w_2)$ | $(g)$ | 0 | 0 |
| 50 | 11 | 33 | $\alpha d_0\overline{g}$ | $(g^2)$ | (C3) | (C8) |
| 52 | $11+i$ | 35 | $h_0^iw_2\overline{h_0^3}$ | $(g)$ | 0 | 0 |
| 53 | 11 | 36 | $\alpha^2\overline{\gamma}$ | $(g^2)$ | (C4) | (C9) |
| 54 | 10 | 35 | $g\overline{\beta^2}$ | $(g^2w_1)$ | (B) | (A) |
| 54 | 10 | 36 | $h_1w_2\overline{h_1}$ | $(g)$ | (B) | (A) |
| 55 | 11 | 38 | $h_0w_2\overline{h_0h_2}$ | – | (F) | (G) |
| 55 | 12 | 38 | $d_0e_0\overline{g}$ | $(g^2)$ | 0 | 0 |
| 56 | 11 | 40 | $i(c_0w_2)$ | $(g)$ | 0 | 0 |
| 59 | 11 | 41 | $\beta g\overline{g}$ | – | (G) | (B) |

Table D.4: Summands in $(E_3(tmf/\nu), d_3)$ (cont.)

| $t-s$ | $s$ | $g$ | $x$ | $\mathrm{Ann}(x)$ | Type$(x)$ | Type$(xw_2^2)$ |
|---|---|---|---|---|---|---|
| 60 | 11 | 42 | $w_2\overline{c_0}$ | $(g)$ | 0 | 0 |
| 60 | $12+i$ | 44 | $i(h_0^{1+i}\alpha w_2)$ | $(g)$ | 0 | 0 |
| 60 | 13 | 44 | $g\overline{\alpha^3}$ | $(g^2)$ | (C5) | (C10) |
| 61 | 12 | 46 | $h_1w_2\overline{c_0}$ | $(g)$ | 0 | 0 |
| 64 | $14+i$ | 51 | $h_0^{1+i}w_2\overline{h_0^2\alpha}$ | $(g)$ | 0 | 0 |
| 72 | 13 | 56 | $h_0w_2\overline{g}$ | $(g)$ | 0 | 0 |
| 72 | 14 | 60 | $h_0^2w_2\overline{g}$ | $(g)$ | 0 | 0 |
| 72 | $15+i$ | 61 | $i(h_0^{1+i}\alpha^2w_2)$ | $(g)$ | 0 | 0 |
| 73 | 13 | 58 | $h_1w_2\overline{g}$ | $(g)$ | (F) | (G) |
| 74 | 14 | 62 | $i(h_1\gamma w_2)$ | $(g)$ | 0 | 0 |
| 76 | $15+i$ | 66 | $h_0^iw_2\overline{h_0\alpha^2}$ | $(g)$ | 0 | 0 |
| 78 | 14 | 65 | $h_1w_2\overline{\gamma}$ | $(g)$ | (A) | (F) |
| 79 | 15 | 69 | $h_0w_2\overline{\alpha\beta}$ | $-$ | (G) | (B) |
| 80 | 15 | 71 | $i(\delta w_2)$ | $(g)$ | 0 | 0 |
| 84 | 15 | 73 | $w_2\overline{\delta}$ | $(g)$ | 0 | 0 |
| 84 | 16 | 77 | $h_0w_2\overline{\delta}$ | $(g)$ | 0 | 0 |
| 84 | 17 | 79 | $h_0^2w_2\overline{\delta}$ | $(g)$ | 0 | 0 |
| 84 | $18+i$ | 80 | $i(h_0^{1+i}\alpha^3w_2)$ | $(g)$ | 0 | 0 |
| 85 | 16 | 79 | $h_1w_2\overline{\delta}$ | $(g)$ | 0 | 0 |
| 88 | $18+i$ | 87 | $h_0^{1+i}w_2\overline{\alpha^3}$ | $(g)$ | 0 | 0 |

Complex (A) is

$$\begin{array}{ccccc}
\langle h_1w_2\overline{\gamma}, gw_2^2\overline{\beta^2}, h_1w_2^3\overline{h_1}\rangle & \xrightarrow{\begin{pmatrix} g^2w_1 & g^6 & 0 \\ 0 & 0 & 0 \\ 0 & 0 & g^2w_1 \end{pmatrix}} & \langle \overline{\gamma}, i(h_1w_2), w_2^2\overline{h_1}\rangle & \xrightarrow{(\,gw_1 \;\; g^2w_1 \;\; g^5\,)} & \langle i(1)\rangle \\
\| & & \| & & \| \\
\frac{R_2}{(g)} \oplus \frac{R_2}{(g^2w_1)} \oplus \frac{R_2}{(g)} & & \frac{R_2}{(g^3w_1)} \oplus \frac{R_2}{(g)} \oplus \frac{R_2}{(g^3w_1)} & & \frac{R_2}{(g^3w_1)}
\end{array}$$

(For typographical reasons we write $\langle h_1w_2\overline{\gamma}, gw_2^2\overline{\beta^2}, h_1w_2^3\overline{h_1}\rangle$ and $\langle\overline{\gamma}, i(h_1w_2), w_2^2\overline{h_1}\rangle$ in place of $\langle h_1w_2\overline{\gamma}\rangle\oplus\langle gw_2^2\overline{\beta^2}\rangle\oplus\langle h_1w_2^3\overline{h_1}\rangle$ and $\langle\overline{\gamma}\rangle\oplus\langle i(h_1w_2)\rangle\oplus\langle w_2^2\overline{h_1}\rangle$, respectively.) The classes $h_1w_2\overline{\gamma}$ and $h_1w_2^3\overline{h_1}$ do not survive, $gw_2^2\overline{\beta^2}$ is replaced by $gw_1w_2^2\overline{\beta^2}$, $w_2^2\overline{h_1}$ is replaced by $w_1w_2^2\overline{h_1}$, and the two classes $\overline{\gamma}$ and $i(h_1w_2)$ are replaced by the single class

$$\gamma\overline{g} = 9_{30} + 9_{31} = g\overline{\gamma} + i(h_1w_2)\,,$$

leaving the direct sum of the four cyclic modules

$$\begin{aligned}
\langle i(1)\rangle &\cong R_2/(g^5, gw_1)\\
\langle \gamma\overline{g}\rangle &\cong R_2/(g^5, gw_1)\\
\langle w_1 w_2^2\overline{h_1}\rangle &\cong R_2/(g^2)\\
\langle g w_1 w_2^2\overline{\beta^2}\rangle &\cong R_2/(g^2)
\end{aligned}$$

at $E_4$. Complex (B) is

$$\begin{array}{ccccc}
\langle \beta g w_2^2\overline{g}, h_0 w_2^3\overline{\alpha\beta}\rangle & \xrightarrow{\begin{pmatrix} g^5 & 0\\ 0 & 0\\ gw_1 & 0\end{pmatrix}} & \langle g\overline{\beta^2}\rangle \oplus \langle h_1 w_2\overline{h_1}\rangle \oplus \langle g w_2^2\overline{h_2^2}\rangle & \xrightarrow{(0\ g^2w_1\ g^6)} & \langle \overline{h_1}\rangle\\
\| & & \| & & \|\\
\dfrac{R_2\oplus R_2}{\langle (gw_1, w_1), (0, g)\rangle} & & R_2/(g^2w_1)\oplus R_2/(g)\oplus R_2/(g^2w_1) & & R_2/(g^3w_1)
\end{array}$$

The classes $\beta g w_2^2\overline{g}$ and $h_1 w_2\overline{h_1}$ do not survive, and $g w_2^2\overline{h_2^2}$ is replaced by $g^5\overline{\beta^2} + g w_1 w_2^2\overline{h_2^2}$, leaving

$$\begin{aligned}
\langle \overline{h_1}\rangle &\cong R_2/(g^6, g^2w_1)\\
\langle g\overline{\beta^2}\rangle &\cong R_2/(g^6, g^2w_1)\\
\langle g^5\overline{\beta^2} + g w_1 w_2^2\overline{h_2^2}\rangle &\cong R_2/(g)\\
\langle h_0 w_2^3\overline{\alpha\beta}\rangle &\cong R_2/(g)\,.
\end{aligned}$$

Type (C) complexes have the form

$$\begin{array}{ccc}
\langle x\rangle & \xrightarrow{gw_1} & \langle y\rangle\\
\| & & \|\\
R_2/(g^2) & & R_2/(g^2)
\end{array}$$

There are ten such summands in $(E_3(tmf/\nu), d_3)$, with $x$ and $y$ as in Table D.5, leaving

$$\begin{aligned}
\langle y\rangle &\cong R_2/(g^2, gw_1)\\
\langle gx\rangle &\cong R_2/(g)
\end{aligned}$$

at $E_4$.

Table D.5: Summands of type (C)

| $n$ | $x$ | $y$ |
|---|---|---|
| 1 | $\alpha\overline{g}$ | $\overline{h_0 h_2}$ |
| 2 | $d_0\overline{\gamma}$ | $i(d_0)$ |
| 3 | $\alpha d_0\overline{g}$ | $d_0\overline{h_0 h_2}$ |
| 4 | $\alpha^2\overline{\gamma}$ | $i(\alpha^2)$ |
| 5 | $g\overline{\alpha^3}$ | $i(d_0 e_0)$ |
| 6 | $\alpha w_2^2\overline{g}$ | $w_2^2\overline{h_0 h_2}$ |

Table D.5: Summands of type (C) (cont.)

| $n$ | $x$ | $y$ |
|---|---|---|
| 7 | $d_0 w_2^2 \overline{\gamma}$ | $i(d_0 w_2^2)$ |
| 8 | $\alpha d_0 w_2^2 \overline{g}$ | $d_0 w_2^2 \overline{h_0 h_2}$ |
| 9 | $\alpha^2 w_2^2 \overline{\gamma}$ | $i(\alpha^2 w_2^2)$ |
| 10 | $g w_2^2 \overline{\alpha^3}$ | $i(d_0 e_0 w_2^2)$ |

Complex (D) is

$$\begin{array}{ccccc} \langle e_0\overline{\gamma}\rangle & \xrightarrow{\binom{gw_1}{w_1}} & \langle i(e_0)\rangle \oplus \langle h_1\overline{\delta}\rangle & \xrightarrow{(w_1\ 0)} & \langle i(c_0)\rangle \\ \| & & \| & & \| \\ R_2/(g^3) & & R_2/(g^3)\oplus R_2/(g) & & R_2/(g) \end{array}$$

The class $i(e_0)$ is replaced by

$$\delta'\overline{h_1} = 8_{20} + 8_{21} = i(e_0 g) + h_1\overline{\delta}\,,$$

and the class $e_0\overline{\gamma}$ is replaced by $e_0 g^2\overline{\gamma}$, leaving

$$\begin{aligned} \langle i(c_0)\rangle &\cong R_2/(g, w_1) \\ \langle \delta'\overline{h_1}\rangle &\cong R_2/(g^2, w_1) \\ \langle h_1\overline{\delta}\rangle &\cong R_2/(g) \\ \langle e_0 g^2\overline{\gamma}\rangle &\cong R_2/(g)\,. \end{aligned}$$

Type (E) complexes have the form

$$\begin{array}{ccc} \langle x\rangle & \xrightarrow{w_1} & \langle y\rangle \\ \| & & \| \\ R_2/(g^3) & & R_2/(g) \end{array}$$

There are four such summands in $(E_3(tmf/\nu), d_3)$, with $x$ and $y$ as in Table D.6, leaving

$$\begin{aligned} \langle y\rangle &\cong R_2/(g, w_1) \\ \langle gx\rangle &\cong R_2/(g^2) \end{aligned}$$

at $E_4$.

Table D.6: Summands of type (E)

| $n$ | $x$ | $y$ |
|---|---|---|
| 1 | $e_0\overline{h_1}$ | $h_1\overline{c_0}$ |
| 2 | $e_0\overline{g}$ | $i(\delta)$ |
| 3 | $e_0 w_2^2\overline{h_1}$ | $h_1 w_2^2\overline{c_0}$ |
| 4 | $e_0 w_2^2\overline{g}$ | $i(\delta w_2^2)$ |

Complex (F) is

$$
\begin{array}{ccc}
\langle h_1 w_2^3 \overline{\gamma} \rangle & = & R_2/(g) \\
\left\downarrow \begin{pmatrix} 0 \\ g^2 w_1 \\ 0 \end{pmatrix} \right. & & \\
\langle h_1 w_2 \overline{g} \rangle \oplus \langle w_2^2 \overline{\gamma} \rangle \oplus \langle i(h_1 w_2^3) \rangle & = & R_2/(g) \oplus R_2/(g^3 w_1) \oplus R_2/(g) \\
\left\downarrow \begin{pmatrix} g^2 w_1 & g^5 & 0 \\ 0 & g w_1 & g^2 w_1 \end{pmatrix} \right. & & \\
\langle \overline{g} \rangle \oplus \langle i(w_2^2) \rangle & = & R_2/(g^3 w_1) \oplus R_2/(g^3 w_1) \\
\left\downarrow \begin{pmatrix} 0 & g^3 \\ 0 & 0 \end{pmatrix} \right. & & \\
\langle i(\beta g), h_0 w_2 \overline{h_0 h_2} \rangle & = & \dfrac{R_2 \oplus R_2}{\langle (g w_1, w_1), (0, g) \rangle}
\end{array}
$$

The classes $h_1 w_2^3 \overline{\gamma}$ and $h_1 w_2 \overline{g}$ do not survive, the two classes $w_2^2 \overline{\gamma}$ and $i(h_1 w_2^3)$ are replaced by

$$\gamma w_1 w_2^2 \overline{g} = 29_{227} + 29_{228} = g w_1 w_2^2 \overline{\gamma} + i(h_1 w_1 w_2^3)\,,$$

and $i(w_2^2)$ is replaced by $g^4 \overline{g} + i(w_1 w_2^2)$. This leaves

$$
\begin{aligned}
\langle \overline{g} \rangle &\cong R_2/(g^6, g^2 w_1) \\
\langle g^4 \overline{g} + i(w_1 w_2^2) \rangle &\cong R_2/(g) \\
\langle \gamma w_1 w_2^2 \overline{g} \rangle &\cong R_2/(g)
\end{aligned}
$$

and the non-cyclic summand

$$\langle i(\beta g), h_0 w_2 \overline{h_0 h_2} \rangle \cong \frac{R_2 \oplus R_2}{\langle (g^3, 0), (g w_1, w_1), (0, g) \rangle}\,.$$

Complex (G) is

$$
\begin{array}{ccc}
\langle h_1 w_2^3 \overline{g} \rangle & = & R_2/(g) \\
\left\downarrow g^2 w_1 \right. & & \\
\langle w_2^2 \overline{g} \rangle & = & R_2/(g^3 w_1) \\
\left\downarrow \begin{pmatrix} g^3 \\ 0 \\ 0 \\ 0 \end{pmatrix} \right. & & \\
\langle \beta g \overline{g}, h_0 w_2 \overline{\alpha \beta} \rangle \oplus \langle i(\beta g w_2^2), h_0 w_2^3 \overline{h_0 h_2} \rangle & = & \dfrac{R_2 \oplus R_2}{\langle (g w_1, w_1), (0, g) \rangle} \oplus \dfrac{R_2 \oplus R_2}{\langle (g w_1, w_1), (0, g) \rangle} \\
\left\downarrow (\, g w_1 \;\, 0 \;\, g^5 \;\, 0 \,) \right. & & \\
\langle g \overline{h_2^2} \rangle & = & R_2/(g^2 w_1)
\end{array}
$$

The class $h_1 w_2^3 \overline{g}$ does not survive, the class $w_2^2 \overline{g}$ is replaced by $w_1 w_2^2 \overline{g}$, the class $\beta g \overline{g}$ is replaced by

$$\gamma^2 \overline{\gamma} = 15_{68} + 15_{69} = \beta g^2 \overline{g} + h_0 w_2 \overline{\alpha \beta}\,,$$

and the class $i(\beta g w_2^2)$ is replaced by $i(\beta g w_1 w_2^2)$. This leaves

$$
\begin{aligned}
\langle g \overline{h_2^2} \rangle &\cong R_2/(g^5, g w_1) \\
\langle \gamma^2 \overline{\gamma} \rangle &\cong R_2/(g^2, w_1)
\end{aligned}
$$

$$\langle h_0 w_2 \overline{\alpha\beta} \rangle \cong R_2/(g)$$
$$\langle w_1 w_2^2 \overline{g} \rangle \cong R_2/(g^2)$$

and

$$\langle i(\beta g w_1 w_2^2), h_0 w_2^3 \overline{h_0 h_2} \rangle \cong \frac{R_2 \oplus R_2}{\langle (g, w_1), (0, g) \rangle} .$$

Complex (H) is $w_2^2$ times complex (D). The class $i(e_0 w_2^2)$ is replaced by

$$\delta' w_2^2 \overline{h_1} = 24_{168} + 24_{169} = i(e_0 g w_2^2) + h_1 w_2^2 \overline{\delta} ,$$

and the class $e_0 w_2^2 \overline{\gamma}$ is replaced by $e_0 g^2 w_2^2 \overline{\gamma}$, leaving

$$\langle i(c_0 w_2^2) \rangle \cong R_2/(g, w_1)$$
$$\langle \delta' w_2^2 \overline{h_1} \rangle \cong R_2/(g^2, w_1)$$
$$\langle h_1 w_2^2 \overline{\delta} \rangle \cong R_2/(g)$$
$$\langle e_0 g^2 w_2^2 \overline{\gamma} \rangle \cong R_2/(g) .$$

## D.3. Calculation of $E_5(tmf/\nu) = H(E_4(tmf/\nu), d_4)$

The $(E_4, d_4)$-term of the Adams spectral sequence for $tmf/\nu$ splits as a direct sum of 28 $R_2$-module complexes of length two, plus 19 complexes of types labeled (A) to (S) and (B2) to (N2) (with some gaps), plus a large summand with trivial differential. The Type-column in Table D.7 gives the label of the complex containing the $R_2$-module generator $x$.

Table D.7: Summands in $(E_4(tmf/\nu), d_4)$

| $t-s$ | $s$ | $g$ | $x$ | $\mathrm{Ann}(x)$ | $\mathrm{Type}(x)$ |
|---|---|---|---|---|---|
| 0 | 0 | 0 | $i(1)$ | $(g^5, gw_1)$ | 0 |
| 0 | $1+i$ | 0 | $i(h_0^{1+i})$ | $(g)$ | 0 |
| 1 | 1 | 1 | $i(h_1)$ | $(g)$ | 0 |
| 2 | 2 | 1 | $i(h_1^2)$ | $(g)$ | 0 |
| 4 | $3+i$ | 1 | $h_0^i \overline{h_0^3}$ | $(g)$ | 0 |
| 5 | 1 | 2 | $\overline{h_1}$ | $(g^6, g^2 w_1)$ | (A) |
| 6 | 2 | 2 | $h_1 \overline{h_1}$ | $(g)$ | 0 |
| 7 | 2 | 3 | $\overline{h_0 h_2}$ | $(g^2, gw_1)$ | (B) |
| 7 | 3 | 2 | $h_0 \overline{h_0 h_2}$ | $(g, w_1)$ | 0 |
| 8 | 3 | 3 | $i(c_0)$ | $(g, w_1)$ | 0 |
| 12 | 3 | 4 | $\overline{c_0}$ | $(g)$ | (C) |
| 12 | 3 | $4+5$ | $i(\alpha)$ | $(g^2)$ | (C) |
| 12 | $4+i$ | 4 | $i(h_0^{1+i} \alpha)$ | $(g)$ | 0 |
| 13 | 4 | 5 | $h_1 \overline{c_0}$ | $(g, w_1)$ | 0 |
| 14 | 4 | 6 | $i(d_0)$ | $(g^2, gw_1)$ | (D) |

Table D.7: Summands in $(E_4(tmf/\nu), d_4)$ (cont.)

| $t-s$ | $s$ | $g$ | $x$ | $\mathrm{Ann}(x)$ | $\mathrm{Type}(x)$ |
|---|---|---|---|---|---|
| 16 | $6+i$ | 7 | $h_0^{1+i}\overline{h_0^2\alpha}$ | $(g)$ | 0 |
| 19 | 5 | 8 | $d_0\overline{h_1}$ | $(g^2)$ | (E) |
| 21 | 6 | 9 | $d_0\overline{h_0h_2}$ | $(g^2, gw_1)$ | (F) |
| 24 | 4 | 9 | $\overline{g}$ | $(g^6, g^2w_1)$ | (G) |
| 24 | 5 | 10 | $h_0\overline{g}$ | $(g)$ | 0 |
| 24 | 6 | $10+11$ | $i(\alpha^2)$ | $(g^2, gw_1)$ | (B) |
| 24 | 6 | 11 | $h_0^2\overline{g}$ | $(g)$ | 0 |
| 24 | $7+i$ | 11 | $i(h_0^{1+i}\alpha^2)$ | $(g)$ | 0 |
| 25 | 5 | 12 | $h_1\overline{g}$ | $(g)$ | 0 |
| 26 | 6 | 12 | $i(h_1\gamma)$ | $(g)$ | 0 |
| 26 | 7 | 12 | $i(\alpha d_0)$ | $(g^2)$ | (H) |
| 28 | $7+i$ | 13 | $h_0^i\overline{h_0\alpha^2}$ | $(g)$ | 0 |
| 29 | 7 | 14 | $i(\alpha e_0)$ | $(g^2)$ | (C) |
| 30 | 6 | 14 | $g\overline{h_2^2}$ | $(g^5, gw_1)$ | (F) |
| 30 | 6 | 15 | $h_1\overline{\gamma}$ | $(g)$ | (F) |
| 31 | 6 | 16 | $\overline{\alpha\beta}$ | $(g^2)$ | (I) |
| 31 | 7 | 15 | $h_0\overline{\alpha\beta}$ | $(g, w_1)$ | 0 |
| 31 | 8 | 15 | $i(d_0e_0)$ | $(g^2, gw_1)$ | (D) |
| 32 | 7 | 17 | $i(\delta)$ | $(g, w_1)$ | 0 |
| 35 | 7 | 18 | $i(\beta g)$ | $-$ | (H) |
| 36 | 7 | 19 | $\overline{\delta}$ | $(g)$ | 0 |
| 36 | 8 | 19 | $h_0\overline{\delta}$ | $(g)$ | 0 |
| 36 | 9 | 19 | $g\overline{h_0^2\alpha}$ | $(g^2)$ | (E) |
| 36 | 9 | 20 | $h_0^2\overline{\delta}$ | $(g)$ | 0 |
| 36 | $10+i$ | 20 | $i(h_0^{1+i}\alpha^3)$ | $(g)$ | 0 |
| 37 | 8 | $20+21$ | $\delta'\overline{h_1}$ | $(g^2, w_1)$ | 0 |
| 37 | 8 | 21 | $h_1\overline{\delta}$ | $(g)$ | 0 |
| 38 | 8 | 22 | $d_0\overline{g}$ | $(g^2)$ | (J) |
| 40 | $10+i$ | 24 | $h_0^{1+i}\overline{\alpha^3}$ | $(g)$ | 0 |
| 42 | 9 | 25 | $e_0g\overline{h_1}$ | $(g^2)$ | (A) |
| 45 | 10 | 28 | $d_0\overline{\alpha\beta}$ | $(g^2)$ | (K) |
| 48 | 9 | 29 | $i(h_0w_2)$ | $(g)$ | (E) |

Table D.7: Summands in $(E_4(tmf/\nu), d_4)$ (cont.)

| $t-s$ | $s$ | $g$ | $x$ | $\mathrm{Ann}(x)$ | $\mathrm{Type}(x)$ |
|---|---|---|---|---|---|
| 48 | 10 | $30+31$ | $\alpha^2\overline{g}$ | $(g^2)$ | (I) |
| 48 | $10+i$ | 31 | $i(h_0^{2+i}w_2)$ | $(g)$ | 0 |
| 49 | 9 | $30+31$ | $\gamma\overline{g}$ | $(g^5, gw_1)$ | 0 |
| 50 | 10 | 33 | $i(h_1^2w_2)$ | $(g)$ | 0 |
| 52 | $11+i$ | 35 | $h_0^iw_2\overline{h_0^3}$ | $(g)$ | 0 |
| 54 | 10 | 35 | $g\overline{\beta^2}$ | $(g^6, g^2w_1)$ | (K) |
| 55 | 11 | 38 | $h_0w_2\overline{h_0h_2}$ | – | (H) |
| 55 | 12 | 38 | $d_0e_0\overline{g}$ | $(g^2)$ | (J) |
| 56 | 11 | 39 | $\alpha g\overline{g}$ | $(g)$ | (L) |
| 56 | 11 | 40 | $i(c_0w_2)$ | $(g)$ | (L) |
| 60 | 11 | 42 | $w_2\overline{c_0}$ | $(g)$ | 0 |
| 60 | $12+i$ | 44 | $i(h_0^{1+i}\alpha w_2)$ | $(g)$ | 0 |
| 61 | 12 | 45 | $e_0g\overline{g}$ | $(g^2)$ | (G) |
| 61 | 12 | 46 | $h_1w_2\overline{c_0}$ | $(g)$ | 0 |
| 63 | 13 | 49 | $d_0g\overline{\gamma}$ | $(g)$ | (M) |
| 64 | $14+i$ | 51 | $h_0^{1+i}w_2\overline{h_0^2\alpha}$ | $(g)$ | 0 |
| 70 | 15 | 58 | $\alpha d_0g\overline{g}$ | $(g)$ | (N) |
| 72 | 13 | 56 | $h_0w_2\overline{g}$ | $(g)$ | (M) |
| 72 | 14 | 60 | $h_0^2w_2\overline{g}$ | $(g)$ | 0 |
| 72 | $15+i$ | 61 | $i(h_0^{1+i}\alpha^2w_2)$ | $(g)$ | 0 |
| 73 | 15 | 62 | $\alpha^2g\overline{\gamma}$ | $(g)$ | (L) |
| 74 | 14 | 62 | $i(h_1\gamma w_2)$ | $(g)$ | (K) |
| 76 | $15+i$ | 66 | $h_0^iw_2\overline{h_0\alpha^2}$ | $(g)$ | 0 |
| 79 | 15 | $68+69$ | $\gamma^2\overline{\gamma}$ | $(g^2, w_1)$ | 0 |
| 79 | 15 | 69 | $h_0w_2\overline{\alpha\beta}$ | $(g)$ | (N) |
| 80 | 15 | 71 | $i(\delta w_2)$ | $(g)$ | 0 |
| 80 | 17 | 72 | $g^2\overline{\alpha^3}$ | $(g)$ | (M) |
| 84 | 15 | 73 | $w_2\overline{\delta}$ | $(g)$ | 0 |
| 84 | 16 | 77 | $h_0w_2\overline{\delta}$ | $(g)$ | 0 |
| 84 | 17 | 79 | $h_0^2w_2\overline{\delta}$ | $(g)$ | 0 |
| 84 | $18+i$ | 80 | $i(h_0^{1+i}\alpha^3w_2)$ | $(g)$ | 0 |
| 85 | 16 | 79 | $h_1w_2\overline{\delta}$ | $(g)$ | 0 |

Table D.7: Summands in $(E_4(tmf/\nu), d_4)$ (cont.)

| $t-s$ | $s$ | $g$ | $x$ | $\mathrm{Ann}(x)$ | $\mathrm{Type}(x)$ |
|---|---|---|---|---|---|
| 86 | 17 | 82 | $e_0 g^2 \overline{\gamma}$ | $(g)$ | 0 |
| 88 | $18+i$ | 87 | $h_0^{1+i} w_2 \overline{\alpha^3}$ | $(g)$ | 0 |
| 96 | $17+i$ | 91 | $i(h_0^{1+i} w_2^2)$ | $(g)$ | 0 |
| 97 | 17 | 93 | $i(h_1 w_2^2)$ | $(g)$ | 0 |
| 98 | 18 | 99 | $i(h_1^2 w_2^2)$ | $(g)$ | 0 |
| 100 | $19+i$ | 105 | $h_0^i w_2^2 \overline{h_0^3}$ | $(g)$ | 0 |
| 102 | 18 | 102 | $h_1 w_2^2 \overline{h_1}$ | $(g)$ | 0 |
| 103 | 18 | 103 | $w_2^2 \overline{h_0 h_2}$ | $(g^2, g w_1)$ | (B2) |
| 103 | 19 | 108 | $h_0 w_2^2 \overline{h_0 h_2}$ | $(g, w_1)$ | 0 |
| 104 | 19 | 110 | $i(c_0 w_2^2)$ | $(g, w_1)$ | 0 |
| 104 | 20 | $112+113$ | $g^4 \overline{g} + i(w_1 w_2^2)$ | $(g)$ | 0 |
| 108 | 19 | 112 | $w_2^2 \overline{c_0}$ | $(g)$ | (C2) |
| 108 | 19 | $112+113$ | $i(\alpha w_2^2)$ | $(g^2)$ | (C2) |
| 108 | $20+i$ | 118 | $i(h_0^{1+i} \alpha w_2^2)$ | $(g)$ | 0 |
| 109 | 20 | 120 | $h_1 w_2^2 \overline{c_0}$ | $(g, w_1)$ | 0 |
| 109 | 21 | 124 | $w_1 w_2^2 \overline{h_1}$ | $(g^2)$ | (O) |
| 110 | 20 | 121 | $i(d_0 w_2^2)$ | $(g^2, g w_1)$ | (D2) |
| 112 | $22+i$ | 132 | $h_0^{1+i} w_2^2 \overline{h_0^2 \alpha}$ | $(g)$ | 0 |
| 115 | 21 | 131 | $d_0 w_2^2 \overline{h_1}$ | $(g^2)$ | (E2) |
| 117 | 22 | 138 | $d_0 w_2^2 \overline{h_0 h_2}$ | $(g^2, g w_1)$ | (P) |
| 120 | 21 | 134 | $h_0 w_2^2 \overline{g}$ | $(g)$ | 0 |
| 120 | 22 | $141+142$ | $i(\alpha^2 w_2^2)$ | $(g^2, g w_1)$ | (B2) |
| 120 | 22 | 142 | $h_0^2 w_2^2 \overline{g}$ | $(g)$ | 0 |
| 120 | $23+i$ | 147 | $i(h_0^{1+i} \alpha^2 w_2^2)$ | $(g)$ | 0 |
| 121 | 21 | 136 | $h_1 w_2^2 \overline{g}$ | $(g)$ | 0 |
| 122 | 22 | 144 | $i(h_1 \gamma w_2^2)$ | $(g)$ | 0 |
| 122 | 23 | 149 | $i(\alpha d_0 w_2^2)$ | $(g^2)$ | (Q) |
| 124 | $23+i$ | 152 | $h_0^i w_2^2 \overline{h_0 \alpha^2}$ | $(g)$ | 0 |
| 125 | 23 | 153 | $i(\alpha e_0 w_2^2)$ | $(g^2)$ | (C2) |
| 126 | 22 | 147 | $h_1 w_2^2 \overline{\gamma}$ | $(g)$ | (P) |
| 127 | 22 | 148 | $w_2^2 \overline{\alpha \beta}$ | $(g^2)$ | (I2) |
| 127 | 23 | 155 | $h_0 w_2^2 \overline{\alpha \beta}$ | $(g, w_1)$ | 0 |

Table D.7: Summands in $(E_4(tmf/\nu), d_4)$ (cont.)

| $t-s$ | $s$ | $g$ | $x$ | Ann$(x)$ | Type$(x)$ |
|---|---|---|---|---|---|
| 127 | 24 | 160 | $i(d_0e_0w_2^2)$ | $(g^2, gw_1)$ | (D2) |
| 128 | 23 | 157 | $i(\delta w_2^2)$ | $(g, w_1)$ | 0 |
| 128 | 24 | 162 | $w_1w_2^2\overline{g}$ | $(g^2)$ | (R) |
| 132 | 23 | 159 | $w_2^2\overline{\delta}$ | $(g)$ | 0 |
| 132 | 24 | 167 | $h_0w_2^2\overline{\delta}$ | $(g)$ | 0 |
| 132 | 25 | 172 | $gw_2^2\overline{h_0^2\alpha}$ | $(g^2)$ | (E2) |
| 132 | 25 | 173 | $h_0^2w_2^2\overline{\delta}$ | $(g)$ | 0 |
| 132 | $26+i$ | 177 | $i(h_0^{1+i}\alpha^3w_2^2)$ | $(g)$ | 0 |
| 133 | 24 | $168+169$ | $\delta' w_2^2\overline{h_1}$ | $(g^2, w_1)$ | 0 |
| 133 | 24 | 169 | $h_1w_2^2\overline{\delta}$ | $(g)$ | 0 |
| 134 | 24 | 170 | $d_0w_2^2\overline{g}$ | $(g^2)$ | (J2) |
| 134 | 26 | $179+180$ | $g^5\overline{\beta^2} + gw_1w_2^2\overline{h_2^2}$ | $(g)$ | (P) |
| 136 | $26+i$ | 185 | $h_0^{1+i}w_2^2\overline{\alpha^3}$ | $(g)$ | 0 |
| 138 | 25 | 181 | $e_0gw_2^2\overline{h_1}$ | $(g^2)$ | (O) |
| 139 | 27 | 193 | $i(\beta gw_1w_2^2)$ | – | (Q) |
| 141 | 26 | 191 | $d_0w_2^2\overline{\alpha\beta}$ | $(g^2)$ | (S) |
| 144 | 25 | 185 | $i(h_0w_2^3)$ | $(g)$ | (E2) |
| 144 | 26 | $194+195$ | $\alpha^2w_2^2\overline{g}$ | $(g^2)$ | (I2) |
| 144 | $26+i$ | 195 | $i(h_0^{2+i}w_2^3)$ | $(g)$ | 0 |
| 146 | 26 | 197 | $i(h_1^2w_2^3)$ | $(g)$ | 0 |
| 148 | $27+i$ | 207 | $h_0^iw_2^3\overline{h_0^3}$ | $(g)$ | 0 |
| 151 | 27 | 210 | $h_0w_2^3\overline{h_0h_2}$ | – | (Q) |
| 151 | 28 | 217 | $d_0e_0w_2^2\overline{g}$ | $(g^2)$ | (J2) |
| 152 | 27 | 211 | $\alpha gw_2^2\overline{g}$ | $(g)$ | (L2) |
| 152 | 27 | 212 | $i(c_0w_2^3)$ | $(g)$ | (L2) |
| 153 | 29 | $227+228$ | $\gamma w_1w_2^2\overline{g}$ | $(g)$ | 0 |
| 156 | 27 | 214 | $w_2^3\overline{c_0}$ | $(g)$ | 0 |
| 156 | $28+i$ | 224 | $i(h_0^{1+i}\alpha w_2^3)$ | $(g)$ | 0 |
| 157 | 28 | 225 | $e_0gw_2^2\overline{g}$ | $(g^2)$ | (R) |
| 157 | 28 | 226 | $h_1w_2^3\overline{c_0}$ | $(g)$ | 0 |
| 158 | 30 | 241 | $gw_1w_2^2\overline{\beta^2}$ | $(g^2)$ | (S) |
| 159 | 29 | 237 | $d_0gw_2^2\overline{\gamma}$ | $(g)$ | (M2) |

Table D.7: Summands in $(E_4(tmf/\nu), d_4)$ (cont.)

| $t-s$ | $s$ | $g$ | $x$ | Ann$(x)$ | Type$(x)$ |
|---|---|---|---|---|---|
| 160 | $30+i$ | 246 | $h_0^{1+i}w_2^3\overline{h_0^2\alpha}$ | $(g)$ | 0 |
| 166 | 31 | 261 | $\alpha d_0 g w_2^2\overline{g}$ | $(g)$ | (N2) |
| 168 | 29 | 244 | $h_0 w_2^3\overline{g}$ | $(g)$ | (M2) |
| 168 | 30 | 256 | $h_0^2 w_2^3\overline{g}$ | $(g)$ | 0 |
| 168 | $31+i$ | 265 | $i(h_0^{1+i}\alpha^2 w_2^3)$ | $(g)$ | 0 |
| 169 | 31 | 266 | $\alpha^2 g w_2^2\overline{\gamma}$ | $(g)$ | (L2) |
| 170 | 30 | 258 | $i(h_1\gamma w_2^3)$ | $(g)$ | (S) |
| 172 | $31+i$ | 270 | $h_0^i w_2^3\overline{h_0\alpha^2}$ | $(g)$ | 0 |
| 175 | 31 | 273 | $h_0 w_2^3\overline{\alpha\beta}$ | $(g)$ | (N2) |
| 176 | 31 | 275 | $i(\delta w_2^3)$ | $(g)$ | 0 |
| 176 | 33 | 291 | $g^2 w_2^2\overline{\alpha^3}$ | $(g)$ | (M2) |
| 180 | 31 | 277 | $w_2^3\overline{\delta}$ | $(g)$ | 0 |
| 180 | 32 | 289 | $h_0 w_2^3\overline{\delta}$ | $(g)$ | 0 |
| 180 | 33 | 299 | $h_0^2 w_2^3\overline{\delta}$ | $(g)$ | 0 |
| 180 | $34+i$ | 307 | $i(h_0^{1+i}\alpha^3 w_2^3)$ | $(g)$ | 0 |
| 181 | 32 | 291 | $h_1 w_2^3\overline{\delta}$ | $(g)$ | 0 |
| 182 | 33 | 302 | $e_0 g^2 w_2^2\overline{\gamma}$ | $(g)$ | 0 |
| 184 | $34+i$ | 315 | $h_0^{1+i}w_2^3\overline{\alpha^3}$ | $(g)$ | 0 |

Complex (A) is

$$\begin{array}{ccc} \langle e_0 g\overline{h_1}\rangle & \xrightarrow{g w_1^2} & \langle\overline{h_1}\rangle \\ \| & & \| \\ R_2/(g^2) & & R_2/(g^6, g^2 w_1) \end{array}$$

The class $e_0 g\overline{h_1}$ is replaced by $e_0 g^2\overline{h_1}$, leaving

$$\langle\overline{h_1}\rangle \cong R_2/(g^6, g^2 w_1, g w_1^2)$$
$$\langle e_0 g^2\overline{h_1}\rangle \cong R_2/(g)$$

at $E_5$. Complex (B) is

$$\begin{array}{ccc} \langle i(\alpha^2)\rangle & \xrightarrow{w_1^2} & \langle\overline{h_0 h_2}\rangle \\ \| & & \| \\ R_2/(g^2, g w_1) & & R_2/(g^2, g w_1) \end{array}$$

The class $i(\alpha^2)$ is replaced by $i(\alpha^2 g)$, leaving

$$\langle\overline{h_0 h_2}\rangle \cong R_2/(g^2, g w_1, w_1^2)$$

$$\langle i(\alpha^2 g)\rangle \cong R_2/(g, w_1)$$

at $E_5$. Complex (B2) is $w_2^2$ times complex (B). The class $i(\alpha^2 w_2^2)$ is replaced by $i(\alpha^2 g w_2^2)$, leaving

$$\langle w_2^2\overline{h_0h_2}\rangle \cong R_2/(g^2, gw_1, w_1^2)$$
$$\langle i(\alpha^2 g w_2^2)\rangle \cong R_2/(g, w_1)$$

at $E_5$. Complex (C) is

$$\begin{array}{ccc} \langle i(\alpha e_0)\rangle & \xrightarrow{\binom{w_1^2}{w_1^2}} & \langle \overline{c_0}\rangle \oplus \langle i(\alpha)\rangle \\ \| & & \| \\ R_2/(g^2) & & R_2/(g) \oplus R_2/(g^2) \end{array}$$

The class $i(\alpha e_0)$ does not survive, and $i(\alpha)$ is replaced by $\overline{c_0} + i(\alpha) = 3_5$. This leaves

$$\langle \overline{c_0}\rangle \cong R_2/(g)$$
$$\langle \overline{c_0} + i(\alpha)\rangle \cong R_2/(g^2, w_1^2)$$

at $E_5$. Complex (C2) is $w_2^2$ times complex (C). The class $i(\alpha e_0 w_2^2)$ does not survive, and $i(\alpha w_2^2)$ is replaced by $w_2^2\overline{c_0} + i(\alpha w_2^2)$. This leaves

$$\langle w_2^2\overline{c_0}\rangle \cong R_2/(g)$$
$$\langle w_2^2\overline{c_0} + i(\alpha w_2^2)\rangle \cong R_2/(g^2, w_1^2)$$

at $E_5$. Complex (D) is

$$\begin{array}{ccc} \langle i(d_0 e_0)\rangle & \xrightarrow{w_1^2} & \langle i(d_0)\rangle \\ \| & & \| \\ R_2/(g^2, gw_1) & & R_2/(g^2, gw_1) \end{array}$$

The class $i(d_0 e_0)$ is replaced by $i(d_0 e_0 g)$, leaving

$$\langle i(d_0)\rangle \cong R_2/(g^2, gw_1, w_1^2)$$
$$\langle i(d_0 e_0 g)\rangle \cong R_2/(g, w_1)\,.$$

Complex (D2) is $w_2^2$ times complex (D). Here $i(d_0 e_0 w_2^2)$ is replaced by $i(d_0 e_0 g w_2^2)$, leaving

$$\langle i(d_0 w_2^2)\rangle \cong R_2/(g^2, gw_1, w_1^2)$$
$$\langle i(d_0 e_0 g w_2^2)\rangle \cong R_2/(g, w_1)\,.$$

Complex (E) is

$$\begin{array}{ccc} \langle g\overline{h_0^2\alpha}\rangle \oplus \langle i(h_0 w_2)\rangle & \xrightarrow{(w_1^2\ \, gw_1)} & \langle d_0\overline{h_1}\rangle \\ \| & & \| \\ R_2/(g^2) \oplus R_2/(g) & & R_2/(g^2) \end{array}$$

The classes $g\overline{h_0^2\alpha}$ and $i(h_0 w_2)$ are replaced by

$$i(\alpha^3 g + h_0 w_1 w_2) = 13_{39} + 13_{40} = g^2\overline{h_0^2\alpha} + i(h_0 w_1 w_2)\,,$$

leaving

$$\begin{aligned} \langle d_0\overline{h_1}\rangle &\cong R_2/(g^2, gw_1, w_1^2) \\ \langle i(\alpha^3 g + h_0w_1w_2)\rangle &\cong R_2/(g)\,. \end{aligned}$$

Complex (E2) is $w_2^2$ times complex (E). The classes $gw_2^2\overline{h_0^2\alpha}$ and $i(h_0w_2^3)$ are replaced by $i(\alpha^3 gw_2^2 + h_0w_1w_2^3)$, leaving

$$\begin{aligned} \langle d_0w_2^2\overline{h_1}\rangle &\cong R_2/(g^2, gw_1, w_1^2) \\ \langle i(\alpha^3 gw_2^2 + h_0w_1w_2^3)\rangle &\cong R_2/(g)\,. \end{aligned}$$

Complex (F) is

$$\begin{array}{ccc} \langle g\overline{h_2^2}\rangle \oplus \langle h_1\overline{\gamma}\rangle & \xrightarrow{(w_1\ w_1)} & \langle d_0\overline{h_0h_2}\rangle \\ \| & & \| \\ R_2/(g^5, gw_1) \oplus R_2/(g) & & R_2/(g^2, gw_1) \end{array}$$

The classes $g\overline{h_2^2}$ and $h_1\overline{\gamma}$ are replaced by

$$\gamma\overline{h_1} = 6_{14} + 6_{15} = g\overline{h_2^2} + h_1\overline{\gamma}\,,$$

leaving

$$\begin{aligned} \langle d_0\overline{h_0h_2}\rangle &\cong R_2/(g^2, w_1) \\ \langle \gamma\overline{h_1}\rangle &\cong R_2/(g^5, gw_1)\,. \end{aligned}$$

Complex (G) is

$$\begin{array}{ccc} \langle e_0g\overline{g}\rangle & \xrightarrow{gw_1^2} & \langle\overline{g}\rangle \\ \| & & \| \\ R_2/(g^2) & & R_2/(g^6, g^2w_1) \end{array}$$

The class $e_0g\overline{g}$ is replaced by $e_0g^2\overline{g}$, leaving

$$\begin{aligned} \langle\overline{g}\rangle &\cong R_2/(g^6, g^2w_1, gw_1^2) \\ \langle e_0g^2\overline{g}\rangle &\cong R_2/(g)\,. \end{aligned}$$

Complex (H) is

$$\begin{array}{ccc} \langle i(\beta g), h_0w_2\overline{h_0h_2}\rangle & \xrightarrow{(w_1\ gw_1)} & \langle i(\alpha d_0)\rangle \\ \| & & \| \\ \dfrac{R_2 \oplus R_2}{\langle (g^3, 0), (gw_1, w_1), (0, g)\rangle} & & R_2/(g^2) \end{array}$$

The classes $i(\beta g)$ and $h_0w_2\overline{h_0h_2}$ are replaced by

$$\gamma^2\overline{h_1} = 11_{37} + 11_{38} = i(\beta g^2) + h_0w_2\overline{h_0h_2}\,,$$

leaving

$$\begin{aligned} \langle i(\alpha d_0)\rangle &\cong R_2/(g^2, w_1) \\ \langle \gamma^2\overline{h_1}\rangle &\cong R_2/(g^2, w_1)\,. \end{aligned}$$

Complex (I) is

$$\begin{array}{ccc} \langle \alpha^2\overline{g}\rangle & \xrightarrow{w_1^2} & \langle\overline{\alpha\beta}\rangle \\ \| & & \| \\ R_2/(g^2) & & R_2/(g^2) \end{array}$$

The class $\alpha^2\overline{g}$ does not survive, leaving

$$\langle\overline{\alpha\beta}\rangle \cong R_2/(g^2, w_1^2)\,.$$

Complex (I2) is $w_2^2$ times complex (I). The class $\alpha^2 w_2^2\overline{g}$ does not survive, leaving

$$\langle w_2^2\overline{\alpha\beta}\rangle \cong R_2/(g^2, w_1^2)\,.$$

Complex (J) is

$$\begin{array}{ccc} \langle d_0e_0\overline{g}\rangle & \xrightarrow{w_1^2} & \langle d_0\overline{g}\rangle \\ \| & & \| \\ R_2/(g^2) & & R_2/(g^2) \end{array}$$

The class $d_0e_0\overline{g}$ does not survive, leaving

$$\langle d_0\overline{g}\rangle \cong R_2/(g^2, w_1^2)\,.$$

Complex (J2) is $w_2^2$ times complex (J). The class $d_0e_0w_2^2\overline{g}$ does not survive, leaving

$$\langle d_0w_2^2\overline{g}\rangle \cong R_2/(g^2, w_1^2)\,.$$

Complex (K) is

$$\begin{array}{ccc} \langle g\overline{\beta^2}\rangle \oplus \langle i(h_1\gamma w_2)\rangle & \xrightarrow{(\,w_1\ \ gw_1\,)} & \langle d_0\overline{\alpha\beta}\rangle \\ \| & & \| \\ R_2/(g^6, g^2w_1) \oplus R_2/(g) & & R_2/(g^2) \end{array}$$

The classes $g\overline{\beta^2}$ and $i(h_1\gamma w_2)$ are replaced by

$$\gamma^2\overline{g} = 14_{61} + 14_{62} = g^2\overline{\beta^2} + i(h_1\gamma w_2)\,.$$

leaving

$$\begin{aligned} \langle d_0\overline{\alpha\beta}\rangle &\cong R_2/(g^2, w_1) \\ \langle\gamma^2\overline{g}\rangle &\cong R_2/(g^5, gw_1)\,. \end{aligned}$$

Complex (L) is

$$\begin{array}{ccc} \langle \alpha^2 g\overline{\gamma}\rangle & \xrightarrow{\binom{w_1^2}{w_1^2}} & \langle\alpha g\overline{g}\rangle \oplus \langle i(c_0w_2)\rangle \\ \| & & \| \\ R_2/(g) & & R_2/(g) \oplus R_2/(g) \end{array}$$

The class $\alpha^2 g\overline{\gamma}$ does not survive, and $\alpha g\overline{g}$ is replaced by

$$\delta'\overline{g} = 11_{39} + 11_{40} = \alpha g\overline{g} + i(c_0w_2)\,.$$

This leaves

$$\begin{aligned} \langle\delta'\overline{g}\rangle &\cong R_2/(g, w_1^2) \\ \langle i(c_0w_2)\rangle &\cong R_2/(g)\,. \end{aligned}$$

Complex (L2) is $w_2^2$ times complex (L). The class $\alpha^2 g w_2^2 \overline{\gamma}$ does not survive, and $\alpha g w_2^2 \overline{g}$ is replaced by $\delta' w_2^2 \overline{g}$. This leaves

$$\begin{aligned} \langle \delta' w_2^2 \overline{g} \rangle &\cong R_2/(g, w_1^2) \\ \langle i(c_0 w_2^3) \rangle &\cong R_2/(g) . \end{aligned}$$

Complex (M) is

$$\begin{array}{ccc} \langle h_0 w_2 \overline{g} \rangle \oplus \langle g^2 \overline{\alpha^3} \rangle & \xrightarrow{\left( w_1 \; w_1^2 \right)} & \langle d_0 g \overline{\gamma} \rangle \\ \| & & \| \\ R_2/(g) \oplus R_2/(g) & & R_2/(g) \end{array}$$

The classes $h_0 w_2 \overline{g}$ and $g^2 \overline{\alpha^3}$ are replaced by

$$(\alpha^3 g + h_0 w_1 w_2) \overline{g} = 17_{72} + 17_{73} = g^2 \overline{\alpha^3} + h_0 w_1 w_2 \overline{g} ,$$

leaving

$$\begin{aligned} \langle d_0 g \overline{\gamma} \rangle &\cong R_2/(g, w_1) \\ \langle (\alpha^3 g + h_0 w_1 w_2) \overline{g} \rangle &\cong R_2/(g) . \end{aligned}$$

Complex (M2) is $w_2^2$ times complex (M). The classes $h_0 w_2^3 \overline{g}$ and $g^2 w_2^2 \overline{\alpha^3}$ are replaced by $(\alpha^3 g w_2^2 + h_0 w_1 w_2^3) \overline{g}$, leaving

$$\begin{aligned} \langle d_0 g w_2^2 \overline{\gamma} \rangle &\cong R_2/(g, w_1) \\ \langle (\alpha^3 g w_2^2 + h_0 w_1 w_2^3) \overline{g} \rangle &\cong R_2/(g) . \end{aligned}$$

Complex (N) is

$$\begin{array}{ccc} \langle h_0 w_2 \overline{\alpha \beta} \rangle & \xrightarrow{w_1} & \langle \alpha d_0 g \overline{g} \rangle \\ \| & & \| \\ R_2/(g) & & R_2/(g) \end{array}$$

The class $h_0 w_2 \overline{\alpha \beta}$ does not survive, leaving

$$\langle \alpha d_0 g \overline{g} \rangle \cong R_2/(g, w_1) .$$

Complex (N2) is $w_2^2$ times complex (N). The class $h_0 w_2^3 \overline{\alpha \beta}$ does not survive, leaving

$$\langle \alpha d_0 g w_2^2 \overline{g} \rangle \cong R_2/(g, w_1) .$$

Complex (O) is

$$\begin{array}{ccc} \langle e_0 g w_2^2 \overline{h_1} \rangle & \xrightarrow{g w_1} & \langle w_1 w_2^2 \overline{h_1} \rangle \\ \| & & \| \\ R_2/(g^2) & & R_2/(g^2) \end{array}$$

The class $e_0 g w_2^2 \overline{h_1}$ is replaced by $e_0 g^2 w_2^2 \overline{h_1}$, leaving

$$\begin{aligned} \langle w_1 w_2^2 \overline{h_1} \rangle &\cong R_2/(g^2, g w_1) \\ \langle e_0 g^2 w_2^2 \overline{h_1} \rangle &\cong R_2/(g) . \end{aligned}$$

Complex (P) is

$$\begin{array}{ccc} \langle h_1 w_2^2\overline{\gamma}\rangle \oplus \langle g^5\overline{\beta^2} + gw_1w_2^2\overline{h_2^2}\rangle & \xrightarrow{(\,w_1\ w_1^2\,)} & \langle d_0 w_2^2\overline{h_0h_2}\rangle \\ \| & & \| \\ R_2/(g)\oplus R_2/(g) & & R_2/(g^2, gw_1) \end{array}$$

The classes $h_1w_2^2\overline{\gamma}$ and $g^5\overline{\beta^2} + gw_1w_2^2\overline{h_2^2}$ are replaced by

$$\gamma^2g^3\overline{g} + \gamma w_1w_2^2\overline{h_1} = 26_{179} + 26_{180} + 26_{181} = g^5\overline{\beta^2} + gw_1w_2^2\overline{h_2^2} + h_1w_1w_2^2\overline{\gamma}\,.$$

This leaves

$$\begin{aligned} \langle d_0w_2^2\overline{h_0h_2}\rangle &\cong R_2/(g^2, w_1) \\ \langle \gamma^2g^3\overline{g} + \gamma w_1w_2^2\overline{h_1}\rangle &\cong R_2/(g)\,. \end{aligned}$$

Complex (Q) is

$$\begin{array}{ccc} \langle i(\beta gw_1w_2^2), h_0w_2^3\overline{h_0h_2}\rangle & \xrightarrow{(\,w_1^2\ gw_1\,)} & \langle i(\alpha d_0w_2^2)\rangle \\ \| & & \| \\ \dfrac{R_2\oplus R_2}{\langle (g, w_1), (0, g)\rangle} & & R_2/(g^2) \end{array}$$

The classes $i(\beta gw_1w_2^2)$ and $h_0w_2^3\overline{h_0h_2}$ do not survive, leaving

$$\langle i(\alpha d_0w_2^2)\rangle \cong R_2/(g^2, gw_1, w_1^2)\,.$$

Complex (R) is

$$\begin{array}{ccc} \langle e_0gw_2^2\overline{g}\rangle & \xrightarrow{gw_1} & \langle w_1w_2^2\overline{g}\rangle \\ \| & & \| \\ R_2/(g^2) & & R_2/(g^2) \end{array}$$

The class $e_0gw_2^2\overline{g}$ is replaced by $e_0g^2w_2^2\overline{g}$, leaving

$$\begin{aligned} \langle w_1w_2^2\overline{g}\rangle &\cong R_2/(g^2, gw_1) \\ \langle e_0g^2w_2^2\overline{g}\rangle &\cong R_2/(g)\,. \end{aligned}$$

Complex (S) is

$$\begin{array}{ccc} \langle gw_1w_2^2\overline{\beta^2}\rangle \oplus \langle i(h_1\gamma w_2^3)\rangle & \xrightarrow{(\,w_1^2\ gw_1\,)} & \langle d_0w_2^2\overline{\alpha\beta}\rangle \\ \| & & \| \\ R_2/(g^2)\oplus R_2/(g) & & R_2/(g^2) \end{array}$$

The classes $gw_1w_2^2\overline{\beta^2}$ and $i(h_1\gamma w_2^3)$ are replaced by

$$\gamma^2w_1w_2^2\overline{g} = 34_{302} + 34_{303} = g^2w_1w_2^2\overline{\beta^2} + i(h_1\gamma w_1w_2^3)\,,$$

leaving

$$\begin{aligned} \langle d_0w_2^2\overline{\alpha\beta}\rangle &\cong R_2/(g^2, gw_1, w_1^2) \\ \langle \gamma^2w_1w_2^2\overline{g}\rangle &\cong R_2/(g) \end{aligned}$$

at $E_5$.

# Bibliography


[1] J. F. Adams, *On the cobar construction*, Proc. Nat. Acad. Sci. U.S.A. **42** (1956), 409–412, DOI 10.1073/pnas.42.7.409. MR79266

[2] J. F. Adams, *On the structure and applications of the Steenrod algebra*, Comment. Math. Helv. **32** (1958), 180–214, DOI 10.1007/BF02564578. MR96219

[3] J. F. Adams, *On the non-existence of elements of Hopf invariant one*, Ann. of Math. (2) **72** (1960), 20–104, DOI 10.2307/1970147. MR141119

[4] J. F. Adams, *On Chern characters and the structure of the unitary group*, Proc. Cambridge Philos. Soc. **57** (1961), 189–199, DOI 10.1017/s0305004100035052. MR121795

[5] J. F. Adams, *Vector fields on spheres*, Ann. of Math. (2) **75** (1962), 603–632, DOI 10.2307/1970213. MR139178

[6] J. F. Adams, *On the groups $J(X)$. I*, Topology **2** (1963), 181–195, DOI 10.1016/0040-9383(63)90001-6. MR159336

[7] J. F. Adams, *A periodicity theorem in homological algebra*, Proc. Cambridge Philos. Soc. **62** (1966), 365–377, DOI 10.1017/s0305004100039955. MR194486

[8] J. F. Adams, *On the groups $J(X)$. IV*, Topology **5** (1966), 21–71, DOI 10.1016/0040-9383(66)90004-8. MR198470

[9] J. F. Adams, *Stable homotopy and generalised homology*, University of Chicago Press, Chicago, Ill.-London, 1974. Chicago Lectures in Mathematics. MR0402720

[10] J. F. Adams, *Operations of the nth kind in $K$-theory, and what we don't know about $RP^\infty$*, New developments in topology (Proc. Sympos. Algebraic Topology, Oxford, 1972), Cambridge Univ. Press, London, 1974, pp. 1–9. London Math. Soc. Lecture Note Ser., No. 11. MR0339178

[11] J. F. Adams and H. R. Margolis, *Modules over the Steenrod algebra*, Topology **10** (1971), 271–282, DOI 10.1016/0040-9383(71)90020-6. MR294450

[12] J. F. Adams and H. R. Margolis, *Sub-Hopf-algebras of the Steenrod algebra*, Proc. Cambridge Philos. Soc. **76** (1974), 45–52, DOI 10.1017/s0305004100048714. MR341487

[13] José Adem, *The iteration of the Steenrod squares in algebraic topology*, Proc. Nat. Acad. Sci. U.S.A. **38** (1952), 720–726, DOI 10.1073/pnas.38.8.720. MR50278

[14] D. W. Anderson, E. H. Brown Jr., and F. P. Peterson, *Pin cobordism and related topics*, Comment. Math. Helv. **44** (1969), 462–468, DOI 10.1007/BF02564545. MR261613

[15] Vigleik Angeltveit and John Rognes, *Hopf algebra structure on topological Hochschild homology*, Algebr. Geom. Topol. **5** (2005), 1223–1290, DOI 10.2140/agt.2005.5.1223. MR2171809

[16] M. F. Atiyah, *Thom complexes*, Proc. London Math. Soc. (3) **11** (1961), 291–310, DOI 10.1112/plms/s3-11.1.291. MR131880

[17] Marc Aubry, *Calculs de groupes d'homotopie stables de la sphère, par la suite spectrale d'Adams-Novikov* (French), Math. Z. **185** (1984), no. 1, 45–91, DOI 10.1007/BF01214973. MR724045

[18] Christian Ausoni, *On the algebraic $K$-theory of the complex $K$-theory spectrum*, Invent. Math. **180** (2010), no. 3, 611–668, DOI 10.1007/s00222-010-0239-x. MR2609252

[19] Christian Ausoni and John Rognes, *Algebraic $K$-theory of topological $K$-theory*, Acta Math. **188** (2002), no. 1, 1–39, DOI 10.1007/BF02392794. MR1947457

[20] Andrew Baker and Andrej Lazarev, *On the Adams spectral sequence for $R$-modules*, Algebr. Geom. Topol. **1** (2001), 173–199, DOI 10.2140/agt.2001.1.173. MR1823498

[21] M. G. Barratt, J. D. S. Jones, and M. E. Mahowald, *Relations amongst Toda brackets and the Kervaire invariant in dimension* 62, J. London Math. Soc. (2) **30** (1984), no. 3, 533–550, DOI 10.1112/jlms/s2-30.3.533. MR810962

[22] M. G. Barratt, M. E. Mahowald, and M. C. Tangora, *Some differentials in the Adams spectral sequence. II*, Topology **9** (1970), 309–316, DOI 10.1016/0040-9383(70)90055-8. MR266215

[23] Tilman Bauer, *Computation of the homotopy of the spectrum* `tmf`, Groups, homotopy and configuration spaces, Geom. Topol. Monogr., vol. 13, Geom. Topol. Publ., Coventry, 2008, pp. 11–40, DOI 10.2140/gtm.2008.13.11. MR2508200

[24] J. C. Becker and D. H. Gottlieb, *The transfer map and fiber bundles*, Topology **14** (1975), 1–12, DOI 10.1016/0040-9383(75)90029-4. MR377873

[25] Mark Behrens, *A modular description of the* $K(2)$*-local sphere at the prime 3*, Topology **45** (2006), no. 2, 343–402, DOI 10.1016/j.top.2005.08.005. MR2193339

[26] M. Behrens, M. Hill, M. J. Hopkins, and M. Mahowald, *On the existence of a* $v_2^{32}$*-self map on* $M(1,4)$ *at the prime 2*, Homology Homotopy Appl. **10** (2008), no. 3, 45–84. MR2475617

[27] Mark Behrens, Mark Mahowald, and J.D. Quigley, *The 2-primary Hurewicz image of tmf*, arXiv:2011.08956.

[28] Mark Behrens and Satya Pemmaraju, *On the existence of the self map* $v_2^9$ *on the Smith-Toda complex* $V(1)$ *at the prime 3*, Homotopy theory: relations with algebraic geometry, group cohomology, and algebraic $K$-theory, Contemp. Math., vol. 346, Amer. Math. Soc., Providence, RI, 2004, pp. 9–49, DOI 10.1090/conm/346/06284. MR2066495

[29] J. Michael Boardman, *Conditionally convergent spectral sequences*, Homotopy invariant algebraic structures (Baltimore, MD, 1998), Contemp. Math., vol. 239, Amer. Math. Soc., Providence, RI, 1999, pp. 49–84, DOI 10.1090/conm/239/03597. MR1718076

[30] M. Bökstedt and I. Madsen, *Topological cyclic homology of the integers*, Astérisque **226** (1994), 7–8, 57–143. $K$-theory (Strasbourg, 1992). MR1317117

[31] M. Bökstedt and I. Madsen, *Algebraic* $K$*-theory of local number fields: the unramified case*, Prospects in topology (Princeton, NJ, 1994), Ann. of Math. Stud., vol. 138, Princeton Univ. Press, Princeton, NJ, 1995, pp. 28–57. MR1368652

[32] Armand Borel, *Sur la cohomologie des espaces fibrés principaux et des espaces homogènes de groupes de Lie compacts* (French), Ann. of Math. (2) **57** (1953), 115–207, DOI 10.2307/1969728. MR51508

[33] A. K. Bousfield, *The localization of spectra with respect to homology*, Topology **18** (1979), no. 4, 257–281, DOI 10.1016/0040-9383(79)90018-1. MR551009

[34] L. E. J. Brouwer, *Über Abbildung von Mannigfaltigkeiten* (German), Math. Ann. **71** (1911), no. 1, 97–115, DOI 10.1007/BF01456931. MR1511644

[35] William Browder, *The Kervaire invariant of framed manifolds and its generalization*, Ann. of Math. (2) **90** (1969), 157–186, DOI 10.2307/1970686. MR251736

[36] Edgar H. Brown Jr. and Michael Comenetz, *Pontrjagin duality for generalized homology and cohomology theories*, Amer. J. Math. **98** (1976), no. 1, 1–27, DOI 10.2307/2373610. MR405403

[37] Robert Ray Bruner, *The Adams spectral sequence of* $H_\infty$ *ring spectra*, ProQuest LLC, Ann Arbor, MI, 1977. Thesis (Ph.D.)–The University of Chicago. MR2611760

[38] R. Bruner, *Algebraic and geometric connecting homomorphisms in the Adams spectral sequence*, Geometric applications of homotopy theory (Proc. Conf., Evanston, Ill., 1977), II, Lecture Notes in Math., vol. 658, Springer, Berlin, 1978, pp. 131–133. MR513570

[39] Robert R. Bruner, *Two generalizations of the Adams spectral sequence*, Current trends in algebraic topology, Part 1 (London, Ont., 1981), CMS Conf. Proc., vol. 2, Amer. Math. Soc., Providence, R.I., 1982, pp. 275–287. MR686121

[40] Robert Bruner, *A new differential in the Adams spectral sequence*, Topology **23** (1984), no. 3, 271–276, DOI 10.1016/0040-9383(84)90010-7. MR770563

[41] Robert R. Bruner, Ext *in the nineties*, Algebraic topology (Oaxtepec, 1991), Contemp. Math., vol. 146, Amer. Math. Soc., Providence, RI, 1993, pp. 71–90, DOI 10.1090/conm/146/01216. MR1224908

[42] Robert R. Bruner, *Extended powers of manifolds and the Adams spectral sequence*, Homotopy methods in algebraic topology (Boulder, CO, 1999), Contemp. Math., vol. 271, Amer. Math. Soc., Providence, RI, 2001, pp. 41–51, DOI 10.1090/conm/271/04349. MR1831346

[43] R. R. Bruner, J. P. C. Greenlees, and J. Rognes, *The local cohomology theorems for tmf and* $H^{*,*}(\mathcal{A}(2))$. In preparation.

[44] R. R. Bruner and J. Rognes, *The Adams spectral sequence for the image-of-J spectrum.* In preparation.

[45] R. R. Bruner, J. P. May, J. E. McClure, and M. Steinberger, *$H_\infty$ ring spectra and their applications*, Lecture Notes in Mathematics, vol. 1176, Springer-Verlag, Berlin, 1986, DOI 10.1007/BFb0075405. MR836132

[46] Robert Bruner, Christian Nassau, and Sean Tilson, *Steenrod operations and $A$-module extensions*, arXiv:1909.03117v3.

[47] Robert Burklund, *An extension in the Adams spectral sequence in dimension* 54, arXiv:2005.08910.

[48] E. Čech, *Höherdimensionale Homotopiegruppen*, Verhandlungen des Internationalen Mathematiker-Kongresses Zürich 1932, II. Band, Orell Füssli Verlag, Zürich und Leipzig, 1932, pp. 203.

[49] Dominic Culver, *The Adams spectral sequence for* 3*-local tmf*, arXiv:1902.04230.

[50] Donald M. Davis, *The cohomology of the spectrum $bJ$*, Bol. Soc. Mat. Mexicana (2) **20** (1975), no. 1, 6–11. MR467749

[51] Donald M. Davis and Mark Mahowald, *$v_1$- and $v_2$-periodicity in stable homotopy theory*, Amer. J. Math. **103** (1981), no. 4, 615–659, DOI 10.2307/2374044. MR623131

[52] Donald M. Davis and Mark Mahowald, *Ext over the subalgebra $A_2$ of the Steenrod algebra for stunted projective spaces*, Current trends in algebraic topology, Part 1 (London, Ont., 1981), CMS Conf. Proc., vol. 2, Amer. Math. Soc., Providence, RI, 1982, pp. 297–342, DOI 10.2307/2374058. MR686123

[53] Donald M. Davis and Mark Mahowald, *The image of the stable $J$-homomorphism*, Topology **28** (1989), no. 1, 39–58, DOI 10.1016/0040-9383(89)90031-1. MR991098

[54] Christopher L. Douglas, John Francis, André G. Henriques, and Michael A. Hill (eds.), *Topological modular forms*, Mathematical Surveys and Monographs, vol. 201, American Mathematical Society, Providence, RI, 2014, DOI 10.1090/surv/201. MR3223024

[55] Daniel Dugger and Daniel C. Isaksen, *The motivic Adams spectral sequence*, Geom. Topol. **14** (2010), no. 2, 967–1014, DOI 10.2140/gt.2010.14.967. MR2629898

[56] Samuel Eilenberg and Saunders MacLane, *Cohomology theory of Abelian groups and homotopy theory. II*, Proc. Nat. Acad. Sci. U.S.A. **36** (1950), 657–663, DOI 10.1073/pnas.36.11.657. MR39252

[57] Samuel Eilenberg and John C. Moore, *Homology and fibrations. I. Coalgebras, cotensor product and its derived functors*, Comment. Math. Helv. **40** (1966), 199–236, DOI 10.1007/BF02564371. MR203730

[58] A. D. Elmendorf, I. Kriz, M. A. Mandell, and J. P. May, *Rings, modules, and algebras in stable homotopy theory*, Mathematical Surveys and Monographs, vol. 47, American Mathematical Society, Providence, RI, 1997. With an appendix by M. Cole, DOI 10.1090/surv/047. MR1417719

[59] A. D. Elmendorf and M. A. Mandell, *Rings, modules, and algebras in infinite loop space theory*, Adv. Math. **205** (2006), no. 1, 163–228, DOI 10.1016/j.aim.2005.07.007. MR2254311

[60] Hans Freudenthal, *Über die Klassen der Sphärenabbildungen I. Große Dimensionen* (German), Compositio Math. **5** (1938), 299–314. MR1556999

[61] Bogdan Gheorghe, Guozhen Wang, and Zhouli Xu, *The special fiber of the motivic deformation of the stable homotopy category is algebraic*, arXiv:1809.09290.

[62] Paul G. Goerss, *Topological modular forms [after Hopkins, Miller and Lurie]*, Astérisque **332** (2010), Exp. No. 1005, viii, 221–255. Séminaire Bourbaki. Volume 2008/2009. Exposés 997–1011. MR2648680

[63] Paul Goerss, Hans-Werner Henn, and Mark Mahowald, *The homotopy of $L_2V(1)$ for the prime 3*, Categorical decomposition techniques in algebraic topology (Isle of Skye, 2001), Progr. Math., vol. 215, Birkhäuser, Basel, 2004, pp. 125–151. MR2039763

[64] P. Goerss, H.-W. Henn, M. Mahowald, and C. Rezk, *A resolution of the $K(2)$-local sphere at the prime 3*, Ann. of Math. (2) **162** (2005), no. 2, 777–822, DOI 10.4007/annals.2005.162.777. MR2183282

[65] P. G. Goerss and M. J. Hopkins, *Moduli spaces of commutative ring spectra*, Structured ring spectra, London Math. Soc. Lecture Note Ser., vol. 315, Cambridge Univ. Press, Cambridge, 2004, pp. 151–200, DOI 10.1017/CBO9780511529955.009. MR2125040

[66] J. P. C. Greenlees and J. P. May, *Completions in algebra and topology*, Handbook of algebraic topology, North-Holland, Amsterdam, 1995, pp. 255–276, DOI 10.1016/B978-044481779-2/50008-0. MR1361892

[67] J. P. C. Greenlees and V. Stojanoska, *Anderson and Gorenstein duality*, Geometric and topological aspects of the representation theory of finite groups, Springer Proc. Math. Stat., vol. 242, Springer, Cham, 2018, pp. 105–130, DOI 10.1007/978-3-319-94033-5_5. MR3901158

[68] Michael A. Hill, *The 3-local* tmf*-homology of* $B\Sigma_3$, Proc. Amer. Math. Soc. **135** (2007), no. 12, 4075–4086, DOI 10.1090/S0002-9939-07-08937-X. MR2341960

[69] M. A. Hill, M. J. Hopkins, and D. C. Ravenel, *On the nonexistence of elements of Kervaire invariant one*, Ann. of Math. (2) **184** (2016), no. 1, 1–262, DOI 10.4007/annals.2016.184.1.1. MR3505179

[70] Michael Hill and Tyler Lawson, *Topological modular forms with level structure*, Invent. Math. **203** (2016), no. 2, 359–416, DOI 10.1007/s00222-015-0589-5. MR3455154

[71] Heinz Hopf, *Abbildungsklassen n-dimensionaler Mannigfaltigkeiten* (German), Math. Ann. **96** (1927), no. 1, 209–224, DOI 10.1007/BF01209163. MR1512315

[72] Heinz Hopf, *Über die Abbildungen der dreidimensionalen Sphäre auf die Kugelfläche* (German), Math. Ann. **104** (1931), no. 1, 637–665, DOI 10.1007/BF01457962. MR1512691

[73] Michael J. Hopkins, *Some problems in topology*, 1984. D.Phil. thesis, Oxford University.

[74] Michael J. Hopkins, *Topological modular forms, the Witten genus, and the theorem of the cube*, Proceedings of the International Congress of Mathematicians, Vol. 1, 2 (Zürich, 1994), Birkhäuser, Basel, 1995, pp. 554–565. MR1403956

[75] M. J. Hopkins, *Algebraic topology and modular forms*, Proceedings of the International Congress of Mathematicians, Vol. I (Beijing, 2002), Higher Ed. Press, Beijing, 2002, pp. 291–317. MR1989190

[76] Michael J. Hopkins and Mark Mahowald, *From elliptic curves to homotopy theory*, Topological modular forms, Math. Surveys Monogr., vol. 201, Amer. Math. Soc., Providence, RI, 2014, pp. 261–285, DOI 10.1090/surv/201/15. MR3328536

[77] Michael J. Hopkins and Haynes R. Miller, *Elliptic curves and stable homotopy I*, Topological modular forms, Math. Surveys Monogr., vol. 201, Amer. Math. Soc., Providence, RI, 2014, pp. 209–260, DOI 10.1090/surv/201/14. MR3328535

[78] Michael J. Hopkins and Jeffrey H. Smith, *Nilpotence and stable homotopy theory. II*, Ann. of Math. (2) **148** (1998), no. 1, 1–49, DOI 10.2307/120991. MR1652975

[79] W. Hurewicz, *Beiträge zur Topologie der Deformationen. I: Höherdimensionale Homotopiegruppen* (German), Proc. Akad. Wet. Amsterdam **38** (1935), 112–119.

[80] Dale Husemoller, John C. Moore, and James Stasheff, *Differential homological algebra and homogeneous spaces*, J. Pure Appl. Algebra **5** (1974), 113–185, DOI 10.1016/0022-4049(74)90045-0. MR365571

[81] Daniel C. Isaksen, *The cohomology of motivic* $A(2)$, Homology Homotopy Appl. **11** (2009), no. 2, 251–274. MR2591921

[82] Daniel C. Isaksen, *Stable stems*, Mem. Amer. Math. Soc. **262** (2019), no. 1269, viii+159, DOI 10.1090/memo/1269. MR4046815

[83] Daniel C. Isaksen, Guozhen Wang, and Zhouli Xu, *More stable stems*, arXiv:2001.04511.

[84] Daniel C. Isaksen and Zhouli Xu, *Motivic stable homotopy and the stable 51 and 52 stems*, Topology Appl. **190** (2015), 31–34, DOI 10.1016/j.topol.2015.04.008. MR3349503

[85] Daniel S. Kahn, Cup $-\,i$ *products and the Adams spectral sequence*, Topology **9** (1970), 1–9, DOI 10.1016/0040-9383(70)90043-1. MR253337

[86] Michel A. Kervaire and John W. Milnor, *Groups of homotopy spheres. I*, Ann. of Math. (2) **77** (1963), 504–537, DOI 10.2307/1970128. MR148075

[87] Stanley O. Kochman, *Stable homotopy groups of spheres*, Lecture Notes in Mathematics, vol. 1423, Springer-Verlag, Berlin, 1990. A computer-assisted approach, DOI 10.1007/BFb0083795. MR1052407

[88] Stanley O. Kochman and Mark E. Mahowald, *On the computation of stable stems*, The Čech centennial (Boston, MA, 1993), Contemp. Math., vol. 181, Amer. Math. Soc., Providence, RI, 1995, pp. 299–316, DOI 10.1090/conm/181/02039. MR1320997

[89] Johan Konter, *The homotopy groups of the spectrum Tmf*, arXiv:1212.3656.

[90] Tyler Lawson and Niko Naumann, *Commutativity conditions for truncated Brown-Peterson spectra of height 2*, J. Topol. **5** (2012), no. 1, 137–168, DOI 10.1112/jtopol/jtr030. MR2897051

[91] Tyler Lawson and Niko Naumann, *Strictly commutative realizations of diagrams over the Steenrod algebra and topological modular forms at the prime 2*, Int. Math. Res. Not. IMRN **10** (2014), 2773–2813, DOI 10.1093/imrn/rnt010. MR3214285

[92] L. G. Lewis Jr., J. P. May, M. Steinberger, and J. E. McClure, *Equivariant stable homotopy theory*, Lecture Notes in Mathematics, vol. 1213, Springer-Verlag, Berlin, 1986. With contributions by J. E. McClure, DOI 10.1007/BFb0075778. MR866482

[93] Wen-Hsiung Lin, $\mathrm{Ext}_A^{4,*}(\mathbb{Z}/2,\mathbb{Z}/2)$ *and* $\mathrm{Ext}_A^{5,*}(\mathbb{Z}/2,\mathbb{Z}/2)$, Topology Appl. **155** (2008), no. 5, 459–496, DOI 10.1016/j.topol.2007.11.003. MR2380930

[94] W. H. Lin, D. M. Davis, M. E. Mahowald, and J. F. Adams, *Calculation of Lin's Ext groups*, Math. Proc. Cambridge Philos. Soc. **87** (1980), no. 3, 459–469, DOI 10.1017/S0305004100056899. MR569195

[95] Arunas Liulevicius, *The factorization of cyclic reduced powers by secondary cohomology operations*, Mem. Amer. Math. Soc. **42** (1962), 112. MR182001

[96] J. Lurie, *A survey of elliptic cohomology*, Algebraic topology, Abel Symp., vol. 4, Springer, Berlin, 2009, pp. 219–277, DOI 10.1007/978-3-642-01200-6_9. MR2597740

[97] Jacob Lurie, *Elliptic Cohomology I: Spectral Abelian Varieties*. https://www.math.ias.edu/~lurie/papers/Elliptic-I.pdf.

[98] Jacob Lurie, *Elliptic Cohomology II: Orientations*. https://www.math.ias.edu/~lurie/papers/Elliptic-II.pdf.

[99] Mark Mahowald, *The metastable homotopy of* $S^n$, Memoirs of the American Mathematical Society, No. 72, American Mathematical Society, Providence, R.I., 1967. MR0236923

[100] Mark Mahowald, *The order of the image of the J-homomorphisms*, Bull. Amer. Math. Soc. **76** (1970), 1310–1313, DOI 10.1090/S0002-9904-1970-12656-8. MR270369

[101] Mark Mahowald, *A new infinite family in* ${}_2\pi_*{}^s$, Topology **16** (1977), no. 3, 249–256, DOI 10.1016/0040-9383(77)90005-2. MR445498

[102] Mark Mahowald, *bo-resolutions*, Pacific J. Math. **92** (1981), no. 2, 365–383. MR618072

[103] Mark Mahowald, *Toward a global understanding of* $\pi_*(S^n)$, Proceedings of the International Congress of Mathematicians, Vol. II (Berlin, 1998), Doc. Math. **Extra Vol. II** (1998), 465–472. MR1648096

[104] M. Mahowald and R. James Milgram, *Operations which detect Sq4 in connective K-theory and their applications*, Quart. J. Math. Oxford Ser. (2) **27** (1976), no. 108, 415–432, DOI 10.1093/qmath/27.4.415. MR433453

[105] Mark Mahowald and Charles Rezk, *Topological modular forms of level 3*, Pure Appl. Math. Q. **5** (2009), no. 2, Special Issue: In honor of Friedrich Hirzebruch., 853–872, DOI 10.4310/PAMQ.2009.v5.n2.a9. MR2508904

[106] Mark Mahowald and Paul Shick, *Periodic phenomena in the classical Adams spectral sequence*, Trans. Amer. Math. Soc. **300** (1987), no. 1, 191–206, DOI 10.2307/2000595. MR871672

[107] Mark Mahowald and Martin Tangora, *Some differentials in the Adams spectral sequence*, Topology **6** (1967), 349–369, DOI 10.1016/0040-9383(67)90023-7. MR214072

[108] Mark Mahowald and Martin Tangora, *An infinite subalgebra of* $\mathrm{Ext}_A(Z_2, Z_2)$, Trans. Amer. Math. Soc. **132** (1968), 263–274, DOI 10.2307/1994893. MR222887

[109] Jukka Mäkinen, *Boundary formulae for reduced powers in the Adams spectral sequence*, Ann. Acad. Sci. Fenn. Ser. A. I. **562** (1973), 42. MR375315

[110] M. A. Mandell and J. P. May, *Equivariant orthogonal spectra and S-modules*, Mem. Amer. Math. Soc. **159** (2002), no. 755, x+108, DOI 10.1090/memo/0755. MR1922205

[111] M. A. Mandell, J. P. May, S. Schwede, and B. Shipley, *Model categories of diagram spectra*, Proc. London Math. Soc. (3) **82** (2001), no. 2, 441–512, DOI 10.1112/S0024611501012692. MR1806878

[112] H. R. Margolis, *Spectra and the Steenrod algebra*, North-Holland Mathematical Library, vol. 29, North-Holland Publishing Co., Amsterdam, 1983. Modules over the Steenrod algebra and the stable homotopy category. MR738973

[113] W. S. Massey, *Products in exact couples*, Ann. of Math. (2) **59** (1954), 558–569, DOI 10.2307/1969719. MR60829

[114] Akhil Mathew, *The homology of tmf*, Homology Homotopy Appl. **18** (2016), no. 2, 1–29, DOI 10.4310/HHA.2016.v18.n2.a1. MR3515195

[115] C. R. F. Maunder, *On the differentials in the Adams spectral sequence for the stable homotopy groups of spheres. I*, Proc. Cambridge Philos. Soc. **61** (1965), 53–60, DOI 10.1017/s0305004100038639. MR175127

[116] C. R. F. Maunder, *On the differentials in the Adams spectral sequence for the stable homotopy groups of spheres. II*, Proc. Cambridge Philos. Soc. **61** (1965), 855–868, DOI 10.1017/s0305004100039219. MR202146

[117] J. Peter May, *The cohomology of the Steenrod algebra; stable homotopy groups of spheres*, Bull. Amer. Math. Soc. **71** (1965), 377–380, DOI 10.1090/S0002-9904-1965-11302-7. MR185596

[118] J. Peter May, *A general algebraic approach to Steenrod operations*, The Steenrod Algebra and its Applications (Proc. Conf. to Celebrate N. E. Steenrod's Sixtieth Birthday, Battelle Memorial Inst., Columbus, Ohio, 1970), Lecture Notes in Mathematics, Vol. 168, Springer, Berlin, 1970, pp. 153–231. MR0281196

[119] J. P. May, *The geometry of iterated loop spaces*, Springer-Verlag, Berlin-New York, 1972. Lectures Notes in Mathematics, Vol. 271. MR0420610

[120] J. P. May, *$E_\infty$ spaces, group completions, and permutative categories*, New developments in topology (Proc. Sympos. Algebraic Topology, Oxford, 1972), Cambridge Univ. Press, London, 1974, pp. 61–93. London Math. Soc. Lecture Note Ser., No. 11. MR0339152

[121] J. Peter May, *$E_\infty$ ring spaces and $E_\infty$ ring spectra*, Lecture Notes in Mathematics, Vol. 577, Springer-Verlag, Berlin-New York, 1977. With contributions by Frank Quinn, Nigel Ray, and Jørgen Tornehave. MR0494077

[122] R. James Milgram, *Group representations and the Adams spectral sequence*, Pacific J. Math. **41** (1972), 157–182. MR304463

[123] Haynes R. Miller, *A localization theorem in homological algebra*, Math. Proc. Cambridge Philos. Soc. **84** (1978), no. 1, 73–84, DOI 10.1017/S0305004100054906. MR494105

[124] Haynes R. Miller, *On relations between Adams spectral sequences, with an application to the stable homotopy of a Moore space*, J. Pure Appl. Algebra **20** (1981), no. 3, 287–312, DOI 10.1016/0022-4049(81)90064-5. MR604321

[125] Haynes R. Miller, Douglas C. Ravenel, and W. Stephen Wilson, *Periodic phenomena in the Adams-Novikov spectral sequence*, Ann. of Math. (2) **106** (1977), no. 3, 469–516, DOI 10.2307/1971064. MR458423

[126] Haynes R. Miller and W. Stephen Wilson, *On Novikov's Ext1 modulo an invariant prime ideal*, Topology **15** (1976), no. 2, 131–141, DOI 10.1016/0040-9383(76)90002-1. MR433434

[127] John Milnor, *The Steenrod algebra and its dual*, Ann. of Math. (2) **67** (1958), 150–171, DOI 10.2307/1969932. MR99653

[128] John W. Milnor and John C. Moore, *On the structure of Hopf algebras*, Ann. of Math. (2) **81** (1965), 211–264, DOI 10.2307/1970615. MR174052

[129] Mamoru Mimura, *On the generalized Hopf homomorphism and the higher composition. II. $\pi_{n+i}(S^n)$ for $i = 21$ and 22*, J. Math. Kyoto Univ. **4** (1965), 301–326, DOI 10.1215/kjm/1250524662. MR177413

[130] Mamoru Mimura and Hirosi Toda, *The $(n+20)$-th homotopy groups of $n$-spheres*, J. Math. Kyoto Univ. **3** (1963), 37–58, DOI 10.1215/kjm/1250524854. MR157384

[131] Stephen A. Mitchell, *On $p$-adic topological $K$-theory*, Algebraic $K$-theory and algebraic topology (Lake Louise, AB, 1991), NATO Adv. Sci. Inst. Ser. C Math. Phys. Sci., vol. 407, Kluwer Acad. Publ., Dordrecht, 1993, pp. 197–204, DOI 10.1007/978-94-017-0695-7_9. MR1367298

[132] R. Michael F. Moss, *Secondary compositions and the Adams spectral sequence*, Math. Z. **115** (1970), 283–310, DOI 10.1007/BF01129978. MR266216

[133] Shunji Mukohda, *On the applications of squaring operations for certain elements in the cohomology of the Steenrod algebra*, 1969. Collected Papers on Natural Science Commemorating the 35th Anniversary of Fukuoka University.

[134] Osamu Nakamura, *Some differentials in the* mod 3 *Adams spectral sequence*, Bull. Sci. Engrg. Div. Univ. Ryukyus Math. Natur. Sci. **19** (1975), 1–25. MR0385852

[135] Christian Nassau, 1997. private communication.

[136] Christian Nassau, 2001. http://nullhomotopie.de/charts/bigpng.png.

[137] Lee S. Nave, *The Smith-Toda complex $V((p+1)/2)$ does not exist*, Ann. of Math. (2) **171** (2010), no. 1, 491–509, DOI 10.4007/annals.2010.171.491. MR2630045

[138] S. P. Novikov, *Cohomology of the Steenrod algebra* (Russian), Dokl. Akad. Nauk SSSR **128** (1959), 893–895. MR0111022

[139] Shichirô Oka, *Ring spectra with few cells*, Japan. J. Math. (N.S.) **5** (1979), no. 1, 81–100, DOI 10.4099/math1924.5.81. MR614695

[140] L. S. Pontryagin, *Homotopy classification of the mappings of an $(n+2)$-dimensional sphere on an $n$-dimensional one* (Russian), Doklady Akad. Nauk SSSR (N.S.) **70** (1950), 957–959. MR0042121

[141] Daniel Quillen, *The Adams conjecture*, Topology **10** (1971), 67–80, DOI 10.1016/0040-9383(71)90018-8. MR279804

[142] Daniel Quillen, *On the cohomology and $K$-theory of the general linear groups over a finite field*, Ann. of Math. (2) **96** (1972), 552–586, DOI 10.2307/1970825. MR315016

[143] Douglas C. Ravenel, *Localization with respect to certain periodic homology theories*, Amer. J. Math. **106** (1984), no. 2, 351–414, DOI 10.2307/2374308. MR737778

[144] Douglas C. Ravenel, *Complex cobordism and stable homotopy groups of spheres*, Pure and Applied Mathematics, vol. 121, Academic Press, Inc., Orlando, FL, 1986. MR860042

[145] David L. Rector, *Steenrod operations in the Eilenberg-Moore spectral sequence*, Comment. Math. Helv. **45** (1970), 540–552, DOI 10.1007/BF02567352. MR278310

[146] Charles Rezk, *Notes on the Hopkins-Miller theorem*, Homotopy theory via algebraic geometry and group representations (Evanston, IL, 1997), Contemp. Math., vol. 220, Amer. Math. Soc., Providence, RI, 1998, pp. 313–366, DOI 10.1090/conm/220/03107. MR1642902

[147] V. A. Rohlin, *New results in the theory of four-dimensional manifolds* (Russian), Doklady Akad. Nauk SSSR (N.S.) **84** (1952), 221–224. MR0052101

[148] John Rognes, *Topological cyclic homology of the integers at two*, J. Pure Appl. Algebra **134** (1999), no. 3, 219–286, DOI 10.1016/S0022-4049(97)00155-2. MR1663390

[149] John Rognes, *Algebraic $K$-theory of the two-adic integers*, J. Pure Appl. Algebra **134** (1999), no. 3, 287–326, DOI 10.1016/S0022-4049(97)00156-4. MR1663391

[150] John Rognes, *Galois extensions of structured ring spectra. Stably dualizable groups*, Mem. Amer. Math. Soc. **192** (2008), no. 898, viii+137, DOI 10.1090/memo/0898. MR2387923

[151] Graeme Segal, *Categories and cohomology theories*, Topology **13** (1974), 293–312, DOI 10.1016/0040-9383(74)90022-6. MR353298

[152] Jean-Pierre Serre, *Sur les groupes d'Eilenberg-MacLane* (French), C. R. Acad. Sci. Paris **234** (1952), 1243–1245. MR46047

[153] Jean-Pierre Serre, *Sur la suspension de Freudenthal* (French), C. R. Acad. Sci. Paris **234** (1952), 1340–1342. MR46048

[154] Jean-Pierre Serre, *Quelques calculs de groupes d'homotopie* (French), C. R. Acad. Sci. Paris **236** (1953), 2475–2477. MR55684

[155] Nobuo Shimada and Akira Iwai, *On the cohomology of some Hopf algebras*, Nagoya Math. J. **30** (1967), 103–111. MR215896

[156] Katsumi Shimomura, *The homotopy groups of the $L_2$-localized Toda-Smith spectrum $V(1)$ at the prime* 3, Trans. Amer. Math. Soc. **349** (1997), no. 5, 1821–1850, DOI 10.1090/S0002-9947-97-01710-8. MR1370651

[157] Joseph H. Silverman, *The arithmetic of elliptic curves*, 2nd ed., Graduate Texts in Mathematics, vol. 106, Springer, Dordrecht, 2009, DOI 10.1007/978-0-387-09494-6. MR2514094

[158] William M. Singer, *Connective fiberings over* BU *and* U, Topology **7** (1968), 271–303, DOI 10.1016/0040-9383(68)90006-2. MR232392

[159] N. E. Steenrod, *Reduced powers of cohomology classes*, Ann. of Math. (2) **56** (1952), 47–67, DOI 10.2307/1969766. MR48026

[160] N. E. Steenrod, *Cohomology operations*, Lectures by N. E. Steenrod written and revised by D. B. A. Epstein. Annals of Mathematics Studies, No. 50, Princeton University Press, Princeton, N.J., 1962. MR0145525

[161] Vesna Stojanoska, *Duality for topological modular forms*, Doc. Math. **17** (2012), 271–311. MR2946825

[162] Vesna Stojanoska, *Calculating descent for 2-primary topological modular forms*, An alpine expedition through algebraic topology, Contemp. Math., vol. 617, Amer. Math. Soc., Providence, RI, 2014, pp. 241–258, DOI 10.1090/conm/617/12286. MR3243402

[163] Robert E. Stong, *Determination of $H^*(\mathrm{BO}(k,\cdots,\infty),Z_2)$ and $H^*(\mathrm{BU}(k,\cdots,\infty),Z_2)$*, Trans. Amer. Math. Soc. **107** (1963), 526–544, DOI 10.2307/1993817. MR151963

[164] Dennis Sullivan, *Genetics of homotopy theory and the Adams conjecture*, Ann. of Math. (2) **100** (1974), 1–79, DOI 10.2307/1970841. MR442930

[165] Martin C. Tangora, *On the cohomology of the Steenrod algebra*, Math. Z. **116** (1970), 18–64, DOI 10.1007/BF01110185. MR266205

[166] Martin Tangora, *Some extension questions in the Adams spectral sequence*, Proceedings of the Advanced Study Institute on the Algebraic Topology (Aarhus Univ., Aarhus, 1970), Mat. Inst., Aarhus Univ., Aarhus, 1970, pp. 578–587. Various Publ. Ser., No. 13. MR0339163

[167] Martin Tangora, *Some homotopy groups mod* 3, Conference on homotopy theory (Evanston, Ill., 1974), Notas Mat. Simpos., vol. 1, Soc. Mat. Mexicana, México, 1975, pp. 227–245. MR761731

[168] Hirosi Toda, *Generalized Whitehead products and homotopy groups of spheres*, J. Inst. Polytech. Osaka City Univ. Ser. A **3** (1952), 43–82. MR60822

[169] Hirosi Toda, *Calcul de groupes d'homotopie des sphères* (French), C. R. Acad. Sci. Paris **240** (1955), 147–149. MR68215

[170] Hirosi Toda, *On exact sequences in Steenrod algebra* mod 2, Mem. Coll. Sci. Univ. Kyoto Ser. A. Math. **31** (1958), 33–64, DOI 10.1215/kjm/1250776948. MR100835

[171] Hirosi Toda, *Composition methods in homotopy groups of spheres*, Annals of Mathematics Studies, No. 49, Princeton University Press, Princeton, N.J., 1962. MR0143217

[172] Hirosi Toda, *An important relation in homotopy groups of spheres*, Proc. Japan Acad. **43** (1967), 839–842. MR230310

[173] Hirosi Toda, *Extended p-th powers of complexes and applications to homotopy theory*, Proc. Japan Acad. **44** (1968), 198–203. MR230311

[174] Guozhen Wang and Zhouli Xu, *The triviality of the 61-stem in the stable homotopy groups of spheres*, Ann. of Math. (2) **186** (2017), no. 2, 501–580, DOI 10.4007/annals.2017.186.2.3. MR3702672

[175] Guozhen Wang and Zhouli Xu, *Some extensions in the Adams spectral sequence and the 51-stem*, Algebr. Geom. Topol. **18** (2018), no. 7, 3887–3906, DOI 10.2140/agt.2018.18.3887. MR3892234

[176] John S. P. Wang, *On the cohomology of the* mod − 2 *Steenrod algebra and the non-existence of elements of Hopf invariant one*, Illinois J. Math. **11** (1967), 480–490. MR214065

[177] George W. Whitehead, *On the homotopy groups of spheres and rotation groups*, Ann. of Math. (2) **43** (1942), 634–640, DOI 10.2307/1968956. MR7107

[178] George W. Whitehead, *The* $(n+2)^{\mathrm{nd}}$ *homotopy group of the n-sphere*, Ann. of Math. (2) **52** (1950), 245–247, DOI 10.2307/1969466. MR37507

[179] Nobuo Yoneda, *Note on products in* Ext, Proc. Amer. Math. Soc. **9** (1958), 873–875, DOI 10.2307/2033320. MR175957

[180] Zen-ichi Yosimura, *Universal coefficient sequences for cohomology theories of CW-spectra. II*, Osaka Math. J. **16** (1979), no. 1, 201–217. MR527026

# Index

# Selected Published Titles in This Series

253 **Robert R. Bruner and John Rognes,** The Adams Spectral Sequence for Topological Modular Forms, 2021

252 **Julie Déserti,** The Cremona Group and Its Subgroups, 2021

251 **David Hoff,** Linear and Quasilinear Parabolic Systems, 2020

250 **Bachir Bekka and Pierre de la Harpe,** Unitary Representations of Groups, Duals, and Characters, 2020

249 **Nikolai M. Adrianov, Fedor Pakovich, and Alexander K. Zvonkin,** Davenport–Zannier Polynomials and Dessins d'Enfants, 2020

248 **Paul B. Larson and Jindrich Zapletal,** Geometric Set Theory, 2020

247 **István Heckenberger and Hans-Jürgen Schneider,** Hopf Algebras and Root Systems, 2020

246 **Matheus C. Bortolan, Alexandre N. Carvalho, and José A. Langa,** Attractors Under Autonomous and Non-autonomous Perturbations, 2020

245 **Aiping Wang and Anton Zettl,** Ordinary Differential Operators, 2019

244 **Nabile Boussaïd and Andrew Comech,** Nonlinear Dirac Equation, 2019

243 **José M. Isidro,** Jordan Triple Systems in Complex and Functional Analysis, 2019

242 **Bhargav Bhatt, Ana Caraiani, Kiran S. Kedlaya, Peter Scholze, and Jared Weinstein,** Perfectoid Spaces, 2019

241 **Dana P. Williams,** A Tool Kit for Groupoid $C^*$-Algebras, 2019

240 **Antonio Fernández López,** Jordan Structures in Lie Algebras, 2019

239 **Nicola Arcozzi, Richard Rochberg, Eric T. Sawyer, and Brett D. Wick,** The Dirichlet Space and Related Function Spaces, 2019

238 **Michael Tsfasman, Serge Vlăduţ, and Dmitry Nogin,** Algebraic Geometry Codes: Advanced Chapters, 2019

237 **Dusa McDuff, Mohammad Tehrani, Kenji Fukaya, and Dominic Joyce,** Virtual Fundamental Cycles in Symplectic Topology, 2019

236 **Bernard Host and Bryna Kra,** Nilpotent Structures in Ergodic Theory, 2018

235 **Habib Ammari, Brian Fitzpatrick, Hyeonbae Kang, Matias Ruiz, Sanghyeon Yu, and Hai Zhang,** Mathematical and Computational Methods in Photonics and Phononics, 2018

234 **Vladimir I. Bogachev,** Weak Convergence of Measures, 2018

233 **N. V. Krylov,** Sobolev and Viscosity Solutions for Fully Nonlinear Elliptic and Parabolic Equations, 2018

232 **Dmitry Khavinson and Erik Lundberg,** Linear Holomorphic Partial Differential Equations and Classical Potential Theory, 2018

231 **Eberhard Kaniuth and Anthony To-Ming Lau,** Fourier and Fourier-Stieltjes Algebras on Locally Compact Groups, 2018

230 **Stephen D. Smith,** Applying the Classification of Finite Simple Groups, 2018

229 **Alexander Molev,** Sugawara Operators for Classical Lie Algebras, 2018

228 **Zhenbo Qin,** Hilbert Schemes of Points and Infinite Dimensional Lie Algebras, 2018

227 **Roberto Frigerio,** Bounded Cohomology of Discrete Groups, 2017

226 **Marcelo Aguiar and Swapneel Mahajan,** Topics in Hyperplane Arrangements, 2017

225 **Mario Bonk and Daniel Meyer,** Expanding Thurston Maps, 2017

224 **Ruy Exel,** Partial Dynamical Systems, Fell Bundles and Applications, 2017

223 **Guillaume Aubrun and Stanisław J. Szarek,** Alice and Bob Meet Banach, 2017

222 **Alexandru Buium,** Foundations of Arithmetic Differential Geometry, 2017

For a complete list of titles in this series, visit the
AMS Bookstore at **www.ams.org/bookstore/survseries/**.